Springer Proceedings in Complexity

For further volumes:
www.springer.com/series/11637

Thomas Gilbert • Markus Kirkilionis • Gregoire Nicolis
Editors

Proceedings of the European Conference on Complex Systems 2012

Editors
Thomas Gilbert
Faculté des Sciences
Université Libre de Bruxelles
Brussels, Belgium

Gregoire Nicolis
Faculté des Sciences
Université Libre de Bruxelles
Brussels, Belgium

Markus Kirkilionis
Mathematics Institute
University of Warwick
Coventry, UK

ISSN 2213-8684 ISSN 2213-8692 (electronic)
Springer Proceedings in Complexity
ISBN 978-3-319-00394-8 ISBN 978-3-319-00395-5 (eBook)
DOI 10.1007/978-3-319-00395-5
Springer Dordrecht Heidelberg New York London

Library of Congress Control Number: 2013944574

Printed on acid-free paper

Springer is part of Springer Science+Business Media (www.springer.com)

Foreword

The present volume contains contributions presented at the ninth *European Conference on Complex Systems*, held at Université Libre de Bruxelles, Brussels, from 2 to 7 September 2012, under the sponsorship of the Complex Systems Society.

The volume is divided into seven parts. The first six parts comprise contributions to the main conference, whether oral or poster, compiled according to the six conference main tracks. The last part includes contributions to some of satellite meetings hosted at the conference.

We are pleased to acknowledge the invaluable help of the colleagues who assisted in the organization of this event, starting with the *Organizing Committee* members, Vincent Blondel, Timoteo Carletti, Enrico Carlon, Anne De Wit, Pierre Gaspard, Albert Goldbeter, Renaud Lambiotte, and Carlo Vanderzande, and the *Steering Committee*, responsible for the development and support of the ECCS conference series, whose members are Fatihcan Atay, Vittoria Colizza, Thomas Gilbert, Janusz Holyst, Jürgen Jost, Markus Kirkilionis (Chair), Kristian Lindgren, Andras Lorincz, Jorge Louçã, Roberto Serra, Mina Teicher, Stefan Thurner, and Jeff Johnson (President of the Complex Systems Society). The six *Track Committees* were skillfully chaired by Claude Baesens, András Lörincz, Eve Mitleton-Kelly, Jacques Demongeot, Peter Allen, and Sorin Solomon, who benefited from the support of Anne De Wit, Pierre Gaspard, Hugues Bersini, Serge Massar, Annick Castiaux, Stéphane Vannitsem, Geneviève Dupont, Tom Lenaerts, Renaud Lambiotte, Nicolas Vandewalle, Vincent Blondel, Timoteo Carletti, Natasa Golo, as well as of many anonymous referees. The eighteen satellite meetings hosted at the conference were masterfully organized by independent committees to whom we are indebted. In addition, we wish to thank the students and staff members at the Université Libre de Bruxelles, without whom the conference could not have been organized.

We wish to express our gratitude to Theo Geisel who delivered the inaugural talk, as well as to the eight invited keynote speakers Charles H. Bennett, Jean-Louis Deneubourg, Manfred Eigen, Santo Fortunato, Peter Grassberger, Jean-Marie Lehn, Raymond Kapral, and Sylvia Walby.

Finally, it is our pleasure to thank the sponsors who enthusiastically supported this conference: the Université Libre de Bruxelles, the Fonds de la Recherche

Scientifique—FNRS, the Belgian Science Policy Office-Belspo, ASSYST—Action for the Science of complex SYstems and Socially intelligent icT, funded under the CORDIS Seventh Framework Programme, Naxys—Namur Center for Complex Systems, Springer Complexity, Oxford University Press, Cambridge University Press, Groupe De Boeck, World Scientific, Wolfram Research, and Star Alliance.

Thomas Gilbert, Conference Chair
Gregoire Nicolis, Program Chair
Markus Kirkilionis, Chair of the Steering Committee

Contents

Part I
Foundations of Complex Systems

Chapter 1
Aggregation and Emergence in Agent-Based Models: A Markov Chain Approach

Sven Banisch, Ricardo Lima, and Tanya Araújo

Abstract We analyze the dynamics of agent-based models (ABMs) from a Markovian perspective and derive explicit statements about the possibility of linking a microscopic agent model to the dynamical processes of macroscopic observables that are useful for a precise understanding of the model dynamics. In this way the dynamics of collective variables may be studied, and a description of macro dynamics as emergent properties of micro dynamics, in particular during transient times, is possible.

1.1 Introduction

Our work is a contribution to interweaving two lines of research that have developed in almost separate ways: Markov chains and agent-based models (ABMs). The former represents the simplest form of a stochastic process while the latter puts a strong emphasis on heterogeneity and social interactions. The usefulness of the Markov chain formalism in the analysis of more sophisticated ABMs has been discussed by [7], who look at 10 well-known social simulation models by representing them as a time-homogeneous Markov chain. Among these models are the Schelling segregation model [11], the Axelrod model of cultural dynamics [1] and the sugarscape model from [6]. The main idea of [7] is to consider all possible con-

This work has benefited from financial support from the Fundação para a Ciência e a Tecnologia (FCT), under the *13 Multi-annual Funding Project of UECE, ISEG, Technical University of Lisbon*. Financial support of the German Federal Ministry of Education and Research (BMBF) through the project *Linguistic Networks* is also gratefully acknowledged (http://project.linguistic-networks.net).

S. Banisch (✉)
Mathematical Physics, Bielefeld University, Bielefeld, Germany

R. Lima
Dream & Science Factory, Marseilles, France

T. Araújo
ISEG—Technical University of Lisbon (TULisbon), Lisbon, Portugal

T. Araújo
Research Unit on Complexity in Economics (UECE), Lisbon, Portugal

T. Gilbert et al. (eds.), *Proceedings of the European Conference on Complex Systems 2012*, Springer Proceedings in Complexity, DOI 10.1007/978-3-319-00395-5_1,

figurations of the system as the state space of the Markov chain. Despite the fact that all the information of the dynamics on the ABM is encoded in a Markov chain, it is difficult to learn directly from this fact, due to the huge dimension of the configuration space and its corresponding Markov transition matrix. The work of Izquierdo and co-workers mainly relies on numerical computations to estimate the stochastic transition matrices of the models.

Consider an ABM defined by a set $\mathbf{N}$ of agents, each one characterized by individual attributes that are taken from a finite list of possibilities. We denote the set of possible attributes by $\mathbf{S}$ and we call the *configuration space* $\boldsymbol{\Sigma}$ the set of all possible combination of attributes of the agents, i.e. $\boldsymbol{\Sigma} = \mathbf{S}^N$. This also incorporates models where agents move on a lattice (e.g. in the sugarscape model) because we can treat the sites as "agents" and use an attribute to encode whether a site is occupied or not. The updating process of the attributes of the agents at each time step typically consists of two parts. First, a random choice of a subset of agents is made according to some probability distribution ω. Then the attributes of the agents are updated according to a rule, which depends on the subset of agents selected at this time. With this specification, ABMs can be represented by a so-called random map representation which may be taken as an equivalent definition of a Markov chain [10]. Hence, ABMs are Markov chains on $\boldsymbol{\Sigma}$ with transition matrix $\hat{P}$. For a class of ABMs we can compute transition probabilities $\hat{P}(x, y)$ for any pair $x, y \in \boldsymbol{\Sigma}$ of agent configurations. We refer to the process $(\boldsymbol{\Sigma}, \hat{P})$ as micro chain.

When performing simulations of an ABM we are actually not interested in all the dynamical details but rather in the behavior of variables at the macroscopic level (such as average opinion, number of communities, etc.). The formulation of an ABM as a Markov chain $(\boldsymbol{\Sigma}, \hat{P})$ enables the development of a mathematical framework for linking the micro-description of an ABM to a macro-description of interest. Namely, from the Markov chain perspective, the transition from the micro to the macro level is a projection of the Markov chain with state space $\boldsymbol{\Sigma}$ onto a new state space $\mathbf{X}$ by means of a (projection) map Π from $\boldsymbol{\Sigma}$ to $\mathbf{X}$. The meaning of the projection Π is to lump sets of micro configurations in $\boldsymbol{\Sigma}$ according to the macro property of interest in such a way that, for each $X \in \mathbf{X}$, all the configurations of $\boldsymbol{\Sigma}$ in $\Pi^{-1}(X)$ share the same property.

The price to pay in passing from the micro to the macrodynamics in this sense [5, 8] is that the projected system is, in general, no longer a Markov chain: long memory (even infinite) may appear in the projected system. In particular, well known conditions for lumpability [8] make it possible to decide whether the macro model is still Markov. Conversely, this setting can also provide a suitable framework to understand how aggregation may lead to the emergence of long range memory effects.

1.2 Application to the Voter Model

We illustrate these ideas at the example of the Voter Model (VM) (see Refs. [4, 9]). In the VM, $\mathbf{S} = \{0, 1\}$ meaning that each agent is characterized by an attribute x_i,

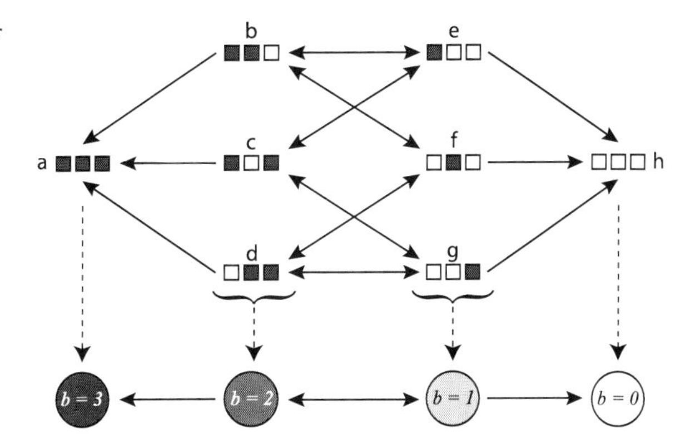

Fig. 1.1 The micro chain for the VM with 3 agents and its projection onto a random walk obtained by agglomeration of states with the same number of black agents b

$i = 1, \ldots, N$ which takes a value among two possible alternatives. The set of all possible combinations of attributes of the agents is $\mathbf{\Sigma} = \{0, 1\}^N$, that is, the set of all bit-strings of length N. At each time step in the iteration process, an agent i is chosen at random along with one of its neighboring agents j. If the states (x_i, x_j) are not equal already, agent i adopts the state of j (by setting $x_i = x_j$). At the microscopic level of all possible configurations of agents the VM corresponds therefore to an absorbing random walk on the N-dimensional hypercube. It is well known that the model has the two absorbing states $(1, \ldots, 1)$ and $(0, \ldots, 0)$. For a system of three agents this is shown in Fig. 1.1.

Opinion models as the VM are a nice examples where our projection construction is particularly meaningful. There, we consider the projection Π_b that maps each $x \in \mathbf{\Sigma}$ into $X_b \in \mathbf{X}$ where b is the number of agents in x with opinion 1. The projected configuration space is then made of the X_b where $0 \leq b \leq N$ (see Fig. 1.1). Markov chain theory, in particular lumpability, allows us to determine conditions for which the macro chain on $\mathbf{X} = (X_0, \ldots, X_b, \ldots, X_N)$ is again a Markov chain. We find that this requires that the probability distribution ω must be invariant under the group $\mathcal{S}_N$ of all the permutations of N agents and therefore uniform. This underlines the theoretical importance of homogeneous or complete mixing in the analysis of the VM and related models.

In this way our method enables the use of Markov chain instruments in the mathematical analysis of ABMs. In Markov chains with absorbing states (and therefore in the ABM) the asymptotic status is quite trivial. As a result, it is the understanding of the transient that becomes the interesting issue. In order to analyze the transient dynamics for the macro dynamics, all that is needed is to compute the fundamental matrix $\mathbf{F}$ of the Markov chain [8]. For the binary VM we are able to derive a closed form expression for the elements in $\mathbf{F}$ for arbitrary N which provides us with all the information needed to compute the mean quantities and variances of the transient dynamics of the model. In addition, we show in the VM with three opinion alternatives ($\mathbf{S} = \{0, 1, 2\}$) how restrictions in communication (bounded confidence) lead to stable co-existence of different opinions because new absorbing states emerge in the macro chain.

1.3 Some Results

The micro chains obtained via the random map representations helps to understand the role of the collection of (deterministic) interaction rules used in the model from one side and of the probability distribution ω governing the sequential choice of the rules used to update the system at each time step from the other side. The importance of this probability distribution is to encode social relations and exchange actions. In our setting it becomes explicit how the symmetries in ω translate into symmetries of the micro chain. If we decide to remain at a Markovian level, then the partition, or equivalently the collective variables to be used to build the macro model must be compatible with the symmetry of the probability distribution ω. In order to account for an increased level of heterogeneity the partition of the configuration space defining the macro-level has to be refined. A first result into this direction is that the symmetry group of agent permutations on ω informs us about ensembles of agent configurations that can be interchanged without affecting the probabilistic structure of micro chain. Consequently, these ensembles can be lumped into the same macro state and the dynamical process projected onto these states is still a Markov chain. It is clear, however, that, in absence of any symmetry, there is no other choice than to stay at the micro-level because no Markovian description at a macro-level is possible in this case.

In our opinion, a well posed mathematical basis for linking a micro-description of an ABM to a macro-description may help the understanding of many of the properties observed in ABMs and therefore provide information about the transition from the interaction of individual actors to the complex macroscopic behaviors observed in social systems. We summarize our main results below:

1. We formulate agent-based models as Markov chains at the micro level with explicit transition probabilities.
2. This allows the use of lumpability arguments to link between the micro and the macro level.
3. In case of a non-lumpable macro description this explains the emergence non-trivial dynamical effects (long memory).
4. In the Voter Model, homogeneous mixing leads to a macroscopic Markov chain which underlines the theoretical importance of homogeneous mixing.
5. This chain can be solved including mean convergence times and variances.
6. The stable co-existence of different opinions with in the bounded confidence model follows from the emergence of new absorbing states in the macro chain.
7. Heterogeneous mixing requires refinement and we show how to exploit the symmetries in the mixing distribution (ω) to obtain a proper refinement.

For further reading, see Refs. [2, 3].

References

1. Axelrod R (1997) The dissemination of culture: a model with local convergence and global polarization. J Confl Resolut 41(2):203–226

2. Banisch S, Lima R, Araújo T (2012) Agent based models and opinion dynamics as Markov chains. Soc Netw. doi:10.1016/j.socnet.2012.06.001
3. Banisch S, Lima R (2012, forthcoming) Markov projections of the Voter model. arXiv:1209.3902 [physics.soc-ph]
4. Castellano C, Fortunato S, Loreto V (2009) Statistical physics of social dynamics. Rev Mod Phys 81(2):591–646
5. Chazottes J-R, Ugalde E (2003) Projection of Markov measures may be Gibbsian. J Stat Phys 111(5–6). doi:10.1023/A:1023056317067
6. Epstein JM, Axtell R (1996) Growing artificial societies: social science from the bottom up. The Brookings Institution, Washington
7. Izquierdo LR, Izquierdo SS, Galán JM, Santos JI (2009) Techniques to understand computer simulations: Markov chain analysis. J Artif Soc Soc Simul 12(1):6
8. Kemeny JG, Snell JL (1976) Finite Markov chains. Springer, Berlin
9. Kimura M, Weiss GH (1964) The stepping stone model of population structure and the decrease of genetic correlation with distance. Genetics 49:561–576
10. Levin DA, Peres Y, Wilmer EL (2009) Markov chains and mixing times. AMS, Providence
11. Schelling T (1971) Dynamic models of segregation. J Math Sociol 1(2):143–186

Chapter 2
Chemically-Driven Miscible Viscous Fingering: How Can a Reaction Destabilize Typically Stable Fluid Displacements?

L.A. Riolfo, Y. Nagatsu, P.M.J. Trevelyan, and A. De Wit

Abstract We experimentally demonstrate that chemical reactions, by producing changes in viscosity at the miscible interface between two fluids, can be the very source of viscous fingering in systems that are otherwise stable in the absence of a reaction. We explain how, depending on whether the reaction product is more or less viscous than the reactants, different patterns develop in the reaction zone.

2.1 Background

Viscous fingering (VF) is the hydrodynamic instability that classically appears when a fluid with a given viscosity displaces another more viscous one in porous media or a Hele-Shaw cell [1]. It has diverse implications in various fields such as hydrology [2], petroleum recovery [1], liquid crystal [3], polymer processing [4], chromatography [5] or CO_2 sequestration to name a few [6].

Experimental [7, 8] and theoretical [9, 10] studies have shown that chemical reactions, by modifying the viscosity of the solutions at hand, can influence miscible VF. Changes in the viscosity profile, induced by a chemical reaction, give rise to variations in the displacement evolution and hence different patterns are observed.

The present work, going further, presents experimental demonstration of reaction-driven viscous fingering of the interface between a more viscous liquid displacing a less viscous one, a displacement that in absence of reaction would typically be stable. It has been theoretically predicted [9, 10] that the necessary condition for such a reactive displacement to undergo fingering is to yield a reaction product with a viscosity either larger or smaller than the viscosity of the reactants. Specifically,

L.A. Riolfo (✉) · A. De Wit
Nonlinear Physical Chemistry Unit, Service de Chimie Physique et Biologie Théorique, Faculté des Sciences, Université Libre de Bruxelles (ULB), CP 231, 1050 Brussels, Belgium

Y. Nagatsu
Graduate School of Engineering, Tokyo University of Agriculture and Technology, Fuchu, Tokyo 183, Japan

P.M.J. Trevelyan
Division of Mathematics & Statistics, University of Glamorgan, Pontypridd CF37 1DL, Wales

T. Gilbert et al. (eds.), *Proceedings of the European Conference on Complex Systems 2012*, Springer Proceedings in Complexity, DOI 10.1007/978-3-319-00395-5_2,

if μ_i and μ_d denote the viscosity of the invading solution and that of the displaced solution respectively, purely chemically-driven VF of the classically stable $\mu_i > \mu_d$ situation should occur provided μ_r, the viscosity in the reaction zone, is either larger than μ_i or smaller than μ_d [9, 10].

We study here both scenarios, viscosity maximum ($\mu_r > \mu_i$) and minimum ($\mu_r < \mu_d$), exploiting the viscosity dependence of polymer solutions on pH. From the experimental findings, the different fingering patterns are analyzed as a function of the viscosity contrast generated by the chemical reaction [11].

The article organizes as follows: In the next section we explain the experimental set up and the chemicals utilized in the experiments. Also in this second section we present our experimental findings. In Sect. 2.3 we discuss and explain the evolution on the displacements presented in the second section. Finally, conclusions are drawn while highlighting the possible impact of this experimental work.

2.2 Methods

Experiments are carried out in a horizontal Hele-Shaw cell consisting of two transparent glass plates 100 mm wide, 500 mm long and 14 mm thick separated by a gap width $b = 0.25$ mm. The fluids are injected linearly at a constant flow rate q. As the displacing more viscous fluid, we use aqueous polymer solutions. When these solutions displace a less viscous dyed non-reactive solution, no instability is observed at the miscible interface between the fluids. However, if the displaced fluid reacts with the polymer, generating a maximum or a minimum in the viscosity profile, the interface can become unstable undergoing fingering. In the displacement experiments where the maximum develops, a more viscous aqueous solution of 0.30 %wt polyacrylic acid (PAA—1250000 MW—Sigma Aldrich) displaces a dyed 0.06M sodium hydroxide (NaOH) aqueous solution. The liquids react at the miscible interface. The reaction product, sodium polyacrylate (SPA), typically presents a viscosity larger than that of both reactants. The chemical reaction at the miscible interface is PAA + NaOH → SPA.

On the other hand, in the case where the minimum in viscosity develops, a sodium polyacrylate (SPA—2100000–6600000 MW—Wako) aqueous solution 0.125 %wt pushes a less viscous 60 %wt glycerol aqueous solution containing 0.5M HCl. In this case the polymer reacts with the acid producing PAA, which here has a viscosity lower than that of both reactants. The reaction is then SPA + HCl → PAA + NaCl.

Figure 2.1 shows the temporal evolution of reaction-driven VF observed in a linear displacement for both cases. When the maximum in viscosity develops, fingers grow behind the reactive interface (Fig. 2.1(a)). On the other hand, in the case of a minimum in viscosity, the interface undergoes fingers that grow towards the displaced fluid (Fig. 2.1(b)).

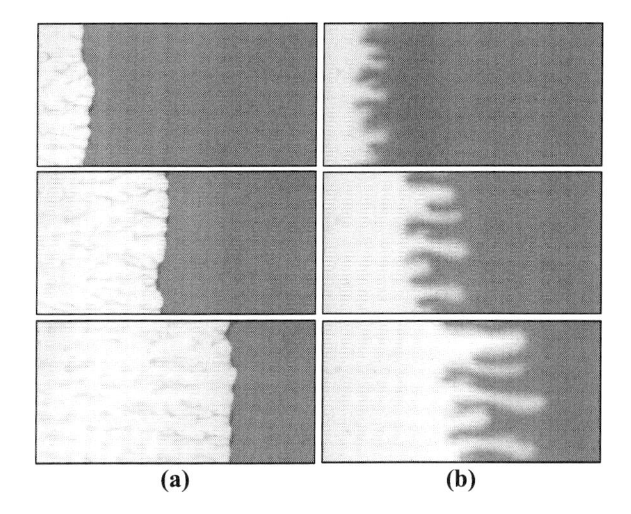

Fig. 2.1 Temporal evolution of reaction-driven VF in a linear displacement. (**a**) A more viscous solution of PAA displaces from left to right a less viscous aqueous dyed solution of NaOH in concentration 0.06M. Flow rate $q = 0.5$ ml/min. Time from top to bottom $t = 75$, 150 and 225 s. (**b**) A more viscous SPA solution displaces from left to right an aqueous dyed solution of 60 %wt glycerol + HCl 0.5M. Flow rate $q = 0.25$ ml/min. Time $t = 140$, 280 and 360 s. Field of view of each image = 4 cm × 8 cm

2.3 Results

In order to understand the systems' evolution we analyze experimentally the viscosity contrasts generated during the displacement experiments. We measure the viscosity of the pure reactants and estimate the viscosity developed in the reaction zone as the viscosity of a mixture of the pure reactants. The respective viscosities are measured with a rotational viscosimeter (Brookfield—Pro Extra II) at the shear rate corresponding to the experimental conditions.

In the displacement experiments with a maximum in viscosity the reactants viscosity are: invading fluid (0.3 %wt PAA) $\mu_i = 870$ cp, displaced fluid (0.06M NaOH) $\mu_d = 1$ cp. Hence, the initial viscosity contrast is stable, because the more viscous fluid displaces the more mobile one. However, in the reaction zone the viscosity developed is approximately $\mu_r = 3880$ cp. Therefore, an unstable contrast of viscosity is developed between the invading fluid and the reaction zone: $\mu_i < \mu_r$ and we have locally a less viscous fluid pushing a more viscous one. As the unstable region is located between the invading fluid and the reaction zone, the fingers should develop in this region. This is consistent with the experiments (Fig. 2.1(a)), where the fingers develop behind the reaction zone toward the invading fluid.

In the displacement with a minimum in viscosity, the viscosities are: invading fluid (0.125 %wt SPA) $\mu_i = 794$ cp, displaced solution (60 % glycerol + 0.5M HCl) $\mu_d = 10$ cp. The viscosity falls to $\mu_r = 5$ cp in the reaction zone. Therefore, even if the initial viscosity contrast is stable, locally an unstable region develops in time between the reaction zone and the displaced fluid ($\mu_r < \mu_d$). The development of the instability is then predicted to occur in the region between the reaction zone and the displaced fluid. This conjecture from the viscosity profiles agrees with the experimental findings exposed in Fig. 2.1(b).

We show here that depending on the unstable viscosity contrast developed during the displacement different patterns develop, and the interface deforms towards opposite directions, either in the displacement direction if a viscosity minimum develops, or against the displacement direction if a maximum in viscosity is chemically induced.

In this way, we have provided experimental evidence of viscous fingering triggered by a chemical reaction at the miscible interface between a more viscous solution displacing a less viscous one in a Hele-Shaw cell. Such a situation is classically stable in the absence of a reaction as we have a fluid with low mobility invading another more mobile one. The chemical reaction, by generating a product either more or less viscous than both reactants, triggers in time a non-monotonic viscosity profile. A locally unstable configuration with adverse mobility gradient develops around the extremum. This leads to fingers developing respectively behind or ahead of the reaction zone depending whether the viscosity profile exhibits a maximum or a minimum.

This results may help to prevent undesirable mixing during fluids displacements, such in the case of waste management in soils [12, 13], but also could lead to control of mixing enhancement in a unique direction in complex scenarios such as in microfluidics [14].

Acknowledgements We acknowledge JSPS, Prodex, the ITN—Marie Curie—Multiflow network and FNRS for financial support.

References

1. Homsy GM (1987) Viscous fingering in porous media. Annu Rev Fluid Mech 19:271
2. De WA, Bertho Y, Martin M (2005) Viscous fingering of miscible slices. Phys Fluids 17:054111
3. Buka A, Kertesz J, Vicsek T (1986) Transitions of viscous fingering patterns in nematic liquid crystals. Nature 323:424
4. Vlad DH, Maher JV (2000) Tip-splitting instabilities in the channel Saffman-Taylor flow of constant viscosity elastic fluids. Phys Rev E 61:5394
5. Broylesa BS, Shalliker RA, Cherrak DE, Guiochon G (1998) Visualization of viscous fingering in chromatographic columns. J Chromatogr A 882:173
6. Lake LW, Schmidt RL, Venuto PB (1992) A niche for enhanced oil recovery in the 1990s. Oilfield Rev 4:55
7. Nagatsu Y, Matsuda K, Kato Y, Tada Y (2007) Experimental study on miscible viscous fingering involving viscosity changes induced by variations in chemical species concentrations due to chemical reactions. J Fluid Mech 571:477
8. Nagatsu Y, Kondo Y, Kato Y, Tada Y (2009) Effects of moderate Damköhler number on miscible viscous fingering involving viscosity decrease due to a chemical reaction. J Fluid Mech 625:97
9. Hejazi SH, Trevelyan PMJ, Azaiez J, De Wit A (2010) Viscous fingering of a miscible reactive $A + B \rightarrow C$ interface: a linear stability analysis. J Fluid Mech 652:501

10. Nagatsu Y, De Wit A (2011) Viscous fingering of a miscible reactive $A + B \rightarrow C$ interface for an infinitely fast chemical reaction: nonlinear simulations. Phys Fluids 23:04310
11. Riolfo LA, Nagatsu Y, Iwata S, Maes R, Trevelyan PMJ, De Wit A (2012) Experimental evidence of chemically-driven miscible viscous fingering. Phys Rev E 85:015304(R)
12. Flowers TC, Hunt JR (2007) Viscous and gravitational contributions to mixing during vertical brine transport in water-saturated porous media. Water Resour Res 43:W01407
13. Maes R, Rousseaux G, Scheid B, Mishra M, Colinet P, De Wit A (2010) Experimental study of dispersion and miscible viscous fingering of initially circular samples in Hele-Shaw cells. Phys Fluids 22:1123104
14. Jha B, Cueto-Felgueroso L, Juanes R (2011) Fluid mixing from viscous fingering. Phys Rev Lett 106:194502

Chapter 3
Dynamical Localization in Kicked Rotator as a Paradigm of Other Systems: Spectral Statistics and the Localization Measure

Thanos Manos and Marko Robnik

Abstract We study the intermediate statistics of the spectrum of quasi-energies and of the eigenfunctions in the kicked rotator, in the case when the corresponding system is fully chaotic while quantally localized. As for the eigenphases, we find clear evidence that the spectral statistics is well described by the Brody distribution, notably better than by the Izrailev's one, which has been proposed and used broadly to describe such cases. We also studied the eigenfunctions of the Floquet operator and their localization. We show the existence of a scaling law between the repulsion parameter with relative localization length, but only as a first order approximation, since another parameter plays a role. We believe and have evidence that a similar analysis applies in time-independent Hamilton systems.

3.1 Introduction

One of the most important manifestations of quantum chaos of low-dimensional classically fully chaotic (ergodic) Hamiltonian systems is the fact that in the (sufficiently deep) semiclassical limit the statistical properties of the discrete energy spectra obey the statistics of Gaussian Random Matrix Theory (RMT). The opposite extreme are classically integrable systems, which quantally exhibit Poissonian spectral statistics (see [1]).

Quantum kicked rotator (QKR) is a typical example in the field of quantum chaos [2]. A typical property of the QKR is the chaos suppression for sufficiently large time scales. The study of the statistical properties of the classical and quantum (semiclassical) parameters in such systems is of great importance. Here we study in detail the semiclassical region where $k > K > 1$, i.e. the regime of full correspondence between quantum and classical diffusion (on the finite time scale $t \leq t_D$) and the manifested quantum dynamical localization for $t > t_D$. Furthermore, we are focused in the probability level spacing distributions in the regime where the system is

T. Manos (✉) · M. Robnik
Center for Applied Mathematics and Theoretical Physics, University of Maribor, Krekova 2, 2000 Maribor, Slovenia

T. Manos
School of Applied Sciences, University of Nova Gorica, Vipavska 11c, 5270 Ajdovščina, Slovenia

T. Gilbert et al. (eds.), *Proceedings of the European Conference on Complex Systems 2012*, Springer Proceedings in Complexity, DOI 10.1007/978-3-319-00395-5_3,

classically strongly chaotic ($K \geq 7$) but quantally localized, i.e in the *intermediate* or *soft* quantum chaos, as it is described in the literature [5].

3.2 The Quantum Kicked Rotator Model

The QKR model [3] is described by the following function

$$\hat{H} = -\frac{\hbar^2}{2I}\frac{\partial^2}{\partial\theta^2} + \varepsilon_0 \cos\theta \sum_{m=-\infty}^{\infty} \delta(t - mT), \tag{3.1}$$

where $\hbar$ is Planck's constant, I is the moment of inertia of the pendulum and ε_0 is the perturbation strength. The motion after one period T of the ψ wave function then can be described by the following mapping

$$\psi(\theta, t+T) = \hat{U}\psi(\theta, t), \tag{3.2}$$

$$\hat{U} = \exp\left(i\frac{T\hbar}{4I}\frac{\partial^2}{\partial\theta^2}\right)\exp\left(-i\frac{\varepsilon_0}{\hbar}\cos\theta\right)\exp\left(i\frac{T\hbar}{4I}\frac{\partial^2}{\partial\theta^2}\right), \tag{3.3}$$

where the ψ function is determined in the middle of the rotation, between two successive kicks. The evolution operator $\hat{U}$ of the system corresponds to one period. Due to the instant action of the perturbation, this evolution can be written as the product of three non-commuting unitary operators, the first and third of which corresponds to the free rotation during half a period $\hat{G}(\tau/2) = \exp(i\frac{T\hbar}{4I}\frac{\partial^2}{\partial\theta^2})$, $\tau \equiv \hbar T/I$, while the second $\hat{B}(k) = \exp(-ik\cos\theta)$, $k \equiv \varepsilon_0/\hbar$ describes the kick. The system's behavior depends only on two parameters, i.e. τ and k and its correspondence with the classical systems is described by the relation $K = k\tau = \varepsilon_0 T/I$. In the case $K \equiv k\tau \gg 1$ the motion is well-known to be strongly chaotic. The transition to classical mechanics is described by the limit $k \to \infty$, $\tau \to 0$ while $K = \text{const}$. In what follows $\hbar = \tau$ and $T = I = 1$. We shall consider mostly the semiclassical regime $k > K$, where $\tau < 1$.

In order to study how the localization affects the statistical properties of the quasienergy spectra we use the model's representation with a finite number N of levels [4, 5]

$$\psi_n(t+T) = \sum_{m=1}^{N} U_{nm}\psi_m(t), \quad n, m = 1, 2, \ldots, N. \tag{3.4}$$

The finite unitary matrix U_{nm} determines the evolution of a N-dimensional vector (Fourier transform of ψ) of the model

$$U_{nm} = \sum_{n'm'} G_{nm'} B_{n'm'} G_{n'm}, \tag{3.5}$$

where $G_{ll'} = \exp(i\tau l^2/4)\delta_{ll'}$ is a diagonal matrix corresponding to free rotation during a half period $T/2$ and the matrix $B_{n'm'}$ describing the one kick has the following form

$$B_{n'm'} = \frac{1}{2N+1}\sum_{l=1}^{2N+1}\left\{\cos\left[(n'-m')\frac{2\pi l}{2N+1}\right] - \cos\left[(n'+m')\frac{2\pi l}{2N+1}\right]\right\}$$
$$\times \exp\left[-ik\cos\left(\frac{2\pi l}{2N+1}\right)\right].$$

The model (3.4) with a finite number of states is considered as the quantum analogue of the classical standard mapping on the torus with closed momentum p and phase θ where U_{mn} describes only the odd states of the systems, i.e. $\psi(\theta) = -\psi(-\theta)$.

3.3 Intermediate Statistics and Comparison of Probability Distributions

Let us first compare the Brody and Izrailev probability distribution functions (PDFs) for the study of the intermediate level statistics. The Brody distribution is defined by the relation

$$P_{BR}(s) = C_1 s^{\beta_{BR}} \exp(-C_2 s^{\beta_{BR}+1}), \tag{3.6}$$

where the two parameters C_1 and C_2 are determined by the normalization conditions $\int_0^\infty P_{BR}(s)ds = 1$ and $\int_0^\infty s P_B(s)ds = 1$. Izrailev suggested the following distribution (see [5] and references there for the details and the argumentation)

$$P_{IZ}(s) = A\left(\frac{1}{2}\pi s\right)^{\beta_{IZ}} \exp\left[-\frac{1}{16}\beta_{IZ}\pi^2 s^2 - \left(B - \frac{1}{4}\pi\beta_{IZ}\right)s\right], \tag{3.7}$$

in order to describe the intermediate statistics, where the parameters A and B are determined again by the two above normalization conditions. Both β parameters, in the strongly localized regime tend asymptotically to 0 with Poissonian statistics while in the chaotic one tend to 1, which excellently approximates the Gaussian Orthogonal Ensemble (GEO) of the RMT. On the other hand, the non-integer β in the PDFs could be associated with the statistics of the quasienergy states with chaotic localized eigenfunctions.

Here, we use $N = 4000$ (which is considerably much larger size compared to the one used in the past studies) and $K = 7$ with $k = 30$. In Fig. 3.1(a) we show the numerical data (histogram) and the two PDFs. Their repulsion parameters have been calculated with best fit procedure independently. The corresponding values are found to be $\beta_{BR} \approx 0.424$ and $\beta_{IZ} \approx 0.419$ for the two PDFs respectively. In the inset figure of Fig. 3.1(a), we may see how the $P_{BR}(s)$ manages to capture and describe better the peak of the distribution where the most significant part of quasienergies ω is concentrated. The dot-dashed gray line indicates the Wigner surmise while

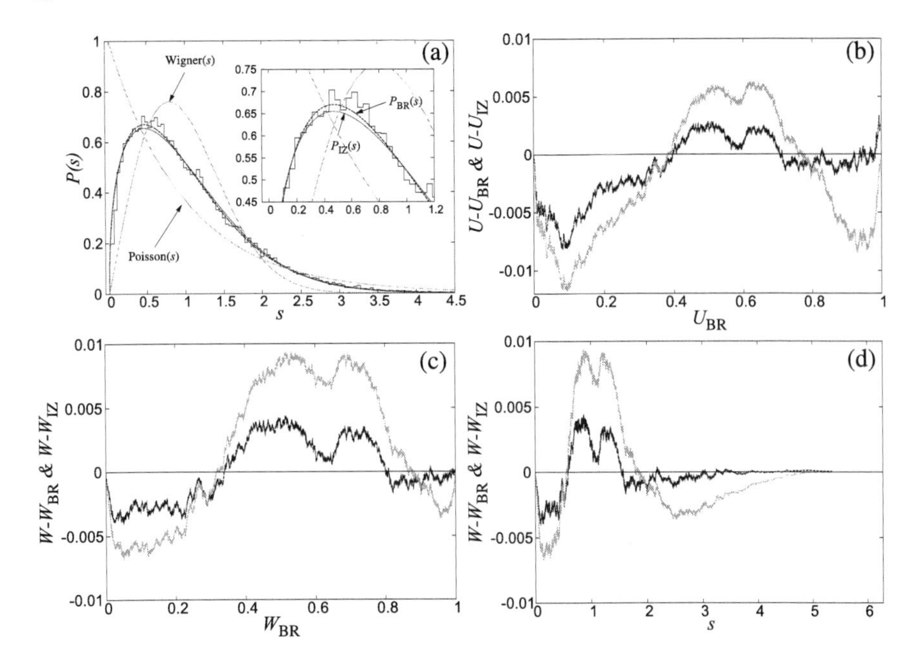

Fig. 3.1 Intermediate statistics (panel (**a**)) for distribution $P(s)$ (histogram—*black solid line*) of the model (3.4)–(3.5) fitted with distribution $P_{BR}(s)$ (*black dashed line*) and $P_{IZ}(s)$ (*black dotted line*) for $N = 4000$, $K = 7$ and $k = 30$ (see text for discussion). *The gray lines* indicate the two extreme distributions, i.e. the Poisson and Wigner. In panels (**b**)–(**d**) we show the comparison of the Brody (*black line*) and Izrailev (*gray line*) PDFs with the numerical using the U-function and W-distribution (see text for discussion)

the dot-dot-dashed one the Poisson distribution. Similar findings have also been found even for smaller sizes of the matrix U_{nm}, where the statistics are improved by sampling more matrices with slightly different values of k as e.g. in [5].

The above statement, regarding the $P_{BR}(s)$ better agreement with the numerical data, becomes more clear when checking the so-called U-functions $U(s) = (2/\pi)\arccos\sqrt{1-W(s)}$ of the two above distributions [6]. The $W(s) = \int_0^s P(x)dx$ is the cumulative (or integrated) level spacing distribution function (CDF). The U-function has the advantage that its expected statistical error δU is independent of s, being constant for each s and equal to $\delta U = 1/(\pi\sqrt{N_s})$, where N_s is the total number of objects in the $W(s)$ distribution. The numerical pre-factor $2/\pi$ is determined in such a way that $U(s) \in [0, 1]$ when $W(s) \in [0, 1]$. We may note here that the β values for the CDFs may be in principle slightly different compared to those found by the PDFs. In Fig. 3.1(b), we show the $U_{BR} - U$ and $U_{IZ} - U$ vs. U_{BR} where one may see that the Brody one is in general *closer* to zero (black line) than the Izrailev one (gray color). This fact indicates that the Brody one fits better the numerical data. This is also evident in Fig. 3.1(c), where the $W_{BR} - W$ and $W_{IZ} - W$ vs. W_{BR} are presented while in Fig. 3.1(d) the $W_{BR} - W$ and $W_{IZ} - W$ vs. s. The horizontal zero line in these panels indicates the complete agreement between the numerical data and theoretical predictions. The repulsion parameters for the CDFs used in panels (b), (c), (d) are $\beta_{BR} \approx 0.396$ and $\beta_{IZ} \approx 0.366$ respectively.

3.4 Scaling Laws and Localized Chaotic Regimes

A number of different ways to measure and estimate the *localization length* of the eigenfunctions have been proposed in the literature. Here, we adopt the well-accepted measure described and justified in e.g. [5]: For each N-dimensional eigenvector of the matrix U_{nm} the *information entropy* is $\mathcal{H}_N(u_1,\ldots,u_N) = -\sum_{n=1}^{N} u_n^2 \ln u_n^2$, where $u_n = \mathrm{Re}\,\varphi_n$ and $\sum_n u_n^2 = 1$. The distribution of u_n^2 for the GOE in the large N-limit tends to the Gaussian distribution and we get $\mathcal{H}_N^{GOE} = \psi(0.5N+1) - \psi(1.5) \simeq \ln(0.5Na) + O(1/N)$, where $a = 4/\exp(2-\gamma) \approx 0.96$, while ψ is the digamma function and γ the Euler constant ($\simeq 0.57$). Then the *entropy localization length* l_H is defined as $l_H = N\exp(\mathcal{H}_N - \mathcal{H}_N^{GOE})$. The fluctuations can be minimized when using the *mean localization length* $\langle l_H\rangle \equiv d$, which is computed by averaging over all eigenvectors of the same matrix (or over an ensemble of similar matrices) $d = N\exp(\langle\mathcal{H}_N\rangle - \mathcal{H}_N^{GOE})$.

The parameter that determines the transition from weak to strong quantum chaos is not the strength parameter k but the *ratio of the localization length* l_∞ to the size N of the system, $\Lambda = l_\infty/N$, where $l_\infty = D_{cl}/2\hbar^2$ and D_{cl} is the classical diffusion constant

$$D_{cl} = \begin{cases} \frac{1}{2}K^2[1 - 2J_2(K)(1 - J_2(K))], & \text{if } K \geq 4.5 \\ 0.30(K - K_{cr})^3, & \text{if } K_{cr} < K \leq 4.5, \end{cases} \tag{3.8}$$

where $K_{cr} \simeq 0.9716$ and $J_2(K)$ the Bessel function. The *localization parameter* is then defined as $\beta_{loc} = d/N$. The scaling law we used is $\beta_{loc}(x) = \gamma x/(1+\gamma x)$, where $x \equiv \Lambda$ and $\gamma \approx 4.2$ which is slightly different (but in agreement) to the one proposed in [7], where $x = k^2/N$ and $K = 5$. In Fig. 3.2(a), we compare β_{BR} repulsion parameter of the $P_{BR}(s)$ with the *localization parameter* β_{loc} through the localization length $d = \langle l_H\rangle$. For the numerical calculations and results regarding the spacing distributions $P(s)$ for the quasienergies, we have considered a wide range of the quantum perturbation parameter k keeping the classical parameter fixed (classically always fully chaotic). In order to ameliorate the statistics, we considered a sample of 161 matrices U_{nm} of size $N = 398$ (≈ 64000 elements), in a similar manner as e.g. in [5]) with slightly different values of k ($\Delta k = \pm 0.00125 \ll k$).

3.5 Summary

We studied aspects of dynamical localization in the kicked rotator, following [4, 5, 7], and largely confirm these results. We here considered the case with $K \geq 7$, where the dynamics is already fully chaotic (ergodic). The fractional power

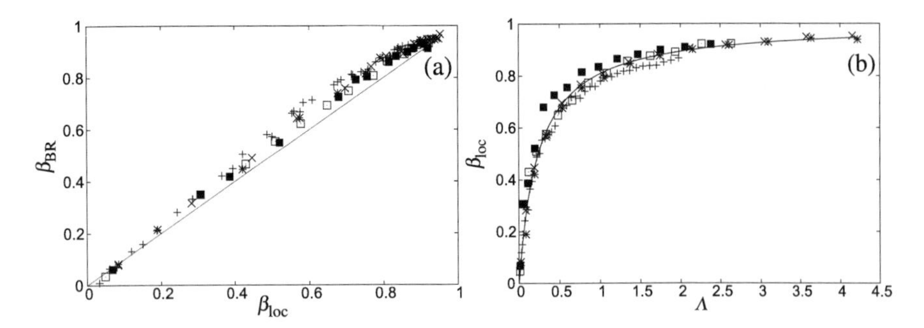

Fig. 3.2 (**a**) The fit parameter β_{BR} as a function of β_{loc} for 161×398 elements, $K = 7$ (+), 14 (×), 20 (∗), 30 (□), 35 (■) for a wide range of k values. (**b**) The parameter β_{loc} vs. Λ where the scaling law (see text) is shown with *the black line*

law level repulsion is clearly manifested, and globally the level spacing distribution is very well described by the Brody or by the Izrailev distribution, with a clear and systematic (although not very large) trend towards Brody rather than Izrailev. We show that the scaling law (β_{loc} vs. Λ) exists, but only as a first order approximation, as we see some scattering of data around the scaling curve. It seems that with increasing dimension of the matrices the scaling curve asymptotes to the limiting curve with only statistical scattering of the data points left. Further research confirms that a similar picture describing the dynamical localization applies also in time-independent systems, like e.g. billiards [8].

Acknowledgements The financial support of the Slovenian Research Agency (ARRS) is gratefully acknowledged.

References

1. Haake F (2001) Quantum signatures of chaos. Springer, Berlin
2. Stöckmann HJ (1999) Quantum chaos—an introduction. Cambridge University Press, Cambridge
3. Casati G, Chirikov BV, Izraelev FM, Ford J (1979) Stochastic behavior of a quantum pendulum under a periodic perturbation. In: Casati eG, Ford J (eds) stochastic behaviour in classical and quantum Hamiltonian systems, Proc Como conf, 1997. Lecture notes in physics, vol 93. Springer, Berlin, pp 334–352
4. Izrailev FM (1988) Quantum localization and statistics of quasienergy spectrum in a classically chaotic system. Phys Lett A 134:13–18
5. Izrailev FM (1990) Simple models of quantum chaos: spectrum and eigenfunctions. Phys Rep 196:299–392
6. Prosen T, Robnik M (1994) Semiclassical energy level statistics in the transition region between integrability and chaos: transition from Brody-like to Berry-Robnik behaviour. J Phys A, Math Gen 27:8059–8077

7. Izrailev FM (1995) Quantum chaos, localization and band random matrices. In: Casati G, Chirikov B (eds) Quantum chaos: between order and disorder. Cambridge University Press, Cambridge, pp 557–576
8. Batistić B, Robnik M (2010) Semiempirical theory of level spacing distribution beyond the Berry-Robnik regime: modeling the localization and the tunneling effects. J Phys A, Math Theor 43:215101

Chapter 4
$A + B \rightarrow C$ Reaction Fronts in Hele-Shaw Cells Under Modulated Gravitational Acceleration

Laurence Rongy, Kerstin Eckert, and Anne De Wit

Abstract We study the dynamics of $A + B \rightarrow C$ reaction fronts propagating under modulated gravitational acceleration by means of parabolic flight experiments and numerical simulations. We observe an accelerated front propagation under hypergravity along with a slowing down of the front under low gravity. By reaction-diffusion-convection simulations of an $A + B \rightarrow C$ front propagating in a thin layer, we can relate this periodic modulation of the front position to the amplification and decay, respectively, of the buoyancy-driven double vortex associated with the front propagation. A correlation between grey-value changes in the experimental shadowgraph images and characteristic changes in the concentration profiles are obtained by a numerical simulation of the imaging process (Eckert et al., Phys. Chem. Chem. Phys., 14:7337–7345, 2012).

4.1 Background

In a variety of chemical systems, buoyancy-driven convection has been shown to remarkably influence the propagation of reaction fronts. For example, the propagation speed of autocatalytic fronts in capillary tubes depends on the angle of inclination of the tube with regard to the vertical [2]. Simpler and more common chemical reaction types than autocatalysis are second-order irreversible reactions of the form $A + B \rightarrow C$. Provided that the two reactants A and B are initially separated in space, a reaction front can also propagate in these chemical systems. It has recently been shown that the reaction-diffusion properties of such fronts [3, 4] are not recovered if they develop in thin horizontal liquid layers where gravity points across the thin layer [5]. This suggests that the dynamics of these fronts can be influenced by chemically-driven buoyancy convection if A, B, and C have different densities. Rongy et al. [6, 7] have shown numerically that the nonlinear dynamics is characterized by one or two flow vortices developing around the front in covered horizontal

L. Rongy (✉) · A. De Wit
Nonlinear Physical Chemistry Unit, Service de Chimie Physique et Biologie Théorique, Faculté des Sciences, Université Libre de Bruxelles (ULB), CP 231, 1050 Brussels, Belgium

K. Eckert
Institute of Fluid Mechanics, Technische Universität Dresden, 01062 Dresden, Germany

T. Gilbert et al. (eds.), *Proceedings of the European Conference on Complex Systems 2012*, Springer Proceedings in Complexity, DOI 10.1007/978-3-319-00395-5_4,

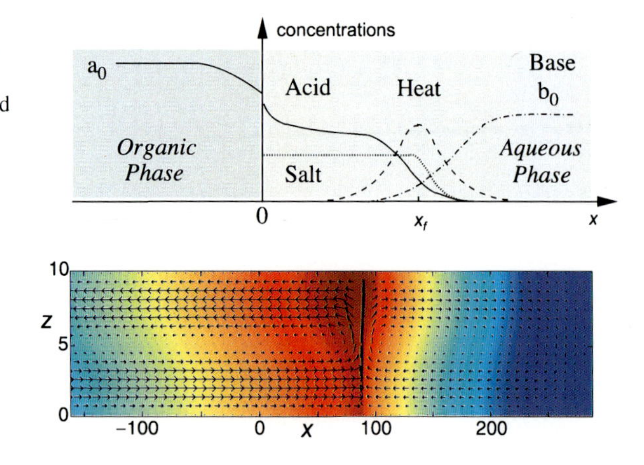

Fig. 4.1 *Top*: Sketch of the reaction-diffusion profiles in the reactive liquid-liquid system. *Bottom*: Velocity field superimposed on salt concentration field ranging from $s = s_{\max}$ in *the middle* (*red*) to $s = 0$ on the sides (*blue*) in the hyper-gravity phase. The maximum intensity of the velocity vectors is 1.1 in dimensionless units

layers. They have furthermore classified the various possible density profiles and related flow properties in a parameter space spanned by the three Rayleigh numbers of the problem. We provide here a direct comparison between experiments and these theoretical predictions. We show moreover that they can be used to understand situations on earth but also more complicated scenarios with modulated gravity, such as the ones observed in parabolic flights.

4.2 Chemical System and Model

The experimental container is a Hele-Shaw (HS) cell where propionic acid (A) undergoes a mass transfer from an organic phase (cyclohexane) into the aqueous layer and is subsequently neutralised by the base TMAH (B), producing a salt, tetramethylammonium propionate (C), according to $A + B \rightarrow C$ (see Fig. 4.1). The reaction front in the HS cell is followed by means of the shadowgraph technique [9]. This technique is sensitive with respect to the Laplacian of the refractive index n and the shadowgraph produces a relative change in the light intensity, from which the reaction front position X_f can be extracted. In ground experiments the resulting hydrodynamic instabilities have been well characterized [5, 8] and we focus here on microgravity situations.

To reduce complexity, the simulations focus entirely on the aqueous phase, instead of treating the complete liquid-liquid system with the mass transfer through the interface. This is justified because, after some minutes only, the reaction front is already sufficiently far from the interface. Thus we consider a two-dimensional (2D) thin aqueous solution layer placed horizontally in the gravity field in which an acid-base reaction, $A + B \rightarrow C$, takes place. The governing equations for the acid, base, and salt concentrations are isothermal reaction-diffusion-convection equations. The evolution of the 2D velocity field, $\vec{v}$, is described by the incompressible Stokes equations and is coupled to the evolution of the concentrations through an equation of state. In the latter we assume a linear dependence between the solution density, ρ,

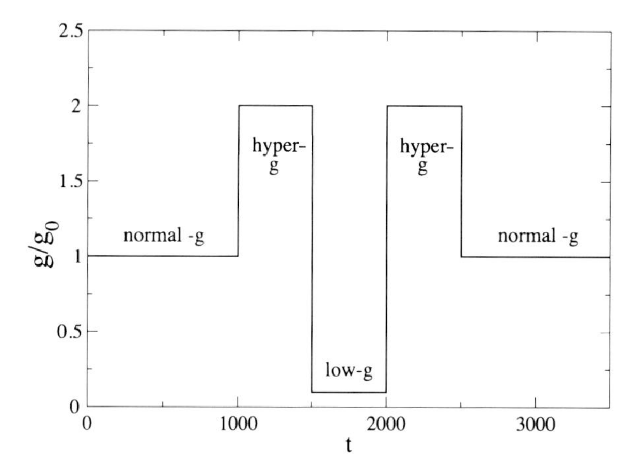

Fig. 4.2 Flight protocol of the normalized magnitude of the gravitational acceleration, g/g_0 with $g_0 = 9.81$ m/s^2, versus time, as used in the simulations

and the concentrations, introducing Rayleigh numbers directly proportional to the experimental contribution of each species to the changes in density. To simulate the gravity modulations of the parabolic flight we let the normalized magnitude of the gravitational acceleration, g/g_0 , with $g_0 = 9.81$ m/s^2, vary with time following the flight protocol (see Fig. 4.2).

4.3 Experimental and Numerical Results

When two miscible solutions containing A and B, respectively, are brought into contact, the dynamics of the $A + B \rightarrow C$ fronts developing in the liquid layer measured experimentally during parabolic flights differ from those obtained on earth [5]. This confirms that, even in thin liquid layers, convective motions can influence the properties of such fronts drastically. When gravitational acceleration varies periodically, we observe an associated periodic change between an accelerated front propagation under hyper-gravity and a slowed down propagation under low gravity. These results are explained by the numerical integration of our reaction-diffusion-convection model. The simulations reproduce the experimental behavior and allow to relate the modulation of the front position to the changes in the buoyancy-driven flow field consisting of a double vortex (see Fig. 4.1). This vortex consists of a rising flow at the reaction front, advecting fresh reactants towards it. The amplification or decay of such a double vortex when increasing or decreasing the g-level, respectively, explains the periodic behavior of the front position.

Figure 4.3 indicates a representation of the reaction front under parabolic flight conditions by two characteristic lines, the leading edge (LE) and the trailing edge (TE) (see caption of Fig. 4.3). The width of the front, defined as the distance between these two lines, is maximum in the low-gravity phase and minimum in the hyper-gravity phase. Thanks to the numerical simulations of the experimental shadowgraph visualization, we are able to interpret the meaning of LE and TE and to establish a correlation between the changes of the grey values in the experimental

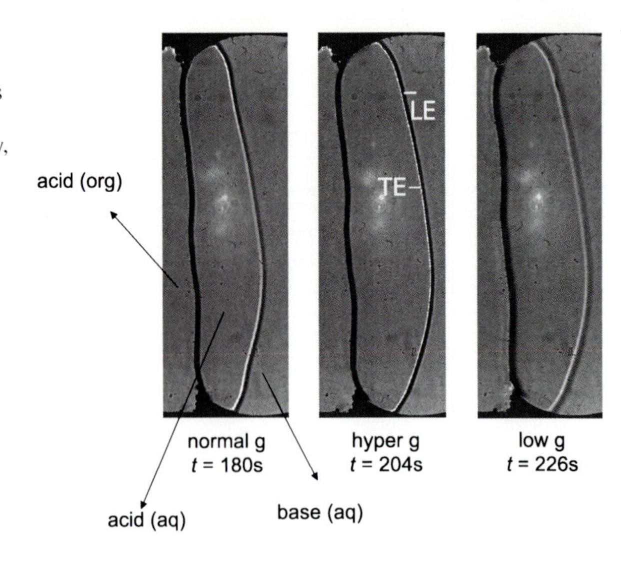

Fig. 4.3 Shadowgraph images of the system during the three characteristic stages of the parabolic flight: normal-gravity, hyper-gravity, and low-gravity phases. The position of the front is followed thanks to two characteristic lines, TE and LE. TE corresponds to the trailing edge, which is the brightest contour of the shadowgraph, and LE is the leading edge, which is the darkest contour and is directed towards the fresh aqueous phase

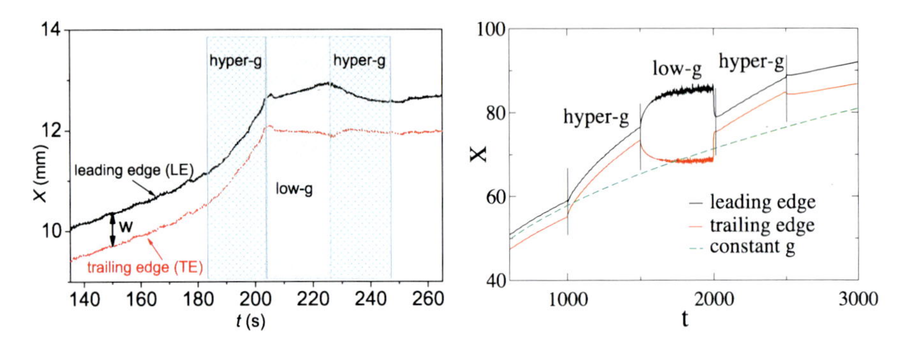

Fig. 4.4 Experimental (*left*) and numerical (*right*) behavior of the positions X_f^{LE} and X_f^{TE} of leading and trailing edges versus time during different stages of one parabola

images with those of the concentration fields. In particular, the observation of an increasing front width at decreased contrast during the low-gravity phase can be related to a change in the curvature of the concentration profiles. Indeed, the positions of the leading and trailing edges are deduced from the positions of minimum and maximum $d^2\langle n\rangle/dx^2$ during one period of g-modulation, where n, the refractive index, is reconstructed from the numerical concentration profiles as a linear combination of the concentration fields of each species.

Figure 4.4 illustrates the behavior of the two characteristic lines, LE and TE, during the different stages of the parabolic flight. There is an excellent qualitative agreement between the experimental and numerical results. In the numerical part of Fig. 4.4 we also evidence the acceleration of the reaction front in the parabolic flight conditions compared to the situation under constant normal gravity $g = 1g_0$.

References

1. Eckert K, Rongy L, De Wit A (2012) $A + B \rightarrow C$ reaction fronts in Hele-Shaw cells under modulated gravitational acceleration. Phys Chem Chem Phys 14:7337–7345
2. Nagypal I, Bazsa G, Epstein IR (1986) Gravity-induced anisotropies in chemical waves. J Am Chem Soc 108:3635–3640
3. Gálfi L, Rácz Z (1988) Properties of the reaction front in an $A + B \rightarrow C$ type reaction-diffusion process. Phys Rev A 38:3151–3154
4. Jiang Z, Ebner C (1990) Simulation study of reaction fronts. Phys Rev A 42:7483–7486
5. Shi Y, Eckert K (2006) Acceleration of reaction fronts by hydrodynamic instabilities in immiscible systems. Chem Eng Sci 61:5523–5533
6. Rongy L, Trevelyan PMJ, De Wit A (2008) Dynamics of $A + B \rightarrow C$ reaction fronts in the presence of buoyancy-driven convection. Phys Rev Lett 101:084503
7. Rongy L, Trevelyan PMJ, De Wit A (2010) Influence of buoyancy-driven convection on the dynamics of $A + B \rightarrow C$ reaction fronts in horizontal solution layers. Chem Eng Sci 65:2382–2391
8. Eckert K, Acker M, Shi Y (2004) Chemical pattern formation driven by a neutralization reaction. I: Mechanism and basic features. Phys Fluids 16:385–399
9. Merzkirch W (1987) Flow visualization. Academic Press, New York

Chapter 5
Effect of Limited Stirring on the Belousov Zhabotinsky Reaction

Florian Wodlei and Mihnea R. Hristea

Abstract The effect of a limited stirring phase on the general behavior of the periodic color change in the Belousov Zhabotinsky was investigated systematically. The effects of stirring on the BZ reaction has other consequences too apart from the pure homogenization of the system. We have investigated the system under different conditions (changing dimensions and volume of the beaker, different stirring rates and times). We found that the stirring can result in the disappearance of an aperiodic phase which is present in the non-stirred case. We suppose that the stirring time plays the role of a 'bifurcation parameter'.

Keywords Belousov Zhabotinsky reaction · Stirring · Rotational flow

5.1 Introduction

The Belousov Zhabotinsky reaction (BZ reaction), named after B.P. Belousov [1] and A.M. Zhabotinsky [2], is a homogeneous chemical reaction system. It performs periodic color changes from red to blue (in case of the ferroin indicator) or from translucent to light yellow (in case of cerium as indicator). In last more than half a century there has been a constant interest in studying BZ reaction because of its rich and complex behaviors ranging from periodic to chaotic nature. Our interest in BZ reaction arose from its high similarity to biological processes. By studying this system we expect to get a better understanding of physicochemical and biological processes such as pattern formation, metabolism, reproduction or adaptability.

As the color oscillations are in visible range one can easily monitor the concentration of ferroin (or cerium) by measuring the transmittance of the light through the solution. The reaction is normally studied in a constantly stirred tank reactor (CSTR) [6]. According to our knowledge until now the effect of a limited stirring

F. Wodlei (✉) · M.R. Hristea
Living Systems Research, Roseggerstr. 27/2, 9020 Klagenfurt, Austria
e-mail: florian.wodlei@ilsr.at

M.R. Hristea
e-mail: mihnea.hristea@ilsr.at

T. Gilbert et al. (eds.), *Proceedings of the European Conference on Complex Systems 2012*, Springer Proceedings in Complexity, DOI 10.1007/978-3-319-00395-5_5,

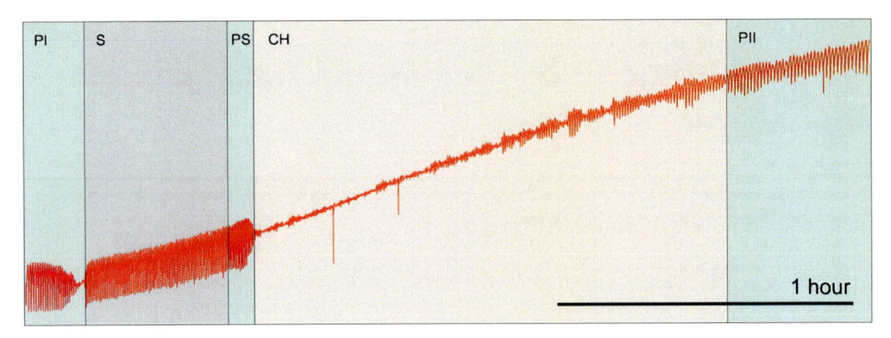

Fig. 5.1 Evolution of periodic color change corresponding to the periodic change of the ferroin concentration. There exist four distinct regions (phases): the first periodic phase (PI), the stirring phase (SP) and a short periodic phase right after (PS), the aperiodic (chaotic) phase (CH) and the second periodic phase (PII). Phases are separated by vertical lines and different background colors for which, except from the stirring phase, the times for the transion from one to the other phase are not that clear

phase has never been investigated systematically although there exists some report where the system was stirred for a limited time [7].

Apart from the experimental investigation of the BZ reaction we are also interested in a theoretical model capable of describing such behavior. For that reason we are also engaged in the physicochemical description of the system in the context of non-equilibrium thermodynamics and hydrodynamics. The mathematical methods needed to describe the system involve functional analysis and nonlinear differential equations. We followed the general ideas of Ilya Prigogine et al. [5] and R.J. Field, E. Körös and R.M. Noyes (FKN model) [3, 4]. A first simplified model able to describe the qualitative behavior we have observed was presented at the European Conference on Complex Systems 2011 [9, 10]. Our model can be understood as a dynamical FKN model as the parameters are time dependent.

In this work we report our investigation on the effect of a limited stirring phase on the behavior of the BZ reaction. The effects of stirring on the BZ reaction has other consequences too apart from the pure homogenization of the system. This can be seen easily from the fact that if the BZ reaction is stirred at a 'high' rate, the color oscillations stop immediately, and when the stirring is stopped, almost immediately the oscillations restart. If the BZ reaction is stirred at a 'low' rate, the color oscillations sustain and moreover the time period of the oscillations becomes regular. If the stirring is done for a limited time the general behavior changes. A typical time evolution of the periodic color change with an stirring phase of 30 minutes is shown in Fig. 5.1.

5.2 Materials and Methods

The experiments were performed with chemicals of analytical quality without further purification. We prepared solutions from the chemicals with the following concentration: Sodium bromate ($NaBrO_3$) 0.5 M, sodium bromide ($NaBr$) 0.18 M, malonic acid 1.5 M, sulfuric acid (H_2SO_4) 3.06 M and ferroin (complex built from o-phenantroline-chloride monohydrate and iron(II)sulfate-7-hydrate) 0.026 M. In an Erlenmeyr flask 7 ml distillated water was poured together with 8 ml $NaBrO_3$ solution, 10 ml from malonic acid solution, 10 ml H_2SO_4 solution, 4 ml NaBr solution and 0.6 ml ferroin solution constituting the total volume to 30 ml. The solution was mixed for 2 minutes after adding sodium bromide and 1 minute after adding ferroin. Right after, the solution was poured into a photometric quartz UV grade spectrophotometric cuvette with a colume of approx. 4 ml. For the measurements, where the system was stirred, a teflon-coated magnetic stirrer of approx. 0.6 mm length was on the bottom of the cuvette.

The time evolution of the transmittance was monitored by a photometric unit consisting of a light dependent resistor (D 9960-23) with the peak wavelength at 600 nm and a light emitting diode (LED 5 mm) with the peak wavelength at 700 nm. The photometric unit was incorporated in an electric circuit such that the color change could be monitored by measuring the voltage from the apparatus constructed. The voltage measurements were performed and recorded using VC820 multimeter from Voltcraft. The measurement was executed in a black box, where the cuvette was inserted such that the photometric unit was measuring the color change approx. 1.5 cm from the bottom of the cuvette and 2.5 cm from the liquid-air interface.

5.3 Results and Discussion

The effect of a limited stirring phase was investigated under different conditions (changing dimensions and volume of the beaker, different stirring rates and times). We found out that the main experimental difference between unstirred and stirred systems is that the 'phase transition' from chaotic phase (CH) to the second periodic phase (PII) takes place earlier in the case where the system is stirred. Furthermore we found out that if the time of stirring is increased up to 60 minutes the chaotic phase disappears completely. In Fig. 5.2 from bottom to top one can see that the time of stirring results in a shortening of the time of the chaotic phase (CH) which can result in a complete disappearance of the chaotic phase (upper panel of Fig. 5.2).

The general behavior is as follows: The color change disappears after some few oscillations (initial phase, periodic phase I) (Fig. 5.1, PI). When stirring is started with a certain stirring rate (approx. 9 Hz) the color oscillations come back and become stable (Fig. 5.1, S). After a time (30 minutes in Fig. 5.1) the stirring is stopped. The periodic behavior continues for some more time (Fig. 5.1, PS) but the oscillations become irregular and smaller. A phase of 'small' irregular oscillations and not always distinguishable from the noise of the measuring apparatus starts and can last up to 1 hour (chaotic phase) (Fig. 5.1, CH). After that a phase of large, regular and ordered oscillations starts which can last up to 10 hours (periodic phase II) (Fig. 5.1, PII).

In general we can distinguish three scenarios. The first scenario, which is shown in the lower panel of Fig. 5.2, appears if the BZ reaction is not stirred. Then the periodic color change evolves from a periodic (PI) over an aperiodic (CH) to an

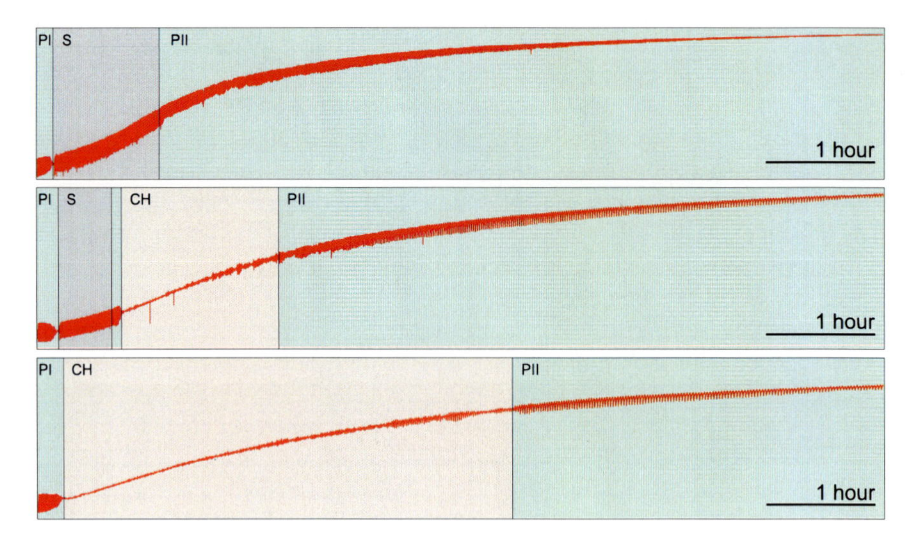

Fig. 5.2 Effect of a limited stirring phase on the evolution of the periodic color change. Lettering as in Fig. 5.1. *Lower panel*: time evolution without stirring phase. *Middle panel*: time evolution with a limiting stirring phase of 30 minutes (*grey area*, letter S) (*the middle panel* of this figure is shown in more detail in Fig. 5.1). *Upper panel*: time evolution with a limiting stirring phase of 60 minutes (*grey area*, letter S). Note that a chaotic phase (CH), as in the cases where the solution is not stirred or only stirred for 30 minutes, is no more visible

again periodic phase (PII) (such a behavior is also reported by Rustici et al. [7]) [8]. The second scenario, which is shown in the middle panel of Fig. 5.2, appears if the BZ reaction is stirred for (at least) 30 minutes right after the first periodic phase (PI). During stirring the oscillations become regular and after stopping they continue for some while. Then a chaotic phase (CH), as in the first scenario, starts and which again after some time changes into a second periodic phase (PII). And the last and most interesting scenario, which is shown in the upper panel of Fig. 5.2 and in more details in Fig. 5.3, appears if the BZ reaction is stirred for (at least) 60 minutes right after the first periodic phase (PI). During stirring the oscillations become regular and after stopping they remain regular and the periodic behavior continues, i.e. the chaotic phase disappears.

In all experiments where we stirred the system for 60 minutes and more there appears an interesting 'shift' where the mean transmittance decreases for a couple of oscillations and then comes back again (inset c in upper panel of Fig. 5.3). We still do not yet understand what that means but it might be connected to the disappearance of the chaotic phase.

5.4 Conclusion

We show here that a limited stirring phase changes the general behavior of the time evolution of the BZ reaction. As long as the stirring rate is not too 'high' the stirring

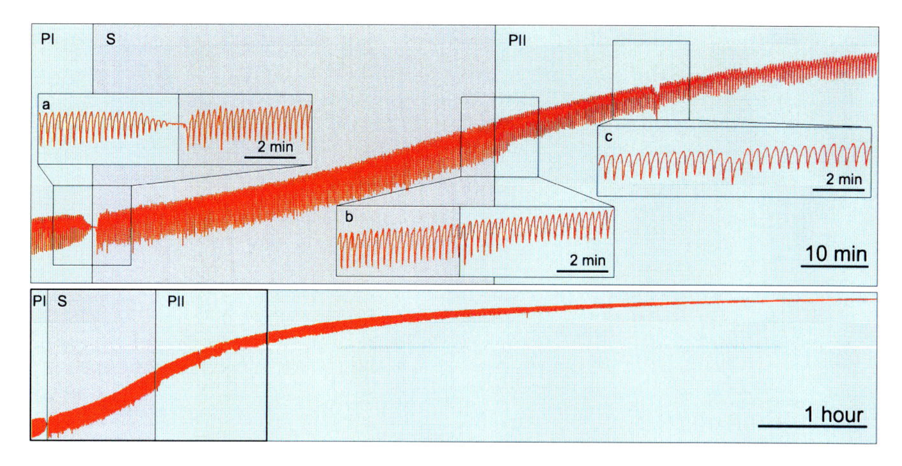

Fig. 5.3 Evolution of periodic color change with a 60 minute stirring phase. Lettering as in Fig. 5.1. *Lower panel*: full record of the periodic color change. *Upper panel*: Magnification of *bold-framed box* from *lower panel* of this figure. Framed boxes (**a**)–(**c**) are magnified in insets of this panel (**a**): Magnification shows the transition from phase PI to S. (**b**): Magnification shows the transition from phase S to PII. Note that there is a smooth transition without the appearance of a chaotic phase. (**c**): Magnification shows a 'shift' where the mean transmittance decreases for a short time (details in Sect. 5.3)

time is more important then the stirring rate for this change. Moreover there exists a threshold value in stirring time (more than 30 minutes and less or equal to 60 minutes) for which the chaotic (inter)phase (CH) disappears. If the stirring time is only 30 minutes the chaotic phase is still present, but compared to the non-stirred case it is shorter (approx. 35 %); if the stirring phase is 60 minutes the phase is not present anymore. In that sense the stirring time plays the role of a 'bifurcation parameter' (see difference between Fig. 5.2 middle and upper panel).

We do not yet have a quantitative model describing the complex behavior we have observed in experiment. However we have developed a first qualitative model, which was presented at the European Conference on Complex Systems 2011 [9, 10] and is still one of our key topic of interest to further investigating the Belousov Zhabotinsky reaction.

Acknowledgements Intensive and fruitful discussions with Christian Wodlei and Oliver Pirker are gratefully acknowledged. We thank Sampurna Chaterjee (Max Planck Institute of Neurological Research, Cologne, Germany) and Debabrata Deb (Department of Chemical and Petroleum Engineering, University of Pittsburgh, USA) for the English language support and also for productive discussions.

References

1. Belousov BP (1959) A periodic reaction and its mechanism. Sb Ref Radiat Med 1:145 (in Russian)

2. Zhabotinsky AM (1964) The periodic course of the oxidation of malonic acid in a solution. Biofizika 9(3):306–311 (in Russian)
3. Noyes RM, Field RJ, Körös E (1972) Oscillations in chemical systems. I. Detailed mechanism in a system showing temporal oscillations. J Am Chem Soc 94(4):1394–1395
4. Field RJ, Körös E, Noyes RM (1972) Oscillations in chemical systems. II. Thorough analysis of temporal oscillation in the bromate-cerium-malonic acid system. J Am Chem Soc 94(25):8649–8664
5. Prigogine I, Glansdorff P (1971) Thermodynamic theory of structure, stability and fluctuations. Wiley, New York
6. Scott SK (1993) Chemical chaos. Clarendon, Oxford
7. Rustici M, Branca M, Caravati C, Marchettini N (1996) Evidence of a chaotic transient in a closed unstirred cerium catalyzed Belousov-Zhabotinsky system. Chem Phys Lett 263:429–434
8. Rossi F, Budroni MA, Marchettini N, Cutietta L, Rustici M, Liveri MLT (2009) Chaotic dynamics in an unstirred ferroin catalyzed Belousov–Zhabotinsky reaction. Chem Phys Lett 480:322–326
9. Wodlei F, Hristea MR (2011) Stirring effect on the Belousov Zhabotinsky reaction. In: Thurner S, Szell M (eds) Book of abstracts ECCS'11, Vienna, p 136
10. Contribution to ECCS'11. http://www.ilsr.at/eccs11_stirring_effect.pdf

Chapter 6
Size Distribution of Barchan Dunes by a Cellular Dune Model

Atsunari Katsuki

Abstract Barchans, which are crescent sand dunes, are observed in desert and on the surface of the Mars. They form barchan field through interaction such as collision processes. In order to investigate dynamics of barchan field, we used cellular dune model. The model includes only saltation and avalanche as the basic sand transport processes. We succeed to reproduce a few hundred of barchans in a numerical simulation. The size of barchans grows more and the number of them is less. Also the size distribution has long-time tail like log-normal distribution.

Keywords Sand dune · Barchan

6.1 Introduction

Sand dunes are found in various places such as desert on the Earth and on the surface of Mars. They are formed through interplay between sand and air flow. Morphological shapes of dunes are determined by the directional variability of flow and the amount of available sand on the ground [1]. In the case of the unidirectional wind flow over a year and the insufficient amount of sand for covering the entire bedrock, barchans are observed. The shape of a barchan is characterized by the two horns that point downwind and the slip face among them [2, 3]. The slip face is a steeps slope formed by an avalanche.

Its dynamics of dune geomorphology has studied by using both laboratory experiments [4, 5] and numerical simulations[6–11]. Most of these researches mainly focused on the morphology and interaction among a few number of barchans. On the other hand, much more still about understanding of barchan field remains to be done. Hastenrath [12] has reported from field measurements that size distribution of dune has a Gaussian distribution. The amount of statistics is not sufficient. Hersen et al. also has reported that size and spacing between barchans are well selected

A. Katsuki (✉)
College of Science and Technology, Nihon University, Chiba 274-8501, Japan
e-mail: katsuki@phys.ge.cst.nihon-u.ac.jp

T. Gilbert et al. (eds.), *Proceedings of the European Conference on Complex Systems 2012*, Springer Proceedings in Complexity, DOI 10.1007/978-3-319-00395-5_6,

[13]. Duran et al. has measured the size distribution of barchan in some places [14]. They found a log-normal distribution of size of barchans. Thus we reproduce a lot of barchans in a numerical field and investigate a size distribution.

6.2 Model

We introduce a numerical model of barchans. It is necessary to keep calculation costs to a minimum in order to reproduce a lot of barchans. Since most of previous dune models have performed on calculation for fluid flow and sand movement, it takes high calculation costs. Thus we simplify these complicated processes down to a few simple steps. Our model is basically a variant of cell models [7, 15] and takes into account only saltation and avalanche as the element processes of a barchan. Although this model has been simplified considerably, it can reproduce many realistic features of dunes [11, 16].

The dune field is divided into square cells. Each cell represents an area of sandy ground that is sufficiently larger than a sand grain. With regard to the basic concept of a cell model, we do not consider a detailed structure inside the cell. A field variable $h(x, y, t)$ that expresses the local surface height is assigned to each cell; t denotes discrete time step and spatial coordinates x and y denote the central position of each cell in the flow direction and in the lateral direction, respectively. The edge length of a cell is considered as a unit length. It should be noted that while x, y, and t are discrete variables, $h(x, y, t)$ assumes a continuous value.

Saltation is the transported process of sand grains by the flow. We model this process by a simple transportation rule without considering fluid dynamical details. The saltation length and saltation mass are denoted by L_S and q_S, respectively. Here, the saltation mass is the volume of sand transferred from one cell to another. Since the area of a cell is unity, q_S represents the change in the height of a cell after saltation. In each time step of the simulation, the saltation mass q_S is shifted from a cell (x, y) to the leeward cell $(x + L_S, y)$. We further assume that the saltation occurs only on an upwind face. The saltation length Ls and the amount of transported sand q_S are modeled by the following rules,

$$L_S = a + bh(x, y, t) - ch^2(x, y, t), \tag{6.1}$$

$$q_S = d, \tag{6.2}$$

where $a = 1.0$, $b = 1.0$, $c = 0.01$ and $d = 0.1$ are phenomenological parameters. In form (6.1), the second term represents that sand are transported farther as the height of sand surface at the take-off cell is higher. The last term is introduced for Ls to saturate at a certain value. Equation (6.1) is used only in the range where L_S is an increasing function of $h(x, y, t)$. The saltation mass q_S is fixed as 0.1 for simplicity. Avalanche is another process in which sand slides down the steepest slope when the angle of the local slope exceeds the angle of repose. This procedure is repeated until

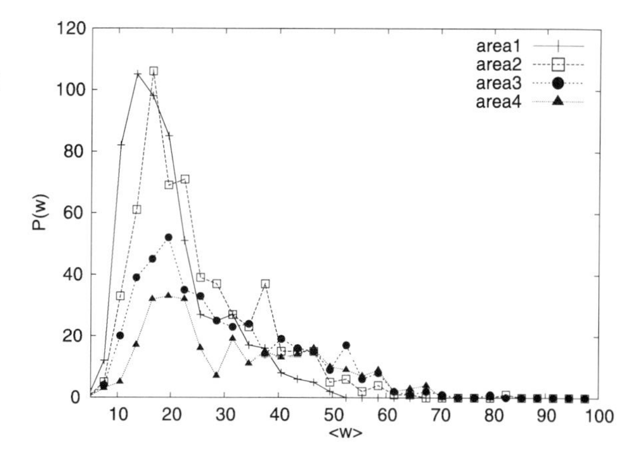

Fig. 6.1 Size distribution of barchan dunes ($\phi = 0.009$). The upwind area (area 1) and the downwind area (area 4) shows "*plus*" and "*triangle*", respectively

all the cells satisfy the stability condition. The angle of repose is fixed at 34° in the present simulations.

6.3 Results and Discussion

We reproduced a few hundred of barchans by a numerical simulation using the above model. The supplied sand from an upwind boundary has constant density ϕ. We investigate about size distribution. The width as the size of the barchan is defined as the distance between two horns. In the numerical field, there are some curious dune like deformed barchans, temporarily. Since it is known that the width and the length of barchan is proportional to height, we exclude the deformed barchans which is far from the scaling 10 %.

In order to investigate time evolution of the size distribution with $\phi = 0.009$, we divides the numerical field to four parts and have studied size distribution in each area (Fig. 6.1). Barchans are formed by self-organization. They are larger through the barchans collisions and the supplied sand, decreasing the number density of them. The large barchan is sometimes divided by collision with small barchan. In the result, small barchans are observed around a large barchan and the size distribution of all barchans in the downwind area become like log-normal distribution. This log-normal distribution is consistent to field measurement by Duran et al. [14].

6.4 Conclusion

We reproduced the barchan field by a coarse-grained dune model, which has only two processes, saltation and avalanche. Next we investigate time evolution of size distribution of barchans. At the upwind area, small barchans emerge and become larger through supplied sand and collision process. The large barchan is divided by collision with small one and becomes smaller. In the result, the size distribution of barchans is like log-normal distribution.

References

1. Wasson RJ, Hyde R (1983) Factors determining desert dune type. Nature 304:337–339
2. Bagnold RA (1941) The physics of blown sand and desert dunes. Methuen, London
3. Cooke R, Warren A, Goudie A (1993) Desert geomorphology. UCL Press, London
4. Hersen P, Douady S, Andreotti B (2002) Relevant length scale of barchan dunes. Phys Rev Lett 89:264301
5. Endo N, Taniguchi K, Katsuki A (2004) Observation of the whole process of interaction between barchans by flume experiments. Geophys Res Lett 31:L12503
6. Wippermann FK, Gross G (1985) The wind-induced shaping and migration of an isolated dune: a numerical experiment. Bound-Layer Meteorol 36:319–334
7. Nishimori H, Ouchi N (1993) Formation of ripple patterns and dunes by wind-blown sand. Phys Rev Lett 71:197–200
8. Lima AR, Sauermann G, Herrmann HJ, Kroy K (2002) Modelling a dune field. Physica A 310:487–500
9. Andreotti B, Claudin P, Douady S (2002) Selection of dune shapes and velocities. Part 2: A two-dimensional modelling. Eur Phys J B 28:341–352
10. Schwämmle V, Herrmann HJ (2003) Solitary wave behavior of sand dunes. Nature 426:619–620
11. Katsuki A, Nishimori H, Endo N, Taniguchi K (2005) Collision dynamics of two barchan dunes simulated by a simple model. J Phys Soc Jpn 74:538
12. Hastenrath SL (1967) The barchans of the Arequipa region, Southern Peru. Z Geomorphol 11:300–311
13. Hersen P, Andersen KH, Elbelrhiti H, Andreotti B, Claudin P, Douady S (2004) Corridors of barchan dunes: stability and size selection. Phys Rev E 69:011304
14. Duran O, Schawaemmle V, Lind PG, Herrmann HJ (2007) How barchan dunes distribute over deserts. arXiv:cond-mat/0701367
15. Werner BT (1995) Eolian dunes: computer simulations and attractor interpretation. Geology 23:1107–1110
16. Katsuki A, Kikuchi M, Endo N (2005) Emergence of a Barchan Belt in a unidirectional flow: experiment and numerical simulation. J Phys Soc Jpn 74:878

Chapter 7
Experimental Study of Buoyancy-Driven Instabilities Around Acid-Base Reaction Fronts

L. Lemaigre, L.A. Riolfo, and A. De Wit

Abstract The interplay between hydrodynamics and chemistry can give rise to complex non linear dynamics. To study how a simple $A + B \rightarrow C$ reaction can affect buoyancy-driven instabilities, we experimentally investigate convective flows appearing at the miscible interface between a solution of a reactant A put on top of a solution of another reactant B in the gravity field when a reaction takes place. The main observation is that the symmetry of the hydrodynamic patterns is drastically modified by the chemistry.

7.1 Introduction

The interplay between hydrodynamics and chemistry can give rise to complex non linear dynamics [1]. For instance, when a hydrodynamic instability appears between two reacting fluids, the flow will bring the reactants in contact and enhance the reaction, which, in turn, will modify the concentrations and affect the flow. In order to better understand the different mechanisms involved in such complex phenomena, it is important to study simple model systems. To study how a simple $A + B \rightarrow C$ reaction can affect buoyancy-driven instabilities, we experimentally investigate convective flows appearing at the miscible interface between a solution of a reactant A put on top of a solution of another reactant B in the gravity field when a reaction takes place.

Let us first review the instabilities taking place in non reactive systems. The convective instabilities affecting a stratification of a fluid A on top of a miscible fluid B in the gravity field are classified in three regimes [2]. The Rayleigh-Taylor (RT) instability occurs when a denser fluid A is put in top of a less dense fluid B. The interface deforms into rising and sinking "fingers" (Fig. 7.1(b)) hence also the name "density fingering" sometimes attributed to this convective mode.

If ρ_A the density of fluid A is lower than ρ_B, the density of fluid B, i.e. if we start from an initially statically stable stratification, instabilities can nevertheless occur because of differential diffusion effects. These can take place if D_A and

L. Lemaigre (✉) · L.A. Riolfo · A. De Wit
Nonlinear Physical Chemistry Unit, Service de Chimie Physique et Biologie Théorique, Faculté des Sciences, Université Libre de Bruxelles (ULB), CP 231, 1050 Brussels, Belgium

T. Gilbert et al. (eds.), *Proceedings of the European Conference on Complex Systems 2012*, Springer Proceedings in Complexity, DOI 10.1007/978-3-319-00395-5_7,

D_B, the diffusion coefficients of A and B are sufficiently different i.e. if their ratio $\delta = D_B/D_A \neq 1$. If $\delta > 1$, the so-called double-diffusive (DD) instability (also commonly named "salt fingering" because of its genericity in ocean dynamics [3]) can be triggered. In this case symmetric ascending and descending fingers also develop across the initial contact line (Fig. 7.1(a)) [2, 4, 5].

In order to study how a chemical reaction can affect these hydrodynamic instabilities, we have experimentally studied the dynamics obtained when the solutes react according to a simple $A + B \rightarrow C$ reaction. The main observation is that the symmetry of the hydrodynamic patterns is drastically modified by the chemistry.

The coupling between an acid-base reaction and buoyancy-driven instabilities has been already partly characterized experimentally for immiscible fluids [6] in a Hele-Shaw cell. This cell is analogous to a porous medium and can be approximated to a 2-dimensional system. The two reactant solutions are injected one on top of the other inside a vertical Hele-Shaw cell in order to obtain a flat contact line. As the cell is composed of two glass plates, the dynamics can easily be visualized by shadowgraphy [7]. This optical technique allows to visualize the variations in refractive index, which are in turn related to the composition of the fluid. It is particularly interesting to visualise colourless solutions. It has indeed been shown that the presence of a colour indicator influences the dynamics [8, 9].

7.2 Results

In a first set of experiments, we start with less dense NaOH on top of a denser solution of HCl (Fig. 7.1(c)). After an induction time of the order of the minute, fingers start to rise from the initial contact line. These fingers merge, then move away from each other and some new fingers appear. The influence of the concentration has been investigated by varying the concentration of the base while keeping the concentration of the acid constant. As the concentration of the base increases, the fingers appear and grow faster, indicating that the system becomes more and more unstable. This result has been confirmed by measuring the length of the fingers by image processing.

As the acid diffuses faster than the base, we are here in the conditions to have a double diffusive instability if there were no chemical reaction. Indeed, this type of instability appears when the upper solute has a lower molecular diffusion coefficient than the lower solute, the lower solution being the denser one [3]. In the non-reactive case, fingers grow symmetrically above and under the contact line (Fig. 7.1(a)). However, in the reactive case the fingers only develop on one side of the contact line. The reaction thus breaks the symmetry of the double diffusive patterns (Compare Figs. 7.1(a) and 7.1(c)).

This breaking of symmetry has also been observed in the case of a Rayleigh-Taylor instability, i.e. when the overlying NaOH solution is denser. Again, a breaking of symmetry is observed (Fig. 7.1(d)). However, when the density of the overlying solution is even more increased, a symmetrical Rayleigh-Taylor is recovered.

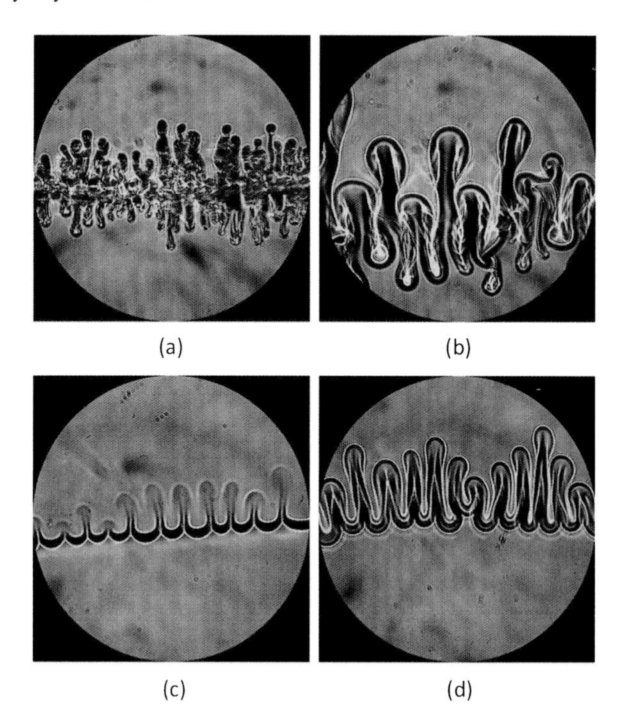

Fig. 7.1 Convective patterns observed for: (**a**) non-reactive double diffusion, (**b**) non-reactive Rayleigh-Taylor, (**c**) reactive double diffusion, (**d**) reactive Rayleigh-Taylor instabilities. Field of view of each image = 2.5 cm

It must be noted that double diffusion and Rayleigh-Taylor are not the only possible buoyancy-driven instabilities. When the overlying solute has a larger molecular diffusion coefficient than the lower solute and the lower solution is denser, fingering appears because of diffusive layer convection. The interplay between this instability and the acid-base reaction involving NaOH and HCl has already been studied in the literature [10–12]. A breaking of symmetry due to the chemical reaction has also been observed in that case.

During this study, the convective patterns induced by the coupling between an acid-base reaction and buoyancy-driven instabilities have been studied experimentally. Future research will focus on performing linear stability analysis and non linear simulations of a related reaction-diffusion-convection model in order to study the stability conditions as well as the nonlinear dynamics at longer times.

Acknowledgements We acknowledge Prodex, the ITN—Marie Curie—Multiflow network and FRS-FNRS for financial support.

References

1. De Wit A (2008) Chemo-hydrodynamic patterns and instabilities. Chim Nouv 99:1–7
2. Trevelyan PMJ, Almarcha C, De Wit A (2011) Buoyancy-driven instabilities of miscible two-layer stratifications in porous media and Hele-Shaw cells. J Fluid Mech 670:38–65
3. Turner JS (1979) Buoyancy effects in fluids. Cambridge University Press, Cambridge

4. Cooper CA, Glass RJ, Tyler SW (1997) Experimental investigation of the stability boundary for double-diffusive finger convection in a Hele-Shaw cell. Water Resour Res 33:517–526
5. Pringle SE, Glass RJ (2002) Double-diffusive finger convection: influence of concentration at fixed buoyancy ratio. J Fluid Mech 462:161–183
6. Shi Y, Eckert K (2006) Acceleration of reaction fronts by hydrodynamic instabilities in immiscible systems. Chem Eng Sci 61:5523–5533
7. Settles GS (2001) Schlieren and shadowgraph techniques. Springer, Berlin
8. Almarcha C, Trevelyan PMJ, Riolfo LA, Zalts A, El Hasi C, D'Onofrio A, De Wit A (2010) Active role of a color indicator in buoyancy-driven instabilities of chemical fronts. J Phys Chem Lett 1:752–757
9. Kuster S, Riolfo LA, Zalts A, El Hasi C, Almarcha C, Trevelyan PMJ, De Wit A, D'Onofrio A (2011) Differential diffusion effects on buoyancy-driven instabilities of acid-base fronts: the case of a color indicator. Phys Chem Chem Phys 13:17295–17303
10. Zalts A, El Hasi C, Rubio D, Ureña A, D'Onofrio A (2008) Pattern formation driven by an acid-base neutralization reaction in aqueous media in a gravitational field. Phys Rev E 77:015304
11. Almarcha C, Trevelyan PMJ, Grosfils P, De Wit A (2010) Chemically driven hydrodynamic instabilities. Phys Rev Lett 104:044501
12. Almarcha C, R'Honi Y, De Decker Y, Trevelyan PMJ, Eckert K, De Wit A (2011) Convective mixing induced by acid-base reactions. J Phys Chem B 115:9739–9744

Chapter 8
Dynamical Trap Effect in Virtual Stick Balancing

Arkady Zgonnikov, Ihor Lubashevsky, and Maxim Mozgovoy

Abstract We present the experimental evidence of the dynamical traps model describing the human fuzzy rationality in the dynamical systems framework. The results of the experiments on virtual stick balancing are compared to the results of the previous studies on the dynamical trap effect. According to the results obtained, we suggest that the dynamical traps model actually captures certain essential features of human fuzzy rationality and therefore may serve as an alternative to the traditional notion of stable equilibrium in describing the behavior of human as an element of complex social systems.

Keywords Mathematical modeling · Emergence mechanism · Human fuzzy rationality · Dynamical traps

8.1 Introduction

Wide variety of physical formalism and notions have been used recently in describing social systems and behavior of human as a part of such systems (e.g., see [1, 2]). Particularly, the basic concepts of Newtonian mechanics are commonly applied in the modeling of traffic flow and motion of groups of animals (fish schools, bird flocks, etc.) [3]. The notions of master equation and Hamiltonian as an energy function were used in the theory of opinion dynamics and the dynamics of culture and language (e.g., [4]). Among other concepts of physics that are widely used in social systems analysis are fluid dynamics, Ginsburg-Landau equations and reaction-diffusion systems (e.g., [5]).

Despite aforementioned advances, one can still note that the mathematical theory of human behavior in social systems is far from being developed well. Apparently, inanimate objects under consideration of Newtonian physics differ substantially in its nature from animate beings, since such features as motivation, morale, memory, learning, etc. are inapplicable to the former. So we may assume that the corre-

A. Zgonnikov (✉) · I. Lubashevsky · M. Mozgovoy
Complex Systems Modeling Lab, University of Aizu, Tsuruga, Ikki-machi, Aizu-Wakamatsu City, Fukushima 965-8560, Japan

T. Gilbert et al. (eds.), *Proceedings of the European Conference on Complex Systems 2012*, Springer Proceedings in Complexity, DOI 10.1007/978-3-319-00395-5_8,

sponding mathematical formalism still should be developed in the domain of social systems in addition to the existing notions derived from the physical ones.

One of such notions widely met throughout probably all branches of physics is a fixed-point attractor, or stable equilibrium point; it is also commonly used in social psychology [6] as like as the notions of periodic attractor and latent attractor. Nevertheless, social objects and systems in the real world demonstrate anomalous dynamics and irregular behavior which often cannot be reduced to established patterns like equilibrium points or limit cycles. The development of individual, specialized notions accounting for the peculiarities of human beings may enable us to better describe and understand complex social systems involving human as a key element.

Let us consider a hypothetical dynamical system controlled by the operator whose purpose is to stabilize the system near an equilibrium point. We assume that the operator does not react on the small deviations from this equilibrium, though these variations are clearly recognized by her perception. In other words, the operator is comfortable with the deviations of a small magnitude. Thus, until the variation becomes large enough, the operator prefers not to intervene the system dynamics. Therefore, any point from a certain neighborhood of the equilibrium one is treated equally by the operator. This assumption is in fact due to the phenomenon of human fuzzy rationality [7]. In the present paper we discuss the notion of dynamical traps which was previously introduced in order to mimic this feature of the bounded capacity of human cognition [8]. Considering the series of virtual experiments we discover the evidence of the dynamical trap effect presence in the human behavior during the stick balancing process. The results obtained may be treated as a step towards understanding the nature of various anomalous phenomena caused by human imperfect rationality in complex social systems.

Dynamical Traps In order to illustrate the dynamical traps concept, let us assume that considered hypothetical dynamical system is described by the following equations

$$\begin{aligned} \dot{x} &= y, \\ \dot{y} &= \Omega(x, y) F(x, y) + \xi(t), \end{aligned} \tag{8.1}$$

and has an equilibrium in the origin of coordinates (0,0). Here $\Omega(x, y)$ stands for the dynamical trap effect, $F(x, y)$ is the sum of the regular forces (implicitly including human control) and $\xi(t)$ is the random factor. $\Omega(x, y)$ could be defined as follows

$$\begin{aligned} \Omega(x, y) &\approx 0 \quad \text{if } (x, y) \in \mathbb{Q}_{\text{tr}}, \\ \Omega(x, y) &= 1 \quad \text{otherwise}, \end{aligned}$$

where $\mathbb{Q}_{\text{tr}}$ is a certain vicinity of the equilibrium point.

In order to explain the meaning of cofactor $\Omega(x, y)$ we consider the behavior of the operator who is approaching the desired phase space position ($x = 0$, $y = 0$). Let us assume that if the current position is far from the origin, the operator perfectly follows the optimal control strategy. If the current position is recognized

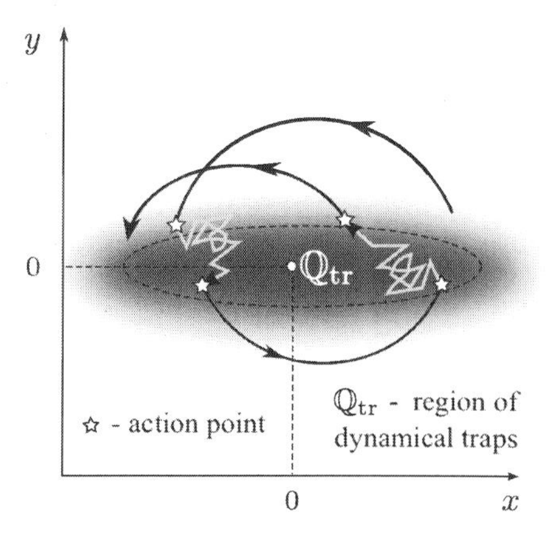

Fig. 8.1 The structure of phase space of system (8.1)

by the operator as "good enough" ($(x, y) \in \mathbb{Q}_{tr}$, i.e. it may not be strictly optimal) due to her fuzzy rationality, she halts active control over the system so that system dynamics is stagnated in certain vicinity of the desired position. Therefore, $\mathbb{Q}_{tr}$ is called the region of dynamical traps. The structure of the described system phase space is presented on Fig. 8.1.

The investigation of the dynamical traps model was originally inspired by a class of intrinsic cooperative phenomena found in the dynamics of vehicle ensembles on highways [9]; later it was shown that the dynamical trap effect could cause emergent phenomena in the chain of oscillators mimicking the interaction of motivated objects [8, 10]. Among other results obtained in [10], it was demonstrated numerically that the "motivated" oscillator from the particle chain under the presence of dynamical trap forms the specific phase space trajectories (see Fig. 8.2). The phase variables distributions were shown to take non-Gaussian forms. The reviewed results demonstrate that the dynamical trap effect could be responsible for establishing of complex patterns of the system motion near the equilibrium point.

Inspite of these achievements up to now there were no experimental evidences of the dynamical trap effect existence in the real world. The purpose of the current work is to provide an experimental background to the theoretical framework developed earlier by comparing the results of previous studies on the dynamical trap effect and the results of the series of experiments aimed at elucidation of some characteristics of human fuzzy rationality in stick balancing task.

In order to exemplify theoretical studies on the dynamical trap effect we consider the process of the inverted pendulum balancing by human. The task of dynamic stick balancing has been investigated widely from various perspectives; studies on both real-world and virtual experiments are available (see, e.g., [11, 12]). However, attention is mainly paid to the in-depth understanding of the mechanical and psychomotor aspects of the human control, while we aim to provide an experimental background to the simple model of human cognition which may be useful in the modeling of complex systems where human decisions play crucial role.

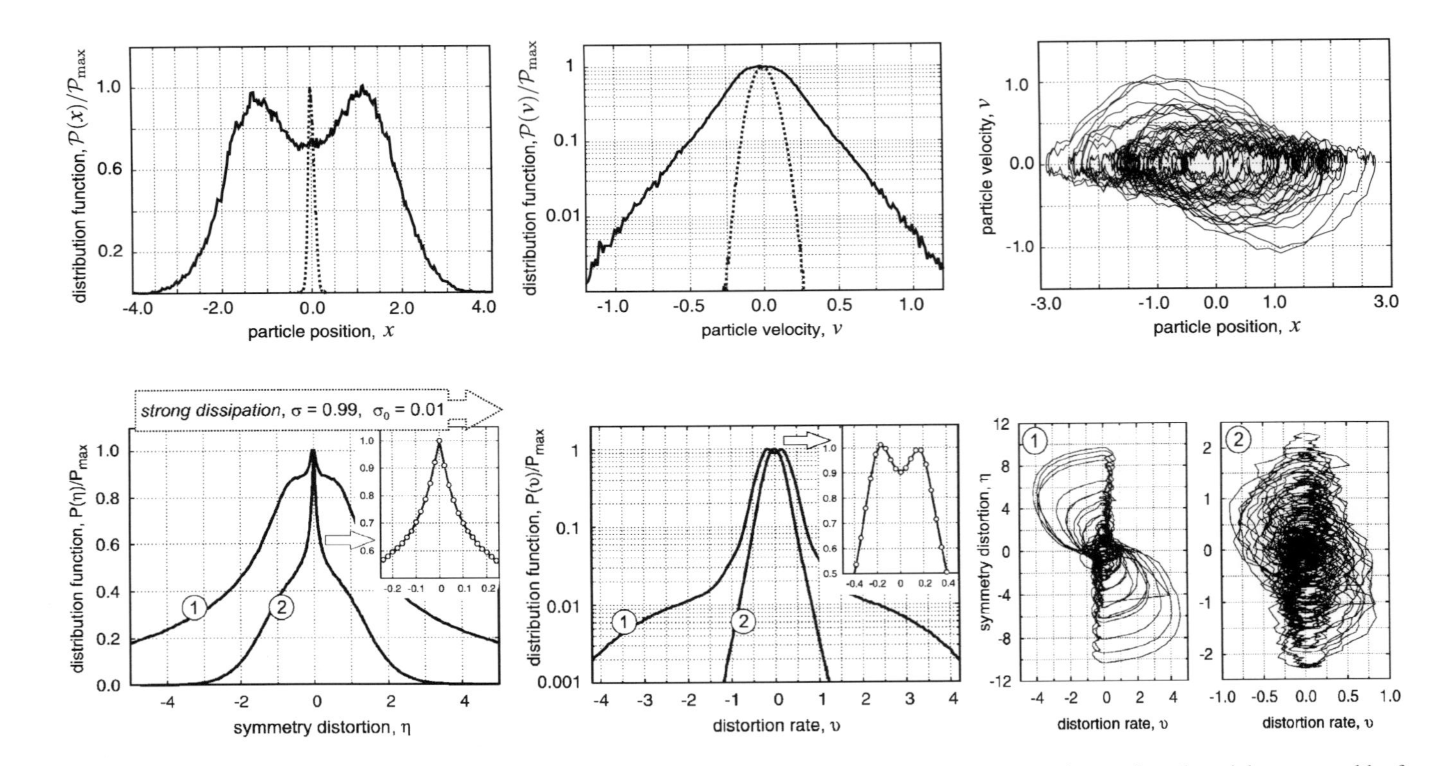

Fig. 8.2 Phase variables density functions and phase trajectories of a single particle from the ensemble of interacting motivated particles governed by fuzzy rationality [10]. *The top three diagrams* correspond to the case of the single moving particle oscillating between two fixed neighbors. On the density diagrams *solid lines* correspond to the case of strong dynamical trap effect and *dotted lines* match the absence of dynamical trap. *Four bottom diagrams* depict phase portraits and density functions of the particle in the middle of 1000-particle chain with low (label 1) and high (label 2) density of particles

8.2 Virtual Experiments

In present work we focus on the computer-based simulation of two-dimensional inverted pendulum motion in viscous environment. Real-world motion capture-based stick balancing experiments were also held, as well as virtual experiments simulating stick balancing in non-viscous environment. Preliminary analysis of the experimental data demonstrated that the corresponding dynamical systems exhibit more complex behavior than the system currently under consideration due to the increased number of phase space variables. Therefore its detailed analysis requires an individual study and does not fall under the scope of present work.

The mechanical system under consideration is described by the following dimensionless mathematical model:

$$\tau\dot{\theta} = \sin\theta - Av(t)\cos\theta. \tag{8.2}$$

Here phase space variable θ is the angle between the stick and the vertical axis, τ is a time scale parameter characterizing the operator perception and the right-hand part of the equation represents the sum of friction and gravity force moments. $v(t)$ stands for the velocity of platform motion which is actually the control parameter affected by system operator while A is constant amplifying coefficient of control effort.

It is notable that the phase space of system (8.2) should comprise not only angle θ but also its derivative $\dot{\theta}$. This assumption is due to the fact that the operator controlling the system evidently perceive the angular velocity of the stick and regulates the value of control effort $v(t)$ based on the current values of both factors. Hence, the system dynamics is determined not only by the stick angle but by the angular velocity as well. The similar approach of the phase space extension was previously proposed in the studies on the car following theory [9] where "position-velocity" phase space was extended by the acceleration as the third independent phase variable.

We developed a simple tool that implements the model described above. The operator has to maintain the angle between the virtual stick and the vertical axis near unstable equilibrium position $\theta_{eq} = 0$ by moving the platform via computer mouse. The total number of subjects participating in the experiment was 12, including both male and female students and professors of different nationalities. Therefore, we achieved participants diversity in nationality, gender and age in order to make the experimental group more or less representative.

A few sessions of the experiments were held. During each session subjects had to control virtual inverted pendulum for the time period of 5 to 20 minutes after 5-minutes adaptation period. To prevent the fatigue effect, sessions were held on the different days. For each participant we have acquired at least three sets of data for various durations of control process.

The numerical data captured from each subject was analyzed separately. It was found that after a short period of adaptation each participant mainly starts to follow the simple strategy of system control:

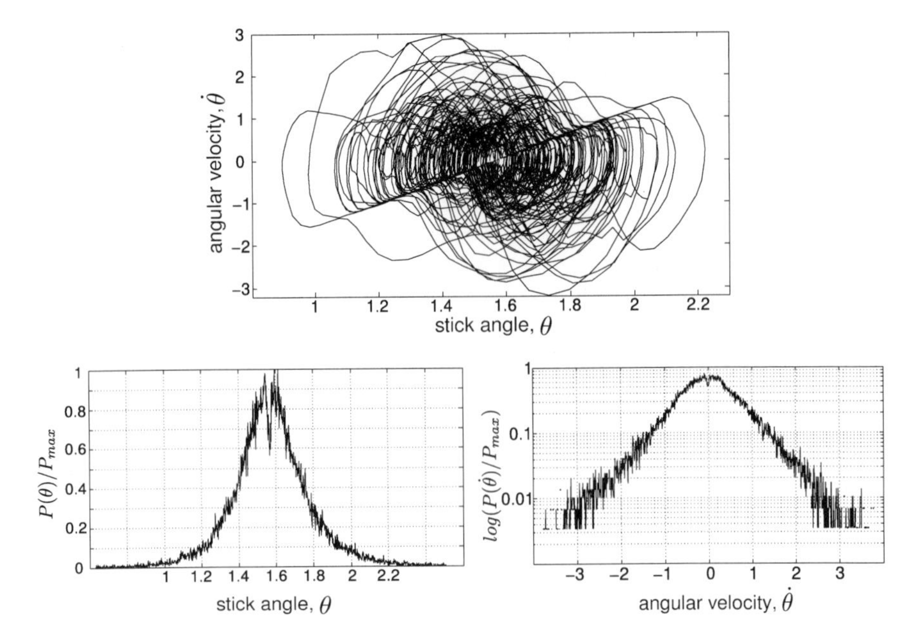

Fig. 8.3 Phase portrait and phase variables distribution (normalized) of the virtual inverted pendulum motion in viscous environment under the control of male student during the time interval of 5 minutes, $\tau = 0.3$, $A = 0.5$. The distribution of the angular velocity is represented in logarithmic scale

- wait until the angle or angular velocity of the stick exceeds certain threshold;
- correct the platform position so that the angular velocity is damped and the stick position is approximately vertical and so on.

From each raw data set obtained we extracted the data required to visualize the phase space trajectories of the inverted pendulum motion in "angle-angular velocity" phase space. It was discovered that the phase portraits of the system under human control are extremely similar in their basic properties for all participants and for any considered process duration. It is notable that though the average magnitude of deviation from the equilibrium and the average time of continuous balancing vary from one subject to another, the structure of phase portrait is stable within the whole group, as like as the probability distribution functions of the phase space variables θ and $\omega = \dot{\theta}$. Figure 8.3 represents the typical system motion trajectory and corresponding distribution functions.

8.3 Discussion

Surprisingly, the structure of the phase trajectory produced by the virtual stick under human control is quite similar to the ones of a single oscillator from the particle ensemble studied in [10] (see Fig. 8.2). The certain dissimilarity of the trajectories in the neighborhood of the equilibrium points is probably due to the fact that these equilibria are of different nature; the oscillators in the chain are artificially exposed to the external white noise disturbance, while the virtual stick is itself unstable at $\theta = 0$.

This comparison may lead us to the assumption that human behavior during the process of two-dimensional inverted pendulum balancing is in some sense analogous to the behavior of the particle balancing its position between two neighbors described by the dynamical trap model. Furthermore, the analysis of the system (8.2) phase variables distributions revealed that the probability density functions for both variables have anomalous bimodal form (see Fig. 8.3), as like as the corresponding functions (Fig. 8.2) found during the analysis of the system of interacting oscillators in [10]. Besides, the form of the angular velocity distribution found is like cusp $\propto \exp(-|\omega|)$, which again is anomalous and highly analogous to the results of the previous studies on the dynamical trap effect. All these facts could be considered as the first experimental evidence of the dynamical trap existence in the real world.

We may therefore expect that the bounded capacity of human cognition could be described by the proposed dynamical trap model, which in turn could be used for modeling of wide variety of complex systems comprising large numbers of interacting human beings. Moreover, one may even speculate that the results obtained give evidence to the fact that the standard notion of fixed-point attractor may not be applicable in dynamical systems where human role is crucial due to the phenomena of fuzzy rationality.

Acknowledgements The work was supported in part by the JSPS "Grants-in-Aid for Scientific Research" Program, Grant no. 245404100001.

References

1. Chakrabarti B, Chakraborti A, Chatterjee A (eds) (2006) Econophysics and sociophysics: trends and perspectives. Wiley, New York
2. Nowak A, Strawinska U (2009) Introduction to applications of physics and mathematics to social science. In: Meyers RA (ed) Encyclopedia of complexity and systems science. Springer, New York, pp 322–326
3. Castellano C, Fortunato S, Loreto V (2009) Statistical physics of social dynamics. Rev Mod Phys 81:591–646
4. Slanina F (2009) Physical models of social processes. In: Meyers RA (ed) Encyclopedia of complexity and systems science. Springer, New York, pp 8379–8405
5. Helbing D (2001) Traffic and related self-driven many-particle systems. Rev Mod Phys 73:1067–1141
6. Vallacher RR (2009) Applications of complexity to social psychology. In: Meyers RA (ed) Encyclopedia of complexity and systems science. Springer, New York, pp 8420–8435
7. Dompere KK (2009) Fuzzy rationality. Springer, Berlin
8. Lubashevsky I (2012) Dynamical traps caused by fuzzy rationality as a new emergence mechanism. Adv Complex Syst 15:1250045. doi:10.1142/S0219525912500452

9. Lubashevsky I, Wagner P, Mahnke R (2003) Bouded rational driver models. Eur Phys J B 32:243–247
10. Lubashevsky IA, Mahnke R, Hajimahmoodzadeh M, Katsnelson A (2005) Long-lived states of oscillator chains with dynamical traps. Eur Phys J B 44:63–70
11. Milton JG, Ohira T, Cabrera JL, Fraiser RM, Gyorffy JB et al (2009) Balancing with vibration: a prelude for "Drift and Act" balance control. PLoS ONE 4(10):e7427. doi:10.1371/journal.pone.0007427
12. Suzuki S, Harashima F, Furuta K (2010) Human control law and brain activity of voluntary motion by utilizing a balancing task with an inverted pendulum. Adv Hum-Comput Interact 2010:215825

Chapter 9
Bounded Capacity of Human Cognition as a New Mechanism of Instability in Dynamical Systems

Ihor Lubashevsky

Abstract A new emergence mechanism related to the bounded capacity of human cognition is considered. It assumes that individuals (operators) governing the dynamics of a certain system try to follow an optimal strategy in controlling its motion but fail to do this perfectly because similar strategies are indistinguishable for them. The main attention is focused on the systems where the optimal dynamics implies the stability of a certain equilibrium point in the corresponding phase space. In such systems the bounded capacity of human cognition gives rise to some neighborhood of the equilibrium point, the region of dynamical traps, wherein each point is regarded as an equilibrium one by the operators. So when a system enters this region and while it is located in it, maybe for a long time, the operator control is suspended. The present work draws on the results obtained previously as well as new ones and is mainly aimed at elucidating the basic principles in constructing a mathematical formalism describing this human feature. In particular, it is demonstrated that oscillator with dynamical traps can be derived within rather general assumptions about human behavior.

In the present *extended description* the main attention is focused on the reasons and motives for developing the concept of dynamical traps.

Keywords Bounded capacity · Human cognition · Dynamical traps · Instability · Emergence

9.1 Introduction

During the last decades there has been a great deal of modeling social systems and behavior of humans as a part of such systems using physical formalism (for a review see, e.g., articles of Encyclopedia [1]). In particular, the basic concepts of Newtonian mechanics are commonly applied to modeling traffic flow, motion of pedestrians, groups of animals (fish schools, bird flocks, etc.) (e.g., Refs. [2–4]), the

I. Lubashevsky (✉)
University of Aizu, Ikki-machi, Aizu-Wakamatsu, Fukushima 965-8560, Japan
e-mail: i-lubash@u-aizu.ac.jp

T. Gilbert et al. (eds.), *Proceedings of the European Conference on Complex Systems 2012*, Springer Proceedings in Complexity, DOI 10.1007/978-3-319-00395-5_9,

phenomenological theory of phase transitions was used to mimic jam formation in congested traffic [5], animal forging was imitated by random walks (for a review see Ref. [6]), probabilistic formalism of statistical physics was demonstrated to be useful in describing opinion dynamics, the dynamics of culture and language (e.g., Refs. [7–9]). The Lotka-Volterra model and the reaction-diffusion systems found their applications in stock market, income distribution, and population dynamics [10], the replicator equations were employed to simulate the moral dynamics [11].

One of basic notions concerned with directly in the present work and widely met throughout probably all the branches of physics is the stationary point, the system dynamics in the vicinity, and emergent phenomena occurring via its instability. The notion of a fixed-point attractor as a stable stationary point in the system dynamics that corresponds to some local minimum in a certain potential relief is widely met in social psychology [12]. The latter is extended even to collections of such fixed point attractors forming a basin. Besides, social psychology uses the notion of latent attractors, periodic attractors representing limit cycles, and deterministic chaos. In addition, the concept of synchronization of interacting oscillators was used to model social coordination [13].

In spite of these achievements we have to note that the mathematical theory of social systems is currently at its initial stage of development. Indeed, animate beings and objects of the inanimate world are highly different in their basic features, in particular, such notions as willingness, learning, prediction, motives for action, moral norms, personal and cultural values are just inapplicable to inanimate objects. This enables us to pose a question as to what *individual* physical notions and mathematical formalism should be developed to describe social systems in addition to the available ones inherited from modern physics. For example, Kerner's hypothesis about the *continuous* multitude of metastable states representing the synchronized phase of traffic flow, on one hand, stimulated developing the three-phase traffic model explaining a number of observed phenomena in congested traffic flow [14, 15]. On the other hand, a *microscopic* mechanism enabling the coexistence of many different metastable states actually at the same point of the corresponding phase space is up to now a challenging problem.

Previously the concept of dynamical traps was introduced to describe the bounded capacity of human cognition in evaluating events, actions, etc. according to their preference and its effects in governing a certain system or entity [16–18], which was partly stimulated by studying the car following dynamics for bounded rational drivers [19, 20]. When, for example, two actions are close to each other in quality from the standpoint of a person (operator) making a decision their choice may be random because he ought to consider them equivalent. The notion of dynamical traps accounts for this feature. In particular, dealing with a dynamical system in the phase space $\mathbb{R}_{xy}$ its stationary point $\{x = 0, y = 0\}$ being initially stable is replaced by a certain neighborhood $\mathbb{Q}_{tr}$ called the dynamical trap region such that when the system goes into $\mathbb{Q}_{tr}$ its dynamics is stagnated. This mimics vain actions of an operator in directing the system motion towards the point $\{0, 0\}$ precisely. Indeed, when the system under the operator control gets any point in $\mathbb{Q}_{tr}$ the operator may consider the current situation perfect because he just does not "see" the point

$\{0, 0\}$ and until the system leaves $\mathbb{Q}_{tr}$ he has no reason to keep the control active. Broadly speaking, it is an alternative to the notion of stationary point in dynamical systems [18]. We note that a concept of dynamical traps but of different nature is met in theory of relaxation oscillations [21] and Hamiltonian systems (see, e.g., Ref. [22]).

9.2 Dynamical Trap Model

The present work is devoted to the general principles in constructing the governing equations allowing for dynamical traps caused by human properties. By way of example, at the starting point of theory development it is assumed that if the operator was able to govern the system perfectly following a certain optimal strategy then its dynamics would be described by the coupled equations

$$\tau \frac{dx}{dt} = F_x(x, y), \qquad \tau \frac{dy}{dt} = F_y(x, y). \tag{9.1}$$

Here τ is a time scale characterizing the operator perception delay, the "forces" $F_x(x, y)$ and $F_y(x, y)$ are determined by *both* the physical regularities of the system mechanics and the active behavior of the operator in controlling the system motion. The origin $\{0, 0\}$ of the coordinate frame is placed at the equilibrium point of system (9.1), i.e., the equalities

$$F_x|_{\substack{x=0\\y=0}} = 0, \qquad F_y|_{\substack{x=0\\y=0}} = 0 \tag{9.2}$$

are assumed to hold. In this context the perfect rationality of the operator means his ability to locate precisely the current position of the system on the phase plane $\mathbb{R}_{xy}$, to predict strictly its further motion, and, then, to correct the current motion continuously. Exactly in this case it is possible to consider that the operator orders the strategies of behavior according to their preference and then chooses the optimal one. As a result the equilibrium point $\{0, 0\}$ must be stable when the aim of operator actions is to keep the system in close vicinity to this point (Fig. 9.1(b)).

The motion of the given system has been presumed to be a cumulative effect of the physical regularities and the operator actions. The notion of dynamical trap kernel implements this feature. Namely, the operator is considered to be able to halt the system motion at a certain multitude $\mathbb{L}$ of points in the phase space $\mathbb{R}_{xy}$ to be called the locus of partial equilibrium $\mathbb{L}_{pe}$ if after getting any point of $\mathbb{L}_{pe}$ the system will stay at it without furthers actions of the operator. In this case the system motion along $\mathbb{L}_{pe}$ is due to random factors or the operator action. If after the operator suspending his control over the system its dynamics becomes unstable with a relatively small increment, the multitude $\mathbb{L}$ is called the unstable kernel $\mathbb{L}_{uk}$. The two cases differ from each other only in the mechanism forcing the system to leave the dynamical trap region $\mathbb{Q}_{tr}$. In the former case it is random or uncontrollable motion of the system in the phase space $\mathbb{R}_{xy}$ (Fig. 9.1(c)), in the latter case it is

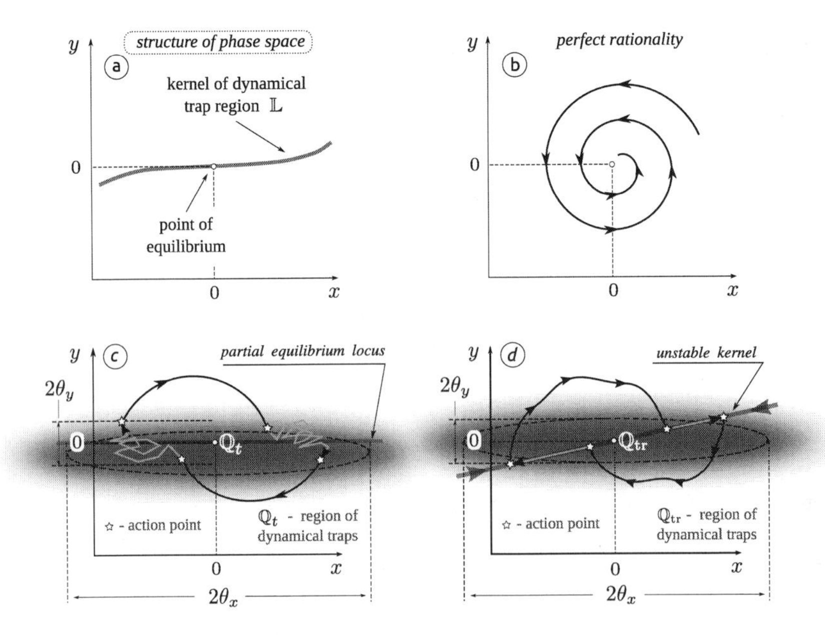

Fig. 9.1 The structure of the phase space $\mathbb{R}_{xy}$ of the system under consideration (**a**); a schematic illustration of its dynamics near the stable equilibrium point $\{0, 0\}$ in the cases of the perfect rationality (**b**), and illustration of system trajectories in the case of dynamical traps with a partial equilibrium locus (**c**) and a unstable kernel (**d**)

the weak system instability without the operator control (Fig. 9.1(d)). The given coordinate frame is chosen so that the partial equilibrium locus $\mathbb{L}_{pe}$ be tangent to the x-axis at the origin or the unstable kernel $\mathbb{L}_{uk}$ correspond to small variations in y. The system motion outside $\mathbb{Q}_{tr}$ is governed by the active behavior of the operator and the physical regularities. The points where the operator control is suspended or resumed are called the action points and treated as random events. In the given work the continuous approximation of action points is used within which the transition from active to passive behavior of the operator is mimicked by introducing noise $\xi(t)$ with amplitude ϵ and the factor

$$\Omega(x, y) = \frac{\Delta^2 + (x/\theta_x)^2 + (y/\theta_y)^2}{1 + (x/\theta_x)^2 + (y/\theta_y)^2} \tag{9.3}$$

such that in the region $\mathbb{Q}_{tr}$ it takes small value $\Delta^2 \ll 1$ whereas outside $\mathbb{Q}_{tr}$ it is about unity. The thresholds θ_x and θ_y determine the dimensions of the region $\mathbb{Q}_{tr}$, namely, $\mathbb{Q}_{tr} = |x| \lesssim \theta_x$ and $|y| \lesssim \theta_y$.

Considering, for example, systems with partial equilibrium locus the following model called the oscillator with dynamical traps

$$\ddot{x} = -\Omega(\dot{x})[x + \sigma\dot{x}] + \epsilon\xi(t). \tag{9.4}$$

is derived under rather general assumptions. Here the cofactor $\Omega(\dot{x})$

$$\Omega(\dot{x}) = \frac{\Delta^2 + (\dot{x})^2}{1 + (\dot{x})^2}, \tag{9.5}$$

describes the dynamical trap effect and σ can be treated as a certain viscous friction. Using model (9.4) it is demonstrated that dynamical traps can cause this oscillator to undergo non-equilibrium phase transitions of a new type where noise plays a constructive role whereas the regular "force" does not exhibit a change in its structure similar to the bifurcation of stationary points. In addition, the work briefly presents available experimental evidence of the dynamical traps obtained in balancing a virtual pole with strong friction.

Then the model of one oscillator with dynamical traps is generalized to a chain of such oscillators to analyze the emergent cooperative phenomena. It is demonstrated that due to the dynamical trap effect complex spatial-temporal patterns can arise, the distribution functions of the particle velocities and positions take anomalous forms with heavy tails and several scales, and it is possible to introduce the notion of dynamical phases which are treated as individual entities arising via self-organization. Special attention is paid to the systems with complex dynamics without noise. In this case a particle can leave the dynamical trap region only because of the mismatch between the operator actions. Besides, a sequence of phase transitions with anomalous properties caused by weak noise is discussed in brief.

9.3 Conclusion

The obtained results enable us to state that the bounded capacity of human cognition in ordering, e.g., events or actions can be regarded as a new mechanism of complex emergent phenomena in systems where human factor plays a crucial role. The mathematical formalism required for the description of this effect draws on the concept of dynamical traps being a certain alternative to the notion of stationary points in dynamical systems. It should pointed out that the dynamical trap effect can cause the system instability giving rise to emergent phenomena even in the case when the initial stationary point was stable.

Acknowledgements The work was supported in part by the JSPS "Grants-in-Aid for Scientific Research" Program, Grant 245404100001.

References

1. Meyers RA (ed) (2009) Encyclopedia of complexity and systems science. Springer, New York
2. Helbing D, Johansson A (2009) In: Encyclopedia of complexity and systems science. Springer, New York, p 6476
3. Schadschneider A, Klingsch W, Klüpfel H, Kretz T, Rogsch C, Seyfried A (2009) In: Encyclopedia of complexity and systems science. Springer, New York, p 3142

4. Helbing H (2001) Rev Mod Phys 73:1067
5. Mahnke R, Kaupužs J, Lubashevsky I (2005) Phys Rep 408:1
6. Bénichou O, Loverdo C, Moreau M, Voituriez R (2011) Rev Mod Phys 83:81
7. Castellano C, Fortunato S, Loreto V (2009) Rev Mod Phys 81:591
8. Slanina F (2009) In: Encyclopedia of complexity and systems science. Springer, New York, p 8379
9. Stauffer D (2009) In: Encyclopedia of complexity and systems science. Springer, New York, p 6380
10. Yaari G, Stauffer D, Solomon S (2009) In: Encyclopedia of complexity and systems science. Springer, New York, p 4920
11. Hegselmann R (2009) In: Encyclopedia of complexity and systems science. Springer, New York, p 5677
12. Vallacher RR (2009) In: Encyclopedia of complexity and systems science. Springer, New York, p 8420
13. Olivier OO, Kelso JAS (2009) In: Encyclopedia of complexity and systems science. Springer, New York, p 8198
14. Kerner BS (2009) In: Encyclopedia of complexity and systems science. Springer, New York, p 9302
15. Kerner BS (2004) The physics of traffic. Springer, Berlin
16. Lubashevsky I, Hajimahmoodzadeh M, Katsnelson A, Wagner P (2003) Eur Phys J B 36:115
17. Lubashevsky I, Mahnke R, Hajimahmoodzadeh M, Katsnelson AA (2005) Eur Phys J B 44:63
18. Lubashevsky I (2012) Adv Complex Syst 15:1250045
19. Lubashevsky I, Mahnke R, Wagner P, Kalenkov S (2002) Phys Rev E 66:016117
20. Lubashevsky I, Wagner P, Mahnke R (2003) Eur Phys J B 32:243
21. Lubashevsky IA, Gafiychuk VV, Demchuk AV (1998) Physica A 255:406
22. Zaslavsky GM (2005) Hamiltonian chaos and fractional dynamics. Oxford University Press, London

Chapter 10
Complex Systems with Trivial Dynamics

Ricardo López-Ruiz

Abstract In this communication, complex systems with a near trivial dynamics are addressed. First, under the hypothesis of equiprobability in the asymptotic equilibrium, it is shown that the (hyper) planar geometry of an N-dimensional multi-agent economic system implies the exponential (Boltzmann-Gibss) wealth distribution and that the spherical geometry of a gas of particles implies the Gaussian (Maxwellian) distribution of velocities. Moreover, two non-linear models are proposed to explain the decay of these statistical systems from an out-of-equilibrium situation toward their asymptotic equilibrium states.

Keywords Statistical models · Equilibrium distributions · Decay toward equilibrium · Nonlinear models

10.1 Introduction

In this paper, different classical results [1–4] are recalled. They are obtained from a geometrical interpretation of different multi-agent systems evolving in phase space under the hypothesis of equiprobability [5, 6]. Two nonlinear models that explain the decay of these statistical systems to their asymptotic equilibrium states are also collected [7, 8].

We sketch in Sect. 10.2 the derivation of the Boltzmann-Gibbs (exponential) distribution [6] by means of the geometrical properties of the volume of an N-dimensional pyramid. The same result is obtained when the calculation is performed over the surface of a such N-dimensional body. In both cases, the motivation is a multi-agent economic system with an open or closed economy, respectively.

Also, a continuous version of an homogeneous economic gas-like model [7] is given in the Sect. 10.2. This model explains the appearance, independently of the initial wealth distribution given to the system, of the exponential (Boltzmann-Gibbs)

R. López-Ruiz (✉)
Dept. of Computer Science, Faculty of Science and Bifi, Universidad de Zaragoza, 50009 Zaragoza, Spain
e-mail: rilopez@unizar.es

T. Gilbert et al. (eds.), *Proceedings of the European Conference on Complex Systems 2012*, Springer Proceedings in Complexity, DOI 10.1007/978-3-319-00395-5_10,

distribution as the asymptotic equilibrium in random markets, and in general in many other natural phenomena with the same type of interactions.

The Maxwellian (Gaussian) distribution is derived in Sect. 10.3 from geometrical arguments over the volume or the surface of an N-sphere [5]. Here, the motivation is a multi-particle gas system in contact with a heat reservoir (non-isolated or open system) or with a fixed energy (isolated or closed system), respectively.

The ubiquity of the Maxwellian velocity distribution in ideal gases is also explained in the Sect. 10.3 with a nonlinear mapping acting in the space of velocity distributions [8]. This mapping is an operator that gives account of the decay of any initial velocity distribution toward the Gaussian (Maxwellian) distribution.

Last section contains the conclusions.

10.2 Systems Showing the Boltzmann-Gibbs Distribution

10.2.1 Multi-agent Economic Open Systems

Here we assume N agents, each one with coordinate x_i, $i = 1, \ldots, N$, with $x_i \geq 0$ representing the wealth or money of the agent i, and a total available amount of money E:

$$x_1 + x_2 + \cdots + x_{N-1} + x_N \leq E. \tag{10.1}$$

Under random or deterministic evolution rules for the exchanging of money among agents, let us suppose that this system evolves in the interior of the N-dimensional pyramid given by Eq. (10.1). The role of a heat reservoir, that in this model supplies money instead of energy, could be played by the state or by the bank system in western societies. The formula for the volume $V_N(E)$ of an equilateral N-dimensional pyramid formed by $N+1$ vertices linked by N perpendicular sides of length E is

$$V_N(E) = \frac{E^N}{N!}. \tag{10.2}$$

We suppose that each point on the N-dimensional pyramid is equiprobable, then the probability $f(x_i)dx_i$ of finding the agent i with money x_i is proportional to the volume formed by all the points into the $(N-1)$-dimensional pyramid having the ith-coordinate equal to x_i. Then, it can be shown that the Boltzmann factor (or the Maxwell-Boltzmann distribution), $f(x_i)$, is given by

$$f(x_i) = \frac{V_{N-1}(E - x_i)}{V_N(E)}, \tag{10.3}$$

that verifies the normalization condition

$$\int_0^E f(x_i)dx_i = 1. \tag{10.4}$$

The final form of $f(x)$, in the asymptotic regime $N \to \infty$ (which implies $E \to \infty$) and taking the mean wealth $\epsilon = E/N$, is:

$$f(x)dx = \frac{1}{\epsilon} e^{-x/\epsilon} dx, \tag{10.5}$$

where the index i has been removed because the distribution is the same for each agent, and thus the wealth distribution can be obtained by averaging over all the agents. This distribution has been found to fit the real distribution of incomes in western societies [9, 10].

10.2.2 Multi-agent Economic Closed Systems

We derive now the Boltzmann-Gibbs distribution by considering the system in isolation, that is, a closed economy. Without loss of generality, let us assume N interacting economic agents, each one with coordinate x_i, $i = 1, \ldots, N$, with $x_i \geq 0$, and where x_i represents an amount of money. If we suppose that the total amount of money E is conserved,

$$x_1 + x_2 + \cdots + x_{N-1} + x_N = E, \tag{10.6}$$

then this isolated system evolves on the positive part of an equilateral N-hyperplane. The surface area $S_N(E)$ of an equilateral N-hyperplane of side E is given by

$$S_N(E) = \frac{\sqrt{N}}{(N-1)!} E^{N-1}. \tag{10.7}$$

If the ergodic hypothesis is assumed, each point on the N-hyperplane is equiprobable. Then the probability $f(x_i)dx_i$ of finding agent i with money x_i is proportional to the surface area formed by all the points on the N-hyperplane having the ith-coordinate equal to x_i. It can be shown that $f(x_i)$ is the Boltzmann factor (Boltzmann-Gibbs distribution), with the normalization condition (10.4). It takes the form,

$$f(x_i) = \frac{1}{S_N(E)} \frac{S_{N-1}(E - x_i)}{\sin\theta_N}, \tag{10.8}$$

where the coordinate θ_N satisfies $\sin\theta_N = \sqrt{\frac{N-1}{N}}$. After some calculation the Boltzmann distribution is newly recovered:

$$f(x)dx = \frac{1}{k\tau} e^{-x/k\tau} dx, \tag{10.9}$$

with $\epsilon = k\tau$, being k the Boltzmann constant and τ the temperature of the statistical system.

10.2.3 The Continuous Economic Gas-Like Model

We consider an ensemble of economic agents trading with money in a random manner [9, 10]. This is one of the simplest gas-like models, in which an initial amount of money is given to each agent, let us suppose the same to each one. Then, pairs of agents are randomly chosen and they exchange their money also in a random way. When the gas evolves under these conditions, the exponential distribution appears as the asymptotic wealth distribution. In this model, the microdynamics is conservative because the local interactions conserve the money. Hence, the macrodynamics is also conservative and the total amount of money is constant in time.

The discrete version of this model is as follows [9, 10]. The trading rules for each interacting pair (m_i, m_j) of the ensemble of N economic agents can be written as

$$\begin{aligned} m_i' &= \sigma(m_i + m_j), \\ m_j' &= (1-\sigma)(m_i + m_j), \\ i, j &= 1 \dots N, \end{aligned} \tag{10.10}$$

where σ is a random number in the interval (0, 1). The agents (i, j) are randomly chosen. Their initial money (m_i, m_j), at time t, is transformed after the interaction in (m_i', m_j') at time $t+1$. The asymptotic distribution $p_f(m)$, obtained by numerical simulations, is the exponential (Boltzmann-Gibbs) distribution,

$$p_f(m) = \beta \exp(-\beta m), \quad \text{with } \beta = 1/\langle m \rangle_{gas}, \tag{10.11}$$

where $p_f(m)dm$ denotes the PDF (*probability density function*), i.e. the probability of finding an agent with money (or energy in a gas system) between m and $m + dm$. Evidently, this PDF is normalized, $\|p_f\| = \int_0^\infty p_f(m)dm = 1$. The mean value of the wealth, $\langle m \rangle_{gas}$, can be easily calculated directly from the gas by $\langle m \rangle_{gas} = \sum_i m_i/N$.

The continuous version of this model [11] considers the evolution of an initial wealth distribution $p_0(m)$ at each time step n under the action of an operator T. Thus, the system evolves from time n to time $n+1$ to asymptotically reach the equilibrium distribution $p_f(m)$, i.e.

$$\lim_{n\to\infty} T^n\big(p_0(m)\big) \to p_f(m). \tag{10.12}$$

In this particular case, $p_f(m)$ is the exponential distribution with the same average value $\langle p_f \rangle$ than the initial one $\langle p_0 \rangle$, due to the local and total richness conservation.

The derivation of the operator T is as follows [11]. Suppose that p_n is the wealth distribution in the ensemble at time n. The probability to have a quantity of money x at time $n+1$ will be the sum of the probabilities of all those pairs of agents (u, v) able to produce the quantity x after their interaction, that is, all the pairs verifying $u + v > x$. Thus, the probability that two of these agents with money (u, v) interact between them is $p_n(u) * p_n(v)$. Their exchange is totally random and then they can give rise with equal probability to any value x comprised in the interval $(0, u+v)$. Therefore, the probability to obtain a particular x (with $x < u+v$) for the interacting pair (u, v) will be $p_n(u) * p_n(v)/(u+v)$. Then, T has the form of a nonlinear integral operator,

$$p_{n+1}(x) = Tp_n(x) = \int\int_{u+v>x} \frac{p_n(u)p_n(v)}{u+v} du dv. \tag{10.13}$$

If we suppose T acting in the PDFs space, it has been proved [7] that T conserves the mean wealth of the system, $\langle Tp \rangle = \langle p \rangle$. It also conserves the norm ($\| \cdot \|$), i.e. T maintains the total number of agents of the system, $\|Tp\| = \|p\| = 1$, that by extension implies the conservation of the total richness of the system. We have also shown that the exponential distribution $p_f(x)$ with the right average value is the only steady state of T, i.e. $Tp_f = p_f$. Computations also seem to suggest that other high period orbits do not exist. In consequence, it can be argued that the relation (10.12) is true. This decaying behavior toward the exponential distribution is essentially maintained in the extension of this model for more general random markets.

10.3 Systems Showing the Maxwellian Distribution

10.3.1 Multi-particle Open Systems

Let us suppose a one-dimensional ideal gas of N non-identical classical particles with masses m_i, with $i = 1, \ldots, N$, and total maximum energy E. If particle i has a momentum $m_i v_i$, we define a kinetic energy:

$$K_i \equiv p_i^2 \equiv \frac{1}{2} m_i v_i^2, \tag{10.14}$$

where p_i is the square root of the kinetic energy K_i. If the total maximum energy is defined as $E \equiv R^2$, we have

$$p_1^2 + p_2^2 + \cdots + p_{N-1}^2 + p_N^2 \leq R^2. \tag{10.15}$$

We see that the system has accessible states with different energy, which can be supplied by a heat reservoir. These states are all those enclosed into the volume of the N-sphere given by Eq. (10.15). The formula for the volume $V_N(R)$ of an N-sphere of radius R is

$$V_N(R) = \frac{\pi^{\frac{N}{2}}}{\Gamma(\frac{N}{2}+1)} R^N, \tag{10.16}$$

where $\Gamma(\cdot)$ is the gamma function. If we suppose that each point into the N-sphere is equiprobable, then the probability $f(p_i)dp_i$ of finding the particle i with coordinate p_i (energy p_i^2) is proportional to the volume formed by all the points on the N-sphere having the ith-coordinate equal to p_i. It can be shown that

$$f(p_i) = \frac{V_{N-1}\left(\sqrt{R^2 - p_i^2}\right)}{V_N(R)}, \tag{10.17}$$

which is normalized, $\int_{-R}^{R} f(p_i)dp_i = 1$. The Maxwellian distribution is obtained in the asymptotic regime $N \to \infty$ (which implies $E \to \infty$):

$$f(p)dp = \sqrt{\frac{1}{2\pi\epsilon}} e^{-p^2/2\epsilon} dp, \tag{10.18}$$

with $\epsilon = E/N$ being the mean energy per particle and where the index i has been removed because the distribution is the same for each particle. Then the equilibrium velocity distribution can also be obtained by averaging over all the particles.

10.3.2 Multi-particle Closed Systems

We start by assuming a one-dimensional ideal gas of N non-identical classical particles with masses m_i, with $i = 1, \ldots, N$, and total energy E. If particle i has a momentum $m_i v_i$, newly we define a kinetic energy K_i given by Eq. (10.14), where p_i is the square root of K_i. If the total energy is defined as $E \equiv R^2$, we have

$$p_1^2 + p_2^2 + \cdots + p_{N-1}^2 + p_N^2 = R^2. \tag{10.19}$$

We see that the isolated system evolves on the surface of an N-sphere. The formula for the surface area $S_N(R)$ of an N-sphere of radius R is

$$S_N(R) = \frac{2\pi^{\frac{N}{2}}}{\Gamma(\frac{N}{2})} R^{N-1}, \tag{10.20}$$

where $\Gamma(\cdot)$ is the gamma function. If the ergodic hypothesis is assumed, that is, each point on the N-sphere is equiprobable, then the probability $f(p_i)dp_i$ of finding the particle i with coordinate p_i (energy p_i^2) is proportional to the surface area formed by all the points on the N-sphere having the ith-coordinate equal to p_i. It can be shown that

$$f(p_i) = \frac{1}{S_N(R)} \frac{S_{N-1}(\sqrt{R^2 - p_i^2})}{(1 - \frac{p_i^2}{R^2})^{1/2}}, \tag{10.21}$$

which is normalized. Replacing p^2 by $\frac{1}{2}mv^2$, $f(p)$ takes the following form $g(v)$ in the asymptotic limit $N \to \infty$,

$$g(v)dv = \sqrt{\frac{m}{2\pi k\tau}} e^{-mv^2/2k\tau} dv. \tag{10.22}$$

This is the typical form of the Maxwellian distribution, with $\epsilon = k\tau/2$ given by the equipartition theorem.

10.3.3 The Continuous Model for Ideal Gases

Here, as we have done in the anterior case of economic systems, we present a new model to explain the Maxwellian distribution as a limit point in the space of velocity distributions for a gas system evolving from any initial condition [8].

Consider an ideal gas with particles of unity mass in the three-dimensional (3D) space. As long as there is not a privileged direction in the equilibrium, we can take any direction in the space and study the discrete time evolution of the velocity distribution in that direction. Let us call this direction U. We can complete a Cartesian system with two additional orthogonal directions V, W. If $p_n(u)du$ represents the probability of finding a particle of the gas with velocity component in the direction U comprised between u and $u + du$ at time n, then the probability to have at this time n a particle with a 3D velocity (u, v, w) will be $p_n(u)p_n(v)p_n(w)$.

The particles of the gas collide between them, and after a number of interactions of the order of system size, a new velocity distribution is attained at time $n+1$. Concerning the interaction of particles with the bulk of the gas, we make two simplistic and realistic assumptions in order to obtain the probability of having a velocity x in the direction U at time $n+1$: (1) Only those particles with an energy bigger than x^2 at time n can contribute to this velocity x in the direction U, that is, all those particles whose velocities (u, v, w) verify $u^2+v^2+w^2 \geq x^2$; (2) The new velocities after collisions are equally distributed in their permitted ranges, that is, particles with velocity (u, v, w) can generate maximal velocities $\pm U_{\max} = \pm\sqrt{u^2+v^2+w^2}$, then the allowed range of velocities $[-U_{\max}, U_{\max}]$ measures $2|U_{\max}|$, and the contributing probability of these particles to the velocity x will be $p_n(u)p_n(v)p_n(w)/(2|U_{\max}|)$. Taking all together we finally get the expression for the evolution operator T. This is:

$$p_{n+1}(x) = Tp_n(x) = \int\int\int_{u^2+v^2+w^2 \geq x^2} \frac{p_n(u)p_n(v)p_n(w)}{2\sqrt{u^2+v^2+w^2}} du dv dw. \quad (10.23)$$

Let us remark that we have not made any supposition about the type of interactions or collisions between the particles and, in some way, the equivalent of the Boltzmann hypothesis of *molecular chaos* [12] would be the two simplistic assumptions we have stated on the interaction of particles with the bulk of the gas. Then, an alternative framework than those usually presented in the literature [13] appears now on the scene. In fact, it is possible to show that the operator T conserves in time the energy and the null momentum of the gas. Moreover, for any initial velocity distribution, the system tends towards its equilibrium, i.e. towards the Maxwellian velocity distribution (1D case). This means that

$$\lim_{n\to\infty} T^n\big(p_0(x)\big) \to p_\alpha(x) = \sqrt{\frac{\alpha}{\pi}} e^{-\alpha x^2} \quad (10.24)$$

with $\alpha = (2\langle x^2, p_0\rangle)^{-1}$. In physical terms, it means that for any initial velocity distribution of the gas, it decays to the Maxwellian distribution, which is just the fixed point of the dynamics. Recalling that in the equilibrium $\langle x^2, p_\alpha\rangle = k\tau$, with k the Boltzmann constant and τ the temperature of the gas, and introducing the mass

m of the particles, let us observe that the Maxwellian velocity distribution can be recovered in its 3D format:

$$p_\alpha(u)p_\alpha(v)p_\alpha(w) = \left(\frac{m\alpha}{\pi}\right)^{\frac{3}{2}} e^{-m\alpha(u^2+v^2+w^2)} \quad \text{with } \alpha = (2k\tau)^{-1}. \qquad (10.25)$$

In general, it is observed that the convergence of the T-iterations of any distribution $p(x)$ to its Gaussian limit $p_\alpha(x)$ is very fast.

10.4 Conclusion

We have shown that the Boltzmann factor describes the general statistical behavior of each small part of a multi-component system whose components or parts are given by a set of random variables that satisfy an additive constraint, in the form of a conservation law (closed systems) or in the form of an upper limit (open systems).

Let us remark that these calculations do not need the knowledge of the exact or microscopic randomization mechanisms of the multi-agent system in order to reach the equiprobability. In some cases, it can be reached by random forces, in other cases by chaotic or deterministic causes. Evidently, the proof that these mechanisms generate equiprobability is not a trivial task and it remains as a typical challenge in this kind of problems.

In order to explain the ubiquity and stability of this type of distributions two models based on discrete mappings in the space of distributions have been proposed. On one hand, the gas-like models interpret economic exchanges of money between agents similarly to collisions in a gas where particles share their energy. The continuous version of a gas-like discrete model where the agents trade in binary collisions has been introduced to explain the stability of the exponential distribution in this kind of economic systems. On the other hand, a nonlinear map acting on the velocity distribution space of ideal gases, which gives account of the decay of an out-of-equilibrium velocity distribution toward the Maxwellian distribution, has been presented. Some properties concerning the dynamical behavior of both operators have also been sketched.

Acknowledgements Several collaborators have participated in the development of different aspects of this line of research. Concretely, X. Calbet, J. Sañudo, J.L. Lopez and E. Shivanian. See the references.

References

1. Huang K (1987) Statistical mechanics. Wiley, New York
2. Munster A (1969) Statistical thermodynamics, vol I. Springer, Berlin
3. Ditlevsen PD (2004) Turbulence and climate dynamics. Frydendal, Copenhagen
4. Yakovenko VM (2009) Econophysics, statistical mechanics approach to. In: Meyers RA (ed) Encyclopedia of complexity and system science. Springer, Berlin, pp 2800–2826

5. Lopez-Ruiz R, Calbet X (2007) Derivation of the Maxwellian distribution from the microcanonical ensemble. Am J Phys 75:752–753
6. Lopez-Ruiz R, Sañudo J, Calbet X (2008) Geometrical derivation of the Boltzmann factor. Am J Phys 76:780–781
7. Lopez JL, Lopez-Ruiz R, Calbet X (2012) Exponential wealth distribution in a random market. A rigorous explanation. J Math Anal Appl 386:195–204
8. Shivanian E, Lopez-Ruiz R (2012) A new model for ideal gases. Decay to the Maxwellian distribution. Physica A 391:2600–2607
9. Dragulescu A, Yakovenko VM (2000) Statistical mechanics of money. Eur Phys J B 17:723–729
10. Dragulescu A, Yakovenko VM (2001) Evidence for the exponential distribution of income in the USA. Eur Phys J B 20:585–589
11. Lopez-Ruiz R, Lopez JL, Calbet X (2012) Exponential wealth distribution: a new approach from functional iteration theory. In: ESAIM: Proceedings of ECIT-2010 conference, vol 36, pp 189–196
12. Boltzmann L (1995) Lectures on gas theory. Dover, New York. Translated by SG Brush
13. Maxwell JC (2003) In: Niven WD (ed) The scientific papers of James Clerk Maxwell, vols I, II. Dover, New York

Chapter 11
Advection of Optical Localized Structures

F. Haudin, R.G. Rojas, U. Bortolozzo, M.G. Clerc, and S. Residori

Abstract We present an experimental and numerical study on the effects of a translational nonlocal coupling induced on localized structures (LS) in the context of nonlinear optics. LS are obtained in a light-valve experiment and made to drift by a mirror tilt in the feedback loop. Phase singularities are detected for small drifts, whereas for large drifts, periodic organizations are observed.

11.1 Background

Advection phenomena are commonly studied in the framework of hydrodynamics where fluid particles or scalar passive quantities can be transported by a flow. One of the most famous examples of the advection phenomenon is the formation of Bénard-von Kármán streets, occurring when a viscous fluid is flowing past an obstacle. Above a critical Reynolds number, organizations with two rows of eddies on either side of its wake can develop [1]. Also in superfluids and Bose-Einstein condensates, vortex shedding from a moving obstacle has been evidenced by numerical simulations of the nonlinear Schrodinger equation [2, 3], showing several analogies with vortex streets in Newtonian fluids [4]. The formation of pairs of vortex-antivortex, also accompanies phase transitions associated with symmetry breaking [5]. Examples can be mentioned in magnets, superfluids [6] or in plasma jets [7].

In optics, vortices have been introduced as the topological defects arising above the laser transition [8]. In this context, they have been identified as the singular

F. Haudin (✉) · U. Bortolozzo · S. Residori
INLN, Université de Nice-Sophia Antipolis, CNRS, 1361 route des Lucioles, 06560 Valbonne, France

F. Haudin (✉)
Nonlinear Physical Chemistry Unit, Service de Chimie Physique et Biologie Théorique, Faculté des Sciences, Université Libre de Bruxelles (ULB), CP 231, 1050 Brussels, Belgium

R.G. Rojas
Instituto de Física, Pontificia Universidad Católica de Valparaíso, Casilla 4059, Valparaíso, Chile

M.G. Clerc
Departamento de Física, FCFM, Universidad de Chile, Casilla 487-3, Santiago, Chile

T. Gilbert et al. (eds.), *Proceedings of the European Conference on Complex Systems 2012*, Springer Proceedings in Complexity, DOI 10.1007/978-3-319-00395-5_11,

points where the field amplitude is zero, while the circulation of the phase gradient on any loop which encloses the vortex core is equal to $\pm 2\pi$, with conservation of the total vorticity.

In nonlinear optics, the first observations of vortices pairs have been reported twenty years ago in photorefractive cavities [9–11] and in Kerr media [12]. The idea of drifting structures was used in a photorefractive cavity, where a wake of alternating vortices has been studied in the transverse field [13], showing analogies with the street of eddies previously mentioned. Concerning localized states, in the LCLV experiment, triangularly shaped localized structures have also shown to present optical vortices [14], where the breaking of the circular symmetry is responsible for the appearance of pairs of oppositely charged phase singularities [15]. In an experiment with two VCSELs facing each other, vortex structures have been recently reported [16]. Furthermore, other studies in the LCLV experiment have evidenced that inducing a translational non local effect can induce secondary instabilities of patterns [17, 18] such as transitions from hexagons to stripes, squares to zigzag [19].

In this context of study, the question we are addressing is how optical localized states can be affected by a translational non local effect. The answer is not obvious a priori since LS are particular objects. Indeed, they have wave properties, described by the amplitude of the electric field, but they are "particle-like" too, if we consider them as elementary pixels that can be switched on and off due to bistability [20]. Recently, the issue of drifting LS guided by using a spatial light modulator has been addressed [21]. However, the deformation of LS due to a translational coupling, TC, and the emission of optical vortices were not studied before. The present results are reported in [22].

11.2 Methods

To characterize the effect of a drift on LS, structures are generated in an experiment comprising a liquid crystal light-valve, LCLV, with optical feedback [23] (Fig. 11.1(a)). The LCLV is made of a thin nematic liquid crystal (LC) layer, with thickness of 15 microns. This layer is sandwiched between a glass plate and a photoconductive wall. The photoconductor is responding like an impedance varying with the intensity of the light it receives and a dielectric mirror is deposited on it. A planar anchoring is imposed to the molecules thanks to a prior treatment of the surfaces in contact with the LC (nematic director parallel to the confining walls). A voltage is applied across the cell. An expanded He-Ne laser beam, with wavelength 632.8 nm, linearly polarized along the vertical direction, is illuminating the cell. More precisely, the optical beam passes through the LC is then reflected back by the dielectric mirror on the rear side of the valve. Finally the light beam is sent in the feedback loop. An optical fiber bundle is used to close the loop and redirect the beam back to the photoconductive side of the LCLV. The nematic director is oriented at 45 degrees and a polarizing cube splitter introduces polarization interference between the ordinary and extraordinary waves. This condition is responsible

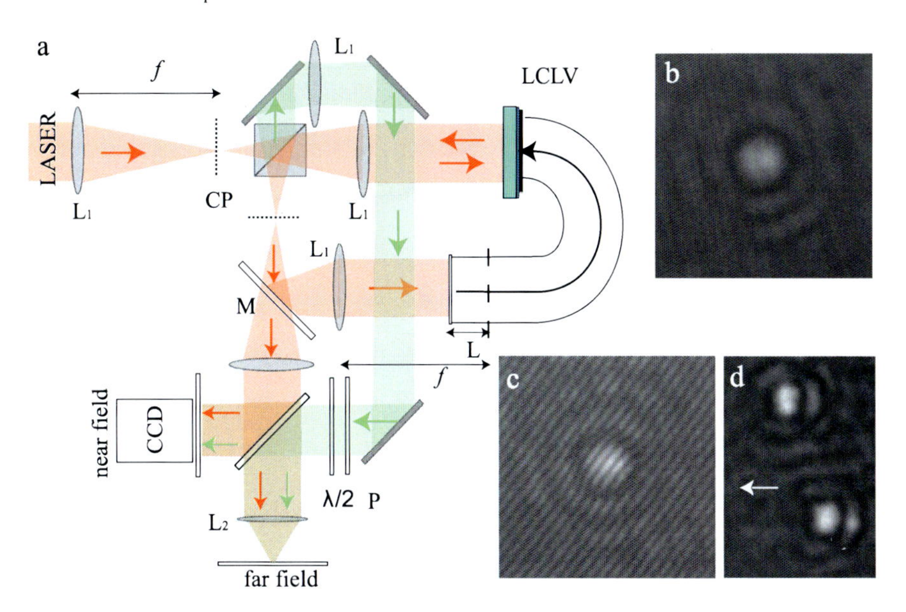

Fig. 11.1 (**a**) Experimental set up with an interferometer. LCLV means Liquid Crystal Light Valve, L_1 is a lens with focal distance $f = 25$ cm; L_2 is a lens with short focal distance, CP a polarisor cube, $\lambda/2$ is a half-wave plate, P a polarisor, L is the free propagation length of light; (**b**) experimental profile of a quasi-motionless LS, (**c**) interferogram of the same structure and (**d**) two experimentally advected LS

for the bistability between differently orientated states of the LC [24]. To observe LS, the second effect needed in the system is diffraction, which is introduced by displacing the fiber bundle over a L distance from this self-imaging configuration: this leads to the selection of a transverse spatial characteristic size for the LS scaling as the square-root of L. A CCD (Charged Coupled Device) camera is used to record the intensity of the LS on an optical plane equivalent to the one of the photoconductor. Starting from a situation with LS in the transverse plane, a mirror of the feedback loop is quickly misaligned, LS start to move in the direction induced by the detuning.

11.3 Results

When they are moving, LS lose their circular symmetry. For relatively small translations, their profile is only slightly deformed. Wavelets are visible in the wake of LS. For larger drifts, the deformation becomes more important.

From the intensity profiles, there is no direct access to the phase of the electric field. Nevertheless, it is possible to detect phase singularities by interferometry, that is, by taking a reference beam and making it to interfere at a given angle with the beam extracted from the feedback loop. Phase singularities are, then, detected as

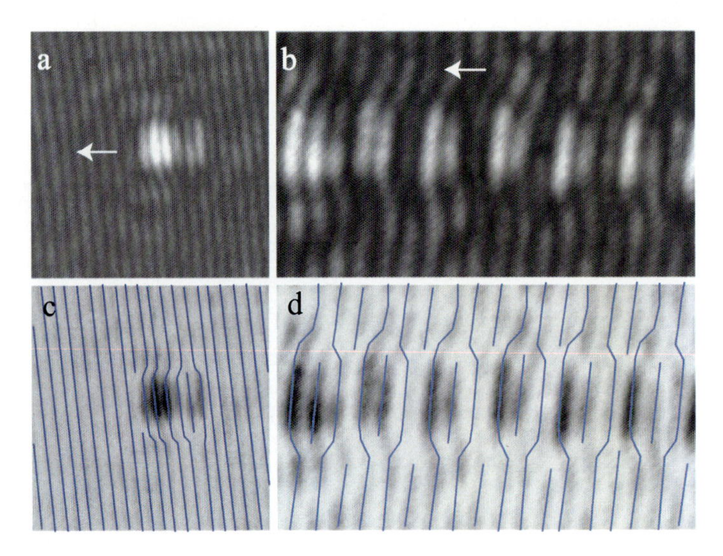

Fig. 11.2 For a small drift: (**a**) interferogram of an advected LS and (**d**) same interogram with inversion of colors and evidence of the fringes to detect phase singularities. For a larger drift: (**c**) interferogram and (**d**) same interogram with inversion of colors and evidence of the fringes to detect phase singularities

fringe dislocations appearing in the interference pattern. To optimize the location of the singularities, we have settled the experiment with a large L in order to have large structures, and the interferometer is adjusted to get a suitable number of fringes per LS. Without using the reference beam, one gets the profile of intensity without the fringes (Fig. 11.1(b)). In an ideal case without any translational effect ($dx = 0$), LS have the usual circular geometry and no phase singularity is detected (Fig. 11.1(c)). When the mirror is detuned, the symmetry of the LS is breaking (Fig. 11.1(d)). In their intensity profile, phase singularities exist, detected by the fringe dislocations, organized by pairs in a symmetric way on each side of the structure and visible either in the front part of the structure or in its wake (Figs. 11.2(a) and (b)). For bigger misalignments of the mirror, periodic organizations develop looking like a necklace of LS (Figs. 11.2(c) and (d)). A remarkable feature is that the characteristic size inside the organization is smaller than the natural size of the LS (for $dx = 0$). Looking at the interference pattern, the presence of dislocations proves the presence of phase singularities periodically distributed in space. All the experimental results show qualitatively two regimes and the existence of phase singularities in both of them.

To better analyze the appearance of these singularities in the profile of drifting LS, numerical simulations are performed. They are done by integrating the equations of the full LCLV model [24] describing the dynamics of the localized structures in the presence of diffraction and a voltage, and by adding a translational coupling. More precisely, the integration of the model is made by using a pseudo spectral method, for which the spatial derivatives and the diffraction operator are solved in Fourier space. The temporal derivate is calculated with an adaptive Runge-Kutta

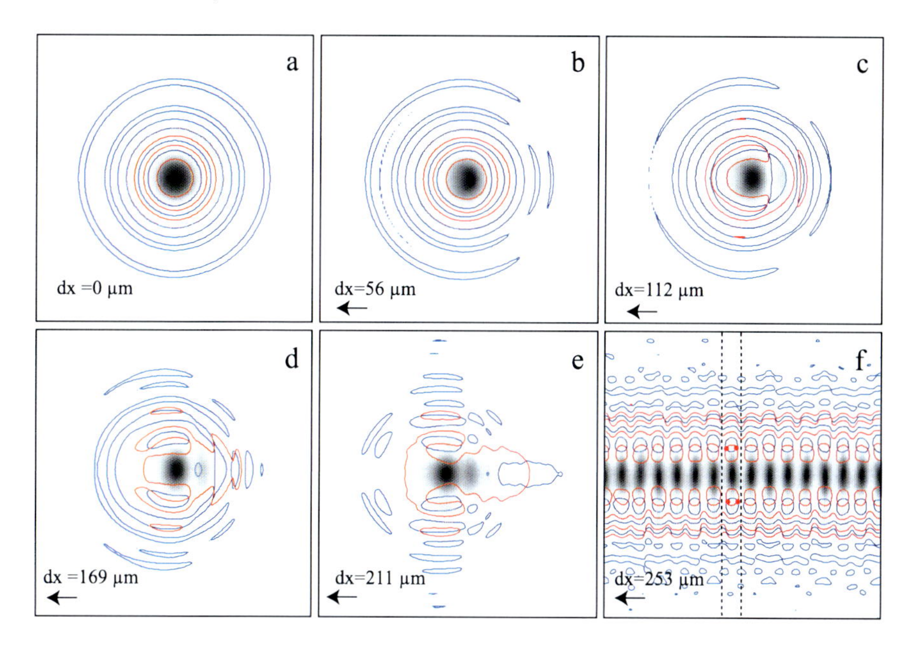

Fig. 11.3 Numerical intensity profiles of LS for growing dx values. In blue (respectively *red*) are represented the null-lines of the real part (respectively imaginary part) of the electric field. Intersections between blue and red lines correspond to phase singularities. The parameters are the following: $L = -16$ cm, the voltage V_0 across the LCLV is 12.9 V, the parameter for the intensity is $\alpha I_{in} = 1.2$ and the different values for the drift parameter are: (**a**) $dx = 0$, (**b**) $dx = 56$, (**c**) $dx = 112$, (**d**) $dx = 169$, (**e**) $dx = 211$ and (**f**) $dx = 253$ μm

algorithm. In the bistable regime, a single LS is generated by applying a Gaussian pulse, then, a translation dx of the intensity on the photoconductor layer is introduced. In the numerical profiles of the electric field of the LS, phase singularities are detected every time a null-line of the real part of the field is crossing a null-line of the imaginary part. When looking at the intensity profiles of LS for increasing value of the translational coupling dx, the numerical results show behaviors similar to the experimental ones. When the LS are motionless, the null-lines of the electric field are circular and never cross each-other. The deformation induced by the translation brings the lines to intersect at multiple points, where optical vortices are nucleated by pairs (Fig. 11.3).

References

1. Tritton DJ (1988) Physical fluid dynamics, 2nd edn. Oxford University Press, London
2. Frisch T, Pomeau Y, Rica S (1992) Transition to dissipation in a model of superflow. Phys Rev Lett 69:1644
3. Huepe C, Brachet ME (2000) Scaling laws for vortical nucleation solutions in a model of superflow. Physica D 140:126

4. Sasaki K, Suzuki N, Saito H (2010) Bénard-von Kármán vortex street in a Bose-Einstein condensate. Phys Rev Lett 104:150404
5. Mermin ND (1979) The topological theory of defects in ordered media. Rev Mod Phys 51:591
6. Nago Y, Inui M, Obara K, Yano H, Ishikawa O, Hata T (2009) Vortex emission by a low-frequency vibrating wire in superfluid ^{3}He-B. J Phys, Conf Ser 150:032071
7. Bulanov SS, Esiev RU, Zharnikov MN, Kamrukov AS, Kozhevnikov IV, Kozlov NP, Morozov MI, Roslyakov IA, Stepanov Yu A (2008) Explosive plasma-vortex source of optical emission. Tech Phys Lett 34:34
8. Coullet P, Gil L, Rocca F (1989) Optical vortices. Opt Commun 73:403
9. Arecchi FT, Giacomelli G, Ramazza PL, Residori S (1991) Vortices and defect statistics in two-dimensional optical chaos. Phys Rev Lett 67:3749
10. Arecchi FT, Boccaletti S, Ramazza PL, Residori S (1993) Transition from boundary- to bulk-controlled regimes in optical pattern formation. Phys Rev Lett 70:2277
11. Ramazza PL, Residori S, Giacomelli G, Arecchi FT (1992) Statistics of topological defects in linear and nonlinear optics. Europhys Lett 19:475
12. Swartzlander GA, Law CT (1992) Optical vortex solitons observed in Kerr nonlinear media. Phys Rev Lett 69:2503
13. Vaupel M, Staliunas K, Weiss CO (1996) Hydrodynamic phenomena in laser physics: modes with flow and vortices behind an obstacle in an optical channel. Phys Rev A 54:880
14. Bortolozzo U, Pastur L, Ramazza PL (2004) Bistability between different localized structures in nonlinear optics. Phys Rev Lett 93:253901
15. Ramazza PL, Bortolozzo U, Pastur L (2004) Phase singularities in triangular dissipative solitons. J Opt A, Pure Appl Opt 6:S266
16. Genevet P, Barland S, Giudici M, Tredicce JR (2010) Bistable and addressable localized vortices in semiconductor lasers. Phys Rev Lett 104:223902
17. Ramazza PL, Bigazzi P, Pampaloni E, Residori S, Arecchi FT (1995) One-dimensional transport-induced instabilities in an optical system with nonlocal feedback. Phys Rev E 52:5524
18. Ramazza PL, Boccaletti S, Giaquinta A, Pampaloni E, Soria S, Arecchi FT (1996) Optical pattern selection by a lateral wave-front shift. Phys Rev A 54:3472
19. Ramazza PL, Ducci S, Arecchi FT (1998) Optical diffraction-free patterns induced by a discrete translational transport. Phys Rev Lett 81:4128
20. Tlidi M, Mandel P, Lefever R (1994) Localized structures and localized patterns in optical bistability. Phys Rev Lett 73:640
21. Cleff C, Gutlich B, Denz C (2008) Gradient induced motion control of drifting solitary structures in a nonlinear optical single feedback experiment. Phys Rev Lett 100:233902
22. Haudin F, Rojas RG, Bortolozzo U, Clerc MG, Residori S (2011) Vortex emission accompanies the advection of optical localized structures. Phys Rev Lett 106:063901
23. Residori S (2005) Phys Rep 416:201
24. Clerc MG, Petrossian A, Residori S (2005) Bouncing localized structures in a liquid-crystal light-valve experiment. Phys Rev E 71:015205(R)

Chapter 12
Comparative Analysis of Buoyancy- and Marangoni-Driven Convective Flows Around Autocatalytic Fronts

M.A. Budroni, L. Rongy, and A. De Wit

Abstract We introduce a reaction-diffusion-convection (RDC) model to study the combined effect of buoyancy- and Marangoni-driven flows around a traveling front. The model allows for a parametric control of the two contributions *via* the solutal Rayleigh number, Ra_c, which rules the buoyancy component and the solutal Marangoni number, Ma_c, governing the intensity of the velocity field at the interface between the reacting solution and air. Complex dynamics may arise when the bulk and the surface flows describe an antagonistic interplay. Typically, spatiotemporal oscillations are observed in the parameter region ($Ra_c < 0, Ma_c > 0$).

12.1 Background

In spatially extended systems, autocatalytic reactions can sustain propagating chemical fronts [1, 2] when coupled to diffusion. The study of such fronts has been traditionally carried out in gels in order to avoid complications due to the onset of convective flows. Convective instabilities inevitably occur in aqueous solutions, where propagating waves generate a self-organised interface between two miscible phases (the products and the fresh reactants respectively) and fluid motions may arise, promoted by density and surface tension gradients across the chemical front. The resulting convective flows feedback with the reaction dynamics, modifying the reaction-diffusion structures. Interestingly, transport phenomena do not act here as homogenising agents, but actively couple with the nonlinear kinetics to induce spatiotemporal reaction-diffusion-convection self-organisation. In this framework, chemical fronts represent nice model systems for studying chemo-hydrodynamic instabilities, which have implications also in practical terms, since emergent behaviours due the coupling between reactions and transport phenomena are widespread in different fields ranging from the atmosphere and the environmental chemistry to the realm of applied processes. In the last years, the dramatic influence of bulk and surface flows on the wave dynamics has been pointed out in

M.A. Budroni (✉) · L. Rongy · A. De Wit
Nonlinear Physical Chemistry Unit, Service de Chimie Physique et Biologie Théorique, Faculté des Sciences, Université Libre de Bruxelles (ULB), CP 231, 1050 Brussels, Belgium

T. Gilbert et al. (eds.), *Proceedings of the European Conference on Complex Systems 2012*, Springer Proceedings in Complexity, DOI 10.1007/978-3-319-00395-5_12,

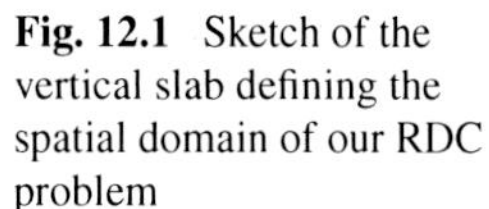

Fig. 12.1 Sketch of the vertical slab defining the spatial domain of our RDC problem

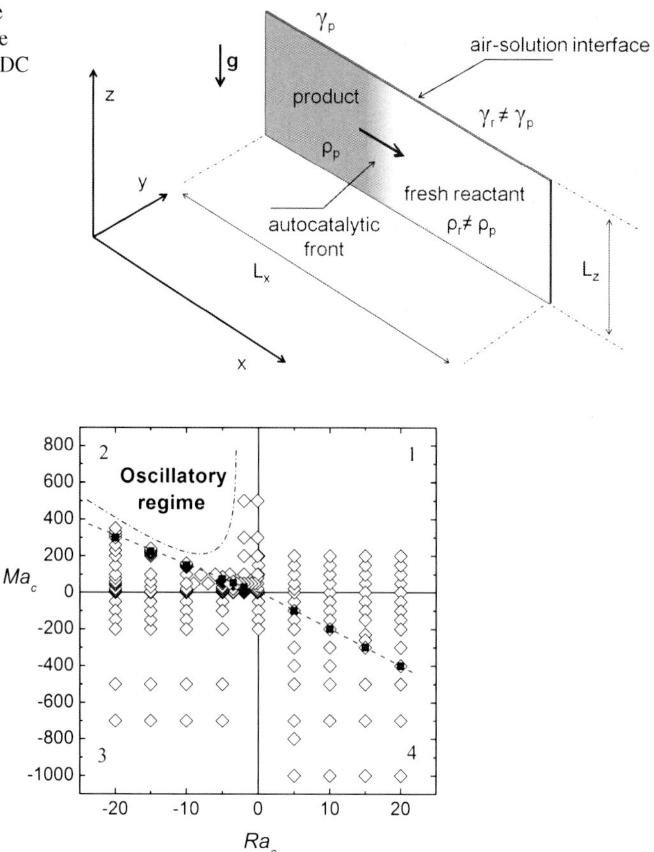

Fig. 12.2 Classification of the dynamical regimes for the asymptotic front mixing length, W, in the parameter space (Ra_C, Ma_C). The first and the third quadrants frame dynamics in which Marangoni and buoyancy effects are cooperative while in the second and the fourth quadrant the two convective contributions act antagonistically. *Empty diamonds* represent stationary solutions. Oscillatory behaviours are located in the second quadrant. When the two effects are antagonist but comparable, W stabilises over a minimum, which scales linearly with Ra_C and Ma_C, as indicated by *black filled squares*. No oscillatory behaviour is observed in the fourth quadrant, even if the chemical front experiences the antagonist influence of the Marangoni- and buoyancy-driven flows

both experimental [3, 4] and theoretical works [4–8], showing that chemical fronts can be distorted, accelerated or even broken by the hydrodynamic feedback. The reaction-diffusion-convection coupling was proved to be also responsible for order-disorder transitions in chemical oscillators, where it controls the route from periodic regimes to spatiotemporal chaos [9].

Several numerical studies were aimed to isolate the role played by pure buoyancy-driven and pure surface-driven flows in chemical travelling structures [5, 6]. While these limit cases were successfully understood, the combined interplay of surface and buoyancy effects still remains unexplored.

12.2 Methods

Here we propose a reaction-diffusion-convection (RDC) model, where both buoyancy and surface solutal contributions to hydrodynamic flows are taken into account. The model is derived by coupling a cubic kinetics, which well describes autocatalytic processes, to Fickian diffusion and to the Navier-Stokes equations governing the velocity of the fluid in the reactor. We consider a 2-dimensional slab of length L_x and height L_z, where the gravity force $\mathbf{g}$ is oriented against the vertical direction z (see Fig. 12.1). The autocatalytic product c, characterized by the density ρ_p and the surface tension γ_p, propagates towards positive x (from left to right in Fig. 12.1) invading the fresh reactant with density ρ_r, surface tension γ_r and initial concentration a_0. The reactor is assumed to be in isothermal conditions. The system equations, conveniently casted into a dimensionless form and written in the *vorticity-stream function* ($\omega - \psi$) form [7],[1] read

$$\frac{\partial c}{\partial t} + \left(\frac{\partial \psi}{\partial z}\frac{\partial c}{\partial x} - \frac{\partial \psi}{\partial x}\frac{\partial c}{\partial z}\right) = \nabla^2 c + c^2(1-c) \tag{12.1}$$

$$\frac{\partial \omega}{\partial t} + \left(\frac{\partial \psi}{\partial z}\frac{\partial \omega}{\partial x} - \frac{\partial \psi}{\partial x}\frac{\partial \omega}{\partial z}\right) = S_c\left(\nabla^2 \omega - Ra\frac{\partial c}{\partial x}\right) \tag{12.2}$$

$$\frac{\partial^2 \psi}{\partial x^2} + \frac{\partial^2 \psi}{\partial z^2} = -\omega. \tag{12.3}$$

The surface effect is introduced by using the Marangoni boundary condition at the upper free surface border:

$$\omega = -Ma_c\frac{\partial c}{\partial x} \quad \text{at } z = L_z. \tag{12.4}$$

This mathematical description provides a mean for a comparative study of the buoyancy and surface-driven effects *via* the change of two key parameters: the solutal Rayleigh number, $Ra_c = -\frac{1}{\rho_0}\frac{\partial \rho}{\partial c}\frac{a_0 L_0^3 g}{D\nu}$, which quantifies the buoyancy influence and the solutal Marangoni number, $Ma_c = -\frac{1}{\mu\sqrt{kD}}\frac{d\gamma}{dc}$, which tunes the intensity of surface effects at the interface between the reacting solution and air.[2] The front dynamics is probed by changing the relative importance of Ra_c and Ma_c and the resulting chemical patterns are quantitatively characterised through the analysis of (i) the asymptotic front mixing length, W, (which measures the distance between the tip and the rear of the front) and (ii) by the topology of the velocity field.

---

[1]The equations are scaled by using the time scale of the chemical process $t_0 = 1/(ka_0^2)$ and the reaction-diffusion characteristic length $L_0 = \sqrt{Dt_0}$; D is the diffusivity of the autocatalytic species; k the kinetic rate constant of the reaction; μ is the water dynamic viscosity, which is related to the kinematic viscosity, ν, via $\nu = \mu/\rho_0$, where ρ_0 is the water density; the *vorticity*, ω, and the *stream function*, ψ, are tied to the velocity field $\mathbf{v} = (u, v)^T$ through the relations $\omega = \nabla \times \mathbf{v}$, ψ, $u = \partial_z \psi$ and $v = -\partial_x \psi$. In Eqs. (12.1)–(12.4), the Schmidt number, $S_c = \nu/D$, gives the balance between momentum and mass diffusion.

[2]See footnote 1.

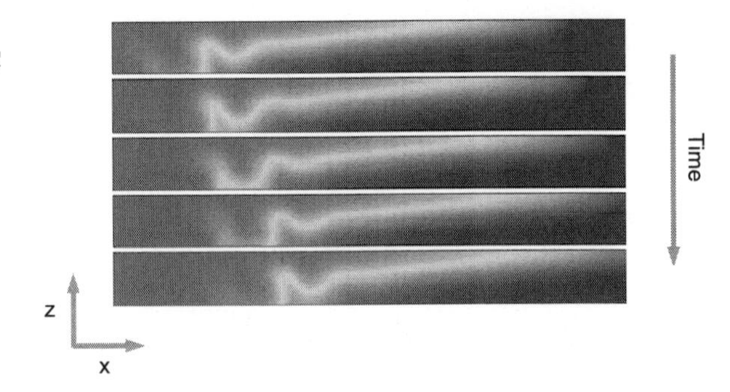

Fig. 12.3 Spatio-temporal oscillations of a typical RDC structure obtained for the antagonist case ($Ra_c = -5, Ma_c = 200$). On the left is the autocatalytic species propagating into the fresh reactants on the right

12.3 Results

An overview of the various possible scenarios can be summarised in the (Ra_c, Ma_c) parameter space, as sketched in Fig. 12.2, where we report the asymptotic dynamics of the front mixing length. Steady asymptotic regimes are found when the bulk and the surface contributions to fluid motions act cooperatively, typically in the regions of the parameter space where the control parameters have the same sign ($Ra_c > 0$, $Ma_c > 0$ or $Ra_c < 0$, $Ma_c < 0$). Complex dynamics may arise when the two effects combine for an antagonistic configuration. An example of these are spatiotemporal oscillations observed as the parameters are set in the region ($Ra_c < 0, Ma_c > 0$) (see Fig. 12.3). Periodic behaviour develop even in the absence of any double-diffusive interplay, which in previous literature was identified as a possible source of complexity [10]. Oscillations at the autocatalytic front occur because the local region of solute denser product brought on top of the less dense non-reacted fluid by the Marangoni surface flow starts sinking under the influence of the gravitational field, forming a finger. The time scale at which this tongue moves down along the z-direction is comparable to that of the front propagation, so that they can annihilate. This mechanism iterates, supporting the oscillatory dynamics. The necessity of an active synergy between the convective contributions for oscillations to occur is confirmed by the strict dependence of the transition upon the reactor height, L_z, which can affect the importance of buoyancy driven flows. Below a critical threshold of L_z, no oscillatory dynamics can be observed, since the buoyancy contribution is suppressed. Nevertheless, we show that a hydrodynamic antagonism is not a sufficient condition for periodic or more complicated oscillations. This is numerically proved by considering the parameters region ($Ra_c > 0$, $Ma_c < 0$), where, even if the system experiences the competing effect of the two convective contributions, it always saturates to an asymptotic steady-state. For both antagonistic cases, the analysis of the front dynamics shows a self-similar architecture which relates the trends of the front observables in the (Ra_c, Ma_c)-space. This study complements earlier efforts to give a significant overview of the possible instability scenarios originated by a chemo-hydrodynamic feedback. Experimental observations in the most common chemical clocks, such as the Iodate Arsenous Acid (IAA) and the Chlorite Tetrationate (CT) reaction, can be framed within this "taxonomy". New experiments are expected to be designed for validating our numerical predictions about the new route to spatiotemporal oscillations found in this work.

References

1. Field R, Burger M (1985) Oscillations and travelling waves in chemical systems. Wiley, New York
2. Kapral R, Showalter K (1994) Chemical waves and patterns. Kluwer Academic, Dordrecht
3. Sebestikova L, Hauser MJ (2012) Buoyancy-driven convection may switch between reactive states in three-dimensional chemical waves. Phys Rev E 85:036303
4. Rossi F, Budroni MA, Marchettini N, Carballido-Landeira J (2012) Segmented waves in a reaction-diffusion-convection system. Chaos 22:037109
5. Rongy L, De Wit A (2006) Steady Marangoni flow traveling with a chemical front. J Chem Phys 124:164705
6. Rongy L, Goyal N, Meiburg E, De Wit A (2007) Buoyancy-driven convection around chemical fronts traveling in covered horizontal solution layers. J Chem Phys 127:114710
7. Budroni M, Rongy L, De Wit A (2012) Dynamics due to combined buoyancy- and Marangoni-driven convective flows around autocatalytic fronts. Phys Chem Chem Phys 14:14619
8. Rongy L, Assemát P, De Wit A (2012) Marangoni-driven convection around exothermic autocatalytic chemical fronts in free-surface solution layers. Chaos 22:037106
9. Budroni MA, Masia M, Rustici M, Marchettini N, Volpert V (2009) Bifurcations in spiral tip dynamics induced by natural convection in the Belousov–Zhabotinsky reaction. J Chem Phys 130:024902
10. Rongy L, Schuszter G, Sinkó Z, Tóth T, Horváth D, Tóth A, De Wit A (2009) Influence of thermal effects on buoyancy-driven convection around autocatalytic chemical fronts propagating horizontally. Chaos 19:023110

Chapter 13
A Field Theory for Self-organised Criticality

Gunnar Pruessner

Abstract Although self-organised criticality has been introduced more than two decades ago, its theoretical foundations remain somewhat elusive: How does it work? What is its link to ordinary critical phenomena? How can exponents be calculated systematically? Does it actually exist at all? In the following a field theory is introduced that addresses these questions. In contrast to previous attempts, this field theory is not phenomenological, or based on symmetry arguments. Rather, it is based on the microscopic dynamics of the Manna Model. Exponents can be calculated in an ϵ-expansion perturbatively in a systematic way. Above the upper critical dimension, the field theory becomes (asymptotically) exact, allowing immediate comparison to numerical results.

13.1 Introduction

Twenty five years ago Bak, Tang and Wiesenfeld [1] conceived the idea of self-organised criticality (SOC), as the phenomenon that some slowly-driven systems dissipate their "load" by sudden outbursts of activity, so called "avalanches", the statistical features of which are scale free [2, 3]. These systems very much behave like ordinary critical phenomena, which somehow mange to tune themselves to the critical point.

Over the last couple of decades, a plethora of models have been developed with the aim to identify the sufficient and necessary conditions for SOC. This has facilitated the development of a range of theories of the mechanism underpinning the SOC phenomenon. Among these models are some that display self-organised criticality in a robust and solid fashion, most notably the Manna Model [4] and the Oslo Model [5]. Numerical evidence suggests that both models belong to the same universality class [6].

Recently, extensive numerical work has been carried out to characterise the Manna model on different (regular) lattices in $d = 1, 2, 3$ dimensions [7, 8], as well

G. Pruessner (✉)
Department of Mathematics, Imperial College London, 180 Queen's Gate, London SW7 2AZ, UK
e-mail: g.pruessner@imperial.ac.uk

T. Gilbert et al. (eds.), *Proceedings of the European Conference on Complex Systems 2012*, Springer Proceedings in Complexity, DOI 10.1007/978-3-319-00395-5_13,

as at and above the upper critical dimension $d = 4$ on hypercubic ones [9, 10]. On that basis, the coefficients of an $\epsilon = 4 - d$ expansion can be extracted [8], suggesting the existence of a field theoretic description. In a convoluted way, a field theory exists already: The equation of motion of the Oslo Model is the quenched Edwards Wilkinson equation [11], which has been painstakingly analysed by Le Doussal, Wiese and Chauve [12]. Based on the link between interface roughness exponent χ and avalanche dimension $D = d + \chi$ [13, 14], one finds $D = 4 - (2/3)\epsilon + 0.04777\epsilon^2$ from [12], which compares acceptably well to the numerical result of $D = 4 - 0.658(5)\epsilon + 0.00962(13)\epsilon^2$ (with a somewhat uncertain second order term), at least to leading order. On the other hand, numerically $z = 2 - 0.239(4)\epsilon$ (with a second order undetermined), compared to $z = 2 - (2/9)\epsilon - 0.0432087\epsilon^2$ from field theory.

A number of SOC models have been analysed using renormalisation group methods in the past. Diaz-Guilera [15, 16] focussed on a Langevin-like description of the Zhang model [17], which highlighted the technical problems associated with the appearance of sharp thresholds [18, 19]. Pietronero, Vespignani and Zapperi [20] presented a physically very appealing real-space renormalisation procedure, which seemed to capture all relevant correlations to produce a very good estimate of $\tau = 1.253\ldots$ for the two-dimensional Manna Model (compared to $\tau = 1.273(2)$, numerically [7]). Although the dynamical exponent $z = 1.234$ was not reproduced as well ($z \approx 1.54$, according to [4, 7, 21]), this procedure was very widely and successfully applied to a number of SOC models [22–25].

On the other hand, based on symmetry and conservation arguments Hwa and Kardar [26] and later Grinstein, Lee and Sachdev [27] were able to determine stochastic equations of motion for SOC models such as the Bak-Tang-Wiesenfeld model [1], which displayed non-trivial scaling. This work has been extended to other models [28–31] focusing more recently on the question whether the universality class of conserved directed percolation, which should contain the Manna Model, is truly distinct from that of ordinary directed percolation [32, 33].

So far, however, none of the established SOC models has been analysed using traditional techniques. Either the method of analysis was developed phenomenologically, for example by ignoring certain correlations, or the equations of motion were determined on the basis of symmetry considerations and expansions rather than the underlying rules. Both approaches were undoubtedly successful, quantitatively as well as qualitatively. Yet, a complete, field-theoretic treatment of SOC based on the microscopic dynamics and using the established methods of critical phenomena remains elusive. What is more, while non-trivial scaling had been found at the relevant fixed points, no dedicated mechanism for the self-organisation to the critical point was identified. Rather, it was observed [20] that SOC models flow to an *attractive, non-trivial* fixed point.

Dickman, Vespignani and Zapperi [28, 34] proposed a more phenomenological explanation of SOC on the basis of absorbing state phase transitions. While this established the link between ordinary (non-equilibrium) phase transitions more firmly (originally suggested by Tang and Bak [35]), the proposed tuning-mechanism has the shortcoming of being essentially linear, thus failing to explain how it triggers a fundamentally non-linear phenomenon. At closer inspection, it remains unclear why the self-organisation occurs at precisely the right pace [36] and how it is linked to the underlying absorbing state phase transition[37].

In summary, to date, there is no theoretical foundation of SOC, which would

- explain how SOC is related to ordinary critical phenomena,
- explain what mechanism triggers the scale invariance observed in SOC systems,
- most importantly, predict exponents and other universal quantities in SOC systems.

In the present manuscript, I present a summary of recent work which tries to overcome these limitations. Using the language of second quantisation, a field theoretic description of the Manna Model can be derived. This field theory is an *exact* representation of its (Poisonian) dynamics. Taking the usual field theoretic steps reduces the number of non-linear vertices to four, resulting in the effective field theory of the Manna Model, which is the first field theory of its kind.

13.2 The Manna Model and Its Field Theory

The Abelian Manna Model [4, 38] is defined as follows. Each site i carries a number of particles z_i. Whenever z_i exceeds 1, site i topples, whereby two particles are removed from that site (thus reducing z_i by 2) and placed at two uniformly, randomly, independently chosen nearest neighbouring sites j (thus increasing their z_j by 1 each). In turn, these sites might topple, as their z_j may exceed the threshold of 1. Some sites i are boundary sites where particles are lost. These sites can be thought of as having "virtual" nearest neighbours "outside" the lattice which can be charged, but which never topple.

The Manna Model is normally initialised with an empty lattice, $z_i = 0$ everywhere, and driven by adding a particle ("charging") at a randomly chosen site (thus increasing z_i by 1). The driving has to cease whenever a sequence of topplings occurs, thereby implementing a separation of time scales. The totality of all topplings between any two driving steps constitutes an avalanche. In particular, the avalanche size s is given by the total number of topplings occurring between any two external charges.

When deriving the field theory, one key step is to capture the seemingly fermionic nature of the Manna Model, as quiescent sites do not support more than one particle. One might be tempted to generalise the Manna Model by stipulating that a site topples after being charged with probability z_j/n, where n is a carrying capacity. For $n = 1$ the Manna Model is recovered, yet, there is no guarantee that the scaling behaviour would remain unaffected by this rather brutal manhandling of the fermionicity.

However, numerical simulations of the modified Manna Model confirm that the resulting model still is in the Manna universality class. What is more, the modified Manna Model with the probabilistic toppling rule is exactly the original Manna

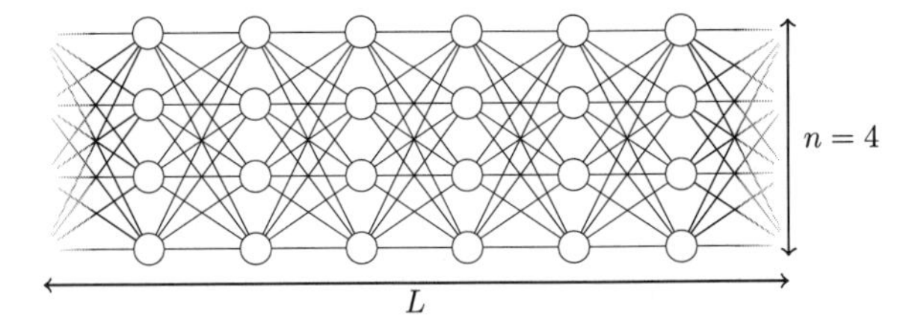

Fig. 13.1 The original Manna Model on the lattice shown above behaves like a modified Manna Model with probabilistic toppling rule, parameterised by the carrying capacity n. Each site in a column is connected to every site in the nearest neighbouring columns. The lattice shown is a snippet of a (decorated) one-dimensional lattice with length L

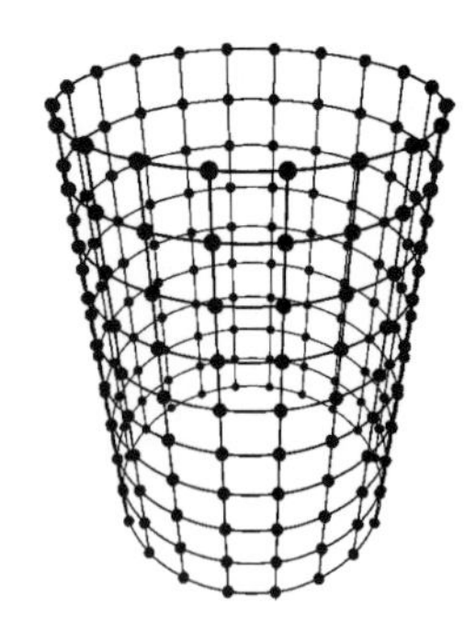

Fig. 13.2 The boundary conditions in the field theory are best chosen to be cylindrical

Model on a lattice like the one shown in Fig. 13.1, if z_i of the modified version is the sum over the z_i in each column in the original version.

The other obstacle to overcome is the nature of the boundary conditions, which are necessarily open at some sites (otherwise stationarity cannot be maintained in the presence of external drive). It turns out that cylindrical boundary conditions are mathematically easiest, see Fig. 13.2. The bare propagator in a d-dimensional system then becomes

$$\frac{1}{-\imath\omega + D(\mathbf{k}^2 + q_n^2)} \tag{13.1}$$

with frequency ω, diffusion constant D and momenta $\mathbf{k} \in \mathbb{R}^{d-1}$ and $q_n \in \mathbb{R}$. The latter are "quantised", $q_n = \frac{\pi}{L} n$ with $n = 1, 2, \ldots$. The eigensystem deployed, $\sqrt{2/L}\sin(q_n z)$, lacks momentum conservation, resulting in a number of complications, which, however, can be overcome and are well known from critical Casimir systems [39].

In the present setup, some exact results are obtained immediately. To start with, the average avalanche size can derived as *exactly* $\langle s \rangle = (d/6)(L+1)(L+2)$ in all dimensions d. The fact that the average avalanche size derived from the bare propagator Eq. (13.1) is identical to the average avalanche size derived on the basis of random walker considerations, indicates that the bare propagator does not renormalise at any order [40].

Table 13.1 Comparison of various moment ratios, derived analytically for the branching random walk on the lattice, which is the (effective) mean field theory of the Manna Model, and the numerical values found for the Abelian Manna Model in $d = 5 > d_c = 4$ dimensions ($d = 5$ avoids the significant logarithmic corrections at $d = 4$)

Observable	Analytical	Numerical (leading order) [10]
$\langle s^3\rangle\langle s\rangle/\langle s^2\rangle^2$	3.08754...	3.061(5)
$\langle s^4\rangle\langle s^2\rangle/\langle s^3\rangle^2$	1.6693...	1.65(2)
$\langle s^5\rangle\langle s^3\rangle/\langle s^4\rangle^2$	1.4005...	1.38(3)

Above the upper critical dimension $d_c = 4$, exponents and universal moment ratios are determined by tree-level diagrams, such as

$$4\langle s^2\rangle = 2\frac{1}{L}\left(\frac{2}{L}\right)^3 \sum_{\substack{n,m,l\\ \text{odd}}} \frac{4}{q_l q_m}$$

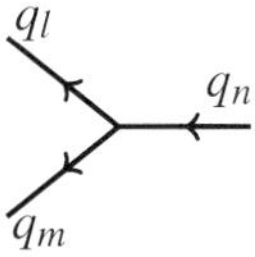

$$\frac{2}{q_n} = \frac{L^6}{560D^2} \tag{13.2}$$

which are easily evaluated numerically, and somewhat more tediously in closed form. These tree-level diagrams give rise to the mean-field theory, which can be extracted and analysed in its own right. The effective mean-field process is a branching random walk, controlled by a branching ratio that self-organises to the critical point. The statistics of that branching random walk is given by the tree-level diagrams. Alternatively, it can be characterised using standard techniques, such as generating functions. The resulting recurrence relations can be solved analytically to give

$$\langle s\rangle = \frac{1}{6}L^2 \tag{13.3}$$

$$\langle s^2\rangle = \frac{1}{140}L^6 + \cdots \tag{13.4}$$

$$\langle s^3\rangle = \frac{131}{138600}L^{10} + \cdots \tag{13.5}$$

$$\langle s^4\rangle = \frac{1129}{5405400}L^{14} + \cdots \tag{13.6}$$

$$\langle s^5\rangle = \frac{18961343}{293318625600}L^{16} + \cdots \tag{13.7}$$

where most of the messy calculations has been performed by means of a computer algebra program [41]. While the (dimensionful) amplitudes are not universal, suitable ratios of the amplitudes are, see Table 13.1.

What is more, the field theory also explains the self-organisation process itself. It turns out that the bare propagator is not *a priori* massless, as the interaction with the lattice and the particles stuck there can hinder or promote spreading of activity. The diagrams contributing to the mass, however, cancel strictly if stationarity is imposed.

This cancellation is caused by a symmetry of two vertices, which in turn is a result of demanding bulk conservation. It is clear, however, that bulk conservation is not in itself a necessary ingredient.

The field theoretical description of the Manna Model also applies to the generalised model with carrying capacity n, where the stationary, spatially averaged particle density fluctuates just above $n/2$. As n increases, the mean value of the particle density approaches $n/2$, corresponding to a branching ratio fluctuating about $1/2$. The fluctuations are irrelevant above the upper critical dimension, but need to be considered in a systematic way, as done in field theoretic renormalisation, below the upper critical dimension. The stationary value itself experiences a non-universal shift away from half-filling, $n/2$, just like the critical temperature is lowered relative to the mean-field value in ferromagnetic phase transitions.

Most importantly, below the upper critical dimension, the field theory gives access, for the first time, to a systematic ϵ-expansion of the critical exponents, which can be compared to the results obtained numerically. As the Manna Model and the quenched Edwards-Wilkinson equation are in the same universality class, the present field theory applies equally to the latter. The quenched noise there might be represented by the substrate configuration in the Manna Model.

13.3 Summary

Using the language of second quantisation, a field theoretic description of the Manna Model can be obtained. This allows, for the first time, a route of analysis of Self-Organised Criticality by means of a renormalised field theory, the most successful method in statistical mechanics. The bare propagator of the field theory incorporates the open boundary conditions which have to be applied in order to maintain a stationary state. A number of exact results, as well as leading order behaviour, can be derived directly from the bare propagator. While a mechanism exists to generate mass, the propagator's mass vanishes due to a certain symmetry between vertices, so that an entire set of diagrams vanishes necessarily in the stationary state. Both mechanism and analytical results can be probed above the upper critical dimension, confirming them by comparison to recent numerical results.

References

1. Bak P, Tang C, Wiesenfeld K (1987) Self-organized criticality: an explanation of $1/f$ noise. Phys Rev Lett 59(4):381–384
2. Jensen HJ (1998) Self-organized criticality. Cambridge University Press, Cambridge
3. Pruessner G (2012) Self-organised criticality. Cambridge University Press, Cambridge
4. Manna SS (1991) Two-state model of self-organized criticality. J Phys A, Math Gen 24(7):L363–L369
5. Christensen K, Corral Á, Frette V, Feder J, Jøssang T (1996) Tracer dispersion in a self-organized critical system. Phys Rev Lett 77(1):107–110

6. Nakanishi H, Sneppen K (1997) Universal versus drive-dependent exponents for sandpile models. Phys Rev E 55(4):4012–4016
7. Huynh HN, Pruessner G, Chew LY (2011) The Abelian manna model on various lattices in one and two dimensions. J Stat Mech 2011(09):P09024
8. Huynh HN, Pruessner G (2012) Universality and a numerical ϵ-expansion of the Abelian manna model below upper critical dimension. Unpubished
9. Lübeck S (2004) Universal scaling behavior of non-equilibrium phase transitions. Int J Mod Phys B 18(31/32):3977–4118
10. Pruessner G, Huynh HN (2012) The manna model on cylinderical, large aspect ratio lattices and above the upper critical dimension. Unpubished
11. Pruessner G (2003) Oslo rice pile model is a quenched Edwards-Wilkinson equation. Phys Rev E 67(3):030301(R)
12. Le Doussal P, Wiese KJ, Chauve P (2002) Two-loop functional renormalization group theory of the depinning transition. Phys Rev B 66(17):174201
13. Paczuski M, Maslov S, Bak P (1996) Avalanche dynamics in evolution, growth, and depinning models. Phys Rev E 53(1):414–443
14. Paczuski M, Boettcher S (1996) Universality in sandpiles, interface depinning, and earthquake models. Phys Rev Lett 77(1):111–114
15. Díaz-Guilera A (1992) Noise and dynamics of self-organized critical phenomena. Phys Rev A 45(2):8551–8558
16. Díaz-Guilera A (1994) Dynamic renormalization group approach to self-organized critical phenomena. Europhys Lett 26(3):177–182
17. Zhang YC (1989) Scaling theory of self-organized criticality. Phys Rev Lett 63(5):470–473
18. Pérez CJ, Corral Á, Díaz-Guilera A, Christensen K, Arenas A (1996) On self-organized criticality and synchronization in lattice models of coupled dynamical systems. Int J Mod Phys B 10(10):1111–1151
19. Corral Á, Díaz-Guilera A (1997) Symmetries and fixed point stability of stochastic differential equations modeling self-organized criticality. Phys Rev E 55(3):2434–2445
20. Pietronero L, Vespignani A, Zapperi S (1994) Renormalization scheme for self-organized criticality in sandpile models. Phys Rev Lett 72(11):1690–1693
21. Chessa A, Vespignani A, Zapperi S (1999) Critical exponents in stochastic sandpile models. In: Proceedings of the Europhysics conference on computational physics CCP 1998, Granada, Spain, September 2–5, 1998. Comp phys commun, vols 121–122, pp 299–302
22. Marsili M (1994) Renormalization group approach to the self-organization of a simple model of biological evolution. Europhys Lett 28(6):385–390
23. Loreto V, Pietronero L, Vespignani A, Zapperi S (1995) Renormalization group approach to the critical behavior of the forest-fire model. Phys Rev Lett 75(3):465–468
24. Hasty J, Wiesenfeld K (1997) Renormalization of one-dimensional avalanche models. J Stat Phys 86(5–6):1179–1201
25. Hasty J, Wiesenfeld K (1998) Renormalization group for directed sandpile models. Phys Rev Lett 81(8):1722–1725
26. Hwa T, Kardar M (1989) Dissipative transport in open systems: an investigation of self-organized criticality. Phys Rev Lett 62(16):1813–1816
27. Grinstein G, Lee DH, Sachdev S (1990) Conservation laws, anisotropy, and "self-organized criticality" in noisy nonequilibrium systems. Phys Rev Lett 64(16):1927–1930
28. Dickman R, Vespignani A, Zapperi S (1998) Self-organized criticality as an absorbing state phase transition. Phys Rev E 57(5):5095–5105
29. Rossi M, Pastor-Satorras R, Vespignani A (2000) Universality class of absorbing phase transitions with a conserved field. Phys Rev Lett 85(9):1803–1806
30. van Wijland F (2002) Universality class of nonequilibrium phase transitions with infinitely many absorbing states. Phys Rev Lett 89(19):190602
31. Bonachela JA, Chaté H, Dornic I, Muñoz MA (2007) Absorbing states and elastic interfaces in random media: two equivalent descriptions of self-organized criticality. Phys Rev Lett 98(15):155702

32. Bonachela JA, Muñoz MA (2007) How to discriminate easily between directed-percolation and manna scaling. Physica A 384(1):89–93. Proceedings of the international conference on statistical physics, Raichak and Kolkata, India, January 5–9, 2007
33. Basu M, Basu U, Bondyopadhyay S, Mohanty PK, Hinrichsen H (2012) Fixed-energy sandpiles belong generically to directed percolation. Phys Rev Lett 109:015702
34. Vespignani A, Dickman R, Muñoz MA, Zapperi S (1998) Driving, conservation and absorbing states in sandpiles. Phys Rev Lett 81(25):5676–5679
35. Tang C, Bak P (1988) Critical exponents and scaling relations for self-organized critical phenomena. Phys Rev Lett 60(23):2347–2350
36. Pruessner G, Peters O (2006) Self-organized criticality and absorbing states: lessons from the Ising model. Phys Rev E 73(2):025106(R)
37. Peters O, Pruessner G (2009) Tuning- and order parameter in the soc ensemble. Unpubished
38. Dhar D (1999) Studying self-organized criticality with exactly solved models. arXiv:cond-mat/9909009
39. Diehl HW, Schmidt FM (2011) The critical Casimir effect in films for generic non-symmetry-breaking boundary conditions. New J Phys 13(12):123025
40. Pruessner G (2012) The average avalanche size in the manna model and other models of self-organised criticality. Preprint. Available from: arXiv:1208.2069
41. Mathematica (2011) Wolfram Research, Inc, Champaign, IL, USA, version 8.0.1.0

Chapter 14
Chaos and Non-linear Tools in Website Visits

Maria Carmela Catone

Abstract The present work is an application of linear and non-linear tools, with particular attention to the chaotic dynamics, in order to analyze the daily visits to the Italian newspaper website "La Repubblica". The series is examined, using the time chart, the recurrence plot and the power spectrum. In the phase space, the detrend series consists of 5 clusters of points, explained by the frequency distribution that is centred on 3 values. The analysis is performed by calculating the embedding dimension, Lyapunov exponents and the correlation dimension that suggests the existence of an attractor. A non-linear forecast of the following values is made. In conclusion, some theoretical issues on the characteristics of a chaotic system emerge.

Keywords Chaos · Recurrence · Websites

14.1 Introduction

This work is a technical application of linear and non-linear tools, with particular attention to chaotic dynamics, in order to understand the trend in the number of visits received by the Italian newspaper website "La Repubblica". Despite the consistency of the influx of visits, it was possible to identify regularities in the behaviour models in the use of the website. Moreover, from this case study, some theoretical issues on the characteristics of a chaotic system emerge. The work is made up of part of a description, an analysis of the number of hits registered over three years (from 31 March 2008 to 21 September 2011) and a prediction; the data is available on the audiweb website and the following software are used: Visual Recurrence Analysis (VRA), Chaos Data Analyzer (CDA) and Excel.

M.C. Catone (✉)
Department of Political Science and Sociology, University of Florence, Florence, Italy
e-mail: mariacarmela.catone@unifi.it

T. Gilbert et al. (eds.), *Proceedings of the European Conference on Complex Systems 2012*, Springer Proceedings in Complexity, DOI 10.1007/978-3-319-00395-5_14,

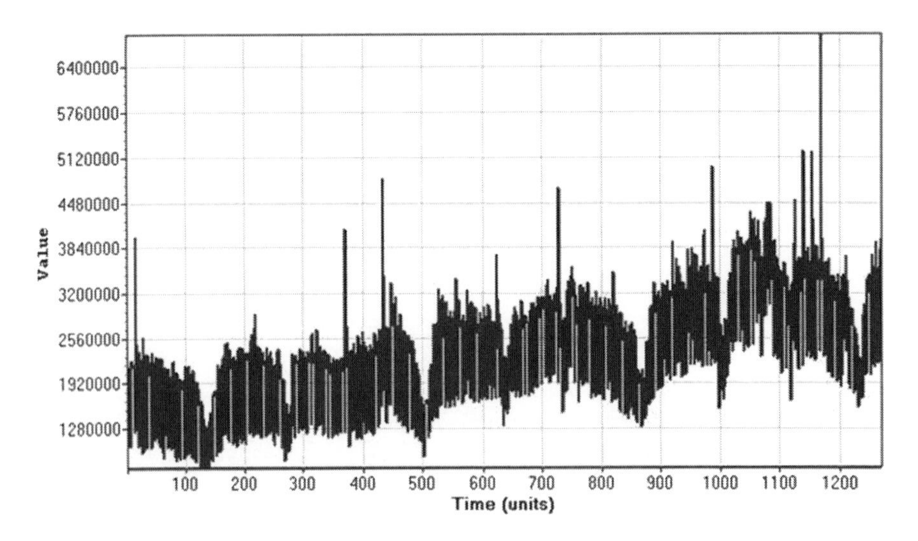

Fig. 14.1 Time series chart

14.2 Development of the Research

At first, the general structure of the series was studied using the VRA [1].

The time series chart (number of visits versus time) is characterized by the presence of a growing trend, peaks caused by exceptional events of everyday life (e.g. the fall of government, earthquakes, etc.) and minimums that occur during the periods August and Christmas (Fig. 14.1).

In particular, the daily visits are on average 2442065 and during the summer holidays they decrease by almost 40 %. Furthermore, during the week there are more visits from Monday to Friday, but they decrease on Saturday and Sunday.

In order to carry out a more thorough examination of this configuration, the number of weekly visits was observed in the phase space with a time delay of 7 days. The graph shows the number of hits on one day in a week versus the number recorded on the same day of the preceding week. A cluster of points emerges around the bisector that extends from 600000 to over 4500000 visits. This distribution means that the values of visits on consecutive weeks tend to be similar. Following this, by carrying out the same analysis with a one-day delay, the occurrences from one day to the next could be examined. In this phase space (Fig. 14.2) the points accumulate along three lines (a bisector and two shorter parallels) and therefore:

- after one day, the number of visits on the bisector tends to remain unchanged. This happens in the transitions Monday–Tuesday, Tuesday–Wednesday, Wednesday–Thursday, Thursday–Friday and Saturday–Sunday. The highest values of the visits are on the upper-middle part of the bisector, and along the lower part of the same line the values of Saturday-Sunday can be found;
- the upper line is associated mainly to the transition Sunday–Monday because the values of the next day are higher than those of the previous day;

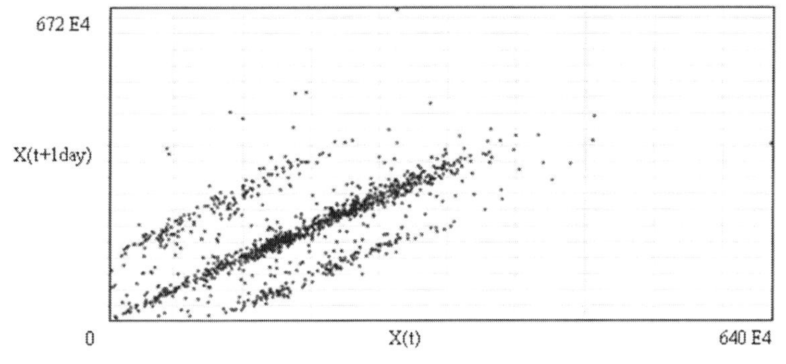

Fig. 14.2 Phase space plot with a one-day delay

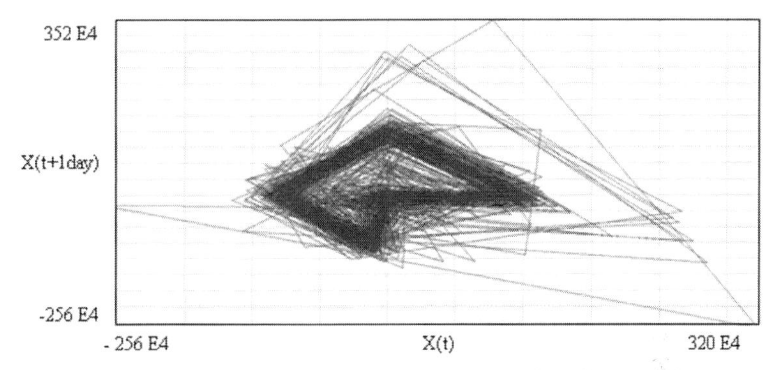

Fig. 14.3 Phase space plot with a one-day delay in the series without linear trend

- the lower line indicates the movement from Friday to Saturday because the values of the successive day are lower than those of the preceding day.

The elimination of the trend is then carried out as some analysis tools require, from a formal point of view, a stationary series, so that the mean and variance are constant in consecutive time periods. The deletion of the linear trend is obtained by subtracting the previous value from its current value. Once the linear trend is removed, stationarity is evident through a simple inspection of its time chart.

Furthermore, in the phase space with a one-day delay there is a configuration (Fig. 14.3) that is completely different from that of the initial series. Actually, it consists of five clusters of points that directly specify the variations of the visits during the week. Joining the consecutive points, paths which are almost cyclic are obtained mainly in a closed area. Representing the data in the phase space with a delay of 2, 3, 4, 5, 6 days, other cyclic diagrams are obtained.

The recurrence plots of the initial and no-trend series are worked out and they put the indexes of the delayed vectors, made up of the data of the series, on the x and y axes. The Euclidean distances between these vectors are represented using a colour scheme: short and long distances are indicated by light and dark colours respectively [2, 3]. The "La Repubblica" recurrence plot is characterized by white/red squared areas around the diagonal in a yellow/green context. The non-homogeneity of the diagram and the white areas around the diagonal show the presence of a non-stationary process. Additionally, there are white line diagonals that denote elements of determinism together with horizontal and vertical lines that are signs of laminarity. The stationarity of the no-trend series emerges, instead, from the greater uniformity of its recurrence plot.

This series was then analyzed with the Chaos Data Analyzer (CDA). A section of the program allows us to establish the possible periodicity of the series. The series shows a periodic behaviour with a fundamental frequency of 0.142 cycles per day. The period (i.e. the inverse of the frequency) equals 7 days, so the series has a weekly periodicity where annual summer and Christmas cycles with minor amplitude are also included. Moreover, the repetition of the development of the signal allows us to make average predictions. Another unit of the program shows that the frequency distribution of the series is centred on three points, characterized by variations of visits: $-B$ (negative), X (almost zero), A (positive). The central population is greater than the almost identical lateral populations.

In an ideal model where weekly visits are on two levels (with a linear growing trend), i.e. upper (Monday-Friday) and lower (Saturday-Sunday), their variations from Monday to Sunday consecutively make a sequence of this type: A, X, X, X, X, $-B$, X. It follows that the variations of the visits on a particular day compared with the next day's visits are identified by these kinds of coordinates: (A, X), (X, X), (X, X), (X, X), (X, $-B$), ($-B$, X), (X, A). This result explains the origin of the five clusters of points in the phase space with a one-day delay, as previously shown.

Next, the series was studied using non-linear analysis tools in order to extract possible chaotic behaviour which occurs in a long run, in an aperiodic way and in a deterministic and bounded system sensitive to the initial conditions [4]. The series lacks the requirement of aperiodicity, but its periodicity is not exact because the signal does not repeat itself perfectly. Thus, I carried out the process of identifying the presence of chaos in the amplitudes of oscillations. The analysis was performed by calculating various coefficients: the correlation dimension, the embedding dimension and Lyapunov exponents [5]. The correlation dimension, which is one of the ways of determining the existence and size of the attractor, was already adequately levelled in the initial series, also because the linear trend was present. In the no-trend series, the correlation dimension tends to stabilize at around the value of 4, suggesting the existence of an attractor. The minimum embedding dimension, i.e. the size of phase space from which we observe the image that preserves the properties of the attractor, equals 4. The Lyapunov exponents are positive, showing a sensitivity to initial conditions.

Following this, the series was surrogated in the phase to validate the presence of determinism in the data. In the new series, the correlation dimension tends to increase without levelling out, Lyapunov exponents usually grow and the points in the phase space form a cloud where the early regularities no longer exist. So the hypothesis that the "La Repubblica" series contains parts of determinism is confirmed as the mixing of Fourier components of the signal produces a more irregular sequence of data.

After extrapolating the data from the series, a forecast of the following values was made, using non-linear analysis tools. Additionally, examining the visits of another Italian newspaper website, "Il Corriere della Sera", a similar behaviour to that of "La Repubblica" was identified.

To explore that a function, on average periodic, can have a chaotic component, I determined the sequence:

$$Zn = Xn + Yn \tag{14.1}$$

where Xn assumes the constant value of 4 and thus a period of 1, and Yn was extracted from the map:

$$Yn + 1 = (1.2\ Yn) \bmod 1 \tag{14.2}$$

which has a chaotic behaviour, being the Lyapunov exponent (log 1.2) positive. The examination of Zn with CDA provided a correlation dimension of around 1 and an embedding (minimum) dimension equal to 1. The Lyapunov exponent of Zn, for proper values of parameters, tends to equal that of Yn. Thus, the Zn is on average periodic, but also chaotic. Therefore, predictions could be made by repeating the averaged function in a period or using non-linear criteria.

In conclusion, it is noted that while the use of the periodicity provides only average values, the non- linear analysis of the series, resuming the determinism of the system, can offer more detailed and precise predictions.

References

1. Marwan N, Romano MC, Thiel M, Khurts J (2007) Recurrence plots for the analysis of complex systems. Phys Rep 438(5–6):237–329
2. Faggini M (2007) Visual recurrence analysis: an application to economic time series. In: Salzano M, Colander D (eds) New economic windows: complexity hints for economic policy, vol 2. Springer, Berlin, pp 69–92
3. Zbilut JP (2005) Use of recurrence quantification analysis in economic time series. In: Salzano M, Kirman A (eds) New economic windows: economics—complex windows. Springer, Berlin, pp 91–104
4. Sprott JC (2003) Chaos and time-series analysis. Oxford University Press, New York
5. Alligood KT, Sauer TD, Yorke JA (2007) Chaos—an introduction to dynamical systems. Springer, New York

Chapter 15
Networks and Cycles: A Persistent Homology Approach to Complex Networks

Giovanni Petri, Martina Scolamiero, Irene Donato, and Francesco Vaccarino

Abstract Persistent homology is an emerging tool to identify robust topological features underlying the structure of high-dimensional data and complex dynamical systems (such as brain dynamics, molecular folding, distributed sensing).

Its central device, the filtration, embodies this by casting the analysis of the system in terms of long-lived (*persistent*) topological properties under the change of a scale parameter.

In the classical case of data clouds in high-dimensional metric spaces, such filtration is uniquely defined by the metric structure of the point space. On networks instead, multiple ways exists to associate a filtration. Far from being a limit, this allows to tailor the construction to the specific analysis, providing multiple perspectives on the same system.

In this work, we introduce and discuss three kinds of network filtrations, based respectively on the intrinsic network metric structure, the hierarchical structure of its cliques and—for weighted networks—the topological properties of the link weights. We show that persistent homology is robust against different choices of network metrics. Moreover, the clique complex on its own turns out to contain little information content about the underlying network. For weighted networks we propose a filtration method based on a progressive thresholding on the link weights, showing that it uncovers a richer structure than the metrical and clique complex approaches.

Keywords Complex networks · Persistent homology · Metrics · Computational topology

15.1 Introduction

Over the last decade complex networks have become one of the prominent tools in the study of social, technological and biological systems. By virtue of their sheer

G. Petri (✉) · F. Vaccarino
ISI Foundation, Via Alassio 11/c, 10126 Torino, Italy

M. Scolamiero · I. Donato · F. Vaccarino
Dipartimento di Scienze Matematiche, Politecnico di Torino, C.so Duca degli Abruzzi n. 24, 10129 Torino, Italy

T. Gilbert et al. (eds.), *Proceedings of the European Conference on Complex Systems 2012*, Springer Proceedings in Complexity, DOI 10.1007/978-3-319-00395-5_15,

sizes and complex interactions, they cannot be meaningfully described and controlled through classical reductionist approaches.

Within this framework, the study of the topology of complex networks, and its implications for dynamical processes on them, has most often focused on the statistical properties of nodes and edges and therefore found a natural and effective description in terms of statistical mechanical models of graph ensembles [1, 2]. These models rely for their formulations on local interactions and become quickly hard to manage when higher correlations are included or one-step approximations are not sufficient, as Schaub et al. [3] pointed out for the case of community detection algorithms for example.

The last few years saw a new perspective emerge that focuses on the very *geometry* of complex network. It was promoted by a large availability of new (typically geosocial) data coming from spatial networks [4], but also by analytical and numerical results on the relations between geometrical properties and global features of complex networks, e.g. the hyperbolic embedding of the Internet with the resulting increased efficiency of greedy routing algorithms [5], stationarity conditions for chemical networks [6] and brain cortex dynamics [7].

In this work, we take on this perspective and study the geometrical properties of networks through the goggles of *persistent homology*, a technique originally introduced by [8, 9] to uncover robust topological information from noisy high-dimensional point clouds. Persistent homology works by extracting from a dataset a growing sequence of simplicial complexes (called *filtration*), indexed by a parameter ϵ, and studying the associated homology groups, which encode the geometrical information (for example, the holes of an n-torus). The robustness of each topological feature is then obtained from the *persistence* of the corresponding generator along the filtration,

For example, in the case of the torus, there will be two persistent generators associated to the two non-equivalent loops on its surface.

Persistent homology has received some attention in the context of networks [10], but there has been no systematic study on its efficiency and sensibility for networks yet. This is of particular importance since, in contrast with the unique natural metric available for point cloud datasets, networks allow various rules to generate the filtration.

Our results will show that the salient features of the homology do not change significantly under different metrics and that there exist a metric scale ϵ_c at which the filtration displays the richest structure.

We will then study a second method to create the filtration, relying only on the network clique structure. Unfortunately, this will turn out to yield little additional information.

In the case of weighted networks it is possible to devise a refined filtration based on the clique structure of the network thresholded by ϵ, which yield a much richer picture than the simple clique complex method.

The rest of this work is organised as follows. In the next section a minimal introduction to homology and its persistent sister is given. The following section will present selected results of simulations and datasets under different choices of metrics for the network filtration.

We conclude then presenting the procedure for the filtration built with the link-thresholded clique structure and briefly discuss the results and implications for future research. In particular, we have discuss the possibility of expanding the method by considering multi-filtrations, that is filtrations indexed by more than one parameter.

15.2 Homology

Formally, homology is an algebraic invariant converting local geometric information of a space into a global descriptor. There are many homology theories, but simplicial homology is the most amenable for computational purposes thanks to its combinatorial structure.

This kind of homology is applied to simplicial complexes, that are combinations of vertices, segments, triangles and higher dimensional analogues, joined according to specific compatibility relations. As we will see in the following, simplicial complexes can be constructed from discrete spaces or networks. Low dimensional homology groups have an intuitive interpretation. Given a simplicial complex X, $H_0(X)$ is the free group generated by the connected components of X, $H_1(X)$ is the free group generated by the cycles in X, $H_2(X)$ is the free group generated by voids—holes bounded by two-dimensional faces. The Betti numbers count the number of generators of such homology groups.

The standard tool to encode this information is the so-called *barcode*, which is a collection of intervals representing the lifespans of such generators. Long-lived topological features can be distinguished in this way from short-lived ones, which can be considered as topological noise. There are various ways of building persistence modules out of a given dataset. The most known are the Rips-Vietoris complex, the Cech complex and the clique complex [8]. The first two require a metric space for the data and are generated by inflating spheres of the same radius around points (or nodes in a network) and associating set of points to simplices according to the overlap of the corresponding spheres. They can also be used to create a filtration out of general network, once a metrical structure is given on the network itself (shortest-path, commute time distance, etc). Besides these two methods, there exist a few methods pertaining to networks only [8], the best known being the *clique complex*, which is generated by associating to each maximal clique the simplex generated by the vertices of the clique.

15.3 Robustness Against Metric Change

Network metrics have been well studied, especially in the context of clustering algorithms [11] and Markov Chain models [12]. In addition to the shortest path and commute time metrics, it is possible to define kernel matrices as functions of the

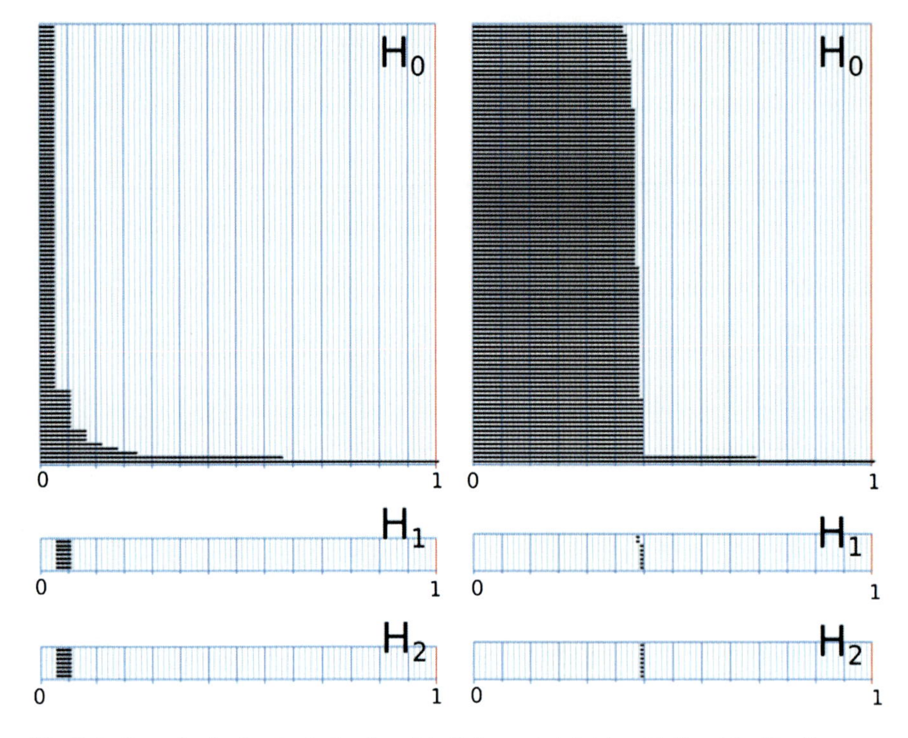

Fig. 15.1 Barcodes for the shortest path metric (*left*, panels (**a**), (**c**) and (**e**)) and the Von Neumann metrics (*right*, panels (**b**), (**d**) and (**f**)) on the C. Elegans brain network. From *top* to *bottom*, we report the intervals of existence of the homological spaces H_0 (panels (**a**) and (**b**)), H_1 (panels (**c**) and (**d**)), H_2 (panels (**e**) and (**f**)). The parameter $\epsilon \in [0, 1]$ increases from *left* to *right*. *Each horizontal line* corresponds to the intervals of existence of a generator of the corresponding homology space. In both cases, the higher homology space are non-trivial only in the vicinity of the merging of a large number of connected components, as highlighted by the drastic reduction in the number of generators of H_0

network's adjacency and Laplacian matrices. From such kernels one obtains a well-defined distance, which effectively turns the network into a metric space.

We analysed the metrics associated to: the shortest paths, the commute time between nodes, exponential diffusion [13] and exponential Laplacian diffusion [11, 14], which emerge as solutions of diffusion processes on the corresponding network, the von Neumann kernel [15],which generalises the hub-authority measures, Markov diffusion [16] and random walks with restart.

For each metric, the filtration was generated and the persistent homology calculated. The analysis was repeated on a range of different networks, spanning different network topologies, sizes and origins (biological, social, technological).

For brevity, in this paper, we show only the comparison of the barcodes obtained using the shortest path and the exponential diffusion (with $\alpha = 0.01$) distances for the *C. elegans* neuronal network (Fig. 15.1). In order to compare the results both metrics have been mapped to the interval [0, 1]. Surprisingly, we found that the higher homology spaces (H_1, $H_2 \ldots$, bottom plots in Fig. 15.1) are trivial for most values of the filtration parameter. They do however show the appearance of generators of higher homology groups in the vicinity of the value of ϵ at which a significant number of connected components merges into few, as shown by the decrease in the number of generators of H_0.

In this respect, our results suggest the existence of a particular value ϵ_c, a *metric scale*, at which one observes the most structure in the metrical representation of the network under study. The same behaviour was found in a number of other networks, ranging from the US air passenger network to the human gene regulatory one. Note moreover that, in general, ϵ_c is different from the average distance between the nodes (in terms of the chosen metric) and therefore cannot be explained as a mere effect of the distances distribution. Moreover, if the appearance of non trivial higher homology groups was only due to the merging of small connected components into a giant component, one would expect to observe the same phenomenon also for the merging of smaller components. However, we did not see any of these signatures, supporting the existence of a characteristic scale ϵ_c.

15.4 Clique Complex and Link Weights Thresholding

Another natural filtration of a network is generated by considering its clique structure. The clique complex is obtained by associated to each maximal k-clique, a completely connected subgraph formed by k nodes, the $(k-1)$-simplex whose vertices are the nodes of the clique. The natural parameter for this filtration is the clique dimension k. Recent work [10] tried to uncover specific signatures of modular and cluster structures in complex networks by making use of this filtration. In our analysis the filtration obtained in this way did not show interesting features in addition to the clique structure itself, which however can be investigated without recurring to homological concepts. However, if we consider weighted networks, it is possible to devise a filtration which combines link weights and clique structure. Given the weighted adjacency matrix ω_{ij}, we let ϵ vary in $(\min \omega_{ij}, \max \omega_{ij})$ and consider a sequence of networks, such that the network at step ϵ contains all links (i, j) with $\omega_{ij} > \epsilon$. As we decrease ϵ from its maximum allowed value, we go from the empty network to the original one. For each step, we build the corresponding clique complex and study the persistent homology of the resulting filtration. Figure 15.2 shows the results of this filtration on a large Facebook-like network of online contacts. It is immediately evident that a very rich topological information is present. Long persistent intervals appear both for some generators of H_1 and H_2. The first implies the existence of chains composed by edges with large weights, whose nodes though are not strongly connected across the chain itself, but only with their two neighbours along the chain. The same reasoning applies to the case of H_2 where the building blocks are not segments but triangles. The presence of long persistent H_2 generators is a signpost for higher ordering in the structure of the online contacts. This means that strong pair interactions organise in long loops without significant triadic closure.

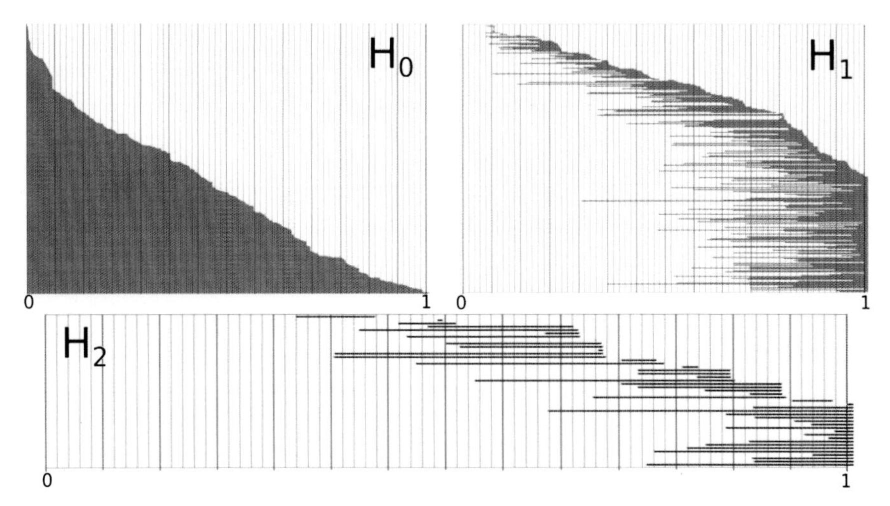

Fig. 15.2 Barcodes obtained from the weighted-clique complex filtration of a network of online contacts for the homology groups H_0 (**a**), H_1 (**b**) and H_2 (**c**). Persistent H_1 and H_2 generators imply that the existence of loops and chains of tethraidra formed by nodes which are weakly interacting with their neighbours in the chain, with the exception of the one directly adjacent *along* the chain. In the case of human contacts, this means that strong pair interactions organise in long loops without significant triadic closure

Finally, we can conclude that this method is able to identify mesoscopic and long-range structures which are present in networks, but would otherwise pass undetected with standard methods, and assigns to them also a measure of robustness in the form of the persistence intervals.

Acknowledgements The authors acknowledge Mario Rasetti for insightful discussions and constant support.

References

1. Newman M (2010) Networks: an introduction. Oxford University Press, New York
2. Albert R, Barabasi A (2002) Statistical mechanics of complex networks. Reviews of Modern Physics 74(1):47–97
3. Schaub MT, Delvenne JC, Yaliraki SN, Barahona M (2012) Markov dynamics as a zooming lens for multiscale community detection: non clique-like communities and the field-of-view limit. PloS One 7(2):e32210
4. Barthlemy M (2011) Spatial networks. Physics Reports 499(13):1–101
5. Boguna M, Papadopoulos F, Krioukov D (2010) Sustaining the internet with hyperbolic mapping. Nature Communications 1:62
6. Conradi C, Flockerzi D, Raisch J, Stelling J (2007) Subnetwork analysis reveals dynamic features of complex (bio)chemical networks. Proceedings of the National Academy of Sciences 104(49):19175–19180
7. Henderson JA, Robinson PA (2011) Geometric effects on complex network structure in the cortex. Phys Rev Lett 107:018102

8. Zomorodian A, Carlsson G (2005) Computing persistent homology. Discrete Comput Geom 33(2):249–274
9. Carlsson G (2009) Topology and data. Bulletin of the American Mathematical Society 46(2):255–308
10. Horak D, Maletic S, Rajkovic M (2009) Persistent homology of complex networks. Journal of Statistical Mechanics: Theory and Experiment 2009(03):P03034
11. Fouss F, Yen L, Pirotte A, Saerens M (2006) An experimental investigation of graph kernels on a collaborative recommendation task. In: Sixth international conference on data mining (ICDM'06), pp 863–868
12. Bolch G, Greiner S, de Meer H, Trivedi KS (1998) Queueing networks and Markov chains: modeling and performance evaluation with computer science applications. Wiley-Interscience, New York
13. Kondor R, Lafferty J (2002) Diffusion kernels on graphs and other discrete input spaces. In: Proceedings of the nineteenth international conference on machine learning (ICML'02), pp 315–322
14. Smola A, Kondor R (2003) Kernels and regularization on graphs. In: Learning theory and kernel machines. Lecture notes in computer science, vol 2777, pp 144–158
15. Kandola J, Shawe-Taylor J, Cristianini N (2002) Learning semantic similarity. Advances in neural information processing systems 15:657–666
16. Yen L, Fouss F, Decaestecker C, Francq P, Saerens M (2007) Graph nodes clustering based on the commute-time kernel. In: Zhou ZH, Li H, Yang Q (eds) Advances in knowledge discovery and data mining. Lecture notes in computer science, vol 4426. Springer, Berlin, pp 1037–1045

Chapter 16
Von Neumann Reproduction: Preliminary Implementation Experience in Coreworlds

Barry McMullin, Declan Baugh, and Tomonori Hasegawa

Abstract We introduce the distinctive, self-referential, logic of self-reproduction originally formulated by John von Neumann, and we present some initial results from novel realisations of this abstract architecture, embedded within two computational worlds: Tierra and Avida. In both cases, the von Neumann architecture proves to be evolutionarily fragile, for unanticipated, but relatively trivial, reasons. We briefly discuss some implications, and sketch prospects for further investigation.

Keywords John von Neumann · Artificial life · Self-reproduction · Coreworlds · Tierra · Avida

16.1 Introduction

As early as 1948, John von Neumann had already formulated and essentially resolved a fundamental paradox in the theory of the evolutionary growth of machine complexity: namely, how any (assumed or "divinely" created) seed machine can, directly or indirectly, give rise to machines arbitrarily more complex than itself [3]. Inspired by Turing's general purpose (programmable) computing machines, his resolution relied on a machine architecture comprising a general purpose programmable *constructor* which could act to decode a symbolic description of an arbitrarily (more) complex target machine and thus construct it. As a special case, this also led to a generic architecture for machine *self-reproduction* (where the description is now a *self*-description, and must be *copied* as well as *decoded*).

This self-reproduction architecture, formulated very abstractly by von Neumann, was subsequently found to be strikingly reminiscent of the biological role of DNA

B. McMullin (✉) · D. Baugh · T. Hasegawa
The Rince Institute, Dublin City University, Dublin 9, Ireland
e-mail: barry.mcmullin@dcu.ie

D. Baugh
e-mail: declan.baugh2@mail.dcu.ie

T. Hasegawa
e-mail: tomonori.hasegawa2@mail.dcu.ie

T. Gilbert et al. (eds.), *Proceedings of the European Conference on Complex Systems 2012*, Springer Proceedings in Complexity, DOI 10.1007/978-3-319-00395-5_16,

(as "symbolic description") and of the molecular machinery of the "genetic code" whereby ribosomes (supported by tRNAs and other enzymes) decode or "translate" symbolic descriptions (presented as mRNAs) into arbitrarily complex protein molecules (and protein machinery). Indeed, the prescient nature of von Neumann's contribution is made clear from the fact that the chemical structure of DNA was not elucidated until 1953, and the programmable "decoding" or "translation" function of the ribosome was not fully formulated until 1960 (the code "proper" only later clarified as being implemented by the aminoacyl-tRNA synthetases).

More generally, von Neumann's architecture gave a concrete mechanical interpretation and implementation of the traditional biological idea that an organism can be decomposed into a set of tacit hereditary "factors" (genome) and a corresponding, manifest, functional, form (phenome). As subsequently emphasised by Pattee [6], however, von Neumann's architecture (and its real-world biological counterparts) also carries with it an intriguing example of *self-reference:* the decoding relationship (the "genotype-phenotype mapping", in biological terms) implemented by the programmable constructor is also represented, in encoded form, within the symbolic description (genome)—and this encoding must be precisely according to, or at least consistent with, the very same mapping that the constructor itself (part of the phenome) implements. This most primitive and original form of *self-reference* has been dubbed "semantic closure" by Pattee, and has also been explicitly discussed by Hofstadter [2]; but the full implications of this self-referential closure for understanding, and fabricating, complex self-organising systems are, as yet, poorly understood.

The present contribution presents a brief summary of some preliminary attempts to revisit and explore this issue afresh, through realising the von Neumann reproduction architecture in two abstract, computational, worlds: *Tierra* and *Avida*.

16.2 Von Neumann Reproduction in the Tierra World

Tierra is an artificial life platform where a population of reproducing computer programs ("organisms") compete with each other for both CPU time and memory space [8]. The platform contains a circular core memory, or *soup*, in which the organisms are embedded. The Tierran instruction set consists of 32 assembler language instructions, each occupying a single word, generally with no operands. However, certain op-codes are overloaded: in isolation, they function as `NOP` instructions, but in certain contexts, sequences of them can be interpreted as immediate operands specifying a biologically inspired pattern-based addressing mechanism. There are various stochastic perturbations in program execution, including: *cosmic rays* (alteration applied to the contents of a random memory location within the soup), *copy errors* (an error occurred when reading and/or writing the contents of a memory location) and *segment deletions* (upon birth, a random organism segment is deleted). Each organism is assigned a (virtual) CPU which includes four general purpose registers, an instruction pointer and a small circular stack. A *slicer* allocates CPU time to each living organism within the soup and a *reaper* limits the population by reaping old and malfunctioning organisms.

The normal protocol for evolutionary experiments in Tierra is for the experimenter to first design "by hand" an organism that is capable of self-reproduction. This is then used as a *seed* organism for one or more specific experimental runs. In each run, the seed gives rise to a population, which then diversifies through mutation, and ultimately evolves through selection and other ecological interactions (potentially including parasitism).

Classically, the reproductive mechanism of Tierran organisms relies on an approach pioneered in predecessor systems such as *Coreworld* [7]. In effect, the parent organism simply inspects and copies its own program directly. We may classify this as reproduction by "self-inspection" or "self-copying". Such a mechanism is not possible for complex organisms in the real world for several practical reasons [4]; but it is closely analogous to the more primitive template replication process underlying *in vitro* RNA evolution, and, indeed, to the DNA replication process that is one component of normal biological reproduction. In particular, it does support inheritable variation and evolutionary exploration of a combinatorially large (for practical purposes, infinite) space of distinct organism strains. Under such a self-copying reproduction architecture, there is no distinction between phenotype and genotype as the entire organism acts as both the template for replication and the active, functional, phenotype (the executing program and any associated data).

In our work we have instead designed a seed organism based on the von Neumann reproductive architecture, and having an explicit genotype-phenotype decomposition. Note that this is entirely specified by the particular configuration of the organism's memory image: it is neither required not prevented by the underlying dynamics of the Tierra world. In the first instance the von Neumann "decoding" (genotype-phenotype mapping) has been simply modelled on the standard biological genetic code. That is, it is a sequential mapping from discrete "codon" symbols in the genome to functional "instructions" and "data" symbols in the phenome, implemented via a lookup table located in the (parental) phenome. This table is therefore directly analogous to the functionality implemented by the aminoacyl-tRNA synthetases in RNA-protein translation. The lookup table is itself, self-referentially, encoded into the genome. While it would be expected that this self-referential mapping would be highly conserved (robust) in evolution, we nonetheless conjecture that some significant long term evolution (either selective or by drift) should be observed.

When all stochastic effects are disabled, our seed organism reproduces effectively and populates the memory to form a stable ecosystem of identical organisms. However, when stochastic effects are switched on, we quickly see the emergence of organisms which we classify as "pathological constructors". These are organisms which are not self-reproducing but which rapidly construct multiple short malfunctioning offspring within their lifetime. Pathological constructors are a hindrance to an ecosystem because their offspring, although sterile, still occupy both memory space and CPU time. When several pathological constructors coincide in time, their production rate can be so high that their non-functional offspring displace the entire population of functional self-reproducing creatures. In the specific initial experiments we have run, this has frequently resulted in complete ecosystem collapse.

We note that this total collapse appears to also rely on a subtle, and presumably unintended, heterogeneity in the Tierran world, whereby an organism located at the "origin" of the memory space (absolute address zero) may preferentially "capture" instruction pointers "lost" by malfunctioning organisms. In particular, a pathological constructor at that location can capture instruction pointers initially associated with its own non-functional offspring. This can dramatically increase its individual fecundity; to the point where, even within a finite, "normal", lifetime it can displace all other organisms in the soup.

Under a series of simulations where each source of random perturbation was individually disabled, the disabling of the segment deletions showed an apparent barrier to the emergence of pathological constructors. We conjecture that the explanation is as follows. For von Neumann style reproducers, all random perturbations which corrupt the copying of the genome will result in an offspring which still has a functional phenome, and which will therefore go on to create at least one further offspring (which may be functional or not). If the corruption of the genome copying is by way of a segment deletion, then the initial offspring will generally be capable of quickly and repeatedly generating further offspring, which are non-functional: quickly because both the copying and the decoding of a short genome will be quick; and generating non-functional offspring because the offspring phenome will be the result of decoding a drastically shortened genome. Thus, segment deletion within a genome can be expected to quite typically result in a pathological constructor as we have already characterised it.

This analysis concludes that the mechanism which results in ecosystem collapse due to pathological constructors appears to depend critically on both the inclusion of segment deletion perturbations and the spatial inhomogeneity affecting memory location zero. By contrast, in order for a pathological constructor to emerge from a classic Tierran self-copier organism, relatively much more specific random perturbations must occur affecting very specific locations, which will alter, but not corrupt the reproductive functionality. This suggests that the probability of pathological constructors emerging within a population of von Neumann reproducers in Tierra is much higher than that of a population of self-copiers.

16.3 Von Neumann Reproduction in the Avida World

Avida is another abstract ("simulated" or "virtual") world which has been extensively used to investigate very general properties of spontaneous evolutionary processes [1, 5]. It is loosely inspired by the structure of a conventional, large scale, cluster computer, with many separate computational nodes, each with one general purpose CPU and a limited local memory. The nodes are sparsely interconnected, typically in a regular two dimensional lattice.[1] The CPU instruction set is config-

[1] The Avida world bears some superficial resemblances to von Neumann's own early formulation of an abstract cellular automaton (CA) world, particularly in its 2D network of discrete computa-

urable on a system wide basis. It is normally reminiscent of a conventional microcontroller, but with some specialised features. A program running on a given node can overwrite the memory of a neighbouring node and in this way replace the program running on that node (effectively re-programming the node). Based on this, a suitably designed program can repeatedly and recursively reproduce itself into neighbouring nodes. Such a program, combined with its associated computational node/hardware, is regarded as an abstraction of a biological organism. If an Avida world is initialised or seeded with a single instance of some such hand-designed organism, a population of organisms will grow to occupy the entire world roughly in the manner of bacteria in a Petri dish. Certain CPU operations in Avida are made unreliable by design. This has the effect that mutant strains of organism can spontaneously arise, multiply, and compete in a Darwinian manner for the finite available "space" (nodes) in the system. Unlike the Petri dish analog, a culture of Avida organisms can be continuously replenished with "nutrients" (analogous to a continuous flow bioreactor) and thus the ongoing evolutionary process can, in principle, be continued indefinitely.

As with Tierra, the "standard" mechanism whereby self-production is achieved in Avida is based on self-copying; but again we have designed a novel organism which incorporates the characteristic genotype-phenotype structure and self-referential genotype-phenotype mapping originally described by von Neumann.

As yet, only preliminary experiments have been run and analysed with this novel self-referential seed organism in the Avida world. However, the consistent experience to date has been that instead of observing either simple conservation or long term evolution in the genotype-phenotype mapping we see relatively rapid degeneration of the entire reproduction mechanism—i.e., emergence of "conventional", non-self-referential, self-copying organisms, comparable to the standard seed organisms. These organisms lack the decomposition into distinct genome and phenome components, lack any genotype-phenotype mapping process, and therefore also lack the characteristic von Neumann self-reference (or Pattee's semantic closure); but they are still capable of self-reproduction. Once such organisms emerge they are selectively favoured in this world (as they avoid the computational load of translation/decoding without incurring any immediate offsetting penalty). It follows that this degeneration is essentially irreversible.

Thus, it is not surprising that self-copiers should selectively displace self-referencing organisms in Avida; nor is it *very* surprising that there might be some available mutational pathways for such degenerative strains to appear, as the Avida world is specifically designed to make reproduction by self-copying extremely easy (it can be achieved with a program as short as 15 instructions under the default instruction set). However, what was surprising was that this degeneration could occur

tional nodes. However, there are also fundamental differences. In the von Neumann CA, each node was a simple finite state automaton with no general purpose memory system (29 states per node, equivalent to less than 5-bits of special purpose memory); whereas each Avida node comprises a general purpose CPU and—by comparison—a substantial general purpose memory system, typically of capacity at least some hundreds or thousands of bits and potentially configurable to be much bigger.

with just a single point mutation in our newly developed seed organism. Given that several aspects of the reproduction cycle need to be well co-ordinated for reproduction to succeed, we had not thought that a single point mutation would be likely to already yield a viable self-copier. Further analysis is ongoing to fully understand the mechanism for this transition.

16.4 Discussion and Future Prospects

The initial results from these experiments show on the one hand that it is, indeed, perfectly possibly to realise the von Neumann reproduction architecture in these coreworld type systems; and, on the other hand, that this reproduction architecture can prove to be unexpectedly brittle. The particular phenomena reported here, though degenerative, are still interesting in their own right. Moreover, we conjecture that it will be possible, through relatively modest modifications to the underlying machinery or "physics" of these worlds, and/or adjustments to the specific detailed implementations of the von Neumann reproduction architecture, to avoid these particular breakdowns and evolutionary collapse. If this can be done, then these platforms should still be fruitful "minimal" environments in which to explore and understand the most basic evolutionary emergence and elaboration of "symbolisation", in the form of spontaneous evolutionary change in genotype-phenotype mapping.

Acknowledgements This work has been supported by the European Complexity Network (Complexity-NET) through the Irish Research Council for Science and Technology (IRCSET) under the collaborative project EvoSym.

References

1. Adami C (1997) Introduction to artificial life. Springer, Berlin
2. Hofstadter DR (1985) The genetic code: arbitrary. In: Metamagical themas: questing for the essence of mind and pattern. Penguin Books, London, pp 671–699
3. McMullin B (2000) John von Neumann and the evolutionary growth of complexity: looking backward, looking forward. Artif Life 6(4):347–361. doi:10.1162/106454600300103674
4. McMullin B, Taylor T, von Kamp A (2001) Who needs genomes? In: Atlantic symposium on computational biology, genome information systems & technology, March 15–17, 2001. Regal University Center, Durham, NC, USA. Available from http://alife.rince.ie/bmcm-cbgi-2001/
5. Ofria C, Wilke CO (2004) Avida: a software platform for research in computational evolutionary biology. Artif Life 10(2):191–229. doi:10.1162/106454604773563612
6. Pattee HH (1982) Cell psychology: an evolutionary approach to the symbol-matter problem. Cogn Brain Theory 5(4):325–341
7. Rasmussen S, Knudsen C, Feldberg R, Hindsholm M (1990) The coreworld: emergence and evolution of cooperative structures in a computational chemistry. Physica D 42(1–3):111–134. Available from http://www.sciencedirect.com/science/article/pii/0167278990900706
8. Tom R (1994) An evolutionary approach to synthetic biology: zen and the art of creating life. Artif Life 1(1/2):179–209. Available from http://life.ou.edu/pubs/zen/

Chapter 17
Modelling Complex Multi-particle Transport: From Smooth Flow to Cluster Formation

Ko van der Weele and Giorgos Kanellopoulos

Abstract One of the major problems with multi-particle flows is their tendency to spontaneously form clusters. This is a hot topic in contemporary science not only because of its fundamental interest but also because of its ubiquity in industrial applications and everyday life. Here we present a clear-cut method to model the clustering, dividing the available space in a grid of discrete cells and describing the dynamics from cell to cell by means of a flux function. The method is illustrated by two representative examples: the onset of clustering in granular flow on a conveyor belt and the formation of traffic jams on the highway. Further insight is gained by studying the continuum limit of the model.

Keywords Pattern formation · Granular transport · Clustering

17.1 Introduction

On the fundamental side, clustering is a paradigmatic example of spontaneous pattern formation in multi-particle systems out of thermodynamic equilibrium [1]. On the side of applications, clustering is encountered in the flow of granular materials such as mining ore, cereals, or rock avalanches [2–4], in traffic flows on the highway (where the spontaneously occurring clusters are known as "phantom jams" [5]), in crowds rushing towards emergency exits (where clustering may lead to trampling and lethal accidents [6]), in ant trails [7, 8], in the dynamics of icebergs in the polar seas [9], and countless other flows that consist of dissipatively interacting entities.

In regions where the particle density is slightly higher than average—due to some external influence or simply by a statistical fluctuation—the particles collectively dissipate more energy than elsewhere, become slower, and hence do not escape so easily from this region anymore. Vice versa, in regions where the density is below

K. van der Weele (✉) · G. Kanellopoulos
Center for Research and Applications of Nonlinear Systems, Department of Mathematics, University of Patras, 26500 Patras, Greece
e-mail: weele@math.upatras.gr

G. Kanellopoulos
e-mail: giorgoskanellopoulos@gmail.com

T. Gilbert et al. (eds.), *Proceedings of the European Conference on Complex Systems 2012*, Springer Proceedings in Complexity, DOI 10.1007/978-3-319-00395-5_17,

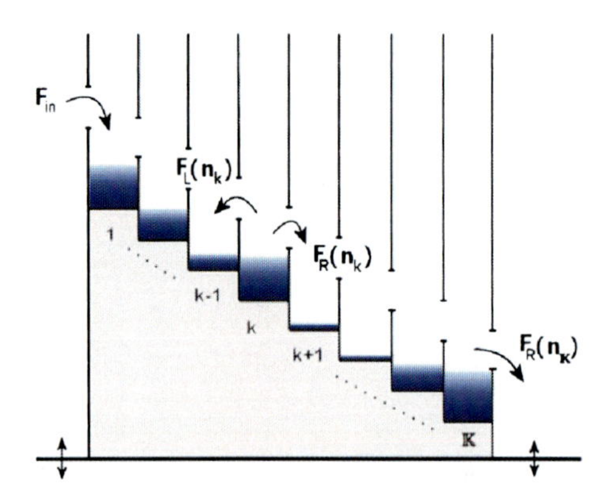

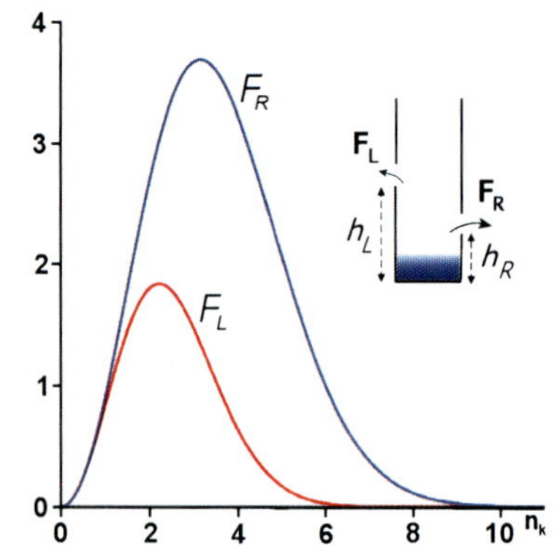

Fig. 17.1 (*Left*): Model transport system consisting of K compartments, vertically vibrated in order to fluidize the granular particles. Material from the kth compartment is able to flow into either of its neighboring compartments: The flow in box $k-1$ is prescribed by a flux function $F_L(n_k)$ and that to box $k+1$ by a larger flux function $F_R(n_k)$. The adjustable inflow rate into the top compartment is denoted by F_{in}. Under standard operating conditions the outflow from the last compartment, $F_R(n_K)$, will be equal to the inflow rate F_{in}. (*Right*): The flux functions given by Eq. (17.2), with $A = 1\ \text{s}^{-1}$, $B_R = 0.1$, and $B_L = 0.2$, i.e., for the case when the height of the barrier to the left is twice as large as that to the right ($h_L = 2h_R$)

average, the particles remain more mobile and move on easily. The separation of dense and dilute regions thus becomes a self-enhancing effect, leading inevitably to the formation of clusters consisting of many slow particles interlaced by diluted regions with only a few, but fast particles [2]. Note that this pattern formation, and the associated decrease in entropy, does not in any way violate the second law of thermodynamics since the transport systems we are dealing with are inherently open with respect to energy (a continuous input of energy is required to keep the particles going, while there is also a continuous dissipation of energy via the particle interactions) and also with respect to mass (due to the in–and outflow of particles).

Here we present a particularly clear-cut and fast method to model the clustering: we divide the available space in a grid of discrete cells and describe the dynamics from cell to cell by means of a flux function. For instance, in a one-dimensional system (such as the conveyor belt for granular materials of Fig. 17.1(a), or a motorway for cars) we divide the total length in K intervals and denote the number density on each of these intervals by $n_k(t), k = 1, \ldots, K$. If the flux per unit time from such an interval to the neighboring interval at the right is given by a flux function $F_R(n_k)$, and the flux towards the left by $F_L(n_k)$, it follows that the dynamics of the system is governed by the following set of K coupled ordinary differential equations:

$$\frac{dn_k}{dt} = F_R(n_{k-1}) - F_L(n_k) - F_R(n_k) + F_L(n_{k+1}), \quad \text{for } k = 1, \ldots, K \qquad (17.1)$$

Each of these equations simply expresses the mass balance for the cell in question: the time rate of change of the density $n_k(t)$ is equal to the influx into cell k from the neighboring cells $k-1$ and $k+1$, minus the outflux from k towards these cells.

We will illustrate the method by means of several representative examples and in doing so learn a good deal about the way in which these flows typically give way to clustering.

17.2 A Model for Granular Transport

One of the main examples we study is the granular transport system depicted in Fig. 17.1, consisting of an array of vibrated connected compartments. A steady inflow $F_{in} = Q$ is applied to the top compartment of the system and under normal operating conditions (i.e., for not too large values of Q) a continuous flow establishes itself all the way down to the last compartment, yielding an outflow that equals the inflow Q.

In the case of granular transport, assuming that the particles are agitated sufficiently vigorously to form a so-called granular gas, the flux functions take the approximate analytic form depicted in Fig. 17.1 [10–13]:

$$F_{R,L}(n_k) = A{n_k}^2 e^{-B_{R,L} n_k^2} \tag{17.2}$$

which contains two parameters, A and $B_{R,L}$. The parameter A sets the absolute flux rate. It is proportional, among other things, to the area of the opening between the boxes and determines how fast (or how slow) the dynamical phenomena take place. It can be incorporated in a dimensionless time scale $\tau = At$. The second (dimensionless) parameter $B_{R,L}$ may be regarded as the "Reynolds number" of the transport process. It is proportional to

$$B_{R,L} \propto \frac{mgh_{R,L}}{m(af)^2} \left(1 - \epsilon^2\right)^2 \left(\frac{N_{\text{tot}} d^2}{\Omega}\right)^2 \tag{17.3}$$

which is the product of three dimensionless numbers representing: (i) the potential energy $mgh_{R,L}$ required to overcome the barrier towards the right or left (respectively at height h_R and h_L above the bottom of the compartment) divided by the kinetic energy $\frac{1}{2}m(af)^2$ inserted by the vibration, where a and f are the amplitude and frequency of the external driving, (ii) the loss of energy involved in each particle collision, with ϵ being the coefficient of normal restitution (equal to 1 for perfectly elastic collisions and about 0.90 for the glass or metal beads used in most laboratory experiments), and (iii) a filling factor, with N_{tot} denoting the total number of particles, d their diameter, and Ω the ground floor area of the system.

The central feature of the flux functions $F_{R,L}(n_k)$ is their one-humped shape, which is directly related to the dissipative nature of the particle interactions. For small values of n_k, the flux from compartment k increases with the density just as in any ordinary gas; beyond a certain value of nk, however, the increasingly frequent interactions make the particles so slow that they are hardly able to overcome the barrier anymore, and hence the flux starts to decrease again. It is thanks to this one-humped shape that clustering is possible, i.e., that a dynamic equilibrium can be established between a well-filled compartment (with a density n_k far to the right of the maximum) and its diluted neighbor compartments (with a density n_k close to 0), since they both have the same flux level.

Also other flow problems with a tendency to cluster can be described by a similar one-humped flux function [14]. The precise shape will differ from one problem to the next, but it will always exhibit a maximum like in Fig. 17.1(b).

17.3 Subcritical Pattern Formation

For sufficiently small inflow rates Q the material flows smoothly through the system and all compartments are filled to the same density level. However, this changes when we keep increasing Q and push it beyond the capacity of the system. At some critical value ($Q = Q_{cr}$) cluster formation becomes inevitable. Interestingly, the clusters are announced in advance (already below Q_{cr}) by the appearance of an oscillatory pattern in the density profile, see Fig. 17.2(a): The compartments now alternately have higher and smaller density. This provides a valuable warning signal in practical applications. If immediate measures are taken, the imminent cluster formation may yet be prevented. If not, a dense cluster is formed in the first compartment and obstructs any further inflow.

The critical flow and the associated wavy density profile can be explained quantitatively in terms of the dynamical flux model introduced above. The appearance of the oscillatory pattern turns out to be connected to a reverse period doubling bifurcation[1] of the previously uniform density profile (see Fig. 17.1(b)) [13].

In order to unravel the *physical* mechanism that lies at the basis of the pattern formation, it is instructive to study the continuum version of the flux model. In this version the compartment numbers $k = 1, \dots, K$ are replaced by a continuous variable x and the number density $n_k(t)$ by $\rho(x,t)\Delta x$, where Δx is the compartment width. The dynamics of the system is now described (instead of by K coupled ordinary differential equations) by a single partial differential equation of the Fokker-Planck type [15]:

$$\frac{\partial \rho}{\partial t} = -P(\rho)\frac{\partial \rho}{\partial x} + \frac{\partial}{\partial x}\left\{D(\rho)\frac{\partial \rho}{\partial x}\right\}, \tag{17.4}$$

where the right hand side contains a drift term and a diffusive term. The drift velocity $P(\rho)$ and the diffusion coefficient $D(\rho)$ both depend on the density:

$$P(\rho) = \Delta x\left(\frac{d\widetilde{F}_R}{d\rho} - \frac{d\widetilde{F}_L}{d\rho}\right) \tag{17.5}$$

$$D(\rho) = \frac{1}{2}\Delta x^2\left(\frac{d\widetilde{F}_R}{d\rho} + \frac{d\widetilde{F}_L}{d\rho}\right) \tag{17.6}$$

[1]The details of the transition (and the associated bifurcation) depend on the exact form of the flux function. For instance, if the one-humped flux function happens to rise from zero not as n^2 (as in Eq. (17.2)) but rather more steeply as $n^{1/2}$ the period doubling bifurcation is no longer reverse but takes place in the forward direction. In that case the system also exhibits oscillatory density profiles of periodicity 4, 8, and so on, before the final clustering sets in.

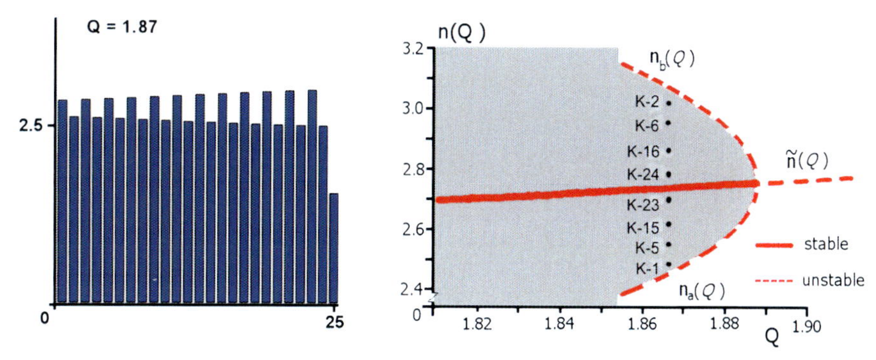

Fig. 17.2 (*Left*): Oscillatory density profile of a 25–compartment conveyor belt, at an inflow rate $Q = 1.87$, just below the critical value Q_{cr} at which clustering sets in. The depicted profile represents a dynamical equilibrium, with a steady flow of $Q = 1.87$ (per time unit) being transported towards the right. (*Right*): The reverse period doubling bifurcation associated with the clustering transition. The basin of attraction of the uniform density $\widetilde{n}(Q)$ is indicated in gray; its boundaries are given (from $Q = 1.855$ onward) by the two elements of an unstable period-2 solution $n_a(Q)$ and $n_b(Q)$. The points $k = K - 1, K - 2, \ldots, K - 23, K - 24$ $(= 24, 23, \ldots, 2, 1)$ denote the densities in the successive compartments from *right* to *left*. The oscillatory profile is the result of the oscillatory convergence towards the value $\widetilde{n}(Q)$

In this continuum version of the model, the clustering (and the whole sequence of events preceding it) is explained as an interplay between drift and diffusion. The drift term turns out to be responsible for the subcritical density oscillations, while the diffusion term is responsible for the eventual clustering. Clustering sets in when the density ρ locally exceeds some critical threshold such that the coefficient $D(\rho)$ becomes negative there; this change of sign gives rise to anti-diffusion, which is the continuum analog of cluster formation [15, 16].

17.4 Connection with Other Transport Problems

17.4.1 Traffic Flow

Even though at first sight they may seem quite unrelated, cars on the highway have much in common with granular particles moving in a preferential direction. The analogy is in fact so strong that it has led to a series of bi-annual conferences "Traffic and Granular Flow" [17]. The engines provide the necessary energy input (the cars are self-driven particles) and just as the particles in a granular gas, they interact dissipatively. They do so without actual collisions but simply because a car that closes in upon another must necessarily reduce its speed; so, just like the granular particles, cars make each other slow and traffic jams are the natural result [5, 6, 18]. Another striking similarity with the granular transport system of Fig. 17.1(a) is the fact that many highways are "compartmentalized" by means of induction loops in the asphalt (usually about 500 m apart), which monitor every car that passes over them, and also its speed.

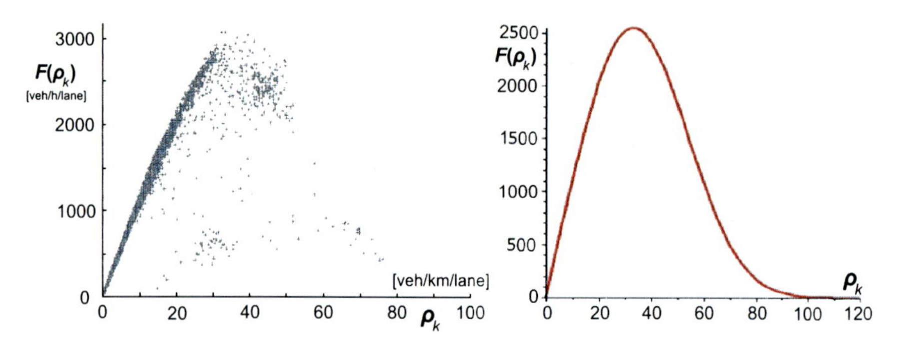

Fig. 17.3 (*Left*): Measured traffic flux (veh/h per lane) as function of the local vehicle density (veh/km per lane) at a certain point along the highway A58 between the cities of Breda and Eindhoven in the Netherlands, during the morning traffic on a working day without exceptional weather conditions (Data courtesy of the Dutch Ministry of Traffic, Public Works, and Water Management). (*Right*): The reconstructed flux function given by Eq. (17.7)

At low densities, up to $\rho_k \approx 30$ veh/km/lane, the cars drive at their desired speed of roughly 110 km/h, but above 30 veh/km/lane the distance between successive cars becomes so small (less than 30 m) that the drivers can no longer maintain this speed. They have to react, brake, and maneuver, and this causes a sudden drop in the mean velocity. The corresponding car flux across the measuring point (density times velocity), widely known as the "fundamental traffic diagram" of traffic analysis [6, 18], is shown in Fig. 17.3(a). It shows the above two regimes very clearly: at low densities the traffic flows freely and the flux function $F(\rho_k)$ shows an upward branch rising to nearly 3000 veh/h/lane, whereas for $\rho_k > 30$ veh/km/lane the traffic becomes congested and the flux goes down dramatically. Just like the granular flux function of Fig. 17.1(b), the car flux $F(\rho_k)$ thus has the one-humped shape which is a prerequisite for clustering. Indeed, on the basis of the measured data in Fig. 17.3(a) one may reconstruct a flux function of the form (see Fig. 17.3(b)):

$$F(\rho_k) = F_R(\rho_k) = A\rho_k e^{-3.32\times 10^{-5}\rho_k^{2.67}} \tag{17.7}$$

with $A = 112$ km/h being the desired velocity at small densities.

Apart from the conspicuous similarities, there are also various interesting differences with respect to the granular flow. For one thing, the traffic flux grows linearly with the density for small ρ_k, in contrast to the granular flux function of Eq. (17.2), and—as discussed in Sect. 17.3—this implies a somewhat different bifurcation structure of the clustering transition. Also at variance with the granular case is the fact that the traffic flux in the direction opposite to the main flow is not just much smaller, but identically zero: $F_L(\rho_k) = 0$. Further one should not forget the in—and outflow of vehicles at the entrances and exits along the highway, which introduces a time—dependent source term $Q_{\text{on/off},k}(t)$ in certain cells k. The set of balance equations for traffic flow then take the following form:

$$\frac{d\rho_k}{dt} = F_R(\rho_{k-1}, \rho_k) - F_R(\rho_k, \rho_{k+1}) \pm Q_{\text{on/off},k}(t), \quad \text{for } k = 1, \ldots, K \tag{17.8}$$

In this equation we have incorporated one further refinement by taking into account that the flux function does not depend solely on the density of the cell k, but also on that of the target cell, meaning that the flux from cell k to $k+1$ will have the general form $F_R(\rho_k, \rho_{k+1})$ instead of $F_R(\rho_k)$. In fact, it will be a decreasing function of the target cell's density ρ_{k+1} since the drivers enter this cell less freely when it is already occupied by a large number of cars than when it is empty [19].

17.4.2 Energy Cascade in Turbulent Fluids

As a third example, and in order to illustrate that the transport need not be restricted to material flows but may also concern the flow of energy or momentum, let us say a few words about the energy cascade in turbulent fluids. As is well known, the energy in three-dimensional turbulence is transported from the large length scales to the smaller ones. One of the models that has been put forward to describe this is the GOY model, named after Gledzer, Ohkitani and Yamada [20–23]. In this model the spectrum of relevant length scales is divided into N discrete shells, with the first shell representing the largest and the Nth shell the smallest length scale. More specifically, the nth shell is determined by a wave number $k_n = k_0\lambda_n$ with $\lambda > 1$ $(n = 1, 2, \ldots, N)$. Each shell is characterized by a complex velocity mode u_n, which is coupled to the modes in the nearest and next nearest shells in a way that mimics the underlying hydrodynamic equations.

Energy is being put into the system in one of the first shells—conventionally in the fourth—and this energy is transferred to the next shells (higher n, smaller length scale) in a way similar to the granular transport studied in the present paper. In dynamic equilibrium all shells contain a certain energy $\frac{1}{2}|u_n|^2$, which is distributed in such a way that the product $u_n k_n^{1/3}$ oscillates around a constant level along most of the cascade [24, 25]. The oscillatory profile of $u_n k_n^{1/3}$ is very reminiscent of that of Fig. 17.2(a), with only one intriguing difference: the periodicity of the oscillations is 3 instead of 2. This period-3 pattern suggests that the oscillatory profile does not always have to be related to a period-doubling bifurcation but that also period-tripling is possible; and in other systems it may well occur via even more elaborate bifurcation schemes.

17.5 Further Extensions and Concluding Remarks

Up to now we have considered one-dimensional transport systems, where the formation of one single cluster automatically implies the breakdown of the flow. If we extend the transport system to a 2D grid, however, the material can stream around this first cluster and the outflow does not come to a halt before the grid is covered by a whole battalion of clusters. The precise positioning of these clusters—another beautiful example of pattern formation—turns out to be sensitively dependent on the special characteristics of the system, its boundary conditions, and also on small random (experimental) density fluctuations.

Naturally, in two spatial dimensions we are dealing with densities $n_{k,l}(t)$ and the right hand side of the generalized balance equation Eq. (17.1) will contain the various fluxes that represent the exchange with all nearest neighbors. If the system under study warrants it, also the exchange with next-nearest neighbors or other cells further away may be included. Boundaries or any obstacles within the system are modeled by forbidden cells. Inflow or outflow of particles, either at the boundaries of the system or in some intermediate cell (as with the on—and off—ramps along the highway), or feedback loops, are equally straightforward to incorporate.

In conclusion, the method outlined above is extremely versatile and can be adapted to a variety of dynamical systems of interest. The model is simple enough to be robustly stable against small uncertainties in the initial conditions and, not unimportant for practical applications, it requires only a minimum of computer time. All these features make the flux method uniquely suited for an accurate phenomenological description of complex multi-particle transport problems.

Acknowledgements This work is part of the research programme of the European Complexity-NET project "Complex Matter", which is financially supported by the Greek General Secretariat for Research and Technology (GSRT).

References

1. Cross MC, Hohenberg PC (1993) Pattern formation outside of equilibrium. Rev Mod Phys 65:851–1112
2. Goldhirsch I, Zanetti G (1993) Clustering instability in dissipative systems. Phys Rev Lett 70:1619–1622
3. Jaeger H, Nagel S, Behringer R (1996) Granular solids, liquids, and gases. Rev Mod Phys 68:1259–1273
4. Van der Weele K (2008) Granular gas dynamics: how Maxwell's demon rules in a non-equilibrium system. Contemp Phys 49:157–178
5. Kerner BS, Konhäuser P (1993) Cluster effect in initially homogeneous traffic flow. Phys Rev E 48:R2335
6. Helbing D (2001) Traffic and related self-driven many-particle systems. Rev Mod Phys 73:1067–1141
7. Nishinari K, Chowdhury D, Schadschneider A (2003) Cluster formation and anomalous fundamental diagram in an ant trail model. Phys Rev E 67:036120
8. John A, Schadschneider A, Chowdhury D, Nishinari K (2009) Traffic-like collective movement of ants on trails: absence of jammed phase. Phys Rev Lett 102:108001
9. Herman A (2011) Molecular-dynamics simulation of clustering processes in sea-ice floes. Phys Rev E 84:056104
10. Eggers J (1999) Sand as Maxwell's demon. Phys Rev Lett 83:5322–5325
11. Van der Weele K, Van der Meer D, Versluis M, Lohse D (2001) Hysteretic clustering in granular gas. Europhys Lett 53:328–334
12. Van der Weele K, Kanellopoulos G, Tsiavos C, Van der Meer D (2009) Transient granular shock waves and upstream motion on a staircase. Phys Rev E 80:011305
13. Kanellopoulos G, Van der Weele K (2011) Subcritical pattern formation in granular flow. Int J Bifurc Chaos Appl Sci Eng 21:2305–2319

14. Van der Weele K, Mikkelson R, Van der Meer D, Lohse D (2004) Cluster formation in compartmentalized granular gases. In: Hinrichsen H, Wolf DE (eds) The physics of granular media. Wiley, New York, pp 117–139
15. Kanellopoulos G, Van der Weele K (2012) Critical flow and clustering in a model of granular transport: the interplay between drift and antidiffusion. Phys Rev E 85:061303
16. Van der Meer D, Van der Weele K, Lohse D (2002) Sudden collapse of a granular cluster. Phys Rev Lett 88:174302
17. Wolf DE, Schreckenberg ME, Bechem A (eds) (1996) Traffic and granular flow. World Scientific, Singapore. See also in: Schreckenberg ME, Wolf DE (eds) (1998) Traffic and granular flow '97. Springer, Berlin; Helbing D et al (eds) (2000) Traffic and granular flow '99. Springer, Berlin; Fukui M et al (eds) (2003) Traffic and granular flow '01. Springer, Berlin; Hoogendoorn SP, Luding S, Bovy PHL, Schreckenberg M, Wolf DE (eds) (2005) Traffic and granular flow '03. Springer, Berlin
18. Chowdhury D, Santen L, Schadschneider A (2000) Statistical physics of vehicular traffic and some related systems. Phys Rep 329:199–329
19. Van der Weele K, Spit W, Mekkes T, Van der Meer D (2005) From granular flux model to traffic flow description. In: Hoogendoorn SP, Luding S, Bovy PHL, Schreckenberg M, Wolf DE (eds) Traffic and granular flow '03. Springer, Berlin, pp 569–578
20. Gledzer EB (1973) System of hydrodynamic type admitting two quadratic integrals of motion. Sov Phys Dokl 18:216–217. English translation from: Dokl Akad Nauk SSSR 209:1046–1048 (1973)
21. Ohkitani K, Yamada M (1989) Temporal intermittency in the energy cascade process and local Lyapunov analysis in fully-developed model turbulence. Prog Theor Phys 81:329–341
22. Yamada M, Ohkitani K (1987) Lyapunov spectrum of a chaotic model of three-dimensional turbulence. J Phys Soc Jpn 56:4210–4213
23. Yamada M, Ohkitani K (1988) The inertial subrange and non-positive Lyapunov exponents in fully-developed turbulence. Prog Theor Phys 79:1265–1268
24. Schorghöfer N, Kadanoff L, Lohse D (1995) How the viscous subrange determines inertial range properties in turbulence shell models. Physica D 88:40–54
25. Kadanoff L, Lohse D, Schorghöfer N (1997) Scaling and linear response in the GOY turbulence model. Physica D 100:165–186

Chapter 18
Out-of-Equilibrium Dynamics in Systems with Long-Range Interactions: Characterizing Quasi-stationary States

Pierre de Buyl

Abstract Systems with long-range interactions (LRI) display unusual thermodynamical and dynamical properties that stem from the non-additive character of the interaction potential. We focus in this work on the lack of relaxation to thermal equilibrium when a LRI system is started out-of-equilibrium. Several attempts have been made at predicting the so-called quasi-stationary state (QSS) reached by the dynamics and at characterizing the resulting transition between magnetized and non-magnetized states. We review in this work recent theories and interpretations about the QSS. Several theories exist but none of them has provided yet a full account of the dynamics found in numerical simulations.

Keywords Vlasov equation · Long-range interactions

18.1 Introduction

Systems with long-range interactions (LRI) display unusual thermodynamical and dynamical properties such as ensemble inequivalence, lack of relaxation to equilibrium or broken ergodicity (see Refs. [1, 2] for a review of the field). These properties stem from the non-additive character of the interparticle interaction potential. Let us mention a few systems belonging to the LRI class: gravitational systems, non-neutral plasmas, 2D fluid dynamics, etc.

We focus in this work on the lack of relaxation to thermal equilibrium in LRI systems when the system is initiated in an out-of-equilibrium state. This phenomenon leaves the system in an intermediary stage of the dynamics that is called a quasi-stationary state (QSS) [3]. This state does not correspond to the equilibrium predicted by statistical mechanics and its lifetime increases algebraically with the number N of interacting particles in the system.

The occurrence of QSS should be taken into account if one is interested in the actual properties of a system. The time needed to reach thermal equilibrium may

P. de Buyl (✉)
Center for Complex Systems and Nonlinear Phenomena, Université libre de Bruxelles, CP 231, Av. F. Roosevelt, 50, 1050 Brussels, Belgium

T. Gilbert et al. (eds.), *Proceedings of the European Conference on Complex Systems 2012*, Springer Proceedings in Complexity, DOI 10.1007/978-3-319-00395-5_18,

prevent a proper observation of equilibrium properties in the available experimental or simulational setting.

In this work, we review the generic steps of the out-of-equilibrium dynamics of LRI systems and use the paradigmatic Hamiltonian Mean-Field (HMF) model and its Vlasov formulation to illustrate those steps. Then, several theories attempting to predict or describe the QSS are reviewed.

18.2 The Hamiltonian Mean-Field Model and the Vlasov Equation

Let us consider the Hamiltonian Mean-Field (HMF) model introduced by Antoni and Ruffo [4]. This model aims at reproducing the collective behavior of more complex models with ferromagnetic or gravitational interactions, for instance.

The particles in the HMF model lie in a 1-dimensional periodic space with position $\theta \in [-\pi : \pi[$. The N-body Hamiltonian is

$$\mathcal{H} = \sum_{i=1}^{N} \frac{p_i^2}{2} + \frac{1}{2N} \sum_{i,j=1}^{N} \left(1 - \cos(\theta_j - \theta_i)\right) \tag{18.1}$$

where θ_i is the position (in $[-\pi : \pi[$) of particle i and p_i is its momentum. N is the total number of particles in the system.

One may also consider the continuum limit of the HMF model. This leads to the Vlasov equation

$$\frac{\partial f}{\partial t} + p \frac{\partial f}{\partial \theta} - \frac{\partial V[f](\theta, t)}{\partial \theta} \frac{\partial f}{\partial p} = 0, \tag{18.2}$$

where

$$V[f](\theta, t) = \int d\theta' dp' f(\theta', p', t) \left(1 - \cos(\theta' - \theta)\right), \tag{18.3}$$

is the interaction potential.

The mean field, or magnetization,

$$\mathbf{m} = \frac{1}{N} \sum_{i=1}^{N} (\cos\theta_i, \sin\theta_i) = (m_x, m_y), \tag{18.4}$$

is used to follow the dynamical evolution of the HMF model. In the continuum limit.

$$\mathbf{m} = \int d\theta dp f(\theta, p)(\cos\theta, \sin\theta) = (m_x, m_y), \tag{18.5}$$

The norm of $\mathbf{m}$ is denoted m.

Equilibrium statistical mechanics allows one to compute the value of m for a given energy or temperature. The authors of Ref. [4] observed a discrepancy in the caloric curve between the theory and their simulations, close to the second order transition separating the magnetized phase ($m > 0$) from the homogeneous phase

($m = 0$). Further investigations revealed the origin of the discrepancy: the system of particles had not reached thermodynamical equilibrium in the simulations.

The evolution of many systems with long-range interactions consists in the following generic steps:

1. An initial condition that does not correspond to equilibrium.
2. Violent relaxation: the observables in the system undergo strong changes. The time scale of this step does not depend on the number of particles N.
3. Quasi-stationary state (QSS). This state may either be stationary or present oscillations. Its lifetime grows algebraically with N.
4. Equilibrium: the state that is predicted by equilibrium statistical mechanics.

Simulations illustrating this dynamical evolution may be found in Refs. [3, 5], for instance. An important consequence of the occurrence of QSS is that in addition to the equilibrium phase transitions that the system may experience, one has to consider an "out-of-equilibrium phase diagram" that is based on the magnetization found in the QSS. In the thermodynamic limit $N \to \infty$, this "out-of-equilibrium phase diagram" is the relevant one.

18.3 Theories for the Quasi-stationary States

Several attempts have been made at predicting the quasi-stationary state reached by the dynamics. We review several of those attempts, namely: Lynden-Bell's theory that is based on an entropy maximization principle [6, 7], the exact stationary regime theory proposed by de Buyl, Mukamel and Ruffo [8] and a dynamical reduction proposed by Levin [9].

None of the aforementioned theories is able to take into account the existence of states whose observables are not constant in time, i.e. they predict a time independent distribution $f(\theta, p)$. It is known that dynamical resonances lead to oscillating regimes [4, 10] and those cannot be predicted.

18.3.1 The Theory of Lynden-Bell

In 1967, Lynden-Bell [6] devised a theory to compute the relaxed state of gravitational systems obeying a Vlasov equation. His theory is based on the maximization of an entropy functional that takes into account the incompressible character of the distribution function in Vlasov dynamics. The computation is based on a coarse graining of phase space but leads to a continuous prediction for the distribution function.

Lynden-Bell's theory has been applied with success to the prediction of the intensity of the Colson-Bonifacio model for the free-electron laser [11] and to provide an out-of-equilibrium phase diagram for the HMF model [6, 7].

18.3.2 BGK Like Theory

Based on the fact that a distribution function that only depends on the energy is stationary in Vlasov dynamics, one may try to construct stationary states. This approach is well know in plasma physics as Bernstein-Greene-Kruskal modes [12]. The authors of Ref. [8] develop this idea while proposing an approximate correspondence between the initial condition and the state that is reached by the system.

The distribution $f(\theta, p)$ is expressed directly as a function of the energy distribution function of the initial condition. For low values of the initial magnetization, the theory fails to predict the final magnetization. Else, it provides good results and predicts a second-order phase transition for $\langle m\rangle$ and $\langle m_x\rangle$. This theory is based purely on dynamical consideration and as such provides interesting complementary information with respect to Lynden-Bell's theory.

18.3.3 Core-Halo and Envelope

The authors of Ref. [9] propose an ansatz for the distribution function f that reproduces the core-halo structure found in the phase space of the HMF model

$$f_S(\theta, p) = \eta_0 \left[\Theta(\epsilon_F - \epsilon) + \chi \Theta(\epsilon_h - \epsilon)\, \Theta(\epsilon - \epsilon_F)\right], \tag{18.6}$$

where $\epsilon(\theta, p) = p^2/2 + (1 - M_S \cos\theta)$, M_S is the value of the magnetization and Θ is the Heaviside function.

This ansatz requires the determination of the energy levels (ϵ_F for the core and ϵ_h for the halo) and of the magnetization M_S. Those values are provided by a reduced dynamical equation. This theory is tested on the transition between magnetized and homogeneous regimes in the HMF model and predicts a first-order like transition for $\langle m\rangle$ and $\langle m_x\rangle$. The order of the transition is confirmed by simulation data for $\langle m_x\rangle$. Simulation data for $\langle m\rangle$ is not given however.

18.4 Discussion and Conclusion

Out of the existing theories aimed at predicting the quasi-stationary states (QSS) that have been applied to the Hamiltonian Mean-Field model, none is able to predict the regimes in which oscillations are found. As is pointed out in Ref. [13], one may relate the time averages of the squared norm of the magnetization to the one of the x component of the magnetization[1] by the following relation:

$$\langle m^2\rangle = \langle m_x^2\rangle = \langle m_x\rangle^2 + \sigma_{m_x}^2, \tag{18.7}$$

[1] Here, m_y can be set equal to zero without loss of generality.

where σ_{m_x} is the time-wise standard deviation of m_x. The choice of an observable thus impacts the results that is found in simulations for non-steady QSS, explaining the different results between Refs. [7] and [9]. As soon as the QSS displays oscillations in the magnetization, $\sigma^2_{m_x} > 0$ and $\langle m^2 \rangle \neq \langle m_x \rangle^2$. The phase diagram provided by Lynden-Bell's theory [7] still represents the most ensemble view of the QSS for the HMF model as well as an actual interpretation in terms of phase transitions.

We have reviewed in this work recent advances in the understanding of the out-of-equilibrium dynamics in systems with long-range interactions. Several theories exist but none of them has provided yet a full account of the dynamics found in numerical simulations. Progress in this direction has been made by the construction of counter-rotation BGK clusters by Yamaguchi [14]. This construction is however not predictive.

Acknowledgements The author would like to acknowledge interesting discussions and collaborations with R. Bachelard, G. De Ninno, D. Fanelli, P. Gaspard, D. Mukamel and S. Ruffo.

References

1. Campa A, Dauxois T, Ruffo S (2009) Statistical mechanics and dynamics of solvable models with long-range interactions. Phys Rep 480:57–159
2. Bouchet F, Gupta S, Mukamel D (2010) Thermodynamics and dynamics of systems with long-range interactions. Physica A 389:4389–4405
3. Yamaguchi YY, Barré J, Bouchet F, Dauxois T, Ruffo S (2004) Stability criteria of the Vlasov equation and quasi-stationary states of the HMF model. Physica A 337:36–66
4. Antoni M, Ruffo S (1995) Clustering and relaxation in Hamiltonian long-range dynamics. Phys Rev E 52:2361–2374
5. Yamaguchi YY (2008) One-dimensional self-gravitating sheet model and Lynden-Bell statistics. Phys Rev E 78:041114
6. Lynden-Bell D (1967) Statistical mechanics of violent relaxation in stellar systems. Mon Not R Astron Soc 136:101–121
7. Antoniazzi A, Fanelli D, Ruffo S, Yamaguchi YY (2007) Nonequilibrium tricritical point in a system with long-range interactions. Phys Rev Lett 99:040601
8. de Buyl P, Mukamel D, Ruffo S (2011) Self-consistent inhomogeneous steady states in Hamiltonian mean-field dynamics. Phys Rev E 84:061151
9. Pakter R, Levin Y (2011) Core-Halo distribution in the Hamiltonian mean-field model. Phys Rev Lett 106:200603
10. Antoniazzi A, Califano F, Fanelli D, Ruffo S (2007) Exploring the thermodynamic limit of Hamiltonian models: convergence to the Vlasov equation. Phys Rev Lett 98:150602
11. Barré J, Dauxois T, De Ninno G, Fanelli D, Ruffo S (2004) Statistical theory of high-gain free-electron laser saturation. Phys Rev E 69:045501(R)
12. Bernstein IB, Greene JM, Kruskal MD (1957) Exact nonlinear plasma oscillations. Phys Rev 108:546–550
13. de Buyl P, Fanelli D, Ruffo S (2012) Phase transitions of quasistationary states in the Hamiltonian mean field model. Cent Eur J Phys 10:652–659
14. Yamaguchi YY (2011) Construction of traveling clusters in the Hamiltonian mean-field model by nonequilibrium statistical mechanics and Bernstein-Greene-Kruskal waves. Phys Rev E 84:016211

Chapter 19
Distance Ratio: An Exploratory Application to Compare Complex Networks

Nuno Caseiro and Paulo Trigo

Abstract This paper describes an experimental application of the distance ratio measure used to compare individual networks among themselves and to analyze the aggregated network representing the group mental model from the field of emergency management.

The data was obtained by surveying a group of Civil Protection graduates and aggregating all the answers (shared mental model). The data allowed us to deepen the analysis of the resulting network in order to research for differences among networks.

Keywords Complex networks · Network comparison · Distance ratio · Emergency management · Mental models

19.1 Introduction

The study of complex networks are an important tool in different fields of knowledge, with applications from social relations to power grids, from genes and proteins to text analysis [1–6].

The usual approaches are centered in the statistical proprieties of one network and its components: the nodes and the links between them (edges). In this case, complex networks are taken as individual structures but they are normally dynamic by nature and it's important to understand the differences or similarities among two or more.

In general, we can compare networks by means of: (i) global property statistics, such as degree distribution, betweenness centrality, assortativity, and clustering

N. Caseiro (✉)
Instituto Politécnico de Castelo Branco, Instituto Universitário de Lisboa—ISCTE, LabMAg, Lisbon, Portugal
e-mail: ncaseiro@ipcb.pt

P. Trigo
Instituto Superior de Engenharia de Lisboa, LabMAg, Lisbon, Portugal
e-mail: ptrigo@deetc.isel.ipl.pt

T. Gilbert et al. (eds.), *Proceedings of the European Conference on Complex Systems 2012*, Springer Proceedings in Complexity, DOI 10.1007/978-3-319-00395-5_19,

coefficient; (ii) subgraph enumerating, such as over- and under-represented motifs than in randomized networks, graphlet degree distribution, joint degree correlations of subgraph, and trained subgraph feature. Each of comparison method has specific characteristics, advantages and disadvantages [7].

When approaching two or more networks that share some elements among them, are taken in several moments in time or in different physical settings the current proposals are limited. Two networks even with different substructures can exhibit some similitude regarding global properties [7].

Some examples of this situation are the network formed by words and phrases in two books in a domain or written by the same author, citations in a field of knowledge in two different years or power grids in different regions (with same equipment). In these cases comparing can be a useful approach to detect differences and similarities between the existing elements and their links.

With the present work we propose to address this gap by applying a measure, Distance Ratio inspired in system analysis [8]. In this context a system is represented by elements connect by causal relations with positive or negative impact within them. It is possible to identify the existence of self-loops, when an element has an impact on itself. But irrespective of the classification, the emerging representation is a different form of network, so our research will test its suitability for complex networks application.

As a practical application the presented indicator (Distance Ratio) is applied to compare different networks representing mental models of emergency managers. These structures are obtained by relating concepts of this field gathered by inquiry from practitioners. It can be used to verify differences among a shared mental model and the individual ones [9, 10] or to test the similarities between professionals of different emergency agents.

In this application, obtained results can be useful for support training and recruitment decisions since they give an indicator of closeness between different mental models.

As a comparative indicator and methodology it can be extended to contexts like plagiarism, when applied to compare the network of words between different documents or other settings where comparison is required.

19.1.1 The Distance Ratio

In the literature there seems to lack a measure that can be used for the purpose of comparing complex networks. We will use an indicator, the Distance ratio (DR), that in its original application, in the domain of system dynamics, was used to calculate two causal maps similarity. Causal maps in systems dynamics are a set of elements that influence each other in a sequence and with a level of intensity. In the language of system dynamics there are more than variables and links. It is possible to identify other constructs such as delayer links, non-linear links, stock variables or rate variables and also feedback loops [11]. The distance ratio is used to compute to what extend two networks are close in terms of their constituents. Thus, an extended matrix A and B is used, where a_{ij} and b_{ij} are their elements. The adjacency matrix, with n variables, will form $an\ xn$ matrix with the strength of the links as cell values [12] or a binary representation (0/1) of the existing links if there is no information about weights.

Since causal maps can be seen as a network where an element can be represented as a node, the influence to other elements are links and the level of intensity is the weight of that link, it is our belief that the DR measure can be extended and applied to complex networks.

The formula for distance ratio is presented below [8, 12]

$$\frac{\sum_{i=1}\sum_{j=1}\text{diff}(i,j)}{(\varepsilon\beta+\delta)v_c^2+\gamma'(2v_c(v_{ua}+v_{ub})+v_{ua}^2+v_{ub}^2)+\alpha((\varepsilon\beta+\delta)v_c^2+\gamma(v_{ua}+v_{ub}))} \tag{19.1}$$

where

$$\text{diff}(i,j)\begin{cases}0, & \text{if } i=j \text{ and } \alpha=1 \text{ (no self-loops)}\\ \Gamma(a_{ij},b_{ij}), & \text{if } i \text{ or } j\notin v_c\wedge i \text{ or } j\in v_a \text{ or } v_b\\ |a_{ij}-b_{ij}|+\delta, & \text{if } a_{ij}*b_{ij}<0\\ |a_{ij}-b_{ij}|, & \text{otherwise}\end{cases} \tag{19.2}$$

and

$$\Gamma(a_{ij},b_{ij})\begin{cases}0, & \text{if } \gamma=0\\ 0, & \text{if } \gamma=1 \text{ and } a_{ij}=b_{ij}=0\\ 1, & \text{otherwise}\end{cases} \tag{19.3}$$

The result obtained by the formula can vary between 0 and 1, where 0 means completely similar and 1 totally different (distant).

A group of parameters are denoted by the Greek letters ($\alpha, \beta, \gamma, \delta, \varepsilon$) and are used to adjust the formula to different contexts. The α parameter can be set to 0 or 1, whether self-loops are allowed or not. To reflect differences in weight between nodes in the networks we pass the max weight to parameter β.

ε accounts for the number of polarities in the matrix and it can take the value 1 or 2 (one polarity—only positive or negative or two polarities—positive and negative).

If $\delta = 0$, we do not differentiate situations where different weights create the same difference values. For instance, a difference between a weight of 4 and 1 is the same as for -2 and 1 as per Eq. (19.2). In the latter, since a negative and a positive value is involved, a value δ is added.

The parameter γ is set for how to interpret matrix cells for which one of the maps does not have an edge because there is a mismatch of nodes. In the context of concept maps the inexistence of a node is a result of the one that creates more difficulties. If we do not want to deduce anything from the absence of nodes, we will set $\gamma = 0$. If we wish to assign meaning to the fact that one person as mentioned one concept (node) and another does not, the value of $\gamma = 1$ signaling that this difference should be taken into account.

19.1.2 Mental Models

Mental models are conceived of as a cognitive structure that forms the basis of reasoning and decision making, and can be seen as a network of associations between concepts in an individual's mind. Mental models have been described as a form of intuitive knowledge that serve as a frame of reference for interpretation of the world which forms the bases for reasoning and working with problems [13].

They are built by individuals based on their personal life experiences, perceptions, and understanding of the world. Mental models provide the mechanism through which new information is filtered and stored. However, the ability to represent the world accurately is always limited and unique to each individual [14].

There a variety of techniques for eliciting mental models, ranging from brainstorming, to interviews or text analysis. They include concept mapping, word association techniques, ordered recall, card sorting procedures, paired-comparison, and the ordered tree technique [12, 15, 16].

They can be applied both individually or to a group of people [10, 15]. Most of procedures used are based on the assumption that an individual's mental model can be represented as a network of concepts and relations [14].

When working together people must share a part of the mental model to deal effectively with each other. Team mental models promote understanding among team members regarding information requirements, the need for communication and coordination [17]. In the case of emergency management cycle the individuals involved need to have a common mental structure to deal with information issues. The lack of a common mental model is a common problem referred to in emergency management literature [18–20].

19.2 Methodology

In order to compute the DR we asked a group of graduate students to identify, organize and relate the concepts they recognized in the emergency management field.

Each element of the survey group was asked to indicate a new level of concepts related to a starting concept. In this situation the initial concept was "Civil Protection". For each of the concepts in the new group, a new sublevel of concepts was requested. By iterating on the above mentioned steps, it is possible to build several levels of inter-related concepts (as represented in Fig. 19.1). An individual structure is elicited representing the mental model of concepts of the respondents with the respective relations. This structure is representable as a network.

As the concepts are words, the responses were verified for major spelling mistakes and typos. This is an important step to ensure that the same concept indicated by two different respondents was not coded as different ones because of a mistake. Finally, individual networks were grouped and processed to create a unique network with all the contributions.

This grouped network can be seen as a shared mental model [10]. In this network, we take into account the number of times a pair of concepts was mentioned by the

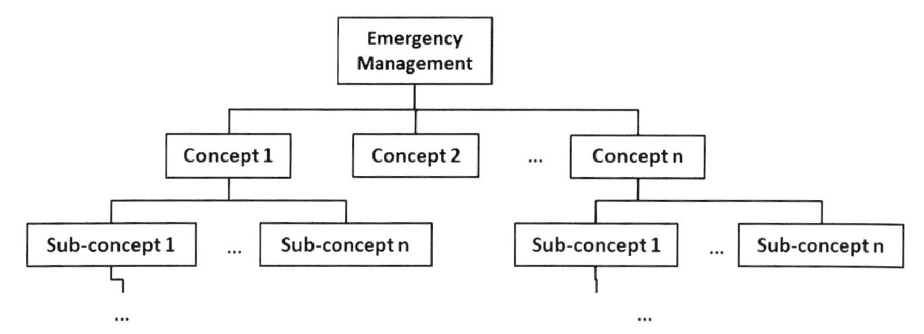

Fig. 19.1 Model of data gathering

respondents, because this stresses the importance of that relation within the group. That total was taken as a weight of the edge between concepts.

The data was processed to transform the concepts identified in a network structure, where a node represents a concept and an edge represents a connection between two concepts. Moreover, for each pair of nodes the respective weight was indicated.

19.3 Results

We compute the DR for each of the networks among subjects. For the current application the parameters in the formula were set to:

- $\alpha = 1$, meaning self-loops were not taken into account, because in the case of complex networks a node do not link to itself directly;[1]
- β =max weight, in the present case the value is one, meaning only that there is a link between two nodes;
- $\gamma = 1$, because we want to take into account the fact that one network has a node and another has not, thus valuing the eventual link that can exist;
- $\delta = 0$, not adding any value to differences;
- $\varepsilon = 1$, because in this case we only have one polarity. All link weights are positive.

The values presented in Table 19.2 point to a low distance ratio since they are very close to 0, the lowest limit, meaning that networks are very similar among them. This can be due to the nature of the sample used. The subjects were all graduate students, with similar training and level of experience.

In spite of this fact, some differences can be perceived between subjects with some of them with more proximity with the others (subject 4) and lower deviation. Or with more relative distance (subject 6).

If we compare the subjects with the group network with the starting concept (the aggregated network mentioned in Table 19.1). We find a mean DR of 0.005 (with

[1] We recall that the Distance Ratio formula is adapted from systems dynamics where self-loops are a normal situation.

Table 19.1 Distance ratio between subjects

		Subjects											Mean	SD
		1	2	3	4	5	6	7	8	9	10	11		
Subjects	1		0.027	0.019	0.013	0.029	0.044	0.02	0.022	0.017	0.023	0.045	0.026	0.0107
	2	0.027		0.018	0.013	0.024	0.029	0.018	0.019	0.016	0.019	0.026	0.021	0.0052
	3	0.019	0.018		0.012	0.018	0.02	0.017	0.017	0.015	0.017	0.017	0.017	0.0023
	4	0.013	0.013	0.012		0.013	0.013	0.014	0.013	0.012	0.013	0.012	0.013	0.0007
	5	0.029	0.024	0.018	0.013		0.032	0.019	0.02	0.016	0.02	0.029	0.022	0.0061
	6	0.044	0.029	0.02	0.013	0.032		0.02	0.022	0.017	0.024	0.07	0.029	0.0169
	7	0.02	0.018	0.017	0.014	0.019	0.02		0.018	0.017	0.019	0.019	0.018	0.0019
	8	0.022	0.019	0.017	0.013	0.02	0.022	0.018		0.015	0.019	0.02	0.019	0.0029
	9	0.017	0.016	0.015	0.012	0.016	0.017	0.017	0.015		0.016	0.015	0.016	0.0015
	10	0.023	0.019	0.017	0.013	0.02	0.024	0.019	0.019	0.016		0.022	0.019	0.0032
	11	0.045	0.026	0.017	0.012	0.029	0.07	0.019	0.02	0.015	0.022		0.028	0.0176

Table 19.2 DR from aggregated network and individual ones

	Subjects											Mean	SD
	1	2	3	4	5	6	7	8	9	10	11		
Grouped Network	0.0055	0.0055	0.0056	0.0056	0.0054	0.0054	0.0059	0.0055	0.0055	0.0056	0.0052	0.0055	0.0001

standard deviation of 0.0001). The value is lower than the values obtained between subjects (with mean of 0.02).

An possible explanation to this is that the aggregated network is constructed by adding all the individual contributions, therefore when we perform the distance ratio between the aggregated network and the individual ones, each one finds a "piece" of itself in the group network thus having a lower distance.

19.4 Conclusions

Network analysis seems to be a valid tool to study mental maps since the concepts and their relationships can be represented by nodes and links.

The application of the Distance Ratio to complex networks seems feasible but requires a broader sample that can increase potential differences. The sample used led to a low Distance Ratio, meaning that the mental models were very similar within the group under study. The distance ratio approach can be refined to take into account the ranking of concepts (node weight) in the case of aggregated networks.

Since this is an exploratory work that used a sample of graduates as the source of concepts, the computed differences between mental models were not very sharp. Future work will try to replicate this approach with professionals in the Civil Protection field, from different agencies and explore if greater dissimilarities exist.

The understanding resulting from that may be helpful to take decisions regarding training improvement and information sharing among individuals or groups in key organizations in the field.

References

1. Butts CT (2009) Revisiting the foundations of network analysis. Science 325(5939):414–416
2. Newman MEJ (2003) The structure and function of complex networks. SIAM Rev 45(2):167
3. Boccaletti S, Latora V, Moreno Y, Chavez M, Hwang D (2006) Complex networks: structure and dynamics. Phys Rep 424(4–5):175–308
4. Dorogovtsev SN, Goltsev AV, Mendes JFF Critical phenomena in complex networks. Networks 1–79. doi:10.1103/RevModPhys.80.1275
5. Dorogovtsev SN, Mendes JF (2001) Language as an evolving word web. Proc - Royal Soc, Biol Sci 268(1485):2603–2606
6. Rubinov M, Sporns O (2010) NeuroImage complex network measures of brain connectivity: uses and interpretations. NeuroImage 52(3):1059–1069
7. Li W, Yang J-Y (2009) Comparing networks from a data analysis perspective. Complex Sci 5:1907–1916
8. Schaffernicht M, Groesser SN (2011) A comprehensive method for comparing mental models of dynamic systems. Eur J Oper Res 210(1):57–67
9. Langfield-Smith K, Wirth A (1998) Measuring differences between cognitive maps. J Oper Res Soc 43(12):1135–1150
10. Langan-Fox J (2001) Analyzing shared and team mental models. Int J Ind Ergon 28(2):99–112
11. Schaffernicht M (2006) Detecting and monitoring change in models. Syst Dyn Rev 22(1):73–88

12. Markóczy L, Goldberg J (1995) A method for eliciting and comparing causal maps. J Manag 21(2):305
13. Muñoz GJ, Glaze RM, Winfred Arthur J, Jarrett S, McDonald JN (2011) Driving mental models as a predictor of crashes and moving violations. In: 26th annual conference of the society for industrial and organizational psychology
14. Jones NA, Ross H, Lynam T, Perez P, Leitch A (2011) Mental models: an interdisciplinary synthesis of theory and methods. Ecol Soc 16(1). http://www.ecologyandsociety.org/vol16/iss1/art46/
15. Carley KM, Palmquist M (1992) Extracting, representing, and analysing mental models. Soc Forces 70(3):601–636
16. Kim MK (2012) Cross-validation study of methods and technologies to assess mental models in a complex problem solving situation. Comput Hum Behav 28(2):703–717
17. Sayama H, Farrell DL, Dionne SD (2010) The effects of mental model formation on group decision making: an agent-based simulation. Complexity 16(3):49–57
18. Kai Y, Qingquan W, Lili R (2008) Emergency Ontology construction in emergency decision support system. In: IEEE international conference on service operations and logistics, and informatics, vol 1, pp 801–805
19. Alexander D (2002) Principles of emergency planning and management. Oxford University Press, London, p 340
20. Griffin BD (2009) Emergency management terms and concepts. High Educ 1–8

Chapter 20
Traveling and Stationary Patterns in Bistable Reaction-Diffusion Systems on Network

Nikos E. Kouvaris, Hiroshi Kori, and Alexander S. Mikhailov

Abstract Traveling and stationary patterns in bistable reaction-diffusion systems have been extensively studied for classical continuous media and regular lattices. Here, we consider analogs of such non-equilibrium patterns in bistable one-component systems on trees and on random networks. As revealed through numerical simulations, traveling fronts exist in network-organized systems. They represent waves of transition from one stable state into another, spreading over the entire network. The fronts can furthermore be pinned, thus forming stationary structures. While pinning of fronts has previously been considered for chains of diffusively coupled bistable elements, the network architecture brings about significant differences. An important role is played by the degree (the number of connections) of a node. For regular trees with a fixed branching factor, the pinning conditions are analytically determined. For large random networks, the mean-field theory for stationary patterns is constructed.

Keywords Self-organization · Pattern formation · Nonlinear dynamics · Bistability · Complex networks · Traveling fronts · Pinning · Stationary patterns

20.1 Introduction

Pattern formation in reaction-diffusion systems for from equilibrium has been widely studied for chemical and biological processes. Within the last years, attention has been turned to self-organization on network systems, where the network nodes are occupied by active elements and the links represent diffusive connections between them. Such systems may correspond to networks of diffusively coupled chemical reactors [1] or biological cells [2]. Detailed studies of synchronization phenomena in oscillatory systems [3] and of infections spreading over networks [4, 5]

N.E. Kouvaris (✉) · A.S. Mikhailov
Department of Physical Chemistry, Fritz Haber Institute of the Max Planck Society, Faradayweg 4-6, 14195 Berlin, Germany

H. Kori
Department of Information Sciences, Ochanomizu University, Tokyo 112-8610, Japan

T. Gilbert et al. (eds.), *Proceedings of the European Conference on Complex Systems 2012*, Springer Proceedings in Complexity, DOI 10.1007/978-3-319-00395-5_20,

have been performed. Turing patterns in activator-inhibitor systems organized on large networks have also been considered [6]. However, systematic research along this directions is still largely missing. To contribute towards such research, we investigate here traveling and pinned fronts in networks formed by diffusively coupled bistable elements.

The study of bistable media is of principal importance for the theory of self-organization in reaction-diffusion systems. Traveling fronts which represent the transition from one stable state into another are providing a typical example of self-organized wave patterns. The velocity and the profile of the front are uniquely determined by the medium properties. Depending on the parameters of the medium, either spreading or retreating fronts can typically found. Stationary fronts, which separate two regions of different stable steady states, are not characteristic for continuous media; they are found only at special parameter values along the border line separating spreading and retreating fronts [7]. When discrete systems, formed by chains of coupled bistable elements, are considered, traveling fronts can however become pinned if diffusion is sufficiently weak, so that, stable stationary fronts which are found within entire parameter regions may arise [8–10].

As shown in our study [11], traveling and pinned fronts are also possible in network-organized systems, but their properties are significantly different. The behavior of the fronts is highly sensitive to network architecture and degrees of network nodes play an important role.

20.1.1 Bistable Systems on Networks

Classical one-component reaction-diffusion systems in continuous media are described by the form

$$\frac{\partial}{\partial t}u(\mathbf{x},t) = f(u,h) + D\nabla^2 u(\mathbf{x},t), \tag{20.1}$$

where $u(\mathbf{x},t)$ is the local activator density, function $f(u,h)$ specifies local bistable dynamics and D is the diffusion coefficient. Depending on the particular context, the activator variable u may represent concentration of a chemical reagent or of biological species which amplifies its own production.

Here we consider analogs of processes described by the model (20.1), which are taking place on networks. In network-organized systems, the activator species occupies nodes of a network and can be transported over network links to other nodes. The connectivity structure of the network can be described in terms of an adjacency matrix $\mathbf{T}$, whose elements are $T_{ij}=1$, if there is a link connecting the nodes i and j $(i,j=1,\dots,N)$, and $T_{ij}=0$ otherwise. Processes in undirected networks, where the adjacency matrix $\mathbf{T}$ is symmetric $(T_{ij}=T_{ji})$, have been considered. Generally, the network analog of system (20.1) is given by

$$\dot{u}_i = f(u_i) + D\left(\sum_{j=1}^{N} T_{ij}u_j - \sum_{j=1}^{N} T_{ji}u_i\right), \tag{20.2}$$

where u_i is the amount of activator in network node i and $f(u_i)$ describes the local bistable dynamics of the activator. The last term in Eq. (20.2) describes diffusive coupling between the nodes. Parameter D characterizes the rate of diffusive transport of the activator over the network links.

Instead of the adjacency matrix, it is convenient to use the Laplacian matrix $\mathbf{L}$ of the network, whose elements are defined as $L_{ij} = T_{ij} - k_i \delta_{ij}$, where $\delta_{ij} = 1$ for $i = j$, and $\delta_{ij} = 0$ otherwise. In this definition k_i is the degree, or the number of connections, of node i given by $k_i = \sum_j T_{ji}$. In the new notations Eq. (20.2) takes the form

$$\dot{u}_i = f(u_i) + D \sum_{j=1}^{N} L_{ij} u_j. \tag{20.3}$$

When the considered network is a lattice, its Laplacian matrix coincides with the finite-difference expression for the Laplacian differential operator after discretization on this lattice.

In our study we use the Schlögl model [12] which is a classical example of a one-component system exhibiting bistable dynamics. This model describes a hypothetical trimolecular chemical reaction which exhibits bistability. In the Schlögl model, the nonlinear function $f(u)$ is a cubic polynomial

$$f(u) = -\frac{\partial V}{\partial u} = -(u - r_1)(u - r_2)(u - r_3), \tag{20.4}$$

so that, $V(u)$ has one maximum at r_2 and two minima at r_1 and r_3. We have performed numerical simulations and analytical investigations of the reaction-diffusion system (20.3) random networks and for irregular trees using the Schlögl model.

20.2 Main Results

As revealed through numerical simulations, analogs of traveling fronts, spreading or retreating, exist in such network-organized systems. Furthermore, stationary patterns, pinned at subsets of network nodes, are found. Our numerical simulations suggest that degrees of the nodes play an important role in such phenomena. The observed behavior is however complex and depends on the architecture of the networks and on how the initial activation was applied [11].

In the special case of regular trees, an approximate analytical theory has been constructed. Our theory, which represents an extension of the respective theory of pinned fronts for the chains [10], reveals that the branching factors of the trees and, thus, the degrees of their nodes, are essential for fronts dynamics [11]. By using this approach, front pinning conditions could be derived and parameter boundaries, which separate pinned and traveling fronts, could be determined. As we have found, propagation conditions are different for the fronts traveling from the tree root to the periphery or in the opposite direction. Generally, all fronts become pinned as the diffusion constant is gradually reduced. While the theory has been developed for regular trees, where the branching factor is fixed, it can be used to interpret the behavior found in irregular trees and also for large Erdös-Rényi networks.

It is known that large random Erdös-Rényi networks can locally be approximated by the trees [13]. Therefore, if the initial perturbation has been applied to a node and starts to spread over the network, its propagation is effectively taking place on a tree formed by the node neighbors. Only when the activation has already covered a sufficiently high fraction of the network nodes, loops start to play a role. When this occurs, the activation may arrive at a node along different pathways and the tree approximation ceases to hold. In this opposite situation, a different theory employing the mean-field approximation can however be applied as has been previously used for spreading-infection problems [14] and the analysis of Turing patterns on the networks [6]. Within the mean-field approximation, statistical properties of the network stationary activity patterns are well reproduced [11].

20.3 Methods

In our study random Erdös-Rényi, scale-free and hierarchical tree networks were considered. Erdös-Rényi networks were constructed by taking a large number N of nodes and randomly connecting any two nodes with some probability p. This construction algorithm yields a Poisson degree distribution with the mean degree $\langle k \rangle = pN$ [15]. In our study we have considered the largest connected component network, namely, we have removed the nodes with the degree $k = 0$.

Tree networks with branching factor $k - 1$ were constructed by a simple iterative method. Starting with a single root node and at each step $k - 1$ nodes are added to each existing node with the degree $k = 1$. After L steps this algorithm leads to a tree network with the size $N = \sum_{l=1}^{L} (k - 1)^{l-1}$, where the root node has degree $k - 1$, the last added nodes have degree 1 and all other nodes have degree k. In our numerical simulations we have also used complex trees consisting of component trees with different fixed branching factors which are connected at their origins.

Scale-free networks were constructed by the preferential attachment algorithm of Barábasi and Albert [15]. Starting with a small number of m nodes with m connections, at each next time step a new node is added, with m links to m different previous nodes. The new node will be connected to a previous node i, which has k_i connections, with the probability $k_i / \sum_j k_j$. After many time steps, this algorithm leads to a network composed by N nodes with the power-law degree distribution $P(k) \sim k^{-3}$ and the mean degree $\langle k \rangle = 2m$.

Another issue in our study was to find the more convenient visualization of the networks, for a better highlighting of self-organized patterns. Thus, network analogs of traveling fronts was illustrated by grouping the nodes according to their distance (the shortest path length) from the first activated node and the average value of the activator density in each group was plotted as a function of the distance [11]. Stationary patterns, were displayed either by ordering the nodes according to their increasing degrees or by using the Fruchterman-Reingold force-directed algorithm which places the nodes with close degrees near one to another in the network projection onto a plane, so that, the localization of the stationary patterns at the subsets of nodes with certain degrees is clearly illustrated.

Acknowledgements Financial support from the DFG Collaborative Research Center SFB910 "Control of Self-Organizing Nonlinear Systems" and from the Volkswagen Foundation in Germany is gratefully acknowledged.

References

1. Karlsson A, Karlsson R, Karlsson M, Cans AS, Strömberg A, Ryttsén F, Orwar O (2001) Molecular engineering: networks of nanotubes and containers. Nature 409(6817):150–152
2. Bignone FA (2001) Structural complexity of early embryos: a study on the nematode caenorhabditis elegans. J Biol Phys 27:257–283
3. Arenas A, Díaz-Guilera A, Kurths J, Moreno Y, Zhou C (2008) Synchronization in complex networks. Phys Rep 469(3):93–153
4. Colizza V, Vespignani A (2008) Epidemic modeling in metapopulation systems with heterogeneous coupling pattern: theory and simulations. J Theor Biol 251(3):450–467
5. Barrat A, Barthelemy M, Vespignani A (2008) Dynamical processes on complex networks. Cambridge University Press, Cambridge
6. Nakao H, Mikhailov AS (2010) Turing patterns in network-organized activator–inhibitor systems. Nat Phys 6(7):544–550
7. Mikhailov AS (1994) Foundations of synergetics I: Distributed active systems, 2nd edn. Springer, Berlin
8. Bootht V, Erneux T (1992) Mechanism for propagation failure in discrete reaction-diffusion systems. Physica A 188:206–209
9. Erneux T, Nicolis G (1993) Propagating waves in discrete bistable reaction-diffusion systems. Physica D 67:237–244
10. Mitkov I, Kladko K, Pearson J (1998) Tunable pinning of burst waves in extended systems with discrete sources. Phys Rev Lett 81(24):5453–5456
11. Kouvaris NE, Kori H, Mikhailov AS (2012) Traveling and pinned fronts in bistable reaction-diffusion systems on networks. PLoS ONE 7(9):e45029
12. Schlögl F (1972) Chemical reaction models for non-equilibrium phase transitions. Z Phys 253(2):147–161
13. Dorogovtsev SN, Mendes JFF (2003) Evolution of networks: from biological nets to the internet and WWW. Oxford University Press, London
14. Pastor-Satorras R, Vázquez A, Vespignani A (2001) Dynamical and correlation properties of the internet. Phys Rev Lett 87(25):3–6
15. Albert R, Barabási AL (2002) Statistical mechanics of complex networks. Rev Mod Phys 74:47–97

Chapter 21
Searching Shortest Paths on Weakly Dynamic Graphs

Jean-Yves Colin, Moustafa Nakechbandi, and A.S. Ould Cheikh

Abstract In this paper, we study weakly dynamic graphs, and we propose an efficient polynomial algorithm that computes in advance shortest paths for all possible configurations. No additional computation is needed after any change in the problem because shortest paths are already known in all cases. We apply this result to a dynamic routing problem. In this problem, messages must be sent from some components (captors for example) to a specific one (a processor for example) as quickly as possible. The actual network is a mesh and the problem can represented by a weighted directed acyclic graph. One known arc has unreliable performances.

Keywords Shortest paths · Dynamic graphs · Route planning

21.1 Introduction

In complex systems, dynamic graphs in which some of their values or even their topologies may change from one moment to another offer new research opportunities.

One of the most famous algorithm, and one of the most used on static graphs, is the shortest-paths algorithm from E.W. Dijkstra [1]. For the more complex dynamic graphs, several models are proposed, probabilistic [2–4] or non-probabilistic [5–7]. In non-probabilistic models, most algorithms update previously computed data after each change in a value or after the addition or removal of an edge [8–11]. In probabilistic models, the optimal paths definition of a shortest (or longest) path is

J.-Y. Colin (✉) · M. Nakechbandi
LITIS, Le Havre University, 5 rue Ph. Lebon, BP 540, 76058 Le Havre cedex, France
e-mail: moustafa.nakechbandi@univ-lehavre.fr

M. Nakechbandi
e-mail: jean-yves.colin@univ-lehavre.fr

A.S. Ould Cheikh
Nouakchott University, Nouakchott, Mauritania
e-mail: ahdsalem@gmx.fr

T. Gilbert et al. (eds.), *Proceedings of the European Conference on Complex Systems 2012*, Springer Proceedings in Complexity, DOI 10.1007/978-3-319-00395-5_21,

different according to the published papers. The most usual definition sets as optimal a path that maximizes the expected value of an utility function chosen by the authors [2, 4]. The problem itself is usually NP-hard.

Among used metaheuristics, one can find ant colony algorithms [9] and other swarm intelligence algorithms. These algorithms are very general and try too to adapt their results following changes in the problem.

In this paper, we study weakly dynamic graphs, and we propose an efficient polynomial algorithm that computes in advance shortest paths for all possible configurations. No computation is needed after any change in the problem because shortest paths are already known in all cases.

We apply this result to one delivery problem for trucks from one regional storehouse to several local stores when one possible point has a variable traversal duration. We apply too this result to the problem of rerouting delivery trucks toward their final destinations when there is a change in the traversal duration of one known point.

21.2 Problem Statement

We first define weakly dynamic graphs.

Definition 21.1 A weakly dynamic graph is a graph in which there is an unstable valuated edge (in an undirected graph) or valuated arc (in a directed graph) between two known vertices x_1 and x_2 of the graph. That edge or arc has an unknown positive value x that may change at any time. All other edges or arcs are stable and their values never change.

In the rest of this paper, we will study a weakly dynamic directed acyclic graph $G(V, E)$ with V being the set of vertices of G and E being the set of arcs. Each arc (i, j) of G is valuated by a positive stable value p_{ij}, except for two known vertices x_1 and x_2 of the graph. This arc (x_1, x_2) from x_1 to x_2 is valuated by an unknown and unstable value x that may change at any time. The length of a path at a given moment is the sum of the values of all the used arcs.

We now want to find in G shortest paths between a given vertex and all other vertices, or alternatively shortest paths from all vertices to a given vertex.

21.3 Main Results

21.3.1 The Proposed Algorithm

In the following, we study only the problem of finding a shortest path between a given vertex s_0 and all other vertices. Finding a shortest path from all vertices to a given vertex is a similar problem that can be solved by using predecessors instead of successors in our algorithm.

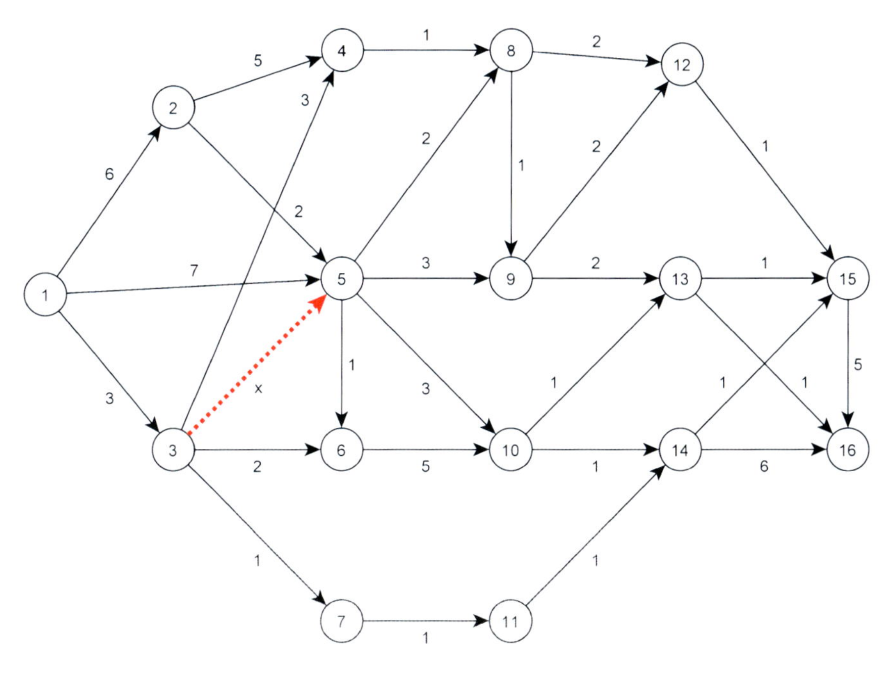

Fig. 21.1 Example of a Weakly Dynamic Graphs. *The arrow* in *dotted lines* represents the non steady bow

The algorithm works in four successive phases:

- it first builds a set of the vertices that can be reached from the starting vertex using arc (x_1, x_2). Only the successors of x_2 are considered because G is a directed graph.
- next, it computes the length $d(x_2, s_n)$ of the shortest path from vertex x_2 to all vertices of the above computed set.
- then it computes the length $ds(s_0, s_n)$ of the shortest path that does not use arc (x_1, x_2), from vertex s_0 to all other vertices of the graph.
- finally, it computes the length $d(s_0, s_n)$ of the shortest path from vertex s_0 to all other vertices of the graph, by comparing
- the length $ds(s_0, x_1) + x + d(x_2, s_n)$ of the shortest path that uses arc (x_1, x_2),
- and the length $ds(s_0, s_n)$ of the shortest path that does not use arc (x_1, x_2). Thus,

  $$d(s_0, s_n) = ds(s_0, x_1) + x + d(x_2, s_n), \quad \text{if } ds(s_0, s_n) > ds(s_0, x_1) + x + d(x_2, s_n)$$

  else $= ds(s_0, s_n)$.

Because the length of any shortest path that uses arc (x_1, x_2) depends on the value of x, the length $d(s_0, s_k)$ of the shortest path from vertex s_0 to another vertex s_k, will also depends on x. In most cases for a vertex s_k, there will be one value $x(s_k)$ of x such that, if x is inferior to this value $x(s_k)$, the shortest path will use arc (x_1, x_2), else it will not use it. Each path itself to each vertex s_n is computed during the lengths computations.

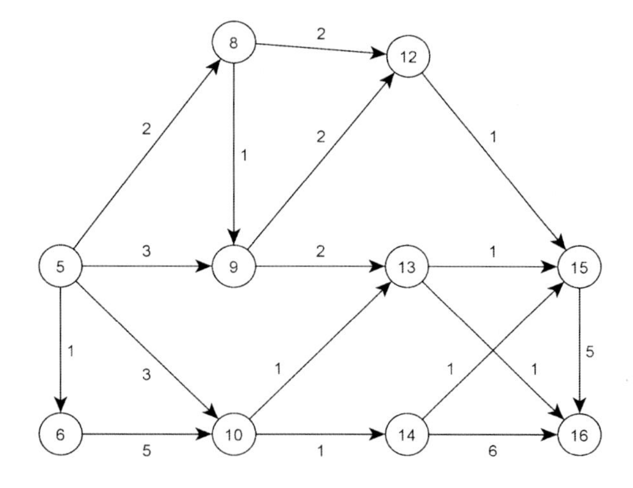

Fig. 21.2 Sub-graph generate by step 1

Example 21.1 We now apply this algorithm on the graph of Fig. 21.1.

Step 1: the set of direct or indirect successors of x_2 is {6, 8, 9, 10, 12, 13, 14, 15, 16}. The corresponding sub-graph is that shown in Fig. 21.2.

Step 2: the computed shortest paths from x_2 to these successors s_n have lengths $d(x_2, s_n)$ of

s_n	5	6	8	9	10	12	13	14	15	16
$d(x_2, s_n)$	0	1	2	3	3	4	5	4	5	6

Step 3: the computed lengths $ds(s_0, s_n)$ of shortest paths that do not use arc (x_1, x_2), from vertex s_0 (with s_0 being vertex 1) to all other vertices of the graph, are

s_n	1	2	3	4	5	6	7	8	9
$ds(x_2, s_n)$	0	6	3	6	7	5	4	7	8

s_n	10	11	12	13	14	15	16
$ds(x_2, s_n)$	10	5	9	10	6	7	11

Step 4: the lengths $d(x_1, x_2)$ of the shortest paths for all vertices is the minimum of the two values not using arc (x_1, x_2) and using arc (x_1, x_2).

s_n	1	2	3	4	5	6	7	8	9
If not using arc (x_1, x_2)	0	6	3	6	7	5	4	7	8
If using arc (x_1, x_2)	–	–	–	–	$3 + x$	$4 + x$	–	$5 + x$	$6 + x$

s_n	10	11	12	13	14	15
If not using arc (x_1, x_2)	10	5	9	10	6	7
If using arc (x_1, x_2)	$6+x$	–	$7+x$	$8+x$	$6+x$	$7+x$

Thus, for some vertices and some values of x, the shortest path uses arc (x_1, x_2), and for these vertices and some larger values of x, the shortest path does not use arc (x_1, x_2).

21.3.2 Some Proprieties of the Algorithm

Theorem 21.1 *The paths computed by the algorithm are shortest paths.*

Theorem 21.2 *The algorithm complexity is* $O(n^2)$.

Definition 21.2 We call critical value of x for a vertex s_k the value $x(s_m)$ such that, if x is inferior to this value, the shortest path from vertex s_0 to vertex s_k will use arc (x_1, x_2) and will have a length that depends on x, else it will not use it and its length will be constant.

Theorem 21.3 *Each vertex* s_m *that is a direct or indirect successor of vertex* x_2 *has* 0 *or* 1 *critical value for the computation of a shortest path from* s_0 *to this vertex.*

A corollary of the last theorem is that the number of critical values is a finite number. Furthermore, if we sort in ascending order the critical values of all vertices of the graph, one can remark that the computed set of shortest paths from s_0 to all other vertices in the graph is the same for all values of x between two consecutive values critical values. So the proposed algorithm can be used to efficiently compute shortest paths for all possible values of x from a given vertex to all other vertices. It can be used too to efficiently compute shortest paths from all vertices of a graph to a given target vertex, by using the predecessors instead of the successors during the computation.

21.4 Application to a Routing Problem

We now apply the algorithm to the following dynamic routing problem. Messages must be sent from some components (captors for example) to a specific one (a processor for example) as quickly as possible. The actual network is a mesh and can represented in this problem by a weighted directed acyclic graph [12]. One arc is known to have unreliable performances for some reason. This problem is thus a weakly dynamic graph problem with vertex s_n being the destination, and arc (x_1, x_2) being the unreliable arc (cf. Fig. 21.3).

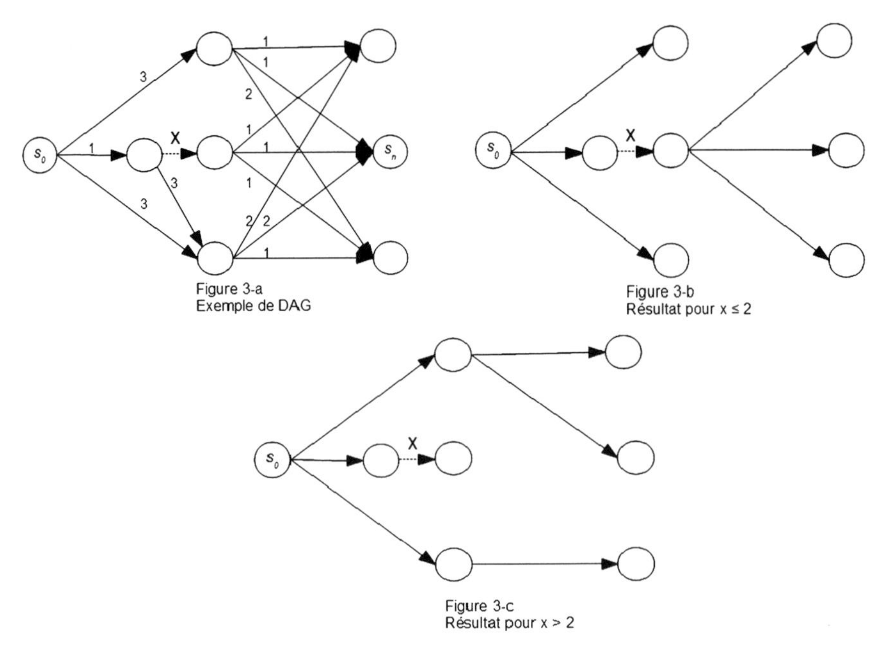

Fig. 21.3 Application 1. Trees of shorter path from vertex s_0 to all others

As said above, we may use a version of the proposed algorithm that uses the predecessors instead of the successors. This allows us to compute shortest paths from all other vertices to vertex s_n. In the example graph of Fig. 21.4, there then are two critical values, 2 and 4, giving 3 intervals. Each interval has its own (inverted tree) of shortest paths spanning the whole graph and leading to vertex s_n (cf. Fig. 21.4). Each gives then a routing policy that is optimal in its interval. These routing policies are then all precalculated and stored in the vertices with their relevant critical values of validity. The current value of x is next used to decide what routing will be used. The unreliable arc is then monitored. When its value changes, the new value is transmitted to all vertices. Each vertex compares it to the critical values and may immediately decide to keep using the currently used local routing policy, or switch to the already computed one best suited to the new value of x, just by checking in which interval the value x is now in. Thus, no time is lost recomputing new optimal paths.

21.5 Conclusions

In this paper, we proposed a new graph model we call weakly dynamic graph, and we presented an algorithm to compute shortest paths for all possible cases in a given graph. This algorithm has a polynomial complexity.

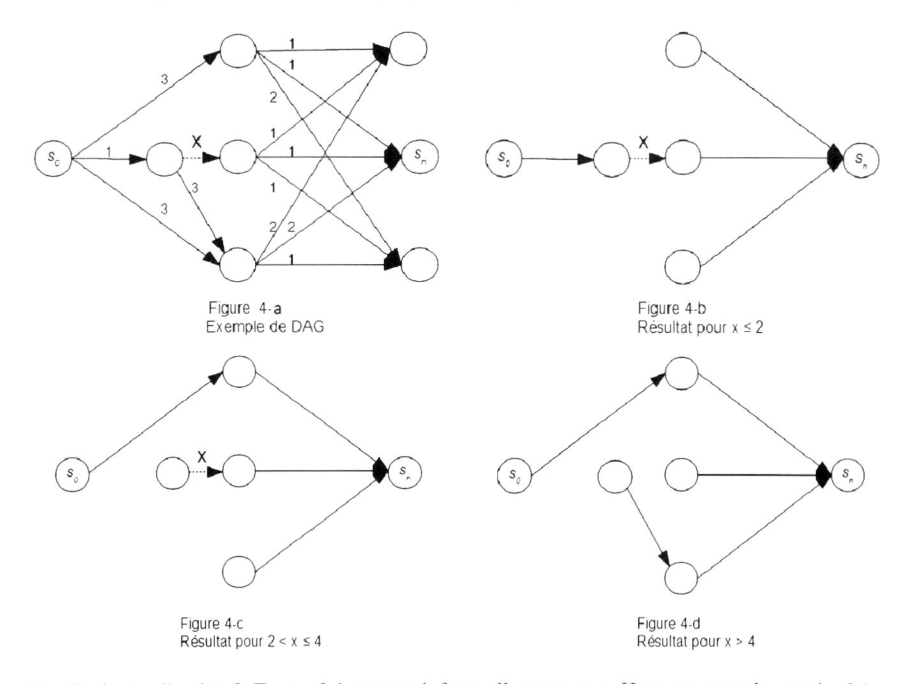

Fig. 21.4 Application 2. Trees of shorter path from all vertex to s_n Here one uses the graph of the previous application (Fig. 21.3(a))

We intend to extend this study to weakly dynamic non oriented graphs, to graphs with 2 or more variable arcs, to logistical routing problems and to the longest paths algorithms used in scheduling and project management.

References

1. Dijkstra EW (1959) A note on two problems in connexion with graphs. Numer Math 1:269–271
2. Fulkerson D (1962) Expected critical path lengths in PERT networks. Oper Res 10:808–817
3. Minoux M (1976) Structures algébriques généralisées des problèmes de cheminement dans les graphes: théorèmes, algorithmes et applications. RAIRO Rech Opér 10(6):33–62
4. Mirchandani PB, Soroush H (1985) Optimal paths in probabilistic networks. A case with temporary preferences. Comput Oper Res 12:365–383. 1985
5. Orda A, Rom R (1990) Shortest-path and minimum-delay algorithms in networks with time dependent edgelength. J ACM 37(3):607–625. 1990
6. Ramalingam G, Reps T (1996) On the computational complexity of dynamic graph problems. Theor Comput Sci 158:233–277. 1996
7. Kuhn F, Oshman R (2011) Dynamic networks: models and algorithms. SIGACT News 42(1):82–96
8. Demetrescu C, Italiano GF (2004) A new approach to dynamic all pairs shortest paths. J ACM 51(6):968–992. 2004

9. Balev S, Guinand F, Pigné Y (2007) Maintaining shortest paths in dynamic graphs. In: Proceedings of the international conference on non-convex programming: local and global approaches theory, algorithms and applications (NCP'07), Rouen, December 17–21, pp 17–21
10. Mao H (2008) Pathfinding algorithms for mutating graphs. Computer Systems Lab
11. Alivand M, Alesheikh AA, Malek MR (2008) New method for finding optimal path in dynamic networks. World Appl Sci J 3:25–33. 2008
12. Bast H, Carlsson E, Eigenwillig A, Geisberger R, Harrelson Ch, Raychev V, Viger F (2008) Fast routing in very large public transportation networks using transfer patterns. In: ESA 2008, vol 1, pp 290–301

Chapter 22
Emergence of Long Range Order in the XY Model on Diluted Small World Networks

Sarah De Nigris and Xavier Leoncini

Abstract We study the XY model on diluted Small World networks, i.e Small World networks whose number of links scales with the system size $N_{links} \sim N^{\gamma}$, $1 < \gamma < 2$. Starting from the regular lattice topology, we first concentrate on the behaviour varying the dilution parameter γ: for low values, the system does not exhibit a phase transition; while for γ approaching 2 a second order transition of the magnetisation arises since the system is near the HMF regime. Hence $\gamma_c = 1.5$ appears to be a critical value: an energy range is observed in which the magnetisation shows important fluctuations and does not reach the equilibrium state. We then take in account the model on a Small World network: for the latter, we have chosen the Watts-Strogatz model, whose topology is parametrized by the rewiring probability p, $0 < p < 1$. We performed microcanonical simulations of the dynamics and we highlight the presence of a second order phase transition appearing even for very low p and γ, when the topology is still near the regular lattice one. Moreover we observe a dependence of the critical energy ϵ_c on the rewiring probability p.

22.1 Introduction

The concept of *Small World network* can be found in systems which spread from sociology and information science to biology and physics. We can take as fundamental and well known examples, among others, the Web [1] or networks of cells in the living, like the neurons [2]. Despite the difference between those systems, some common features arise in these real world networks like an high level of clustering or the "Small World" effect itself, i.e. little average distance between two nodes of the network. These shared properties are, looking at the heterogeneity of the examples, quite independent from the punctual nature of the agents interacting on networks and thus it represents a challenge to describe how common *topological* features stem from simple assumptions which can be taken regardless to the particular model considered. Moreover the dynamical processes taking place on the top of

S. De Nigris (✉) · X. Leoncini
CPT-CNRS, UMR 7332, Aix-Marseille Université, Campus de Luminy, Case 907, 13288 Marseille Cedex 9, France
e-mail: denigris.sarah@gmail.com

T. Gilbert et al. (eds.), *Proceedings of the European Conference on Complex Systems 2012*, Springer Proceedings in Complexity, DOI 10.1007/978-3-319-00395-5_22,

networks are profoundly influenced by the underlying structure, pointing out a non trivial interplay between the topology and the dynamics [3]. The statistical physics approach could hence turn the *qualitative* matching between network structure and dynamical processes into *quantitative* and the present work aims to inscribe itself on this line. In this purpose, we chose a model for the network, the Watts and Strogatz model [4], and a model for the interaction, the XY model for rotors. The latter, in spite of its simple formulation, displays a very rich phenomenology and it has been applied in various physical contexts: from solid state physics [5] to superconductors and, recently, in more peculiar systems like bird flocks [6]. On the other hand, the Watts-Strogatz networks model catches the two aforementioned features of Small World networks and, at the same time, it provides naturally a control parameter on the topology, the rewiring probability.

In Sect. 22.2 we present the XY model and we introduce the order parameter used to characterize the phase transition of the system. Then, in Sect. 22.3, we first concentrate on the behaviour of the model on a regular lattice, considering how its thermodynamic properties vary changing the number of neighbours k connected to each spin. Further on, in Sect. 22.4, we introduce the algorithm to create a Small World network and we analyze the influence of this network structure on the phase transition displayed by the XY model. We conclude, in Sect. 22.5, with a concise summary of the main results.

22.2 The XY Model

In general the XY model describes a system of N pairwise interacting units. At each unit i is assigned a real number θ_i, called the *spin* of i. In the following, we will consider the XY model from the point of view of Hamiltonian dynamical systems by adding a kinetic energy term to the XY Hamiltonian. The total Hamiltonian H takes hence the form:

$$H = \sum_{i=1}^{N} \frac{p_i^2}{2} + \frac{J}{2k} \sum_{i,j=1}^{N} \epsilon_{i,j} \big(1 - \cos(\theta_i - \theta_j)\big). \tag{22.1}$$

Because of the periodicity of the cosine function in the Eq. (22.1), the phase space for θ_i is restricted to the interval $(-\pi, \pi]$. We associate to each spin i a canonical momentum p_i whose coupled dynamics with the $\{\theta_i\}$ will be given by the set of Hamilton equations:

$$\begin{cases} \dot{\theta}_i = \frac{\partial H}{\partial p_i} = p_i \\ \dot{p}_i = -\frac{\partial H}{\partial \theta_i} = -\frac{J}{k}(\sum_j \epsilon_{i,j} \cos\theta_j \sin\theta_i - \sum_j \epsilon_{i,j} \sin\theta_j \cos\theta_i). \end{cases} \tag{22.2}$$

The coupling constant J in Eqs. (22.1)–(22.2) is chosen positive in order to obtain a ferromagnetic behaviour and in the following it will be set at 1 without loss of generality. We encode the information about the links connecting the units in the *adjacency matrix* $\epsilon_{i,j}$:

$$\epsilon_{i,j} = \begin{cases} 1 & \text{if } i, j \text{ are connected} \\ 0 & \text{otherwise.} \end{cases} \tag{22.3}$$

By construction, $\epsilon_{i,j}$ is a symmetric matrix with null trace since we do not consider directed links. The system possesses two constants of motion preserved by the dynamics: the energy $H = E$ and the total angular momentum $P = \sum_i p_i$ which are set by the initial conditions. We chose to start the system with a Gaussian distribution for both for the spins and the momenta. The numerical integration of Eq. (22.2) is performed using a symplectic integrator [7], which ensures the conservation of the momenta E and P and the symplectic structure. The thermodynamic quantities are hence calculated averaging over time and over different network realisations. The key quantity of our study is the order parameter $\mathbf{M} = (m_x, m_y)$:

$$\mathbf{M} = \begin{cases} m_x = \frac{1}{N} \sum_j \cos(\theta_j) \\ m_y = \frac{1}{N} \sum_j \sin(\theta_j). \end{cases} \tag{22.4}$$

A well known limit of this model is recovered when the spins are fully coupled, the *Hamiltonian Mean Field (HMF)* model. It will be our paradigm of confrontation investigating the behaviour of the magnetisation $\mathbf{M}$ in Eq. (22.4) since the *HMF* model displays a second order phase transition of the order parameter $\mathbf{M}$. This transition has been widely studied, both numerically and analytically, and in this work we aim to understand if it also arises when the topology of the spins connections becomes non trivial. In fact, in one and two dimensions the Mermin-Wagner theorem predicts that the XY model with only local interactions should not possess long-range order at any finite temperature. But including more and more long-range interactions in the XY model we argue that it leads to true long-range order since we are approaching the mean field regime. In the following section, we address this issue investigating the transition in the XY model from short-range to long-range interactions as the number of connections is increased.

22.3 Regular Lattice Topology

We introduce the *dilution* γ as our parameter of interest to shift continuously from the short-range to the long-range regime. It is defined as follows [8]:

$$\gamma = \frac{\log(N_L)}{\log(N)},$$

where N_L is the total number of links and $\gamma \in (1, 2]$. Hence the configuration corresponding to the case $\gamma = 1$ is the linear chain with only nearest neighbours coupling and, on the other hand, $\gamma = 2$ corresponds to the full coupling of the spins, the *HMF* case. In Eq. (22.1) the normalisation constant k corresponds to the number of links per unit, called the *degree*, and it is imposed by γ:

$$k = \frac{2^{2-\gamma}(N-1)^{\gamma}}{N}. \tag{22.5}$$

The prefactor $2^{2-\gamma}$ accounts for the case of a linear chain ($\gamma = 1$) in which we set 2 links per unit. We construct this way a lattice in which each spin is connected to $k/2$ neighbours on each side and the width of this neighbourhood is imposed by our choice of the dilution. Having set the structure of the lattice, we performed simulations in the microcanonical ensemble and we studied the evolution of the total equilibrium magnetisation $\langle M \rangle$ (Eq. (22.4)) where $\langle \ldots \rangle$ denotes the time average and $M = |\mathbf{M}|$. We first concentrated on low dilution values ($\gamma < 1.5$) : as expected, the system doesn't show a phase transition of the order parameter since low dilution implies the existence of just short range interactions. In Fig. 22.1 the magnetisation vanishes with the system size, so that in the thermodynamic limit we expect the residual magnetisation to be zero. Nevertheless, quasi-long-range order could still arise at finite temperatures like in the 2D XY model which displays the Kosterlitz-Thouless phase transition [9]. This particular phase transition is characterized by the change in behaviour of the correlation function, which decays as a power law at low temperatures and exponentially in the high temperature phase. Hence to test the eventual presence of a K–T transition, we consider the correlation function:

$$c(j) = \frac{1}{N}\sum_{i=1}^{N} \cos(\theta_i - \theta_{i+j[N]}). \tag{22.6}$$

At equilibrium, the correlation decays exponentially fast (Fig. 22.1) at any temperature in the physical range, confirming the absence of the aforementioned phase transition. For those values of γ, we can conclude that the number of links is still too low to entail a change in the 1-D behaviour. Symmetrically, the other important limit to consider is $\gamma > 1.5$ when we approach the full coupling of the spins. As shown in Fig. 22.2, the mean field transition of the order parameter is recovered in this dilution regime: it is worth stressing here that we recover the mean field result even for γ significantly lower than 2, e.g. for $\gamma = 1.6$, implying that global coherence is still reachable with a weaker condition than the full coupling.

In both cases, $\gamma \to 1$ and $\gamma \to 2$, the variance of the magnetisation $\langle \sigma^2 \rangle = \langle (M - \langle M \rangle)^2 \rangle$ has the expected linear scaling with the system size, ensuring the reaching of equilibrium in our simulations.

The transition between the 1-D behaviour and the mean field phase appears to be critical for $\gamma_c = 1.5$: for low energies $0.45 \lesssim E \leq 0.75$ the magnetisation is affected by important fluctuations and it does not reach the equilibrium state (Fig. 22.3) on the timescales considered. Moreover the correlation function in Eq. (22.6) does not prove helpful in characterizing this peculiar state: it acquires the exponential behaviour only for energies higher than $\epsilon = 0.7$, while in the interesting interval

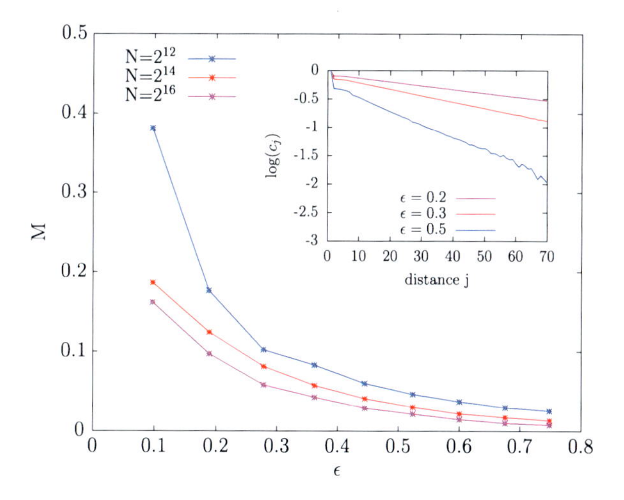

Fig. 22.1 Equilibrium magnetisation versus the energy density $\epsilon = H/N$. (*inset*) Correlation function c_j for $\gamma = 1.25$

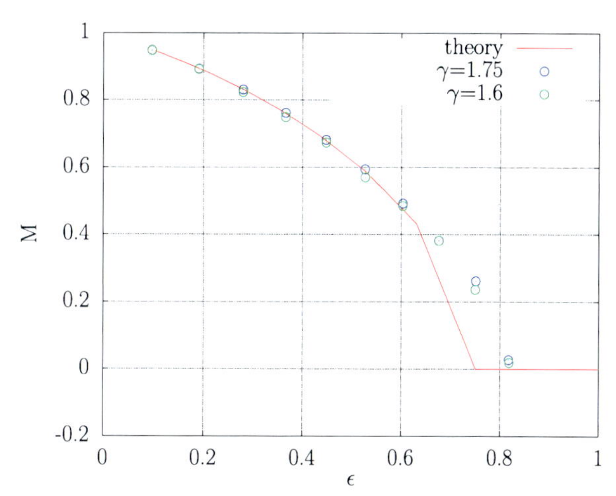

Fig. 22.2 Mean field phase transition of the magnetisation for $N = 2^{18}$. The shift in the energy of the phase transition is due to the size finiteness of the simulations, while *the solid curve* represents the analytical calculation for the HMF model in the TD limit

of energies it is heavily affected by the fluctuations and it is impossible to properly determine its behaviour. We observed these effects on several sizes from $N = 2^{12}$ up to $N = 2^{18}$ and, when considering the scaling of $\langle \sigma^2 \rangle$ with the size (Fig. 22.3), it is evident that the variance is not affected by the increasing system size. We argue that at $\gamma = 1.5$ the number of links is at its lowest value to allow the arising of long range order: at the moment, a more complete characterization of this state is ongoing as well as a theoretical effort to explain the particular nature of this dilution value.

22.4 Small World Network

The algorithm chosen to produce networks is issued from the seminal paper of Watts and Strogatz [4] and it acts on a regular network rewiring randomly the links. In practice we start from the previous regular lattice in which each vertex is connected

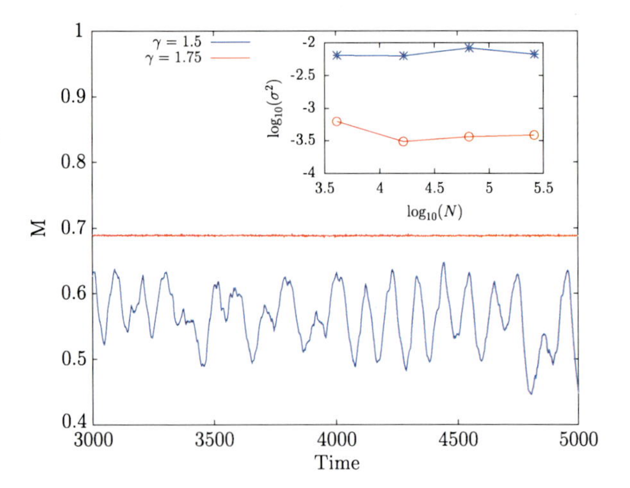

Fig. 22.3 Temporal evolution for the magnetisation for $N = 2^{18}$ and $E = 0.6$. (*inset*) Scaling of the magnetisation variance $\langle \sigma^2 \rangle$ for $\gamma = 1.5$ for $\epsilon = 0.6$ (*stars*) and $\epsilon = 0.74$ (*dots*)

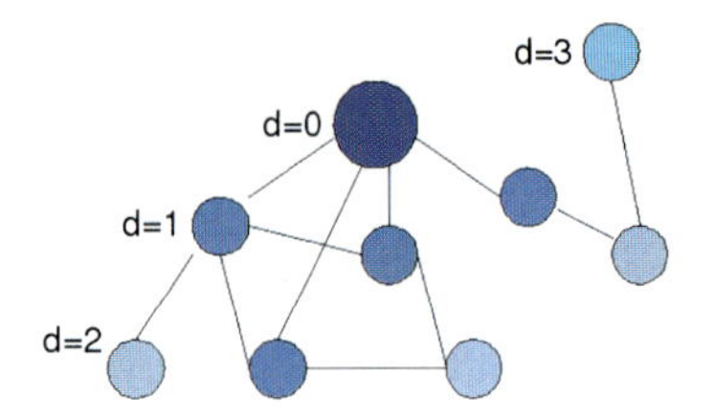

Fig. 22.4 Path lengths starting from the blue vertex

to his k neighbours and according to a fixed probability p each link is either left untouched either it is rewired. Hence we have a parameter, the *probability* p, to tune the level of rewiring: for low p values, the network is almost regular; on the other hand, for high p values, almost all the links are rewired and the network is random. With this parameter p, we can pass continuously between these two limit cases. Depending on the system size, a parameter region can be found in which the network has high connectivity and little average distance, being a Small World network. In order to quantify this interval of interest, we define two parameters, the *average path length* and the *connectivity*. The first is related to the "Small World effect" itself, i.e. the property of some networks to have a logarithmic growth of the average distance between two vertices $\langle l \rangle$ with the system size:

$$\langle l \rangle \sim \log(N). \tag{22.7}$$

This scaling indicates that $\langle l \rangle$ grows slower than linear with the system size which, on the contrary, is the behaviour of a regular network. This slow growth is the signature of *shortcuts* and, from the point of view of statistical physics, these shortcuts can imply the emergence of *global coherence*, as we shall see later. We define $\langle l \rangle$ as:

$$\langle l \rangle = \frac{1}{N} \sum_i l_i, \tag{22.8}$$

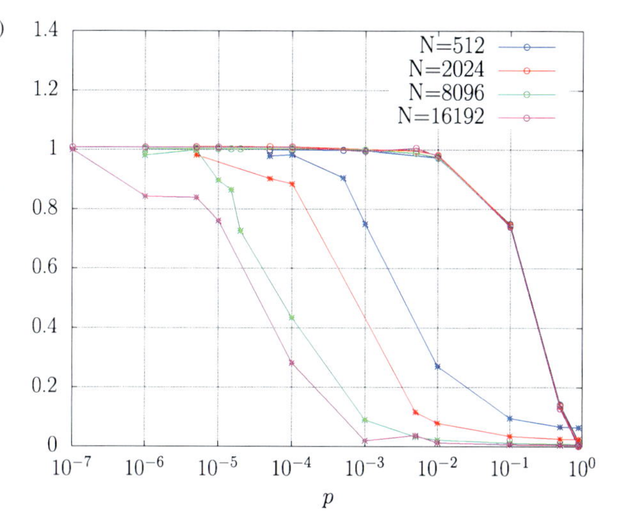

Fig. 22.5 Connectivity (*dots*) and average path length (*stars*) versus rewiring probability for $\gamma = 1.25$ and several sizes. The connectivity and the average path length are normalised to the corresponding value for the regular lattice

with l_i being the longest path attached to the i vertex. To quantify these paths, since the network lacks a metric, we count the number of edges between two vertices. Starting for instance from the i vertex, we have that his neighbours are at distance $d = 1$, the neighbours of the neighbours are at $d = 2$ and so on for the successive generations of neighbours (Fig. 22.4). Finally, to ensure that the path taken is the shortest possible, we impose to consider each vertex only once avoiding this way to come back on links already explored. For the *connectivity* we have:

$$C = \frac{1}{N} \sum_i c_i,$$

where

$$c_i = \frac{e_i}{\frac{1}{2} k_i (k_i - 1)}. \tag{22.9}$$

In Eq. (22.9), k_i stands for the *degree* of the vertex i while e_i is the number of links existing between the k_i neighbours of the vertex: the maximal number of couples between the neighbours is $\frac{1}{2} k_i (k_i - 1)$ and we count how many of these triangles effectively exist, e_i. In practice the connectivity quantifies the average amount of clustering per vertex and it is, by definition, a local parameter. We measured the topological quantities C and $\langle l \rangle$ varying the rewiring parameter p and averaging on different network realisations per each p value. As shown in Fig. 22.5, p changes the topology of the network: for low p values, C and $\langle l \rangle$ are both high and we are thus in the regular network region. On the other side, for high p values, the network has low values of C and $\langle l \rangle$ since it is completely random. In the intermediate zone the network is a Small World one, having high connectivity and low average path length. This region varies with the system size since it is delimited by the fall of the average path length which is a global parameter. Figure 22.5 provides thus a “map” indicating when the network is in the Small World regime. Using this knowledge,

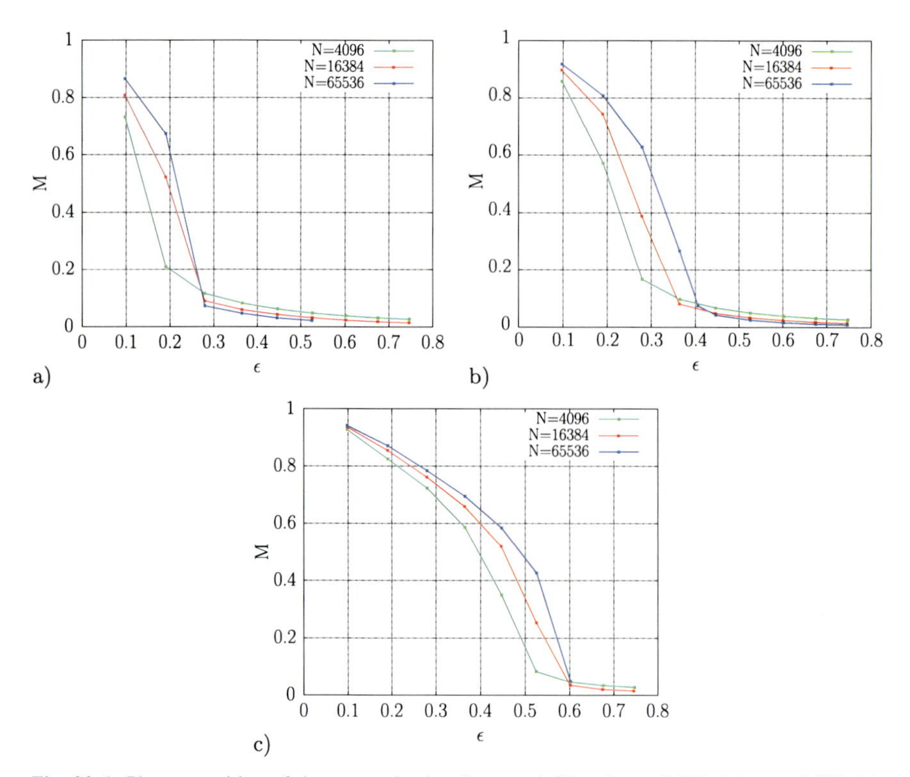

Fig. 22.6 Phase transition of the magnetisation for $\gamma = 1.25$ and $p = 0.001$ (**a**), $p = 0.005$ (**b**), $p = 0.05$ (**c**)

we were able to investigate the XY model on this particular topology as we aimed to highlight the interplay between the progressive introduction of rewiring and the emergence of long range order. Consequently we took $\gamma = 1.25$ for the dilution: in this case the regular network does not show a phase transition as discussed in Sect. 22.3 and, on the other hand, it has been shown that for *random networks* the mean field transition is recovered at any value of γ in the thermodynamic limit [8]. In between, in the Small World regime, the presence of the transition is strongly dependent on the system size and on p (Fig. 22.6). The reason for this behaviour is encoded in Fig. 22.5: the critical probability p_c to have the breakdown of $\langle l \rangle$ scales as $1/N$ and low values of $\langle l \rangle$ imply that enough shortcuts have been created to lead a shift from the 1-D topology. Hence we expect that, in the thermodynamic limit, $p_c \to 0$ with the increasing system size, as argued also in [10], but this limit proves more and more numerically expensive and hence difficult to evaluate. In Fig. 22.6 we show the transition for three low values of probability $p = 0.001$, $p = 0.005$ and for $p = 0.05$.

We remark first that the lowest size considered $N = 4096$ does not show the mean field transition implying that the probability is too low to produce enough shortcuts for this particular system size. Moreover we observe a shift of the transition energy

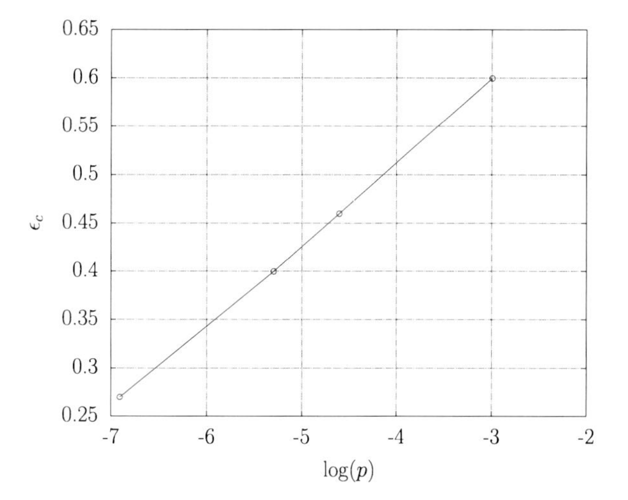

Fig. 22.7 Dependence of the critical energy ϵ_c on the rewiring probability p

ϵ_c with p: from Fig. 22.7 we observe that the phase boundary in well described by the form $\epsilon_c \approx C \ln(p)$. Our results appear hence in coherence with [11] and [10], but in that case the simulations were performed in the canonical ensemble while we deal with totally different dynamics since we use the microcanonical picture. This difference in not negligible since ensemble equivalence, which exists for the HMF model [12], cannot be taken for granted in long-range systems [13].

22.5 Conclusion

In this last section, we would like to resume the logical steps of our work and to give a perspective of further developments. In Sect. 22.2, we first introduced our model for the interaction, the XY model for rotors. We then recalled that a limit case of our model ($\gamma = 2$) is the *Hamiltonian Mean Field (HMF)* model and we stressed the presence of a second order phase transition in the *HMF* which we retrieve, with some differences, in the more general XY model on networks. In Sect. 22.3, we first focused on the regular lattice topology in which we controlled the degree of each spin via the dilution parameter γ. We showed to limit cases: the low dilution regime, where the long-range order is absent, and the high dilution phase in which the global coherence is recovered when the dilution overcomes the threshold of $\gamma = 1.5$. Interestingly, we highlighted that the phase transition is not a direct consequence of the full coupling of the spins, like in the *HMF* model, but it can still arise even for $\gamma = 1.6$, quite far hence from the extremal configuration of $\gamma = 2$. The main result of Sect. 22.3 is the evidence of a non trivial behaviour when $\gamma = 1.5$: the important fluctuations affecting the order parameter and the invariance of these effects on the system size in a whole interval of energies suggest the need of an enhanced analysis to characterize this state, probably with methods coming from the field out-of-equilibrium critical phenomena. Section 22.4 is devoted to Small World networks with particular attention to their topological parameters, the connectivity and the average path length. In this section we described operatively a model, the Watts-Strogatz model, which via the rewiring probability p, allowed us to explore the topologies of the three main configurations taken in account: random, regular and Small World networks. Finally, we presented the numerical results of simulations of the XY model on Small World networks with low dilution. We highlighted the effect of the average path length l in giving the system *global coherence:* we observe the arising of phase transitions only in regimes of low l which imply the fundamental presence of *shortcuts*. These shortcuts are responsible for the efficiency of information transmission throughout the network and they allow the emergence of a collective behaviour in a 1-D network even surprisingly when the low dilution would imply the absence of long-range order. This result indicates that it exists a complex interplay between the number of links, given by γ, and their distribution, influenced by p, and this issue is object of ongoing investigations. Moreover a non trivial effect induced by the rewiring gives the logarithmic scaling of the critical energy with the probability p and the mechanism underneath this effect is, at our knowledge, still unexplained.

Acknowledgements The authors would like to thank A. Barrat for fruitful discussions and S. De Nigris is grateful to A. Machens for informing her of the ECCS 2012 conference. S. De Nigris is financially supported by DGA/DS/MRIS.

References

1. Albert R, Jeong H, Barabasi AL (1999) The diameter of world wide web. Nature 401:130–131
2. Lago-Fernandez LF, Huerta R, Corbacho F, Siguenza JA (2000) Fast response and temporal coherent oscillations in small-world networks. Phys Rev Lett 84:2758–2761
3. Dorogovtsev AV, Goltsev SN, Mendes JFF (2008) Critical phenomena in complex networks. Rev Mod Phys 80. doi:10.1103/RevModPhys.80.1275
4. Watts DJ, Strogatz SH (1998) Collective dynamics of 'small-world' networks. Nature 393:440–442
5. Halsey TC (1985) Topological defects in the fully frustrated xy model and in 3he-a films. J Phys C, Solid State Phys 18:2437
6. Toner J, Tu Y (1995) Long-range in a two-dimensional dynamical XY model: how birds fly together. Phys Rev Lett 75:4326–4329
7. McLachlan RI, Atela P (1992) The accuracy of symplectic integrators. Nonlinearity 5:541–562
8. Ciani A, Ruffo S, Fanelli D (2011) Long-range interactions and diluted networks. Springer, Berlin
9. Kosterlitz JM, Thouless DJ (1973) Ordering, metastability and phase transitions in two-dimensional systems. J Phys C, Solid State Phys 6:1181–1203
10. Kim BJ, Hong H, Holme P, Jeon GS, Minnhagen P, Choi MY (2001) XY model in small-world networks. Phys Rev E 64:056135
11. Medvedyeva K, Holme P, Minnhagen P, Kim BJ (2003) Dynamical critical behaviour of the XY model in small-world networks. Phys Rev E 67:036118
12. Campa A, Dauxois T, Ruffo S (2009) Statistical mechanics and dynamics of solvable models with long-range interactions. Phys Rep 480:57–159
13. Leyvraz F, Ruffo S (2002) Ensemble inequivalence in systems with long-range interactions. J Phys A, Math Gen 35:285–294

Chapter 23
Role Detection: Network Partitioning and Optimal Model of the Lumped Markov Chain

Maguy Trefois and Jean-Charles Delvenne

Abstract Nowadays, complex networks are present in many fields (social science, chemistry, biology, ...) as they allow to model systems with interacting agents. In many cases, the number of interacting agents is large (from hundreds to millions of nodes). In order to get information about the functionality of the underlying system, we are interested in studying the structure of the network. One way to do that is by partitioning the network. In this paper, we present a method to detect a partition of the network such that the dynamics of a random walker on the lumped network is a good model of the dynamics of a random walker in the original network.

23.1 Introduction

Nowadays, complex networks are present in many fields (social science, information theory, chemistry, biology, computer science, ...) as they allow to model systems with interacting agents. In many cases, the number of interacting agents is large (from hundreds to millions of nodes). In order to get information about the functionality of the underlying system, we are interested in studying the structure of the network. One way to do that is by partitioning the network into communities (many links within the clusters and few links between them). In the last decade, this community detection problem has attracted many interest in research [1–3, 6, 8–10, 12, 13].

In this paper, we present a method to detect a partition of the network such that the dynamics of a random walker on the lumped network is a good model of the dynamics of a random walker in the original network. In particular, our strategy

M. Trefois (✉) · J.-C. Delvenne
Department of Applied Mathematics, Université catholique de Louvain, Louvain-la-Neuve, Belgium

J.-C. Delvenne
Facultés Universitaires Notre-Dame de la Paix, Namur Complex Systems Center (NAXYS), Namur, Belgium

J.-C. Delvenne
Center for Operations Research and Econometrics (CORE), Université catholique de Louvain, Louvain-la-Neuve, Belgium

T. Gilbert et al. (eds.), *Proceedings of the European Conference on Complex Systems 2012*, Springer Proceedings in Complexity, DOI 10.1007/978-3-319-00395-5_23,

allows to find the communities in a well clustered network, or to discover if the network is multipartite. Moreover, in the case of a lumpable Markov chain, this strategy provides the partition with respect to which the chain is lumpable [7].

23.2 The Partitioning Problem

Consider an undirected and unweighted network. The dynamics defined on the network is the following: being at node i, the probability of jumping to node j is

$$p_{ij} = \begin{cases} \frac{A_{ij}}{k_i} & \text{if } k_i \neq 0 \\ 0 & \text{otherwise} \end{cases}$$

where k_i is the degree of node i and A is the adjacency matrix of the network. This dynamical process is a Markov chain on the network.

We are interested in partitioning the network so that the dynamics defined on the blocks is a good model of the dynamics in the original network. More precisely, we look for a partition $S = \{S_1, \ldots, S_n\}$ such that for any blocks S_k, S_l and for any nodes $i, j \in S_k$,

$$\sum_{m \in S_l} p_{im} = \sum_{m \in S_l} p_{jm}. \quad (23.1)$$

Notice this partitioning problem exactly corresponds to the lumpability of the Markov chain defined on the original network.

In general, the Markov chain defined on the network in not lumpable, which means that there does not exist a relevant partition S having exactly property (23.1). That is why we are interested in the most relevant partition whose dynamics on the blocks is a good model of the dynamics in the original network. The blocks of this partition will be called "roles" (this role definition differs from those proposed in [4, 5, 11]). In next section, we present our strategy to find such a partition.

23.3 The Objective Function

In [7], E et al. suggest a method in order to partition the network as defined in previous section. However, in their method they have to fix in advance the number of roles to detect. As this number is a priori unknown, this seems to be a big disadvantage of their strategy. That is why we present another strategy in which the relevant number of roles is provided by the method itself.

The role partition will be represented by a lumped network. The nodes of the lumped network correspond to the different roles and the weight of the directed edge from node n_k to node n_l in the lumped network represents the probability of jumping from node n_k to node n_l.

Given a role partition $S = \{S_1, \ldots, S_n\}$ of the original network, the weight m_{kl} of the edge from node n_k to node n_l in the corresponding lumped network is given by the arithmetic mean of the probabilities of jumping from a node of role S_k to any node of role S_l, that is

$$m_{kl} = \frac{1}{|S_k|} \sum_{i \in S_k} p(i, S_l),$$

where $p(i, S_l) = \sum_{j \in S_l} p_{ij}$ is the probability of jumping from node i to any node of role S_l.

We would like to find a partition $S = \{S_1, \ldots, S_n\}$ such that for any nodes i and j belonging to a same block and for any block S_l, the probabilities $p(i, S_l)$ and $p(j, S_l)$ are very similar, that is we would like to find a partition S which minimizes the expression:

$$\sum_{k,l=1}^{n} \sum_{i \in S_k} \frac{(p(i, S_l) - m_{kl})^2}{|S_k|}.$$

However, the partition with only one block and the partition with the maximum number of blocks (that is, any node of the original network is a block) are trivial solutions. So, minimizing previous expression does not provide a relevant partition. To deal with this problem, we compare the observed "variance" $e_{kl} := \sum_{i \in S_k} \frac{(p(i,S_l)-m_{kl})^2}{|S_k|}$ with its expected value $E(e_{kl})$ in a null model (e.g., the Erdos-Rényi model). Then, we compute the mean of these differences on all pairs of blocks.

Consequently, we would like to find a partition minimizing the function:

$$f(S = \{S_1, \ldots, S_n\}) = \frac{1}{n^2} \sum_{k,l=1}^{n} e_{kl} - E(e_{kl}).$$

Notice that the partition with only one block and the partition with the maximum number of blocks are not trivial minimizers of f.

We will show the efficiency of this objective function through several examples.

Acknowledgements We acknowledge support from the Belgian Programme of Interuniversity Attraction Poles and an Action de Recherche Concertée (ARC) of the French Community of Belgium.

References

1. Aynaud T, Guillaume JL (2011) Multi-step community detection and hierarchical time segmentation in evolving networks. In: Proceedings of the 5th SNA-KDD workshop
2. Blondel V, Guillaume J-L, Lambiotte R, Lefèvre E (2008) Fast unfolding of communities in large networks. J Stat Mech Theory Exp. doi:10.1088/1742-5468/2008/10/P10008
3. Browet A, Absil PA, Van Dooren P (2011) Community detection for hierarchical image segmentation. In: Proceedings of the 14th international conference on combinatorial image analysis, IWCIA'11, pp 358–371
4. Cason T (2012) Node-to-node similarity measures and role extraction in networks. PhD thesis, Université catholique de Louvain, Belgium

5. Cooper K, Barahona M (2010) Role-based similarity in directed networks. E-print. arXiv:1012.2726
6. Delvenne JC, Yaliraki SN, Barahona M (2010) Stability of graph communities across time scales. Proc Natl Acad Sci USA 107:12755–12760
7. E W, Li T, Vanden-Eijnden E (2008) Optimal partition and effective dynamics of complex networks. Proc Natl Acad Sci USA 105:7907–7912
8. Fortunato S (2010) Community detection in graphs. Phys Rep 486:75–174
9. Lambiotte R, Delvenne J-C, Barahona M (2009) Laplacian dynamics and multiscale modular structure in networks. arXiv:0812.1770
10. Mucha P et al (2010) Community structure in time-dependent, multiscale, and multiplex networks. Science 328:876
11. Reichardt J, White DR (2007) Role models for complex networks. Eur Phys J B 60:217–224
12. Schaub M, Delvenne J-C, Yaliraki SN, Barahona M (2012) Markov dynamics as a zooming lens for multiscale community detection: non clique-like communities and the field-of-view limit. PLoS ONE 7:e32210
13. Traag VA, Van Dooren P, Nesterov Y (2011) Narrow scope for resolution-limit-free community detection. Phys Rev E 84:016114

Chapter 24
Kinetic Limit of Dynamical Description of Wave-Particle Self-consistent Interaction in an Open Domain

Bruno Vieira Ribeiro and Yves Elskens

Abstract In a closed domain Ω of space, we consider a system of N particles $\sigma^N = (x_1, v_1, \ldots, x_N, v_N)$ interacting via a pair potential U. In this region, particles also interact self-consistently with a wave $Z = A\exp(\mathrm{i}\phi)$. We consider injection of particles in Ω, so N varies in time.

Given initial data $(Z^N(0), \sigma^N(0))$ and a boundary source/sink, the system evolves according to a Hamiltonian dynamics to $(Z^N(t), \sigma^N(t))$. In the limit of infinitely many particles (kinetic limit), this generates a Vlasov-like kinetic equation for the distribution function $f(x, v, t)$ coupled to an envelope equation for $Z(t) = Z^\infty(t)$. The solution (Z^∞, f) exists and is unique for any initial data with finite energy, provided that Ω has smooth enough boundaries.

Further, for any finite time t, given a sequence of initial data such that $\sigma^N(0) \to f(0)$ weakly and $Z^N(0) \to Z(0)$ as $N \to \infty$, the states generated by the Hamiltonian dynamics $(Z^N(t), \sigma^N(t))$ are such that $\lim_{N\to\infty}(Z^N(t), \sigma^N(t)) = (Z^\infty(t), f(x, v, t))$.

24.1 Introduction

With the development of theories for the dynamics of wave-particle interaction, the N-body Hamiltonian description of plasma systems (and alikes) has often been

B. Vieira Ribeiro (✉)
Instituto de Física, Universidade de Brasília, CP 04455, 70919-970 Brasília, DF, Brazil
e-mail: brunovr@fis.unb.br

B. Vieira Ribeiro · Y. Elskens
Equipe Turbulence Plasma, case 321, PIIM, UMR 7345 CNRS, Aix-Marseille Université, Campus Saint-Jérôme, 13397 Marseille Cedex 13, France

B. Vieira Ribeiro
e-mail: bruno.vieiraribeiro@univ-amu.fr

Y. Elskens
e-mail: yves.elskens@univ-amu.fr

B. Vieira Ribeiro
CAPES Foundation, Ministry of Education of Brazil, 70040-020 Brasília, DF, Brazil

T. Gilbert et al. (eds.), *Proceedings of the European Conference on Complex Systems 2012*, Springer Proceedings in Complexity, DOI 10.1007/978-3-319-00395-5_24,

used alongside the well known Vlasovian model. Major developments have been achieved in the study of the agreement between these descriptions in the $N \to \infty$ limit, in which the dynamics formally reduces to the kinetic theory. More precisely, for long-range forces, and mean-field particle-particle interactions in the absence of waves, it has been shown that the $N \to \infty$ limit commutes with the time evolution of the system, see [1, 2]. For the wave-particle interaction case, Firpo and Elskens [3] have shown that the mean-field methods are also applicable and have proven the equivalence of descriptions. Similar techniques also apply when the field obeys a wave equation, see [4].

So far, these works assume periodic boundary conditions on an infinite domain. Our goal is to consider open systems with finite extension, which can account for injection of particles and reveal the possible importance of boundary terms, and study their kinetic limit. We start with an open system of particles interacting via a pair-wise smooth potential. In this case, the potential is assumed to be twice differentiable and bounded, the potential vanishes outside the interaction region defining the finite open system, outside which the particles are free. The injection of particles is accounted for by giving "fake" initial data outside the interaction region of space. Moreover, particles are coupled to wave-like degrees of freedom, much as a wavefield.

24.2 Dynamics

Consider a one-dimensional system of particles interacting via a pair potential in a closed region defined by position coordinates $x \in \Omega \equiv [0, L]$. Inside this region, particles also interact with waves with given natural frequencies ω_{0j}, wave numbers k_j, phases θ_j and intensities I_j.

The Hamiltonian describing this system is given by

$$H(x, p, X, Y) = \sum_r \frac{p_r^2}{2} + \sum_j H_{0j}(X_j, Y_j) + \varepsilon' \sum_{r,r'} U(x_r, x_r') R(x_r) R(x_r') \\ + \varepsilon \sum_{r,j} k_j^{-1} \beta_j (Y_j \sin k_j x_r - X_j \cos k_j x_r) R(x_r), \tag{24.1}$$

where (x_r, p_r) are the canonical position and momentum of the rth particle and (X_j, Y_j) are canonical variables for the *Cartesian* components of the complex mode, which is related to the intensity-phase components of the wave by

$$Z_j = X_j + \mathrm{i}Y_j = \sqrt{2I_j}\,\mathrm{e}^{-\mathrm{i}\theta_j}. \tag{24.2}$$

This Hamiltonian is a modification of that of [5], plus a particle-particle interaction term as that of [2], limited to the region where $R(x) > 0$. ε and ε' are coupling constants chosen to avoid divergences in the $N \to \infty$ limit (in this limit, we expect $\varepsilon' N = 1$, for example). The first two terms of Eq. (24.1) correspond, respectively, to free particles and free waves, and the last term corresponds to wave-particle coupling in Ω. The third term accounts for the particle-particle interaction, with a pair potential U bounded, Lipschitz continuous, and symmetrical w.r.t. its two arguments. As we consider all interactions to take place only in Ω, we introduce the function R in the Hamiltonian. It is a Lipschitz continuous function with value 0 for all $x \in]-\infty, -\delta[$ and $x \in]L+\delta, \infty[$, and value 1 for $x \in [\delta, L-\delta]$, for a given small positive constant δ. Therefore, outside Ω, the Hamiltonian just expresses free motion of particles and waves.

The dynamical equations of motion for the system are

$$\dot{x}_r = p_r \tag{24.3}$$

$$\begin{aligned} \dot{p}_r = {} & R(x_r)\sum_j \varepsilon\beta_j \Im\left(Z_j e^{ik_j x_r}\right) - \sum_{r'} \varepsilon' \partial_{x_r} U\left(x_r, x_{r}'\right) R(x_r) R\left(x_{r}'\right) \\ & + \sum_j \varepsilon\beta_j k_j^{-1} \Re\left(Z_j e^{ik_j x_r}\right) \partial_{x_r} R(x_r) \\ & - \sum_{r'} \varepsilon' U\left(x_r, x_{r}'\right) \partial_{x_r} R(x_r) R\left(x_{r}'\right) \end{aligned} \tag{24.4}$$

$$\dot{Z}_j = -i\omega_{0j} Z_j + \sum_r \varepsilon\beta_j k_j^{-1} i e^{-ik_j x_r} R(x_r), \tag{24.5}$$

where we use for H_{0j} a harmonic oscillator term

$$H_{0j} = \sum_j \omega_{0j} \frac{X_j^2 + Y_j^2}{2}. \tag{24.6}$$

We now introduce the velocity $v_r = p_r$ as all particles have unit mass. We introduce wave envelopes

$$a_j(t) = C^{-1} Z_j(t) e^{i\omega_{0j} t}, \tag{24.7}$$

with an appropriate constant C. For simplicity, here, we work with only one mode, dropping the subscript j, and let $\beta_j' = \varepsilon\beta_j C$ and . Then,

$$\dot{x}_r = v_r \tag{24.8}$$

$$\begin{aligned} \dot{v}_r = {} & R(x_r)\beta' \Im\left(a e^{ikx_r - i\omega_0 t}\right) - \sum_{r'} \varepsilon' \partial_{x_r} U\left(x_r, x_{r}'\right) R(x_r) R\left(x_{r}'\right) \\ & + \varepsilon \frac{\beta_j'}{k} \Re\left(a e^{ikx_r - i\omega_0 t}\right) \partial_{x_r} R(x_r) - \sum_{r'} \varepsilon' U\left(x_r, x_{r}'\right) \partial_{x_r} R(x_r) R\left(x_{r}'\right) \end{aligned} \tag{24.9}$$

$$\dot{a} = \frac{i\beta_j'}{C^2 k} \sum_r e^{-ikx_r + i\omega_0 t} R(x_r). \tag{24.10}$$

The positions and velocities of particles determine an empirical sum σ^R of point measures on $\Gamma \equiv \Omega \times \mathbb{R}$ space,

$$\sigma^R(x, v, t) = \eta \sum_r \delta\left(x - x_r(t)\right) \delta\left(v - v_r(t)\right) R(x_r), \tag{24.11}$$

counting particles in Γ. The prefactor η is chosen as to keep a finite mass in the kinetic limit $N \to \infty$.

The space Γ is equipped with the distance

$$\left\|(x,v)-\left(x',v'\right)\right\| = \alpha\left(|x-x'|+\tau|v-v'|\right), \tag{24.12}$$

where α^{-1} and τ are, respectively, length and time scales. In the mode space, $\mathcal{Z}$, we use the distance

$$\|a-a'\| = \zeta|a-a'|, \tag{24.13}$$

with a real positive coefficient ζ.

24.3 Kinetic Limit

The kinetic limit we are interested in corresponds to the sequence of point measures σ^R converging to a continuous measure σ, defined by a positive density $f(x,v,t)$ in Γ, where the density f is a (weak) solution of the Vlasov-like system, dual to (24.8)–(24.10), given by

$$\partial_t f + v\partial_x f + F[f,a]\partial_v f = 0 \tag{24.14}$$

$$\dot{a} = \frac{i\beta'}{Ck}\int_\Gamma f(x,v,t)e^{-ikx+i\omega_0 t}R(x)dxdv, \tag{24.15}$$

with the force field

$$\begin{aligned} F(x,v,t) = R(x)\bigg(&\beta'\Im\left(a(t)e^{ikx-i\omega_0 t}\right) - \varepsilon'\int_\Gamma \partial_x U\left(x,x'\right)R\left(x'\right)df'\bigg) \\ &+ \partial_x R(x)\bigg(\frac{\beta'}{k}\Re\left(a(t)e^{ikx-i\omega_0 t}\right) \\ &- \varepsilon'\int_\Gamma U\left(x,x'\right)R\left(x'\right)df'\bigg), \end{aligned} \tag{24.16}$$

where

$$df' = f\left(x',v',t\right)dx'dv'. \tag{24.17}$$

We want to account for particles being injected in Ω. So, beside the mass measure σ (or σ^R), we introduce a *boundary flux* measure ν, in (t,v) space, that counts particles being injected in Ω through $x_1 = -\delta$ or $x_2 = L+\delta$. Thus, ultimately, we have two types of trajectories to account for: (a) that of particles with initial conditions in Γ which remain in Γ; and (b) that of particles with initial conditions outside Γ which are injected into Γ at a finite time (note that we are *not* interested in particles after they leave Γ, nor in those that do not enter it in a finite time).[1] Therefore, the evolution of trajectories in one-particle (x,v) space is governed by a flow T as follows. For case (a)

$$\text{(a)}\quad \left(x_r(t),v_r(t)\right) = T_{t,s}\left(x_r(s),v_r(s)\right), \tag{24.18}$$

[1] We are relying on the assumption that particles never re-enter Ω once they leave, which is reasonable because the force vanishes outside Γ.

describing particles that evolve from a time s to time t inside Γ. And, for case (b)

$$\text{(b)} \quad \big(x_r(t), v_r(t)\big) = T_{t,t'}\big[T^0_{t',s}\big(x_l - v_r(s)(s-t'), v_r(s)\big)\big], \qquad (24.19)$$

describing injected particles that, at some time $t' < s$, were outside Γ a distance $|v_r(s)t'|$ from the boundary, where T^0 is the free motion map. Clearly, $T_{t,s}$ depends on the wave envelope and particle distribution history during $[s,t]$.

By duality, any measure μ_s of the system is transported by the flow as

$$\mu_t = \big(\mu_s + \nu_. \circ T^0_{.,s}\big) \circ T_{s,t}\big[a(.), \mu_.\big], \qquad (24.20)$$

where we recall the proper parameters in the flow T.

Let $\mathcal{M}_+$ be the space of positive measures μ on Γ and $\mathcal{N}_+$ be the space of positive boundary flow measures ν. In these spaces, define the distance

$$\begin{aligned} d_{\Gamma,[0,T]}\big(\mu,\mu';\nu,\nu'\big) = \sup_{\phi\in\mathcal{D}} \bigg| &\int_{\Gamma_c} \phi(x,v)d\big(\mu-\mu'\big)(x,v) \\ &+ \sum_{l=1}^{2} \int_{\Lambda} R(x)\phi\big(x_l - vt', v\big)d\big(\nu_. - \nu'_.\big)\big(t',v\big)\bigg| \\ &+ \sup_{\phi\in\mathcal{D}} \bigg| \int_{\Gamma_\delta} R(x_l)\phi(x,v)d\big(\mu-\mu'\big)(x,v)\bigg| \end{aligned} \qquad (24.21)$$

$$\mathcal{D} = \big\{\phi | \phi \in C^{0,1}_{\mathrm{b}}(\mathbb{R}\times\mathbb{R});\ \|\phi\|_{\mathrm{uL}} \le 1\big\}, \qquad (24.22)$$

where $C^{0,1}_{\mathrm{b}}(\mathbb{R}\times\mathbb{R})$ stands for the space of bounded, Lipschitz continuous functions on $\mathbb{R}\times\mathbb{R}$ and $\|\cdot\|_{\mathrm{uL}} \equiv \max\big\{\|\cdot\|_{\mathrm{u}}, \lambda\mathrm{Lip}(\cdot)\big\}$, with a scaling constant λ. We decompose the space Γ into two regions $\Gamma_c \cup \Gamma_\delta$,

$$\Gamma_c = \big\{(x,v) : x \in \Omega,\ \mathrm{dist}(x,\partial\Omega) > \delta;\ v \in \mathbb{R}\big\}, \qquad (24.23)$$

$$\Gamma_\delta = \big\{(x,v) : \mathrm{dist}(x,\partial\Omega) \le \delta;\ v\cdot\mathbf{n} < 0\big\}, \qquad (24.24)$$

where $\mathbf{n}$ is the normal outward to Ω. Finally, we also introduced the space

$$\Lambda = \big\{\big(t',v\big) : t' \in [0,T];\ v\cdot\mathbf{n} < 0\big\}. \qquad (24.25)$$

For any ν, we may use the simpler expression $d_\Gamma(\mu,\mu') = d_{\Gamma,[0,T]}(\mu_t,\mu'_t;\nu,\nu)$. Then, our distance in $\mathcal{M}_+ \times \mathcal{Z}$ is given by

$$\big\|\big(\mu_t, a(t)\big) - \big(\mu'_t, a'(t)\big)\big\| \equiv d_\Gamma\big(\mu_t,\mu'_t\big) + \big\|a(t) - a'(t)\big\|. \qquad (24.26)$$

24.4 Results

Our main results are

Theorem 24.1 *Let* $|\partial U| \le B_1$ *and* $|\partial_x U(x,x') - \partial_y U(y,x)| \le B_2|x-y|$, *for positive and real constants* B_1 *and* B_2. *Given two different initial data* $(\mu_0, a(0))$ *and*

$(\mu'_0, a'(0))$ *in* $\mathcal{M}_+ \times \mathcal{Z}$, *and same boundary fluxes, the kinetic evolution equations generate, for any positive time* t, *unique states* $(\mu_t, a(t))$ *and* $(\mu'_t, a'(t))$ *from the initial data, respectively. Furthermore, for any* $t > 0$,

$$\left\| \left(\mu_t, a(t)\right) - \left(\mu'_t, a'(t)\right) \right\| \leq e^{\xi t} \left\| \left(\mu_0, a(0)\right) - \left(\mu'_0, a'(0)\right) \right\| \qquad (24.27)$$

for a strictly positive constant ξ.

Theorem 24.2 *Given a continuous measure* $\sigma_0 \in \mathcal{M}_+$ *and a sequence of point measures* $\sigma^R_{0N} \in \mathcal{M}_+$ *defining the initial distribution of N particles in* (x, v) *space, such that* $\lim_{N\to\infty} d_\Gamma(\sigma^R_{0N}, \sigma_0) = 0$, *and given an initial wave envelope* $a(0) \in \mathcal{Z}$, *for all times* $0 \leq t \leq T$ *consider the resulting measure and envelope* $(\sigma^R_{tN}, a^R(t))$ *generated by* H *and the kinetic solution* $(\sigma_t = f(x, v, t)dxdv, a(t))$ *of* (24.14)–(24.15). *Then,* $\lim_{N\to\infty} d_\Gamma(\sigma^R_{tN}, \sigma_t) = 0$ *and* $\lim_{N\to\infty} a^R_N(t) = a(t)$.

The theorems are proven by standard arguments (see [6]).

Acknowledgements The authors thank N. Dubuit for fruitful discussion and Marco A. Amato for initiating and collaborating in the ongoing researches.

B. Vieira Ribeiro is supported by a grant from CAPES Foundation through the PDSE program, process number: 8510/11-3.

References

1. Neunzert H (1984) An introduction to the nonlinear Boltzmann-Vlasov equation. In: Cercignani C (ed) Kinetic theories and the Boltzmann equation. Lect notes math, vol 1048. Springer, Berlin, pp 60–110
2. Spohn H (1991) Large scale dynamics of interacting particles. Springer, Berlin
3. Firpo M-C, Elskens Y (1998) Kinetic limit of N-body description of wave-particle self-consistent interaction. J Stat Phys 93:193–209
4. Elskens Y, Kiessling MK-H, Ricci V (2008) The Vlasov limit for a system of particles which interact with a wave field. Commun Math Phys 285:673–712
5. Elskens Y, Escande D (2003) Microscopic dynamics of plasmas and chaos. IOP Publishing, Bristol
6. Kiessling MK-H (2008) Microscopic derivations of Vlasov equations. Commun Nonlinear Sci Numer Simul 13:106–113

Chapter 25
The Emergence of Pathological Constructors when Implementing the Von Neumann Architecture for Self-reproduction in Tierra

Declan Baugh and Barry Mc Mullin

Abstract John von Neumann's architecture for *genetic reproduction* provides an explanation in principle for how arbitrarily complex machines can construct other ("offspring") machines of equal or even greater complexity. We designed a von Neumann style self-reproducing ancestor within the framework of the Tierra platform, which implements a (mutable) genotype-phenotype mapping during reproduction. However, we have consistently observed a particular phenomenon where what we call *pathological constructors* quickly emerge, which ultimately lead to catastrophic ecosystem collapse. Pathological constructors are creatures which rapidly construct multiple short malfunctioning offspring within their lifetime. Pathological constructors are a hindrance to an ecosystem because their offspring, although sterile, still occupy both memory space and CPU time. When several pathological constructors coincide in time, their production rate can be so high that their non-functional offspring displace the entire population of functional self-reproducing creatures, resulting in ecosystem collapse. We investigate the origin of pathological constructors, and consider how a more mutational robust architecture which is less susceptible to the emergence of these creatures can be created.

Keywords Von Neumann · Genetic reproduction · Tierra · Artificial life · Genotype-phenotype mapping · Evolutionary growth of complexity · Pathological constructors

25.1 Introduction

As early as 1948, John von Neumann had formulated his theory of the evolutionary growth of machine complexity [1, 2]. This theory provides a proof-of-principle demonstration that machines can directly, or indirectly, give rise to machines arbi-

D. Baugh (✉) · B. Mc Mullin
The Rince Institute, Dublin City University, Dublin, Ireland
e-mail: declan.baugh2@mail.dcu.ie

B. Mc Mullin
e-mail: barry.mcmullin@dcu.ie

T. Gilbert et al. (eds.), *Proceedings of the European Conference on Complex Systems 2012*, Springer Proceedings in Complexity, DOI 10.1007/978-3-319-00395-5_25,

trarily more complex than themselves. This machine architecture is comprised of two specific parts, the phenotype and the genotype.

The phenotype is the functional, active section of the machine, and the genotype is the passive section, dedicated to information storage. For genetic reproduction, under some arbitrary genotype-phenotype mapping the genotype must contain an encoded description of the phenotype. Conversely, the phenotype must include the functionality to both decode the genotype and construct an offspring phenotype.

Previous work with evolutionary systems where the agents are responsible for their own self reproduction has been based exclusively on machine architectures which reproduce via *template-reproduction*, where there is no division of labour between genotype and phenotype. In this case, self reproduction is performed by self inspection, and no explicit mutable genotype-phenotype mapping is implemented.

Within the platform of Tierra, we designed an ancestor that reproduces via genetic reproduction. More importantly, this design implemented a mutable genotype-phenotype mapping as described by the von Neumann architecture, where the arbitrary mapping between genotype and phenotype is subject to heritable mutations. We aim to explore if alternative, viable mutational pathways are introduced while implementing this architecture. However, during implementation within the Tierra platform, several unanticipated phenomena emerged, which are examined and documented here.

25.2 Implementation of the von Neumann Architecture for Machine Self-reproduction Within Tierra

Classically, the reproductive mechanism of Tierran creatures rely on self-copying,[1] which involves the creatures activating a reproduction mechanism which incrementally copies the contents of each memory location of the parent creature to an available space in memory which will become the offspring. There is no distinction between phenotype and genotype for a self-copying creature as the entire creature acts as both the template for replication, and the implementation of the reproduction cycle and all other functionality.

In order to achieve von Neumann style reproduction we must first devise a method of mapping inert numbers within the genotype to active instructions to be executed within the phenotype. There exists an infinite number of possible mappings which we could implement, however, we chose to implement ours via the inclusion of a *look-up table*. The look-up table provides a method of translating the inert numbers within the memory locations of a parent genotype to functional instructions which can be executed as part of the offspring phenotype. For this particular genotype-phenotype mapping we chose a 1:1 mapping where a single number within the genotype is translated to a single instruction within the phenotype, therefore, the look-up table consists of 32 memory locations, each containing a value

[1] Analogous to RNA template replication.

which corresponds to an instruction within the Tierran instruction set. The look-up table of the seed ancestor will therefore represent a random permutation of the all possible instructions.

Prior to reproduction, this seed creature must first allocate space for an offspring. During constructing of an offspring phenotype, the parents Ax register incrementally steps through each memory location within its genotype, and the number stored at each address is inspected. A second register, Bx which initially points to the start of the look-up table, is displaced by the number which was inspected by Ax. The number within the updated Bx memory location (which lies within the look-up table), is now written to the offspring phenotype, where it will subsequently function as an instruction. This activity facilitates the mapping of numbers which are stored within the parent genotype, to instructions incorporated in the offspring phenotype. Furthermore, random perturbations within the look-up table facilitate the alteration of the genotype-phenotype mapping. This may have the effect of introducing new mutational pathways for the creature, which was not possible under the previously unaltered look-up table.

Upon construction of the offspring phenotype, the parent's genotype is incrementally copied to the offspring space and the connection between parent and offspring is severed. At this point, the parent loses write access to the offspring's memory block, and a new CPU is created and allocated to the offspring. While copying the genotype, should a random perturbation occur which affects the encoded description of the look-up table (or otherwise modify the decoding process), then the creature's offspring will incorporate a mutated genotype-phenotype mapping. This is the particular phenomenon which we initially set out to investigate.

25.3 The Emergence of Pathological Constructors from Genetic Reproducers

When all random perturbations are disabled our seed creature reproduces effectively and populates the memory to form a stable ecosystem of identical creatures. However, when all random perturbations are switched on we immediately see a large emergence of pathological constructors which saturate available CPU time and memory space. Under a series of simulations where each source of random perturbation was individually disabled, the disabling of the segment deletions showed an apparent prevention against the emergence of pathological constructors. When a large segment deletion occurs while copying the genotype from parent to offspring, the resultant creature will typically consist of a functional phenotype, assigned to a partial genotype. This creature continues to rapidly produce offspring, (due to the short genotype), but these offspring are non-functional as they consist of a corrupt phenotype, assigned to a corrupt genotype. When several such pathological constructors coincide in time, their production rate can be so high that their non-functional offspring displace the entire population of functional self reproducing creatures, i.e., ecosystem collapse.

For von Neumann style genetic reproducers, all random perturbations which corrupt the genotype will result in a constructor which will create at least one functional or non functional offspring. A genotype which experiences a segment deletion will result in a pathological constructor which can construct many non-functional offspring before it is killed by the reaper.

This analysis concludes that the mechanism which results in ecosystem collapse due to pathological constructors appears to depend critically on both the one generation delay from when a random perturbation occurs in a genotype and when it is expressed in the phenotype, and the inclusion of segment deletions. The combination of these factors results in a high level of ease in which segment deletions can lead to corrupt genotypes, while still leaving a functioning phenotype.

By contrast, in order for a pathological constructor to emerge from a self-copier, relatively much more specific random perturbations must occur upon very specific locations which will alter, but not corrupt the reproductive functionality. This suggests that the probability of emerging pathological constructors within a population of genetic-reproducers is much higher than that of a population of self-copiers.

25.4 Conclusions and Future Work

The highlighted intricate properties of the von Neumann self reproducing automata, implemented in Tierra suggest that this may not be mutationally robust architecture to support genetic reproduction. A combination of the effects of the segment deletions and the generation delay in expressing random perturbations contribute to the abundant emergence of pathological constructors, hence increasing the ecosystem's susceptibility to catastrophic collapse.

It is worth noting that in the typical reproduction cycle of complex (multicellular) biological organisms, most of the "decoding" of the genotype takes place as development of the offspring, i.e., it is under the direction of the (embryonic) offspring phenotype rather than the parental phenotype [3]. If we incorporate this concept within the von Neumann architecture, where the offspring phenotype is decoded from the offspring genotype (as opposed to the parent genotype which is the case with von Neumann's architecture), then this design may not exhibit the one generation delay from when a random perturbation occurs in a genotype, and when it is expressed in the phenotype. A corrupt genotype will immediately be assigned a corrupt phenotype, and hence will not reproduce. It seems likely that such an architecture, implemented in Tierra, would be more evolutionary stable and much less vulnerable to emergence of pathological constructors.

Acknowledgements This work has been supported by the European Complexity Network (Complexity-NET) through the Irish Research Council for Science and Technology (IRCSET) under the collaborative project EvoSym.

Appendix

Source code to reproduce results in this paper can be accessed at: http://alife.rince.ie/evosym/sab-2012-db.zip.

References

1. von Neumann J (1948) The general and logical theory of automata. In: Cerebral Mechanisms in Behaviour, pp 1–32
2. McMullin B (2000) John von Neumann and the evolutionary growth of complexity: looking backward, looking forward. Artif Life 6(4):347–361
3. Buckley W (2008) Computational ontogeny. Biol Theory 3(1):3–6

Part II
Complexity, Information and Computation

Chapter 26
A Preferential Attachment Model for Efficient Resources Selection in Distributed Computing Environments

María Botón Fernández, Francisco Prieto Castrillo, and Miguel A. Vega-Rodríguez

Abstract In the last decade, Complex Network theory has been applied in many disciplines to solve a wide range of problems. Most social, biological and technological networks are modelled as complex networks from their topology point of view.

In this regard, an Efficient Resources Selection (ERS) model was proposed in a previous work to solve the resources selection problem in grid environment (i.e. to find a suitable resource set for grid applications). In this model, the infrastructure resources are considered nodes of a complex network that evolves during application execution. On the other hand, the edges represent the interaction between resources during the tasks execution. Besides, within the selection process the Preferential Attachment technique (Barabási and Réka, Science, 286(5439):509–512, 1999) is applied to determine the most efficient resources. This efficiency parameter is calculated using both resources degree and fitness values.

In the present contribution, a summary of this ERS model along with an analysis of its relevance parameters is exposed. The obtained results are also discussed.

Keywords Complex systems · Self-adaptive applications · Grid computing · Optimization

26.1 Introduction

In recent years, Grid computing [2, 3] has become a powerful environment enabling researchers to execute massive computing applications. This is due to the fact that

---

M.B. Fernández (✉) · F.P. Castrillo
Dept. Science and Technology, Ceta-Ciemat, Trujillo, Spain
e-mail: maria.boton@ciemat.es

F.P. Castrillo
e-mail: francisco.prieto@ciemat.es

M.A. Vega-Rodríguez
Dept. Technologies of Computers and Communications, University of Extremadura, Caceres, Spain
e-mail: mavega@unex.es

T. Gilbert et al. (eds.), *Proceedings of the European Conference on Complex Systems 2012*, Springer Proceedings in Complexity, DOI 10.1007/978-3-319-00395-5_26,

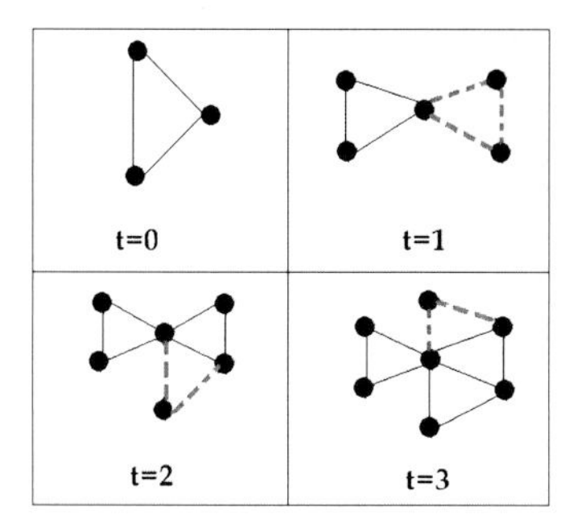

Fig. 26.1 Evolution of the complex network built at runtime. *The broken lines* represent new links

Grid infrastructures are composed by an unlimited amount of heterogeneous resources geographically dispersed.

This infrastructure has been applied successfully in a wide range of projects and domains: for analyzing extraterrestrial intelligent signals for patterns, in the high performance data mining services, in medical imaging applications, in high energy physics experiments developed at CERN as well as in biological projects.

However, the heterogeneous and changing characteristics of Grid resources along with the dynamic nature of such infrastructure lead to non-trivial task scheduling limitations. Hence, a challenge topic of this type of infrastructure is the resources selection, i.e., finding a suitable resource set for the application deployment. In a previous work [4], we focused on solving this problem applying a complex network algorithm known as Preferential Attachment [1], i.e., *the most popular/efficient* resources are selected at every application execution cycle. In this regard, we established the following rules:

- The infrastructure resources are considered nodes of a complex network built during application deployment (Fig. 26.1).
- At every application execution cycle a resource set is selected to perform the corresponding tasks set. This resource set composes a complete subgraph in the complex network.
- The edges of the complex network represent the constraint *executing tasks from the same task set.*

To accomplish this goal, an Efficient Resources Selection (*ERS*) model was proposed. This optimization strategy is defined at the user level, which means that the methodology to choose the grid resources applies only both basic grid concepts and operations. In a grid infrastructure, users interact with elements through the User Interface (*UI*) by using certain command set. As it is shown in Fig. 26.2, users submit their application tasks through the *UI*. These tasks are managed by the metascheduler called Resource Broker (*RB*) which sends them to a specific site (known in grid terminology as Resources Centres). Then, tasks are finally handled by the Computing Element (*CE*) in the corresponding site. The worker nodes (*WN*) are the compute nodes where tasks are executed.

Once the model was implemented, a test set was defined and a real grid was chosen as testbed. From the obtained results we concluded that the proposed model benefits grid applications improving infrastructure throughput. However, we consider

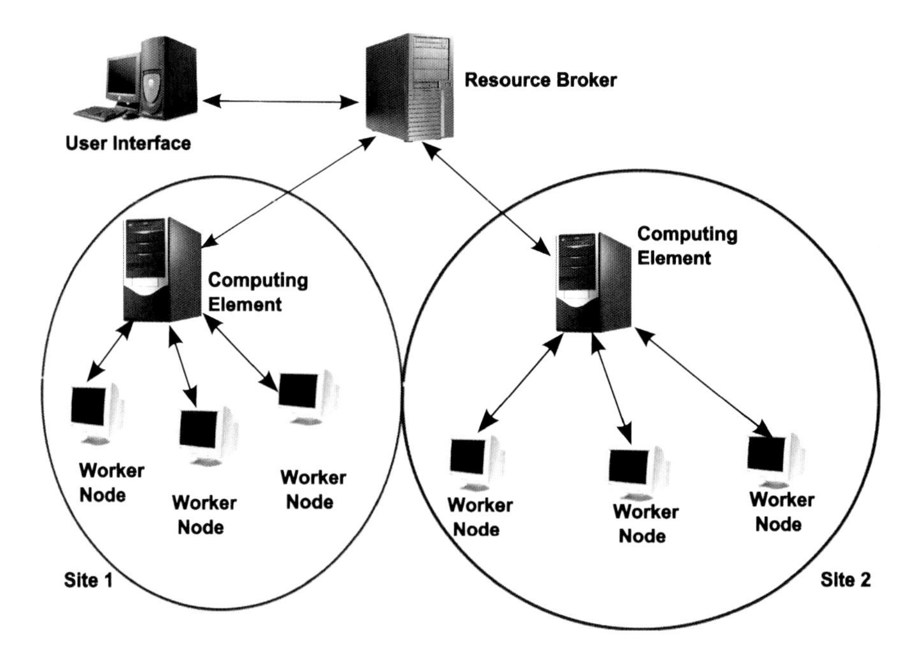

Fig. 26.2 Representation of a basic grid infrastructure

that more insight into model behaviour is needed. For that reason, in the present contribution we analyse the effects of the model relevance parameters in the application execution.

26.2 Aim

1. Modelling the Grid Infrastructure as a Complex Network to select the most efficient resources.
2. Provide a self-adaptive capability to Grid applications.
3. Determine the influence of the model relevance parameters in the application behaviour.

26.3 Model Definition

The *ERS* model is based on the mapping between two spaces: a task space denoted as J and a heterogeneous resource space R. As stated, the resources handled by the model are the grid schedulers known as Computing Elements (*CE*). During the application execution, the resources efficiency is continuously monitored. This way we ensure that the best resources are used.

The proposed model is composed by three modules as shown in Fig. 26.3. All the actions related to the efficient selection are encapsulated in the *Intelligence Module*;

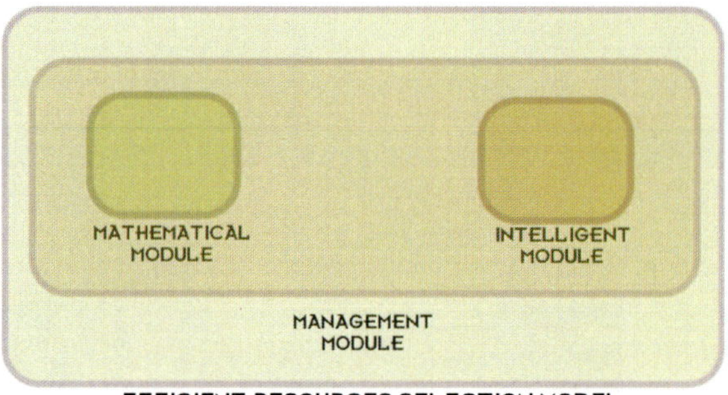

Fig. 26.3 Modules that compose the proposed ERS model

the mathematical formulation composes the *Mathematical Module* and, finally, the *Management Module* is responsible for preparing the environment and invoking the other modules as needed.

The model execution flow has four main steps that are presented as follows. Firstly, the space J is partitioned into equally sized task sets. Next, a resource set is randomly chosen for the initial task set. These operations are encapsulated within the *Management Module*. Once the tasks are executed, the corresponding efficiency metrics for the involved resources are calculated (the *Mathematical Modules* is invoked). This performance information allows the model to classify the resources and to select the best ones at every execution cycle. The way to choose these resources is by invoking the *Intelligence Module* where the particular designed *PA* (described in the following section) is implemented.

As a summary, the present strategy is modelled from the complex network perspective. The *CEs* are represented as nodes in a complex network which grows at runtime. The task space is divided in several subsets, all of them with the same size. At every application execution cycle a task subset is launched. The links between nodes in our resulting network represent the constraint *executing tasks for the same subset*.

26.4 Heterogeneous Preferential Attachment

As stated, the algorithm used in the *Intelligence Module* is inspired on the Preferential Attachment (*PA*) technique [1] for selecting the most *popular* resources at every application execution cycle. The main idea of this technique is that new nodes in the complex network will connect more likely with those which have a higher degree.

Since every resource (*CE*) is considered as a node, taking into account the PA rules, each node has associated two parameters: the degree and the fitness. The fitness Fi (Eq. (26.2)) indicates how the node has performed the tasks assigned to it.

Consequently, using both parameters (degree and fitness) it can be determined how efficient the resource has been. This new metric, denoted as efficiency (Eq. (26.1)) will determine how new nodes will connect to the existing ones, i.e., the efficiency E_i determines the link probability. Hence, in this new PA version additional node features are superposed to the link probability.

$$E_i(k, F) = (k_i \cdot F_i)/k_{\max}, \tag{26.1}$$

where $k_{\max}$ is the maximum degree value for a specific resources set, k_i is the resource degree value and F_i is the obtained fitness value of the resource.

$$F_i = (a \cdot \epsilon_i + b\Delta T_i)/(a + b). \tag{26.2}$$

The fitness F_i of a particular resource is calculated by using the increment of the processing time ΔT_i (Eq. (26.3)) along with other three parameters: on one hand, the percentage of successfully completed task ϵ_i. On the other hand, two relevance parameters a and b specified by users. Thus, the users can decide the highest priority condition for their specific needs.

$$\Delta T_i = (T_{\max} - T_i)/(T_{\max} - T_{\min}). \tag{26.3}$$

The increment of processing time depends on maximum and minimum time values ($T_{\max}$ and $T_{\min}$ respectively) for a specific resource set. Ti is the processing time for a concrete resource and it is calculated as shown in Eq. (26.4).

$$T_i = T_{\text{comm}_i} + \sum_{j \in NT_i} T_{\text{comp}_{j,i}}. \tag{26.4}$$

26.5 Analysis and Results

The tests were performed in a real grid infrastructure belonging to the Spanish Grid Initiative (*ES-NGI*) project [5]. The emergence of such infrastructure is based on the growing demand by scientists for more computational resources. In addition, the *ES-NGI* proposal encourages collaboration and data sharing in the scientific community.

In a previous work, a test set was designed to verify that the proposed model performs an efficient selection. It must also be highlighted that two additional goals were fixed during the model definition phase: to reduce the application execution time and to increase the successfully completed tasks rate. For that reason, the proposed strategy is compared with the traditional or standard resources selection in grid environment (based on a process called match-making in which the *RB* chooses among all available *CEs* those with a higher rank that fulfill the task requirements).

In those preliminary tests, relevance parameters values a and b were fixed at 60 % and 40 % respectively. Although these values are specified by users we decided to begin with slightly higher importance to tasks rate, because we suppose this is an important issue for scientists. From the obtained results we concluded that the model performs well and reaches the established goals with this pair of values.

However, in the present contribution we consider to analyse how the model behaves for the possible pair of values. These results would be useful for a deeper understanding of the model behaviour. Furthermore, researchers may use this information to decide how to specify their experiments by using our strategy.

For this evaluation, we have chosen the six most significant pairs of values within the range of possibilities. By observing the results it can be deduced that the model performs well with a remarkable exception; the overall application execution is higher for the pair of values were execution time has a minimal relevance. Moreover, it is interesting that is in these two cases where the model reaches the worst rate of successful completed tasks. In conclusion, the ERS model based on the PA technique is a favourable strategy for grid application deployment.

26.6 Summary

The resources selection problem and the application adaptation in Grid infrastructures have been investigated. Furthermore, an efficient resources selection model was proposed in a previous work by monitoring the resources efficiency.

Based on the data obtained in a previous work, we analyse the influence of the relevance parameters in the grid applications deployment for a further knowledge of the model behaviour.

From the exposed results it is possible conclude that the ERS-PA version gets a good time reduction and an appropriate task rate in the different tests. The values of this pair of parameters do not have a significant influence in the model performance.

In conclusion, the proposed approach is beneficial for scientific applications in Grid environments.

Acknowledgements María Botón-Fernández is supported by the PhD research grant of the Spanish Ministry of Economy and Competitiveness at the Research Centre for Energy, Environment and Technology (CIEMAT). The authors would also like to acknowledge the support of the European Funds for Regional Development.

References

1. Barabási A-L, Réka A (1999) Emergence of scaling in random networks. Science 286(5439):509–512
2. Foster I (2002) What is the grid? A three point checklist. GRIDtoday 1(6):22–25
3. Foster I, Kesselman C, Tuecke S (2001) The anatomy of the grid. Enabling scalable virtual organizations. In: Sakellariou R, Keane JA, Gurd JR, Freeman L (eds) Euro-Par 2001. LNCS, vol 2150. Springer, Heidelberg, pp 1–4
4. Botón-Fernández M, Prieto Castrillo F, Vega-Rodríguez MA (2011) Self-adaptive deployment of parametric sweep applications through a complex networks perspective. In: Proceedings of the 2011 international conference on computational science and its applications: Part II. ICCSA'11. LNCS, vol 6783. Springer, Heidelberg, pp 475–489
5. The National Grid Initiative for Spain. http://www.es-ngi.es/

Chapter 27
The Challenge of Software Complexity

Kevin Moore and Michel Wermelinger

Abstract Given the interdisciplinary nature of complex network studies, there is a practical need for dialogue between theorists proposing graph measurements and those seeking to apply them into a domain. We consider this in the domain of software complexity by highlighting the distinctive nature of networks representing software's internal structure and also by describing the application of one such proposal, the offdiagonal complexity, against two examples of software. The results showed the promise of using complex networks to measure software complexity but also demonstrated the confounding effects of size. Based on that application we make proposals to improve the dialogue between theory and experiment.

Keywords Software complexity · Software evolution · Graph theory · Software metrics · Offdiagonal

27.1 Importance and Properties

Today's society is heavily dependant on software. It runs our computers, our phones and the internet, while managing economies and communications. This pervasiveness means that any improvement in understanding software has a potentially enormous payback from better project management, control of costs and increased quality. Past practice of software development could be seen as a chimera of art-form and engineering with success or failure in projects seemingly dependent on anecdotal wisdoms. While a comprehensive theoretical framework seems elusive, current practice has become increasingly evidence-based and draws from a wide range of disciplines such as psychology, sociology, data-mining and complexity theories.

K. Moore (✉) · M. Wermelinger
Computing Department, The Open University, Milton Keynes, UK
e-mail: ou@kevin.moore.name

M. Wermelinger
e-mail: m.a.wermelinger@open.ac.uk

T. Gilbert et al. (eds.), *Proceedings of the European Conference on Complex Systems 2012*, Springer Proceedings in Complexity, DOI 10.1007/978-3-319-00395-5_27,

That software is complex is also largely self-evident. Brooks [8] (of "The Mythical Man-Month" fame) argues that complexity is one of the fundamental essences associated with software. As such, understanding this inherent property would make great inroads into understanding software overall.

While there are several viewpoints into software such as its cognitive, computational, problem or solution complexity [9], this paper focuses on the structural complexity of the code, arguing that it provides the most direct understanding of the product.

27.1.1 Software as a Complex Network

The variety of coding languages, styles and paradigms makes processing and quantifying code hard to generalise. One solution is to abstract the code into a network graph, with vertices representing a chosen unit of code and edges representing an arbitrary relationship between those units. By representing the interconnections between collaborating modules, objects, classes, methods, and subroutines with a network graph, software becomes another domain capable of investigation with the interdisciplinary toolset of complex networks.

The basic technique is well established [20, 28] and while more recent developments have for instance considered graphing the entire socio-technical system [7], obtaining a measurement that represents the complexity of source code's basic structure and that can be connected to software development practice remains desirable.

27.1.2 Software as a Typical Network

Software networks appear as typical complex networks exhibiting both small-world behaviour and having a long and fat-tailed degree distribution obeying a power law. If they are constructed as directed graphs, the degree distributions of the inward and outward links differ, with the exponent for incoming edges being less than that of the outgoing and showing a better fit to the power law [12, 18, 23, 28].

Solé and Valverde [24] identify software networks as heterogeneous, scale-free and with some modular structure; a characterisation that also includes a wide range of biological and technical systems. Based on earlier work [27] they suggest this commonality is due to such systems being shaped through a processes of optimisation, a suggestion that reflects software development well. Technical and biological networks are typically disassortative, i.e. vertices with a high degree preferentially attach to those with low degree, as opposed to social networks which typically show assortative mixing [22]. Perhaps unsurprisingly, software networks have been empirically confirmed as disassortative [15, 24].

Software networks can therefore be recognised as typical examples of complex graphs, but some atypical aspects of software create distinctive challenges and opportunities.

Table 27.1 Example sizes of real-world networks with software networks in *bold*

Network	Nodes
Les Miserables character co-appearance	77
American football games	115
Tomcat 4.1.40[a]	181
C. elegans neural net	302
Netbeans 6.8[a]	1532
S. cerevisiae protein-protein interaction	1870
Tomcat 4.1.40[b]	2699
Netbeans 6.8[b]	14378
AS internet topology	22963
BEA Weblogic 8.1 middleware platform[b]	80095

[a]Nodes represent packages

[b]Nodes represent classes

27.1.3 Software as an Atypical Network

Software networks demonstrate a wide variation of size, reflecting the range of available software from small tools to major applications, but are often large in comparison with other networks commonly used in complexity research [18, 19, 21] as shown in Table 27.1.

The same software network can be considered at different resolutions, i.e. by considering different code units as vertices. For example, in code written in Java, a popular programming language, one can consider classes (which group related functions) and packages (which group related classes). While any scale-free network could be considered in the same way, in software these two 'granularity levels' (or equivalent ones for other programming languages) are particularly significant and represent meaningful and deliberate constructs to software developers. It is possible that the complexity of software networks behaves differently at different resolutions while remaining the same coherent network.

Software evolves through multiple *versions* as the code is modified in response to fault fixing and feature requests, but also as a result of refactoring activity. This activity occurs when developers attempt to rework the code structure while preserving functionality. While refactoring is tricky to isolate from other coding activity, this offers a network that has been changed, hopefully simplified, and yet remains functionally the same. Software networks can also evolve by widespread deletion, as functionality is split out of the main product in a sort of software cell division'. The reverse can also happen as existing external products are absorbed wholesale. Even under more routine development it is uncertain what growth models are being applied; as a designed product it is clearly neither stochastic nor perfectly deterministic. The earlier suggestion that an optimisation process is at work seems likely, but it is unclear exactly what developers are optimising for.

Despite this apparent chaos, the evolution of software size is well described with an inverse square model that results in a decaying growth curve. In this model S_t is the size of version t and E is a model parameter [26]:

$$S_t = S_{t-1} + E/(S_{t-1})^2 \tag{27.1}$$

The evolution of software complexity is not as well described, although it is argued that complexity will increase as software evolves [17]. Directly measuring software complexity by measuring its representation as a complex network firstly requires identifying a proposed measure and then applying it to example software.

27.2 Experiments

27.2.1 Offdiagonal Complexity

Proposed by J.C. Claussen [10] following earlier discussions and preprints, this measure is capable of distinguishing complex networks from those with a regular or random structure. Its basis is the observation that for complex networks the values in a node-node degree correlation matrix are more evenly spread along the offdiagonals. Such correlations between the degrees of pairs of nodes allows the construction of an approximative complexity estimator from the entropy of the normalised distribution.

We computed the offdiagonal complexity (OdC) of two medium-sized software networks through their evolution [19]. This required the development of software implementing OdC, a process that encountered practical difficulties such as interpreting the mathematical notations, which appeared to vary between the original and citing authors, limited examples and apparent errors in the examples given. While these issues were neither insurmountable nor unexpected they did cause uncertainty in validating the software implementation.

Two major free and open source software projects, the integrated development environment Netbeans [4] and the Apache webserver component Tomcat [1], were used as datasets. The available stable releases of each software project were converted into network graphs and their OdC values taken alongside established size measures, such as the number of Java classes, using a custom toolset christened netMetric [5]. For each release two network graphs were created, giving views of the software at different granularities: one to represent the dependencies between Java packages (referred to as 'p2p') and another to represent dependencies between Java classes ('c2c' and considered the more detailed).

Netbeans showed nearly a fourfold increase in size, supporting previous understandings of software evolution such as Lehman's 6th law of continuing growth [17]. However the evolution of OdC behaved differently, challenging Lehman's 2nd law of increasing complexity.

The change of OdC behaviour after release 5.5.1 appears to be due to the removal of the Java Enterprise Edition (J2EE) functionality into a separate product and suggests that removal allowed the product to continue growing in size significantly

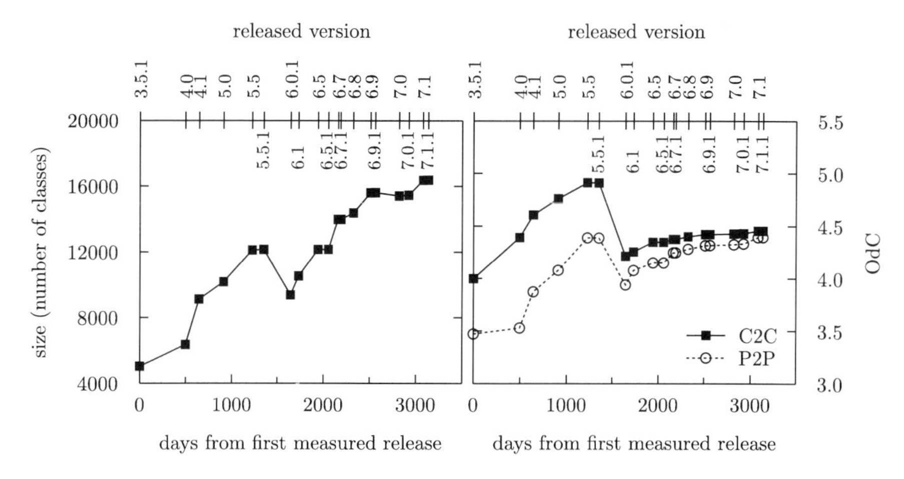

Fig. 27.1 Netbeans size and OdC evolution for packages named org.netbeans.* and their classes

without comparable OdC increases. The releases studied were the major stable versions (and not for instance the developers' in-progress snapshots) which can be categorised as 'new' or 'maintenance' releases. As can be seen in Figure 27.1, there is no discernible difference between new releases and their corresponding maintenance releases (e.g. 5.5 and 5.5.1). Normally, maintenance releases correct defects of the previous release by changing the code within code units instead of changing the software's higher-level structure.

A similar pattern of 'punctuated equilibrium', in which sharp changes are followed by a stable period, has been observed in the evolution of other systems, e.g. in Eclipse [30] (a similar product to Netbeans). The most drastic change was observed when the Rich Client Platform was added, causing a major restructuring of Eclipse's software architecture.

The major versions of Tomcat showed far less distinctive evolution in either size or OdC. This is understandable as a consequence of Tomcat implementing a fixed specification meaning that beyond defect fixes the software changes little.

As well as measuring the entire software system, selected subsystems were investigated in the same manner. In Netbeans, each subsystem demonstrated its own evolutionary pattern for both size and OdC in agreement with other works showing that software evolution proceeds differently in different areas of the codebase [14, 16]. Tomcat again showed little evolution within subsystems. These observations on software networks suggest that growth and perhaps system complexity arise from localised changes in the network.

Offdiagonal complexity was shown to be realistically computable and to show informative behaviour as the software evolved through its releases. However a strong correlation with size (Pearson's $r = 0.86$) limits its usefulness in evaluating software complexity since size is easier and quicker to measure. However, with refinement, the use of degree correlations in an entropy measure could still provide a measurement distinct from size. Claussen [11] offers the "full OdC" as a way of comparing

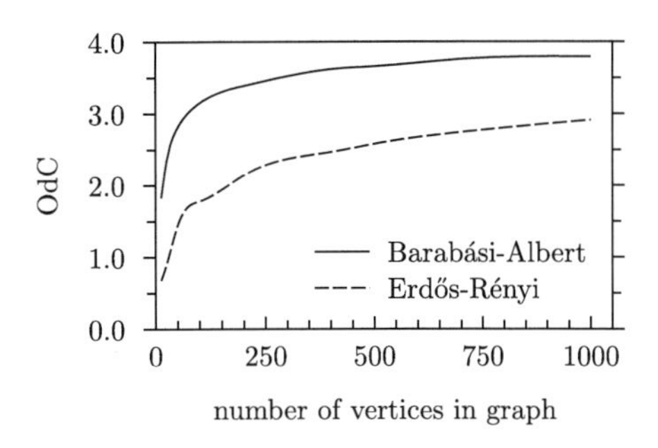

Fig. 27.2 Examples of OdC behaviour in synthetic networks

networks of different size, and Anastasiadis et al. [6] replaced the Boltzman-Gibbs entropy in OdC with the generalised Tsallis, and suggested that changing the parameter involved in Tsallis' entropy could make OdC sensitive to particular structures. Unfortunately there was no suggestion as to what those structures might be. Building the correlation matrix using the idea of a remaining degree distribution à la Newman [22] might also improve sensitivity to structural complexity.

As a way of examining its null-model behaviour, we also computed the OdC on synthetic Barabási-Albert and Erdős-Rényi graphs, observing a rapidly decreasing sensitivity as the number of vertices increased, see Fig. 27.2. This suggests that the OdC is most useful for smaller graphs with less than ∼300 vertices. These scaling properties demonstrate that measures that appear promising when applied to graphs with tens of vertices lose their practicality applied to the typically much larger software networks. Indeed it suggests that OdC is reflecting a complexity arising from size and not just from structure. This confounding effect of size when measuring complexity is a significant practical issue.

27.2.2 Practical Issues

Based on the experience with OdC we make several suggestions for proposed graph measurements that would be helpful for experimentalists, e.g. software engineering researchers like us, interested in complexity metrics.

- The scaling properties should be described. Ideally a proposal should be insensitive to size, but a linear or monotonic relationship with size would still be of practical use since software size can be measured and thus accounted for.
- Describing the computability of the metric with a 'big O' notation would allow an assessment of practicality. The availability of this was instrumental in choosing to experiment with OdC.
- Providing a reference algorithm in any coding language, including pseudo-code, could improve understanding, especially for non-mathematicians.
- Offering downloadable example networks with correct values published would help in verifying software implementations.

– A discussion on how the proposal behaves (if at all) against network properties such as diameter or average degree, and what type of network it is relevant for, would help in assessing its suitability to measure software networks. A proposal that for instance focused on polytrees would be unsuitable since they don't represent software networks.
– Any suggestions as to what structural features it may be sensitive to would also support the assessment of usefulness.

Ideally this information could be curated into a repository allowing the easy selection of proposals for experiment. While admittedly creating more work for the theorists, the advantage is the increased visibility of their proposal with a faster take up and feedback against real world networks. The nature and form of that feedback should be suggested by theorists as part of establishing a dialogue between theorists and those wanting to apply measurement proposals.

The availability of multiple datasets such as the Qualitas Corpus [25], Helix [29] and the Software-artifact Infrastructure Repository [13], alongside toolsets for creating call graphs such as netMetric [5], Dependency Finder [2] and Doxygen [3], provide a ready and extensive source of graphs for analysis. Software is a dynamic process with large amounts of ancillary information (such as changelogs) creating software networks whose evolution is potentially observable step-by-step. Measuring complexity in the structure of software remains elusive, but approached through complex networks it is a potentially rich field for study.

27.3 Conclusions

In this paper we have shown how software networks offer some distinctive challenges and opportunities when measuring complexity which could be of interest to theorists, particularly in terms of how complex networks evolve. The application of the offdiagonal complexity to a software network has been described and shown to be of interest but limited practical use for measuring software complexity. Based on that, proposals are made in the anticipation of fostering a positive dialogue between theorists proposing graph measures and those investigating their practical application.

Acknowledgements We thank Jim Hague, from the Physics Department, and Jozef Siran, from the Mathematics Department, for comments on a draft of this paper.

References

1. Apache tomcat. Open source software implementation of java servlet and javaserver pages. http://tomcat.apache.org/
2. DependencyFinder. A suite of tools for analyzing compiled java code. http://depfind.sourceforge.net/

3. Doxygen. A documentation system. http://www.doxygen.org/
4. Netbeans. The smarter and faster way to code. http://netbeans.org/
5. netMetric. A tool for analysing large java codebases. http://sourceforge.net/p/netmetric
6. Anastasiadis A, Costa L, Gonzáles C, Honey C, Széliga M, Terhesiu D (2005) Measures of structural complexity in networks. In: Complex systems summer school, Santa Fe
7. Bird C, Nagappan N, Gall H, Murphy B, Devanbu P (2009) Putting it all together: Using socio-technical networks to predict failures. In: The 20th international symposium on software reliability engineering, ISSRE'09. IEEE Press, New York, pp 109–119
8. Brooks FJ (1987) No silver bullet essence and accidents of software engineering. IEEE Comput 20(4):10–19
9. Cardoso A, Crespo R, Kokol P (2000) Two different views about software complexity. In: Proceedings of European software conference on metrics (ESCOM). Elsevier, Amsterdam, pp 433–438
10. Claussen JC (2007) Offdiagonal complexity: a computationally quick complexity measure for graphs and networks. Physica A 375(1):365–373
11. Claussen JC (2008) Offdiagonal complexity: a computationally quick network complexity measure—application to protein networks and cell division. In: Mathematical modeling of biological systems, vol II, pp 279–287. arXiv:0712.4216
12. Concas G, Marchesi M, Pinna S, Serra N (2007) Power-laws in a large object-oriented software system. IEEE Trans Softw Eng 33(10):687–708
13. Do H, Elbaum S, Rothermel G (2005) Supporting controlled experimentation with testing techniques: an infrastructure and its potential impact. Empir Softw Eng 10(4):405–435. http://sir.unl.edu/portal/index.php
14. Gall H, Jazayeri M, Klosch R, Trausmuth G (1997) Software evolution observations based on product release history. In: Proceedings on the international conference on software maintenance. IEEE Comput Soc, Los Alamitos, p 166
15. Gao Y, Xu G, Yang Y, Liu J, Guo S (2010) Disassortativity and degree distribution of software coupling networks in object-oriented software systems. In: IEEE international conference on progress in informatics and computing (PIC), vol 2, pp 1000–1004
16. Godfrey M, Tu Q (2000) Evolution in open source software: a case study. In: Proceedings of the international conference on software maintenance, pp 131–142
17. Lehman MM, Fernández-Ramil JC (2006) Rules and tools for software evolution planning and management. In: Madhavji N, Fernández-Ramil J, Perry D (eds) Software evolution and feedback: theory and practice. Wiley, New York, pp 539–563
18. Louridas P, Spinellis D, Vlachos V (2008) Power laws in software. ACM Trans Softw Eng Methodol 18(1):2:1–2:26
19. Moore K (2011) Evaluating offdiagonal complexity as a metric of software evolution. Master's thesis, Open University
20. Myers C (2003) Software systems as complex networks: structure, function, and evolvability of software collaboration graphs. Phys Rev E 68:046116
21. Newman M Network data. http://www-personal.umich.edu/~mejn/netdata/
22. Newman M (2002) Assortative mixing in networks. Phys Rev Lett 89(20):208701
23. Potanin A, Noble J, Frean M, Biddle R (2005) Scale-free geometry in oo programs. Commun ACM 48(5):99–103
24. Solé R, Valverde S (2004) Information theory of complex networks: on evolution and architectural constraints. In: Complex networks, pp 189–207
25. Tempero E, Anslow C, Dietrich J, Han T, Li J, Lumpe M, Melton H, Noble J (2010) Qualitas corpus: a curated collection of java code for empirical studies. In: Asia pacific software engineering conference, APSEC2010. http://qualitascorpus.com/
26. Turski W (2006) A simple model of software system evolutionary growth. In: Madhavji N, Fernández-Ramil J, Perry D (eds) Software evolution and feedback: theory and practice. Wiley, New York, pp 131–141
27. Valverde S, Cancho R, Sole R (2002) Scale-free networks from optimal design. Europhys Lett 60:512

28. Valverde S, Solé R (2003) Hierarchical small worlds in software architecture. Dyn Contin Discrete Impuls Syst 14(suppl):1–11
29. Vasa R, Lumpe M, Jones A (2010) Helix—software evolution data set. http://www.ict.swin.edu.au/research/projects/helix
30. Wermelinger M, Yu Y, Lozano A, Capiluppi A (2011) Assessing architectural evolution: a case study. Empir Softw Eng 16(5):623–666. http://oro.open.ac.uk/28753/

Chapter 28
The Internet Geographical PoP Level Maps

Yuval Shavitt and Noa Zilberman

Abstract Inferring the Internet PoP level maps is gaining interest due to its importance to many areas, e.g., for tracking the Internet evolution and studying its properties. We introduce DIMES's Internet PoP-level connectivity maps, annotated with geographical information and created using a structural approach to automatically generate large scale PoP level maps. The generated PoP level maps dataset is presented and a detailed analysis of a map is provided. PoP level maps have a wide range of applications, introduced in this work. We survey some of these applications and propose further opportunities for future research.

28.1 Introduction

The Internet is one of the most interesting networks to study. It is a man-made network, used by billions of people in their everyday life. The structure of this network is of a special interest, as every service provider applies his own policies and design rules to his portion of the network, called an Autonomous System (AS). The AS level is most commonly used to draw Internet maps, as it is relatively small (tens of thousands of ASes) and therefore relatively easy to handle. An AS may represent a local ISP as well as a large company spanning across continents. The connectivity and growth of this network is driven by a large number of factors: from business agreements between service providers, local population growth, technological trends and more. Looking at the Internet topology from the AS level is thus coarse: it does not indicate the size of the AS nor local aspects and does not provide any geographic notion. IP and Router level maps represent the other extreme: they contain too many details to suit practical purposes, and the large number of entities makes them very hard to handle.

This work was partially funded by the Israeli Science Foundation's center of knowledge grant 1685/07.

Y. Shavitt (✉) · N. Zilberman
School of Electrical Engineering, Tel-Aviv University, Tel Aviv, Israel
e-mail: shavitt@eng.tau.ac.il

N. Zilberman
e-mail: noa@eng.tau.ac.il

T. Gilbert et al. (eds.), *Proceedings of the European Conference on Complex Systems 2012*, Springer Proceedings in Complexity, DOI 10.1007/978-3-319-00395-5_28,

Service providers tend to place multiple routers and other networking equipment in a single location called a Point of Presence (PoP). The equipment placed in a PoP is used to serve a certain area and to connect it to higher hierarchies within the AS. A PoP is owned by one AS, however several PoPs owned by different vendors many times reside within the same campus, that provides them the infrastructure they need. For studying the Internet evolution and for many other tasks, PoP level maps give a better level of aggregation than router level maps with a minimal loss of information. PoP level graphs allow to examine the size of each AS network by the number of physical co-locations and their connectivity instead of by the number of its routers and IP links, which is an important contribution. The points of presence are not only counted, but also provided with a geographical location and information about the size of the PoP. Using PoP level graphs one can detect important nodes of the network, understand network dynamics, examine types of relationships between service providers as well as routing policies and more.

While aggregating IPs to AS-level is a fairly simple task, PoP level maps are more difficult to create. Andersen et al. [2] used BGP messages for clustering IPs and validated their PoP extraction based on DNS. Rocketfuel's [16] generated PoP maps using tracers and DNS names. The iPlane project also generates PoP level maps and their connectivity [9] by first clustering IP interfaces into routers and then clustering routers into PoPs. They did so by estimating the length of the reverse path, with the assumption that reverse path length of routers in the same PoP will be similar.

Assigning a location to an IP address, let alone a PoP, is a complicated task. The most common way to do so is using a geolocation service. Geolocation services use DNS resolution [16], hand-labeled hostnames [1], user's information provided by partners [3], and more. Geolocation services are not highly accurate, as we showed in [14]. Thus a measurement based approach was suggested to approximate the geographical distance of network hosts [7, 8, 10].

This work presents PoP level connectivity maps generation and analysis, based on an algorithm described in [5]. The traceroute measurements used in this work were generated by DIMES, a highly-distributed Internet measurements infrastructure [13]. DIMES achieves high distribution of vantage points by employing a community based distribution methodology that uses Internet users' PCs for measurements.

28.2 PoP Level Maps Construction

A PoP is a group of routers which belong to a single AS and are physically located at the same building or campus. In most cases [6, 11] the PoP consists of two or more backbone/core routers and a number of client/access routers. The client/access routers are connected redundantly to more than one core router, while core routers are connected to the core network of the ISP. The algorithm we use for PoP extraction looks for bi-partite subgraphs with delay constraints in the IP interface graph of an AS; no aliasing to routers is needed [5]. The bi-partites serve as cores of the PoPs and are extended with other nearby interfaces.

To identify the geographical location of a PoP, we use the geographic location of each of its IPs. As all the PoP IP addresses should be located within the same campus, the location confidence of a PoP is significantly higher than the confidence that can be gained from locating each of its IP addresses separately. The location of an IP address is obtained from numerous geolocation databases, and the PoP's location is set to the median of all PoP's IP locations. Every PoP location is assigned a range of convergence, representing the expected location error range based on the information received from the geolocation databases. Further discussion of the extraction and geolocation algorithms is provided in our previous works [5, 14].

The connectivity between PoPs is an important part of PoP level maps [15]. We generate PoPs connectivity graph using unidirectional links. We define a link L_{SD} as a the aggregation of all unidirectional edges originating from an IP address included in a PoP S and arriving to an IP address included in a PoP D. Each of the IP level links has an estimate of the median delay measured along it, with the median calculated on the minimal delay of a basic DIMES operation. A basic DIMES operation is comprised of four consecutive measurements and all measured values are roundtrip delays [5].

28.3 Data Set

The collected dataset for PoP level maps is taken from DIMES [4]. We use all traceroute measurements taken during weeks 42 and 43 of 2010, totaling 33 million, which is an average of 2.35 million measurements a day. The measurements were collected from over 1308 vantage points, which are located in 49 countries around the world.

The 33 million measurements produced 9.1 million distinct IP level edges (no IP level aliasing was performed). Out of these, 258K edges had less than the median delay threshold, and had sufficient number of measurements to be considered by the PoP extraction algorithm. A total of 4098 PoPs where discovered, containing 67422 IP addresses. The geographic spread of these PoPs around the world is shown on Figure 28.1(left). Although the number of discovered PoPs is not large, as the algorithm currently tends to discover mainly large PoPs while missing many access PoPs, the large number of IP addresses and the spread around the world allow a large scale and meaningful PoP level connectivity evaluation.

The PoP level connectivity map generated from the data set [15] contains 86760 links, which are an aggregation of 1.65 million edges. Out of the 4098 discovered PoPs in week 42, 2010, 4091 have at least one PoP level link. 2405 PoPs have outgoing links, and 4073 PoPs have incoming links. Out of those, 18 PoPs have only outgoing links and 1686 have only incoming links. Note that a PoP without any PoP level links, or a PoP with only incoming or outgoing links still have additional IP-level connecting edges. As the full map is too detailed to display, a partial map is

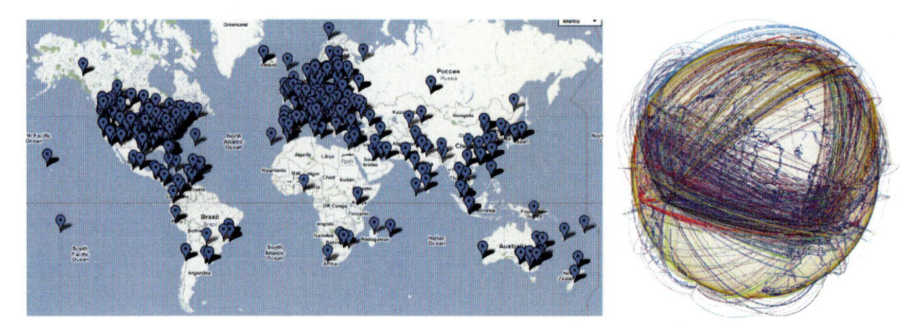

Fig. 28.1 An Internet PoP Level Location Map (*left*) and a Partial Connectivity Map (*right*)—Week 42, 2010

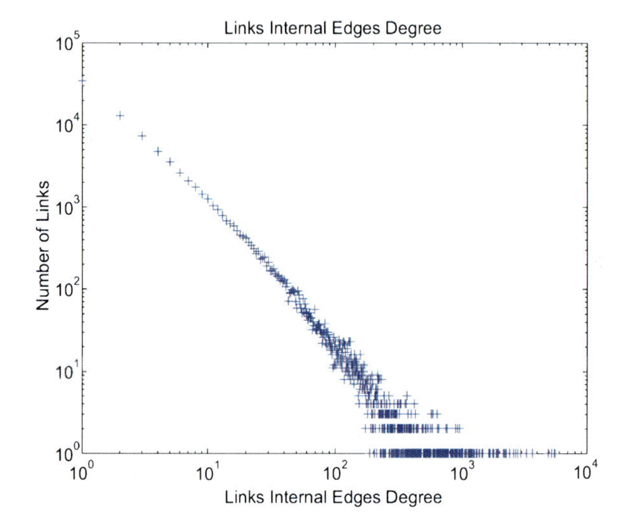

Fig. 28.2 Number of edges within a link vs. number of PoP level links

shown in Figure 28.1(right), demonstrating the connectivity between 430 ASes on PoP level.

Almost all the IP edges that are aggregated into links are unidirectional: 99.2 %. This is a characteristic of active measurements: the number of vantage points is limited in number and location, thus most of the edges can be measured only one way. However, at PoP links level, 6.5 % of the links are bi-directional: eight times more than the bi-directional edges. This demonstrates one of the PoPs strengths, as it provides a more comprehensive view of the networks' connectivity without additional resources. The average number of edges within a unidirectional link is 7.5, and the average number of edges within a bidirectional link is 44.7. This is not surprising, as it is likely that most of the bidirectional links will connect major PoPs within the Internet's core and thus be easily detected.

An additional view of edges aggregation into links is given by Figure 28.2. The X-axis shows the number of edges aggregated into a link, while the Y-Axis is the number of PoP-level links. The graph shows a Zipf's law relation between the two, as 82.6 % of the links aggregate ten edges or less, and less than 2 % aggregate 100 edges or more. The large number of edges per link is explained by the fact that a measured edge is not a point-to-point physical connection: Take two routers, A & B, connected by a single fiber; If one of the routers has 48 ports, and we measure through each one of them, we detect 48 edges between the two routers (incoming port i on router A and the single connected incoming port of router B). We find that the number of links per PoP also behaves according to Zipf's law.

Looking at the number of links by destination PoP, 46 % of the PoPs are connected by 10 links or less and the average number of links per PoP is 21. Most of the PoPs are connected to PoPs outside their AS: 71.5 % of the source PoPs and 99 % of the destination PoPs. Interestingly, only 62.2 % of the destination PoPs have links within their AS, which indicates that many PoPs are detected only thanks to inter-AS measurements, and thus that the detected PoPs are probably large ones and not small local access PoPs. We believe this is also the reason why few destination PoPs have a small number of links: PoPs with a large number of links to other ASes are more likely to be discovered by our algorithm.

28.4 Applications of PoP Level Maps

PoP level maps can be leveraged for a large number of research interests. The most obvious area is the study of Internet network topology, as it represents a level of the network that was barely considered in the past. Tying a geographic location and a size to a PoP, PoP level maps offer an opportunity to investigate service providers' actual presence and influence on the network. An additional benefit is the ability to study types of relationships (ToR) between service provider on different locations around the globe.

An additional aspect of PoP level maps relates to cyber security research. As shown by Schneider et al. [12], DIMES's PoP level map can be leveraged to study the robustness of a network. The Map can also be used for several Geolocation purposes, such as improving the accuracy of Geolocation databases [14] and for distance estimation.

Another application of PoP level maps is the study of Internet's evolution. By adding the maps as a new indicator, on top of economic, geographic and demographic parameters, a better understanding of the network's growth can be achieved.

28.5 Conclusion and Future Work

We presented here DIMES's PoP-level connectivity maps. The PoP level connectivity maps provide a new look at the Internet's topology with a better level of aggregation than router level maps and more information than AS level maps. The maps provide network topology information, annotated with geographic location and link delay, thus providing a large-scale look on the Internet using a light data set.

The PoP level links maps are now available through the DIMES website [4] for download, and can be useful to researchers in the fields of complex networks, Internet topology, Geolocation, and more.

References

1. Quova (2010). http://www.quova.com
2. Andersen DG, Feamster N, Bauer S, Balakrishnan H (2002) Topology inference from BGP routing dynamics. In: Internet measurement workshop, pp 243–248
3. Digital Envoy (2010). NetAcuity Edge. http://www.digital-element.com/our_technology/edge.html
4. DIMES. Distributed internet measurements and simulations. http://www.netdimes.org/
5. Feldman D, Shavitt Y, Zilberman N (2011) A structural approach for PoP geolocation. Comput Netw. doi:10.1016/j.comnet.2011.10.029
6. Greene BR, Smith P (2002) Cisco ISP essentials. Cisco Press, Indianapolis
7. Gueye B, Ziviani A, Crovella M, Fdida S (2006) Constraint-based geolocation of internet hosts. IEEE/ACM Trans Netw 14(6). doi:10.1109/TNET.2006.886332
8. Laki S, Matray P, Haga P, Sebok T, Csabai I, Vattay G (2011) Spotter: a model based active geolocation service. In: IEEE INFOCOM 2011, Shanghai, China
9. Madhyastha HV, Anderson T, Krishnamurthy A, Spring N, Venkatara- mani A (2006) A structural approach to latency prediction. In: Proceedings of the 6th ACM SIGCOMM conference on internet measurement, IMC'06, pp 99–104
10. Padmanabhan VN, Subramanian L (2001) An investigation of geographic mapping techniques for internet hosts. In: Proceedings of the 2001 conference on applications, technologies, architectures, and protocols for computer communications, SIGCOMM'01, pp 173–185
11. Sardella A (2006) Building next-gen points of presence, cost-effective PoP consolidation with juniper routers. White paper, Juniper Networks
12. Schneider CM, Moreira AA, Andrade JS, Havlin S, Herrmann HJ (2011) Mitigation of malicious attacks on networks. Proc Natl Acad Sci USA 108(10). doi:10.1073/pnas.1009440108
13. Shavitt Y, Shir E (2005) DIMES: let the internet measure itself. In: ACM SIGCOMM computer communication review, vol 35
14. Shavitt Y, Zilberman N (2011) A geolocation databases study. IEEE J Sel Areas Commun 29(9). doi:10.1109/JSAC.2011.111214
15. Shavitt Y, Zilberman N (2012) Geographical internet PoP level maps. In: TMA workshop, pp 121–124
16. Spring N, Mahajan R, Wetherall D (2002) Measuring ISP topologies with rocketfuel. In: ACM SIGCOMM, pp 133–145

Chapter 29
Practical Approach to Construction of Internal Variables of Complex Self-organized Systems and Its Theoretical Foundation

Dalibor Štys, Petr Jizba, Tomáš Náhlík, Karina Romanova, Anna Zhyrova, and Petr Císař

Abstract We propose a method for characterizing the image—multidimensional projection—of complex, self-organising, system. The method is general and may be used for characterisation of any structured, experimentally observable, complex self-organized systems. The method is based on calculation of information gain by which a point contributes to the total information in the image, the point information gain, PIG. We have also derived related variables, the point information gain entropy PIE and point information gain entropy density PIED. The later values are unique to a structured information and may be used for analysis of similarity by clustering, identification of states etc. We illustrate our key results using the example of living cell. We discuss practical limits of the analogy between this observable self-organising system and its possible theoretical model using an example of chaotic attractor.

Keywords Rényi entropy · Point information gain · Principal component analysis · Principal manifold · Chaotic attractor

D. Štys (✉) · T. Náhlík · K. Romanova · A. Zhyrova · P. Císař
School of Complex Systems, FFPW, University of South Bohemia, Zámek 136, 373 33 Nové Hrady, Czech Republic
e-mail: stys@frov.jcu.cz

T. Náhlík
e-mail: nahlik@frov.jcu.cz

K. Romanova
e-mail: romanova@frov.jcu.cz

A. Zhyrova
e-mail: zhyrova@frov.jcu.cz

P. Císař
e-mail: cisar@frov.jcu.cz

P. Jizba
FNSPE, Czech Technical University in Prague, Břehová 7, 115 19 Prague, Czech Republic
e-mail: p.jizba@fjfi.cvut.cz

T. Gilbert et al. (eds.), *Proceedings of the European Conference on Complex Systems 2012*, Springer Proceedings in Complexity, DOI 10.1007/978-3-319-00395-5_29,

29.1 The Method

Complexity and self-organisation leads to formation of structures objects. Highly discussed examples are living organisms which range from simple cells through herds, flocks, insect colonies to humans and their herding behavior or cities. We focused on analysis and interpretation of information on these structures [7–10]. We devised the method of calculation of the point information gain (PIG) $\mathrm{PIG}_{\alpha,x,y}$, the information contribution to the image using the Rényi entropy concept.[1,2]

$$\mathrm{PIG}_{\alpha,x,y} = \frac{1}{1-\alpha}\ln\left(\sum_{i=1}^{n} p_{i,x,y}^{\alpha}\right) - \frac{1}{1-\alpha}\ln\left(\sum_{i=1}^{n} p_{i}^{\alpha}\right), \tag{29.1}$$

where $p_{i,x,y}$ and p_i are probabilities of occurrence of given intensity in the image without and with the examined point, n is number of intensity levels and α is a dimensionless coefficient.

By summation of PIG we obtain derived quantities, the point information gain entropy PIE, and point information gain entropy density PIED, the sum of all PIG levels at given alpha.

$$\mathrm{PIE/\ points}_{\alpha} = \sum_{i=1}^{n} \mathrm{PIG}_{\alpha,i}. \tag{29.2}$$

PIE and PIED for infinite number of α create a state space in which the point is unique for each image differing in properties or position of any of the points. In practice, we use only 13 different α values for each of the camera colour channels. Values of α were selected on the basis of previous experience of Petr Jizba with other types of dataset and their usefulness was confirmed in many experiments. Yet, we do not exclude that an experimental dataset may be obtained, for whose discrimination we shall need to calculate PIG for other α values.

In [9] we show that PIE/points may be successfully used for multivariate statistical analyses such as principal component analysis [6]. There may be derived clusters in the state space and transitions between them. We recently found out that the information gain calculation has been independently developed as a very successful optimisation criterion in widely distant field of self-organising semantics, here it is used in construction of decisive trees [1]. The point information gain may thus be a natural feature of self organising system whose significance has not been fully understood. At Fig. 29.1 we show an example how a trajectory of a living cell described in a PIE/points space may be used for objective identification of cell states during the cell cycle.

---

[1] http://expertomica.eu/software/eec/.

[2] http://expertomica.eu/software/pie/.

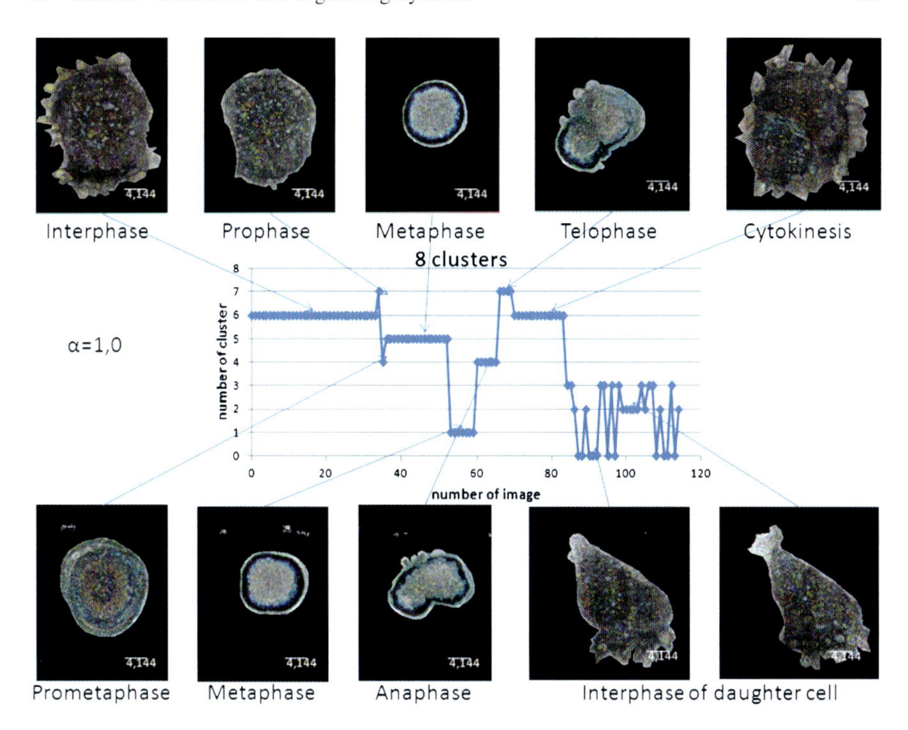

Fig. 29.1 Objective analysis of the cell cycle: images of the cell were abstracted from the microscopic image and values PIE/points were calculated for 13α + values for each color channel. Resulting 39 image variables were subjected to principal component analysis and clustered. In the initial phase for the cell cycle results of clustering closely resembled those of manual annotation. In the later phase the objective analysis indicated that the structure of cell interior still undergoes significant changes which remain undetected by expert analysis. The analysis was made using the Unscrambler software (http://www.camo.com/)

29.2 Limits of the PIED Metrics

A obvious limit of any measured system lies in the fact that we measure only in distinct times and with distinct precision. As comprehensively delineated by Žampa [11] (see also [7]), this fact has important influence on causal relations in the dynamic system. Essentially, determinism is turned into stochastic causality. In the state space containing several basins of attraction, such stochastic causality may have much deeper origin. The probability density function of transition between two attractor zones of a chaotic attractor may have very complicated shapes [2]. The assumptions of the general stochastic systems theory are still valid, but the origin of the probabilistic behaviour in self—organised systems is much deeper, truly fundamental.

Key question in the application of combination the stochastic systems theory and theory of chaotic attractor for interpretation of observable results in self-organising systems is the question whether the image which is analysed is more a reflection of the group (ball) of points in the phase space which evolved during the interval of the measurement or rather a portrait of the whole self organising system. This depends on properties of the system and has to be determined experimentally. In the case that we observe just an average of small number of points, we may use the same approach as we use in measurement of standard technical systems. We must study its evolution in time, determine the best approximation to internal orthogonal variables and interpret them in terms of a model described by differential equations and error functions.

In case of, for example, living cells, we may well assume that decisive elements of the structure passed the trajectory sufficiently often in the time interval of the data capture to account for probabilistic interpretation of the result. Theoretically very well substantiated probabilistic approach is calculation of the generalised dimension D_α [4, 5] which includes the calculation of Rényi entropy at the limit to infinitely small yardstick ϵ

$$D_\alpha = \lim_{\epsilon \to 0} \frac{\frac{1}{1-\alpha} \log(\sum_i p_i^\alpha)}{\log \frac{1}{\epsilon}} \tag{29.3}$$

The similarity of our approach of characterisation of the image by PIED and calculation of the generalised dimension was discussed earlier [9]. Any real application will always be complicated by distortion of the image by instrument transmission function (i.e. microscope point spread function), standard electromechanical sources of error and, with the same importance, by existence of several self organising systems in one observed system, often organised in a hierarchical manner. For example in the case of living cell we must consider self organisation and asymptotic stability of folded proteins and multiprotein complexes, membranes, organelles etc. We most probably in all cases encounter a mixed situation, when part of the observable objects passes within the time of measurement the state trajectory sufficiently often to account for the calculation of the D_α while others do not.

At Fig. 29.2 we show real example of analysis of several cell cycles. The cell cycle may be considered as a sequence of changes in parameters of the internal self-organising system of the cell which are expressed as different observable structures. In fact we should also consider that for each of the cell states we have completely different system. As may be seen, the time of duration of each of the phases of the cell cycle differs widely. For prometaphase we see time ranging from two measurement intervals (60 s) up to 45 (1450 s). The dataset is sparse although the measurement took 24 hours and the manual analysis in the e-cells program[3] several working days.

29.3 Conclusions and Outlook: Multivariate Analysis of Datasets from Dynamic Self-organising Systems

For objective interpretation and analysis of general experimental data it is good to have a model of expected system dynamics. Only in such case data about the

[3]http://expertomica.eu/software/ecell/.

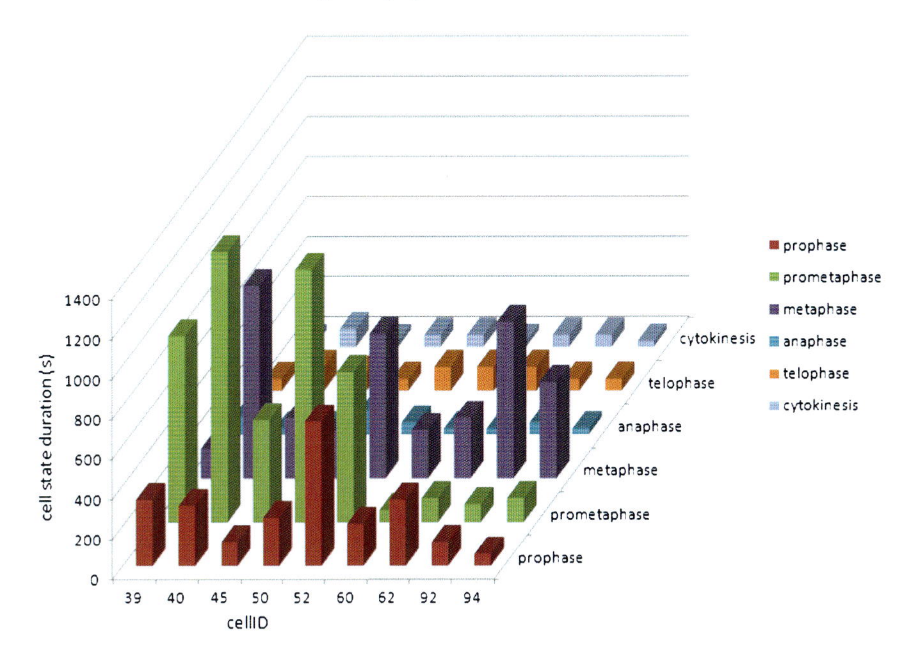

Fig. 29.2 Manually annotated phases (state) in the cell cycle of a single cell. Clearly, the duration of individual states differs significantly for individual cells and no obvious rule seems to be detectable. Certainly the amount of data is very small, yet the dataset was large (over 2000 images) and time resolution was clearly insufficient. Automation of the annotation process is needed, but to achieve it, the system model must be created

system may be compared between various datasets characterising the same system. Such models are available for most technical systems, but are difficult to be achieved for natural self-organising systems. Stochastic models for technical systems are created to account mainly for measurement uncertainities. The case of complex, namely self-organising, systems may include, besides experimental error, also jumps between two distinct regions of the chaotic attractor. At events such as changes between cell states during the cell cycle, we may assume also transitions between two attractors with significantly different parameters or, equally well, two completely different attractors, possibly with different dimensionality. Such events have not been studied in detail. As we have shown at Fig. 29.1, the use of principal component analysis is well possible for clustering of similar states. But it does not bring any information about the possible models. We may potentially utilise various methods of multivariate non-linear analysis of which the principal manifold approach [3] is the most instructive one. Similarly as the principal component analysis assumes that the manifold to which the model should be fitted—the equivalent of the mechanical model—is the plane in multidimensional space, we may assume that the resulting principal manifold is the best approximation to the manifold of the attractor responsible for stability of an observed system. And the choice of chaotic attractor which project to similar manifold may be made. This approach is in many aspects similar to the approach of mechanical engineer who approximates movement of the device by equations of mechanics. Experiments in this direction are in progress and will be reported at the conference.

Acknowledgements This work was partly supported and co-financed by the South Bohemian Research Center of Aquaculture and Biodiversity of Hydrocenoses (CZ.1.05/2.1.00/01.0024) and by the Grant Agency of the University of South Bohemia (152/2010/Z 2012).

References

1. Aberer K (2011) Peer-to-peer data management. Synthesis lectures on data management, vol 3. Morgan and Claypool Publishers, San Rafael, pp 1–150
2. Cvitanović P, Artuso R, Mainieri R, Tanner G, Vattay G (2009) Chaos: classical and quantum. Niels Bohr Institute, Copenhagen
3. Gorban A, Zinovyev A (2010) Principal manifolds and graphs in practice: from molecular biology to dynamical systems. arXiv:1001.1122 [cs.NE]
4. Grassberger P, Procaccia P (1983) Characterization of strange attractors. Phys Rev Lett 50:346–349
5. Grassberger P, Procaccia P (1983) Measuring the strangeness of strange attractors. Physica D 9:189–208
6. Pearson K (1901) On lines and planes of closest fit to systems of points in space. Philos Mag 2:559–572
7. Stys D, Vanek J, Nahlik T, Urban J, Cisar P (2011) The cell monolayer trajectory from the system state point of view. Mol BioSyst 7:2824–2833
8. Stys D, Urban J, Vanek J, Cisar P (2010) Analysis of biological time-lapse microscopic experiment from the point of view of the information theory. Micron 41:478–483
9. Stys D, Jizba P, Papacek S, Nahlik T, Cisar P (2012) On measurement of internal variables of complex self-organized systems and their relation to multifractal spectra. In: Kuipers A, Heegaard PE (eds) IWSOS 2012. LCNS, vol 7166. Springer, Berlin, pp 36–47. ISBN 978-3-642-28582-0
10. Urban J, Vanek J, Stys D (2009) Preprocessing of microscopy images via Shannon's entropy. Pattern Recognit Inf Process 283:187
11. Zampa P, Arnost R (2004) In: 4th WSEAS conference

Chapter 30
An Efficient Simulator for Boolean Network Models

Stefano Benedettini and Andrea Roli

Abstract Boolean networks (BNs), first introduced by Kauffman as genetic regulatory network models, are the subject of notable works in complex systems biology literature. BN models lately garnered much attention because it has been shown that BNs can capture important phenomena in genetics and biology in general. In this work, we illustrate the main properties and design principles of a new efficient, flexible and extensible BN simulator, named the Boolean Network Toolkit. This simulator makes it possible to easily set up experiments and analyse the most relevant features of BN's dynamics.

30.1 Introduction

Boolean networks (BNs) have been firstly introduced by Kauffman [1] and subsequently received considerable attention in the composite community of complex systems. Recent advances in this research field can be mainly found in works addressing themes in gene regulatory networks (GRNs) or investigating properties of BNs themselves. These models are often studied by means of simulation processes and an issue arises concerning the efficiency and usability of the simulator. In fact, this tool plays a crucial role for the results which can be attained because it should enable the investigator to analyse large size systems, in order to avoid conclusions that might be wrong because of the limited size of the simulated networks or undersampling of some dynamic properties. In addition, a specific research stream in this area is the *ensemble approach* [2], that aims at finding classes of GRN models (such as BNs) which match statistical features of genes, such as the number of cell types of an organism or cell dynamics in case of perturbation. In recent works, [1] it

[1] See, e.g., the work by Benedettini et al. [3].

S. Benedettini (✉)
European Centre for Living Technology, Venice, Italy
e-mail: s.benedettini@unive.it

A. Roli
DEIS-Cesena, Alma Mater Studiorum, University of Bologna, Bologna, Italy
e-mail: andrea.roli@unibo.it

T. Gilbert et al. (eds.), *Proceedings of the European Conference on Complex Systems 2012*, Springer Proceedings in Complexity, DOI 10.1007/978-3-319-00395-5_30,

has been shown that the application of stochastic local search (SLS) techniques is an effective approach to automatically design BNs to match certain desiderata. This approach requires efficient and easily reconfigurable BN simulation tools, as well.

In this work, we illustrate the main properties and design principles of a new simulator, named the Boolean Network Toolkit (BNTK) [4].

30.1.1 Simulator Requirements

BN models simulation has two main requirements: (i) efficiency and (ii) easy experiment configuration. The first requirement addresses the issues of simulating large size BNs and of sampling with sufficient precision level dynamic properties of the networks, such as number of attractors, length of cycles and size of basins of attraction. A further issue adds upon simple simulation of a model when the investigator aims at finding an model instance satisfying given requirements, for example a BN with a given attractors landscape. Hence the requirement of exploring an enormous search space. It appears clear that, in order to explore such a large search space effectively, we both need sophisticated search algorithms and an efficient simulator. In fact, the search method will likely have to sample the search space many times and each of these samples, a BN, will have to be evaluated, i.e., simulated; the simulation being the most computationally expensive operation in this process.

Other important aspects are that of modularity and flexibility. The simulator should be flexible enough to allow the researcher to quickly set up experiments of different kind which go beyond the usual basic tasks, like computing a BN's trajectory or finding a set of attractors. For instance, the experiment described in [5] to find Threshold Ergodic Sets is rather complex and requires a high degree of customisability; the simulator should provide all that. As a further prerequisite, the simulator should not also be limited to synchronous BNs and handling also other different models of networks, such as BNs with stochastic dynamics.

To summarise, the simulator should satisfy the following requirements:

Flexibility. The simulator should enable the researcher to implement different kinds of experiments in diverse contexts and under different constraints. It should also provide a way to easily modify a BN, a fundamental operation required by design and training algorithms.

Ease of integration. The simulator is going to be integrated with other software libraries, typically written in `C++` for greater efficiency.

Free and Open Source Software. We strive to maintain our software free so as to foster adoption and extension.

Modularity and separation of concerns. It should be able to support different BN update schemes and possibly different network models. Also, modularity is a key to achieve flexibility by allowing the researcher to modify and extend the tool itself.

Efficiency. The simulator should be as efficient as possible.

30.2 Design of the Boolean Network Toolkit

This section briefly explains the design choices made to implement the BNTK. First we will introduce the key abstractions in the simulator pertinent to BNs. Afterwards, we describe the architecture of the simulator and we will see how these abstractions can be extended to encompass different kinds of update schemes and possibly new network models. The requirements listed at the end of the introduction push us to choose C++ as our implementation language; the main programming paradigm adopted is, therefore, Object Oriented, but, as we will see, we will also employ elements of functional programming.

Simulator Architecture The main entity in the simulator is, of course, the BN. The principal design choice that can help us to achieve good modularity and flexibility is illustrated in the following. In the BNTK, network state, dynamics and topology are separate concepts. Network state can be any indexable data structure whose elements are Booleans, but in principle also integers or double precision floats depending on the domain of node values; this is a crucial characteristic if we want to extend the simulator with other network models. For BNs we employ the compact bitset structure provided by the Boost Dynamic Bitset Library. Network topology is a stateless entity, structurally represented by a directed graph data structure whose nodes, indexed by integers, containing their update function. A network topology exposes methods to access its graph structure and, most importantly, it provides a method to compute the next value of a node given the current state vector. In order to implement network topology, we use the excellent Boost Graph Library [6]. Network dynamics is modeled by a specific entity that realises a particular node update scheme.

Fundamental Abstractions The main abstraction in our design is a "synchronous BN". In the Object Oriented paradigm, it would be intuitive to model a BN as a mutable object composed of a state vector and a topology. On the other hand, such abstraction is rather distant to the original mathematical concept that sees BNs as discrete maps, i.e., functions. In formal term, a BN is a function F that maps the current state to the next state $s_{t+1} = F(s_t)$. Notice that the form of F is neither dependent on the state nor the time parameter. Moreover, from a computational point of view, a function is an immutable object. The functional programming paradigm is the preferred approach to model BNs in the BNTK.

The most basic simulation task is to evolve a BN a certain number m of steps starting from initial condition s_0. With a functional approach this is straightforward. Formally, we just need to take the first m elements from the sequence $F^i(s_0), i \in \mathbb{N}$; what we need is a way to *lazily*[2] represent such sequence, or, in computer science parlance, a *stream*. Lazily evaluated sequences, such as lists, are typically available in all major functional programming languages.

[2]"Lazy" roughly means "computed on demand".

Almost all computational tasks on BNs involve computing a trajectory, therefore the concept of stream, or lazy list, is pervasive in the simulator. Such abstraction is fundamental because, as we will see, it makes it possible to easily decompose complex experiments on BNs into simpler and reusable components.

With a trajectory, it is easy to find an attractor: it is sufficient to apply one of the well-known cycle finding algorithms. In the BNTK we implemented two algorithms. One is a naïve search that stores every state encountered and looks for repeated values. This algorithm does not scale well, in terms of both time and space complexity, with long transients, therefore we also provide an alternative, Brent's cycle detection algorithm [7] (see also [8]), which scales much better for long transients (chaotic RBNs, for instance), but is slower for short sequences. Finding an attractor starting from an initial condition can be in turn interpreted as a higher-order function which takes a BN (a function in our interpretation), an initial state and returns an attractor object. Let us write the type of such function as $State \rightarrow Attractor$.

Now that we have encapsulated the computation of an attractor into a function, we can easily calculate a set of attractors starting from a sequence of initial conditions. To do so we can use another important higher-order function on lists called `map`, that, roughly, is an abstraction of a 'for' loop. `map` is a function that takes a function from a type A to another type B ($A \rightarrow B$ for short), a list of element of type A ($[A]$ for short) and returns a list of elements of type B (in symbols `map` $: (A \rightarrow B), [A] \rightarrow [B]$).

30.3 BNTK Functionalities

The BNTK contains also the implementation of some functionalities representing common use cases. Among them, we mention finding the set of attractors, their length and basin of attraction and computing dynamic measures such as the Derrida plots. In the following, we show a couple of use case examples.

BN Trajectory Let us start with simply simulating the evolution in time of a BN until an attractor is found.

```
1 BooleanNetwork bn = // read BN from file
2 BooleanDynamics* dyn = synchronous_dynamics(bn); // get an
      updater
3 State s0 = // an initial state vector
4 State s = (*dyn)(s0); // state at t=1
5 s = (*dyn)(s); // state at t=2
6 s = (*dyn)(s); // state at t=3
7 BrentCycleFinder finder(dyn);
```

where `finder` is a function object, implementing Brent's cycle detection algorithm, that takes a state vector and returns an attractor. Notice that neither `BooleanNetwork` or `BooleanDynamics` are stateful objects and that network state is externally maintained. This way, we are able to run several network evolutions in parallel without copying the same topology over and over.

Attractor Set Let us find a set of attractors in an N node network starting from m random initial conditions. Implementation is in the following snippet:

```
1 RandomStateGen states(N, m);
2 BooleanNetwork bn = // read BN from file
3 BooleanDynamics* dyn = synchronous_dynamics(bn); // get an
      updater
4 NaiveFinder finder(dyn);
5 AttractorRange ar = states | finder;
6 std::set<Attractor> attractors(ar.begin(), ar.end());
```

where `RandomStateGen` is a *range*[3] that yields m random Boolean vectors and `NaiveFinder` is a function object that implements the naïve cycle detection algorithm. In Line 5 we show how new ranges can be constructed by using the pipe operator and, in Line 6, we demonstrate how ranges interoperate with STL classes. We also notice that `Attractor` objects provide a comparison operator, so they can be inserted into sets.

Basins of Attraction A small extension to the previous use case is to enumerate the whole configuration space of a network and count the multiplicity of each attractor[4], so that we obtain their basin size. To do so, we use a `std::map` to count the occurrences and a `StateEnumerator` object, a range, like `RandomStateGen`, which yields all state vectors of a certain dimension N.

```
1    StateEnumerator allStates(N);
2    BooleanNetwork bn = // read BN from file
3    BooleanDynamics* dyn = synchronous_dynamics(bn); // get an
         updater
4    NaiveFinder finder(dyn);
5    AttractorRange ar = allStates | finder;
6    std::map<Attractor,int> m;
7    for(AttractorRange::iterator it = ar.begin(); it != ar.end
         (); ++it) {
8      if(m.count(*it) == 0) // new attractor
9        m[*it] = 1; // also inserts attractor into m
10     else
11       m[*it] += 1; // increment occurrences
12   }
```

where the `C++` programmer can recognise the familiar STL iteration idiom. We also recall that in the first case of the if statement (Line 9), the attractor pointed by iterator `it` is also inserted into the map. In the end, map `m` contains, for each attractor, the number of states in its basin.

[3]A range is a stream encapsulated in a `C++` object

[4]Of course, such experiment is feasible only for small networks.

Derrida Plot A useful mathematical tool to analyse BN dynamics is the Derrida plot [9], which is a graphical way to visualise the sensitivity of a network to perturbations. The brute-force computation of a Derrida plot is, of course, prohibitive for all but the smallest networks. Nevertheless, we can resort to approximation and, for each data point in the plot, take the average on a sample of network states. Let us show how we can employ the BNTK to calculate the data points in a Derrida plot, supposing that we set the number of sample for each data point to M.

```
1  BooleanNetwork bn = // read BN from file
2  BooleanDynamics* dyn = synchronous_dynamics(bn); // get an
       updater
3  std::vector<double> derrida(N);
4  for (int x = 0; x < N; ++x) {
5    std::vector<int> distances(M);
6    for (int i = 0; i < M; ++i) {
7      State s = random_state(N);
8      State sPrime = random_flips(s, x); // x random flips
9      State s1 = (*dyn)(s);
10     State sPrime1 = (*dyn)(sPrime);
11     distances[i] = hamming_distance(s1, sPrime1);
12   }
13   derrida[x] = average(distances);
14 }
```

Aside from `average`, all other functions are implemented in the toolkit; specifically, `random_state` generates a state vector whose element have value 1 with probability 0.5 while `random_flips` returns a state with x randomly chosen bits flipped (`hamming_distance` has intuitive meaning). This shows how the toolkit can be easily extended with new experiments. As a matter of fact, all use cases presented in this section are already implemented in the toolkit by ready-to-use functions.

30.4 Ongoing Work

The BNTK can be downloaded from: http://booleannetwork.sourceforge.net. The tool is actively used and maintained and important extensions are also planned. Having encapsulated network dynamics in a class allows us to freely define different update schemes without modifying other code. It is sufficient to implement the interface `BooleanDynamics` for all update schemes desired. Although not currently available, implementations of different update strategies, such as asynchronous and random, are on their way.

Another extension regards the integration of other network types, in particular, we plan on adding two alternative BN models, namely, Boolean Threshold Networks [10] and Glass Networks [11].

References

1. Kauffman S (1993) The origins of order: self-organization and selection in evolution. Oxford University Press, London
2. Kauffman S (2004) A proposal for using the ensemble approach to understand genetic regulatory networks. J Theor Biol 230:581–590
3. Benedettini S, Roli A, Serra R, Villani M (2011) Stochastic local search to automatically design Boolean networks with maximally distant attractors. In: Di Chio C, Cagnoni S, Cotta C, Ebner M, Ekárt A, Esparcia-Alcázar A, Merelo J, Neri F, Preuss M, Richter H, Togelius J, Yannakakis G (eds) Applications of evolutionary computation. Lecture notes in computer science. Springer, Heidelberg, pp 22–31
4. Benedettini S The Boolean network toolkit. http://sourceforge.net/projects/booleannetwork/
5. Serra R, Villani M, Barbieri A, Kauffman S, Colacci A (2010) On the dynamics of random Boolean networks subject to noise: attractors, ergodic sets and cell types. J Theor Biol 265(2):185–193
6. Siek JG, Lee LQ, Lumsdaine A (2002) The Boost graph library: user guide and reference manual. Addison-Wesley, Reading
7. Brent RP (1980) An improved Monte Carlo factorization algorithm. BIT Numer Math 20:176–184
8. Knuth D (1998) The art of computer programming. Volume 2: Seminumerical algorithms, 3rd edn. Pearson Education, Upper Saddle River
9. Derrida B, Pomeau Y (1986) Random networks of automata: a simple annealed approximation. Europhys Lett 1(2):45–49
10. Heckel R, Schober S, Bossert M (2010) On random boolean threshold networks. In: International ITG conference on source and channel coding (SCC), pp 1–6
11. Kappler K, Edwards R, Glass L (2003) Dynamics in high-dimensional model gene networks. Signal Process 83:789–798

Chapter 31
Inferring Information Across Scales in Acquired Complex Signals

Suman Kumar Maji, Oriol Pont, Hussein Yahia, and Joel Sudre

Abstract Transmission of information across the scales of a complex signal has some interesting potential, notably in the derivation of sub-pixel information, cross-scale inference and data fusion. It follows the structure of complex signals themselves, when they are considered as acquisitions of complex systems. In this work we contemplate the problem of cross-scale information inference through the determination of appropriate multiscale decomposition. Our goal is to derive a generic methodology that can be applied to propagate information across the scales in a wide variety of complex signals. Consequently, we first focus on the determination of appropriate multiscale characteristics, and we show that singularity exponents computed in microcanonical formulations are much better candidates for the characterization of transitions in complex signals: they outperform the classical "linear filtering" approach of the state-of-the-art edge detectors (for the case of 2D signals). This is a fundamental topic as edges are usually considered as important multiscale features in an image. The comparison is done within the formalism of reconstructible systems. Critical exponents, naturally associated to phase transitions and used in complex systems methods in the framework of criticality are key notions in Statistical Physics that can lead to the complete determination of the geometrical cascade properties in complex signals. We study optimal multiresolution analysis associated to critical exponents through the concept of "optimal wavelet". We demonstrate the usefulness of multiresolution analysis associated to critical exponents in two decisive examples: the reconstruction of perturbated optical phase in

S.K. Maji (✉) · O. Pont · H. Yahia
Geostat Team, INRIA Bordeaux Sud-Ouest, Bordeaux, France
e-mail: suman-kumar.maji@inria.fr

O. Pont
e-mail: oriol.pont@inria.fr

H. Yahia
e-mail: hussein.yahia@inria.fr

J. Sudre
Dynbio Team, LEGOS, UMR CNRS 5556, Toulouse, France
e-mail: joel.sudre@legos.obs-mip.fr

T. Gilbert et al. (eds.), *Proceedings of the European Conference on Complex Systems 2012*, Springer Proceedings in Complexity, DOI 10.1007/978-3-319-00395-5_31,

Adaptive Optics (AO) and the generation of high resolution ocean dynamics from low resolution altimetry data.

31.1 Introduction

Most real-world signals are complex signals, usually difficult to describe but possessing a high degree of redundancy [1]. In particular, in the case of Fully Developed Turbulence (FDT), there is a relation between the spectrum of singularity exponents associated to structure functions and the existence of a multiscale hierarchy [2]. Turbulent flows, although chaotic in nature, possess a complex arrangement of geometrical structures related to the cascading properties of physical variables [3]. The same type of conclusion can be inferred from multiscale analysis of most natural complex signals [4]. As a consequence, the paradigm of understanding natural signals as acquisitions of complex systems with unknown phase space is a useful one [5]. The properties of physical cascading variables reflect the transfer of energy, or more generally information, taking place from larger scales to smaller ones. Recent developments in microcanonical framework for the computation of *singularity exponents* and the derivation of singularity spectra have lead to a sensible improvement in the numerical techniques for the determination of multiscale characteristics of real signals [6, 7]. Experimental analysis on different real world signals, ranging from stock market time series to atmospheric perturbated optical path shows that these systems are not only found to be multiscale, but their singularity spectra are also coincident. Consequently, the precise numerical computation of geometrically localized singularity exponents in single acquisitions of complex systems, without the averages taken on grand ensembles, unveils the determination of their universality class [6]. The statistical characteristics of information in these signals can be described from the localisation and precise value of singularity exponents. As a consequence, it should be possible to transfer across the scales extra physical information from lower scale to higher resolution, a procedure which unveils considerable enhancements of high resolution mapping of natural phenomena.

In this paper, we demonstrate that microcanonical formulations for understanding and evaluating the mechanisms that govern the evolution of dynamical systems lead to accurate inference schemes across the scales in complex signals. We show that the singularity exponents can be used in multiresolution analysis for accurate inference of information across the scales. The profound reason for this fact comes from the observation that geometrically localized singularity exponents encode transitions in complex signals in a much more accurate manner than done with linear filtering processing techniques [8, 9], as will be demonstrated in this work, in particular in the case of 2D images and the accurate determination of edges (which are typically multiscale characteristics on an image). Consequently, we study the notion of optimal wavelet for inferring information across the scales. Our fundamental contribution in this work is to show that multiresolution analysis associated to geometrically localized singularity exponents is a very good candidate for inferring information across the scales in complex signals. We take two specific examples: the reconstruction of the optical phase shift perturbated by atmospheric turbulence (Adaptive Optics) and the high resolution mapping of ocean dynamics using sea surface temperature maps. In the first example, we derive a radically new and non-linear approach for reconstructing the perturbated optical phase; while in the second, we show that oceanic dynamical information acquired at low resolution (pixel size: 22 km s) from altimetry can be transferred across the scales at high resolution sea surface temperature data (pixel size: 4 km s) to produce high resolution mapping of oceanic currents. In both the cases we use a proper wavelet decomposition technique on the signal of the singularity exponents to help us inferring information along the scales of the signal.

The paper is organised as follows: in Sect. 31.2 we present a brief discussion on the evolution of the theory of singularity exponents, in Sect. 31.3 we present the numerical analysis for the singularity exponents and the idea of the most informative set within a signal. Theory behind the reconstruction of the whole signal from the most informative set is explained in Sect. 31.4. Notion of optimal wavelet, for inferring information pointwise in a cascade, is introduced in Sect. 31.5. The experimental data used is discussed in Sect. 31.6 and the results are shown in Sect. 31.7. Finally, we conclude in Sect. 31.8.

31.2 Universality Class and Multiscale Organisation

A power-law behaviour in the thermodynamical variables, and also time and spatial correlation functions, is commonly observed in systems with high order transitions. The underlying dynamics of such systems can be observed, at the macroscopic scale, in the form of a power-law [10]. It was soon realized that the exponents of the power laws define different classes: systems characterized by same values of singularity exponents belong to the same universality class, which implies the presence of a common macroscopic behaviour independent of the microscopic dynamics of each system [11]. Different singularity spectra of very different physical systems can match a same curve. Such a correspondance can be explained by the existence of a common underlying dynamical system, the universality class, responsible for similar statistical properties of information at macroscopic scale [11].

Previous works attempt to relate the general organisation of a multiscale structure with the existence of cascade process [10]. In these works, a multiscale signal s is characterized by the power-law scaling in the order p moments of some related variable $\mathbb{T}_r$, in the way:

$$\left\langle |\mathbb{T}_r s|^p \right\rangle = A_p r^{\tau_p} + o\left(r^{\tau_p}\right) \quad (r \rightarrow 0) \tag{31.1}$$

The existence of multplicative cascade process was first justified by Kolmogorov in his theory on turbulence [3]. Kolmogorov proposed the following: given two scales r and L, $0 < r < L$, we can characterize the distribution of the velocity field by an injection parameter $\eta_{r/L}$ as:

$$|\mathbb{T}_r s| \doteq \eta_{r/L} |\mathbb{T}_L s| \tag{31.2}$$

where the symbol '$\doteq$' means that both sides are equally distributed and $\eta_{r/L} = [r/L]^{\alpha}$. From this relation, the order p moments have the following relation:

$$\langle |\mathbb{T}_r s|^p \rangle = [r/L]^{\alpha p} \langle |\mathbb{T}_L s|^p \rangle \tag{31.3}$$

Comparing Eqs. (31.1) and (31.3) we get, $\tau_p = \alpha p$. However, experiments show that in the case of FDT, the relationship between τ_p and p is not linear rather it is a convex bell-shaped curve, a condition known as 'anomalous scaling' [7]. To apply Kolmogorov's decomposition in anomalous scaling, certain assumptions have to be made:

- $\eta_{r/L}$ has to be interpreted as a random variable, independent of L.
- The variables $\eta_{r/L}$ has to be infinitely divisible to ensure downward process from scale L to r is verified directly or in several stages giving rise to the famous cascade process.

It has been verified [3] that an injection mechanism as the one proposed by Kolmogorov leads to the understanding of a underlying geometrical structure in a multiplicative cascade process, together with the knowledge of the exponents τ_p for inferring information along the scales of the signal. This experimental outcome of self-similarity led researchers to propose a different model for the generation of exponents.

31.3 Singularity Analysis in the Microcanonical Framework

Criticality, and the associated *critical exponents* are key notions in Statistical Physics to understand phase transitions, which are prototypes of scale invariant phenomena [10]. The spectrum of singularity exponents in a system determine its multiscale properties which are accessible statistically. We will say that a signal s is multiscale in a microcanonical sense, if for at least one multiscale functional dependant on scale r, it is assumed that for any point $\boldsymbol{x}$ the following equation holds:

$$\mathbb{T}_r s(\boldsymbol{x}) = \alpha(\boldsymbol{x}) r^{h(\boldsymbol{x})} + o\big(r^{h(\boldsymbol{x})}\big) \quad (r \to 0) \tag{31.4}$$

for some functions $\alpha(\boldsymbol{x})$ and $h(\boldsymbol{x})$. The exponent $h(\boldsymbol{x})$, which is a function of the point $\boldsymbol{x}$, is called a *singularity exponent* or *Local predictability exponent (LPE)* of the point [7]. The central problem is to compute at high numerical precision the value of $h(\boldsymbol{x})$ at point $\boldsymbol{x}$: bad approximations of singularity exponents lead to poor reconstructions.

31.3.1 Local Predictability Exponents

According to microcanonical formulations [10], a multiscale signal is supposed to satisfy Eq. (31.4) for a family of functionals $\mathbb{T}_r$, at any point $\boldsymbol{x}$ in the signal domain, and have a singularity spectrum computed from singularity exponents as a convex curve function of h [10]. Equation (31.4) is a pointwise and localized version of the definition used in introducing singularity spectrum [6, 12]: we do not make use of statistical averages and grand ensembles as in practice such an ensemble average is not accessible. Rather, we seek to evaluate $h(\boldsymbol{x})$ at point $\boldsymbol{x}$. We denote $\mathcal{F}_h$ the component in the signal's domain associated to singularity exponent value h as:

$$\mathcal{F}_h = \{\boldsymbol{x} : h(\boldsymbol{x}) = h\} \tag{31.5}$$

In other words, we can say that each point $\boldsymbol{x}$ in the signal is characterized by a singularity exponent $h(\boldsymbol{x})$ which is typical to one component $\mathcal{F}_h$, i.e., the components are level sets of the function $h(\boldsymbol{x})$. This family of sets is naturally associated to the multiscale hierarchy in a signal and in the case of natural images, it is expected that there exists a particular set which comprises the point where sharp transitions within the signal are well recorded. We will call this set as the *Most Singular Manifold* or MSM. Geometrically speaking, it is the singularity component associated with the smallest possible value h_∞ , finite for signals corresponding to physical variables that cannot diverge. We will denote this set by $\mathcal{F}_\infty$ and can be expressed as:

$$\mathcal{F}_\infty = \{\boldsymbol{x} : h(\boldsymbol{x}) = h_\infty = \min(h(\boldsymbol{x}))\} \tag{31.6}$$

The MSM plays a fundamental role in the multiscale geometrical hierarchy of natural images. Visual inspection of this set reveals a structure which is characterized by the presence of 'edges' or contours in natural images [1]. It will be understood hereafter that the MSM contains the most informative set in an image so that the whole signal can be reconstructed from the restriction of its gradient to the MSM. Moreover we will see that singularity exponents lead to a notion of edge that matches much better across the scales than the edges computed from classical filtering methods. Before we go deeper into the subject of MSM and its application to reconstructible systems, we give a brief overview for the determination of the singularity exponents.

31.3.2 Singularity Analysis

The singularity exponents for experimental, discretized data can be calculated using different methods [7], but for our case we will use the *Unpredictable Points Manifold* (herein referred to as UPM) [13, 14]. According to this method, we make point estimates of the singularity exponent, namely:

$$h(\boldsymbol{x}) = \frac{\log(\tau_\Psi s(\boldsymbol{x}, r_0))/\langle \tau_\Psi s(., r_0)\rangle}{\log r_0} + o\left(\frac{1}{\log r_0}\right) \tag{31.7}$$

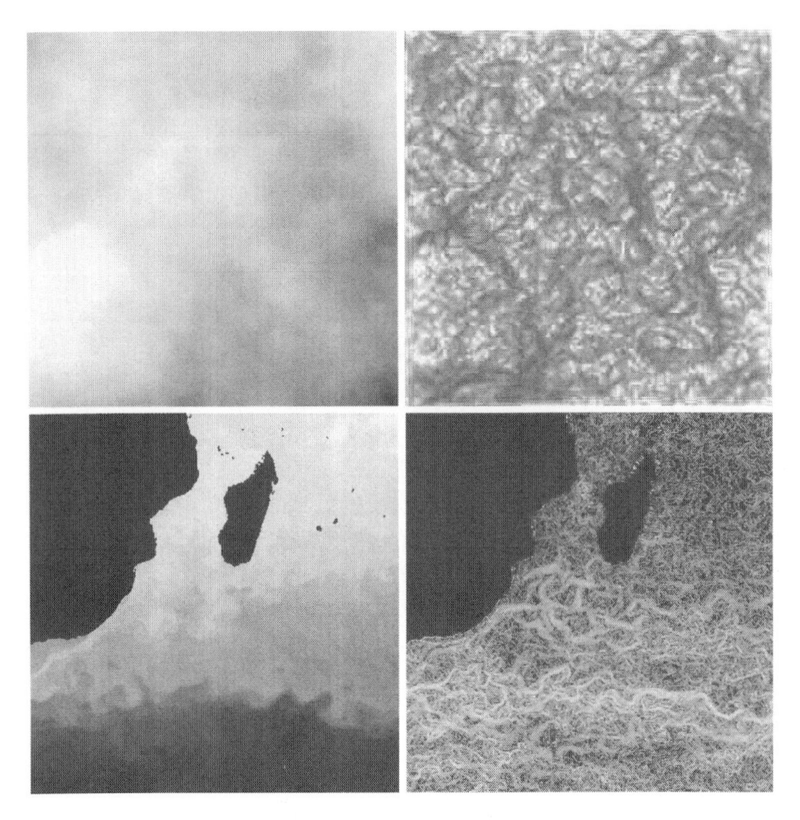

Fig. 31.1 *Top left*: Image of a simulated optical phase perturbated by atmospheric turbulence. The image corresponds to a 128 × 128 pixels sub-image extracted from an original 256 × 256 pixels image to avoid the pupil boundary. *Top right*: Image of the singularity exponents computed from the phase data. *Bottom left*: Excerpt of the Agulhas current below the coast of South Africa (sea surface temperature image: each pixel record the temperature of the upper layer of the sea). *Bottom right*: Singularity exponents of the Agulhas current

where $\langle \tau_\Psi s(., r_0) \rangle$ is the average value of the wavelet projection over the whole signal and $o(\frac{1}{\log r_0})$ is a diminishing quantity and r_0 is the minimum scale. If the signal s is an image of size $M \times N$, then we choose $r_0 = 1/\sqrt{M \times N}$. The singularity exponents computed on our experimental dataset are shown in Fig. 31.1.

The values $h(\boldsymbol{x})$ are computed for all points $\boldsymbol{x}$ within the signal domain. Now, coming back to MSM, sorting of these singularity exponent values based on a typical threshold value 0 defines the standard reconstruction set in the MSM method. Such a set often provides a robust and accurate reconstruction and is defined by:

$$\mathcal{E}_{\text{MSM}} = \bigcup_{h_\infty \le h \le 0} \mathcal{F}_h \tag{31.8}$$

31.4 Reconstructible Systems

In this section, we are led to find mathematically a functional $\mathcal{G}$ which permits the reconstruction of the signal's gradient from its restriction to the MSM. The functional must satisfy the properties of being deterministic, linear, translationally-invariant, isotropic and yield correct power spectrum of natural images. We consider the gradient measure of the signal $s = \nabla s(\boldsymbol{x})$ and integrate it over the multifractal set of most unpredictable points $\mathcal{F}_\infty$. A deterministic representation of the gradient measure for the signal can be:

$$\nabla s(\boldsymbol{x}) = \mathcal{G}(\nabla s|_{\mathcal{F}_\infty})(\boldsymbol{x}) \tag{31.9}$$

Considering the fact of $\mathcal{G}$ being linear, an integral representation can be given by:

$$\nabla s(\boldsymbol{x}) = \int_{\mathcal{F}_\infty} \nabla s(\boldsymbol{y}) G(\boldsymbol{x}, \boldsymbol{y}) \mathrm{d}\boldsymbol{y} \tag{31.10}$$

where $G(\boldsymbol{x}, \boldsymbol{y})$ is a density measure of the function $\mathcal{G}$ and is a 2×2 matrix. Using isotropy, standard power spectrum (in the form $1/\|\boldsymbol{f}\|^2$) for the associated spectral measures of natural images, one obtains the following formula [1, 10] expressed in Fourier space:

$$\hat{s}(\boldsymbol{f}) = \frac{\boldsymbol{f}.\widehat{\nabla s|_{\mathcal{F}_\infty}}(\boldsymbol{f})}{i\|\boldsymbol{f}\|^2} \tag{31.11}$$

where i is the imaginary unit, $i \equiv \sqrt{-1}$ and $\widehat{\ }$ denotes the Fourier transform. We normalize the result by taking the vector field $\boldsymbol{v_0}$ unitary and normal to the MSM instead of $\nabla s|_{\mathcal{F}_\infty}$; where $\boldsymbol{v_0}(\boldsymbol{x}) = \nabla s(\boldsymbol{x})\delta_{\mathcal{F}_\infty}$, $\delta_{\mathcal{F}_\infty}$ being the density of the proper Hausdorff measure restricted to the set $\mathcal{F}_\infty$ [1]. We therefore perform integration over all the space (the restriction is still present, but now introduced by $\boldsymbol{v_0}$):

$$\hat{s}(\boldsymbol{f}) = \frac{\boldsymbol{f}.\widehat{\boldsymbol{v_0}|_{\mathcal{E}}}(\boldsymbol{f})}{i\|\boldsymbol{f}\|^2} \tag{31.12}$$

Fourier inversion of this formula gives the reconstruction of the image from the restriction of the gradient field to the MSM. Results of reconstruction on the MSM of experimental datasets, and their performance over classical edge detection algorithms, are shown in Tables 31.2 and 31.1. It is seen that in the case of acquisitions of turbulent signals, the reconstruction based on the MSM (we call it MSM in Tables 31.2 and 31.1) performs significantly better among the algorithms tested. In fact, when it comes to the case of turbulent signals, the classical edge detectors like Sobel [15], Prewitt [16], Roberts [17], Laplacian of Gaussian (LoG) [18, 19], Zerocross [20, 21], Canny [8] to a more recent non-linear approach called NLFS [22], dedicated to the computation of edges in digital images, are systematically outperformed by MSM in terms of reconstruction from a compact representation of its edge pixels. As a consequence, the fundamental notion of edge, which is a basic multiscale feature, is much more well encoded by the set $\mathcal{E}_{\mathrm{MSM}}$ defined in Eq. (31.8). This tends to show that singularity exponents are good candidates for an accurate multiresolution analysis. In the next section, we develop the notion of optimal wavelet.

31.5 Inferring Information Across the Scales Using Microcanonical Analysis

To infer the cascading properties pointwise (called microcanonical cascade) we introduce the concept of optimal wavelet. Let $s(\boldsymbol{x})$ be a multiscale signal and let $\Psi(\boldsymbol{x})$ be a wavelet. We define the wavelet projection of s on Ψ at position $\boldsymbol{x}$ and resolution r as:

$$T_{\Psi}|\nabla s|(\boldsymbol{x}, r) = \int |\nabla s|(\boldsymbol{y})\Psi\left(\frac{\boldsymbol{x}-\boldsymbol{y}}{r}\right)d\boldsymbol{y} \tag{31.13}$$

We can now define a random variable $\zeta_{r/L}(\boldsymbol{x})$ as

$$T_{\Psi}|\nabla s|(\boldsymbol{x}, r) = \zeta_{r/L}(\boldsymbol{x})T_{\Psi}|\nabla s|(\boldsymbol{x}, L) \tag{31.14}$$

Now, we can talk about a wavelet Ψ which, if determined, will make $\zeta_{r/L}(\boldsymbol{x})$ independent of $T_{\Psi}|\nabla s|(\boldsymbol{x}, L)$. Such a wavelet is called an *optimal wavelet*. In subsection *Optimal Wavelet Analysis*, we propose a new algorithm for a very robust detection of the optimal wavelet in 2D signals. The new methodology helps us to detect the presence of an optimal wavelet, in a totally unconstrained way, from the signal itself. Once determined, the optimal wavelet has the potential of unlocking the signal's microcanonical cascading properties through simple wavelet multiresolution analysis [23].

31.5.1 Multiresolution Analysis & Fast Wavelet Transform

Multiresolution analysis is a mathematical formalism that deals with the phenomenon of detail-structured viewing of objects [23]. Data redundancy is minimized by use of dyadic wavelet sequences which are Hilbertian frames associated to dyadic partition of the space/frequency domain.

Any signal $|s\rangle$ can be represented in a dyadic wavelet basis of mother wavelet $|\Psi\rangle$ [24] as follows (from now on we use the notation $|s\rangle$ for the signal):

$$|s\rangle = \sum_{j=-\infty}^{\infty}\sum_{k}\alpha_{j,k}|\Psi_{j,k}\rangle \tag{31.15}$$

where

$$|\Psi_{j,k}\rangle(\boldsymbol{x}) = 2^{j/2}\Psi\left(2^{j}\boldsymbol{x}-k\right) \tag{31.16}$$

and $\alpha_{j,k}$, are called *wavelet coefficients*. The wavelet coefficients $\alpha_{j,k}$ can be obtained by a simple projection of the signal $|s\rangle$ onto the basis function $\Psi_{j,k}$, namely:

$$\alpha_{j,k} = \langle s|\Psi_{j,k}\rangle \tag{31.17}$$

The decomposition process using multiresolution analysis gives rise to an image fourth smaller than the previous one. Therefore, each *parent coefficient* $\alpha_p \equiv \alpha_{j-1,\lfloor k/2 \rfloor}$, at the coarser scale $j-1$, covers the same spatial extent of four *children coefficients* $\alpha_c \equiv \alpha_{j,k}$ at the finer scale j.

31.5.2 Approximation of Microcanonical Cascade

In the wavelet analysis of 2D signals, persistence along the scales implies a relation of the form between the wavelet coefficients:

$$\alpha_c = \eta_1 \alpha_p + \eta_2 \tag{31.18}$$

with η_1, η_2: random variables independant of α_c and α_p and also independant of each other. For an optimal wavelet the above equation takes the form $\alpha_c = \eta_1 \alpha_p$.

31.5.3 Optimal Wavelet Analysis

Any given signal $|s\rangle$ can be represented in terms of their cascade variables η and wavelet coefficients α as:

$$|s\rangle = \sum_{j \neq 0,k} \prod_{j',k} \eta_{j',\lfloor k/2^{j-j'} \rfloor} \alpha_{0,0} |\Phi_{j,k}\rangle + \alpha_{0,0} |\Phi_{0,0}\rangle \tag{31.19}$$

where $\Phi_{j,k}$ is the wavelet basis for the optimal wavelet. Experimental observations show that the expectation of the signal $\langle |s\rangle \rangle = 0$ as $\langle \alpha_{0,0} \rangle = 0$ due to symmetry.

We multiply the sign of $\alpha_{0,0}$ i,e, $\sigma_{0,0}$ with the signal s and then compute the average. Since, in our case we have an ensemble of dynamically equivalent signals, we compute the average over $|s^i\rangle$ to get the expected value for all these signals; where i is the index of an ordering of the signals. Equation (31.19) can be generalized to:

$$\left\langle \sigma^i_{0,0} |s^i\rangle \right\rangle = \langle |\alpha_{0,0}| \rangle |\Phi_{0,0}\rangle \tag{31.20}$$

We try to estimate the sign of $\alpha_{0,0}$. Let $\epsilon_{0,0}$ be the estimation, we then have

$$\left\langle \epsilon^i_{0,0} |s^i\rangle \right\rangle = \langle \epsilon_{0,0} \sigma_{0,0} |\alpha_{0,0}| \rangle |\Phi_{0,0}\rangle \propto \Phi_{0,0} \tag{31.21}$$

Principle application of finding the optimal wavelet on a given set of images is quite simple. The procedure is as follows:

- We subdivide a given image $|s^i\rangle$ over small areas of equal sizes and normalize individually every sub-image.
- We find the correlation between the sub-images of $|s^i\rangle$: $C_i = \langle s^i | s^k \rangle$.
- For every i we find the average of the correlation.
- We then find for which l, the average correlation is maximum. Let it be i^*.
- We call $|s^{i^*}\rangle$ as the *most central element* (MCE).

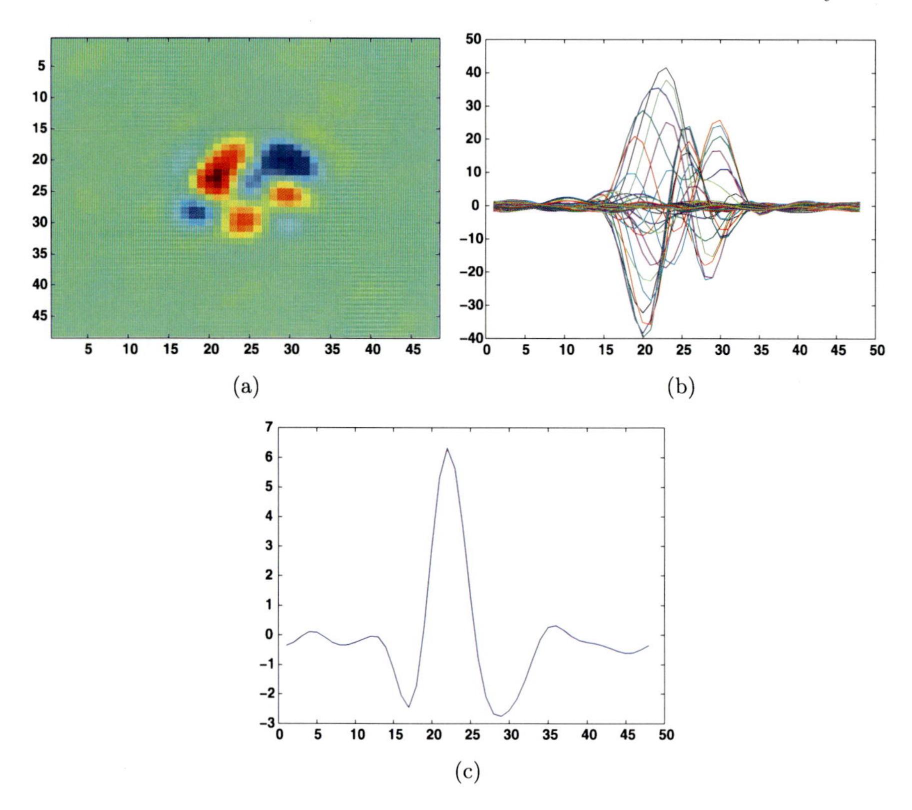

Fig. 31.2 (**a**) Sub-image at 48 × 48 pixels resolution obtained after orientation with the sign with the MCE (**b**) 2-D plot of the sub-image (**c**) 2-D plot for the sum over the columns

Since we don't know $\Phi_{0,0}$, we make the wavelet projection of the signal on the element which has the most dependancy with all the other elements (dominant presence of the term $\langle|\alpha_{0,0}|\rangle|\Phi_{0,0}\rangle$); i,e, the MCE. So, we have

$$\begin{aligned}\epsilon^i_{0,0} &= \sigma(C_{i,i*}) = \sigma\left(\langle s^i|s^{i*}\rangle\right)\\ &= |\alpha^*_{0,0}|\sigma_{0,0}\sigma^*_{0,0}\langle s|\Phi_{0,0}\rangle \end{aligned} \tag{31.22}$$

If we have a correct estimate of the sign, we can say $\langle\epsilon^i_{0,0}|s^i\rangle\rangle \propto \langle\sigma^i_{0,0}|s^i\rangle\rangle$. And we estimate the wavelet from Eq. (31.21). Since wavelets are normalized by definition, we can cancel the proportionality factor in Eq. (31.21):

$$\left\langle\epsilon^i_{0,0}|s^i\rangle\right\rangle = \left\langle\sigma(C_{i,i*})|s^i\rangle\right\rangle = \left\langle\sigma\left(\langle s^i|s^{i*}\rangle\right)|s^i\rangle\right\rangle \tag{31.23}$$

We have tested this algorithm on Benzi model [25] to construct multiaffine fields based on an order 2 Gaussian wavelet decomposition. The preliminary results are shown in Fig. 31.2. Since, the process of finding an optimal wavelet is still under review and subjected to constant experimentation, we approximate the optimal wavelet by a Battle-Lemarié wavelet which is found to give an acceptable approximation of the optimal wavelet.

31.6 DATA

31.6.1 Atmospheric Phase

The data is shown in Fig. 31.1(top). The datasets consists in simulated optical phase perturbated by the Earth's upper layer turbulent atmosphere. The optical phase corresponds to the acquisition of a point source (representing a star far away enough in outer space so that the optical phase reaching the telescope is in the form of planar wavefronts). These data are provided by the French Aerospace Lab-ONERA, and they have the following imaging characteristics:

- diameter of the telescope: 8 m,
- seeing at 5 microns: 0.85 arcseconds,
- wind's speed: 12.5 m/s,
- acquisition frequency: 250 Hz,
- pupil size: 256×256 pixels, but for our experimental purpose we take 128×128 pixels from the centre to eliminate boundary effects.

We have the Hartmann-Shack (HS) acquisition of the x and y slopes for the phase data provided by Onera given by 208 effective HS sub-pupils (size 16×16) which samples the pupil of the telescope. The distribution of the sub-pupils within the telescope is shown in Fig. 31.4(a). Figures 31.4(b) and 31.4(c) shows the x and y low resolution acquisition of the phase data, which gives us an approximation of low resolution x and y components of the gradients for the phase, by the HS sensor.

31.6.2 Sea Surface Temperature

Sea surface temperature data (SST) are global acquisitions of the temperature of the ocean's upper layer. Data is radiometrically corrected so that pixels values represent celsius degrees. Data is acquired by the MODIS instrument orbiting around earth, pixel resolution is 4 kms, data is acquired on 2 August 2007. In our experiment, we also use low resolution products representing geostrophy and Ekman currents deduced from altimetry data according to method exposed in [26]. Pixel size of altimetry products is 22 kms. Figure 31.3(a) shows the altimetry, Fig. 31.3(b) the MODIS data and Fig. 31.3(c) shows the low-resolution motion field derived from altimetry according to [26].

31.7 Results

31.7.1 Edge Detection and Singularity Exponents

First we detail the comparison results on edge detection using classical linear filtering and the set provided by Eq. (31.8). Reconstruction has been performed on the

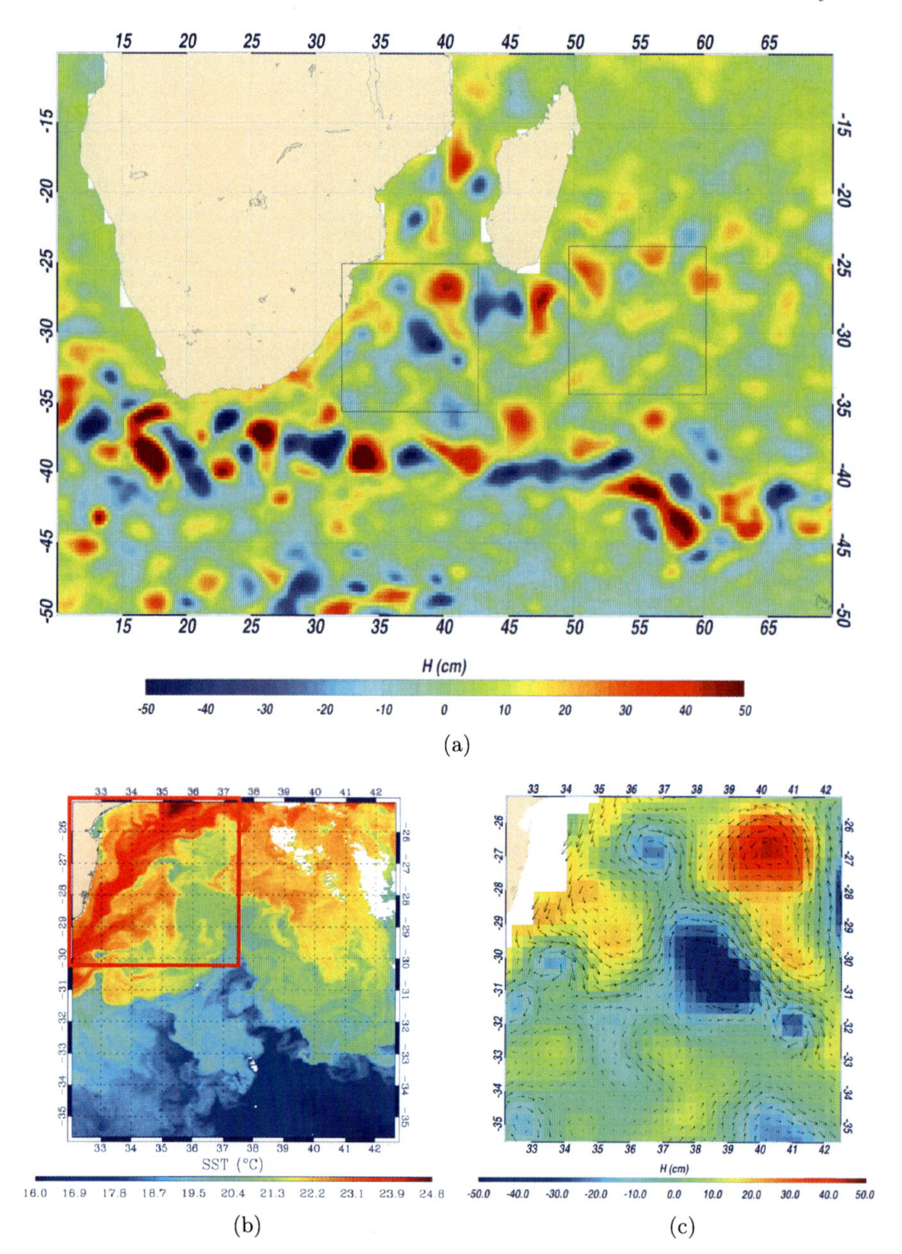

Fig. 31.3 (**a**) Altimetry data (**b**) Sea Surface Temperature (SST) acquired by MODIS satellite on August 2, 2007 (**c**) low-resolution motion field derived from altimetry

edge files computed on the phase and sea surface temperature images. Performance of the reconstruction on classical edge detectors to a more recent nonlinear derivative approach (called NLFS) [22] and MSM has been presented in Table 31.1. Also,

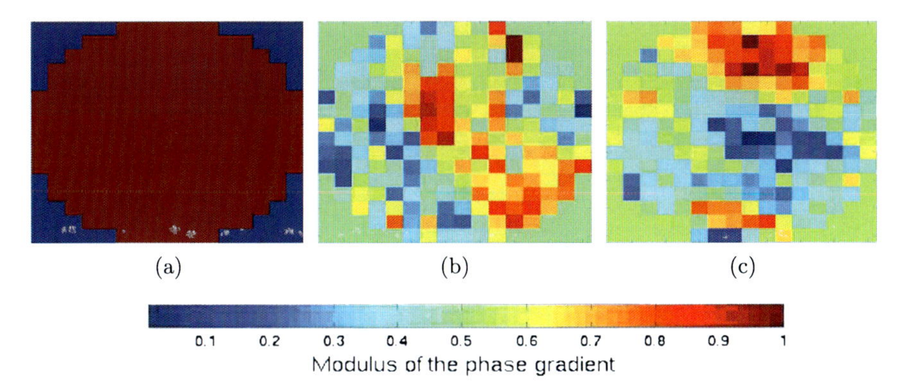

Fig. 31.4 (**a**) Distribution of the sub-pupils within the telescope (**b**) HS acquisition of the x slope for the phase data (**c**) HS acquisition of the y slope for the phase data

we evaluate the quality of the reconstruction using the *peak signal to noise ratio* (psnr, expressed in decibels dB) defined by:

$$\mathrm{psnr} = -20.0 \times \log_{10}\left(\frac{1}{\lambda(\Omega)} \frac{(\int_{\Omega}(s(\boldsymbol{x}) - s_r(\boldsymbol{x}))^2 \mathrm{d}\boldsymbol{x})^{1/2}}{\Delta_s} \right) \tag{31.24}$$

where Ω is the image domain, $\lambda(\Omega)$ its Lebesgue measure (image size), s the original image, s_r the reconstructed image, and Δ_s is the dynamical range of s, i.e. the difference between the maximal and minimal values. Better reconstructions tend to have a higher psnr. A quantitative evaluation of the results are presented in Table 31.2.

31.7.2 Reconstruction of Optical Phase

Also, we show that the application of wavelet multiresolution analysis technique on the signal (optical phase perturbated by atmospheric turbulence) of the singularity exponents provided by a simple approximation of an optimal wavelet (here a third order Battle-Lemarié wavelet) help us to infer information along the scales of the signal which in turn can be used to properly reconstruct the signal from low resolution to high resolution. The process is summarized below:

- We first compute the third order Battle-Lemarié wavelet coefficients associated to the signal of the singularity exponents computed on the perturbated phase signal,
- for each component (x and y) of the phase gradient at low resolution (16×16 sub-image, see Sect. 31.6.1), back project the component from low to high resolution to get a phase's gradient at higher spatial resolution of 128×128.

Consequently, we reconstruct the phase by performing inverse gradient operation on the norms of the gradients. We also check the robustness of our reconstruction

Table 31.1 Performance of different edge detection algorithms. *row 1*: Atmospheric phase data. *row 2*: SST data of the Agulhas current below the coast of South Africa

Original	MSM	NLFS	Canny	LoG	Sobel	Zero-Cross	Roberts	Prewitt

Table 31.2 Evaluation of edge detection algorithms

Image	Algorithm	Parameter(s)	Density	Reconstruction (psnr)
Sea temp.	NLFS [22]	$\sigma = 0.2$	22.24 %	10.15 dB
Sea temp.	Canny	$\sigma = 0.001, \alpha = 01$	13.094 %	9.65 dB
Sea temp.	Laplacian	$\sigma = 0.001, \alpha = 01$	24.47 %	10.16 dB
Sea temp.	Sobel	$\sigma = 0.001$	24.58 %	9.58 dB
Sea temp.	Zero-crossing	$\sigma = 0.001$	13.95 %	9.60 dB
Sea temp.	Roberts	$\sigma = 0.001$	27.97 %	10.22 dB
Sea temp.	Prewitt	$\sigma = 0.001$	24.83 %	9.83 dB
Sea temp.	MSM	parameter free	17.24 %	11.30 dB
Phase	NLFS [22]	$\sigma = 0.25$	24.92 %	8.30 dB
Phase	Canny	$\sigma = 0.001, \alpha = 01$	14.11 %	7.25 dB
Phase	Laplacian	$\sigma = 0.001, \alpha = 01$	28.48 %	7.24 dB
Phase	Sobel	$\sigma = 0.3$	5.83 %	6.48 dB
Phase	Zero-crossing	$\sigma = 0.001$	15.88 %	6.61 dB
Phase	Roberts	$\sigma = 0.001$	34.74 %	7.77 dB
Phase	Prewitt	$\sigma = 0.001$	26.72 %	6.96 dB
Phase	MSM	parameter free	15.75 %	13.18 dB

algorithm by adding different proportions of Gaussian white noise to the data. Results obtained, as shown in Fig. 31.6, show visual resemblance of the reconstructed signal with the original one.

31.7.3 High Resolution Ocean Dynamics

In this experiment, the low resolution vector field shown in Fig. 31.3(c) and derived from altimetry data is used to generate a high resolution vector field corresponding to SST data. First, the singularity exponents are computed on SST data. Then a multiresolution analysis is performed on the resulting singularity exponents from SST spatial resolution (4 km s) down to altimetry resolution (22 km s). The low resolution vector field shown in Fig. 31.3(c) is propagated, componentwise, up to SST resolution and the resulting vector field is prefiltered using an $1/\|f\|$ filter. The results are shown in Fig. 31.5: a high resolution vector field representing the ocean dynamics at resolution 4 km s is obtained from the multiresolution analysis of the singularity exponents. Validation has been performed on the outputs of a 3D simulation model, and shows proper reconstruction of the high resolution vector field both in norm and direction: 80 % of vectors are correctly computed. This method provides a very interesting alternative to classical motion computation techniques that use conservation hypothesis (optical flow) or Maximum Correlation methods.

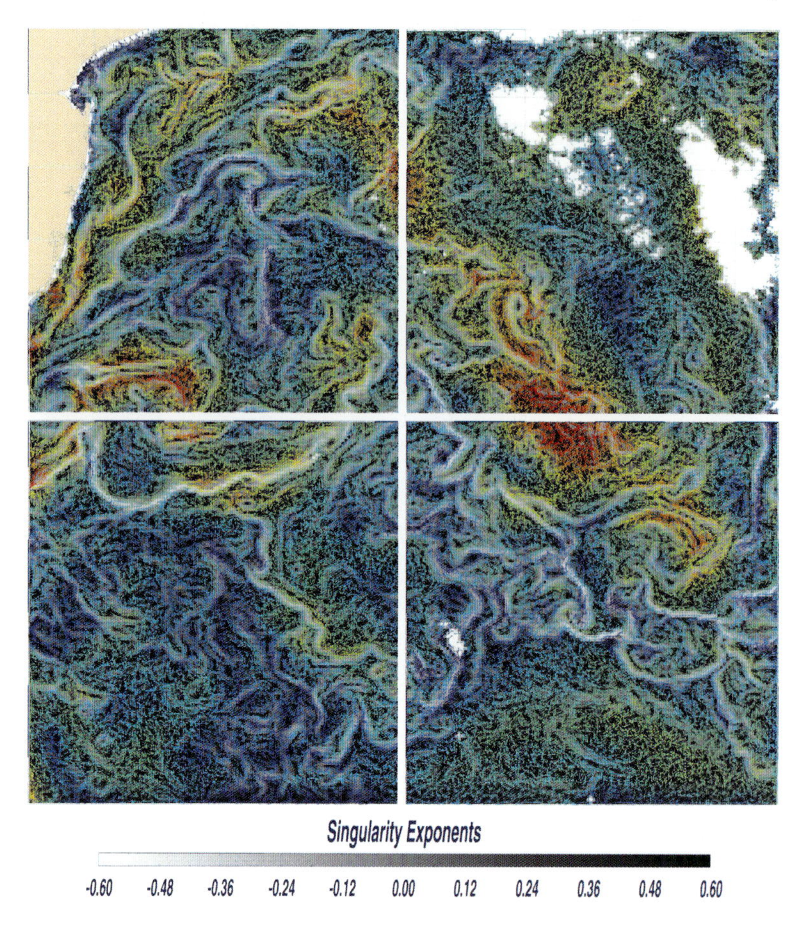

Fig. 31.5 Vector field computed at high resolution SST MODIS data using the low resolution altimetry of Fig. 31.3(c) and the multiresolution analysis of the SST singularity exponents as explained in Sect. 31.7. The color of the vectors indicate their norm from 0.0 $\mathrm{cm\,s^{-1}}$ (*blue*) to 83.9 $\mathrm{cm\,s^{-1}}$ (*red*). In the background we also display the singularity exponents

31.8 Conclusion

In this work we set up and study a multiresolution analysis scheme general enough to suit the case of acquisitions of general complex systems. We first study geometrically localized singularity exponents in natural signals, computed in a microcanonical framework, from which singularity spectra can be derived. We study their relations with high order transitions in associated phase spaces, and conclude that they unlock a notion of transition that outperforms all classical "linear filtering" approaches for edge detection in the case of 2D images. Edges are typical multiscale features, which should maximize information content in natural signals. We study the performance of reconstructible systems both with transitions associated to singularity exponents and the edge pixels provided by standard edge detection tech-

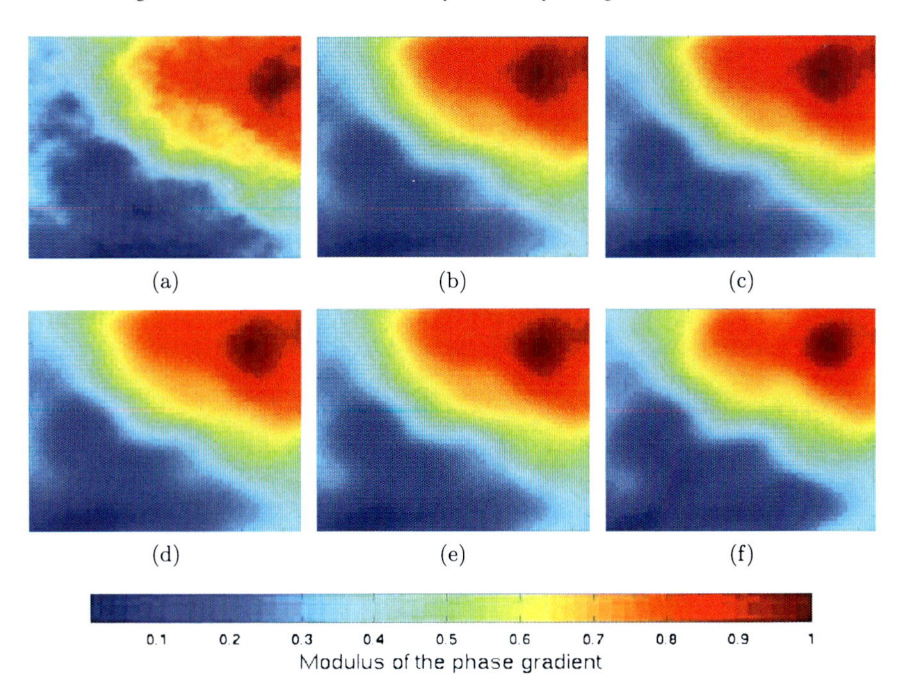

Fig. 31.6 (**a**) The original phase. Reconstructed phase (**b**) without noise. (**c**) with an input SNR of 40 dB. (**d**) with an input SNR of 26 dB. (**e**) with an input SNR of 14 dB. (**f**) with an input SNR of 6 dB

niques. Examples are chosen among the most difficult natural signals: acquisition of turbulent phenomena (perturbated optical phase and ocean dynamics acquired from space). We study a multiresolution analysis scheme associated to the signal of singularity exponents, and in doing so we provide an effective determination of optimal wavelets, which are wavelets whose associated multiresolution analysis is optimal w.r.t inference across the scales. We show the power of the approach by studying two specific examples: the reconstruction of the phase perturbated by atmospheric turbulence applied to adaptive optics and the generation of high resolution ocean dynamics from low resolution acquired altimetry signals. The method is general enough to provide an effective approach to infer sub-pixel information in most natural complex signals.

Acknowledgements Suman Kumar Maji's Ph.D. is funded by a CORDIS grant and Région Aquitaine OPTAD research project grant.

References

1. Turiel A, del Pozo A (2002) Reconstructing images from their most singular fractal manifold. IEEE Trans Image Process 11:345–350

2. Parisi G, Frisch U (1985) On the singularity structure of fully developed turbulence and predictability in geophysical fluid dynamics. In: Ghil M, Benzi R, Parisi G (eds) Proc intl school of physics E Fermi. North-Holland, Amsterdam, pp 84–87
3. Frisch U (1995) Turbulence. Cambridge University Press, Cambridge
4. Turiel A, Parga N (2000) The multi-fractal structure of contrast changes in natural images: from sharp edges to textures. Neural Comput 12:763–793
5. Boffetta G, Cencini M, Falcioni M et al (2002) Predictability: a way to characterize complexity. Phys Rep 356(6):367–474
6. Pont O, Turiel A, Perez-Vicente C (2009) Empirical evidences of a common multifractal signature in economic, biological and physical systems. Physica A 388:2025–2035
7. Turiel A, Perez-Vicente C, Grazzini J (2006) Numerical methods for the estimation of multifractal singularity spectra on sampled data: a comparative study. J Comput Phys 216:362–390
8. Canny J (1986) A computational approach to edge detection. IEEE Trans Pattern Anal Mach Intell 8:679–698
9. Faugeras O (1993) Three-dimensional computer vision: a geometric viewpoint. MIT Press, Cambridge. ISBN 0-262-06158-9
10. Turiel A, Yahia H, Perez-Vicente C (2008) Microcanonical multifractal formalism: a geometrical approach to multifractal systems. Part I: singularity analysis. J Phys A, Math Theor 41:015501. doi:10.1088/1751-8113/41/1/015501
11. Lovejoy S, Schertzer D (1990) Multifractals, universality classes, satellite and radar measurements of clouds and rain. J Geophys Res 95:2021–2034
12. Arneodo A, Bacry E, Muzy J (1995) The thermodynamics of fractals revisited with wavelets. Physica A 213:232–275
13. Pont O, Turiel A, Perez-Vicente C (2009) Description, modelling and forecasting of data with optimal wavelets. J Econ Interact Coord 4:39–54. doi:10.1007/s11403-009-0046-x
14. Pont O, Turiel A, Yahia H (2011) An optimized algorithm for the evaluation of local singularity exponents in digital signals. In: IWCIA, vol 6636, pp 346–357
15. Sobel I (1978) Neighbourhood coding of binary images fast contour following and general array binary processing. Comput Graph Image Process 8:127–135
16. Prewitt J (1970) Object enhancement and extraction. In: Picture process psychopict, pp 75–149
17. Roberts LG (1965) Machine perception of three dimensional solids. In: Tippett JT et al (eds) Optical and electro-optical information processing. MIT Press, Cambridge
18. Rosenfeld A (1969) Picture processing by computer. Academic Press, New York
19. Marr D, Hildreth E (1980) Theory of edge detection. Proc R Soc Lond B, Biol Sci 207:187–217
20. Haralick RM (1984) Digital step edges from zero crossing of second directional derivatives. IEEE Trans Pattern Anal Mach Intell 6:58–68
21. Torre V, Poggio TA (1986) On edge detection. IEEE Trans Pattern Anal Mach Intell 8:147–163
22. Laligant O, Truchetet F (2010) A nonlinear derivative scheme applied to edge detection. IEEE Trans Pattern Anal Mach Intell 32:242–257
23. Pottier C, Turiel A, Garçon V (2008) Inferring missing data in satellite chlorophyll maps using turbulent cascading. Remote Sens Environ 112:4242–4260
24. Mallat S (1999) A wavelet tour of signal processing, 2nd edn. Academic Press, San Diego
25. Benzi R, Biferale L, Crisanti A, Paladin G, Vergassola M, Vulpiani A (1993) A random process for the construction of multiaffine fields. Physica D 65(4):352–358
26. Sudre J, Morrow R (2008) Global surface currents, a high resolution product for investigating ocean dynamics. Ocean Dyn 58:101–118. doi:10.1007/s10236-008-0134-9

Chapter 32
On the α-Shiner–Davison–Landsberg Complexity Measure

Thomas L. Toulias and Christos P. Kitsos

Abstract Shannon entropy is essential for the study of complex continuous systems as it forms various complexity measures. Shiner–Davison–Landsberg (SDL) complexity is such a measure used for the characterization of complex bio-systems, especially in the study of the EEG signals on epileptic seizures. We consider a continuous system whose various states can be described by a wide range of distributions provided by the family of the γ-ordered Normal distributions. Moreover, for the construction of the SDL measure of complexity we consider the generalized Shannon entropy derived via the generalized Fisher's entropy type measure of information J_α. The obtained α-SDL complexity is evaluated and studied with regards to the absolute complexity state, which is important in bio-systems.

Keywords Fisher's information measure · Shannon entropy · SDL complexity measure · γ-Ordered normal distribution

32.1 Introduction

The role of complexity, in the behavioral study of bio-systems, attracts increasingly attention in applications, [6, 8]. For the study of EEG signals on epileptic seizures, through complexity measures, see [9, 10] and [7] among others.

The Shiner–Davison–Landsberg (SDL) complexity measure K_{SDL} is introduced and extended in Sect. 32.3. This measure of complexity is a "universal" measure in the sense that its dependence on "disorder" is the same between systems with significant structural differences and is used for the study of systems to determine their degree of "order" or "disorder". An extension of the SDL complexity applied on systems described by the family of γ-ordered Normal distributions is adopted and evaluated in Sect. 32.3.

T.L. Toulias (✉) · C.P. Kitsos
Technological Educational Institute of Athens, 12210 Egaleo, Athens, Greece
e-mail: t.toulias@teiath.gr

C.P. Kitsos
e-mail: xkitsos@teiath.gr

T. Gilbert et al. (eds.), *Proceedings of the European Conference on Complex Systems 2012*, Springer Proceedings in Complexity, DOI 10.1007/978-3-319-00395-5_32,

The multivariate and elliptically contoured γ-ordered Normal distribution (emerged through the generalized Fisher's information J_α) [5] is defined as follows:

Definition 32.1 The p-dimensional random variable X follows the γ-ordered Normal $\mathcal{N}_\gamma^p(\mu, \Sigma)$, $\gamma \in \mathbb{R} \setminus [0, 1]$, with position parameter $\mu \in \mathbb{R}^p$ and scale matrix $\Sigma \in \mathbb{R}^{p\times p}$, when the density function f_X is of the form

$$f_X(x; \mu, \Sigma, \gamma) = C_\gamma^p |\det \Sigma|^{-1/2} \exp\left\{-\frac{\gamma-1}{\gamma} Q(x)^{\frac{\gamma}{2(\gamma-1)}}\right\}, \quad x \in \mathbb{R}^p, \tag{32.1}$$

with Q the quadratic form $Q(x) = (x-\mu)\Sigma^{-1}(x-\mu)^T$. The normality factor C_γ^p is defined as

$$C_\gamma^p = \pi^{-p/2} \frac{\Gamma(\frac{p}{2}+1)}{\Gamma(p\frac{\gamma-1}{\gamma}+1)} \left(\frac{\gamma-1}{\gamma}\right)^{p\frac{\gamma-1}{\gamma}}. \tag{32.2}$$

The position parameters μ is the mean of X, i.e. $\mu = \mathrm{E}(X)$. The family of the γ-ordered Normal distributions $\mathcal{N}_\gamma^p(\mu, \Sigma)$ provides a "smooth bridging" between the elliptically countered multivariate Uniform $\mathcal{U}^p(\mu, \Sigma)$, Normal $\mathcal{N}^p(\mu, \Sigma)$, Laplace $\mathcal{L}^p(\mu, \Sigma)$ and the degenerate Dirac $\mathcal{D}^p(\mu)$ distributions. That is, all the above significant distributions are members of the $\mathcal{N}_\gamma^p(\mu, \Sigma)$ family of distributions for certain values of order γ due to the following Theorem, see [2] and [3] for details.

Theorem 32.1 *The multivariate γ-ordered Normal distribution $\mathcal{N}_\gamma^p(\mu, \Sigma)$, for order values of $\gamma = 0, 1, 2$ and $\pm\infty$ coincides with*

$$\mathcal{N}_\gamma^p(\mu, \Sigma) = \begin{cases} \mathcal{D}^p(\mu), & \gamma = 0,\ p = 1, 2, \\ 0, & \gamma = 0,\ p \geq 3, \\ \mathcal{U}^p(\mu, \Sigma), & \gamma = 1, \\ \mathcal{N}^p(\mu, \Sigma), & \gamma = 2, \\ \mathcal{L}^p(\mu, \Sigma), & \gamma = \pm\infty. \end{cases} \tag{32.3}$$

32.2 Generalized Fisher's Information Measure

The entropy type information measure $J_\alpha(X)$ for a random variable X, defined through the probability density $f_X(x)$ of X, is given by

$$J_\alpha(X) = \int_{\mathbb{R}^p} f_X(x) \left\| \nabla \log f_X(x) \right\|^\alpha dx, \tag{32.4}$$

see [5] and [4]. This is a (power) generalization of the known Fischer's entropy type information measure $J(X)$, since $J_2(X) = J(X)$ for every random variable X.

The generalized entropy power $N_\alpha(X)$ is of the form [4]

$$N_\alpha(X) = \nu_\alpha e^{\frac{\alpha}{p} H(X)}, \tag{32.5}$$

with normalizing factor ν_α given by the appropriate generalization of $(2\pi e)^{-1}$, namely

$$\nu_\alpha = \left(\frac{\alpha - 1}{\alpha e}\right)^{\alpha - 1} \pi^{-\frac{\alpha}{2}} \left[\frac{\Gamma(\frac{p}{2} + 1)}{\Gamma(p\frac{\alpha-1}{\alpha} + 1)}\right]^{\frac{\alpha}{p}}, \quad \alpha \in \mathbb{R} \setminus [0, 1]. \tag{32.6}$$

Theorem 32.2 *The Shannon entropy for the multivariate and elliptically countered Uniform, Normal and Laplace distributed X (for $\gamma = 1, 2, \pm\infty$ respectively) is given by*

$$H(X) = \begin{cases} \log \frac{\pi^{p/2} |\sqrt{\det \Sigma}|}{\Gamma(\frac{p}{2}+1)}, & X \sim \mathcal{N}_1^p(\mu, \Sigma), \\ \log \sqrt{(2\pi e)^p |\det \Sigma|}, & X \sim \mathcal{N}_2^p(\mu, \Sigma), \\ \log \frac{p! e \pi^{p/2} \sqrt{|\det \Sigma|}}{\Gamma(\frac{p}{2}+1)}, & X \sim \mathcal{N}_{\pm\infty}^p(\mu, \Sigma), \end{cases} \tag{32.7}$$

while $H(X)$ is infinite when $X \sim \mathcal{N}_0^p(\mu, \sigma^2 \mathbb{I}_p)$.

Through the generalized entropy power N_α as in (32.5), we can work to obtain an extended form of the Shannon entropy. This is due to the following Definition.

Definition 32.2 The generalized Shannon entropy of a random variable X, denoted by $H_\alpha(X)$, is a J_α-related Shannon entropy, i.e. the known Shannon entropy of X, where its corresponding entropy power is considered to be the generalized entropy power of X, i.e.

$$N_\alpha(X) = \nu e^{\frac{2}{p} H_\alpha(X)}. \tag{32.8}$$

Therefore, from (32.8) we derive the relation between the generalized Shannon entropy $H_\alpha(X)$ and the usual Shannon entropy $H(X)$,

$$H_\alpha(X) = \frac{p}{2} \log \frac{\nu_\alpha}{\nu} + \frac{\alpha}{2} H(X). \tag{32.9}$$

Practically (32.9) presents a linear (affine) transformation of $H(X)$ depending on parameter α and dimension $p \in \mathbb{N}$, i.e. for fixed p a class of generalized Shannon entropy is obtained for a p-variate random variable X.

Theorem 32.3 *The J_α-related Shannon entropy of the multivariate $X_\gamma \sim \mathcal{N}_\gamma(\mu, \Sigma)$ is given by*

$$H_\alpha(X_\gamma) = \frac{2\gamma - \alpha}{2\gamma} p + \frac{p}{2} \log \left\{ 2\pi \left(\frac{\alpha - 1}{\alpha}\right)^{\alpha - 1} \left(\frac{\gamma}{\gamma - 1}\right)^{\alpha \frac{\gamma - 1}{\gamma}} \right.$$
$$\left. \times \left[\frac{\Gamma(p\frac{\gamma-1}{\gamma} + 1)}{\Gamma(p\frac{\alpha-1}{\alpha} + 1)}\right]^{\frac{\alpha}{p}} |\det \Sigma|^{\frac{\alpha}{2p}} \right\}. \tag{32.10}$$

For $\alpha = \gamma$ it is $H_\alpha(X_\alpha) = \frac{1}{2}\log\{(2\pi e)^p|\det\Sigma|^{\alpha/2}\}$. Moreover for a random variable X following the multivariate Uniform, Normal and Laplace distributions ($\gamma = 1, 2, \pm\infty$ respectively), it is

$$H_\alpha(X) = \begin{cases} \frac{2-\alpha}{2} + h^p_{\gamma,a}, & X \sim \mathcal{N}^p_1(\mu,\Sigma), \\ p + \frac{\alpha}{2}\log\{(2/e)^{p/2}\Gamma(\frac{p}{2}+1)\} + h^p_{\gamma,\alpha}, & X \sim \mathcal{N}^p_2(\mu,\Sigma), \\ p + \frac{p}{2}\log p! + h^p_{\gamma,a}, & X \sim \mathcal{N}^p_{\pm\infty}(\mu,\Sigma), \end{cases} \quad (32.11)$$

where $h^p_{\gamma,\alpha} = \frac{\alpha}{2}\log\{(2\pi)^{p/\alpha}(\frac{\alpha-1}{\alpha})^{p(\alpha-1)/\alpha}[\Gamma(p\frac{\alpha-1}{\alpha}+1)]^{-1}|\sqrt{\det\Sigma}|\}$. For the limiting degenerate case of $\gamma = 0$ we obtain $H_\alpha(X_0) = (\operatorname{sgn}\alpha)(+\infty)$.

Notice that despite the rather complicated form of the $H_\alpha(X_\gamma)$ with $\alpha \neq \gamma$, the J_α-related Shannon entropy of an α-ordered Normal distributed X_γ has a very compact relation.

32.3 The α-SDL Complexity Measure

The Shannon entropy in a continuous system is defined over a random variable X as the expected value of the *information content*, $I(X)$ say, of X, i.e. $H(X) = E\{I(X)\}$. This can be considered as a measure of the "disorder" of a system. However in applied sciences, the normalized Shannon entropy $H^* = H/\max\{H\}$, is usually considered as a measure of "disorder" because H^* is independent of all the various states that the system can adopt. Respectively, the quantity $\Omega = 1 - H^*$ is considered as a measure of "order".

A quantitative measure of complexity with the simple possible expression is the "order-disorder" product $K^{\omega,h}$ given by [1]

$$K^{\omega,h} = \Omega^\omega H^{*h} = H^{*h}(1 - H^*)^\omega = \Omega^\omega(1-\Omega)^h, \quad \omega, h \in \mathbb{R}_+, \quad (32.12)$$

which is usually called as *simple complexity with "order" power ω and "disorder" power h*.

The Shiner–Davison–Landsberg (SDL) [10] measure of complexity K_{SDL} is an important measure in bio-sciences, defined as a special case of (32.12) with $\omega = h = 1$, i.e.

$$K_{SDL} = 4K^{1,1} = 4H^*(1 - H^*) = 4\Omega(1-\Omega). \quad (32.13)$$

We can extend the notion of K_{SDL} not by power-parametrization of order/disorder, as in $K^{\omega,h}$, but with the use of the generalized Shannon entropy defined described in Sect. 32.2. The new parametrized SDL complexity, called the *α-SDL complexity measure*, is defined by $K^\alpha_{SDL} = 4H^*_\alpha(1 - H^*_\alpha)$, with H^*_α being the normalized form of H_α. Therefore, the α-SDL complexity preserves the "qualities" of the usual SDL complexity mentioned above.

For evaluation purposes, consider a continuous system where its various states can be described by a wide-range of distributions such as the family of the univariate γ-ordered Normal distribution with normalized Shannon entropy $H_\alpha^*(X_\gamma) = H_\alpha(X_\gamma)/H_\alpha(Z)$ where $X_\gamma \sim \mathcal{N}_\gamma(\mu, \sigma^2)$ as in (32.1), and $Z \sim \mathcal{N}(\mu, \sigma_Z^2)$ with $\sigma_Z^2 = \operatorname{Var}(X_\gamma)$. The linear expression of H_α through H, as obtained in (32.9), is now essential for the normalization of H_α^* because the Normal distribution (also a member of the $\mathcal{N}_\gamma$ family with $\gamma = 2$) provides maximum entropy not only for H, but also for the generalized (Shannon) entropy H_α, for every $\mathcal{N}_\gamma$ distributions with equally given variances, i.e. $\sigma_Z^2 = \operatorname{Var} X_\gamma$. Hence, $\max_\gamma\{H_\alpha(X_\gamma)\} = H_\alpha(X_2) = H_\alpha(Z)$, and the normalization is valid.

Theorem 32.4 *The α-SDL complexity of a random variable $X_\gamma \sim \mathcal{N}_\gamma(\mu, \sigma^2)$ is given by*

$$K_{SDL}^\alpha(X_\gamma) = \frac{8\log\{(2\pi)^{1/\alpha}\sigma e^{\frac{2\gamma-\alpha}{\alpha\gamma}}(\frac{\gamma}{\gamma-1})^{\frac{\gamma-1}{\gamma}}(\frac{\alpha-1}{\alpha})^{\frac{\alpha-1}{\alpha}}\frac{\Gamma(\frac{\gamma-1}{\gamma}+1)}{\Gamma(\frac{\alpha-1}{\alpha}+1)}\}}{\log^2\{(\pi\sigma)^2 e^{\frac{4-\alpha}{2}}(\frac{\gamma}{\gamma-1})^{2\frac{\gamma-1}{\gamma}}(\frac{\alpha-1}{\alpha})^{2\frac{\alpha-1}{\alpha}}\frac{\Gamma(3\frac{\gamma-1}{\gamma})}{\Gamma^2(\frac{\alpha-1}{\alpha}+1)\Gamma(\frac{\gamma-1}{\gamma})}\}}$$

$$\times \log\left\{\pi e^{\frac{2-\gamma}{\gamma}}\left(\frac{\gamma}{\gamma-1}\right)^2\frac{\Gamma(3\frac{\gamma-1}{\gamma})}{2\Gamma^3(\frac{\gamma-1}{\gamma})}\right\}, \tag{32.14}$$

while the α-SDL absolute complexity, $\max\{K_{SDL}^\alpha(X_\gamma)\} = 1$, *is obtained for*

$$\sigma_*^2 = 2^{-\frac{\alpha+2}{\alpha}}\pi^{\frac{\alpha-2}{\alpha}}e^{4\frac{\alpha-\gamma}{\alpha\gamma}-1}\left(\frac{\gamma}{\gamma-1}\right)^{2\frac{\gamma+1}{\gamma}}\left(\frac{\alpha-1}{\alpha}\right)^{\frac{2}{\alpha}}\frac{\Gamma(3\frac{\gamma-1}{\gamma})\Gamma^2(\frac{\alpha-1}{\alpha})}{\Gamma^5(\frac{\gamma-1}{\gamma})}, \quad \gamma \neq 2. \tag{32.15}$$

Notice that $K_{SDL}^\alpha(X_2) = 0$ for every α, i.e. the normal distribution provides absolute complexity on every α-SDL complexity measure. Thus, K_{SDL}^α extends the (expected) result that Normal distributions (as maximum entropy distributions) vanishes the SDL complexity. Moreover, it can be proved, through (32.14), that there is no Uniform or Laplace distribution that provides absolute complexity on all α-SDL complexity measures (including the usual SDL complexity).

32.4 Discussion

A computational overview of the SDL complexity clarifies the theoretical study of Sect. 32.3.

Figure 32.1 illustrates the behavior of the 2-SDL complexity, i.e. the usual SDL complexity $K_{SDL}(X_\gamma) = K_{SDL}^2(X_\gamma)$ with $X_\gamma \sim \mathcal{N}_\gamma(\mu, \sigma^2)$ for various scale parameters σ^2.

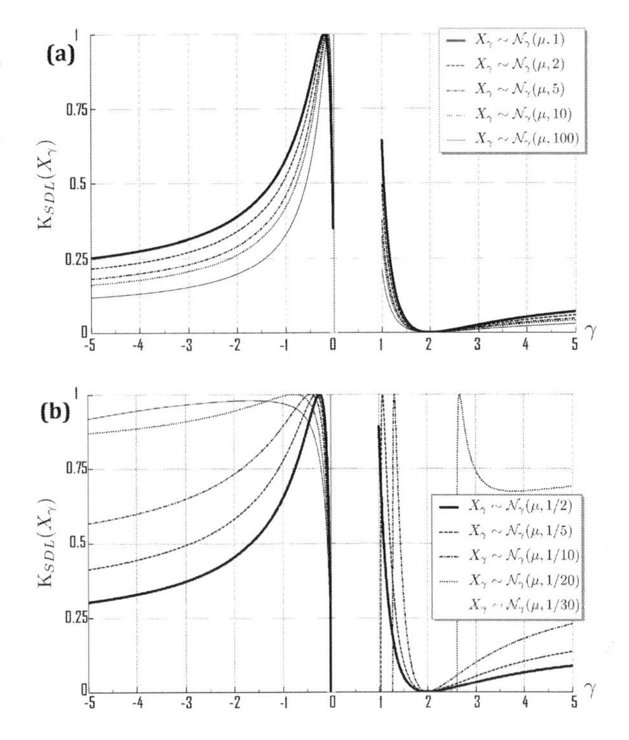

Fig. 32.1 Graphs of the 2-SDL complexity $K^2_{SDL}(X_\gamma)$ along γ for various values of σ^2 (*the left-side graph* obtained for $\sigma^2 \geq 1$ while *the right-side graph* for $\sigma^2 < 1$) with $X_\gamma \sim \mathcal{N}_\gamma(\mu, \sigma^2)$

Table 32.1 Evaluation of $K^2_{SDL}(X_\gamma)$, $X_\gamma \sim \mathcal{N}_\gamma(\mu, \sigma^2)$ for various γ and σ^2 as well as the corresponding $\gamma^* = \gamma^*(\sigma^2)$ values for each σ^2 that provide "absolute" 2-SDL complexity

γ	σ^2									
	1/30	1/20	1/10	1/5	1/2	1	2	5	10	100
−5	0.917	0.869	0.567	0.413	0.302	0.251	0.215	0.180	0.161	0.118
−4	0.944	0.884	0.599	0.444	0.329	0.275	0.236	0.199	0.177	0.131
−3	0.965	0.907	0.649	0.493	0.372	0.314	0.271	0.229	0.205	0.153
−2	0.978	0.943	0.732	0.582	0.454	0.388	0.339	0.290	0.261	0.197
−1	0.969	0.993	0.890	0.775	0.652	0.579	0.520	0.458	0.420	0.328
0	0.000	0.000	0.000	0.000	0.000	0.000	0.000	0.000	0.000	0.000
1	−1.03	−1.44	−4.08	−18.7	0.894	0.647	0.496	0.377	0.319	0.210
2	0.000	0.000	0.000	0.000	0.000	0.000	0.000	0.000	0.000	0.000
3	−0.26	0.746	0.099	0.053	0.033	0.025	0.029	0.019	0.015	0.010
4	−0.96	0.675	0.180	0.103	0.066	0.052	0.043	0.038	0.030	0.021
5	−2.34	0.690	0.223	0.137	0.089	0.071	0.058	0.047	0.042	0.030
$\pm\infty$	−0.51	0.789	0.416	0.279	0.194	0.157	0.132	0.109	0.097	0.070
γ_*	–	2.678	1.383	1.079	−0.23	−0.20	−0.17	−0.15	−0.14	−0.11

Evaluations of the 2-SDL complexity $K_{SDL}^{2}(X_{\gamma})$ for values of $\gamma = -5, -4, \ldots, 5$, and $\gamma = \pm\infty$ for various σ^2 are given in the following Table 32.1 (as these values form, in some extent, Fig. 32.1).

The last row of Table 32.1 presents the values $\gamma_{*} = \gamma_{*}(\sigma^2)$ (for each given σ^2 of the first row of the Table 32.1) which provide "absolute" 2-SDL complexity, i.e. for $K_{SDL}^{2}(X_{\gamma_{*}}) = 1$.

References

1. Feldman DP, Crutchfield JP (1998) Measures of statistical complexity: why? Phys Lett A 238:244–252
2. Kitsos CP, Toulias TL (2012) On the multivariate γ-ordered normal distribution. Far East J Theor Stat 38(1):49–73
3. Kitsos CP, Toulias TL (2011) On the family of the γ-ordered normal distributions. Far East J Theor Stat 35(2):95–114
4. Kitsos CP, Toulias TL (2010) New information measures for the generalized normal distribution. Information 1:13–27
5. Kitsos CP, Tavoularis NK (2009) Logarithmic Sobolev inequalities for information measures. IEEE Trans Inf Theory 55(6):2554–2561
6. Lopez–Ruiz R, Mancini HL, Calbet X (1991) A statistical measure of complexity. Phys Lett A 209:321–326
7. Lamberti PW, Martin MT, Plastino A, Rosso OA (2004) Intensive entropic non-triviality measure. Physica A 334:119–131
8. Martin MT, Plastino A, Rosso OA (2003) Statistical complexity and disequilibrium. Phys Lett A 311:126–132
9. Rosso OA, Martin MT, Plastino A (2003) Brain electrical activity analysis using wavelet-based informational tools (II): Tsallis non-extensivity and complexity measures. Physica A 320:497–511
10. Shiner JS, Davison M, Landsberg PT (1999) Simple measure for complexity. Phys Rev E 59(2):1459–1464

Chapter 33
State Space Properties of Boolean Networks Trained for Sequence Tasks

Andrea Roli, Matteo Amaducci, Lorenzo Garattoni, Carlo Pinciroli, and Mauro Birattari

Abstract In a recent work, it has been shown that Boolean networks (BN), a well-known genetic regulatory network model, can be utilised to control robots. In this work, we use a genetic algorithm to train robots controlled by a BN so as to accomplish a sequence learning task. We analyse the robots' dynamics by studying the corresponding BNs' phase space. Our results show that a phase space structure emerges enabling the robot to have memory of the past and to exploit this piece of information to choose the next action to perform. This finding is in accordance with previous results on minimally cognitive behaviours and shows that the phase space of Boolean networks can be shaped by the learning process in such a way that the robot can accomplish non-trivial tasks requiring the use of memory.

33.1 Introduction

Dynamical systems provide metaphors and tools which can be effectively used to analyse artificial agents, such as robots [3, 15]. The dynamical systems metaphor has also been advocated as a powerful source of design principles for robotics [8]. The core idea supporting this viewpoint is that information processing can be seen as the evolution in time of a dynamical system [12]. In this paper, we show that a dynamical systems perspective makes it possible to analyse the behaviour of a robot controlled by Boolean networks and explain it in terms of trajectories in the Boolean network's state space.

Boolean networks (BNs) have been introduced by Kauffman [4] as a gene regulatory network (GRN) model. BNs have been proven to reproduce very important phenomena in genetics and they have also received considerable attention in the research communities on complex systems [1, 4]. A BN is a discrete-state and

Mauro Birattari acknowledges support from the Belgian F.R.S.-FNRS.

A. Roli (✉) · M. Amaducci
Dipartimento di Informatica: Scienza e Ingegneria, DISI, Alma Mater Studiorum, Università di Bologna, Cesena, Italy

L. Garattoni · C. Pinciroli · M. Birattari
IRIDIA-CoDE, Université libre de Bruxelles, Brussels, Belgium

T. Gilbert et al. (eds.), *Proceedings of the European Conference on Complex Systems 2012*, Springer Proceedings in Complexity, DOI 10.1007/978-3-319-00395-5_33,

discrete-time dynamical system whose structure is defined by a directed graph of N nodes, each associated to a Boolean variable x_i, $i = 1, \dots, N$, and a Boolean function $f_i(x_{i_1}, \dots, x_{i_{K_i}})$, where K_i is the number of inputs of node i. The arguments of the Boolean function f_i are the values of the nodes whose outgoing arcs are connected to node i. The state of the system at time t, $t \in \mathbb{N}$, is defined by the array of the N Boolean variable values at time t: $s(t) \equiv (x_1(t), \dots, x_N(t))$. The most studied BN models are characterised by having a *synchronous* dynamics—i.e., nodes update their states at the same instant—and *deterministic* functions. However, many variants exist, including asynchronous and probabilistic update rules [13]. BN models' dynamics can be studied by means of usual dynamical systems methods [2, 12], hence the usage of concepts such as state (or phase) space, trajectories, attractors and basins of attraction. BNs can exhibit complex dynamics and some special ensembles have been deeply investigated, such as that of Random BNs [4, 11].

In a recent work, it has been shown that BNs can be utilised to control robots [10]. A BN is coupled with a robot by defining a set of input nodes, whose values are imposed by the robot's sensor readings, and a set of output nodes, which are used to maneuver the robot's actuators. The BN is trained by means of a learning algorithm that manipulates the Boolean functions. The algorithm employs as learning feedback a measure of the performance of the BN-controlled robot (in the following, BN-robot) on the task to perform.

In this work, we use a genetic algorithm to train a BN-robot so as to accomplish a task concerning *sequence learning* and we analyse their dynamics by studying the characteristics of the corresponding BNs' state space. Our results show that a state space structure emerges enabling the robot to have memory of the past and to exploit this piece of information to choose the next action to perform. In the following of this brief contribution, we outline the task to accomplish and we illustrate the main results achieved. For completeness, we include a description of materials and methods.

33.2 The Task

Sequence learning is one of the most prominent activities in humans, animals, as well as artificial agents and systems [14]. Sequence tasks involve the use of some kind of *memory* which enables the agent to choose the next action depending on the past. The main kinds of sequence tasks are: sequence prediction, generation, recognition and sequential decision making. Sequence learning is clearly a difficult task, due to the fact that forms of memory structures are needed. Several techniques exist to tackle the problem, including recurrent neural networks, hidden Markov model, dynamic programming, reinforcement learning and evolutionary computation techniques, such as the ones used in this work.

In our experiment, the BN-robot must learn to recognise a sequence of colours, by performing certain actions. The environment in which the BN-robot operates is a straight corridor. Along the corridor, the ground is painted in three different

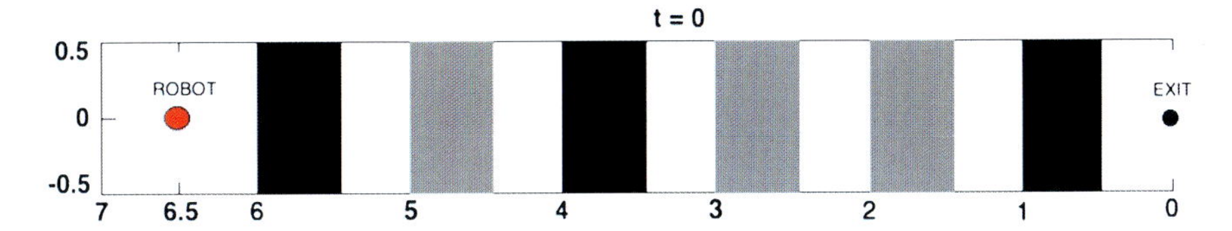

Fig. 33.1 An example of the BN-robot's working environment. The order of *black* and *grey stripes* is randomly chosen at each trial

colours: white (W) represents the background, while black (B) and grey (G) denote the symbols of a sequence to be recognised. See Fig. 33.1 for an example of the environment. The BN-robot, placed at the beginning of the corridor, moves along it, turning its LEDs on when it encounters a black or grey stripe in the right sequence and keeping the LEDs off when the colour is not in the right order or it is the background colour. In our case, the sequence to be recognised is a cyclic repetition of black and grey. For example:

```
Colours along the corridor:  W   B  W   G  W   G   W   B  W   B
BN-robot's LEDs status:     OFF ON OFF ON OFF OFF OFF ON OFF OFF
```

This task is dynamic, in that the robot needs to decide whether to switch on or off the LED, on the basis of information concerning the past. To carry out this task, the robot needs to exploit some sort of memory.

33.3 Results

A successful BN-robot is one which correctly switches its LEDs on and off according to the desired sequence, when encountering different colours on the ground. Since this task requires some kind of memory structure to be constructed, we analysed the state space traversed by the BN controlling a robot with the aim to understand its operation and dynamics. A similar approach has also been used in previous works in evolutionary robotics [6, 15]. To analyse a BN, we extracted a sample of 1000 trajectories in the state space by simulating the robot in corridors with colours in random order. We gathered such trajectories and generated a graph of the observed state transitions. The first relevant observation we derived from this analysis is that the size of the state space traversed by the BN is a very tiny fraction of the whole potential state space, which is of size 2^{30}. Indeed, the number of states in the collected trajectories is about 200, on average; hence, the learning process shapes the BN in such a way that its dynamics is confined in a limited portion of the state space. A further notable property of the BN dynamics of the robot is that memory is implicitly represented by connecting different areas of the state space, each devoted to a specific set of actions. A compact view of the state space can be provided in the form of a finite state automaton (FSA), in which states represent clusters of connected states in the BN phase space. Indeed, the BN phase space can be clustered in sets of states which encode the memory of the previous colour encountered

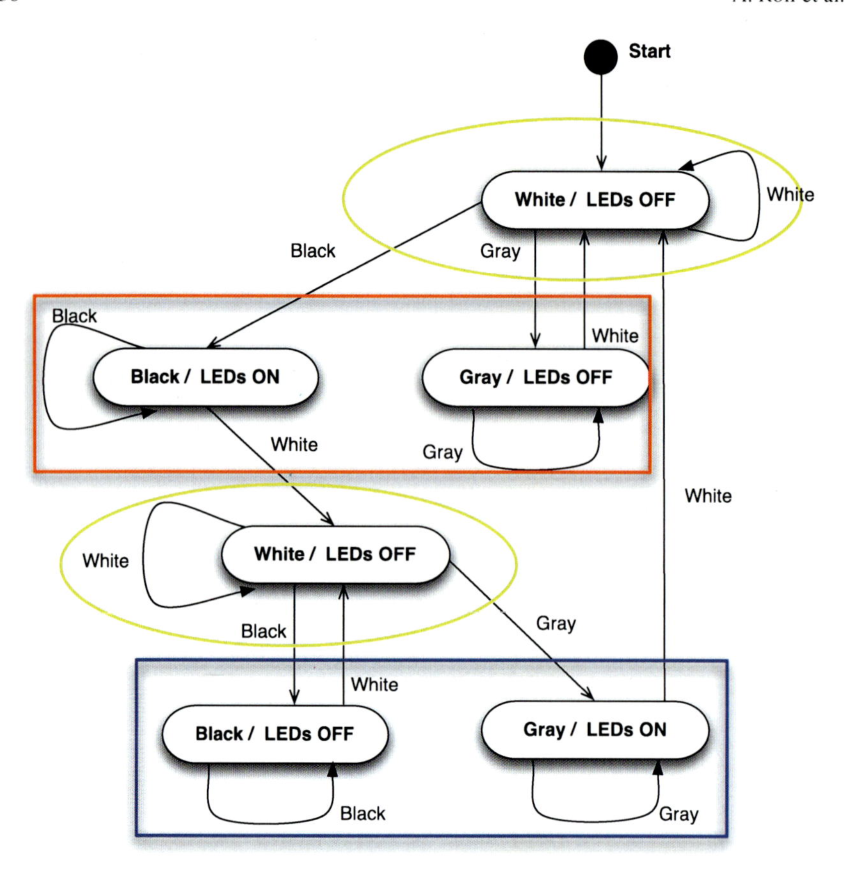

Fig. 33.2 Finite state automaton representation of the trajectory graph over the state space for a typical BN. A state in the automaton represents a cluster of BN's states in which the BN remains until a specific input is received

by the robot. In Fig. 33.2, we report the FSA of a typical successful BN-robot. At the beginning of the trial, the BN is in a state space area in which the values of the output nodes are such that the BN-robot goes straight and keeps its LEDs off until a coloured stripe is found. Then, depending on the detected colour, the BN goes into either of two regions, which we will denote as the upper (red) and lower (blue) rectangles. As we can see, the mechanisms in the two clusters is dual: in the upper one, the robot switches its LEDs on when it encounters a black stripe either if it is the first non-background colour it detects or if a grey stripe has been previously found; conversely, the second state space cluster is devoted to recognising grey stripes, after a black one has been encountered.

In summary, we can assert that the information concerning the last seen colour is implicitly stored in the state space area in which the BN operates. This finding is in accordance with previous results on minimally cognitive behaviours [15] and shows that the state space of BNs can be shaped by the learning process in such a way that the BN-robot can accomplish non-trivial tasks requiring the use of memory.

33.4 Materials and Methods

In this experiment, we control an *e-puck* robot [5] by means of a BN. The robots are simulated with the ARGoS simulator [9]. The values of a set of network nodes (BN input nodes) are imposed by the robot's sensor readings, and the values of another set of nodes (BN output nodes) are observed and used to encode the signals for maneuvering the robot's actuators. The BN controlling the robot has 30 nodes in total and the function of each node depends on the value of 3 other nodes, chosen at random. Four nodes are used as inputs and encode the proximity sensors (North, South, East, West)—the node is set to 1 if an obstacle is near the robot—and two encode the ground colours (00 $\leftrightarrow$ black, 01 $\leftrightarrow$ grey, 11 $\leftrightarrow$ white). Two nodes are used to control the robot wheels, which can be individually either set to a constant non-zero speed or stopped.

The BNs controlling the robots are designed by means of a genetic algorithm (GA),[1] according to the evolutionary robotics approach [7]. The genetic algorithm adopts a proportional selection and applies mutation and crossover operators. Mutation is implemented by randomly choosing a node and an entry in its Boolean function truth table and flipping it.[2] The crossover operator is a single point crossover, operating on the binary string given by the concatenation of the nodes' Boolean functions. The training process changes the Boolean functions, while the BN topology is kept constant (it is generated according to the random BN model, as described by Kauffman [4]). The fitness function is computed as the average of the performance across 10 trials, in which the sequence of colours is randomly generated. The performance of the BN-robot in a trial is computed as the distance it can walk along the corridor by correctly switching its LEDs and avoiding the walls. The GA is run with the following parameter setting: population size equal to 20, elitism set to 2, $p_{\mathrm{mut}} = 0.02$, $p_{\mathrm{cx}} = 0.1$, the number of generations is set to 5000. The GA is run 60 times, starting from randomly generated BNs. The successful runs, i.e., those returning BN-robots correctly performing the sequence task, were 10 out of 60.

33.5 Conclusion

In this brief contribution, we have outlined the results of the analysis of the behaviour of BN-robots trained to accomplish a sequential task. The behaviour of a BN-robot is studied by means of the phase space analysis of the corresponding BN.

---

[1]Other search techniques have also been used and we obtained the same qualitative results.

[2]Details can be found in [10].

The results show that the training process shapes the phase space so as to restrict the BN dynamics to few, relatively small, areas. Phase space areas play the role of memory, as they implicitly store the information concerning the past which is relevant for the BN-robot to choose the next action. In addition, we observe that the analysis of the BN phase space can be simplified by clustering the set of states and studying the corresponding finite state automata. This method may also be subject to formal verification, making it possible to validate the robot's behaviour.

References

1. Aldana M, Coppersmith S, Kadanoff L (2003) Boolean dynamics with random couplings. In: Kaplan E, Marsden J, Sreenivasan K (eds) Perspectives and problems in nonlinear science. A celebratory volume in honor of Lawrence Sirovich. Springer applied mathematical sciences series. Springer, Berlin
2. Bar–Yam Y (1997) Dynamics of complex systems. Studies in nonlinearity. Addison-Wesley, Reading
3. Beer R (1995) A dynamical systems perspective on agent-environment interaction. Artif Intell 72:173–215
4. Kauffman S (1993) The origins of order: self-organization and selection in evolution. Oxford University Press, London
5. Mondada F, Bonani M, Raemy X, Pugh J, Cianci C, Klaptocz A, Magnenat S, Zufferey JC, Floreano D, Martinoli A (2009) The e-puck, a robot designed for education in engineering. In: Gonçalves P, Torres P, Alves C (eds) Proceedings of the 9th conference on autonomous robot systems and competitions, vol 1, pp 59–65
6. Nolfi S (2005) Categories formation in self-organizing embodied agents. In: Cohen H, Lefebvre C (eds) Handbook of categorization in cognitive science. Elsevier, Amsterdam, pp 869–889
7. Nolfi S, Floreano D (2000) Evolutionary robotics. MIT Press, Cambridge
8. Pfeifer R, Bongard J (2006) How the body shapes the way we think: a new view of intelligence. MIT Press, Cambridge
9. Pinciroli C, Trianni V, O'Grady R, Pini G, Brutschy A, Brambilla M, Mathews N, Ferrante E, Di Caro G, Ducatelle F, Stirling T, Gutiérrez A, Gambardella L, Dorigo M (2011) ARGoS: a modular, multi-engine simulator for heterogeneous swarm robotics. In: Proceedings of the IEEE/RSJ international conference on intelligent robots and systems (IROS 2011). IEEE Comput Soc, Los Alamitos, pp 5027–5034
10. Roli A, Manfroni M, Pinciroli C, Birattari M (2011) On the design of Boolean network robots. In: Di Chio C, Cagnoni S, Cotta C, Ebner M, Ekárt A, Esparcia-Alcázar A, Merelo J, Neri F, Preuss M, Richter H, Togelius J, Yannakakis G (eds) Applications of evolutionary computation. Lecture notes in computer science, vol 6624. Springer, Heidelberg, pp 43–52
11. Serra R, Villani M, Graudenzi A, Kauffman S (2007) Why a simple model of genetic regulatory networks describes the distribution of avalanches in gene expression data. J Theor Biol 246:449–460
12. Serra R, Zanarini G (1990) Complex systems and cognitive processes. Springer, Berlin
13. Shmulevich I, Dougherty E (2009) Probabilistic Boolean networks: the modeling and control of gene regulatory networks. SIAM, Philadelphia
14. Sun R, Giles C (2001) Sequence learning: from recognition and prediction to sequential decision making. IEEE Intell Syst 16(4):67–70
15. Yamauchi B, Beer R (1994) Sequential behavior and learning in evolved dynamical neural networks. Adapt Behav 2(3):219–246

Chapter 34
Towards a Deeper Understanding of the Complex Behaviour Observed in the Distribution of Words in Written Texts

Concepción Carretero-Campos, Marcelo A. Montemurro, Pedro Bernaola-Galván, Ana V. Coronado, and Pedro Carpena

Abstract Here we show that the recently reported presence of long-range correlations in the distribution of words along texts is due to the complex distribution of the keywords, while common words are not correlated. Indeed we prove that the degree of long-range correlations of a word at long scales is a good measure of its relevance to the text. Additionally, we develop a model able to reproduce the spatial distribution of a word in a text, based on the long-range correlations observed for the word. The model not only reproduces the complex behaviour characterized by the presence of correlations at long scales and the degree of relevance of the word, but also the probability distribution of the inter-occurrences distances in the whole range of scales.

Keywords Long-range correlations · Keyword detection · Complex structure of words in texts · Word relevance and complexity

34.1 Introduction and Background

In the last years there has been growing interest for the study of human language in the context of complex systems. Written texts are good candidates for such approach, since they are composed of single elements (words) which can interact among them in complex forms and at different levels, controlled by grammatical rules of the particular language, the literary genre, the writer's style and the information the text aims to communicate.

The approaches to text analysis have been focused on three main topics:

(i) The automatic detection of relevant words. The idea is to detect keywords in a text (i.e. words related to the main topics of the text) without the use of external information. One successful strategy proposed to tackle this problem [1] uses the fact that relevant words attracts themselves, and are concentrated in certain regions

C. Carretero-Campos (✉) · P. Bernaola-Galván · A.V. Coronado · P. Carpena
Departamento de Física Aplicada II, Universidad de Málaga, 29071 Málaga, Spain

M.A. Montemurro
Faculty of Life Sciences, The University of Manchester, M13 9PT Manchester, UK

T. Gilbert et al. (eds.), *Proceedings of the European Conference on Complex Systems 2012*, Springer Proceedings in Complexity, DOI 10.1007/978-3-319-00395-5_34,

of the text forming clusters and giving rise to large frequency fluctuations. However, common words are distributed more homogeneously along the text. Therefore, the larger the clustering, the larger the relevance of the word. Thus, by quantifying properly the clustering of each word, a ranking of relevance can be obtained. That connection between clustering and relevance has also been observed with a different approach [2].

(ii) Long-range correlations in written texts. The complex interactions between words occurring at many levels that we commented above produce an also complex non-trivial spatial structure in written texts, that can be quantified through the analysis of the long-range correlations present in the text. The results of such studies conclude that written texts present long-range structures that give raise to long-range correlations which have been quantified for different texts and languages [3–5].

(iii) Models for the spatial distribution of words in texts, and specifically for the probability distribution of the inter-occurrence distances for a given word, $p(d)$. Some results show [6] that in general $p(d)$ for any word is given by a stretched-exponential distribution. Although this result agrees fairly well with experimental observations for large d, fails at short distances d (see methods and results), where $p(d)$ is systematically overestimated.

Here we show that the three main topics described above are not independent and in contrast, are deeply related and can be considered as different faces of the same problem. First, we show that the long-range correlations observed in texts are due to their keywords: we quantify the long-range correlation of any word in a text, and show that relevant words present strong long-range correlations and are responsible for the correlations observed in the whole text, while common words are not correlated and do not contribute to the text correlations. Indeed, we present results indicating that the degree of correlations of a word is also a good measure for its relevance to the text. Second, and the main result of our paper, we present a model able to reproduce the spatial distribution of a word in a text. Starting from the correlations of the word and its frequency in the text, the model predicts the position of consecutive appearances of the word and is able to reproduce all the interesting properties of the word: its long-range correlations, its spatial structure characterized by the distribution of distances $p(d)$ in the whole range of d (solving the problem of previous results) and also the degree of relevance of the word quantified by the clustering approach referred above.

34.2 Methods

34.2.1 Keyword Detection: The Clustering Measure C

Carpena et al. [1] proposed the measure C. It is an improvement of the method developed by Ortuño et al. [7], which is based on the statistical analysis of the distributions of distances between successive occurrences of a word. It follows the hypothesis that the most informative words of a text have a strong self-attraction and tend to form clusters, while common words are placed randomly everywhere in the text giving raise to a more homogeneous distribution.

Given a word with frequency n in a text, the inter-occurrences distances are denoted by $\{d_i\}$. Ortuño et al. [7] proposed the normalized standard deviation $\sigma = s/\langle d\rangle$ as a measure of the relevance of the word. The larger the value of σ, the larger the relevance. To avoid the dependence of σ on the probability p of the word in the text, it needs to be normalized to $\sigma_{nor} = \frac{\sigma}{\sqrt{1-p}}$, where $\sqrt{1-p}$ is the expected value of σ for a word with probability p randomly distributed. But both σ and σ_{nor} strongly depend on the frequency of occurrence n. In order to take into account the information provided by the clustering σ_{nor} and by the frequency n, the measure C is defined as the deviation of σ_{nor} with respect to the expected value in a random text $\langle\sigma_{nor}\rangle(n)$ in units of the expected standard deviation $sd(\sigma_{nor})(n)$ [1]:

$$C(\sigma_{nor}, n) = \frac{\sigma_{nor} - \langle\sigma_{nor}\rangle(n)}{sd(\sigma_{nor})(n)}. \tag{34.1}$$

$C = 0$ indicates that the word is randomly distributed. The larger the value of C, the larger the statistically significant clustering, and thus the relevance of the word. Negative values of C indicate repulsion.

34.2.2 Quantification of Long-Range Correlations in Words: Detrended Fluctuation Analysis

The occurrences of every word of a text are represented by a binary sequence. Given a text of length N, for each word we generate a binary sequence $x(i)$ by assigning the value 1 on all positions i where the word appears in the text and 0 in all other positions. In this way, we have a different binary sequence for every word. We choose a mapping of the text into a binary sequence in order to avoid the introduction of spurious correlations due to the numeric assignment [8]. The correlations of those sequences are quantified by the scaling exponent α obtained by Detrended Fluctuation Analysis (DFA).

DFA [9] is a method widely used to estimate long-range power-law correlations. It measures the average of the fluctuations $F(l)$ of the sequence at different scales l. A power-law relation,

$$F(l) \propto l^{\alpha}. \tag{34.2}$$

indicates the presence of scaling. The degree of correlations are quantified by α, which can be estimated fitting the slope of $\log F(l)$ versus $\log l$. If $\alpha = 0.5$ the sequence is uncorrelated, while if $\alpha < 0.5$ or $\alpha > 0.5$ there are negative or positive correlations, respectively.

Note that in our computation we have considered l from 8 to $N/10$, a range of scales in which it has been shown that DFA provides accurate results [10].

34.2.3 A Model for Word Appearances

The goal of the model developed here is to reproduce the distribution of a word along the text as well as its long-range correlations. To achieve this objective, we propose to generate a binary sequence with long-range correlations that represents the occurrences of every word throughout the text (1 when the word appears and 0 otherwise).

In a first step, by Fourier Filtering Method (FFM) [11], we generate a sequence of random real numbers $x(i)$ with long-range power-law correlations characterized by the scaling exponent α. In a second step, to obtain from $x(i)$ a binary sequence with the desired correlation exponent, we assume that the word appears along the text in the positions i in which the sequence exceeds a threshold (r), so the occurrences of the word can be represented by a binary sequence obtained by mapping $x(i)$ to 1 if $x(i) \geq r$, and to 0 otherwise. The threshold r is obtained numerically by imposing that the probability of having values of the sequence above the threshold is just the frequency of the word we want to model. We denote the distribution of inter-occurrences distances obtained by this mapping as $p_0(d)$.

The analytical form of $p_0(d)$ is in general unknown. By means of numerical simulations we have obtained that the cumulative distribution $P_0(d)$ (defined as $P_0(d) \equiv \int_d^{\infty} p_0(x)dx$) is of stretched exponential form [12], i.e.

$$P_0(d) \sim e^{-(d/c)^{\beta}}, \tag{34.3}$$

where the constants β and c depend on the word considered. This functional form (corresponding to the dashed line in Fig. 34.5) has been proposed recently [6] to characterize the distribution of distances between successive occurrences of the same word. Such expression agrees fairly well with experimental observations for large d but systematically overestimate the distribution at short distances d (see Fig. 34.5).

Nevertheless, we notice that real words have repulsion at short distances, due to restrictions imposed by the grammar that does not allow the occurrence of the same word at very short distances. The previous mapping leads to Eq. (34.3) and does not account for this feature. To incorporate this phenomenon we propose a repulsion factor $f(d)$ that modifies the distribution of inter occurrences distances obtained by the mapping described above. We accept a distance d (and therefore we accept a '1' at a distance d from the previous one in the binary sequence) with probability $f(d)$ given by

$$f(d) = 1 - e^{-\frac{(d-(d_{\min}-1))}{a}}. \tag{34.4}$$

if $d \geq d_{\min}$ and $f(d) = 0$ if $d < d_{\min}$, where $d_{\min}$ is the observed minimum inter-occurrence distance of the real word along the text. The parameter a gives information about the characteristic scale of the repulsion, and depends on the word considered. With this repulsion factor, the final $p(d)$ provided by our model is of the form:

$$p(d) \sim f(d)p_0(d). \tag{34.5}$$

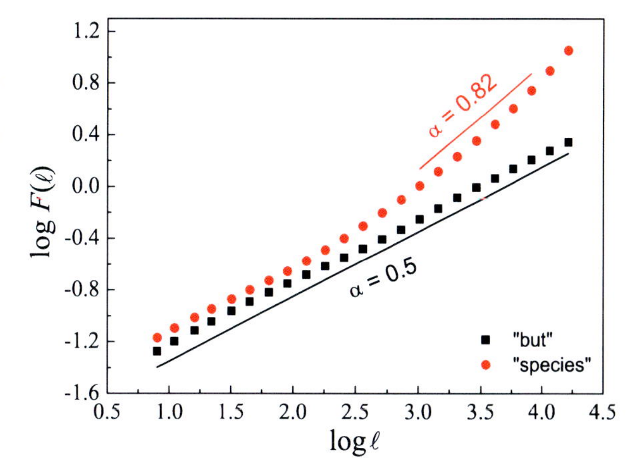

Fig. 34.1 Detrended Fluctuation Analysis applied to the binary sequences which represent the occurrences of the words "species" (relevant) and "but" (non-relevant) along the book "The Origin Of The Species" by Charles Darwin

Note that the repulsion $f(d)$ modifies $p_0(d)$ lowering it only at short distances, but keeps unaffected the large distances regime: for long enough distances d, $f(d) \simeq 1$ and, consequently, $p(d) \simeq p_0(d)$. In this way, $p(d)$ presents the correct behaviour at all scales.

The three parameters of the model (the correlation exponent α, the threshold r and the repulsion scale parameter a) can be initially estimated from the text analyzed: α is calculated using DFA in the binary sequence obtained from the text for the word considered, r is obtained from the frequency of the word in the text, and a is estimated as the distance d at which the distribution of distances of the real word departs from the stretched exponential (see Fig. 34.5).

34.3 Results and Discussion

34.3.1 A Link Between Relevance and Long-Range Correlations in Words

The results we show here are obtained using the book "The Origin of Species" by Charles Darwin (6th Edition).[1] It has a length $N = 193786$ words and contains a vocabulary of 8186 word types.

Firstly we calculate the relevance measure C for those 8186 word types to obtain a ranking of relevance. Additionally, for each word with frequency al least 100 (a total of 252 words), we map its occurrences to a binary sequence (see methods) and quantify the correlations of this sequence by DFA. We consider a cut-off for the frequency to avoid results affected by low statistics.

Figure 34.1 shows the results of the DFA applied to a relevant word, "species" (position 4 in the ranking), and to a non-relevant word, "but" (position 8185 in the

[1] It has been downloaded from the Project Gutenberg web page. http://www.gutenberg.org.

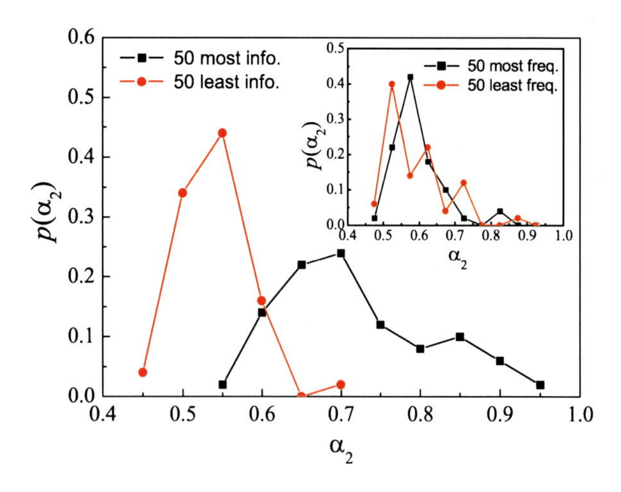

Fig. 34.2 Probability distribution of the scaling exponent α_2 obtained by a linear least squares fit of $\log F(l)$ versus $\log l$ in a range of distances l from 1000 to 10000. For the 50 most informative words versus the 50 least informative words from the book "The Origin Of The Species" by Charles Darwin. Inset: the same for the 50 most frequent words versus the 50 least frequent words

Table 34.1 The first 10 words and the last 10 words extracted from the book "The Origin Of The Species" by the correlation exponent at long scales α_2

Word	α_2	Word	α_2
seeds	0.945	case	0.500
islands	0.905	also	0.499
young	0.901	whether	0.498
water	0.889	hence	0.495
flowers	0.859	so	0.493
forms	0.849	either	0.491
varieties	0.843	both	0.484
organs	0.836	give	0.484
breeds	0.833	then	0.471
species	0.816	us	0.464

ranking). We observe that at short distances both words have a random distribution characterized by a scaling exponent α close to 0.5. However, at a distance beyond the effects of grammatical rules, we observe a crossover in the scaling exponent of the relevant word towards a value of α significantly greater than 0.5, which indicates strong long-range correlations.

The existence of a crossover in the correlations at intermediate scales appears in general only for relevant words (words with high C), while non-relevant words have a correlation exponent close to 0.5 at all scales and do not have a crossover to a second regime. To illustrate this behavior, for each word we compute a linear least squares fit of $\log F(l)$ versus $\log l$ in a range of distances l from 1000 to 10000 to have a value for the correlation exponent at long scales, which we denote α_2. In Fig. 34.2 we show the probability distribution of α_2 for the 50 most informative words and for the 50 least informative words. The separation between the two distributions suggests that α_2 can be used to discriminate the relevant words of a text, the larger the value of α_2 the larger the relevance, as we can see in Table 34.1. Note

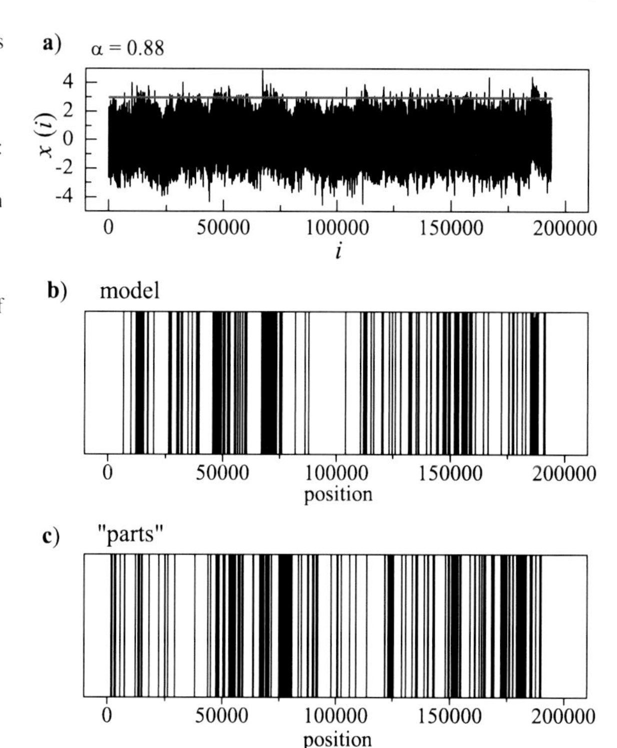

Fig. 34.3 Generation process of an artificial word that models the distribution of the word "parts" along the book "The Origin Of The Species": (**a**) sequence of random numbers $x(i)$ with correlation exponent α (*black line*) and a threshold r (*red line*) used to map the real sequence to a binary one, (**b**) occurrences of the modeled word along the text, (**c**) occurrences of the real word along the text

that the probability distributions of α_2 for the 50 most frequent words and for the 50 least frequent words almost do not differ one from another (inset in Fig. 34.2), so these results are independent on the frequency of the words.

The results we have shown here allow us to conclude that: (i) The degree of long-range correlations of a word at long scales is directly related to its relevance and (ii) The long-range correlations that have been observed in texts are due to the complex distribution of the relevant words along the text. Common words, that have a homogeneous distribution, do not contribute to the existence of long-range correlations in texts.

The link between long-range correlations and relevance have also been found for other books and in different languages, suggesting that it is an universal feature.

34.3.2 Complex Properties of Words Reproduced by the Model

Here we present some results of the model we propose to reproduce the correlation properties of the words. Figure 34.3 displays the generation process of an artificial word that models the distribution of the word "parts" along the book "The Origin Of The Species" (for a suitable election of the parameters, see methods). As we can see in Figs. 34.3(b) and 34.3(c) the spatial distribution of the modeled word agrees

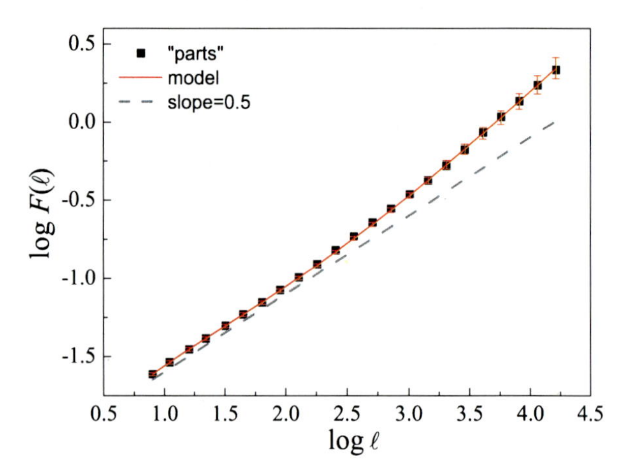

Fig. 34.4 Detrended Fluctuation Analysis applied to the binary sequence which represents the occurrences of the word "parts" (*solid black squares*) along the book "The Origin Of The Species" by Charles Darwin, and the fitting obtained by the modeled word (*red solid line*). A straight line with slope 0.5 (*gray dashed line*) is plotted for comparison

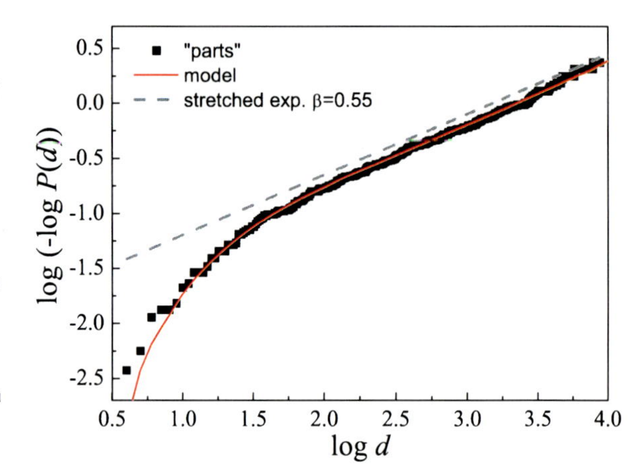

Fig. 34.5 Cumulative distribution $P(d)$ of the inter-occurrences distances of the word "parts" (*solid black squares*) in the book "The Origin Of The Species" (plotted in a scale in which the stretched exponential is a straight line, see Eq. (34.3)), and the fitting obtained by the modeled word (*red solid line*). A straight line with slope $\beta = 0.55$ (*gray dashed line*) is plotted for comparison

with the real word, giving raise to a similar cluster structure. If now we apply the DFA method to the binary sequences generated by 256 iterations of the model and then average the curves, we prove that the model accurately retains the correlations structure of the word (Fig. 34.4). The average curve of the model provides a very good fitting of the correlations of the word "parts" at all scales. In this way we prove that the model reproduces the distribution of a word along the text as well as its long-range correlations.

In addition, we focus on the probability distribution of the inter-occurrence distances of the word "parts", in order to know if the model is also able to reproduce it. Motivated by the fact that the function used in the literature to fit this distribution is the stretched exponential [6], in Fig. 34.5 we represent $\log(-\log(P(d)))$ versus $\log d$, where $P(d)$ is the cumulative distribution of the inter-occurrence distances of the word. In this scale, the stretched exponential (see Eq. (34.3)) would appear as a straight line of slope β, but this behaviour is only observed at long distances. At short distances, which are affected by the grammatical rules, the real distribution is overestimated. However, if we plot the distribution of distances provided by our model, we are able to faithfully reproduce the behaviour at all distances since we have had into account the repulsion that real words present at short distances.

Finally, we compute the average value of the relevance measure C over all the iterations of the model, and compare it with the value obtained for the real word. For the specific word "parts", for which $C = 9.71$, we have an average $\langle C \rangle = 12.55$ with a standard deviation $sd_C = 3.9$, and therefore the real C value is clearly within the confidence interval.

The results of the model are completely general: It is possible to find suitable parameters which provide similar fittings for others words of this book, for other books and for different languages (as we will detail in a forthcoming paper). Therefore, we can conclude that the model we have developed is able to retain all the properties of interest of words in written texts: it reproduces the complex behaviour characterized by the presence of correlations at long scales, the spatial structure characterized by the distribution of distances, and the degree of relevance quantified by the clustering measure C.

Acknowledgements This work has been supported by Grant no. P07-FQM03163 from Spanish Junta de Andalucía.

References

1. Carpena P, Bernaola-Galván P, Hackenberg M, Coronado AV, Oliver JL (2009) Level statistics of words: finding keywords in literary texts and symbolic sequences. Phys Rev E 79:035102(R)
2. Montemurro MA, Zanette DH (2010) Towards the quantification of the semantic information encoded in written language. Adv Complex Syst 13(2):135–153
3. Montemurro MA, Pury PA (2002) Long-range fractal correlations in literary corpora. Fractals 10:451–461
4. Bhan J, Kim S, Kim J, Kwon Y, Yang S, Lee K (2006) Long-range correlations in Korean literary corpora. Chaos Solitons Fractals 29:69–81
5. Şahin G, Erentürk M, Hacinliyan A (2009) Detrended fluctuation analysis in natural languages using non-corpus parametrization. Chaos Solitons Fractals 41:198–205
6. Altmann EG, Pierrehumbert JB, Motter AE (2009) Beyond word frequency: bursts, lulls, and scaling in the temporal distributions of words. PLoS ONE 4(11):e7678
7. Ortuño M, Carpena P, Bernaola-Galván P, Muñoz E, Somoza AM (2002) Keyword detection in natural languages and DNA. Europhys Lett 57(5):759–764
8. Voss RF (1992) Evolution of long-range fractal correlations and $1/f$ noise in DNA base sequences. Phys Rev Lett 68:3805–3808
9. Peng C-K, Buldyrev SV, Havlin S, Simons M, Stanley HE, Goldberger AL (1994) Mosaic organization of DNA nucleotides. Phys Rev E 49:1685–1689
10. Hu K, Ivanov PC, Chen Z, Carpena P, Stanley HE (2001) Effect of trends on detrended fluctuation analysis. Phys Rev E 64:011114
11. Makse HA, Havlin S, Schwartz M, Stanley HE (1996) Method for generating long-range correlations for large systems. Phys Rev E 53:5445–5449
12. Carretero-Campos C, Bernaola-Galván P, Ivanov PC, Carpena P (2012) Phase transitions in the first-passage time of scale-invariant correlated processes. Phys Rev E 85:011139

Chapter 35
Shared Information—New Insights and Problems in Decomposing Information in Complex Systems

Nils Bertschinger, Johannes Rauh, Eckehard Olbrich, and Jürgen Jost

Abstract How can the information that a set $\{X_1, \dots, X_n\}$ of random variables contains about another random variable S be decomposed? To what extent do different subgroups provide the same, i.e. shared or redundant, information, carry unique information or interact for the emergence of synergistic information?

Recently Williams and Beer proposed such a decomposition based on natural properties for shared information. While these properties fix the structure of the decomposition, they do not uniquely specify the values of the different terms. Therefore, we investigate additional properties such as strong symmetry and left monotonicity. We find that strong symmetry is incompatible with the properties proposed by Williams and Beer. Although left monotonicity is a very natural property for an information measure it is not fulfilled by any of the proposed measures.

We also study a geometric framework for information decompositions and ask whether it is possible to represent shared information by a family of posterior distributions.

Finally, we draw connections to the notions of shared knowledge and common knowledge in game theory. While many people believe that independent variables cannot share information, we show that in game theory independent agents can have shared knowledge, but not common knowledge. We conclude that intuition and heuristic arguments do not suffice when arguing about information.

N. Bertschinger (✉) · J. Rauh · E. Olbrich · J. Jost
MPI für Mathematik in den Naturwissenschaften, Inselstraße 22, 04109 Leipzig, Germany
e-mail: bertschi@mis.mpg.de

J. Rauh
e-mail: jrauh@mis.mpg.de

E. Olbrich
e-mail: olbrich@mis.mpg.de

J. Jost
e-mail: jost@mis.mpg.de

T. Gilbert et al. (eds.), *Proceedings of the European Conference on Complex Systems 2012*, Springer Proceedings in Complexity, DOI 10.1007/978-3-319-00395-5_35,

35.1 Introduction

The field of complex systems investigates systems which are composed of many components or sub-systems. Such a system is considered as complex if these components interact in intricate ways and exhibit dependencies at all scales. Informally, complex systems are often described in terms of information that is exchanged between components. Thus, information theory is a natural tool to study complex systems.

As an example from neural coding, consider two neurons which provide information about some stimulus. Many scientists have tried to uncover whether both neurons provide redundant information about the stimulus or act synergetically, i.e. provide information which can only be recovered when the joint response of both cells is recorded simultaneously [1, 2]. Similarly, one could ask for the unique information of each response, i.e. information that can be obtained from one of the cells, but not the other. For example, the brain separates visual information into the where and what pathways [3] which potentially provide unique information with respect to each other. Another way to explain the intuition on how information can be decomposed, is to consider two agents which are interrogated about certain topics. For example, assume that one agent is an expert in physics and biology, whereas the other one has studied art and biology. In this case, both agents could answer questions about biology being their shared topic. Furthermore, each agent has additional unique information about physics and art, respectively. Considering their joint responses an interrogator might be able to draw interesting connections between art and physics none of the agents is aware of. This would correspond to the synergetic information in this case.

In general, when considering more than two random variables, there may be different combinations of shared, unique and synergistic information, depending on how the information is distributed among the random variables. The total mutual information $I(S : X_1, \dots, X_n)$ should then be a sum of different terms with a well-defined interpretation. At the moment, it is not clear how many such terms are necessary in the general case of n interacting elements. Williams and Beer recently proposed one such decomposition, which they call *partial information (PI) decomposition* [4]. This decomposition is naturally derived from simple intuitive properties that such a decomposition should satisfy.

Before explaining the construction of Williams and Beer, we first have a look at the case of $n = 2$ explanatory variables in Sect. 35.2. In Sect. 35.3 we discuss natural properties that such a decomposition should satisfy and, following Williams and Beer, use these properties to derive the PI decomposition. In Sect. 35.4 we propose additional properties that relate the values of shared information in situations where we ask for information about different variables. In Sect. 35.5 we discuss the measure $I_{\min}$ proposed by Williams and Beer and compare it to another function I_I, i.e. the minimum of the pairwise mutual informations. We show that the function $I_{\min}$ may decrease when we ask for information about a larger variable. In Sect. 35.6, we study the case for three variables. We show that it is difficult to assign intuitively plausible values to all partial information terms, even in the simple XOR-example.

Using this example we show that the structure of the PI lattice is incompatible with a symmetry property which we call strong symmetry.

In Sect. 35.7 we propose a geometric picture for information decomposition [5]. This view provides an appealing mathematical structure and provides additional insights into the structure of information. Within this geometric framework, we compare our ideas to the measures proposed in [4] and [6]. Then, in Sect. 35.8, we study the game theoretic notions of shared and common knowledge that are used to describe epistemic states of multi-agent systems, and we discuss how these notions are related to the problem of decomposing information. We conclude with an outlook on the possibility of a general decomposition of information.

35.2 The Case of Two Variables

First, we fix the notation and recall some basic definitions from information theory [7]. We assume that a system consists of N components $X_1, \dots, X_N$. For simplicity we assume that the set of possible states $\mathcal{X}_i$ that a component X_i can be in is finite. Thus, the set of all possible states for the whole system is given by $\mathcal{X}_1^N = \times_{i=1}^N \mathcal{X}_i$.

Given a probability distribution p on $\mathcal{X}_1^N$, the X_i become random variables. Mutual information between two random variables X and Y quantifies the information about Y that is gained by knowing X and vice versa. It can be defined as

$$I(X:Y) = \sum_{y \in \mathcal{Y}} p(y) D\big(p(X|y) || p(X)\big) \tag{35.1}$$

where $D(p(X|y)||p(X)) = \sum_{x \in \mathcal{X}} p(x|y) \log_2 \frac{p(x|y)}{p(x)}$ is the Kullback-Leibler (KL) divergence between $p(X|y)$ and $p(X)$.[1] The KL divergence is often considered as a distance between probability distributions even though it is not a metric. But, like a metric, it vanishes if and only if the two distributions are identical. It can also be interpreted as an information gain: if one finds out that $Y = y$ then $D(p(x|y)||p(x))$ bits of information are gained about X. It is well known that the mutual information is symmetric and vanishes if and only if X and Y are independent.

Consider now three random variables X_1, X_2 and S. The (total) mutual information $I(S; (X_1, X_2))$ quantifies the total information that is gained about S if the outcome of X_1 and X_2 is known. How do X_1 and X_2 contribute to this information?

For two explanatory variables, we expect four contributions to $I(S:X_1X_2)$:

$$\begin{aligned} I(S:X_1X_2) = {} & SI(S:X_1;X_2) + UI(S:X_1 \setminus X_2) + UI(S:X_2 \setminus X_1) \\ & + CI(S:X_1;X_2) \end{aligned} \tag{35.2}$$

The shared (redundant) information $SI(S:X_1;X_2)$, the unique informations UI and the complementary (synergistic) information $CI(S:X_1;X_2)$. Intuition tells us

[1] Here, $p(X)$ denotes the probability distribution of the random variable X. When referring to the probability of a particular outcome $x \in \mathcal{X}$ of this random variable, we write $p(x)$.

that the individual mutual informations that are provided by X_1 and X_2 should decompose as

$$\begin{aligned} I(S:X_1) &= SI(S:X_1;X_2) + UI(S:X_1 \setminus X_2) \\ I(S:X_2) &= SI(S:X_1;X_2) + UI(S:X_2 \setminus X_1). \end{aligned} \tag{35.3}$$

Using the full decomposition (35.2) and the chain rule of mutual information [7] we find that the conditional informations correspond to unique and complementary information, e.g. $I(S:X_1|X_2) = UI(S:X_1 \setminus X_2) + CI(S:X_1;X_2)$. Furthermore, we recover the fact that the co-information I_{Co} [8] contemplates shared and complementary information, i.e.

$$\begin{aligned} I_{Co}(S:X_1:X_2) &:= I(S:X_1|X_2) - I(S:X_1) \\ &= CI(S:X_1;X_2) - SI(S:X_1;X_2) \end{aligned} \tag{35.4}$$

Unfortunately, the three linear equations (35.2) and (35.3) do not completely specify the four functions on the right hand side of (35.2). To determine the decomposition (35.2) it is sufficient to define one of the functions SI, UI and CI. It seems to be a difficult task to come up with a reasonable and well-motivated definition of SI such that the induced definitions of UI and CI via Eqs. (35.2) and (35.3) are non-negative. The same is true when trying to find formulas for UI or CI. Note that any definition of the unique information fixes two of the terms in (35.2). This leads to the consistency condition

$$I(S:X_1) + UI(S:X_2 \setminus X_1) = I(S:X_2) + UI(S:X_1 \setminus X_2), \tag{35.5}$$

which resembles the chain rule. Indeed, $UI(S:X_1 \setminus X_2)$ can be considered as a version of conditional information which does not contain the complementary information.[2]

Apart from the problem of finding formulas for SI, UI and CI, a second problem is how to generalize the decomposition (35.2) to more than two explanatory variables. A possible solution to both problems was recently proposed by Williams and Beer.

35.3 Natural Properties of Shared Information and the Partial Information Lattice

Williams and Beer [4] base their construction of a non-negative decomposition of $I(S:X_1 \cdots X_n)$ on the notion of redundancy or shared information. Let

[2] A related notion has been developed in the context of cryptography to quantify the secret information. Although the secret information has a clear operational interpretation it cannot be computed directly, but is upper bounded by the *intrinsic mutual information* $I(S:X_1 \downarrow X_2)$ [9, 10]. Unfortunately, the intrinsic mutual information does not obey the consistency condition (35.5), and hence it cannot be interpreted as unique information in our sense.

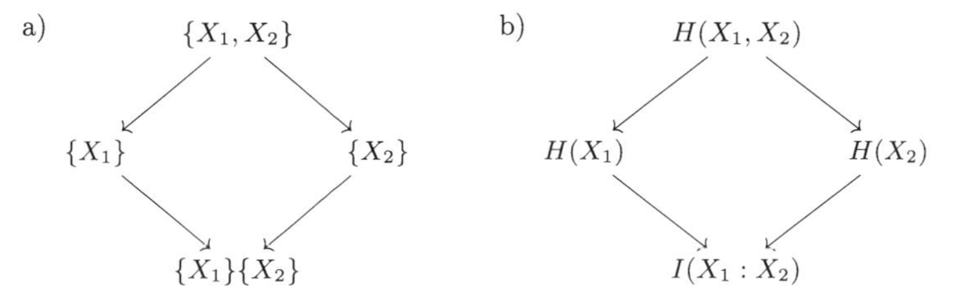

Fig. 35.1 The PI lattice for two random variables. (**a**) The sets corresponding to the nodes in the lattice. (**b**) The redundancies at the nodes for $S = \{X_1, X_2\}$, assuming strong symmetry (see ($\mathbf{S}_1$) in Sect. 35.4)

$\mathbf{A}_1, \ldots, \mathbf{A}_k \subseteq \{X_1, \ldots, X_n\}$, and denote by $I_\cap(S : \mathbf{A}_1; \ldots; \mathbf{A}_k)$ the information about S that is shared among the random variables in the sets $\mathbf{A}_1, \ldots, \mathbf{A}_k$. It is natural to demand that $I_\cap$ satisfy the following properties:

(**GP**) $I_\cap(S : \mathbf{A}_1; \ldots; \mathbf{A}_k) \geq 0$ *(global positivity)*.
($\mathbf{S}_0$) $I_\cap(S : \mathbf{A}_1; \ldots; \mathbf{A}_k)$ is symmetric in $\mathbf{A}_1, \ldots, \mathbf{A}_k$ *(weak symmetry)*.
(**I**) $I_\cap(S : \mathbf{A}) = I(S : \mathbf{A})$ equals the mutual information of S and $\mathbf{A}$ *(self-redundancy)*.
(**M**) $I_\cap(S : \mathbf{A}_1; \ldots; \mathbf{A}_k) \leq I_\cap(S : \mathbf{A}_1; \ldots; \mathbf{A}_{k-1})$, with equality if $\mathbf{A}_{k-1}$ is a subset of $\mathbf{A}_k$ *(monotonicity)*.

The properties ($\mathbf{S}_0$), (**I**) and (**M**) have been proposed as axioms of shared information by Williams and Beer in [4]. As Williams and Beer observe, (**GP**) is a consequence of the other properties. Here we like to state it as a separate property, since we want to discuss what happens if we drop or relax some of these properties.

The properties ($\mathbf{S}_0$) and (**M**) imply that it is sufficient to define the function $I_\cap(S : \mathbf{A}_1; \ldots; \mathbf{A}_k)$ in the case that $\mathbf{A}_i \not\subseteq \mathbf{A}_j$ for all $i \neq j$. A family of sets $\mathbf{A}_1, \ldots, \mathbf{A}_k$ with this property is called an *anti-chain*. The anti-chains form a lattice with respect to the partial order defined by $(\mathbf{B}_1, \ldots, \mathbf{B}_k) \leq (\mathbf{A}_1, \ldots, \mathbf{A}_l)$ if and only if for each $i = 1, \ldots, l$ there exists $j \in \{1, \ldots, k\}$ such that $\mathbf{B}_j \subseteq \mathbf{A}_i$. If S is fixed, then ($\mathbf{S}_0$) and (**M**) imply that $I_\cap(S : \cdot)$ is a monotone function on the lattice of anti-chains of $\{X_1, \ldots, X_n\}$: If $(\mathbf{B}_1, \ldots, \mathbf{B}_k) \leq (\mathbf{A}_1, \ldots, \mathbf{A}_l)$, then

$$I_\cap(S : \mathbf{B}_1, \ldots, \mathbf{B}_k) = I_\cap(S : \mathbf{B}_1, \ldots, \mathbf{B}_k, \mathbf{A}_1, \ldots, \mathbf{A}_k) \leq I_\cap(S : \mathbf{A}_1, \ldots, \mathbf{A}_l).$$

This lattice is also called the *partial information (PI) lattice*. In this paper, we focus on the case of two or three random variables, and the corresponding lattices are depicted in Figs. 35.1 and 35.2.

Properties (**M**) and (**I**) imply $I_\cap(S : \mathbf{A}_1; \ldots; \mathbf{A}_k) \leq I_\cap(S : \mathbf{A}_1) = I(S : \mathbf{A}_1) \leq I(S : X_1 \ldots X_n)$. To obtain a decomposition of this total mutual information, we need to associate to each element of the PI lattice a "local quantity" I_∂ in such a way that

$$I_\cap(S : \mathbf{A}_1; \ldots; \mathbf{A}_k) = \sum_{(\mathbf{B}_1, \ldots, \mathbf{B}_l) \leq (\mathbf{A}_1, \ldots, \mathbf{A}_k)} I_\partial(S : \mathbf{B}_1, \ldots, \mathbf{B}_l).$$

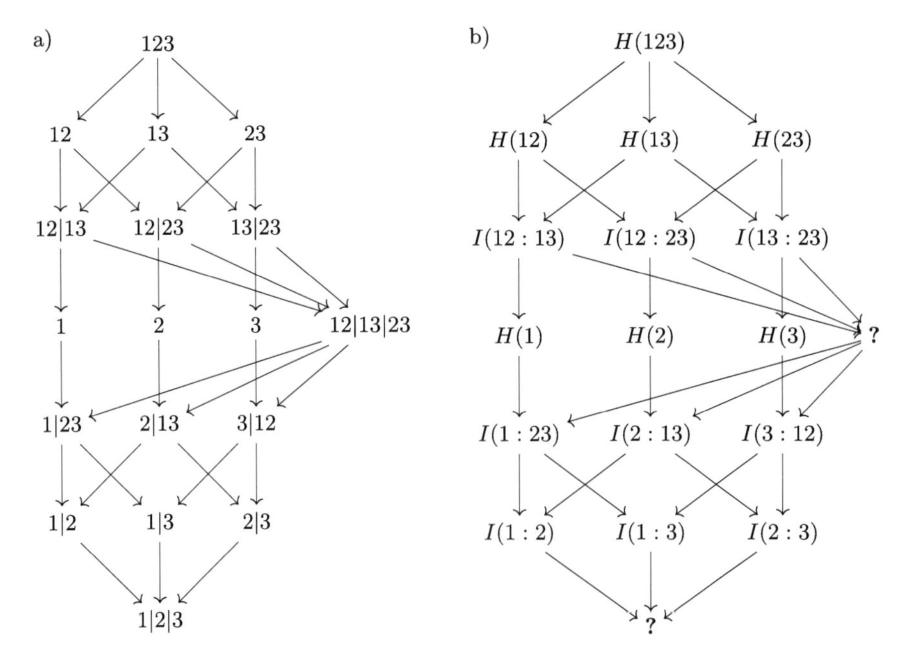

Fig. 35.2 The PI lattice for $n = 3$. For simplicity the sets are abbreviated by juxtaposing the indices of the corresponding variables. For example, 12|13 corresponds to $\{X_1, X_2\}\{X_1, X_3\}$. (**a**) The PI lattice. (**b**) The redundancies at the nodes, assuming strong symmetry and $S = \{X_1, X_2, X_3\}$

One can show, using the notion of a Möbius inversion, that such a function I_∂ always exists, and I_∂ is uniquely determined from $I_\cap$.

As an example consider again the case of two variables (Fig. 35.1). When S is given, then the upper three terms in the lattice correspond to the mutual informations $I(S : X_1)$, $I(S : X_2)$ and $I(S : X_1 X_2)$. The lowest term, $I_\cap(S : X_1; X_2)$ is the shared information $SI(S : X_1; X_2)$. The PI decomposition is

$$\begin{aligned} I_\cap\big(S : \{X_1 X_2\}\big) &= I_\partial\big(S : \{X_1\}\{X_2\}\big) + I_\partial\big(S : \{X_1\}\big) \\ &\quad + I_\partial\big(S : \{X_2\}\big) + I_\partial\big(S : \{X_1 X_2\}\big), \\ I_\cap\big(S : \{X_1\}\big) &= I_\partial\big(S : \{X_1\}\{X_2\}\big) + I_\partial\big(S : \{X_1\}\big), \\ I_\cap\big(S : \{X_2\}\big) &= I_\partial\big(S : \{X_1\}\{X_2\}\big) + I_\partial\big(S : \{X_2\}\big), \\ I_\cap\big(S : \{X_1\}\{X_2\}\big) &= I_\partial\big(S : \{X_1\}\{X_2\}\big). \end{aligned}$$

A comparison with (35.2) and (35.3) shows that

$$\begin{aligned} I_\partial\big(S : \{X_1 X_2\}\big) &= CI(S : X_1; X_2), \\ I_\partial\big(S : \{X_1\}\big) &= UI(S : X_1 \setminus X_2), \end{aligned}$$

$$I_\partial\big(S:\{X_2\}\big) = UI(S:X_2 \setminus X_1),$$

$$I_\partial\big(S:\{X_1\}\{X_2\}\big) = SI\big(S:\{X_1\}\{X_2\}\big).$$

As stated above, when $I_\cap$ is known, then I_∂ can be computed uniquely using a Möbius inversion. In general, I_∂ may have negative values. In order to have a natural interpretation of the PI decomposition, we need to require:

(LP) $I_\partial \geq 0$ *(local positivity).*

Local positivity can also be expressed as a condition on $I_\cap$, see [4].

35.4 Further Natural Properties of Shared Information

The properties presented in the preceding section were identified by Williams and Beer and are naturally related to the notion of the PI lattice. Unfortunately, they are not enough to specify the function $I_\cap$ uniquely. The properties are incomplete for mainly two reasons: First, they do not tell us much about the left hand side apart from the normalization condition (**I**). Second, they do not tell us enough about what happens when we add another argument on the right.

In this section we propose natural properties that describe the role of the left-hand side. Our first proposal is the following property:

($\mathbf{S}_1$) $I_\cap(S:\mathbf{A}_1;\dots;\mathbf{A}_k)$ is symmetric in $S,\mathbf{A}_1,\dots,\mathbf{A}_k$ *(strong symmetry).*

In the following, we mostly consider the case that $S=\{X_1,\dots,X_n\}$, and in this case (**M**) and ($\mathbf{S}_1$) together imply that $I_\cap(S:\mathbf{A}_1;\dots;\mathbf{A}_k) = I_\cap(\mathbf{A}_1:\mathbf{A}_2;\dots;\mathbf{A}_k)$, and hence we may omit the first argument S.

Unfortunately, strong symmetry is not satisfied by many information theoretic quantities that are used to quantify shared information or synergy, but nevertheless we think that it is natural: If $I_\cap$ has just two arguments, then strong symmetry does hold, since the mutual information is symmetric. In other words, the amount of information that one random variable X_1 contains about another variable X_2 is the same as the amount of information that X_2 carries about X_1. It is natural to assume that an analogous statement should hold if $I_\cap$ has more than two arguments. Note that the co-information I_{Co} is symmetric in all its arguments.

Under the strong symmetry assumption, if we consider two variables X_1 and X_2 and set $S=\{X_1,X_2\}$, then all functions are fixed. The corresponding lattice is depicted in Fig. 35.1(b). We will see later that, given the other properties, strong symmetry contradicts the local positivity in the case of three random variables X_1,X_2,X_3. The implications of this will be discussed later.

A weaker property restricting the dependence on the first argument is the following:

(LM) $I_\cap(S:\mathbf{A}_1;\dots;\mathbf{A}_k) \leq I_\cap(SS':\mathbf{A}_1;\dots;\mathbf{A}_k)$ *(left monotonicity).*

This property captures the intuition that if $\mathbf{A}_1, \dots, \mathbf{A}_k$ share some information about S, then at least the same amount of information is available to reduce the uncertainty about the joint outcome of S and S'. Left monotonicity follows, of course, from monotonicity and strong symmetry.

Another property, which is independent from strong symmetry and which also implies (**LM**), is the following:

(**LC**) $I_\cap(SS' : \mathbf{A}_1; \dots; \mathbf{A}_k) = I_\cap(S : \mathbf{A}_1; \dots; \mathbf{A}_k) + I_\cap(S' : \mathbf{A}_1; \dots; \mathbf{A}_k|S)$ *(left chain rule).*

where $I_\cap(S' : \mathbf{A}_1; \dots; \mathbf{A}_k|S)$ is given by $\sum_{s \in \mathcal{S}} p(s) I_\cap(S' : \mathbf{A}_1; \dots; \mathbf{A}_k|s)$, i.e. all distributions are conditioned on s and then the average is taken to obtain a conditional information. This property is a natural generalization of the chain rule of mutual information. Moreover, a similar property is used in Shannon's axiomatic characterization of entropy.

Unfortunately, the left chain rule is not fulfilled by any of the proposed measures for shared information that we discuss later. Nevertheless, we state it here, since we find it mathematically appealing. The same is true for left monotonicity: Most measures do not satisfy (**LM**), see Sect. 35.5.

The left chain rule together with local positivity also implies the following property which has recently been proposed by [6]:

(**Id**$_2$) $I_\cap(\mathbf{A}_1 \cup \mathbf{A}_2 : \mathbf{A}_1; \mathbf{A}_2) = I(\mathbf{A}_1 : \mathbf{A}_2)$ *(identity).*

The identity property implies that $I_\cap(\{X_1, X_2\} : X_1; X_2)$ vanishes if X_1 and X_2 are independent. At first sight it seems natural that independent random variables cannot share information. However, in Sect. 35.8 we will argue that they may indeed share information in this case.

35.5 The Functions $I_{\min}$ and I_I

Williams and Beer define a function $I_{\min}(S, \mathbf{A}_1, \dots, \mathbf{A}_k)$ which satisfies all their properties (**GP**), (**S**$_0$), (**I**) and (**M**) as follows:

$$
\begin{aligned}
I_{\min}(S : \mathbf{A}_1; \dots; \mathbf{A}_k) &= \sum_s p(s) \min_i \sum_{a_i} p(a_i|s) \log \frac{p(s|a_i)}{p(s)} \\
&= \sum_s p(s) \min_i \sum_{a_i} p(a_i|s) \log \frac{p(a_i|s)}{p(a_i)} \\
&= \sum_s \min_i \sum_{a_i} p(a_i, s) \log \frac{p(a_i, s)}{p(a_i) p(s)}.
\end{aligned}
$$

The idea is the following: For each i compare the prediction $p(s|a_i)$ of S by $\mathbf{A}_i$ with the prior distribution $p(s)$ of S. Then combine a minimization over i with a suitable average using the joint distribution of $\mathbf{A}_i$ and S.

The order of the minimization and the averaging plays a crucial role. If we interchange it, we obtain another function

$$I_I(S:A_1;\ldots;A_k)=\min_i \sum_s p(s)\sum_{a_i} p(a_i|s)\log\frac{p(s|a_i)}{p(s)}=\min_i\bigl\{I(S:A_i)\bigr\}.$$

This function I_I satisfies the same properties, including local positivity (**LP**) (the proof of [4] that proves (**LP**) for $I_{\min}$ applies). Of course, I_I does not at all capture the intuition behind the notion of shared information: I_I just compares absolute values of mutual informations, without caring whether different variables contain "the same information." We will later argue that $I_{\min}$ suffers from a similar flaw (in particular, $I_\cap = I_I$ in the examples considered below). Note that any function $I_\cap$ satisfying the properties (**GP**), (**S**$_0$), (**I**) and (**M**) satisfies $I_\cap \le I_I$. In particular, $I_{\min} \le I_I$.

The function I_I satisfies left monotonicity. However, $I_{\min}$ does not: For example, the following joint probability distribution

X_1	X_2	S	S'	
0	0	0	0	1/6
0	1	0	0	1/6
0	1	0	1	1/6
1	1	0	1	1/6
1	0	1	1	2/6

satisfies $I_{\min}(S:X_1;X_2)=\frac{1}{3}+\frac{2}{3}(\frac{3}{4}\log_2 3-1) > I_{\min}(SS':X_1;X_2)=\frac{1}{3}$. This example can be understood as follows: If $S=0$, then both X_1 and X_2 have some information about S and thus contribute $\frac{3}{4}\log_2 3-1$ bits to $I_{\min}$ in this case. However, if we additionally condition on S', then in any case one of X_1 or X_2 carries no information: To be precise, if $(S,S')=(0,0)$, then X_2 is uniformly distributed, and if $(S,S')=(0,1)$, then X_1 is uniformly distributed. Thus, in both cases the minimization contributes zero bits to $I_{\min}$. The remaining case $(S,S')=(1,1)$ is equivalent to the case $S=1$, where both X_1 and X_2 are fixed, and contributes one bit with weight $\frac{1}{3}$.

Omitting the calculations we mention that the redundancy measure proposed by [6] (and denoted by I_{HSP} in Sect. 35.7) also violates left monotonicity in the same example.

35.6 The Case of Three Variables

For three variables, the PI lattice is depicted in Fig. 35.2(a). Under the assumption of strong symmetry all but two values in this lattice are fixed, see Fig. 35.2(b). The unknown values correspond to the information shared by three random variables.

In the following, we discuss an example with three random variables X_1, X_2, X_3: Assume that X_1 and X_2 are independent binary random variables, and let

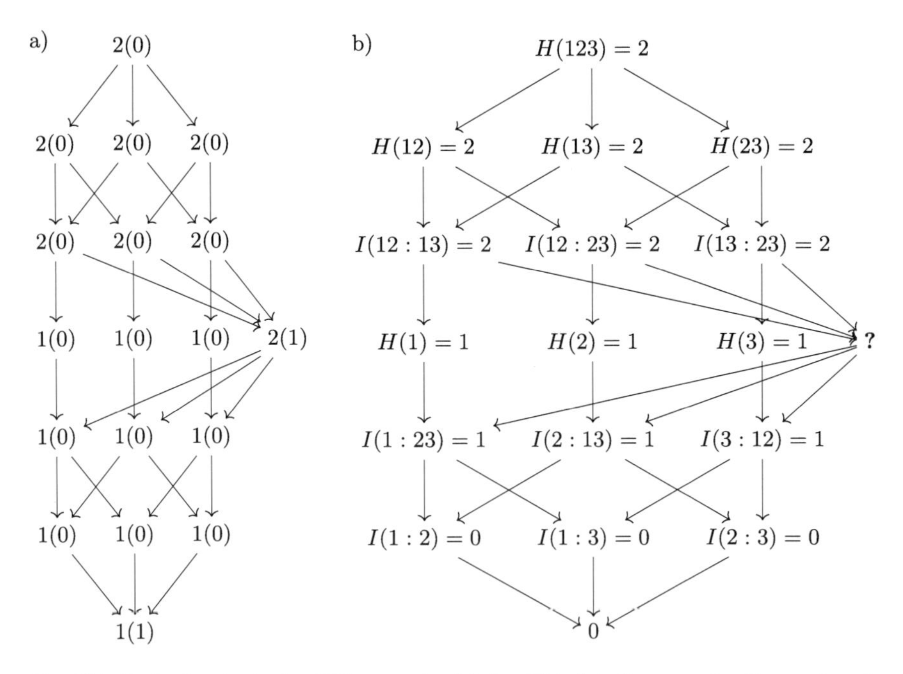

Fig. 35.3 Redundancies in the XOR-example: (**a**) $I_{min}(123, \cdot)$ in the example. The numbers in parentheses are $I_\partial(123, \cdot)$. (**b**) The shared information assuming strong symmetry

$X_3 = X_1 \oplus X_2$, where $\oplus$ denotes the sum modulo 2 or the XOR-function. Note that this example is symmetric in X_1, X_2 and X_3. Figure 35.3(a) shows the values of I_{min} and I_∂ in this example for $S = \{X_1, X_2, X_3\}$; in other words, we decompose the information that the system has about itself. What is striking is that the lowest entry in this lattice does not vanish: According to I_{min}, X_1, X_2 and X_3 share one bit of information, although they are pair-wise independent. This fact that independent variables may share information according to I_{min} has also been observed and criticized in [6]. We will later give an argument from game theory that explains how independent variables can share information. Nevertheless, in our opinion one bit of shared information is too much in this situation: The absolute value of one bit of shared information needs to be compared to the fact that each of X_1, X_2, X_3 does not carry more than one bit of information. Note that in the XOR-example $I_{min} = I_I$.

A close analysis of this also reveals that strong positivity is incompatible with the PI lattice:

Theorem 35.1 *There is no measure of shared information that satisfies* (**S**$_1$), (**M**), (**I**) *and* (**LP**).

Proof Assume that $I_\cap$ is a monotone function on the PI lattice that satisfies strong symmetry (**S**$_1$). In the PI lattice for the XOR-example we can express all values on the lattice in terms of entropies and mutual informations, with one exception, see Fig. 35.3(b). Note that, by strong symmetry, $I_\cap(X_1X_2X_3 : \mathbf{A}_1; \dots; \mathbf{A}_k) = I_\cap(\mathbf{A}_1; \dots; \mathbf{A}_k)$ whenever $\mathbf{A}_1 \cup \cdots \cup \mathbf{A}_k \subseteq \{X_1, X_2, X_3\}$. Comparing with Fig. 35.2(b) we see that the information shared by X_1, X_2 and X_3 must vanish by monotonicity, since the terms on the next layer also vanish, $I(X_i, X_j) = 0$ for $i \neq j$. Only the information shared by the pairs $\{X_1, X_2\}$, $\{X_1, X_3\}$ and $\{X_2, X_3\}$ is not determined. However, we can bound these terms by the monotonicity. Similarly, we can compute bounds on I_∂. Namely,

$$\begin{aligned} &I_\partial\big(\{X_1, X_2\}; \{X_1, X_3\}; \{X_2, X_3\}\big) \\ &\quad = I_\cap\big(\{X_1, X_2\}; \{X_1, X_3\}; \{X_2, X_3\}\big) - I_\cap\big(\{X_1\}; \{X_2, X_3\}\big) \\ &\qquad - I_\cap\big(\{X_2\}; \{X_1, X_3\}\big) - I_\cap\big(\{X_3\}; \{X_1, X_2\}\big) \pm 0 \le 2 - 3 = -1, \end{aligned}$$

where ± 0 represents a sum of terms belonging to the lowest two layers of the PI diagram, and these terms all vanish. This calculation shows that local positivity is not possible. □

To resolve this problem, one of the properties mentioned in Theorem 35.1 has to be dropped. The easiest solution is to drop strong symmetry. What are the alternatives? We have to keep self-redundancy (**I**) and local positivity (**LP**), since we want to find a decomposition of mutual-information into positive terms. Therefore, if we want to keep strong symmetry, we need to replace monotonicity (**M**). It is probably a good idea to keep the inequality condition in (**M**), but it is conceivable to replace the equality condition. However, one must keep in mind that the equality condition is essential in justifying the use of the PI lattice: Without this condition the values of the function $I_\cap$ on arbitrary collections of subsets are not determined by its values on the antichains, and so the PI lattice is not any more the natural domain of shared information. Therefore, without the equality condition in (**M**) we need to compute many more terms to completely specify $I_\cap$. In turn, this means that there are many more local terms I_∂. With these additional terms it may be possible to obtain local positivity and strong symmetry at the same time.

Heuristically, what happens in the XOR-example is the following: The term $I(X_1 : X_2X_3)$ on the third layer in Fig. 35.2 (counted from below) is equal to one bit, since we can compute X_1 from X_2 and X_3, and hence $I(X_1 : X_2X_3) = H(X_1)$. Intuitively, the information shared between X_1 and $\{X_2, X_3\}$ is precisely the information contained in X_1. However, the three terms $I(X_1 : X_2X_3)$, $I(X_2 : X_1X_3)$ and $I(X_3 : X_1X_2)$ on the third layer are not independent, since X_1, X_2 and X_3 are not completely independent, but only pairwise independent. Hence, if we compute the information shared by all three pairs, we cannot just add up these three bits: We have to subtract (at least) one bit, which we overcounted. Somehow this one bit that we overcounted does not have a place in the PI lattice.

If we drop strong symmetry and keep the PI lattice, it is still the question how to distribute the information over the PI lattice in the XOR-example. In any case, monotonicity implies that $I_\cap(S : X_1X_2; X_1X_3; X_2X_3) \le I(S : X_1X_2X_3) = 2$. On the other hand, the other three values on the third layer, the three mutual informations $I(S : X_i)$, are all equal to one bit. These values restrict the possible values of I_∂, and it is not easy to motivate a non-negative assignment on intuitive grounds, even for this simple example.

35.7 A Geometric Picture of Shared Information

One problem that makes it difficult to define shared information is that there is no known experimental way to extract shared information. In this section we want to assume that shared information can be extracted or modelled concretely. We not only search for a number that measures the amount of shared information, but we want to represent the information itself.

As a motivation consider the case of two random variables X, Y from the perspective of coding theory. Suppose that we want to transmit information about X and Y over some channel. Then the capacity that we need must exceed the amount of information that we want to transmit. To transmit a single variable X, we need a capacity of $H(X)$. To be precise, this statement only becomes true asymptotically: When we want to transmit a string of n values of n independent copies of X, then, for large n, if we have a channel with a capacity of $H(X)$ per time unit ΔT, then the time needed to transmit X is roughly $n\Delta T$. In the same sense, to transmit X and Y together, we need a channel of capacity $H(\{X, Y\})$. Suppose that X was already transmitted, i.e. both sender and receiver know the value of X. As Shannon showed, in this case a channel of capacity $H(Y|X) = H(\{Y, X\}) - H(X)$ is sufficient to transmit the remaining information, such that the receiver knows both X and Y. Hence, $H(Y|X)$ has the natural interpretation of unique information of Y with respect to X, and as Shannon's theorem shows, the unique information can be isolated and transmitted separately. The question is: Which other parts of information can be isolated?

As before, we consider information about a random variable S. We follow the paradigm that our information or belief about S can be encoded in a probability distribution $p(S)$. Suppose that X is another random variable. If S is not independent of X, then a measurement of X gives us further information about S. For example, if we know that $X = x$, then our belief about S can be encoded in the conditional probability distribution $p(S|x)$. Thus, the information that X carries about S can be encoded in a family $\{p(S|x)\}_{x\in\mathcal{X}}$ of probability distributions for S. These distributions encode the *posterior* beliefs about S conditioned on each outcome of X.

As motivated by Shannon, information can be quantified by logarithms of probabilities: The information that the state of the variable S is equal to the specific value s is worth $-\log_2(p(S = s))$. Our uncertainty about S, when our knowledge is encoded in the distribution $p(S)$, is then equal to the expected information gain when we learn the value of S:

$$\sum_s p(s)\big(-\log\big(p(s)\big)\big) =: H(S).$$

Similarly, the information that we gain when we learn that $X = x$ is equal to the conditional entropy $H(S|X = x) = -\sum_s p(s|x)\log(p(s|x))$. The (expected) information that X brings us about S is obtained by averaging $H(S|X = x)$ and comparing the value with $H(S)$; this agrees with the mutual information:

$$\sum_x p(x)\sum_s p(s|x)\log\bigl(p(s|x)\bigr) - \sum_s p(s)\log\bigl(p(s)\bigr)$$
$$= \sum_x p(x)\sum_s p(s|x)\log\left(\frac{p(s|x)}{p(s)}\right) = I(X:S).$$

The situation can be pictured geometrically. Let $\mathcal{P}_S$ be the set of all probability distributions for S. Geometrically,

$$\mathcal{P}_S = \left\{p : \mathcal{S} \to \mathbb{R} : p(s) \geq 0, \sum_{s\in\mathcal{S}} p(s) = 1\right\}$$

is a simplex. The family $\{p(S|x)\}_x$ is a point configuration in $\mathcal{P}_S$, indexed by the outcomes x of the random variable X. The information gain is then the mean reduction of uncertainty (in the sense of Shannon information) when replacing the prior $p(S)$ with the family $\{p(S|x)\}_x$.

According to our geometric interpretation of information, the shared information that $X_1, \ldots, X_k$ carry about S should also be representable as a weighted family of probability distributions for S. The question is how to construct this weighted family from the posteriors $\{p(S|x_i)\}_{x_i}$ and the joint distribution of $X_1, \ldots, X_k$ and S. Suppose that we have found such a family representing the shared information, and denote it by $\{p_{x_1|x_2|\cdots|x_k}(S)\}_{x_1,\ldots,x_k}$. Then we want to quantify the shared information. There are two natural possibilities:

$$SI_{lr}(S : X_1; \ldots; X_k) := \sum_{x_1,\ldots,x_k}\sum_s p(s, x_1, \ldots, x_k)\log\left(\frac{p_{x_1|x_2|\cdots|x_k}(s)}{p(s)}\right)$$
$$SI_{KL}(S : X_1; \ldots; X_k) := \sum_{x_1,\ldots,x_k} p(x_1, \ldots, x_k)D(p_{x_1|x_2|\cdots|x_k}||p).$$

The function SI_{KL} has the advantage that it always satisfies global positivity, regardless of how we construct $p_{x_1|x_2|\ldots|x_k}$. By contrast, the function SI_{lr} directly measures the change of surprise when we replace the prior distribution $p(s)$ with the distribution $p_{x_1|x_2|\cdots|x_k}$. Depending on how we construct $p_{x_1|x_2|\cdots|x_k}$ the value of SI_{lr} may become negative.

We would like to have the following properties:

1. The construction should be symmetric in $x_1, \ldots, x_k$.
2. If $k = 1$, then we obtain the posterior: $p_{x_1}(S) = p(S|x_1)$.
3. More variables share less information:
   $D(p_{x_1|\cdots|x_k}(S)||p(S)) \leq D(p_{x_1|\cdots|x_{k-1}}(S)||p(S))$.

These properties are related to the properties (**S**), (**I**) and (**M**) as stated above, but re-formulated to hold point wise for each joint outcome $x_1, \dots, x_k$.

A natural candidate satisfying the above properties is given by

$$p_{x_1|\cdots|x_k}(S) = \operatorname{argmin}\left\{ D\left(\sum_{i=1}^{k} \lambda_i p(S|x_i) || p(S) \right) : \lambda_i > 0, \sum_i \lambda_i = 1 \right\}.$$

Since the KL divergence is convex, the function $p \mapsto D(p||p(S))$ has a unique minimum on any closed convex set. This shows that the above definition is well-defined. Moreover, the definition ensures that $p_{x_1|\cdots|x_k}(S)$ belongs to the convex hull of the posteriors $p(S|x_i)$ for $i = 1, \dots, k$. This models the fact that $p_{x_1|\cdots|x_k}(S)$ only involves information that is present in these posteriors. In fact, $p_{x_1|\cdots|x_k}$ is the least informative distribution from this convex set.

The construction of $p_{x_1|\cdots|x_k}(S)$ implies the following property, which gives an idea in which sense $p_{x_1|\cdots|x_k}(S)$ summarizes information shared among all the posteriors $p(S|x_i)$:

Lemma 35.1 *If all $p(S|x_i)$ satisfy some linear inequality, then $p_{x_1|\cdots|x_k}(s_2)$ satisfies the same inequality. In particular:*

1. *If $p(s_1|x_i) \leq p(s_2|x_i)$ for all i, then $p_{x_1|\cdots|x_k}(s_1) \leq p_{x_1|\cdots|x_k}(s_2)$.*
2. *If $p(s|x_i) = 0$ for all i, then $p_{x_1|\cdots|x_k}(s) = 0$.*

Unfortunately, SI_{lr} violates monotonicity, and with SI_{KL} the synergy can become negative. Both facts can be illustrated with the same example:

From (35.4) we find that

$$CI(S : X_1; X_2) = I(S : X_1|X_2) - I(S : X_1) + SI(S : X_1; X_2)$$

and thus the non-negativity of CI requires that

$$SI(S : X_1; X_2) \geq I(S : X_1) - I(S : X_1|X_2) = I_{Co}(S : X_1 : X_2). \tag{35.6}$$

Now, if S is a function of X_2, then $I(S : X_1|X_2)$ vanishes, and therefore (35.6) implies $SI(S : X_1; X_2) \geq I(S : X_1)$. Together with (**M**) we obtain

$$SI(S : X_1; X_2) = I(S : X_1), \quad \text{if } S \text{ is a function of } X_2. \tag{35.7}$$

Consider the following distribution

s	x_1	x_2	$p(s, x_1, x_2)$
0	0	0	2/6
0	1	0	1/6
1	0	1	1/6
1	1	1	2/6

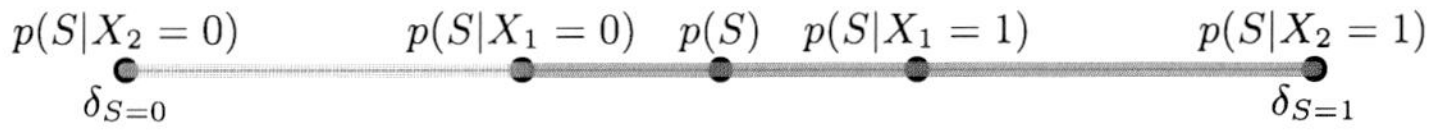

$p(S|X_2=0)$ $p(S|X_1=0)$ $p(S)$ $p(S|X_1=1)$ $p(S|X_2=1)$
$\delta_{S=0}$ $\delta_{S=1}$

Fig. 35.4 The construction of $p_{x_1|x_2}$ for the example to SI_{KL} and SI_{lr}. The set of probability distributions of the binary variable S is the interval between the two point measures $\delta_{S=0}$ and $\delta_{S=1}$. The convex hull of $p(S|X_1=0)$ and $p(S|X_2=0)$ is marked in *green*. The closest point to the prior is $p(S|X_1=0)$. The convex hull of $p(S|X_1=0)$ and $p(S|X_2=1)$ is marked in *red*; it contains the prior

The relative location of $p(S)$ and the posteriors of S given one or two of X_1 and X_2 is visualized in Fig. 35.4. Under this distribution S and X_1 are positively correlated, while $S=X_2$, and thus $I(S:X_1|X_2)=0$. Consider the case $X_1=x_1\neq x_2=X_2$ in which X_1 and X_2 have conflicting posterior about S, i.e. $p(S|x_2)$ assigns probability one to $S=x_2$, whereas $p(S|x_1)$ assigns a higher probability to $S=x_1\neq x_2$. Thus, $p_{x_1|x_2}(S)$ is equal to the prior $p(S)$ in this case. On the other hand, if $X_1=X_2=x$, then both posteriors favor $S=x$. The convex hull of $p(S|x_1)$ and $p(S|x_2)$ is an interval, and the posterior $p(S|x_1)$ is the closest point to the prior $p(S)$. Therefore, $p_{x_1|x_2}(S)=p(S|x_1)$. In total,

$$\begin{aligned}
&I(S:X_1)-SI_{KL}(S:X_1;X_2)\\
&\quad=\sum_{x_1,x_2}p(s,x_1,x_2)\big(D_{KL}\big(p(S|x_1)||p(S)\big)-D_{KL}\big(p_{x_1|x_2}(S)||p(S)\big)\big)\\
&\quad=\sum_{x_1\neq x_2}p(s,x_1,x_2)D_{KL}\big(p(S|x_1)||p(S)\big)>0,
\end{aligned}$$

and therefore (35.7) is violated. One can check that in this case $SI_{lr}(S:X_1;X_2)$ also violates (35.7), but in the other direction. Therefore, SI_{lr} violates monotonicity.

The geometric strategy pursued in this section can be compare with the strategy by Williams and Beer in [4] that leads to the definition of $I_{\min}$. The formula

$$\begin{aligned}
I_{\min}(S:\mathbf{A}_1;\dots;\mathbf{A}_k)&=\sum_s p(s)\min_i\sum_{a_i}p(a_i|s)\log\frac{p(a_i|s)}{p(a_i)}\\
&=\sum_s p(s)\min_i D\big(p(\mathbf{A_i}|S)||p(\mathbf{A_i})\big)
\end{aligned}$$

defining $I_{\min}(S;\mathbf{A_1};\dots;\mathbf{A_n})$ is similar to the defining equation of SI_{KL}, but involves the conditional distributions $p(a_i|s)$ of the input given the output S. In our opinion it is much more natural to work with distributions over the output variable S, since, after all, we are interested in information about S. Of course, the defining equation of $I_{\min}$ can be rewritten in the form

$$I_{\min}(S:\mathbf{A}_1;\dots;\mathbf{A}_k)=\sum_s p(s)\min_i\sum_{a_i}p(a_i|s)\log\frac{p(s|a_i)}{p(s)},$$

which resembles the definition of SI_{lr}, but involves minimizing over the inputs.

The proposed definition of the posteriors $p_{x_1|\cdots|x_k}(s)$ involves similar ideas as the definition of shared information in [6]. We only sketch these connections and refer to the manuscript [6] for the precise definitions. To distinguish their function from other functions we call it I_{HSP}. The definition of $I_{HSP}(S : X_1; X_2)$ involves approximating the posteriors $p(s|x_1)$ by the convex hull family of posteriors $p(s|x_2)$ for all possible values x_2 of X_2. However, as defined in [6] this approximation, denoted by $p_{(x_1 \searrow X_2)}(s)$, is not unique. Then

$$I_{HSP}(S : X_1; X_2) = \min\left\{\sum_{s,x_1} p(s, x_1) \log \frac{p_{(x_1 \searrow X_2)}(s)}{p(s)}, \right.$$

$$\left. \sum_{s,x_2} p(s, x_2) \log \frac{p_{(x_2 \searrow X_1)}(s)}{p(s)} \right\}.$$

Note that in both definitions of $p_{(x_1 \searrow X_2)}(s)$ and $p_{x_1|\cdots|x_k}(s)$ the notion of the convex hull is used as a means to describe the set of distributions that involve information contained in a set of posterior distributions. The difference between both approaches is that [6] do not try to extract and represent the joint information pointwise, but they try to model the information contained in X_1 using the posterior distributions of X_2. This breaks the symmetry, and therefore, in the end, one has to take a minimum. Furthermore, this definition is only meaningful in the case of two random variables and violates the left monotonicity (see Sect. 35.5).

35.8 Game Theoretic Intuitions

Without an operational definition it is hard to decide which of the above properties and geometric structures are best suited to capture the concept of shared information. In order to get a better idea of what is actually meant when talking about shared information, we highlight some aspects from the perspective of game theory.

Scientists in both game theory [11] and computer science [12] have studied how knowledge is distributed among a group of agents. Since knowledge can be regarded as certain information, results from these disciplines can provide additional insights into shared information. The basic formalism of epistemic agents considers a set $\mathcal{S}$ of possible states of the world or situations. The knowledge of an agent i is represented as a partition X_i on $\mathcal{S}$. Such a partition can be considered as a function $X_i : \mathcal{S} \to \mathcal{X}_i$ mapping states of the world to possible observations $\mathcal{X}_i$ that are available to the agent.[3] Thus, each agent i might not be able to observe the actual state s of the world, but given an observation x_i he considers all situations in $X_i^{-1}(x_i) = \{s \in \mathcal{S} \mid X_i(s) = x_i\}$ to be possible.

[3]Note the similarity to the definition of a random variable as a measurable map from a probability space to outcomes. In fact, if we choose an arbitrary probability distribution on $\mathcal{S}$, then the partition X_i, considered as a function $\mathcal{S} \to \mathcal{X}_i$, becomes a random variable.

Suppose that agent i observes $x_i \in \mathcal{X}_i$. Then i is said to know an event, corresponding to a subset $E \subset \mathcal{S}$, if the event occurs in all situations that the agent holds possible given x_i, i.e.

$$X_i^{-1}(x_i) \subseteq E.$$

This gives rise to the knowledge operators $K_i : 2^{\mathcal{S}} \to 2^{\mathcal{S}}$ taking an event E to all situations where agent i knows this event:

$$K_i(E) = \{s \in S \mid \text{agent } i \text{ knows } E \text{ given the observation } X_i(s)\}. \tag{35.8}$$

$K_i(E)$ can itself be considered as an event. Using this operator K_i, we can compute the situations where an event E is shared knowledge between agents $1, \dots, n$, i.e. where every agent knows E:

$$SK(E) = \bigcap_{i=1}^{n} K_i(E)$$

Note that this does not imply that every agent knows that every agents know E. The much stronger requirement that everyone knows E, and everyone knows that everyone knows this, and so on, is formalized by iterating the above construction and referred to as *common knowledge*:

$$CK(E) = \bigcap_{k=1}^{\infty} SK^k(E), \quad \text{where } SK^k(E) = \big(SK\big(\cdots SK(E)\cdots\big)\big) \quad (k \text{ iterations}).$$

As an example consider the case of three binary random variables X_1, X_2 and S, where X_1 and X_2 are independent and S consists of a copy of both of them. Then, the set of possible situations, i.e. the support of the joint distribution $p(x_1, x_2, s)$, consists of four possible states:

X_1	X_2	S
0	0	00
0	1	01
1	0	10
1	1	11

The information partitions correspond to the projections on the respective components of the joint state, e.g.

$$X_1^{-1}(0) = \{(0, 0, 00), (0, 1, 01)\},$$
$$X_2^{-1}(1) = \{(0, 1, 01), (1, 1, 11)\}.$$

For the event $E = \{(0, 0, 00), (0, 1, 01), (1, 0, 10)\}$ we find that

$$K_1(E) = \{(0, 0, 00), (0, 1, 01)\}$$
$$K_2(E) = \{(0, 0, 00), (1, 0, 10)\},$$

and therefore $SK_{1,2}(E) = \{(0, 0, 00)\}$ since both agents 1 and 2 can exclude the state $(1, 1, 11)$ in this case. Thus, we conclude that there exists non-trivial shared information between X_1 and X_2, namely that $S \neq 11$, even though X_1 and X_2 are independent of each other and neither of them knows the state of the other. On the other hand, there is no common knowledge between X_1 and X_2, since $SK_{1,2}(SK_{1,2}(E)) = \emptyset$.

Note that $I_{\min}(S : X_1; X_2) = I_I(S : X_1; X_2) = 1$ bit in this example, if we assume that X_1 and X_2 are independent and uniformly distributed. If we say that $I_{\min}$ measures the shared information, then this implies that X_1 and X_2 have no unique information. This is surprising, given that X_1 and X_2 are independent. Regarding the game theoretic analysis we see that the shared knowledge only rules out one state. Thus, a reasonable definition of shared information might give a positive value to $I_\cap(S : X_1; X_2)$ even if $I(X_1 : X_2) = 0$, but should certainly stay below 1 bit. Maybe a value of $\log(4/3)$ would be a good idea, since the number of possibilities is reduced from four to three. Note that (**Id**$_2$), as proposed in [6], would require that $I_\cap(S : X_1; X_2) = 0$ whenever $I(X_1 : X_2)$ vanishes.

At present, it is not clear how the difference between shared and common information could be formulated in information theoretic terms. One may also ask, whether a desired decomposition of information, should take into account shared information or rather refer to common information. It would probably be easier to use shared information in a decomposition, because otherwise one needs to decompose the information into terms describing the information that X_1 knows that X_2 knows, but X_2 does not know whether it is known by X_1, and so on. On the other hand, common knowledge is represented as a partition (see [11]), and hence corresponds to a random variable after introducing a probability measure on $\mathcal{S}$. In contrast, shared knowledge cannot be represented as a partition. Maybe this explains why it is difficult, and may even be impossible, to represent shared information as a random variable.

Note that the condition (**Id**$_2$) takes into account the mutual information between elements $\mathbf{A_i}$ of the right hand side. Their relationship is not considered in the definition of shared knowledge, but only appears in the higher-order terms which are iterated in the case of common knowledge. Therefore, the property (**Id**$_2$) is more natural for common information than for shared information. The same holds true for (**LC**), since (**LC**) implies (**Id**$_2$).

35.9 Conclusions

We have discussed natural and intuitive properties that a measure of shared information should have. We have shown that some of these properties contradict each other. This shows that intuition and heuristic arguments have to be used with great care when arguing about information.

In particular, we discussed the partial information decomposition and lattice introduced by Williams and Beer. We have shown that a positive decomposition according to the PI lattice contradicts another desirable property, called strong symmetry. We are unsure whether this is an argument against strong symmetry, or whether the PI lattice has to be refined, since it is difficult to assign plausible values to the PI decomposition for the XOR-example.

Williams and Beer also proposed a concrete measure $I_{\min}$ of shared information. We show that in some examples this measure yields unreasonably large values. The problem is that $I_{\min}$ does not distinguish whether different random variables carry *the same* information or just *the same amount* of information. This phenomenon has also been observed by others. However, most people focussed on the property that independent variables may share information about themselves. We argue, using ideas from game theory, that this fact in itself does not speak against $I_{\min}$; but we agree that the absolute value that $I_{\min}$ assigns to the shared information is too large. In our opinion, what is more striking, is that $I_{\min}$ is not monotone in its left argument: Random variables share less information about more.

We expect that further progress requires a more precise, operational idea of what shared information should be. We believe that our results provide additional insights, even thought we have mainly revealed pitfalls regarding the notion of shared information. Thus, despite some recent progress, the quest for a general decomposition of multi-variate information is still open.

Acknowledgements This work was supported by the VW Foundation (J. Rauh) and has received funding from the European Community's Seventh Framework Programme (FP7/2007-2013) under grant agreement no. 258749 (to E. Olbrich).

References

1. Schneidman E, Bialek W, Berry MJ (2003) Synergy, redundancy, and independence in population codes. J Neurosci 23(37):11539–11553. 17
2. Latham PE, Nirenberg S (2005) Synergy, redundancy, and independence in population codes, revisited. J Neurosci 25(21):5195–5206. 25
3. Pessoa LG, Ungerleider L (2008) What and where pathways. Scholarpedia 3(10):5342
4. Williams P, Beer R (2010) Nonnegative decomposition of multivariate information. arXiv: 1004.2515v1
5. Ay N, Olbrich E, Bertschinger N, Jost J (2011) A geometric approach to complexity. Chaos 21(3):037103
6. Harder M, Salge C, Polani D (2012) A bivariate measure of redundant information. CoRR. arXiv:1207.2080 [cs.IT]
7. Cover T, Thomas J (1991) Elements of information theory, 1st edn. Wiley, New York
8. Bell AJ (2003) The co-information lattice. In: Proc fourth int symp independent component analysis and blind signal separation (ICA 03)
9. Maurer U, Wolf S (1997) The intrinsic conditional mutual information and perfect secrecy. In: IEEE international symposium on information theory
10. Christandl M, Renner R, Wolf S (2003) A property of the intrinsic mutual information. In: IEEE international symposium on information theory
11. Aumann RJ (1976) Agreeing to disagree. Ann Stat 4(6):1236–1239
12. Halpern JY (1995) Reasoning about knowledge: a survey. In: Gabbay D, Hogger CJ, Robinson JA (eds) Handbook of logic in artificial intelligence and logic programming, vol 4. Oxford University Press, London, pp 1–34

Chapter 36
Probabilistic Real Swarm Logical Gate

Yuta Nishiyama, Yukio-Pegio Gunji, and Andrew Adamatzky

Abstract Computation can be implemented by natural entities. We established that a swarm of soldier crabs probabilistically performed basic logical operations through geometrically constrained channels. Two inputs three outputs logical gate G simultaneously performed NOT x AND y, x AND y and x AND NOT y. The logical gate G also could be treated as a logical gate G′ that performed NOT x. Next we proposed a logical gate G″ that performed x OR y. Moreover we illustrated how three fundamental gates could be assembled into circuits. Then we discuss about an integration of gates and a circuit as a relationship between part and wholeness.

Keywords Biological computing · Part and wholeness · Soldier crab · Swarm

36.1 Introduction

Colliding real materials can implement logical operations [1]. A mutual interplay between a computational device and an environment is revealed by such implementation.

A nondissipative and reversible logical computation is implemented by colliding elastic balls, namely the billiard ball model (BBM) [2]. Correct configurations of balls and massive reflectors on frictionless plane have a logical universality. In the case of BBM, any perturbations of the system directly influence the trajectory of ball because billiard balls and a container can be separated from each other.

In reaction-diffusive systems, a computation is implemented by collision of traveling waves in the Belousov-Zhabotinsky medium [3]. Localized wave-fragments

Y. Nishiyama (✉) · Y.-P. Gunji
Kobe University, Kobe, Japan
e-mail: y_nishiyama@hotmail.co.jp

Y.-P. Gunji
e-mail: pegioyukio@gmail.com

A. Adamatzky
University of the West England, Bristol, UK
e-mail: adamatzky@gmail.com

T. Gilbert et al. (eds.), *Proceedings of the European Conference on Complex Systems 2012*, Springer Proceedings in Complexity, DOI 10.1007/978-3-319-00395-5_36,

that the BZ reaction at subexcitable threshold exhibits when projected light is controlled are regarded as signals for a computation. In the case of computation in a dissipative medium, perturbations are utilized as computational resources. It is too difficult to maintain the computational domain where an experimenter is required to distinguish controlled computational resources from unintended perturbations.

Natural phenomena can be regarded as a computation that involves not only a sequence of operations of objects but ambiguousness about each of them. Actually living organism as a self-organizing closure can always maintain a boundary with respect to itself in an open environment. Using organisms as computational devices enable a logical operation in which any perturbations are utilized as computational resources and its domains are maintained. For example, slime molds enable to implement a logical collision gate when two of them are fused or avoided with each other dependent on the gradient and the path to propagate in architecture-based computing devices [4, 5]. It was reported that *Physarum* logical gate was robust against external perturbations [4]. The robustness of computation implemented by living organisms is important and remarkable property because such computer can respond to indefinite environments [6].

A swarm of soldier crabs, *Mictyris guinotae*, has a clear boundary. The inherent individual movements, however, are intricate. It seems that the nature of swarm simultaneously involves two aspects of an individual freedom and a collective unity. We have proposed a swarm model based on a mutual anticipation in which each agent had a velocity vector and some potential transitions regarded as inherent noise [7]. This model showed that inherent noise contributed cohesion of swarm and the noise was not distinguished from external one. Then it predicted a robust swarm logical gate [8]. Moreover we confirmed the feasibility of swarm logical gate implemented by real soldier crabs [9]. In this paper, we report that a probabilistic swarm logical gate performs logical conjunction, negation and disjunction. We can also construct the other probabilistic operations by assembling the three fundamental gates.

36.2 Materials and Methods

Boolean logical gates has X- and Y-input channels and A-, B- and C-output channels (Fig. 36.1). The X-input connects to C-output in a straight line, B-output at an angle of forty-five degrees and A-output at an angle of ninety ones. Y-input connects to A-output, B-output and C-output in a similar way. The gate has two gradients with respect to make crabs move. One gradient is implemented by utilizing escaping behavior motivated by intimidation plate just behind each input channel. Crabs are sensitive to sudden standing or moving big objects because such objects ordinarily are likely to be enemies in the flat. Thus the intimidation plate popped up can cause them to escape away from itself. The other gradient is implemented by physical gravitational stimulus produced by slope immediately following each input channel. A slope facilitates straightforward movement of swarms and prevents individuals coming back starting gate to some extent.

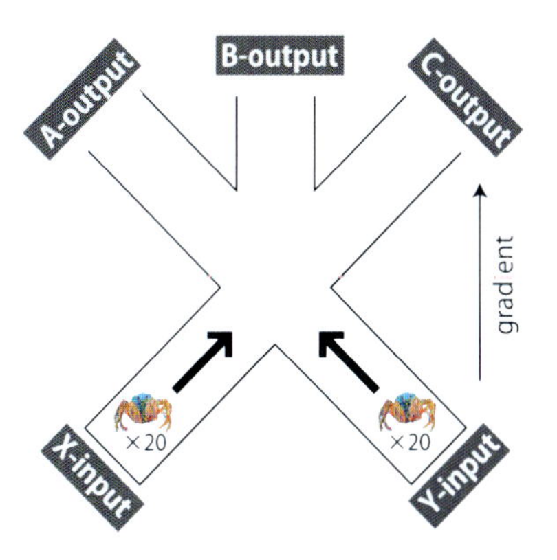

Fig. 36.1 Diagram of two inputs (X and Y) three outputs (a, b and c) Boolean logical gate G. Soldier crabs move from south to north

We set a swarm of 20 subjects in closed X- and/or Y-input channels just prior to each trial. The swarm represented value of 1. Crabs were left about two minutes for adaptation. After it, input channels were opened and intimidation plates were simultaneously popped up, and then crabs go ahead to the output channels.

36.3 Results

36.3.1 Three Fundamental Gates

Two-inputs three-outputs Boolean gate G

$$\langle x, y\rangle \rightarrow \langle \mathit{NOT}\ x\ \mathit{AND}\ y,\ x\ \mathit{AND}\ y,\ x\ \mathit{AND}\ \mathit{NOT}\ y\rangle$$

is realized with swarms of solider crabs, where $x, y \in \{0, 1\}$.

Usually, presence or absence of entities represents True or False values of logical variables in collision-based gates [1]. In our experimental crab gate, we had to define threshold values of population in an output channel because not all crabs within a swarm arrive at one exit. We assume that if number of crabs in a channel exceed a threshold value then the channel's variable takes value True, otherwise False. If the threshold value is too low then the swarm collision gate with inputs (0, 1), (1, 0) or (1, 1) is almost certain to output 1 in all output channels, conversely too high output 0. Thus, the threshold value is a control parameter responsible for the type of operation implemented by the swarm logical gate.

We estimate an output value in each output channel by a proportion, in which output values a, b, or c are 1 if the proportion of crabs in A-, B-, or C-channels exceeds a threshold, otherwise values a, b, or c are set to 0.

First we make A-channel implement NOT x AND y. Let A_{01} an event in which A-channel produces output 1 for a pair of inputs $\langle 0, 1\rangle$ and let A_{10} and A_{11} events in which A-channel produces output 0 for pairs of inputs $\langle 1, 0\rangle$ and $\langle 1, 1\rangle$, respectively. Figure 36.2(a) illustrates a probability that A_{xy} happen under a given threshold value of proportion for a pair of inputs $\langle x, y\rangle$. It shows that A-channel is able to implement

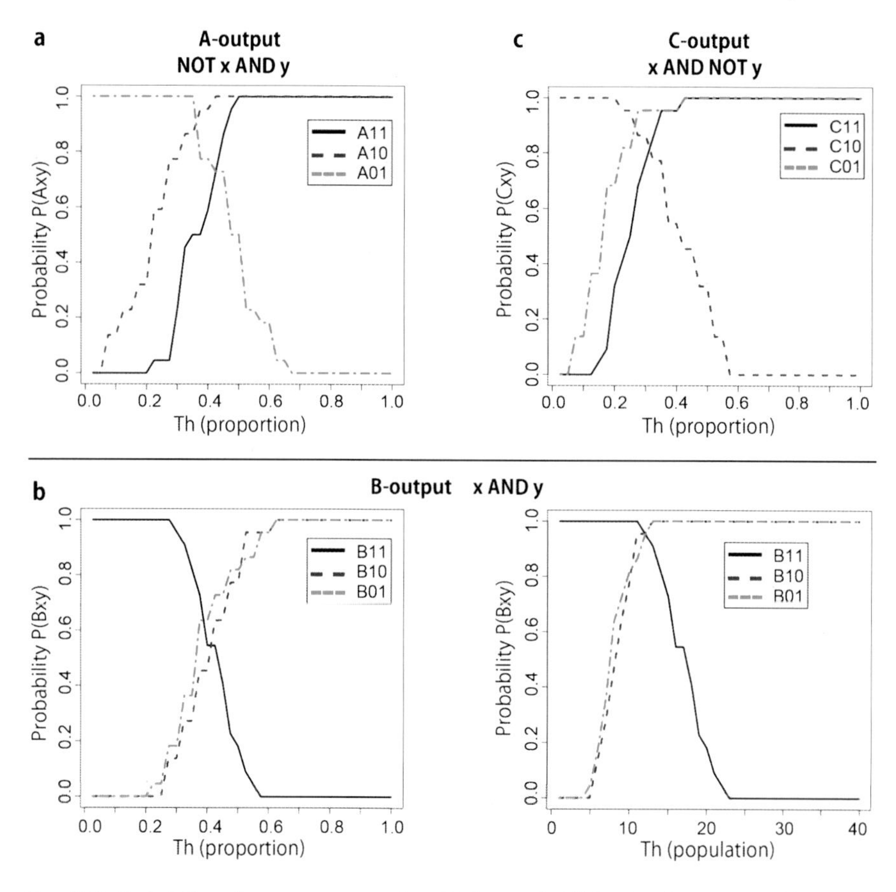

Fig. 36.2 Probability of correct computation. Higher probabilities for three types of inputs are simultaneously, more reliable the computation in an output channel is. In a case of B-output, the reliability is quite low when we use the threshold value of proportion, see *right top*. But we can use it of population in order to make it correctly compute

NOT x AND y with probabilities $P(A_{01}) = 0.73$, $P(A_{10}) = 1$ and $P(A_{11}) = 0.86$ under the threshold value $Th = 0.45$.

Similarly, we make C-channel implement x AND NOT y. Let C_{10} an event in which C-channel produces output 1 for a pair of inputs $\langle 1, 0 \rangle$ and let C_{01} and C_{11} events in which C-channel produces output 0 for pairs of inputs $\langle 0, 1 \rangle$ and $\langle 1, 1 \rangle$, respectively. C-channel is able to implement x AND NOT y with probabilities $P(C_{01}) = 0.95$, $P(C_{10}) = 0.77$ and $P(C_{11}) = 0.86$ under the threshold value $Th = 0.33$ (Fig. 36.2(c)).

Next we make B-channel implement x AND y. Let B_{11} an event in which B-channel produces output 1 for a pair of inputs $\langle 1, 1 \rangle$ and let B_{01} and B_{10} events in which B-channel produces output 0 for pairs of inputs $\langle 0, 1 \rangle$ and $\langle 1, 0 \rangle$, respectively. Unfortunately, the probability $P(B_{xy})$ is unreasonable (Fig. 36.2(b), left). It suggested that there was no preference of B-channel when swarms were set in both

Table 36.1 Probabilities and number of correct computation regarding Boolean gate G

	$P(A_{xy})$	$P(B_{xy})$	$P(C_{xy})$	$P(G_{xy}) = (= P(A_{xy}, B_{xy}, C_{xy}))$
$x = 0, y = 1$	0.73 (16/22)	0.95 (21/22)	0.95 (21/22)	0.73 (16/22)
$x = 1, y = 0$	1 (22/22)	0.95 (21/22)	0.77 (17/22)	0.77 (17/22)
$x = 1, y = 1$	0.86 (19/22)	0.95 (21/22)	0.86 (19/22)	0.73 (16/22)

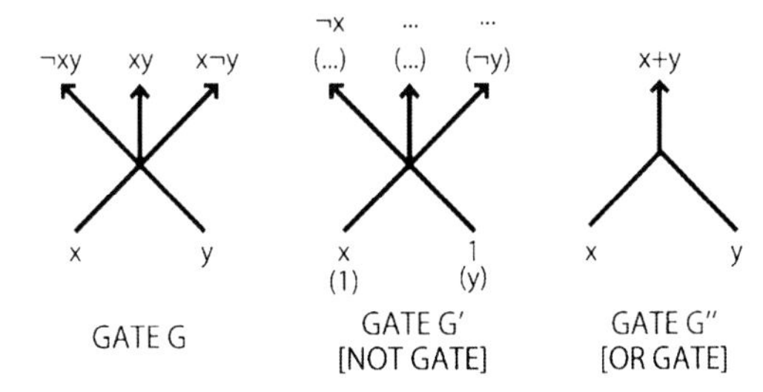

Fig. 36.3 Schematic diagrams of Boolean gate G, G′ and G″. Note that the gate G′ can be gotten by regarding either input of gate G as a part of architecture

X- and Y-input channels. We can, however, make the B-channel implement x AND y by using threshold value regarding population. Figure 36.2(b) right illustrates a probability that B_{xy} happens under a given threshold value of population. It shows that B-channel is able to implement x AND y with probabilities $P(B_{01}) = 0.95$, $P(B_{10}) = 0.95$ and $P(B_{11}) = 0.95$ under the threshold value $Th = 12$.

The threshold values could be chosen in such a way that the A-, B- and C-channels implement NOT x AND y, x AND y, and x AND NOT y, respectively. The Boolean gate G simultaneously satisfies three events with same inputs. So the probability $P(G_{xy})$ is equivalent to a joint probability $P(A_{xy}, B_{xy}, C_{xy})$. Thus we can find probabilities $P(G_{01}) = 0.73$, $P(G_{10}) = 0.77$ and $P(G_{11}) = 0.73$ (Table 36.1).

Boolean gate G′

$$x \rightarrow NOT\, x$$

is realized with swarms of solider crabs.

We can obtain Boolean gate G′ that performs logical negation when a swarm that leaves either input in Boolean gate G is regarded as a part of computer (Fig. 36.3). The output value of G′ can be obtained in A-output channel when a swarm that leaves Y-input channel is a part of computer. Then the probabilities $P(G'_0)$ and $P(G'_1)$ are equivalent to $P(A_{01})$ and $P(A_{11})$ respectively, where G'_x represents an event that the output value is 1 or 0 against absence or presence of a swarm in the input channel. Similarly, C-output channel gives an output value when X-input is a part of computer, then $P(G'_0) = P(C_{10})$ and $P(G'_1) = P(C_{11})$. Thus it must be concluded that $P(G'_0) = P(A_{01}) = P(C_{10})$ and $P(G'_1) = P(A_{11}) = P(C_{11})$. Though the number of occurrences of A_{01} (16/22) empirically differs from that of C_{10} (17/22), the difference is no more than one time. To avoid confusion, we discriminate the use of A-output from that of C-output as G'_{Ax} and G'_{Cx}. Consequently $P(G'_{A0}) = 0.73$, $P(G'_{C0}) = 0.77$, $P(G'_{A1}) = 0.86$ and $P(G'_{C1}) = 0.86$ (Table 36.2).

Table 36.2 Probability regarding Boolean gate G′

	$P(G'_{Ax})$ (= $P(A_{x1})$)	$P(G'_{Cx})$ (= $P(C_{1x})$)
x = 0	0.73	0.77
x = 1	0.86	0.86

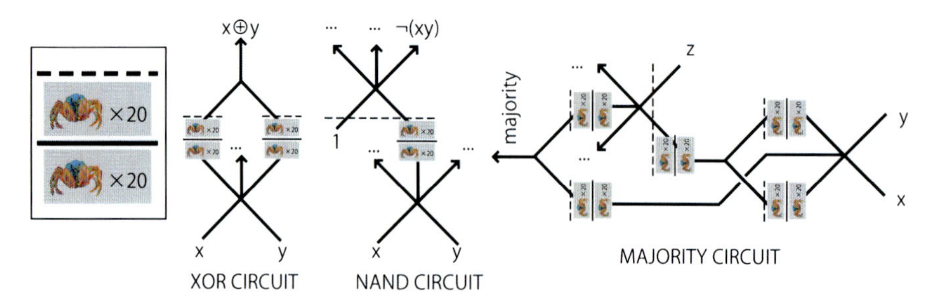

Fig. 36.4 Examples of various circuits. A joint has two reservoirs that one corresponds to output channel of previous gate and another has a swarm in order to discharge it when the first one has output value 1. *Dotted line* represents a door to discharge

Boolean gate G″

$$\langle x, y \rangle \rightarrow \langle x\ OR\ y \rangle$$

is realized with swarms of solider crabs.

Boolean gate G″ can perform logical disjunction when it is constructed by two input channels and one output channel with the gradients like Boolean gate G (Fig. 36.3). It requires no threshold value because presence of a crab in an output channel is sufficient to know entry. There were no cases that no crabs arrive at output channels in the case of Boolean gate G. So we assume $P(G''_{xy}) = 1$.

36.3.2 Examples of Logical Circuits

Boolean gate G, G′ and G″ can be assembled into XOR, NAND and MAJORITY circuits.

We introduce a joint to connect a few logical gates (Fig. 36.4, box). The joint consists of two rooms. One room can keep crabs that have arrived at an output channel of previous logical gate and also functions as an output channel of the gate. Another one discharges a swarm when the first one has output value 1.

Figure 36.4 illustrates a few circuits, for instance. In principle, success probabilities of each circuit are calculated from probabilities mentioned above (Table 36.3).

Table 36.3 Probability of correct computation regarding a few circuits. They are joint probability or conditional probability dependent on constructions of G, G′ and G″

x	0	0	1	1
y	0	1	0	1
P(XOR) = P(A, C)		0.73	0.77	0.73
P(NAND) = P(B) ∗ P(G′)	0.73(0.73)	0.69(0.73)	0.69(0.73)	0.81(0.81)

x	0	0	0	1	1	1	1
y	0	1	1	0	0	1	1
z	1	0	1	0	1	0	1
P(MAJORITY)	0.95	0.69	0.69	0.73	0.73	0.73	0.73

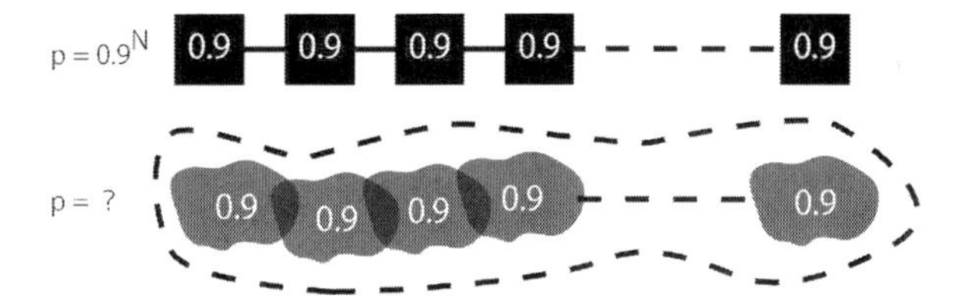

Fig. 36.5 Predicted probability of assembled probabilistic logical gate in series when we assumed that uncertain elements are certain (*top*) and that living organism autonomously response to their environments

36.4 Discussion

We introduced three fundamental logical gates implemented by soldier crabs and a few examples of circuits made from them. The logical gates correctly performed with probabilities of over 0.70. Such lowness implies that crabs autonomously response to environments including the other individuals [10]. It is, therefore, expected that crabs logical gate can response to a change of environment. A concept of circuit was simple assembly of probabilistic gates. Assembling the gates decrease the probability of correct output in principle. The calculated probabilities were based on that of each fundamental gate. A program is constructed by assembling fundamental gates. When we utilize the probabilistic logical gates to construct the larger program, its probability gradually become low as well as the circuits. For instance, consider a gate working with probability of 0.9. If a program would be series of N of the gate, we might expect that it performed correct operation with probability of 0.9^N (Fig. 36.5, top). In this time, an uncertain element, namely the probabilistic gate, would be regarded as a certain one. Our computational devices, however, are living organisms. Their behavior is not able to separate from their environment. It is difficult to decide a frame of environment that living organisms response. A behavior of gate, therefore, may differ from that of gate integrated into program. Consequently, the probability of performance of program will differ from

0.9^N (Fig. 36.5, bottom). Such uncertainty means no absolute inefficient computation but rather should be introduced in a positive manner.

What is a relationship between local and global properties in a computation? A logical gate implemented by slime mould can correctly perform logical operation due to change of local behavior even if a part of architecture is broken [4]. The logical operation is global behavior. The designer should understand that the global computation causes from the local behaviors. If do so, the logical gate is broken when slime mould unfortunately behave. Nevertheless, the broken gate is correctly working. Thus local and global properties are confused owing to mediated by living computational devices. Such mediation emerges as robust computation in which the function is maintained even if the structure changes.

Our swarm model demonstrated robust computation in which a diversity of individual movement was confused with external perturbation [8]. It is predicted that the logical gate with robustness is constructed by implementing velocity-matching between two swarms. The velocity-matching phenomenon is often found in nature. We, however, hardly demonstrate them in current empirical settings [9]. Our experimental arena had degree of freedom of which individuals bring easy walking due to cork-made floor and their behaviors might be self-centered than usual. Future works requires mobile and cohesive swarm in experimental environments.

References

1. Adamatzky A (ed) (2002) Collision-based computing. Springer, Berlin
2. Fredkin E, Toffoli T (1982) Conservative logic. Int J Theor Phys 21:219–253
3. Costello BDL, Adamatzky A (2005) Experimental implementation of collision-based gates in Belousov–Zhabotinsky medium. Chaos Solitons Fractals 25:535–544
4. Tsuda S, Aono M, Gunji YP (2004) Robust and emergent Physarum logical computing. Biosystems 73:45–55
5. Adamatzky A (2010) Slime mould logical gates: exploring ballistic approach. arXiv:1005.2301v1 [nlin.PS]
6. Tsuda S, Zauner KP, Gunji YP (2007) Robot control with biological cells. Biosystems 87:2–3. 215–223
7. Gunji YP, Murakami H, Niizato T, Adamatzky A, Nishiyama Y, Enomoto K, Toda M, Moriyama T, Matsui T, Iizuka K (2011) Embodied swarming based on back propagation through time shows water-crossing, hour glass and logic-gate behavior. In: Lenaerts T et al (eds) Advances in artificial life, pp 294–301
8. Gunji YP, Nishiyama Y, Adamatzky A (2011) Robust soldier crab ball gate. Complex Syst 20:94–104
9. Nishiyama Y, Gunji YP, Adamatzky A (2012) Collision-based computing implemented by soldier crab swarms. Int J Parallel Emerg Distrib Syst 1–8. doi:10.1080/17445760.2012.662682
10. Nishiyama Y, Enomoto K, Toda M, Moriyama T, Gunji YP (2011) Autonomous oscillations of soldier crabs. In: Proc ICMC2011, pp 117–119

Chapter 37
The Role of Complex Systems in Public-Private Service Networks

Ameneh Deljoo, Marijn Janssen, and Y.-H. Tan

Abstract The complex systems field provides powerful instruments and concepts for understanding evolution and developments. Public Private Service Networks (PPSN) are increasingly advocated as beneficial for the delivery of public services. Yet how these networks evolve and adopt new policies is ill-understood. In this paper the characteristics of PPSN and Complex Adaptive Systems (CAS) are compared and it is suggested that PPSN can be viewed as a particular type of CAS. We argue that CAS research methods and tools and agent based simulation of PPSN can help to increase our understanding of the evolutionary nature of PPSN.

Keywords Public private networks · Complex adaptive systems · Agent based modeling

37.1 Introduction

Public Private Service Networks (PPSN) are constellations of public and private organizations who have to collaborate with each other to make services available to businesses or citizens. An example is the social security in which disabled persons can get a personal budget from the government and use this to buy medicare that best suits their purposes. This results in complex relationships and dependencies among the parties involved.

The rapid evolution and expansion of Information and Communication Technologies (ICT) capabilities, capacity and pervasiveness gives rise to opportunities and challenges for the sustainability and transformation of society and economy. Yet taking advantage of these opportunities is often difficult and innovation projects

A. Deljoo (✉) · M. Janssen · Y.-H. Tan
Faculty of Technology, Policy and Management, Delft University of Technology, Delft, The Netherlands
e-mail: A.deljoo@tudelft.nl

M. Janssen
e-mail: M.F.W.H.A.Janssen@tudelft.nl

Y.-H. Tan
e-mail: Y.Tan@tudelft.nl

T. Gilbert et al. (eds.), *Proceedings of the European Conference on Complex Systems 2012*, Springer Proceedings in Complexity, DOI 10.1007/978-3-319-00395-5_37,

often fail. However, organizations continue to be designed to achieve stability of operations in an environment of increasing instability so traditional engineering approaches are difficult. Evolutionary approaches need to be taken because there are many dependencies and the organizations are autonomous. Meanwhile it is necessary to understand the development of PPSN which is influenced by the interactions and dependencies among stakeholders.

In this work, we will argue that PPSN can be conceptualized as a particular type of Complex Adaptive Systems (CAS). We use the general definition of CAS as "a system that emerges over time into a coherent form, and adapts and organizes itself without any singular entity deliberately managing or controlling it" [1]. In this paper we discuss the relationship between PPSN and the CAS theory and make the argument why CAS is a suitable lens for conceptualizing PPSN. This paper is structured as follows. Next, we introduce and describe PPSN, CAS and agent based simulation (ABM) and their benefits. ABM is one of the ways to simulate the PPSN. Finally, we compare the CAS and PPSN characteristics. Based on the similarities we will argue that PPSN can be conceptualized as a type of CAS. Finally we will discuss the benefits of conceptualizing PPSN as CAS, draw conclusions and provide our future research directions.

37.2 Background

37.2.1 Public-Private Service Network

PPSNs refer to arrangements where the private sector is used that traditionally have been provided by public organizations. In addition to private execution and financing of public investment, the most important characteristics of PPSNs is an emphasis on service delivery to satisfy the needs and expectations of stakeholders groups.

PPSN can be typified as having multiple stakeholders [2]. These stakeholders each have their own goals and interests. Not only organization are a particular type of stakeholder, also single organizations typically consist of multiple stakeholders, sometimes with competing views. PPSN consist of many players having their own responsibilities which are free in their decision making and have a variety of installed systems which are developed independently. There are both vertical or hierarchical relationship (often within organizations). As such there is no central authority and organizations employ a wide variety of technologies. As public-private service networks are a relatively new phenomenon, there is little understanding of their evolution.

37.2.2 Complex Adaptive Systems

Government officials and other decision makers increasingly encounter a daunting class of problems that involve systems composed of very large numbers of diverse interacting parts. These systems are prone to seemingly uncontrollable behaviors. CAS is used in a various areas including social systems, ecologies, economies, cultures, politics, technologies, traffic jam, weather [1]. These traits are the hallmarks of what scientists call complex systems. A complex system is composed of many parts that interact with and adapt to each other and, in so doing, affect their own individual environments [3]. The combined system level behavior arises from the interactions of parts that are, in turn, influenced by the overall state of the system. Global patterns emerge from the autonomous but interdependent mutual adjustments of the components.

According to John Holland CAS is "a dynamic network of many agents (which may represent cells, species, individuals, firms, nations) acting in parallel, constantly acting and reacting to what the other agents are doing. The control of a CAS tends to be highly dispersed and decentralized. If there is to be any coherent behavior in the system, it will have to arise from competition and cooperation among the agents themselves. The overall behavior of the system is the result of a huge number of decisions made every moment by many individual agents" [4].

Most work in CAS has been conduct in highly abstract and artificial system, like cellular automata and genetic algorithm, and in this fields of economics, medical. Nowadays, other fields like management and organization has attracted to using CAS in their research [5]. It has been successful in understanding phenomena and provide guidelines for development.

37.2.3 Agent-Based Modeling

Understanding CAS requires tools that themselves are complex to create and understand. Often agent-based modeling (ABM) are used for modeling CAS [6]. Shalizi defines ABM as "An agent is a persistent thing which has some state we find worth representing, and which interacts with other agents, mutually modifying each other's states. The components of an agent based model are a collection of agents and their states, the rules governing the interactions of the agents and the environment within which they live" [7].

ABM is a one of simulation modeling tools, including applications to real world business problems and complexity [8]. ABM provides a modeling and simulation approach which can be beneficial for a CAS approach and is useful in creating understandable results for managers and business men. Its four areas of application are discussed by using real world applications: flow simulation, organizational simulation, market simulation, and diffusion simulation. For each category, one or several business applications are described and analyzed. In ABM, a system is defined as a collection of autonomous decision making entities called agents [9]. Each agent individually assesses its situation and makes decisions on the basis of a set of rules and each agent has an own feature. Agents may execute various behaviors appropriate for the system they represent for example, producing, consuming, or selling. Repetitive competitive interactions between agents are a feature of ABM.

ABM does not smooth over nonlinearities and can embrace major transitions and catastrophes. In particular, "ABM has provided good results in situations in which the behavior of individuals cannot be defined clearly, when the description of this behavior cannot be adequately expressed by equations, or when the complexity of differential equations becomes too much, and when the system is more appropriately built on activities that in aggregate and predictable processes, and when the behavior has a more stochastic" [10].

37.3 PPSN as CAS

37.3.1 Can PPSN be Viewed as Complex Systems?

In ABM agents pursue their own interests. This is often true for organizations and stakeholders having their own interests, requirements and objectives as explained in the part about PPSN. That may run contrary to those of the principal (the state). These assumptions are very similar to the basics of a PPSNs, with several public organizations involved and the government acting as the principal and the private sector as the agent. When the government instructs the private sector to perform a certain task on its behalf, it tries to negotiate a contract that will stipulate the relevant parameters, including the nature and quantity of output; specific benchmarks and timing; and the tools by which the government will control the performance of the private sector operator.

New ICT enables dependencies at a global scale adding to a higher complexity. When actors want to achieve anything, they find that there are multiple actors with various interests that are dependent on each other to achieve their goals. The problems that (government) organizations face are more complex than before as the act of governing also has to deal with such interdependencies because most acts of government transcend organizational boundaries [11]. In PPSN organizations operating in a network have different goals. In addition, as the situation and structure of the various organizations differs, any arrangement in the network has to deal with these differences. When it comes to public-private networks, the difference between the goals, resource and values makes it potentially conflicting. In order to realize anything in such complex multi actor network, governments have to acknowledge these networks and collaborate with others [12–14]. Changes in objectives, in resources, or in the environment may require the adaptation of a single organization which might influence other organizations in the PPSN. In the public sector, networks present an alternative to (bureaucratic) policy and decision making processes and service delivery [2]. Other potential issues specific to public private cooperation include accountability, transparency and privacy[11]. These types of potential conflicts of interests cannot always be avoided and should be taken into account and modeled when developing a ABM of PPSN.

The similarities between the characteristics of CAS and PPSN are outlined in Table 37.1. This table shows that CAS can be a suitable lens for conceptualizing the

Table 37.1 CAS and PPSN characteristics

	CAS	PPSN
Content	Autonomous, intelligent and goal-driven agents	Independent actors pursuing their own goals
Platform	Emerging behavior by interacting agents	Actor behavior influences network evolution
Central core	Decentralization (with some degree of control)	Interacting private and public actors based on vertical but mostly horizontal relationship
Behavior	Non-linearities, emergent behavior	Non-linearity's, evolutionary nature
Simulation & modeling	Agent-based models, complex network-based models	Discrete-event modelling, agent-based modeling

complexity in PPSN and for understanding issues in development and evolution of PPSN. The perspective of a CAS as an empirical lens to understand the public sector developments have been used before which proved useful for identifying principles guiding its development [5]. In a similar vein it will likely suitable for understanding PPSN. Through the CAS lens, our intention is that by conceptualizing PPSN as CAS, public managers and policy makers can acquire better understanding of interaction and network governance. Our future research aims to understand the behavior of PPSN and use the CAS to derive simple rules to guide its development.

37.3.2 Can PPSN be Modeled Using Agent-Based Modeling?

Each organization or business consist of dozens or hundreds of actors which can be conceptualized as "agents", which are acting autonomously and which interactions make up complex behavior of the system as a whole. It is possible to simulate this complex behavior by programming software agents with relatively simple rules and letting them interact with one another. By changing agents' activities and behaviors at a local level, it is possible to improve the performance of the system as a whole. ABM captures the complex network of interactions and connections that make up real systems and make it possible to see emergent patterns and unexpected changes and events [6, 10]. Providing this we expect that conceptualizing PPSN as CAS will result into the following advantages.

- *ABM can provide insight into causes of emergent phenomena of CAS.*

Emergent phenomena results from the interactions of individual entities [15]. An emergent phenomenon can have properties that are decoupled from the properties of the part. For example, in a PPSN the service provision can be delayed due to a lack of innovation of one party or by an update of functionality by all parties. A traffic jam, which results from the behavior of and interactions between individual vehicle drivers, may be moving in the direction opposite that of the cars that cause it [8]. This characteristic of emergent phenomena makes it difficult to understand and predict CAS.

- *ABM uses a more natural description of a PPSN.*

Whether one is attempting to describe a traffic jam, the stock market, voters, or how an organization works, ABM makes the model seem closer to reality, as PPSN consist of autonomous parties having their own goals and interests.

- *ABM can provide a framework for testing development strategies.*

The flexibility of ABM can be observed along multiple dimensions. For example, it is easy to add or remove agents to an agent based model like organizations enter or leave a PPSN. ABM also provides a natural framework for tuning the complexity of the agents: behavior, degree of rationality, ability to learn and evolve, and rules of interactions [10]. Another dimension of flexibility is the ability to change levels of description and aggregation: one can easily play with aggregate agents, subgroups of agents, and single agents, with different levels of description coexisting in a given model. One may want to use ABM when the appropriate level of description or complexity is not known ahead of time and finding it requires some tinkering [16].

37.4 Conclusion

In this paper we argued that CAS provides a suitable lens for conceptualizing and understanding PPSNs. We demonstrated that CAS and PPSN characteristics are quiet similar and that the evolutionary nature of PPSN can likely be understood by taking a CAS lens. A PPSN is made up of interacting organizations and organizations consist of coalitions of stakeholders having their own interests. There is many relationships and interactions which determines the system behavior.

Our future work will focus on using ABM and CAS for understanding PPSN and its evolution. A multi agent simulation model will be developed capturing socio and technical elements. Agents will be used to model the behavior of each individual entity and the aggregate behavior of individuals can show the dynamic and emergent behavior over time. Once an ABM is created, various parameters can be manipulated and rules could be modified in order to study the emergent outcomes and to study the adoption of new developments. In this way the implications of design principles can be evaluated and used to modify the design principles.

References

1. Dooley K (1996) Complex adaptive systems: a nominal definition, vol 8, pp 2–3
2. Provan KG, Milward HB (2001) Do networks really work? A framework for evaluating public-sector organizational networks. Public Adm Rev 61:414–423

3. Levin SA (2002) Complex adaptive system: exploring the known, the unknown and the unknowable, vol 40. AMS, Providence, pp 3–19
4. Holland JH (1992) Complex adaptive systems. Daedalus 121:17–30
5. Janssen M, Kuk G (2006) A complex adaptive system perspective of enterprise architecture in electronic government. In: Conference on a complex adaptive system perspective of enterprise architecture in electronic government, p 71
6. Tcsfatsion L (2003) Agent-based computational economics: modeling economies as complex adaptive systems. Inf Sci 149:262–268
7. Shalizi CR (2006) In: Deisboeck TS, Kresh JY (eds) Methods and techniques of complex systems science: an overview complex systems science in biomedicine. Springer, New York, pp 33–114
8. Bonabeau E (2002) Agent-based modeling: methods and techniques for simulating human systems. Proc Natl Acad Sci USA 99(3):14
9. Brown DG (2006) Agent-based models. In: Geist H (ed) The Earth's changing land: an encyclopedia of land-use and land-cover change. Greenwood Publishing Group, Westport, pp 7–13
10. Nicholas RJ (2001) An agent-based approach for building complex software systems. Commun ACM 44:35–41
11. Nutt PC (2006) Comparing public and private sector decision-making practices. J Public Adm Res Theory 16:289–318
12. Klievink B (2011) Unravelling interdependence coordinating public-private service networks. Faculty of Technology, Policy and Management, TU Delft, Delft
13. Walter PW (1990) Neither market nor hierarchy: network forms of organization, vol 12. JAI Press, London, pp 295–336
14. Arase D (1994) Public-private sector interest coordination in Japan's ODA. Pac Aff 67:171–199
15. Anderson PW (1972) More Is different, vol 177. AAAS, Washington, pp 393–396
16. Rounsevell MDA, Robinson DT, Murray-Rust D (2012) From actors to agents in socio-ecological systems models. Philos Trans R Soc Lond B, Biol Sci 367:259–269

Chapter 38
Revisiting von Neumann's Architecture of Machine Self-reproduction Using *Avida*

Tomonori Hasegawa and Barry McMullin

Abstract In an attempt to explore potential for the evolutionary growth of complexity and distinctive mutational pathways, we revisit the machine self-reproduction originally proposed by John von Neumann. Preliminary experiments have been run using a designed von Neumann style self-reproducer novelly implemented within the Avida world. Among initial results, we have observed ones where the designed self-reproducer in this particular world degenerates into a self-copier with surprising ease. In this paper, we briefly report the result and discuss implications and future works.

Keywords Evolutionary growth of complexity · Genotype-phenotype mapping · Self-reproduction · Avida · Von Neumann

38.1 Introduction

38.1.1 Von Neumann's Architecture of Self-reproduction

Machine self-reproduction was substantially theorised by John von Neumann in the late 1940s and early 1950s [1]. His theory implicates that there can be potential for the evolutionary growth of complexity within the world of machines, as observed in a real biological evolution, where the degree of complexity can not only decrease but also maintain or increase [2]. The proposed architecture of machine reproduction characteristically comprises active and passive components (see Fig. 38.1). The active component includes a constructor and a tape copier, which are controlled by a controller to collaboratively reproduce an arbitrary machine that is described by the passive component called a description tape. When the tape describes the active component itself, the reproduction amounts to machine self-reproduction. Once

T. Hasegawa (✉) · B. McMullin
The Rince Institute, Dublin City University, Dublin, Ireland
e-mail: tomonori.hasegawa2@mail.dcu.ie

B. McMullin
e-mail: barry.mcmullin@dcu.ie

T. Gilbert et al. (eds.), *Proceedings of the European Conference on Complex Systems 2012*, Springer Proceedings in Complexity, DOI 10.1007/978-3-319-00395-5_38,

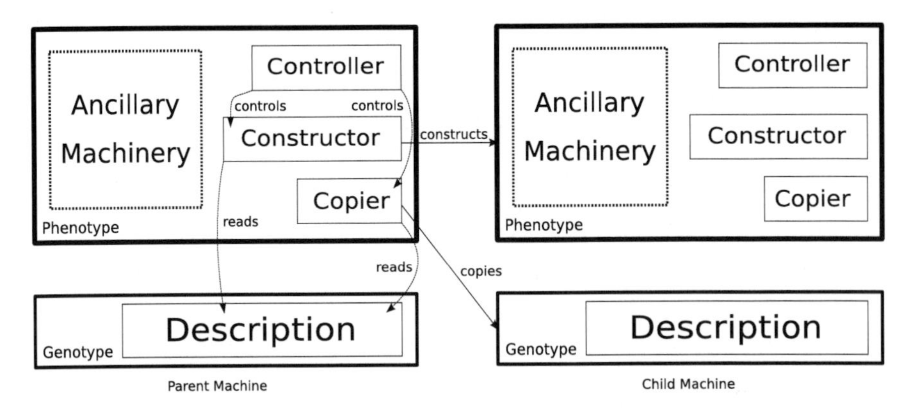

Fig. 38.1 Von Neumann architecture of machine self-reproduction: In this schema, given a machine description, the controller takes the initiative to manoeuvre (i) the constructor to decode the description, according to a specific procedure, and to create an offspring machine's active components, as well as (ii) the copier to copy the description into an offspring machine. The ancillary machinery is a part that is not directly associated with reproduction

such machines repeatedly self-reproduce, they will increase exponentially in number until they encounter some resource limit.

38.1.2 Genotype-Phenotype Mapping

Insights in molecular biology, including the discovery of the DNA structure, have shown how von Neumann's architecture of self-reproduction based on copying and decoding resonates with the biological counterpart based on transcription and translation. Translation is underpinned by a genotype-phenotype mapping, which basically refers to the relationship between genome and its expression, such as physical or behavioural traits (or, the relationship between DNA and resulting proteins). In computational systems it may be described as decoding a sequence of numbers, according to some predefined rule, into another sequence of numbers that serve as particular functions and/or properties. In terms of the von Neumann's architecture, the active and passive components are analogous to phenotype and genotype, respectively. Likewise, the decoding rule is implemented by the constructor and it must be represented in the description tape in some encoded form, in a totally self-consistent way.

Von Neumann's architecture of machine self-reproduction reasonably deserves to be revisited, since it has not been implemented and investigated in depth within such an evolutionary platform as *Avida* (despite the significance as pointed out in [3]). The present paper briefly summarises a preliminary attempt and its initial result of our research which aims to characterise the von Neumann style self-reproducer in the exploration of its potential for the evolutionary growth of complexity.

38.2 Implementation

38.2.1 The Avida *World*

The Avida platform is a simulated world which has been widely used in evolutionary research [4, 5]. Inspired by its predecessors including Tierra [6], the system is designed to observe spontaneous evolutionary dynamics of a population of virtual organisms. The world is represented as a two-dimensional lattice of computational nodes, which seemingly resembles cellular automata. Each node in the Avida world, in contrast, is equipped with a virtual CPU (rather than a finite state machine) and a local memory, and can represent an organism which "metabolises" by running the program. Organisms can interact locally by overwriting or replacing another program on a neighbouring node. A successfully programmed organism can repeatedly self-reproduce, placing its offspring into the neighbourhood, in the manner of bacteria in a Petri dish.

A typical Avida simulation run is initiated with a single seed program, called an *ancestor*, designed by the experimenter. The standard self-reproduction in the Avida world is achieved by self-copying based on self-inspection (roughly analogous to the template replication of RNA or DNA). Over time, one can observe that a population grows, possibly diversifying and filling up the world. While certain errors designed into CPU operations serve as inheritable variation, the limited resources such as CPU time and memory space cause competition among variant organisms, thereby giving rise to the Darwinian evolution.

38.2.2 Ancestor *Design*

Based on the architecture of machine self-reproduction, we have designed a novel ancestor to seed the Avida world (see Fig. 38.2). The novel ancestor may be decomposed into *genome* and *phenome*, corresponding to the passive machinery and the active machinery of von Neumann's self-reproductive architecture, respectively. The self-reproduction of the ancestor is designed as follows: the *phenome* supports the active process of *decoding* and *copying* of the description (*genome*), including allocation of memory in the parental memory where the memory image of a prospective offspring is created, and also dividing off that part as an offspring.[1]

Decoding represents the genotype-phenotype mapping, as it signifies what kind of transformation of numbers is applied to the description. We have designed decoding such that utilises a *lookup table* incorporated in the phenome, on the basis

[1]We have included no explicit ancillary machinery in our design of the ancestor, that is, every instruction of its phenome somehow commits to the reproductive process. If, through accumulated mutations, there should prove to be any part of the program that does not necessarily engage in the reproductive process, then that part may be regarded as ancillary machinery in von Neumann's terms.

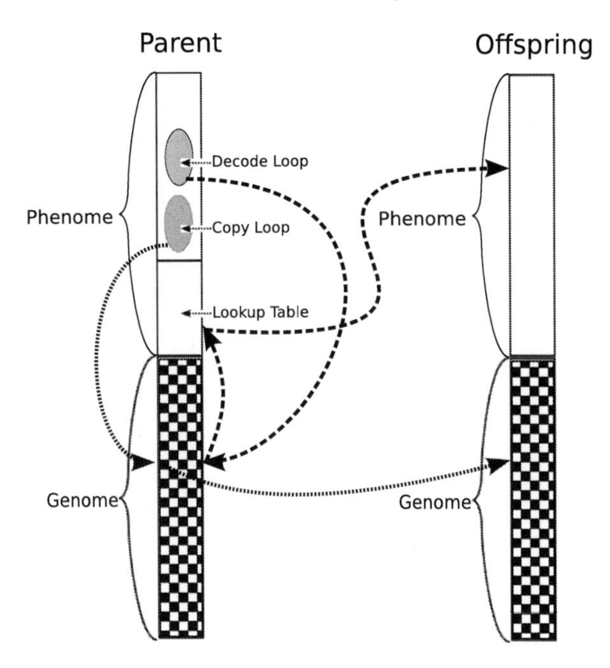

Fig. 38.2 Designed self-reproduction of the novel ancestor

of mapping one number into another, where the parent phenome part reads a number from its genome part, looks up what is at the lookup table location represented by the read number, and then writes the looked-up new number into the offspring phenome part. The mechanism hinges on the fact that a memory content may be interpreted as an instruction, in the way predefined in an *instruction set* in the system configuration setting, or, alternatively, as an uninterpreted numerical, usually a distinct integer (in the sense that opcodes are numbers assigned to operations).

The initial choice of the mapping is quite arbitrary, but in practice, we have picked up a mapping that is invertible, with one-to-one correspondence (in a way similar to codons translated into proteins), so that with the designed lookup table it is possible to reverse-engineer the ancestor. As a lookup table, we use a list of numbers which is a permutation of the instruction set. With it, each number representing an instruction, as defined in the instruction set, finds a number that it translates into from the lookup table. Given the designed phenome, therefore, we could obtain its corresponding genome to complement the ancestor using the lookup table incorporated in the phenome, so that the phenome is the decoded version of the genome and the genome is the encoded version of the phenome.

The genotype-phenotype mapping in this novel program is analogous to the mapping from a string of words (i.e. genome) into another string of words (i.e. phenome) underlain by the mapping between words defined by the lookup table. Therefore, it is reasonable to expect the genotype-phenotype mapping to be mutable (and thus presumably evolvable), with the lookup table itself and the usage of it being open to change evolutionarily.

The ancestor obtained in this way is much larger in size and spends more steps to self-reproduce than the standard shortest (self-copying) ancestor does, due not only to the more complex procedure of self-reproduction but also to the primitiveness of the instructions that resembles assembly language, which makes the program structure substantially cumbersome.

38.3 Experimental Results and Discussion

38.3.1 Behaviour of the Designed Ancestor

Our naïve expectation was that the novel ancestor would be able to give rise to a "traditional" Avidian evolutionary process. More specifically, we expected that the ancestor could reproduce reliably enough to increase in population, under appropriate rates for the stochastic effects, where the total population growth is limited by the size of the Avida world. Not only that, some mutant capable of breeding true and still maintaining von Neumann's architecture was expected to emerge, through stochastic effects applied on the genome, and to undergo selection and evolution accordingly.

When seeded in the Avida world, the novel ancestor turns out to self-reproduce successfully, by decoding and copying its description as designed. This indicates the implementation of a von Neumann style ancestor incorporating genotype-phenotype mapping in this system is realisable.

With this seed program, we ran Avida simulations with mutation repeatedly, selectively allowing only single-point copy mutation to occur in the experiment, and singled out runs where any mutant becomes dominant. For each of these runs, we traced mutational changes and identified the transitional path from the ancestor to the final dominant strain. Analysis of them establishes that there are several cases of takeover by self-copiers, which are degenerated from the ancestor with the von Neumann style architecture. This marks the loss of the decomposition into genome and phenome, and also the loss of the mutable (or evolvable) genotype-phenotype mapping.

38.3.2 Degenerative Displacement

Surprisingly enough, it is found that one step of single-point mutation is sufficient to cause this degenerative transition. The unanticipated mechanism of the degeneration is as follows: the "decode" cycle is destroyed by a mutational change and starts to malfunction, but the "copy" cycle remains unharmed and happens to function to copy the entire memory image, overwriting a few mistakes made by the decode loop, and successfully makes both ends meet so as to precisely self-reproduce.

Once a self-copying strain arises, it is not at all surprising that it displaces the von Neumann self-reproducers because it is selectively favoured by the Avida system due to its high reproduction rate. In this particular case, the self-copier, being the same length as the hand-designed ancestor, takes less than half of the CPU cycles the ancestor requires. This is because for a memory content to be written into one location of the offspring, the decoding needs to execute more instructions than the copying, the cost of which self-copiers avoid.

Determining what style of self-reproduction an organism employs can be rather cumbersome because even when the program of the organism looks the same, the use, or interpretation, of the numbers in the program can be completely different. Instead, we have examined the gestation time since it indicates, even if crudely, the style of self-reproduction. Out of the organisms obtained throughout the run, there are a number of offspring whose self-reproduction takes approximately half of the gestation time that the ancestor requires, implying they are not von Neumann style self-reproducers, but highly likely to be self-copiers. In terms of gestation time, no organism was found that was viable, self-reproducing and mutated from the ancestral organism and that still preserved von Neumann style self-reproduction.

38.4 Conclusion and Future Work

Preliminary experiments have shown that there are some lineages where the reproduction mechanism degenerates back into the self-copying mode with considerable ease. That is, a mutation is expressed, causing the dysfunction of the decoding cycle, and, as a result, the decomposition into genome and phenome as well as the genotype-phenotype mapping are lost. The emerged self-copiers reproduce themselves rather quickly because they do not need intricate procedures and thus they are able to become dominant over the hand-designed ancestor. The displacement is not surprising per se, given the Avida system that favours faster self-reproducers; however, it is surprising that this degeneration can occur only with one step of single point mutation in the novel ancestor.

Further analysis of the mechanism for this degenerative transition is ongoing. Although it failed to demonstrate the potential that von Neumann style self-reproducers may have with a particular ancestral design and a particular system setting, the result shows how evolutionary search can discover an unanticipated mutational pathway. As an important next step, we intend to introduce a new preventive mechanism to effectively offset the intrinsic advantage of self-copiers, so that we can refocus on the characterisation of the von Neumann style self-reproducer and the investigation of the evolvability of genotype-phenotype mapping in the Avida world.

Resources including the ancestral program and the Avida configuration file we used in the experiment can be downloaded at http://alife.rince.ie/evosym/eccs_2012_th.zip.

Acknowledgements This work has been supported by the European Complexity Network (Complexity-NET) through the Irish Research Council for Science and Technology (IRCSET) under the collaborative project EvoSym. We would also like to thank the anonymous reviewers for the original paper.

References

1. Von Neumann J (1951) The general and logical theory of automata. In: Cerebral mechanisms in behavior, pp 1–41
2. McMullin B (2000) John von Neumann and the evolutionary growth of complexity: looking backward, looking forward. Artif Life 6(4):347–361
3. McMullin B, Taylor T, von Kamp A (2001) Who needs genomes?
4. Adami C (1997) Introduction to artificial life. Springer, Berlin
5. Ofria C, Wilke CO (2004) Avida: a software platform for research in computational evolutionary biology. Artif Life 10(2):191–229
6. Ray T (1994) An evolutionary approach to synthetic biology: Zen and the art of creating life. Artif Life 1(1/2):179–209

Chapter 39
Decimation of Fast States and Weak Nodes: Topological Variation via Persistent Homology

Irene Donato, Giovanni Petri, Martina Scolamiero, Lamberto Rondoni, and Francesco Vaccarino

Abstract We study the topological variation in Markov processes and networks due to a coarse-graining procedure able to preserve the Markovian property. Such coarse-graining method simplifies master equation by neglecting the *fast* states and significantly reduces the network size by decimating weak nodes.

We use persistent homology to identify the robust topological structure which survive after the coarse-graining.

Keywords Markov processes · Complex networks · Coarse-graining (theory) · Persistent homology · Computational topology

39.1 Introduction

Networks have received large attention over the last decade [5, 6] because of their ability to encode complex behaviours in simple ways, namely through the topology of their connectivity and the type of interactions between their elements. Furthermore, such interactions often evolve according to stochastic rules, characteristic that tightly connects complex networks to Markov processes. However, the salient features of large networks can be hard to identify due to the sheer size of the systems and heterogeneity in linking and weight distributions. In this context, effective techniques to highlight dominant structures within large networks are of extreme importance both for control and understanding [3]. In particular, such techniques have to preserve chosen network properties while reducing the complexity of the system.

We analyse a coarse-graining procedure inspired by the method recently developed for continuous Markov processes [1, 2]. This procedure simplifies the numer-

I. Donato (✉) · G. Petri · M. Scolamiero · F. Vaccarino
ISI Foundation, Via Alassio 11/c, 10126 Torino, Italy

I. Donato (✉) · L. Rondoni · F. Vaccarino
Dipartimento di Scienze Matematiche, Politecnico di Torino, C.so Duca degli Abruzzi n. 24, 10129 Torino, Italy

L. Rondoni
INFN, Sezione di Torino, Via P. Giuria 1, 10125 Torino, Italy

T. Gilbert et al. (eds.), *Proceedings of the European Conference on Complex Systems 2012*, Springer Proceedings in Complexity, DOI 10.1007/978-3-319-00395-5_39,

ical treatment of chemical and biological processes in terms of master equations reducing the number of variables with a decimation of *fast* states and a subsequent renormalization of the weights of all the surviving states in order to preserve the Markovian property.

In particular, we slightly modify the method of [1] in order to extend it to the case of complex networks. Whereas, in the case of Markov processes, the faster states are removed, in the case of networks we consider as candidates for decimation the nodes displaying smaller strength i.e. weak nodes. For every decimated node i, we introduce edges between all pairs of i's neighbours, effectively substituting the node itself with a *clique* composed by its former neighbours. The renormalization procedure of assigning weights to these new edges is described in the following section.

Importantly, we show on a range of different networks that this decimation technique is able to preserve the robust homological properties of the system, obtained through a persistent homology approach, while at the same time reducing significantly the complexity of the computation. We see also that the topological features depend crucially on the variation of coarse-graining level.

We report the example of *C. elegans* brain neural network of the type of results obtained for networks under the procedure that decimates the weak nodes and an asymmetric scale-free graph associated to an irreversible Markov process, as example of the procedure that decimates fast states.

39.2 Decimation of Fast States and Weak Nodes

The coarse-graining procedure, considered here, allows to advance the understanding of a certain Markov process with very different time-scale by neglecting the fast dynamic. Indeed, if the process is described by Master equation (here ω_{ij} indicates the rate of going from state i to state j whereas P_i is the probability of being in state i)

$$\frac{dP_i}{dt} = \sum_{i \neq j} (P_i \omega_{ij} - P_j \omega_{ji}) \tag{39.1}$$

and we are interested in the slow dynamics, it is usually not necessary but computationally demanding to exactly integrate the fast dynamics.

In this method the coarse-graining is parametrized by some threshold i.e. the coarse-graining level. Particularly, we can decimate all states having an average permanence time smaller than a prescribed threshold $\Delta\tau$ (*fast* states), where the time spent in a generic state n is exponentially distributed with average $\tau_n = 1/\omega_n^{out}$ and ω_n^{out} is the sum of the outgoing rates from n.

Similar considerations can be made in the context of network analysis, where typically the main interest is the description of the underlying backbone of the system, which is usually defined in terms of link weights [3] or connectivity [4]. From this point of view, weak nodes, that is nodes with low outgoing strength, are good candidates for removal, since they do not contribute significantly to the main structure of the network. In particular, given a strength threshold s and a node i, we decimate all the nodes with total outgoing strength $s_{ij} = \sum_{j \in \Gamma_i} \omega_{ij} < s$ (*weak* nodes).

It is important to replace the decimated nodes with effective interactions between the nodes they used to bridge. Namely, this means that a disappearing node i is substituted by links creating a fully connected *clique* among i's former neighbours. The weights of the *clique* edges are setted in such a way to preserve the Markovian propriety that leads to the following renormalization [1]:

$$\tilde{\omega}_i^j = \omega_i^j + \frac{\omega_i^n \omega_n^j}{\omega_n^{out}} \tag{39.2}$$

Note that this procedure corresponds to neglect the time spend in the removed state. We use the same renormalization (39.2) in the case of the network although the decimated nodes are the weak ones.

Interestingly, it was proved that, given a set of candidates for decimation, the result of this procedure does not depend on the order in which the nodes are removed, i.e. the procedure is commutative, widening its applicability and generality.

39.3 Persistent Homology

Formally, homology is an algebraic invariant converting local geometric information of a space into a global descriptor. There are many homology theories, but simplicial homology is the most amenable for computational purposes thanks to its combinatorial structure.

This kind of homology is applied to simplicial complexes, that are combinations of vertices, segments, triangles and higher dimensional analogues, joined according to specific compatibility relations. Low dimensional homology groups have an intuitive interpretation. Given a simplicial complex X, $H_0(X)$ is the free group generated by the connected components of X, $H_1(X)$ is the free group generated by the cycles in X, $H_2(X)$ is the free group generated by voids—holes bounded by two-dimensional faces. The Betti numbers count the number of generators of such homology groups.

The standard tool to encode this information is the so-called *barcode*, which is a collection of intervals representing the lifespans of such generators. Long-lived topological features can be distinguished in this way from short-lived ones, which can be considered as topological noise. Moreover, the value of the filtration parameter at which a certain generator provides important information about its role within the network.

There are various ways of building persistence modules out of a given dataset. The most known are the Rips-Vietoris complex, the Cech complex and the clique complex [7]. Here, we exploit the network's weighted clique structure. The clique complex is obtained by associating to each maximal k-clique, a completely connected subgraph formed by k nodes, the $(k - 1)$-simplex whose vertices are the clique's nodes. In this way however one produces a single simplicial complex, which

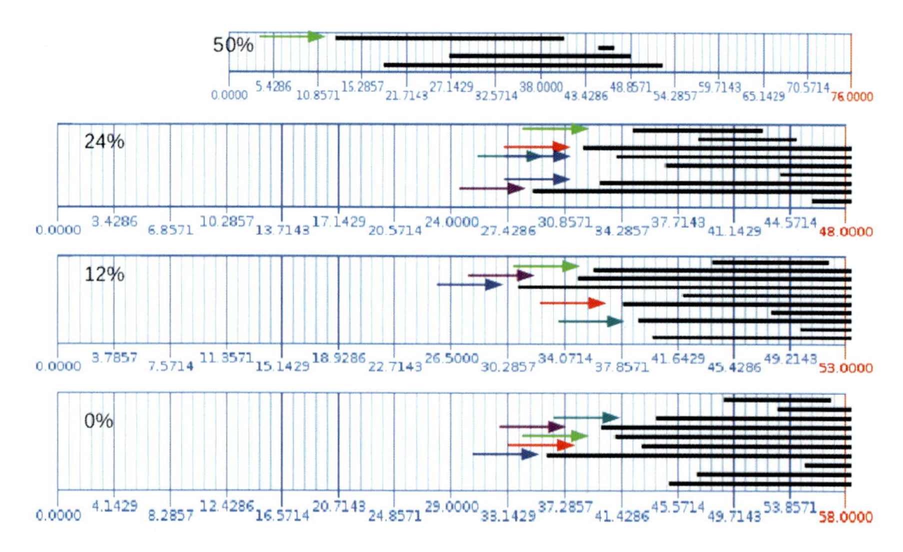

Fig. 39.1 The barcodes are shown with the percentage of the decimated nodes. *The arrows* indicates the persistence of the same loops. *The overlapped arrows* mean that two cycles have merged. We see the disappearance of the persistent cycles only when the decimation becomes massive

does not convey in itself any information about the robustness of the topological features it displays.

It is possible to devise a filtration, i.e. a sequence of simplicial complexes, combining link weights and clique structure. Given the weighted adjacency matrix ω_{ij}, we let ϵ vary in $\big(\min(\omega_{ij}), \max(\omega_{ij})\big)$ and consider a sequence of networks, such that the network at step ϵ contains all links (i, j) with $\omega_{ij} > \epsilon$. As we decrease ϵ from its maximum allowed value, we go from the empty network to the original one. For each step, we build the corresponding clique complex and study the persistent homology of the resulting filtration.

39.4 Topological Variation Due to the Decimation Procedure

We applied the decimation procedure to different weighted graphs representing real networks and than we have studied the relevant topological features to investigate if they are destroyed by the coarse-graining. We report two examples, one is the *C. elegans* neural network, where we decimated *weak* nodes, and the other one is an asymmetric weighted scale-free graph where we decimated *fast* states because of its interpretation as an irreversible Markov process.

We show the barcodes for the H_1 generators at different percentages of nodes removed for both the *C. elegans* neural network (Fig. 39.2) and the scale-free graph (Fig. 39.1). The barcodes summarize the topological information dictated by the dynamic of the network. Indeed, they are a collection of horizontal lines representing the homology generators in an arbitrary order. The starting and final points of these segment indicate the filtration level where they have appeared and disappeared. In particular, the H_1 generators are the cycles i.e. edges set in circular way with weight stronger then the near ones. The moment in the filtration in which they appeared gives information about the scale of the cycles (or voids if we are taking in consideration the H_2 generators). Indeed, persistent homology is a tool able to see also the structures (cycles, voids etc. ...) in the meso-scale and, furthermore, able to show the edges and nodes constituting such structures. In Figs. 39.2 and 39.1, for example, arrows of the same color highlight cycles that are present both in the initial graph and in the decimated ones. Hence, we can distinguish between graphs displaying the same homology generators distribution or having the same generators.

Decimation and Persistent Homology Applied to a Random Weighted Scale-Free Markov Process We have applied the coarse-graining procedure to a 50 states continuous Markov process, obtained from a directed scale-free graph. The motivation for applying this method to a scale-free graph is that it is recognized to reproduce observed properties of the world-wide web [9] and that this coarse-graining procedure is especially useful to integrate master equation when there are a lot of different scales.

The model used to create the initial data gives a graph that grows with preferential attachment so that the in- and out-degrees distributions follow power laws [8]. After we added random weights to each edge and divided the weights from i to j by the number of neighbours of i in order to prevent that the lifetime of state i would be the shorter the higher is its degree centrality. This step permit also to increase the separation of time scales. The H_1 generators for the full network and the ones corresponding to 12 and 24 percent of the weaker states removed (Fig. 39.1) evidence a good preservation of the relevant cycles. If a node of a certain cycle is removed we observe the shortening of that cycle in the coarse-grained graph. There could also be the merging of two cycles when a node to be decimated belong to both structures.

Importantly, the decimation procedure was also applied to a graph obtained in the same way that the previous one but without dividing the weights by the number of neighbours. We observed in this case that the cycles after the coarse-graining are not so long preserved because of the higher degrees of the fast states.

Decimation and Persistent Homology Applied to C. Elegans Neural Network Fig. 39.2 reports the barcode for the H_1 generators, which correspond to loops composed by strong links, obtained from the weighted clique persistent homology after a series of decimation. The plots refer in particular to respectively 40, 60, 80 and 85 percent of the network nodes being removed. Interestingly, we find that a set of generators, characterized by an early appearance in the barcode and a long persistence, that is emerging at large weight thresholds, survive through the decimation. This is particularly interesting especially when considering the high percentages of node removal, hence highlighting the extreme robustness and importance of the topological network features identified through their persistence. The other smaller cycles instead disappear under the decimation results, confirming their role as topological noise.

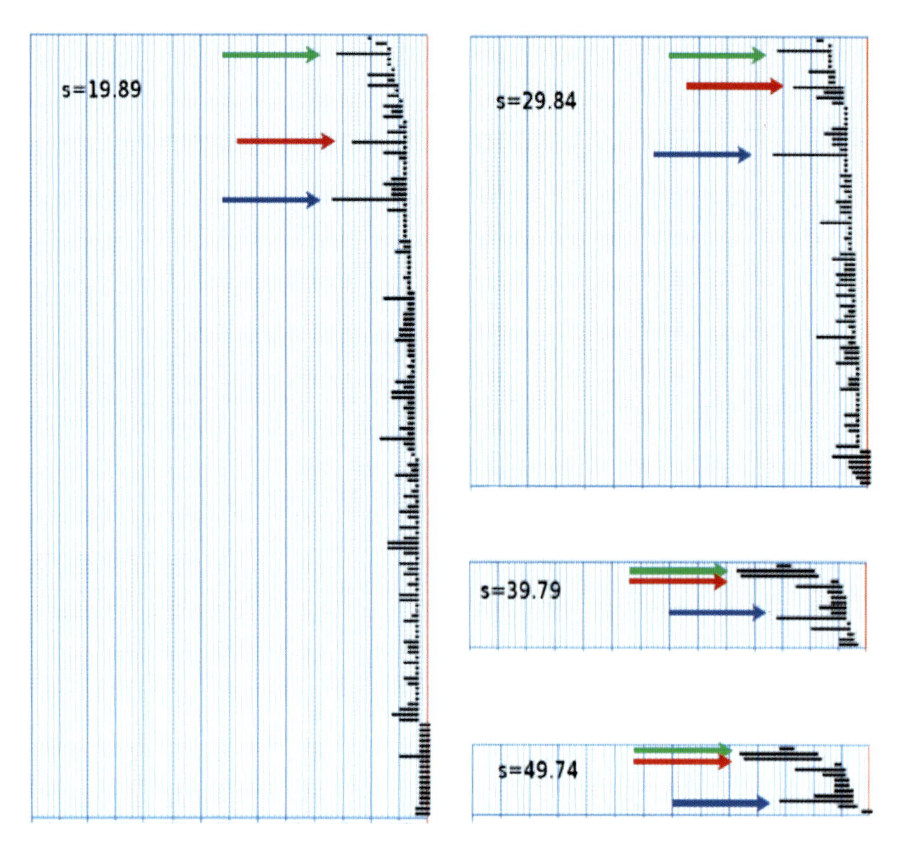

Fig. 39.2 In the first two figures (the ones corresponding to a threshold values $s = 19.89$ and $s = 29.84$) the generators are the same although decreased because of a reduction of the 34 per cent of the nodes between the two figures. The cycles start to change for a threshold value $s = 39.79$, where the reduction of the nodes become very large (more then 50 per cent with respect to *the above figure*), but an accurate study bring us to the conclusion that the cycles are only slightly varied. For example, the first stripe on the figure with $s = 29.84$ corresponds to the persistence cycle between the nodes 71, 74, 216, 207, whereas the first stripe on the figure with $s = 19.89$ corresponds to the cycles 71, 74, 216, 305 i.e. 305 takes the place of 207 that, effectively, was decimated

39.5 Conclusion

We used the persistent homology to explore topological features dictated by the dynamics on networks and Markov processes.

We conclude that the relevant topology features are essentially preserved under a coarse-graining based on decimation of weak nodes and fast states. The dynamics, before and after the decimation procedure, concentrates on the same cycles provided that the nodes even if fast should not be of the higher degree.

Acknowledgements L. Rondoni gratefully acknowledges financial support from the European Research Council under the European Community's Seventh Framework Programme (FP7/2007-

2013)/ERC Grant agreement No. 202680. The EC is not liable for any use that can be made on the information contained herein.

References

1. Pigolotti S, Vulpiani A (2008) J Chem Phys 128:154114
2. Puglisi A, Pigolotti S, Rondoni L, Vulpiani A (2010) J Stat Mech 2010:P05015
3. Serrano M, Boguñá M, Vespignani A (2009) Extracting the multiscale backbone of complex weighted networks. Proc Natl Acad Sci USA 106(16):6483
4. Grady D, Thiemann C, Brockmann D (2011) Parameter-free identification of salient features in complex networks. arXiv:1110.3864
5. Albert R, Barabasi A (2002) Statistical mechanics of complex networks. Rev Mod Phys 74(1):47–97
6. Dorogovtsev SN, Goltsev AV, Mendes JFF (2008) Critical phenomena in complex networks. Rev Mod Phys 80(4):1275–1335. doi:10.1103/RevModPhys.80.1275
7. Zomorodian A, Carlsson G (2005) Computing persistent homology. Discrete Comput Geom 33(2):249–274
8. Bollobás B, Borgs C, Chayes J, Riordan O (2003) Directed scale-free graphs. In: Proceedings of the fourteenth annual ACM-SIAM symposium on discrete algorithms, pp 132–139
9. Barabasi A, Albert R (1999) Emergence of scaling in random networks. Scienze 286:509–512. Also in Physica A 281:69–77 (2000)

Part III
Prediction, Policy and Planning, Environment

Chapter 40
Characteristics of Seismic Networks in Spatial Scales

D.D. Kang, D.I. Lee, and K. Kim

Abstract We investigate the seismic network by considering the cell resolution and the temporal causality of the earthquake from seismic data taken in southern California of USA. The topological properties of our seismic network are simulated and analyzed. We mainly estimate the global network metrics such as the mean degree, the probability distribution of degree, the clustering coefficient, the characteristic path length, and the global efficiencies. Particularly, Our results in random networks are compared to other findings which estimate the global network metrics.

Keywords Seismic network · Random network · Clustering coefficient · Local efficiency · Global efficiency

40.1 Introduction

Over the last two decades, research of complex systems [1] has been applied to new methods and techniques of studying the intermittent nature of turbulence, various financial time series, wavelet transform approaches, growing and non-growing networks [2], and seismic phenomena [3], amongst others. In network theory, small-world and scale-free network models have been studied widely in various applications of these scientific fields. These two network models have played a crucial role in understanding complex phenomena. Of the many systems of current interest, the degree distribution for scale-free networks is interesting, because it follows a power law, and for random networks it decays faster than exponentially.

The seismic network is a phenomenon of dynamical behavior in complex seismic time series [4–6]. Abe and Suzuki have analyzed the spatio-temporal properties of seismicity from the viewpoint of the Tsallis entropy under appropriate constraints.

D.D. Kang · D.I. Lee
Department of Environmental Atmospheric Sciences, Pukyong National University, Busan 608-737, Republic of Korea

K. Kim (✉)
Department of Physics, Pukyong National University, Busan 608-737, Republic of Korea
e-mail: kskim@pknu.ac.kr

T. Gilbert et al. (eds.), *Proceedings of the European Conference on Complex Systems 2012*, Springer Proceedings in Complexity, DOI 10.1007/978-3-319-00395-5_40,

In particular, the correlation function has been a main issue in theoretical and numerical investigations of aftershock phenomena. Several theoretical formulae have been used to carry out the computational simulation of earthquakes.

In this paper, Our purpose of this work is to simulate and analyze the topological robustness of the seismic network against the spatial shift and the scale in a volume, which is located on a tectonic plate without boundaries, from seismic data collected on the southern California of USA. Section 40.2 introduces in detail the theoretical method and the numerical calculation of complex networks. Our main results are summarized in Sect. 40.3.

40.2 Theoretical Method and Numerical Calculation

First of all, the mean degree is calculated as $\langle k\rangle = \frac{1}{N}\sum_{i=1}^{N} k_i$, where k_i is the number of degrees on the i-th vertex. The probability distribution of degree [7] is represented in terms of

$$p(k) \sim k^{-\gamma}, \tag{40.1}$$

where γ is the degree exponent.

We introduce the characteristic path length L_c given by

$$L_c = \frac{1}{N(N-1)} \sum_{i,j\in N} d_{ij}, \tag{40.2}$$

where Eq. (40.2) is the average of the shortest paths between each node and the other nodes in the network. To examine the cliquishness of earth networks, we can calculate the mean clustering coefficient as

$$C = \frac{1}{N}\sum_{i=1}^{N} C_i, \tag{40.3}$$

where $C_i = 2E_i/k_i(k_i-1)$, and E_i denotes the number of links among k_i neighbors of a node i. Two networks in this work show a similar clustering property.

Finally, the global efficiency E_g [1] is defined by

$$E_g = \frac{\sum_{i\neq j\in G} d_{ij}^{-1}}{\sum_{i\neq j\in G_{id}} l_{ij}^{-1}}, \tag{40.4}$$

where d_{ij} is the shortest path length between two vertices i and j, and l_{ij} is the spatial distance between two stations i and j when they are directly connected. For the various graphs it is obtained that $0 \le E \le 1$, and $E = 1$ for the ideal case ($G = G_{id}$). Through the above statistical quantities, we can simulate and analyze the topological measures in the complex network, and these mathematical techniques have been confirmed in the empirical and realistic investigation of diverse models.

In order to examine the robustness of a network topology against spatial shifts and scales in an volume, we select three kinds of width with 6.88 km, 10.31 km,

Table 40.1 Our estimated measures compared with other network in previous published works

Networks	Nodes	Edges	$\langle k\rangle$	L_c	C
California earthquake (6.88 km)	6945	84874	12.22	1.26	0.58
California earthquake (10.31 km)	4271	64185	15.03	1.19	0.64
California earthquake (17.19 km)	2741	50675	18.49	1.14	0.70
Random network	6945	84757	12.20	1.52	0.003
Random network	4271	64077	15.00	1.40	0.007
Random network	2741	50646	18.48	1.29	0.013

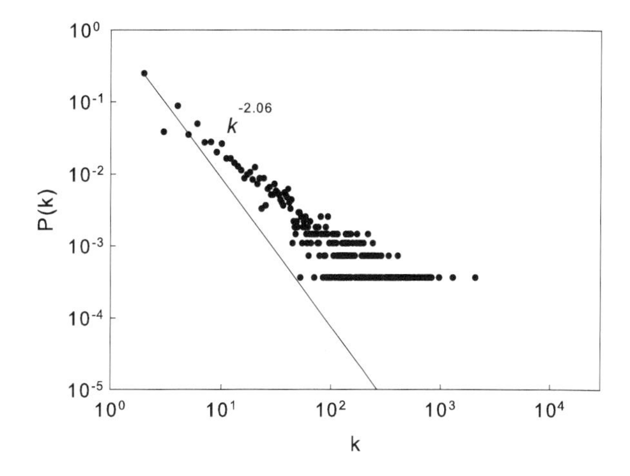

Fig. 40.1 Probability distribution of the degree for the cell width 17.19 km, where *the solid line* has a slope of −2.06

and 17.19 km. We mainly perform the numerical computations for four statistical quantities given by Eqs. (40.1)–(40.4).

Our values are compared to those of other networks in Table 40.1. From the result of the numerical analysis, Fig. 40.1 shows the probability distribution of degree as a function of the degree for the volume width with 17.19 km. As the width increases, the value L_C increases from our result. Our result is found that the global coefficients are given by 1.0×10^{-3}, 1.6×10^{-3}, and 2.5×10^{-3}, respectively, and these values are similar to those of the random network [8]. As the volume width increases, the clustering coefficient for the earthquake network increases, similar to the result of Ref. [9].

40.3 Conclusions

In summary, we have examined the topological properties of the earthquake network such as the mean degree, the probability distribution of degree, the characteristic path length, the clustering coefficient, and the global efficiency from the seismic data of southern California in USA.

In this paper, an estimate of network metrics is found using network theory. We believe that the idea by which this result is obtained may be a useful starting point for a new approach to understanding the topological properties of network theory in earthquake. We suggest that a recent network approach to earthquakes analysis is very reliable in three-dimensional cells. Further work is needed for the case with boundaries of more than two tectonic plates. Our hope is that the formalism of our analysis will be extended to both the discrimination and the characterization of various earthquakes in other nations.

Acknowledgements This work was supported by Center for Atmospheric Sciences and Earthquake Research (CATER 2012-6110) and by the National Research Foundation of Korea through a grant provided by the Korean Ministry of Education, Science & Technology in 2011 (No. K1663000201107900).

References

1. Boccaletti S, Latora V, Moreno Y, Chavez M, Hwang DU (2006) Complex networks: structure and dynamics. Phys Rep 424:175–308
2. Kim Y, Choi W, Yook SH (2012) Modified Penna bit-string network evolution model for scale-free networks with assortative mixing. J Korean Phys Soc 60:621–624
3. Abe S, Suzuki N (2006) Complex-network description of seismicity. Nonlinear Process Geophys 13:145–150
4. Abe S, Pastén D, Muñoz V, Suzuki N (2011) Universalities of earthquake-network characteristics. Chin Sci Bull 56:3697–3701
5. Telesca L, Lapenna V, Macchiato M (2005) Multifractal fluctuations in seismic interspike series. Physica A 354:629–640
6. Abe S, Suzuki N (2012) Universal law for waiting internal time in seismicity and its implication to earthquake network. Europhys Lett 97:1–5
7. Amaral LAN, Scala A, arthelemy M, Stanley HE (2000) Classes of small-world networks. Proc Natl Acad Sci USA 97:11149–11152
8. Kang DD, Lee D-I, Chang K-H, Park J-K, Kim K (2012) Characteristics of network metrics in seismic phenomena. J Korean Phys Soc 61:1163–1166
9. Baek WH, Lim G, Kim K, Choi YJ, Chang KH, Jung JW, Yi M, Lee DI, Ha DH (2012) Analysis of topological properties in a seismic network. Physica A 391:2279–2285

Chapter 41
You Are Who Knows You: Predicting Links Between Non-members of Facebook

Emöke-Ágnes Horvát, Michael Hanselmann, Fred A. Hamprecht, and Katharina A. Zweig

Abstract Could online social networks like Facebook be used to infer relationships between non-members? We show that the combination of relationships between members and their e-mail contacts to *non*-members provides enough information to deduce a substantial proportion of the relationships between non-members. Using structural features we are able to predict relationship patterns that are stable over independent social networks of the same type. Our findings are not specific to Facebook and can be applied to other platforms involving online invitations.

Keywords Online social network · Privacy · Link prediction · Machine learning · Random forest classifier

41.1 Introduction

Inference of user attributes and link prediction in online social networks is a challenging task that has attracted the attention of many researchers in the past few years. They showed that characteristics of a given user, such as its political preference or its sexual orientation, can be accurately inferred based on the attributes of its

E.Á. Horvát (✉) · M. Hanselmann · F.A. Hamprecht
Interdisciplinary Center for Scientific Computing (IWR), Heidelberg Collaboratory for Image Processing (HCI), University of Heidelberg, Speyererstr. 6, 69115 Heidelberg, Germany
e-mail: agnes.horvat@iwr.uni-heidelberg.de

F.A. Hamprecht
e-mail: fred.hamprecht@iwr.uni-heidelberg.de

F.A. Hamprecht · K.A. Zweig
Marsilius Kolleg, University of Heidelberg, Hauptstr. 232, 69117 Heidelberg, Germany

K.A. Zweig
e-mail: zweig@cs.uni-kl.de

E.Á. Horvát · K.A. Zweig
Technical University of Kaiserslautern, Gottlieb-Daimler-Strasse 48, 67663 Kaiserslautern, Germany

T. Gilbert et al. (eds.), *Proceedings of the European Conference on Complex Systems 2012*, Springer Proceedings in Complexity, DOI 10.1007/978-3-319-00395-5_41,

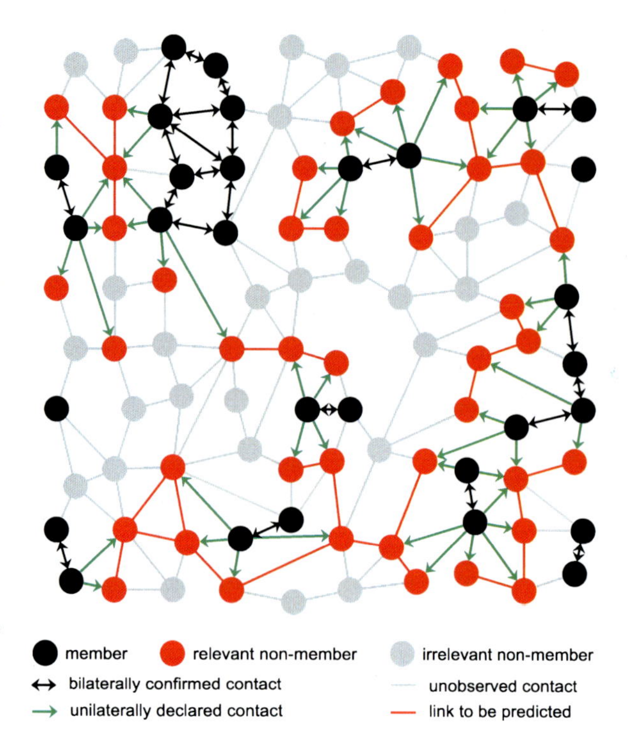

Fig. 41.1 Division of the social network into members (*black nodes*) and non-members. In the depicted toy example, a fraction of $\rho = 0.3$ (30 out of 100) individuals are members. The relevant subset of non-members consists of those who are in contact with at least one member (*red nodes*). A fraction of $\alpha = 0.5$ (15 out of 30) members have disclosed their e-mail contacts to non-members. The edges between members (*black*, bi-directed arrows) and to non-members (*green arrows*) are used to predict edges between non-members (*red lines*). For clarity, the values of ρ and α are exaggerated and the weak ties between individuals are omitted. Figure reprinted from [2] with permission

friends [3, 7, 9, 14]. Also, previously unrevealed or future relationships have been predicted with high precision using both supervised [6, 13] and unsupervised [5] learning methods. As an extension of these insights into the transparency of the members of such platforms, we asked a novel question: How many of the relationships between *non*-members can online social networks infer [2]? Besides revealing the potential of predicting relationships *outside* the network, we present a challenging link prediction task where learning and testing were performed on distinct Facebook networks.

41.2 Ground Truth Imputation

Given an online social network, a society is divided into a fraction ρ of members and $1 - \rho$ of non-members (see Fig. 41.1). Members are linked through mutually confirmed friendship relationships. Furthermore, aiming at expanding their circle of friends, a fraction α of the platform members import their whole e-mail address-book, thereby sharing also their contacts to non-members. Based on the seemingly innocuous combination of these two information types, we infer links between the non-members.

Data comprising the whole social network is unattainable (i.e., friendships via online social networks, e-mail contacts to non-members, and acquaintances between them). Thus, in order to provide the ground truth for a machine learning approach, we impute the missing information. We assume that the available real Facebook friendship networks of students from five different US universities (`UNC`, `Princeton`, `Georgetown`, `Oklahoma`, and `Caltech`) represent their complete online social network [11]. We then partition the students into members and non-members. We do not have a clear understanding of how people decide upon joining online social networks. On one hand, a recent analysis of the growth of Facebook showed that the probability of a non-member joining the platform increases with the structural diversity of its acquaintances who are members, i.e., with the number of connected components in its Facebook neighbourhood [12]. On the other hand, there is indication that platforms recruit their members through a mixture of online mediated invitations by friends who are already members and through independent decisions by individuals who are not yet friends of a member [4]. In line with the latter investigation, we cover a wide range of possible mechanisms: we use a series of different member recruitment models and show that our results are stable for all of them. The considered models are the following: (1) the *breadth-first search* model (BFS): once we have a starting member, all her friends become members followed by all their friends and so on, (2) the *depth first search* model (DFS): a randomly chosen friend of a member joins, followed by a randomly chosen friend of the new member and so on, (3) we start a *random walk* (RW) from a member and restart the walk as soon as we would choose someone who is already a member, (4) the *ego networks selection* model (EN): we select the members randomly and together with them, all their friends join the platform, and (5) the *random selection* model (RS): people decide independently from their friends whether to become a member or not, i.e., we choose each member randomly from the remaining non-members. Accordingly, the ground truth imputation for our inference problem consists in fixing the fraction of members ρ, partitioning the community into members and non-members by using one of the member recruitment models above described, and finally choosing the disclosure parameter α, thereby controlling for the percentage of contacts that are made public. Having devised the ground truth, we use the following approach.

41.3 Link Prediction

The available Facebook networks are anonymized. In the absence of user attributes, we base our predictions solely on topological graph features. For each pair of non-members, we compute a set of features deduced from the known structural properties of (online) social networks [8, 10]. For example, based on the recognition that people sharing a friend are usually friends themselves, we include a feature that counts the number of neighbours two non-members share. Other features weight this number in several ways (e.g., by the popularity of the common neighbour) or count the number of paths of length 3 between the two non-members. We use the feature vectors to employ a standard supervised learning method called random forest classifier [1]. We adjust the parameters of the classifier on a training set and then apply it to a test set. We predict that those pairs of non-members are linked for which the edge probability determined by the algorithm is higher than some threshold value. In a final step, we validate our predictions by comparing them with the ground truth. We use two measures to quantify the accuracy of the algorithm:

(1) the *Area Under the Curve* (*AUC*) which is a standard machine learning measure that quantifies the probability that the classifier algorithm assigns higher prediction values to true positives than to true negatives. Thus, a perfect classifier has an *AUC* of 1 while random guessing results in an *AUC* of 0.5.
(2) the *positive predicted value* of the k top-ranked predictions (PPV_k), introduced by [5], is defined as the percentage of correctly classified edges among the first k pairs in the ranking and is thus equal to the sensitivity achieved by predicting these k samples to be edges.

Instead of training and testing within the same network, we assure the independence of these two sets by learning and testing on different networks. We do so by devising two training schemes. (1) In the $4 \rightarrow 1$ cross-prediction scenario the classifier is trained on samples from *four* data sets and tested on samples from a fifth data set. With this scheme we avoid overfitting. (2) In the $1 \rightarrow 1$ cross-prediction setting the classifier is trained on *one* university data set and evaluated on another. The goal here is to evaluate whether a single network contains enough characteristic patterns to obtain high-quality predictions for an entirely different network.

41.4 Results

Imputing the ground truth required introducing two parameters (the membership ρ and the disclosure α) as well as a member recruitment model (BFS, DFS, RW, EN, or RS). We investigate the prediction accuracies for a wide range of their combinations using two measures (*AUC* and PPV_k) and two training schemes.

First, we examine the performance of our algorithm with $4 \rightarrow 1$ cross-prediction for each combination of ρ and α, all member recruitment models and all five university data sets (see Fig. 41.2). Based on the minimal (lower triangle) and maximal (upper triangle) *AUC* and PPV_k values, we see that the differences between the member recruitment models are small in most cases. The *AUC* values are above 0.85 for all combinations with $\rho \geq 0.5$ and $\alpha \geq 0.4$ in the case of `UNC`, `Princeton`, `Georgetown` and `Oklahoma`, for all member recruitment models except the BFS. This implies that in most cases the prediction is considerably better than random guessing. The PPV_k is at least 0.4 for the same range of ρ and α and in the case of `UNC`, `Georgetown`, and `Oklahoma`, and for all member recruitment models except the BFS and the DFS. A value of 0.4 means that when selecting the k samples with the highest prediction values, at least 40 % of them indeed represent two non-members that know each other. To interpret this value correctly, we have to note that our data set shows a striking imbalance which makes prediction difficult. While there is a huge number of node pairs that could be linked by an edge, there

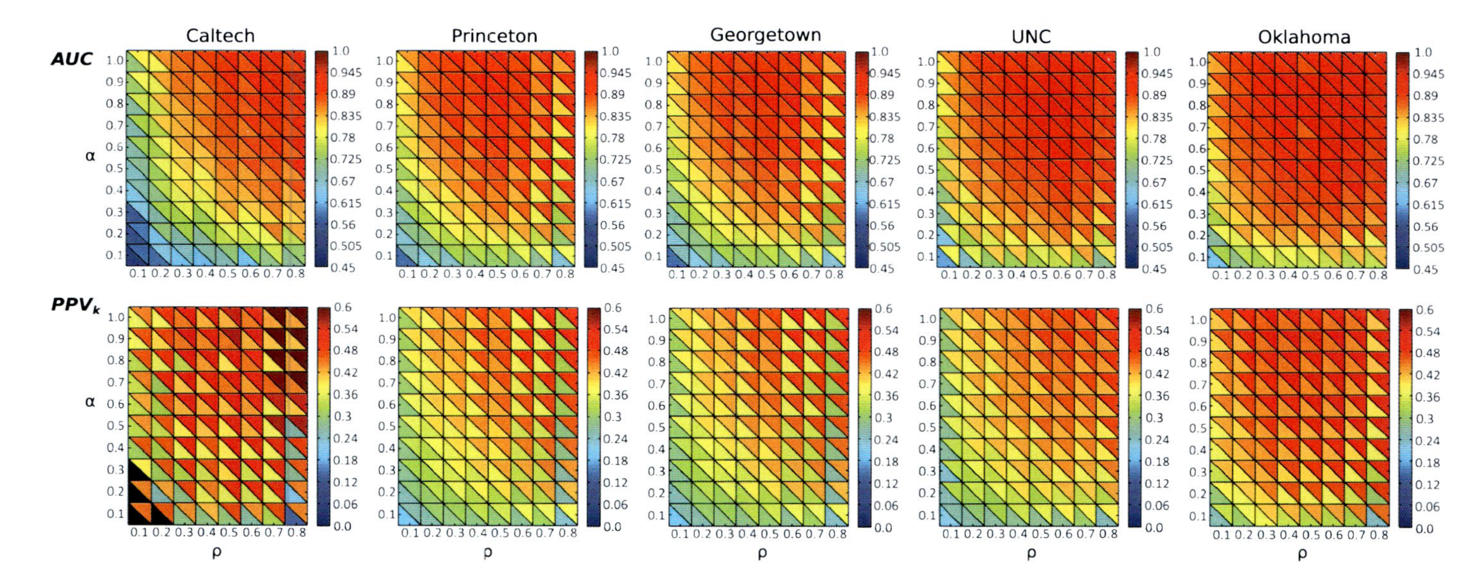

Fig. 41.2 Minimal (*lower triangle*) and maximal (*upper triangle*) prediction accuracy in the $4 \rightarrow 1$ training scheme for all five member recruitment models as a function of the membership parameter ρ and the disclosure α. *Upper row*: *AUC*; *lower row*: PPV_k; *black triangles* denote data points where PPV_k was smaller than the according fraction of positive samples among all samples, i.e., it was worse than expected by chance. Figure reprinted from [2] with permission

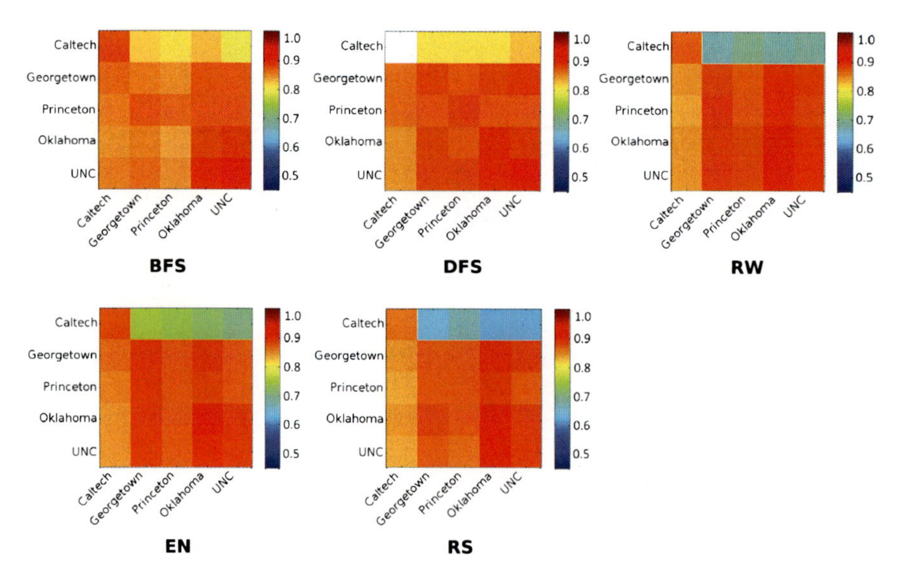

Fig. 41.3 $1 \to 1$ cross-prediction accuracy: *AUC* values for each of the five member recruitment models when $\rho = \alpha = 0.5$. The y and x-axis show on which network the random forest was trained and tested, respectively. *The white field* indicates that there were too few edge samples to reasonably train the classifier. Figure reprinted from [2] with permission

are only a few pairs which are truly linked. More precisely, depending on the chosen member recruitment model and on the ρ membership and α disclosure parameters, the ratio between the number of edges and non-edges lies between 0.0002 and 0.03 for four out of five university networks.

Second, in the $1 \to 1$ cross-prediction setting, we evaluate how reliable the predictions are if the random forest is trained on only one network at $\rho = \alpha = 0.5$. Given the coverage of Facebook and the heavy usage of the friend finder application by both novice members and experienced users of the platform, these estimates of ρ and α are rather conservative. Figure 41.3 shows the corresponding prediction accuracy. On the diagonal, we plot as reference the prediction accuracy when we train and test on the same network, while the off-diagonal elements correspond to the cross-prediction case. It can be seen that some data sets are easy to predict, namely `Oklahoma` and `UNC`, while `Caltech` is hard to predict based on any of the four other data sets. Furthermore, if the classifier is trained on `Caltech` data, the predictions are consistently the worst among all cross-predictions. The intuition behind this observation is that `Caltech` is a clear outlier among the used data sets because it is by far the smallest and the densest.

41.5 Conclusions

Our work reveals the potential that social network platforms have in predicting links between non-members based only on the connection patterns of the befriended members and their e-mail contacts to non-members. Accordingly, people without a profile on an online social network—such as Facebook—are not immune against data mining based on data available to the given platform. This finding is based solely on topological features, i.e. we used purely contact data and no user attributes. If we had access to more comprehensive data including details about the members like their age, location, or occupation, then our inference could be considerably more accurate.

Acknowledgements E.Á. Horvát and K.A. Zweig were supported by the Heidelberg Graduate School of Mathematical and Computational Methods for the Sciences, University of Heidelberg, Germany, which is funded by the German Excellence Initiative (GSC 220). F.A. Hamprecht and K.A. Zweig were supported by a fellowship of the Marsilius Kolleg, University of Heidelberg, Germany.

References

1. Breiman L (2001) Random forests. Mach Learn 45:5–32
2. Horvát E-Á, Hanselmann M, Hamprecht F, Zweig KA (2012) One plus one makes three (for social networks). PLoS ONE 7(4):e34740
3. Jernigan C, Mistree B (2009) Gaydar: Facebook friendships expose sexual orientation. First Monday 14:10. Online
4. Kumar R, Novak J, Tomkins A (2006) Structure and evolution of online social networks. In: Proceedings of the 12th ACM SIGKDD international conference on knowledge discovery and data mining, pp 611–617
5. Liben-Nowell D, Kleinberg J (2007) The link-prediction problem for social networks. J Am Soc Inf Sci Technol 58:1019–1031
6. Lichtenwalter RN, Lussier JT, Chawla NV (2010) New perspectives and methods in link prediction. In: Proceedings of the 16th ACM SIGKDD international conference on knowledge discovery and data mining, pp 243–252
7. Lindamood J, Heatherly R, Kantarcioglu M, Thuraisingham B (2009) Inferring private information using social network data. In: Proceedings of the 18th international conference on world wide web, pp 1145–1146
8. Mislove A, Marcon M, Gummadi KP, Druschel P, Bhattarcharjee B (2007) Measurement and analysis of online social networks. In: Proceedings of the 7th ACM SIGCOMM conference on internet measurement, pp 29–42
9. Mislove A, Viswanath B, Gummadi KP, Druschel P (2010) You are who you know: inferring user profiles in online social networks. In: Proceedings of the 3rd ACM international conference on web search and data mining, pp 251–260
10. Newman MEJ, Park J (2003) Why social networks are different from other types of networks. Phys Rev E 68:036122
11. Traud A, Kelsic E, Mucha P, Porter M (2011) Comparing community structure to characteristics in online collegiate social networks. SIAM Rev 53:526–543
12. Ugander J, Backstrom L, Marlow C, Kleinberg J (2012) Structural diversity in social contagion. Proc Natl Acad Sci USA 109(16):5962–5966
13. Wang C, Satuluri V, Parthasarathy S (2007) Local probabilistic models for link prediction. In: 7th IEEE international conference on data mining, pp 322–331
14. Zheleva E, Getoor L (2009) To join or not to join: the illusion of privacy in social networks with mixed public and private user profiles. In: Proceedings of the 18th international conference on world wide web, pp 531–540

Chapter 42
Vulnerability Analysis of Interdependent Infrastructure Systems

Gaihua Fu, Mehdi Khoury, Richard Dawson, and Seth Bullock

Abstract Complex network approaches have been used to analyse physical or social networks. Previous research has tended to focus on studying single, isolated systems and ignores the fact that many of these systems are developing into a "network of networks". Over the years these systems have become increasingly interconnected. Due to this interdependence the failure of one network component may propagate across the system of systems, resulting in cascading failure. This research studies interdependent networks from a "system-of-systems" perspective, and a suite of tools have been developed to analyse infrastructure networks and their associated interdependencies. The aim of this research is to facilitate system-scale understanding of infrastructure networks and the implications of interdependencies between them. This understanding could be used to improve the design of infrastructure that is more resilient and adaptable to natural and manmade hazards.

Keywords Complex networks · Interdependencies · Vulnerability · Cascading failure

42.1 Introduction

Complex system approaches have been used to analyse the robustness of physical or social networks [1, 6]. Previous research has tended to focus on studying single, iso-

G. Fu (✉) · R. Dawson
Newcastle University, Newcastle upon Tyne, UK
e-mail: gaihua.fu@ncl.ac.uk

R. Dawson
e-mail: richard.dawson@ncl.ac.uk

M. Khoury · S. Bullock
University of Southampton, Southampton, UK

M. Khoury
e-mail: mk7@ecs.soton.ac.uk

S. Bullock
e-mail: sgb@ecs.soton.ac.uk

T. Gilbert et al. (eds.), *Proceedings of the European Conference on Complex Systems 2012*, Springer Proceedings in Complexity, DOI 10.1007/978-3-319-00395-5_42,

lated systems and ignores the fact that many of these systems are developing into a "network of networks" [3, 5]. For example, in the infrastructure domain, successful operation of the energy distribution system requires water for cooling, transport to supply fuel, and ICT systems for control and management. Over the years these systems have become increasingly interconnected and mutually dependent in complex ways. Due to this interdependence the failure of one network component may propagate across the system of systems, resulting in cascading failure across multiple sectors. As the extent and complexity of interdependencies increases, so does the risk of such failure [7]. Some advances have been made in investigating relatively simple, spatially constrained representations of systems and their interdependencies [2]. However, this simplification often results in network models, analytical methods and modelling outputs often being far removed from those required for real-world cases. Consequently, more extensive analysis and modelling of interdependent networks is required in order to identify and understand the role of interdependencies and the risks associated with them.

In the *Resilient Futures* project (http://r-futures.ecs.soton.ac.uk), we study interdependent infrastructure networks from a "system-of-systems" perspective, and a suite of tools have been developed to represent and analyse infrastructure networks and their associated interdependencies. The aim of this research is to facilitate system-scale understanding of infrastructure networks and the implications of interdependencies between them. This understanding could be used to improve the design of infrastructure that is more resilient and adaptable to future socio-economic developments and climate change. In this paper, we present the infrastructure network modelling work that we have conducted to date and we report some preliminary results of the research.

42.2 Infrastructure Network Model

We have developed a modelling framework for exploring cascading failure of interdependent networks [4]. Here, we first establish a number of isolated networks, each representing an infrastructure system exhibiting some topology. Such a network can either be spatial or aspatial in our modelling framework. In a spatial network a node has attributes that indicate its location in a predefined geographical space. In an aspatial network these attributes are absent. Interdependencies between networks are represented by a number of edges, each connecting a node in one network with a node in another. Figure 42.1 shows a simple spatial interdependent network system that is generated with our network modelling tool.

Interdependencies can be customised along a few dimensions so as to provide the capacity to model various complex coupling modes of interdependent systems. Firstly, inter-network connections can be generated according to different criteria, including random connections, or co-related connections according to spatial proximity or node degree. Secondly, inter-network connections can be configured to generate networks with different interdependent directionality and density. The interde-

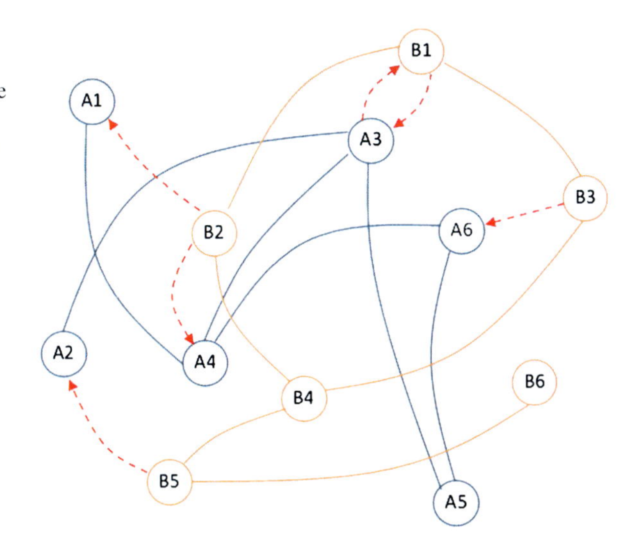

Fig. 42.1 A spatial interdependent system that couples two networks, where A_i nodes are from one network, B_j nodes are from another network, an undirectred intra-network connection is represented with a *solid edge*, and a directed inter-network connection (that links an A_i node with $a B_j$ node) is represented with a *dashed edge*

pendent system shown in Fig. 42.1 couples two networks according to the spatial proximity of network nodes.

We model an attack on one network that disables some proportion of the network nodes directly and indirectly brings about a cascade of additional node failures as a consequence of compromised interdependencies. Such additional node failures happen recursively and may result in system failure extending far beyond the original attack footprint. We show that the post-attack performance of both networks is mediated by

- network topology (e.g., regular, small world, centralised, decentralised)
- the nature and extent of network interdependency (e.g., directed versus undirected dependencies, their density and correlation structure)
- the type of attack (i.e., random, targeted, spreading, or spatial)
- the rule for establishing post-attack viability (e.g., a network component might be viable only if it is above some critical size, or if it is sufficiently connected to a viable component another network, etc.)

42.3 Experimental Results

42.3.1 Impacts of Interdependencies on Network Performance

We studied the vulnerability of interdependent network systems by performing experiments over systems that couple two Erdős–Rényi networks, with each network having 10000 nodes and average degree of 4. We initiated cascading failure by removing a fraction, q, of randomly chosen nodes from one network. In line with other percolation based research [1], we use the relative size, P, of the largest connected

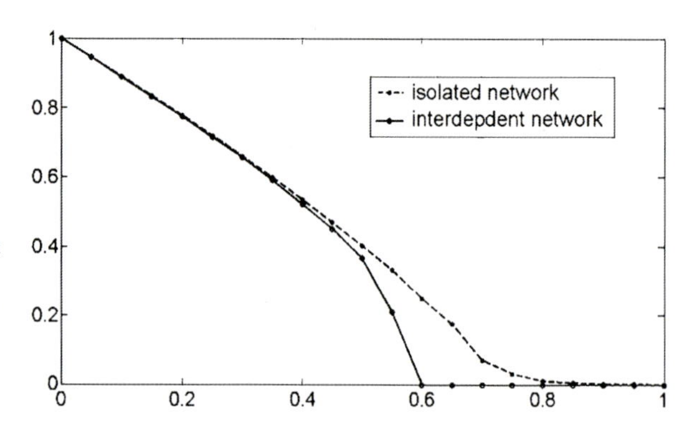

Fig. 42.2 Performance comparison of an interdependent network against that of a single network. Each curve represents the mean performance of 100 simulations of interdependent networks that couple two fully connected Erdős–Rényi networks, with average interdependent degree of 2

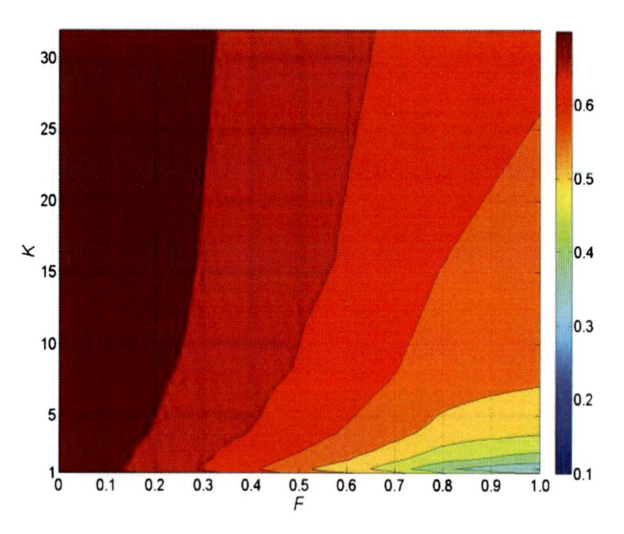

Fig. 42.3 Aggregate performance of interdependent network when K and F are varied. Each map represents the mean performance of 100 simulations of interdependent networks that couple two 10000-node Erdős–Rényi networks with average degree of 4

component as a measure of system performance. Some key findings are presented below and readers are referred to [4] for more information on our experimental results.

As shown in Fig. 42.2, an interdependent network system is more vulnerable than an individual, isolated network, typically exhibiting a lower failure threshold than a single, isolated network. The disruption in an interdependent network is disproportionate to the attack size, experiencing abrupt system collapse at some failure threshold, q_c.

The performance of an interdependent network varies with the configuration of the inter-network dependencies. We use F to denote the portion of dependent nodes that a network has, and use K to denote the average interdependent degree (the number of supporting nodes that a dependent node is directly connected to). Figure 42.3 shows the aggregate performance of an interdependent network when K and F are varied. The aggregate performance measure is calculated as the integral, IP, of the relative size, P, of the largest component in the form of:

$$IP = \int_0^1 P(q) \tag{42.1}$$

It demonstrates that a system with lower interdependent degree and larger portion of dependent nodes is more vulnerable than a system with smaller portion of dependent nodes but higher interdependent degree.

42.3.2 Maintaining Interdependence

Until now, we have followed previous studies in evaluating the post-attack performance of a pair of interdependent networks, A and B, by measuring the size of the largest remaining connected component in each network. This calculation has ignored the extent to which either of these two largest components is connected to surviving nodes in the other network. Implicitly, this approach assumes that a cluster of surviving B nodes can exist as a viable network in the absence of any connection to the A network, and vice versa. As a consequence, our analysis above shows that reducing the number of interdependencies between A and B is a good way of improving system resilience, since when one network is attacked the other is shielded from the consequences.

However, this approach to measuring post-attack viability does not take into account the initially symbiotic nature of two infrastructure systems which is an important property of modern infrastructure, e.g., a railway network cannot survive in isolation from a road transport network because it needs the road network to deliver passengers, goods, and personnel in order to operate.

Consequently, here we evaluate the post-attack viability of a network by measuring the size of the largest component that meets an interdependency threshold. This threshold is expressed in terms of the proportion of nodes within the component that are connected to at least one surviving node in the other network. Setting the threshold at 10 %, for instance, demands that in order for a network B component to be viable, at least 10 % of its nodes must be connected to a surviving A node, i.e., an A node that itself is within a component that meets the 10 % interdependency threshold.

Figure 42.4 depicts how the post-attack viability of a particular pair of weakly coupled networks varies with the size of an attack that removes some proportion of randomly selected nodes in A. When the size of the largest component in B is used as a measure of post-attack viability (dashed line), around 80 % of the network can survive even after large attacks. With the introduction of the interdependency threshold, a catastrophic failure occurs once the attack reaches a critical size.

Figure 42.5 compares the two approaches to measuring post-attack viability for a wide range of coupled Erdős–Rényi networks, depicting how the measures vary with the average degree of the two networks and the extent of the interdependency

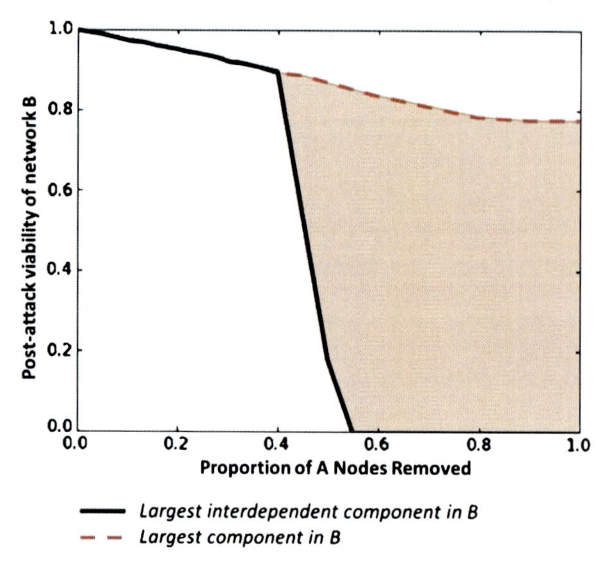

Fig. 42.4 Two measures of post-attack viability for a pair of networks A and B: largest remaining component (*dashed line*) vs. largest remaining component with at least 10 % interdependence (*solid line*). Each network comprises $N = 500$ nodes connected as an Erdős–Rényi graph with average degree 4, and with 20 % of A nodes interdependent with randomly chosen unique B nodes. These dependencies are undirected

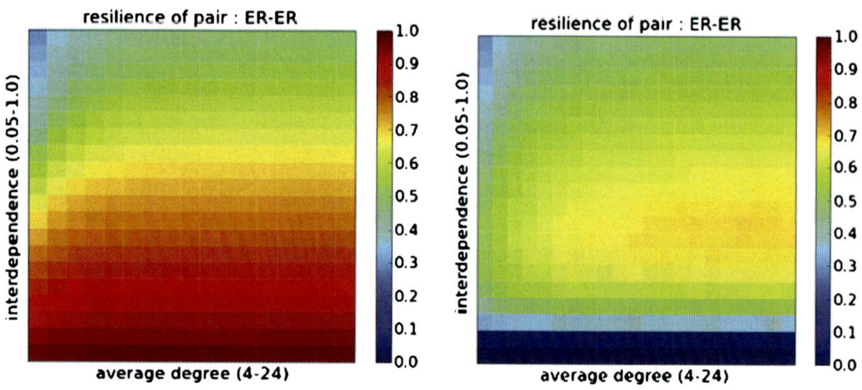

Fig. 42.5 Two measures of post-attack viability: (*left*) the size of the largest connected component, and (*right*) the size of the largest connected component that meets an interdependency threshold of 10 %. Each heat-map cell represents the mean aggregate post-attack viability of 25 pairs of random 500-node Erdős–Rényi networks. For each pair of networks, aggregate post-attack viability was again measured in terms of IP, the area under an attack-size vs. post-attack viability curve

between them. Here the introduction of the interdependency threshold qualitatively changes the relationship between degree, interdependency and post-attack viability. The nature of this change is sensitive to the initial topology of the networks (Erdős–Rényi, Barabási-Albert, spatial, etc.) and the nature of the attack (random, spatial, spreading).

The fact that modern infrastructure networks are highly specialised and need to be constantly connected to each other in order to survive ensures that the loss of inter-network connectivity may be as destructive as cascading failure when modelling the resilience of interdependent networks.

42.4 Conclusions

This paper presents a network modelling and analytical approach for studying the vulnerability of complex interdependent systems. We demonstrate that disruption in interdependent networks can be disproportionate to attack size, and that the magnitude of cascading failure can be significantly increased when interdependencies are sub-optimal. The approach described here has value for infrastructure stakeholders, providing hitherto unavailable analysis of how to: (1) maximise reliability of interconnected infrastructures subject to disruption; (2) adapt existing infrastructure systems to meet the challenges imposed by natural and malicious threats and hazards. Ongoing work is extending this analysis to consider issues around capacity, lag and latency in network connections, and systems with more than two networks. A case study using the UK electricity transmission grid and railway network is also underway.

Acknowledgements This research is sponsored under the UK EPSRC grant no EP/I005943/1.

References

1. Albert R, Barabási AL (2002) Statistical mechanics of complex networks. Rev Mod Phys 74(1):47–97
2. Buldyrev SV, Parshani R, Paul G, Stanley HE, Havlin S (2010) Catastrophic cascade of failures in interdependent networks. Nature 464(7291):1025–1028
3. Pederson P, Dudenhoeffer D, Hartley S, Permann M (2006) Critical infrastructure interdependency modeling: a survey of US and international research. Idaho National Lab, USA
4. Fu G, Dawson R, Khoury M, Bullock S (2012, submitted) Cascading failure in networks of networks: impact of redundancy and directionality
5. Gao J, Buldyrev SV, Havlin S, Stanley HE (2011) Robustness of a network of networks. Phys Rev Lett 107:195701
6. Newman M, Barabasi AL, Watts DJ (2006) The structure and dynamics of networks, 1st edn. Princeton University Press, Princeton
7. Rinaldi SM, Peerenboom JP, Kelly TK (2001) Identifying, understanding and analyzing critical infrastructure interdependencies. IEEE Control Syst 21(6):11–25

Chapter 43
Human Security—A View Through the Lens of Complexity

Anthony J. Masys

Abstract Figures compiled by the Department for International Development (DfID) suggest that between 50000 and 100000 people, more than half of them children under five, died in the 2011 Horn of Africa crisis that affected Somalia, Ethiopia and Kenya. The US government estimates separately that more than 29000 children under five died in the space of 90 days from May to July last year. The accompanying destruction of livelihoods, livestock and local market systems affected 13 million people overall. Hundreds of thousands remain at continuing risk of malnutrition (http://www.oxfam.org/sites/www.oxfam.org/files/bp-dangerous-delay-horn-africa-drought-180112-en.pdf, 2012). The threats to human security are multiple, complex and interrelated and often mutually reinforcing. The complexity view of the human security domain facilitated by Actor Network Theory (ANT) is supported by methods of network analysis and computational simulation that highlight how dynamic networked strategies associated with human security require a continuing process of inquiry, adaptation and learning. Through this analysis emerge critical insights regarding human security policy and the shaping of interventions.

Keywords Human security · Complexity · Actor network theory

43.1 Introduction

Human security is far more than the absence of conflict. It is multidimensional, encompassing education and health, democracy and human rights, protection against environmental degradation and the proliferation of deadly weapons [2]. Stemming from the 1994 Human Development Report by the United Nations Development Programme (UNDP), human security is broadly defined as 'freedom from fear and freedom from want' highlighting seven key components (economic, food, health, environmental, personal, community and political security).

Several events over the past number of years have exposed serious weaknesses pertaining to human security with respect to prevention and emergency-response ca-

A.J. Masys (✉)
Centre for Security Science, Defence R&D Canada, Ottawa, Canada
e-mail: Anthony.masys@drdc-rddc.gc.ca

T. Gilbert et al. (eds.), *Proceedings of the European Conference on Complex Systems 2012*, Springer Proceedings in Complexity, DOI 10.1007/978-3-319-00395-5_43,

pabilities in various countries. In general, these problems were not the result of specific conditions, but rather were the product of complex processes involving more fundamental issues. To address these complex human security challenges, it is argued by the Commission on Human Security (CHS), that a new paradigm of security is required. It is recognized that:

First, human security is needed in response to the complexity and the interrelatedness of both old and new security threats—from chronic and persistent poverty to ethnic violence, human trafficking, climate change, health pandemics, international terrorism, and sudden economic and financial downturns. Such threats tend to acquire transnational dimensions and move beyond traditional notions of security that focus on external military aggressions alone.

Second, human security is required as a comprehensive approach that utilizes the wide range of new opportunities to tackle such threats in an integrated manner. Human security threats cannot be tackled through conventional mechanisms alone. Instead, they require a new consensus that acknowledges the linkages and the interdependencies between development, human rights and national security [3].

Given the interdependencies, uncertainty and complexity associated with human security, the traditional linear approach to problem framing/solving and policy formulation is an inadequate way to work with socio-technical/economic/political systems. Systems thinking, characterized by seeing wholes and interconnections is critical to understanding these complex systems. As described in Masys (2010, 2012), the systems lens can enable decision makers to see beyond events and detect underlying patterns [4, 5]. The systems lens 'acknowledges that knowledge is multiple, temporary and dependent on context—with different points of view providing a constant challenge to any existing viewpoint or system' [6].

43.2 Actor Network Theory

Applied to the social sciences in general and in this case human security domain, complexity theory and systems thinking provides a perspective of the 'social world' that reveals emergent properties, nonlinearity, consideration of the 'dynamic system', interactions, interrelations that is transforming the traditional views of the social [7]. Addressing issues that lie at the foundation of sociological theory, complexity theory facilitates '... a re-conceptualization and re-thinking regarding the nature of systems reflecting dynamic inter-relationships between phenomenon. The new theorizations of system within complexity theory radically transform the concept making it applicable to the most dynamic and uneven changing phenomena' [8]. Important features that characterize complex systems and their behaviour include the ability to produce properties at the collective level that are not present when the components are considered individually as well as their sensitivity to small perturbations. As an integrating element, complexity theory provides not a methodology per se, but rather 'a conceptual framework, a way of thinking, and a way of seeing the world' [9], and presents itself as a powerful way to view the human security domain.

As described in detail in Masys (2010) [4], sociology offers an interesting approach for looking at complex systems through the application of Actor Network Theory (ANT). The complexity lens of ANT examines the inter-connectedness and the inherent relationality of the heterogeneous elements characterized by the technological and non-technological (human, social, organizational) elements. Actor Network Theory is a theoretical perspective that has evolved to address the socio-technical paradigm and in particular the conceptualization of the 'social'.

The network space of the actor network provides the domain of analysis that presents human security as a network of heterogeneous elements that shape and are shaped by the network space. Yeung (2002) notes that much of the work that draws on actor network theory places its analytical focus on unearthing the complex web of relations between humans and non-humans [10]. The interaction of non-human actors with the human actors gives shape and definition to identity and action.

The complexity lens of ANT challenges the binary distinction that leads one to a priori designate an actor as human or non human artefact. In place of this, ANT presents a schema of a network that is characterised by relations, fluidity and dynamics. This network schema has far reaching implications beyond the visual representations to include how we understand the temporal and spatial heterogeneity that is resident within the actor network. As such, this challenges our notions of far/close, small scale/ large scale and inside/outside [11] and to think in terms of associations and relations. This inherent relationality is central to our understanding of ANT and is the hallmark of complexity thinking. The spatial and temporal implications are profound. The actor network recognizes that 'what is acting at the same moment in any place is coming from many other places, many distant materials, and many faraway actors' [12] which is particularly relevant for the human security domain.

43.3 Actor Network Theory in Practice

ANT in practice explores the ways that the networks of relations are constructed, interact, how they compete with other networks, and how they are made more durable over time [13].

In so doing, ANT presents all entities (people, objects, concepts and actions) as taking form and attributes as a function of their relation with other entities (Law, 1999). One may say that the relationality is brought about '...through a wide array of networked or circulating relationships that are implicated within different overlapping and increasingly convergent material worlds' [14].

ANT treats both human and machine (non-human) elements in a symmetrical manner, thereby facilitating the examination of a situation where Callon (1999) argues, "... it is difficult to separate humans and non-humans", and where "the actors have variable forms and competencies" [15].

43.4 Translation/Inscription

As described in detail in Masys (2010) [4] fundamental processes within ANT are inscription and translation. Inscription refers to the way technical artifacts embody patterns of use: Technical objects thus simultaneously embody and measure a set of relations between heterogeneous elements (Akrich, 1992:205). Monterio (2000:77) argues that although 'inscription' might sound deterministic, '...the artifact is always interpreted and appropriated flexible, the notion of inscription may be used to describe how concrete anticipations and restrictions of future patterns of use are involved in the development and use of a technology' [16]. Inscriptions enable action at a distance by creating 'technical artefacts' that ensure the establishment of an actor's interests such that it can travel across space and time and thereby influence other work (Latour, 1987) [17]. Inscripted artifacts such as texts and images are central to knowledge work [18] and thereby can shape sensemaking, decision making and action.

The process of translation has been described as pivotal in any analysis of how different elements in an actor network interact [19]. As a transformative process, translation emphasizes '...the continuous displacements, alignments and transformations occurring in the actor network' [20]. Translation rests on the idea that actors within a network will try to enroll (manipulate or force) the other actors into positions that suit their purposes. When an actor's strategy is successful and it has organized other actors for its own benefit, it can be said to have translated them. As articulated by Yeung (2002:6), 'Actors in these relational geometries are not static "things" fixed in time and space, but rather agencies whose relational practices unleash power inscribed in relational geometries and whose identities, subjectivities, and experiences are always (re)constituted by such practices' [10].

Within the context of the case study (2011 Horn of Africa Drought), an examination of actors such as those characterized from technologies to policies, facilitates an exploration of how these "actors" mediate action and how they are entangled in local socio-technical/economic/political configurations. Thereby inscription and translation processes inherent within the actor network are seen to be implicated in the evolving crisis.

The lens of ANT facilitates the view of the world in terms of heterogeneous elements, thereby employing a "systems thinking" perspective of the problem space. Network analysis and agent based modeling support the actor network lens by providing insights into the processes of translation and inscription as well as providing greater insights into the interdependencies inherent within the human security domain.

43.5 Scope/Methodology

This study examines the human security domain from an all hazards perspective, encompassing both man-made and natural disasters. Such disasters as the Hurricane

Katrina (2005), Deepwater Horizon (2011) oil spill, Fukushima tsunami (2011) as well as recent humanitarian crises (2011 Horn of Africa Drought) reveal the complex landscape associated with human security characterized by interdependencies and interrelationships in a complex networked topology. This complexity view of the human security domain facilitated by ANT (follow the actor) is supported by methods of network analysis and computational simulation, that highlight the 'hybrid collectif' (the intersection of human, informational and physical domains) and how dynamic networked strategies associated with human security require a continuing process of inquiry, adaptation and learning [21]. As noted in Styhre (2002), the complexity lens of ANT reveals that changes result from a multiplicity of interconnected causes and effects [22]. Within the context of understanding human security, drawing upon the complexity lens of ANT supported by network analysis and agent based modeling facilitates a break from '...mechanistic, linear, and causal methods of analysis towards viewing interdependence and interrelation rather than linearity and exclusion' [23].

43.6 Discussion

As described by the CHS [3], 'new ways of thinking' are required (if not essential) to manage the complex problems associated with human security. Cited in Barnes, Matka, and Sullivan, (2003: 277), Cilliers (2000) highlights the 'rich interaction' and 'abundance' of feedback which means that 'any activity in the system reverberates throughout the system, and can have effects that are very difficult to predict' [24]. Through the complexity lens of ANT, we recognize that systems do not have elements, but it is the interdependencies and interactions among the elements that create the whole. Thus complexity theory suggests that studying the interdependencies and interactions among the elements, as well as the unity of the system itself will provide critical insights for understanding an organization and its systems properties [25]. Complexity moves away from linear thinking about cause and effect to nonlinear models that seek to understand the association and interaction between factors. Here the argument with regards to human security is that a fully relational view of the interrelations and interdependencies associated with the human, physical and informational domains is necessary. Such a perspective reveals how changes to the socio-political-economic-technical system become enrolled into complex and subtle blendings of human actors and technical, political and economic artifacts, to form actor networks (hybrid collectifs). Integrating concepts from complexity theory supports ANT perspective by facilitating an '... explanatory framework of interrelationships: of how individuals and organizations interact, relate and evolve... Change happens in the context of this intricate intertwining at all scales' [26]. Human security, through the complexity lens of ANT thereby emerges as a network of heterogeneous actors. Through this lens it highlights that a solution space must focus on understanding the complex interdependencies and nonlinear temporal and spatial realities.

Human security domain 'architects' are regularly faced with the prospect of having to anticipate the consequences of their actions, and avoid unintended consequences, without comprehensive information about the system. The actor network analysis reveals underlying complexity pertaining to human security issues. For example:

- Structural complexity—the actor network emerges as a network of heterogeneous actors interrelated in a rhizomal 'network' topology.
- Network evolution—the actor network changes over time
- Connection diversity—relations between different actors are heterogeneous and are shaped by translation and inscription processes.
- Dynamic complexity—the actor themselves are actor networks, adding to the overall dynamic complexity of the system
- Actor diversity—actors are not homogeneous, they are heterogeneous
- Meta-complication—the factors above are all interrelated thereby adding another level of complexity in understanding the dynamic complexity of actor networks as it relates to human security (derived from Strogatz (2001:269)) [27].

Understanding the structure and function of complex networks has recently become the foundation for explaining many different real-world complex biological, technological and informal social phenomena. The analysis methodology of ANT, 'Following the actors', revealed the notion of a complex co-evolving system characterized by '... intricate and multiple intertwined interactions and relationships. Connectivity and interdependence propagates the effects of actions, decisions and behaviours..., but that propagation or influence is not uniform as it depends on the degree of connectedness' [26]. Through the complexity lens of ANT, a 'network mapping' informed by agent-based modeling highlights the complex spatial and temporal interdependencies that reside within the actor network (Fig. 43.1). Moreover, it reveals the importance of interoperability across actors (technical, social, political and economic) and the notion of uncertainty and assumptions that can shape human security.

As described in the ODI report [28], the outbreak of contagious livestock disease and limited intervention against the outbreak of Foot and Mouth Disease (FMD) has been exacerbated by increased cross-border population migrations in the region. The scarce resources available contribute to heightened tensions and violence resulting in a significant number of deaths at feeding points and cross border areas. Migration has resulted in eroding sustainable livelihood by affecting education programs. For example, '... In Ethiopia more than 280 schools, affecting at least 58,000 students in Somali and Oromia regions, remain closed as a result of the drought' [28].

The reports (Levine et al., 2011) argue that '...Humanitarian response in pastoral areas in the Horn of Africa has consistently been late' [28]. Although technical interventions, such as early warning mechanisms were in place and operational, they failed to trigger effective response. This compounded by a weakness in livelihood analysis and lack of coordination [28], reflects a fundamental dysfunctionality within the actor network. The actors to support early warning interpretation, sensemaking, decision making and action were not aligned. This inscribed dysfunctionality was 'translated' within the actor network thereby resulting in a contagion

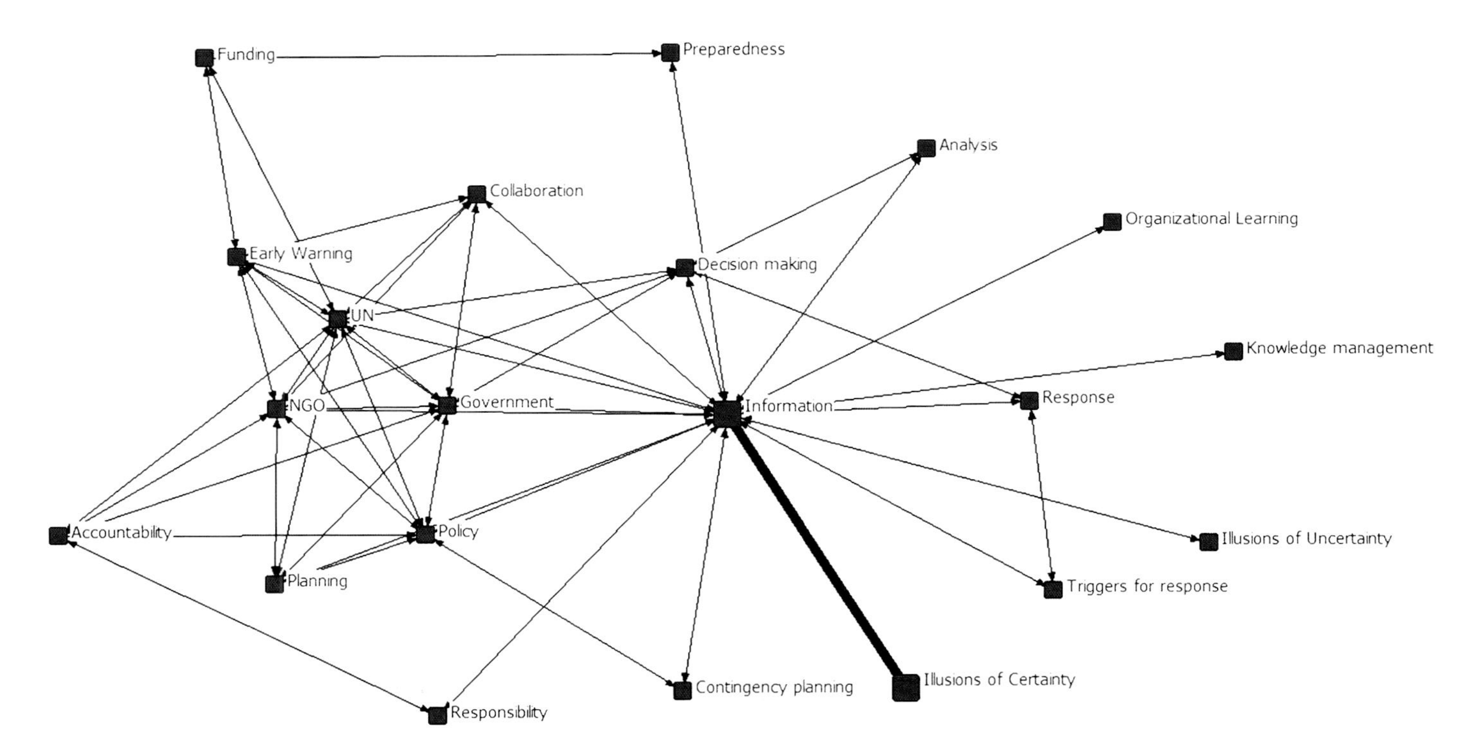

Fig. 43.1 Actor network visualization—abridged high level view

like effect whereby ineffective decision making emerged. As part of the intervention strategy, it was noted by Levine et al (2011:3) '... many contingency plans across Kenya, Somalia and Ethiopia are long documents with a great deal of information, but they rarely include the elements that would actually help speed up response if the contingency did occur. Most plans lacked many or even all of the following:

- an overall strategy to which the various planned interventions were to contribute;
- rationale for the interventions;
- justification or rationale for the proposed scale of intervention;
- impact targets;
- clear triggers for deciding when to implement;
- anticipated calendar months for implementation;
- what needed to be monitored to know when to implement;
- a clear link to likely budgets—are the plans realistic?
- specific actors given specific responsibilities for which they could be held accountable;
- a link to the prevailing situation, or to what was most expected or feared at that time, including issues such as conflict, freedom of movement and food prices;
- a situational analysis that included predictions about what would be going on outside the area, e.g. movement of livestock in or out—without which most pastoral livelihood interventions would make little sense;
- discussion of specific locations within the area—given mobile livelihoods, which strategy would be needed in which location? Which areas would be likely to need most/least help? Where could conflict issues be a problem?;
- a link to an assessment of the degree of help needed (how much livelihood support did people need to protect their herds or to survive?); and
- (most important of all) a link to preparedness: to actions to be taken before the contingency arose in order to be ready to implement the contingency plan on time [28].

Illusion of certainty emerge from the actor network comprised of early warning systems, contingency plans, government guidelines and policies, planning guidelines, mental models. When modeled as an actor network, these inscripted dysfuctionalities created a mindset that focused on 'siloed' thinking. Epidemiological (SIR) network modeling describes well how inscription and translation processes can result in ineffective solutions. Outdated and misunderstood mental models of risk, crisis and disaster management shaped sense making, decision making action and inaction.

Levine et al (2011:7) argue that '... a system perspective can often reveal how behaviour that is competent from the standpoint of each individual actor does not contribute to achieving the overall goals which collectively all the actors in the 'system' say they are working towards, in different ways. System problems often result when different actors do not really share the objectives, or when they do not agree on which elements contribute to a single system' [28]. Actors such as text (policy, legislation) can be understood as a network aligning heterogeneous elements (people, other texts, equipment, procedures, institutions) to achieve a particular objective. A weak alignment of policy with actors (NGOs) creates a dysfunctional actor network, one in which is affected by shock to the system. This is a function of the influence of the actor (translation). These actors may emerge as a black box which appears to be closed thereby assuming to be reliable and stable themselves (i.e. Contingency plans).

The complexity lens of ANT supports this through viewing the dynamics of the problem space in terms of translation and inscription processes. Resilience emerges from managing the risks not the crisis and this requires understanding the need for heterogeneous engineering (aligning actors that cross the physical, human and informational domains) and understanding the interdependencies that reside within them. This resonates with Levine et al (2011:18) comment that: 'The most challenging was the 'system dimension', meaning that none of the changes involved one single actor improving its own performance in isolation' [28].

43.7 Findings

Complex problems are, by their very nature, difficult to predict. Complex issues pertaining to human security, risk, crisis and disaster management are not amenable to detailed forecasting. Rather than fixing the shape of policy responses in advance, responses need the flexibility to adapt to emerging insights. However, traditional approaches to human security issues often assume that causality is well-established, linear and predictable. This 'mindset' supports an inherent rigidity within the planning and operational phases.

The reality is that complex problems involve conflicting goals. Divergent policy interpretations and mental models arise from and contain different assumptions and understanding of resident uncertainties. From the actor network analysis it is argued that agendas and goals are not value neutral and have defining affects on the actor network dynamics. To better understand the processes of translation and inscription, a network analysis was conducted drawing upon the SIR model. The visualization of this SIR model within the context of ANT revealed inherent qualities of the actor network. Showing how conflicts and alignment affect the network and are rooted in the 'mindset' that permeates the system through inscription and translation.

Resilience thereby emerges from the analysis as a key quality of human security. The concept of resilience encompasses a capacity to anticipate and manage risks and the ability to survive threats and respond to challenges. Hollnagel et al. (2006:316) asks the question:

'How do people detect that problems are emerging or changing when information is subtle, fragmented, incomplete or distributed across different groups involved in production processes and in safety management. Many studies have shown how decision makers in evolving situations can get stuck in a single problem frame and miss or misinterpret new information that should force re-evaluation and revision of the situation assessment (Johnson et al., 2001; Patterson et al., 2001)' [29].

To cope with the complexity, resilience thereby requires a continuous monitoring system and supporting actor network to act and react proactively and reactively.

Resilience becomes an emergent property of the complex system and the actor network. Resilience requires an approach to viewing the complex problem space to reveal the interdependencies within the hybrid collectif in order to make it proactive; flexible; adaptive; and prepared. It must be aware of the impact of actions, as well as the failure to take action [29].

Through the complexity lens of actor network theory supported by network analysis and agent based modeling, it is understood that problems such as human security cannot be understood in isolation. They are systemic problems that require systemic solutions.

43.8 Conclusions

The study of complex network topologies across many domains has generated greater understanding of the network properties and has facilitated strategies to optimize the given topologies for specific problem solving opportunities. For example, the approach of targeting high-degree nodes has been suggested to be an effective disease and computer virus prevention strategy (nodal vaccination). As described in Braha and Bar-Yam (2006:63), agile strategies can be implemented to facilitate '… monitoring and vaccinating nodes based on centrality over time' [30]. By means of the actor network worldview with network analysis and simulation reveal how small changes in the state of some system components can cause large effects with respect to human security.

This exploratory and interpretive research highlights the necessity for a complexity view of the 'wicked' problem space associated with the human security. Hence managers, leaders and those involved in humanitarian or relief operations should recognize the complexity inherent in human security operations and be responsive to the emergence of weak signals and recognize potential opportunities for constructive action through an integrated comprehensive approach.

Acknowledgements Grateful acknowledgement is given to Dan Braha, Hiroki Sayama and Yaneer Bar-Yam for their fruitful insights and instruction during the 2011 Winter School on Complex Systems- New England Complex Systems Institute.

References

1. Save the Children (2012) A dangerous delay: the cost of late response to early warnings in the 2011 drought in the Horn of Africa. Joint Agency briefing paper. Available at http://www.oxfam.org/sites/www.oxfam.org/files/bp-dangerous-delay-horn-africa-drought-180112-en.pdf
2. Sommaruga C (2004) The global challenge of human security. Foresight 6(4):208–211
3. United Nations Trust Fund for Human Security (2009) Human security in theory and practice: applications of the human security concept and the United Nations trust fund for human security, pp 1–79. United Nations, New York
4. Masys AJ (2010) Fratricide in air operations: opening the black box- revealing the social. PhD dissertation, University of Leicester, UK

5. Masys AJ (2012) Black swans to grey swans—revealing the uncertainty. Disaster Prev Manag 21(3):320–335
6. Wilkinson A, Eidinow E (2008) Evolving practices in environmental scenarios: a new scenario typology. Environ Res Lett 3:1–11
7. Dooley KJ, Corman SR, McPhee RD, Kuhn T (2003) Modeling high-resolution broadbrand discourse in complex adaptive systems. Nonlinear Dyn Psychol Life Sci 7(1):61–85. 2003
8. Walby S (2003) Complexity theory, globalisation and diversity. Presented to the conference of the British Sociological Association, University of York
9. Mitleton-Kelly E (2004) Complex systems and evolutionary perspectives on organisations: the application of complexity theory to organisations
10. Yeung HWC (2003) Towards a relational economic geography: old wine in new bottles? Paper presented at the 98th annual meeting of the association of American geographers, Los Angeles, CA, March 19–23, 2003. Available at http://courses.nus.edu.sg/course/geoywc/publication/Yeung_AAG.pdf
11. Latour B (1996) On actor-network theory. A few clarifications. Soz Welt 47:369–381
12. Latour B (2005) Reassembling the social: an introduction to actor network theory. Oxford University Press, London
13. Powell JL, Owen T (2011) Actor network theory and social science: possibilities and implications. J Public Adm Gov 1(2):140–157. www.macrothink.org/jpag
14. Urry J (2005) The complexities of the global. Theory Cult Soc 22(5):235–254
15. Callon M (1999) Actor-network theory: the market test. In: Law J, Hassard J (eds) Actor network and after. Blackwell Sci./The Sociological Review, Oxford/Keele, pp 181–195
16. Monteiro E (2000) Actor network theory and information infrastructure. In: Ciborra C (ed) From control to drift. The dynamics of corporate information infrastructures. Oxford University Press, London, pp 71–83. www.idi.ntnu.no/~ericm/ant.FINAL.htm
17. Latour B (1987) Science in action: how to follow scientists and engineers through society. Open University Press, Milton Keynes
18. Wickramasinghe N, Tumu S, Bali RK, Tatnall A (2007) Using actor network theory (ANT) as an analytic tool in order to effect superior PACS implementation. Int J Netw Virtual Organ 4(3):257–279
19. Somerville I (1997) Actor network theory: a useful paradigm for the analysis of the UK cable/online socio-technical ensemble? Available at http://hsb/baylor.edu/eamsower/ais.ac.97/papers/somervil.html
20. Viseu AAB (2005) Augmented bodies: the visions and realities of wearable computers. PhD dissertation, University of Toronto
21. Comfort LK (2005) Risk, security and disaster management. Annu Rev Pol Sci 8:335–356
22. Styhre A (2002) Non-linear change in organizations: organization change management informed by complexity theory. Leadersh Organ Dev J 23(6):343–351
23. Dennis K (2007) Time in the age of complexity. Time Soc 16:139–155
24. Barnes M, Matka E, Sullivan H (2003) Evidence, understanding and complexity: evaluation in non-linear systems. Evaluation 9(3):265–284
25. Anderson R, Crabtree B, Steele D, McDaniel R (2005) Case study research: the view from complexity science. Qual Health Res 15(5):669–685
26. Mitleton-Kelly E, Papaefthimiou MC (2000) Co-evolution of diverse elements interacting within a social ecosystem. In: Proceedings of feast 2000 international workshop on feedback and evolution in software and business processes, Imperial College, London, pp 10–12
27. Strogatz SH (2001) Exploring complex networks. Nature 410:268–276
28. Levine S, Crosskey A, Abdinoor M (2011) System failure? Revisiting the problems of timely response to crises in the Horn of Africa. Network paper number 71. Humanitarian Policy Group, Overseas Development Institute
29. Hollnagel E, Woods DD, Leveson N (2006) Resilience engineering: concepts and precepts. Ashgate Publishing, Hampshire
30. Braha D, Bar-Yam Y (2006) From centrality to temporary fame: dynamic centrality in complex networks. Complexity 12(2):59–63

Chapter 44
Mitigating Risks of Event Avalanches Caused by Climate Change

Lessons for Sustainable Urban Design

Ljubomir Jankovic

Abstract Development of the human society and its technological, economic and financial systems, coupled with the population growth, has resulted in high interconnectivity between individual and corporate entities. These entities form networks of co-dependent agents which operate under critical connectivity. Climate change has brought about an increased frequency of extreme events, such as heat waves, droughts, floods and hurricanes, which can easily set off event avalanches that propagate throughout these networks. This paper looks into event propagation characteristics of production and consumption networks and into how these characteristics can be designed and managed so as to prevent such extreme events from becoming event avalanches that sweep through the network and result in considerable human and material costs. It draws conclusions on how sustainability of an urban environment can be maintained at the time of occurrence of extreme events.

Keywords Climate change · Event avalanche · Economic network · Critical connectivity · Percolation · Risk · Mitigation · Sustainability · Urban design

44.1 Introduction

Development and prosperity of the human society depends on essential materials such as energy, food, water, and on essential activities such as movement, communication and trade. Heat waves, droughts, floods and hurricanes are just some examples of recent events caused by the climate change which can affect availability of, or access to these essential materials, or disrupt the essential activities. As the population growth, movement, communication and trade are constantly shaping an interdependent network of organizations and individuals, the question is how much or how far an extreme event can propagate through such network? This paper looks into the notion of percolation as event carrier through the network, and into the ways

L. Jankovic (✉)
Birmingham School of Architecture, Birmingham Institute of Art and Design, Birmingham City University, Gosta Green, Corporation Street, Birmingham B4 7DX, UK
e-mail: Lubo.Jankovic@bcu.ac.uk

T. Gilbert et al. (eds.), *Proceedings of the European Conference on Complex Systems 2012*, Springer Proceedings in Complexity, DOI 10.1007/978-3-319-00395-5_44,

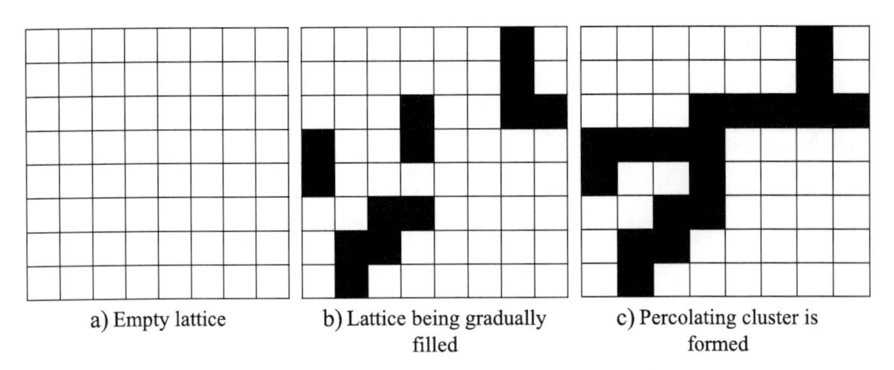

Fig. 44.1 Percolation in a square lattice

of controlling percolation as means of risk mitigation. The subject of network percolation has been investigated theoretically in diverse fields, such as interdependent networks [1], insect nests [2], epidemics [3], small world networks [4] and others. Conversely, the work in this paper is based on computational experiments with an economic network model, in order to increase the understanding of the metrics of complex behavior of networks influenced by extreme events arising from climate change. The work presented here would also fit well to other kinds of large-scale impacts on socio-economic systems. However, impacts from climate change can be expected to be regular and therefore require permanent design interventions in order to mitigate its effects.

44.2 Percolation and Event Avalanches in a Square Lattice

In this section we define an event avalanche and analyze conditions for its occurrence using a simplified network in the form of a square lattice with a neighbor to neighbor connectivity, as shown in Fig. 44.1.

In this configuration, an event avalanche will be defined as a continuous path for event propagation between two sides of the system. At first, the lattice is empty (Fig. 44.1(a)), and as we gradually fill in the empty sites (Fig. 44.1(b)), a percolating cluster is formed (Fig. 44.1(c)). This cluster spans the entire lattice from one end to another.

Weisstein [5] defines percolation threshold pc as a fraction of the points of the lattice that, when filled, creates a continuous path between two sides. Site percolation is defined as a continuous path through the cells, and bond percolation as continuous path through links between the cells. There are different percolation thresholds for different lattice topologies, such as diamond, honeycomb, triangular, square and other configurations. For an infinite square lattice, site percolation threshold is calculated as ~59.3 %, and bond percolation threshold as 50.0 % [5]. Percolation thresholds for different lattice topologies are different from one topology to another.

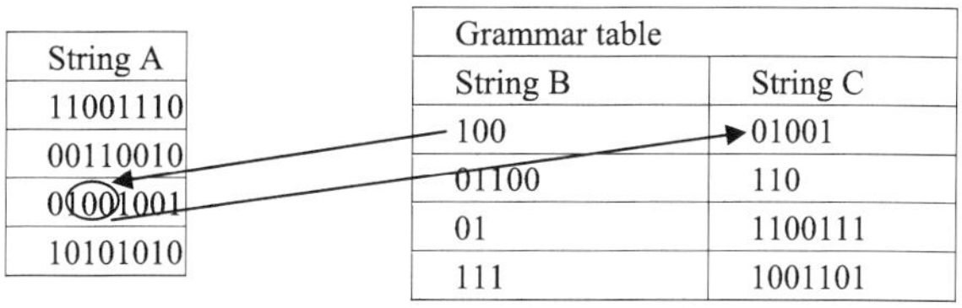

Fig. 44.2 Random grammars illustration—if string A is picked at random and string B matches any substring of A then this substring of A is replaced with string C

String A
11001110
00110010
01001001
10101010

Grammar table	
String B	String C
100	01001
01100	110
01	1100111
111	1001101

From the above definition and analysis it follows that an event avalanche will occur at either site percolation threshold or bond percolation threshold or above. It also follows that an event avalanche will be prevented from occurring when the lattice is below the percolation threshold pc. We therefore conclude that an event avalanche can be mitigated by reducing the lattice density below the percolation threshold.

In this section we discussed a simplified network in the form of a square lattice with neighbor to neighbor connectivity, in order to explain the conditions for occurrence of event avalanches and their mitigation. The next section will discuss how these concepts apply to an economic network with a radically different topology.

44.3 An Economic Network Model Based on Random Grammars

In this section we introduce a simple model of an economic network and analyze its percolation threshold as a mitigating method for event avalanches. Whereas the simple lattice described in the previous section relied on a neighbor to neighbor connectivity to establish the network, the model in this section represents a network of spatially distributed agents with long range connections, analogous to those in real economic networks.

This model uses bit strings to represent natural resources and products created from those resources. Agents, which represent companies, or types of companies, operate on these strings using random grammar rules, as those reported by Kauffman [6]. A grammar rule consists of an input/output pair of bit strings. When a grammar rule is applied to a bit string A, then if the input string B matches any substring of A, then this substring is replaced with the output string C (Fig. 44.2). Only one instance of the substring of A will be replaced. This process simulates the supply-demand match between pairs of agents.

The model maintains two 'pools' of bit strings: the resource pool and the product pool. The resource pool contains a set of randomly generated initial strings which are always available. When a grammar rule acts on a resource string to change it into a new string, the original resource string is still preserved. The product pool is initially empty, and any new strings created from resource strings are placed in it. When a grammar rule acts on a product string, the original is removed and replaced with the new string. Thus resource strings are in infinite supply and product strings are not.

During the model operation, either a resource string or a product string is selected at random, each time step, with equal probability, and passed to an agent chosen at random, to apply its grammar rules. Whenever an agent uses a resource string, its color, which represents fitness, moves towards red, and whenever it uses a product string, its color moves towards green. Whenever an agent successfully processes any string, its size increases. Agent size also continuously decreases over time. Lines are drawn between an agent that created the string and the agent that used it. The thickness of such lines increases each time when the link between the same two agents is reinforced. Thus thicker lines indicate more frequent transfer of products between the same pairs of agents.

The next section describes computational experiments and results obtained using this model, in order to test network sensitivity to parameter changes and consequences of these changes on percolation properties.

44.4 Main Results

Due to the location independence of the network agents, only the bond percolation rather than site percolation will be considered in the economic network model. For the same reason, there are no edges in this system, and therefore a new definition for percolation is needed. This will be investigated through three computational experiments (Figs. 44.3, 44.4 and 44.5) documented in this section.

In addition to measuring cluster sizes, we also calculate Shannon entropy [7] in order to assess event transmission properties of the network, as follows:

$$H = -k\Sigma p(i)\log\big(p(i)\big)$$

where $p(i)$ is the probability that ith agent will form a link with another agent and k is a constant that accounts for duplication of links and scaling. The summary of inputs and outputs of the experiments, with 100 agents in each experiment, is shown in Table 44.1. After a number of experiments with the same input parameters, the results always converged towards similar values. The "largest local cluster" in Table 44.1 means a cluster of agents linked through processing strings of the same origin, and the "largest overall cluster" is a cluster of clusters. We discuss the results in the next section.

44.5 Discussion

The results show that with high connectivity of over 1.8 in Experiment 1 (Fig. 44.3 and Table 44.1) a very large overall cluster of 60 (out of 100) agents is formed, representing 60 % of the network (Table 44.1). Shannon entropy, which represents the network's transmission capacity, is in the upper quartile of the [0, 1] interval, showing high event transmission through the network. An extreme event will therefore propagate through the majority of this network. In Experiment 2 (Fig. 44.4), with

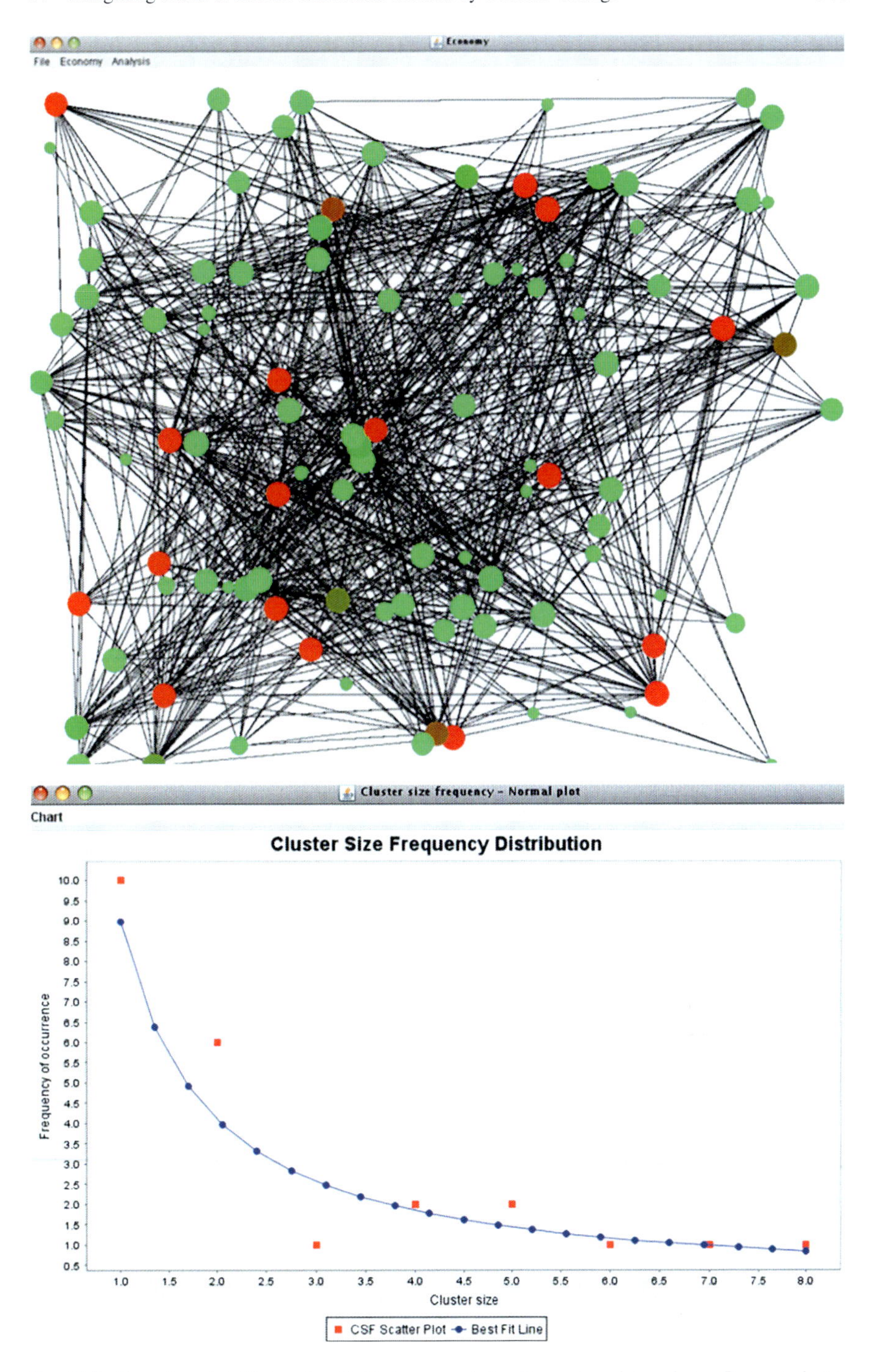

Fig. 44.3 Experiment 1: Network with high connectivity. *Top*: network topology; *Bottom*: cluster size frequency distribution

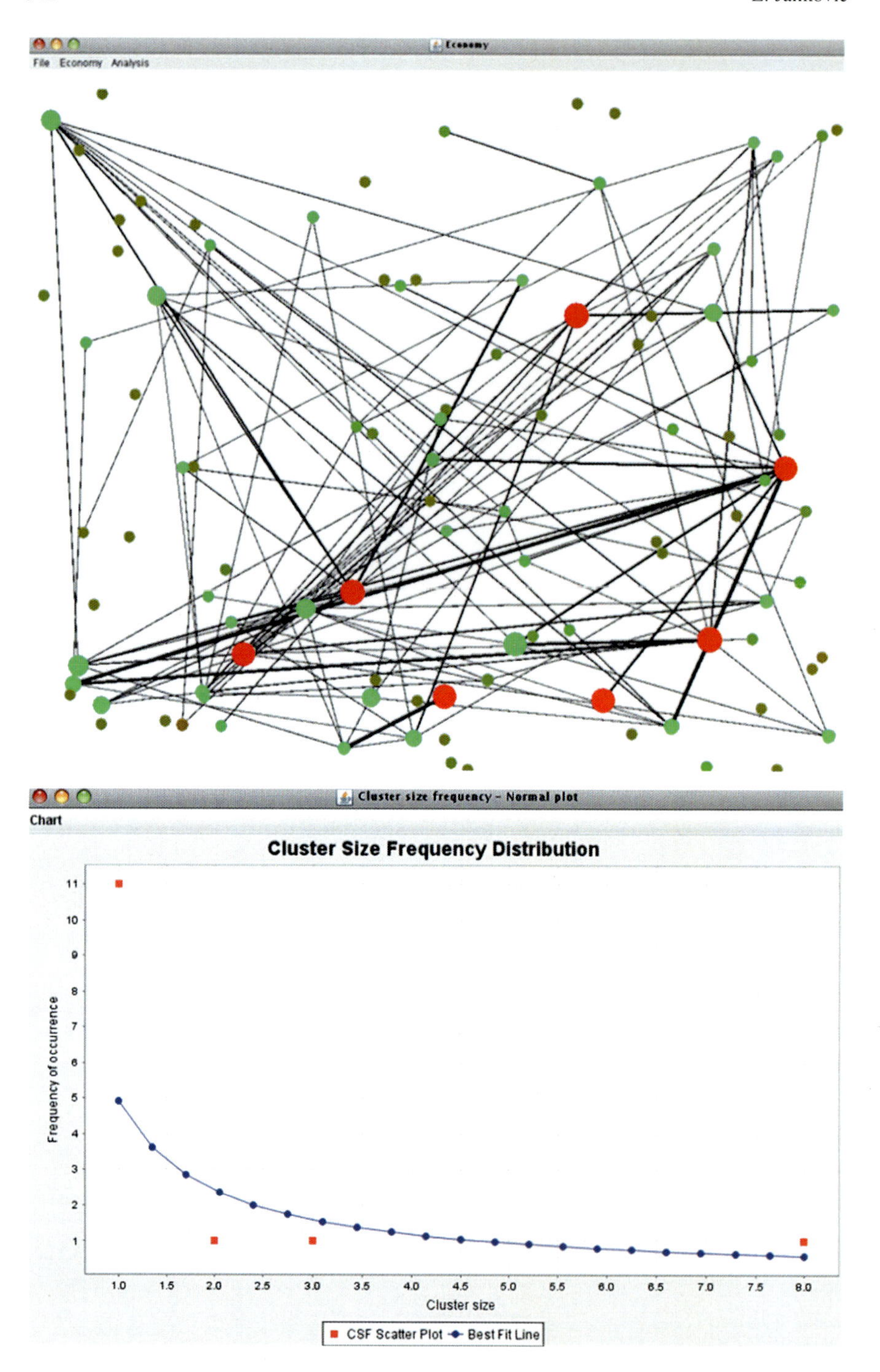

Fig. 44.4 Experiment 2: Network with medium connectivity. *Top*: network topology; *Bottom*: cluster size frequency distribution

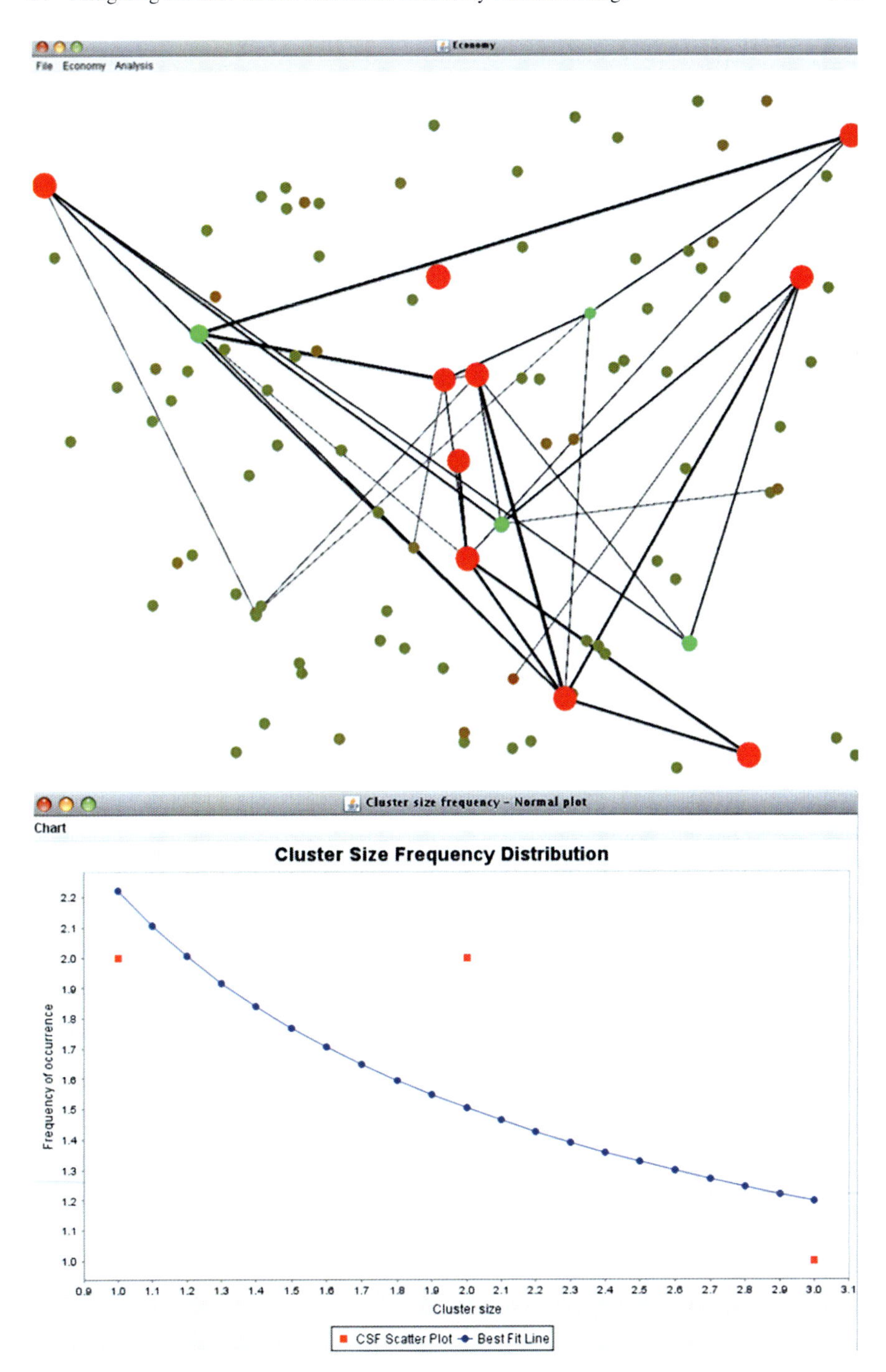

Fig. 44.5 Experiment 3: Network with low connectivity. *Top*: network topology; *Bottom*: cluster size frequency distribution

Table 44.1 Summary of experiment inputs and outputs

Inputs	Experiment 1	Experiment 2	Experiment 3
String A: Number of resources	10	10	10
String A: Resource string length	10	5	5
String B: Rule input length	5	5	5
String B: Rule input length variation	1	1	2
String C: Rule output length	6	4	1
String C: Rule output length variation	3	0	0
Outputs	**Experiment 1**	**Experiment 2**	**Experiment 3**
Total number of clusters	24	14	5
Average Shannon entropy H	0.8589	0.0577	0.0671
Average connectivity	1.8419	0.4081	0.3792
Largest local cluster	8	8	3
Largest overall cluster	60	24	9

the grammar rules altered to achieve lower connectivity (Table 44.1), the largest overall cluster of 24 agents is formed, representing 24 % of the network, and in Experiment 3 (Fig. 44.5), with even lower connectivity, the size of the overall cluster represents 9 % of the network agents. In the two latter cases, the connectivity and Shannon entropy are low, showing that the majority of network is not transmitting. The connectivity is therefore correlated with the cluster size and transmission properties of the network, and can be used as a control mechanism.

In Experiment 3 (Table 44.1), Shannon entropy is slightly higher than in Experiment 2, (0.0671 vs. 0.0577), even though there are fewer number of links and therefore lower transmission capability. Although counterintuitive, this is believed to be perfectly in order, as it is the result of operation of the random grammar. Considering that the difference is small, it is unlikely to be significant.

The percolation threshold in this type of network can therefore be redefined as percentage of agents bonded by the largest overall cluster that can affect the majority of the network. This is not a fixed value but it merely depends on the judgment of the acceptable risk, whilst still permitting normal network operation. There was no evidence of a sharp threshold, such as in the Erdős–Rényi random graph model and its variants.

44.6 Lessons for Sustainable Urban Design

What is the meaning of this analysis in terms of lessons for sustainable urban design? Networks at or above the percolation threshold will transmit extreme events resulting from the climate change throughout the system, with possibly devastating effects. However, the network model presented in this paper does not have a clear percolation threshold. Instead, the extent of percolation is determined as a percentage of agents bonded by the largest interlinked cluster (cluster of clusters) or by the magnitude of Shannon entropy. Using connectivity as a control mechanism, the size of the largest cluster and the magnitude of Shannon entropy can be altered, thus altering the transmission properties, and possibly protecting the system from devastation.

However, networks require percolation in order to operate efficiently. Propagation of "good" events should not be limited by keeping the network permanently below the percolation threshold. We therefore infer that flexibility in network connectivity will increase network's resilience. This effectively means that the co-dependence between the network agents needs to be controlled up or down by the agents, depending on external conditions. If conditions are favorable, agents should be able to increase their connectivity, and if extreme events are influencing the network, agents need to be able to reduce their connectivity and thus limit the extent of event propagation.

This analysis indicates that processes in an urban environment can be maintained without interruption only if the underlying network architecture has essential local resources, such as local food and energy production, local fuel, food and medicine storage, or secure long-range connectivity to the sites from which these resources can be drawn, and the ability to deploy these resources at times when undesirable events start propagating through the network. Design of urban environment should incorporate such facilities that make the deployment of local resources easily possible at times when needed.

In an urban environment that solely depends on long-range unsecure connections to essential resources, it will not be possible to change the network connectivity very easily, thus making such network prone to sweeping changes, either desirable or undesirable.

But does this mean that these local facilities might not be used the majority of time, and that there would be some redundancy in the urban environment? Indeed, a robust urban network must contain a degree of redundancy if it is to survive a large event caused by climate change, and this will be the price to pay for a sustainable urban design.

Therefore, adaptable connectivity and easily deployable local resources are the key to sustaining an urban environment at the time of occurrence of extreme events.

44.7 Conclusions and Future Work

The work presented in this paper focused on the analysis of event propagation in a network of production and consumption agents. Percolation properties of the network were defined and analyzed using Shannon entropy, agent connectivity, the number and size of local clusters, and the size of the overall cluster. It was found that Shannon entropy was generally proportional to all other indicators and that it could be used for determining the overall percolation properties of the network. Whilst there was no clear percolation threshold for this type of network, and the percolation was merely the matter of judgment of the proportion of the network that can be allowed to propagate an extreme event, it was possible to control the propagation by changing the network connectivity.

As the network presented in the paper is reminiscent of underlying networks in urban environments, the paper discussed the implications of propagation of extreme events arising from climate change on the sustainable operation of an urban environment. The key lessons for sustainable design of an urban environment arising from this analysis are: the provision of easily deployable local resources, or secure long-range connectivity to resources, and adaptable connectivity in response to extreme events. A sustainable urban environment must contain a degree of redundancy in the form of local facilities for energy production, storage of food, fuel and other necessities that can be deployed quickly at times of extreme events.

Although the presented study would also fit well to other kinds of large-scale impacts on socio-economic systems, impacts from climate change can be expected to be regular and hence a need for permanent design interventions that will mitigate its effects.

The economic model presented here is an initial step in the invcstigation of event propagation properties in an economic network. Future work will involve the addition of several different markets in the model, implementation of limited budgets for individual agents, and an ability to add or remove agents dynamically.

Acknowledgements The author acknowledges the use of a complexity source code library from InteSys Ltd.

References

1. Buldyrev SV, Parshani R, Paul g, Stanley HE, Havlin S (2010) Catastrophic cascade of failures in interdependent networks. Nature 464:1025–1028
2. Valverde S, Corominas-Murtra B, Perna A, Kuntz P, Theraulaz G, Sol´e RV Percolation in insect nest networks: evidence for optimal wiring. Working paper no 09-03-07, Santa Fe Institute
3. Moore C, Newman MEJ Epidemics and percolation in small-world networks. Working paper no 00-01-002, Santa Fe Institute
4. Watts D (1999) Small worlds—the dynamics of networks between order and randomness. Princeton University Press, Princeton
5. Weisstein EW Percolation threshold. From MathWorld—A Wolfram web resource. http://mathworld.wolfram.com/PercolationThreshold.html
6. Kauffman S (1996) At home in the universe—the search for laws of complexity. Penguin, Baltimore
7. Bar-Yam Y (1997) Dynamics of complex systems. Pegasus Books, Oakland

Chapter 45
Reliable Probabilities Through Statistical Post-processing of Ensemble Forecasts

Bert Van Schaeybroeck and Stéphane Vannitsem

Abstract We develop post-processing approaches based on linear regression that make ensemble forecasts more reliable. First of all we enforce climatological reliability (CR) in the sense that the total variability of the forecast is equal the variability of the observations. Second, we impose ensemble reliability (ER) such that the spread around the ensemble mean of the observation coincides with the one of the ensemble members. Since, generally, different ensembles have different sizes, standard post-processing methods tend to overcorrect ensembles with large spreads. By taking variable values of the error variances, our forecast becomes more reliable at short lead times as reflected by a flatter rank histogram. We illustrate our findings using the Lorenz 1963 model.

45.1 Introduction

The atmosphere (and its climate) is a complicated system involving multiple components, each with their own time and spatial scales, in constant interaction with one another. This system displays the property of sensitivity to initial conditions drastically limiting its predictability horizon [3]. This property is also shared by many deterministic detailed atmospheric models, reflecting the chaotic nature of their dynamics.

Operational numerical weather and climate models suffer from the presence of both initial-condition and modeling errors. In particular, even though during last decades the amount of satellite observations has increased tremendously, a substantial lack of observations above the oceans remains allowing room for improvement on the quality of initial conditions. In addition, it is nowadays also realized that model errors (errors due to unresolved scales, ill-tuned model parameters and neglected interactions between different components of the climate system) play a major role in the deterioration of forecasts as a function of lead time [6].

The recognition of these intrinsic uncertainties on the initial conditions and model physics forms the starting point of current-day weather forecast practice,

B. Van Schaeybroeck (✉) · S. Vannitsem
Royal Meteorological Institute, Ringlaan 3, 1180 Brussels, Belgium
e-mail: bertvs@meteo.be

T. Gilbert et al. (eds.), *Proceedings of the European Conference on Complex Systems 2012*, Springer Proceedings in Complexity, DOI 10.1007/978-3-319-00395-5_45,

through the development of an ensemble system in which a set of trajectories are integrated in time starting from different initial conditions, and/or using different model physics. The aim of using ensembles is to produce a probabilistic instead of a deterministic forecast. The spread of the ensemble around its mean can be viewed as an uncertainty measure of the current forecast and provides the possibility to evaluate the different potential scenarios that could be followed by the atmosphere. This uncertainty information is of crucial importance as the reliability of the forecast may be strongly flow dependent. However, despite huge efforts at constructing ensembles which produce a good estimate of uncertainty, all experiments show that ensemble forecasts are consistently under-dispersive or overconfident for longer lead times [5]. This also means that outliers occur more frequently than expected. The reasons of this ensemble feature include the immense complication of the natural system in comparison with the finite dimensional model (inducing the presence of model errors) and the finite number of ensemble members. It requires corrections either by modifying the model itself or by performing post-processing of the forecasts.

It is common to correct forecasts by the simplest post-processing method, the bias correction. More generally post-processing (also called Model Output Statistics, MOS) is an approach used for correcting certain aspects of the new forecasts, based on statistical features of prior comparisons between model outputs and observations. The most widespread approach applies ordinary least-squares regression (LMOS) to relate observations and model predictions, thereby assuming the sole presence of errors in the observations [1]. As the bias correction method, LMOS corrects the mean but it tends to degrade the variability, especially at long lead times. This depletion of the ensemble variance is related to the progressive decorrelation of the forecast, associated with the chaotic nature of its dynamics. As a consequence, when applying LMOS on each member of an ensemble forecast, all members will be mapped to a constant value for long lead times.

Different authors proposed an approach based on linear regression that introduces ensemble and climatological reliability [2, 4]. We outline a general framework based on maximum likelihood estimation and Lagrange multipliers for imposing an arbitrary number of constraints and generalize their set-up for use with arbitrarily many predictors. Moreover we discuss how to use the ensemble spread for estimating the error variances in order to avoid overcorrection of ensembles with large spreads.

45.2 Notation

Consider the meteorological variable X for which we have N observations $(X_{O,1}, \dots, X_{O,N})$. Corresponding to each observation n, the m-th member of our ensemble forecast produces values $(V_{1n}^m, \dots, V_{Pn}^m)$ for the P different meteorological variables or predictors. The first predictor V_1 is the one corresponding to the variable X. We also define the ensemble mean values as $(\bar{V}_{1n}, \dots, \bar{V}_{pn})$.

Given a training set of observations and corresponding predictors, we want to improve the bare uncorrected forecast V_1 by using all model predictors. More specifically for each member m of an ensemble n we construct a corrected forecast or predictand:

$$X^m_{C,n} = \alpha + \sum_p \beta_p \bar{V}_{pn} + \gamma \epsilon^m_n, \tag{45.1}$$

where we defined the deviation from the ensemble mean $\epsilon^m_n = V^m_{1n} - \bar{V}_{1n}$. This relation implies that a bias correction is realized using α, the ensemble mean variability is corrected by adjusting the β's and the ensemble spread by adjusting γ.

45.3 Constrained Maximum Likelihood Estimation

Basic climatological constraints for forecasts are the equality of mean and variance of the corrected forecast with the mean and variance of the observations. Note that even for perfect model output the associated variance may differ from the variance of the observations due to the presence of measurement errors and representativity errors. The constraint for *climatological reliability* (CR) is:

$$\sigma^2_C = \sigma^2_O,$$

where σ^2_C is the variance of the corrected forecast and σ^2_O the variance of the observations.

A reliable forecast is characterized by the fact that the observation may be considered as a member of the ensemble forecast and hence has the same statistical properties. Defining the error variance of the ensemble n as $\sigma^2_{\epsilon,n}$, this implies that the quadratic error divided by $\sigma^2_{\epsilon,n}$ is equal to one, once averaged over all observations and forecasts. The condition for *ensemble reliability* (ER) is therefore:

$$1 = \frac{1}{N} \sum_{n=1}^{N} \frac{(\bar{X}_{C,n} - X_{O,n})^2}{\sigma^2_{\epsilon,n}}. \tag{45.2}$$

A mathematical tool for introducing the ER and CR constraints is by introducing a constrained likelihood. Assuming the errors on the ensemble mean of the corrected forecast are normally distributed with mean zero and variance $\sigma^2_{\epsilon,n}$, the constrained log-likelihood becomes:

$$\begin{aligned} \ln \mathcal{L}(\alpha, \boldsymbol{\beta}, \gamma, \mu, \lambda) = & -\frac{1}{2} \sum_{n=1}^{N} \ln\left(2\pi \sigma^2_{\epsilon,n}\right) - \frac{1}{2} \sum_{n=1}^{N} \frac{(\bar{X}_{C,n} - X_{O,n})^2}{\sigma^2_{\epsilon,n}} \\ & + \lambda\left(\sigma^2_C - \sigma^2_O\right) + \mu\left(1 - \frac{1}{N} \sum_{n=1}^{N} \frac{(\bar{X}_{C,n} - X_{O,n})^2}{\sigma^2_{\epsilon,n}}\right). \end{aligned}$$

Upon maximization of this functional with respect to the different parameters we obtain the maximum likelihood estimate (MLE). The ER forecast is obtained by setting $\lambda = 0$, the CR forecast by setting $\mu = 0$, the unconstrained MLE forecast by assuming $\lambda = \mu = 0$ while the ER+CR assumes both to be nonzero.

45.3.1 Discussion

The error variance $\sigma^2_{\epsilon,n}$ of an ensemble n is assumed to be given, but has to be specified. We adopt two approaches: a variable and a constant error variance. We start by assuming a constant error variance in the sense that it is independent on n and the obvious choice is the average ensemble variance. Equation (45.2) then simply states that the mean square error of the ensemble mean equals the average ensemble variance [5]. We find that the CR forecast is equivalent to Errors-in-Variables MOS or EVMOS [7–9] which appropriately takes into account the presence of errors in both the forecasts and the observations.[1] Enforcing only ER and using a constant error variance we find that the CR constraint is automatically satisfied as shown before by different authors for the case of one predictor [2, 4].

For a skillful ensemble forecast, the magnitude of the error, which is the squared difference between ensemble mean and the observation, may be estimated by the ensemble variance. Therefore, for an inhomogeneous chaotic system there is a dependence of the error variance $\sigma^2_{\epsilon,n}$ on the ensemble n itself. Applying in that case the unconstrained MLE, we find that the ER constraint is automatically satisfied.

45.4 Verification

We test the usefulness of the calibration methods in the context of the well-known Lorenz 63 model by focussing on ensemble scores at short lead times. The system describes thermal convection and involves three coupled first-order differential equations in time for the variables x, y and z:

$$\begin{aligned} \dot{x} &= \sigma(-x + y), \\ \dot{y} &= rx - y - xz, \\ \dot{z} &= xy - bz. \end{aligned}$$

We adopt the conventional parameter choice $(\sigma, r, b) = (10, 28, 8/3)$ such that the system exhibits chaotic behavior. For generating the observations we assume a slightly biased parameter $r' = r + 0.001$ from the one used for generating the model data. The system is perfectly reliable at time zero in the sense that the observation is randomly sampled from the forecast ensemble. Our training and verification sets include 50.000 ensembles of each 51 members.

45.4.1 Ensemble Forecast Skill

Figure 45.1(a) shows the average continuous ranked probability skill score (CRPSS) of the different forecasts for the z-variable against lead time. The CRPS score is the

[1] In fact, it is equivalent to EVMOS applied on the ensemble mean.

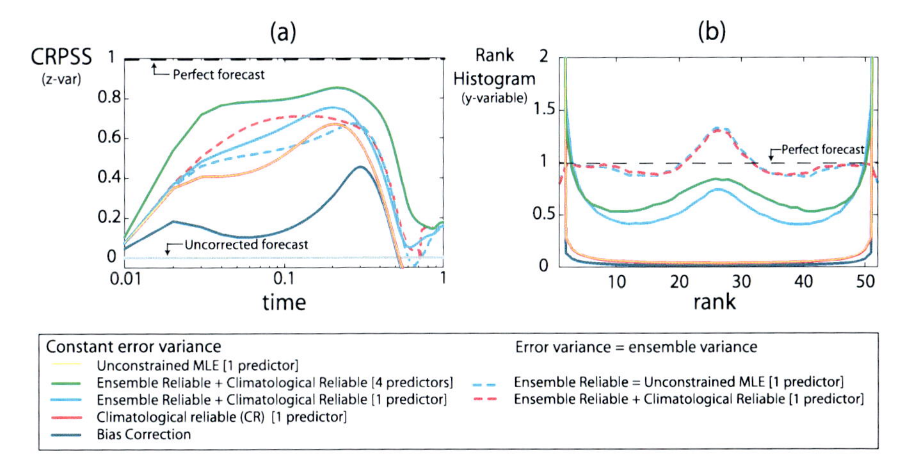

Fig. 45.1 (**a**) CRPS skill scores for different corrected forecasts applied on the z-variable as a function of the lead time. (**b**) Rank histogram as a function of the 52 ranks for different corrected forecasts applied on the y-variable. Both plots are generated using the Lorenz 63 model with a model error and averaged over 50.000 ensemble forecasts with 51 ensemble members each

integrated squared difference between the cumulative distribution functions of the ensemble forecast and the observation [10]. The CRPS skill score of a forecast is then equal to $\mathrm{CRPSS} = 1 - \mathrm{CRPS}/\mathrm{CRPS}_u$ where CRPS_u is the value associated with the uncorrected forecast. We distinguish between corrections produced assuming a constant error variance (full lines) and a variable one (dashed lines). For a forecast worse (better) than the uncorrected forecast, the CRPSS is negative (positive) while a perfect forecast has a CRPSS value equal to one. The bias correction (black line) gives rise to the smallest correction while the unconstrained MLE (yellow line) and the CR forecast (red line) improve the forecast but degrade its quality after some time. The ER+CR forecast (blue full line) and all the approaches using a variable error variance (dashed lines) are consistently better. By far the best approach is the ER+CR approach (green full line) with constant error variances but using four predictors as opposed to one.

45.4.2 Calibration

Figure 45.1(b) evaluates the calibration as expressed by the rank or Talagrand diagram for the different forecasts for the y-variable integrated over lead times between zero and one.

The rank histogram shows the frequency with which the observation lies at a specific place in the set of ranked ensemble forecasts. Ideally the rank histogram is a straight line at the value one such that the observation can be viewed as a member of the ensemble. Departure from uniformity may, among others, indicate over-dispersion or under-dispersion of the ensemble forecast. It is clearly seen that the bias-corrected forecast, the CR forecast and the unconstrained MLE forecast (all with a constant error variances) are badly calibrated. The ER+CR forecast with constant error variances performs better but still includes many outliers. These outliers can be attributed to ensembles with small spreads. The best calibrated forecasts are the ones using a variable error variance since these approaches correct small ensembles as well as large ensembles.

45.5 Conclusions

In summary, we introduced a post-processing technique based on linear regression enforcing the forecast to become climatologically reliable (CR) and/or to have reliable ensembles (ER). We have tested our methods using a low-order chaotic system with model error. For short lead times our constrained approaches are superior to the uncorrected forecast, the bias-corrected forecast and the forecast obtained by an unconstrained maximum likelihood estimation. Moreover the use of the ensemble spread for estimating the error variance of the ensemble mean leads to a better calibration and avoids the undercorrection of ensembles with small spreads. Note also that, as opposed to most statistical post-processing techniques, our new approaches are computationally cheap.

References

1. Glahn HR, Lowry DA (1972) The use of model output statistics (MOS) in objective weather forecasting. J Appl Meteorol 11:1203
2. Johnson C, Bowler N (2009) On the reliability and calibration of ensemble forecasts. Mon Weather Rev 137:1717
3. Kalnay E (2002) Atmospheric modeling, data assimilation and predictability. Cambridge University Press, Cambridge
4. Kharin VV, Zwiers FW (2003) Improved seasonal probability forecasts. J Climate 16:1684
5. Leutbecher M, Palmer TN (2008) Ensemble forecasting. J Comput Phys 227:3515
6. Nicolis C, Perdigao RAP, Vannitsem S (2009) Dynamics of prediction errors under the combined effect of initial condition and model errors. J Atmos Sci 66:766
7. Vannitsem S (2009) A unified linear model output statistics scheme for both deterministic and ensemble forecasts. Q J R Meteorol Soc 135:1801
8. Van Schaeybroeck B, Vannitsem S (2011) Post-processing through linear regression. Nonlinear Process Geophys 18:147
9. Van Schaeybroeck B, Vannitsem S (2012) Toward post-processing ensemble forecasts based on hindcasts, Scientific et Technical Publications of Royal Meteorological Institute Belgium, 61
10. Wilks DS (1995) Statistical methods in the atmospheric sciences. Academic Press, San Diego

Chapter 46
CoenoSense: A Framework for Real-Time Detection and Visualization of Collective Behaviors in Human Crowds by Tracking Mobile Devices

Martin Wirz, Tobias Franke, Eve Mitleton-Kelly, Daniel Roggen, Paul Lukowicz, and Gerhard Tröster

Abstract There is a need for event organizers and emergency response personnel to detect emerging, potentially critical crowd situations at an early stage during city-wide mass gatherings. In this work, we present a framework to infer and visualize crowd behavior patterns in real-time from pedestrians' GPS location traces. We deployed and tested our framework during the 2011 Lord Mayor's Show in London. To collection location updates from festival visitors, a mobile phone app that supplies the user with event-related information and periodically logs the device's location was distributed. We collected around four million location updates from over 800 visitors. The City of London Police consulted the crowd condition visualization to monitor the event. We learned from the police officers that our framework helps to assess occurring crowd conditions and to spot critical situations faster compared to the traditional video-based methods. With that, appropriate measure can be deployed quickly helping to resolve a critical situation at an early stage.

46.1 Introduction

Understanding the behavior of pedestrian crowds in physical spaces is important in many areas ranging for urban planning, policy making at community and state level as well as design, and management of pedestrian facilities and transportation systems. Therefore, in the past few years, efforts have increased to study how human crowds form and how specific collective behavior patterns among the involved individuals emerge.

M. Wirz (✉) · D. Roggen · G. Tröster
Wearable Computing Laboratory, ETH Zürich, Zürich, Switzerland

T. Franke · P. Lukowicz
Embedded Intelligence, DFKI Kaiserslautern, Kaiserslautern, Germany

E. Mitleton-Kelly
Complexity Research Programme, LSE, London, UK

T. Gilbert et al. (eds.), *Proceedings of the European Conference on Complex Systems 2012*, Springer Proceedings in Complexity, DOI 10.1007/978-3-319-00395-5_46,

To do so, simulation tools have been used to study the self-organizing effects of large groups of pedestrians. Different models exist to simulate pedestrian dynamics. The most popular models include: Physical models that model pedestrians based on the analogy to gases or fluids; the social force model together with its extensions [1]; and cellular automata [2]. To perform a calibration of the model parameters, experiments under controlled conditions can be performed [3, 4] or video footage capturing pedestrian dynamics can be evaluated [5].

Simulations can then be used to study the effect of architectural configurations on crowd dynamics, the emergence of collective behavior patterns in urban spaces, the influence of commuting patterns in a subway system, etc. and help to derive data-backed recommendations useful to advise policy makers and other stake holders. One area where knowledge of crowd behaviors and pedestrian dynamics is of special importance is in the management and monitoring of large-scale mass events and city-wide festivals. It is a top priority for every organizer of such an event to be able to maintain a high standard of safety and to minimize the risk of incidents. Hence, establishing adequate safety measures is important.

However, deriving and deploying optimized emergency strategies based on simulations remains challenging due to many unpredictable factors inherent with the nature of such events. A key issue is that the actual number of attendees of such events may greatly deviate from estimations as it depends on factors like weather conditions, alternative events, and the program etc. Yet, the biggest challenge of all is that the behavior of the crowd during an event remains highly unpredictable. These challenges foster the need to detect critical crowd situations like overcrowding at an early stage in order to rapidly deploy adequate safety measures to mitigate the impact of a potentially dangerous situation. To do so, real-time information about the behavior of the crowd is required. At present, mostly video-based monitoring systems come into operation for this task. Recent research has focused on developing computer-based methods to automatically analyze the recorded scenes and to detect abnormal and potentially dangerous crowd situations [6–9].

Vision-based approaches face several limitations: Cameras can not capture elements outside their fields of view or occluded by other obstacles [8] and it is still difficult to fuse information from many cameras to obtain global situational awareness [10]. Another drawback is the need for good lighting conditions. Furthermore, as many events happen during the night, the application of a vision based approach is limited.

Some countries, such as the UK may also use helicopters to gain an overview either in daylight or at night with thermal imaging. The use of helicopters, however, is expensive and requires highly trained personnel; they are therefore usually only used for events, which are expected to be problematic.

As an alternative to traditional approaches, we see a big potential in monitoring crowd behaviors by tracking the locations of the attendees via their mobile phone. We believe that the high distribution of location-aware mobile phones in our society and the acceptance to share personal context information enables such an approach. In this work we present CoenoSense, a platform able to infer crowd conditions in real-time from location information collected on mobile phones of attendees of a mass gathering. This information can be made instantaneously available to event organizers and emergency response personnel through intuitive visualizations. Our approach relies on participatory sensing paradigms by offering incentives to users to deliberately share their location information. As a system trial, we deployed and tested our framework during the Lord Mayor's Show in London in November 2011.

46.2 CoenoSense Data Collection Platform

To infer crowd conditions, we require the location of festival attendees. We use attendees' mobile phones to obtain their location as most of today's mobile phones are situation and location aware. Methods to obtain location information include GPS positioning and WiFi/GSM-fingerprinting [11]. Collecting location updates on user's mobile devices requires users to install and run a dedicated application on their mobile phones. At first sight, such an approach may appear undesirable, as it can be assumed that people are not willing to install such an application. In the case of a mass gathering, this may mean that only a fraction of all attendees would run such an application and many would opt for not having their location tracked for various reasons, including privacy concerns and energy considerations. Nevertheless, we believe this approach is still viable and promising by following a participatory sensing approach where users themselves are motivated to deliberately share their location information by offering them a set of attractive incentives.

In a preceding study, we have verified that people are willing to share location information if they receive some benefits or if they realize that sharing such information is for their own good and safety [12]. By following such an approach, we offer users full control of the recordings and allow them to disable it at anytime. In addition strict ethical guidelines are observed, and all data provided is anonymous. A clear statement explaining how an individual user's privacy is safeguarded is given before the app is downloaded. To collect location information of festival attendees, we developed a generic festival app for mobile devices which can be tailored to a specific event and provides the users with relevant, event-related information such as the festival program, a map indicating points of interest and background information, as well as travel information. These features are designed to be attractive and useful during the event to reach a large user base. While a user is running the app, GPS location is regularly sampled with a frequency of 1 Hz on the device and periodically sent to our servers running the CoenoSense (www.coenosense.com) framework. CoenoSense acts as a centralized repository to store the location updates received from many mobile devices simultaneously and allows for real-time processing of the collected data. Our app offers users full control over the shared data and at anytime, data recording can be disabled. CoenoSense can be used to visualize occurring crowd behaviors from the aggregated location updates. The safety aspect becomes operational through a combination of the visualization feature, which shows potential overcrowding or turbulence, and the geographically targeted messaging function, which enables the organisers or emergency services to advise users in that particular location.

46.3 Inferring and Visualizing Crowd Conditions

To improve pedestrians' safety, much research has been devoted to understanding crowd behaviors and to identify critical crowd conditions by conducting lab experiments and evaluating empirical data from real mass gatherings. An obvious, yet important crowd characteristic to assess the criticality of a situation is the density of a crowd. For example, most stampedes occur in high-density crowds [3]. Different methods to measure crowd density and to identify dangerous overcrowding have been proposed [9]. Density, however, is not the only relevant characteristic; movement velocity, the flow direction of a crowd and turbulent movement have also been identified as important indicators of critical situations [4, 5]. In general, Helbing and Johansson [5, 13] suggest quantifying the hazard to the crowd (and with this the criticality of a situation in the crowd) by a measure they call *crowd pressure*, defined as the local pedestrian density multiplied by the variance of the local velocity of the crowd.

We want to provide emergency response personnel with the means to instantaneously assess the current crowd situation during mass gatherings. Displaying information as an overlay over a map is a common approach to present spatial information and allows for a quick assessment of the situation. In this section we elaborate on our methods to infer crowd conditions. We are going to show how, given the location updates from the users, an estimation of *crowd pressure* can be obtained and how this information can be visualized. Other measures can be derived and visualized in a similar fashion.

We use heat map visualization to display crowd pressure. A heat map is a graphical representation of spatial data where regions are colored according to measurement values found at the specific location. Heat map visualizations have been used in different applications to convey various types of spatial information.

It is in the nature of our data that we have areas with many active app users and by contrast, areas with very few; also large parts with no users. We are only able to infer crowd conditions in areas where location updates have been recorded. By calculating crowd conditions in an area, the amount of active users (and hence available location updates) is relevant: Information from areas where many location updates were recorded might be more important as this region seems to be more popular than in regions with only a few individuals. To visualize the crowd conditions and to also include the importance, we adjust the heat map generation method in the following way: We use a color gradient to indicate the crowd conditions at a location with a varying opacity level that corresponds to density at the location. We calculate the density by performing a Kernel Density Estimation (KDE) [14] of the active users' location updates at a given time.

KDE is a non-parametric way of creating a smooth map of density values in which the density at each location reflects the concentration of sample points. Hereby, each sample point contributes to the density estimation based on the distance from it. By using a Gaussian kernel K, the density estimation $\hat{d}$ at each location X is given by:

$$\hat{d}(X,t) = \frac{1}{N \cdot h} \sum_{i=1}^{N} K\left(\frac{X - X_{i,t}}{h}\right) \tag{46.1}$$

with h the Bandwidth (an application dependent smoothing parameter), and $X_{1,t}, \dots, X_{N,t}$ the users' current location. The Gaussian kernel function K is given by

$$K(u) = \frac{1}{\sqrt{2\pi}} \exp\left(-\frac{1}{2}u^2\right) \tag{46.2}$$

By determining the density values $\hat{d}(X,t)$ for each location and mapping each density value to a color using a color gradient, a heat map representing the participant density estimation is obtained.

The obtained density estimation for a location is then directly mapped to the opacity value of the point. A very low density value will result in an almost transparent point, while a high density value will result in a fully opaque point. Hence, the more users are situated around a location, the more intense the location color. Regions where no data is available remain transparent. The coloring is then determined by calculating the crowd conditions at the specific location and mapping this value to a color value using a color gradient. With this, for each point in space, we obtain an opacity value representing the density together with a color value for the crowd condition at that location.

According to [13], the crowd pressure is given as

$$P(X) = \rho(X) \cdot \mathrm{Var}_X(\mathbf{v}) \tag{46.3}$$

where ρ is the local pedestrian density and $\mathrm{Var}_X(\mathbf{v})$ the local velocity variance. In our case, we can obtain a density measure using Formula (46.1) and can calculate the velocity variance as

$$\widehat{\mathrm{Var}}_{X,t}(\mathbf{v}) = \frac{\sum_{i=1}^{N} |\mathbf{v}_{X_{i,t}} - \langle \mathbf{v} \rangle_{X,t}|^2 \cdot K(\frac{X - X_{i,t}}{h})}{\sum_{i=1}^{N} K(\frac{X - X_{i,t}}{h})} \tag{46.4}$$

where $K(u)$ is the Gaussian kernel according to Eq. (46.2) to determine the weight and $X_1, X_2, \dots, X_N$ are the locations of the active users. $\langle \mathbf{v} \rangle_{X,t}$ is $\hat{v}(X,t)$, the local crowd velocity at a given location X and can be seen as the weighted average velocity of each user at a given location by weighting the speed values of each user depending on the distance to that location with a Gaussian weighting scheme via

$$\hat{v}(X,t) = \frac{\sum_{i=1}^{N} v_{i,t} K(\frac{X - X_{i,t}}{h})}{\sum_{i=1}^{N} K(\frac{X - X_{i,t}}{h})} \tag{46.5}$$

With this, the formula to calculate the crowd pressure estimation $\hat{P}(X,t)$ at the location X at time t is given by:

$$\hat{P}(X,t) = \hat{d}(X,t) * \widehat{\mathrm{Var}}_{X,t}(\mathbf{v}) \tag{46.6}$$

Fig. 46.1 Heat map visualization of crowd pressure

The heat map is generated analogously to the other crowd conditions by mapping the crowd pressure to a color and combining it with the opacity obtained from the crowd density. Figure 46.1 shows an example of such a heat map.

46.4 Exemplary Use Case

To understand the usefulness of a real-time visualization of crowd conditions during mass gatherings, we deployed the system during the Lord Mayor's Show 2011 in London on November 12th. The Lord Mayor's Show is a street parade in the City of London, the historic core of London. A new Lord Mayor, mayor of the City of London, is appointed every year and this public parade is organized to celebrate his inauguration. The annual one-day event attracts about half a million spectators each year and is one of the City's longest established and best known annual events dating back to 1535.

The event starts at 11:00 am and the processional route goes from the Mansion House via Bank, St. Paul's Cathedral and Fleet Street to the Aldwych. Then the whole procession sets off again at 1 pm to take the new Lord Mayor back to Mansion House. The procession finally ends at about 2:30 pm when the last floats reach the City.

In collaboration with the event organizers, we tailored our festival app to the event and distributed it for free as the festival's official app. It was advertised on the Lord Mayor's Show website and available through Apple's iTunes app store.

Data collection was active between 00:01 am and 11:59 pm on November 12, but only if the user was in a specific geographical area around the festival venue. Over the whole day, we collected a total of 3′903′425 location updates from 827 different users. During the parade, location updates from up to 244 users were received simultaneously, at any one time.

Figure 46.1 shows the heat map visualization of the crowd pressure at 5:21 pm. It is right after the end of the firework display as people are leaving the festival venue. It is clearly visible that there are high densities of users around train, Subway and bus stations as there are opaque regions to be found. These are locations pedestrians find public transportation to travel home and where naturally potentially dangerous situations may emerge.

46.5 Conclusion

Understanding the behavior of pedestrian crowds in physical spaces in real-time is important for many fields of application. In this work, we introduce a framework to infer real-time crowd conditions by tracking people's movement traces via their mobile phone. By aggregating and visualizing this information as heat maps, we can offer an intuitive way to obtain a global view of the crowd situation and to assess different crowd conditions instantaneously. There are also some challenges coming along with our approach. Mainly, due to the nature of our participatory sensing approach, our system can only collect data from the users of the mobile phone app. This is usually only a subset of all attendees. Keeping this in mind, two aspects are crucial:

– *Ensuring a large user base*: Providing attractive incentives is important to reach a large user base. We reach this by offering festival attendees an enhanced experience by using our festival app. Hereby, a user study has revealed a set of attractive features and helped us to design the app accordingly [12]. Advertising the app in an appropriate way is key. Recently, big events started to offer their own app. Our tools can easily be integrated into their existing solution.
– *Seeing users as probes*: By designing robust crowd condition measures that are robust with respect to the ratio of the app users, it is possible to extract accurate crowd condition measures, even when not all attendees are being tracked. To do so, we have to consider the users as probes and conclude from their behavior the overall crowd situation. This can be achieved e.g. through calibration of the data. While the ratio of app users to festival attendees remains unknown, we assume that the spatio-temporal distribution of users reflects the distribution of attendees at any one time during the event. With this assumption in mind, an actual crowd density estimation can still be obtained by determining the ratio of mobile app users to festival users in a given area e.g. by inspecting CCTV recordings. This ratio has to be updated periodically throughout the event.

The framework can provide the following features of value to policy makers: (a) an overview, not available by the usual means of crowd monitoring including

CCTV, as it can cover a larger area at any one time, for longer; (b) it is cheaper than an helicopter. Helicopters do have thermal imagery technology, but they are expensive and need highly trained personnel to fly them and on the ground; (c) especially valuable at night, (e.g. during the Fireworks display at the LMS), when CCTV cameras are not effective; (d) can be used to plan future events and to position barriers, ambulance stations, loos, etc. more accurately; (e) using the heatmap is intuitive and does not require any training, although it does need a trained officer to identify potential critical issues and take appropriate action. Overall the framework was found by the organisers and emergency services to be a valuable tool in taking appropriate action quickly to avoid a potential incident, thus increasing safety. From a complexity theory perspective it shows evolving emergent crowd dynamics; it can be used to illustrate self-organising behavior of groups;

During the Lord Mayor's Show 2011, only the emergency response personnel and security personnel had access to the real-time visualization of the crowd conditions. It will be of further interest from a complexity science point of view to investigate the dynamics evolving when festival attendees themselves are given access to such crowd information. It would then be of interest to study how the available information is considered in their decision making process and what kind of co-evolutionary dynamics will emerge. Ultimately, we would like to understand if such information can help to lower the number of overcrowded situations, while decreasing turbulence and crowd pressure.

Acknowledgements This work is supported under the FP7 ICT Future Enabling Technologies Programme under grant agreement No. 231288 (SOCIONICAL).

References

1. Helbing D, Molnar P (1995) Social force model for pedestrian dynamics. Phys Rev E 51:4282–4286
2. Burstedde C, Klauck K, Schadschneider A, Zittartz J (2001) Simulation of pedestrian dynamics using a two-dimensional cellular automaton. Physica A 295(3–4):507–525
3. Helbing D, Buzna L, Johansson A, Werner T (2005) Self-organized pedestrian crowd dynamics: experiments, simulations, and design solutions. Transp Sci 39(1):1–24
4. Steffen B, Seyfried A (2010) Methods for measuring pedestrian density, flow, speed and direction with minimal scatter. Physica A 389(9):1902–1910
5. Johansson A, Helbing D, Al-Abideen HZ, Al-Bosta S (2008) From crowd dynamics to crowd safety: a video-based analysis. Adv Complex Syst 11:479–527
6. Krausz B, Bauckhage C (2012) Loveparade 2010: automatic video analysis of a crowd disaster. Comput Vis Image Underst 116(3):307–319
7. Mehran R, Oyama A, Shah M (2009) Abnormal crowd behavior detection using social force model. In: IEEE computer vision and pattern recognition
8. Zhan B, Monekosso D, Remagnino P, Velastin S, Xu L (2008) Crowd analysis: a survey. Mach Vis Appl 19(5–6):345–357
9. Gong S, Loy CC, Xiang T (2011) Security and surveillance. In: Moeslund TB, Hilton A, Krüger V, Sigal L (eds) Visual analysis of humans. Springer, London
10. Song B, Sethi RJ, Roy-Chowdhury AK (2011) Wide area tracking in single and multiple views. In: Visual analysis of humans. Springer, London

11. Kim D, Kim Y, Estrin D, Srivastava M (2010) Sensloc: sensing everyday places and paths using less energy. In: Proc of the 8th ACM conference on embedded networked sensor systems. ACM, New York
12. Wirz M, Roggen D, Troster G (2010) User acceptance study of a mobile system for assistance during emergency situations at large-scale events. In: 3rd international conference on human-centric computing (HumanCom 2010). IEEE Press, New York
13. Helbing D, Johansson A, Al-Abideen H (2007) Dynamics of crowd disasters: an empirical study. Phys Rev E 75:046109
14. Scott D (1992) Multivariate density estimation. Wiley, New York

Chapter 47
An Agent-Based Model for the Analysis of the Energy Sources Diffusion Dynamics

Alessandro Filisetti, Stefano Bontempi, and Marco Setti

Abstract A novel model devoted to the characterization of the diffusion dynamics of three energy sources, traditional, bioenergy-like and solar-like, in a socio-economic energy system composed of general industries is presented. During the simulation each industry defines its strategy about the implementation of new technologies for the procurement of the energy needed to fulfill its internal activities. The research focuses on two different socio-economic energy systems descriptions: the first description is characterized by industries operating only by means of economic assessments, while in the second case imitation phenomena are introduced so that industries define their strategies not only by cost-benefits analysis but observing the behavior of the neighborhood as well.

47.1 Introduction

In the recent years the attention toward a sustainable society is increasing. From the scientific point of view the analysis of the complex dynamics characterizing the transition of the socio-economic systems toward a eco-compatible future turn to be of paramount importance to address decision makers in the formulation of suitable initiatives, especially with regard to the guidelines traced by the Kyoto Protocol for the year 2020.[1]

To this aim, this preliminary work is based on the analysis of the dynamics emerging from the diffusion of new energy sources in a virtual socio-economic energy system composed of general industries assessing the investment in new technologies to produce energy in order to perform their internal activities.

Although the study is focused on energy sources only, the model essentially deal with the diffusion of new technologies in a dynamic environment characterized by different agents performing each one its own strategy.

[1] http://unfccc.int/kyoto_protocol/status_of_ratification/items/2613.php.

A. Filisetti (✉) · S. Bontempi · M. Setti
C.I.R.I., Energy and Environment Alma Mater Studiorum, University of Bologna, via F.lli Rosselli, 107, 42123 Reggio Emilia, Italy

T. Gilbert et al. (eds.), *Proceedings of the European Conference on Complex Systems 2012*, Springer Proceedings in Complexity, DOI 10.1007/978-3-319-00395-5_47,

To observe the dynamics emerging from the interactions of the actors, an agent-based description of the socio-economic system has been chosen. The interest around this kind of representation has been growing and many different models have been developed, as for example in the technology optimization and diffusion studies [2, 5, 6] or in policies planning supporting decision analysis [1, 3, 8, 9].

A central aspect of the agent-based models regards the non-linear way by which agents can interact, showing those behaviors typical of complex systems.

The technologies considered in this work represent three different ways of collecting energy. The idea is to represent a traditional source of energy distributed in the same way all over the territory, a bioenergy-like source localized only in a small part of the environment and a solar-like energy source available all over the territory with different irradiation degrees according to the possible orientations of the existing industries and different expositions in different places.

With regard to the technology diffusion, policy makers can tune parameters of the system in different ways in order to push the diffusion of a technology all over the territory. To this aim the research focuses on the influence that solar investment prices could have on the diffusion of such a technology. Furthermore, since an important aspect of new technologies diffusion concerns imitation and emulation phenomena typical of socio-economic systems, results obtained in a system composed of industries defining their own strategies only by economic analysis are compared with those of a system composed of industries with different levels of imitation contributing to the overall assessment of the success of the investment.

47.2 Description of the Model

The main entities of the model are general industries having different energy consumptions. The internal destination of the provided energy is not taken into account. It is assumed that industries need a certain amount of energy each month to perform their activities, hence they have different costs and revenues associated with the specific energy technology adopted. At this level of abstraction it is also assumed that industries dimension are proportional to energy needs.

Energy can be provided by means of three different sources: traditional, E_T from now on, bioenergy-like, E_B, and a solar-like, E_S. In accordance with the different energy prices industries may adopt a hybrid recipe of energy sources, so that the total amount of necessary energy may be provided by different technologies, according to the most suitable combination.

If energy is provided under the E_T form, industries will pay an overall cost C_T^j, proportional to the energy need, that is equal all over the environment, Eq. (47.1).

$$C_T^j = c_T k_j, \tag{47.1}$$

where c_T is the unitary cost (€/kWh) of E_T and k_j is the energy need (kWh/month) of the j-th industry.

In the case of E_B the part of energy produced by biomass technologies is sold, so that revenues contribute to decrease traditional energy procurement costs. Nevertheless adopting E_B industries incur additional costs proportional to the distance from the bio-energy sources, Eq. (47.2).

$$C_B^j = C_T^j + D_j c_d k_j^B - c_B k_j^B, \tag{47.2}$$

where D_j is the distance from the bio-energy source, c_d represents transportation costs (€/km/kWh), k_j^B stands for the portion of energy produced by means of E_B and c_B is the price at which energy is sold. Accordingly, the third energy source E_S does not represent a cost but a value to subtract from the cost of the traditional furniture, hence:

$$C_S^j = C_T^j - P_S I_j k_j^S, \tag{47.3}$$

where P_S are the unitary earnings for selling the energy produced to the energy supplier, I_j is the irradiation constant of the place where industry j is placed and k_j^S stands for the portion of energy produced by means of E_S.

The model is developed using Netlogo [7], a programming framework tailored for the simulation of complex systems.

In this work an agent-based description is adopted. By means of the agent-based modeling each industry is able to perform a personal strategy based on its own peculiarities and knowledge about the environment.

Figures 47.1(a) and 47.1(b) represent the two different layers of the system, the layer representing the bio-energy availability and the layer representing the solar irradiation.

Each step of the simulation represents one month, hence at each month industries define their own strategies.

In this preliminary work two different strategies are adopted in order to assess the energetic issue. The first strategy is based on a simple economic analysis of the investment necessary to be equipped with the potential new plants. The second strategy takes into account the imitation process so that industries define their behavior not only considering the mere economic estimation but also looking at the behavior of the industries present in their neighborhood.

At the beginning of each simulation all industries are equipped with traditional technologies, then once a year (on average) they evaluate the possibility to change the energy source. The economic analysis is made computing the net present value (NPV) and the payback period (PBP) associated with the assessed investment [4]. The imitation dimension affects each industry differently according to the propensity P_j, with $0 \leqslant P_j \leqslant 1$, to the imitation of the behavior of the neighbors. In accordance with P_j the economic evaluation of the incoming deriving from the new technology is distorted. Neighbors adopting the technology under investigation will increase the perceived revenues from that technology while neighbors adopting different technologies will tend to increase the costs associated.

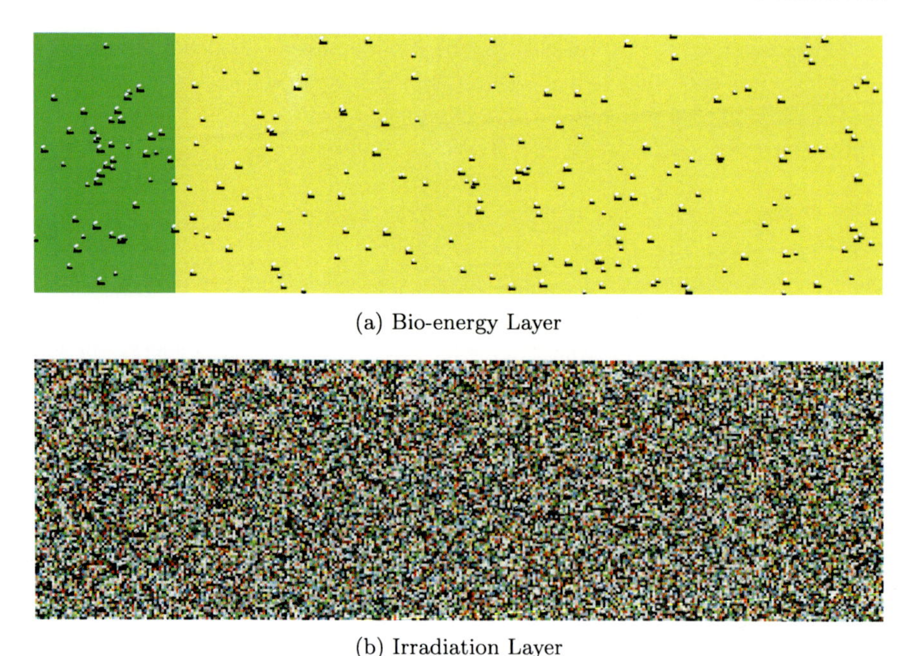

(a) Bio-energy Layer

(b) Irradiation Layer

Fig. 47.1 The figure shows the virtual environment of the agent-based simulation. (**a**) *The green zone* represents the E_B source that is not available all over the environment. *The yellow zone* is only the negative part of the *green one*, hence no bio-energy is present on *the yellow zone*. The size of the agents is proportional to its energy need. (**b**) In this figure the solar-like irradiation is represented. At this stage the irradiation is randomly distributed all over the territory representing different possible solar expositions of the industries

The total attractiveness of each technology is proportional to the energy dimension of the neighbors and inversely proportional to their distance, Eq. (47.4).

$$A_E = \sum_{i=1}^{N_E^{(D)}} k_i d_{ij}^{-1}, \tag{47.4}$$

where i stands for the different industries using energy E in the neighborhood, D is the maximum radius of the neighborhood, k_i is the dimension, i.e. the energy need, of the i-th industry, d_{ij} stands for the distance between industry j and industry i and E stands for the specific energy source, E_T, E_B and E_S. Then the relative attractiveness of each energy source is computed dividing the attractiveness of each technology for the overall attractiveness such that $[A_T] = A_T/(A_T + A_B + A_S)$, $[A_B] = A_B/(A_T + A_B + A_S)$ and $[A_S] = A_S/(A_T + A_B + A_S)$.

The relative attractivenesses are then used to skew the economic analysis of the investment overestimating the potential of a technology in case of several neighbors adopting that technology, and underestimating such potential otherwise.

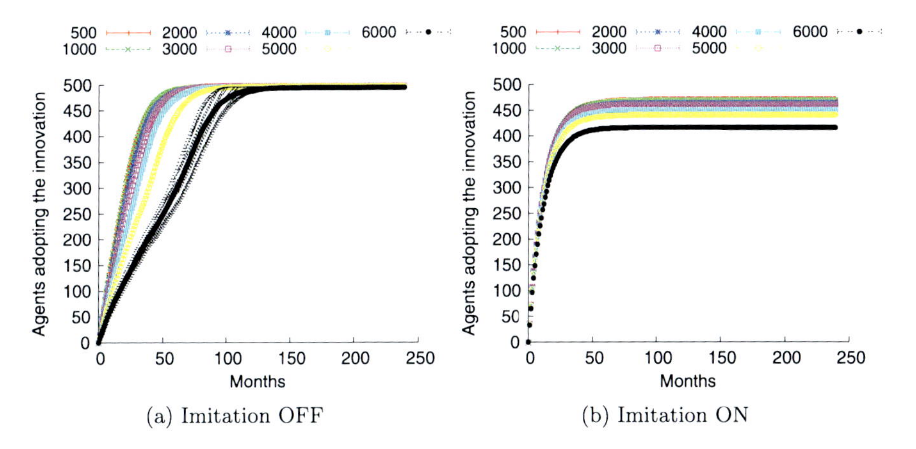

(a) Imitation OFF

(b) Imitation ON

Fig. 47.2 The figures show the number of industries adopting the innovation in time. On the Y axis the average number of industries with error bars is shown while on the X axis the months elapsed from the beginning of the simulation is shown. Panel (**a**) shows the behavior of the socio-economic energy system when imitation is turned on while panel (**b**) shows the behavior of the socio-economic system without imitation phenomena. Parameters: Overall number of industries: 500, E_T cost: 0.25 €/kWh, E_B price: 0.3 €/kWh, E_S price: 0.35 €/kWh, E_B new plant investment cost: 5000 €/KW, investment interest rate: 0.04

47.3 Results

In this work a particular behavior has been investigated. In order to characterize the influence of the cost of new solar-like plants on the diffusion dynamics of such a technology 25 different runs for each price value, 500, 1000, 2000, 3000, 4000, 5000, 6000 €/kWh are performed for a total number of 175 simulations.[2]

Figures 47.2(a) and 47.2(b) show the number of industries adopting a new solar-like technology to produce their energy. When imitation is not considered, it is straightforward that increasing the price of the investment, i.e. the price of the new plant installation, the number of industries adopting the new technology decreases, Fig. 47.2(b). Nevertheless it is possible to observe that in view of the increment of the price the decrease in the number of industries is not too pronounced.

On the other hand, if imitation phenomena is considered the situation changes dramatically, Figs. 47.2(a) and 47.3. The choice of each industry is not longer based on a mere economic estimation but it is based on the behavior of the other industries as well. In Fig. 47.2(a) is possible to observe that a after an initial transient the overall behavior of the system is channeled toward a diffuse adoption of the new technology. Also with higher investment costs, i.e. 6000 €/kWh, the attractiveness of the first adopters tend to convince the other industries to overestimate the perceived revenues triggering the diffusion phenomenon.

[2]Please refer to the caption of the images for the values of the other parameters used.

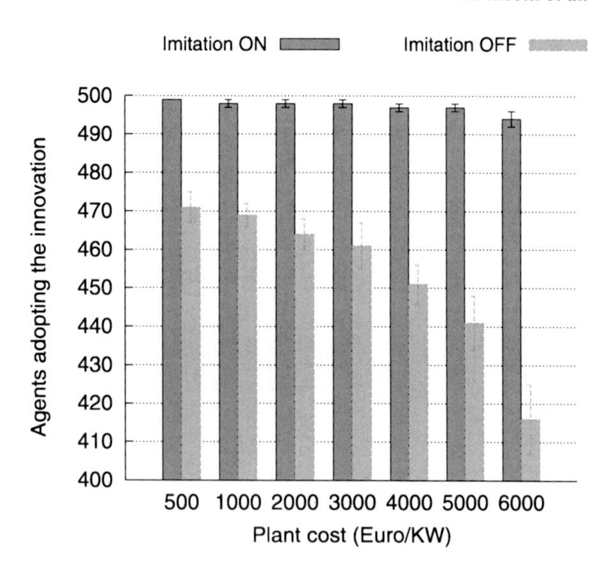

Fig. 47.3 The figure shows the total average number of industries adoption the innovation at end of the simulations (25 runs for each parameter setting). Overall number of industries: 500, E_T cost: 0.25 €/kWh, E_B price: 0.3 €/kWh, E_S price: 0.35 €/kWh, E_B new plant investment cost: 5000 €/KW, investment interest rate: 0.04

47.4 Conclusions and Further Activities

In this work a novel model for the study and the characterization of the complex dynamics emerging during the processes of diffusion of a new technologies in socio-economic energy systems is presented.

In particular the influence of the investment costs associated with a solar-like energy production technology on the number of industries investing on it is investigated. Moreover, two different systems, the first one composed of industries defining their own strategies on economic assessment only and the other one composed of industries which strategies are based on both economic evaluations and imitation processes (with different weights) are compared. Results show that the technology diffusion rate is higher in systems characterized by imitation processes.

Although this is a very simple novel model, an unexpected behavior emerging from the interaction of the industries in a virtual socio-economic energy system has been clearly shown. Actual world shows indeed more complex dynamics, nevertheless it is sufficient deal with such a simple representation of the reality to get the importance of imitation phenomena in fostering diffusion dynamics in socio-economic systems. Finally policy makers, and decision makers in general, should have clearly kept in mind the role that very attractive actors play in speeding up the diffusion of new technologies as well as new sustainable initiatives in general.

To this aim next analysis will be focused on hybrid systems in which some industries tend to imitate while other industries adopt strategies based only on economic analysis of the system.

An other important aspect will be the introduction of policies, e.g. new carbon taxes on old technologies or variation in the investment interest rates, during the simulation in order to observe how the system re-organize itself in response to that policies.

Acknowledgements We would thank prof. Roberto Serra, prof. Marco Villani, prof. Cesare Zanasi, Cosimo Rota, Claudia Severi and Matteo Vittuari for their helpful comments and suggestions.

References

1. Beck J, Kempener R, Cohen B, Petrie J (2008) A complex systems approach to planning, optimization and decision making for energy networks. Energy Policy 36(8):2795–2805
2. Chappin EJL (2010) Agent-based modelling of energy infrastructure transitions. Int J Crit Infrastructures 6(2):106
3. Cormio C, Dicorato M, Minoia A, Trovato M (2003) A regional energy planning methodology including renewable energy sources and environmental constraints. Renew Sustain Energy Rev 7(2):99–130
4. Khan MY, Jain PK (1999) Theory and problems in financial management. Tata McGraw-Hill, New Delhi
5. Nikolic I, Dijkema GPJ (2010) On the development of agent-based models for infrastructure evolution. Int J Crit Infrastructures 6(2):148–167
6. Schwarz N, Ernst A (2009) Agent-based modeling of the diffusion of environmental innovations—an empirical approach. Technol Forecast Soc Change 76(4):497–511
7. Wilensky U (1999) NetLogo http://ccl.northwestern.edu/netlogo/. Center for Connected Learning and Computer-Based Modeling, Northwestern University, Evanston, IL
8. Zellner ML (2008) Embracing complexity and uncertainty: the potential of agent-based modeling for environmental planning and policy. Plan Theory Pract 9(4):437–457
9. Zeng Y, Cai Y, Huang G, Dai J (2011) A review on optimization modeling of energy systems planning and GHG emission mitigation under uncertainty. Energies 4(10):1624–1656

Chapter 48
Complexity and Standards—Programming Innovation

Anna Andreyevna Zaytseva

Abstract This work intends to explain step-by-step the special role of standards and their complexity in exercising innovation policy in the European Union. Standards might be an important policy tool for intentional diffusion of market sectors beyond textually available official policy documents/guidelines. This diffusion, having been initiated intentionally, continues in a self-regulated way as the interest driven for emergence of innovation. I study standardization in the context of the present Lead Market Initiatives, i.e. the official innovation policy areas of 2012 and put my main analytic focus on the most complex area eHealth with standard IEEE 11073 Personal Health Data as example. I also use the taxonomy of standards in Information and Communication Technologies by David (Economic policy and technological performance, Cambridge University Press, Cambridge, 1987) and Krechmer (Technical communications standards: new directions in innovation, 1999) for doing my policy analysis of the European innovation policy. Moreover, the research has suggestions for other paths of complex analysis.

Keywords eHealth · Diffusion · Innovation · Standard · Policy analysis · Information and communication technologies (ICT) · Complexity science · ISO IEEE 11073 personal health data

48.1 Introduction

This work is a policy analysis of the European innovation policy performed with the use of technical knowledge in Information and Communication Technologies (ICT)

A.A. Zaytseva (✉)
Faculty of Mathematics and Natural Sciences, University of Oslo, Oslo, Norway
e-mail: annza944@gmail.com

A.A. Zaytseva
Faculty of Arts and Sciences, Linköping University, Linköping, Sweden

A.A. Zaytseva
Faculty of Law, State University—Higher School of Economics, Moscow, Russian Federation

T. Gilbert et al. (eds.), *Proceedings of the European Conference on Complex Systems 2012*, Springer Proceedings in Complexity, DOI 10.1007/978-3-319-00395-5_48,

and legal knowledge for analyzing documents on standardization and innovation issued by competitive European institutions. This is a contribution to the scientific discussion on standardization versus innovation. However, the policy analysis in this work does not address any special type of problematics in the classical scientific understanding, since the main object of the analysis is standardization in the context of the present innovation policy exercised at the moment. Thus, the purpose of this work is to find and explain hidden codes of behavior of stakeholders and institutions involved in the studied area of relations. The target audience of my analysis is policy-makers, policy-analysts and stakeholders using standards within the framework of activities aimed to innovation. I also determine other ways of analytical development of research for developing the thought process in Complexity Science depending on the synergic knowledge and further purposes for closer studying the involved fields.

48.2 Background and Research Questions

The European Union encourages the emergence and development of new technological areas and innovation by a variety of measures, e.g. grants within policy programmes, procuring innovation at regional levels, establishing special policy areas for innovation, etc. Let me choose for my policy analysis the latter, i.e. Lead Market Initiatives, which may be found in the Communication "A lead market initiative for Europe" COM (2007) 860, and other documents (see [17]). The chosen field belongs to official innovation policy tools for the stimulation and emergence of innovation in the European internal market. Standardization, however, goes hand in hand with the industrial, managerial, social, environmental and technological processes of almost all possible areas of the European internal market. Thus, in this work I go far beyond analysis of standards as "codified technical information for macroeconomic growth" [1]. We find the basic implementation concept of standardization for innovation encouragement in the Communication "Towards an increased contribution from standardisation to innovation in Europe" COM (2008) 133 and the Communication "A strategic vision for European standards: Moving forward to enhance and accelerate the sustainable growth of the European economy by 2020" COM (2011) 311, where standardization as a tool is considerably developed. It is especially exciting to find basic regulations regarding standardization in non-binding guidelines first, such as Communications, whereas documents specifying the standardization exercising are Directives of the European institutions for each particular innovation policy area come second. It says that the basic guidelines may change, but why?

In the Europe 2020 Strategy the European Union declares three major priorities to reach in the nearest future, i.e. "smart growth" based on knowledge and innovation, "sustainable growth" oriented to a greener economy and "inclusive growth" for social cohesion and high employment. The Lead Market Initiatives as policy measures are described as follows: they determine demand driven areas out of any technology push; have broad market segmentation; pursue strategic social and economic interest with major in environmental challenges; imply use of flexible policy instruments (which we can refer standards to); do not pick of "champions" but rather encourage the overall development of markets as such [12]. Standards are expected to stimulate high performance both in encouragement of smart, sustainable and inclusive growth [10], as well as to facilitate the development of the Lead Market Initiatives [12].

Standards may be of different complexities and may be used for the development of products, services, processes, or may not exist in some areas at all, which is an issue of particular situations. Having noticed that standards have high expectation in the respective policy guidelines regarding encouragement of innovation, I have the following questions: What does the European Union mean with giving standards the role of innovation "encouragers"? What makes standards "strategic assets" for innovation [11]? May there be any other policy ideas behind the encouragement to use standards in innovation areas?

48.3 Methodologies

The aim of my research is to do policy analysis within the framework of standardization for innovation encouragement and clarify the concrete role of standards there for the engineering purpose of the European Union as a supranational entity. The research is qualitative and meanwhile analytic with overlapping knowledge of policy analysis in the sphere of innovation and knowledge of standardization taxonomy by Paul David [3] and Ken Krechmer [6] from the area of Telecommunication Studies. I combine this with text analysis of the relevant legal documents about each particular Lead Market Initiative in order to figure out answers to my research questions, extracting knowledge basically from these official innovation policy resources.

I do it step-by-step, algorithmically by use of steps from (1) to (8) in the text, which helps me find meaning by each degree of complexity embedded in standards, since I guess this is an important factor for determining the complexity of technologies and required actions contained in standards, especially for the purpose of studying synergies for emerging innovation. I would be glad if my analytic work will raise the discussion in socio-technical research and get contributions in overlapping knowledge areas. The research consists of two main parts to make the analysis coherent and further developed depending on the emphasis on a particular step from (1) to (8), so that my readers are sent to one when needed in order to follow up the analysis and discoveries.

48.4 Research Design and Research Flow

48.4.1 Part One

The part "Research design" is directly connected with the part "Discoveries". I recommend my readers to read first a step in research design and then go to the respective step in "Discoveries". Between the steps number 5 and 6 there are explanations regarding interconnectivity of each step in the part.

The following steps are done in the first stage of the analysis in order to collect information and organize it:

(1) The Lead Market Initiatives are cases to understand and conclude on the standardization "library". By "library" of standards I mean the degree of formal standardization in each of these official innovation policy areas. The areas may overlap with other market or public policy sectors and have a different variety of sub-areas. Thus, the term "library" is used as a concept regarding standards as mature knowledge, or "building blocks of infrastructures" [4], within innovative areas. At the official pages of the European Standards Bodies and other organizations authorized for standardization we may always find the required updated information on formal standardization and ongoing standardization process in relevant sub-areas of the Lead Market Initiatives. At this step I do exploratory research of the legal acts, policy guidelines and other formal documentation (such as Communications, White Papers, Working Documents, Directives regarding each Lead market Initiative, etc.).[1]

(2) I break down the areas of the Lead Market Initiatives into synergies of different market sectors/activities, which shows me what sectors each Lead Market Initiative is all about. The purpose of this step is to find out possible combinations for technologies and space for new possible policy development by clarifying sectoral contents of each Lead Market Initiative. Exploratory and comparative research based on conceptualization of the innovative technologies is done to map the synergies. The basic principles of the Lead Market Initiatives, mentioned in "Introduction" are taken as comparative criteria to verify the synergies.

(3) The taxonomy of standards in Information and Communication Technologies (ICT) by Paul David [3] and Ken Krechmer [6] is introduced with the purpose of guiding our suggestion regarding internal complexity of standards and hence their architecture: from reference standards to similarity, compatibility and etiquette standards. The aim is not to test the theory of standards, but apply it to the ongoing policy analysis.

(4) The ICT area is put in focus for the limitation of research as an innovation priority in the European Union. By the method of deduction we do exploratory research of the ICT role in the innovation policy and analyze documentary materials of areas synergic to ICT. As example, these areas may be found in "2010–2013 Standardisation Work Programme for industrial innovation", e.g. eHealth, eInclusion, eBusiness, eGovernment, ePublishing, etc. [16] Such materials are documentation of legal and non-legal power containing policy guidelines by the European institutions. eHealth is an ICT priority and a Lead Market Initiative, and is synergic to medicine and medical practices, healthcare, education, information systems and devices, etc.

[1] More available information on formal standardization in the Lead Market Initiatives may be found by opening up pages on respective Lead Market Initiative here: [18].

(5) ISO IEEE 11073 Personal Health Data is taken as example of a complex advanced standard belonging to the fourth category of the standards' taxonomy, i.e. the most advanced—etiquette standards. The architecture of ISO IEEE 11073 Personal Health Data and its synergies are analyzed for possible technological outcomes for innovation in the area of eHealth [2]. The methods of the Systems Theory are implemented in the sense that we determine possible elements and their constellations by the architecture of the object-oriented programming.

Below brief explanations for the interconnectivity will follow.

For the purpose of our analysis, it is important how many standards exist in an area and at what stage of the standardization process the standards find themselves, i.e. whether they are under standard-setting or re-setting. These are important qualities of the "library" of standards in an area (1), since even an initiative to formal standard-setting shows policy intentions by the European institutions. Such resources may also be standardization mandates or reports by the Commission. That helps figure out synergies and sub-synergies of the studied innovation policy areas by the method of deduction (2). Since standards as such are in focus, their embedded knowledge is classified in the taxonomy of standards by David [3] and Krechmer [6]. This taxonomy contains four types of standards from the easiest to the most advanced with space for innovation (3). As an example, we take the ISO IEEE 11073 Personal Health Data standard [2] in the context of the Ambient Assisted Living Programme [13] in the area of eInclusion within eHealth, which is one of the key areas in ICT for Europe (4). eHealth is one of the European Lead Market Initiatives as of year 2012 and has the highest level of organizational and regulatory complexity, which is encouraging for different scales and proportions of synergies. The most interesting aspect is that eHealth has the lowest degree of formal standardization within and no common standards as a Lead Market Initiative. That conclusion comes from the analysis in part (1).

Since we are interested in the internal complexity of standards in order to find out why standards are a "strategic asset" enabling innovation [11], we shall clarify in a particular example a complex standard's architecture to find out sub-systems responsible for particular functions and determine open spaces allowing potential synergies for improvement of the technology using the standard. In order to do that, we apply to the object-oriented programming useful for assessment of the chosen above ISO IEEE 11073 Personal Health Data, since it gives descriptions and classifications of the classes of objects in the system of this standard (5). Analytically we can suggest new patterns of technologies for innovation, based on combinations of synergies. Thus, we can assume patterns of behavior of stakeholders performing particular actions in making innovations. That research may be further developed in the analysis of any particular sphere of life expecting innovations of particular type—technological, administrative or procedural—to emerge. However, our focus is targeted mainly at technological innovations with their division into thematic spheres, such as ICT, environment, healthcare, etc. Which actors, why and how they may prioritize particular innovations there and what is the purpose of such behavior expected by the European institutions in the policy instruments of standardization will be our contribution to the advanced policy analysis of the European innovation policy context.

48.4.2 Part Two

At the second stage, i.e. the stage of analytic co-relations, we analyze the following:

(6) Co-relation of the results gained at part (1) with those gained in part (3) is done by projection of the documentary analysis to the standards' taxonomy. Thus, we correlate standards of the Lead Market Initiatives with types of standards in the theory of standardization. This is possible because of the description of standards in each area, analyzed in part (1).

(7) The theory of successful advanced standards by Fomin et al. [5] is introduced for the purpose of studying an "open space" of advanced standards. This is done as a contribution to part (3) for the purpose of widening the issue. Here we discover a base for the assumption that the behavior to innovate may be programmed. It may depend on the correlation of the result of part (7) with part (6). This part may be developed in the frame of another research with socio-technological bias: the results of part (7) may be used as a ground for modeling synergic, smart technological constellations by Small and Medium Economies (SME) in practice or work on projects on smart solutions (smart houses, smart cities, etc.)

(8) Projection of the results of the parts (1), (2), (5) and (6) to figure out possible responses to challenges by doing thematic innovations of technological, social and environmental character—for the purpose of the social market economy as a main policy of the European union, the single market policy—for the understanding of the division of stakeholders' roles there depending on the thematic synergy for innovation. These conclusions may clarify programming performance of stakeholders discovered in the part (7).

Explanations about connectivity follow here:

Let us comment what's been mentioned above. The gained legal analysis of the documents regarding standardization of the Lead Market Initiatives (1) is compared with the complexity of standards (6) described in the taxonomy of standards by David [3] and Krechmer [6] in part (3) to figure out synonymous classes in the taxonomy through their functional descriptions and nature. In the perspective of Systems Theory we are interested in the intensity of the actions taken by the stakeholders on the process of their implementation of standards of the spheres of official innovation areas, and hence—the standards of particular complexity, as it is established in the part (6). Here we come closer to the programming character of standards.

A complement to the theory of standardization is the theory of successful/advanced standards by Fomin et al. [5]. It is based on the role of negotiation, sense-making and design for standard-setting in the following sense: negotiation is information transfer among areas of the Lead Market Initiatives and their synergic areas (the latter are potentially highly possible by complex standards); sense-making is the result of the information transfer which may take form of matter-energy and lead to changes in the very architecture/system of standards, which influences the synergic pattern of the Lead Market Initiatives; design is a potential formation of new constellations of technologies as outcomes of interactions among technologies, stakeholders, ideas and other elements allowed by the highest complexity of advanced standards.

Our research implies multilevel analysis and contextual focus, which is useful for structuring the gained data and assumptions for testing by simulation as an advanced standardization policy analysis. We move from the Lead Market Initiatives to smaller technological areas (micro-context) where innovation may emerge notwithstanding its size and the status of the discoverer (including meso-contexts) up to the European level of the social market economy (macro-context). According to the Europe 2020 Strategy [10], the single market policy shall be transformed into a social market economy with smart, inclusive and sustainable growth by year 2020. The social market economy has at least three big areas of policies—a social (exercised mainly at local and national levels of the Member States), an economic (exercised mainly at the level of Member States, the European supranational level and the global level for competitiveness) and an environmental sphere (exercised at all levels in different proportions by stakeholders depending on their technological and economic potential) [9].

The next step is to look at synergies of areas depending on complexity of standards, the context of technological relations and the level of stakeholders' performance. Open spaces in advanced standards of particular combinations of technologies are expected to exist in a strategic way founded in the documents on official innovation areas. A potential for negotiations, sense-making and design may be developed there at an advanced level (8). Thus, in our research we determine diffusion of markets to predict what outcomes the new technologic areas may give to the European Union due to the course of normal social development. Synergies of standards in complexity allocate new technologies at all levels of the European policy system depending on the scale and number of the open spaces for innovation in standards.

48.5 Discoveries

What we discovered appears in all parts of the analysis from (1) to (8), which has already been briefly commented. Below are examples picked out from the whole analysis in accordance with the steps showed in the part "Research Design and Research Flow".

Step 1. To summarize, we find that the Communication "Towards an increased contribution from standardisation to innovation in Europe" [11] defines standardization as "a voluntary cooperation among industry, consumers, public authorities and other interested parties for the development of *technical specifications* based on consensus" (*italics* supplied). At the same time, we find that David defines standard as "a set of *technical specifications* adhered to by a producer, either tacitly or as a result of a formal agreement" (*italics* supplied) [3]. It is interesting to notice that Council Directive of 21 December 1989 89/686/EEC on the approximation of the laws relating to personal protective equipment (i.e. on Protective Textiles) defines harmonized standard as "a text containing *technical specifications*" (italics supplied), which makes us believe that we may use the taxonomy of standards by David [3] and Krechmer [6] for the legal and policy analysis of standardization in the European innovation policy. Based on the "library" of standards, i.e. the amount of formal standards in the Lead Market Initiatives, we allocate the latter in a "rainbow" from most formally standardized to least standardized: Protective Textiles, Recycling, Sustainable Construction, Renewable Energies, Bio-based Products and eHealth. This means that we move from predictability to unpredictability. By the next step we define synergies, which may take place in the scope of this unpredictability.

Step 2. eHealth is synergistic with ICT, public health infrastructure, medicine and economics and social inclusion; Sustainable Construction is synergic with social infrastructures, materials, energy and environmental issues; Bio-based Products are synergic with bio-systems, chemistry, physics, materials, energy and other environmental issues; Recycling is related to energy, chemistry, waste issues and regional policies; Renewable Energies is in synergy mainly with carbon technologies, biosphere, resource issues and are perspective for the development of smart city concept; whereas Protective Textiles are synergic mainly with physics and chemistry. Thus, eHealth is much more complex than Protective Textiles. By the next step we apply to the complexity of the very standards.

Step 3. As regards the taxonomy of standards, we have the following: reference Standards are the easiest regarding complexity; Standards for Minimal Admissible Attributes (or Similarity Standards) contain the allowed deviations from what is required; Standards for Interface Compatibility (Compatibility Standards) contain descriptive features of two or more items to interact and qualities of the lower standards [3]; Etiquette Standards are open-ended, i.e. they maintain communication among many items and contain spaces for compatibility with inventions [6]. This theory comes from ICT Studies.

Step 4. According to Horizon 2020, innovation priorities are ICT, nanotechnologies, advanced materials, biotechnology, advanced manufacturing and processing, space and multi-disciplinary approach [15], which may also be considered as an innovation policy instrument. This is Complexity Science and it may be found in the context of "Future and Emerging Technologies" (FET). The outputs of the FET-projects are formed into a bigger initiative called "Coordination and Support Actions" containing the best Complexity Science methodologies for practical policies improvements [14]. The users of our analysis may be found there or, for example, it may be public or private structures doing policy analysis by simulation programs. The complexity and the sectoral use of standards may be criteria for such work.

Step 5. According to the Fifth Call of the Ambient Assisted Living Programme, the implementation of standard ISO IEEE 11073 shall be explained in projects applications [13]. Due to its architecture, the standard is capable of synergies with neuroscience, cognition, self-expression (due to the Enumeration element), networks and different types of e-services (due to Agent-Manager interactions), smart compact devices (due to Medical Device System), etc. [2] Further implementation is possible for human-computer interaction in healthcare (plus social integration at meso and even macro levels) and the development of other e-services related to healthcare, etc. This is a fruitful environment for complex standards, where their complexity plays a role in terms of the space/gap in the standards' architectures to innovation through synergies and beyond.

Step 6. From the documents on standards of the Lead Market Initiatives the complexities of standards are approximately determined by a projection of the gained textual analysis to the standardization taxonomy by David [3] and Krechmer [6]. For example, standards of Protective Textiles are likely to be close to Similarity Standards (encouraging mainly high quality; recourses e.g. [7, 8]), standards of Bio-based Products—to Compatibility and Similarity Standards (encouraging mainly innovation; recourses e.g. [19, 20]), while standards of synergies in eHealth—to Etiquette Standards (directly oriented to innovation, whereas the very standards of eHealth are still in the process of formulation by CEN as of year 2012).

Step 7. The theory of successful/advanced standards [5] suggests three conceptual elements in the open space of standards encouraging innovation, namely negotiation, sense-making and design (see explanations to the part (7) in part two, "Research Design and Research Flow"). These categories may be further analyzed by use of the parameters of the contexts (micro, meso and macro) and the levels of interactions among stakeholders (see explanations to the second part of the "Research Design and Research Flow", paragraph 3). The Systems Theory is useful to work out categories of stakeholders, the nature of interaction and environments. The standards closer to etiquette standards are observed to be in less formally standardized Lead Market Initiatives with high technologies and complex technological challenges. Much negotiation is required for sense-making and design, and this leads to a high diffusion of sectors. ISO IEEE 11073 Personal Health Data as one example and standardization initiatives according to "2010–2013 ICT Standardisation Work Programme for industrial innovation" actively supports synergies for eHealth and eInclusion [16]. Common standards seem to be avoided, making stakeholders search for synergic solutions and be more responsible for making decisions in cooperation and networking. However, other responsibilities may also be kind of algorithmically "programmed" depending on the needed outcome from the stakeholders' performance.

Step 8. Guidelines or algorithms regarding the stakeholders' behavior becomes very evident in regard to the type of standard, its synergy with environment, the context and the level of the stakeholders main activity. The complexity of standards corresponds to the complexity of actions as outcome and hence, the degree of diffusion of market sectors. Even if a standard is only planned, the relations around its expected implementation may reflect is planned complexity, since standards contains guidelines or algorithms regarding its synergies. The more advanced a standard is, the higher the performance of the stakeholder, the deeper the diffusion of the synergic sectors and more alternatives allowed in the open space of standards, the higher the possibility for innovation, the higher economic and social expectation by European institutions from the stakeholder and finally the lower the European Union's responsibility is for innovation emergence and social integration stability. The less advanced a standard is, the lower the performance of the stakeholder, the lesser the sectoral diffusion, the lower innovation expectations, but the higher the product quality, and the higher the European Union's responsibility in stimulation of innovation by other means. This is the answer to the last research question.

48.6 Conclusion

Complex standards with high complexity and spaces for innovation may be called a "strategic asset" for innovation encouragement [11], since it has strategic "gaps" for introduction of new synergies and ideas into the technology using the standard. Standards of lower complexity have mainly other functions, such as high quality support. Standards used in ICT (as example of ISO IEEE 11073 Personal Health Data), especially in the context of FET, are likely to have an advanced architecture for synergies. Results of successful projects and their complexity may be further analyzed by methodologies in a cross-disciplinary way to find models of perspective responses for the social market economy and targets of the Europe 2020 Strategy. The contribution may be done for the development of the concept "intelligent policy" at example of standardization, which may be considered as such. The possibility to program innovation by synergic algorithms in advanced standards is a tool for smart diffusion of the single market. The trends are described in Step 8. Our analysis and approaches may be used for analysis of policies where standardization may take place through the method of simulation, but also for the modeling of scenarios in Complexity Science for policy analysis regarding the further development of the European policy system beyond 2020 through standardization and innovation policy technologies.

References

1. Blind K (2004) The economics of standards: theory, evidence, policy. Edward Elgar, Cheltenham Glos
2. Clarke M (2008) Developing a standard for personal health devices based on 11073. In: Andersen SK et al (eds) eHealth beyond the horizon—get it there. IOS Press, Amsterdam. http://person.hst.aau.dk/ska/MIE2008/ParalleSessions/PapersForDownloads/10.Sta/SHTI136-0717.pdf
3. David PA (1987) Some new standards for the economics of standardization in the information age. In: Dasgupta P, Stoneman P (eds) Economic policy and technological performance. Cambridge University Press, Cambridge
4. Fomin V (2003) The role of standards in the information infrastructure development, revisited. In: Standard marking: a critical research frontier for information systems, MISQ special issue workshop, international conference on information systems, Seattle, Washington, December 12–14, pp 302–313. http://www.google.ru/url?sa=t&rct=j&q=&esrc=s&frm=1&source=web&cd=3&ved=0CC4QFjAC&url=http%3A%2F%2Fciteseerx.ist.psu.edu%2Fviewdoc%2Fdownload%3Fdoi%3D10.1.1.4.253%26rep%3Drep1%26type%3Dpdf&ei=lVmBUO2OJMio4gS3n4CgDQ&usg=AFQjCNFnnoWJsgTK69zRgetAg6-TVJBX_g&cad=rjt

5. Fomin V, Keil T, Lyytinen K (2003) Theorizing about standardization: integrating fragments of process theory in light of telecommunication standardization wars. Sprouts: working papers on information systems. Press of Case Western Reserve, Cleveland. http://sprouts.aisnet.org/186/1/030102.pdf
6. Krechmer K (1999) Technical communications standards: new directions in innovation. International Center for Standards Research University of Colorado at Boulder (1999). Online article in MS-Word http://www.csrstds.com/siit.html
7. Council Directive 89/686/EEC on personal protective equipment. http://eur-lex.europa.eu/LexUriServ/LexUriServ.do?uri=CONSLEG:1989L0686:20031120:EN:PDF
8. Council Directive 89/656/EEC on the minimum health and safety requirements for the use by workers of personal protective equipment at the workplace. http://eur-lex.europa.eu/LexUriServ/LexUriServ.do?uri=CONSLEG:1989L0656:20070627:EN:PDF
9. Communication from the Commission (2010) Towards a single market act: for a highly competitive social market economy: 50 proposals for improving our work, business and exchanges with one another, COM 608 final. http://ec.europa.eu/internal_market/smact/docs/single-market-act_en.pdf
10. Communication from the Commission (2010) Europe 2020: a strategy for smart, sustainable and inclusive growth, COM 2020. http://ec.europa.eu/eu2020/pdf/COMPLET%20EN%20BARROSO%20%20%20007%20-%20Europe%202020%20-%20EN%20version.pdf
11. Communication from the Commission (2008) Towards an increased contribution from standardisation to innovation in Europe, COM 133 final. http://eur-lex.europa.eu/LexUriServ/LexUriServ.do?uri=COM:2008:0133:FIN:en:PDF
12. Communication from the Commission (2007) A lead market initiative for Europe, COM 860 final. http://eur-lex.europa.eu/LexUriServ/LexUriServ.do?uri=COM:2007:0860:FIN:en:PDF
13. Ambient Assisted Living Programme, ICT-based solutions for (self-)management of daily life activities of older adults at home. http://www.aal-europe.eu/get-involved/calls/call-5-daily-life-activities
14. Future and Emerging Technologies (FET), CORDIS, FP7, ICT. http://cordis.europa.eu/fp7/ict/programme/fet_en.html
15. Horizon 2020, Research and innovation. http://ec.europa.eu/research/horizon2020/index_en.cfm?pg=home&video=none
16. ICT Standardisation Work Programme for industrial innovation, 1st update—January 2011. European Commission, Enterprise for Industry Directorate-General. http://ec.europa.eu/enterprise/sectors/ict/files/ict-policies/2010-2013_ict_standardisation_work_programme_1st_update_en.pdf
17. European Commission, Enterprise and Industry, Industrial Innovation, Innovation policy. http://ec.europa.eu/enterprise/policies/innovation/policy
18. European Commission, Enterprise and Industry, Industrial Innovation, Demand-side policies. http://ec.europa.eu/enterprise/policies/innovation/policy/lead-market-initiative/index_en.htm
19. Mandate M/492 addressed to CEN, CENELEC, and ETSI for the development of horizontal European standards and other standardisation deliverables for bio-based products, 7th March 2011. ftp.cen.eu/CEN/Sectors/List/bio_basedproducts/M_492.pdf
20. Mandate M/429 addressed to CEN, CENELEC, and ETSI for the elaboration of a standardisation programme for bio-based products, 10th October 2008. ftp.cen.eu/CEN/Sectors/List/bio_basedproducts/M_429.pdf

Chapter 49
The Right to a Due Deliberation, Mental Models of Judicial Reasoning and Complex Systems

Enrique Cáceres Nieto

Abstract This work aims at proposing a theory on mental models of legal reasoning in Roman Law Tradition. The essay integrates the approaches of legal epistemology, complex sciences, corporate governance, neural networks and mental models in a coherent conceptual theory whose goals are: to explain the way in which legal institutions, and particularly the judicial ones, participate in the construction of social reality (including corruption problems) and highlight some ideas about strategies to intervene in its correct functioning. It is an important step in developing a constructivist legal theory.

Keywords Legal constructivism · law and complexity · Legal epistemology · Legal theory · Legal philosophy · Mental models · Cognitive sciences

49.1 Introduction

One of the most relevant issues in contemporary legal theory is to define the conditions that determine the respect to the human right to a due deliberation by the judges.

Nowadays responses come from different theoretical traditions: analytical jurisprudence, applied legal epistemology and artificial intelligence applied to law. Unfortunately, they share the following problems: (1) they are product of mere speculations and not the result of empirical research; (2) assume an individualistic perspective of judicial reasoning.

In my ongoing research I assume a different point of departure and goal. I assume that: (1) It is necessary a descriptive model of judicial reasoning before a normative one; (2) The mental models theory developed by cognitive psychology provides a good framework to face the problems of knowledge elicitation and representation of judicial reasoning presupposed by the descriptive model recounted above; (3) In real world mental models of judicial reasoning emerge from self-organizing processes

E. Cáceres Nieto (✉)
Legal Research Institute and C3-Science Complexity Center, National Autonomous University of México, Mexico City, Mexico
e-mail: encacer@hotmail.com

T. Gilbert et al. (eds.), *Proceedings of the European Conference on Complex Systems 2012*, Springer Proceedings in Complexity, DOI 10.1007/978-3-319-00395-5_49,

which take place as the result of the connexions and interdependence of judges and lawyers in courts and can became dysfunctional (as happens in the case of corruption or negligence) (4) Therefore the understanding of judicial reasoning requires to be studied from the approach of complex sciences and the advances of corporate governance. The goal will be to develop a legal epistemic engineering that help judges to respect the right to a due deliberation.

In my research, I justify the use of neural networks as a theoretical metaphor to be further developed the mental model that I propose. I have placed legal reasoning in the broadest context corresponding to the emerging epistemological paradigms of situated cognition and extended mind. I maintain that legal institutions are self-organizing social networks that comprise the environment from which arise presupposed mental models in judicial judgments. In this sense, mental models of judicial reasoning are found in the dimension of situated cognition and the extended mind.

Along with Francisco Varela I assume that scientific knowledge goes from a cloudy to a crystal state.

As we all know Complex Systems Theory has strongly impacted on different scientific domains with different degrees of crystallization. Its influence has not been the same in physics or biology as in social science. One field in which the complexity is virtually absent is legal theory.

My main goal in this talk is to propose some advances of my efforts to integrate complexity and cognitive sciences in a new kind of legal theory that I proposed to denote with the expression: 'Legal Constructivism'. Even if this is much closer to the cloudy than to the crystal state, I think it could be considered the first step to a future integration of the theory of complexity in the legal theory domain.

The type of mental model of judicial reasoning in criminal law that I propose was obtained by the expansion of the explanatory power of a theoretical framework developed in order to construct an expert system to assist in reaching judicial judgments in family law.

The qualitative research techniques used to stimulate judges' knowledge were the following: (1) unstructured interviews; (2) shadowing self; (3) interruption analysis and (4) mind maps.

The learning curve implies different changes in the network structure along a diachronic axis, which represents legal proceedings. This involves dynamics of states that corresponds to the activation of connectivity patters in the subject's memory on coming in contact with the medium (the case) and the dynamics of parameters derived from the modifications to the values assigned to the different concepts throughout the process.

Traditional legal thought is linear and consequentialist: it assumes implication relationships between what legal principles prescribe and the social reality that arises. However, this reality is very often completely different, if not contradictory, to the ultimate aims of the law.

Replacing linear thinking with complex thinking can offer new ways of understanding the impact law can have on building social reality. Specifically, the theory of mental models show great potential in defining teaching strategies based on complex learning and situated learning. Likewise, mental models empower judges' extended mind as a result of the computational formalization of these models by means of expert systems that assist in legal judgment.

49.2 Ubis Societas Ubi Ius

This very old Latin expression means that wherever there is a human society there is Law. It explicitly refers to the main assumption: Legal norms regulate social behavior.

However, the reality frequently shows that legal norms do not have any impact on the change of social dynamics, and that they even produce opposite results to their explicit goals.

For instance, the explicit goal of prisons is the social rehabilitation of criminals, but it is a widespread true that they usually function as schools of crime.

As in the case of prisons, other legal institutions present an amazing phenomenon: they show the same dynamics even if they come from different legal systems, countries and cultures. Traditional legal theory cannot respond to this question.

49.3 The Role of Cognitive Science in Legal Constructivism: Situated Cognition and Extended Mind [7]

The theory of situated cognition, holds that cognition involves not just the brain, but also an individual's physical self: nervous, endocrine, musculoskeletal and other systems. In other words, cognition is embodied. To a large extent, these processes take place without an individual's conscious control, but in a self-organizing manner due to the interaction with the environment.

The theory of extended mind holds that cognition is not restricted to an individual's biological "packaging" but can extend to natural elements, as well as technological and cultural ones (placing law in this last category).

In this paper, I maintain that legal institutions are self-organizing social networks that comprise the environment from which presupposed mental models arise. In this sense, mental models of judicial reasoning are found in the dimension of situated cognition and the extended mind.

49.4 The Role of Complexity: The Lucifer Effect

Social psychologist Philip Zimbardo is best known as the father of the 1971 Stanford Prison Experiment. He used a mock prison populated with student volunteers to illustrate the extent to which identity is situated within a social setting; student volunteers randomly chosen to play guards became cruel, while those playing inmates became rebellious and depressed.

His troubling finding was that almost anyone, given the right "situational" influences, can be induced to abandon moral scruples and cooperate in violence and oppression. In other words Zimbardo provides an important experimental support for the situated cognition thesis and demonstrates that certain variables can induce self organizing processes in legal institutions from which new social dynamics emerge that include the transformation of the embodied cognition of human agents without conscious control.

One of the most amazing results of Zimbardo's research was to find the same kind of variables introduced in his university experiment when he studied the transformations occurred in real institutions as the Abu Ghraib prison.

The variables in the Lucifer Effect experiment were:

- Power
- Conformity and obedience
- Deindividuation
- Dehumanization
- Evil for inactivity
- Banality of evil

49.5 Lucifer Effect in Criminal Justice in Mexico

Corruption and inefficiency are the main manifestation of the Lucifer effect in many Mexican legal institutions, including criminal courts. The situation is so bad that even if the constitution establishes the presumption of innocence as the main principle in criminal trials, real practices function in the opposite way: there is a presumption of guilt which is very difficult to defeat.

"In general terms one can affirm that the probability that a person who has been recorded before the district attorney to be absolved is minimal. The judicial statistics published every year by INEGI situate in about 90 % the percentage of condemnatory judgments in the whole country" [6].

It is amazing to realize that:

> "...the description of the criminal processes revealed by the research seems not to be exclusive of Mexico".

In the Colombian case, it has been observed that:

> "The Judge's task is limited to confirm the decision made by the district attorney and to impose the punishment (see [6, p. 68])".

Even if there are no comparative empirical researches that let us to know what happened in all the criminal courts in Latin America, informal comments in international legal congresses let us suppose that all of them function in a similar way.

49.6 The Recipe for Evil in Criminal Courts (Hypothesis 1)

I assume that the recipe for evil in criminal courts is:

1. Lack of an acceptable epistemic normativity in order to determine when the standard of proof accepted by the legal system has been satisfied and therefore somebody must be declared innocent or guilty.
2. Lack of transparency of the cognitive process that occurs in the black box of the judges minds.
3. Invulnerability to corrective interactions within the systemic environment (civil society).

49.7 From The Lucifer Effect to the "Saint George" Effect (Hypothesis 2)

The main hypothesis of this talk is that if it is possible to induce the emergence of a certain kind of dynamics with the variables of the receipt of the evil (The Lucifer Effect), then, it must be possible to induce a virtual dynamics in an institutional system if other variables are introduced. I propose to refer to this change as The Saint George Effect. In other words: "The saint George Effect" seeks to be the antidote of The Lucifer Effect.

The receipt of The Saint George Effect.

1. Epistemic normativity to determine when the standard of proof accepted by the legal system has been or not achieved and therefore to decide if somebody must be declared innocent or guilty.
2. This is the most important variable because it makes possible the rest of the variables of the receipt for the Saint George Effect.
3. To have epistemic artefacts that make transparent the cognitive processes followed by judges when they are making legal decisions.
4. Inter-subjective control of the cognitive processes of judges by the systemic environment (civil society).

Following on I will focus my attention in the two first variables.

49.8 Properties of the Model of Epistemic Normativity of Judicial Reasoning (The First Variable)

49.8.1 Naturalization

One of the main problems in contemporary legal theory is the construction of suitable models of epistemic normativity that determine the cognitive processes that can justify the declaration of innocence or guilt in a trial.

Up to the moment, several models have been proposed based on different theoretical frameworks: frequentist probability, bayesian induction, deductive logic, etc. Nevertheless all of them suffer from a lack or naturalized base, that is to say, they have not taken into consideration the way in which judges processes the evidence when they make decisions.

The model that I propose seeks to replace this deficiency using the theory of mental models developed by contemporary cognitive science and a theoretical metaphor based on the dynamics of artificial neural networks.

49.8.2 Epistemic Artefacts

A biological property of different species consists of adapting their environment in order to satisfy their needs.

The man is not only a producer of physical artefacts, but also of what has been called "cultural artefacts" or "epistemic artefacts" which are the result of exteriorizing mental representations. For instance: the cave paintings, the maps, the mathematical equations, the language, the Law.

Once epistemic artefacts have been exteriorized, they go on the public dimension where they act as scaffolds on which other members of the community can generate new epistemic artefacts by adaptive processes. Hereby, the aforementioned artefacts exceed the mere mental dimension to turn into guides for the action, as happens with the planes of an architectural project, or the performance of a symphony. Certainly, they also make possible the coordination of collective actions, their evaluation and their intersubjective control.

In spite of possessing a symbolic equivalent content, some manifestations of epistemic artefacts are more efficient than others. For this reason we prefer reports with histograms to those which have only text, and we prefer maps to oral explanations about how to get somewhere.

The model (epistemic artifact) that I propose takes into consideration the preponderance of visual representations and will be presented in a graphical way.

49.9 Mental Models of Judicial Reasoning as Epistemic Artefacts

Mental models are:

> "… declarative representations of how the world is organized and may contain both general, abstract knowledge and concrete cases that exemplify this knowledge. So, strong models allow for both abstract and case-based reasoning [9]."

The purpose of this work is to present a representation of the mental models judges use in their reasoning, in which explicit knowledge corresponds to their training in law school and implicit knowledge, to their interactions in court [4].

49.10 Cognitive Self-organization and Legal Reasoning

The researcher Dan Simon [8] has carried out in the field of experimental psychology and my own in artificial intelligence [2] present very similar results in terms of the cognitive self-organization that takes place in both juries (the Common Law system) and professional judges (the Romano-Germanic system). The most relevant characteristics are: (1) the activation and adaptability of previous cognitive patterns to solve new cases; (2) unconscious coherentist displacement to organize apparently unconnected information to, for instance, link different kinds of evidence; (3) on-going changes in the state of the mental model throughout the process, according to the arguments and counterarguments, and proof and counterproof presented by the parties; (4) the interdependent way value is assigned to the different pieces of evidence; (5) the influence of elements that are secondary to the case (as may occur with racial prejudice) in changing the state of the system; and (6) the fact that the same case can produce different mental models despite applying the same law.

49.11 Mental Models, Elicitation and Knowledge Representation [4]

The type of mental model (epistemic artefact) presented was obtained while developing an expert system to assist in reaching judicial judgments.

The qualitative research techniques used to stimulate judges' knowledge were the following: (1) unstructured interviews; (2) shadowing self; (3) interruption analysis and (4) mind maps.

49.12 Architecture and Dynamics of the Mental Model of Judicial Reasoning [4]

The final model was a heterogeneous complex network whose various elements fall under the following categories: (1) descriptive statements of the facts in question; (2) regulatory statements set forth by law; (3) evidence; (4) concepts of the general theory of crime; and (5) the final ruling.

The interconnection between these components was modelled as a neural network with different layers for each of the mentioned elements. The structure of this multi-layered network is as follows: (1) an entrance layer that corresponds to the legally relevant terms in the description/portrayal/account of the case; (2) and exit layer consisting of two activating neurons that are binarily exclusive; (3) a hidden layer hierarchically higher than the layer of evidence that corresponds to the defining characteristics of a criminal nature; (4) a hidden layer hierarchically higher than that of the criminal layer that corresponds to concepts of exculpatory circumstances consistent with the type of crime in question; and (5) a hidden layer hierarchically higher to that regarding the exculpatory circumstances that correspond to concepts of the general theory of crime.

The connections are bidirectional, inter-layer, intra-layer and reiterative. The connectivity density is high and operates between all the layers.

The dynamics of the system assumes that the activation of the different neurons works according to the level of reaching the threshold of the active state that corresponds to a given epistemic value (believing that "p", considering "p" proven, etc.). It is supposed that the intensity of neuronal connections does not work deterministically, but fuzzily. In the model, three degrees of intensity are represented by different colours: low (yellow), medium (orange) and red (equivalent to reaching the threshold and the corresponding activation of the neuron or neurons). Synaptic connections can be excitatory or inhibitory [5]. This characteristic is very important since legal reasoning is dialogical and defeasible. In other words, it implies the opposition between the various components of the network whose processing (to a large extent cognitively self-organizing) should arise from a coherent structure that corresponds to the final decision of the sentence.

49.12.1 The Hattori Case [1]

Hattori—a 17 year old Japanese student that was a guest at family Haymaker's home—and Haymaker Junior (also 17 years old) were invited to a Halloween party organized by their senior high classmates.

On the afternoon of Saturday October the 17th of 1992 the 2 teenagers headed for Baton Rouge (where the party was going to take place). Hattori was characterized as "John Travolta". For his part Haymaker pretended he had an injured arm and wounds in his face (This was all "make believe").

They hadn't realized until it was too late, that they had the wrong address. So they ended up at Peairs' family's home thinking that the party was to be celebrated there.

Peairs' family members were Rodney (the husband), Bonnie (the wife), and their daughters Brittany and Stacey of 11 and 7 years old respectively.

Haymaker Junior rang the doorbell. When Bonnie opened the door, she could only see Haymaker, but when she turned her head and saw Hattori, with an expression of intense fear she suddenly shot the door on the boys' faces without letting them to express the reason that had brought them there. She ask come to Rodney because somebody want to enter in the house. He takes a gun, opens the door and

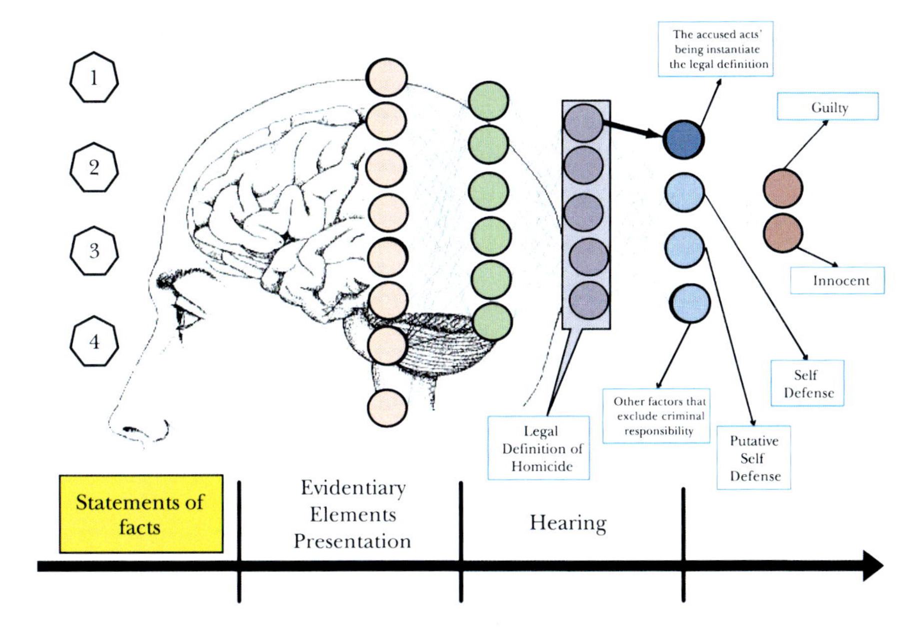

Fig. 49.1 Structure of the defining homicide layer [3, p. 244]

cry the expression Freeze but Hattori continues walking. Rodney shoots and kills the boy.

Each party declares the following:

The legally relevant elements of the reconstructed narratives that activate the input layer are the following:

Relevant Discursive Elements

1. The accused shoots and causes the Hattori's death.
2. Hattori had trespassed private property,
3. With a menacing aspect
4. And didn't pay attention to the discouraging instruction.

As a result of the activation of the input layer (legally relevant facts), a connectivity pattern gets activated between the input layer and the layers corresponding to the legal definition of the crime in question (possible homicide), and to the factors that exclude criminal responsibility layer (possible self defense) related to legally relevant facts 2–4. The intensity of the connection gets a low plausibility value.

The legal definition of homicide instantiates the theoretical concept of a crime, so, the respective neuron gets activated at the corresponding layer (see Figs. 49.1, 49.2 and 49.3).

Figures 49.4 and 49.5 display the inputs offered during the probatory stage.

These evidentiary items activate the second hidden layer. They correspond to the confession and the two expert testimonies.

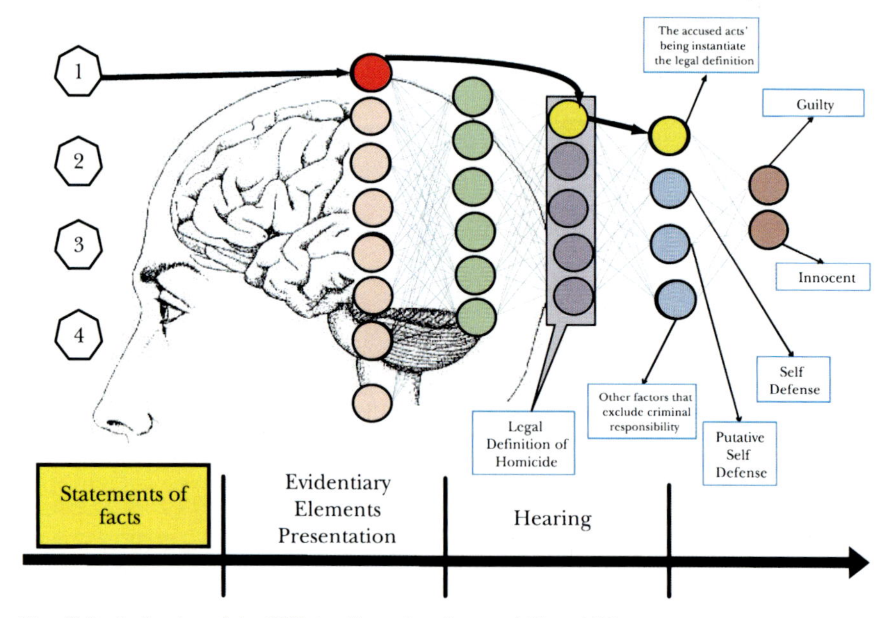

Fig. 49.2 Activation of the DHL by discursive element 1 [3, p. 245]

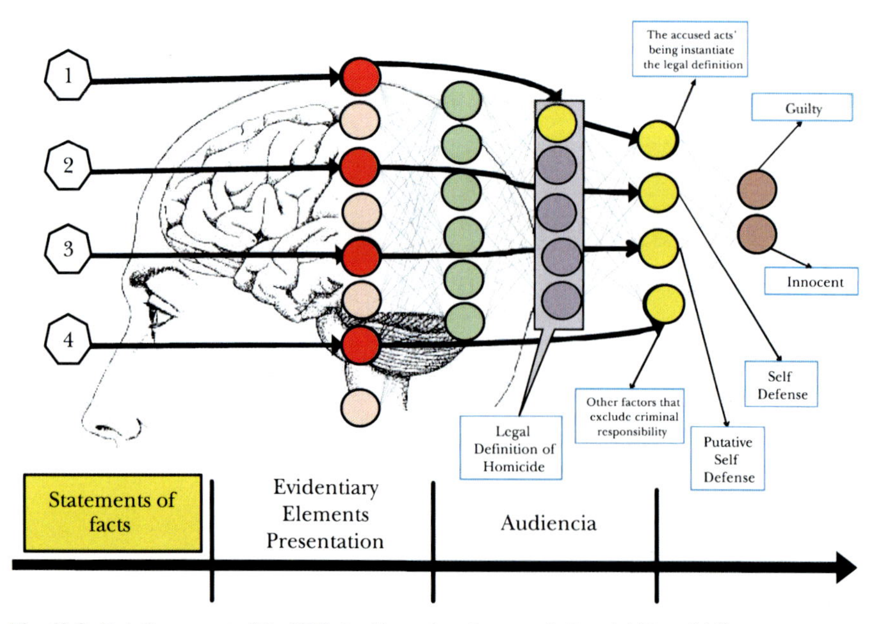

Fig. 49.3 Reinforcement of the DHL by discursive elements 2, 3 and 4 [3, p. 245]

The reinforcement of the evidentiary items (excitative weight) increases the intensity of the connection in the second hidden layer; the corresponding thresh-

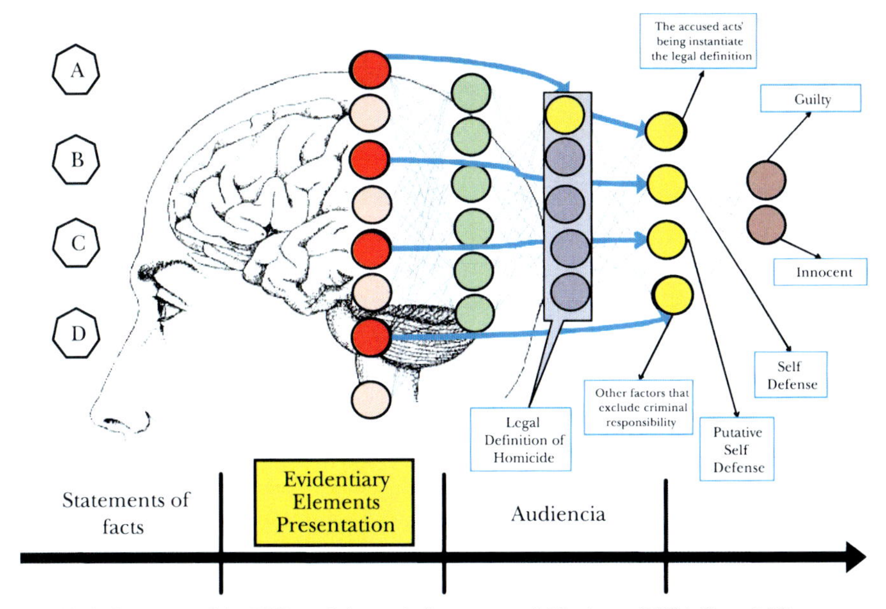

Fig. 49.4 Structure of the DHL and the excluding responsibility layer (ERL) [3, p. 246]

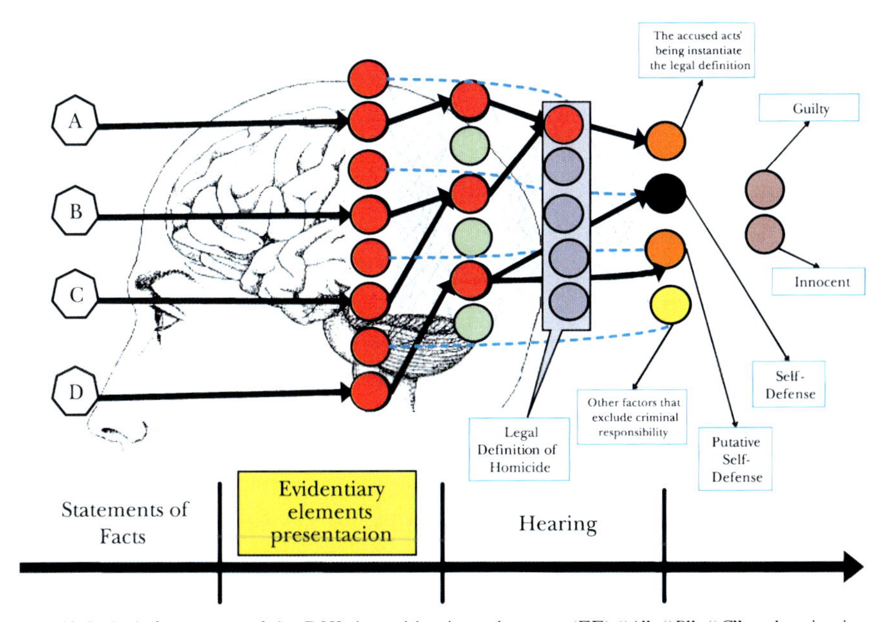

Fig. 49.5 Reinforcement of the DHL by evidentiary elements (EE) "*A*", "*B*", "*C*" and activation of the ERL by EE "*D*" [3, p. 246]

old is satisfied; and the neuron corresponding to the justified belief that Rodney committed homicide, gets activated. Nonetheless, it must be pointed out that this belief still does not satisfies the required threshold in order to activate the layer that corresponds to the general theory of criminal deeds, due to the fact that other neurons corresponding to the factors that exclude criminal responsibility such as self defense are still activated, even though their weight may be "low" (inhibiting).

But we should keep in mind that there is also an evidentiary item D which corresponds to Hattori's friend. The content of this testimony indicates first, that the victim was a Japanese student (which neutralizes the hypothesis of Hattori being dangerous); second, that he did not master the language (which explains why he kept walking toward Rodney despite the warning expressed in the form of "freeze"); and third it indicates that they had the wrong address (which explains why Hattori had entered Peairs' property with such confidence).

These elements create an inter-layer connectivity pattern towards the neuron corresponding to self defense. The pattern has an inhibitory weight, which is why the belief in a possible objective self defense gets deactivated.

Nonetheless the layer corresponding to the factors that exclude criminal responsibility remains active with a low plausibility value, due to a possible subjective self defense (putative self defense). The putative self defense implies that the homicide had been the result of Bonnie and Rodney experiencing an undefeatable subjective belief that they were facing a real threat with a plausibility value of "Medium".

But there are two other evidentiary items:

"E" is a testimony whose content indicates that Hattori did not walked towards the Peairs in a threatening way but more like in a joyful fashion. It also indicates that Hattori's characterization as John Travolta would not lead someone to think that he was dangerous.

"F" is another testimony whose content indicates that Hattori had only given two steps towards Bonnie and Rodney when he got shot (see Figs. 49.6 and 49.7).

The neurons activated by these elements generate a connectivity pattern with inhibitory weight with respect to the neuron that corresponds to the factor that excludes criminal responsibility of subjective self defense. This last neuron of subjective self defense along with the layer of all the factors that exclude criminal responsibility gets deactivated. In this state, an inhibitory weight gets transferred to the layer that corresponds to unjustified crime. With this transference of inhibitory weight the threshold to change to an active state is reached. This state generates a last and new connectivity pattern with the output layer that activates the neuron that corresponds to "criminally responsible". The latter state activates the effectors of the system and its corresponding motor behavior which in this case amounts to pronouncing a performative declaration which has the effect of changing the state of affairs of the operational environment of the legal operator. This has the consequence of a new legal reality being constructed.

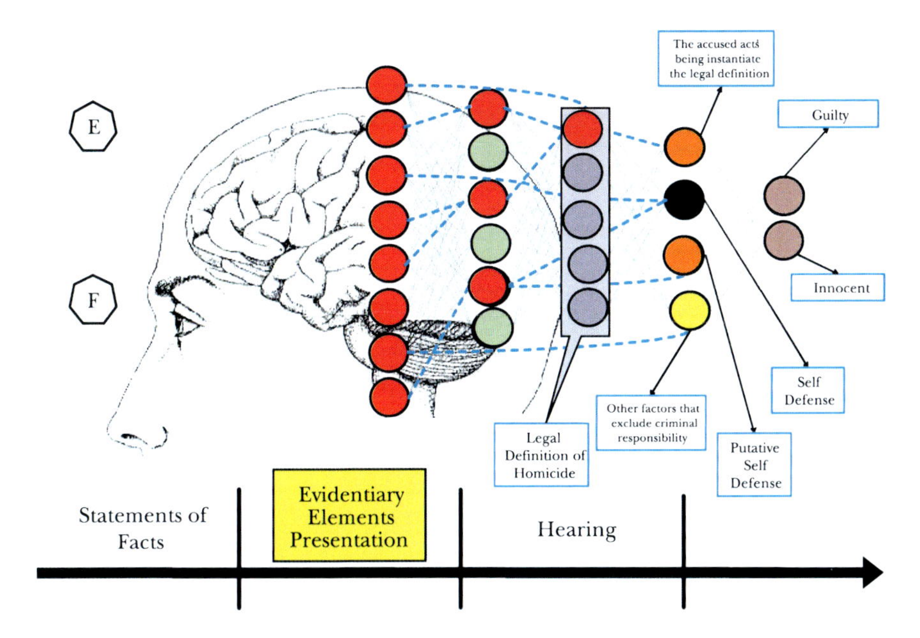

Fig. 49.6 Activation of ERL with plausibility value of "medium" [3, p. 248]

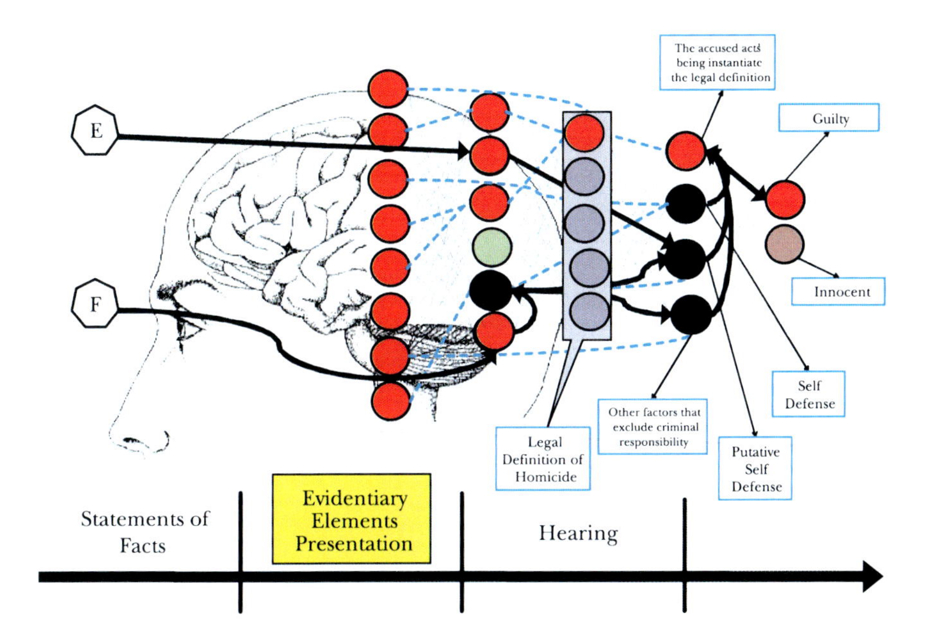

Fig. 49.7 Deactivation of ERL by EE "*E*" and "*F*", and final state of the system (Guilty) [3, p. 248]

49.13 The Saint George Effect Revisited

As it was previously claimed, mental models are epistemic artefacts.

They are the must important ingredient of the Saint George effect recipe because they are the support of the other ones:

1. The graphic representation of mental models transforms the black boxes of judges in transparent ones.
2. The public character of mental models make them accessible to the systemic environment (Law schools, research institutes, NGO[s, layers [s bars, etc) and therefore also make possible the intersubjective control of judges-s cognitive processes by sharing the same epistemic normativity.
3. Mental models show great potential in defining teaching strategies based on complex learning and situated learning.
4. Likewise, mental models empower judges' extended mind as a result of the computational formalization of these models by means of expert systems that assist in legal judgment.

49.14 Final Remark

Traditional legal thought is linear and consequentialist: it assumes implication relationships between what legal principles prescribe and the social reality that arises. However, this reality is very often completely different, if not contradictory, to the ultimate aims of the law.

Replacing legal thinking with complex thinking can offer new ways of understanding the impact law can have on building social reality.

References

1. Cáceres E (2010) Steps towards a constructivist and coherentist theory of judicial reasoning in civil law tradition, law and neuroscience. Oxford University Press, London
2. Cáceres E (2008) EXPERTIUS: a Mexican judicial decision-support system in the field of family law. Proceedings of the 21st international conference on legal knowledge and information systems, JURIX
3. Cáceres E (2009) Pasos hacia una teoría constructivista y conexionista del razonamiento judicial en la tradición del derecho romano-germánico. Problema Anuario de Teoría y Filosofía del Derecho, México, número 3
4. Cáceres E (2010) Judicial reasoning, mental models and complex networks. Advances in analysis and decision-making for complex and uncertain systems, The International Institute for Advanced Studies in Systems Research and Cybernetics.
5. Crespo A (2002) Cognición humana, Spain, Editorial Centro de Estudios Ramón Areces, SA, p 125
6. Pásara L (2006) Cómo sentencian los jueces en el distrito federal en materia penal, Instituto de Investigaciones Jurídicas, UNAM, México, p 50

7. Robbins A (2008) Short primer of the situated cognition. The Cambridge handbook of the situated cognition. Cambridge University Press, Cambridge
8. Simon D (2004) A third view of the black box: cognitive coherence in legal decision making. The University of Chicago Law Review
9. Van Merriënboer JJG et al (2002) Blueprints for complex learning: the 4C/ID-model Educ Technol Res Dev 50(2):48

Chapter 50
MOSIPS Agent-Based Model for Predicting and Simulating the Impact of Public Policies on SMEs

Federico Pablo-Martí, Antonio García-Tabuenca, María Teresa Gallo, Juan Luis Santos, María Teresa del Val, and Tomás Mancha

Abstract This paper presents MOSIPS (MOdel of Simulation of Impacts of Public Policies on SMEs), a multi-agent model of an open economy. It is part of a user-friendly object-oriented interactive intelligent policy simulation system allowing forecasting and visualizing the socio-economic potential impact of public policies for supporting SMEs. MOSIPS model specifies the characteristics of every agent, and its particular feature is that it locates accurately agents in the space, using this information to precisely determine the interaction network. It represents the dynamic behaviour of people and enterprises, analysing their decisions and interactions in the social networks. It can be used to model macro-economic features of a system and permits focusing on a specific part of the economy, both at sector and spatial level.

Keywords Agent-based model · Prediction and simulation · Policy evaluation

50.1 Introduction

Recently, agent-based models (ABM) have increased their importance in economics. In particular, the last financial crisis was not predicted by standard macroeconomic models. Due to several of their assumptions, they were not able to represent that significant deviation from the equilibrium growth path predicted. In contrast, if the approach is bottom-up, starting with the specification of the agents involved in the economy, it appears and emergent behaviour of the system which cannot be explained from the behaviour of the representative agent. This allows the appearance of bubbles, followed by a sharp reduction in prices and a lowering in expectations. Multi-agent models have been used to study economic systems in several ways: we can find examples of conceptual works on agent-based economic models [22] a variety of agent-based models focused on a part of the economy, for example, leverage effects in financial markets [10]. Multi-agent models of the economy as a whole are infrequent, examples are the models by [9, 12, 14] and the EURACE model [7].

MOSIPS model includes a number of features of the previous referred models, but it represents the economy making the emphasis in Small and Medium Enter-

F. Pablo-Martí (✉) · A. García-Tabuenca · M.T. Gallo · J.L. Santos · M.T. del Val · T. Mancha
Institute of Economic and Social Analysis—IAES, University of Alcala, Madrid, Spain

T. Gilbert et al. (eds.), *Proceedings of the European Conference on Complex Systems 2012*, Springer Proceedings in Complexity, DOI 10.1007/978-3-319-00395-5_50,

prises (SMEs) and the factors faced in their creation and growth. Thus, entrepreneurship and access to finance appears as two major issues. These enterprises choose their location not only optimizing the place, but taking into account the residence of the owner. This characteristic makes necessary to locate the entrepreneurs. Moreover, the demand SMEs face is determined by their location and their size, which conditions their visibility. SMEs' suppliers, workers and consumers tend to be near their location. Then, it is crucial to locate every agent in its real place to allow determining realistic interaction networks and the correct performance of every firm. For our purposes, in order to truthfully represent a local, regional or national economy, both firms and people should be placed with their individual characteristics. In addition, public administration, financial sector and the external sector are represented as well, as they interact with SMEs establishing policies, giving access to finance, competing with them or allow selling part of their production abroad.

What is described here is the abstract model. The description follows partially the agent-based model documentation guidelines developed at the 100th Dahlem Conference "New Approaches in Economics after the Financial Crisis" [23]. As specified by these guidelines, the first part provides an overview (Sect. 50.2) and the second one explains general concepts underlying the model's design (Sect. 50.3). Section 50.4 provides the specification of data needing and its treatment, as it requires a much more complex process than the majority of macroeconomic models. Finally, a brief conclusion looks at the further developments of the model and summarizes its major characteristics.

50.2 Overview

The rationale behind MOSIPS model (Fig. 50.1) is closely related to the purpose of the project for which is designed: to develop a policy simulation system allowing forecasting and visualizing the socio-economic potential impact of public policies for supporting SMEs.

50.2.1 Rationale

MOSIPS model represents the dynamics of behaviour and decisions of agents, and their interactions. It forecasts the evolution of an economic system over a time horizon of one quarter to several years. It is based on a multi-agent approach at the micro-economic level. It can be used to model macro-economic features of a system and allow focusing in a specific part of the economy, at sector and spatial level, evaluating the effects of a policy over the firms and the individuals, depending on their initial characteristics.

MOSIPS model provides the framework to test the accuracy of micro-foundations specified outside the scope of the representative agent paradigm reproducing a virtual reality to evaluate the effects of economic policy. The obtained results have a range of error due to the randomness of individual processes and the building of

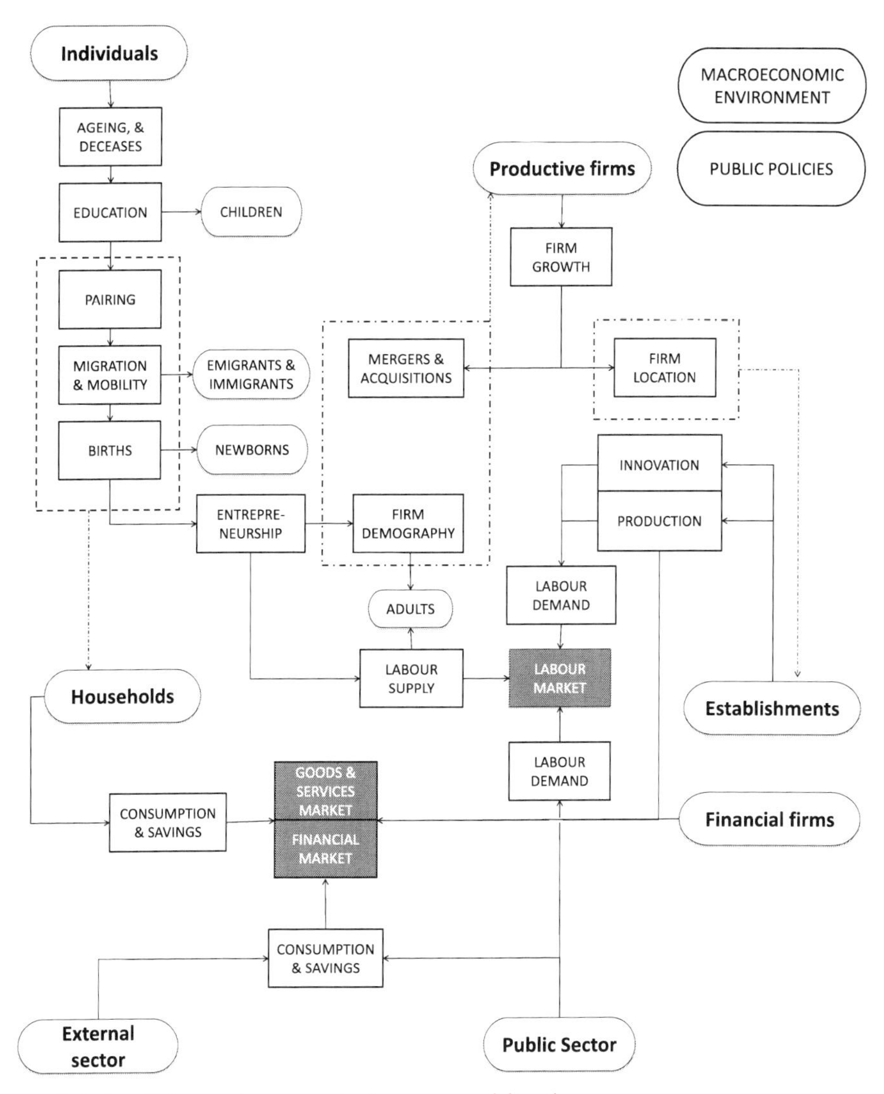

Fig. 50.1 MOSIPS model components. Source: own elaboration

the database. This approach can be seen as an extension to Arrow-Debreu general equilibrium theory [4], as the unique result computed by standard models is one of the possible outcomes: the optimal trajectory excluding part of the heterogeneity of the agents, and not having into account spatial issues with a sufficient degree of accuracy.

The object of study is a local economic system, disaggregated by branches, which represents the economic system under consideration. Then, the effects of a policy are studied both at sector and at spatial level. These effects also can be observed according to other characteristics such as the size of the firms, their innovative behaviour or their financial situation. The effects of policies can also affect to the population, and they can also be studied taking into account their location or the individual characteristics.

50.2.2 Agents

The model includes two main kinds of agents: firms and households. These agents, by means of commercial or social mechanisms, interact among themselves and with other entities within their environment but which are external to the agents' identity and decisions.

MOSIPS concerns particularly these two types of agents and their scope from the moment that a part of individuals chooses to become entrepreneurs with the intention to start up a firm. While such a choice is clearly influenced by the decisions they have made previously [2, 16, 20] (e.g., their type of education, family influence, or acceptance of a firm as inheritance), as well as by the environment created by other entities or markets [3] (i.e., government regulations or rate of interest required on a loan to an entrepreneur), the analysis underscores the behaviours and decisions of 'entrepreneurs-firms' agents.

Partnerships among firms are also possible, so that sometimes groups or conglomerates of firms aiming at achieving common goals (for innovation or export activities for example) can be formed. Therefore, in the proposed model, groups of agents are or may be relevant [17, 21].

In addition, each 'firm' agent can be considered from the perspective of the 'individuals' agents comprising such firm, that is, from the different degrees of responsibility and the ability to make decisions that individuals belonging to a firm have. In this way, one can include workers, technicians, managers, directors, and owners. According to this view, which rests upon the theories on human resources and knowledge management as well as the agency problem or theory, each individual who is part of the firm takes initiatives based on simple, or sometimes complex, management options, which may even be opportunistic or contradictory to the objectives of the firm or the interests of the owner [5]. These behaviours, dealt with individually (each member of the firm is an individual), would lead to a deepening or specialisation of the proposed model, which eventually would bring new ideas and approaches on firm development and growth in the territory analysed (with the information and data warehouse used), as well as any possible imbalances.

Every agent has the characteristics pointed by [24]. They decide their characteristics autonomously trying to maximize their expected profits/utility. The communication among agents takes place by market prices and social networks. Agents react to the changes in their environment, but sometimes anticipate these changes in order to define their decisions, showing a proactive behaviour.

50.2.3 Other Entities

In addition to these two basic types of agents, there appear other complementary entities for their activities involved to a higher or lesser extent in the modelling process but which are pivotal in the composition of agents. These entities do not make decisions directly in the process, but the evolution of their behaviours in time clearly impacts on the creation of the expectations and decisions of firms (and individuals). Specifically, these entities are the public sector, the financial system, and the local environment.

50.2.4 Boundaries

The model faces a major constraint in its development: the existence and behaviour of the agents which comprise the external sector. Agents and institutions that compose it (companies, public administrations, households...) incorporate permanent or sequential new inputs to the model in its running time, interacting with businesses in the analysed territory. For example, in trade relations between import and export companies, decisions are produced on both sides, in the country and abroad. Often, in business practice are taken decisions on capital investment out of the country. However, such decisions are not fully considered in the model. In turn, these decisions are based on the behaviour of other agents or external entities that have their own behaviours and different rules (for example, labour legislation, taxation of companies, the price of industrial land, or the difference productivity in the tradable good or service).

These external effects modify the behaviour of agents in the area to be analysed and, therefore, may be measured at least indirectly or by a method of approximation. Then, all the consumers of the exported goods and services, and firms that produce the imports are considered as several entities (one for each economic region), and their behaviour is only predicted in an aggregate way making use of several macroeconomic indicators. For the purposes of international trade, the model uses international exchange rates, which express fairly accurately the strengths and weaknesses of different economies, or what is the same level of competitiveness of domestic and foreign companies. Likewise, in the case of investment in the country by foreign companies, there are several difficulties in the determination of the behaviour of foreign agents not individually and explicitly modelled.

50.2.5 Relations

The types of relationships that structure the interactions among agents are of a diverse nature (Fig. 50.2). These are developed based on the various activities conducted by entrepreneurs and firms in their processes of recruitment and procure-

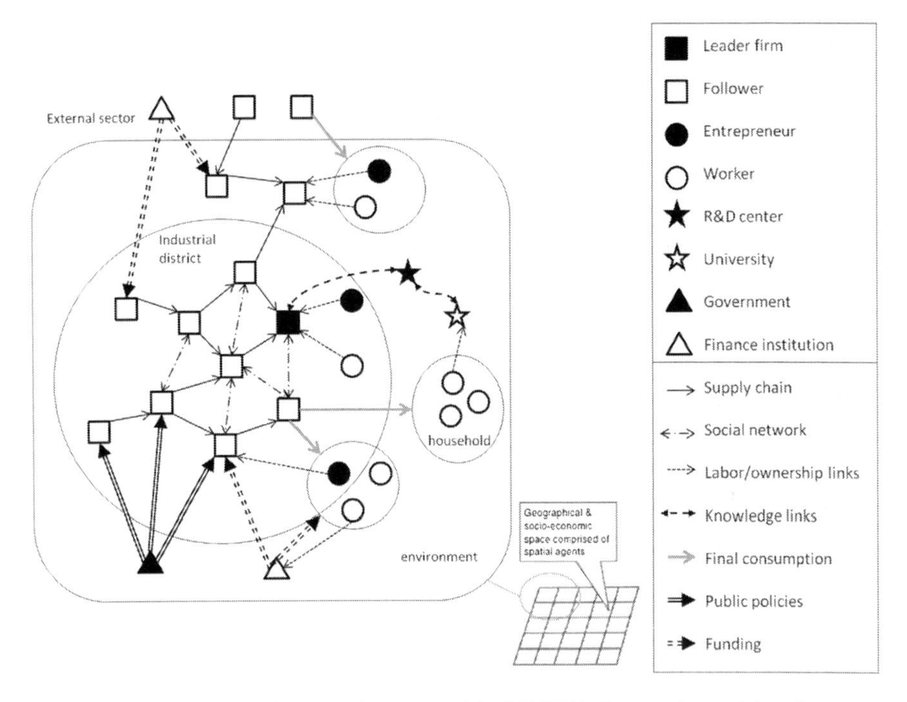

Fig. 50.2 Agents and environment incorporated in MOSIPS. Source: Own elaboration based on [1]

ment of inputs, human resources management, production, innovation and technology management, product development, financial management, and marketing and sales strategy. Also influential are the possible strategies for growth and territorial expansion of investments (within or outside the territory). Individuals on themselves or grouped in households have relations with other individuals and with firms resulting from their labour, consumption and investment activities.

In general, the market itself sets a 'virtual' kind of network relationships among firms in their pursuit of needs/opportunities for personnel, intermediate inputs, production equipment and technology, and sales niches, which are provided by the different types of markets (labour, goods and services...) [6, 19]. But the general market also establishes the relationship of rivalry and competition between them. In sum, these are network relations that provide information for cooperation or competition as appropriate [13, 15].

Contractual relationships on the labour market are structured between individuals and firms. They are based on search processes, where firms in a context of imperfect information choose the best candidates and individuals offer their work to firms that provide the most attractive terms. In general, these relationships are normative since they are based on labour legislation.

50.2.6 Activities

Individuals are born into households where they grow, consume, pursue an education, and, eventually, die. Households may change their location and increase or decrease the number of their members. Upon reaching working age, individuals choose in each period whether to join the workforce and, if so, become employees or entrepreneurs. Job seekers offer their work to firms in their environment recruiting workers, which take the decision regarding who to hire. Individuals who choose to become entrepreneurs create firms and choose the location.

Firms produce and sell their products on the market. They also choose the cheapest and most reliable suppliers. Additionally, they change in terms of size through internal growth or by acquiring other firms, provided they have adequate funding. They can apply for funding from the financial system, based on their repayment capacity and on the financial market conditions. Further, firms decide the level of their commitment to innovation, both in terms of processes and products. They modify their workforce by hiring or laying off workers according to their labour skills or to production needs. Firms disappear if they go bankrupt or if the businessperson so chooses.

The public sector incorporates its rules and policies by modifying the attributes and behaviours of agents. Banks procure funds from agents with a funding capacity, and they provide funds to those who require them. The financial resources available for firms and households can differ from the level of savings of agents due to the financial flows with other countries and the circumstances of the financial system. The local environment includes aspects that affect the agents from a territorial perspective such as closeness from infrastructures, the existence of firm clusters, or congestion problems.

In the modelling process, the properties of the protocols that govern the interaction between individuals and companies are based on relations of production and consumption, employment and lending, adopting a microeconomic perspective. In this sense, the price of competitors' products in relation to themselves constitutes one of the main signals received by the agents. They act taking the suitable decisions for the acquisition of inputs, recruitment of factors, production and sales. These relations take place in markets.

With respect to these factors and product markets, the model provides an approximation to local environments, on one side. But on the other side it also assumes the existence of other broader environments, at national or international level. That is, the analysis of interaction of agents is focused on their interest in a defined territorial space, such as a region. Relationships with agents from other areas are analysed in a more simplified approach. For example, the majority of goods are bought by large retailers in international markets, and then they sell them to small retailers. Final customers do not have access to a high number of sellers due to informational costs. Then, the appropriate scale of the first market is international, small retailers only have access to the regional market and final costumers tend to purchase goods at local level.

The same assumptions are made in the relation between firms and workers, who are unemployed or employed in other companies and they want to change their jobs. It could be considered that both companies and workers face the regional employment supply and demand, respectively. In most of the approaches, it is assumed that all the agents act against the market, the aggregate behaviour of the rest of agents, looking to optimize their interest. However, in most of the cases, every agent creates its own behaviour associated to the decisions of its neighbours. It arises from the information and expectations generated by the rest, weighted depending on their spatial and relational proximity. For example, a company located in a municipality is able to produce and sell its production with a slightly different price from a competitor of a neighbouring municipality, while in other part of the region prices can be lower. Then, agent actions and decisions are highly affected by the behaviours of agents in the proximity, but it also depends on the aggregate behaviour, emerged from the decisions of every agent.

Thus, all companies are somehow interconnected, but these links are stronger in environments which are closer. In any case, those behaviours associated with the environment may also depend on the sector, the concentration of supply and demand or the degree of public promotion of a product (e.g., which is derived from the impact of advertising).

Individuals face the same interaction protocols and information flows, but applied to their decisions. They obtain most of the information from firms which they are linked, but also from the aggregate behaviour (e.g. the unemployment rate, GDP growth, price index). Individuals also condition their decisions taking into account the performance of other agents who are linked with. Then, a potential entrepreneur will decide to create his own enterprise with a higher probability if both their acquaintances and the information she has about the general performance of the economy is promising for her success.

50.3 Design Concepts

This section of the paper presents some of the highlights of the general approach of the model

50.3.1 Time, Activity Patterns and Activation Schemes

Time is modelled discretely. Each period consists of several steps. The length of the period determines the temporal resolution of the model and is determined largely by the characteristics and temporal reference of the data used. The model can consider any time interval without affecting its characteristics. However, the quarter has been taken as the primary reference since it is considered that most of the decisions of the agents have a maturation period around this length.

Actions are triggered instantly at the time when the 'central clock' determines each period. They, however, do not need to be carried out in each period. The user can choose a different periodicity for some of them.

By observing this basic temporal sequence of events, the model is fed with information and data proposed in the system architecture for the years 2007 to 2011. This means that the simulation system starts from 2007 and forecasts of the modelling can be developed from 2012.

50.3.2 Interaction Protocols and Information Flows

Matching interactions and business activities are bilateral. These are gravitational interactions where intensity depends on "visibility", which, for an agent, means the expected relevance of its interaction with the counterpart. In the case of matching of individuals, each individual selects a group of people with whom s/he interacts and who s/he subjectively evaluates based on its attributes. In the case of firms, sellers offer their product to the market and buyers choose their supplier from a group of sellers who are selected according to their closeness and to the size of their firms.

Each firm demands workers featuring certain characteristics. Among the firms seeking a worker's profile, workers choose the most "attractive" ones in terms of salary and distance. Matching occurs when the best possible combination for both parties is achieved.

50.3.3 Forecasting

Agents base their forecasts on their past experience, within a context of incomplete information. Households determine their levels of consumption and savings from their income experience in prior periods following a scheme inspired by the life cycle and permanent income hypotheses. Firms make their decisions based on the experience gained with clients and competitors.

50.3.4 Behavioural Assumptions and Decision Making

Agents have bounded rationality and act in an environment of imperfect information. Interactions take place predominantly in the close environment of the agents. The chances of interaction among agents depend on their visibility, understood as indicated in Sect. 50.3.2.

Firms select their suppliers and their workers. Consumers choose the firm in which they work and their consumptions elections. In all the situations the process is the same. Firstly, the agent evaluates its performance. For this purpose, it examines not only its results, but also receives information about other agents. It can be biased

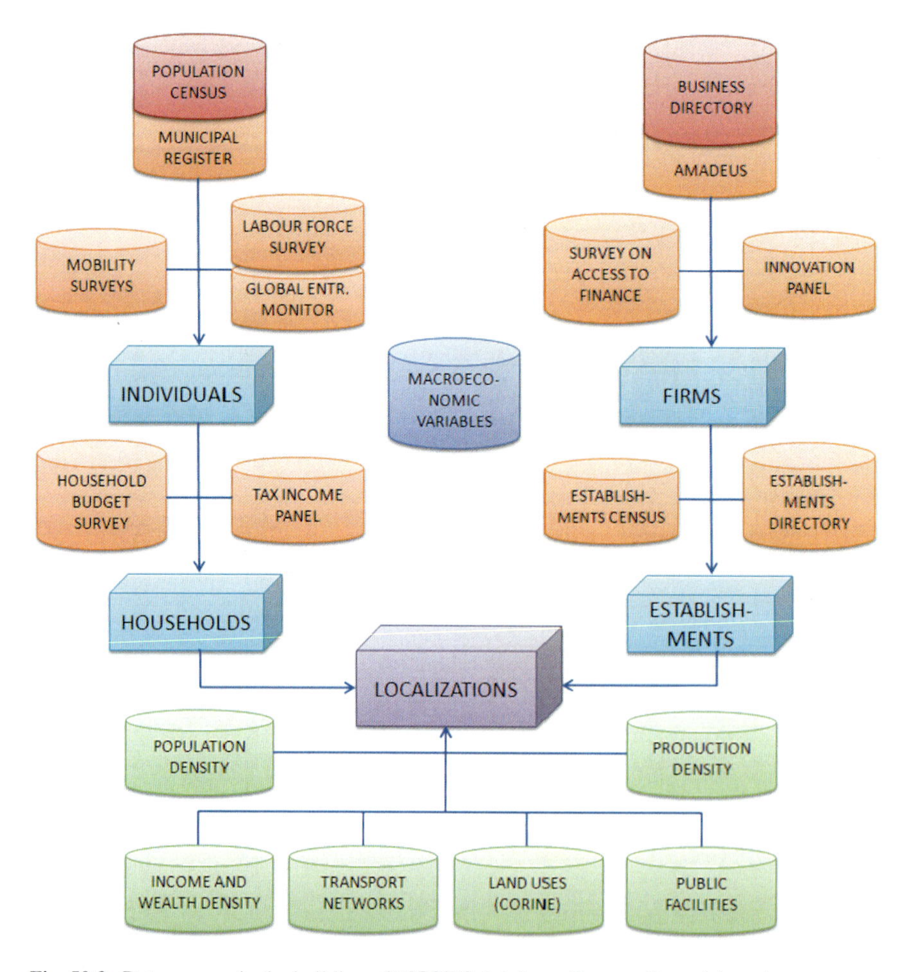

Fig. 50.3 Data sources in the building of MOSIPS data base. Source: Own elaboration

and incomplete. If the performance is sufficiently good, the agent does not search for new agents (suppliers, workers, etc.), whereas if its performance is not sufficiently good, according to the expectations of the agent, or worse than the average perceived performance, the firm will look for better agents to work with (Fig. 50.3).

50.3.5 Learning

The structural characteristics of households and firms evolve through learning. This is done by imitation and mutation procedures. The structural characteristics of individuals are modified in each period either due to random factors or by imitation of the behaviour of the agents regarded as displaying more appropriateness (like benchmarking).Thus, by means of a selection evolution process, agents get adapted to the circumstances through learning, as it happens, for instance, in the case of the initial reservation wage, or the choice of distance when seeking suppliers.

50.3.6 Population Demography

In the model, entries and exits of households and firms occur in each period. Within households, there are births and deaths. Births depend on the location and personal circumstances of the mother, while deaths hinge on the individual's age as well as other factors. Individuals can change their location when appropriate in a cost-benefit scheme.

Some individuals are entrepreneurs, and they emerge as such when, from a subjective point of view, it is convenient for them to be entrepreneurs. Similarly, they stop being entrepreneurs when it no longer suits them. Entrepreneurs with one or more firms are businessmen. The birth and death of firms follow the decisions taken by businessmen, who are dependent on economic or personal factors. They also make decisions about the location of their firms.

50.3.7 Level of Randomness

There are two main sources of randomness. The major one is generated in the creation of agents. There is available a certain information, but it is incomplete, and agents are created following random rules, accordingly and conditioned to their known characteristics. For example, an employed person who works in a given sector is matched to a firm of that sector with an establishment near his residence, but there is a possibility of make incorrectly this process. Then, if the actual firm has a good performance and the matched firm goes bankrupt, this individual will lose his job, while in the reality, he continues working. The counterpart is an actual unemployed will continue be working in the modelled reality. As we can see, part of these random possibilities cancel at aggregate level, but can have huge effects at the micro level, making unfeasible to obtain accurate micro information.

Another significant source of randomness is the creation of networks, as this is the part of the model for which there is a lower level of data available. Networks are extremely important for learning processes and changing behaviours. Thus, the model increases its randomness along periods of simulation, leading to stochastic dynamics for a large number of variables (household characteristics, prices, amounts, innovation, location, etc.).

50.3.8 Miscellaneous

The MOSIPS model is inspired by the circular pattern of income where the financial field is explicitly integrated. This is crucial in the current crisis process given the serious financial constraints of firms.

Additionally, MOSIPS provides a highly precise spatial outlook since agents are located individually using GIS techniques, which makes it possible to observe the impact of policies at a micro-spatial level.

50.4 Data Bases for MOSIPS Model

In order to represent the society it is necessary to build two databases, one for individuals and families and other for firms and establishments. They are complemented with the macroeconomic environment, and public, financial and foreign sector, taken as a whole.

The variables for the macroeconomic environment and public policies, allow the agents information creating their expectations and it is added to the information they have from themselves and other individuals and firms whom they have connections.

We make use of two techniques that allow achieve a great grade of accuracy in the process: statistical matching and downscaling. Thus, the model has a high degree of scalability, and allows the user focusing from individual effects up to total variation or putting the focus in the agents of a specific area or a group of SMEs with similar characteristics.

50.4.1 Statistical Matching

Statistical matching is used to fusion information from different microdata sources [8]. The different data sources incorporated in the building process of MOSIPS data base are shown in Fig. 50.3.

In the model the main one for individuals is the Population Census, which informs about the number of the total population and the main individual characteristics: age, gender, location at regional level, family composition, level of studies, etc. It is complemented firstly with the municipal register to update individual data. The scope of this data base is enlarged with the Labour Force Survey to become acquainted labour status of individuals. The EU Labour Force Survey is a large household sample survey providing quarterly results on labour participation of people aged 15 and over as well as on persons outside the labour force. All definitions apply to persons aged 15 years and over living in private households. The Global Entrepreneurship Monitor explores the role of entrepreneurship in national economic growth, unveiling detailed national features and characteristics associated with entrepreneurial activity. It is used to value the chances of developing new enterprises. Mobility surveys make possible to know patterns of mobility of individuals, and allow creating demand functions spatially defined [18]. Tax Income Panel is the best indicator of income and Household Budget Survey permits identify the consumption in every sector and the savings as the remaining part of the income. It allows us to ascertain the consumption expenditure of households residing, as well as the distribution of said expenditure among the different consumption divisions. Other sources are used to acquire information not present in the previous presented sources. The main one is mortgage duration and amount, present in the Annual Report of Property Registers.

Enterprises data base is built in a similar way, starting from the Business Directory, which contains information about the number of enterprises, the sector they belong and their size. However, is a poor source with respect to the Population Census and only includes aggregate information. Then, it must be fulfilled with the microdata exhaustive information enclosed in other statistical source called Amadeus.

This data base should be enlarged including characteristics from the Survey on Access to Finance and the Innovation Panel. The Survey on Access to Finance covers micro, small, medium-sized and large firms and it provides evidence on the financing conditions faced by SMEs compared with those of large firms every six months. In addition to a breakdown into firm size classes, it provides evidence across branches of economic activity, euro area countries, firm age, financial autonomy of the firms, and ownership of the firms. The Innovation Panel is a statistical instrument for studying the innovation activities of firms over time. It takes into account the heterogeneity in the firms' decisions (such as different shares of intramural R&D and external R&D in total innovation expenditures) or in the effects (such as the different impacts on productivity). Other sources are used to create the proper number and size of establishments for every firm and to locate them in the space.

The statistical matching process is taken in several stages. For individuals is necessary to start from the Population Census, and then add labour information though a microsimulation model that is built in order to obtain accurate data. Then, we are sure that every individual is coherent with himself across the time periods. For example, a civil servant cannot be fired. Then, it is possible to link information about income and consumption at household level, due to the available microdata information. Finally, entrepreneurial activity is compute for every adult. Enterprises data base is constructed following the same scheme, starting with the census (business directory) and adding information about their characteristics (number of workers, financial statements...) and access to finance, which is a major issue for the growth and overall performance of SMEs.

50.4.2 Downscalling

In order to accurately define the placement of agents, a raster of locations must be included, with information of land price, demography, uses of land and transport networks. This raster is built making use of downscaling techniques. From information at local level or other (e.g. per Ha or per km^2) we arrive to the level of each cell included in the raster [11].

After making the statistical matching process, both firms and households are located into the raster. Finally, networks are built between individuals, firms and among individuals and firms. These relations are related to the information provided by markets of factors, goods and services.

50.5 Conclusions

MOSIPS model is an agent-based model (ABM) which consists of a set of agents (firms/establishment and individuals/households) with its own attributes which interact each other according to a set of appropriate rules. Although the scope of the application of model is wide, the main simulation object is the impact of public policies on the small and medium enterprises (SMEs) from a bottom-up approach.

This kind of model makes possible to tackle the analysis of complex phenomena, capturing complex processes not fully described by traditional techniques. The masterpiece is the agent, instead of the whole system. It is also explicitly considered and modelled the heterogeneity of agents, their social interactions and decision-making processes. The interaction between the agents is non-linear, and they can adapt themselves, learn, evolve, and even develop some self-organization mechanisms that allow them to acquire collective properties or characteristics that do not have individually.

Consequently there will be outcomes arising from the interaction between agents and between them and the environment in which they operate. In other words, aggregate macroeconomic results will emerge from the behaviour of individuals that take part in these complex environments.

The building blocks of the MOSIPS includes the agents' population and other entities; the interaction paradigm among agents; the activities the agents develop as well as their adaptive capability—e.g. the degree of re-activeness and pro-activeness—and finally the object of the simulation.

Acknowledgements This work has been conducted in the context of MOSIPS project (Modeling and Simulation of the Impact of public Policies on SMEs) of the EU 7th Framework Programme with Grant Agreement 288833.

References

1. Albino V, Carbonara N, Giannoccaro I (2007) Supply chain cooperation in industrial districts: a simulation analysis. Eur J Oper Res 177:261–280
2. Aldrich HE, Cliff JE (2003) The pervasive effects of family on entrepreneurship: toward a family embeddedness perspective. J Bus Venturing 18(5):573–596
3. Ardagna S, Lusardi A (2010) Explaining international differences in entrepreneurship: the role of individual characteristics and regulatory constrains. In: Lerner J, Schoar A (eds) International differences in entrepreneurship. University of Chicago Press, Chicago, pp 17–62
4. Arrow KJ, Debreu G (1954) The existence of an equilibrium for a competitive economy. Econometrica 22:265–290
5. Brunet I, Alarcón A (2004) Teorías sobre la figura del emprendedor. Rev Sociol 73:81–103
6. Coviello N, Munro H (1995) Growing the entrepreneurial firm: networking for international market development. Eur J Mark 29(7):49–61
7. Dawid H, Gemkow S, Harting P, van der Hoog S, Neugart M (2011) The Eurace@Unibi model: an agent-based macroeconomic model for economic policy analysis. Available from www.wiwi.uni-bielefeld.de/vpl1/projects/eurace/eurace-unibi.html
8. DíOrazio M, DiZio M, Scanu M (2006) Statistical matching: theory and practice. Wiley, New York

9. Dosi G, Fagiolo G, Roventini A (2008) Schumpeter meeting Keynes: a policy-friendly model of endogenous growth and business cycles. Laboratory of Economics and Management, SantAnna School of Advanced Studies. LEM working paper series 21
10. Farmer J, Foley D (2009) The economy needs agent-based modelling. Nature 460:685–686
11. Gallego FJ (2010) A population density grid of the European Union. Popul Environ 31:460–473
12. Gintis H (2006) The emergence of a price system from decentralized bilateral exchange. BE J Theor Econ 6:1302–1322
13. Gulati R (1999) Network location and learning: the influence of network resources and firm capabilities. Strateg Manag J 20(5):397–420
14. Mandel A, Fürst S, Lass W, Meissner F, Jaeger CC (2009) Lagom generiC: an agent-based model of growing economies. ECF working paper 1. Available from www.europeanclimate-forum.net/index.php?id=ecfworkingpapers
15. Meyer M, Aderhold J, Duschek S (2004) Organizing social complexity in production networks. J Acad Bus Econ 3(1):1
16. Nielsen K, Sarasvathy S (2011) Who re-enters entrepreneurship? And who ought to? An empirical study of success after failure. In: Dime-Druid Academy winter conference in Aalborg, January 20–22, Aalborg, Denmark
17. Roessl D (2005) Family businesses and interfirm cooperation. Fam Bus Rev 18:203–214
18. Schenk T, Löffler G, Rauh J (2007) Agent-based simulation of consumer behaviour in grocery shopping on a regional level. J Bus Res 60:894–903
19. Slotte-Kock S, Coviello N (2010) Entrepreneurship research on network processes: a review and ways forward. Entrep Theory Pract 34(1):31–57
20. Stam E, Audretsch D, Meijard J (2008) Renascent entrepreneurship. J Evol Econ 18(3–4):493–507
21. Street CT, Cameron A-F (2007) External relationships and the small business: a review of small business alliance and network research. J Small Bus Manag 45(2):239–266
22. Tesfatsion L, Judd K (2006) Agent-based computational economics. II. In: Handbook of computational economics. Elsevier, Amsterdam
23. Wolf S, Bouchaud J-P, Cecconi F, Cincotti S, Dawid D, Gintis H, Hoog S, Jaeger CC, Kovalevsky DV, Mandel A, Paroussos L (2011) Describing economic agent-based models, Dahlem ABM documentation guidelines. In: Proceedings of the 100th Dahlem conference on new approaches in economics after the financial crisis, August 28–31, 2010
24. Wooldridge M, Jennings NR (1995) Intelligent agents: theory and practice. Knowl Eng Rev 10:115–152

Chapter 51
Integrating Collective Decision-Making Models and Agent-Based Simulation

Pablo Lucas and Diane Payne

Abstract Collective Decision-Making Models (henceforth CDMM) are mathematically deterministic formulations (i.e. without probabilistic inputs or outputs) aimed at explaining the behaviour of individuals in dynamic negotiations given any number of issues, in which the participants attempt to influence the outcome of a final and binding decision. Albeit different CDMM have produced acceptable predictions to actual final collective outcomes, both the data collection process and the interpretation of CDMM results require attention to the rather strict underlying assumptions in each of these models. Our contribution is thus twofold: (I) replication for systematic testing of the Challenge and Exchange CDMM assumptions, along with their requirements consisting of the Compromise, Mean and Median models, using an agent-based framework; and (II) insights gained from these tests regarding the dynamics of CDMM runs and their combinations using input from three datasets.

Keywords Replication · Evidence · Development · Validation

51.1 Introduction

Collective Decision-Making Models (CDMM) have been developed and applied in the social sciences as tentative explanatory approaches as to how individuals attempt to influence each another in a negotiation process. Despite the relative success of CDMM at producing generally acceptable predictions of actual outcomes, the underlying assumptions of CDMM are strict and this has adverse effects to understand the dynamics of each model. As these implementations are not publicly available online, four of these models have been replicated in an agent-based model (ABM) framework in order to systematically test their assumptions under different precision floating-points. This process resulted in:

P. Lucas (✉) · D. Payne
Geary Institute, University College Dublin, Dublin, Ireland
e-mail: pablo.lucas@ucd.ie

D. Payne
e-mail: diane.payne@ucd.ie

T. Gilbert et al. (eds.), *Proceedings of the European Conference on Complex Systems 2012*, Springer Proceedings in Complexity, DOI 10.1007/978-3-319-00395-5_51,

- a controlled environment for reproducible and flexible testing of CDMM;
- an understanding of which CDMM are most stable, in terms of predictions;
- insights into the assumptions and internal dynamics of the Exchange Model.

The replication of CDMM is reviewed next, and then the findings are discussed.

51.2 Collective Decision-Making Models

Collective decision-making models (CDMM) are mathematically deterministic formulations (i.e. without probabilistic inputs or outputs) that are aimed at explaining the behaviour of individuals in dynamic negotiations given any number of issues, in which all the participants attempt to influence the outcome of a final and binding decision. CDMM take into account, per issue, three normalised values (PSP) that represent: the initial position,[1] the salience[2] and the power[3] of every agent (an individual or organisation) in a decision-making process. CDMM are aimed at producing a prediction of the outcome of an actual negotiation by mathematically processing, according to a CDMM algorithm, all PSP values belonging to each agent per issue. Thus understanding the assumptions and dynamics of each CDMM is crucial for insights about their dynamics, including:

- amount of conflict, backing and acceptance between participants;
- nature of the influence and implementation (i.e. compliance) processes;
- stability of individual and collective outcomes, given the initial conditions;
- strategic information about how agents agree across any number of issues.

Both the data collection process and the interpretation of CDMM results require particular attention to the rather strict underlying assumptions in each of these models. This is important as one can process the same empirical dataset with different CDMM and analyse which result approximates best to the actual outcome, suggesting thus how individuals may have behaved. The model input data involves data about the PSP values from, ideally three, expert stakeholders by decomposing a collective decision into independent issues, provided that each is:

- an important element of the actual collective decision-making process;
- and a continuum, on which every agent PSP value can be justifiably allocated.

51.3 Integrating Replications of CDMM into an ABM

The following CDMM have been replicated and integrated into a single ABM framework: Challenge, Compromise, Exchange, Mean and Median model. Repli-

[1]Initial position is also referred to as "preferred or voting position", per issue.

[2]Salience is also referred to as one's "importance attached to", per issue.

[3]Power is also referred to in the literature as one's "resources" or "capabilities", being immutable across issues and potentially mobilised by the agent in question, per issue.

cation is important for model verification and validation, as this allows a thorough inspection of how these are specified and, ultimately, provides insights into their internal dynamics and how they were originally implemented. The replications were time-consuming due to the unavailability of the original source codes, lack of published details regarding all of the assumptions and the algorithmic procedures necessary for the implementations. Another required effort was to design a process to standardise the input data to be used in the ABM, as the original formats had to be reorganised for tabulation using relational indexes for actors, issues and PSP values. That also included the removal of various special characters (such as trailing spaces and other invisible text markers) found in the original format and normalisation of values. Once all data entries have been standardised, the ABM takes as input the indexes of agents, issues and PSP values. The output consists of: the predicted collective outcome (including final PSP values and losses), simulated potential challenges and simulated actual challenges (including a rank based on relative individual power and outcome) and matches within and across issues. The setup used for this paper involved in running the following independent sequence of models (i.e. each run loaded the original PSP values, per dataset): Mean, Median, Challenge, Exchange, Challenge-Exchange and Exchange-Challenge. The last two are composites, in that the output of the former model is the input of the latter. The intention of testing this is twofold: observe whether a decision-making process is best characterised as a shift from one to another, instead of being fully based on the assumptions of one CDMM, and analyse the effects of one CDMM model has on another in terms of PSP.

51.4 Obtained Findings from the Integrated Replication

The integration of all the aforementioned CDMM into an ABM framework allowed for the flexible and systematic testing of assumptions and dynamics underlying each model. This provided key insights into understanding how these deterministic models operate, particularly with regards to whether the precision of calculations affects the assumptions, internal dynamics and predicted outcomes per model. The Mean, Median and Challenge models remained largely unaffected by these tests. This finding, in terms of robustness, reinforces the analysis by [1] that the most successful models are those mainly based on computing some type of mean amongst agent PSP values. However the Exchange model is impacted by these changes in terms of how many pairwise exchanges can actually occur and how much the final predicted outcome diverges from the actual outcome—both per issue and across issues. One can observe a clear decrease of exchanges with the increase of precision in this model. The same effect occurs regarding when combining models, i.e. the setups where the output of one model becomes the input of another model. Both the Challenge-Exchange (CE) and the Exchange-Challenge (EC) tests incur differences driven by the precision issue observed in the Exchange model. An exception to that is the CE setup, where the Challenge model can cause an increase in the number of exchanges by realigning the PSP values of all agents per dataset. No exchanges can happen at all if the precision of calculations is set between 2 and 4, depending on the dataset. This is a relevant, previously unknown finding about the dynamics of the Exchange model. It suggests that the number of exchanges depends directly on how strict are the requirements for both agents to accept equal gains, and this corroborates to highlight further the rigidness of the original assumption regarding actual exchanges. For most models, lower precisions yield greater differences in deviations. Dataset A is the only one that yields a different dynamic when processed with the CE setup. All other models, for all datasets, yield stationary trends converging to either one value or a range. For dataset C, the results can be summarised as: (a) the EC setup performs almost identically to the Challenge model, with prediction errors varying equally between -0.1 and -0.05; (b) all other models performances are similar, with prediction errors varying between 0.20 and 0.27. The precision issue in the Exchange model is significant as, depending on the dataset, it can lead to deviations from the actual outcome of up to 0.85. This large difference in deviations is rather noticeable as most predicted outcomes deviate below 1 and that these variations arise due to precision issues.

51.5 Discussion and Final Remarks

The relevance of implementing CDMM in an ABM framework allowed for systematic testing of the Challenge and Exchange Models in a controlled environment that allows reproducible and flexible testing of hypotheses. The replication of CDMM is itself a worthwhile exercise, as during this process it was possible to identify issues of accuracy regarding the specification of CDMM algorithms and calculations that can impact the interpretation of obtained results. Moreover, in doing so, a process for standardising (empirical and/or simulated) datasets for use in the ABM has been designed, so that the same implementation can be used to process other data sources that follow the specifications for normalising values and flagging of missing entries. The increasing demand to understand the processes driving collective decisions, including for instance policy-making and various socio-economic phenomena, is turning ABM simulations into a form of surrogate reasoning about the actual phenomenon in question. In this sense, our contribution through the ABM lens is the further understanding of how two CDMM work in terms of their assumptions and internal dynamics. The replication allowed the identification and interpretation of differences in robustness between the propositions of the Challenge and Exchange models, along with how these models interplay once combined in a setup where the output of one is the input of another. These findings provide insights that are helpful to improve the modelling of collective decision-making and the understanding as to why fundamentally different CDMM (i.e. strategic and non-strategic ones) can perform similarly. The nature of an ABM requires rigorously correct implementation standards; as researchers using this framework must design clear computational experiments for testing hypotheses, which are formally specified in terms of a computerised simulation model, to gain insights by analysing the results of computational experiments. Thus one must ensure that observed unexpected results are indeed due to unforeseen relationships in the model itself, and not due to an artefact (i.e. an observation results from a particular implementation) or error (i.e. a mismatch between the design and the implementation) [2, 3]. The replications of all CDMM required a detailed scrutiny of their assumptions and procedural specifications, so that the difference in results observed due to calculation precision seem to be an artefact in the original model specification, as either assumptions indeed did not account for issues related to precision or there was no clear specification regarding it.

References

1. Achen C (2006) Evaluating political decision-making models. In: Thomson R, Stokman F, Achen C, König T (eds) The European Union decides. Cambridge University Press, Cambridge
2. Polhill G, Izquierdo LR, Gotts NM (2005) The ghost in the model (and other effects of floating point arithmetic). J Artif Soc Soc Simul 8(1):31
3. Galán JM, Izquierdo LR, Izquierdo SS, Ignacio Santos J, del Olmo R, López-Paredes A, Edmonds B (2009) Errors and artefacts in agent-based modelling. J Artif Soc Soc Simul 12(1):1

Chapter 52
Agent-Based Simulation for Complex Social Systems: Support for the Developer

Amineh Ghorbani and Virginia Dignum

The successful implementation of policies in complex social environments, require a deep understanding of interdependencies between many actors with different perspectives. In order to understand, analyse and design such complex systems, advanced modelling tools are required.

Models of many types have been extensively used for researching complex social systems. Modelling is especially suitable when prototyping or experimenting with the real system is expensive or impossible. In one extreme of the scale, text-based models are well suitable to support argumentation and decision-making by groups of people, but are not computable. On the other extreme, mathematical models, including dynamical systems, statistical models, differential equations, or game theoretic models, enable a precise and computable representation of the system, but their complexity rises exponentially as the complexity of behaviours grow, so that describing complex individual behaviour with equations often becomes an intractable task. In [11], simulation is described as a third way to represent social models, being a powerful alternative to these types. Simulation has a high descriptive power and can easily be run on a computer. Moreover, simulation can represent non-linear relationships, which are often tough problems for the mathematical approach.

In particular, Agent-Based Modelling and Simulation (ABMS) is a powerful approach for the analysis of complex social systems. This comes from the fact that agent related concepts allow the representation of organizational, social and behavioural aspects of individuals in a society and their interactions. In ABMS, system behaviour emerges as the result of the combined activity of many (tens, hundreds, thousands, millions) individuals, each following its own behaviour rules, living together in some environment and communicating with each other and with the environment. ABMS can support both qualitative and quantitative evaluation methodologies. With the advent of software platforms in recent years, agent-based

A. Ghorbani (✉) · V. Dignum
Faculty of Technology, Policy and Management, Delft University of Technology, Delft, The Netherlands
e-mail: a.ghorbani@tudelft.nl

T. Gilbert et al. (eds.), *Proceedings of the European Conference on Complex Systems 2012*, Springer Proceedings in Complexity, DOI 10.1007/978-3-319-00395-5_52,

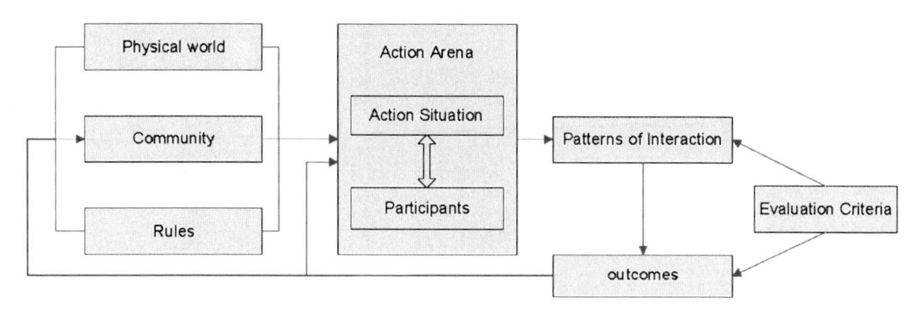

Fig. 52.1 The IAD framework

modelling (ABM) has increased in popularity among social scientists (e.g. Repast [9], Netlogo [15], Swarm [8]).

To aid model development, some researchers have provided guidelines on how to build agent-based models (e.g. [1, 3, 5]). The general steps that may be more or less refined by different scientists include model conceptualization/design, implementation, validation/verification, and analysis of data. Nevertheless, existing methodologies, require the modeller to possess substantial programming knowledge. Explicit model conceptualization, which entails describing the set of concepts that will constitute the "building blocks of the model", is generally recognized to be a crucial step in building software models because it leads modellers to better capture, analyse and understand what they are actually modelling [16]. Moreover, while understanding and explaining individual behaviour is extremely complex, social rules or *institutions* are more elicitable [13] and hence more readily identified and captured by modellers. Therefore, we propose to support model development by grounding it on institutional concepts. In the social sciences, the institutional analysis and development framework (IAD) proposed by Ostrom [10] has been used successfully for many years for the conceptualization and analysis of complex social systems.

The IAD framework (cf. Fig. 52.1 focuses the analyst's attention on individuals who make decisions over some course of action. In IAD, the action arena is the unit of analysis and focus of investigation. An action situation is the "social space where individuals interact, exchange goods and services, engage in appropriation and provision activities, solve problems, or fight" [10]. Policy processes and outcomes are assumed to be affected, to some degree, by four types of variables external to individuals: (1) attributes of the physical world, (2) attributes of the community within which actors are embedded, (3) rules that create incentives and constraints for certain actions, and (4) interactions with other individuals.

The IAD framework helps policy makers to organize diagnostic, analytical, and prescriptive capabilities. However, the development and use of simulations involve making precise assumptions about a limited set of variables and parameters to derive precise predictions about the results of combining these variables using a particular theory. In order to support the systematic design of agent-based simulations for complex social systems based on the IAD framework, we have developed a framework, MAIA (*M*odelling *A*gent systems based on *I*nstitutional *A*nalysis), which ex-

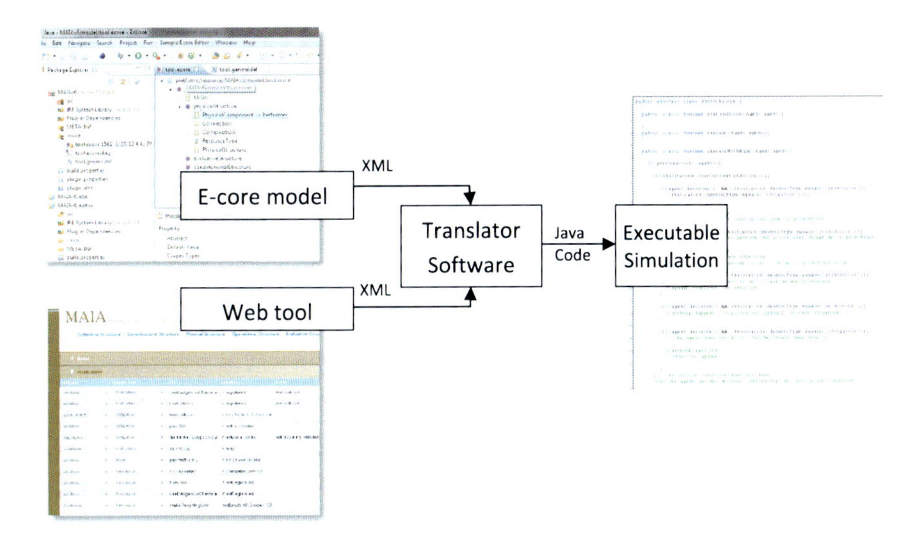

Fig. 52.2 The MAIA simulation package consist of a meta-model presented as an E-core specification, a web-based application to guide conceptualization and a translator code that produces executable code from a MAIA-based model

tends and formalizes IAD with the required modelling constructs necessary to build executable simulations.

MAIA provides an extensive set of modelling concepts rich enough to capture a large range of complex social phenomena. This set includes social concepts such as norms, culture, personal values and preferences, social roles, responsibility and dependency. The MAIA simulation package is presented in Fig. 52.2. To support the development of MAIA based models, a web-based application, takes the user through a step by step procedure, checks for consistency among the concepts and produces XML code. The user of this software application does not need any computer science background. Furthermore, MAIA provides formalized rules on the translation of social concepts to programming code, using a Model-Driven Engineering (MDE) approach that provides a transformation between the MAIA meta-model and a Java program. MDE approaches have been advocated when constructing agent-based models of social systems [4, 6, 12].

The application of MAIA in various real world cases (consumer lighting transition [7], woodfuel market in Switzerland [14], e-waste recycling sector in Bangalore [2] and bio-gas energy production in the Netherlands) shows the usefulness of the MAIA simulation package in several aspects. First, the concepts (e.g. role, personal value) are more easy to grasp by the social scientist because they are taken from the social science theories. Furthermore, compared to other tools, because the web-based application is based on the MAIA framework, it provides a rich set of social concepts. Second, the MAIA web-based tool, appears to be efficient and reliable due to various consistency checks and automatic completion of many fields. Third, with the provided set of rules, it is theoretically and practically possible to implement a translator that automatically generates code from the set of social concepts.

Even without the use of a mediator software that would automatically generate code from the set of concepts, as we saw in our case studies, it is fairly straightforward to develop a Java program from the information provided in the web-based application. This would however, require programming knowledge. In this respect, an added value of using the tool is the possibility of having a team of modellers to develop an agent-based model: system analysts who decompose the system into MAIA concepts using the tool and do not necessarily need programming knowledge and programmers who use this information to develop a simulation program. This practice was conducted in three of our case studies. The transfer of the knowledge between the model developers was through the tool outputs.

In short, the MAIA platform supports the development of agent-based models for policy making by providing (1) a methodology that provides guidelines on how to produce executable code from a conceptualized model, (2) a web-based application that supports the conceptualization process, and (3) a (semi) automatic transformation to generate executable simulations.

References

1. Drogoul A, Vanbergue D, Meurisse T (2003) Multi-agent based simulation: where are the agents? In: Multi-agent-based simulation II, pp 43–49
2. Ghorbani A, Dignum V, Sheoratan S, Dijkema G (2011) Applying the Maia methodology to model the informal E-waste recycling sector. In: The seventh conference of the European social simulation association (ESSA)
3. Gilbert GN, Troitzsch KG (2005) Simulation for the social scientist. Open University Press, Milton Keynes. ISBN 0335216005
4. Hassan S, Fuentes-Fernández R, Galán JM, López-Paredes A, Pavón J (2009) Reducing the modeling gap: On the use of metamodels in agent-based simulation. In: 6th conference of the European social simulation association (ESSA 2009), pp 1–13
5. Heath B, Hill R, Ciarallo F (1998) A survey of agent-based modeling practices (January 1998 to July 2008). Journal of Artificial Societies and Social Simulation 12(4):9
6. Janssen MA, Alessa LN, Barton M, Bergin S, Lee A (2008) Towards a community framework for agent-based modelling. Journal of Artificial Societies and Social Simulation 11(2):6
7. Ligtvoet A, Ghorbani A, Chappin EJ (2011) A methodology for agent-based modeling using institutional analysis applied to consumer lighting. In: Agent technologies for energy systems, Tenth international conference on autonomous agents and multi agent systems (AAMAS), Taipei, Taiwan
8. Minar N (1996) The swarm simulation system: a toolkit for building multi-agent simulations. Santa Fe Institute, Santa Fe
9. North MJ, Collier NT, Vos JR (2006) Experiences creating three implementations of the repast agent modeling toolkit. ACM Transactions on Modeling and Computer Simulation (TOMACS) 16(1):1–25
10. Ostrom E (2005) Understanding institutional diversity. Princeton University Press, Princeton
11. Ostrom TM (1988) Computer simulation: the third symbol system. Journal of Experimental Social Psychology 24(5):381–392
12. Sansores C, Pavón J (2005) Agent-based simulation replication: a model driven architecture approach. In: MICAI 2005: Advances in artificial intelligence, pp 244–253

13. Scharpf FW (1997) Games real actors play: actor-centered institutionalism in policy research. Westview Press, Boulder
14. Steubing B, Kostadinov F, Ghorbani A, Wager P, Zaha R, Thees C, Ludwig C (2011) Agent-based modeling of a woodfuel market factors that affect the availability of woodfuel. In: ISIE'11, Berkeley, US
15. Tisue S (2004) Netlogo: design and implementation of a multi-agent modeling environment. In: Proceedings of agent
16. Winograd T, Bennett J, De Young L, Hartfield B (1996) Bringing design to software. ACM Press, New York

Chapter 53
Coping with the Complexity of Cognitive Decision-Making: The TOGA Meta-Theory Approach

Marta Weronika Wronikowska

Abstract One of the more complex systems of the real-world is the intelligent agent reasoning. The key elements of this mental activity is cognitive decision-making (CDM). The aim of this work is to recognize the initial situations of depended types of CDM which are realized by an intelligent agents, such as a human beings, their organizations and intelligent human-machine systems. The assumed framework of this classification is based on Gadomski's TOGA meta-theory (Top-down Object-based Goal-oriented Approach). The model of the information, preferences and knowledge (IPK) enables classification and recognition of the human cognitive errors in CDM. In parallel, the dependence of IPK on the assumed agent ontology is underlined.

Keywords Abstract intelligent agent · Cognitive decision-making · Complex systems · Meta-theory TOGA · Model IPK

53.1 Introduction and Theory

Independently of whether at professional work or in private life, we always feel constrained to make some choices. Whereas making choice is ending usually that what commonly and intuitively is called decision making. A variety of many situations leads to necessity for change our state and in consequence to choose the right action. A number of models of such decision processes particularly in the fields of engineering and economics was analyzed in many books. However their models mainly focused on the properties of the problem and not on individual abilities and preferences. It can be said, that many computational (this means possible to computer simulation) came into being, domain dependent type of decision making models, but not dependent on cognitive property so called decision-maker (policy-maker). On the other hand these models usually assumed its total rationality, not allowing either to take into account the emotional factors or biological limitations in inference process nor to consideration the actual individual knowledge and range of conceptual individual ontology of people who make decision. However cognitive

M.W. Wronikowska (✉)
ENEA: Italian National Agency for New Technologies, Energy and Sustainable Economic Development, UTFISST, Casaccia Research Centre, Via Anguillarese 301, Rome, Italy
e-mail: Marta.Wronikowska@yahoo.com

T. Gilbert et al. (eds.), *Proceedings of the European Conference on Complex Systems 2012*, Springer Proceedings in Complexity, DOI 10.1007/978-3-319-00395-5_53,

decision making is such a class of functional models, which assume independence from the application domain, and focused on specific functions of mind and the role of possession of ontology in the decision making process. A fundamental property of cognitive decision making is also the fact that this process can begin in a situation of insufficient data to make the expected decision (i.e. the final choice) [14].

In recent decades the cognitive decision-making (CDM) has always been a domain of more research activities in different application fields, see for example [3, 5, 13]. In parallel, for the above reason the authors used different comparisons or definitions of this concept. A few of them have been presented below:

- Cognitive decision-making process itself is based on a complex human psychological process made up of cognitive and motivational elements [7].
- Cognitive decision making remains a species of decision making and like other types of decision making, it should be the outcome of intelligent or rational means-ends deliberation [6].
- Cognitive (perspective) decision-making is a dynamic process in which temporally varying the noisy information is integrated across multiple time scales with a decision resulting when the information stream relevant of one of the alternative actions crosses a threshold [1].
- Cognitive decision making is a term popularly used to describe different situations. Very often, authors use this concept without defining it, or without quoting the definition constructed by someone else. Frequently authors try construct own definition but sometimes their results are poor. Critique leave involved in content of cognitive decision making. One can only add, that using terms without defining make many confusion; misunderstandings what every scientist should know.
- For the purposes of this work the following definition has been accepted:
- Cognitive decision making (CDM) is a complex decision-making patterned on the functioning of human mind with its biological constraints, and based on current information, knowledge and preferences of the decision maker.
- The above mentioned definition is congruent with a general frame of the TOGA meta-theory. Such an assumption enables to apply the TOGA's IPK (Information Preferences Knowledge) model as a basic conceptual framework for the modeling of cognitive decision making. Hence, CDM is described through two conceptualization layers model. One of them (information processing layer) is based on the IPK model and the second (decision making layer) on the criteria or alternatives searching in the simple decision-making model. Both layers have been described below.

53.1.1 Information Processing Layer

In the TOGA meta-theory, the IPK model [2] consists an elementary repetitive unit/cell of an intelligent agent.[1] The IPK cell is composed of 3 interacting key

[1]Intelligent agent (IA)—is composed of "Carrier of IA" (CIA—*physical system which is the carrier of an Abstract Intelligent Agent* (*AIA*) for example human body or computer hardware) and

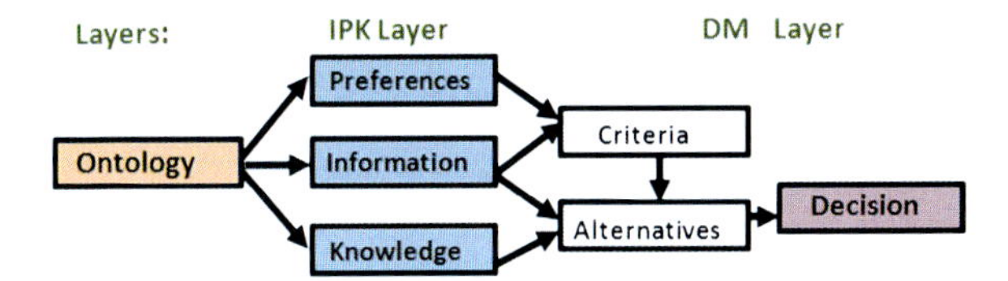

Fig. 53.1 The influence of ontology on the components of the two mentioned layers of CDM

systems. They are shortly represented by the concepts: information, preferences, knowledge. Where, the following definitions are assumed:

- Information (I)—the data which have the meaning and represent a specific property of a preselected domain of human or artificial agent's activity [12].
- Preferences (P)—are ordered relations among two states of the domain of the agent's activity [12].
- Knowledge (K)—is every abstract system, which is able to transform the information into the other information or knowledge [2].

For every IPK cell, all of them are related to one preselected domain of activity (D) of intelligent agent. The basic functional scheme [11] of the reasoning interaction between I, P and K can be presented (using known data processing: data → processing system → results) in four simplified sequences:

New Information (about D) → Information System → Information about state D.

Information about state D → Preferences System → Goal state of D. (Information about state D ∧ Goal state of D) → Knowledge System → selected Knowledge X.

Information about state D → Knowledge X → Information A → Action in D.

Where, *New Information* is an information portion/entity which arrives from D and modifies the agent's mental representation of the state of D domain in its Information System.

53.1.2 Decision-Making Layer

One can accept that the decision-making of an intelligent agent is based on the application of criteria to the alternatives, what by a procedural choice, produces a decision. In order to be used, criteria and alternatives have to be specialized, what usually requires information which specify different aspects of the current state of the agent domain of activity.

In the systemic CDM ontology, the definitions of its four key canonic concepts are the following:

- Criteria—are a set of rules which are used for a choice from a specified set of alternatives. For example: a maximal acceptable travelling time or a number of the operation steps. Those criteria depend on policy-maker's preferences.
- Alternatives—are actions available for realization by decision-maker, and resulting from this agent's knowledge base. whose realization can or should lead to the expected preferred state of the domain of activity (a goal state).

"Abstract Intelligent Agent" (AIA—*complex property of the dynamic structure of a system which has a capability to be intelligent*, for example human mind or computer software) see more [9].

– Decision—it is such alternative from an alternatives set, which satisfy previously accepted criteria.

Information has been defined before, but in the context of alternatives and criteria, it can be any part of the description of the state of a domain Q which can be involved in agent decision. Formally one can write:

$$D = A_D = Cr\{A\},$$

it means that decision D is a result of the choice performed by criteria operator on the set of alternatives. More precisely:

$$D(Q, I, P, K) = Cr[P, Ic]\{A_1(K_1, Ia_1), A_2(K_2, Ia_2) \cdots A_N(K_N, Ia_N)\} \quad (53.1)$$

where: D—is a decision related to the state of a domain Q. Cr—is a criteria system obtained from the agent preferences bases. P—the set of intelligent agent preferences. Ic—is an information being parameters of the criteria, and $Ic \in I_{sin}$, where I_{sin} is information about an initial decisional situation. K_n—the intelligent agent knowledge. Ia_n—the information being an parameter of the alternatives, and $Ia_n \in I_{sin}$, where $n = 1, \ldots, N$. $A_1 \cdots A_2 \cdots A_N$—a set of alternatives dependent on IA operational knowledge set K and their specializations Ia_n.

From the above perspective, the CDM of Y intelligent agent is also based on a local specific and the personal ontology Ω of Y.

Definition Every ontology is a such set of concepts and relations between them, which are necessary to describe a selected domain for a given class of goals.

In each case, before the initialization of the CDM process of an intelligent agent, we have to know the IA domain of activity and consequently, its goal-oriented IPK (Ω) bases.

If we intend to analyze the CDM of a few intelligent agents which cooperate for a common goal, then they have to have a common problem dependent ontology Ωp.

$$\Omega p = \text{Min. } \Omega 1 \cap \Omega 2 \cap \cdots \cap \Omega N \neq 0, \quad \text{for } N \text{ cooperating agents.}$$

General influence of the ontology choice on the a final decision in a goal-oriented problem is presented on the Fig. 53.1.

53.2 Modeling of Cognitive Decision Making

According to the TOGA meta-theory a decision-making starts with the agent situation which is not congruent with its maximal preferences and may rely either on the development of criteria while the alternatives are known or on the development of alternatives when the criteria have been established before (what has been illustrated by expression (53.1). Through taking into account the state of choice criteria, alternatives of a decision-maker and available information, one can distinguish eight cognitive situations (Si) presented in the Table 53.1. Some of them lead to the initialization of the CDM.

The situation S1 makes impossible the performing of decision-making because its goal is impossible to determine. On the other hand, the situation S2 activates the realization of just known behavioral procedure, because the possessed criteria, alternatives and information enable only the realization of a unique mental process

Table 53.1 Types of cognitive decision-making initial situations depending on criteria, alternatives and information [14]

CDM SITUATIONS	S1	S2	S3	S4	S5	S6	S7	S8
CRITERIA	−	+	+	−	+	−	+	−
ALTERNATIVES	−	+	−	+	−	+	+	−
INFORMATION	−	+	−	+	+	−	−	+

− not sufficient, + sufficient

and this is not considered CDM. Finally, S7 is logically impossible, when the situation S3, S4, S5, S6 and S8 are typical situations from which the cognitive decision making can start.

For example, the CDM starting from S3 requires to construct an operational knowledge, and at the consequence, requires information Ia or Ic, see (53.1). Starting from S4, an IA requires only new criteria building from an available do-main preference base.

Figure 53.2 illustrates only three characteristic CDM functional procedures which can require the repetition of information acquisitions.

The first type (CDM I) appears when in the initial state of decision-making lacks only alternatives (S5).

The second (CDM II) type occurs when there is the lack of information in the initial states of the decision-making (S7), which may enable to use available criteria and alternatives, initially, not yet specialized.

The third type (CDM III) is when in the initial state of CDM lack only criteria (S4). They can be acquired from a domain expert.

Remark In the case S1, the CDM is abandoned or can lead to randomly selected decision or so called meta-decision [10, 12].

53.3 Application and Results

The presented model of the cognitive decision making is congruent with the Universal Management Paradigm of TOGA [8], which should enable to use the developed CDM for the programming of decision support systems applied for the large scale emergency as well as autonomous robots. The CDM model can provide the information acquisition scheme to a client who intends to buy a car by top-down and goal-oriented restricting his domain of information search. This was presented in Wronikowska's master thesis [14]. The developed model frames of CDM are planned to be applied in the prototype of the intelligent decision support system for nuclear power plant operators.

53.4 Concluding Remarks

The recognized CDM models can be specialized in any area and for every type of intelligent agent, i.e. for a human being, a computer or a robot. Its basic invari-

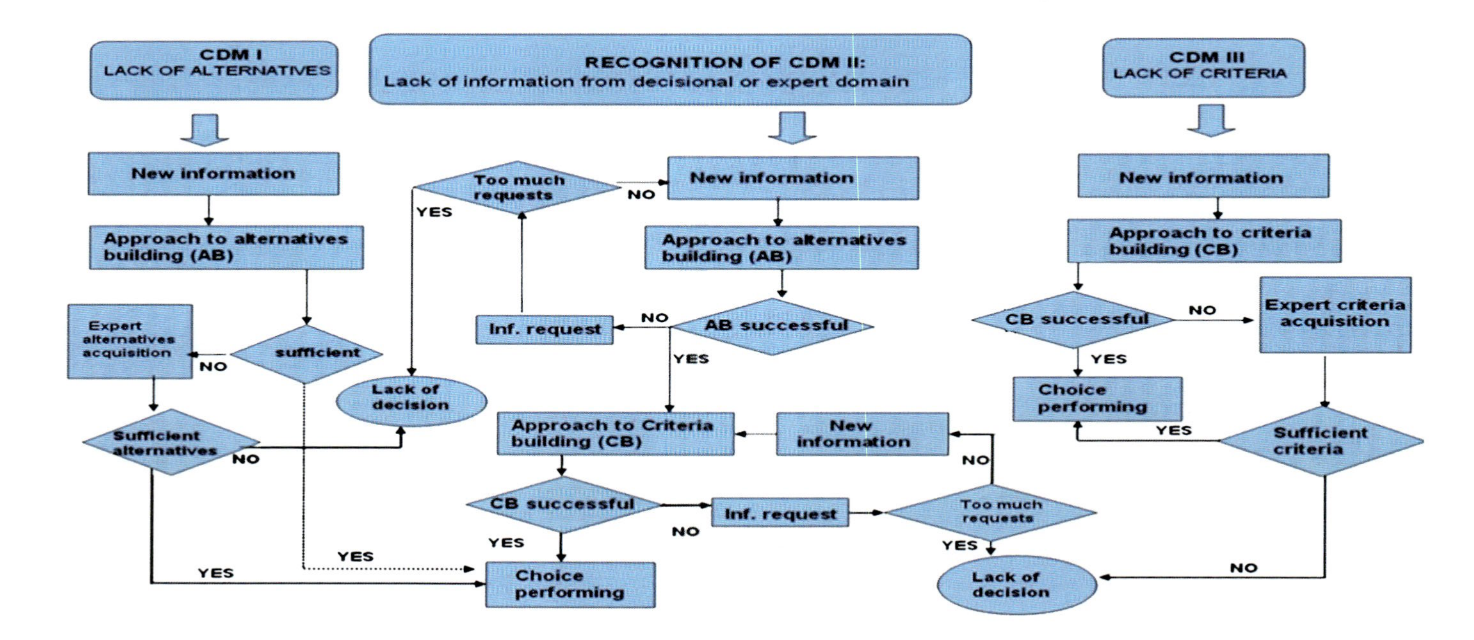

Fig. 53.2 An example of general scheme of the realization of second type of cognitive decision making which is driven by information acquired during the decision-making process

ant properties form the so called cognitive architecture of the cognitive decision-making system [4] and TOGA [8, 10, 12]. From the psychological perspective, the top-down specialization of the presented model enables to distinguish 2 types of preferences, knowledge, as well as, ontologies. One is rational and the second is emotional. The last can be caused by different sensual perception, and different psycho-somatic states of decision maker. The complexity and importance of such modeling and eventually future computer simulation seems to be important for the future research. The presented work is included in the Wronikowska research study performed subsequently to her master thesis (2010) at the Adam Mickiewicz University, the fellowship at the Italian Research Agency related to nuclear power plant operator's error, as well as, the initial stage of her PhD study at Sapienza University of Rome, 2012.

Acknowledgements I would like to thank Adam Maria Gadomski for a precious comments which I received during the edition of the last version of this article.

References

1. Anderson JJ (2008) Cognitive aspects of decision making. Workshop, Washington DC Crystal City Hiltonn
2. Dipoppa G, Gadomski AM, Vicoli G (2002) Intelligent advisor: cognitive user friendly interaction. WP1 technical report. The ITEA-SOPHOCLES project documentation, Rome
3. Duncan TR (2002) IMC using advertising and promotion to build brands. McGraw-Hill, Boston
4. Langley P, Laird JE, Rogers S (2009) Cognitive architectures: research issues and challenges. Cogn Syst Res 10(2):141–160
5. Levi I (1980) The enterprise of knowledge. An essay on knowledge. Credal probability and chance. MIT Press, Cambridge
6. Levi I (1997) Rationality and commitment. In: The covenant of reason: rationality and the commitments of thought. Cambridge University Press, Cambridge
7. Ogiela L, Ogiela MR (2009) Cognitive techniques in visual data interpretation. Studies in computational intelligence, vol 228. Springer, Berlin
8. Gadomski AM (1994) TOGA: a methodological and conceptual pattern for modeling of abstract intelligent agent. In: Proceeding of the first international round-table on abstract intelligent agent, Rome
9. Gadomski AM (1994) Intelligent-agent's worlds in the TOGA conceptualization. In: Proceedings of the Workshop on Intelligent information systems, Wigry, Poland, June 6–10, 1994, Instytut Podstaw Informatyki PAN, Warszaw
10. Gadomski AM (1998) Risk based reasoning in decision-making for emergency management. In: SRA Europe annual conference risk analysis: opening the process, Paris
11. Gadomski AM, Salvatore A, Di Giulio A (2003) Case study analysis of disturbs in spatial cognition: unified TOGA approach. In: Proceeding of 2nd international conference on spatial cognition (ICSC2003), Rome
12. Gadomski AM, Zimny TA (2009) Application of IPK (information, preferences, knowledge) paradigm for the modelling of precautionary principle based decision-making. In: Critical information infrastructure security. Springer, Berlin
13. Massaro DW (1998) Perceiving talking faces: from speech perception to a behavioral principle. MIT Press, Cambridge
14. Wronikowska MW (2010) The use of ontologies in cognitive decision-making. Master thesis (unpublished), Institute of Psychology, Adam Mickiewicz University, Poznan

Part IV
Biological Complexity

Chapter 54
Computing Birth-Death Fixation Probabilities for Structured Populations

Burton Voorhees

54.1 Introduction

Evolutionary dynamics is a broad and rapidly developing field, subsuming a wide variety of topics and directions of research. A relatively recent development has been the use of edge-weighted graphs to model interaction patterns in heterogeneous populations. While early studies suggested that at least some common population structures have no influence on fixation probability [1, 2], it is now clear that there are other population structures that can exert a strong influence, either suppressing or enhancing the effects of selection relative to drift [3, 4]. In this paper, an approach to studies of evolutionary processes on a graph G with N vertices is developed in terms of the full population state space, consisting of 2^N binary valued vectors, and a Markov process on this state space with transition matrix defined in terms of the edge weight matrix of G. Solution for the steady state of this process yields two vectors spanning the null space of the graph Laplacian of the state transition diagram. Components of the first of these vectors give extinction probabilities for specified distributions of mutants and components of the second vector give fixation probabilities. A parameter called graph determinacy is introduced, measuring the spread of fixation probability across the state space. A family of graphs called circular flows is defined, which includes cycles of arbitrary width and length, funnel and cascade graphs, star graphs, layered networks, and generalizations of these. The fixation probability for complete bipartite graphs is given for the first time. Comparison of several examples to the Moran process shows that a graph may enhance selection for only limited values of the fitness parameter.

54.2 Fixation Processes on Graphs

Consider a homogeneous population of N individuals consisting of m mutants with fitness r and $N - m$ normals with relative fitness 1, evolving in discrete time. At

B. Voorhees (✉)
Center for Science, Athabasca University, 1 University Dr., Athabasca, AB T9S 3A3, Canada

T. Gilbert et al. (eds.), *Proceedings of the European Conference on Complex Systems 2012*, Springer Proceedings in Complexity, DOI 10.1007/978-3-319-00395-5_54,

each iteration a random individual is chosen to reproduce and another (or the same) individual is chosen to die and be replaced by a clonal copy of the reproducing individual. The reproductive choices are biases by fitness: for $r > 1$, mutants are somewhat more likely to be chosen for reproduction. The population state at any time is represented by a length N binary vector $\vec{v} = (v_1, \ldots, v_N)$ where v_i is 0 or 1 respectively as vertex i is occupied by a normal or a mutant. This is the birth-death Moran process [5]. Mathematically, it is represented by a complete graph with N vertices, with all edges (including loops at each vertex) given a weight $1/N$, indicating the probability of an individual at any given vertex replacing an individual at an adjacent vertex. Much of the interest in this model has been in computing the probability of a single mutation going to fixation in the population. For the Moran process, this fixation probability ρ is well-known:

$$\rho = \frac{1 - \frac{1}{r}}{1 - \frac{1}{r^N}} \tag{54.1}$$

For an N-vertex directed graph G with edge-weight matrix W a population state is the N-dimensional binary vector $\vec{v} = (v_1, \ldots, v_N)$ with v_k equal to 0 or 1 as vertex k is occupied respectively by a normal or by a mutant. Thus, the full state space is $V(H^N)$, the vertex set of the N-hypercube. The matrix W defines a state transition matrix T on $V(H^N)$, and a corresponding state transition diagram $STD(G)$. The matrix T is easily constructed: For each state $\vec{v} = (v_1, \ldots, v_N)$ define vectors $\vec{a}(\vec{v})$, $\vec{b}(\vec{v})$ with components

$$a_j(\vec{v}) = \sum_{i=1}^{N} v_i w_{ij}, \qquad b_j(\vec{v}) = \sum_{i=1}^{N} (1 - v_i) w_{ij} \tag{54.2}$$

Thus, a_j is the probability that an edge from a mutant vertex terminates at vertex j and b_j is the probability that an edge from a normal vertex terminates at vertex j.

Theorem 54.1 *For a birth-death process defined on a graph G with edge-weight matrix W, let vectors $\vec{a}(\vec{v})$ and $\vec{b}(\vec{v})$ be defined as in Eq.* (54.2), *where $\vec{v}$ ranges over the entire state space. Then for all j, $1 \le j \le N$:*

1. *The probability of a transition from the state $(v_1, \ldots, v_{j-1}, 0, v_{j+1}, \ldots, v_N)$ to the state $(v_1, \ldots, v_{j-1}, 1, v_{j+1}, \ldots, v_N)$ is equal to*
$$\frac{r a_j(\vec{v})}{N - m + rm}$$
2. *The probability of a transition from the state $(v_1, \ldots, v_{j-1}, 1, v_{j+1}, \ldots, v_N)$ to the state $(v_1, \ldots, v_{j-1}, 0, v_{j+1}, \ldots, v_N)$ is equal to*
$$\frac{b_j(\vec{v})}{N - m + rm}$$
3. *The probability that $\vec{v}$ remains in the same state is equal to*
$$\frac{r\vec{a}(\vec{v}) \cdot \vec{v} + \vec{b}(\vec{v}) \cdot \vec{v}'}{N - m + rm}$$

The state transition diagram is an edge-weighted directed graph on 2^N vertices. The transition matrix T derived from Theorem 54.1 is row stochastic and has an eigenvalue 1 with corresponding eigenvector $\vec{x} = \underline{1}$ where $\underline{1}$ is the 2^N-dimensional vector of all ones.

Since the introduced mutation will either go extinct or go to fixation, the birth-death process has two absorbing states, represented by the N-dimensional vectors $\underline{0}$ and $\underline{1}$. The row T_{0j} is chosen to correspond to the extinction state $\underline{0}$ and the final row $T_{2^N-1,j}$ is chosen to correspond to the fixation state $\underline{1}$. Since T_{ij}^k is the probability of a k-step transition from state i to state j, the matrix $T^* = \lim_{k\to\infty} T^k$ consists of initial and final non-zero columns with all other entries equal to zero and the solution of $(I - T)\vec{x} = 0$ is

$$x_i = uT_{i0}^* + vT_{i,2^N-1}^* \equiv u\eta_i + v\mu_i \tag{54.3}$$

where u and v are free parameters. With this definition, η_i is the probability that state i goes to extinction and μ_i is the probability that it goes to fixation. If S_m is the subset of states containing m mutants, the probability of fixation from a single mutant vertex is

$$\rho = \frac{1}{N} \sum_{\vec{v} \in S_1} \mu_i \tag{54.4}$$

In general, solution of $(I - T)\vec{x} = 0$ involves the need to solve $2^N - 2$ linear equations. The number of equations can be reduced, however, if the graph G has a large automorphism group allowing partition of states into equivalence classes [6]. For a Moran process, for example, the equivalence classes are just the sets S_m for $0 \le m \le N$ and the state space is isomorphic to $\{0, 1, \ldots, N\}$. For any state with m mutants, let x_m be the probability that a member of S_m goes to fixation. For the Moran process with self-replacement $w_{ij} = 1/N$ for all i, j. Hence $\vec{a}(\vec{v}) = (\frac{m}{N})\underline{1}, \vec{b}(\vec{v}) = (\frac{N-m}{N})\underline{1}$ and the probabilities in Eqs. (54.2)–(54.4) are

$$\frac{rm}{N(N-m+rm)}, \qquad \frac{N-m}{N(N-m+rm)}, \qquad \frac{r(1+r)m^2 - 2Nm + N^2}{N(N-m+rm)} \tag{54.5}$$

To derive an equation for x_m, note that there are $N - m$ zeros that can become ones, and m ones that can become zeros, hence:

$$\left[1 - \frac{(1+r)m^2 - 2Nm + N^2}{N(N-m+rm)}\right] x_m - \frac{rm(N-m)}{N(N-m+rm)} x_{m+1} - \frac{m(N-m)}{N(N-m+rm)} x_{m-1} = 0$$

which reduces to

$$(1+r)x_m - rx_{m+1} - x_{m-1} = 0 \quad 1 \le m \le N-1 \tag{54.6}$$

with $x_0 = u$, $x_N = v$ as boundary conditions. These equations can be solved recursively, for the Moran process the solution of Eq. (54.8) is

$$x_m = \frac{1}{\sum_{k=0}^{N-1} r^k} \left(u \sum_{k=0}^{N-m-1} r^k + v \sum_{k=1}^{m} r^{N-k} \right) \tag{54.7}$$

Thus, for each μ_i such that the corresponding state is in S_m, the fixation probability is

$$\mu_i = \frac{\sum_{k=1}^{m} r^{N-k}}{\sum_{k=0}^{N-1} r^k} \tag{54.8}$$

and the overall single vertex fixation probability for S_1 is $N\mu_1/N$, or

$$\rho = \frac{r^{N-1}}{\sum_{k=0}^{N-1} r^k} \tag{54.9}$$

which is equivalent to Eq. (54.1).

For the general case, with $\vec{v} = (v_1, \ldots, v_N)$, $v'_j = 1 - v_j$, define s_v as the denary form of the binary number $(v_1, \ldots, v_N) : s_v = \sum_{k=1}^{N} v_k 2^{N-k}$. Then, for $\vec{v} \in V(H^N)$ the system of equations for $\vec{x}$ becomes

$$\begin{aligned} &\big[N + (r-1)m - r\vec{a}(\vec{v}) \cdot \vec{v} - b(\vec{v}) \cdot \vec{v}'\big] x_{s_v} - r \sum_{j=1}^{N} a_j(\vec{v}) v'_j x_{s_v + 2^{N-j}} \\ &\quad - \sum_{j=1}^{N} b_j(\vec{v}) v_j x_{s_v - 2^{N-j}} = 0 \end{aligned} \tag{54.10}$$

Since T is row stochastic, $I - T$ is just the graph Laplacian L of the state transition diagram and the vectors $\vec{\eta}$ and $\vec{\mu}$ span the null space of L.

The *determinacy* $\delta(G)$ of the population graph is defined as

$$\delta(G) = 1 - \frac{\vec{\eta} \cdot \vec{\mu}}{|\vec{\eta}||\vec{\mu}|} \tag{54.11}$$

Clearly $0 \le \delta(G) \le 1$. This quantity measures the degree to which extinction or fixation is pre-determined, starting from a randomly chosen initial state.

54.3 Suppressing and Enhancing Selection

If ρ_M is the fixation probability for a Moran process then any graph G for which $1/N \le \rho_G < \rho_M$ suppresses selection in favor of drift while if $\rho_M < \rho_G$ it enhances selection.

Let $\{G_s | 0 \le s \le k\}$ be a family of $k+1$ directed graphs with $|G_s| = n_s$ and with each G_s having a strictly sub-stochastic weight matrix W_s. For $N = \sum_{s=0}^{k} n_s$, let

$\{M_{s+1,s}\}$ be a set of $n_{s+1} \times n_s$ matrices, with $s+1$ evaluated mod$(k+1)$, such that the $N \times N$ matrix W with form

$$W = \begin{pmatrix} W_0 & 0 & 0 & 0 & \dots & 0 & M_{1,0} \\ M_{2,1} & W_1 & 0 & 0 & \dots & 0 & 0 \\ 0 & M_{3,2} & W_2 & 0 & \dots & 0 & 0 \\ 0 & 0 & M_{4,3} & W_3 & \dots & 0 & 0 \\ & & & & \vdots & & \\ 0 & 0 & 0 & 0 & \dots & W_{k-1} & 0 \\ M_{0,k} & 0 & 0 & 0 & \dots & 0 & W_k \end{pmatrix} \tag{54.12}$$

is row stochastic. The graph $G = \bigcup_{s=0}^{k} G_s$ with weighted edge matrix W will be called a circular flow. If (G, W) is a circular flow such that $n_{s-1} < n_s$ for $1 \leq s \leq k$ then (G, W) is a funnel while if this inequality is reversed it is a cascade. If all n_s are equal it is a cycle. A circular flow is simple and homogeneous if (a) $W_s = 0$ for all s, and (b) the matrix $M_{s,s+1}$ consists entirely of ones for all s (with $s+1$ evaluated mod$(k+1)$).

Since analytic treatment of circular flows in general is difficult, attention is restricted to simple homogeneous case, in which all level s states with the same number of mutants are equivalent and the full state space vector can be partitioned into blocks of length n_s. Population members of the resulting set of equivalence classes can be represented by a vector $\vec{m} = (m_0, m_1, \dots, m_k)$ where m_s indicates the number of mutants located at level s. Given a class label $(m_0, \dots, m_{s-1}, m_s, m_{s+1}, \dots, m_k)$ define

$$\iota(\vec{m}) = m_k + \sum_{s=0}^{k-1} m_s \prod_{j=1}^{k-s} (n_{s+j} + 1) \tag{54.13}$$

Using the notation $(m_0, \dots, m_{s-1}, m_s \pm 1, m_{s+1}, \dots, m_k) \rightarrow \iota(\vec{m}, m_s \pm 1)$ Eq. (54.10) becomes

$$\left[\sum_{s-0}^{k} \frac{m_s(n_{s+1} - m_{s+1}) + r m_{s+1}(n_s - m_s)}{n_s} \right] x_{\iota(\vec{m})} - \sum_{s=0}^{k} \frac{m_s(n_{s+1} - m_{s+1})}{n_s} x_{\iota(\vec{m}, m_s - 1)} - r \sum_{s=0}^{k} \frac{m_{s+1}(n_s - m_s)}{n_s} x_{\iota(\vec{m}, m_s + 1)} = 0 \tag{54.14}$$

The single vertex fixation probability is

$$\rho_F = \frac{1}{N} \left\{ n_k x_1 + \sum_{s=1}^{k} n_{k-s} x_{\iota(s)} \right\}, \quad \iota(s) = \prod_{j=k-s}^{k-1} (n_{j+1}) + 1 \tag{54.15}$$

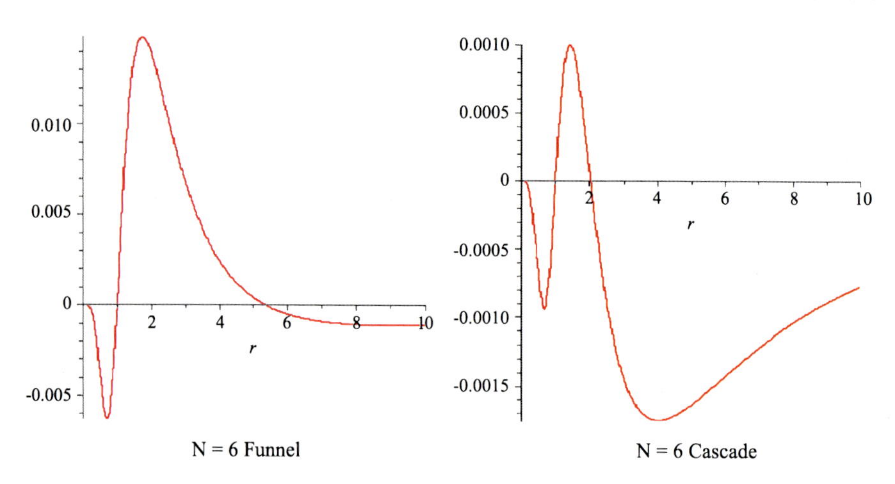

Fig. 54.1 Difference between fixation probabilities for funnel and cascade graphs and complete graphs on six vertices

Circular flows include cycles of arbitrary width, funnel and cascade graphs, and star graphs. With respect to star graphs, the general results obtained in [6] are reproduced. Inspection of the form of that result, together with examination of a number of examples of complete bipartite graphs $K_{s,n}$ $(n_0 = s, n_1 = n)$ leads to the following:

Conjecture The fixation probability for the complete bipartite graph $K_{s,n}$ is

$$\rho_S(s,n) = \left(\frac{r^{n+s-1}}{sr+n}\right)\left[\frac{(snr+n^2-sn+s^2)(nr+s)^{n-s-1}}{P(s,n)}\right] \tag{54.16}$$

where

$$P(s,n) = \frac{r^{n+s}(nr+s)^{n-s}-(sr+n)^{n-s}}{r^2-1}. \tag{54.17}$$

Equations (54.16) and (54.17) reduce to the n-star fixation probability for $s = 1$. The polynomial $P(s,n)$ also exhibits a number of interesting symmetry properties.

What is of particular interest in the case of funnel and cascade graphs, as well as for cycles with constrictions, is that selection is enhanced only for limited ranges of the fitness parameter. This result is new and unexpected. Figure 54.1 shows plots of the fixation probability minus the corresponding Moran fixation probability for an $N = 6$ graph that is a funnel or cascade depending on the direction of flow. The graphs of Fig. 54.1 enhance selection for only a limited range of fitness values. The funnel range for enhanced selection is $1 < r < 5.369515496$, for the cascade it is $1 < r < 2.030404551$.

54.4 Discussion

Suppression of selection is particularly significant as a defense against rapidly reproducing deleterious mutations, as in cancers [7], and may have value in attempts to control the spread of infectious diseases. Structures that enhance selection may prove valuable for models of sensory neural networks in which it is important to quickly identify a stimulus and produce an appropriate response. Mathematically, the fact that the matrix $I - T$ is a graph Laplacian provides a link to spectral graph theory (e.g., [8, 9]).

A point of interest in the cases studied in this paper is that funnels, cascades, and cycles with constriction provide enhancement of selection only for limited ranges of the fitness parameter. In neural networks, where interactions can be both excitatory and inhibitory, this suggests that inhibition and disinhibition of particular edges or vertices can act to control the degree of enhancement or suppression of selection. In order to begin addressing such possibilities, however, further work on networks with heterogeneous weights, as well as time depending weights is required.

Other directions for continued research are study of other simple graphs, graphs with non-uniform edge weight distributions, attempts to develop approximation methods for analysis of more complex graphs, and application to problems in evolutionary biology, neurology, and sociology.

References

1. Maruyama T (1974) A simple proof that certain quantities are independent of the geographical structure of populations. Theor Popul Biol 5(2):148
2. Slatkin M (1981) Fixation probabilities and fixation times in a subdivided population. Evolution 35:477
3. Shakarian P, Roos P, Johnson A (2011) A review of evolutionary graph theory with applications to game theory. Biosystems. doi:10.1016/j.biosystems.2011.09.006
4. Liberman E, Hauert C, Nowak MA (2005) Evolutionary dynamics on graphs. Nature 433:312. 7023
5. Moran P (1958) Random processes in genetics. Math Proc Camb Philos Soc 54(01):60
6. Broom M, Rychtár J (2008) An analysis of the fixation probability of a mutant on special classes of non-directed graphs. Proc R Soc A 464:2609
7. Nowak MA, Tarnita CE, Antal T (2010) Evolutionary dynamics in structured populations. Philos Trans R Soc B 365:19
8. Nowak MA, Michor F, Iwasa Y (2003) The linear process of somatic evolution. Proc Natl Acad Sci USA 100:14966
9. Chung FRK (1997) Spectral graph theory. CBMS regional conference series in mathematics. AMS, Providence

Chapter 55
Modeling of Spatially Extended Delay-Induced Circadian Oscillations Synchronized by Cell-to-Cell Communications

Dmitry A. Bratsun and Andrey P. Zakharov

Abstract We propose a spatially extended deterministic model with time delay for the circadian oscillations of protein concentrations. Our model is based on the non-linear interplay between two proteins forming a time-delayed feedback loop comprised both positive and negative elements. In order to study spatio-temporal dynamics of the system, a novel algorithm of the numerical simulation of time-delayed reaction-diffusion systems is proposed. The algorithm based on finite difference method involves storing in a computer memory not all, but some selected nodal data, and the subsequent interpolation to determine intermediate values. Spatio-temporal protein patterns excited in complete darkness are studied numerically. It is shown that the synchronization of biorhythms can be produced by either of two mechanisms: (i) basal transcription factors and (ii) cell-to-cell communication.

Keywords Time-delay · Circadian rhythms · Pattern formation

Circadian rhythms are biological rhythms that are common to almost all living organisms. A remarkable feature of these rhythms is that they are not simply a response to 24 hours environmental cycles imposed by the Earth's rotation, but instead are generated internally by cell autonomous biological clocks. After the decades of research, the genetic mechanism of circadian oscillations has been widely recognized as a core of this phenomenon. Thus, the transcription/translation processes should be taken into account seriously when one starts to model the circadian rhythms. As it is known now, a feedback influence of protein on its own expression can be delayed which leads to non-Markovian phenomena in this system [1]. It is evident that the delay prevents the system from achieving equilibrium, and results instead in the familiar limit cycle oscillations. The deterministic and stochastic properties of gene regulation taking into account the non-Markovian character of gene transcription/translation was studied in [2, 3]. We have shown that time delay in the protein production or degradation may change the behavior of the system from sta-

D.A. Bratsun (✉) · A.P. Zakharov
Theoretical Physics Department, Perm State Pedagogical University, Perm, Russia
e-mail: dmitribratsun@rambler.ru

T. Gilbert et al. (eds.), *Proceedings of the European Conference on Complex Systems 2012*, Springer Proceedings in Complexity, DOI 10.1007/978-3-319-00395-5_55,

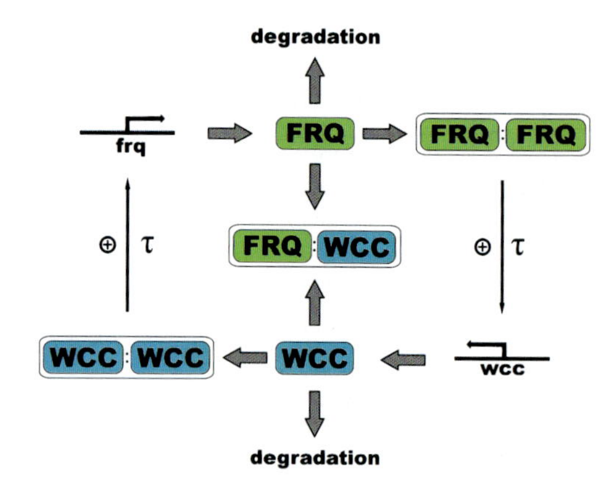

Fig. 55.1 Network architecture of the circadian rhythm molecular components in *N.crassa*

tionary to oscillatory even when a deterministic counterpart of the stochastic system exhibits no oscillations.

The filamentous fungus Neurospora crassa is an excellent model system for investigating the mechanism of circadian rhythmicity because of the wealth of genetic and biochemical techniques available. It is easy to grow and has a haploid life cycle that makes genetic analysis simple since recessive traits will show up in the offspring. The genome of N. crassa was recently reported as completely sequenced and all data are freely available online. The genome is about 43 megabases long organized in 7 chromosomes and includes approximately 10000 genes. With advances in molecular biology, understanding of the Neurospora circadian clock has improved, and main genetic components of this clock have been determined. Please see Fig. 55.1 for a simplified graphical depiction of this network architecture.

The primary molecular components of the circadian oscillator are the frequency and white collar genes (white collar 1 (wc-1) and white collar 2 (wc-2)) which form a feedback loop comprised of both positive and negative elements [4]. The corresponding white-collars proteins WC-1 and WC-2 are transcription factors which form a heterodimeric complex known as the white collar complex WCC. The WCC acts as a positive regulator of FRQ by activating its transcription in the dark and in response to blue light (WC-1 is a photoreceptor), while the frequency protein dimerizes and then acts as a negative regulatory element by binding to and inhibiting the function of WCC. As the circadian cycle progresses the FRQ protein is phosphorylated and degraded which allows the cycle to begin anew. Also, these species can be removed by association with WCC to form FRQ/WCC complexes. Furthermore the production of WC-1 and FRQ proteins are subject to a delay on the order of several hours after the mRNAs are made again mediated through an unknown post transcriptional mechanism. The previous experimental efforts have highlighted also the importance of degradation of the core clock components, particularly that of FRQ, plays in establishing the period of the circadian rhythm. However quantitative information about the magnitude of the degradation rates are still lacking for this core clock component.

In this work, we propose the model of temporal circadian dynamics of the Neurospora which is work is a further simplification of models proposed by Smolen et al. [4] and Sriram and Gopinathan [5]. The main difference to earlier approaches is that we derive the dynamic equations directly from a set of biochemical reactions. In the part of spatially extended delay-induced circadian oscillations in Neurospora our modeling seems to be first in the literature. Perhaps this can be explained by a prevailing tradition in the study of circadian oscillations. It is possible also that this is due to computational difficulties that arise when studying spatially extended reaction-diffusion systems with time delay. To overcome these difficulties, we propose a new method for the numerical study of such systems and focus on the deterministic spatio-temporal dynamics neglecting the stochasticity of the system.

In the model of the N. crassa oscillator we assume there are two primary components: the heterodimeric WCC complex and the FRQ protein. We start our analysis from a set of biochemical reactions constituting the mechanism of bioclock and finally arrive to the following two-variable spatially extended systems [3]:

$$\frac{\partial F}{\partial t} = \frac{1}{1+4K_1^F F}\left(A_F + k_F \frac{K_1^W K_2^F \varphi(t) W^2(t-\tau)}{1+K_1^W K_2^F \varphi(t) W^2(t-\tau)} - B_F F - kFW\right) + D\left(\frac{\partial^2 F}{\partial x^2} + \frac{\partial^2 F}{\partial y^2}\right), \tag{55.1}$$

$$\frac{\partial W}{\partial t} = \frac{1}{1+4K_1^W W}\left(A_W + k_W \frac{K_1^F K_2^W F^2(t-\tau)}{1+K_1^F K_2^W F^2(t-\tau)} - B_W W - kFW\right) + D\left(\frac{\partial^2 W}{\partial x^2} + \frac{\partial^2 W}{\partial y^2}\right). \tag{55.2}$$

Here F and W stand for number of isolated monomers of FRQ and WCC respectively and D is the coefficient of protein diffusion in the cell. For simplicity, we assume that the diffusion coefficients of FRQ and WCC proteins are equal. Even supposing that the delay is defined by the length of the path traveled by RNA polymerase along the gene, one obtains different values since the wc-1 gene (it is part of the locus NCU02356.5) is a one and a half times longer than the frq gene (it is in the locus NCU02265). But the exact values of the delays are currently unknown, and for simplicity we assume that time delays have the same values: $\tau_F = \tau_W = \tau$. The model (55.1)–(55.2) includes a positive feedback loop in which activation of FRQ production by WCC increases the level of FRQ, leading to an increase in the level of WCC itself. The negative feedback loop in which FRQ represses the frq gene transcription by binding to the WCC is also modeled.

As it is known, the Neurospora not only has the advantage that powerful genetics and molecular techniques are able to be performed on it, but it has another advantage—circadian rhythms of conidiation that is easily monitored on Petri dishes. To observe the phenotypic expression of the Neurospora clock, conidia are inoculated at some place of a Petri dish. After growth for a day in a constant light, the position of the growth front is marked and the culture is transferred to constant dark. The light-dark transfer synchronizes the cells in the culture and sets the clock running from subjective dusk. Following transfer, the growth front is marked every 24 hours with the aid of the red light, which has no effect on the clock. Growth rate is constant and the positions of the readily visualized orange conidial bands (separated by undifferentiated mycelia) allow determination of both period and phase of the rhythm. Thus, the computational domain $\Sigma \in (x, y)$ where the protein fields are solved numerically can be interpreted as a flat area of two-dimensional physical space of a Petri dish occupied by the mycelium of Neurospora. In fact, N. crassa is multicellular organism, and the translation of proteins occurs within individual cells. But we can consider the mycelium of the fungus as a whole due to the important feature of Neurospora: the mycelium of the organism consists of branched hyphae which show apical polar growth. The fungal hyphae are typically composed of multiple cells or compartments demarcated by septa with the central pore sometimes up to 0.5 microns in diameter. Thus, the protein produced in the separate cells of Neurospora seems to be able to cross the intercellular walls, and we can assume an existence of joint molecular cloud of protein inside a whole organism.

In order to perform two-dimensional simulations of delayed-induced circadian oscillations governed by Eqs. (55.1)–(55.2), we define a two-dimensional domain $\Sigma : (0 < x < 200, 0 < y < 200)$ with the following conditions for concentrations of FRQ and WCC proteins imposed at the boundary of the domain:

$$\left.\frac{\partial F}{\partial x}\right|_{x=0,200} = 0, \quad \left.\frac{\partial F}{\partial y}\right|_{y=0,200} = 0, \quad \left.\frac{\partial W}{\partial x}\right|_{x=0,200} = 0, \quad \left.\frac{\partial W}{\partial y}\right|_{y=0,200} = 0. \tag{55.3}$$

The initial-boundary value problem (55.1)–(55.3) has been solved by finite difference method described in the previous section. The explicit scheme was adopted to discretize equations. The equations and boundary conditions have been approximated on a rectangular uniform mesh 400×400 using a second order approximation for the spatial coordinates. In some calculations, we introduce the boundary Γ separating the region where the Reactions (55.1)–(55.2) take place and the nonreactive area.

If the protein can easily pass through this boundary, we consider it as a free interface between immiscible fluids. Then the concentration of reactant which diffuses through the surface subjects to the boundary condition

$$F_1|_\Gamma = F_2|_\Gamma, \qquad D_1 \left.\frac{\partial F_1}{\partial \mathbf{n}}\right|_\Gamma = D_2 \left.\frac{\partial F_2}{\partial \mathbf{n}}\right|_\Gamma, \tag{55.4}$$

where $\mathbf{n}$ is the vector normal to the interface, $D_{1,2}$ are the coefficients of protein diffusion in the areas on opposite sides of Γ.

The spatial phase synchroniation is the process when spatially distributed cyclic signals tend to oscillate with a repeating sequence of relative phase angles. We have noticed above that some external stimuli can synchronize the spatio-temporal behaviour of the system. The external control of this active medium can be performed, for example, via the basal transcription factors. We found the system is particularly sensitive to the basal transcription of the WCC protein governed by the parameter A_W. With increase of WCC produced via the basal transcription machinery, the spatio-temporal structure of the system becomes more ordered. In contrast to the distinct chaotic pattern at $A_F = A_W = 0$ formed due to break up of spiral waves,

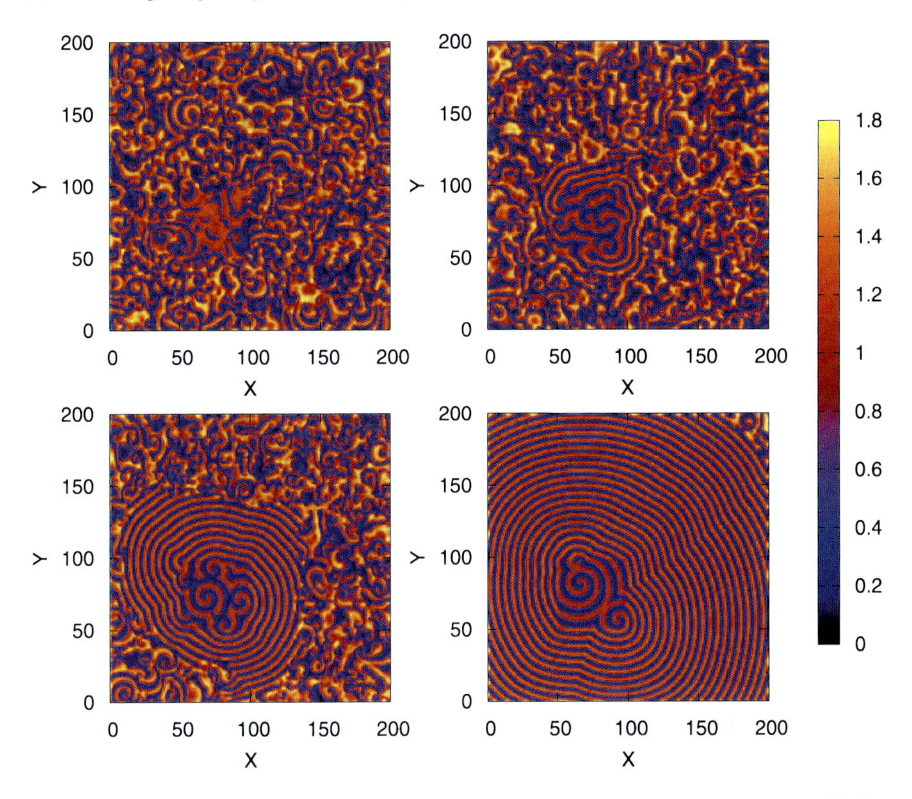

Fig. 55.2 The spatial synchronization of oscillations of the FRQ protein in the system with the basal transcription machinery, acting locally in the square area $50 < x < 100$, $50 < y < 100$ in the left lower quadrant of the domain Σ. In the rest of the domain the basal transcription is inhibited. *The frames* from *left* to *right* and from *up* to *down* correspond to times $t = 5000$, 7200, 11000, 20000 respectively. The coefficient of diffusion is $D = 0.01$

the structures for large values A_W looks more ordered. Since the basal transcription can be repressed, it can be switched on only in some places of the organism. So it would be interesting to see what happens when the basal transcription factors have been activated locally. Figure 55.2 gives an example of such numerical simulation. It is assumed that the basal transcription machinery ($A_F = 0$, $A_W = 2$) acts only in the area $50 < x < 100$, $50 < y < 100$ in the left lower quadrant of the domain Σ. It becomes effective at time $t = 5000$. We see that local basal transcription results in the global effect: it produces spatially synchronized oscillations over the whole domain Σ. From the point of view of nonlinear dynamics, the spiral traveling wave has recovered its structure. So, the process shown in figure is reverse to the core break up of spiral wave discussed above. It should be noted that the process of self-organization goes far beyond the area where the basal transcription works.

Finally we present the results of the numerical simulations showing that the synchronization of the circadian oscillations can occur due to intercellular communications via chemical signals (Figs. 55.3 and 55.4). We have attempted to simulate this

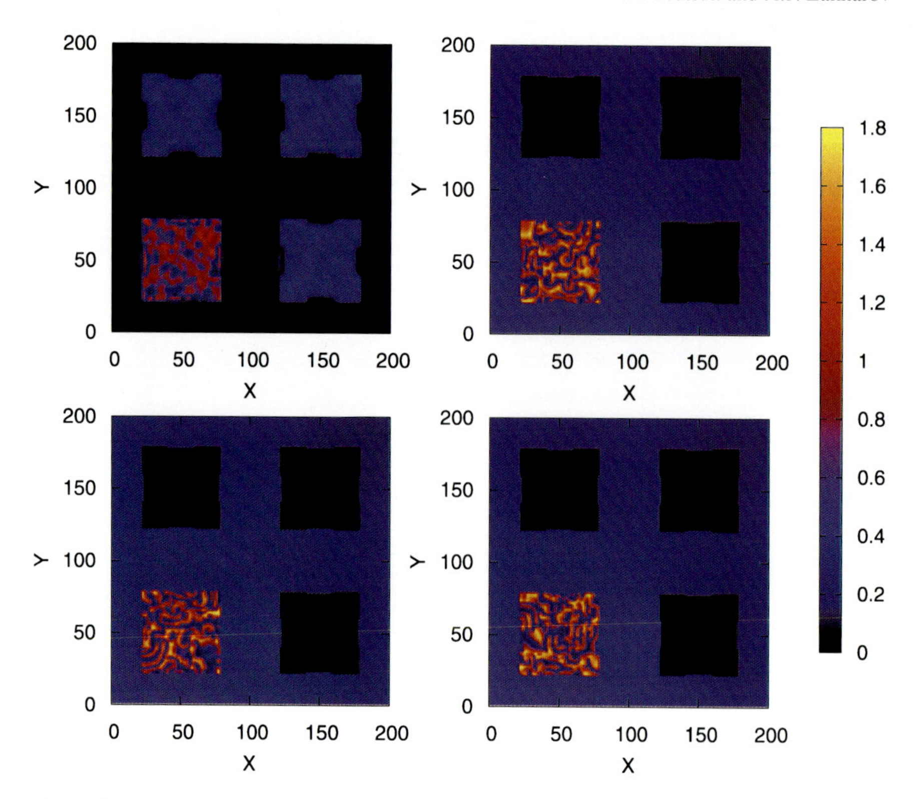

Fig. 55.3 Evolution of the concentration of the FRQ protein in complete darkness. There are four "cells" in the domain of integration, but only one of them is switched on. *The frames* from *left* to *right* and from *up* to *down* correspond to times $t = 200, 5000, 10000, 20000$ respectively

process by considering the four square areas of 60 to 60 referred to below as "cells", within which the system of Eqs. (55.1)–(55.2) for circadian oscillations has been integrated. The mixed boundary conditions have been imposed on the border of the area. There are four segments of the open border ("membranes") with the condition (55.4) of free diffusive penetration of the reagent species through fluid-fluid interface. The rest of the border is considered to be impermeable in accordance with the formula (55.3). Thus, each closed area is a rough model of living cell with its own circadian oscillations occurring inside the cell. Each area has four membranes and can communicate with the extracellular world via exchange of a certain protein.

We have supposed that such carrier of signals connecting the cell to the extracellular world is the FRQ protein. At the same time, we assume that the complex WCC cannot overcome the membrane and exists only inside the cell. Being in the extracellular space the protein does not react, but it can diffuse. Its dynamics obeys the standard diffusion equation.

Figure 55.3 presents the evolution of the concentration of FRQ in complete darkness and zero basal transcription, when only one cell (#3) is switched on. At these parameters the spiral wave develops into phase turbulence due to mechanism of the

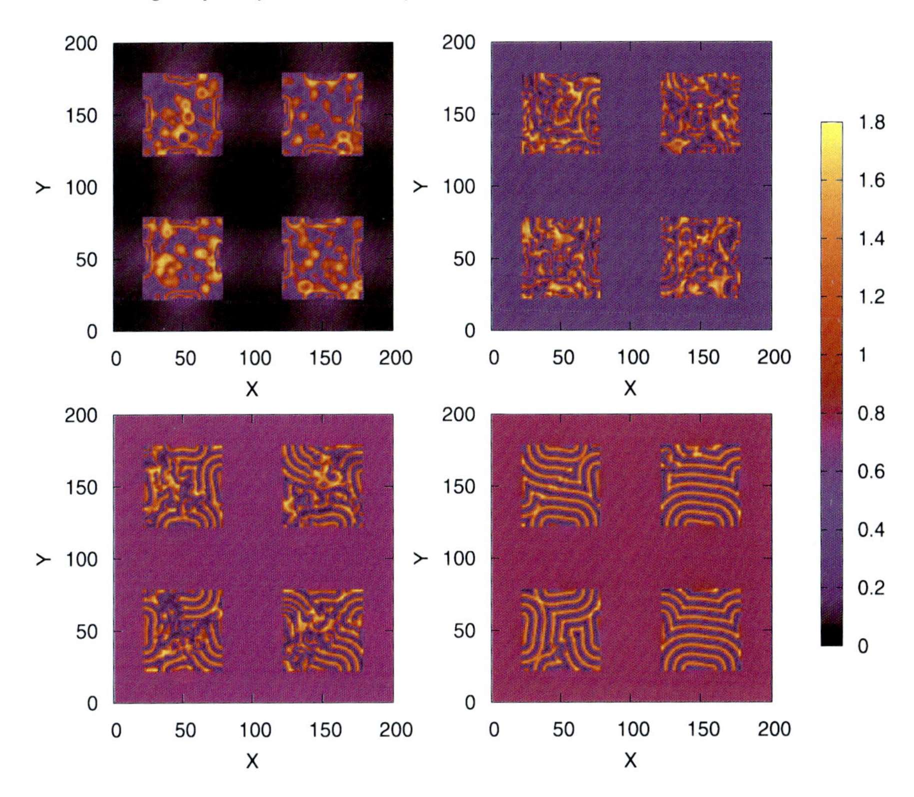

Fig. 55.4 The spatial synchronization of oscillations of the FRQ protein due to intercellular communications. *The frames* from *left* to *right* and from *up* to *down* correspond to times $t = 200, 1000, 5000, 15000$ respectively

core breakup. If the system does not receive any external signals, the chaotic state can be maintained for indefinitely long time periods. In the case shown in Fig. 55.3, there is the flow of FRQ from the cell #3 outward. The protein gradually fills the extracellular space due to diffusion. As long as the cell is the only cell in the system, a state of chaos continues to persist. The spatial synchronization of circadian oscillations occurs when there are a whole group of functioning cells in the system (Fig. 55.4). We found that the state of spatially synchronized bioclocks working in the different cells is achieved, when the concentration of transmitted protein in the extracellular space reaches a certain average value. The oscillations in the cell group are synchronized, when the level of concentration is about $F \approx 0.8$. Thus, in the group consisting of a large number of cells, the spatial synchronization seems to be attained more quickly between the cells that intensively produce the protein and exchange more vigorously the signals with their neighbors. This conclusion is supported by the recent experimental findings demonstrating spatio-temporal synchronization of circadian oscillations in a population of E. Coli [6]. The authors have noticed that oscillations arise because the acyl-homoserine lactone (this is a small molecule that can diffuse across the cell membrane) has a dual role, both enabling activation of the genes necessary for intracellular oscillations and mediating the coupling between cells.

Acknowledgements The work was supported by the Department of Science and Education of Perm region (project C26/244), the Ministry of Science and Education of Russia (project 1.3103.2011) and Perm State Pedagogical University (project 031-F).

References

1. Liu Y, Loros JJ, Dunlap JC (2000) Phosphorylation of the Neurospora clock protein frequency determines its degradation rate and strongly influences the period length of the circadian clock. Proc Natl Acad Sci USA 97:234–239
2. Bratsun D, Volfson D, Hasty J, Tsimring LS (2005) Delay-induced stochastic oscillations in gene regulation. Proc Natl Acad Sci USA 102:14593–14598
3. Bratsun DA, Zakharov AP (2011) Modelling spatio-temporal dynamics of Circadian rhythms in Neurospora crassa. Comput Res Model 3:191–213 (Russian)
4. Smolen P, Baxter DA, Byrne JH (2001) Modeling circadian oscillations with interlocking positive and negative feedback loops. J Neurosci 21:6644–6656
5. Sriram K, Gopinathan MS (2004) A two variable delay model for the circadian rhythm of Neurospora crassa. J Theor Biol 231:23–38
6. Danino T, Mondragon-Palomino O, Tsimring L, Hasty J (2010) A synchronized quorum of genetic clocks. Nature 423:326–330

Chapter 56
Topology Drives Calcium Wave Propagation in 3D Astrocyte Networks

Jules Lallouette and Hugues Berry

Abstract Glial cells are non-neuronal cells that constitute the majority of cells in the human brain and significantly modulate information processing via permanent cross-talk with the neurons. Astrocytes are also themselves inter-connected as networks and communicate via chemical wave propagation. How astrocyte wave propagation depends on the local properties of the astrocyte networks is however unknown. In the present work, we investigate the influence of the characteristics of the network topology on wave propagation. Using a model of realistic astrocyte networks (>1000 cells embedded in a 3D space), we show that the major classes of propagations reported experimentally can be emulated by a mere variation of the topology. Our study indicates that calcium wave propagation is favored when astrocyte connections are limited by the distance between the cells, which means that propagation is better when the mean-shortest path of the network is larger. This unusual property sheds new light on consistent reports that astrocytes *in vivo* tend to restrict their connections to their nearest neighbors.

56.1 Introduction

More than half of the cells in the human brain are glial cells. These non-neuronal cells have recently been evidenced to play a direct active role in information transfer in the brain. Indeed astrocytes (the main subtype of glial cells) not only react to but can also modulate synaptic communication between neurons [8, 16]. Astrocytes are also themselves inter-connected as networks on which chemical waves propagate [9]. Understanding astrocyte-astrocyte and astrocyte-neuron cross-talks is thus crucial to understanding the brain [8]. Chemical communication within astrocyte networks is characterized as elevations of intracellular calcium that propagate from cell to cell though protein channels called gap-junctions channels (GJC). However,

J. Lallouette (✉) · H. Berry
BEAGLE, INRIA RhôneAlpes, Université de Lyon, LIRIS, UMR5205, Villeurbanne, France
e-mail: jules.lallouette@inria.fr

H. Berry
e-mail: hugues.berry@inria.fr

T. Gilbert et al. (eds.), *Proceedings of the European Conference on Complex Systems 2012*, Springer Proceedings in Complexity, DOI 10.1007/978-3-319-00395-5_56,

depending on the experimental conditions, the reported speed and extent of the propagation (the number of cells that participate in the waves) vary over a large range. These discrepancies can be interpreted as many different intracellular biochemistry conditions or as different topologies of the astrocyte networks. Here, we questioned the latter hypothesis, that the various propagation ranges observed experimentally may be explained on the sole basis of the topology of astrocyte networks.

In the brain, astrocytes are believed to occupy separate spatial territories, connecting to their nearest astrocyte neighbors only [2, 15]. Beyond this general setting, the structure of astrocyte networks is still unknown. Recent experimental evidence however suggests that astrocytes networks display different topologies depending on the brain region [18] and even that neuronal activity can modify these topologies by regulating inter-astrocyte GJC [9]. In contrast, in the modeling literature, most articles consider astrocyte networks embedded in a two dimensional space [7, 13] and connected using regular lattices [10, 11, 13]. The effect of more complex or more realistic topologies has been restricted to small networks (5–10 cells) [6, 12], that do not allow comparison of propagation ranges with biological data.

In the present work, we study a model of realistic astrocyte networks (>1000 cells embedded in a 3d space) and investigate how the characteristics of the network topology affect calcium wave propagation at a network level. We show that indeed, the major classes of observed propagations can be emulated by a mere variation of the topology. Our study indicates that calcium wave propagation is favored when astrocyte connections are limited by the distance between the cells, which means that propagation is better when the mean-shortest path of the network is larger. Altogether, our findings offer a sound theoretical understanding of calcium wave propagations in astrocyte networks.

56.2 The Model

Our model consists of two main parts: a model describing calcium dynamics inside each astrocyte on the one hand, and the network topology models to connect astrocytes, on the other hand.

56.2.1 Calcium Dynamics Model

Albeit experimental protocols monitor wave propagation as variations of intracellular calcium, the molecule that is transmitted through GJC to connected astrocytes is not calcium but another messenger, called IP_3. When IP_3 level in the cell cytoplasm is large enough, a calcium surge is released from intracellular calcium stores to the cell cytoplasm. In turn, since the IP_3 producing (and degrading) enzymes are activated by cytoplasmic calcium, this calcium elevation will lead to increased IP_3 levels, some of which can be transported through a GJC to a connected astrocyte.

The IP_3 entering the connected cell can then regenerate the original calcium signal if the transferred IP_3 amount is large enough.

To model the dynamics of this system we used the ChI model, developed and studied at the single-cell level in [10, 17]. This model uses three coupled nonlinear equations to describe calcium dynamics in each astrocyte i:

$$\frac{dC^i}{dt} = J_{chan}(C^i, h^i, \mathrm{IP}_3^i) + J_{leak}(C^i) - J_{pump}(C^i)$$

$$\frac{dh^i}{dt} = (h_\infty(C^i, \mathrm{IP}_3^i) - h^i)/\tau_h(C^i, \mathrm{IP}_3^i)$$

$$\frac{d\mathrm{IP}_3^i}{dt} = P_{PLC\delta}(C^i, \mathrm{IP}_3^i) - D_{3K}(C^i, \mathrm{IP}_3^i) - D_{5P}(\mathrm{IP}_3^i) + J_{net}^i$$

where C^i, h^i and IP_3^i are respectively the cell-averaged calcium concentration, the fraction of open intracellular IP_3 receptors at the calcium stores and the cell-averaged concentration of IP_3 messenger of cell i. The terms in the RHS represent calcium flux between intracellular compartments or internal enzyme kinetics and can be found in [10, 17]. $J_{net}^i = \sum_{j \in N(i)} J_{j \to i}$ represents the total flux of IP_3 from neighbouring cells j to cell i. To model the IP_3 flux $J_{j \to i}$ though GJC between cells j and i, we use a non-linear coupling (see [10] for justification):

$$J_{j \to i} = \frac{F}{2}\left(1 + \tanh\left(\frac{|\Delta_{ji}\mathrm{IP}_3| - \mathrm{IP}_3^{thr}}{\mathrm{IP}_3^{scale}}\right)\right)\frac{\Delta_{ji}\mathrm{IP}_3}{|\Delta_{ji}\mathrm{IP}_3|}$$

where F is the coupling strength, $\Delta_{ji}\mathrm{IP}_3$ is the IP_3 difference between cells j and i and IP_3^{thr} is a threshold below which the flux is small. Note that $J_{j \to i}$ can be positive or negative, depending on the sign of the IP_3 gradient.

Except when explicitly indicated below, all parameters were taken from [10] (FM encoding conditions). All the resulting coupled ODEs were numerically integrated using a 4th order Runge-Kutta scheme with a time step of 10 ms. The model was implemented in C++ and was run on the IN2P3 Computing Center which we thank for providing the computer resources.

56.2.2 Network Topologies

Spatial Structure One major goal of our study is to obtain propagation measures in the model that can be compared to experimental data (propagation extent, speed, etc.). A recent experimental study [19] has exhibited the distribution of the nearest astrocyte-astrocyte distance in mice brains (mean distance $\mu_{exp} = 50$ μm, coefficient of variation $cv_{exp} = 0.25$, minimum distance $m_{exp} = 20$ μm). In accordance with these results, we reproduced this distribution by first placing N astrocytes at the nodes of a regular 3-dimensional cubic grid (interspacing distance a_{grid}). The position of each cell was then jittered by adding to its coordinates a random normal variable (zero mean, variance σ). From a simple grid search, the best match between this model and the distribution obtained in [19] was found with $a_{grid} = 70$ μm, $\sigma = 55$ μm.

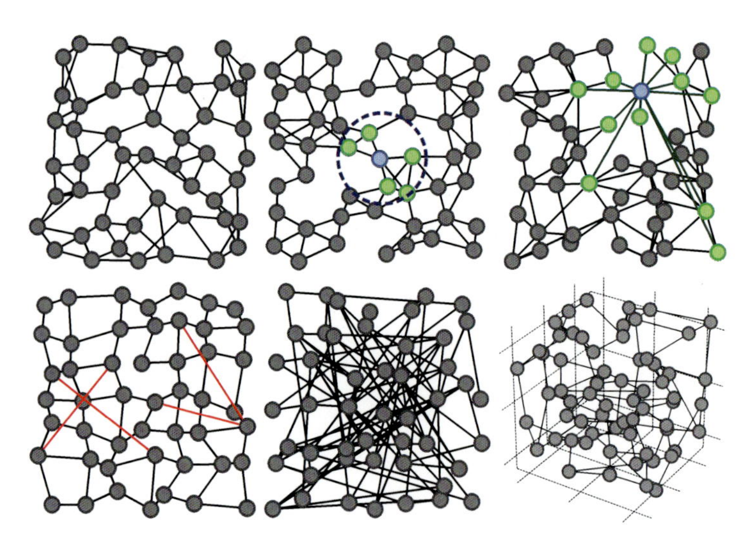

Fig. 56.1 Representations of networks obtained with each model. From *left* to *right* and *top* to *bottom*: regular degree, link radius, spatial scale free, shortcut, Erdős-Rényi and three dimensional version of a regular degree network. Note that most of these illustrations are two-dimensional for readability, but the networks used in the present study are all three-dimensional

Topological Structure We investigated five kinds of network topologies:

- *Regular degree* networks are constructed by linking each astrocyte to its k_{reg} nearest neighbors in space.
- *Link radius* networks are spatial networks in which each astrocyte is linked to all the astrocytes found within distance r_{link}.
- *Spatial scale free* networks are built incrementally by the classical preferential attachment rule, but taking spatial distances into account (detailed description in [1]). In short, a parameter (r_c) controls the trade-off between scale-free structure and the link restriction to short intercell distances.
- *Shortcut* networks are based on 3d cubic lattices. Each edge is then rewired with probability p_s: one of the edge end is replaced by a (uniformly) randomly chosen cell. Those networks are similar to the classical *small-world* topologies.
- *Erdős-Rényi* networks are the only non-spatial networks: each pair of node is linked with a probability p_{er} independently of distance.

Two-dimensional illustrations of the networks created with each of these models are shown in Fig. 56.1. Each of these topologies present specific features that are not necessarily biologically realistic but will provide us insights into what kind of features networks should or should not have in order to support realistic wave propagation. In each case, we varied the parameters specific to the topology so that the mean degree of the network $2 \leq \langle k \rangle \leq 17$. For each parameter value, the measurements were averaged over 20 realizations of the topology with the given parameters. All networks had $N = 1331$ astrocytes.

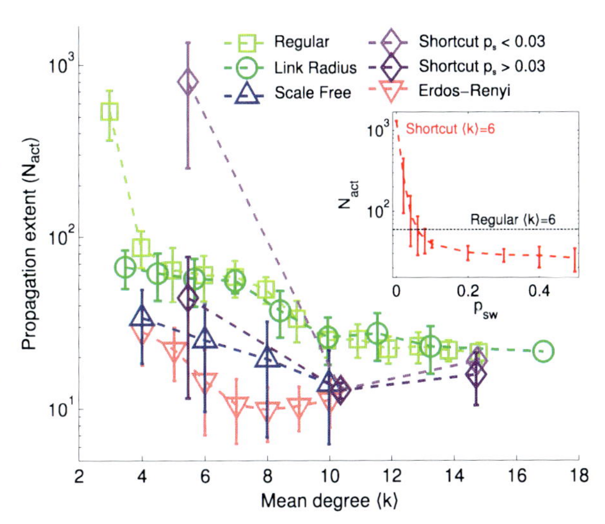

Fig. 56.2 Propagation extent N_{act}, as a function of the mean degree of the networks $\langle k \rangle$. Values are grouped in classes of network models and error bars plot the standard deviation in log scale. *The inset* contains the propagation extent for shortcut networks with $\langle k \rangle = 6$ and as a function of p_s (*red curve*), and the mean propagation extent for regular degree networks, with parameter $k_{reg} = 6$

56.2.3 Stimulation

All the experiments are conducted using the same stimulation procedure: we stimulate one astrocyte in the center of the network (to avoid boundary effects on spatial networks) by coupling it during a stimulation time t_{stim} via GJC to a virtual astrocyte whose IP_3 cell concentration is kept constant at IP_3^{bias}. All simulations are carried on during $t_{sim} = 200$ s, giving enough time for a calcium wave to fully propagate to its maximum extent. The total number of activated cells during the simulation was found to be independent of t_{stim} and IP_3^{bias} for $t_{stim} \geq 100$ s and $IP_3^{bias} \geq 2$ μM, so we fixed during all simulations $t_{stim} = t_{sim}$ and $IP_3^{bias} = 2$ μM.

56.3 Results

In order to distinguish topological effects from the effects introduced by GJC or biophysical parameters, we first carried out series of simulations with fixed GJC parameters ($F = 2$ μM s^{-1} and $IP_3^{thr} = 0.3$ μM) while varying only the network topology.

56.3.1 Topological Influences

To ease comparison with experimental studies, we quantified the extent of wave propagation by measuring the total number of activated cells during the simulation, N_{act}, where a cell is considered activated at time t if the amount of cytoplasmic calcium it contains is larger than 0.7 μM. Fig. 56.2 presents the variations of N_{act} when the mean degree $\langle k \rangle$ varies; whatever the network model, the higher $\langle k \rangle$, the lower N_{act}. This is a first surprising result of our simulations: *whatever the network type,*

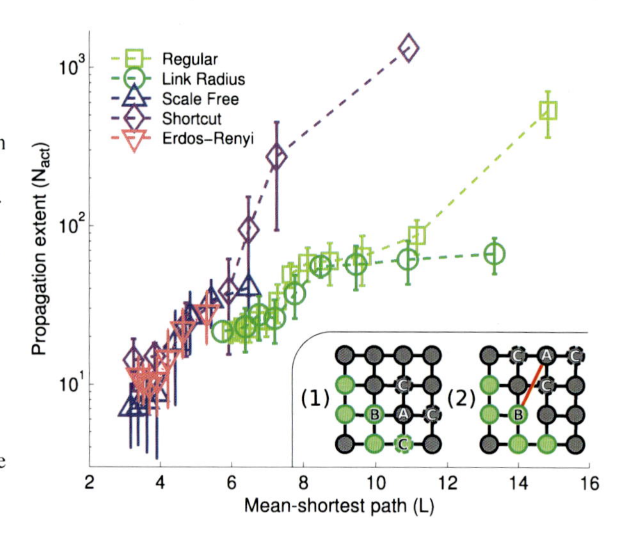

Fig. 56.3 Relations between propagation extent N_{act} and the mean-shortest path L for each network model; error bars plot standard deviation in log scale. All models display the same qualitative behavior. Inset (*1*) represents a small example of propagation in an network strongly influenced by space while inset (*2*) uses a network weakly influenced by space. *Green cells* are activated cells that form the wave front; the wave is propagating from the *bottom left* corner to the *top right* one

wave propagation is generically improved for networks with small mean degree. This result is however coherent with an intuitive understanding of the way IP_3 diffuses in the network: in order to activate the astrocyte it arrives in (say astrocyte A), transferred IP_3 must overcome a threshold value; therefore if A is connected to a lot of inactive neighbors, most of this IP_3 influx will leak to these inactive neighbors before it can activate A. Moreover, if it becomes activated, A will give less IP_3 to each of its unactive neighbors if it has a lot of unactive neighbors. Consequently, high mean-degree tends to impair calcium wave propagation.

However, for a given value of $\langle k \rangle$, N_{act} can be down to 6-folds smaller in Erdős-Rényi networks than in regular or link radius networks. Therefore, the topology class still has an important effect on the propagation and $\langle k \rangle$ cannot alone explain the observed differences. Indeed, as we can see from the inset of Fig. 56.2, the shortcut network model goes from very large (spanning almost all the network) to very small extents, as shortcuts are more frequent while its mean degree is conserved. This behavior could hence be linked to the influence of space on topology: when links are restricted to a spatial neighborhood, the propagation extent is greater than when long distance links (or shortcuts) are added. Additionally, heavy space limitation also prevents the existence of highly connected nodes, influencing the mean degree $\langle k \rangle$. Furthermore, experimental findings indicate that real astrocyte networks are connected via a distance-controlled connectivity, i.e. large distance connexions are not likely (see [8, 9]).

This influence of space on topology can be quantified by the mean-shortest path L of our networks as heavy space limitation yields high L values and the addition of shortcuts reduces it.

Fig. 56.3 presents the relation between the propagation extent, N_{act} and mean-shortest path L. Again, all network models follow the same qualitative behavior: the larger the mean-shortest path, the better the propagation. This is a second surprising finding since it implies that, *contrary to the intuition that shortcuts improve signal propagation in networks, the addition of shortcuts, decreasing L, actually prevents long range propagation in our model of astrocytic networks.* But, just like for the conditions on $\langle k \rangle$ values, this relation between L and the extent of propagation can be seen as the influence of space on networks.

Indeed, reliable activation of an astrocyte can be ensured if several of its neighbors are simultaneously active (its IP_3 influx is then high enough to trigger a calcium spike) and heavy space constraints on topology is a way to ensure this. The rationale behind this affirmation is exposed by the two insets of Fig. 56.3; inset (1) presents a case in which space restriction is strong while inset (2) presents a case in which it is a bit relaxed, because of a "long distance" link (in red). The wave is propagating from the bottom left corner of the networks to the top right corner. An unactivated cell A is linked to an activated cell B in the "wavefront" (green cells); when space influence is high (inset (1)), cell A is spatially close to the wave front, its neighbors (marked C) are hence likely to be activated. Conversely, when space influence is low (inset (2)), the neighbors of A (marked C) are less likely to be in the wavefront because A can be spatially distant from the wavefront. Consequently, in our example, cell A is more likely to get activated when space influence is high (inset (1)) because it has two activated neighbors while it only has one when space influence is low (inset (2)).

However, this space influence cannot be captured by the mean clustering coefficient $\langle C \rangle$ of our networks. Indeed, as it is shown in the inset of Fig. 56.2, the propagation extent in 3D lattices ($\langle k \rangle = 6$, $p_s = 0$ and $\langle C \rangle = 0$) is larger than in regular networks ($\langle k \rangle = 6$ and $\langle C \rangle = 0.27 \pm 0.1$). This particular behavior on lattices (also visible on Fig. 56.3) can be explained by considering the set of nodes that are within a given topological distance l from a reference node (i.e. the shell l of this reference node). Lattices have no links inside each shell; all the links are between shells. Hence a node in the shell $l+1$ will have a mean value of $\langle k \rangle /2$ links coming from the shell l. In other network models, there are intra-shell links so the number of links between shell l and shell $l+1$ is comparatively lower and activation is less easily transmited. This will not however be detailed here and is left to a future publication.

Therefore, according to our simulations, calcium wave propagation is expected to be favored in networks with large mean-shortest path, small mean degree and/or strong preference to connect only to nearest neighbors. In vivo, large regenerative waves spanning all the observable cells are sometimes (though rarely) observed (cf. [14]). Our results indicate that this could be linked to a transient decrease of gap-junction coupling levels, effectively leading to a reduction of the network mean degree, or to a restriction of gap-junction coupling to a few neighbors.

56.3.2 Biophysical Influences

In order to determine the influence of biophysical and GJC properties on propagation, we now fix the topology to the arguably more biologically realistic one,

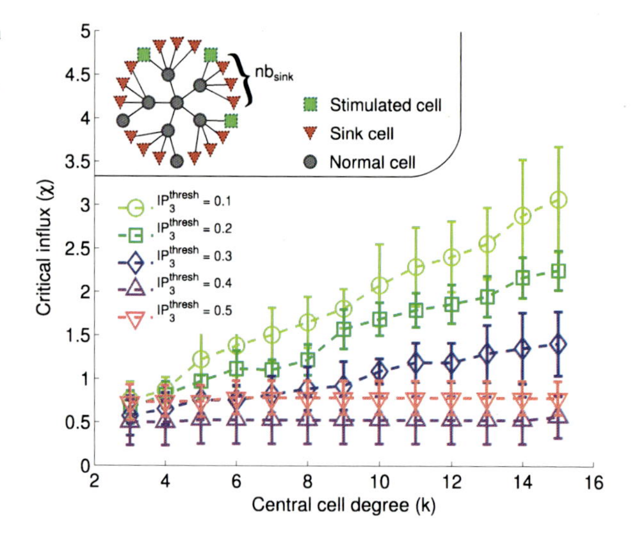

Fig. 56.4 Relations between χ and k for different values of IP_3^{thresh}; error bars plot the standard deviation. *The inset* represents the simple architecture that we used in order to determine these values

the link radius model, as links are expected to be restricted to nearest neighbors. We moreover fix the mean degree value to a realistic biological value of $k \approx 12$ (cf. [8]). We next studied the influence of the GJC parameters (F and IP_3^{thresh}). The extent of propagation was not significantly affected by F but was found to depend strongly on IP_3^{thresh}, the minimum IP_3 gradient above which IP_3 is significantly transferred between the cells via GJC: the higher this parameter, the more individual astrocytes can *store* IP_3 before diffusing it to neighboring unactivated astrocytes. Consequently, the propagation range N_{act} was low (less than 50 cells) up to $\mathrm{IP}_3^{thresh} = 0.3$ μM and for $\mathrm{IP}_3^{thresh} \geq 0.6$ μM. However, for a restricted range of values $0.4 \leq \mathrm{IP}_3^{thresh} \leq 0.5$ μM, propagation was strongly enhanced, reaching $N_{act} \sim 300$ activated astrocytes (compared to the ~ 30 activated astrocytes in previous simulations for the same topology).

In order to understand the strong effects of IP_3^{thresh} on the propagation extent, we conducted a systematic analysis of the activation conditions in the small network displayed at the top of Fig. 56.4. This network is made of one central cell, linked to k neighbors, that are in turn each linked to one "normal" (stimulable) cell and nb_{sinks} "sink" cells, that are artificially maintained in non activated state (red triangles on the figure). For each test, we chose a number of nb_{stim} "normal" cells, among the k ones that can be stimulated, and stimulate them (green squares on the figure). We then run the simulation and observe whether this number nb_{stim} of stimulated cells is enough to activate the central cell. This simple network was designed so as to test the hypothesis that we formulated hitherto in order to explain how wave propagate in our networks: an activated cell (here the neighbors of the central cell) will more easily activate its neighbors (here we focus on the central cell) if it has few unactivated neighbors (the sink cells) to which IP_3 can leak.

For fixed values of GJC parameters, k and nb_{sinks}, we found that the central cell became activated only if nb_{stim} overcomes a threshold nb_{stim}^{crit}. In stimulated branches of our network, when the cell connected to the central cell becomes activated, it passes a given total IP_3 outflux Q to its nb_{sink} neighbor sinks and to the central cell (there is no flux to the stimulated cell as its IP_3 level is already high). The partial flux received by the central cell from one of the k branches is then $Q/(nb_{sink}+1)$ and the total flux received by the central cell is $nb_{stim} \times Q/(nb_{sink}+1)$. We define the normalized critical total influx needed in order to activate the central cell as:

$$\chi = \frac{nb_{stim}^{crit}}{nb_{sink}+1}$$

Figure 56.4 presents the values of χ as k changes and for several values of IP_3^{thresh}. It is apparent from this figure that the IP_3^{thresh} range for which we witnessed high propagation above ($IP_3^{thresh} \in [0.4, 0.5]$) are remarkable here as well, since the values of the threshold χ do not depend on k in this range of IP_3^{thresh} values. This means that, for these IP_3^{thresh} values, the central cell gets activated by a constant number of activating cells (a constant total influx χ), regardless of its degree; IP_3 leaking from the central cell does not prevent its activation. Conversely, for low values of IP_3^{thresh}, the total number of activating cells needed in order to activate the central cell increases linearly with the number of its neighbors. This behavior is not, however, to be confused with a standard threshold model of propagation (cf. [3, 5, 21]) where a node gets activated if a fixed fraction of its neighbors are activated; in our case, one has also to take into account the neighborhood of the activated nodes: a given node is less *activating* if its own neighbors are unactivated (because IP_3 will leak to them). Hence, in the specific case of calcium wave propagation in astrocyte networks, one needs to take into consideration the two-hop neighborhood of a node in order to determine whether it will get activated. Interestingly, one can define a generalized version of a threshold model that accounts for these two-hops neighborhood effects; presenting such a generalized model is beyond the scope of the present paper. We just note here that its behavior was found similar to those presented in Figs. 56.2 and 56.3.

Taken together, these results explain why the propagation extent N_{act} depended on $\langle k \rangle$ in the above results. Indeed, all the above results have been obtained using $IP_3^{thresh} = 0.3$ for which χ increases with k. In this case, the influx threshold is increasingly high, rendering the propagation increasingly difficult, with increasing mean degrees. These results also show that the influence of topology on propagation is itself regulated by biophysical parameters (here GJC parameters); the resulting propagation is a complex interplay between these parameters and the network topology.

56.4 Conclusion

A major point in the results showed above is that we could obtain various extents of wave propagation by sole variations of the network topology; more precisely, two macroscopic measures on the network topology, its mean degree $\langle k \rangle$ and its mean-shortest path L were enough to determine the extent of propagation.

Table 56.1 Different types of waves can be reproduced by varying only the network topology

Wave type [exper. ref.]	N_{act}	$\langle k \rangle$	L
Locally synchronized [19]	~ 10	~ 8	~ 4
Spatially restricted [4]	~ 40	~ 7	~ 8
Regenerative [14, 20]	~ 500	~ 3	~ 14

Table 56.1 presents the three major types of waves reported in experimental articles with their respective extent. In our model, each type of wave can be reproduced using different values of $\langle k \rangle$ and L. Very small extents (locally synchronized event) can be obtained with relatively high degree networks and short mean-shortest path while regenerative waves (propagating to most of the cells) are triggered on small degree networks with high mean-shortest path. Therefore, large scale propagation is achieved when astrocytes are linked to nearest neighbors while linking to more (or more distant) astrocytes decreased propagation extent. Hence, while differences in propagation extent in the biological literature are usually explained by differences in biophysical parameters or in signaling pathways, we showed here that, in our model, topology alone was enough to explain these differences. As it is known that the structure of astrocytic networks varies throughout the brain and could even be shaped by neuronal activity [9], the influence of topology on wave propagation *in vivo* may be a fruitful hypothesis in order to understand the bidirectional relations between neurons and astrocytes at a network level.

For instance, our results could be evoked to formulate an hypothesis concerning the observation that regenerative waves usually occur in cultured astrocytes networks, where space has a very restrictive influence on the links, as these in vitro networks are constrained to a 2D surface. This would strongly reduce the number of close neighbors thus, according to our model, favor long-range propagation .

References

1. Barthélemy M (2011) Spatial networks. Phys Rep 499(1–101):2010
2. Bushong EA, Martone ME, Jones YZ, Ellisman MH (2002) Protoplasmic astrocytes in ca1 stratum radiatum occupy separate anatomical domains. J Neurosci 22(1):183–192
3. Centola D, Eguíluz VM, Macy MW (2007) Cascade dynamics of complex propagation. Physica A 374(1):449–456
4. Charles A (1998) Intercellular calcium waves in glia. GLIA 24(1):39–49
5. Dodds PS, Watts DJ (2004) Universal behavior in a generalized model of contagion. Phys Rev Lett 92(21):218701
6. Dokukina I, Gracheva M, Grachev E, Gunton J (2008) Role of network connectivity in intercellular calcium signaling. Physica D 237(6):745–754
7. Edwards JR, Gibson WG (2010) A model for ca2+ waves in networks of glial cells incorporating both intercellular and extracellular communication pathways. J Theor Biol 263(1):45–58
8. Giaume C (2010) Astroglial wiring is adding complexity to neuroglial networking. Front Neuroenerg 2:129
9. Giaume C, Koulakoff A, Roux L, Holcman D, Rouach N (2010) Astroglial networks: a step further in neuroglial and gliovascular interactions. Nat Rev Neurosci 11(2):87–99

10. Goldberg M, Pittà MD, Volman V, Berry H, Ben-Jacob E (2010) Nonlinear gap junctions enable long-distance propagation of pulsating calcium waves in astrocyte networks. PLoS Comput Biol 6(8):e1000909
11. Höfer T, Venance L, Giaume C (2002) Control and plasticity of intercellular calcium waves in astrocytes: a modeling approach. J Neurosci 22(12):4850–4859
12. Kang M, Othmer HG (2009) Spatiotemporal characteristics of calcium dynamics in astrocytes. Chaos 19(3):037116
13. Kazantsev VB (2009) Spontaneous calcium signals induced by gap junctions in a network model of astrocytes. Phys Rev E, Stat Nonlinear Soft Matter Phys 79(1):010901
14. Kuga N, Sasaki T, Takahara Y, Matsuki N, Ikegaya Y (2011) Large-scale calcium waves traveling through astrocytic networks in vivo. J Neurosci 31(7):2607–2614
15. Ogata K, Kosaka T (2002) Structural and quantitative analysis of astrocytes in the mouse hippocampus. Neuroscience 113(1):221–233
16. Perea G, Araque A (2010) Glia modulates synaptic transmission. Brains Res Rev 63(1–2):93–102
17. Pittà MD, Volman V, Levine H, Pioggia G, Rossi DD, Ben-Jacob E (2008) Coexistence of amplitude and frequency modulations in intracellular calcium dynamics. Phys Rev E, Stat Nonlinear Soft Matter Phys 77(3):030903
18. Roux L, Benchenane K, Rothstein JD, Bonvento G, Giaume C (2011) Plasticity of astroglial networks in olfactory glomeruli. Proc Natl Acad Sci USA. doi:10.1073/pnas.1107386108
19. Sasaki T, Kuga N, Namiki S, Matsuki N, Ikegaya Y (2011) Locally synchronized astrocytes. Cereb Cortex. doi:10.1093/cercor/bhq256
20. Scemes E, Giaume C (2006) Astrocyte calcium waves: what they are and what they do. GLIA 54(7):716–725
21. Watts DJ (2002) A simple model of global cascades on random networks. Proc Natl Acad Sci USA 99(9):5766–5771

Chapter 57
Modelling Spatial Dynamics of Plant Coastal Invasions

James T. Murphy and Mark P. Johnson

Abstract Biological invasion refers to the introduction of non-native species of plants or animals which adversely affect local ecosystems and transform their structure and species composition. As a result, costly control efforts often have to be put in place to protect habitats. An example of an invasive problem on a global scale is the plant species *Spartina anglica* which is a salt marsh grass found in the inter-tidal zones of coastal habitats. In this study, an agent-based modelling approach was taken to analyse the emergent dynamics of Spartina populations in a simulated coastal environment. The model was used to analyse the impact of various factors such as the shape and pattern of colony development and seedling placement on invasion dynamics in order to be able to devise efficient control strategies.

Keywords Agent-based model · Individual-based model · Spatial dynamics · Invasive species · Coastal ecosystem

57.1 Introduction

Biological invasion refers to the introduction of non-native species of plants or animals which adversely affect local ecosystems and transform their structure and species composition. It has been identified in the Millennium Ecosystem Assessment as one of the principal environmental problems influencing future economic and social development in the world [1]. As a result, costly control efforts often have to be put in place to protect habitats. An example of an invasive problem on a global scale is the plant species *Spartina anglica* (Common Cordgrass) which is a salt marsh grass found in the inter-tidal zones of coastal habitats. It is characterised by its ability to trap large amounts of sediments and thus over time replace mudflats with badly drained marshes. This can endanger the habitats of many species (such as invertebrates and shorebirds) which depend on the mudflats and reduces overall biodiversity in coastal regions.

J.T. Murphy (✉) · M.P. Johnson
Ryan Institute, National University of Ireland Galway, Galway, Ireland
e-mail: james.murphy@nuigalway.ie

T. Gilbert et al. (eds.), *Proceedings of the European Conference on Complex Systems 2012*, Springer Proceedings in Complexity, DOI 10.1007/978-3-319-00395-5_57,

The patterns of invasion of Spartina are amenable to spatially-explicit modelling strategies that take into account both temporal and spatio-temporal processes. In the case of invasive plant species, spatially explicit modelling techniques are necessary to take into account how neighbourhood interactions, such as the presence and spatial arrangement of nascent foci, can influence invasion pattern [2]. In this study, an agent-based (or individual-based) modelling approach was taken to analyse the emergent dynamics of Spartina populations in a simulated coastal environment. The model was used to analyse the impact of various factors such as the shape and pattern of colony development and seedling placement on invasion dynamics in order to be able to devise efficient control/eradication strategies. An agent-based modelling tool called CoastGEN was developed specifically for this project in the C++ programming language. The model was built upon a revised and expanded version of an agent-based modelling framework called Micro-Gen, which has been used previously to study bacterial growth dynamics [3]. The model is fully parallelised to take advantage of distributed computing architectures and it represents a robust and adaptable framework to simulate spatially and temporally heterogeneous phenomena.

The agent-based approach differs from traditional mathematical population models that commonly use global parameters or state variables to describe the growth and development of a biological population [4]. A high-level mathematical approach has many advantages in terms of computational efficiency and can give good insights at the population-level, but it is sometimes difficult to trace back the system behaviour to that of the individual organisms. For example, it does not explicitly identify the underlying factors determining a particular growth rate or carrying capacity associated with a population. Cellular automata theory is another more low-level technique that has been used to explain pattern formation in colonies [5].

57.2 Model Overview

The model represents an inter-tidal environment as a discrete, two-dimensional grid with each grid element (or "cell") corresponding to 1 m^2 of surface area. This allows for heterogeneity in the environmental conditions and spatial distribution of organisms, as opposed to assuming a completely homogeneous, mixed environment. Each *Spartina* plant is represented in the model as a group of agents that together represent a "rhizome network". Each *Spartina* agent has a growth rate parameter associated with it (which is fitted to field data) that determines the increase in biomass per time step of the simulation. Each grid location may be occupied by a maximum of one *Spartina* agent. Once an agent has grown to fill that grid location (i.e. the biomass exceeds a specified cell threshold) then either growth ceases, if all neighbouring cells are occupied, or else a new agent is created and added to a randomly chosen free neighbouring cell. This means that growth is limited to peripheral agents of the plant, thus resulting in a constant radial growth rate over time for circular patches, as observed in nature [6].

Table 57.1 Input parameters for CoastGEN simulations of Spartina sp. on inter-tidal mud flat environment. The "growth rate" parameter was estimated by fitting to an observed radial expansion rate of 0.77 my^{-1}) from the literature [7].
b.u. = simulation biomass units

Input parameter	Value
Length of simulation loop (days)	1
Grid size (no. of cells)	1000^2
Cell size (m^2)	1
Initial no. of seedlings	256
Growth rate (b.u. $loop^{-1}$ $agent^{-1}$)	15.8
Biomass limit (b.u. $cell^{-1}$)	10000
Loop start eradication	20000

The model is constructed using the object-oriented programing paradigm of C++. The initial phase of the program involves the creation and initialisation of an array of *Spartina* agents which are stored in an array data structure. The input parameters for the simulation are entered via a text input file. These specify physical parameters such as the size and scale of the environment as well as parameters for the *Spartina* agents such as the initial number of seedlings and their average growth rate (Table 57.1).

The main program loop consists of a series of steps representing the main biotic and abiotic functions of the system. Each loop represents a discrete day of real-time during which the agents grow and expand into neighbouring cells in the environment.

57.3 Results and Discussion

The model was initially validated by comparing the growth curve of a single simulated *Spartina* clone with results from field studies of Spartina clones in Willapa Bay, USA [7]. The growth rate was fitted to these studies and the characteristic circular pattern of growth with a linear increase in radius length over time was confirmed. This is because the rate of surface area expansion of a *Spartina* plant is limited by the length of the growing perimeter of the patch. The ratio of the circumference/area of a growing circular *Spartina* patch approaches zero over time, due to the fundamental geometric relationship between the circumference and the area of a circle.

Following this, a number of theoretical studies were carried out to explore the effects of the spatial distribution of initial seedling placement on invasion dynamics. An important strength of the agent-based modelling approach is that it can be used to explicitly model complex shapes and calculate the impact of their spatial distribution on the effective growth rate. Plants such as Spartina are limited by their available growing surface and the impact of the spatial distribution on this parameter was quantitatively analysed to predict their relative growth rate.

One simulation was run with 256 initial seedlings randomly dispersed across the environment of 1000^2 grid positions and a second control run with the initial seedling placement concentrated near the top boundary of the environment.

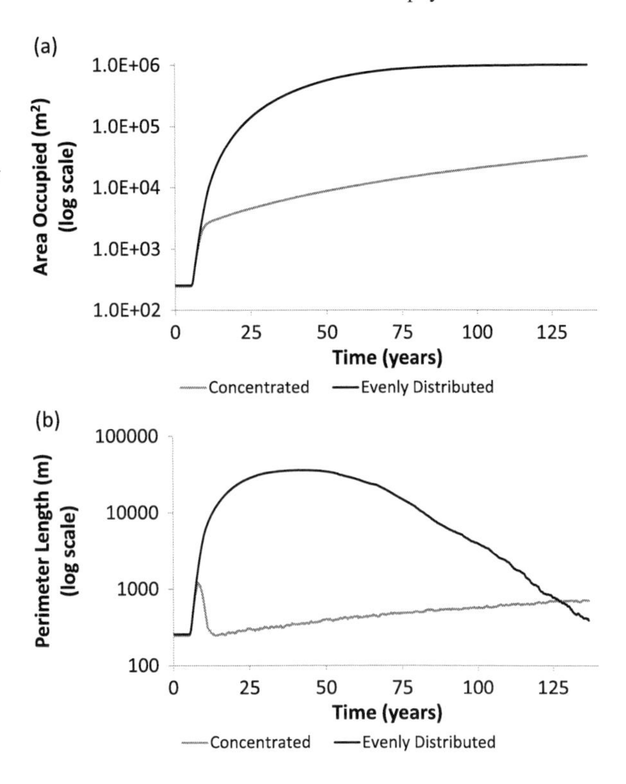

Fig. 57.1 Comparison of predicted (**a**) growth curves and (**b**) perimeter lengths of populations of *Spartina sp.* when initial positions of seedlings are either (i) evenly distributed across the environment, or (ii) concentrated in *top-left corner*. Total environment size = 1 km^2, initial number of seedlings = 256

When the seedlings are placed close to one another they quickly merge to form a single colony with a reduced perimeter/area ratio (Fig. 57.1(a)). However, when the seedlings are more evenly dispersed it results in a higher growth rate as the length of the growing perimeter relative to the total area covered is much greater (Fig. 57.1(b)).

A number of simulations were also carried out to explore different eradication/control strategies for Spartina infestations on an open mudflat environment (Fig. 57.2). Firstly, a completely random eradication strategy was implemented whereby grid positions were chosen at random to remove *Spartina* agents from. This would be analogous to a worst-case scenario where aerial application of an herbicide resulted in incomplete or patchy coverage, for example due to wind dispersal. The model predicted a rapid recovery of the population within 10 years following the eradication event. An alternative, more co-ordinated, approach to eradication involves removing *Spartina* plants at the periphery of a meadow first before gradually moving to the interior, in an analogous way to peeling the layers off of an onion. The population does not recover to normal levels until 50 years after the initial intervention, which represents a 5-fold increase over the purely random approach. The most optimal solution tested was to eradicate the infestation in a linear fashion parallel with the edge of the habitat. There is at least a couple of orders of magnitude difference in the effectiveness of the linear row-by-row eradication strategy compared to the random scenario (Fig. 57.3).

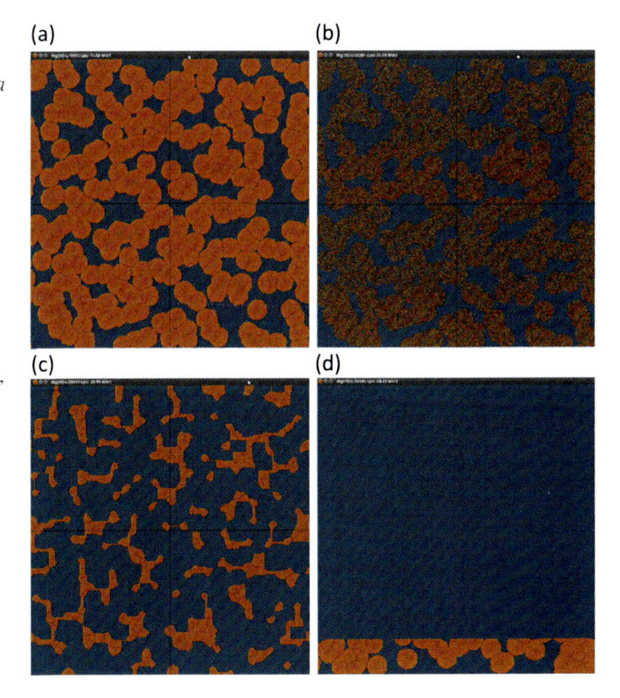

Fig. 57.2 (**a**) Graphical display from CoastGEN of 256 growing circular *Spartina* patches (*orange*) growing on uniform mudflat (*blue*) environment prior to eradication event (54.5 years into simulation). Simulation was run in parallel on four processors with overlapping boundary conditions. (**b**)–(**d**): Spartina populations after (**b**) random eradication, (**c**) perimeter-first eradication, and (**d**) row-by-row eradication strategies were implemented

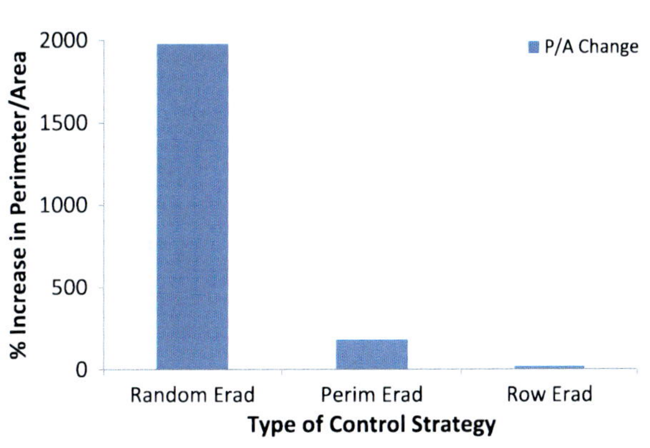

Fig. 57.3 Relative increase in total perimeter/area (P/A) ratio of randomly distributed *Spartina* population in response to each control strategy: Random eradication, Perimeter Eradication and Row-by-Row Eradication strategies. Values for P/A expressed relative to the normal uncontrolled population

The importance of removing smaller foci as a more effective strategy than attacking large meadows has been investigated mathematically for perfectly circular plant colonies [2]. However, the agent-based modelling framework implemented here allows the study of more complex meadow shapes as found in nature. It is clear from this that an inadequate spatial strategy of control can lead to increased potential for re-invasion, for example when a large meadow is broken up into a number of smaller patches as a result of control efforts [4]. This highlights the importance of modelling efforts to assess the potential impact of proposed strategies and highlight emergent dynamics that may occur as a result of interventions due to the specific morphology of an invasion.

The spatial pattern of an invasion plays a key role in the rate of spread of the species and understanding this can lead to significant cost savings when implementing control strategies. Our model framework can be used to explicitly represent complex spatial and temporal patterns of invasion in order to be able to quantitatively predict the impact of these factors on the invasion dynamics. This would be a useful tool for assessing eradication strategies and choosing optimal control solutions in order to be able to minimise control costs.

References

1. Gewin V (2005) Industry lured by the gains of going green. Nature 436(7048):173
2. Moody ME, Mack RN (1988) Controlling the spread of plant invasions—the importance of nascent foci. J Appl Ecol 25(3):1009–1021
3. Murphy JT, Walshe R, Devocelle M (2008) A computational model of antibiotic-resistance mechanisms in methicillin-resistant staphylococcus aureus (mrsa). J Theor Biol 254(2):284–293
4. Taylor CM, Hastings A (2004) Finding optimal control strategies for invasive species: a density-structured model for spartina alterniflora. J Appl Ecol 41(6):1049–1057
5. Huang HM, Zhang LQ, Guan YJ, Wang DH (2008) A cellular automata model for population expansion of spartina alterniflora at Jiuduansha shoals, Shanghai, China. Estuar Coast Shelf Sci 77(1):47–55
6. Dennis B, Civille JC, Strong DR (2011) Lateral spread of invasive spartina alterniflora in uncrowded environments. Biol Invasions 13(2):401–411
7. Grevstad FS (2005) Simulating control strategies for a spatially structured weed invasion: Spartina alterniflora (loisel) in pacific coast estuaries. Biol Invasions 7(4):665–677

Chapter 58
Dynamical Aspects of Information in Copolymerization Processes

Pierre Gaspard

Abstract Natural supports of information are given by random copolymers such as DNA or RNA where information is coded in the sequence of covalent bonds. At the molecular scale, the stochastic growth of a single copolymer with or without a template proceeds by successive random attachments or detachments of monomers continuously supplied by the surrounding solution. The thermodynamics of copolymerization shows that fundamental links already exist between information and thermodynamics at the molecular scale, which opens new perspectives to understand the dynamical aspects of information in biology.

Keywords Thermodynamics of copolymerization · Entropy production · Stochastic processes · Information theory · Shannon disorder · Mutual information · Mutations · DNA replication · DNA sequencing

58.1 Introduction

Under nonequilibrium conditions, the emergence of dynamical order is already in action at the molecular scale during copolymerization processes. Copolymers are special because they constitute the smallest physico-chemical supports of information. Little is known about the thermodynamics and kinetics of information processing in copolymerizations although such reactions play an essential role in many complex systems, e.g. in biology. In this context, recent advances have been performed which shed a new light on the nonequilibrium constraints required to generate information rich copolymers [1–4].

Natural supports of information are given by random copolymers where information is coded in the sequence of covalent bonds, as already suggested by Schrödinger with his concept of aperiodic crystal [5]. Random copolymers exist in chemical and biological systems. Examples are styrene-butadiene rubber, proteins, RNA, and DNA, this latter playing the role of information support in biology.

P. Gaspard (✉)
Center for Nonlinear Phenomena and Complex Systems, Université Libre de Bruxelles, Campus Plaine, CP 231, 1050 Brussels, Belgium
e-mail: gaspard@ulb.ac.be

T. Gilbert et al. (eds.), *Proceedings of the European Conference on Complex Systems 2012*, Springer Proceedings in Complexity, DOI 10.1007/978-3-319-00395-5_58,

At the molecular scale, the stochastic growth of a single copolymer proceeds by successive random attachments or detachments of monomers $\{m\}$ continuously supplied by the surrounding solution:

$$m_1 m_2 \cdots m_{l-1} + m_l \rightleftharpoons m_1 m_2 \cdots m_{l-1} m_l \tag{58.1}$$

The solution is supposed to be sufficiently large to play the role of a reservoir where the concentrations of monomers are kept constant. In this regard, the stochastic growth of a single copolymer is modeled by a Markovian process with transition rates depending on the fixed concentrations of monomers in the surrounding solution [1–4].

According to local detailed balancing, the rates of forward and reversed transitions have ratios that are determined by the free energies of the copolymers $m_1 m_2 \cdots m_l$ in physical equilibrium with the surrounding solution. Thermodynamic quantities can thus be defined and their time evolution studied during the copolymerization process. In this way, fundamental relationships can be established between the thermodynamics of copolymerization processes and the information content encoded in growing copolymers [1–4]. The purpose of this communication is to present the latest results obtained in this framework.

58.2 Results

As shown in Ref. [1] for copolymerization with or without a template, the thermodynamic entropy production is related not only to the average value of the free energy per monomer in the grown sequence, but also to the Shannon disorder of the sequence itself. This result is at the origin of dissipation-error tradeoff during copolymer growth [6].

Two growth regimes are identified:

(1) A regime close to the thermodynamic equilibrium where the copolymer can grow in an adverse free-energy landscape by the entropic effect of its Shannon disorder. In this regime, the disorder of the grown sequence dominates the process even in the presence of a template, in which case the copying process generates a lot of errors.

(2) A regime farther away from equilibrium where the growth proceeds because the free energy of monomer attachment is favorable. In this regime, the error rate drops to low values.

In Refs. [1, 3], these regimes were studied as a function of the free energy driving force. In Ref. [4], results have been reported on a model of free copolymerization where the attachment and detachment rates are controlled by the concentrations of monomers in the surrounding solution. In the present communication, this study of the dependence on monomeric concentrations is extended to a model of copolymerization with a template. The template as well as the growing copy are composed of two monomers $m = 1$ and $m = 2$. The pairs 1–1 and 2–2 are favored between the template α and the copy ω. The pairs 1–2 and 2–1 are considered as errors during

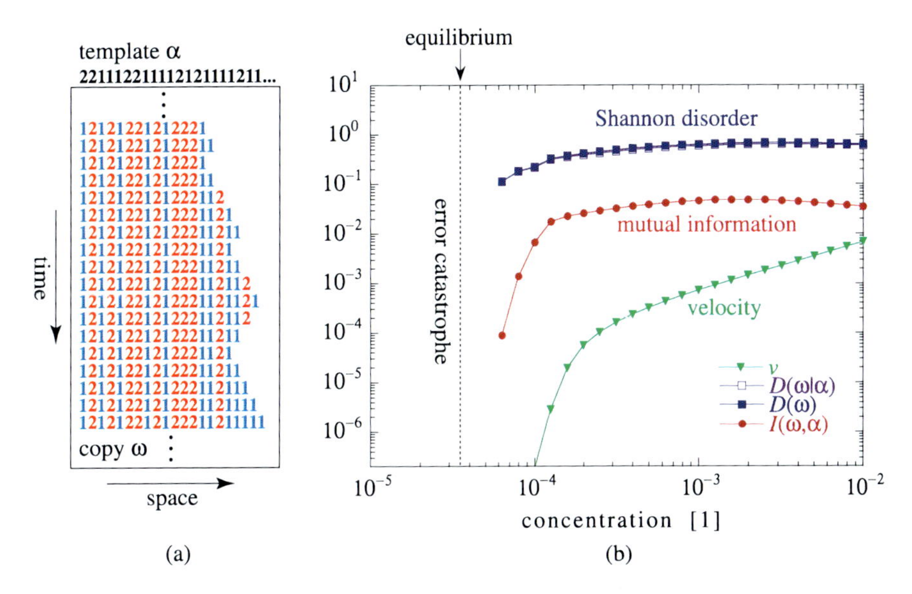

Fig. 58.1 Stochastic growth of the copolymer ω on the template α, as simulated by Gillespie's algorithm with the parameter values $k_{\mathrm{correct}} = 1$, $k_{\mathrm{error}} = 0.5$, $k_{\mathrm{off}} = 10^{-3}$, and $[2] = 1.3 \times 10^{-3}$. The template is generated by a Bernoulli process of probabilities $(\frac{1}{2}, \frac{1}{2})$. (**a**) Space-time plot at the concentration $[1] = 2 \times 10^{-3}$; (**b**) The mean growth velocity v, the Shannon disorder $D(\omega)$ of the copy ω, the Shannon disorder $D(\omega|\alpha)$ of the copy ω conditioned to the template sequence α, and the mutual information $I(\omega, \alpha)$ between the copy ω and the template α versus the concentration $[1]$ of the monomers of species 1

information transmission from the template to the copy. The kinetic mechanism of elongation is the following:

$$\begin{array}{llcl} \alpha: & n_1 n_2 \cdots n_{l-1} n_l n_{l+1} \cdots & & n_1 n_2 \cdots n_{l-1} n_l n_{l+1} \cdots \\ \omega: & m_1 m_2 \cdots m_{l-1} \qquad + m_l & \rightleftharpoons & m_1 m_2 \cdots m_{l-1} m_l \end{array} \tag{58.2}$$

The attachment rates are given by $w_{+m|n} = k_{+m|n}[m]$ and the detachment rates by $w_{-m|n} = k_{-m|n}$. The attachement rates are proportional to the monomeric concentrations $[m]$, while the detachment rates are not since detachments do not need the presence of monomers in the surrounding solution. The rates are supposed to be independent of the end m_{l-1} of the copy ω, which is a simplifying assumption. The rate of formation of correct pairs is defined as $k_{\mathrm{correct}} \equiv k_{+1|1} = k_{+2|2}$ and the error rate as $k_{\mathrm{error}} \equiv k_{+1|2} = k_{+2|1}$. All the detachment rates are assumed to take the same value: $k_{\mathrm{off}} \equiv k_{-1|1} = k_{-1|2} = k_{-2|1} = k_{-2|2}$.

The stochastic process is simulated with Gillespie's algorithm. Figure 58.1(a) illustrates the fluctuating growth of a copy in space and time. In this example, the error rate is larger because of the smallness of the ratio $k_{\mathrm{correct}}/k_{\mathrm{error}} = 2$. This ratio is determined by the strength of the pairing bonds. Figure 58.1(b) depicts the mean growth velocity as well as the quantities characterizing the information content of the copy ω compared to the template α versus the concentration of monomers 1 in the surrounding solution. The mean growth velocity vanishes at equilibrium, which exists at the concentration $[1]_{\rm eq} \simeq 0.3 \times 10^{-4}$ if $[2] = 1.3 \times 10^{-3}$. Both the Shannon disorder $D(\omega)$ of the copy and the conditional disorder $D(\omega|\alpha)$ of the copy with respect to the template are larger than the mutual information $I(\omega, \alpha)$ between the copy and the template. In any case, the three quantities are related to each other by the well-known formula

$$I(\omega, \alpha) = D(\omega) - D(\omega|\alpha) \tag{58.3}$$

from information theory [1, 3]. The mutual information characterizes the fidelity of information transmission between the template and the copy. Figure 58.1(b) shows that this fidelity decreases close to equilibrium. The reason is that the out-of-equilibrium directionality is lost close to equilibrium where fluctuations go either forward or backward because the principle of detailed balancing prevails at equilibrium. Consequently, there is a multiplication of errors close to equilibrium. This error catastrophe is avoided by maintaining the system far enough from equilibrium.

58.3 Conclusions and perspectives

The results show that fidelity in copying a copolymer requires the supply of enough free energy from the attachment of monomers. In this respect, the nonequilibrium driving should exceed a critical value in order to transmit information in copolymerization processes with a template, such as DNA replication [1]. The statement by Manfred Eigen that "information cannot originate in a system that is at equilibrium" [10] is rigorously proved in the present framework. The thermodynamics of copolymerization thus shows that fundamental links already exist at the molecular scale between information and thermodynamics [1–4].

The transition between the two growth regimes could be experimentally investigated in chemical or biological copolymerizations. In polymer science, methods have not yet been much developed to perform the synthesis and sequencing of copolymers for the information they may support. However, such methods are already well developed for DNA and under development for single-molecule DNA or RNA sequencing [7–9]. These methods could be used to test experimentally the predictions of copolymerization thermodynamics by varying NTP and pyrophosphate concentrations to approach the regime near equilibrium where the mutation rate increases.

These considerations open new perspectives to understand the dynamical aspects of information in biology. During copolymerization processes with a template (as it is the case for replication, transcription or translation in biological systems), information is transmitted although errors may occur due to molecular fluctuations, which are sources of mutations. The two main features of biological systems—namely, metabolism and self-reproduction—turn out to be related in a fundamental way since information processing is constrained by energy dissipation during copolymerizations. Moreover, the error threshold for the emergence of quasi-species in the hypercycle theory by Eigen and Schuster [11] could be induced at the molecular scale by the transition towards high fidelity replication beyond the transition between the two growth regimes [12]. In this way, prebiotic chemistry could be more closely linked to the first steps of biological evolution.

Acknowledgements This research is financially supported by the Belgian Federal Government (BELSPO), the Communauté française de Belgique, and the Université Libre de Bruxelles.

References

1. Andrieux D, Gaspard P (2008) Proc Natl Acad Sci USA 105:9516
2. Jarzynski C (2008) Proc Natl Acad Sci USA 105:9451
3. Andrieux D, Gaspard P (2009) J Chem Phys 130:014901
4. Gaspard P (2012) Self-organization at the nanoscale scale in far-from-equilibrium surface reactions and copolymerizations. In: Mikhailov AS, Ertl G (eds) Proceedings of the International Conference "Engineering of Chemical Complexity", Berlin Center for Studies of Complex Chemical Systems, July 4–8, 2011. World Scientific, Singapore, pp 4–8. doi:10.1142/9789814390460_0003. ISBN 978-981-4390-45-3
5. Schrödinger E (1944) What is life? Cambridge University Press, Cambridge
6. Bennett CH (1979) Biosystems 11:85
7. Greenleaf WJ, Block SM (2006) Science 313:801
8. Eid J et al (2009) Science 323:133
9. Uemura S, Echeverría Aitken C, Korlach J, Flusberg BA, Turner SW, Puglisi JD (2010) Nature 464:1012
10. Eigen M (1992) Steps towards life: a perspective on evolution. Oxford University Press, Oxford
11. Eigen M, Schuster P (1977) Naturwissenschaften 64:541. Ibid, 65:7 (1978). Ibid, 65:341 (1978)
12. Woo H-J, Wallqvist A (2011) Phys Rev Lett 106:060601

Chapter 59
Emergence of Gene Regulatory Networks Under Functional Constraints

Marcin Zagórski

Abstract Gene regulatory networks allow the control of gene expression patterns in living cells. We ask here to what extent the network architecture is determined by the output patterns of gene regulatory networks. Given a framework for describing regulatory interactions and dynamics (Burda et al., Proc. Natl. Acad. Sci. USA, 108:17263, 2011), we consider in the space of all regulatory networks those that have prescribed functional capabilities. Markov Chain Monte Carlo sampling is then used to determine how these functional constraints lead to specific structures of the interactions. Particularly, we generate ensemble of regulatory networks with yeast cell-cycle (Li et al., Proc. Natl. Acad. Sci. USA, 101:4781, 2004) biological trajectory imposed. As a result, we find that on average 55 % of interactions are well reproduced, and concerning the whole ensemble almost all networks have from 40 % to 70 % of links in common with yeast cell-cycle network.

Keywords Gene regulatory networks · Yeast cell-cycle · Transcription factors · Essential interactions

59.1 Introduction

After billions of years of evolution Earth's life is a very diverse phenomenon, yet all the living organisms are made of simple building blocks called cells. The single cell is a device designed to interpret internal or external signals in order to enhance its survival prospects. We focus here on gene regulatory networks (GRN), the set of interactions between genes. These interactions along with the gene expression machinery allow all living cells to control their gene expression patterns. In the last decade, our knowledge how any given gene can affect another's expression has been significantly extended through various experiments. For example, small gene networks have been constructed to implement simple functions *in vivo* [3, 4], and much larger sets of interactions have been derived from a number of organisms

M. Zagórski (✉)
Institute of Physics, Jagiellonian University, Reymonta 4, 30-059 Kraków, Poland
e-mail: marcin.zagorski@uj.edu.pl

T. Gilbert et al. (eds.), *Proceedings of the European Conference on Complex Systems 2012*, Springer Proceedings in Complexity, DOI 10.1007/978-3-319-00395-5_59,

[5–7]. Therefore it has been possible to show that several subgraphs of interactions ("motifs") arise more frequently than might be expected [8–11]. Since we know the structure of GRNs better and better, a question arises whether the constraints associated with network functionality are major determinants of network architecture?

In paper [1] with Z. Burda, A. Krzywicki, O.C. Martin we have shown that motifs can emerge in network architecture due to functional constraints (output patterns) imposed on GRNs. In the case where the regulatory networks are constrained to exhibit multistability, we found a high frequency of gene pairs that are mutually inhibitory and self-activating. In contrast, networks constrained to have periodic gene expression patterns had a high frequency of bifan-like motifs involving four genes with at least one activating and one inhibitory interaction. However, this results were obtained with idealised gene expression patterns, and now we impose yeast cell-cycle [2] pathway, to see to what extent we can reproduce a biological network within our model [12].

59.2 Methods

59.2.1 *General Framework*

Compared to the well-known model of Boolean networks (see [13] and references therein) in which a given gene can be either *on* or *off*, here we allow gene expression to have intermediate values. We consider a system of N genes where each gene produces a corresponding protein. In our previous works [1, 14] we restricted gene products to be only transcription factors (TFs), but in order to study cell cycling systems we need to include other molecular species (e.g. cyclins). We include these molecular species by assuming that similarly to TF-DNA binding, if we have two molecules, their facing elements (atoms, bases, amino acids, ...) have to "match" for the two molecules to bind. For simplicity from this point onwards we refer to all gene products as TFs.

Particularly, for ith gene its normalized expression level S_i is a continuous variable ranging from 0 to 1, where zero means no production of TF and one corresponds to maximal production rate. Since we have N genes we can define a vector variable $\mathbf{S} = (S_1, S_2, \ldots, S_N)$ which we call a phenotype. In our approach, we assume that every gene can be influenced by any of N types of TFs. As a result we obtain a $N \times N$ weight matrix $\mathbf{W}$ where a given entry W_{ij} corresponds to the strength of interaction between ith gene and jth TF. Hereafter we refer to $\mathbf{W}$ as the genotype and a formula to determine the values of W_{ij} will be given later.

To find gene expression pattern $\mathbf{S}(t)$ at any given time t, we propose a deterministic dynamics described by a map $\mathbf{S}(t+1) = G(\mathbf{S}(t), \mathbf{W})$, where we call initial phenotype $\mathbf{S}(0)$. This discrete dynamics can be represented by a sequence of steps leading to an attractor which is either a cycle or a fixed point.

If we had only one target phenotype, as it is in [14], we would sample the space of all genotypes leading from $\mathbf{S}^{(initial)}$ to the "vicinity" of some fixed $\mathbf{S}^{(target)}$. In order to quantify, how close a given phenotype is to the target one, we define a fitness function:

$$F(\mathbf{S}) = \exp\left(-f D\left(\mathbf{S}, \mathbf{S}^{(target)}\right)\right), \tag{59.1}$$

where $D(\mathbf{S}, \mathbf{S}') = \sum_i |S_i - S_i'|$ is the difference of expression levels for each gene, and $f \in \mathbb{R}$ is a control parameter.

In case of multiple target phenotypes, we can still use Eq. (59.1), but the "total" distance used to calculate fitness should be a sum of distances from n fixed point phenotypes and corresponding target phenotypes; for cycling behavior target phenotypes are consecutive steps of the imposed cycle.

59.2.2 Microscopic Interactions

In order to determine the strength of interaction between TF and DNA strand, we represent each TF as well as each binding site by a character string of length L with characters belonging to a 4 letter alphabet. Following the standard practice [15], we assume that the free energy of one TF molecule bound to its target site is, up to an additive constant, equal εd_{ij}, where ε is the single mismatch energy and d_{ij} is a number of mismatches between ith binding site and jth TF. Furthermore, one can define the "interaction strengths" W_{ij} via Boltzmann factor

$$W_{ij} = e^{-\varepsilon d_{ij}}, \tag{59.2}$$

with normalizing constant set to 1 (cf. [16]). In case of n_j TFs of jth type one can derive [16, 17] the probability p_{ij} that precisely one of them is bound to the binding site of ith gene

$$p_{ij} = \frac{1}{1 + 1/(W_{ij} n_j)} = \frac{1}{1 + \exp(\varepsilon d_{ij} - \ln(n S_j))}, \tag{59.3}$$

which dependence is known to physicists as Fermi function. In the above formula, for the sake of simplicity, we assume that $n_j = n S_j$ where n is a model parameter representing the number of TFs. For the current work we use $N = 11$ (this is the number of genes in simplified yeast cell-cycle network [2]), $n = 1000$, $L = 12$ and $\varepsilon = 1.75$, though we have checked that for biologically relevant parameters the model findings are qualitatively the same [14]. To keep the framework simple, we assume that gene i transcription is "on" whenever at least one TF is bound within its regulatory region and otherwise it is "off". For inhibitory interaction the TF of type j bound to its binding site is assumed to stop the transcription. The (normalized) mean expression level of a gene is then identified with the probability that transcription is "on". Hence, for gene i with both activators and repressors we have

$$S_i(t+1) = \left[1 - \prod_j \left(1 - p_{ij}(t)\right)\right] \prod_{j'} \left(1 - p_{ij'}(t)\right), \tag{59.4}$$

where j runs over activating interactions and j' over inhibitory interactions, and $p_{ij}(t)$ is given by Eq. (59.3) with S_j replaced by $S_j(t)$. Additionally, to avoid situation where gene expression products with very small concentrations, which are not welled modeled by our mean-field approach, cause acceleration of the genes expression we introduce a phenomenological correction: a small threshold expression H, such that if $S_i(t) < H$ than $S_i(t) = 1/n$ (by default for $n = 1000$ we set $H = 0.01$). In Eq. (59.4), just like in many other modeling frameworks, we use discrete time [2, 13, 18, 19].

59.2.3 *Mutation-Selection Balance*

Since the space of viable GRNs is only a tiny fraction of the space of all regulatory networks, we need to introduce some effective sampling method. In particular, we use Markov Chain Monte Carlo with the Metropolis rule to explore this ensemble. The procedure is following, we start with random genotype that is we draw all the characters representing gene's regulatory regions and TF molecules randomly, and calculate the corresponding weight matrix $\mathbf{W}$. Next, with each step we apply a point mutation to characters representing DNA binding sites (alternatively we change the character of interaction from activatory to inhibitory or vice-versa), recalculate $\mathbf{W}$, and according to Eq. (59.4) the associated fixed point phenotypes or cycling expression patterns $\mathbf{S}$. Afterwards, having $\mathbf{S}$ we compute fitness of the genotype and accept or reject the attempted move according to Metropolis acceptance probability. This way by applying mutation-selection balance we obtain, after some initial period, an ensemble of viable genotypes constrained to have particular function.

59.2.4 *Essential Interactions*

Having obtained an ensemble of viable genotypes one would like to know which of the interactions between DNA regulatory regions and TFs are essential for network function. In order to get this information, we remove one of the interactions from the genotype and check if the gene expression pattern corresponding to this modified genotype is still close to the target phenotype. If the removal of interaction from GRN leads to loss of its functional capabilities we refer to this interaction as *essential*. Furthermore, the set of all essential interactions for a given genotype defines *essential network* for that GRN.

59.3 Results and Discussion

The simplified yeast cell-cycle pathway [2] consists of 13 consecutive gene expression patterns where the last target pattern is additionally a stationary state. In order to sample all possible GRNs having the same target phenotypes as the biological pathway, we impose on our system a sequence of 12 target patterns with additional condition that the last state is the fixed point of our dynamics. Furthermore, through MCMC sampling we generate an ensemble of genotypes which stay close to the predefined trajectory, and check for the essential interactions. As a result, we have a set of about 10000 independent networks for which we study similarities with biological network. Here we only list initial results, for more comprehensive description please refer to [12].

If we check for the similarities with the biological network having 29 links (see simplified yeast cell-cycle network in [2]), we get on average 16 links matching the biological network. This corresponds to roughly 55 % agreement, and if we check for other GRNs within ensemble almost all of them reproduce from 40 % to 70 % of interactions found in biological network. An important point is that generated networks have on average a bit less than 27 links, so in terms of fraction of biological links (L_{bio}) to the total number of interactions in a network (L_{tot}) the average L_{bio}/L_{tot} is equal to 60 % and almost all values of L_{bio}/L_{tot} are between 50 % and 75 %. Moreover, if we consider all interactions which are present in generated GRNs, only one link from yeast network is not found in any of networks in the ensemble, and in 97 % of networks there are the same 11 interactions which are found in yeast network.

These results are very striking if we realize that no bias toward biological network is incorporated inside our framework on any level, apart from the imposed target expression pattern. By using MCMC sampling procedure we produce many regulatory networks which are evolvable and a given target expression pattern can be realized through different topologies, yet all this GRNs have common features with the corresponding biological network. Since in our model the network architecture emerges from purely random background due to imposed functional patterns and selection pressure, it makes an interesting connection with the evolution of real biological networks.

References

1. Burda Z, Krzywicki A, Martin OC, Zagorski M (2011) Proc Natl Acad Sci USA 108:17263
2. Li F, Long T, Lu Y, Ouyang Q, Tang C (2004) Proc Natl Acad Sci USA 101:4781
3. Elowitz M, Leibler S (2000) Nature 403:335
4. Gardner T, Cantor C, Collins J (2000) Nature 403:339
5. Herrgard M, Covert M, Palsson B (2004) Curr Opin Biotechnol 15:70
6. Salgado H et al (2006) Nucleic Acids Res 34:D394
7. Hu Z, Killion P, Iyer V (2007) Nat Genet 39:683
8. Shen-Orr S, Milo R, Mangan S, Alon U (2002) Nat Genet 31:64
9. Ma H et al (2004) Nucleic Acids Res 32:6643
10. Lee T et al (2002) Science 298:799
11. Zhu J et al (2008) Nat Genet 40:854
12. Zagorski M, Krzywicki A, Martin OC Edge usage, motifs and regulatory logic for cell cycling genetic networks. Phys Rev E (2013). doi:10.1103/PhysRevE.87.012727

13. Kauffman S (1993) Origins of order: self-organization and selection in evolution. Oxford University Press, Oxford
14. Burda Z, Krzywicki A, Martin OC, Zagorski M (2010) Phys Rev E 82:011908
15. von Hippel P, Berg O (1986) Proc Natl Acad Sci USA 83:1608
16. Gerland U, Moroz J, Hwa T (2002) Proc Natl Acad Sci USA 99:12015
17. Lässig M (2007) BMC Bioinform 8:S7
18. Wagner A (1996) Evolution 50:1008
19. Bornholdt S, Rohlf T (2000) Phys Rev Lett 84:6114

Chapter 60
Numerical Continuation of Equilibria of Cell Population Models with Internal Cell Cycle

Charlotte Sonck, Markus Kirkilionis, and Willy Govaerts

Abstract Mathematical modelling of the cell cycle has been a subject of study for a few decades. J.J. Tyson and B. Novák have developed several models for the cell cycle of budding yeast, fission yeast and other organisms. Our goal is to incorporate these realistic models in structured cell population ODE models to study the behaviour of the cells at population level.

Our approach is to consider a chemostat, i.e. a container with a constant influx of nutrient and a constant outflow of organism-nutrient mixture. In this chemostat the growth of the organisms is dependent on the nutrient level, the organisms divide according to the cell cycle model and are washed away by the dynamics of the chemostat. The goal is to study the influence on the overall behaviour of the cell population of natural parameters such as the growth rate of the cells and the nutrient concentration.

An equilibrium state of this cell population model requires a constant distribution of the mass of cells born per unit of time. Numerically, the idea is to obtain the equilibrium as the fixed point of a map. We implement this map in our code as the output of a large collection of integrations over age for cells born with a given initial mass, followed by their implications for the consumption of nutrient. A found equilibrium can then be continued under parameter variation.

60.1 Introduction

The cell cycle is a key element in life and has been studied for decades, but some open questions remain. The mathematical modelling of the cell cycle can help to uncover the underlying mechanisms and to solve some of these open questions. J.J. Tyson and B. Novák are prominent leaders in this research and have developed

C. Sonck (✉) · W. Govaerts
Department of Applied Mathematics and Computer Science, Ghent University, Ghent, Belgium
e-mail: Charlotte.Sonck@UGent.be

M. Kirkilionis
Mathematics Department, University of Warwick, Coventry, UK

T. Gilbert et al. (eds.), *Proceedings of the European Conference on Complex Systems 2012*, Springer Proceedings in Complexity, DOI 10.1007/978-3-319-00395-5_60,

several widely studied models for the cell cycle of budding yeast, fission yeast and other organisms (see for example [2]).

Our goal is to incorporate these realistic models for the cell cycle in structured cell population models describing unicellular organisms living in a continuous culture. This way, we can study the behaviour of the cells at the population level and look into its dependence on the nutrient level (for more information on structured consumer resource models, see [1]). We use a chemostat model and attach a physiological structure to the cells, describing their internal cell cycle (which we base on the models of Tyson and Novák) and assume that the progression through the cell cycle depends on the nutrient concentration in the environment of the cells.

60.2 Cell Cycle Mechanism and Change of Nutrient

As a starting point, we have based our cell cycle mechanism on the Toy model of Tyson and Novák [3]. This model consists of 4 state components: the mass m, the concentration of cyclin/Cdk dimers X, the concentration of active Cdh1/APC complexes Y and the concentration of a protein that activates Cdh1 at Finish A. In this model, it is assumed that a cell divides if the concentration X crosses 0.1 from above. The internal state of the cell x in this case thus corresponds to a vector with m, X, Y and A as components. The equations are

$$\frac{dm}{dt} = g_m(x, S) = \mu m\left(1 - \frac{m}{m_{\max}}\right)\frac{S}{\zeta_1 + S},$$

$$\frac{dX}{dt} = g_X(x, S) = k_1 - \left(k_2' + k_2'' Y\right)X,$$

$$\frac{dY}{dt} = g_Y(x, S) = \frac{(k_3' + k_3'' A)(1 - Y)}{J_3 + 1 - Y} - \frac{k_4 m X Y}{J_4 + Y},$$

$$\frac{dA}{dt} = g_A(x, S) = k_5' + k_5'' \frac{(mX)^n}{J_5^n + (mX)^n} - k_6 A,$$

where we have adjusted the equation for the mass by incorporating a dependency on the nutrient concentration S in the chemostat. All the parameters in these equations are positive. We assume that a cell must have a minimal size $m_{\min}$ before it can divide and that, when dividing with mass m, it splits into two cells: a cell with mass ϕm and a cell with mass $(1 - \phi)m$ $(0 < \phi \leq 0.5)$ and with the X, Y and A of the original cell. We furthermore assume that a cell cannot reach a maximal size $m_{\max}$ before having divided or died. It follows that the possible birth masses are in $\Omega_b|_m = [\phi m_{\min}, (1 - \phi)m_{\max}]$.

The nutrient concentration S fluctuates according to the nutrient consumption by the cells in the population and the intrinsic rate of change of the resource $f(S)$ (the rate of change in absence of the consumer). We assume the so-called chemostat condition:

$$f(S) = D\left(S^0 - S\right),$$

where D is the dilution rate and $S^0 > 0$ the concentration of the limiting nutrient contained in the feeding bottle of the chemostat. As a first step towards the construction of an algorithm to find an equilibrium of this cell population model, we assume that the nutrient concentration S is constant and equal to a parameter $\bar{S}$ and calculate the evolution of the cells for this fixed $\bar{S}$. This will result in an equilibrium condition for $\bar{S}$, see further.

We define $\mathcal{F}(a, x_0, \bar{S})$ as the probability for an individual to reach age a given that it had state x_0 at birth and that it has experienced a constant resource concentration $\bar{S}$. Further we denote by $\beta(x, \bar{S})$ the rate of division of a cell at state x under constant nutrient concentration $\bar{S}$ and by $\gamma(x, \bar{S})$ the rate of food consumption of such an individual of state x, also under constant nutrient concentration $\bar{S}$. Let μ_0 denote the constant individual mortality rate, which corresponds to the outflow from the chemostat.

In the Toy model, it is assumed that a cell divides when X crosses 0.1 from above. We refine this by introducing the following rate of division:

$$\beta(m, X) := \begin{cases} 0 & \text{if } m < m_{\min} \text{ or } X \notin (0.1 - \epsilon_\beta, 0.1 + \epsilon_\beta) \text{ or } \frac{dX}{dt} \geq 0, \\ \frac{2\epsilon_\beta}{X - (0.1 - \epsilon_\beta)} - 1 & \text{if } m \geq m_{\min} \text{ and } X \in (0.1 - \epsilon_\beta, 0.1 + \epsilon_\beta) \text{ and } \frac{dX}{dt} < 0. \end{cases}$$

We note that $\beta(m, X)$ has a singularity at $X = 0.1 - \epsilon_\beta$ so that a cell necessarily divides when X crosses $0.1 - \epsilon_\beta$ from above.

60.3 Discretisation of the Birth State and Equilibrium Equations

We will consider cohorts of cells, that each correspond to cells that were born with a certain birth state x_0, and follow these cohorts over age untill the survival probability $\mathcal{F}(a, x_0, \bar{S})$ in the cohort is negligible. As a starting point, we use a fixed number of cohorts N and only discretise the m-component of x_0 since we assume that all cells will divide when X is close to 0.1. We discretise $\Omega_b|_m$ in the simplest possible way, with a uniform meshing and choose 0.1, 0.5 and 0.5 as start values for respectively X, Y and A in every cohort. So we have the following discretisation points of Ω_b:

$$x_{0i} = \begin{pmatrix} d_{0i} := \phi m_{\min} + (i - \frac{1}{2}) \frac{((1-\phi) m_{\max} - \phi m_{\min})}{N} \\ 0.1 \\ 0.5 \\ 0.5 \end{pmatrix} \quad \text{for } i = 1, \ldots, N.$$

We then have the following equilibrium condition equations for the evolution of the state x and the survival probability $\mathcal{F}$:

$$m'(a, x_{0i}, \bar{S}) = g_m\big(m(a, x_{0i}, \bar{S}), X(a, x_{0i}, \bar{S}), Y(a, x_{0i}, \bar{S}), A(a, x_{0i}, \bar{S}), \bar{S}\big),$$

$$m(0, x_{0i}, \bar{S}) = d_{0i},$$

$$X'(a, x_{0i}, \bar{S}) = g_X\big(m(a, x_{0i}, \bar{S}), X(a, x_{0i}, \bar{S}), Y(a, x_{0i}, \bar{S}), A(a, x_{0i}, \bar{S}), \bar{S}\big),$$
$$X(0, x_{0i}, \bar{S}) = 0.1,$$
$$Y'(a, x_{0i}, \bar{S}) = g_Y\big(m(a, x_{0i}, \bar{S}), X(a, x_{0i}, \bar{S}), Y(a, x_{0i}, \bar{S}), A(a, x_{0i}, \bar{S}), \bar{S}\big),$$
$$Y(0, x_{0i}, \bar{S}) = 0.5,$$
$$A'(a, x_{0i}, \bar{S}) = g_A\big(m(a, x_{0i}, \bar{S}), X(a, x_{0i}, \bar{S}), Y(a, x_{0i}, \bar{S}), A(a, x_{0i}, \bar{S}), \bar{S}\big),$$
$$A(0, x_{0i}, \bar{S}) = 0.5,$$
$$\mathcal{F}'(a, x_{0i}, \bar{S}) = -\big(\mu_0 + \beta\big(m(a, x_{0i}, \bar{S}), X(a, x_{0i}, \bar{S})\big)\big)\mathcal{F}(a, x_{0i}, \bar{S}),$$
$$\mathcal{F}(0, x_{0i}, \bar{S}) = 1.$$

We define $\theta(a, x_{0i}, \bar{S})$ as the amount of nutrient consumed per unit of time by a cell of age a in cohort i when it experienced a constant nutrient concentration $\bar{S}$. The corresponding equations are:

$$\theta'(a, x_{0i}, \bar{S}) = \gamma\big(m(a, x_{0i}, \bar{S}), X(a, x_{0i}, \bar{S}), Y(a, x_{0i}, \bar{S}), A(a, x_{0i}, \bar{S}), \bar{S}\big) \times \mathcal{F}(a, x_{0i}, \bar{S}),$$
$$\theta(0, x_{0i}, \bar{S}) = 0.$$

Finally we define $r_0(a, x_{0i}, \bar{S})$ as the number of cells born per unit of time with initial state x_{0i} that originate from mother cells with age a or less, that have experienced a constant nutrient concentration $\bar{S}$ during their life-time. We have to keep in mind that these newborn cells can be the smaller or bigger part of the divided cell, so this gives us the following equations:

$$r_0'(a, x_{0i}, \bar{S}) = \sum_{j \in I_{i1}} \beta\big(m(a, x_{0j}, \bar{S}), X(a, x_{0j}, \bar{S})\big)\mathcal{F}(a, x_{0j}, \bar{S})\bar{b}(x_{0j}) + \sum_{k \in I_{i2}} \beta\big(m(a, x_{0k}, \bar{S}), X(a, x_{0k}, \bar{S})\big)\mathcal{F}(a, x_{0k}, \bar{S})\bar{b}(x_{0k}), \quad \text{with}$$
$$I_{i1} = \left\{ j | 1 \le j \le N \wedge \phi m(a, x_{0j}, \bar{S}) \in \left[d_{0i} - \frac{\Delta m}{2}, d_{0i} + \frac{\Delta m}{2} \right[\right\},$$
$$I_{i2} = \left\{ k | 1 \le k \le N \wedge (1-\phi) m(a, x_{0k}, \bar{S}) \in \left[d_{0i} - \frac{\Delta m}{2}, d_{0i} + \frac{\Delta m}{2} \right[\right\},$$
$$r_0(0, x_{0i}, \bar{S}) = 0.$$

These equations are solved until age $\bar{a}$, when for every $x_{0i} \in \Omega_b$ we have that $\mathcal{F}(\bar{a}, x_{0i}, \bar{S}) < \epsilon$ for a given small positive ϵ.

An equilibrium of this cell population model corresponds to a constant distribution of the mass of cells born per unit of time $\bar{b}(x_{0i})$ ($i = 1, \dots, N$) and a constant nutrient concentration $\bar{S}$. We can calculate the amount of nutrient consumed per unit

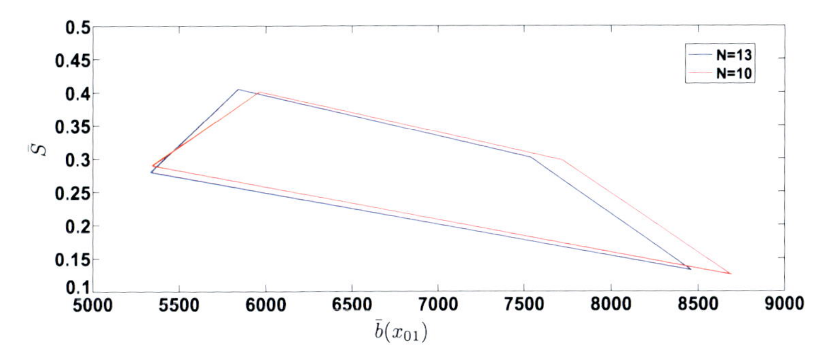

Fig. 60.1 The observed 4-cycle for $N = 10$ and $N = 13$ in $(\bar{b}(x_{01}), \bar{S})$-space

of time by, the originally $\bar{b}(x_{0i})$, cells in cohort i when having experienced a constant nutrient concentration $\bar{S}$ as $\theta(\bar{a}, x_{0i}, \bar{S}) \times \bar{b}(x_{0i})$. The equilibrium $(\bar{S}, \bar{b}(x_{0i}))$ is then the $N + 1$ vector that fulfills the following $N + 1$ equations:

$$f(\bar{S}) - \sum_{i=1}^{N} \Theta(x_{0i}, \bar{S})\bar{b}(x_{0i}) = 0,$$

$$r_0(\bar{a}, x_{0i}, \bar{S}) - \bar{b}(x_{0i}) = 0, \quad \forall i = 1, \dots, N.$$

This corresponds to the fixed point of the following map:

$$\begin{pmatrix} \bar{S} \\ \bar{b}(x_{0i}) \end{pmatrix} \longrightarrow \begin{pmatrix} S^0 - \frac{1}{D} \sum_{i=1}^{N} \Theta(x_{0i}, \bar{S})\bar{b}(x_{0i}) \\ r_0(\bar{a}, x_{0i}, \bar{S}) \end{pmatrix}. \tag{60.1}$$

When the equilibrium is found, it can then be continued under parameter variation. Natural parameters are the growth rates of the cells and the concentration of influx nutrient S^0.

60.4 Results

We implemented the map (60.1) using C++ and the CVODE solver. The results we obtained so far, for simple choices for the functions and parameter values, include several 4-cycles.

For example, for $N = 10$, we start the map with initially 1 cell in every cohort and evaluate the map repeatedly. After some iterations of the map, it is clear that there are only cells born in cohort 1 and in cohort 4. The amount of cells born in these two cohorts is equal, as they correspond respectively to the small and large part of the mother cell after division. If we repeatedly evaluate the map, we observe a 4-cycle after a transient time, see Fig. 60.1. To check if this was reflecting some inherent oscillating behaviour of the cell population model and to rule out possible numerical ghost effects as the cause, we repeated the same calculation for $N = 13$. After some iterations of the map, the same 4-cycle is observed (see Fig. 60.1).

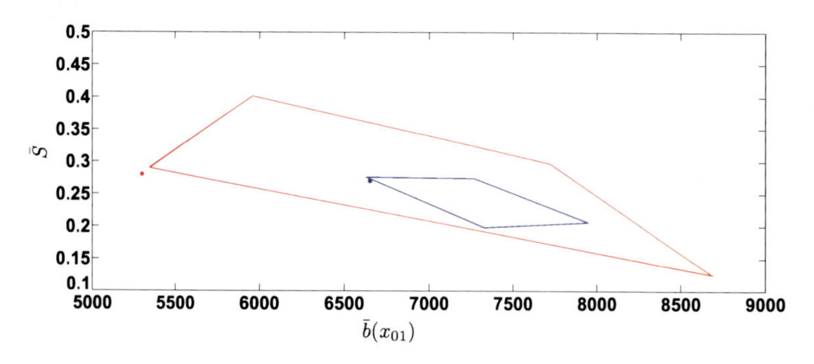

Fig. 60.2 Two sTable 4-cycles for $N = 10$ with corresponding start points in $(\bar{b}(x_{01}), \bar{S})$-space

We also observed that the map has at least 2 different sTable 4-cycles (an observation we made for both $N = 10$ and $N = 13$). In Fig. 60.2 the 4-cycles are depicted for $N = 10$ with the corresponding start point of the calculations.

References

1. de Roos AM, Diekmann O, Getto Ph, Kirkilionis MA (2010) Numerical equilibrium analysis for structured consumer resource models. Bull Math Biol 72:259–297
2. Tyson JJ, Novák B (2001) Regulation of the eukaryotic cell cycle: molecular antagonism, hysteresis, and irreversible transitions. J Theor Biol 210:249–263
3. Tyson JJ, Novák B (2002) Cell cycle controls. In: Fall CP, Marland ES, Wagner JM, Tyson JJ (eds) Computational cell biology. Springer, New York, pp 261–284

Chapter 61
Bistability and Oscillations in a Skeleton Model for the Cyclin/Cdk Network Driving the Mammalian Cell Cycle

Claude Gérard and Albert Goldbeter

Abstract A network of cyclin-dependent kinases (Cdks) based on intertwined negative and positive feedback loops regulates the mammalian cell cycle. We have recently proposed a skeleton model for this Cdk network, which incorporates Cdk regulation through phosphorylation-dephosphorylation and includes the positive feedback (PF) loops that underlie the dynamics of the G1/S and G2/M transitions of the cell cycle (Gérard et al., FEBS J., 279:3411–3431, 2012). We showed that the multiplicity of PF loops promotes the occurrence of bistability and increases the amplitude of oscillations in the various cyclin/Cdk complexes. Stochastic simulations further indicated that the presence of multiple PF loops enhances the robustness of Cdk oscillations with respect to molecular noise. Here we show that this skeleton model can produce complex modes of oscillatory behavior, which are due to the interaction between the multiple oscillatory circuits contained in the Cdk network driving the cell cycle.

Keywords Mammalian cell cycle · Cdk network · Bistability · Cdk oscillations · Positive feedback loops · Robustness to molecular noise · Deterministic and stochastic simulations

61.1 Models for the Cell Cycle

A network of cyclin-dependent kinases (Cdks) controls progression along the four successive phases G1, S (DNA replication), G2, and M (mitosis) of the mammalian cell cycle [1, 2]. When cells are not proliferating, they remain in a quiescent phase, denoted G0. The Cdk network is regulated by intertwined negative and positive feedback loops. Negative feedback loops play a key role in generating self-sustained oscillations in the network. Since positive feedback (PF) loops were shown to participate in the mechanism of biological oscillations in a number of cellular systems, the question arises as to their role in the oscillatory dynamics of the Cdk network.

C. Gérard (✉) · A. Goldbeter
Faculté des Sciences, Université Libre de Bruxelles (ULB), Campus Plaine, CP 231, 1050 Brussels, Belgium
e-mail: cgerard1@ulb.ac.be

T. Gilbert et al. (eds.), *Proceedings of the European Conference on Complex Systems 2012*, Springer Proceedings in Complexity, DOI 10.1007/978-3-319-00395-5_61,

A number of experimental and theoretical studies, mostly devoted to the early cell cycles in amphibian embryos [3–7] and to the yeast cell cycle [8], showed that positive feedback contributes to the robustness of oscillatory behavior. Building on these studies, we return to this issue by using a mathematical model for the Cdk network that drives the mammalian cell cycle.

Several models were proposed to account for parts of the mammalian cell cycle, especially the G1/S transition [9, 10], the restriction point in G1 [11], or the G2/M transition [12]. We recently proposed a detailed model describing the dynamics of the global Cdk network driving the mammalian cell cycle [13]. This model consists of four Cdk modules, each centered around one cyclin/Cdk complex. Cyclin D/Cdk4-6 and cyclin E/Cdk2 ensure progression in G1 and elicit the G1/S transition, respectively; cyclin A/Cdk2 promotes progression in S and the transition S/G2, while the activity of cyclin B/Cdk1 brings about the G2/M transition. This detailed model for the Cdk network, which contains 39 variables, includes both negative and positive feedback loops. We used this model to show that in the presence of sufficient amounts of growth factor the Cdk network is capable of temporal self-organization in the form of sustained oscillations [13]. The latter correspond to the ordered, sequential activation of the various cyclin/Cdk complexes that control the successive phases of the cell cycle. These sustained oscillations of the various cyclin/Cdk complexes that drive progression in the cell cycle suggest that the cell division cycle can be viewed as a true cellular rhythm [14].

We previously showed that we may relinquish many biochemical details in building a skeleton, 5-variable model for the mammalian cell cycle, without losing the key dynamical properties of the Cdk network [15]. Thus, sustained oscillations in the various cyclin/Cdk complexes occur in the skeleton model in the presence of sufficient amounts of growth factor. The skeleton model also accounts for the existence of a restriction point in G1 beyond which the presence of the growth factor is not needed to complete a cycle.

61.2 Effect of Positive Feedback Loops on the Dynamics of the Cell Cycle

We extended the skeleton model by incorporating the regulation of Cdk2 and Cdk1 by the phosphatase Cdc25 and the kinase Wee1 (see Fig. 61.1). This allowed us to assess the role of positive feedback on the dynamics of the Cdk network [16]. Multiple PF loops are indeed associated with the regulation of cyclin E/Cdk2 and cyclin B/Cdk1 through phosphorylation-dephosphorylation. Whereas cyclin E/Cdk2 is subjected to a single PF via its mutual activation with the phosphatase Cdc25, cyclin B/Cdk1 is regulated by two PF loops via its activation of Cdc25 and its inhibition, through phosphorylation, of its inhibitory kinase Wee1. Negative feedback loops are also present in the Cdk network and play a key role in its oscillatory dynamics: this network, based on four cyclin/Cdk modules controlling the successive phases of the cell cycle, is indeed regulated in such a manner that each Cdk module activates the next Cdk module while inhibiting the preceding one (Fig. 61.1).

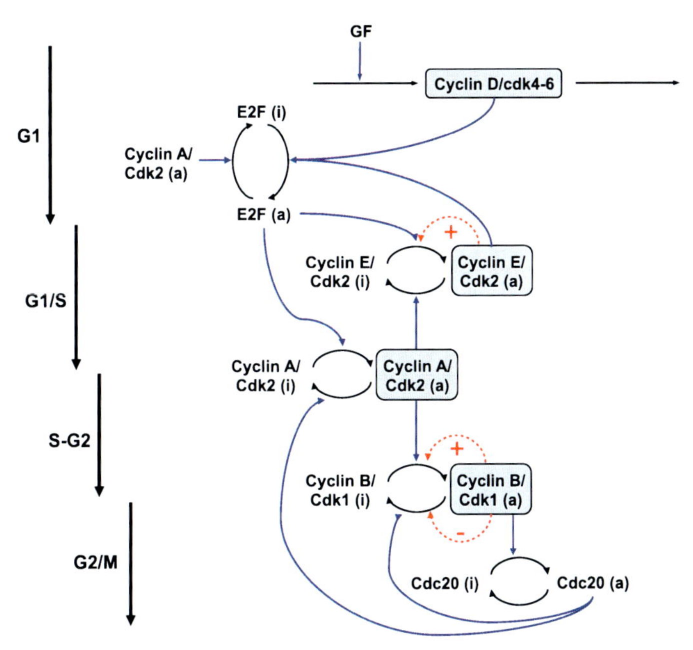

Fig. 61.1 Scheme of the extended skeleton model for the mammalian cell cycle (see [16]). The model contains the four main cyclin/Cdk complex, the transcription factor E2F, and the protein Cdc20. The presence of growth factor (GF) induces the entry in the G1 phase of the cell cycle by promoting the synthesis of cyclin D/Cdk4-6 complex. This complex allows the activation of the transcription factor E2F, which will elicit the synthesis of cyclin E/Cdk2, at the G1/S transition, and cyclin A/Cdk2, during the S phase of DNA replication. During G2, cyclin A/Cdk2 also activates the synthesis of cyclin B/Cdk1, which will permit the peak of activity of cyclin B/Cdk1 at the G2/M transition. During mitosis, cyclin B/Cdk1 activates by phosphorylation the protein Cdc20. This creates a negative feedback loop in the activity of cyclin A/Cdk2 and cyclin B/Cdk1 by promoting the degradation of these complexes. The regulations exerted by Cdc20 allow the cell to complete mitosis, and to start a new cycle if the growth factor is present in sufficient amount. Moreover, positive feedback loops can be added in the regulation of cyclin E/Cdk2 and cyclin B/Cdk1, which control the G1/S and G2/M transitions (*dashed arrows in red*). These positive feedback loops are due to the mutual activation between Cdk2, Cdk1 and their phosphatase Cdc25 and to the mutual inhibition between Cdk1 and the kinase Wee1

We showed [16] that multiple PF loops give rise to bistability in the presence of sufficient zero-order ultrasensitivity [17]. Negative feedback loops built into the structure of the Cdk network are critical in allowing the system to base its oscillatory behavior on repetitive, bistable transitions. The range of bistability increases with the number of PF loops. Oscillations are then characterized by a plateau in Cdk activity, associated with bistability (see Fig. 61.2(A)). Stochastic simulations show that the robustness of Cdk oscillations with respect to molecular noise increases in such conditions [16].

Here we show that the extended version of the skeleton model for the Cdk network can also produce complex periodic oscillations or quasiperiodic oscillations (see Fig. 61.2). Such complex oscillations were previously found in the detailed

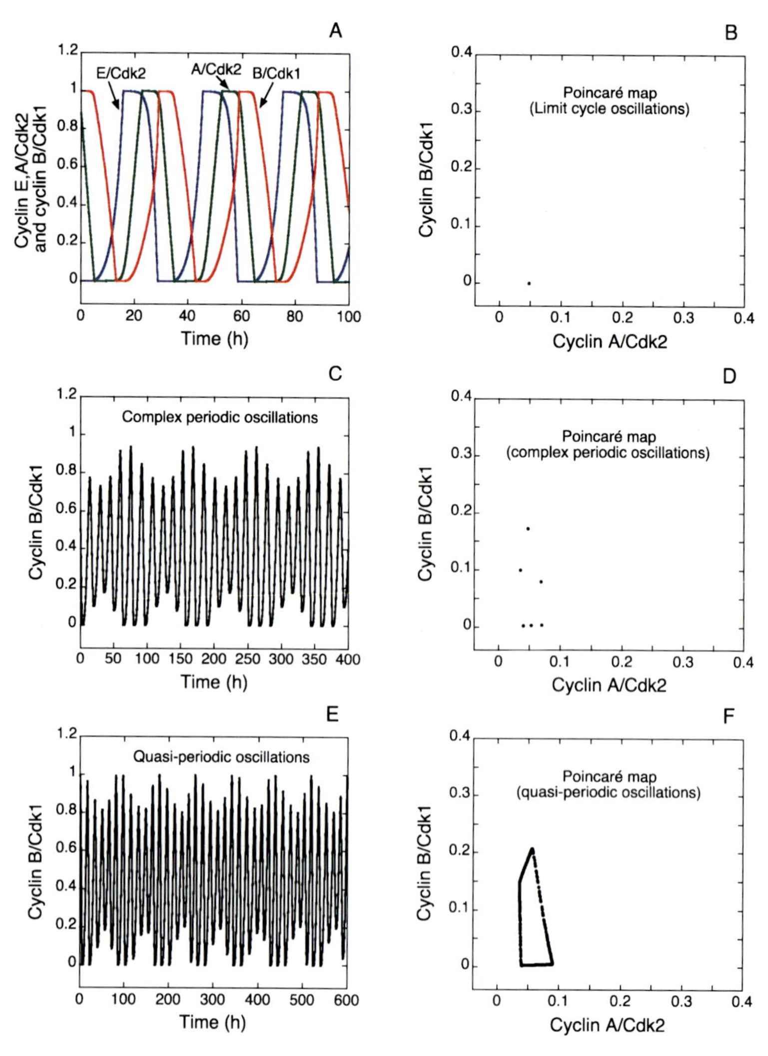

Fig. 61.2 Simple vs. complex oscillatory behaviors in the extended version of the skeleton model for the Cdk network driving the mammalian cell cycle. (**A**) Time evolution of cyclin E/Cdk2, cyclin A/Cdk2 and cyclin B/Cdk1 corresponding to simple periodic, limit cycle oscillations of the cyclin/Cdk network. The time evolution of cyclin B/Cdk1 corresponding to complex periodic or quasi-periodic oscillations is shown in C and E, respectively. Poincaré sections established by plotting the levels of cyclin B/Cdk1 versus cyclin A/Cdk2 corresponding to the passage through a maximum in the level of cyclin E/Cdk2 are shown for limit cycle oscillations (**B**), complex periodic oscillations (**D**), and quasi-periodic behavior (**F**). Numerical values of the parameters for the simple periodic oscillations (**A**), (**B**) are those of Table 2 in Ref. [16] in the presence of strong zero-order ultrasensitivity (ZOU) and in the presence of positive feedback loops ($b_1 = 1$, $b_2 = 1$, $K_{ib} = 0.5$ μM). For complex periodic oscillations (**C**), (**D**), values of parameters are as in (**A**), (**B**) with $V_{1cdc20} = 1.1\ \text{h}^{-1}$, $V_{2Me} = 8\ \text{h}^{-1}$, $V_{2Ma} = 3\ \text{h}^{-1}$, and $V_{2Mb} = 4\ \text{h}^{-1}$. Parameter values for quasi-periodic oscillations (**E**), (**F**) are as in (**C**), (**D**) with $V_{1cdc20} = 1\ \text{h}^{-1}$

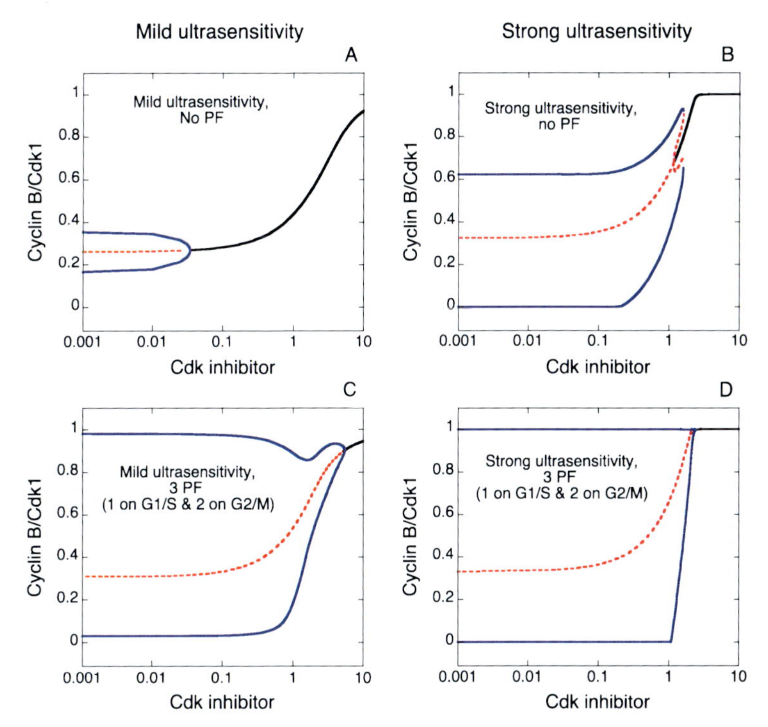

Fig. 61.3 Dynamical behavior of the Cdk network as a function of the level of Cdk inhibitor. Bifurcation diagrams of cyclin B/Cdk1 versus the level of Cdk inhibitor are shown in the absence (**A**), (**B**) or in the presence (**C**), (**D**) of positive feedback (PF) loops in the Cdk network. Simulations are performed in the presence of mild (**A**), (**C**) or strong ultrasensitivity (**B**), (**D**) characterizing the activation/inactivation cycle of the cyclin/Cdk complexes through phosphorylation-dephosphorylation. *Black curves* represent stable steady states; *red dashed curves* represent unstable steady states, while *blue curves* represent the envelope, i.e. maxima and minima of sustained oscillations. We have incorporated the action of a Cdk inhibitor in the skeleton model [16] by considering the effective concentrations of the various cyclin/Cdk complexes when they act as protein kinases in the model, which are equal to their concentrations multiplied by the factor $(1/(1 + \text{inhibitor}))$

model for the mammalian cell cycle and in the skeleton model without positive feedback regulation (see [15, 18]). Much as in these models, complex modes of oscillatory behavior in the extended skeleton model incorporating positive feedback loops originate from the interaction between the multiple oscillatory circuits contained in the Cdk network. However, based on numerous simulations, it seems that these complex modes of oscillatory behavior are less frequent than simple periodic oscillations, which corroborates the view that the ordered progression in the different cell cycle phases is controlled by simple periodic oscillations, and that such oscillations correspond to the physiological mode of oscillations in the Cdk network.

We further investigate the combined effect of positive feedback loops and of Cdk inhibitors on sustained oscillations of the various cyclin/Cdk complexes. Bifurcation diagrams of the Cdk network established as a function of the level of Cdk inhibitor show that the domain of sustained oscillations is much larger in the presence of positive feedback loops (see Fig. 61.3). The level of Cdk inhibitor must then be much larger to suppress the oscillatory behavior of the Cdk network. This result bears on the use of Cdk inhibitors as anticancer drugs [19] since sustained oscillations of the cyclin/Cdk complexes correspond to active cell proliferation. Computational models for the Cdk network driving the cell cycle might thus prove useful for studying the effect of Cdk inhibitors on the dynamics of the cell cycle in the framework of cancer therapy.

Acknowledgements This work was supported by grant no. 3.4607.99 from the Fonds de la Recherche Scientifique Médicale (F.R.S.M., Belgium), the Belgian Federal Science Policy Office (IAP P6/25 "BioMaGNet": "Bioinformatics and Modeling—From Genomes to Networks"), and the F.R.S.-FNRS (Belgium) in conjunction with the ErasysBio+ project C5Sys, "Circadian and Cell Cycle Clock Systems in Cancer". C. Gérard currently holds a postdoctoral fellowship from the Foundation Philippe Wiener—Maurice Anspach in the Department of Biochemistry at the University of Oxford.

References

1. Morgan DO (1995) Principles of Cdk regulation. Nature 374:131–134
2. Morgan DO (2006) The cell cycle: principles of control. Oxford University Press, London
3. Novak B, Tyson JJ (1993) Numerical analysis of a comprehensive model of M-phase control in Xenopus oocyte extracts and intact embryos. J Cell Sci 106:1153–1168
4. Goldbeter A (1993) Modeling the mitotic oscillator driving the cell division cycle. Comment Theor Biol 3:75–107
5. Ferrell JE Jr., Machleder EM (1998) The biochemical basis of an all-or-none cell fate switch in Xenopus oocytes. Science 280:895–898
6. Pomerening JR, Sontag ED, Ferrel JE Jr (2003) Building a cell cycle oscillator: hysteresis and bistability in the activation of Cdc2. Nat Cell Biol 5:346–351
7. Sha W, Moore J, Chen K, Lassaleta AD, Yi C-S, Tyson JJ, Sible JC (2003) Hysteresis drives cell-cycle transitions in Xenopus laevis egg extracts. Proc Natl Acad Sci USA 100:975–980
8. Chen KC, Calzone L, Csikasz-Nagy A, Cross FR, Novak B, Tyson JJ (2004) Integrative analysis of cell cycle control in budding yeast. Mol Biol Cell 15:3841–3862
9. Qu Z, Weiss JN, MacLellan WR (2003) Regulation of the mammalian cell cycle: a model of the G1-to-S transition. Am J Physiol, Cell Physiol 284:349–364
10. Swat M, Kel A, Herzel H (2004) Bifurcation analysis of the regulatory modules of the mammalian G1/S transition. Bioinformatics 20:1506–1511
11. Novak B, Tyson JJ (2004) A model for restriction point control of the mammalian cell cycle. J Theor Biol 230:563–579
12. Aguda BD (1999) A quantitative analysis of the kinetics of the G2 DNA damage checkpoint system. Proc Natl Acad Sci USA 96:11352–11357
13. Gérard C, Goldbeter A (2009) Temporal self-organization of the cyclin/Cdk network driving the mammalian cell cycle. Proc Natl Acad Sci USA 106:21643–21648
14. Goldbeter A, Gérard C, Gonze D, Leloup JC, Dupont G (2012) Systems biology of cellular rhythms. FEBS Lett 586:2955–2965
15. Gérard C, Goldbeter A (2011) A skeleton model for the network of cyclin-dependent kinases driving the mammalian cell cycle. Interface Focus 1:24–35
16. Gérard C, Gonze D, Goldbeter A (2012) Effect of positive feedback loops on the robustness of oscillations in the network of cyclin-dependent kinases driving the mammalian cell cycle. FEBS J 279:3411–3431

17. Goldbeter A, Koshland DE Jr (1981) An amplified sensitivity arising from covalent modification in biological systems. Proc Natl Acad Sci USA 78:6840–6844
18. Gérard C, Goldbeter A (2010) From simple to complex patterns of oscillatory behavior in a model for the mammalian cell cycle containing multiple oscillatory circuits. Chaos 20:045109
19. Cicenas J, Valius M (2011) The Cdk inhibitors in cancer research and therapy. J Cancer Res Clin Oncol 137:1409–1418

Chapter 62
Centrality Clubs and Concepts of the Core: Decoding the Communicative Organisation of the Brain

Emma K. Towlson, Petra E. Vértes, Sebastian E. Ahnert, and Edward T. Bullmore

Abstract Ideas from graph theory have facilitated many exciting advances in the understanding of connectivity in the human brain. It is known that the brain exhibits the 'rich club' phenomenon, characterised by a densely interconnected set of hub nodes. We generalise the definition of a rich club to other centrality measures, demonstrating the possibility of a family of 'rich clubs'. We compare the fMRI network architectures for rich clubs in healthy and schizophrenic individuals and apply a similar analysis to the C. elegans neural network.

Keywords Complexity · Networks · Brain · fMRI · Rich-club · Centrality · Organisation

62.1 Introduction

Graph theory provides a tremendously useful approach to understanding the organisation of complex systems. As a network, the human brain is especially interesting due to its extraordinary complexity and the further constraints that being spatially embedded brings [1]. This work applies selected concepts to fMRI data to move towards identifying a functional 'core' of the brain and describing the related hierarchical organisation, hinting at the way it processes information. The exhibition of the 'rich-club' phenomenon has already been explored in complex systems such as

E.K. Towlson (✉) · S.E. Ahnert
Theory of Condensed Matter Group, Department of Physics, Cavendish Laboratory, University of Cambridge, Cambridge, UK

P.E. Vértes · E.T. Bullmore
Behavioural & Clinical Neuroscience Institute, Department of Psychiatry, University of Cambridge, Cambridge, UK

E.T. Bullmore
Cambridgeshire & Peterborough NHS Foundation Trust, Cambridge, UK

E.T. Bullmore
GlaxoSmithKline, Clinical Unit Cambridge, Addenbrooke's Hospital, Cambridge, UK

T. Gilbert et al. (eds.), *Proceedings of the European Conference on Complex Systems 2012*, Springer Proceedings in Complexity, DOI 10.1007/978-3-319-00395-5_62,

the Internet [2] the power grid and protein interactions [3] and more recently the human connectome [4]. This rich-club phenomenon is characterised by a much greater density of connections between the highest degree nodes of a network than to or between the nodes of a lower degree, thus forming an elite densely interconnected 'club' of hub nodes. The existence of such a regime in a system can shed light on its nature, providing information regarding the robustness, efficiency and functional specialisation of the network. The rich club has often been thought to add value to the overall performance of the system; for example, the rich club of the electrical power grid enhances its resilience to failure of a single hub and therefore reduces the probability of a systemic blackout.

We have further explored the *C. elegans* nervous system, with a special focus on its "rich club". We predicted that the cellular connectome of the *C. elegans* nervous system would have a rich club organization that is similar, in terms of its economical trade-off between topological value and physical cost, to the rich clubs of human brain networks and other physically-embedded complex systems.

62.2 Materials and Methods

62.2.1 Rich-Clubs

To quantify the rich club effect, the degree of each node in the network (i.e. how many other nodes it is connected to) must first be calculated and all nodes with degree $\leq k$ removed. The *rich club coefficient* for the remaining sub-graph, $\Phi(k)$, is then the ratio of the number of existing connections to the number that would be expected if the network was fully connected, and formally is given by [2, 3]:

$$\Phi(k) = \frac{2M_{>k}}{N_{>k}(N_{>k} - 1)} \tag{62.1}$$

where $N_{>k}$ is the number of nodes with degree $> k$ and $M_{>k}$ is the number of edges between them. The computation of $\Phi(k)$ for all values of k in the network of interest yields a rich club curve.

However, the higher degree nodes in a network have a higher probability of sharing connections with each other simply by chance, so even random networks generate increasing rich club coefficients as a function of increasing degree threshold, k. To control for this effect, the rich club curve for the human brains and the *C. elegans* neuronal network were normalised relative to the rich club curves of 1000 comparable random networks. The random networks were generated by performing multiple $(100 \times M)$ double edge swaps or permutations on the original graphs. A double edge swap removes two randomly selected edges a–b and c–d and replaces them with the edges a–c and b–d (assuming they do not already exist, in which case a new edge pair must be selected). This permutation procedure ensures that the number of nodes and edges, and the degree distribution, of the nematode network are all conserved in the random networks. The normalised rich club coefficient is then given by:

$$\Phi_{norm}(k) = \frac{\Phi(k)}{\Phi_{random}(k)} \tag{62.2}$$

where $\Phi_{random}(k)$ is the average value of $\Phi(k)$ across the random networks.

The existence of rich club organisation is defined by $\Phi_{norm}(k) > 1$ over some range of values of threshold degree k. We used a probabilistic approach to define the threshold criteria for a rich club more precisely. At every different threshold degree, we estimated $\Phi_{random}(k)$ for 1000 realisations of the random networks, and estimated the standard deviation of $\Phi_{random}(k)$, denoted σ. The threshold range of the rich club regime was then specified by those values of k for which $\Phi(k) \geq \Phi_{random}(k) + 1\sigma$. Thus a rich club could be said to exist in the subgroup of network nodes defined by an arbitrary degree threshold if $\Phi_{norm}(k) = 1 + 1\sigma$; but we also defined rich clubs by the more stringent criterion of the points of greatest z-score.

Metric calculations and network manipulations were carried out using the Python networkx library [6] and Matlab.

62.2.2 Functional Brain Networks

An exploration based on these rich-clubs was carried out on complex networks generated from fMRI data [5] of 3 cohorts, each comprising of 15 healthy individuals and 12 with diagnosed chronic schizophrenia. The groups were matched for age, premorbid IQ and years of education. Each group had either been taking placebo or antipsychotic medication. Resting-state fMRI time series were acquired at 82 cerebral regions (later subdivided into 471 homogeneously sized regions) over 17 minutes and interregional associations were quantified with wavelet correlations in the frequency range 0.061–0.125 Hz. Complex networks were then found by thresholding the strength of correlation to acquire the top 10 % of connections. Both individual and averaged networks (by persistence of the presence of an edge across subjects) were extracted and rich-clubs identified by calculating the rich-club coefficient and locating the greatest z-score within the regimes.

62.3 Results

62.3.1 The **C. elegans** *Neuronal Network*

The rich-club criterion was satisfied for the *C. elegans* connectome when the threshold value for degree k, was in the range $35 < k < 73$ [7]. In what follows, we will focus on the rich club for degree thresholds in the range $39 \leq k < 45$, where we have that $\Phi(k) \geq \Phi_{random}(k) + 3\sigma$.

There were 11 neurons in this rich club: 8 were located anteriorly in the lateral ganglia of the head (AVAR/L, AVBR/L, AVDR/L, AVER/L); and 3 were located posteriorly in the lumbar (PVCR/L) and dorsorectal (DVA) ganglia. These are precisely the interneuronal components of the locomotor circuit plus DVA (known to participate in locomotory regulation). There was very high efficiency of connectivity between rich club neurons: $E_{Rich} = 0.92$. By way of comparison, the efficiency of connections between the 268 "poor periphery" neurons that were *not* in the rich club was much lower: $E_{Poor} = 0.38$. The rich club was also distinguished by high betweenness centrality, indicating that rich club neurons were often on the shortest paths between all pairs of neurons in the system; nine of the 11 rich club neurons (AVAR/L, AVBR/L, AVER/L, DVA, PVCR/L) were ranked in the top 10 of all neurons in terms of their betweenness centrality (with values ranging from 0.0277 to 0.103). A motif analysis [7] gave further evidence to the centrality of the rich-club to integrating communication.

On examination of the development of the *C. elegans* network, we found that the rich-club is formed in its entirety before the nematode is capable of even twitching movements. The probability of this happening by chance (all 11 neurons being born by this time) is 0.02.

62.3.2 Healthy and Schizophrenic Functional Networks

In agreement with previous findings from structural data [4], we found that all healthy subjects display degree rich-club behaviour amongst consistent brain regions; an average of 14 % of the network participates in the rich-club. Notably, these are also the regions which are commonly found to be activated in task-related studies. Whilst those with chronic schizophrenia do show this rich-club behaviour, it is far less pronounced and not at all consistent across individuals (with an average size of 9 % of the network participating in the rich-club). A bias towards inter-hemispheric connections is also observed. When the schizophrenic patients are administered anti-psychotic medication, however, we observed that the rich-club structures become considerably more like the 'healthy' dense rich-clubs. Interestingly, the drug sulpride appears to more successfully restore this organisation than aripiprazol (the most recent of the two medications to be developed), though both give rise to rich-clubs which contain on average 11 % of the network.

62.4 Centrality Clubs

This behaviour is exposed through the examination of degree, but there are numerous other measures which attempt to quantify how 'central' a node is, such as betweenness, closeness or eigenvector centrality, giving rise to the possibility of a family of 'rich-clubs'. Together, they may expose much about the topological importance of individual and groups of regions in the brain, and offer many suggestions towards the nature of information flow. The extension to different kinds of rich-club is analogously made by the same algorithm. The centrality metric must be calculated for each node in the network and for a threshold k a subgraph is found by only considering those nodes with a centrality $> k$. The measure specific rich-club coefficient is then found by calculating the ratio of the total centrality of the subgraph to the maximum total centrality a graph of that size could possibly have. So for the example of betweenness centrality (BC), a measure of how many shortest paths between pairs a node lies on, we find the maximal total betweenness centrality when the graph is organised into a straight line (forcing most nodes to lie on many shortest paths). Thus the betweenness rich-club coefficient, $\Phi_{BC}(k)$, for a network with N nodes, is given by:

$$\Phi_{BC}(k) = \frac{3\sum_i (BC_i)_{>k}}{N_{>k}(N_{>k}-1)(N_{>k}-2)} \tag{62.3}$$

Once this is done for all k, the curve is normalized as above.

Notably, whilst there is a large correlation between relative values of centrality measures, betweenness centrality is found to be most unlike the others in the functional human brain networks, making this rich-club and the classical degree centrality based rich-club the ones of most interest. Betweenness centrality based rich-clubs find densely connected 'clusters' which can be very loosely connected to each other (this is akin to '*knotty-centrality*' [8]). Preliminary results hint at interesting differences between healthy and schizophrenic individuals and a link to the robustness of the original network (examined by fragmentation with targeted and random attack). Persistently topologically significant regions/groups of regions hint at both the architecture and the nature of control of information flow, or functional 'cores', in the brain.

References

1. Achard S, Bullmore E (2007) Efficiency and cost of economical brain functional networks. PLoS Comput Biol 3:e17
2. Zhou S, Mondragon RJ (2004) The rich-club phenomenon in the internet topology. IEEE Commun Lett 8:180–182
3. Colizza V et al (2006) Detecting rich-club ordering in complex networks. Nat Phys 2:110–115
4. Van den Heuvel M, Sporns O (2011) Rich-club organization of the human connectome. J Neurosci 31(44):15775–15786
5. Lynall M et al (2011) Functional connectivity and brain networks in schizophrenia. J Neurosci 30(28):9477–9487
6. Hagberg AA, Schult SA, Swart PJ (2008) Exploring network structure, dynamics, and function using networkx. In: Proceedings of the 7th Python in science conference (SciPy2008), pp 11–15
7. Towlson EK, Vértes PE, Ahnert SE, Schafer WR, Bullmore ET (2012) The rich club of the C. elegans neuronal connectome. Preprint. doi:10.1523/JNEUROSCI.3784-12.2013
8. Shanahan M, Knotty-Centrality WM (2012) Finding the connective core of a complex network. PLoS ONE 7(5):e36579

Chapter 63
A Broader Perspective About Organization and Coherence in Biological Systems

Martin Robert

Abstract The implications of large-scale coherence in biological systems and possible links to quantum theory are only beginning to be explored. Whether quantum-like coherent phenomena are relevant, or even possible at all, at the high temperatures of biological systems remains unsettled. Here, we discuss a broader perspective on biological organization and how quantum-like dynamics and coherence might shape the very fabric from which complex biological systems are organized. Regardless of its exact nature, a unique form of coherence seems apparent at multiple scales in biology and its better characterization may have broad consequences for the understanding of living organisms as complex systems.

Keywords Complex systems · Emergence · Biological systems · Cellular dynamics · Quantum biology

63.1 Background

The spatio-temporal organization, coherence and complexity found in biological systems appear overwhelming and the underlying principles are only beginning to be deciphered. It is widely accepted that these principles must be based on more fundamental ones at the sub-molecular or molecular level. Since life depends on complex networks of chemical reactions, there are likely deep connections to the underlying quantum theory, which defines chemistry so well. However, quantum phenomena are often associated with paradoxical states characterizing the microscopic world of subatomic particles or other fundamental levels of matter that do not easily fit our perception of the reality of larger systems, including biological ones. In addition, while quantum theory is considered to be one of the most successful scientific theories, its implications beyond the microscopic level are only beginning to gain attention. Whether quantum-like phenomena are relevant or even possible at all, at the high temperatures of biological systems remains somewhat controversial.

M. Robert (✉)
Institute for Advanced Biosciences, Keio University, Tsuruoka, Yamagata, Japan
e-mail: mrobert@ttck.keio.ac.jp

T. Gilbert et al. (eds.), *Proceedings of the European Conference on Complex Systems 2012*, Springer Proceedings in Complexity, DOI 10.1007/978-3-319-00395-5_63,

Here, based on existing findings and proposals we explore its possible relevance and utility for providing insight into the remarkable dynamics and coherence found in living systems. We highlight some specific examples of organization and large-scale coherence in biological systems that might be connected or display properties better described by non-classical or quantum-like features including quantum coherence.

63.2 Proposal/Results

63.2.1 Mechanisms to Maintain Coherence

The standard reasoning suggests that quantum coherence can't be maintained at high temperature in biological systems for any significant period of time due to strong coupling to the environment. Prior studies have shown that under conditions relevant in biology, systems will decohere very rapidly [1]. However, some recent studies provide evidence that quantum coherence can be maintained, in some systems of biological relevance, for much longer time-scales than expected [2]. Perhaps the best established example supporting the presence of quantum coherence in biological systems include electron transfer machinery in photosynthesis [3, 4]. Other suggestive examples where direct evidence may still be lacking include the avian magnetoreception and orientation system, enzyme catalysis and reaction dynamics [5], and some effects in ligand-receptor interactions in cellular signaling processes [6]. In addition, a quantum mechanical model has been proposed as a possible mechanism to explain some forms of adaptive mutation [7]. However, most of these examples of possible quantum coherence in biological systems appear to be extensions of widely accepted quantum chemistry principles, albeit occurring within the warm environment of the cell. As such, these processes may not be as surprising. Here, we aim to reflect mainly on whether quantum-like phenomena might play a role or have explanatory power at much higher levels of organization including the cellular, organ, and even whole organism level.

(1) *Open Confinement.* Arguments for ruling-out any form of quantum-like coherence at high temperature may often ignore some important and unique properties of biological systems. These include prominently, the confined, protected, and controlled cellular environment provided by the insulating nature of the cell membrane. This complex structure provides spatial confinement of internal elements and activities while at the same time allowing the necessary channels to operate two-way exchanges with the environment. The cell thus constitutes a unique spatially confined system with an open architecture that facilitates controlled exchanges with the environment. This "open confinement" appears critical in maintaining cellular coherence. It may seem paradoxical that coherence could be maintained through exchanges with the environment, the very phenomenon that is usually considered the main factor in decoherence. However, the type of exchanges occurring across the living cell membrane, for example, are highly specific and are controlled by a tight

molecular filter through which only certain exchanges are allowed, in contrast to other systems. In addition, some specific forms of interactions and coupling between a system and its environment have been shown to be coherence-promoting [1].

(2) *Metabolic Activity and Negentropy.* In addition, biological systems are known to operate in a thermodynamic state that is far from equilibrium my making use of energy consuming/producing metabolic activities. These allow constant renewal of cellular components and can be conceptualized as an important error correction mechanism and as countering decoherence. Without metabolism, cellular activities would soon come to a halt due among other things to the inability to replace defective cellular components and maintain a constant energy supply. Schrodinger introduced the term negative entropy [8] or negentropy to describe this generation and maintenance of order in biological systems that are otherwise immersed in decoherence-inducing environments. This concept of negentropy therefore appears to describe the constant renewal of damaged cellular constituents linked to metabolic activity and how it acts to resist the natural tendency toward decoherence.

(3) *Multi-level Communication and Downward Causation.* Finally, the complex and multi-scale systems of information exchange and signaling within biological systems further contributes to the maintenance of long-term coherence, in biological systems, whatever the exact description that coherence may fit. As argued by Noble [9, 10], organization in biological systems cannot be simply described from bottom-up principles but is strongly linked to downward causation. In this form of interaction, elements that are the product of the whole system or that are generated from the higher levels of organization feedback and regulate lower level elements in a self-referential manner. Accordingly, there seems to be no privileged level of causation in biological systems and these levels interact at all scales. The self-generated feedback loop may be seen as strongly coherence inducing since the coherence-inducing information is self-generated, self-reinforcing and not dependent on external factors. It may thus conceptually also represent another form of system insulation against decoherence. Together these unique properties, and likely others not described here, provide a unique infrastructure that allows the maintenance of a high level of organization and possibly unique form of coherence in biological systems.

63.2.2 On Complex Systems and Organization

Defining exactly what a complex system is remains a somewhat fuzzy issue. Key features appear to be system openness with fuzzy boundaries, a nested architecture, memory or hysteresis, non-linearity, feedback loops, and emergence. Here, we mainly refer to the organized complexity connected with emergence. One consensus appears to be that complexity originates from systems with a large number of interacting elements. Moreover, as described above, it is likely that complexity is to some extent the result of downward causation, where the higher level emerging functionality and the associated information—which can't be intuitively deduced from the lower level component and its interaction—can feedback and regulate the systems elements at the lower level. This multi-level and intertwined bottom-up and top-down exchange of information may be at the root of complexity.

63.2.3 Coherence, Unity, and the Dynamics of Complex Systems

A high level of coherence is apparent in living systems and cellular processes in the form of naturally occurring oscillatory phenomena. These can be linked to self-organized collective behavior and to the temporally organized expression of genes, proteins, and metabolites in single cells as well as in coherent cell populations. Striking examples of such high coherence include the oscillatory dynamics of most intracellular components observable in synchronized continuous cultures of yeast [11, 12]. Such continuous cultures of yeast and also *E. coli* have been shown to self-organize and display synchronized respiratory oscillations and display high temporal coherence at the level of gene, protein, and metabolite expression. While the exact mechanisms of synchronization have not been clarified, the cellular redox state as well as H_2S and acetaldehyde levels have been found to be important [12–14]. In *E. coli*, valine and other metabolite exchanges might be responsible for similar synchronization events [15]. Such results suggest that most, if not all, cellular components oscillate, claims supported by other studies [16] and in other systems [17]. The phase and frequency of oscillations depends on conditions and can be manipulated. Analysis of their dynamics upon perturbations and over extended time periods show multi time-scale fractal-like structure and chaotic attractors [18]. Moreover, it is interesting that other dynamical cellular states have been characterized as chaotic attractors and proposed to be connected with cellular pluripotency and differentiation [19]. Moreover, support for the relevance of quantum-like coherence in biological systems operating at the edge of quantum chaos [20] suggests there might be important analogies worth exploring further. Whether experimental systems such as synchronized populations of yeast, and the like, might represent tools to explore these chaotic dynamical states seems like a possibility to consider. Extending the same arguments suggests that similar phenomena may also be present at higher cellular and even whole organism level albeit at different spatio-temporal scales.

Another example may be a phenomenon reminiscent of quantum-like interference recently observed in bacterial gene expression [21]. These findings show how quantum-like statistical models may be useful, without being constrained by the absolute definitions derived from quantum physics. Basically, the authors argued that the dynamics of biological systems might require statistical descriptions where quantum-like probability amplitudes are more appropriate than classical probabilities. They also mentioned that this process might arise from the complexity of information processing in adaptive biological systems. Although the authors did not suggest that these quantum statistical models imply any underlying quantum processes, one could argue that it might be a likely consequence.

Information is deeply connected with complexity, emergence and self-organization at multiple scales [22]. Recently, Wiesner et al. [23] reported that some quantum computation models can result in descriptions of systems that have lower entropy than classical models. Since biological systems display apparently complex but highly organized dynamics, such quantum-like models properties could play a role in minimizing entropy and maximizing information content. These authors also suggested that correlations between elements in complex systems may be connected to quantum effects and quantum information [24]. While their work was not presented or derived for this purpose, it may not be farfetched to try and extend this idea and to apply it to the highly correlated and coherent examples found in living systems, such as those described above. These principles might be connected with the encoding and compression of biological information and with the phenomenon of emergence. Is it possible that by extending existing concepts of chaos theory and complex systems, we might postulate that the coherent oscillatory phenomena that are omnipresent in biological systems constitute an informational infrastructure that make quantum-like information processing and systems organization possible within cells, organs, and even whole organisms? The need to refer to non-classical behavior would then arise when the complexity of a system is such that its behavior and interactions cannot be described by classical measures of coherence.

63.2.4 Limitations and Need for Further Developments

It has been argued that classical principles are sufficient to describe the form of coherence and non-linear dynamics that describe biological systems and many such classical approaches have been used successfully. This remains an important argument against the need and justification to use quantum mechanics to describe those systems, in addition to the fact that our actual experience of biological system does not seem to display the weirdness, non-locality and entanglement properties that typically characterize quantum mechanics. In addition, the non-linear causal structure in complex biological systems that may facilitate the maintenance of coherence may however be incompatible with quantum superposition, for example [25]. However, in spite of such sensible arguments it remains interesting to ask whether the enormous phase space and computation-like tasks explored or performed by biological systems might require information processing that go beyond classical systems. Quantum search algorithms can allow to sense and explore multiple states simultaneously, a property that can provide efficiency that is beyond that made possible by classical systems [1, 3]. As we have described, the use of quantum mechanical principles at higher levels of biological organization may require further conceptual refinements that will account for both its relevance and limitations at this level of complexity.

At some level, the concept of unity and non-locality emerging in quantum states of matter like Bose-Einstein condensates and superconductors/superfluids, might be an important link or analogy to represent coherent organization in living systems, rather than more rigid definitions of quantum processes, based on entanglement and superposition. Order in biological systems, such as within the cell, might produce functional analogs of such supercooled states of matter where a large number of particles are correlated and result in collective effects and emergence of quantum phenomena at the macroscopic level. Such ideas about very low effective temperatures possibly existing within the confined cellular environment have previously been explored for their relevance in biological systems [8, 26, 27]. In any case, some properties of biological systems may not easily be described or explained by classical theory, and suggest there may be some need to extend our view of quantum-like processes or other non-classical properties.

To this day, most evidence of quantum phenomena in biological systems is to be found at the microscopic, physical chemistry level [2, 6]. However as we suggest here, it may also be at the higher level of cellular and organismal levels that quantum-like phenomena might play some unsuspected and significant role.

63.2.5 Testing a Possible Role of Quantum-Like Processes in Higher-Level Biological Systems

Overall, it seems there may be already some support both in existing experimental and theoretical studies that quantum-like phenomena have an unsuspected relevance in biological systems. Were more clear demonstrations of fundamental quantum behavior coherence and even entanglement or superposition in biological systems to surface in the coming years, this would represent an exciting development. It may be therefore be worth investigating further whether quantum-like coherent interactions may not only be relevant but might form an important element of the very fabric around which biological systems are organized. Because of the deep connections between non-linear dynamics and quantum properties in physical systems it should be possible to design experiments on dynamical cellular systems and determine, through their dynamical properties, whether quantum processes are at work at all. This could include further details and more definite measurements about the models of coherence in photosynthesis and enzyme kinetics, as described above. Alternately, these could also take the form of accurate measurements on the dynamics of coherent cellular systems similar to those described here, or even the analysis of thermodynamic exchanges of cellular/organism with their environment. Further research in this direction is therefore warranted.

63.3 Conclusion and Significance

The issue of coherence and its mechanisms in biological systems remains unresolved and the debate about the relevance of quantum mechanics in biology continues. However, there is some accumulating evidence that at some level, biological organization and the non-linear dynamics of cellular and organism may display properties highly reminiscent of quantum-like coherence that is not readily explained by classical means. It is also possible that the type of coherence displayed by biological systems fits neither the classical nor the quantum-like definition of coherence and might represent yet another form of coherence unique to living organisms. In any case, the significance of the presented view is not in trying to redefine what quantum-like coherent states may be in biological systems, but rather about the consequences of that coherence, whatever its nature. Such a perspective suggests that the key to the understanding of cellular and biological systems as inseparable wholes resides intrinsically in their non-linear dynamics, non-equilibrium dynamics and highly coherent interactions. Ignoring these principles that are linked to self-organization and emergence is bound to result in failures in attempts to understand and model living systems. Finally, this perspective on biological systems suggests a need for placing more emphasis, when observing, experimenting, and studying living and other complex systems, on the maintenance and proper analysis of the interactions between the system constituents at any level of complexity. Fortunately, such an objective seems to be at the core of a growing multidisciplinary field emphasizing systems approaches in biology and whose findings will likely have important consequences for biological sciences.

Epilogue This work is meant to be conceptual and the author acknowledges that some of the ideas expressed remain hypothetical, especially with respect to the proposal of the relevance of quantum coherence at higher levels of biological organization. The main objective of this work is to integrate existing concepts/knowledge and possibly apparently disparate points of view and reflect on whether they may be more connected than expected. If so, how they might be useful for understanding biological complexity is another objective. The aim is therefore, first and foremost, to favor a productive interdisciplinary discussion on the issue of complexity and coherence in biology.

Acknowledgements The author is grateful Johnjoe McFadden, Jaime Gómez Ramírez, and Takuya Morozumi for stimulating discussions and feedback.

References

1. Davies PCW (2004) Does quantum mechanics play a non-trivial role in life? Biosystems 78:69–79
2. Lloyd S (2011) Quantum coherence in biological systems. J Phys Conf Ser 302:012037
3. Engel GS, Calhoun TR, Read EL, Ahn T-K, Mancal T, Cheng Y-C, Blankenship RE, Fleming GR (2007) Evidence for wavelike energy transfer through quantum coherence in photosynthetic systems. Nature 446:782–786
4. Scholes GD (2011) Quantum biology: coherence in photosynthesis. Nat Phys 7:448–449
5. Hay S, Scrutton NS (2012) Good vibrations in enzyme-catalysed reactions. Nat Chem 4:161–168
6. Arndt M, Juffmann T, Vedral V (2009) Quantum physics meets biology. HFSP J 3:386
7. McFadden J, Al-Khalili J (1999) A quantum mechanical model of adaptive mutation. Biosystems 50:203–211

8. Schrödinger E (1944) What is life—the physical aspect of the living cell. Cambridge University Press, Cambridge
9. Noble D (2008) Claude Bernard, the first systems biologist, and the future of physiology. Exp Physiol 93:16–26
10. Noble D (2008) The music of life: biology beyond genes. Oxford University Press, London
11. Murray DB, Beckmann M, Kitano H (2007) Regulation of yeast oscillatory dynamics. Proc Natl Acad Sci USA 104:2241–2246
12. Murray DB, Klevecz RR, Lloyd D (2003) Generation and maintenance of synchrony in Saccharomyces cerevisiae continuous culture. Exp Cell Res 287:10–15
13. Sohn H, Murray DB, Kuriyama H (2000) Ultradian oscillation of Saccharomyces cerevisiae during aerobic continuous culture: hydrogen sulphide mediates population synchrony. Yeast 16:1185–1190
14. Murray DB, Engelen F, Lloyd D, Kuriyama H (1999) Involvement of glutathione in the regulation of respiratory oscillation during a continuous culture of Saccharomyces cerevisiae. Microbiology 145:2739–2745
15. Robert M, Murray D, Honma M, Nakahigashi K, Soga T, Tomita M (2012) Extracellular metabolite dynamics and temporal organization of metabolic function in E. coli. In: Proceedings of IEEE/ICME international conference on complex medical engineering (CME 2012), pp 197–202
16. Ptitsyn AA, Gimble JM (2011) True or false: all genes are rhythmic. Ann Med 43:1–12
17. Papatsenko I, Levine M, Papatsenko D (2010) Temporal waves of coherent gene expression during Drosophila embryogenesis. Bioinformatics 26:2731–2736
18. Murray DB, Lloyd D (2007) A tuneable attractor underlies yeast respiratory dynamics. Biosystems 90:287–294
19. Furusawa C, Kaneko K (2009) Chaotic expression dynamics implies pluripotency: when theory and experiment meet. Biol Direct 4:17
20. Vattay G, Kauffman S, Niiranen S (2012) Quantum biology on the edge of quantum chaos. arXiv:1202.6433
21. Basieva I, Khrennikov A, Ohya M, Yamato I (2011) Quantum-like interference effect in gene expression: glucose-lactose destructive interference. Syst Synth Biol 5:59–68
22. Gershenson C, Fernández N (2012) Complexity and information: measuring emergence, self-organization, and homeostasis at multiple scales. Complexity 18(2):29–44
23. Wiesner K (2010) Nature computes: information processing in quantum dynamical systems. Chaos 20:037114
24. Anders J, Wiesner K (2011) Increasing complexity with quantum physics. Chaos 21:037102
25. Ellis GFR (2011) On the limits of quantum theory: contextuality and the quantum-classical cut. arXiv:1108.5261
26. Matsuno K (1999) Cell motility as an entangled quantum coherence. Biosystems 51:15–19
27. Matsuno K (2006) Forming and maintaining a heat engine for quantum biology. Biosystems 85:23–29

Chapter 64
Modelling Biological Form

Rebecca Cotton-Barratt and Markus Kirkilionis

Abstract Computer-based models of biological evolution have typically ignored morphological form. Yet there are compelling reasons to believe that form has an important role to play in the long-term development of the system. We present a novel computer model of form-driven evolution, modelling form abstractly via hierarchical structures represented by directed acyclic graphs. This model incorporates elements from a wide variety of recent advances in evolutionary modelling, including adaptive dynamics, genetic algorithms, population dynamics, hypercycles and graph theory. We see several non-trivial phenomena emerge from this minimal imposed structure.

64.1 Introduction

Morphological form has long been vital to taxonomy, and although advances in genetics have provided phylogenetic clues for extant organisms, palaeontologists must still use form as their prime data. Charles Darwin's important observations on biological form and, in particular, modifications in form were critical to the development of his theory of natural selection. However, form has been under utilised in the mathematical modelling of evolution (e.g. the areas of population genetics and adaptive dynamics), relegating it as a preserve of palaeontologists and therefore interesting only as a product of evolutionary processes [1]. Form has fared somewhat better in evolutionary pattern formation and self-assembly models [2, 3], and evolutionary pattern prediction in morphospace analysis [4–6]. However, whilst such models describe the results of selection on form, and may even hint at the underlying dynamics, they fail at closing the feedback loop of the constraint form imposes on evolution and vice versa.

The exception to this are hypercycles which model cyclically coupled self-reproducing networks [7]. For much of the population dynamics of our model

R. Cotton-Barratt (✉) · M. Kirkilionis
Centre for Complexity Science, University of Warwick, Zeeman Building, Coventry CV4 7AL, UK

T. Gilbert et al. (eds.), *Proceedings of the European Conference on Complex Systems 2012*, Springer Proceedings in Complexity, DOI 10.1007/978-3-319-00395-5_64,

we take inspiration from this model, however, our current model is solely competitive and does not have the additional complexity of cooperation enforced by the hypercycles. This is a further extension to our model which we hope to explore.

Of course, it could be argued that if evolutionary processes can be modelled sufficiently without form there is no need to over-complicate models by incorporating it, especially since a lack of clear understanding of the phenotype-genotype map means large assumptions must be made to do so. Nevertheless, macroecological observations over deep time suggest that, since the initial radiation of the Cambrian explosion (approximately 550 mya), life on earth has undergone an evolutionary shift in both tempo and mode linked to the rise of animals [8]. This "explosion" of diversification and niche construction has resulted in the meta-stable biosphere we see today [9]. There are two potential explanations for this shift in macroevolutionary and macroecological behaviour—the new heterotrophic dynamics of the system or the evolvability of modular forms. As always with palaeontological data, it is difficult to separate pattern from process.

Consider the possibility that this shift resulted from a fundamental change in evolvability. It has been noted that there is a distinct lack of metric for comparing developmental innovations, in terms of their degree of difference [10]. Similarly, whilst there exists the Hamming distance for mutations in genetic code, there is no morphological difference measurement, except perhaps to calculate how "derived" one character is from another. Yet, the distribution of evolutionary novelties is non-random over time [10]. What causes these periods of innovation and (relative) stasis (first formally described as punctuated equilibrium [11])? Recent mathematical models such as adaptive dynamics and the Tangled Nature model both exhibit stasis followed by rapid reorganisation. The latter's behaviour emerges from community wide destabilisation and subsequent reconfiguration, controlled by the interaction strength of member's of the ecosystem [12]. The behaviour of the former is seen against the backdrop of mutant-resident invasions, and results from population dynamic assumptions about mutant frequency and fitness (and therefore the probability of establishment) [13].

We present a model that explores the evolvability of forms within the framework of evolutionary dynamics. The model is constructed in such a way so as to explore structure and its affect on evolutionary processes. Hierarchical structures are found in all biological systems, from the trophic interactions between species (e.g. food webs), to regulatory gene networks. In general, it is possible to split such structures into endogenous and exogenous classes—that is, those structures within individuals and those individuals form. The primary structures this model aims to investigate are endogenous. Whilst much work has been done on hierarchical structures in genetics, relatively little work has been done on the phenotypic results of such structures [14]. Such phenotypic structures naturally fall within the classification of Baupläne or body plans, which despite having a genetic basis (through Homeobox genes) are first and foremost morphological features.

In our basic model species are represented by their unique body plans, which are formed as hierarchical graphs. This hierarchical depiction is an elegant way of intuitively displaying the different levels of biological organisation, which could be called the informational content of the body plan. Note that it is essentially the developed information content, it does not include any of the development machinery (nor the genotype information content) for translating the genotype into the phenotype, which could be called the development information.

Each vertex within a species' body plan graph is assigned to a level. Levels represent the vertex's hierarchical position within the species graph. Although this is a convenient mathematical abstraction, it nevertheless has a basis in biology via the levels of organisation typically seen in organisms (for example the proteins which cause cells to specialise which can then form conglomerates such as organs). Hierarchical structures are evolved through the mutation of individual vertices, which may copy individual nodes or subgraphs, or create new nodes or subgraphs. Competition between mutants and residents occurs based on ability scores which are tied individual nodes, and which represent how fit a species is with that particular mutation. This paper explores some of the preliminary results from this model.

64.2 Model

64.2.1 Ability, Maintenance, and Fitness

We begin by defining the graph structure associated to a species. Let n_{ijk} be a node i in level j of species k, which has an ability score a_{ijk} assigned to it. Ability scores are randomly generated on level 0 l_0, but are the sum of ability scores of nodes they are connected to at higher levels. Edges can only exist between nodes of different levels, and are directed such that higher nodes depend on lower nodes (i.e. all edges flow "downwards" from higher levels to lower levels). Note that this results in species which are easily analysable by the criteria defined in [15].

Fitness is directly related to the ability scores of nodes. In the simplest form of the model, fitness is equal to the total (summed) ability scores of a species $\phi(i) = A(i) = \sum a_{ijk}$. If normalisation is present in the model, then fitness can be defined relative to the mean fitness of the population, that is $\phi(i) = \hat{A}(i) = \frac{A(i)}{\bar{A}}$ where $\bar{A} = \frac{\sum A(i)}{N(t)}$ and $N(t)$ is the number of species extant in the system at that particular time step. Under normalisation therefore species with better than average total ability have a fitness greater than 1, whilst those with lower than average total ability have a fitness lower than 1.

Note that so far, the species that would be pre-eminent amongst all others would be a fully connected graph with a large number of nodes. As a counter to this artificially peaked fitness landscape, we can introduce the concept of node maintenance into the model. Maintenance works analogously to ability scores, in that higher node maintenances are the sum of "downstream" node maintenances. However, the two are independently generated and normalised. Therefore, to calculate fitness it becomes important to know the ratio of ability to maintenance, $\phi(i) = \frac{\hat{A}(i)}{\hat{M}(i)}$.

64.2.2 Mutation

We assume that mutations which are detrimental to the ability score of the species can never be established, and are therefore not modelled (following the arguments presented in [13]). We use this argument, as well as the results presented in [16], as justification for not implementing any mutation steps which involve deletion of nodes. This is in contrast to many evolutionary models of protein interaction networks [2].

Mutation occurs as a Poisson process. The mean waiting time is inversely proportional to the total system size—that is, for larger system sizes the mean waiting time is shorter (assuming that the mutation rate is taken to be proportional to the number of births in a system). The mutation rate is also assumed to be proportional to the number of births in a species; therefore species with high frequency have a higher probability of being chosen for mutation. However, all nodes within a species may be chosen with equal probability. Further modifications of the model will investigate the effects of a probability distribution of nodes, where nodes on lower levels have a higher probability of mutating.

There are two initial possibilities for mutation—an existing node is copied or a new node is created. These occur with probability p and $1 - p$ respectively. If a new node is created within a species, there is a probability η of it being connected to any other node, provided that node is on a lower level. Therefore all subgraphs of a species' body plan are random graphs. At node creation a level is "attached" to the node which represents its place in the hierarchical network. This level can be thought of as a measure of the complexity required to form such a structure. When a new node is created, its level is designated as one higher than the highest level node it is connected to.

The new node's ability score is equal to the sum of all the ability scores of its downstream connected nodes. If an existing node is copied, there are two further possible outcomes. The node may be copied and its existing connections maintained (thus duplicating the subgraph's edges) or a new subgraph may be constructed (following the construction rules detailed above). These occur with probability q and $1 - q$ respectively. Note that if the chosen node is a level 0 node, the node is simply copied and assigned a new ability score (since level 0 nodes have no subgraphs of their own). Level 0 node ability scores are drawn from a normal distribution, with $\mu = 0$, $\sigma = 0.2$. Therefore, mutations at this level statistically cause much lower changes in total ability score than evolving more hierarchical structures. Note also that level 0 nodes (which contribute to the ability scores of higher level nodes they are connected to) will contribute multiple times to the total ability score of the species, in direct proportion to the number of distinct subgraphs the node is a member of.

64.2.3 Level Scaling

Allometric scaling laws in biology are well-known empirically, and well-studied theoretically. The most frequently used scaling law states that the metabolic rate of an organism scales as the mass of the organism to a power of between $\frac{2}{3}$ and $\frac{3}{4}$. Various hypotheses have been proposed to explain the observed difference in scaling exponent [17, 18]. A good comparison of the two models can be found in [19]. In Dynamic Energy Budget models, maintenance is also proportional to effective organism size, this time in the form of volume [20]. Thus there is a clear link between the number of cells in an organism (the biomass) and associated biological processes. What is unclear is whether this relationship scales with other properties of organisms, such as object complexity or hierarchical complexity.

Given that there exists a relationship between organism size and biological processes, it is unsurprising to find that other correlations exist between other physical properties of organisms. Bell and Mooers [21] found that organismal complexity scaled with organism size, where organismal complexity is defined as object complexity using the biological complexity categories defined in [15]. Similarly, Changizi [22] found that an increase in expression complexity (the number of distinct expressions in a system) was created exclusively by increasing the number of component types. In other words, the hierarchical object complexity increases solely because of an increase in object complexity, which intuitively follows. Therefore object complexity scales with organism size, and hierarchical object complexity scales with object complexity.

It should be noted that the process scaling laws in [21, 22] are based on physiological data from extant species. Even assuming that there exists an underlying mechanical basis for such laws, it is possible that they are not universal so much as strongly selected for. Extinct mutations may well have partly explored the parameter space of scaling exponents, but that such exponents may be dependent on environmental conditions such as oxygen concentration. In an evolutionary model, it is therefore debatable whether such laws should be hard coded in.

Given that maintenance scales with size, and that hierarchical complexity scales with size, we propose to use in our model a maintenance that scales with hierarchical complexity. We construct this using the following metric: let m_{ijk} be the maintenance cost of node i in level j of species k. Then the scaled maintenance cost $\hat{m}_{ijk}$ for a node of level j is

$$\hat{m}_{ijk} = L^{\gamma} m_{ijk} \tag{64.1}$$

where L is a value assigned to the level j and γ is the scaling coefficient.

64.2.4 Genetic Algorithm

Our genetic algorithm differs from many instantiations by lacking a genetic component. As such, coarse body scale mutations (as defined above) are the only way for the algorithm to explore graph morphospace. As one of the aims of this project is to explore the effects of different formal models of evolution, the mutation options (namely, creation of new nodes and subgraphs or copying nodes and subgraphs) are kept the same for both the genetic algorithm version of the model and the population dynamics version. However recombination is not considered in this model, as all reproduction is clonal.

The genetic algorithm operates on a simple set of instructions. The two major steps of the process are mutation and selection. Mutation has already been discussed, however selection is sufficiently different in the two evolutionary processes of the model that it is sensible it consider it separately in each case. Selection in the genetic algorithm works as a hard lower bound on the fitness of species, such that any species with a fitness lower than the threshold is deleted. Considered from another perspective, this is the same as saying that a varying percentage of species (ordered by fitness) go on to "reproduce" in the next step of the algorithm. One important point to note is that selection is checked at all levels present in an organism.

64.2.5 Population Dynamics

The environment is taken to be the resident population or sub-population. Because mutation occurs on evolutionary time, there must be some separation of time scales between the mutation dynamics and the population dynamics. Between mutation steps, therefore, censuses are taken of the abilities of all species in the population and population frequencies are shifted accordingly. Note that because of the way the model is constructed, total system size (if static) is effectively irrelevant as all that concerns us are the changing, normalised, frequencies of the different species. However, a system size dependent on the level of complexity reached by species present, representing the exploration and discovery of new ecological niches is a potential extension of the model.

Regardless of the type of mutation, all mutants are deemed to be separate species. They are assigned a frequency (which can also be considered a copy number) which is some small fraction of the parent species' frequency. Thus mutants act as perturbations of the stable ecological system. Note that, as argued above, those mutants which exist in the model are already chosen from the ones that could exist and have the potential to establish themselves. A further extension of the model is to explore the effect of different splitting frequencies, and for the splitting to be proportional to the difference of ability scores between resident and mutant.

64.3 Results

The aim of this model is to investigate the conditions under which evolvability is increased, conserved, or decreased. How do we define evolvability? Evolvability, as understood within the framework of this model, is defined to be the future potential for change that a body plan may have. Note that we can directly relate this to the concept of ESSes and CSSes as defined in adaptive dynamics. It is easy to imagine that, in a fixed environment, there should exist at least one perfectly adapted

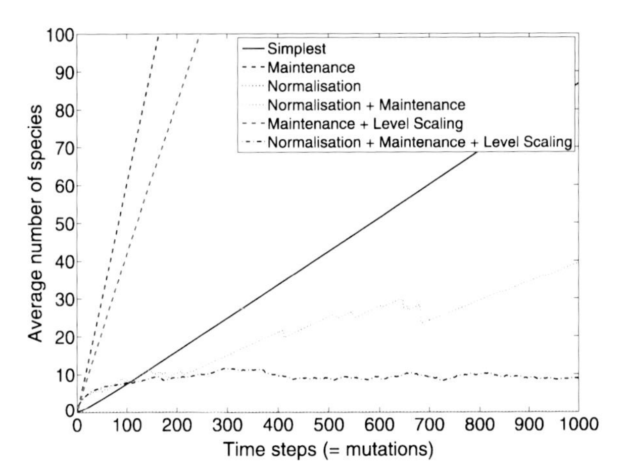

Fig. 64.1 Genetic algorithm framework comparing the effect of different combinations of parameters, including normalisation, maintenance, and level scaling, on number of extant species. The selection regime is constant at $\chi = 0.9$ independent of level. Level scaling, if present, is $\gamma = \frac{3}{4}$. *Each line* is the result of averaging 100 independent runs

organism. It is less obvious that biological evolution would ever find such an organism in finite time. A strong result would be to show that either there exists, for a given parameter range, such an uninvadable strategy (body plan) or that all such strategies are invadable. It appears that in nature such strategies, once established, are very hard to displace (since there has been no extinction or innovation since the end of the Cambrian explosion). An analogous result would provide an insight into the conditions under which such body plans are stable, and suggest whether a profound faunal turnover could ever occur (for example, if all that it required was a sufficiently large environmental perturbation).

Genetic Algorithm Initially, we wish to consider the apparent effects of changing the model parameters. The two main measures of perturbation are average standing diversity (that is, the number of distinct species extant at any time point) and average degree of nodes in the system (again averaged over the number of species extant at any particular time point). In Fig. 64.1, we can see the independent and combined effects of normalisation, maintenance, and level scaling. This was calculated under a fixed selection regime, where selection was constant for all levels present in an organism.

There are three distinct ecosystem modes seen in Fig. 64.1. The first mode is where the "birth" rate of new mutants exceeds the extinction or "death" rate. This is characterised by a generally straight line (normalisation and maintenance being the exception), whose gradient gives the ratio of births of new species to extinctions. The second mode is where the birth rate of new mutants is approximately equal to the extinction rate. Data produced by the model when normalisation, maintenance and level scaling is the only form of the genetic algorithm model to produce this behaviour. The third mode is when the extinction rate exceeds the birth rate of new mutants, as seen in the case where normalisation is the only modification present. In this case, this is an example of the selection function being too restrictive for that simplistic case.

Average degree distribution allows for a crude measure of the average connectedness of species in the system. We calculate the average degree distribution

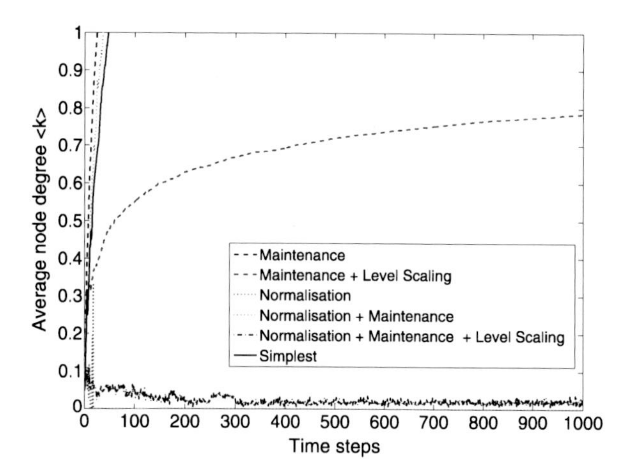

Fig. 64.2 Genetic algorithm framework comparing the effect of different combinations of parameters, including normalisation, maintenance, and level scaling on average degree. The selection regime is constant at $\chi = 0.9$ independent of level. Level scaling, if present, is $\gamma = \frac{3}{4}$. *Each line* is the result of averaging 100 independent runs

of a species, $S(i)$, as $\langle k_{S(i)} \rangle = \frac{2e(S(i))}{v(S(i))}$. To calculate the average degree distribution in the system as a whole, we average over the number of species, such that $\langle k_t \rangle = \frac{\sum_{i=1}^{N} \langle k_{S(i)} \rangle}{N(t)}$. A high average degree implies a fully connected graph in the majority of species, such that $e(S) \gg v(S)$. A low average degree implies an ecosystem of very low connected graphs, where $v(S) \gg e(S)$.

Similarly to the diversity measure, we can again categorise the results seen in Fig. 64.2 into three distinct modes. The first is characterised by a large population of (close to) fully connected graphs. This can be seen in the simplest (no modifications), normalisation and maintenance and maintenance versions of the model. The second is characterised by a smooth curve evolution of the average degree, seen for maintenance and level scaling in Fig. 64.2. If we consider the diversity for the model version which has maintenance and level scaling, we can see that the population is in the regime of effectively infinite diversity, and so the population is large. However, the degree distribution appears to evolve towards a constant (≈ 0.75). Given this, it appears that new mutations that evolve in this system evolve towards a particularly robust set of graphs—since there is no reason to expect the system to converge upon a particular average degree. Further work will explore the average degree dependence on mutation, following the methods in [2, 23].

A further measure of interest is whole system extinction—runs where the system enters into an unsustainable state. This is obviously highly dependent on the selection function chosen. Figure 64.3 shows the percentage of runs with all model parameters present (normalisation, maintenance, and level scaling) which end in whole system extinction for different values of χ in the constant selection regime. The dependence is strongly non-linear, suggesting a tipping point. However, this phenomena was only found in the first 100–200 time steps of a run. Long running systems (> 200 time steps) were therefore somewhat self-selecting, as unstable resident/mutant system configurations were already eliminated. There is no reason why these systems should be self-stabilising and further work is required to determine a mechanism—though initial investigation suggests a dependence on the magnitude of the difference in ability between residents and their mutants.

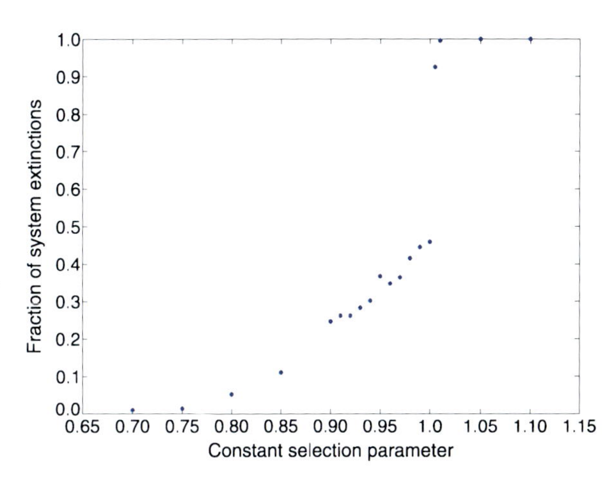

Fig. 64.3 Genetic algorithm framework comparing the effect of different combinations of parameters, including normalisation, maintenance, and level scaling on percentage of whole system extinctions. The selection regime is constant independent of level. Level scaling, if present, is $\gamma = \frac{3}{4}$. *Each line* is the result of averaging 1000 independent runs

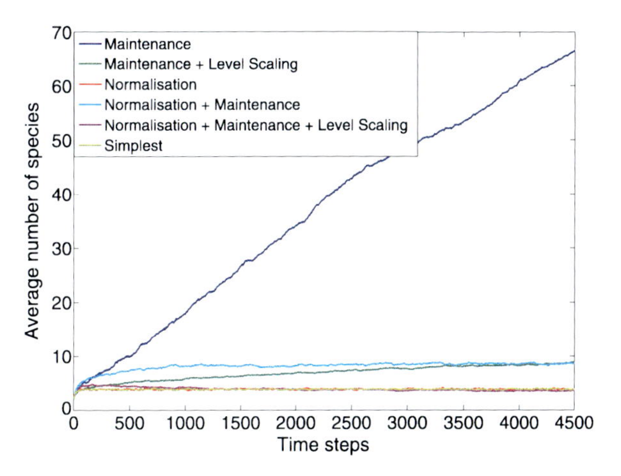

Fig. 64.4 Population dynamics framework comparing the effect of different combinations of parameters, including normalisation, maintenance, and level scaling, on number of extant species. Level scaling, if present, is $\gamma = \frac{3}{4}$. Poisson process parameter $\lambda = 0.1$. Note that this results in the same expected number of mutations as in the genetic algorithm results described above. *Each line* is the result of averaging 500 independent runs

Population Dynamics As with the previous section, two measures of perturbation are average standing diversity and average node degree. The same model modifications can be added within the population dynamics framework, except for the selection function which is no longer appropriate (since all calculations involve the changing frequencies of graph species, depending on their fitness). Species are, however, deemed extinct if their frequency is less than 0.001. A new parameter of mutation rate is introduced, λ, where $\frac{1}{\lambda}$ is the mean waiting time of the Poisson process. Note that at one extreme ($\lambda \gg 1$), we recover a mutation rate that approximates the genetic algorithm (i.e. one mutation per time step).

Considering firstly the diversity over time (Fig. 64.4), the most obvious difference between this framework and the genetic algorithm framework is the tendency for systems to average "steady state" diversities much more than the genetic algorithm. In particular, we see several stable standing system sizes (shown more clearly in Fig. 64.5). Unlike in the genetic algorithm framework, normalisation on its own

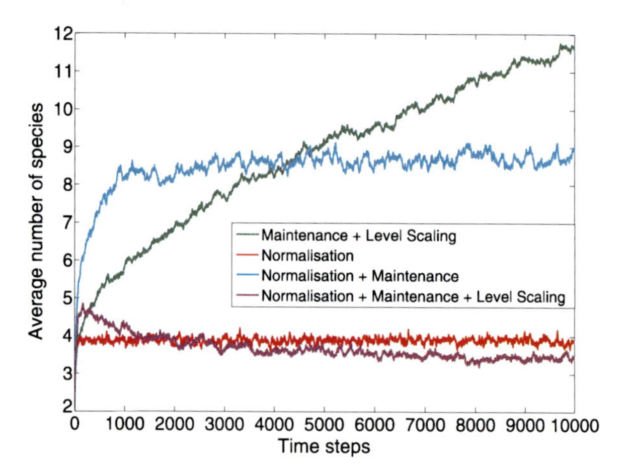

Fig. 64.5 As Fig. 64.4 but considered over longer time. Note that this data has the same expected number of mutations as the genetic algorithm results (seen in Fig. 64.1) and is therefore directly comparable

has no effect on the model. This is expected, as the very nature of population dynamics requires a degree of internal normalisation (i.e. that the frequencies of species are updated according to the fitness of a species as compared to the overall fitnesses of the other species in the system at that time). As a result, normalisation only has an effect when combined with maintenance, since it then normalises maintenance across species which alters the ratio of ability to maintenance independently, as opposed to maintenance without normalisation.

We see also that the main effect of normalisation in the population dynamics framework is to stabilise the system and create this standing diversity. This is distinct to the effect seen in the genetic algorithm framework, where stabilisation required normalisation and maintenance and level scaling. A counter-intuitive result is that maintenance increases the effective carrying capacity of the system. A possible explanation is that maintenance increases the amount of variability in fitness, since there are now two variables (ability and maintenance scores) that make up the fitness score, rather than one (ability scores). If we consider fitness as the real line, $\mathbb{R}$, this allows species to occupy closer positions on this line thereby reducing the rate of replacement (since fitness scores may be closer, and rate of change in frequencies is dependent on the absolute magnitude difference of fitnesses). A model with no maintenance has mutations which change fitness solely on the basis of altering the structure of the graph through new ability scores of nodes, rather than topology combined with the interplay of cost-benefit ratios.

Finally, considering the evolution of average degree of nodes in the system over time (Fig. 64.6 we see a validation of the claim that normalisation does not alter the system when maintenance is not present. As expected, level scaling dramatically influences the average degree of nodes, suppressing highly connected "monster" graphs. What should be noted is that average degrees in population dynamics framework are much higher than in the genetic algorithm framework for similar combinations of parameters.

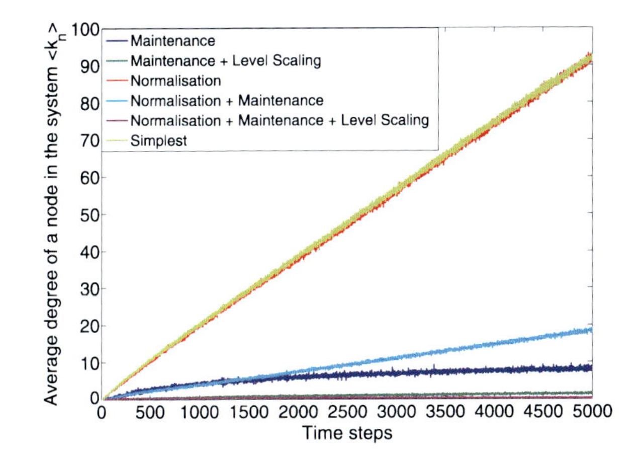

Fig. 64.6 Population dynamics framework comparing the effect of different combinations of parameters, including normalisation, maintenance, and level scaling, on the average degree of a node in the system. Level scaling, if present, is $\gamma = \frac{3}{4}$. Poisson process parameter $\lambda = 0.1$. Note that this results in the same expected number of mutations as in the genetic algorithm results described above. Each line is the result of averaging 500 independent runs

64.4 Conclusion

We present the preliminary results from a model examining the importance of evolutionary modelling frameworks and parameter choices on the evolution of form. The model evolves graph structures using a set of simple mutation rules, and looks at the diversity and average node degree under differing parameter sets.

We conclude that under identical starting conditions, genetic algorithm and population dynamic frameworks produce radically different results, even with the same parameter options. This has important implications for modelling in those subject areas which have no intrinsic reason for choosing either framework. It also requires much greater justification for choosing particular evolutionary modelling frameworks, since results are so model dependent.

In further work, we hope to quantify further the parameter space of these models, and also investigate more thoroughly the evolutionary trajectories of different graphs. We propose to do this by implementing a genotype and an external environment to asses the sensitivity of systems of external perturbations.

Acknowledgements We thank EPSRC for funding this research, and Hans Metz for his valuable insights.

References

1. Amundson R (2005) The changing role of the embryo in evolutionary thought: roots of evo-devo. Cambridge University Press, Cambridge

2. Solé R, Fernandez P, Kauffman S (2003) Adaptive walks in a gene network model of morphogenesis: insights into the Cambrian explosion. Int J Dev Biol 47:685–693
3. Johnston I, Ahnert S, Doye J, Louis A (2011) Evolutionary dynamics in a simple model of self-assembly. Phys Rev E 83(6):066105
4. Pie M, Weitz J (2005) A null model of morphospace occupation. Am Nat 166(1):E1–E13
5. Niklas K (1999) Evolutionary walks through a land plant morphospace. J Exp Bot 50(330):39
6. Costello J, Colin S, Dabiri J (2008) Medusan morphospace: phylogenetic constraints, biomechanical solutions, and ecological consequences. Invertebr Biol 127(3):265–290
7. Eigen M, Schuster P (1977) The hypercycle. A principle of natural self-organization. Part A: Emergence of the hypercycle. Naturwissenschaften 64(11):541–565
8. Butterfield N (2007) Macroevolution and macroecology through deep time. Palaeontology 50(1):41–55
9. Butterfield N (2011) Animals and the invention of the phanerozoic earth system. Trends Ecol Evol 26(2):81–87
10. Erwin DH (2000) Macroevolution is more than repeated rounds of microevolution. Evolut Develop 2(2):78–84
11. Eldredge N, Gould S (1972) Punctuated equilibria: an alternative to phyletic gradualism. Model Paleobiol 82:115
12. Christensen K, Di Collobiano S, Hall M, Jensen H (2002) Tangled nature: a model of evolutionary ecology. J Theor Biol 216(1):73–84
13. Metz J, Geritz S, Meszéna G, Jacobs F, Van Heerwaarden J (1996) Adaptive dynamics, a geometrical study of the consequences of nearly faithful reproduction. In: Stochastic and spatial structures of dynamical systems, pp 183–231
14. Erwin D, Davidson E (2009) The evolution of hierarchical gene regulatory networks. Nat Rev Genet 10(2):141–148
15. McShea D (1996) Perspective: Metazoan complexity and evolution: is there a trend? Evolution 50(2):477–492
16. Anderson J, Tataru P, Staines J, Hein J, Lyngsø R (2012) Evolving stochastic context-free grammars for RNA secondary structure prediction. BMC Bioinform 13(1):78
17. Demetrius L (2006) The origin of allometric scaling laws in biology. J Theor Biol 243(4):455–467
18. West G, Brown J, Enquist B (1997) A general model for the origin of allometric scaling laws in biology. Science 276(5309):122–126
19. van der Meer J (2006) Metabolic theories in ecology. Trends Ecol Evol 21(3):136–140
20. Nisbet R, Muller E, Lika K, Kooijman S (2000) From molecules to ecosystems through dynamic energy budget models. J Anim Ecol 69(6):913–926
21. Bell G, Mooers A (1997) Size and complexity among multicellular organisms. Biol J Linn Soc 60(3):345–363
22. Changizi M (2001) Universal scaling laws for hierarchical complexity in languages, organisms, behaviors and other combinatorial systems. J Theor Biol 211(3):277–295
23. Vázquez A, Flammini A, Maritan A, Vespignani A (2001) Modeling of protein interaction networks. Preprint. arXiv:cond-mat/0108043

Chapter 65
A Novel Approach to Analysing Fixed Points in Complex Systems

Iain S. Weaver and James G. Dyke

Abstract Complex systems are frequently characterised as systems of many components whose interactions drive a plethora of emergent phenomena. Understanding the history and future behaviour of planet Earth, arguably the most complex known system in the universe, is an ambitious goal and remains at the core of complexity science. From the establishment of the planet's magnetic dipole, to the interplay between life and it's environment. The dynamics across all scales are characterised by their numerous interacting components.

Of particular interest is how such a system may be stable at all, and the role of life in establishing this apparent stability. We present a novel analytic approach to a model of a coupled life-environment system. The model demonstrates that even random couplings between many species, and a multidimensional environment can produce stable, and robust configurations. The extent to which this observation is general, rather than being unique to the intricacies of the model may only be revealed by thorough analysis. The model is found to be invariant with the number of biotic components past a lower limit. Additionally, rather than increases in environmental complexity leading to a reduction in the possibility of steady states, it is proven that the converse is true, suggesting that the proposed mechanism may be applicable to even high dimensional complexity.

65.1 Introduction

The Gaia hypothesis proposes the idea that life on Earth and it's abiotic environment are tightly coupled in such a way as to maintain conditions to be within a range essential for life to exist [1]. This phenomena is commonly referred to as homeostasis; in the face of external and internal perturbations, factors such as surface temperature, and atmospheric composition appear to be tightly reigned. Indeed, catastrophic perturbations have occurred in Earth's past, characterised by mass extinction events and climatic shifts. In response, the system stabilises rather than being condemned to a lifeless state.

I.S. Weaver (✉) · J.G. Dyke
School of Electronics and Computer Science, University of Southampton, Southampton, UK
e-mail: isw1g10@soton.ac.uk

T. Gilbert et al. (eds.), *Proceedings of the European Conference on Complex Systems 2012*, Springer Proceedings in Complexity, DOI 10.1007/978-3-319-00395-5_65,

A number of mechanisms have been proposed for these phenomena [2], perhaps most notably by the original proponent of the Gaia hypothesis himself. In the original "Daisyworld" model, [3] seek to dispel the idea that the Gaia hypothesis relies on any sort of global control, rather that natural selection is sufficient. The model describes a grey planet, seeded by black and white daisies, orbiting a star of gradually increasing luminosity. Being abundant in all other factors required for the daisies to flourish, the model's principle variable is surface temperature. With the right choice of feedback the model exhibits homeostasis in that the temperature of the planet is maintained roughly constant in the face of an increasingly bright star, and life is maintained across a much greater range than might naïvly be predicted. While an important proof of concept for the Gaia hypothesis, spurring a great deal of additional research (see [4] for an review of the developments in this area), it is difficult to see the Daisyworld mechanism as being generally applicable; the Earth system is characterised by numerous principle variables, and rather than a pair of competing species, a multitude of forms of life exist across a vast range of conditions. Furthermore, there is no reason to believe *a priori* that life should organise into homeostatic states, as opposed to run-away positive feedback [5].

The reorganisation of the Earth system in response to destabilising perturbations can be seen as an example of Ashby's [6] notion of ultrastability. Ashby's "Homeostat" model was, in contrast to Daisyworld, a physical device which was able to respond to this type of perturbation with spontaneous reorganisation, through a random search mechanism. The result was that the systems variables are constrained within some limited range; a potential likeness with the Earth system.

Dyke [7] proposes a model in which a large number of biotic components interact through a shared environment. The model is able to reproduce the salient features of both Ashby's [6] Homeostat, and Watson and Lovelock's [3] original Daisyworld model, without the need to prescribe a tendency for positive or negative feedback. This model of many biotic elements, interacting through their multidimensional environment serves as the starting point of this paper. In Sect. 65.2, we present the model in a simplified state, along with some characteristic results in Sect. 65.3. We then pose a number of potential criticisms which would conflict with the proposition of a general model in Sect. 65.4. We systematically address the model treatment of biotic complexity in Sect. 65.4.1, along with the environmental complexity in Sect. 65.4.2. These results are discussed along with the possibility of application to a broader class of complex system in Sect. 65.5.

65.2 Model Formulation

Throughout this article, we consistently used the boldface notation to identify vectors of many components, and subscript to identify individual elements,

$$\boldsymbol{X} = \begin{bmatrix} X_1 \\ X_2 \\ \vdots \\ X_n \end{bmatrix}.$$

Dykes's [7] "Daisystat" model expresses life as K biotic elements whose overall activity is influenced by the state of their shared environment, represented the N variables in the vector $\boldsymbol{E}$. This environment is itself influenced by the biota; it's variables may be decreased or increased by the individual biotic elements, through consumption, excretion or some other process with no bias towards positive or negative feedback. In essence, the model consists of two principle assumptions:

(i) *Environment Affects Life* Each biotic element in the system occupies a niche, the relatively narrow envelope of possible conditions in which it can respire, proliferate, or otherwise maintain activity. As we depart from the optima, the biotic element will tolerate the change to some extent, remaining active outside it's optimal conditions. Eventually we depart the niche of this biotic element, and it may die, become dormant, or otherwise inactive, while other biotic elements may be better suited to the new environment, and increase in activity. However, there are limits to this process. The environment must remain within some *essential range* for the biota to be active at all. While species may prefer a wide range of different ambient temperatures, it is hard to imagine life flourishing in ice, or steam.

We are primarily concerned with the fixed points of this model where the activity of each biotic element is given by it's steady state value. If the biota is able to reorganise sufficiently quickly compared to changes in the environment, the biotic elements can be said to exist at this environment-dependant steady-state activity, removing the need to explicitly incorporate time dependence. The steady-state activity, $\alpha(\boldsymbol{E})$, is maximised at a point in the space of environmental variables, $\boldsymbol{\mu}$, it's niche. As $\boldsymbol{E}$ departs from $\boldsymbol{\mu}$, the steady-state activity of this biotic element decreases. There are a range of defensible options for representing this behaviour, and we later investigate the extent to which this choice is important for the model dynamics. For simplicity, we choose the activity of an individual biotic element, $\alpha_i(\boldsymbol{E})$, to be a Gaussian, centred at $\boldsymbol{\mu}_i$ with characteristic width σ_E, that is

$$\alpha_i(\boldsymbol{E}) = \exp\left(-\frac{|\boldsymbol{E} - \boldsymbol{\mu}_i|^2}{2\sigma_E^2}\right) \tag{65.1}$$

where $\boldsymbol{\mu}$ is the optimum for an individual biotic element chosen randomly in the interval $[0 : R]$, the essential range. For the purposes of this section, we use $\sigma_E = 5$, $R = 100$ and $K = 10^4$ biotic elements, distributed across this range.

(ii) *Life Affects Environment* The impact of life on the environment is hard to generalise. As stated, the Earth system is composed of numerous principle variables. Some of these variables are considered resources, consumed by some forms of life, and possibly excreted by others, such as oxygen and carbon dioxide in the atmosphere, or phosphorous in the soil. Other variables such as surface temperature are not consumed in this way, although species of foliage may have an effect on this variable indirectly by modifying the planetary albedo locally. Rather than make this distinction, we opt simply to impose random, and unbiased couplings between life, and it's environment.

The biotic elements may feedback negatively or positively, and strongly or weakly (or not at all), depending on the value of a weighting term. Whatever the weighting, we say their effect is directly proportional to their activity; an abundant, or highly active element is likely to have a more significant impact. The total biotic effect is found by summing the contributions of each element

$$F_i(\boldsymbol{E}) = \sum_{j=1}^{K} \omega_{j,i} \alpha_j(\boldsymbol{E}) \tag{65.2}$$

where $\omega_{j,i}$ is the weight of biotic element j on environmental variable i, chosen randomly from the interval ± 1. Additionally, each of the environmental variables may be affected by some external perturbing force, $\boldsymbol{P}$. The net change in the environment is the sum of these features

$$\tau_{E_i} \frac{\mathrm{d}E_i}{\mathrm{d}t} = P_i + F_i \tag{65.3}$$

where τ_{E_i} is the characteristic timescale for changes in environmental variable i, chosen to be equal between variables for convenience, and F_i is the sum of effects from the biota.

Model Behaviour Equation (65.3) describes the time evolution of the environmental variables towards its fixed points, which occur where the sum of effects of the biotic elements on each resource exactly opposes the external perturbation on that resource, that is

$$F_i = -P_i \quad \text{for } i = 1 \cdots N. \tag{65.4}$$

If they exist, these points may further be stable or unstable. If a change in E_i results in an opposing change in F_i, the point is stable in this direction. A fixed point in the model must be stable in all directions, that is

$$\frac{\mathrm{d}F_i}{\mathrm{d}E_i} < 0 \quad \text{for } i = 1 \cdots N. \tag{65.5}$$

65.3 Model Results

Despite it's very general formulation, and largely random parameters, the model exhibits a range of interesting behaviour. The sum of uncorrelated Gaussians results in homeostatic behaviour in response to increasing perturbing force. This also leads to hysteresis loops illustrated for the case of a single environmental variable in Fig. 65.1. The behaviour is almost identical to Watson and Lovelock's Daisyworld model, except rather than homeostasis being designed, such points have emerged spontaneously from the biological complexity of the model.

A further criticism to the Daisyworld model is that it hinges on a single environmental variable, and therefore necessarily exhibits simple behaviour. On the other

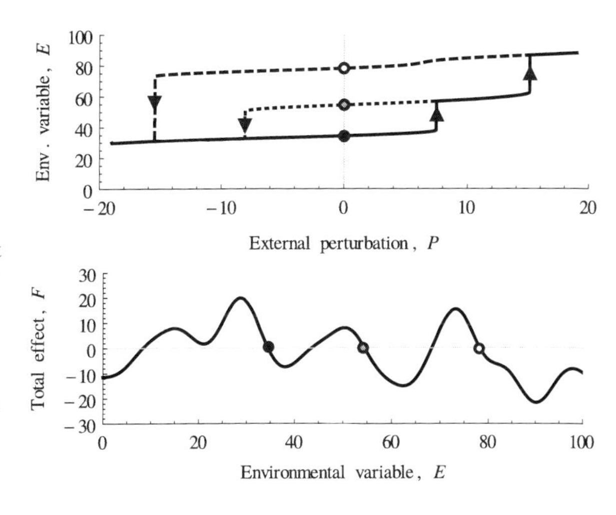

Fig. 65.1 The model has homeostatic fixed points, shown here for a single environmental variable. Population effects oppose increasing (*solid line*) or decreasing (*dashed line*) perturbations, maintaining a roughly constant environment (*top*). Hysteresis is caused by the existence of multiple solutions, shown for the case of $P = 0$ (*bottom*). Sharp transitions into new stable states occur when such points vanish

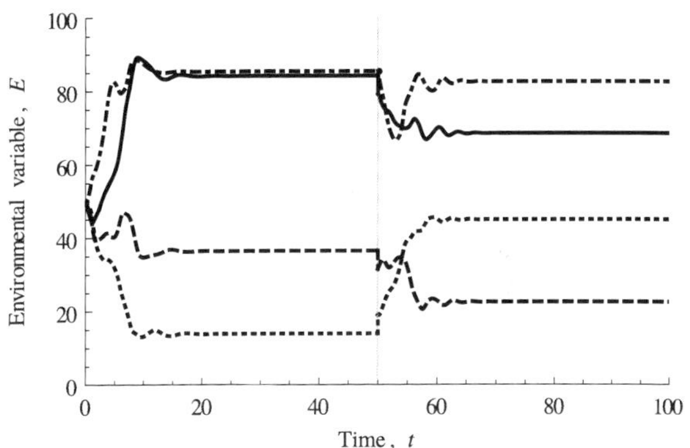

Fig. 65.2 The $N = 4$ model, initialised with all environmental variables at the centre of the essential range. After reaching a steady-state, a perturbation is applied to the variables at time 50 which is sufficient for the model to enter a new attractor

hand the Earth system could not possibly be reduced in this way [8]. Initial intuitions may suggest that with increasing environmental variables the likelihood of finding a point stable in all dimensions simultaneously would vanish exponentially. However, transitions between neighbouring attractive points can be seen in higher dimensional systems. A model of four environmental variables is shown to settle into a stable fixed point in Fig. 65.2, and further illustrates that stability can be re-established after a perturbation removes the system from the basin of attraction corresponding to this point, qualitatively similar to the four unit Homeostat. Figure 65.3 directly examines the basins of attraction for a two dimensional model and illustrates the complicated structure of underlying basins of attraction.

65.4 Analysis

While the model may give insights to key features of the Earth system, it is difficult to establish to what extent this may represent a general mechanism where homeosta-

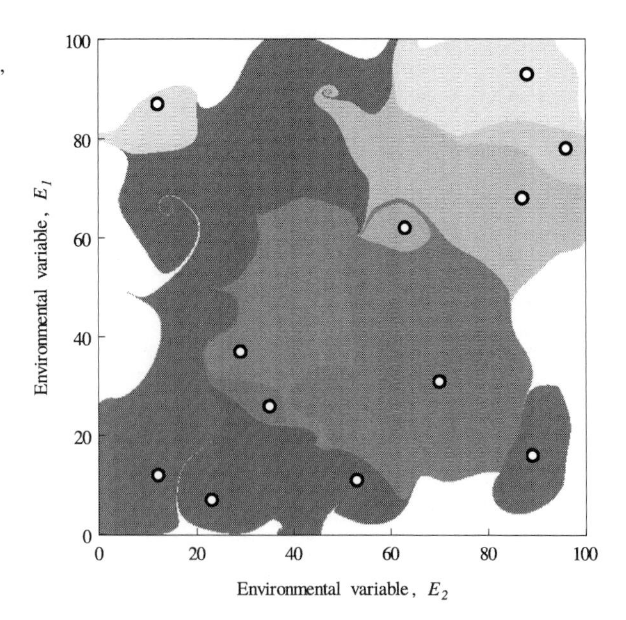

Fig. 65.3 The $N = 2$ model shows many stationary points, indicated by points. The basins of attraction which lead to these points are indicated by *the shaded enclosing regions*, while initial conditions which would leave the essential range are *coloured white*

sis is inevitable, and if or how it is constrained by assumptions and implementation. Three of the key matters to address are;

- Does the model behaviour change with increasing numbers of biotic components?
- How is it affected by increasing environmental complexity?
- To what extent is the choice of underlying functions important?

One may intuit that with very many biotic elements, there is a tendency towards uniformity in $\boldsymbol{F}$, reducing the likelihood of finding stationary points. While early intuitions were that increasingly complex systems enjoyed increased stability [9], the work of [10] and [11] contended this, showing that in a network interpretation of complexity, increased numbers and strength of connections ultimately led to instability. In our model, fixed points must be stable in all environmental variables simultaneously, suggesting that with an increasingly complex environment stability becomes impossible.

65.4.1 Behaviour with Number of Biotic Elements, K

We introduce the covariance function as a means to characterise the nature of the sum of biotic effects $\boldsymbol{F}$. The covariance function encodes the degree of correlation between points in $\boldsymbol{E}$. We write $k_i(\boldsymbol{E}, \boldsymbol{E}')$ as shorthand for the covariance of F_i, $\langle F_i(\boldsymbol{E}) F_i(\boldsymbol{E}')\rangle$, at two arbitrary points in the space of environmental variables,

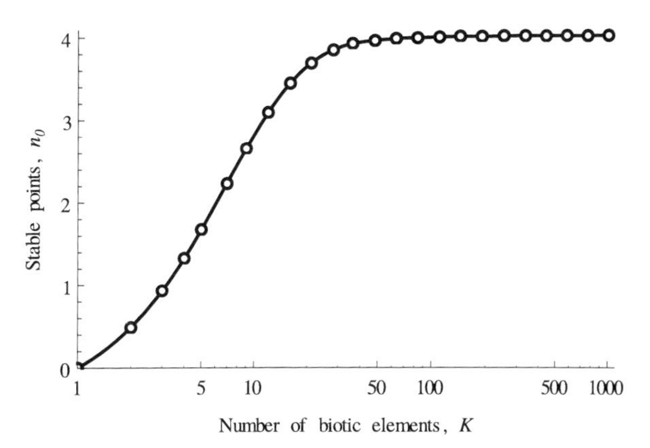

Fig. 65.4 The number of fixed points increases and then saturates with increasing number of biotic components K. The number of fixed points linearly decreases with increasing the width of the biotic component abundance function σ_E and linearly increases with increases in the width of the essential range R

$\boldsymbol{E}$ and $\boldsymbol{E}'$.

$$k_i(\boldsymbol{E}, \boldsymbol{E}') = \left\langle \sum_{n,m=1}^{K} \omega_{i,n}\omega_{i,m}\alpha_n(\boldsymbol{E})\alpha_m(\boldsymbol{E}') \right\rangle. \tag{65.6}$$

At this point, we can exploit the absence of correlations first between individual biotic elements, and then between the weights ω, and the biotic activity. The first observation leads us to conclude the off-diagonal terms, where $i \neq j$, do not contribute to the covariance. The second enables us to separate the expectation values of ω and α, giving

$$k_i(\boldsymbol{E}, \boldsymbol{E}') = K\sigma_\omega^2\langle\alpha(\boldsymbol{E})\alpha(\boldsymbol{E}')\rangle. \tag{65.7}$$

where σ_ω^2 is the variance of the random variable ω. The right side of this equation can be identified simply as the covariance of the individual biotic activity functions. This result illustrates that the covariance of the summed functions share the functional form of the individual functions of which it is comprised. The characteristic length, and therefore the propensity for $\boldsymbol{F}$ to form attractive fixed points, is *independent* of the biotic complexity of the model. This can be verified numerically by examining how the expected number of fixed points in a single variable model varies with the number of biotic elements, K. Figure 65.4 illustrates that the number of fixed points saturates quickly, and that further increasing K does not modify the behaviour of the model.

65.4.2 Behaviour with Number of Environmental Variables, N

A fixed point in $F_i(\boldsymbol{E})$ occurs in a small interval of $\boldsymbol{E}$ if its sign changes across the interval. Labelling the interval $\boldsymbol{\epsilon}$, this condition can be expressed

$$F_i(\boldsymbol{E})F_i(\boldsymbol{E} + \boldsymbol{\epsilon}) < 0 \tag{65.8}$$

and the expected number of such points in the unit interval, n_0, is found from a product of indicator functions of the form of Eq. (65.8)

$$n_0 = \left\langle \prod_{i=1}^{N} \frac{1}{\epsilon} \left[F_i(\boldsymbol{E}) F_i(\boldsymbol{E} + \boldsymbol{\epsilon}) < 0 \right] \right\rangle. \tag{65.9}$$

We have used [...] to represent an indicator function, returning one if the expression true, and zero otherwise. The expectation of an indicator function may be interpreted as the *probability* of it's contents being true, and the product of several therefore gives the probability of many conditions being met simultaneously. Each term in the product may be treated independently due to the independence between the biotic effects on the different environmental variables F_i and F_j. The problem is therefore reduced to finding the value of the series of N expectation values. Expanding for small ϵ gives

$$p = \left\langle \left[F_i(\boldsymbol{E}) < -\epsilon F_i'(\boldsymbol{E}) \right] \right\rangle \tag{65.10}$$

where $F_i'(\boldsymbol{E})$ is used to indicate the derivative of $F_i(\boldsymbol{E})$ in the $\hat{\boldsymbol{\epsilon}}$ direction (commonly written as $\nabla_{\hat{\epsilon}} F_i(\boldsymbol{E})$). To find the expectation value of this indicator function, we need to know how $F_i(\boldsymbol{E})$ and $F_i'(\boldsymbol{E})$ are distributed. Rather than suffer any loss of generality, we make three important observations. Firstly, at any point in $\boldsymbol{E}$ within the essential range, $F_i(\boldsymbol{E})$ is a sum of *independent* contributions from the biotic elements. Therefore, by the central limit theorem, each point follows a Gaussian distribution. Additionally, this distribution has a mean of zero as previously stated. There is no tendency for positive or negative feedback between the biota and environment. Finally, we note $F_i(\boldsymbol{E})$ and $F_i'(\boldsymbol{E})$ to be uncorrelated as a consequence of our independent parameters $\boldsymbol{\mu}$ and ω.

The problem is now dramatically reduced, we need only find the variance of the Gaussian random variables $F_i(\boldsymbol{E})$ and $F_i'(\boldsymbol{E})$, labelled σ_F^2 and $\sigma_{F'}^2$ respectively. Here it is useful to observe that providing the width of the biotic activity functions are small compared to the essential range, the covariance $k(\boldsymbol{E}, \boldsymbol{E}')$ is *stationary*; k depends only on the distance $|\boldsymbol{E} - \boldsymbol{E}'|$. Having already determined the covariance of $F_i(\boldsymbol{E})$ in Eq. (65.7), the variance therefore may be written as

$$\sigma_F^2 = k_i(\boldsymbol{0}). \tag{65.11}$$

We can write a similar expression for $F_i(\boldsymbol{E})$, and remove the directional derivative from the expectation value to give

$$\begin{aligned} \sigma_{F'}^2 &= \left\langle F'(\boldsymbol{E}) F'(\boldsymbol{E}') \right\rangle \big|_{\boldsymbol{E}=\boldsymbol{E}'} \\ &= \nabla_{\boldsymbol{E},\hat{\epsilon}} \nabla_{\boldsymbol{E}',\hat{\epsilon}} k_i\left(\boldsymbol{E} - \boldsymbol{E}'\right)\big|_{\boldsymbol{E}=\boldsymbol{E}'} \\ &= -k_i''(\boldsymbol{0}). \end{aligned} \tag{65.12}$$

Next, we substitute Eqs. (65.11) and (65.12) into Eq. (65.10)

$$p = \iint \left[F_i < -\epsilon F_i' \right] P(F_i) P\left(F_i'\right) \mathrm{d}F_i \mathrm{d}F_i' \tag{65.13}$$

where $P(F_i)$ and $P(F_i')$ are the Gaussian distributions

$$P(F_i) = \frac{1}{\sqrt{2\pi}\sigma_F} \exp\left(-\frac{F_i^2}{2\sigma_F^2}\right), \tag{65.14}$$

$$P(F_i') = \frac{1}{\sqrt{2\pi}\sigma_{F'}} \exp\left(-\frac{F_i'^2}{2\sigma_{F'}^2}\right). \tag{65.15}$$

After a change of variable, $\frac{F_i}{\sigma_F} \to x$ and $\frac{F_i'}{\sigma_{F'}} \to x'$, we can exploit spherical symmetry in x and x' to find the expectation of the indicator function to be

$$\begin{aligned} p &= \iint [x\sigma_F < -\epsilon x'\sigma_{F'}] \frac{e^{-\frac{x^2}{2}}}{\sqrt{2\pi}} \frac{e^{-\frac{x'^2}{2}}}{\sqrt{2\pi}} \mathrm{d}x \mathrm{d}x' \\ &= \frac{1}{\pi} \operatorname{atan}\left(\epsilon \frac{\sigma_{F'}}{\sigma_F}\right) \end{aligned} \tag{65.16}$$

which can be expanded to first order for small ϵ, and substituted into Eq. (65.9) to give

$$n_0 = \left(\frac{p}{\epsilon}\right)^N = \left(\frac{1}{\pi}\sqrt{-\frac{k''(\mathbf{0})}{k(\mathbf{0})}}\right)^N, \tag{65.17}$$

which is consistent with [12] for the case of a one-dimensional model (Theorem 4.1.1). Counter to intuition, the number of stable fixed points within the essential range of the model may *increase* exponentially, rather than vanishing to zero providing there is a sufficiently wide essential range. For the simple example of Gaussian functions the expected number of fixed points is

$$n = \left(\frac{R}{\sqrt{2\pi}\sigma_E}\right)^N, \tag{65.18}$$

of which the fraction 2^{-N} are attractive. Eq. (65.18) makes clear the role of the width of biotic activity functions in guiding the model, while the specific function chosen is unimportant. Indeed, skewed, bimodal and to some extent, long-tailed functions can be shown to produce similar behaviour. Not only may very high dimensional systems exist in stable, stationary states, but the number of such states may be exponentially great. As before, we can verify this relationship numerically up to four environmental variables, shown in Fig. 65.5.

65.5 Conclusion

It has been demonstrated that collections of biotic elements interacting with a complex environment in randomly parametrised ways can not only reach stable configurations which are robust to external perturbations, but display a plethora of complex

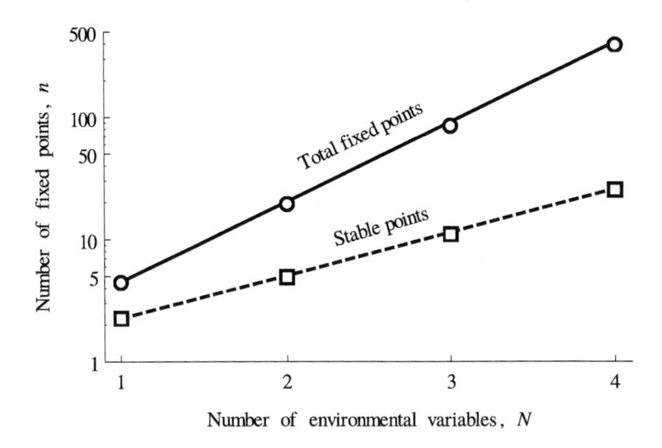

Fig. 65.5 Equation (65.17) can be used to find the expected number of fixed points across the essential range with an increasingly complex environment. Numerical simulations are plotted as points, where statistical errors are at most the size of plot points. Simulation confirms the exponential increase in fixed points with the addition of environmental variables

phenomena. The extent to which this is a generally applicable principle is addressed by an analytic investigation into the model behaviour with increasing biotic, and environmental complexity.

By observing that the sum of many functions shares the covariance function of it's components, we find that rather than a tendency towards uniformity, increasing the number of biotic elements does not hinder the ability of the model to form attractive fixed points. A relatively sparse biota display almost identical properties to one containing many times more elements.

Furthermore, we exploit the Gaussian nature of sums of random biotic effects to derive expressions for the expected number of model fixed points for given dimensionality. In doing so, we find the choice of biotic activity function to be largely arbitrary. Rather than multidimensional systems lacking attractive fixed points, we find the number of such points can increase exponentially with the dimensionality, at a rate determined by the characteristic width of the individual biotic activity functions.

Through this analysis, we have preserved generality where possible, though a number of significant assumptions are made. The covariance of the total biotic effects is approximated by a stationary function, a condition that requires the width of individual activity functions to be small compared to the essential range for life. Additionally it is assumed that the number of biotic elements is sufficiently large that our reliance on the central limit theorem is justified. While numerical validation is provided here, the extent to which these conditions are met by real systems may be the main limitation of this approach.

Acknowledgements This work was supported by an EPSRC Doctoral Training Centre grant (EP/G03690X/1).

References

1. Lovelock J (1979) A new look at life on earth. Oxford University Press, Oxford
2. Lenton T et al (1998) Gaia and natural selection. Nature 394(6692):439–447
3. Watson A, Lovelock J (1983) Biological homeostasis of the global environment: the parable of daisyworld. Tellus B 35(4):284–289
4. Wood AJ, Ackland GJ, Dyke JG, Williams HTP, Lenton TM (2008) Daisyworld: a review. Rev Geophys 46:RG1001
5. Kirchner J (1989) The Gaia hypothesis: can it be tested. Rev Geophys 27(2):230
6. Ashby W (1952) Design for a brain. Wiley, New York
7. Dyke JG (2010) The daisystat: a model to explore multidimensional homeostasis. In: Proceedings of the eleventh international conference on the simulation and synthesis of living systems. Artificial life XI. MIT Press, Cambridge, pp 349–359
8. Kirchner J (2003) The Gaia hypothesis: conjectures and refutations. Clim Change 58(1):21–45
9. Odum E (1971) Fundamentals of ecology. Saunders, Philadelphia
10. Gardner M, Ashby W (1970) Connectance of large dynamic (cybernetic) systems: critical values for stability. Nature 228:784
11. May R (1972) Will a large complex system be stable? Nature 238:413–414
12. Alder A, Strassen V (1981) On the algorithmic complexity of associative algebras. Theor Comput Sci 15(2):201–211

Chapter 66
Inquiring Protein Thermostability: Is Resistance to Temperature Stress a Rigidity/Flexibility Trade-off?

Maria Kalimeri, Simone Melchionna, and Fabio Sterpone

Abstract In this work, we are studying the behavior of two homologous hyperthermophilic and mesophilic proteins via molecular dynamics simulations at different temperatures and at timescales that reach up to hundreds of nanoseconds. A multidisciplinary conformational analysis on the resulting trajectories, supports the idea that thermostability is induced by an interesting partition of flexible and rigid parts along the protein matrix and points out, to that end, the crucial role of electrostatic interactions.

Keywords Thermal stability · Protein flexibility · Molecular dynamics simulations

66.1 Introduction

Understanding how proteins maintain a stable fold in different thermodynamic conditions and how their motion at different length and time scales correlates to biological functions are of principal importance in biophysics. Here we are interested in a special class of proteins: thermophiles. These proteins can resist thermal stress being able to function at temperatures as high as 100 °C. The molecular origin of such special stability is still unknown and extremely appealing for technological applications, e.g. biotechnology or industrial chemical catalysis.

From the point of view of physics, thermophiles are a privileged case study for gaining insight on the main forces that keep a protein folded and functional. In particular it is key to inquire whether or not thermal resistance of thermophilic proteins

M. Kalimeri (✉) · F. Sterpone
Sorbonne Paris Cité, Laboratoire de Biochimie Théorique, Univ. Paris Diderot, UPR9080 CNRS, Paris, France
e-mail: maria.kalimeri@ibpc.fr

F. Sterpone
e-mail: fabio.sterpone@ibpc.fr

S. Melchionna
CNR-IPCF, Consiglio Nazionale delle Ricerche, Rome, Italy
e-mail: simone.melchionna@roma1.infn.it

T. Gilbert et al. (eds.), *Proceedings of the European Conference on Complex Systems 2012*, Springer Proceedings in Complexity, DOI 10.1007/978-3-319-00395-5_66,

is caused by a special rigidity of the protein matrix as assumed by common belief. In order to tackle the complexity of this challenge we have designed a multi-scale strategy based on atomistic and coarse-grained Molecular Dynamics simulations as well as on the use of advanced techniques for sampling rare events and enhanced protein conformational changes.

Here we report the results of a preliminary, multidisciplinary analysis that supports the idea that thermostability is induced by an interesting partition of flexible and rigid parts along the protein matrix and points out, to that end, the crucial role of electrostatic interactions. We note that we quantify flexibility here in terms of atomistic deviations around a mean position.

66.1.1 Theoretical Background

Thermodynamically, protein stability relates to the energetics of the transition from the native state F to the unfolded state U :

$$\Delta G^{f\to u} = G^{u} - G^{f} = -k_b T \ln\left(\frac{\langle U\rangle}{\langle F\rangle}\right) \tag{66.1}$$

where G stands for the free energy, and $\langle U\rangle$ and $\langle F\rangle$ are the sizes of the populations occupying the unfolded and folded states, respectively. By increasing temperature the relative population of the unfolded state is favored. Thermophilic proteins are characterized by a high melting temperature or in other words the unfolded state is favored only at very high temperatures (80–90 °C).[1]

The molecular origin of such stability shift is not clear since, thermodynamically, several scenarios are plausible. In the simple two-state model [2] the higher melting temperature may result from (i) a larger $\Delta G^{f\to u}$ difference characterizing the $\langle U\rangle$ and $\langle F\rangle$ populations at a comparable temperature, (ii) a slow variation of such difference as temperature increases, and finally (iii) from a shift of the temperature associated to the stability state (where the folded state is preferential) [13].

At the same time the free energy of unfolding is a result of a fine interplay between enthalpic and entropic forces:

$$\Delta G^{f\to u} = \Delta H^{f\to u} - T\Delta S^{f\to u} \tag{66.2}$$

The enhanced stability of thermophiles can be rationalized either by considering a favoring enthalpic contribution, or by being entropic in nature. In the former case the special packing of residues in the protein matrix or the higher internal connectivity (H-bonds, salt-bridges) have being invoked, while for the latter residual secondary structure in the unfolded state or enhanced flexibility of the folded state have both been proposed.

Molecular Dynamics is a powerful technique to explore the protein behavior in atomistic resolution via the direct integration of the classical equations of motion

[1] The melting temperature or mesophilic proteins is around 40 °C.

[5, 10, 11, 13–15]. Moreover it can be used in combination with advanced techniques or simplified coarse-grained models in order to explore the conformational many-fold landscape, the folding/unfolding process and gather information on the kinetics and thermodynamics of the system. In this respect it is also worth mentioning the strategic use of tools borrowed from the theory complex networks and systems to analyze protein dynamics and conformational landscape, e.g. protein internal contacts and h-bonds networks, Markov State Model for conformational transitions [4, 8, 10, 11, 15].

66.2 Results

In the following we present the preliminary results of our research on the flexibility/rigidity response of protein to thermal stress at different levels of spatial resolution (See also methodology section at the end).

Stability vs. Unfolding We first stress that by performing simulations in the hundred-nanosecond timescale and longer, in a range of physical temperature (25–100 °C) we verify a lower stability for a mesophilic versus a hyperthermophilic protein. At the working temperature of the hyperthermophilic homologue (85 °C) the former explores the early steps of the unfolding process while the latter maintains its fold stable (e.g. high secondary structure conservation and low deviation from crystallographic native state). This finding shows that the present Force Field for biomolecular simulation contains all the ingredients necessary to distinguish the different temperature-related stability of proteins.

Atomistic Fluctuations A first insight on how the flexibility/rigidity of the protein matrix changes upon thermal stress is recovered by a detailed and rigorous analysis of atomistic fluctuations. This can be performed routinely by computing the root mean square fluctuation (RMSF) of atomic positions along a trajectory after removing rigid body motions,

$$\mathrm{RMSF}_i = \sqrt{\frac{1}{T}\sum_{t_j=1}^{T}\left(x_i(t_j) - \tilde{x}_i\right)} \tag{66.3}$$

where T is the total time and $\tilde{x}$ is a reference position for particle i, usually the time-averaged one. In general such analysis is performed rather blindly without special care on the effective meaning of the observable. In particular RMSF measures the second moment of the distribution of atomic positions; this parameter is meaningful only if this distribution is approximately unimodal. At a long time-scale simulation since the protein experiences large conformational changes the above condition breaks down. It is then necessary to use a precise procedure for individuating the maximal length scale that—in an average sense—allows to compute correctly the atomistic fluctuations.

To that end, we follow the rigorous procedure introduced by Maragliano et al. [5]. The time window on which we perform our block sampling is about 350 ps. At longer time scales mean atomic positions start to experience many-fold localization. We compute the RMSF for backbone C-alpha atoms and perform block averages on several fragments of the trajectory. We find that, despite the fact that the magnitude of the observable is comparable for both proteins, there is a very intriguing difference in the partitioning of flexible (high RMSF) and rigid (low RMSF) fragments along the sequence among the two systems. In particular the latter shows a remarkable anti-correlating behavior in the RMSF between groups of neighboring residues that seems to be independent of the temperature, a sort of caging effect borrowing a concept from liquid state theory. For the hyperthermophilic protein flexible and rigid parts of the sequence alternate more frequently and regularly than in its mesophilic homologue. Thus a more regular distribution of rigid fragments possibly stops the energy flow along the protein matrix preventing progressive unfolding.

Electrostatics The above results are complemented by an analysis of electrostatics interactions which due to a surplus of charged amino acids for the hyperthemophile are considered to play a crucial role for thermal stability. We begin by observing, as expected, a substantially larger number of salt-bridges for the aforementioned system as well as a higher number of possible ionic pair combinations and thus extended salt-bridge clusters.

Furthermore, a big fraction of salt-bridges present in the crystal structure of the hyperthemophile show an exceptional stability which is not the case for the mesophile. These salt-bridges have been verified to be related with the less flexible parts of the matrix, thus it is possible that they act as clamps or stopping points, that way enhancing if not organizing the anti-correlating behavior of atomistic fluctuations.

We further continue with the calculation of the electrostatic characteristic path length (CPL) [17] which is defined as the average number of contacts needed to connect, along the shortest path, two randomly chosen nodes [15]. We follow the same technique as in Ref. [15]. The nodes of our system are the Ca-atoms and a connection exists if there exists an attractive electrostatic interaction between the respective residues (salt-bridge or hydrogen bond). Our results agree with the respective ones therein. Namely, the hyperthermophile reacts to the temperature increase by decreasing its CPL. The same is not true for the mesophilic protein. CPL is inversely proportional to the degree of electrostatic connectivity of the fold, thus this finding suggests a positive correlation between electrostatic connectivity and thermal stability.

Collective Variables The flexibility/rigidity of the protein matrix is also investigated via the construction of 2D free-energy landscapes representations over a set of collective variables such as the radius of gyration, the fraction of native contacts or the deviation of instantaneous protein configurations from the native state. For the folded state both proteins show a rather harmonic basin with comparable width. This finding supports again the idea that the thermophiles do not show a special rigidity.

Clearly a strong deviation is observed at high temperature when the thermophile is rather stable and the mesophile starts to unfold.

Compressibility Previous work has drawn the attention to an existing correlation between protein compressibility and stability [1]. Low compressibility generally correlates to an increased enthalpic stability and a suggested uniform core-to-surface distribution of charged amino-acids. We rigorously compute the compressibility of the protein as a function of temperature [6] and find that while at ambient temperature the two proteins show similar compressibility this is not true at higher temperature where the hyperthermophile shows a smaller one. We stress here that since compressibility relates to the fluctuations of protein volume with respect to that of the simulation cell, we take caution in extending the calculation only to the steady part of the trajectory. In other words at higher temperature we exclude the unfolding process of the mesophile. The volume of protein and its fluctuation are computed via the Voronoi tessellation of the space. We also point out that this precise evaluation of the atomistic volume allow us to check whether or not the mesophile and hyperthermophile are characterized by a different packing behavior, and this is not the case.

66.3 Conclusions

In this work we present a case for which a hyperthermophilic protein exhibits, in general, a comparable degree of flexibility in comparison to its mesophilic counterpart. The difference between the two systems however, concerns how flexibility/rigidity is partitioned in the protein matrix at the atomistic fluctuations timescale and how it relates to the distribution and number of key interactions (e.g. salt-bridge between charged amino acids). Moreover we report a clear difference in the response to thermal stress (as expected due to the different thermal stability), with the hyperthermophilic variant showing a systematic lower compressibility and increased electrostatic connectivity. In the coming months we will explore in more detail the configurational landscape of the two proteins considering the kinetics between local stable conformational states, hence gaining information on the relative distribution of free energy barriers separating local clusters of similar configuration.

Our present results are based on atomistic simulations. We are now refining a coarse-grain model in order to account in an effective way the specific ion-pair interactions that are key for thermostability. The coarse-grained potential has been extracted from atomistic simulations of charged amino acid pairs in dilute solution. The iterative Boltzmann inversion procedure allows the construction of an effective interaction that has been merged in the existing coarse-grained model for protein simulation OPEP [7]. After concluding tests on a small reference system, we will soon be able to investigate in detail the unfolding/folding process of thermophiles via this simplified, hence low-time consuming model.

Methodology We realize molecular dynamics simulations of two homologue proteins, hyperthermophile and mesophile gdomains of elongation factor Tu using the NAMD package [9]. The employed force field is Charmm22 [3] with the TIP3P model for water molecules. Initial coordinates for both systems were obtained from the crystal structures found in Protein Data Bank (PDB) after isolating the amino acid stretches of each protein's gdomain. The crystallographic PDB codes for the mesophilic and the hyperthemophilic species are 1EFC and 1SKQ, respectively [12, 16].

Acknowledgements The research leading to these results has received funding from the European Research Council under the European Community's Seventh Framework Programme (FP7/2007-2013 Grant Agreement no. 258748).

References

1. Dadarlat VM, Post CB (2003) Adhesive-cohesive model for protein compressibility: an alternative perspective on stability. Proc Natl Acad Sci USA 100:14778–14783
2. Feler G (2010) Protein stability and enzyme activity at extreme biological temperatures. J Phys Condens Matter 22:323101
3. MacKerell AD Jr., Brooks B, Brooks CL III, Nilsson L, Roux B, Won Y, Karplus M (1998) CHARMM: the energy function and its parameterization with an overview of the program. Encycl Comput Chem 1:271–277
4. Mann M, Klemm K (2011) Efficient exploration of discrete energy landscapes. Phys Rev E 83:011113
5. Maragliano L, Cottone G, Cordone L, Ciccotti G (2004) Atomic mean-square displacements in proteins by molecular dynamics: a case for analysis of variance. Biophys J 86(5):2765–2772
6. Marchi M (2003) Compressibility of cavities and biological water from Voronoi volumes in hydrated proteins. J Phys Chem B 107(27):6598–6602
7. Maupetit J, Tuffery P, Derreumaux P (2007) A coarse-grained protein force field for folding and structure prediction. Proteins 69(2):394–408
8. Numata J, Wan M, Knapp EW (2007) Conformational entropy of biomolecules: beyond the quasi-harmonic approximation. Genome Inf 1:192–205
9. Phillips JC, Braun R, Wang W, Gumbart J, Tajkhorshid E, Villa E, Chipot C, Skeel RD, Kale L, Schulten K (2005) Scalable molecular dynamics with NAMD. J Comput Chem 26:1781–1802
10. Prada-Gracia D, Gómez-Gardeñes J, Echenique P, Falo F (2009) Exploring the free energy landscape: from dynamics to networks and back. PLoS Comput Biol 5(6):e1000415
11. Rao F, Garrett-Roe S, Hamm P (2010) Structural inhomogeneity of water by complex network analysis. J Phys Chem B 114:15598–15604
12. Song H, Parsons MR, Rowsell S, Leonard G, Phillips SE (1999) Crystal structure of intact elongation factor EF-Tu from Escherichia coli in GDP conformation at 2.05 A resolution. J Mol Biol 285:1245–1256
13. Sterpone F, Melchionna S (2012) Thermophilic proteins: insight and perspective from in silico experiments. Chem Soc Rev 41:1665–1676
14. Sterpone F, Bertonati C, Briganti G, Melchionna S (2009) Key role of proximal water in regulating thermostable proteins. J Phys Chem B 113(1):131–137

15. Tavernelli I, Cotesta S, Di Iorio EE (2003) Protein dynamics, thermal stability, and free-energy landscapes: a molecular dynamics investigation. Biophys J 85(4):2641–2649
16. Vitagliano L, Ruggiero A, Masullo M, Cantiello P, Arcari P, Zagari A (2004) The crystal structure of Sulfolobus solfataricus elongation factor 1alpha in complex with magnesium and GDP. Biochemistry 43:6630–6636
17. Watts DJ, Strogatz SH (1998) Collective dynamics of "small world" networks. Nature (London) 393:440–442

Chapter 67
Finding Missing Interactions in Gene Regulatory Networks Using Boolean Models

Eugenio Azpeitia, Nathan Weinstein, Mariana Benítez, Elena R. Alvarez-Buylla, and Luis Mendoza

Abstract Gene regulatory networks (GRNs) play a fundamental role in development and cellular behavior. However, due to a lack of experimental information, there are missing interactions in the GRNs inferred from published data. It is not a trivial task to predict the position and nature of such interactions. We propose a set of procedures for detecting and predicting missing interactions in Boolean networks that are biologically meaningful and maintain previous experimental information. We tested the utility of our procedures using the GRN of the *Arabidopsis thaliana* root stem-cell niche (RSCN). With our approach we were able to identify some missing interactions necessary to recover the reported gene stable state configurations experimentally uncovered for the different cell types within the RSCN.

67.1 Background

Dynamical modeling is one of the most commonly used approaches for studying gene regulatory networks (GRNs), which has provided key insights of system-level

E. Azpeitia · M. Benítez · E.R. Alvarez-Buylla (✉)
Instituto de Ecología, Universidad Nacional Autónoma de México, Cd. Universitaria, Mexico DF 04510, Mexico
e-mail: eabuylla@gmail.com

E. Azpeitia
e-mail: emazpeitia@gmail.com

M. Benítez
e-mail: marianabk@gmail.com

E. Azpeitia · E.R. Alvarez-Buylla
Centro de Ciencias de la Complejidad (C3), Universidad Nacional Autónoma de México, Cd. Universitaria, Mexico DF 04510, Mexico

N. Weinstein · L. Mendoza (✉)
Instituto de Investigaciones Biomédicas, Universidad Nacional Autónoma de México, Cd. Universitaria, Mexico DF 04510, Mexico
e-mail: lmendoza@biomedicas.unam.mx

N. Weinstein
e-mail: nathan.weinstein4@gmail.com

T. Gilbert et al. (eds.), *Proceedings of the European Conference on Complex Systems 2012*, Springer Proceedings in Complexity, DOI 10.1007/978-3-319-00395-5_67,

properties such as robustness and modularity [2, 6]. However, the construction of dynamic models is usually done with a limited amount of experimental data. This often results in incomplete models due to missing information. Nevertheless, the formalization that underlies dynamical modeling allows for the prediction of some missing interactions, though this is a non-trivial task.

Boolean networks (BNs) are one of the simplest dynamical modeling approaches. BNs consist in a set of nodes (usually representing genes), where each node can only have two values, 0 if the gene is OFF and 1 if the gene is ON. The state of each node at a given time is determined by a Boolean function (BF) of the activation states of its regulatory inputs. Despite their simplicity, BN models have a rich behavior that yields meaningful information about the network under study. Hence, BN models have been successfully used for the analysis of diverse GRNs, including *A. thaliana* flower organ determination [7], and *Drosophila melanogaster* segment polarity [1], among others.

In deterministic Boolean GRNs, the system eventually attains activation patterns that are stationary or that cycle through several configurations of gene activation. These patterns are known as fixed-point and cyclic attractors, respectively. Kauffman [13] proposed that the attractors of BNs could represent the experimentally observed gene expression patterns or configurations that characterize different cell types in biological systems. Usually, when the attractors do not coincide with the reported multigene activation configurations, it is assumed that there are some missing nodes or interactions.

In BNs the number of possible BFs of a node increase as a double exponential function (2^{2^i} where i represents the number of inputs). Thus, the number of possible BFs describing a BN quickly exploits, making impossible to test all the possibilities. However, not all BFs are biologically meaningful [17]. Moreover, the use experimental information could greatly reduce the number of BFs to tests. Hence, a set of procedures that allow us to generate only biologically meaningful BFs that at the same time do not contradict previous experimental information, could help us to generate a reduced number of BFs to test which should be experimentally testable. We developed here a set of procedures capable to do this.

Recently we developed a Boolean GRN of the root stem-cell niche (RSCN) [4]. Our study revealed that the inferred RSCN GRN still lacks some important information because the set of expected attractors did not coincide with the set of expected attractors. In order to test the utility of our procedures, we use them to generate the BFs of all possible missing interactions of the RSCN GRN. We analyze the effect in the set of attractors of including one by one each BF. Then, we examined in further detail the BFs whose inclusion allowed us to recover the attractors observed experimentally. Our procedures narrowed down the nature and number of missing interactions in the RSCN GRN, produce biologically meaningful BF that did not contradicted experimentally reported data and produced testable prediction. Importantly, the procedures are general enough to be used in any BN, thus making it suitable to explore more generic theoretical questions.

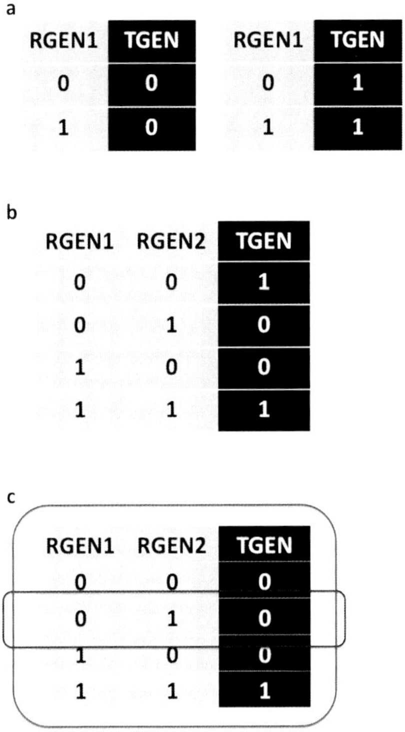

a

RGEN1	TGEN
0	0
1	0

RGEN1	TGEN
0	1
1	1

b

RGEN1	RGEN2	TGEN
0	0	1
0	1	0
1	0	0
1	1	1

c

RGEN1	RGEN2	TGEN
0	0	0
0	1	0
1	0	0
1	1	1

Fig. 67.1 Truth tables with examples of the procedures. In (**a**) two examples of a target gene (TGEN) whose expression is independent of its regulatory gene (RGEN1). As observed, TGEN value remains constant despite the expression value of its RGEN1. In (**b**) TGEN is negatively regulated by RGEN2 in the first pair of rows of the truth table, and positively in the second pair of rows of the truth tables. In (**c**) is highlighted how a loss-of-function mutant of RGEN1 may be represented in a single line of the truth table (enclosed in *purple*), while a yeast two hybrid analysis with a chromatin immunoprecipitation is represented by the whole truth table (enclosed in *grey*)

67.2 Methods

67.2.1 Procedures

We designed a set of procedures that omitted the generation of BFs in which: (1) one or more of the regulatory genes did not have any influence over the regulated gene, (2) the proposed Boolean function was not consistent with experimentally reported data, or (3) a gene could act as both a positive and negative regulator under different conditions. Is important to note that genes with positive and negative activity have been reported, however, they appear to be rare in biomolecular systems.

When a regulatory gene does not affect its target gene value, the expression value of the target gene must remain unchanged despite the expression value of the regulatory gene (Fig. 67.1(a)). When a regulatory gene acts as a positive and a negative regulator of its target gene, the expression value of the target gene must pass from 0 to 1 under some conditions when the regulatory gene changes from 0 to 1, and the expression value of the target gene must pass from 1 to 0 under other conditions when the regulatory gene changes from 0 to 1 (Fig. 67.1(b)). Finally, maintaining consistency with experimental data is more complicated and requires several procedures that depend on the quality and quantity of information available. For example, some experiments, like loss-of-function mutants, provide information that is represented in a single row of a BFs' truth table, while other, like a combination of a yeast two hybrid analysis with a chromatin immunoprecipitation can provide information that is represented with the complete truth table of a BFs (Fig. 67.1(c)). Thus, we designed four different procedures that allow us to maintain consistency with experimental information when: (1) the information is represented by single row in the BFs' truth tables, (2) to maintain the sign of regulation (positive or negative) of a regulatory gene, (3) to maintain regulatory genes interactions (e.g., dimer formations) and (4) when the experimental information is represented by the whole BFs.

To test the utility of the procedure we designed an algorithm to use them in a real GRN. Our algorithm generated the BFs of all possible missing interaction in a GRN and then test the effect of including one by one each BF on the set of attractors. The interaction that most improved the model was incorporated into the model, and then this new model was tested in the same way. The criteria to asses if the addition of a regulatory interaction was an improvement are, in order of relevance: (1) the number of expected attractors obtained, (2) the number of non-expected attractors obtained, and (3) the number of total fixed-point attractors in the model. If more than one interaction equally improved the model, one of them was randomly selected and added to the BN model. After the inclusion of an interaction, we continued adding interactions with the same criteria until: (1) the model reached only the expected attractors, or (2) the inclusion of three consecutive interactions did not improve the model by increasing the number of expected attractors obtained, or reducing the number of non-expected attractors.

67.3 Results

To test our procedures we updated the RSCN GRN [4]. Because the objective of this research was to detect missing interactions, to update the GRN first we omitted the interactions predicted by our previous work. Then, even though it is well documented that PLT genes are essential for RSCN maintenance [8], we removed them, since *PLT* genes acted only as an output node in the model. Third, we included in the updated version of the model nodes for miRNA165/6, the transcription factor PHABULOSA (PHB), and the receptor kinase ACR4 [5, 18]. Fourth, because the auxin signaling pathway describe a unidirectional pathway we were able to reduced it to only two nodes. Fifth, we included novel regulatory interactions reported in the literature [15, 16]. Also, because our model does not incorporate space explicitly, to simulate molecular diffusion, we include a positive self-regulatory edge in nodes whose products diffuse (i.e. SHR, CLE, and miRNA165/6) (Fig. 67.2). Finally, some nodes in the network already have four inputs, and the addition of a 5th regulator over any node would be computationally very demanding, since the number of possible BFs increases from $\approx 6.5 \times 10^4$ to $\approx 4 \times 10^9$. For this reason, we created intermediary nodes that integrate the influence of two regulators over any gene with 4 regulators. For instance, *WOX5* expression is repressed when *CLE40* is perceived by the membrane receptor *ACR4*. Because *WOX5* had 4 regulators, we

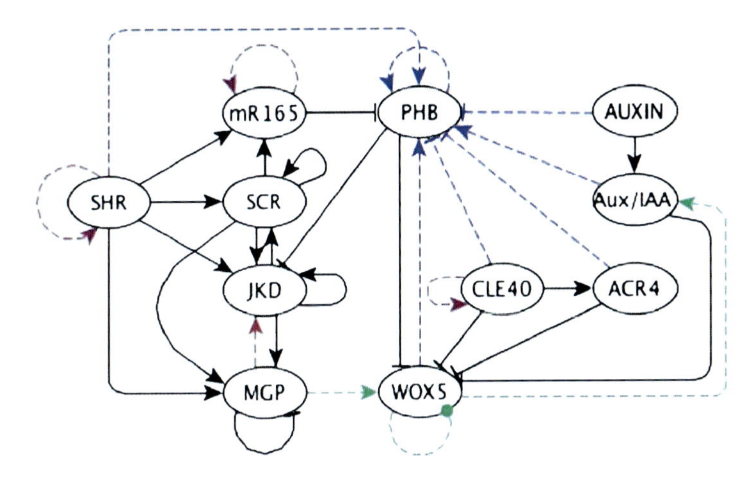

Fig. 67.2 The updated RSCN GRN with predicted missing interactions. RSCN GRN with predicted missing interactions. For clarity intermediary nodes were not included here. *Pink*, *green* and *blue edges* are the three predicted interactions required to recover the expected attractors, and are grouped according to the nodes functions. *Blue edges* are always a regulation over *PHB*. *The pink edge* is a positive regulation of MGP over *JKD*. *The green edges* are always a regulation over *WOX5*. *The doted green edge* can be a negative or a positive regulation of WOX5 over itself

Table 67.1 Expected attractors

CT/G	SHR	miR	JKD	MGP	PHB	SCR	IAA	A/I	WOX	CLE	ACR
CVC	1	0	0	0	1	0	0	1	0	0	0
PVC	1	1	0	0	0	0	0	1	0	0	0
End	1	1	1	1	0	1	0	1	0	0	0
Cor	0	1	1	0	0	0	0	1	0	0	0
LCC	0	0	0	0	0	0	1	0	0	1	1
VI	1	1	0	0	0	0	1	0	0	0	0
CEI	1	1	1	1	0	1	1	0	0	0	0
CLEI	0	1	0	0	0	0	1	0	0	1	1
QC	1	1	1	0	0	1	1	0	1	0	0

CT = Cell type, G = Gene, CVC = Central Vascular cells, PVC = Periferal vascular cells, End = Endodermis, Cor = Cortex, LCC = Lateral root-cap and columella cells, VI = Vascular initials, CEI = Cortex-endodermis initials, CLEI = Columella and lateral root-cap-epidermis initials, QC = Quiescent center, miR = miRNA165/6, IAA = Auxin, A/I = Aux/IAA, WOX = WOX5, CLE = CLE40 and ACR = ACR4

reduced *CLE40* and *ACR4* activity to a single node. This strategy allows for the exhaustive testing of BFs.

Based on available experimental data we expected 9 fixed-point attractors (Table 67.1). Some attractors represented more than one cell type due to lack of experimental information that is still required to distinguish among them (Fig. 67.3).

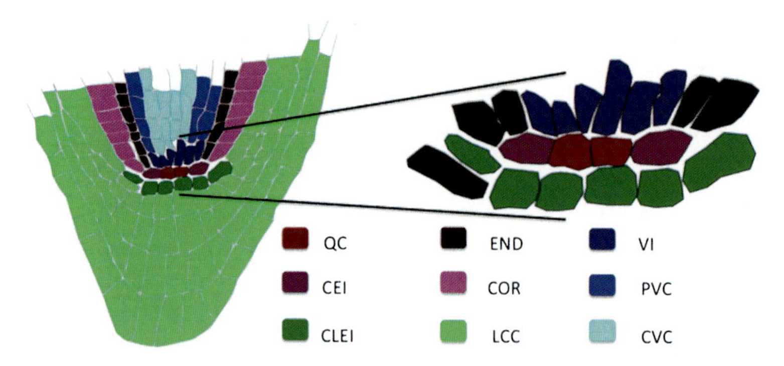

Fig. 67.3 The *A. thaliana* root tip and RSCN. Here, the expected attractors, which characterize each cell type stable gene configuration, are distinguished with different colors. As observed, some of the expected attractors represent more than one cellular type. QC = Quiescent center; END = Endodermis; VI = Vascular initials; CEI = Cortex-endodermis initials; COR = Cortex; PVC = Peripheral vascular cells; CLEI = Collumela-epidermis-lateral-root-cap initials; LCC = Collumela and lateral root cap; CVC = Central vascular cells

The BN model was not able to recover the expected attractors showed in Table 67.1. Thus, we algorithm to test our procedures utility with this GRN 10 times, ending with 10 different models that predicted different putative interactions. Finally, we analyzed the biological significance of the included interactions.

With the RSCN GRN model based exclusively on experimental information, we obtained 7 of the 9 expected attractors, 21 attractors without meaning in the RSCN context, and 4 cyclic attractors. Hence, as explained above, we included all possible interactions and their associated Boolean functions one by one, then we selected those changes that improved the consistency between the model and the experimental data. The inclusion of three types of interactions was sufficient to recover the expected attractors. However, the addition of these interactions did not eliminate cyclic attractors nor attractors without meaning in the RSCN context. In fact, the inclusion of these three interactions always increased the number of cyclic and/or unexpected attractors.

Interestingly, the three interactions mentioned above were functionally similar in the 10 replicas of the search process (Fig. 67.2). The first interaction is a regulation that *restricts PHB* expression domain to the vascular cells. This regulation was accomplished through positive regulation by those nodes with a similar expression domain (e.g. *SHR*) or through negative regulation by those genes with a complementary expression pattern (e.g. *CLE* and *ACR4*). Biologically, we believe that the likely regulator of *PHB* is a member of the *KANADI* (*KAN*) gene family. *KAN* genes were not included in this GRN model, because no connections with any node of the RSCN GRN in the root has been described yet, but *KAN* have antagonistic roles with *PHB* in the shoot and have a complementary expression pattern with *PHB* in the root [11, 12]. The second interaction is a *WOX5* self-regulatory loop. *WOX5* loop could be direct or indirect, and positive or negative (Figure 67.2). Interestingly, there is some experimental and theoretical evidence suggesting the existence of this loop through the auxin signaling pathway [4, 9]. The third interaction is a positive regulation of MAGPIE (MGP) over *JACKDAW* (*JKD*). This is contrary to the proposed antagonistic relation between *JKD* and *MGP* through a negative regulation of MGP over *JKD* [19]. The interplay between *JKD*, *MGP*, *SCR* and *SHR* is complex [16, 19] and there is no consensus on its mechanism. Our simulations suggest that it is necessary to consider other possible regulatory mechanisms.

After the inclusion of 11 to 15 interactions, the performance of the resulting GRN models no longer improved. After this point, almost all models reduced both the number of cyclic and biologically meaningless attractors to 3. Interestingly, some interactions were present in several of the 10 final models. Specifically, the most common interactions were: (1) inhibition of *SHR*, (2) activation of *SHR* by PHB, (3) negative regulation of PHB over auxin, and (4) negative regulation of Aux/IAA or SHR over *CLE*. Our results emphasize the lack of data concerning the regulation of key nodes of the RSCN GRN. Unraveling how these genes are regulated will be fundamental to our understanding of how the RSCN is maintained.

As expected, the use of our procedures producer only experimentally testable predictions that did not contradicted previously reported experimental data. The procedures were capable to greatly reduce the number of BFs of putative missing interactions. For example, to predict the first putative missing interactions, sing our procedures we only tested ~ 3000 out of $\approx 8 \times 10^9$ possible BFs. Anyhow, to produce each of the RSCN GRN that recovered the expected attractors, we needed to test around 100000 BFs, which is a highly demanding computational process. Moreover, if we consider that the number of BFs increase as a double exponential function, there is still an important constrain that needs to be tackled in the future.

67.4 Conclusion

In GRNs it is possible to test the effect of adding or modifying all possible interactions. However, even for small BNs, the number of possible BFs is overwhelming. Nonetheless, we presented in this work a set of procedures to reduce the number of BFs of putative missing interactions. Our procedures produce only experimentally testable BFs that do not contradict experimental data. To systematically predict possible missing interactions, and we have applied our procedures to the *A. thaliana* RSCN GRN. Importantly, the procedures were capable to greatly reduce the number of BFs generated. However the total reduction is dependent on the quality of the experimental information and because the number BFs of a node increase as a double exponential function, its utility is constrained for network with low connectivity.

For the specific case of the RSCN GRN we could not recover a network topology that yielded the observed configurations alone, without additional unobserved attractors. However, our work provides important predictions concerning additional interactions and a novel RSCN GRN architecture that could be experimentally tested. One limitation of this approach is the fact that additional missing nodes may be required to recover the observed set of configurations without unobserved ones.

Interestingly, some of the genes involved in RSCN maintenance are also involved in other aspects of plant development such as epidermis differentiation [10] and vascular development [20]. This challenges the stem cell pedigree idea, and as proposed before, suggests that the stem cell state is not independent of the local cellular micro-environments characteristic of the stem cell niche [14].

We believe our method would be improved by its incorporation into an existent dynamical network analyzer (e.g. [3]). Another possible improvement for our methodology would be the inclusion of a genetic algorithm, which would allow us to search for additional missing interactions. Given that the methodology used in this study is very general, we believe that this kind of exploration could help guide experimental research of any system amenable to BN analyses, as well as theoretical questions. For instance, this methodology can be used to study the constraints that a given network topology imposes on attractor evolvability.

Acknowledgements This work constitutes a partial fulfillment of the doctorado en Ciencias Biomédicas of the Universidad Nacional Autónoma de México (UNAM). E. Azpeitia acknowledges the scholarship and financial support provided by the National Council of Science and Technology (CONACyT), and UNAM. E.R. Alvarez-Buylla thanks financial support from Conacyt (81433, 81542, 1667705,180098,180380) and PAPIIT (IN229003-3, IN226510-3, IN204011-3, IB201212-2). E.R. Alvarez-Buylla is currently sponsored by the Miller Institute for Basic Research in Science, University of California, Berkeley, USA. We thank Rigoberto V. Perez-Ruız and Diana Romo for technical and logistical assistance.

References

1. Albert R, Othmer HG (2003) The topology of the regulatory interactions predicts the expression pattern of the segment polarity genes in Drosophila melanogaster. J Theor Biol 223:1–18
2. Alvarez-Buylla ER, Benítez M, Dávila EB, Chaos A, Espinosa-Soto C, Padilla-Longoria P (2007) Gene regulatory network models for plant development. Curr Opin Plant Biol 10:83–91
3. Arellano G, Argil J, Azpeitia E, Benítez M, Carrillo M, Góngora P, Rosenblueth DA, Alvarez-Buylla ER (2011) "Antelope": a hybrid-logic model checker for branching-time Boolean GRN analysis. BMC Bioinform 12:490
4. Azpeitia E, Benítez M, Vega I, Villarreal C, Alvarez-Buylla ER (2010) Single-cell and coupled GRN models of cell patterning in the Arabidopsis thaliana root stem cell niche. BMC Syst Biol 4:134
5. Carlsbecker A, Lee JY, Roberts CJ, Dettmer J, Lehesranta S, Zhou J, Lindgren O, Moreno-Risueno MA, Vatén A, Thitamadee S, Campilho A, Sebastian J, Bowman JL, Helariutta Y, Benfey PN (2010) Cell signalling by microRNA165/6 directs gene dose-dependent root cell fate. Nature 465:316–321
6. de Jong H (2002) Modeling and simulation of genetic regulatory systems: a literature review. J Comput Biol 9:67–103
7. Espinosa-Soto C, Padilla-Longoria P, Alvarez-Buylla ER (2004) A gene regulatory network model for cell-fate determination during Arabidopsis thaliana flower development that is robust and recovers experimental gene expression profiles. Plant Cell 16:2923–2939
8. Galinha C, Hofhuis H, Luijten M, Willemsen V, Blilou I, Heidstra R, Scheres B (2007) PLETHORA proteins as dose-dependent master regulators of Arabidopsis root development. Nature 449:1053–1057

9. Gonzali S, Novi G, Loreti E, Paolicchi F, Poggi A, Alpi A, Perata P (2005) A turanose-insensitive mutant suggests a role for WOX5 in auxin homeostasis in Arabidopsis thaliana. Plant J 44:633–645
10. Hassan H, Scheres B, Blilou I (2010) JACKDAW controls epidermal patterning in the Arabidopsis root meristem through a non-cell-autonomous mechanism. Development 137:1523–1529
11. Hawker NP, Bowman JL (2004) Roles for class III HD-Zip and KANADI genes in Arabidopsis root development. Plant Physiol 135:2261–2270
12. Izhaki A, Bowman JL (2007) KANADI and class III HD-Zip gene families regulate embryo patterning and modulate auxin flow during embryogenesis in Arabidopsis. Plant Cell 19:495–508
13. Kauffman SA (1969) Metabolic stability and epigenesis in randomly constructed genetic nets. J Theor Biol 22:437–467
14. Loeffler M, Roeder I (2004) Conceptual models to understand tissue stem cell organization. Curr Opin Hematol 11:81–87
15. Miyashima S, Koi S, Hashimoto T, Nakajima K (2011) Non-cell-autonomous microRNA165 acts in a dose-dependent manner to regulate multiple differentiation status in the Arabidopsis root. Development 138:2303–2313
16. Ogasawara H, Kaimi R, Colasanti J, Kozaki A (2011) Activity of transcription factor JACKDAW is essential for SHR/SCR-dependent activation of SCARECROW and MAGPIE and is modulated by reciprocal interactions with MAGPIE, SCARECROW and SHORT ROOT. Plant Mol Biol 77:489–499
17. Raeymaekers L (2002) Dynamics of Boolean networks controlled by biologically meaningful functions. J Theor Biol 218:331–341
18. Stahl Y, Wink RH, Ingram GC, Simon R (2009) A signaling module controlling the stem cell niche in Arabidopsis root meristems. Curr Biol 19:909–914
19. Welch D, Hassan H, Blilou I, Immink R, Heidstra R, Scheres B (2007) Arabidopsis JACKDAW and MAGPIE zinc finger proteins delimit asymmetric cell division and stabilize tissue boundaries by restricting SHORT-ROOT action. Genes Dev 21:2196–2204
20. Zhou GK, Kubo M, Zhong R, Demura T, Ye ZH (2007) Overexpression of miR165 affects apical meristem formation, organ polarity establishment and vascular development in Arabidopsis. Plant Cell Physiol 48:391–404

Chapter 68
Can Hermit Crabs Perceive Affordance for Aperture Crossing?

Kohei Sonoda, Toru Moriyama, Akira Asakura, Nobuhiro Furuyama, and Yukio-P. Gunji

Abstract An animal's perception of its body size is modified when it adapts its body to changes in a complex environment. This ability is essential for animals that use tools or cross apertures. Here, we show that terrestrial hermit crabs, *Coenobita rugosus*, which frequently change shells, can perceive the width of the aperture to be crossed, dependent on the shape of their shells. Hermit crabs walked in a corridor that had two different size apertures; most of the crabs with a large shell did not cross the narrow aperture, indicating an awareness of aperture width. Moreover, most of the crabs with a small shell with an attachment did not select the narrow aperture, either. These results are the first demonstration of animals perceiving affordance while carrying objects.

Keywords Body · Hermit crab · Tool use · Affordance · Aperture crossing

68.1 Introduction

Animals must perceive an extended body when they use tools [1–4]. This idea has been tested in humans in aperture crossing tasks involving hand-held objects [5]. In the experiments, participants judged whether they would be able to carry an object through an aperture of a particular size by perceiving its width. They can also perceive the width of the object by utilizing its inertial variables while wielding it.

K. Sonoda (✉) · N. Furuyama
Information and Society Research Division, National Institute of Informatics, Tokyo, Japan
e-mail: koheisonoda@gmail.com

T. Moriyama
Department of Bioengineering, Shinshu University, Nagano, Japan

A. Asakura
Seto Marine Biological Laboratory, Kyoto University, Yakayama, Japan

Y.-P. Gunji
Department of Earth and Planetary Sciences, Kobe University, Kobe, Japan

T. Gilbert et al. (eds.), *Proceedings of the European Conference on Complex Systems 2012*, Springer Proceedings in Complexity, DOI 10.1007/978-3-319-00395-5_68,

Fig. 68.1 Crab with extension; Scale bar, 1 cm

People in wheelchairs or drivers of automobiles can perceive the width of objects and the affordance needed for aperture crossing [6, 7]. In the case of driving a car, this sense, called "car body sense", enables us to conjecture what the distance is between, say, a street gutter and the car. In this sense, we can also feel when we have come close to scratching the car against a wall or other car. A human driver can gain car body sense during the early stages of learning to drive. Thereafter, the driver can easily control cars that differ in size. We can therefore assert that the body image, as typified by car body sense, can be extended, reduced or changed so that the driver can adapt to a variety of individual cars. Therefore, the user of a machine is assumed to assimilate his/her own specific position relative to that of the machine. The user gains the knowledge of the width and the length of the machine without the experience of a crash that reveals the true dimensions.

Here, we replace the situation of a user and a machine with a hermit crab, *Coenobita rugosus*, and its gastropod shell. Hermit crabs often change shells of various shapes and sizes [8–10]; some marine species also attach sea anemones to their shells [11]. The hermit crab must then adjust to each new shell and/or anemone. This adjustment provides a simple model for investigating the dynamical relationship between a new component and the whole body size perception in non-human subjects. We tested this idea by attaching new small shells to the shells of terrestrial hermit crabs and assessing their ability to perceive affordance for aperture crossing.

68.2 Materials and Methods

Small *Coenobita rugosus, in Nerita Linnaeus* shells (major axis length of 20–40 mm), were collected from Iriomote Island (Uehara, Taketomi-cho, Yaeyama-gun, Okinawa), in August 2011. Specimens were collected at a rate of about 20 a day and kept in plastic containers, fed *ad libitum* on popcorn and water and left in isolation at 28 °C for 24 hours. We attached small *Neritidae* shells to some shells with instant glue mixed with sea sand (Fig. 68.1). The experiments were recorded using a video

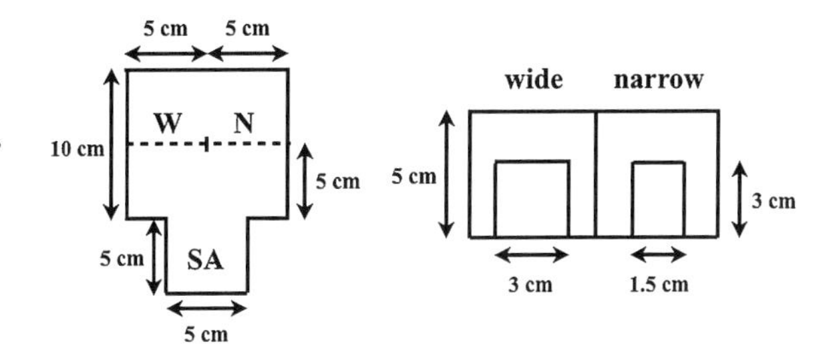

Fig. 68.2 Corridor (*left*) and apertures (*right*) for the experiments: SA, starting area; W, wide aperture and N, narrow

Table 68.1 Results of experiment 1

	Wide	Narrow
Small	11	9
Big	19	1

camera (HDC-TM700, Panasonic). The camera was placed horizontally 1 m above the course.

The corridor was 10 cm long, 10 cm wide, and 5 cm high, and had a center partition that had two apertures and a starting area (Fig. 68.2). The two apertures were shuffled according to a coin toss in each trial. In experiment 1, the shell was unmodified. In experiment 2, three small shells were attached to each crab's shell. The attached shells were 0.6–0.9 cm height and fixed along the minor axis of each crab's shell (Fig. 68.1). The crabs were held in a plastic container for 10 minutes after the attachment.

Crabs of both sexes were used, ($n = 45$: 26 males; 19 females), and sex did not affect the behavior in the experiments. In the experiments, we divided the crabs into two groups according to crab's body size. Big crabs with their original shells could not pass through the narrow aperture and small crabs could pass the narrow aperture. Twenty small crabs and twenty big crabs were used in experiment 1 and five small crabs in experiment 2. Each crab was not pre-trained and only participated in a single trial. We counted which aperture each crab passed through (Fig. 68.2). We did not use data from trials in which the crab moved within 3 seconds of the start. Furthermore, we did not evaluate cases in which crabs touched the external walls (except those in the start area) with body parts such as antennae or legs before passing through the aperture. Each crab could move freely in any direction until it passed through either aperture. We counted only the first trial in the suitable manner for each crab. If the crab did not start, the trial was canceled. We compared the number of passages with Fisher's exact test in experiment 1 with an alpha level of 0.05.

68.3 Results

In experiment 1, with no extension, the number of passages through the wide aperture by big crabs (19/20) was greater than the number of passages by small crabs (11/20) ($p = 0.00836$; Table 68.1). Thus, the crab's original body size affected its aperture crossing behavior.

Before experiment 2, crabs were held in a plastic container for 10 minutes after attaching the new shell. In the experiment with extension, most of the small crabs did not select the narrow aperture. Some crabs approached the narrow aperture, but ended up with the wide aperture.

68.4 Discussion

Without extended shells, the rate of crossing the wide aperture for big crabs was greater than that for small crabs (Table 68.1), showing that crabs preferred the wide aperture when they could not cross the narrow aperture with their shells. With extended shells, all crabs crossed the wide aperture even if their original size was small. In particular, some crabs with extensions changed direction to the wide aperture on their way to the narrow aperture. Therefore, the crabs appeared to detect the width of the apertures relative to the width of their extended shells.

Hermit crabs depend on empty gastropod shells and gather information about shells [8–10]. They can attach an anemone to their shells for the sake of balance [11]. When the shell was extended, its weight and gravity center changed. However, the crabs quickly managed to balance the shell presumably by altering their leg positions and posture within the shell. They also managed to balance asymmetric plastic plate attachments [12]. Through these adjustments of balance, the crabs may be able to detect the width of their body and compare it with the width of the aperture before crossing. This is similar to the perceptions observed by humans wielding objects or controlling machines [5–7]. In the present study, we have shown that hermit crabs can rapidly embody extended shells and obtain adequate perceptions for crossing through apertures.

Acknowledgements The Murata Science Foundation supported this work.

References

1. Shumaler RW, Walkup KR, Beck BB (2011) Animal tool behavior: the use and manufacture of tools by animals. Johns Hopkins University Press, Baltimore
2. Iriki A, Tanaka M, Iwamura Y (1996) Coding of modified body schema during tool use by macaque postcentral neurons. NeuroReport 7:2325–2330
3. Botvinick M, Cohen J (1998) Rubber hands 'feel' touch that eyes see. Nature 391:756
4. de Vignemont F (2011) Embodiment, ownership and disownership. Conscious Cogn 20:82–93
5. Wagman JB, Taylor KR (2005) Perceiving affordances for aperture crossing for the person-plus-object system. Ecol Psychol 17(2):105–130
6. Higuchi T, Hatano N, Soma K, Imanaka K (2009) Perception of spatial requirements for wheelchair locomotion in experienced users with tetraplegia. J Physiol Anthropol 28:15–21
7. Shaw RE, Flascher OM, Kadar EE (1995) Dimensionless invariants for intentional systems: measuring the fit of vehicular activities to environmental layout. In: Flanch J, Hancock P, Cairo J, Vicente K (eds) Global perspective on the ecology of human-machine systems. CRC Press, Boca Raton, pp 293–358

8. Elwood RW (1995) Motivational change during resource assessment in hermit crabs. J Exp Mar Biol Ecol 193:41–55
9. Osorno JL, Fernández-Casillas L, Rodríguez-Juárez C (1998) Are hermit crabs looking for light and large shells? Evidence from natural and field induced shell exchanges. J Exp Mar Biol Ecol 222:163–173
10. Hazlett BA (1981) The behavioral ecology of hermit crabs. Annu Rev Ecol Syst 12:1–22
11. Brooks WR (1989) Hermit crabs alter sea anemone placement patterns for shell balance and reduced predation. J Exp Mar Biol Ecol 132:109–121
12. Sonoda K, Asakura A, Minoura M, Elwood RW, Gunji Y–P (2012) Hermit crabs perceive the extent of their virtual bodies. Biol Lett 8(4):495–497

Chapter 69
A Framework for Scalable Cognition

Towards the Implementation of Global Brain Models

David R. Weinbaum

Abstract The human brain is still the most competent problem solving system we know. However, as a biological construct, it is impossible to expand because of many developmental constraints such as its confinement to the cranium. How can we create a general problem solving system inspired by the brain but not so constrained? In functional terms, how can we scale up cognition? We introduce a framework for scalable cognition, where complex cognitive functions emerge from coordinated coalitions of simple cognitive agents. This is a step towards the Global Brain, a scalable intelligent system with the potential ability to tackle planetary level challenges and beyond.

Keywords Attention · Cognition · Coalition · Challenge · Cognitive · Agent · Cooperation · Emergence · Global brain · Influence · Relevance · Scalable cognition · Self-organization

69.1 Introduction—Brains and the Global Brain

This paper aims to introduce a conceptual framework for a scalable model of a brain. In biology, a brain is an organ that evolved as a specialized means of survival [1]. The brain's general function can be summarized as the 3C function, namely: the combined Communication, Command, and Control functions of a complex organism. Brain, the central nervous system of an organism, also became a metaphor to the 3C structure of systems in general but more in specific of autonomous agents, that is, systems capable of purposeful autonomous behavior. The general investigation of brains as complex control structures falls under the disciplines of general system theory and cybernetics on one hand and cognitive science on the other hand [1, 26].

Our work is based on a fundamental premise inspired by knowledge of biological systems: complex control structures emerge from the coordinated interaction of simple control structures to create hierarchies of control [13]. For example: populations

D.R. Weinbaum (✉)
The Global Brain Institute, Vrije Universiteit Brussel Pleinlaan 2, 1050 Brussels, Belgium
e-mail: David.Weinbaum@vub.ac.be

T. Gilbert et al. (eds.), *Proceedings of the European Conference on Complex Systems 2012*, Springer Proceedings in Complexity, DOI 10.1007/978-3-319-00395-5_69,

of cells organize into multi cellular organisms that in turn organize into societies, hives, schools and other super-organism formations. The emergence of a new level of control is called a meta-system transition (MST) [28, 29] because it brings forth a new kind of autonomous agency. It is our working assumption that the emergence of control hierarchies [25] and subsequent metasystem transitions is, under the proper conditions, a scalable process bound only by physical constraints such as the availability of matter/energy or the speed of light. A scalable model of brains aims to provide a viable theoretical framework of how highly complex control structures emerge in populations of relatively simple interacting agents and what are the necessary conditions and specific mechanisms that may bring about such emergence. Our primary subject matter is therefore the scaling up of cognition as a viable means to achieve the emergence of the Global Brain.

A Global Brain is the projected product of the next meta-system transition of life on planet earth and possibly beyond [14–17, 21]. Such transition will be the outcome of the emergence of a new control structure from the coordinated interactions of human and machine agents (and possibly other biological agents as well) mediated by the internet. The Global Brain will facilitate communication, control and command on the planetary level and will be in fact a new kind of autonomous agency with as yet unpredictable intelligent competences.

It is argued that the prospect of the emergence of a Global Brain as an open ended system capable of demonstrating general intelligence depends on realizing a scalable cognitive process. We describe an agent based framework for scalable cognition by first defining cognition as the combination of two selective processes: *Selection for relevance* (attention mechanism) and *selection for effective action* (intention mechanism). These selective processes are context sensitive and operate on events that mediate differences in the state of the agent's environment. The structure of cognitive agents and the relevant structure of the environment co-define each other and therefore co-emerge.

Our framework suggests that the scaling up of the cognitive process is realizable by embedding in agents the tendency to form cooperative coalitions. Every such coalition is in fact a super-agent constructed from simpler constituent agents operating together in a collective cognitive process. Coalitions are self-similar dynamic structures formed and dismantled according to their relevance.

The relevance and 'survivability' of a coalition depends: (1) The existence of sufficient triggers from the environment to which they respond effectively according to a context sensitive set of criteria. (2) The extent by which they influence other coalitions and participate in higher level coalitions. (3) A decay factor that basically implements the tendency of coalitions to disintegrate and release resources if not used for a long time.

The concept of challenge [18] is defined in our framework as designating the context sensitive items of relevance that are selected by the attention mechanism of agents and suggest the selection of an appropriate action. These items are analogous to the items 'brought to consciousness' in Baars' global workspace theory [2, 3]. As attention is spreading among agents, it recruits their resources to a coordinated action. We say that challenges propagate within the population of agents along paths of influence that together form a network of influence. The propagation of challenges is analogous, in many aspects, to the monetary flow in a market system. This is based on the understanding that the currency of influence 'buys' the attention of the agent which is necessary for mobilizing them to action. While attention represents the value of the agent's actions, its (successful) actions, in turn, gain influence that can buy the attention of other agents. Finally, central to our framework, is the concept of vertical propagation of challenges. Vertical propagation is associated with the emergence of higher scales of cognitive processing and takes place as challenges at a certain level are combined through the interactions of agents to a challenge at a higher level of cognition.

The ultimate test of implementing the framework is the demonstration of general intelligence i.e. the spontaneous discovery of problems in the environment and the emergence of specific problem solving capabilities without the guidance of a designer. This is of course a very hard problem to begin with but this paper makes some conceptual headway in figuring how to get there.

69.2 Cognitive Agents

A cognitive agent is an agent characterized by displaying cognitive activity. Cognitive activity in the broadest sense may be defined as a non-trivial derivation of actions in response to events in the agent's environment [10, 11]. Non-trivial here means that the derivation of actions is influenced by the environment, by the situation of the agent and follows a goal or a fitness criteria. Cognitive activity may also include adaptation/learning of future derivations of actions based on the success or failure of previous actions.

Generally, an event is any difference in the environment that affects the situation of at least one agent. An action is any effect an agent may produce in its environment. Actions therefore produce events in the environment. An agent is identified by the events that affect it, by the events it is capable to produce and by the manner the latter are associated with the first. Of course the manner of association encodes a semantic structure; though it may include random or probabilistic elements, it cannot be entirely random. In other words, an agent must have structure. Similarly, the environment of the agent must have some structure otherwise there is no meaningful way for the agent to associate its actions to events since an environment without structure will necessarily respond randomly to the agent's actions. The realization of cognitive activity must assume a structural coupling [20] between the agent and the environment. Such coupling has consequences on the dynamic structures of both the agent and the environment.

A simple yet a very general working definition of cognition can be given now: cognition is the iterative coordinated processes of:

1. Selecting from the incoming stream of events which events are relevant and which are not. Relevance need not necessarily be a binary value. Events can be prioritized with varying levels of relevance according to the selection mechanism

involved. The mechanism responsible for selection for relevance will be referred from here on as the *attention mechanism*. Attention as an elementary cognitive function is exactly the singling out of relevance.

2. Given the current (most) relevant event, selecting from the available options of response what is the most effective action to execute next. An action may produce an event, change the agent's state or do nothing. The mechanism responsible for selection for effective action will be referred from here on as the *intention mechanism*. Intention as an elementary cognitive function is exactly the singling out of action.

According to this working definition cognition is basically a selective process. Implicit in this definition is that selection is made according to some set of criteria and possibly according to an internal state that encodes goals, drives, representations of the environment, memory of past interactions, predictions of future events and more. These implicit elements constitute together what may be called the context of the cognitive process. In the absence of context there is no cognition. Relevance, the mark of the agent's intelligent interaction with its environment is context sensitive. It is the agent's dynamic situation which guides its cognitive activity. Consequently the agent's own actions affect its situation closing a cybernetic loop through the environment.

69.3 A Framework for Scalable Cognition

The working definition of cognition suggested above is inspired by Bernard Baars' global workspace theory of consciousness [2, 3] and Stan Franklin's application of the theory in his work on the ontology of cognition [9, 10] and artificial minds [8]. Yet, our definition aims to highlight different aspects of the cognitive process in order to prepare the ground for a scalable framework for cognition. In essence, the global workspace model of consciousness operates as follows: many highly specialist and relatively simple cognitive functional modules are working in parallel, processing incoming information and competing on grabbing the central stage of the agent's cognitive process. Once an item of information wins the competition it is globally broadcast to all modules, recruiting a great portion of the computational resources of the agent to further attend to the relevant piece of information while other items are being suppressed. This grabbing of the central stage means the item was 'brought to consciousness'. But the glory of each such item is fleeting as importance decays in time and soon the whole sequence of competition, and global broadcast repeats itself.

Our starting point is fundamentally different. First, it is synthetic and not analytic i.e. it does not aim to explain an existing system (i.e. human cognition) but to construct a general framework for an artificial cognitive process. There is no a priori given shape to the system. In fact we aim for open ended emergence of complex cognitive functions. Second, as we aim to describe a scalable cognitive process, we need to address structures which are self-similar at various scales. This is not a requirement of the global workspace model. Third, our framework aims to describe cognition as distributed within a diverse population of agents, while the original global workspace model, though utilizing massive parallelism at early stages, basically converges to a single stream of processing—the stream of functional consciousness.

We adapt from the global workspace model the basic idea that the selection for relevance executed by the *attention mechanism* is the key for accessing resources in cognition. The simple competition and global broadcast model in Baars' model is replaced by a more general concept of *ad hoc* workspaces called coalitions. Coalitions are groups of interacting agents analogous to Baars' specialized modules. A coalition implies coordination and sharing of information among its participants that facilitates a collective cognitive function i.e. specialized selection for relevance (attention) and specialized selection for action (intention) preformed collectively. Coalitions are consolidated by means of spreading activation and are constituted from the resources and know-how of the participating agents. Coalitions are a product of self-organization within populations of interacting agents and possess emergent capabilities. In this self-organizing process items of relevance and the coalitions that attend to them co-emerge. This is reminiscent of the way bacteria colonies coordinate feats of collective cognition [4–6]. In our global brain framework the workspace(s) from which actions ensue will always be multiple, distributed and dynamic. The framework will not be confined anymore to the constraints imposed by how brains evolved and developed in higher animals with central nervous system.

With this framework we aim to achieve the following characteristics:

Scalability: Hierarchies of agents that emerge from populations of simpler agents.

Plasticity: The tendency to consolidate coalitions is balanced by the freedom of every agent to form *ad hoc* opportunistic coalitions and the 'forgetting' of infrequent coalitions.

Self-organization: No top-down design is involved. The self-organizing nature of distributed cognitive activity is actually necessary to allow any kind of general problem solving capability. Also, such system is highly adaptive: if the nature of the environment changes radically, the consolidated coalitions that are not relevant will tend to fade out and new coalitions will consolidate in response to the new conditions.

Heterogeneity: Agents of different kinds and function are capable of interaction and coordination.

The realization of such a framework involves of course many difficult problems but its basic prospects of success involve finding solutions for two conceptual problems: (1) How to realize a general mechanism of attention? And (2) How to realize a scalable version of the mechanism in 1? I.e. how groups of simple cognitive agents align their local attention mechanisms into the collective attention[1] of the group?

[1]For example: the touch, the scent and sight of a flower are communicated via different agencies. To construct a concept of 'flower' these agencies must somehow interact and exchange information.

69.4 The Distributed Attention Mechanism and Propagation of Challenges

According to the challenge propagation paradigm [18], a challenge is any event that invites an agent to act. A difference in sugar gradient for a bacteria, the sight of a predator for a deer, the ring of a telephone for someone waiting for an important call, a significant continuous increase in human produced greenhouse effecting the global climate (for whom is this a challenge?), these are all events that can be considered as challenges. It is clear that a challenge must be a challenge for someone or something and therefore not every event is automatically a challenge. For an event to become a challenge it is necessary that it will be selected by an agent to be acted upon. An event may be a challenge for one agent and not for another. Similarly an event may be a challenge for an agent at a given point in time and not be a challenge for it at another point etc. Events that are challenges at a given scale (bacterial) are not even registered at other scales (human).

The concept of challenge clearly fits the framework developed above. We propose that challenges are the products of attention mechanisms. As such, a challenge is an item of relevance; it is not a mere difference, not even a difference that merely triggers an action. A challenge is the bringing forth of a *context sensitive relevance* which sets the ground for the selection of action. The notion of challenge is quite abstract: agents need not have the same goals or the same attention discrimination (what they decide to be relevant) in order to respond to challenges in a cooperative manner or even coordinated manner. Yet if challenges are context sensitive and therefore particular to an agent, how are we to understand the notion of the propagation of challenges among agents?

69.4.1 Influence Networks (Horizontal Propagation)

We first consider the meaning of propagation of challenges between two agents. We define the concept of influence as follows: given a network of interacting agents, let us consider two distinct cognitive agents A and B. We will say that B is influenced by A if and only if the selection of items of relevance at the locality of agent B are caused, directly or indirectly, by the actions of agent A.

In a population of interacting cognitive agents, we can conceive of deriving an influence network [12, 19] that represents how challenges are propagated among agents. In terms of attention mechanism, the influence network embeds the information of how local challenges may or may not propagate among agents to produce widespread or global items of relevance. In other words, it may indicate trends of interest and relevance that characterize the attention dynamics of the whole population (see also [22, Chap. 8: Influence Networks]).

For another important example see [24] describing a decision making process in swarms where the relevant information is distributed among many agents.

Influence precisely means what events attract the attention of agents and consequently recruit their resources towards action. It seems that in the large picture, the flow of influence is highly predictive (and instructive) to the flow of resources and actions. Agents can adapt their attention mechanisms and/or their intention mechanisms, and base such adaptations on the information embedded in the influence network. Conceivably such adaptations may increase cognitive fitness.

To achieve a global increase of cognitive fitness, we need to assume that cognitive agents operate with tendency to regulate the usage of their own resources (self-preservation), and the resources available in their environment (exploiting other agents' resources). Such regulative activity can be modeled as economic activity in an economic system.[2] In such system actions are products, attention represents the value of the agent's actions (i.e. its products) and influence is the money. The higher the attention threshold, the more difficult it is to mobilize the agent to action. Agents engage in 'bidding' on the attention of other agents in order to 'buy' their attention and by that mobilize them to action but for that they need enough influence at their disposal to gain such attention. An agent gains influence (money) by effective action. It can be said therefore that influence as money buys actions as products and actions gain influence to the agents that produces them. The relations between influence and attention (representing action value) can define the economic dynamics of such economic model.

69.4.2 Vertical Propagation

In vertical propagation of challenges we mean to indicate a process by which one or more challenges bring about the emergence of a compound higher scale challenge which is attended by the coordinated attention mechanism of many agents. Since every challenge is the outcome of selection for relevance by an agent, vertical propagation is a concept dealing with how local items of relevance may coalesce into a global item of relevance within a group of interacting cognitive agents. In our framework, vertical propagation is the first stage of the formation and the consolidation of a coalition of agents and the coordination of their actions. Vertical propagation is therefore critical to the whole framework of scalable cognition. Clearly, vertical propagation must be facilitated by influence networks. It is the flow of influence among agents that may result in their coordinated activity which is necessary for vertical propagation. Two compelling examples of actual mechanisms of vertical propagation are quorum sensing [23, 24] and stigmergy [27] both appear primarily in populations of micro-organisms and social insects but can also arise in populations of higher animals.

A complementary aspect of vertical propagation exists in the way that already established coalitions activate their constituent agents and recruit their resources

[2]The general idea of attention as a commodity that needs to be managed appears in [7], though in a much more specific and business oriented context.

towards a shared goal. Such activation can recruit an agent even if it was not directly activated by an event from the environment. In contrast to the bottom-up propagation described above, this is a top-down propagation from the super-agent to the agents that constitute it. Notice that there is no specific activating entity, but an emergent mutual activation in response to a certain set of events. The emergence of such top-down control is associated with the above mentioned meta-system transitions.

References

1. Ashby WR (1960) Design for a brain: the origin of adaptive behavior. Wiley, New York
2. Baars BJ (1993) A cognitive theory of consciousness. Cambridge University Press, Cambridge
3. Baars BJ (2005) Global workspace theory of consciousness: toward a cognitive neuroscience of human experience. Prog Brain Res 150:45–53
4. Ben-Jacob E (2003) Bacterial self-organization: co-enhancement of complexification and adaptability in a dynamic environment. Philos Trans R Soc, Math Phys Eng Sci 361(1807):1283–1312
5. Ben-Jacob E, Levine H (2006) Self-engineering capabilities of bacteria. J R Soc Interface 3(6):197–214
6. Ben-Jacob E, Aharonov Y, Shapira Y (2004) Bacteria harnessing complexity. Biofilms 1(04):239–263
7. Davenport TH, Beck JC (2001) The attention economy: understanding the new currency of business. Harvard Business School Press, Cambridge. ISBN 1-57851-441-X
8. Franklin S (1995) Artificial minds. MIT Press, Cambridge. ISBN 0-262-06178-3
9. Franklin S (2006) A foundational architecture for artificial general intelligence. In: Proceedings of the AGI workshop on advances in artificial general intelligence: concepts, architectures and algorithms. IOS Press, Amsterdam, pp 36–57
10. Franklin S (2008) An ontology for cognition. Institute for Intelligent Systems, The University of Memphis. http://ccrg.cs.memphis.edu/tutorial/PDFs/AnOntologyforCognition.pdf
11. Franklin S, Kelemen A, McCauley L (1998) IDA: a cognitive agent architecture. In: IEEE international conference on systems, man, and cybernetics, vol 3, pp 2646–2651
12. Friedkin NE, Johnsen EC (1999) Social influence networks and opinion change. Adv Group Process 16(1):1–29
13. Heylighen F (1991) Cognitive levels of evolution: pre-rational to meta-rational. In: The cybernetics of complex systems—self-organization, evolution and social change, pp 75–91
14. Heylighen F (1997) Towards a global brain: integrating individuals into the world-wide electronic network. In: Der Sinn der Sinne. Steidl Verlag, Göttingen, pp 302–318. http://pespmc1.vub.ac.be/papers/GBrain-Bonn.html
15. Heylighen F (2008) Accelerating socio-technological evolution: from ephemeralization and stigmergy to the global brain. In: Globalization as evolutionary process: modeling global change, vol 10, p 284. http://pespmc1.vub.ac.be/Papers/AcceleratingEvolution.pdf
16. Heylighen F (2011) Conceptions of a global brain: an historical review. In: Grinin LE, Carneiro RL, Korotayev AV, Spier F (eds) Evolution: Cosmic, biological, and social. Uchitel Publishing, Moscow, pp 274–289. http://pcp.vub.ac.be/papers/GBconceptions.pdf
17. Heylighen F (2011) The GBI vision: past, present and future context of global brain research. GBI working paper. http://pespmc1.vub.ac.be/Papers/GBI-Vision.pdf
18. Heylighen F (2012) Challenge propagation: a new paradigm for modeling distributed intelligence. Technical report 2012-01
19. Kempe D, Kleinberg J, Tardos É (2003) Maximizing the spread of influence through a social network. In: Proceedings of the ninth ACM SIGKDD international conference on Knowledge discovery and data mining, pp 137–146

20. Maturana HR (1975) The organization of the living: a theory of the living organization. Int J Man-Mach Stud 7(3):313–332
21. Mayer-Kress G, Barczys C (1995) The global brain as an emergent structure from the Worldwide Computing Network, and its implications for modeling. Inf Soc 11(1):1–27. http://www.ccsr.uiuc.edu/web/Techreports/1990-94/CCSR-94-22.pdf
22. Montgomery J (2011) Mathematical models of social systems. http://www.ssc.wisc.edu/~jmontgom/influencenets.pdf
23. Seeley TD (2010) Honeybee democracy. Princeton University Press, Princeton. http://books.google.com/books?hl=en&lr=&id=txMkdG9G5acC&oi=fnd&pg=PA1&dq=honeybee+democracy&ots=rCs2hCZ5qN&sig=0wzPoaPbe1lOdN1DACyyZFfVFbk
24. Seeley TD, Visscher PK, Schlegel T, Hogan PM, Franks NR, Marshall JAR (2012) Stop signals provide cross inhibition in collective decision-making by honeybee swarms. Science 335(6064):108–111. doi:10.1126/science.1210361
25. Simon HA (1962) The architecture of complexity. Proc Am Philos Soc 106(6):467–482
26. Simon HA (1980) Cognitive science: the newest science of the artificial. Cogn Sci 4(1):33–46
27. Theraulaz G, Bonabeau E (1999) A brief history of stigmergy. Artif Life 5(2):97–116. doi:10.1162/106454699568700
28. Turchin V (1977) The phenomenon of science. A cybernetic approach to human evolution. Columbia University, New York. http://pespmc1.vub.ac.be/pos/TurPOS-prev.pdf
29. Turchin V (1995) A dialogue on metasystem transition. World Futures 45:5–57. http://en.scientificcommons.org/42814465

Chapter 70
Multi-agent Simulation for Enzyme Kinetics

Viviane Galvão, Rafaela Galante, José G.V. Miranda, and Sandra A. Assis

Abstract In this work we have developed a three-dimensional multi-agent-based model to investigate the enzyme kinetics. In our model, we have four types of agents: enzyme, substrate, buffer and product. There are void sites to simulate the mobility of the molecules. The molecules obey a Maxwell-Boltzmann velocity distribution and move by using the Moore neighborhood. In our rules, buffer does not change its state, enzyme can denature, and substrate can cleavage in two products. This model has been validated with experiments for kinetics of invertase. The model results reproduce the Michaelis-Menten kinetics. Also, they show that the velocity of cleavage is related to the initial fraction of substrate, initial fraction of enzyme, and depend on the temperature.

Keywords Multi-agent · Enzyme kinetics · Computational model

V. Galvão (✉) · R. Galante
Departamento de Ciências Biológicas, Universidade Estadual de Feira de Santana, 44036-900 Feira de Santana, BA, Brazil
e-mail: vivgalvao@gmail.com

R. Galante
e-mail: rsgalante@gmail.com

V. Galvão · J.G.V. Miranda
Instituto de Física, Universidade Federal da Bahia, 40210-340 Salvador, BA, Brazil

J.G.V. Miranda
e-mail: vivas@ufba.br

V. Galvão
Faculdade de Tecnologia e Ciências, Núcleo de Biotecnologia (NuBiotec), 41741-590 Salvador, BA, Brazil

S.A. Assis
Departamento de Saúde, Universidade Estadual de Feira de Santana, 44036-900 Feira de Santana, BA, Brazil
e-mail: sandrinhaassis@yahoo.com.br

T. Gilbert et al. (eds.), *Proceedings of the European Conference on Complex Systems 2012*, Springer Proceedings in Complexity, DOI 10.1007/978-3-319-00395-5_70,

70.1 Introduction

Enzymes are biological catalysts that increase the rates of chemical reactions. A classic enzyme has an active site in which a substrate can bind. The principal parameters affecting the enzyme performance are substrate and enzyme concentration, temperature, and pH [1]. For instance, in a well-known degradation reaction, invertase enzyme (EC 3.2.1.26) can catalyze the sucrose cleavage [2]. Invertase catalyzes the irreversible sucrose cleavage into glucose and fructose. These molecules are visualized indirectly through the quantity of reducing sugar [1].

Enzymatic reactions are dynamic, but the information about the experiments is not dynamic due to the large time interval separating observations [1]. A suitable approach to know the concentration for each type of molecules at any time can be obtained by the development of computational models. The models developed for enzyme kinetics are based on differential equations [3], Petri networks [4], lattice gas [5], and cellular automata [6–10]. However, we can also represent molecules by using autonomous agents. Additionally, to mimic a more realistic molecular motion we have developed a three-dimensional multi-agent-based model for enzymatic reaction including the Maxwell-Boltzmann velocity distribution. Also, we have performed a set of experiments for kinetics of invertase to validate our parameters.

70.2 Computational Model

Our computational model description follows the ODD protocol [11].

70.2.1 Purpose

The aim is to explore the possibility of an agent-based model to simulate the enzyme kinetics, *in silico*, giving the exact concentration of the formed products.

70.2.2 State Variables and Scales

The three-dimensional regular lattice consists of a grid with $100 \times 100 \times 100$ sites. This lattice possesses periodic boundary conditions and each site can be a void site or can be occupied by a single agent type. Time and space are discrete and all actions occur at constant intervals called time steps. In our model, 100 time steps correspond to the experimental data of 50 minutes. This model includes four different types of agents: enzyme, substrate, buffer and product. The only parameter is the total number of agents. The quantity of enzyme and buffer is constant during the simulation. The quantity of buffer is given by the difference between the total number of agents and the quantity of substrate and enzyme.

70.2.3 Process Overview and Scheduling

All types of agents have an initial random distribution. Each agent is chosen randomly and moved to void site. These agents have a Maxwell-Boltzmann velocity distribution and move by using the three-dimensional Moore neighborhood [12]. The inclusion of a velocity distribution is important because the temperature and denaturation effect can be taken into account. Also, there are molecules groups with different velocities. Thus, we use the Maxwell-Boltzmann velocity distribution defined as:

$$N_v = 4\pi N \left(\frac{1}{2\pi T} \right)^{3/2} v^2 e^{-v^2/2T} \tag{70.1}$$

Eq. (70.1) was simplified to represent N molecules with unit mass and the Boltzmann constant was set to 1.

70.2.4 Initialization

The spatial distribution pattern of enzyme, substrate and buffer is random. Initially, the solution does not possess product and enzyme denatured; these agent types only appears during the time evolution. The temperature and the initial fractions of enzyme, substrate and buffer are given by the experimental data.

70.2.5 Submodels

The time evolution is equally run in the complete lattice and it is obtained when the transition rules are applied to all agents randomly selected. Each agent changes its state according to a local rule, which depends only on the adjacent neighbors. The transition rules are uniformly applied to each molecule. A representation to explain the rules for substrate cleavage and enzyme denaturation is shown in Fig. 70.1.

70.2.6 Experiment

For invertase extraction, 1 g dry yeast cells (*S. cerevisiae*) was macerated in 30 mL of ethyl eter and silica. The cells were centrifuged on 10 g $\times$ 10 min. The enzyme purification was done through the application of supernatant on Sephadex G-25 column and washed with 50 mM acetate buffer (pH 4.5). The elution was done with flow rate of 2 mL/min. Sucrose hydrolysis was catalyzed by invertase in sodium acetate buffer (50 mM, pH = 4.5). Reducing sugars were measured according to the 3,5-dinitrosalicylic acid method [13]. Enzyme concentration was measured by the

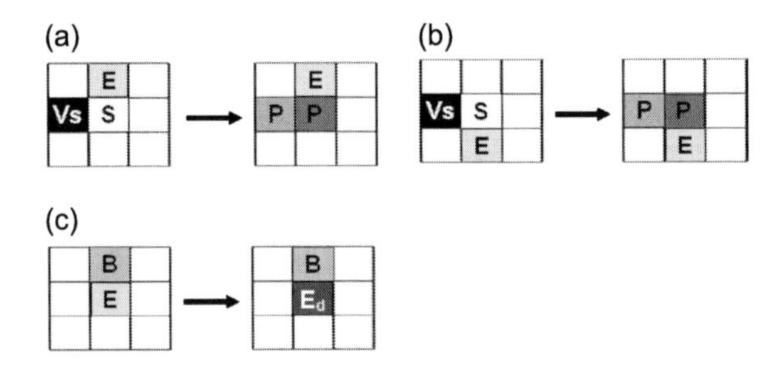

Fig. 70.1 Schematic representation of the transition rules for our enzyme kinetics model. The Moore neighborhood is represented in two-dimensions for clarity. (**a**) If there is a substrate (S) in the central site; if there is a Void site (Vs) in the left of the S, and if the site up of the S is an enzyme (E); the S site changes into product (P) and the Es site changes into another product (P) [1]. This rule is specific due to the lock and key model of enzyme-substrate interaction. (**b**) If there is a S in the central site; if there is an Es in the left of the S, and if the site down of the S is an E; the S site changes into P and the Es site changes into a P [1]. (**c**) In the neighborhood of an E; if there is a molecule with the velocity greater than a maximum velocity at 50 °C, the E changes its configuration and becomes denaturated (E_d). This occurs because the structure of most enzymes can unfold by increase of temperature

Bradford method [14]. The effect of substrate concentration was studied by determining the initial velocities (V_o) at different concentrations. The Michaelis–Menten constant [15] (k_M) and the maximum velocity (V_{max}) were determined using the Lineweaver-Burk method [16]. Average values of triplicates (differing about 5 %) were calculated.

70.3 Results and Discussion

For each set of parameters, the average value of 20 simulation runs was taken. The parameters optimization was done by using the fit to our experiment. In the experiment the total volume of solution is 0.75 ml. The quantity of buffer is given by the difference between the total volume of solution and the quantity invertase and sucrose. In all experiments, the quantity of sucrose is 0.25 ml. However, we can vary the molarity of sucrose. The larger molarity of sucrose that we measure was 0.03 M. Hence, in our computational data, the 0.03 M sucrose corresponds to fraction of sucrose 0.333, the 0.02 M sucrose corresponds to fraction 0.222 and so forth.

The relation between experimental absorbance (A) and the number of products (N_p) is given by $A = \alpha N_p$, where α represents the dimensional parameter. To determine the α parameter, the experimental value of the absorbance at 50 minutes was divided by the final quantity of enzyme. Figure 70.2 shows that these values has a linear behavior. The curve is given by $\alpha = \beta \mathrm{E} + \gamma$, where E is the quantity of enzyme (ml), β is the fraction of enzyme affecting the absorbance, and γ is the fraction of reducing sugar affecting the absorbance. The values of β and γ are 7.8×10^{-5} and 2.4×10^{-6}, respectively, with $R^2 = 0.999$.

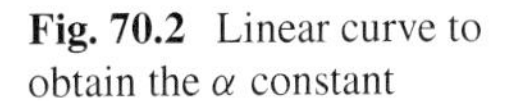
Fig. 70.2 Linear curve to obtain the α constant

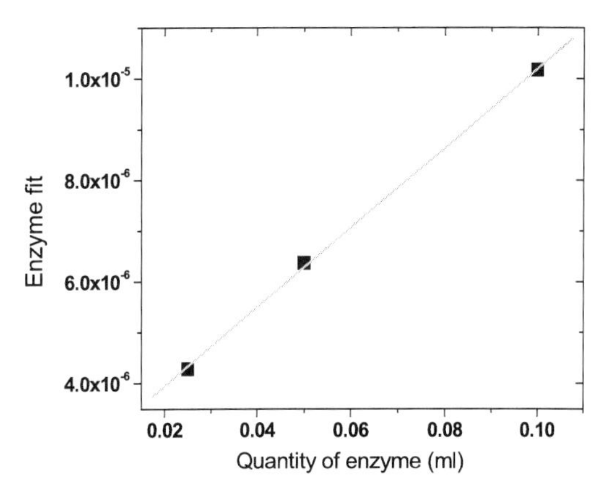

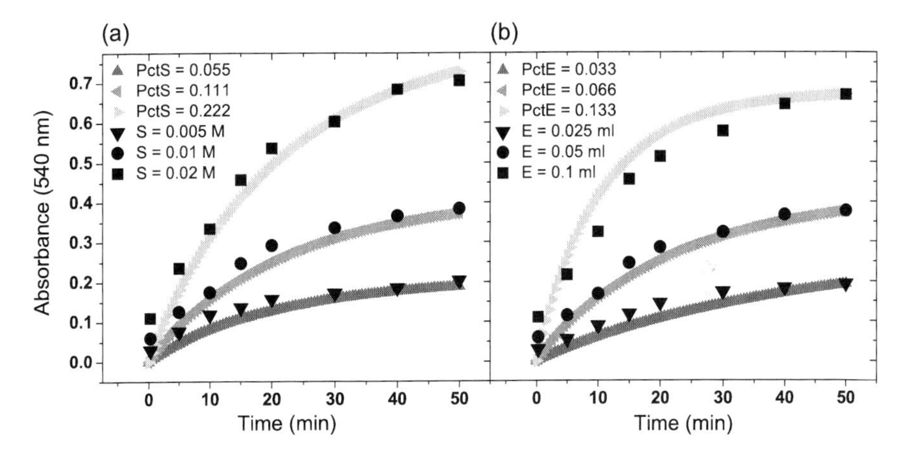

Fig. 70.3 Absorbance as a function of time for different fractions of substrate (S) and enzyme (E). In the experimental and computational data, temperature (T) is 50 °C, pH is 4.5, and total volume of solution (VS) is 0.75 ml. In the computational data the fraction of lattice-site occupied (So) is 0.6. (**a**) Substrate stability. In this experiment we use E = 0.05 ml; then, in the simulation we use PctE = 0.0666. (**b**) Enzyme stability. In this experiment we use 0.01 M sucrose; then, in the simulation we use PctS = 0.111

The absorbance for different time and fractions of enzyme and substrate is shown in Fig. 70.3. We can observe that the quantity of product formed tends to stabilize during the time evolution. Also, the larger the concentration of enzyme and substrate increases, the larger the first absorbance measure. However, our computational data start on zero because initially the formation of products is not present. Thus, the larger the time, the better the concordance between computational and experimental data.

Figure 70.4 shows the Lineweaver-Burk representation of the experimental and computational data. The experimental $v_{\max}$ and k_M are 0.66 μmols $\min^{-1}$ and 0.04 mM, respectively. The computational $v_{\max}$ and k_M are 0.79 μmols $\min^{-1}$ and

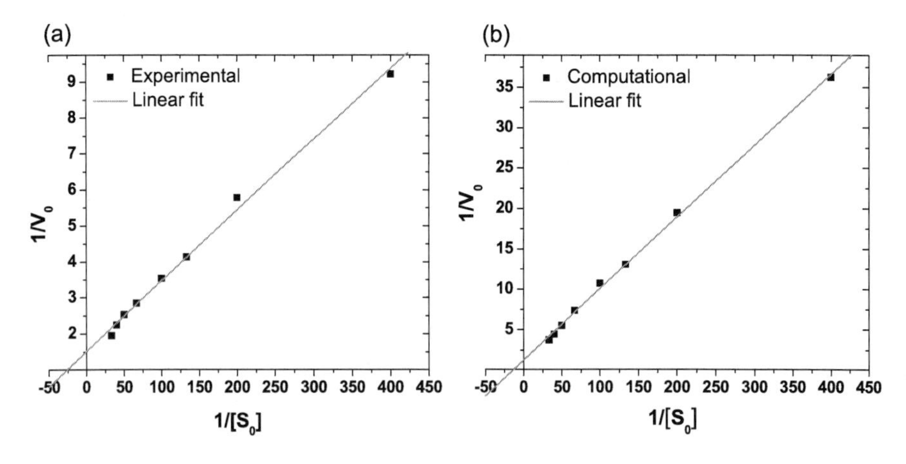

Fig. 70.4 Lineweaver-Burk representation of substrate concentration effect on the activity of enzyme. The initial velocity (V_0) is given by μmols min^{-1}. The others parameters used were I = 0.05 ml, T = 50 °C, pH = 4.5, and VS = 0.75. (a) Experimental data. (b) Computational data. In this simulation we use $S_0 = 0.6$. (See Fig. 70.3 for the meaning of the labels)

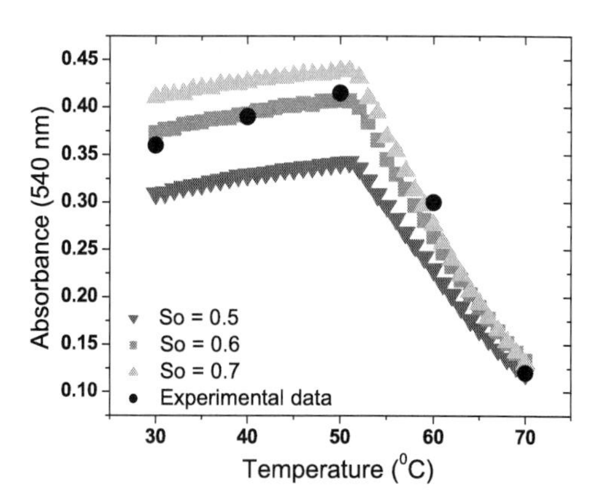

Fig. 70.5 Absorbance as a function of temperature for different fractions of lattice-site occupation. The others parameters used were I = 0.05 ml, 0.02 M sucrose, pH = 4.5, and VS = 0.75. (See Fig. 70.3 for the meaning of the labels)

0.07 mM, respectively. The equation y = 0.02x + 1.52 with $R^2 = 0.99$ was estimate for the experimental data and y = 0.09x + 1.26 with $R^2 = 0.99$ for the computational data.

To plot Fig. 70.5, we have obtained the absorbance at 15 minutes of reaction as a function of temperature for different fractions of lattice-site occupation. The enzyme exhibited maximum activity at 50 °C and it becomes denatured when the temperature was superior to 50 °C in both experimental and computational data. Therefore, the absorbance of reducing sugar increases up to 50 °C and after this temperature the absorbance decreases due to the invertase denaturation.

70.4 Conclusion

We have demonstrated that our model can reproduce some of the useful characteristics of enzyme kinetics. Also, it gives the possibility to know the concentrations of all molecules at each time step. As far as we know this paper presents the first computational model for enzymatic reaction based on regular lattice that includes in a three-dimensional lattice, effect of temperature and denaturation. The inclusion of temperature and denaturation was only possible by using the Maxwell-Boltzmann velocity distribution. The computational and experimental data were compared for the evolution of reducing sugar formation, initial velocity and temperature effect, giving a close agreement. The most important is that our results reproduce the Michaelis-Menten kinetics and give a good fitting using the Lineweaver-Burk method.

Acknowledgements This work was supported by the Brazilian agencies FAPESB and CNPq.

References

1. Koolman J, Roehm K-H (2006) Color atlas of biochemistry, 2nd edn. Thieme, Stuttgart
2. Gomes E, Guez MAU, Martin N, Silva R (2007) Enzimas termoestáveis: fontes, produção e aplicação industrial. Quím Nova 30:136–145
3. Santos AMP, Oliveira MG, Maugeri F (2007) Modelling thermal stability and activity of free and immobilized enzymes as a novel tool for enzyme reactor design. Bioresour Technol 98:3142–3148
4. Dobrescu R, Popa SA, Purcarea V, Vasilescu C (2009) Model for enzyme kinetics using Petri networks. Farmacia 57:691–702
5. Mejdani R (1994) A lattice gas model on a tangled chain for enzyme kinetics. Physica A 206:332
6. García-Olivares A, Villarroel M, Marijuán PC (2000) Enzymes as molecular automata: a stochastic model of self-oscillatory glycolytic cycles in cellular metabolism. Biosystems 56:121–129
7. Kier LB, Cheng C-K, Testa B, Carrupt P-A (1996) A cellular automata model of enzyme kinetics. J Mol Graph Model 14:227–231
8. Kier LB (2000) A cellular automata model of bond interactions among molecules. J Chem Inf Comput Sci 40:1285–1288
9. Seybold PG, Kier LB, Cheng C-K (1997) Simulation of first-order chemical kinetics using cellular automata. J Chem Inf Comput Sci 37:386–391
10. Weimar JR (2002) Cellular automata approaches to enzymatic reaction networks. Lect Notes Comput Sci 2493:294–303
11. Grimm V et al (2006) A standard protocol for describing individual-based and agent-based models. Ecol Model 198:115–126
12. Wolfram S (1986) Theory and applications of cellular automata. World Scientific, Singapore
13. Miller G (1959) Use of dinitrosalicylic acid reagent for detection of reducing sugar. Anal Chem 31:426–428
14. Bradford MM (1976) A rapid and sensitive method for the quantization of microgram quantities of protein utilizing the principle of protein-dye binding. Anal Biochem 72:248–254
15. Michaelis L, Menten ML (1913) Die kinetik der invertinwirkung. Biochem Z 49:333–369
16. Lineweaver H, Burk D (1934) The determination of enzyme dissociation constants. J Am Chem Soc 56:658–666

Part V
Interacting Populations, Collective Behavior

Chapter 71
Fast and Accurate Decisions as a Result of Scale-Free Network Properties in Two Primate Species

Cédric Sueur, Andrew J. King, Marie Pelé, and Odile Petit

Abstract The influence of particular individuals on others opinions and behaviours has long been studied by social and political scientists, and it is often suggested that certain individuals can act as leaders because they are socially connected, and have more 'influence' over others. However, this idea is difficult to test in a real-world (human or non-human) setting. Here, we present a study that describes the collective movements of two primate species: *Macaca tonkeana* and *Macaca mulatta* faced with the decision of when to stop resting and start foraging. We show that individuals that are central to the group's social network elicit stronger follower behaviour and are crucial to the achievement of consensus decisions. This 'embedded' leader-follower dynamic improves the efficiency of the decision-making process, enabling faster decision times. Our data additionally suggest that a behavioural rule-of-thumb 'follow my close affiliate' can result in the most central individual leading decisions by virtue of the scale-free properties of the network. This may allow groups to utilise the knowledge of elder, dominant, or natal individuals (who are often central in social networks) whilst simultaneously maintaining bonds with highly social individuals which may bring indirect fitness benefits itself.

Keywords Speed-accuracy trade-off · *Macaca* · Eigenvector · Centrality · Optimality · Decision-making

C. Sueur (✉) · O. Petit
Unit of Social Ecology, Université Libre de Bruxelles, CP231, Campus Plaine, Bd du triomphe, 1050 Brussels, Belgium
e-mail: cedric.sueur@iphc.cnrs.fr

C. Sueur · M. Pelé · O. Petit
Département Ecologie, Physiologie et Ethologie, Centre National de la Recherche Scientifique, Strasbourg, France

C. Sueur · M. Pelé · O. Petit
Institut Pluridisciplinaire Hubert Curien, Université de Strasbourg, Strasbourg, France

A.J. King
Structure and Motion Laboratory, Royal Veterinary College, University of London, Hertfordshire AL9 7DY, UK

T. Gilbert et al. (eds.), *Proceedings of the European Conference on Complex Systems 2012*, Springer Proceedings in Complexity, DOI 10.1007/978-3-319-00395-5_71,

71.1 Introduction

Sociologists and political scientists study the influence of particular individuals on others opinions and behaviour with the aim of understanding how specific individuals can have a disproportionate influence on group activities and decisions [1]. In non-human animals, certain individuals are similarly observed to steer the behaviour of group-mates [2]. In primate societies for example, a few key individuals 'police' behavioural interaction [3], initiate changes in behaviour [4], or choose the direction in which groups travel [5]. The behaviour of these 'leaders' confer benefits on all individuals in the group; stabilising within-group behavioural interactions and maintaining group cohesion [2, 3]. But group-mates can incur short-term costs by following leaders [5], and when individuals are faced with making a collective decision, aggregated knowledge is predicted to be superior (in terms of time and accuracy [6]) to that of a single leader [7, 8]. How such leader-follower social dynamics evolve is therefore somewhat of a puzzle.

Researchers investigating spatial associations or behavioural interactions in animal groups often report that the structure of the social network is 'scale-free' meaning that a small proportion of individuals have more numerous and stronger relationships than their group-mates [2, 9]. These individuals can be categorised as 'central' to the network by robust network statistics [9]. Such centrality might be correlated to individuals' dominance rank, age or number of relatives, and represents the sum of influences of measurable and non-measurable factors defining an individual [2, 4]. As a consequence, individual centrality offers a useful tool for quantifying the connectivity of an individual within a network, and may be a useful way to quantify an individual's social 'influence'.

We propose that variability in social relationships—that result in highly structured social networks that display scale-free properties [2, 9]—may offer an explanation for both how, and why, we consistently see specific leaders having a pivotal role in determining the activities of primate (and other) groups. We explored decision-making of two primate species: *Macaca tonkeana* and *Macaca mulatta* when faced with the collective decision to stop resting and begin foraging. It is known from previous analyses that all individuals can initiate a change in activity (i.e. make a move) but have varying success in eliciting follower behaviour [4]. Here we show that both *Macaca tonkeana* and *Macaca mulatta* social networks are scale free, and test the hypothesis that central individuals are elevated to leadership roles, resulting in an efficient decision-making process [2, 6, 10]. Specifically, we predicted that central individuals would attract a more enthusiastic following when making an initiation, which would result in group-mates responding to, and joining an initiation sooner (prediction 1), and thus reducing time taken to reach the decision to move (prediction 2). If we assume a positive feedback, that is, an individual will be more likely to initiate a movement the more it is followed, we also expected to see central individuals making more initiations (prediction 3).

71.2 Material and Methods

Two study groups, *Macaca tonkeana* ($n = 10$), and *M. mulatta* ($n = 15$) living in semi-free ranging conditions at the Center of Primatology, Strasbourg University, France, were studied for a period of five months. *M. tonkeana* were studied from November 2005 to March 2006, and *M. mulatta* were studied from May 2006 to August 2006. Each group lived in a park (fenced field of 0.5 ha) with trees, bushes and grassy areas. Animals had free access to an inside shelter (20 m^2) where commercial pellets and water were provided *ad libitum*. Fruit and vegetables were distributed once a week, outside of observation sessions. Group members used the parkland in a heterogeneous way and moved collectively between areas devoted to specific activities. Their environment was thus large enough to study collective movements as shown in several previous studies on these groups [4, 11].

Prior to, and during collective movements ($n = 146, 113$ respectively), groups were observed continuously and simultaneously by two observers. The beginning of a collective movement was defined by the departure of a single first individual (initiator or leader) walking more than 10 m in < 40 s. Any individual walking for more than 5 m in a direction that formed an angle smaller than 45° with the direction of the initiator and within 5 min after its departure was labelled a joiner. The latency of the first joiner is the lag between the initiator departure and the one of the first joiner whilst the latency of joining of all participants is the lag between the initiator departure and the one of the last joiner. Identities of all individuals taking part in a movement were recorded. For more details of the methods employed, and operational definitions of terms used, see Sueur and colleagues' studies [4] and our supplementary material.

Affiliative relationships among individuals were measured by body contact (either grooming or more general body contact) outside of collective movement contexts using instantaneous sampling at 5 minute intervals. We collected 298 scans for *M. tonkeana* and 219 for *M. mulatta*. All scans were incorporated in a sociomatrix to calculate centrality of each group member, and we measured the Eigenvector centrality coefficients with Socprog 2.4 [2, 12]. The distribution of these coefficients follows a power law in *M. tonkeana* (curve estimation test: $R^2 = 0.94$, $F1,8 = 138$, $p < 0.0001$) and *M. mulatta* (curve estimation test: $R^2 = 0.94$, $F1,9 = 143$, $p < 0.0001$) showing that networks are scale-free. Graphs of these networks were drawn with Ucinet 6.0 [13] with positions of individuals on the graph determined by multi-dimensional scaling [12, 13]. We also assessed dominance in a competitive foraging situation which was found to be highly linear. Analyses of dominance can be found in the supplementary material.

To test whether central individuals are elevated to leadership roles in these scale-free networks, which we hypothesise results in an efficient decision-making process, we conducted three Generalized Linear Models (GLMs) with Poisson error structure. In each GLM we tested whether an individual's eigenvector centrality coefficient was related the number of followers an individual attracted (prediction 1), the latency of follower behaviour (prediction 2), and to the number of initiations an individual made (prediction 3). The effect of eigenvector centrality was measured

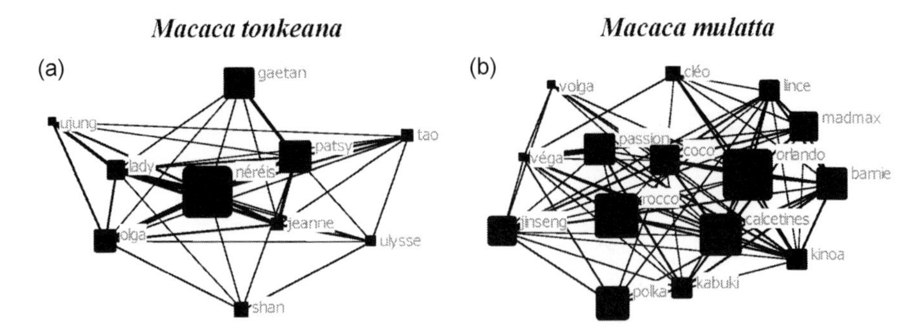

Fig. 71.1 Relationship between social network centrality and leadership in (**a**) *M. tonkeana* and (**b**) *M. mulatta*. *Squares* (*nodes*) correspond to individuals. Links between nodes represent dyadic social relationships, and the most central individuals are situated at the centre of the graph. The size of the node represents the number of initiations an individual makes in the context of collective movements

using z scores and statistical significance assessed by a Wald test. Models and tests were performed in R 2.12 (R Foundation for Statistical Computing).

71.3 Results

Individuals central to the social network attracted a more enthusiastic following (i.e. more joiners) when initiating a movement (*M. tonkeana*, $z = 3.79$, $p = 0.0001$, Fig. 71.1(a); *M. mulatta*, $z = 4.92$, $p < 0.0001$, Fig. 71.1(b)) in support of our first prediction.

We also found that following a central individual resulted in faster decision times, reducing both the latency to the first group-mate joining initiator (*M. tonkeana*, $z = -20.96$, $p < 0.0001$; *M. mulatta*, $z = -4.48$, $p < 0.0001$; Fig. 71.1), and the latency for all individuals to join the movement (*M. tonkeana*, $z = -10.08$, $p < 0.0001$, Fig. 71.2(a); *M. mulatta*, $z = -8.59$, $p < 0.0001$; Fig. 71.2(b)) supporting our second prediction. Finally, individuals with larger centrality coefficients initiated more movements (*M. tonkeana*, $z = 4.79$, $p = 0.0001$; *M. mulatta*, $z = 2.68$, $p = 0.007$, Fig. 71.2), as expected if a positive feedback occurs based on success of past initiations (prediction three). Overall, our findings suggest that following central individuals results in an efficient process (i.e. more individuals agreed with the initiator) with faster decision times.

71.4 Discussion

The way individual decision-making scales to a decision at the group level is crucial, since the combined outcome of these decisions will ultimately determine individual

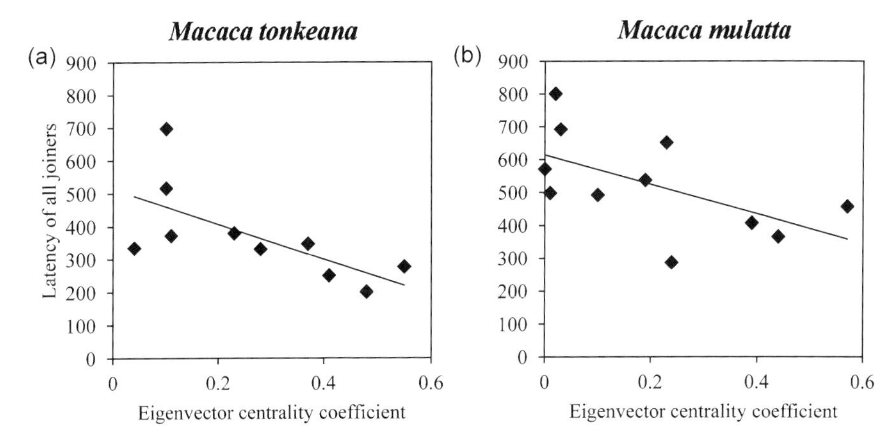

Fig. 71.2 Mean latency (sec) of all joiners according to the eigenvector centrality of initiator in (**a**) *M. tonkeana* and (**b**) *M. mulatta*. *The line* represents the linear fit

fitness. The mechanisms underlying a decision should therefore be selected to 'improve' decisions in terms of the speed taken to reach a decision, and the accuracy of the outcome [6]. Here, we demonstrate for the first time that the specific leader-follower dynamics we observed across two macaque species improve the efficiency of the decision-making process. What is more, this leader-follower dynamic is directly dependent on the social network properties of the group. This finding mirrors recent research on homing pigeons (*Columbia livia*), where a specific hierarchical structure of leader-follower behaviour appears to be more efficient than an egalitarian structure (i.e. everyone following everyone else) [14].

It is already known that following leaders with advanced knowledge can be beneficial for all members of a group [8, 15]. These knowledgeable individuals are often elder, dominant, or natal individuals [2, 4, 5], and are also central to social networks (we do not yet know the causality here) [2, 10]. Thus, we propose that following central individuals not only results in groups utilising knowledge of elder, dominant, or natal individuals (which may represent more accurate decisions), but simultaneously enables faster decision times, and also allows individuals to maintain social bonds with popular individuals which may bring indirect fitness benefits [16]. Thus, following conspecifics which have more numerous and stronger relationships represents a successful strategy for both leaders and followers. We must also emphasise that this strategy does not require that individuals know which is the most central or social individual inside their group. Instead, a straight-forward rule-of-thumb 'follow my close affiliate's behaviour' will result in individuals more often following the most central individual, by virtue of the scale-free network properties [17]. In summary, our findings imply that individuals are elevated to leadership roles as a consequence of the properties of the social network, and this process appears to be self-reinforcing, since successful individuals initiate more collective movements.

Acknowledgements We are grateful to A. Jacobs for his help on statistics. A.J. King was supported by a NERC Fellowship.

References

1. Fowler JH, Schreiber D (2008) Biology, politics, and the emerging science of human nature. Science 322:912–914
2. King AJ, Johnson DDP, Van Vugt M (2009) The origins and evolution of leadership. Current Biology 19:R911–R916
3. Flack JC, Girvan M, de Waal FBM, Krakauer DC (2006) Policing stabilizes construction of social niches in primates. Nature 439:426–429
4. Sueur C, Petit O (2008) Organization of group members at departure is driven by social structure in Macaca. International Journal of Primatology 29:1085–1098
5. King A, Douglas C, Huchard E, Isaac N, Cowlishaw G (2008) Dominance and affiliation mediate despotism in a social primate. Current Biology 18:1833–1838
6. Ward AJW, Herbert-Read JE, Sumpter DJT, Krause J (2011) Fast and accurate decisions through collective vigilance in fish shoals. Proceedings of the National Academy of Sciences 108:2312–2315
7. Conradt L, Roper TJ (2003) Group decision-making in animals. Nature 421:155–158
8. Katsikopoulos KV, King AJ (2010) Swarm intelligence in animal groups: when can a collective out-perform an expert? PLoS ONE 5:e15505
9. Lusseau D, Newman MEJ (2004) Identifying the role that animals play in their social networks. Proceedings of the Royal Society B: Biological Sciences 271:S477–S481
10. Lusseau D, Conradt L (2009) The emergence of unshared consensus decisions in bottlenose dolphins. Behavioral Ecology and Sociobiology 63:1067–1077
11. Sueur C, Salze P, Weber C, Petit O (2011) Land use in semi-free ranging Tonkean macaques Macaca tonkeana depends on environmental conditions: a geo-graphical information system approach. Current Zoology 57:8–17
12. Whitehead H (2009) SOCPROG programs: analysing animal social structures. Behavioral Ecology and Sociobiology 63:765–778
13. Borgatti S, Everett M, Freeman L UCINET 6 for Windows: software for social network analysis. http://www.analytictech.com/
14. Nagy M, Akos Z, Biro D, Vicsek T (2010) Hierarchical group dynamics in pigeon flocks. Nature 464:890–893
15. Couzin ID, Krause J, Franks NR, Levin SA (2005) Effective leadership and decision-making in animal groups on the move. Nature 433:513–516
16. Silk JB, Alberts SC, Altmann J (2003) Social bonds of female baboons enhance infant survival. Science 302:1231–1234
17. Borgatti SP, Mehra A, Brass DJ, Labianca G (2009) Network analysis in the social sciences. Science 323:892–895

Chapter 72
How to Turn an Available Data-Warehouse into Interactive Visualization Tools for Stakeholder's Empowerment

Giuseppe Roccasalva and Andrea Valente

Abstract The essay presents some results of a research work carried out at Polytechnic of Turin in collaboration with CSI Piemonte (in-house company). CSI Piemonte offered the data warehouse collected during the last decades for Public Bodies for different purpose. The objective was to find new data visualization tools in order to create new exploitation of digital data and possibly new businesses. The intelligent use of public property information becomes an important player for the information business, the research activities, for the governance and the democracy. Mostly, in the field of Business Intelligence becomes increasingly important the discipline of Data Visualization and this essay tries to translate and widen up some concepts for other discipline and debates.

Keywords Data driven journalism · Story telling · Community empowerment · Big data and data visualization

72.1 Introduction

The research center LAQ-tip (High Quality Lab-Territorial Integrated Project) of Polytechnic of Turin and CSI Piemonte[1] decided to start studding different data visualization tools (as Gapminder from Google Data Explorer, ManyEyes, Open eXplorer and Fineo). CSI Piemonte offered to investigate the data warehouse in order to create new exploitation of digital data. The work started intuitively and incrementally by testing and questioning the basic and most common data.

[1] www.csi.it (in-house ICT company of Piedmont Region).

G. Roccasalva (✉)
Laq-TIP (High Quality Laq-Territorial Integrated Project), Faculty of Architecture, Politecnico di Torino, Via P. C. Boggio 61, 10138 Torino, Italy
e-mail: giuseppe.roccasalva@polito.it

A. Valente
Faculty of Engineering and Management, Politecnico di Torino, Via Fratelli Carle 9, 10121 Torino, Italy
e-mail: val.andrea@hotmail.it

T. Gilbert et al. (eds.), *Proceedings of the European Conference on Complex Systems 2012*, Springer Proceedings in Complexity, DOI 10.1007/978-3-319-00395-5_72,

In order to give a fare picture of the big data it is necessary to start by the available local data-warehouse. CSI Piemonte manages for its customers 1300 alphanumeric databases, 1400 spatial databases and 160 databases to support public policies of local administrations [1]. The intelligent use of public property information becomes an important player for their information business, the research but also for the democracy and for the governance [3].

72.2 Visualization and Learning

Statistical information are becoming more understandable and accessible in the last years. "We discover the world through our eyes" says Stephan Few, one of the most important manager of visual communication [4]. According to Few, forms of communication such as the graphics are still largely primitive. Charts were invented to bring to light the significant data that would be impossible to interpret with a simple table. Graphs are extremely important cause they add to simple tables the visual dimension, which is more intuitive creating relation between values, colors and shapes. Eyes are holding most of human sensors (nearly 70 %) giving a crucial role to visible perception. Recently, in the field of Business Intelligence becomes crucially important the discipline of Data Visualization [5]. Few highlights three common visualization techniques: dashboards, geo-spatial visualization and dynamic visualization. Next it will be simply explained the basic tools studied.

72.3 Tools

The research questioned how to gather a share knowledge of strategic tools for data visualization. In order to start a white book of profitable visualization tools, the research project was implementing and critically studding four tools.

72.3.1 Google Public Data Explorer [6]

This is a tool that facilitates the exploration, visualization and communication of large data sets. This tool is useful to create views of public data. It currently has four display format: line chart, bar graph animated, animated map, animated bubble chart.

Figure 72.1 shows the dynamic graph that this tool allows to generate.

With the work done with the Data Explorer application it is possible to get three views: a line chart, an animated bar chart and an animated bubble chart. To upload data was necessary to create a dataset DSPL (Dataset Publishing Language). It is a .ZIP file that contains a set of XML files and CSV files. The XML file contains metadata, the information about the data used, and CSV files are simple tables that contain data to be interrogated and visualized. For the implementation of a database with Data Explorer is therefore required knowledge of XML (Google guarantees

Fig. 72.1 Screen shot of latest experiment; visualizing the history rate of foreigner population between some Community of the Metropolitan Area of Turin

a tutorial). The interface with which the user interacts, once implemented the data, offers the possibility to observe the indicators that can be chosen for analysis. Clicking on a marker will be possible to proceed with the display. Loaded this page the user can interact with charts available choosing the menu at the top left. The line chart shows metric values over time in a traditional, line chart format. The animated bar chart shows a bar for a single metric at each point in time. The bar colors can be linked to a different metric. As time advances, the bars show the new values in an animated fashion. The animated bubble chart plots two metrics against each other, one on the x-axis, and one on the y-axis. Additional metrics can be linked to the bubble sizes and colors. As time advances, the bubble positions, sizes, and styles change according to their new values in an animated fashion.

72.3.2 ManyEyes [7]

This is a social network visualization designed by IBM. It allows you to create and share charts and infographics. Once you choose the data set to be transformed into visual form, you can upload and select the output type chosen. The capacity to manage big size data lad data and the real time high quality graphics offer to users a wide range of structured data coming from unstructured information. With a few clicks you can get very pleasant bubble charts, cloud charts, bar charts and pie charts, with the ability to perform drill-down operations (which is a segmentation and creation of hypercubes in the tassonomy). It is also possible to download data sources and use them directly on Many Eyes platform to build other types of views.

The research analyzed a matrix chart. It divides the screen into a grid. Rows represent the values in one text column (e.g., education level) and columns represent another text column (e.g., index of knowledge of English). Each cell then shows a circle or bar that represents the value for its row/column combination. The two modes of the matrix, circles and bars, are useful in different situations. Bars, whose height represents numeric values, are better for exact comparisons and allow space for more columns. Circles, which show values via area, are good for showing non-negative values that vary greatly, and allow space for more rows. It can flip between the two modes in the "Expert Options" menu to the left of the visualizations. It can be displayed a third dimension of data using color. When you select a column to be represented by color, circles will turn into miniature pie charts and bars will be broken into differently-colored pieces. This option should be used with caution, since it's possible to create a cluttered display, but it can often be effective. A matrix chart takes a table with at least two text columns, for the x- and y-categories. If there is a third text column it can be used to determine colors. If there are no numeric columns, then the bubbles or bars simply show the count of each category combination.

72.3.3 Open eXplorer [8]

This is a geo-statistical tool that was designed by prof. Michael Jern at Linkoping University. Geospatial data are joined to a dynamic display. The main user interface is divided into four boxes: the thematic map, scatter plot, histogram (or parallel coordinate graph) and the area of path analysis. Each box become as a story-telling experience. Moreover, the bar of "play" can offer time component in a dynamic exploration.

The display of thematic map shows, in the upper left navigation tools, analysis tools and the legend of the chromatic scale used. The navigation tools allow you to intervene on location and zoom of the map. The Scatter Plot is a graph that allows the simultaneous display of four sizes for each country represented: the first one combined with a map (color of the bubble), the other two are represented on the axis of scatter plot (X axis, Y axis) and the last one is related to the width of the bubble. The parallel coordinate (PCP) graph allows you to simultaneously view with lines and histograms, different sizes. The PCP is a geo- visualization technique used to identify trends and data sets (clusters) to support in-depth analysis on the relationship between territories and indicators. With this tool is possible to represent also a dynamic time graph. The data file to be imported can be done using Excel and must be well designed with respect to the position of the values and required metadata. The Excel file containing the spreadsheet with the data must be saved in Unicode format .TXT.

72.3.4 Fineo [9]

This is a web-application which uses patterns of Sankey diagrams charts. The tool represents the continuous data streams such as money, energy or materials in a system, in order to represent the relationships between the size of categorical data. This innovative app manages to make sense of a set of multidimensional data through an interactive approach. This application has been created by Density Design, a research laboratory of the Department of Design at the Polytechnic of Milan. The research objective is to exploit the potential of information visualization through the construction of various design-computer app.

The flow chart that this application allows to create, in addition to being visually effective, it demonstrates a simple implementation of good data. To charge the collection of information desired, the user needs to upload a file.TSV, an extension that Excel can create. In this case was wanted to show, for the two indicators 'Using ICT' and 'importance levels', information relating to ICT questionnaire given by CSI Piemonte. The graph allows you to analyze the various branches of business, how important it is for employees to use ICT. On the left are the areas where employees work ordered by decreasing values (from top to bottom), bringing the mouse pointer over one of these variables (e.g. Immigrants) we note that the streams converge on the variables of the other indicator on the right (in this case 'significance levels'). Conversely, selecting one of the variables on the right side (e.g. unemployed), will perform an analysis regarding the level of importance for the various sectors. The range of variables is calculated according to the percentage present in the dataset. Although the magnitude of the flow is proportional to the value assigned.

In all of these case data includes demographic information about Turin and the cities around it, and data concerning the use of ICT in the firms of Piedmont.

72.4 Concluding Remarks

Four application tools that provide new graphic techniques than those currently existing in the CSI Piemonte, were examined and the concept of story-telling was introduced. The study is part of an on going program, some choice has been done and possibly some shallow judgment. The principal goal of this work is to demonstrate that data visualizations are become more accessible for users and this application made data been discussable during many meetings with professionals from different background and discipline. Among the main results [2], it is our belief that it is possible to interact with big dataset making animated or dynamic graphs that assure more simplicity and profitability. Moreover, big data are "deaf stories" which wait to be amplified and extended to all. The crucial role of visualizing and learning is passing through many different disciplines and it is increasing as the size of data collected and available.

Acknowledgements Special gratitude goes to Sylvie Occelli, manager at IRES Piemonte (Economical and Social Research Center) for the support at Future ICT conference, for hosting this initial results of the research project at ECCS and for the future interest and collaboration. A truthful thanks goes to Giuliana Bonello (referent for CSI Piemonte) for improving the collaboration with this research.

References

1. CSI piemonte (2009) Bilancio sociale
2. Roccasalva G et al (2012) The future of cities and regions. Springer, Berlin
3. Il Sole 24ore, Il Big Bang dei dati (2011). Maggio
4. Few S (2007) Data visualization, past, present and future. Cognos Innovation Center, Gennaio
5. Few S (2006) Visual communication, IBM Cognos Innovation Center
6. http://www.google.com/publicdata/home
7. http://www-958.ibm.com/software/analytics/manyeyes/
8. http://www.ncomva.com/wp-content/uploads/2012/02/Publishing-stories-created-in-Statistics-eXplorer-in-HTML5-March-2012.pdf
9. http://www.densitydesign.org/research/fineo/

Chapter 73
How Do Fish Use the Movement of Other Fish to Make Decisions?

From Individual Movement to Collective Decision Making

Arianna Bottinelli, Andrea Perna, Ashley Ward, and David Sumpter

Abstract Recent experiments by Ward et al. have shown that fish a moving fish group detects hidden predators faster and more accurately than isolated individuals. The increase in speed, in particular, seems to be a consequence of the movement-mediated nature of the interactions used by fish to share information. The present work aims at investigating the link between movement and information transfer underlying collective decisions in fish. We define an individual-based self-propelled particle (SPP) model of the decision-making process analyzed by Ward et al. We fit it to data in order to deduce the smallest set of interaction rules consistent with the experimentally observed behaviour. We infer the relative weight of different social forces on fish movement during the decision-making process. We find that, in order to reproduce the observed experimental trends, both the social forces of alignment and attraction have to be introduced in the model, alignment playing a more important role than attraction. We finally apply this model to make theoretical predictions about fish ability to detect and avoid a moving predator in a natural environment such as open water.

Keywords Collective animal behaviour · Decision making · SPP models · Fish

73.1 Introduction

Animals living in groups are required to make collective decisions about where to collect food, the timing and direction of group travel, the choice of a new shelter or the detection and avoidance of predators [1]. The decision-making accuracy of groups is typically predicted to be greater than that of the single group members, initially increasing with group size before leveling off [2]. This phenomenon can be explained by the fact that larger groups of animals are more effective than smaller

A. Bottinelli (✉) · A. Perna · D. Sumpter
Uppsala University, Uppsala, Sweden

A. Ward
University of Sidney, Sidney, Australia

T. Gilbert et al. (eds.), *Proceedings of the European Conference on Complex Systems 2012*, Springer Proceedings in Complexity, DOI 10.1007/978-3-319-00395-5_73,

groups or solitary individuals at gathering information, whereas single group members can exploit the informations collected by other members of the group and integrate them to take better decisions [3]. The "many eyes" hypothesis, that ability to detect predators increases with group size, could be one of the main evolutionary drives for the formation of animal groups [4, 5].

While pooling information from different group members increases accuracy, this improvement may come at a cost in terms of decision speed. Under this scenario, the speed-accuracy trade-off would be due to the additional time spent by different individuals for pooling their individual preferences and converge to a unitary decision. Interestingly, however, in experiments the integration of informations possessed by different group members in a fish school was found not have a great cost in terms of speed, allowing not only for more accurate, but also for faster decisions [6]. While the increase in accuracy with group size is successfully predicted by a model of optimal information transfer among group members, the increase of speed seems to be a consequence of the specific nature of the interactions used by fish to share information. Interactions are encoded into fish response to their neighbours' behaviour and mediated by movement, allowing information to spread quickly through the group [7]. Subsequently, information is filtered and integrated according to some simple local heuristic rule [8, 9]. Understanding these heuristics, as well as the relevant cues related to fish movement that support information transfer, is an important step towards a comprehensive understanding of decision-making's experimental outcomes and, in general, towards a mechanistic-based explanation of group behaviour.

In the present work we aim at investigating the specific nature of movement interactions underlying fast and accurate collective decision in fish as observed in [6]. In order to describe fish motion and the decision making process together we took a modeling-numerical approach where fish are represented as interacting self-propelled particles (SPPs). These particles move in a two dimensional space by updating their position according to their own driving force (represented as vectors), but also according to social forces related to the motion of other fish. The whole ensemble of forces driving the particle's movement are called their "rules of motion", and the relative influence of different forces on the motion process is represented by the length of the corresponding vectors, these lengths being parameters of the model. We numerically reproduce the decision-making experiment analyzed by Ward et al. in [6] by setting the simulation in an environment that reproduces the experimental set-up. Our main goal here is to infer the relative importance of the different contributions to fish movement and, in particular, the relevance of social interactions during the collective choice of a group direction in order to escape an hidden predator.

Once obtained the relevant interactions and their relative importance, a further step in our work is to make theoretical predictions about their effectiveness in a different environment such as open water. We generalized the model in order to simulate groups of fish swimming according to the inferred rules of motion but free of the constraints of a tank and under the threat of a moving predator. Through a numerical approach we tested whether the same rules of motion could lead to increases in the success of predator avoidance with group size also in this more general set-up.

73.2 Experimental Background: The Y-Maze Experiment

Our analysis is based on the empirical data collected by Ward et al. in [6]. In these experiments, the authors put groups of different size of mosquitofish, *Gambusia holbrooki*, in a Y shaped tank. A plastic model of a predator was allocated to one of the arms of the Y-maze at random and suspended in midwater to simulate a real predator. In pilot trials, the fish showed a strong aversive response to the predator once they detected it. During the experiment, five different group sizes of fish (1, 2, 4, 8, and 16) were added to a container set in the stem of the Y, then the box was raised, releasing the fish. In all cases, the fish made their way down the Y and into one of the arms. All trials were filmed and the fish were subsequently tracked.

Ward et al. were able to define two zones for their analysis: the area immediately before the bifurcation point of the tank, where the decision-making process takes place, called "decision zone" and the area crossed before reaching the decision zone, called "approach zone". The boundary between the two zones corresponded to the changing point in the behavior of the experimental animals before and during the decision-making process, and is situated around 16 cm from the bifurcation point of the Y-maze. For a picture of the experimental setup see Fig. 73.2.

In both the approach and the decision zones, each fish was characterized in terms of speed, path tortuosity (defined as the ratio of the path taken by the fish to the straight line distance between the beginning and the end of that path), time spent in the considered zone and the accuracy of its decision. The fish is considered to have made an accurate decision if enters into the arm of the Y-maze that does not contain the replica of the predator.

Ward et al. observed that, while a single fish is able to avoid the predator only in the 55.6 % of trials, which is not significantly different from random choice, the proportion of fish making an accurate decision increased with group size: individuals in groups of 8 or 16 fish were significantly more likely to make accurate decisions (i.e., to avoid the replica predator) than fish tested in isolation. The rate of increase in accuracy with group size was compatible with a perfect many eyes theory, stating that, for small groups of animals, the probability of all individuals avoiding the predator is equal to that of at least one individual detecting it. In particular, given the probability that a single fish spots the predator is p^1_{spot}, the probability that a group of size n avoids the predator is given by

$$P_n = 1 - \frac{1}{2}\left(1 - p^1_{spot}\right)^n \tag{73.1}$$

In [6], p^1_{spot} was computed from the one-fish experiment and resulted equal to 0.11 (under the assumption that when the fish fail to spot the predator they take a random branch of the Y-maze). The factor 1/2 in Eq. (73.1) accounts for the fact that even if the group does not spot the predator it has a 50 % chance of choosing the correct branch.

Ward et al. also measured that in the decision-making zone swimming speed is an increasing function of group size, while path tortuosity decreased with increasing group size. In particular, solitary fish and those in pairs decreased their swimming speed in the decision zone compared with the approach zone, whereas those in larger groups did not. This result is inconsistent with and expectation that integrating information among a larger number of individuals would require a longer time to converge to a collective decision.

73.3 Model and Methods

73.3.1 Fish as Interacting Self-Propelled Particles (SPPs)

One of the simplest ways of modeling animal movement and interaction is the Self-Propelled Particles approach introduced in 1995 by Vicsek and collaborators [10–13]. Animals are described as point particles moving with a constant speed and updating their direction at discrete time increments by adopting the average direction of motion of the particles in their local neighbourhood, plus a random perturbation [10]. This approach allows the study of the different global behaviours emerging from the introduction of different rules of interaction between particles (see [14] for a recent review), and therefore is particularly suited for our purpose of finding the minimal set of social interactions allowing for fast and accurate decisions.

Here, we consider the self-propelled particle model used in [15], and adapted from [16]. Each fish is a particle characterized by a direction of movement, a constant speed, an interaction radius R and a blind angle (see Fig. 73.1). At each time step, every fish will interact with all the neighbours within distance R, except for the ones in the region behind them, corresponding to the blind angle. The interaction radius together with the blind angle define an interaction zone whose size is equal for all the simulated fish and deduced from empirical data. We set the interaction radius to around 16 cm, which is also compatible with the size of the decision zone and the blind angle is fixed at 60 degrees from experimental considerations on the visual system of fish [17].

The direction of motion of each simulated fish is determined by the combined effect of different forces that act on the single particle. These contributions are described through vectors whose magnitude is proportional to their relative importance on the particle's motion, while the direction of the vectors can change at each time step, depending, for example, on the neighborhood of each fish. Forces contributing to the fish movement can be divided in two main subgroups: "individual forces" and "social forces". Individual forces represent basic characteristics of the fish movement, they are independent of the local neighborhood surrounding a particular fish, and describe its behaviour when it is alone. Conversely, social forces represent the tendency to align with or to join conspecifics within the local interaction zone of

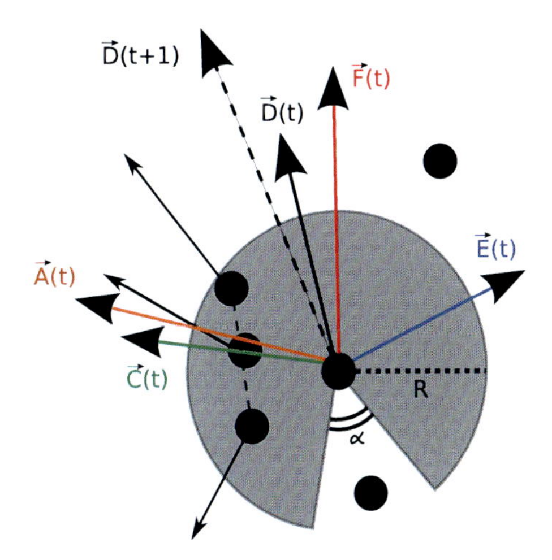

Fig. 73.1 Illustration of the SPP model. In the model fish are described as particles, each characterized by an interaction zone (*grey area* around the fish) defined by the interaction radius R and a blind angle α. While updating its position, the focal fish will take into account only the neighbours within this zone. The contributions to the movement of the fish are represented through forces acting on the particle (*arrows*), and we chose them to be the inertia in the current direction of movement $\vec{D}$, a force towards the favoured direction $\vec{F}$, an angular noise $\vec{E}$, the attraction towards the mass center of the interacting neighbours $\vec{C}$, and the alignment with their direction $\vec{A}$. The length of each vector represents the relative importance of the corresponding force, and is a parameter of the model. During simulations, at each time step all these contributions are computed and summed to give the new direction of movement of each fish $\vec{D}(t+1)$ (*dashed*)

each fish. As each fish experiences different interacting neighbourhood, these social forces are different for different individuals of the group.

The individual forces are inertia, aiming and a random force. The inertia $\vec{D}$ is the tendency to maintain the previous direction of motion. The "aiming force" $\vec{F}$ points towards the particles favoured direction. In the case of the y-shaped maze this point is the top shelter area of the maze. Then angular perturbation $\vec{E}$ represents uncertainty in the path. Lastly, at each time step the constant value of speed is perturbed by a Gaussian random error that is different for each fish and independent of neighbours. Note that this last addition does not introduce a further parameter since the value of speed is fixed according to the average speed measured from empirical data with the experimental standard deviation. These three forces together with the updating rule for speed are enough to describe the basic behaviour of a fish swimming alone [14].

Many options are available when it comes to introduce social forces, allowing to choose the level of detail in the description of individual behaviour [10–15]. Since our aim is to find the minimal set of interaction rules explaining experimental results, in the present model we introduce just two of them: the attraction towards the mass center of the interacting neighbours $\vec{C}$, and a force of alignment with their direction $\vec{A}$.

Given the above forces, the updating rule determining the actual motion of each fish is that at each time step all the described contributions are computed and summed to give the new direction of movement:

$$\vec{D}(t+1) = d\hat{D}(t) + f\hat{F}(t) + e\hat{E}(t) + c\hat{C}(t) + a\hat{A}(t) \tag{73.2}$$

Where all the forces have been decomposed in their modulus (small letter), which is the relevance of the force, as well as the unknown parameter of the model, and their unitary direction (capital, hatted letter). The resulting direction $\vec{D}(t+1)$ is then normalized, and each particle moves in the direction given by $\hat{D}(t+1)$ with speed taken from a Gaussian distributed around the mean experimental value.

73.3.2 Numerical Simulation of the Y-Maze Experiment

We simulated the decision-making process by running repeated simulations of groups of 1, 2, 4, 8 and 16 fish swimming in a Y-shaped environment of the same size of the experimental tank. Figure 73.2 shows the comparison between a typical run of the numerical simulations and a frame in an experimental trial with eight fish. In our simulated environment we distinguish three main zones, the approach zone, the decision zone and the zone after the bifurcation. The border between approach and decision zones is placed 16 cm from the bifurcation point, as in the experiments. At each run of the simulations the predator is randomly set in one of the two branches.

In a typical run of the simulation, the fish start from the beginning of the main branch of the Y maze, in the approach zone, with initial speeds, positions and entrance delays chosen according to experimental data. These delays are introduced to account for the fact that real fish do not all start moving through the maze at the same time. Since this is likely to have a consequence on the number of interacting neighbours per fish, we decided to reproduce such delays, as well as the other initial conditions, by randomly choosing a set of experimental initial conditions observed for a group of fish of the same size as the simulated one, and applying them to our simulations.

Once in the approach zone, fish move according to the rules of motions described in the previous section with, the force $\vec{F}$ pointing towards the top of the tank where the bifurcation is (see Fig. 73.2). This choice causes simulated fish to swim towards the decision zone and the branching point and reproduces the preference of real fish for deeper water and darker areas, as the two arms of the Y-shaped tank in experiments were [6].

The predator-avoidance task is numerically reproduced by assigning to each fish an individual probability $p^1_{spot} = 0.11$ of spotting the predator in the moment it enters in the decision zone. This probability is set according to experimental data so that on average a single fish has a final probability of 0.55 to avoid the predator, according to Eq. (73.1) [6]. In the case a fish spots the predator, it stops behaving according to the rules of motion and moves straight towards the safe branch of the tank with the speed it had when entering the decision zone. Depending on the strength of the social forces compared to the individual ones, the influence of a spotting fish on its neighbours leads to different global outcomes, from the likely avoidance of the predator by the whole group in the case of strong social interactions, to the noninteracting case, in which the probability of avoiding the predator is 0.55 independent of the size of the group.

To fit the model, we ran repeated simulations of the decision-making process for a wide range of parameters' value, with the scale of interaction fixed to $d = 1$. The fitting procedure is divided in two steps: we first obtain the strength of the aiming direction f and of the noise e by matching data and simulations for one fish. We then used these values in simulations of larger groups (2, 4, 8, 16 fish) and compared them with the corresponding data sets to infer the values of the social forces of alignment a and attraction c.

In the first part of the fitting process we ran 108 numerical realizations of the decision-making process for one fish and for each couple (f, e) of the values representing the strength of the force towards the bifurcation point and of the angular noise. The number of realizations is the same number of experimental trials by Ward et al., and one realization is constituted of the average of 100 runs from the same randomly extracted initial condition. For each set of parameters and group size we computed two global observables in the decision zone, the path tortuosity described by Ward et al., and the circular standard deviation of fish turning. Circular standard deviation can be thought of as a measure of the entity of directional changes. It is given by

$$CStD(t) = \sqrt{-2\log\bigl(r(t)\bigr)}, \tag{73.3}$$

where, if $\theta_i(t)$ is the angular turning of the i-th fish at time t and N the group size, then

$$r(t) = \sqrt{\left(\frac{\sum \sin\theta_i(t)}{N}\right)^2 + \left(\frac{\sum \cos\theta_i(t)}{N}\right)^2}. \tag{73.4}$$

Once we obtained the average tortuosity (T) and circular standard deviation (CStD) for each couple of parameters (f, e), we selected only the values compatible with the empirical ones, i.e. the values that were within one standard deviation, and the corresponding sets of "good" parameters $(f, e)_T$ and $(f, e)_{CStD}$. We then chose the best values $(f, e)_{best}$ in the intersection of the above sets $(f, e)_\cap = (f, e)_T \cap (f, e)_{CStD}$ as the values for which the Euclidean distance between numerical and experimental observables was minimized:

$$(f, e)_{best} = \min_{(f,e)_\cap} \sqrt{(T_{sim} - T_{exp})^2 + (CStD_{sim} - CStD_{exp})^2}. \tag{73.5}$$

Note that the accuracy for a single fish was not calculated at this stage, since it was fixed by $p^1_{spot} = 0.11$.

In the second part of the fitting process, for each couple (a, c) of the values of the social forces we ran 16 numerical realizations for groups of 2, 4, 8 and 16 fish with the individual parameters $(f, e)_{best}$ inferred in the analysis described above.

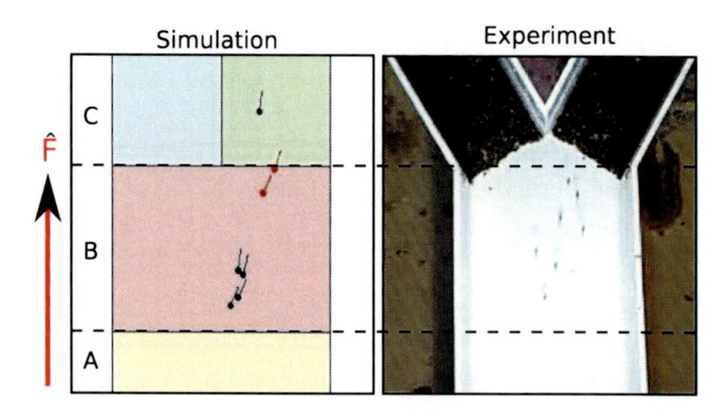

Fig. 73.2 Comparison between the visualization of our numerical set up and the experimental one. Following empirical results from [6], we distinguished three zones in our simulations: the approach zone (A), the decision zone (B) and the zone after bifurcation (C). Simulated fish started from the bottom of the tank, in the approach zone, with initial speeds, positions and entrance delays chosen according to experimental data. Fish swim across the tank due to the aiming force $\vec{F}$ that is set to point upwards, and have an individual probability $p^1_{spot} = 0.11$ of spotting the predator in the moment they enter the decision zone. This reproduces the experimental observed change in fish behaviour when crossing the threshold between zones A and B. The predator is randomly set in one of the two branches at the beginning of each run and a fish spotting it will stop behaving according to the rules of motion, swimming straight and with constant speed towards the safe branch

Again the number of numerical realizations is equal to the number of experimental trials in [6] and one realization is given by the average of 100 runs from the same randomly extracted initial condition. The process that allowed us to find the best values for alignment and attraction to neigbours follows a similar procedure as the one adopted to find the values of the individual forces. For each group size N, we computed the average accuracy (A) and tortuosity (T) and selected the couple of parameters $(a,c)^N$ corresponding to the values compatible with empirical results. We then intersected the couples of values obtained for tortuosity and accuracy $(a,c)^N_\cap = (a,c)^N_A \cap (a,c)^N_T$ at each group size, and subsequently took the union of these sets: $(a,c)_\cup = \cup_N (a,c)^N_\cap$. The best values $(a,c)_{best}$ are finally chosen from the union set $(a,c)_\cup$ as the values minimizing the Euclidean distance between experimental and numerical average for both accuracy and tortuosity:

$$(a,c)_{best} = \min_{(a,c)_\cup} \sqrt{\sum_N \big((A_{sim} - A_{exp})^2 + (T_{sim} - T_{exp})^2\big)_N}. \qquad (73.6)$$

73.3.3 Simulations in Open Water

After fitting the parameters for the specific predator-avoidance task in the Y-maze, we adapted the model to numerically test the ability of the same fish to spot and escape a predator in different environmental conditions. Since the relevance of social interactions has been obtained for a very specific situation, we are interested in whether the same rules of motion could lead to an increasing in the success of predator avoidance with group size also in a more general set-up. In particular, we simulate the model in open water for groups of shoaling fish of different size under the threat of a moving predator chasing them.

The first modification to the original model regards the set-up. Simulations take place in a squared tank centered in the axis origin and with a side that is one meter long. The predator is now free to move in the tank, and we now define some simple rules of motion. For the sake of simplicity, the only contributions to the predator's movement are inertia, which is set to be unitary, and a force of attraction towards the closest fish, which can be thought as the favourite direction of the predator. We chose this force to have the same strength as the aiming direction of fish $\vec{F}$ that we fitted from the Y-maze experiment. The speed of the predator is chosen to be constant and equal to 6 mm/frame, which is smaller than the lower average speed registered in experiments (that is, $v = 7.8$ mm/fr for one fish [6]). This choice, although arbitrary, is done in order to detect the advantage of the observed increasing in decision-making speed as a function of the group size. In the simulation, the predator starts from the upper-right corner of the tank, with an initial direction towards the origin of the axes.

The simulated fish are initially distributed with random positions and directions in a square of 10×10 centimeters, placed 30 centimeters away from the center of the tank. The initial speed of fish, as well as their speed throughout the whole simulations, is now extracted from the empirical distribution characterizing the corresponding group size in the Y-shaped tank experiment [6]. As regards the rules of motion of shoaling fish, the generalization of the model requires only a change in the favoured direction of the fish, i.e. in the direction of the aiming force $\vec{F}$, that in this set of simulations points towards the center of the tank. The effect of this choice is that the group will maintain a circular motion around the center of the tank without hitting the borders, preventing the definition of an interaction with walls.

In order to simulate the predator-avoidance task in open water, we assigned a spotting probability to each fish that is again $p^1_{spot} = 0.11$, but in this new environment a fish can spot the predator when their mutual distance is equal to the length of the decision zone. If this happens, the fish stops behaving according to the rules of motion, swimming straight and at a constant speed in the opposite direction with respect to the position of the predator, until it hits the walls of the tank. Once a fish reaches the border of the tank and exits it, it cannot go back and is considered to be "safe". As in the simulations of the Y-maze presented in the previous section, fish that do not detect the predator are influenced by the behaviour of the neighbours who did, and through social interactions has the chance to reach the border of the tank and avoid the predator. Conversely, a fish might not manage to escape the predator, whether it spots it or not, indeed if it gets closer than 6 cm from the point representing the predator[1] it is considered to be "eaten" and is canceled from simulations.

[1]This is chosen according to the fact that the predator replica used in [6] was 12 cm long.

In this setting, we ran 50 realizations for groups of fish of the same size as the ones analyzed in the Y-maze (1, 2, 4, 8 and 16), each realization being the average over 100 runs starting from the same initial condition and ending when all the fish are either "safe" or "eaten". For each group size we measured the proportion of fish avoiding the predator and the proportion of eaten fish. Among the "safe" fish, we then distinguished between these who avoided the predator because of detection and the ones escaping due to social interactions. Indeed, the latter quantity is the more relevant observable in order to assess the relevance of the social forces inferred in the Y-maze experiment in a predator-avoidance task.

73.4 Results

Initially we tried to describe fish as self-propelled particles interacting only by mean of attraction towards the mass center of neighbours, as in [16]. This was not sufficient to fit data. In fact, for each given group size the averaged observables assumed a value that was almost independent of the strength of attraction. Increasing with the value of the parameter c made little difference to the outcome, and even the best values did not compare favorably with experimental results. A similar situation was found when we tried to fit parameters by introducing only alignment with neighbours, as in [10]. Only the combination of these two social forces together provided significant variation in outcome as a function of parameters to allow us to successfully fit the data.

From the first step of the inference process, that is by matching the average simulated tortuosity and circular standard deviation with the corresponding experimental averages for one fish, we found the best values for the individual forces to be $(f, e)_{best} = (1.7, 2.7)$. Figure 73.3 shows the inference process leading to the obtained result. Here we plot the difference between the simulated observables (circular standard deviation and tortuosity respectively) and their empirical value for each couple (f, e) of the parameters. Both circular standard deviation and tortuosity increase with increasing angular noise and with decreasing aiming force, giving a regular surface whose difference with experimental values intersects zero in both cases. The black central band corresponds to the simulated quantities that are near to those found in experiments. The parameters (f, e) matching these values are reported in Fig. 73.3(C). This allows us to visualize the set $(f, e)_{\cap}$. This result is an intermediate step towards the fitting of social forces, but still it shows that randomness (e) is larger than the aiming force (f) and than inertia (d)

The second step was to look at simulations of groups of 2, 4, 8 and 16 fish. Varying the strength of social forces we finally fitted tortuosity and accuracy against data. Comparing simulated and experimental values at different values of the parameters (a, c) revealed that accuracy is an increasing function of both attraction and alignment. Tortuosity increases with attraction only, being a (slightly) decreasing function of alignment. Despite this trade-off between attraction and alignment in determining the trend of tortuosity, for all the group sizes it was possible to find

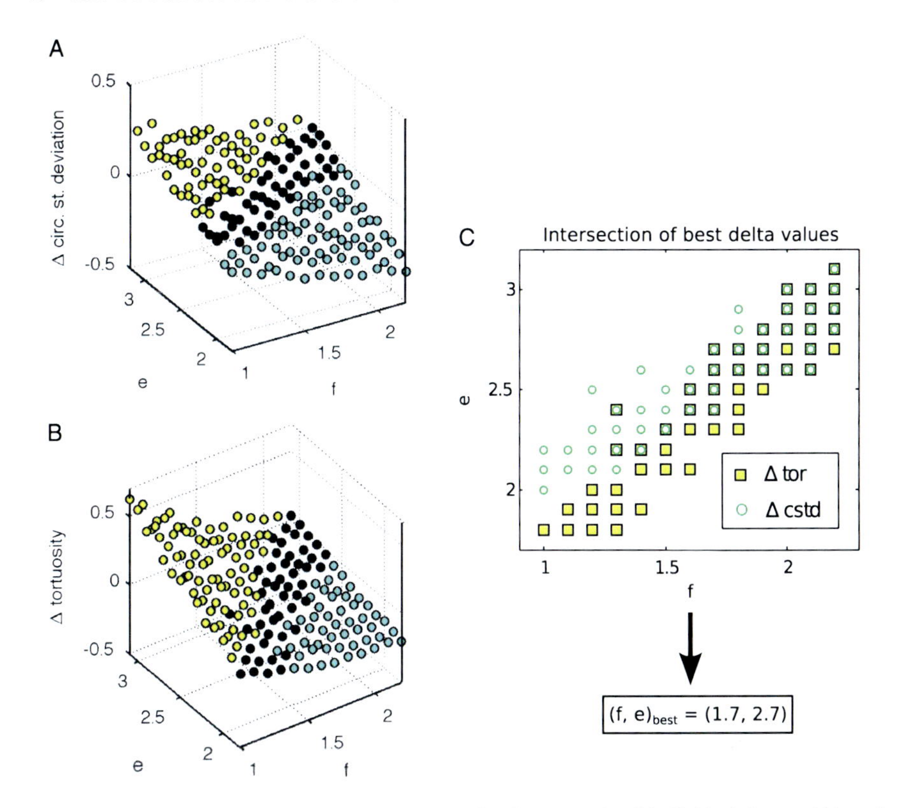

Fig. 73.3 Visualization of the inference process for the strength of individual forces. (**A**) and (**B**) The difference between the average of simulated observables (circular standard deviation and tortuosity respectively) for a single fish and their empirical value is plotted for each couple (f, e) of parameters. *The black central band* corresponds to the simulated quantities differing at most one standard deviation from the experimental values. (**C**) The couples of parameters (f, e) matching the black strip are reported in the same plane, square corresponding to the compatible values for tortuosity and circles for circular standard deviation. The couples of parameters corresponding to both a square and a circle constitute the set $(f, e)_{\cap}$, and a minimization on this intersection gave us $(f, e)_{best} = (1.7, 2.7)$

a set of parameters values for which simulations are compatible with empirical results. The final values of alignment and attraction resulting from the minimization process on the union of the compatible sets of parameters are $a = 4.2$ and $c = 3.1$. We do not show here the plots showing the variation of accuracy and tortuosity with the parameters (a, c), since they would not be particularly informative, and the visualization of the union set $(a, c)_{\cup}$ would be quite difficult due to the large number of subsets involved. Instead in Fig. 73.4 we show the match between our simulations and the data set from Ward's experiment. The simulations are in a good qualitative agreement with empirical data, reproducing quite closely the trends originally observed by Ward et al. These results suggest that alignment has a relevant role in the increasing of decision-making efficiency with group size. Notice that, despite circular standard deviation has been fitted to data only to retrieve the values of indi-

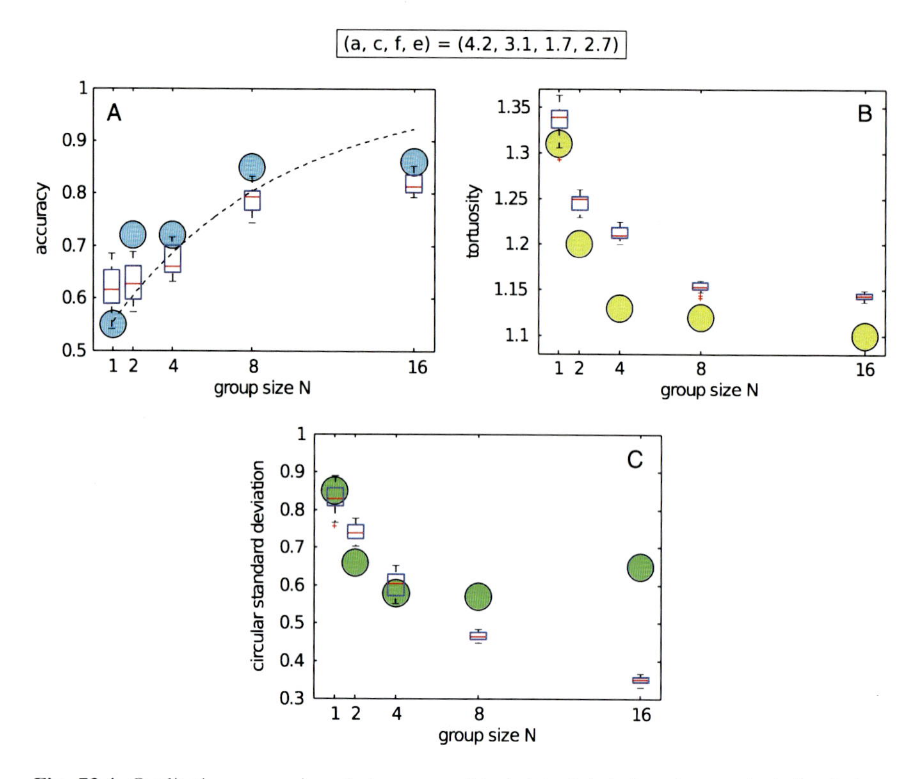

Fig. 73.4 Qualitative comparison between empirical data (*circles*) and numerical simulations (*boxes*). The model has been simulated with the parameters fixed to $(a, c, f, e) = (4.2, 3.1, 1.7, 2.7)$. Accuracy, tortuosity and circular standard deviation have been computed for each group size and plotted together with the corresponding experimental data. All the simulations are in a good qualitative agreement with empirical data, reproducing quite closely the trends originally observed by Ward et al. (**A**) Simulated accuracy is compatible with both empirical data and theoretical predictions at all group sizes. (**B**) Simulated tortuosity well resembles data for small group sizes, while at $N = 8$ and 16 decreasing is more difficult to achieve due to the trade-off between alignment and attraction in controlling this quantity. (**C**) Also the trend of circular standard deviation is reproduced, despite this observables has not been considered during the fitting of social interactions

vidual forces, the values of alignment and attraction obtained by fitting accuracy and tortuosity allow to also reproduce the trend of this third observable (Fig. 73.4(C)). Finally the robustness of the presented results have been tested by changing of the size of blind angle α from 0 to 180 degrees, showing no significant variation in the considered range.

We then tested the predator avoidance as a function of group size also in a more general set-up of open water. The test was performed by fixing the parameters according to the values obtained by fitting the Y-maze experiment, i.e. $(a, c, f, e) = (4.2, 3.1, 1.7, 2.7)$. The quantities that are relevant to assess the success in predator avoidance are the proportion of "safe" fish and the proportion of fish avoiding the predator due to social interaction only and not because they spot-

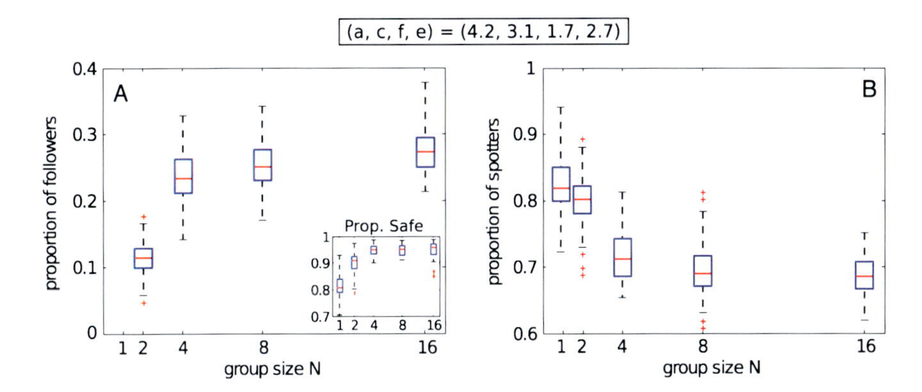

Fig. 73.5 Results of the simulations of the model in open water with $(a, c, f, e) = (4.2, 3.1, 1.7, 2.7)$. (**A**) The proportion of fish escaping due to social interactions only is an increasing function of group size, as well as the total proportion of safe fish (*inset*), meaning that the rules of motion fitted in the Y-maze make larger group more successful in predator avoidance also in different environment. (**B**) The proportion of spotting fish decreases with increasing group size, suggesting an important role of social interactions

ted it. The second quantity is particularly interesting, since it reveals how the actual relevance of social interactions relates with group size. In our simulations we found that both these observables are an increasing function of group size, (Fig. 73.5(A)), despite the surprising fact that the proportion of fish spotting the predator decreases with group size (Fig. 73.5(B)).

73.5 Conclusions

We have presented a first attempt to relate a collective decision-making process to the explicit rules of motion for fish movement. We have shown that a self-propelled particle model can reproduce the increasing speed and accuracy with group size. In doing so we inferred the explicit values of the parameters providing the best match to experimental data. Finally, we investigated the possibility of extending the obtained results to a more general setting by testing the fitted model in a different environment.

When defining the model we initially considered just one social force: the attraction towards the center of mass of the interacting neighbours. In a previous work on the same fish species (*Gambusia holbrooki*) as in the Ward et al. experiment, Herbert-Read et al. find only a weak role of alignment and a strong role for attraction while analyzing the rules of motion of groups of swimming in a square tank [18]. Furthermore, Strömbom has shown that a rich and complex range of behaviour can be achieved from only local attraction and a blind angle [16]. However, we found that it was not possible to fit data by mean of attraction alone. A similar situation was found when we tried to fit parameters by introducing only alignment. Only the combined contribution of these two social forces allowed us to fit the data.

It is therefore clear that fitting a basic self-propelled particle model requires both the social forces of alignment and attraction.

Not only is alignment important, but it is actually more important than attraction in information transfer mediating. This result is consistent with the suggestion that alignment with neighbors can allow information to be transmitted rapidly through fish schools [19]. At first sight, it appears however to be in contrast with experimental observations on the same fish [18]. There are however fundamental differences in the two analyzed settings. Herbert-Read and co-workers inferred the rules of motion for groups of mosquitofish swimming freely in a square tank after a period of acquaintance, where no decision making nor predation risk are involved. In such a circumstance it is reasonable to assume that the fish aim primarily to maintain cohesion. Attraction to conspecifics is then the most relevant force to achieve this goal by, for example, preserving a common speed. However, in a situation where fish have to collectively decide between the two arms of the tank, one of which is occupied by the predator replica, the goals of the fish change. Furthermore, the experimental set up adopted by Ward et al. does not allow fish to spread too much, already ensuring that the group will be quite packed. In this context, one explanation of the importance of alignment is that a sharp direction change provides a social cue about which of two available branches to take. A clear turning by the spotting fish communicates to conspecifics that it has some extra knowledge about which path to take, and the easiest way for other fish to be sure to take the safe branch of the tank is to assume its same direction. That is to align with the turning fish.

It is possible that our results and interpretation are strictly linked to the modeling choices we made in fitting data. The SPP approach introduces a strict distinction between forces and makes it necessary to choose which ones to introduce for representing interactions between fish. Our decision of prioritizing alignment and attraction over other social forces is arbitrary, and further work could be done by exploring if it would be possible to fit Ward's data by using different forces within an SPP model or even by trying different models. For example, introducing repulsion along with attraction instead of alignment could have lead to results more compatible with observations emphasizing the role of attraction [18]. Models have shown that highly polarized groups can be obtained from attraction and repulsion without involving alignment [16, 20]. Another alternative approach would be to fit different models than the ones involving social forces acting on self-propelled particles. Good candidates could be models where individual rules of motion are based on retinal information processing [21] or on simple heuristics integrating information about the surrounding environment [22]. Models involving speed variation, which appears to be an important feature characterizing fish motion [18, 23], could also provide a better fit. An interesting question in this direction is whether the rules of motion found by Herbert-Read and co-workers alone can explain the main features of the decision-making process observed by Ward et al. This could be investigated by defining a data-driven model based on the rules of motion found in [18], assigning a spotting probability to individual fish and simulating it in a Y-maze like environment. Despite the preliminary nature of the work presented here, it is the first attempt we know of to relate the empirical outcome of a collective decision-making process to the explicit rules of motion of animals. There is much additional research which can be done in this area.

The last step in our work has been to ask about the possible generalizations of our model. By simulating the model in open water with the parameters fitted from the Y-maze experiment, we were able to observe that both the proportion of "safe" fish and the proportion of fish escaping the predator due to interactions are an increasing function of group size. These results support the hypothesis that the decision-making accuracy of groups is typically greater than that of the single group members [2]. Our analysis suggests that the interactions involved in the decision-making process in the Y-shaped tank should be the same social forces leading to a successful predator avoidance also in open water, and therefore they might turn out to be environment-independent.

A final intriguing hypothesis emerging from our analysis of the model in open water is that increased accuracy with group size may be accompanied with a decreased probability of individuals detecting predators. We found that the proportion of fish spotting the predator decreases with group size, while the proportion of safe fish increases. This result is consistent with an apparent decreased vigilance with group size. It is however inconsistent with the idea that it is "many eyes" which makes the group safer. It is rather fewer eyes and more efficient transfer of information that allows groups to outperform individuals.

References

1. Dall SRX, Giraldeau LA, Olsson O, McNamara JM, Stephens DW (2005) Information and its use by animals in evolutionary ecology. Trends Ecol Evol 20(4):187–193
2. King AJ, Cowlishaw G (2007) When to use social information: the advantage of large group size in individual decision making. Biol Lett 3(2):137–139
3. Couzin ID (2009) Collective cognition in animal groups. Trends Cogn Sci 13(1):36–43
4. Treherne J, Foster W (1980) The effects of group size on predator avoidance in a marine insect. Anim Behav 28(4):1119–1122
5. Lima SL (1995) Back to the basics of anti-predatory vigilance: the group-size effect. Anim Behav 49(1):11–20
6. Ward AJW, Herbert-Read JE, Sumpter DJT, Krause J (2011) Fast and accurate decisions through collective vigilance in fish shoals. Proc Natl Acad Sci USA 108(6):2312–2315
7. Sumpter D, Buhl J, Biro D, Couzin I (2008) Information transfer in moving animal groups. Theory Biosci 127(2):177–186
8. Sumpter DJT, Krause J, James R, Couzin ID, Ward AJW (2008) Consensus decision making by fish. Curr Biol 18(22):1773–1777
9. Ward AJW, Sumpter DJT, Couzin ID, Hart PJB, Krause J (2008) Quorum decision-making facilitates information transfer in fish shoals. Proc Natl Acad Sci USA 105(19):6948–6953
10. Vicsek T, András Czirók EBJ, Cohen I (1995) Novel type of phase transition in a system of self-driven particles. Phys Rev Lett 75:1226
11. Czirók A, Vicsek T (2000) Collective behavior of interacting self-propelled particles. Physica A 281:17–29
12. Czirok A, Barabasi A, Vicsek T (1999) Collective motion of self-propelled particles: kinetic phase transition in one dimension. Phys Rev Lett 82(1):209–212
13. Couzin ID, Krause J, James R, Ruxton GD, Franks NR (2002) Collective memory and spatial sorting in animal groups. J Theor Biol 218(1):1–11

14. Vicsek T, Zafeiris A (2012) Collective motion. Phys Rep. doi:10.1016/j.physrep.2012.03.004
15. Mann RP (2011) Bayesian inference for identifying interaction rules in moving animal groups. PLoS ONE 6(8):e22827
16. Strombom D (2011) Collective motion from local attraction. J Theor Biol 283(1):145–151
17. Rountree RA, Sedberry GR (2009) A theoretical model of shoaling behavior based on a consideration of patterns of overlap among the visual fields of individual members. Acta Ethol 12(2):61–70
18. Herbert-Read JE, Perna A, Mann RP, Schaerf TM, Sumpter DJT, Ward AJW (2011) Inferring the rules of interaction of shoaling fish. Proc Natl Acad Sci USA 108(46):18726–18731
19. Radakov D (1973) Schooling in the ecology of fish. Wiley, New York
20. Romey WL (1996) Individual differences make a difference in the trajectories of simulated schools of fish. Ecol Model 92(1):65–77
21. Lemasson B, Anderson J, Goodwin R (2009) Collective motion in animal groups from a neurobiological perspective: the adaptive benefits of dynamic sensory loads and selective attention. J Theor Biol 261(4):501–510
22. Moussaïd M, Guillot EG, Moreau M, Fehrenbach J, Chabiron O, Lemercier S, Pettré J, Appert-Rolland C, Degond P, Theraulaz G (2012) Traffic instabilities in self-organized pedestrian crowds. PLoS Comput Biol 8(3):e1002442
23. Hemelrijk CK, Hildenbrandt H, Reinders J, Stamhuis EJ (2010) Emergence of oblong school shape: models and empirical data of fish. Ethology 116(11):1099–1112

Chapter 74
Self-organized Flocking with Conflicting Goal Directions

E. Ferrante, W. Sun, A.E. Turgut, M. Dorigo, M. Birattari, and T. Wenseleers

In flocking, a large number of individuals move cohesively in a common direction. Many examples can be found in nature: from simple organisms such as crickets and locusts to more complex ones such as birds, fish and quadrupeds.

Reynolds was the first to propose a computational model of flocking [1]. The behavior of each individual is made of three parts: separation, cohesion, alignment. Separation means that the individual moves away from its neighbors. Cohesion means that the individual stays close to its neighbors. Alignment means that the individual matches the velocity of its neighbors.

This paper studies flocking in the robotics setting. One of the earliest attempt to realize flocking in robotics was done by Matariç [2]. She created a set of "basic behaviors": safe-wandering, aggregation, dispersion and homing. Turgut et al. [3] implemented flocking on real robots using two behaviors: proximal control and alignment control. Proximal control combines the separation and cohesion components and was realized using the framework of artificial physics as done by Spears et al. [4]. Alignment control is realized through a novel sensing system which they called virtual heading sensor.

More recent research in biology showed that only a small group of informed individuals who have information about a desired goal direction is sufficient lead the whole group in that direction [5]. These leaders are implicit, in the sense that the rest of the swarm is not aware of their presence. Inspired by [5], Çelikkanat [6] extended [3] by providing the goal direction to only a proportion of the robots, which they referred to as informed robots. They found that, similarly to [5], only a minority of informed robots is enough to guide the whole group.

In this paper, we study flocking of a swarm of robots when information about two distinct goal directions is present in the swarm. This case can be instantiated

E. Ferrante (✉) · W. Sun · A.E. Turgut · M. Dorigo · M. Birattari
Université Libre de Bruxelles, Brussels, Belgium

E. Ferrante · A.E. Turgut · T. Wenseleers
Katholieke Universiteit Leuven, Leuven, Belgium

T. Gilbert et al. (eds.), *Proceedings of the European Conference on Complex Systems 2012*, Springer Proceedings in Complexity, DOI 10.1007/978-3-319-00395-5_74,

in many practical examples: a swarm that has to go in one direction while avoiding an obstacle; a swarm that has to avoid a dangerous locations while going to a target location; or a swarm that has to execute, in parallel, two tasks in two different locations. In general, we can identify three different macroscopic objectives that we might want to attain: (a) a swarm that moves to the average direction among the two (for example to avoid the obstacle) without splitting; (b) a swarm that selects the most important of the two directions (for example the direction to avoid danger) and follows it without splitting; (c) a swarm that splits in a controlled fashion in the two directions (for example, in the parallel task execution case).

This paper proposes a solution for the first objective: a method for moving the swarm in the average between the two conflicting goal directions. We show that this objective can be attained using a similar methodology as the one proposed in [3] and [6]. We execute systematic experiments using a realistic robotics simulator. In the experiments, a small proportion of robots is informed about one goal direction, another small proportion about the other goal direction, and the rest of the swarm is non-informed. We study the effect of what we believe are the critical parameters: the overall proportion of informed robots, the difference between the size of the two groups of informed robots and the difference between the two goal direction.

74.1 Method

We use a similar method as the one used in [3]. At each time step, a flocking control vector is calculated as $\mathbf{f} = \alpha\mathbf{p} + \beta\mathbf{h} + \gamma\mathbf{g_i}$, where $\mathbf{p}$ denotes the proximal control vector, $\mathbf{h}$ denotes the alignment control vector, $\mathbf{g_i}$ with $i = \{1, 2\}$ denotes the vector that indicates the two goal directions denoted with θ_1 and θ_2. For informed robots $\gamma = 1$, whereas $\gamma = 0$ for uninformed robots. The values of the other parameters are fixed to $\alpha = 1$, $\beta = 4$ for all the robots.

Using proximal control, each robot keeps a desired distance (d_{des}) with its neighbors to avoid collisions and to achieve cohesion. To do this, the robot only needs to know the relative distance d_i and bearing ϕ_i of each neighbor i. The formula $p_i(d_i) = 12\epsilon[\frac{{d_{des}}^{12}}{d_i^{13}} - \frac{{d_{des}}^{6}}{d_i^{7}}]$, based on the Lennard-Jones potential [7], encodes attraction and repulsion rules. If the actual distance d_i is smaller than $d_{des} = 0.6$ m, $p_i(d_i)$ is negative and the rule is repulsive, otherwise it is attractive. The parameter $\epsilon = 0.5$ controls the strength of the attraction/repulsion rule. After computing $p_i(d_i)$ for each neighbor, the proximal control vector is computed as $\mathbf{p} = \sum_{i=1}^{k} p_i(d_i)e^{j\phi_i}$, where k is the number of neighbors.

Using alignment control, each robot aligns to the average orientation of its neighbors. Each robot detects its own orientation θ_0 and sends it to its neighbors. The robot receives an angle θ_i from its ith neighbor that represent the neighbor's orientation. In this way, it is as if the robot can sense the orientation of its neighbors. The robot then calculates the alignment control vector as: $\mathbf{h} = \frac{\sum_{i=0}^{k} e^{j\theta_i}}{\| \sum_{i=0}^{k} e^{j\theta_i} \|}$, where $\| \cdot \|$ denotes the norm of a vector and k denotes the number of neighbors. Given the flocking control vector **f**, the robot's forward and angular speed are computed as the projection of **f** on x-axis and y-axis of the robot as in [8]. The forward speed u is directly proportional to x component of force, and the angular speed ω is directly proportional to y component of force: $u = K_1 f_x, \omega = K_2 f_y$. $K_1 = 2$ and $K_2 = 2$ are the forward and angular gains, respectively. We also limit forward speed and angular speed to $u \in [0, U_{\max}]$ and $w \in [-\Omega_{\max}, \Omega_{\max}]$, with $U = 20$ cm/s and $\Omega_{\max} = \frac{\pi}{2}$ rad/s.

We use the simulated version of the foot-bot robot developed in [9]. We use the following sensors and actuators: (i) A light sensor able to detect the bearing of a distant light source, that is used by the robot to measure its orientation; (ii) A range and bearing sensing and communication device (RAB), that is used by the robot to obtain range and bearing of the neighbors in proximal control and to communicate its orientation in alignment control; (iii) Two wheels actuators that is used by the robot to move. For motion, we use the differential drive model as in [3] to convert the forward speed u and the angular speed ω into the linear speed of left and right wheels: $N_L = u + \frac{\omega}{2}l, N_R = u - \frac{\omega}{2}l$, where $l = 5$ cm is the distance of two wheels.

74.2 Experiments

We use the ARGoS simulator developed in [10]. It is a modular, multi-engine, open-source simulator for heterogeneous swarm robotics.

74.2.1 Experimental Setup

A swarm of $N = 100$ foot-bots is placed, with random orientations, in an arena of 12×12 m. A remote light source is positioned far from the robots to provide common reference frame. A proportion of ρ_1, ρ_2 robots are informed about the goal direction θ_1, θ_2, respectively. Thus, $N_A = N\rho_1$ is the number of robots informed of goal direction θ_1 and $N_B = N\rho_2$ is the number of robots informed of goal direction θ_2.

We study the capability of the swarm to follow the theoretical average direction in different parameter conditions. In particular, we are interested in determining (i) the impact of the total proportion of informed robots, (ii) the impact of the difference between N_A and N_B and (iii) the impact of the difference between two goal directions $\theta_2 - \theta_1$. For the second case, we classify the experiments in three sets: no difference between N_A and N_B ($N_A = N_B$), small difference ($N_A - N_B = 2$) and high difference ($N_A - N_B > 2$). To study the impact of $\theta_2 - \theta_1$ and to reduce the parameter space, we fix the $\theta_1 = 0$, and we only vary $\theta_2 \in \{10, 20, \ldots, 170, 179, 180\}$. We consider the following values of ρ_1 and ρ_2 (proportion of informed robots): $\{(0.01, 0.01), (0.01, 0.05), (0.02, 0.04), (0.1, 0.1), (0.09, 0.11), (0.01, 0.19)\}$. Each experiment is repeated $R = 100$ times and lasts $T = 500$ simulated seconds.

74.2.2 Metrics

In this paper we are interested in having a swarm that is cohesive and moves along the theoretical average direction between the two given goal directions. We use two metrics to evaluate the degree of attainment of these two objective: split probability and average group direction.

Split Probability To measure split probability, we first compute the number of groups g at the end of each experiment as suggested in [5]. g_i denotes the number of groups at the end of the ith run. After executing R independent runs, the split probability is calculated as: $p = \frac{\sum_{i=1}^{R}(\min\{2, g_i\}-1)}{R}$.

Average Direction The average group direction is simply the vectorial average of all robots orientations: $\bar{\theta} = \angle \sum_{i=1}^{N} e^{j\theta_i}$. We plot the average group direction against the theoretical average direction, that takes into account the number of robots informed about each goal direction: $\hat{\theta} = \angle(N_A e^{j\theta_1} + N_B e^{j\theta_2})$.

74.3 Results

According to the results (not shown), no matter the value of N_A, N_B and $|\theta_2 - \theta_1|$, the swarm does not split.

We now report and discuss the average group direction in the three cases: $N_A = N_B$ (no difference), $N_B - N_A = 2$ (small difference) and $N_B - N_A > 2$ (large difference).

No Difference ($N_A = N_B$) Figure 74.1(left column) shows that, in most of the cases, the average direction strictly follows the theoretical average direction $\hat{\theta}$. The most noticeable exception is the $\rho_1 = \rho_2 = 0.01$ case (Fig. 74.1(a)). In this case, the robots are not able to follow the theoretical average direction when $\theta_2 - \theta_1$ is too high, and the distribution of the average group direction has high standard deviation. This can be explained by the fact that the total proportion of informed robots is not high enough to drive the swarm in the desired direction, as argued in [6]. In fact, Fig. 74.1(d) shows that, with higher total proportion of informed robots, the swarm follows the theoretical average direction more precisely for most configurations of $\theta_2 - \theta_1$.

Small Difference ($N_B - N_A = 2$) Figure 74.1(central column) shows that, no matter $\theta_2 - \theta_1$, the group follows the theoretical average direction. When $\theta_2 - \theta_1$ is high, larger total proportion of informed robots (Fig. 74.1(e)) correspond to lower spread in the distribution. $\theta_2 - \theta_1 = 180$ degrees is a special case as it presents many outliers. Otherwise, the swarm is able to follow the theoretical average even for $\theta_2 - \theta_1 = 179$ degrees.

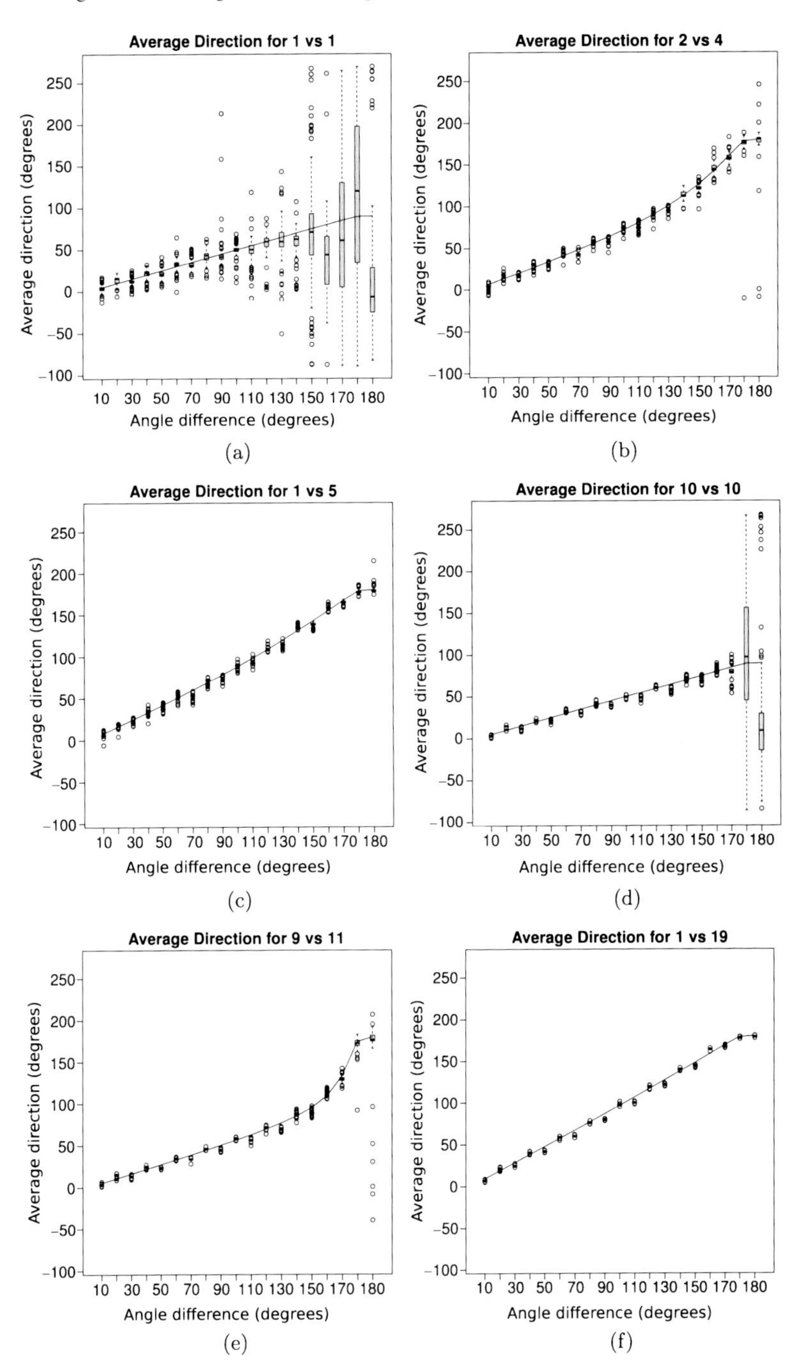

Fig. 74.1 Distribution of average for $N_A = N_B$ (*left column*), $N_B - N_A = 2$ (*central column*) and $N_B - N_A > 2$ (*right column*). *Solid black line* represents the theoretical average direction in total informed robots. The above and nether edges of the box indicate first and third quartiles. *The black center line* indicates the median for each dataset

Large Difference ($N_B - N_A > 2$) Figure 74.1(right column) shows that the theoretical average direction represents the goal direction that is known by the majority. Here, a precise following of the theoretical average direction always takes place, even when $\theta_2 - \theta_1 = 180$. Additionally, when the total proportion is small, there are a few outliers (Fig. 74.1(c)). These outliers are otherwise not present for higher total proportion of informed (Fig. 74.1(f)).

74.4 Conclusion and Future Work

We studied flocking of swarm of mobile robots where information about two conflicting goal directions is present in the swarm. In this setting, we believe three possible macroscopic objectives can be identified: (a) making the swarm follow the average direction between the two without splitting; (b) making the swarm follow one (the most important) direction between the two without splitting; (c) making the swarm split in a controlled fashion and allocate to the two goal directions. In this paper, we propose a method based on [3, 6] and we show that it is capable to attain the first objective. This result presents some difference with the results in [5], where they also studied the conflicting goal direction case but showed that the resulting average direction strongly depends on the difference between the two goal directions. This lack of agreement might be due either to the different methodology or to the different level of detail in the simulations.

This work open many doors for possible future extension. In fact, in a parallel on-going work [11], we are studying how to attain the second objective. This objective was attained by using a special communication strategy called self-adaptive communication strategy (SCS). However, in that work we assumed that the priority of the goal directions are known by the informed robots.

We believe that information transfer and communication are the key to attain the desired macroscopic objectives in self-organized flocking, even in presence of conflicting goal direction. Future work will deal with how to design direct or indirect communication strategies to make the swarm split in a controlled fashion or to deal with the second objective under the case where the priority of the goal directions are not known in the swarm.

Acknowledgements This work was supported by the European Union (ERC Advanced Grant "E-SWARM: Engineering Swarm Intelligence Systems" (contract 246939) and FET project ASCENS) and by the Vlaanderen Research Foundation Flanders (H2Swarm project). Mauro Birattari, and Marco Dorigo acknowledge support from the F.R.S.-FNRS of Belgium's French Community.

References

1. Reynolds C (1987) Flocks, herds and schools: a distributed behavioral model. In: Stone MC (ed) Proc of the 14th annual conference on computer graphics and interactive techniques, SIGGRAPH'87. ACM Press, New York, pp 25–34

2. Mataric MJ (1994) Interaction and intelligent behavior
3. Turgut AE, Çelikkanat H, Gökçe F, Şahin E (2008) Self-organized flocking in mobile robot swarms. Swarm Intell 2(2):97–120
4. Spears WM, Spears DF, Hamann JC, Heil R (2004) Distributed, physics-based control of swarms of vehicles. Auton Robots 17:137–162
5. Couzin ID, Krause J, Franks NR, Levin SA (2005) Effective leadership and decision-making in animal groups on the move. Nature 433:513–516
6. Çelikkanat H, Turgut A, Şahin E (2008) Guiding a robot flock via informed robots. In: Asama H, Kurokawa H, Ota J, Sekiyama K (eds) Distributed autonomous robotic systems (DARS 2008). Springer, Berlin, pp 215–225
7. Suranga Hettiarachchi WMS (2009) Distributed adaptive swarm for obstacle avoidance. Int J Intell Comput Cybern 2(4):644–671
8. Ferrante E, Turgut AE, Huepe C, Stranieri A, Pinciroli C, Dorigo M (2012) Self-organized flocking with a mobile robot swarm: a novel motion control method. Adapt Behav. doi:10.1177/1059712312462248
9. Bonani M, Longchamp V, Magnenat S, Rétornaz P, Burnier D, Roulet G, Vaussard F, Bleuler H, Mondada F (2010) The marxbot, a miniature mobile robot opening new perspectives for the collectiverobotic research. In: Proceedings of the IEEE/RSJ international conference on intelligent robots and systems (IROS). IEEE Comput Soc, Washington, pp 4187–4193
10. Pinciroli C, Trianni V, O'Grady R, Pini G, Brutschy A, Brambilla M, Mathews N, Ferrante E, Di Caro G, Ducatelle F, Birattari M, Gambardella LM, Dorigo M (2012) ARGoS: a modular, parallel, multi-engine simulator for multi-robot systems. Swarm Intell 6(4). doi:10.1007/s11721-012-0072-5
11. Ferrante E, Turgut AE, Stranieri A, Pinciroli C, Birattari M, Dorigo M (2011) A self-adaptive communication strategy for flocking in stationary and non-stationary environments. IRIDIA Tech-rep IridiaTr2012-002

Chapter 75
Garden Ants Lasius Niger Perceive a Rotating Landmark

Mai Minoura, Kohei Sonoda, Tomoko Sakiyama, and Yukio-P. Gunji

Abstract Garden ant (*Lasius niger*) workers are well known to use visual landmarks for navigation, which are gained by stopping to memory the view like a snap shot, as comparing with the view in front of them. Here, we researched whether ants could perceive rotating landmark or not. We let ants to walk on the rotating road with a sheet attached trail pheromone. Since we discovered the answer is yes, we tried to approach the reason as excluding the confliction between two types of information, pheromone and visual landmark.

Keywords Garden ant foraging · Navigation · Rotating landmark · Flexible landscape information · Pheromone information

75.1 Introduction

Animals navigate themselves in using various information, and that can encourage efficient exploration and exploitation. Since foraging and homing behavior of ants based on landmark navigation is taken as a minimal mechanism for navigation [1, 2], the view-dependent template models are proposed, compared to the field observation [3, 4]. Navigation toolkits of desert ants, *Cataglyphis* which is investigated as a model animal for navigation, consists of skylight compass, path integration, view-dependent ways of place recognizing and landmarks [5]. Ants use various combinations of toolkits, which is ubiquitous and inevitable strategy for animal navigation, e.g., skylight cannot be used as a kind of compass till it is linked with a particular criteria such as an ant's familiar view.

Therefore, different modes of information are confronted, negotiated and knitted together to produce coherent motor outputs [6], which could accomplish flexible use

M. Minoura (✉) · T. Sakiyama · Y.-P. Gunji
Department of Earth and Planetary Sciences, Kobe University, Rokkodai-cho 1-1, Nada, Kobe 657-8501, Japan
e-mail: maichi_mino@yahoo.co.jp

K. Sonoda
Information and Society Research Division, National Institute of Informatics, Hitotsubashi 2-1-2, Chiyoda, Tokyo 101-8430, Japan

T. Gilbert et al. (eds.), *Proceedings of the European Conference on Complex Systems 2012*, Springer Proceedings in Complexity, DOI 10.1007/978-3-319-00395-5_75,

of information such as view and/or landmark. Ants acquire and use information in navigation in context-dependent way [1] (Collett et al. 1998). Moving vector is used dependent on landmark. Cataglyphis exhibits considerable flexibility in coupling and decoupling different modules and referential systems to and from each other [7]. Garden ant, *Lasius* also use hybrid flexible combination of visual and chemical cues in navigation [8, 9].

The question arises regarding flexibility of landscape information. The first one is, how flexibly landscape view is used in navigation. The question whether the insects learn landscape image continuously or at particular moments can still not be definitely answered. This seems to be related to flexibility and ambiguity of landscape information. The second question is, how the flexible landscape information is used in a definite way at each moment. Since the flexibility sometimes reveals vagueness, It is possible for insects to confront indecision in navigation frequently. Insects, however, usually make a decision to walk even in a conflict condition with respect to different mode of information. In other words, vagueness of landscape information is efficiently used as navigation tool.

To answer the two questions, we conducted a mobile-landscape experiment for garden ants, *Lasius niger*, to give rise to dynamic conflict condition between visual and odor information. Foraging box containing a nest box is connected to a lane on a turn table which is linked to a feeder area. A lane is secreted by trail pheromone and all area of turn table is covered by a sheet of paper. Each ant walks through a lane to the feeder in straight forward. After feeding each ant turn back to the lane. Soon after he or she enters the lane, it is rotated by an electric motor. Because ants appear to form associative links between specific figure and the visual background [10], the rotation increase the vagueness of landscape image. Moreover, the rotation dynamically increase the degree of conflict between pheromone trail and visual cues. We here show that ants can use landscape image in a definite way, and switches the preferred information at a critical point of visual and odor information.

75.2 Materials and Methods

75.2.1 Study Species

We studied three *Lasius niger* colonies collected on the University of Kobe campus. The colonies were queenless with 900–1200 workers. Queenless colonies forage, make trails and are frequently used in foraging experiments. Experimental colonies were housed in plastic foraging boxes (35 × 25 × 4 cm) containing an antistatic plastic nest box (12 × 8 × 1 cm) covered with clear red sheets. The walls of each plastic boxes were coated with Fluon® to prevent ants from escaping. Colonies were kept at room temperature (21–22 °C). We supplied the colonies with honey and water ad libitum.

75.2.2 Experimental Setup

Colonies were deprived of the food for 5–6 days prior to testing to ensure that foragers were motivated to forage and recruit nest mates to a food. The food was sucrose solution syrup (50 % w/w) placed on a feed plat (2 × 8 cm). Foraging boxes were connected to a feed plat via a plastic bridge and a foraging trail (a lane or a turn table). The foraging box, the foraging trail and the feed plat were arranged linearly. The lane was 3 cm wide and 30 cm long and covered with a same size tracing paper to get a pheromone sheet. The turning table was 30 cm in diameter and covered with a paper. We attached the pheromone sheet (3 × 30 cm) to the turning table along the diameter (Fig. 75.1). The turning table was rotated by a synchronous motor (typeD-5N, NIDEC SERVO CORPORATION). The experiments were performed in a room that had artificial room lights. The room contained different 3 kinds of lab equipment and furniture and so provided visual landmarks of different sizes and colors. Experiments were recorded using a video camera (GZ-MG740, Victor). The camera was placed horizontally 80 cm above the center of the turn table. Angle data were acquired using avidemux2 and the GIMP toolbox.

Experiments were designed to conflict between pheromone information and visual one by rotating a pheromone sheet. Ants were expected to turn when they perceived that they headed in the opposite direction of the nest box on the rotating pheromone sheet. 90 ants were used in experiments and 30 ants were used in each experiment. Rotation speed of the turning table were 3, 1 and 0 r/min in experiment 1, 2 and 3, respectively. There was no significant difference in rotational directions and hence we fixed clockwise rotation.

Before experiments, to obtain a pheromone sheet, a colony was connected to a lane covered with a tracing sheet (Fig. 75.1). The colony was allowed to explore the lane and access to the turning table on a platform at the end of the lane for 20 min. Then the lane was replaced with a turn table covered with a circle sheet and the pheromone sheet was attached on the desk. The pheromone sheet corresponded to the colony used in each experiment. The same pheromone sheet was used for 30 min in 10 trials. The circle sheet under the pheromone sheet was changed to new one in each trial.

In experiments, ants were used one-by-one in each trial. An ant was allowed to go through the pheromone sheet on a turn table via bridge and access the food. The desk did not rotate in this time. After the ant climbed the bridge, it was separated from a foraging box to prevent the other ants from invading. The turn table started rotating when the ant got onto it after feeding. The trial was concluded when the ant reached an edge of the desk on the nest side.

For analysis, we defined the starting end (S) where the ants got onto the pheromone sheet, the goal end (G) that was the opposite side of S on the sheet, the feed side (F) and the nest side (N) (Fig. 75.1). Before the desk rotated, SG vector was coincident with FN vector. We measured the directions (the nest side or the feed side) of the ants and the pheromone sheet (SG vector) every 0.5 s. We also measured the angle between the FN vector and the SG vector when the ants reached the farthest points (the ant's turning points) from S in experiment 1.

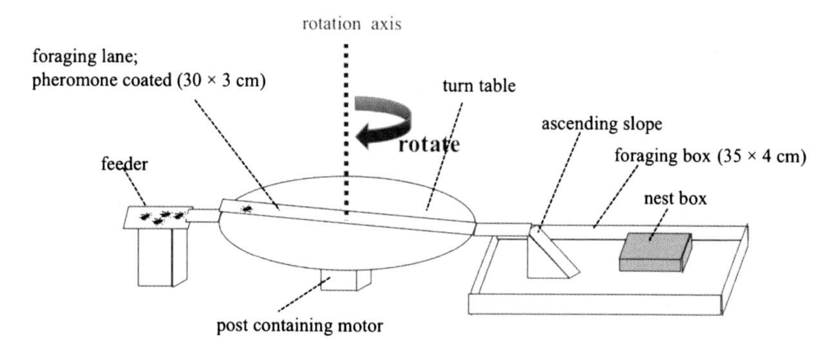

Fig. 75.1 Schematic diagram for apparatus

We observed the nest side direction rate (R) for the ants and the pheromone sheets from the directional data. We also observed, under the condition that the direction of the pheromone sheet was feed side, the conditional nest side direction rate (CR) for the ants. The sets of R were compared using Wilcox rank sum test with continuity correction in each experiment. The sets of CR were compared using Kruskal-Wallis rank sum test. We used Steel-Dwass post-hoc comparisons to explore significant main factors. Circularly distributed data were treated via circular statistics. Mean orientation vectors are given in polar coordinates with the angle ϕ and the length r describing mean orientation of the pheromone sheet at turning points and the amount of scatter about the mean, respectively. The mean and the variance were given in numerical value with the maximum likelihood method, using a circular normal distribution. The Rayleigh test was used to check for uniformity. We use an alpha level 0.0.

75.3 Results

In experiment 1, at 3 r/min, the distribution of the nest side direction rate (R) of the ants was different from that of the pheromone sheet (SG vector) ($W = 802$, $p < 0.0001$; Fig. 75.2(a)). The angles at turning points were not distributed randomly over the entire 360° range ($\phi = 98.5 \pm 29.4°$, $r = 0.881$, $p < 0.0001$; Fig. 75.2(c)). In experiment 2, at 1 r/min, there was no significant difference between the distribution of R of the ants and that of the pheromone sheet ($W = 416$, $p = 0.602$ ns; Fig. 75.2(a)). In experiment 3, at 0 r/min, the distribution of R of the ants was different from that of the pheromone sheet ($W = 345$, $p < 0.01$; Fig. 75.2(a)).

To examine how much the ants conformed to the pheromone in homing under the condition that the pheromone sheet pointed the feed side, we compared the conditional nest side direction rate (CR) for the three experiments. There was a significant difference among the experiments ($\chi_2 = 56.5$, $p < 0.0001$; Fig. 75.2(b)). With the Steel-Dwass post-hoc comparisons, we observed that CR of EXP 1 was greater than

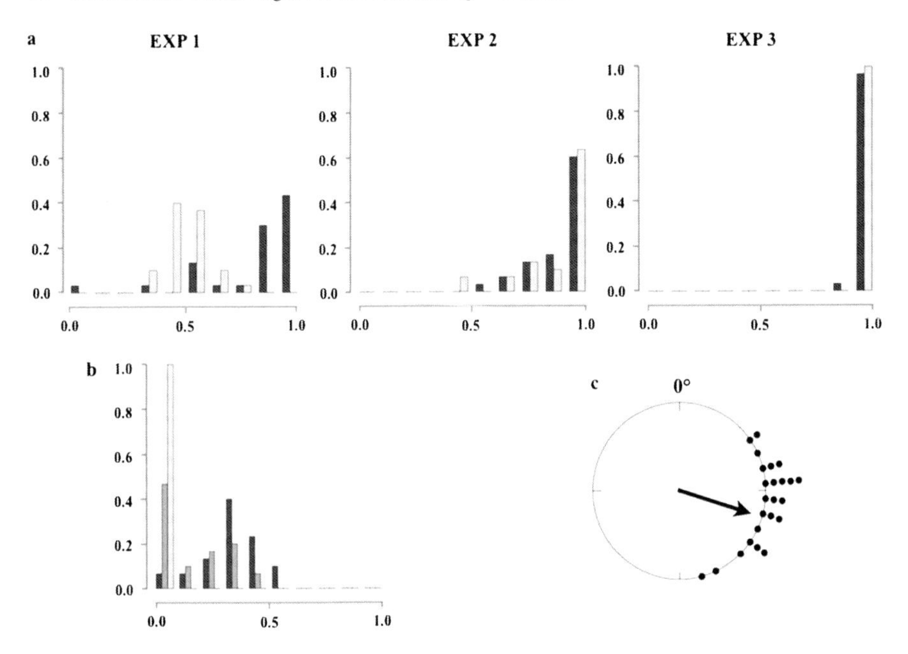

Fig. 75.2 Results: (**a**) Distributions of the nest side direction rates for the ants (*black*) and the pheromone (*white*). (**b**) Distributions of the conditional nest side direction rates for the ants, under the conditions that the pheromone sheet pointed the feed side, in EXP 1 (*black*), EXP 2 (*gray*) and EXP 3 (*white*). (**c**) Angles of the pheromone sheet when the ants turned are given relative to the direction of the feed-nest line

that of EXP 2 ($t = 4.26$, $p < 0.0001$; Fig. 75.2(b)) and that of EXP 3 ($t = 6.93$, $p < 0.0001$; Fig. 75.2(b)) and that of EXP 2 was also greater than that of EXP 3 ($t = 5.11$, $p < 0.0001$; Fig. 75.2(b)).

In this work, the adjustment was considerably quick and correct. In the experiment, flexibly landscape view and conflicts between pheromone and visual information occurred when the desk's rotational angle became 90°. The ant's adjustments (U-turn) (Fig. 75.3) were observed around 90°. This result shows that ants observed conflicting between the information and immediately changed the priority between them.

75.4 Discussion

Adjustments between two information in navigation are not restricted in ant's behavior. Presumably, these phenomena will be observed in a lot of animals. These adjustments are based on distinguishing two different informational types (such as social and private information) while confusing them. This viewpoint is very important to consider intelligence in animals.

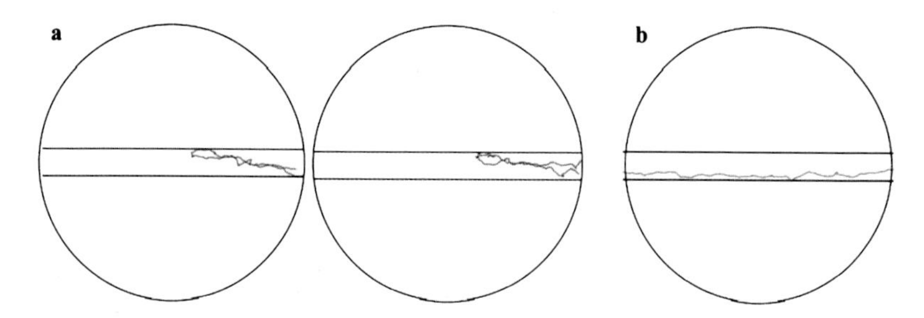

Fig. 75.3 Path Records: Actual path records of ants in (**a**) experiment 1 (3 r/min), and (**b**) experiment 3 (0 r/min)

References

1. Collett M, Collett TS, Bisch S, Wehner R (1998) Local and global vectors in desert ant navigation. Nature 394:269–272
2. Collett M, Collett TS, Wehner R (1999) Calibration of vector navigation in desert ants. Curr Biol 9:1031–1034
3. Cartwright BA, Collett TS (1983) Landmark learning in bees: experiments and models. J Comp Physiol 151:521–543
4. Basten K, Mallot HA (2010) Simulated visual homing in desert ant natural environments: efficiency of skyline cues. Biol Cybern 102:413–425
5. Wehner R (2003) Desert ant navigation: how miniature brains solve complex tasks. J Comp Physiol A 189:579–588
6. Zeil J (2012) Visual homing: an insect perspective. Neurobiology. doi:10.1016/j.conb.2011.12.008
7. Collett TS, Collett M, Wehner R (2001) The guidance of desert ants by extended landmarks. J Exp Biol 204:1635–1639
8. Aron S, Beckers R, Deneubourg JL, Pasteels JM (1993) Memory and chemical communication in the orientation of two mass-recruiting ant species. Insectes Soc 40:369–380
9. Grüter C, Czaczkes T, Ratnieks FL (2011) Decision making in ant foragers (Lasius niger) facing conflicting private and social information. Behav Ecol Sociobiol 65:141–148
10. Graham P, Durier V, Collett TS (2001) The binding and recall of snapshot memories in wood ants (*Formica rufa* L.). J Exp Biol 207:393–398

Chapter 76
In vivo, in silico, in machina: Ants and Robots Balance Memory and Communication to Collectively Exploit Information

Melanie E. Moses, Kenneth Letendre, Joshua P. Hecker, and Tatiana P. Flanagan

Abstract Ants balance the use of remembered private information and communicated public information to maximally exploit resources. This work determines how the strategy that best balances these two sources of information, and the performance of that best strategy, depend on the information in the distribution that is available to be exploited, and the number of ants in the colony. We answer this question by (1) measuring the rates at which ants foraging for seeds in manipulative field studies, (2) simulating ant foraging strategies and measuring resulting foraging performance, and (3) implementing foraging strategies as algorithms for search behaviors in teams of cooperatively searching robots.

Keywords Ants · Swarm robotics · Agent based modeling · Evolutionary algorithms

76.1 Introduction

The behavior of the largest ant colonies emerges from the individual behaviors of millions of ants, where behaviors can be influenced by distributed communication among nest mates. Ant colonies are canonical distributed systems. Without central control, the interactions among millions of communicating individuals enable ant colonies to search and respond to complex, dynamic landscapes effectively.

We hypothesize that ant colonies, immune systems and other complex biological systems use similar strategies to accomplish effective decentralized search.

M.E. Moses (✉) · K. Letendre · J.P. Hecker · T.P. Flanagan
Department of Computer Science, University of New Mexico, Albuquerque, NM 87131, USA
e-mail: melaniem@unm.edu

K. Letendre
e-mail: kletendre@unm.edu

J.P. Hecker
e-mail: jhecker@unm.edu

T.P. Flanagan
e-mail: tpaz@unm.edu

T. Gilbert et al. (eds.), *Proceedings of the European Conference on Complex Systems 2012*, Springer Proceedings in Complexity, DOI 10.1007/978-3-319-00395-5_76,

By demonstrating how behavioral rules of individual ants result in colony-level responses to changing food distributions, we elucidate principles that underlie emergent behavior of other complex systems in biology, computation and societies.

In this work, we aim to understand how ants balance the use of remembered private information and communicated public information to achieve efficient, robust, and scalable search. We use evolutionary optimization algorithms to find the balance between communication and memory that maximizes seed intake in simulated foraging ants. Those behaviors are then encoded as algorithms in physical robots that search for RFID tags individually or in teams of three.

In vivo. Our field studies are conducted on three species of *Pogonomyrmex* seed harvesters whose colony size varies from dozens to thousands [6]. Seeds are hard to find, so the duration of a foraging trip, which includes travel time and search time, is dominated by search time. These ants often use site fidelity, a process in which an ant remembers and returns to the last site in which it found a seed [1, 2]. When resources are clumped, site fidelity reduces an individual ant's search times for other nearby seeds [8]. Seed harvesters appear to lay pheromone trails to recruit nest mates to large piles of food, but this may be rare under natural circumstances [10]. We ask how colony size (the number of foragers in the colony) and food distribution affect the rate at which seeds are collected.

In silico. We simulate ant foraging using a set of agent-based models (ABMs) of 100 foragers on a grid, with parameters describing individual ant behavior optimized by a genetic algorithm (GA). GAs enable multi-parameter optimization by simulating evolutionary processes. GAs have been successfully used to evolve parameters for use in swarm robotics [5]. Our GA selects parameters that specify how ants travel from the nest, search, and balance use of site fidelity and pheromone communication to maximize seed collection rates in simulations.

In machina. Swarm robotics is necessitated by problems that are inherently too complex or difficult for a single robot, and by the need to develop systems that are cheaper, more adaptive, and robust to failures, errors and dynamic environments [4]. Like ant colonies and other complex systems, robotic swarms have potential to utilize efficient, robust, distributed approaches to physical tasks. Effective algorithms for swarm robotics must extend beyond simulation to intelligently deal with the complexities of navigating in real environments.

We build low cost robots based on the open source Arduino platform equipped with ultrasound, wifi to allow communication with a central server, a compass, and ability navigate via dead reckoning. We test how quickly individual robots and teams of three robots collect RFID tags distributed in ways that mimic our field studies and simulations. The ant foraging ABM was modified to model our swarm robots and our experimental setup. Thus, simulations provide both a theoretical benchmark and a basic architecture for using GAs to optimize real world parameters.

76.2 Methods

In vivo. We conducted manipulative field experiments on three sympatric species of *Pogonomyrmex* seed-harvesters in the summers of 2008 and 2009 in Albuquerque, New Mexico [6, 7]. We baited each colony with dyed seeds arranged in a ring around the colony entrance (Fig. 76.1(a)). Equal numbers of seeds were placed in four distributions varying pile number and size. For the largest colonies we used 1 pile of 256 red seeds; 4 piles of 64 purple seeds; 16 piles of 16 green seeds; and a random scattering of 256 blue seeds. After placing the baits, an observer time stamped the color of each seed brought into the nest. Data are reported for 27 experiments, nine on each of the three species.

For every experiment we calculated a normalized foraging rate: first we calculated the rate at which seeds from each distribution were collected by dividing the number of seeds collected from a distribution by the time between collecting the first and last seed of that color; the normalized foraging rate was calculated by dividing the seed rate for a piled distribution by the seed rate for randomly scattered seeds. These three normalized rates (one each for red, purple, and green seeds) measures how much faster clumped seeds are collected compared to randomly distributed seeds. The normalized measure allows us to meaningfully compare across variable colony sizes and activity levels, and to compare results from the field to those from the model.

In silico. Simulations are derived from the model described in [6, 11]. We use GAs to optimize the behavior of simulated ants foraging in three different food environments (clumped, power law and random), using site fidelity alone, recruitment alone, neither strategy, or both foraging strategies together.

Within each simulated colony, every ant shares the same set of parameters that determines its behavior. 100 ants forage on a grid of 4000×4000 cells, with 25600 seeds placed in one of the three distributions. At model initialization, all ants begin at a nest located at the center of the grid. Upon picking up food, an ant decides whether to leave a pheromone trail on the return trip to the nest, or remember the location and return to it, or abandon that search site. This decision is based on the number of other seeds in neighboring grid cells, as actual ants might use smell or briefly handle seeds nearby to gauge their density. An ant laying a trail deposits pheromone on each cell it walks over during its trip back to the nest. This pheromone evaporates from the grid over time.

The ABM requires estimating 12 floating point parameters that are not known from field studies. These parameters determine degree of turning during the correlated random walk of a searching ant; the probability that an ant will remember the site at which the seed was found or lay a trail to that site; the dependence of the probability of remembering or trail laying as a function of local seed density; evaporation rate of the pheromone trails; and probability that ants abandon pheromone trails. We used a GA to find a set of parameters for every ant in a simulated colony that maximized foraging rate by that colony on a particular seed distribution.

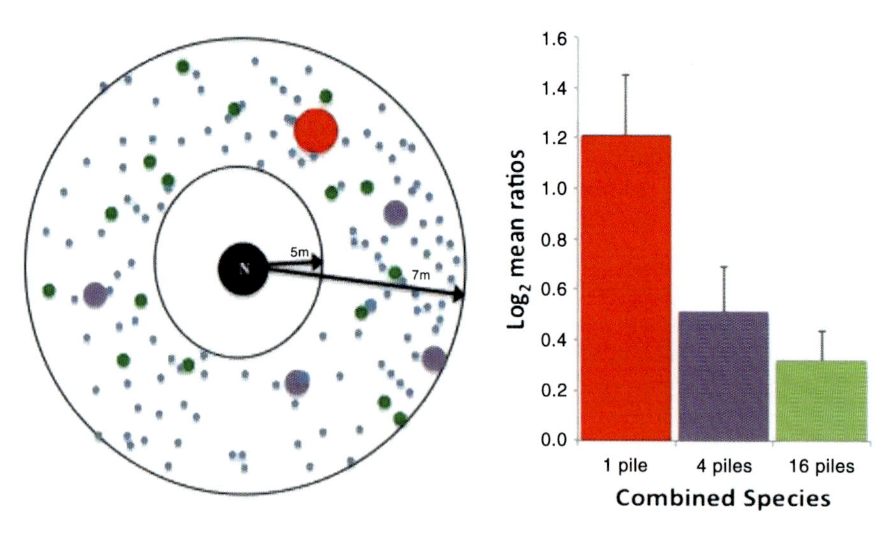

Fig. 76.1 (**A**) Experimental seed distribution around the nest entrance of a *P. rugosus* colony. *Each colored circle* is a pile of millet seeds dyed to that color. The size of each circle represents the relative number of seeds in that pile. All four seed colors are provided simultaneously for each experiment. (**B**) *Bars* indicate $\log_2$ transformed *normalized rates* (foraging rate of piled seeds divided by foraging rate of random seeds) for three seed distributions averaged over all 27 experiments. *Error bars* are standard errors

In machina. We adapted behaviors identified in simulations into algorithms that determine the behavior of robots searching for RFID tags [9]. In each hour-long experiment, robots begin at a 'nest' to which they return once they have located a tag. At the nest, robots communicate with a laptop for localizing and error correction by the robots' ultra- sonic sensors, and for managing the communication of pheromone trails (encoded as x, y coordinate pairs where the nest represents 0, 0). We programmed each robot to stay within a 3 m radius 'virtual fence' to deter drift outside of the experimental area. In every experiment, 32 RFID tags were arranged in one of three different patterns: random, clustered, or power law. Each was distributed in a ring between 50 cm and 200 cm from the nest. The clustered layout has four piles of eight tags. The power law layout mimics the seed distribution in Fig. 76.1, but with 32 tags rather than 1024 total seeds. Experiments are replicated 5 times each under identical conditions for individual robots and for groups of three bots.

A simulation system was adapted from the ant foraging simulation to precisely replicate the behavior of the robots' movements, interactions and their experimental area. In addition to simulating the 3-m radius area to which the physical robots were restricted, we also simulated the behavior of the robots in a much larger unbounded area, with tags distributed in the same density, but in such large numbers that even large swarms of robots collect only a small fraction of the available tags. We simulated 1- and 3-robot swarms, and also scaled up to 30 and 100 robot swarms to observe the scaling properties of the system.

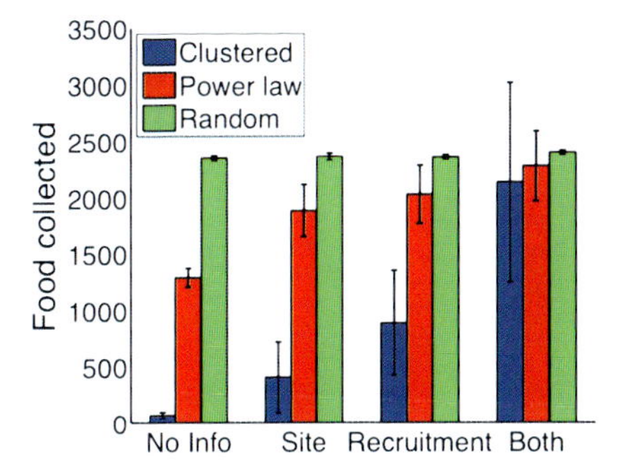

Fig. 76.2 Bars represent number of seeds collected during simulated foraging trials by colonies of 100 foragers. Colonies forage on clustered, random, and power law distributed food, after optimization by GA to maximize food collection rate on those distributions. Results show the effect of using site fidelity, pheromone recruitment, both methods together, or neither (no information use)

76.3 Results

Figure 76.1(b) shows the normalized foraging rates—the rates at which piled seeds are collected divided by rate at which random seeds are collected when seeds are distributed as in Fig. 76.1(a). A value of 0 indicates that seeds from a piled distribution are collected at the same rate as randomly distributed seeds. All piled distributions are collected significantly faster than random seeds. The $\log_2$-transformed normalized rates are 0.3, 0.5 and 1.2 for seeds distributed in 16, 4 and 1 pile. We note that the foraging rates decrease in proportion to the information required to find additional seeds once a seed of a given color is found (4 bits for the 16 piles, 2 bits for the 4 piles and 0 bits for the single pile).

The foraging success of virtual ants evolved by the GA is shown in Fig. 76.2. Foragers collect the most food when it is distributed at random (green bars). In this case, the GA evolves parameters that distribute foragers evenly across the grid. Little benefit accrues from memory or communication when seeds are not clumped. In both power law and clustered distributions, seeds are collected faster using site fidelity than communication alone, but both together are most effective. Foraging rate on the clustered distribution (blue bars) is affected by foraging strategy more than the other two distributions.

Figure 76.3 shows how individual physical robots search for RFID tags. Simulations replicate these behaviors. Figure 76.4 shows how long it takes a single robot and a team of 3 robots to collect 8 of 32 tags (A) in real robots and (B) in simulations designed. Results are shown when robots use only site fidelity, and no pheromone communication. 3 robots collect tags twice as fast as a single robot, in real- world and simulated experiments.

Figure 76.5(A) shows an example of how quickly tags are collected by a single robot (real and virtual) when robots combine site fidelity and pheromone-like communication. Virtual robots outperform real robots because real robots sometimes get lost and then communicate incorrect locations to team-mates. Figures 76.5(B)

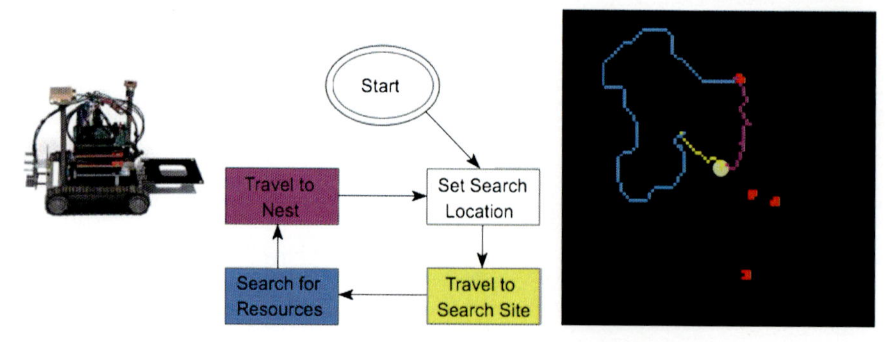

Fig. 76.3 A robot begins its search at a globally shared central nest site (*center circle*) and sets a search location. The robot then travels to the search site (*yellow line*). Upon reaching the search location, the robot searches for tags via a biased random walk (*blue line*) until tags (*red squares*) are found or a probabilistic timeout occurs. The robot returns to the nest (*purple line*), possibly laying a pheromone trail and/or remembering the previous location and returning to it

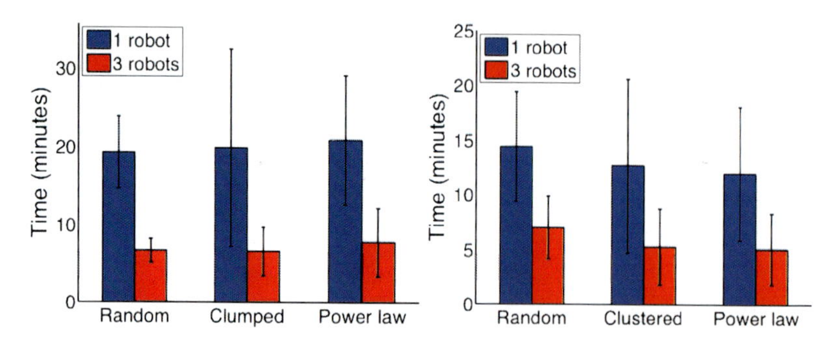

Fig. 76.4 Time to collect 25 % of the tags from three different distributions for one and three robots (**A**) in simulation and (**B**) in physical robots. *Each bar* is an average of 5 experiments

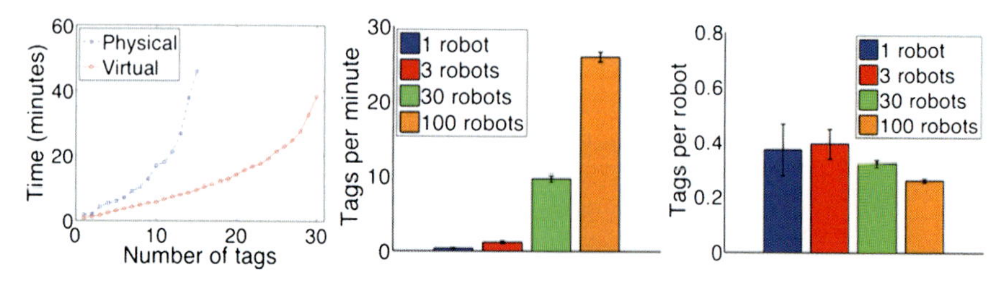

Fig. 76.5 Tags collected using pheromones and site fidelity in combination (**A**) example RFID collection curves by teams of 3 real and virtual robots. (**B**) Rate tags are collected and (**C**) minutes to collect a tag per individual robot, in different simulated team sizes

and 76.5(C) show that large simulated teams of 100 robots collect seeds approximately 70 times faster than individual robots. Each robot in a larger team collects tags slightly slower because robots on larger teams have to travel a further average distances to collect tags (a larger number of tags is necessarily, on average, further from the nest).

76.4 Discussion

Collective search depends on the balance of individual memory and communicated information. We are particularly interested in how search strategies change with the size of the collective and the distribution of the resource that is being collected. Figure 76.2 shows that the most clumped distributions of seeds are collected substantially faster when individual memory is supplemented with communication. Centralized systems are characterized by diminishing returns—each task takes longer to complete in larger systems [12]. However, we see only very modest declines in per-robot foraging rates in large teams when communication is distributed–teams of 100 robots collect tags 70 times faster than single robots. Similarly, we saw no significant difference in seed collection rate across colony sizes in our field study—large colonies collect seeds as fast as small colonies, even though the average distance traveled from the nest to a seed is longer for larger colonies [7].

Information theory quantifies information as the amount of randomness in a distribution, but it says nothing about how animals make use of that information in terms relevant to fitness [3]. Our experiments and simulations allow us to quantify how different strategies that exploit information about the distribution of resources improves search. Figure 76.2 shows that memory and communication improve foraging success the most on the most clustered distributions. This is because each bit of information about the location of a seed is of greater value when the entire seed distribution can be described with fewer bits. Thus, memory and information exchange helps the colony exploit clumped distributions but not random distributions.

We use simple robots to test real-world implementation of swarm foraging algorithms based on site fidelity and pheromone-like communication. Thus far, in our experiments with physical robots, pheromones actually hamper foraging success because lost robots miscommunicate resource locations. However, when we replicate robot behavior in simulation (in which robots are never lost) we find that combining memory and communication is an effective strategy for teams of up to 100 robots, suggesting that by improving robot localization, our architecture and algorithms are scalable to large robotic swarms.

Understanding effective decentralized search in ant colonies provides design principles for engineered robotic swarms. Moreover, many other complex systems search effectively without centralized control—immune systems find pathogens, market economies find efficient pricing mechanisms, and evolution finds strategies that enable populations to survive. By elucidating how effective search strategies emerge from behaviors of individual components, this work lends insight into complex systems more generally.

References

1. Beverly BD, McLendon H, Nacu S, Holmes S, Gordon DM (2009) How site fidelity leads to individual differences in the foraging activity of harvester ants. Behav Ecol 20(3):633–638
2. Crist TO, MacMahon JA (1991) Individual foraging components of harvester ants: movement patterns and seed patch fidelity. Insectes Soc 38(4):379–396
3. Dall SRX, Giraldeau L, Olsson O, McNamara JM, Stephens DW (2005) Information and its use by animals in evolutionary ecology. Trends Ecol Evol 20(4):187–193
4. Dorigo M, Sahin E (2004) Swarm robotics—special issue editorial. Auton Robots 17(2–3):111–113
5. Dorigo MV et al (2004) Evolving self-organizing behaviors for a swarm-bot. Auton Robots 17(2):223–245
6. Flanagan TP, Letendre K, Burnside W, Fricke GM, Moses M (2011) How ants turn information into food. In: IEEE symposium on artificial life (ALIFE), pp 178–185
7. Flanagan TP, Letendre K, Moses ME (2012) Quantifying the effect of colony size and food distribution on harvester ant foraging. PLoS ONE 7(7):e39427
8. Haefner JW, Crist TO (1994) Spatial model of movement and foraging in harvester ants (Pogonomyrmex) (I): the roles of memory and communication. J Theor Biol 166:299–313
9. Hecker JP, Letendre K, Stolleis K, Washington D, Moses ME (2012) Formica ex machina: ant swarm foraging from physical to virtual and back again. In: Proceedings of the 8th international conference on swarm intelligence, Brussels. Lecture Notes in Computer Science, vol 7461
10. Holldobler B (1976) Recruitment behavior, home range orientation and territoriality in harvester ants, *Pogonomyrmex*. Behav Ecol Sociobiol 1(1):3–44
11. Letendre K, Moses ME (2012, in review) Synergy in ant foraging strategies: memory and communication alone and in combination. Unpublished
12. Banavar JR, Moses ME, Brown JH, Damuth J, Rinaldo A, Sibly RM, Maritan A (2010) A general basis for quarter power scaling in animals. Proc Natl Acad Sci USA 107(36):15816–158120

Chapter 77
Popularity and Similarity Among Friends: An Agent-Based Model for Friendship Development

Sma Abbas

Abstract This work investigates how local preferences, social structural constraints and randomness might affect the development of the friendship network in Facebook. We do this by analyzing a snapshot Facebook dataset of Princeton University's students, and by building an agent-based simulation for comparison. Several different, but plausible, processes of friendship network development are proposed in which the structural information of the growing network and the student preferences are taken into account and then compared with the data. 'Network formation based on personal preference and social structure with some randomness' matches the data best, and is thus the preferred hypothesis for the way that students add "friends" on Facebook.

Keywords Facebook · Community structure · Agent-based modelling · Social network analysis (SNA)

77.1 Introduction

In the last few years, internet has embraced active participation of its users by becoming social. One of the main services to make the web social, 2.0, is Social Networking Systems (SNS), such as Facebook and MySpace. Social Networking Systems (SNS) have changed the outlook and the behavior of millions of people worldwide. With the abundance of enormous data, which was impossible before, a lot of research has been done to understand how and what engages people to use an SNS in their day to day lives. Our aim is similar, but mainly focused on how people develop their social network. And also, how segregated, if any, the social network is; and how we might work towards a better integration in a society. The aim of this paper is to reconstruct the development of the social network, so that an understanding of it could be developed.

This work investigates how local preferences and social structural constraints might affect the development of the friendship network in Facebook. We do this by

S. Abbas (✉)
Centre for Policy Modelling, Manchester Metropolitan University Business School, Manchester, UK
e-mail: ali@cfpm.org

T. Gilbert et al. (eds.), *Proceedings of the European Conference on Complex Systems 2012*, Springer Proceedings in Complexity, DOI 10.1007/978-3-319-00395-5_77,

analyzing a snapshot Facebook dataset of Princeton University, and by building an agent-based simulation for comparison. This is an extension of our work [1] for a larger and diverse dataset. Several different, but plausible, processes of friendship network development are proposed in which the structural information of the growing network and the student preferences are taken into account and then compared with the data. 'Network formation based on personal preference and social structure' matches the data best, and is thus the preferred hypothesis for the way that students add "friends" on Facebook.

A lot of social network based models have been proposed. From a general but realistic social network (e.g. see [13, 14]) to a data-driven students' social network [30], but they do not address how such a network might develop within an online environment. This paper attempts to address this concern. First, we simulate some possible strategies of how students meet and develop their social network. Then we compare the obtained results with the underlying dataset we have used and in this way are able to make some inferences as to the probable strategies that the students used.

The magnitude of the data present in the online SNS is enormous, and presents itself as a rich source of social information for analysis. According to studies, most of the online social networks act as a representation of the offline, or real social networks [8, 9]. So it could be assumed as an approximation or a proxy of a real world social network. Not only does an SNS capture the social network, but also the activity between users. Mainly due to privacy concerns and also due to its vast commercial value, this data even by the research community is quite difficult to acquire. So we are left with either a snapshot with limited information, or an activity log without any social network. A huge data set of longitudinal nature of Facebook has been collected, but is available with a limited access [20]. The aim of this paper is to reconstruct the development of the social network with the help of an agent-based methodology, so that a possible history of the social network and a better insight of local processes could be developed.

The paper is divided in different sections. In Sect. 77.2, we define the reference data on which our agent-based model is based—its characteristics and network structure. After that, in Sect. 77.3, we define our model and the strategies of interaction it offers. Simulation results and their comparison with the dataset are presented in Sect. 77.4. Related work is summarized in Sect. 77.5. At the end, in Sect. 77.6, we summarize our findings and present the future outlook of our research by concluding the paper.

77.2 The Reference Dataset

We have used the data of students and faculty members of Princeton who use Facebook. This was provided to us by Mason A. Porter of Oxford University, and has been studied by him and others in [31]. The dataset includes both the attributes and social structure for 6596 people. For each person, it contains eight attributes, which

Table 77.1 Attribute spread

Attributes	Dorm	Major	Year	High school
Missing (%)	33.76	24.86	11.77	20.7
Unique	57	41	26	2235
Average	115.72	160.88	244.30	2.95
St. Dev.	293.06	268.68	399.13	29.21

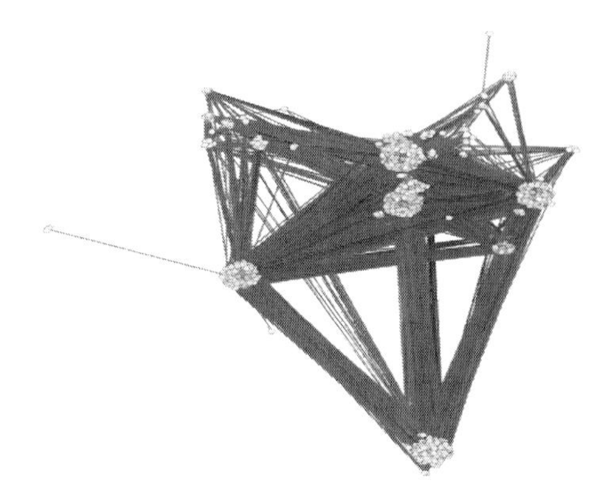

Fig. 77.1 Princeton social graph

are: ID, student/faculty status, gender, major, second major/minor, dorm/house, year, high school; and also friendship links for each student. On average, each person has almost 88 friends. We note that it is a snapshot—it represents only links and attributes present at one single point of time. The data is completely anonymized where simple integer values represent each attribute. The underlying anonymous dataset of Facebook includes both the attributes and social structure for 6575 student of Princeton University. In total there are 293307 links—averaging to 89.2 friends. Each person has four attributes, which are: major course of study (major); their place of living (dorm); year they joined the university, and their high school information. As for the spread of each attributes and their missing values, we have summarized it in Table 77.1.

Since it is a relatively bigger university as compared to Caltech covered in [1], we see a very diverse population when we see the number of various high schools. Missing information in the dataset has been coded by 0. We have dealt it carefully in our model. As for the network structure, we have shown it in Fig. 77.1. We can clearly see groups/communities in the network structure.

77.3 Model Outline

In order to understand the dynamics of this social network, we simulate it using an agent-based simulation. The main aim of the work is to understand the interplay of social processes and their impact on the network structure as a whole. Thus the key focus is on analyzing how students interact and build their social network over time. We can then see which strategy of interaction seems to produce the best representation of a social network as judged by a comparison with the reference dataset. In this section, the term agent will be used to refer to a student.

77.3.1 Simulation Setup, Execution and Termination

The number of agents in all simulation runs is 6575, based on the underlying dataset of Princeton University students. Each individual in the dataset provided the attributes for one agent in the simulation. All agents are created at the start. While initializing a simulation run, the agents are chosen in a random order. Interaction strategy for all the agents is set once in the beginning. It does not change. Each simulation runs until the number of links made is the same as in the reference dataset—293320. No link is dropped or modified once it is created.

77.3.2 Rules for Adding Friends

In this section, we discuss how the agents might interact with each other, in terms of making friends in real life. It is assumed that, by and large, these real life social links will then be duplicated within Facebook. We do not claim that we present an exhaustive list of possible strategies; rather the idea is to explore some plausible ways that depend on the micro-level preference of agents and then evaluate them.

$$S_a = \frac{|\{(i, j)E : \text{s.t. } a_i = a_j\}|}{(|E|)}$$

In order to identify the significance of attributes of the four attributes we have considered in social network development, we relied on affinity [26] to guide us. It measures the ratio of the fraction of links between attribute-sharing nodes, relative to what would be expected if attributes were random. It ranges from 0 to infinity. Values greater than 1 indicate positive correlation; whereas values less than but greater than 0 have negative correlation. For an attribute a, such as dormitory, we first calculate the fraction of links having the same dormitory, for instance. It is represented by:

where a_i represents the value a for a node i. In other words, we are identifying the total number of matched nodes with the same attribute values for an attribute a. E represents total number of links. And then we calculate E_a which represents the expected value when attributes are randomly assigned. It is calculated by:

$$E_a = \frac{\sum_{i=0}^{k} T_i(T_i - 1)}{|U|(|U| - 1)},$$

Table 77.2 Affinity values of the four attributes for all the strategies of interactions

Dorm affinity	Major affinity	Year affinity	High school affinity
1.48	1.32	4.07	0.89

Table 77.3 Values of the four attributes for all the strategies of interactions

Dorm preference	Major preference	Year preference	High school preference
60	60	80	30

where T_i represents the number of nodes with each of the possible k attribute values and U is the sum of all T_i nodes, i.e., $U = \sum_{i=0}^{k} T_i$. The ratio of the two is known as *affinity* : $A_a = \frac{S_a}{E_a}$ [26]. Here are the affinity measures of the four attributes in Table 77.2:

We see that year is the most important attribute. This result matches previously published work [31]. Each agent is initialized with the four attributes (major, dorm etc.) of a corresponding individual recorded in the Princeton data set. We have just used these four attributes because of the conformity in the earlier studies done on students. And also, we found them, using the affinity measure, of the utmost importance. The values for each of the four attributes can be seen in Table 77.3. These values have been found to be the best fitted values when compared with the reference dataset.

All agents have a preference for each of the four attributes we have selected which is known as "Personal Preference". The idea has been inspired from homophily—the love of the similar [24]. It is a probabilistic match of attributes between the source and the target agents. We have shown the illustration in Table 77.4 for the year attribute. A chance out of 100 is randomly selected in a uniform fashion. If it is under the predefined preference value (80 in case of year preference) and the attribute values of both the source and target agents are known (non-zero) and match with each other, then the dormitory preference is satisfied; and we set the dormitory flag to true. Also, if the chance is greater than the preference value, it is satisfied as well. We repeat the same process for the remaining attributes. If all the four attributes' conditions are satisfied, we make a friendship link between the source and the target agents.

We have devised four different plausible strategies for agent interaction—each involves matching students using their attributes, but in different ways. Personal preference is not taken into account when there are missing values for all the four attributes. Hence, in this case, we totally neglect the preference of both the source and the target agent. All of the four interaction strategies use the attribute values defined in Table 77.3.

Table 77.4 Algorithm to calculate "*Personal Preference*"

```
1.  Agent Source = getSourceAgent()
2.  Agent Target = getTargetAgent()
3.  Integer YP = getYPValue() // Get year preference value which
    is fixed as 80
4.  Boolean sameYear = False
5.  Integer chance = get_random_integer(100)
6.  IF (chance < YP){ // 0 < chance <= YP
7.  IF (Source.getYear() == Target.getYear()) AND (Source.
    getYear() != 0 And Target getYear()!= 0)){ sameYear = True }
8.  }ELSE{ sameYear = True }
9.  ...
10. //repeat the same evaluation for the rest of the attributes
    (Dorm, Major etc.)
11. IF (sameYear AND sameMajor AND sameDorm AND sameHighSchool)
12. // If all conditions satisfy
13. form_a_link(Source, Target) //create a friendship link
    between the two
```

77.3.3 Random Strategy—Strategy 1

Each source agent selects a randomly chosen target agent after every time step or simulation tick. The target agent is selected using a uniform probability distribution. After the selection, the source agent determines if the target agent satisfies its personal preference. If it does, an undirected link is created among them, which shows that they are friends.

77.3.4 Friend of a Friend (FOAF) Strategy—Strategy 2

In this strategy, there are two phases for each agent. In the first phase, all agents, on their own, are asked to make only limited random friends selected in a uniform distribution. This should satisfy both the source and target agents' preferences. If these are not satisfied, they do not form a link. After this first phase, personal preferences are not taken into account. From then on, in the second phase, new friends are selected in a "friends-of-friends" manner. During this phase, starting from the first friend of a friend—whose degree is consider as the reference point, in chronological order, we search its friends (FOAF) and continue searching till we find a suitable agent. As soon as we find an available FOAF which has a greater degree than the reference FOAF—showing the popularity, we select it and then form a friendship link between the two.

77.3.5 *Party Strategy—Strategy 3*

In this strategy the personal preferences are also not taken into account. All students arrange a small party which is held on a regular basis. The number of participants in a party is 100. The selection of the party participants is totally independent and unbiased towards any attribute. At each party, a maximum of 300 new (random) friendships are made. Due to the random selection of party participants, there is a chance of selecting nodes which are already connected to each other. In that case, no new link is established.

77.3.6 *Hybrid Strategy—Strategy 4*

This strategy is a combination of the above three strategies. At every simulation time step, a simulation strategy between random and FOAF is chosen on a uniform basis. In order not to overwhelm the randomness, the party strategy is run in every 20th time step. Also, unlike the original model [1], the party mode is executed on a local level, and not on a global level.

77.4 Results

In this Section, we compare the simulation results with the reference dataset. First we compare the global or overall results in Sect. 77.4.1 and then in Sect. 77.4.2, we discuss the attribute level comparison.

77.4.1 *Global Results*

In this Section, we compare the structure based on the overall network of the reference dataset with the various simulation strategies. In Table 77.5, we have summarized the basic Social Network Analysis (SNA), over the reference dataset and the simulation results of the four interaction strategies.

Random and Party modes are the most deviant ones when compared with the underlying reference dataset. In terms of standard deviation in number of degrees (# of friends), they are not even close. Also the distribution of degree is *normal*—bell shaped, as opposed to exponential. The FOAF mode has good results in terms of assortativity, transitivity and even the best fitted distribution. In standard deviation, however, the difference is quite large.

Hybrid mode captures the standard deviation, assortativity and connectedness and also best fitted distribution, quite well, when compared with the reference dataset. In terms of transitivity, it is almost half as the reference dataset. In order

Table 77.5 Reference dataset and simulation output comparison

Dataset/Model	St. Dev. degree	Assortativity	Transitivity	Best fitted distribution
Ref.	78.55	0.09	0.16	Exponential (Alpha = 1.98)
Random	18.12	0.08	0.019	Normal
FOAF	93.76	0.11	0.09	Exponential (Alpha = 1.84)
Party	19.64	−0.002	0.03	Normal
Hybrid	79.97	0.105	0.07	Exponential (Alpha = 1.97)

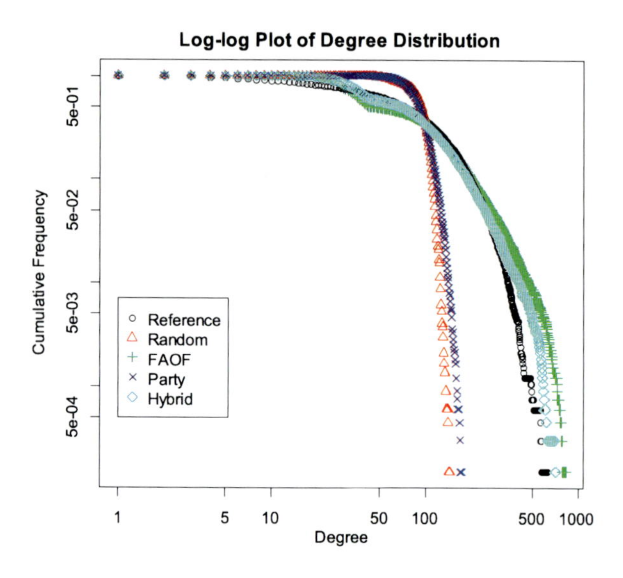

Fig. 77.2 Log-log plot of Total Degree Distribution of all the four simulation strategies and the reference dataset

to align it with that of the reference dataset, we ran a sensitivity analysis over the parameter space. We did find better results when the parameters were changed, but that hampered the standard deviation and assortativity. Hence we focused more on the overall degree fitting and assortativity. The parameter values for the reference dataset, FOAF and Hybrid mode are also mentioned, where Hybrid mode has almost the same alpha value as the reference dataset, for the fitted distribution. The fitting of the degrees have been calculated by setting the minimum degree to 40.

We have summarized in Fig. 77.2, the degree distribution of the reference and the four interaction strategies. This only shows the final node degrees after the simulation has been finished. The reference, the FOAF and the hybrid strategy's degree distributions show a power law effect which suggests that most of the nodes have few links while only a few nodes have a lot of links. The other two strategies, random and party seem normally distributed in nature. Their links are more or less uniformly distributed.

If we consider various studies on number of friends in Facebook (see [25, 28, 32]), mostly all have found that it does have a power law outlook, but there was a seminal work which proved this common belief wrong. According to [11], Facebook has not one but two power-law regimes: one for node degrees less than 300 and one for greater degrees. We found the similar pattern in [2] too. In this case, however, we don't see that for two reasons: firstly, the dataset is too small and secondly the dataset just contains inter-school links. Hence we see just one power-law outlook.

In Table 77.5 we can clearly identify that Hybrid strategy remains the best candidate when it is compared with the reference dataset. The underlying distribution of both the reference and Hybrid strategy can be identified by such a huge standard deviation; which in turn reflects our earlier finding that both of these are in fact power law distribution.

77.4.2 Attribute Level Results

In this Section, we compare the results of our simulation runs of all the four strategies for each of the attributes with the reference dataset. We measured the results in terms of the Silo Index. This is an index which identifies the degree of inter-links between nodes with a particular attribute value in a (social) network. If a set of nodes having a value Y for an attribute X, has all the links to itself, and not to any other values of attribute X, that means a very strong community exists, which is totally disconnected from the rest of the network. In short, this index helps us identify how cohesive inter-attribute links are. It ranges from -1 to 1, representing the extreme cases (no in-group links to only in-group links respectively). It can be written as:

$$\frac{I-E}{I+E}$$

where I represents the number of internal links and E the number of external links. In other words, it is the ratio of the difference in internal and external links, to the total links. It is quite similar to $E-I$ index [17], but with the opposite sign. In $E-I$ index we have value 1 when all links are external, while Silo Index has value 1 when all are internal. Hence, the Silo Index could be written as an $I-E$ index. Since the Hybrid strategy has shown the best results, and also due to space limitations, we are presenting Silo Indices comparison of this strategy alone, with that of reference dataset, in Fig. 77.3.

Apart from High School Silo Index, the rest has quite a high (> 0.83) correlation with that of the reference dataset. Hence it is the preferred mode of interaction. The number of High Schools in the dataset is quite high as shown earlier in Table 77.1. We could not find better correlation of it with varying parameters values. As for the degree mixing [27] which determines how nodes of similar degrees are connected with each other, we have plotted it in Fig. 77.4.

In lower degrees (< 400), there is a high rate of similarities between the Hybrid and the reference dataset. In high degrees, the Hybrid mode is slightly different.

After comparing all the four attributes, Hybrid strategy takes the lead when compared with the reference dataset; it presents itself as a good candidate for describing how students might have developed their social network.

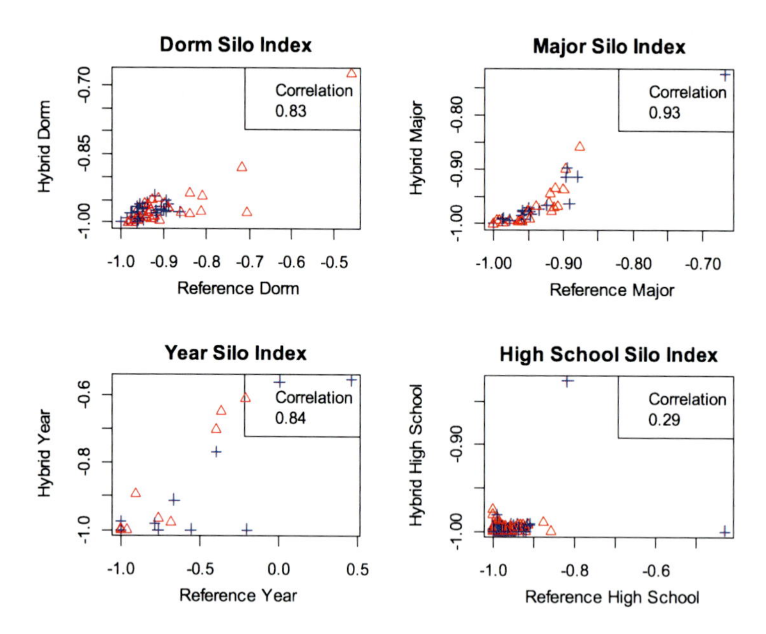

Fig. 77.3 Silo index for dorm, major, year and high school attributes for hybrid strategy and the reference network

77.5 Related Work

A plethora of research in SNS has been done over the last five years. It is impossible to cover all of it; hence some of the relevant work is being mentioned here. The major focus of such work has been the identification of the static nature of SNS, such as [25]. One of the early works before the popular SNS came into being, was a study done on Club Nexus, a Stanford students online environment in [3] back in 2001. They found that people having similar attributes are more likely to form a friendship link. Based on the various classification of users, an SNS growth model has been presented [19]. Instead of just a snapshot of a social network, but interactions among users, Golder et. el found close social circles in [12] which categorizes the general notion of online "friends" into a broader spectrum. A very detailed quantitative study on students to identify their cultural preferences was done in [20].

To understand the behavior of students' real social network development, a function of contact frequency and shared interests has been used in to make a model. Jackson et el. in [15] developed a model in which a neighbourhood search is done to develop a social network; this can result in many of the characteristics of observed networks.

Adalbert studied Facebook from an economist's point of view [22]. The data which he collected and then studied showed that race plays the most significant role in student friendship development—especially in the case of minorities. In his pre-

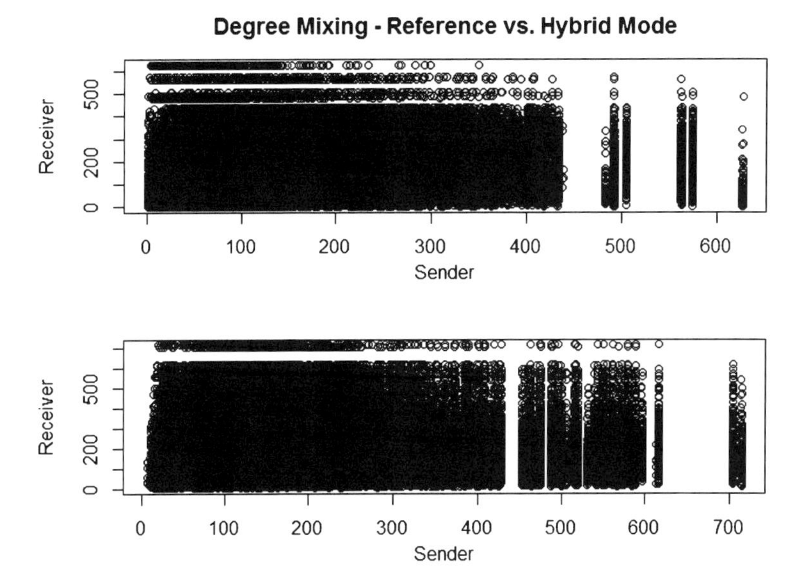

Fig. 77.4 Degree mixing of hybrid mode and the reference dataset

vious study [23], out of students of Taxes A&M, he found that majority of meeting new friends (26 %), were driven by members of the same school organizations. In an another study carried out on students' network [21], race and local proximity, such as dorm were determined to play the most important role, followed by common interests such as major and similar social standing, which in turn were followed by common characteristics such as same year. In our data, however, we could not verify the race factor, as this information is not present in the dataset that we have used.

In case of SNS growth, unlike our model, there are some studies that identify the different classes of users [18]. And also, based on the activity of users, a couple of studies show their social network development [12]. Based on only the structure of an SNS, a couple of exploration techniques have also been devised to predict what new links users are going to make [4, 6], but they usually do not take into account the rich information of attributes of users [10].

In mainstream computer science, there are many "mechanistic and yet tractable" [16] network models, such as Preferential Attachment [7] which specifies an edge creation mechanism, resulting in a network with power-law degree distribution. These models, however, do not take node attributes into account. And in machine learning and social network analysis, where the emphases has been more focused on in the development of statistically sound models that consider the structure of the network as well as the features of nodes and edges in the network [16]. Examples of such models include the Exponential Random Graphs [29] and Stochastic Block Model [5]. These network models are generally intractable and do not offer emergence [16].

77.6 Conclusion and Future Work

An agent-based simulation has been described that attempts to explain how students make SNS links, taking into account both endogenous and exogenous factors.

This is an extension of [1] for a larger and diverse university dataset, in which we tried to understand how local preferences and the structural factors might help develop a social network. Unlike the original model, in order to control randomness of Party Mode, we introduced it at an individual level, rather than at the university level. We tailored the model according to the underlying structure of the reference dataset. We have devised and explored a limited number of strategies for student interaction. We compared our simulation results to data gathered from students' Facebook network of Princeton University. We relied on both structural aspects using SNA and semantic using Silo Index for comparison. The strategies of interaction varied from preferential attachment—based on the attribute values, to complete random interactions.

After analyzing the results and comparing them with the reference dataset, we determined that Hybrid strategy, which is a combination of all three strategies: Random, FOAF and Party does the best. It captures the basic essence of the underlying network. From network level measures to the attribute level comparison, it presents itself as a good candidate for the understanding of students' interactions and social network development. Also, FOAF mode captured most of the aspect, apart from the standard deviation in number of friends, which resulted in a different slope for power law outlook. The initial setting of highly similar friends leads to a cohesive community structure and also the friends-of-a-friend process with a power law outlook. Random and Party strategies which are dominated by the random meeting of friends at events did not explain the data well. We do not claim that we presented an exhaustive list of possible social processes, but rather analyzed a few plausible variations. Focusing on personal preference, social structure with some randomness, presents itself as a promising strategy of interaction. While only pre-simulation statistics based on the underlying data, such as correlation, do not necessarily present the best parameter values. For the initial friendship links, the parameter space has to be explored to find the best match.

In future work, we would like to make a more general model, which captures both local and global aspects of a social network. This model will be based on several datasets and on the findings of this model. Also, with the aid of the earlier studies on social network—specifically online social network, we will try to design and understand the processes involved. We will focus both on internal and environmental aspects.

Acknowledgements We would thank our colleagues at the Centre for Policy Modelling for their helpful comments and feedback. We also would like to thank Mason Porter of Oxford University, who shared the underlying dataset with us.

References

1. Abbas S (2011) An agent-based model of the development of friendship links within Facebook. In: 7th European social simulation association conference, Montpellier, France
2. Abbas S (2011) Ethnic diversity in Facebook. Methodology. Manchester. Retrieved from http://www.cfpm.org/~ali/FacebookReport/FacebookOnomapGroupReport.pdf
3. Adamic L, Buyukkokten O, Adar E (2003) A social network caught in the web. First Monday 8:1–22
4. Agarwal A, Chakrabarti S (2007) Learning random walks to rank nodes in graphs. In: Proceedings of the 24th international conference on Machine learning (ICML'07), pp 9–16. doi:10.1145/1273496.1273498
5. Airoldi EM, Blei DM, Fienberg SE, Xing EP (2008) Mixed membership stochastic blockmodels. J Mach Learn Res 9:1981–2014. Retrieved from: http://www.pubmedcentral.nih.gov/articlerender.fcgi?artid=3119541&tool=pmcentrez&rendertype=abstract
6. Backstrom L, Leskovec J (2011) Supervised random walks: predicting and recommending links in social networks. In: Proceedings of the fourth ACM international conference on Web search and data mining, pp 635–644. doi:10.1145/1935826.1935914
7. Barabási AL, Albert R (1999) Emergence of scaling in random networks. Science 286(5439):509–512. doi:10.1126/science.286.5439.509
8. Boyd DM, Ellison NB (2007) Social network sites: definition, history, and scholarship. J Comput-Mediat Commun 13(1):210–230. doi:10.1111/j.1083-6101.2007.00393.x
9. Ellison NB, Steinfield C, Lampe C (2007) The benefits of Facebook "Friends": social capital and college students' use of online social network sites. J Comput-Mediat Commun 12(4):1143–1168. doi:10.1111/j.1083-6101.2007.00367.x
10. Gao B, Liu T, Wei W, Wang T (2011) Semi-supervised ranking on very large graphs with rich metadata. In: Proceedings of the 17th ACM SIGKDD international conference on knowledge discovery and data mining, vol 49, pp 96–104. doi:10.1145/2020408.2020430
11. Gjoka M, Kurant M, Butts C (2009) Unbiased sampling of Facebook. Search, pp 1–15. Retrieved from http://scholar.google.com/scholar?hl=en&btnG=Search&q=intitle:Unbiased+Sampling+of+Facebook#0
12. Golder SA, Wilkinson D, Huberman BA (2007) Rhythms of social interaction: messaging within a massive online network. In: Steinfield C, Pentland B, Lewis AK et al (eds) Social networks, vol 30, pp 330–342
13. Hamill L (2010) Communications, travel and social networks since 1840: a study using agent-based models. In: Social networks
14. Hamill L, Gilbert N (2008) A simple but more realistic agent-based model of a social network. In: Proceedings of European social simulation association annual conference, Brescia, Italy
15. Jackson MO, Rogers BW (2007) Meeting strangers and friends of friends: how random are social networks? Am Econ Rev 97(3):890–915. doi:10.1257/aer.97.3.890
16. Kim M (2010) Multiplicative attribute graph model of real-world networks. Internet Math 8(1–2):113–160. doi:10.1080/15427951.2012.625257
17. Krackhardt D, Stern RN (2011) Informal networks and organizational crises: an experimental simulation ∗. Soc Psychol Q 51(2):123–140
18. Kumar R, Novak J (2010) Structure and evolution of online social networks. In: Link mining: models, algorithms, and applications, pp 337–357. Retrieved from http://www.springerlink.com/index/X246586QW2477685.pdf
19. Kurant M, Gjoka M, Butts C (2011) Walking on a graph with a magnifying glass: stratified sampling via weighted random walks. ACM SIGMETRICS Perform Eval Rev 39:241–252. doi:10.1145/2007116.2007145
20. Lewis K, Kaufman J, Gonzalez M, Wimmer A, Christakis N (2008) Tastes, ties, and time: a new social network dataset using Facebook.com. Soc Netw 30(4):330–342. doi:10.1016/j.socnet.2008.07.002
21. Marmaros D, Sacerdote B (2006) How do friendships form? Q J Econ 121(1):79–119. doi:10.1162/003355306776083563

22. Mayer A (2009) Online social networks in economics. Decis Support Syst 47(3):169–184. doi:10.1016/j.dss.2009.02.009
23. Mayer A, Puller SL (2008) The old boy (and girl) network: social network formation on university campuses. J Public Econ 92:329–347
24. McPherson M, Smith-Lovin L, Cook JM (2001) Birds of a feather: homophily in social networks. Annu Rev Sociol 27(1):415–444. doi:10.1146/annurev.soc.27.1.415
25. Mislove A, Marcon M, Gummadi KP, Druschel P, Bhattacharjee S, Bhattacharjee B (2007) Measurement and analysis of online social networks. In: Proceedings of the 7th ACM SIGCOMM conference on internet measurement (IMC'07), p 29. doi:10.1145/1298306.1298311
26. Mislove A, Viswanath B, Gummadi KP, Druschel P (2010) You are who you know: inferring user profiles in online social networks. In: Proceedings of the third ACM international conference on web search and data mining, pp 251–260
27. Newman M (2003) Mixing patterns in networks. Phys Rev E 67:026126. Retrieved from http://pre.aps.org/abstract/PRE/v67/i2/e026126
28. Panzarasa P, Opsahl T, Carley KM (2009) Patterns and dynamics of users' behavior and interaction: network analysis of an online community. J Am Soc Inf Sci 60(5):911–932. doi:10.1002/asi.21015
29. Pattison P (1996) Logit models and logistic regressions for social networks. Psychometrika 61(3):401–425. Retrieved from http://www.springerlink.com/index/T2W46715636R2H11.pdf
30. Singer HM, Singer I, Herrmann HJ (2009) Agent-based model for friendship in social networks. Phys Rev E, Stat Nonlinear Soft Matter Phys 80(2):026113. Retrieved from http://www.ncbi.nlm.nih.gov/pubmed/19792206
31. Traud A, Kelsic E, Mucha P (2008) Community structure in online collegiate social networks. Organization, pp 1–15. Retrieved from http://www.uvm.edu/~pdodds/files/papers/others/2008/traud2008a.pdf
32. Wilson C, Boe B, Sala A, Puttaswamy KPN, Zhao BY (2009) User interactions in social networks and their implications. In: Proceedings of the fourth ACM European conference on Computer systems (EuroSys'09), p 205. doi:10.1145/1519065.1519089

Chapter 78
Characterizing and Modeling Collective Behavior in Complex Events on Twitter

A.J. Morales, J. Borondo, J.C. Losada, and R.M. Benito

Abstract All around the world people are increasingly using Internet and online social networks to relate among each other. This fact is bringing a unprecedented amount of user generated data, which is certainly attracting research on several fields. In this work we analyze the user interactions in Twitter around two politically motivated events, like a Venezuelan protest and the 2011 Spanish Presidential electoral campaign. We found that users participated quite heterogeneously, as a tiny fraction of them concentrates much of the activity or collective attention. This heterogeneity gives place to critical features, like interaction networks with power law distributions and modular structure. Although online social networks appear to be a pure social environment, we found traditional agents, such as well known politicians and media hold loads of influence among the participants.

Over the past years, new technologies and specially online social networks have penetrated into the world's population at an accelerated pace. An important feature of these communication tools is that they provide a large amount of user generated content, useful for research on political activism [1, 2], marketing techniques [3] and social influence dynamics [4]. In this study, we use data available from Twitter, to unveil and analyze the structural and dynamical patterns from the user interactions, in order to characterize the emergent collective behavior. We have focused our study around two relevant events: a Venezuelan political protest, that took place exclusively online, and the 2011 Spanish Presidential electoral process. On these events, users posted messages identified with special keywords, such as #SOSInternetVE for the Venezuelan protest and #20N for the electoral campaign. These special keywords identified the topics which messages we downloaded, using the Twitter API. The properties of these datasets are presented in Table 78.1.

A.J. Morales (✉) · J. Borondo · J.C. Losada · R.M. Benito
Grupo de Sistemas Complejos, Universidad Politécnica de Madrid, ETSI Agrónomos, 28040 Madrid, Spain
e-mail: alfredo.moralesg@alumnos.upm.es

R.M. Benito
e-mail: rosamaria.benito@upm.es

T. Gilbert et al. (eds.), *Proceedings of the European Conference on Complex Systems 2012*, Springer Proceedings in Complexity, DOI 10.1007/978-3-319-00395-5_78,

Table 78.1 Datasets properties

Topic	Period	Messages	Participants
20N	Nov. 5–20, 2011	370000	100000
SOSInternetVE	Dec. 14–19, 2010	420000	77700

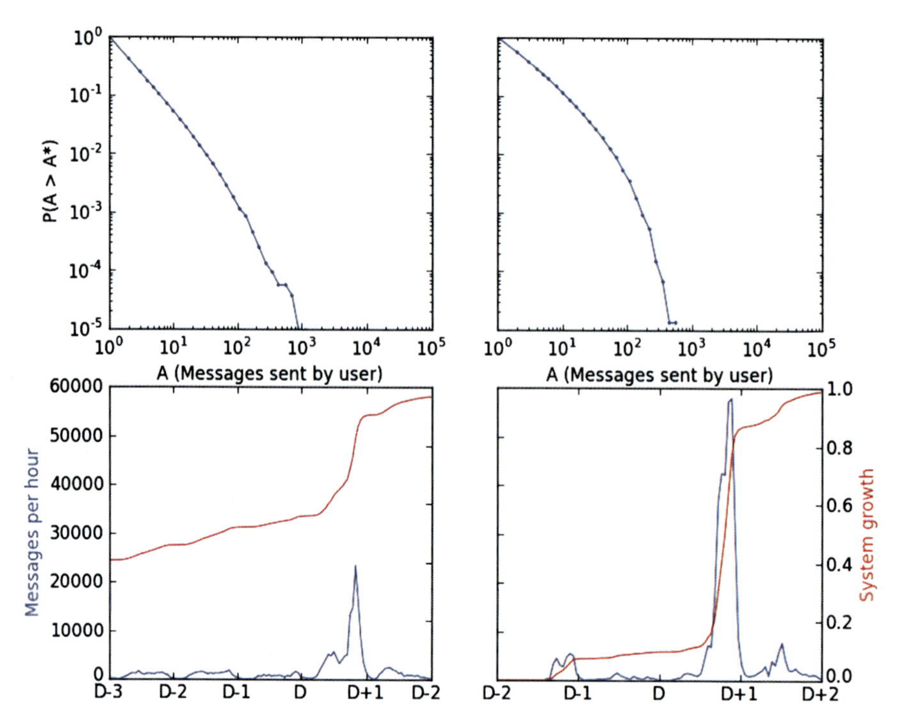

Fig. 78.1 Complementary cumulative distribution of messages sent by user (*top*) and Message rate through time (*bottom*). *The left* and *right panels* correspond to 20N and SOSInternetVE topics respectively

The distribution of the user activity, in terms of messages sent by user, as well as the message ratio through time, are shown in Fig. 78.1. The results indicate that users participated in an extremely heterogeneous way. In both cases, we found that the large majority of users (over 90 %) posted only a few messages each, while half of the messages were posted by less than 10 % of the participant population. This fact implies that the conversations were actually fed by a small portion of very active participants. Besides, both topics grew in a bursty manner, as can be seen in the bottom of Fig. 78.1, where the activity is heterogeneously concentrated in time. It can be noticed that a single conversation may grow up to 60 % of its final size in less than 8 hours.

However, not everybody's messages (or activity) have the same impact on the development of the event, since it remarkably depends on the source's connectivity inside the social substratum. To analyze this matter, we have constructed networks

Table 78.2 Followers, Retweet and Mention networks degree Pearson correlation (r)

Topic	$r_{F,R}$	$r_{F,M}$	$r_{M,R}$	$r_{A,R}$
20N	0.55	0.44	0.35	0.30
SOSInternetVE	0.57	0.70	0.85	0.15

based on "who follows who", which are subgraphs of the Twitter's global followers network, made with the events participants. On Twitter, when a user posts a message, this is instantaneously delivered to his/her own followers. Therefore, these networks represent the social substratum and available channels through which the information may flow along an event. In Fig. 78.2 we present the in and out degree distributions, which illustrate the heterogeneous connectivity found among the participants. In fact, while the large majority of users are followed by less than 20 users each, half of the social links are targeted to less than 2 % of the users. This means that the messages written by these hubs are delivered (and probably read) by half of the participants.

This heterogeneous connectivity gives place to an heterogeneous collective attention. To study so, we have also built other networks, linking the participants according to "who retweeted (retransmitted) who" and "who mentioned who". These interaction mechanisms display effective links where messages were propagated and selectively delivered, respectively. These networks have directed and weighted edges, and the in and out strength distributions are also presented in Fig. 78.2. It can be appreciated that both mechanisms are scale-free at the incoming links, which are the result of the aggregation of individual efforts, reflected in the out strength distribution. In fact, while the large majority is hardly mentioned or retweeted, less than 1 % of the participants, concentrate half of the mentions and retweets. Such an exclusive elite concentrates the largest part of the collective attention in both mechanisms.

Influence in Twitter has been considered to depend not only on the user's topological features in the followers graph, but also on the user's topological features in the retweet and mention graphs [4]. In the two considered cases, these measures are remarkably correlated, as may be seen in Table 78.2, where we present the Pearson correlation for the in degree and in strength values across the three networks, which resulted to be positive at all cases. We detected that the small fraction of hubs (who are influencers among the participants) act like information producers, posting messages widely delivered and retransmitted throughout the network. On the other hand, we found that the large majority of users act like information consumers, either actively or passively. Nevertheless, in order to gain influence, the regular users must play an active part in the conversation, as we demonstrated in a previous study [1], where we detected several cases of regular users who equaled the retransmission levels gained by popular accounts, by means of increasing their activity several order above. This is also supported by the positive Pearson coefficient between the retweets in strength and the user activity, also presented in Table 78.2.

In order to unveil how such heterogeneous users interacted with each other, we calculated the assortativity by degree coefficient [5, 6] for all networks. The results presented in Table 78.3, show that the emergent networks from Twitter are disassor-

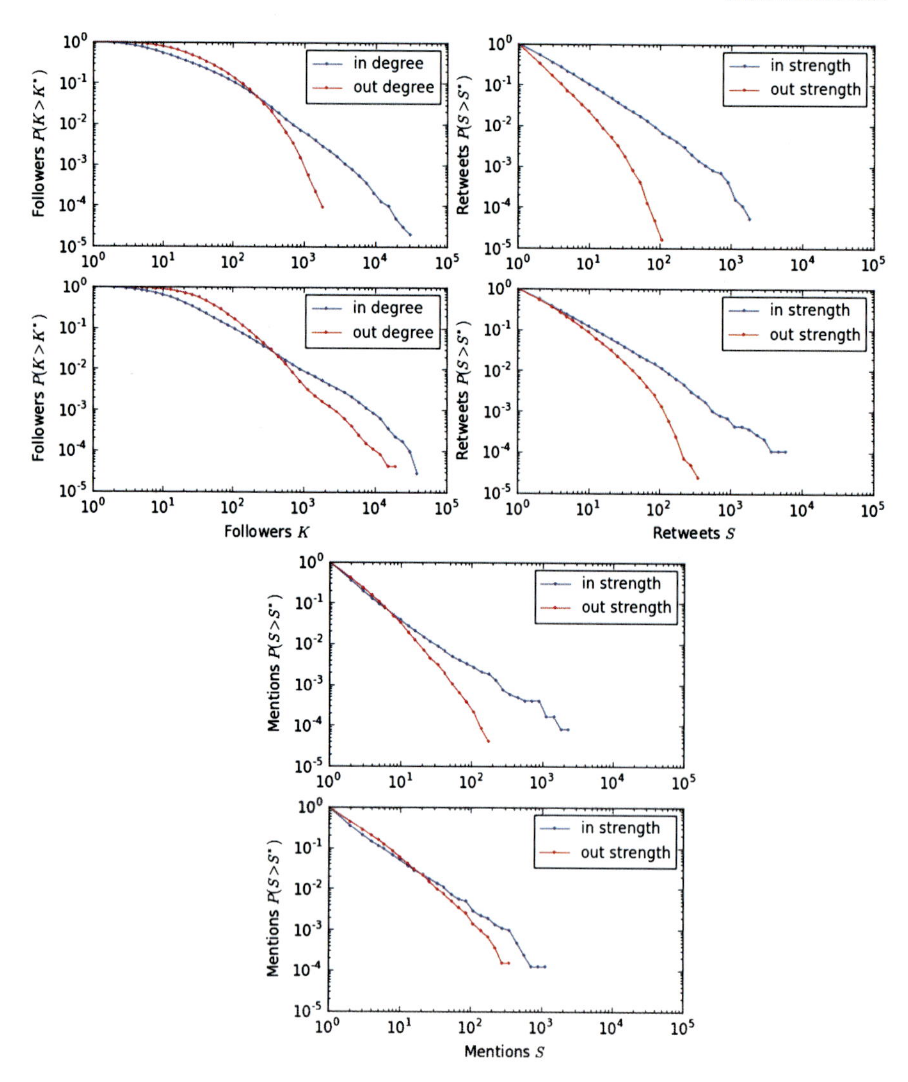

Fig. 78.2 Degree distribution of the Follower Network (*top left*), strength distribution of the Retweet Network (*top right*) and strength distribution of the Mention Network (*bottom*). *The top* and *bottom panels* correspond to 20N and SOSInternetVE topics respectively

tative. This result displays the asymmetric shape of these networks, where the hubs that concentrate much of the incoming links, are often targeted by regular users, who neither mention nor retweet too much, and receive few of the collective attention. Previous works on network assortativity [5], state that social networks tend to be assortative, as popular people want to be friend with popular people, and regular people are usually friends among the regular people. However our measures indicate something different. Hu and Wang [7] reported that other online social networks are also disassortative. The reason for this observations, relies on the difference between

Table 78.3 Assortativity by degree of the followers, retweet and mention networks

Topic	Followers	Retweets	Mentions
20N	−0.09	−0.06	−0.09
SOSInternetVE	−0.10	−0.15	−0.14

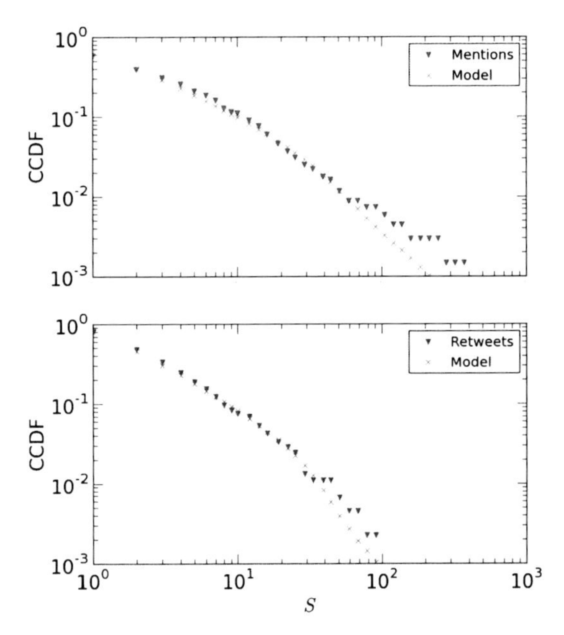

Fig. 78.3 Strength distributions of the political filtered Mention (*top*) and Retweet (*bottom*) networks and model results after 500 realizations

the online and offline world. For example, in Twitter regular people are now able to relate and communicate with popular accounts, either by following, mentioning or retweeting their messages. These new kinds of interactions are responsible for the changes in the structural and dynamical patterns previously reported on social networks.

Such different profiles also give place to the emergence of community structures, as the information consumers usually participate around their preferred information producers. In order to unveil such structures, we have performed community detection analysis based on modularity optimization [8] and random walks [9]. We have found that the retransmission and mention graphs present a higher modular structure than the followers one, being the retransmission graph even more segregative than the mentions map. Such structural differences reinforce the idea that the retransmissions and mentions channels are a substructure of the social substratum that endorses the individual preferences, and also indicate that people are more selective when taking action, than when just receiving the information [1].

On top of this, we have also found that the information producers, at the core of each community, are usually related to mainstream, celebrities or politicians accounts. This lead us to state that even though online social networks appear to be a pure social environment, traditional media agents hold loads of influence inside the network, that they use to boost their messages. However, according to the nature of the event and the interaction mechanism, some collectives may play a more influential role than others among the users. For example, in the 2011 Spanish electoral process [2], mentions are mostly targeted to politicians, while retransmissions are dominated by mainstream, since the first mechanism is used to send personal opinions and the other one is used to rapidly propagate information like news.

Finally, we have been able to model the modular and segregative structure of the mention and retransmission graphs, based on the formalism of heterogeneous preferential attachment [10], by designing connection rules for both micro and mesoscale. The idea behind this model is that the probability of a node i interacting with a node j not only depends on their respective degree, but also on an affinity value between them. This affinity value comes from a function that allow us to tune the mesoscale, independently from the microscale connectivity rules.

We tested this model with the mention and retweet networks of the Spanish electoral process, filtered by official politicians accounts. We found these subgraphs to be highly segregative, since the Pearson coefficient across parties are very close to 1 ($r_M = 0.905$ and $r_R = 0.990$), indicating a considerable lack of debate between the politicians. To model the mesoscale, we first calculated the affinity value across political parties, as the relative flux of interactions among them. In Fig. 78.3 we present the real and modeled strength function for both networks. It can be noticed that the model reproduces very well these distributions, as well as the Pearson coefficient across parties ($r_M = 0.86 \pm 0.03$ and $r_R = 0.989 \pm 0.005$).

In summary, our study reveals the complexity behind the interactions among users and the information diffusion process during particular events on Twitter. These interactions allow us to characterize and model the user's individual and collective behavior. We found that these topics were fed by a small portion of very active participants and driven by a smaller portion of very noticed influencers. These influncers are mostly related to main stream and celebrities, who use the social network to boost the importance of their messages. However, we found that influence might always be boosted by any participant when the activity is remarkably increased. The results obtained bring new insights into how people relate with each other in these communication tools and may serve as frameworks for professionals who use them, in order to maximize the network's potential.

References

1. Morales AJ, Losada JC, Benito RM (2012) Users structure and behavior on an online social network during a political protest. Physica A 391:5244–5253. doi:10.1016/j.physa.2012.05.015
2. Borondo J, Morales AJ, Losada JC, Benito RM (2012) Characterizing and modeling an electoral campaign in the context of Twitter: 2011 Spanish presidential election as a case study. Chaos 22:023138. doi:10.1063/1.4729139
3. Wu S, Hoffman JM, Mason WA, Watts DJ (2011) Who says what to whom on Twitter. In: 20th annual world wide web conference. ACM, New York

4. Cha M, Haddadi H, Benevenuto F, Gummad KP (2010) Measuring user influence on twitter: the million follower fallacy. In: Proceedings of the fourth international AAAI conference on weblogs and social media, Washington
5. Newman MEJ (2003) Mixing patterns in networks. Phys Rev E 67:026126
6. Foster JG, Foster DV, Grassberger P, Paczuski M (2010) Edge direction and the structure of networks. Proc Natl Acad Sci USA 107:10815
7. Hu HB, Wang XF (2009) Disassortative mixing in online social networks. Europhys Lett 86:18003
8. Blondel VD, Guillaume J-L, Lambiotte R, Lefebvre E (2008) Fast unfolding of communities in large networks. J Stat Mech 2008:P10008
9. Rosvall M, Bergstrom CT (2008) Maps of random walks on complex networks reveal community structure. Proc Natl Acad Sci USA 105:1118
10. Santiago A, Benito RM (2008) An extended formalism for preferential attachment in heterogeneous complex networks. Europhys Lett 82:58004

Chapter 79
Majority Rule with Differential Latency: An Absorbing Markov Chain to Model Consensus

Gabriele Valentini, Mauro Birattari, and Marco Dorigo

We study a collective decision-making mechanism for a swarm of robots. Swarm robotics [9] is a novel approach in robotics that takes inspiration from social insects to deal with large groups of relatively simple robots. Robots in a swarm act only on the basis of local knowledge, without any centralized director, hence, in a completely self-organized and distributed manner. The behavior of the swarm is a result of the interaction of its components, the robots. The analysis of collective decision-making mechanisms plays a crucial role in the design of swarm behaviors.

We analyze a swarm robotics system originally proposed by Montes de Oca et al. [7]. Robots in the swarm need to collectively decide between two possible actions to perform, henceforth referred to as action A and action B. Actions have the same outcome but different execution times. The goal of the swarm is to reach consensus on the action with the shortest execution time. In particular, Montes the Oca et al. studied this system in a collective transport scenario where robots in the swarm need to transport objects from a *source* area to a *destination* area. To this end, robots can choose between two possible paths. This corresponds to perform action A or action B. The two paths differ in length and thus in the traversal time. Each robot in the swarm has an *opinion* for a particular path. Moreover, an object is too heavy for a single robot to be transported. A team of 3 robots is needed. The team collectively decides which path to take considering the opinion favored by the majority.

Opinion formation models, such as the majority-rule model by Galam [2], allow us to study and analyze this kind of systems. Krapivsky and Redner [4] provided an

G. Valentini (✉) · M. Birattari · M. Dorigo
IRIDIA, CoDE, Université Libre de Bruxelles, 50, Av. F. Roosevelt, CP 194/6, 1050 Brussels, Belgium
e-mail: gvalenti@ulb.ac.be

M. Birattari
e-mail: mbiro@ulb.ac.be

M. Dorigo
e-mail: mdorigo@ulb.ac.be

T. Gilbert et al. (eds.), *Proceedings of the European Conference on Complex Systems 2012*, Springer Proceedings in Complexity, DOI 10.1007/978-3-319-00395-5_79,

analytical study of the majority-rule model under the assumption of a well mixed[1] population of agents. Later, Lambiotte et al. [5] extended the work of Krapivsky and Redner introducing the concept of *latency*. In the model of Lambiotte et al., when an agent switches opinion as a consequence of the application of the majority rule, it turns in a latent state for a latency period that has stochastic duration. A latent agent may still participates in voting, thus influencing other agents, but its opinion does not change as a result of the decision. This extension gives rise to a richer dynamics depending on the duration of the latency period. Based on these works, Montes de Oca et al. [7] proposed the *differential latency* model where the duration of the latency period depends on the particular opinion adopted. After a decision, differently from the model of Lambiotte et al., the team of agents become latent with a common latency period and is not involved in further voting until the end of the latency period. Montes de Oca et al. showed that the differential latency in the majority-rule model steers the agents towards consensus on the opinion associated to the shortest latency. Montes de Oca et al. applied these results to the study of the swarm robotics system described above by modeling actions of the robots as opinions and their execution times as the latency periods of different duration.

In the context of swarm robotics, a number of works has been devoted to the differential latency model. Montes de Oca et al. [7] first proposed a fluid-flow analysis of this model—using a system of ODEs—aimed at studying the dynamics leading to consensus. This analysis, derived in the limit of an infinite population, deterministically predicts consensus as a function of the initial configuration of the system. However, in a finite population, random fluctuations may drive the system to converge to the long path, even when the fluid-flow model predicts that it should converge to the short one. Later, Scheidler [10] extended the previous analysis using methods from statistical physics—e.g., master equation and Fokker-Planck equation—to derive continuous approximations of a system with a finite population size. With this approach, Scheidler was able to study the exit probability, i.e., the probability that the system eventually reaches consensus on the opinion associated to the shortest latency, and the expected time necessary to reach consensus. Finally, Massink et al. [6] provided a specification of the system using a stochastic process algebra. On the basis of this specification, the authors obtained a statistical model checking and a fluid-flow analysis.

Continuous approximations provide reliable predictions only when the number of robots is relatively large—e.g., thousands of robots. However, swarm robotics aims to design scalable control policies that operate for swarms of any size, ranging from tens to millions of robots. These models cover only the upper part of this range. Besides, from a continuous approximation model, it is usually hard to derive statistics different from the expected value, which in turn, often gives a poor representation of the underlying distribution—e.g., when the variance is large compared to the expected value or when the distribution is not symmetric.

[1] In a well mixed population each agent has the same probability to interact with each other agent [8].

The aim of this work is to study the majority rule with differential latency with an approach able to cope with the limitations of previous approaches. Inspired by the work of Banish et al. [1], we use the formalism of time homogeneous Markov chains with finite state space [3]. In particular, it results that the Markov chain describing the system is absorbing [3]. This approach allows us to consider systems of any finite size and to derive reliable estimations of both the exit probability and the distribution of the number of decisions necessary to reach consensus.

79.1 Markov Chain Model

We model the majority rule with differential latency in a system of M robots as an absorbing Markov chain [3]. Robots can be *latent* or *non-latent*. Only non-latent robots, once grouped in a team of 3 members, take part to the decision-making mechanism. As in Montes de Oca et al. [7], we consider a scenario where the number k of latent teams is constant, and where the latency period follows an exponential distribution whose expected value depends on the team's opinion. Without loss of generality, we consider the expected latency periods to be 1 for opinion A and $1/\lambda$ with $0 \leqslant \lambda \leqslant 1$ for opinion B. Moreover, we are interested in the number ϑ of applications of the majority rule, thus, we consider each application of the decision-making mechanism as one step of the process along the chain. At each step ϑ we consider 3 stages:

(1) A latent team becomes non-latent (it finishes its latency period).
(2) A new team of 3 robots is randomly formed out of the set of non-latent robots.
(3) The team applies the majority rule to decide the team's opinion. Next, it turns in a latent state.

We are interested in the evolution over ϑ of the number of robots with opinion A—the opinion associated to the shortest latency. Let $\mathbb{N}$ be the set of naturals. The state of the Markov chain is a vector $s = (s^l, s^n)$, where $s^l \in \{l : l \in \mathbb{N}, 0 \leqslant l \leqslant k\}$ is the number of latent teams with opinion A and $s^n \in \{n : n \in \mathbb{N}, 0 \leqslant n \leqslant M - 3k\}$ is the number of non-latent robots with opinion A. The state space of the Markov chain consists of m states, where $m = (k+1)(M-3k+1)$ is the cardinality of the Cartesian product of the domains of s^l and s^n. By s_i and s_j we refer to two generic states. By s_a and s_b we refer to the consensus states in which the whole swarm agrees on opinion A and B, respectively. Notice that s_a and s_b are the *absorbing* states of the chain, that is, states that once reached can never be left [3].

At the generic step ϑ, the process moves from $s(\vartheta) = s_i$ to $s(\vartheta+1) = s_j$ following the aforementioned 3 stages. At stage (1), a latent team finishes its latency period, becomes non-latent and disbands. The probability p_i that this team has opinion A is:

$$p_i = \frac{s_i^l}{s_i^l + \lambda(k - s_i^l)}. \tag{79.1}$$

The set of non-latent robots with opinion A increases of $c=3$ units, if the disbanding team has opinion A; and of $c=0$, otherwise. At stage (2), 3 random robots form a new team in the set of non-latent robots. We are interested in the probability q_i that the new team has a number $0 \leqslant d \leqslant 3$ of preferences for opinion A. This probability is given by the hyper-geometric distribution

$$q_i(d;c) = \frac{\binom{s_i^n+c}{d}\binom{M-3k-s_i^n+3-c}{3-d}}{\binom{M-3k+3}{3}} \tag{79.2}$$

of the $M-3k+3$ preferences in the current set of non-latent robots, composed of s_i^n+c votes for opinion A and $M-3k-s_i^n+3-c$ votes for opinion B. At stage (3), the majority rule is applied and the outcome is determined by the value of d. Eventually, the process moves to the next state $s(\vartheta+1)=s_j$.

Equations (79.1) and (79.2) allow us to define the transition probabilities between each possible pair of states s_i and s_j. These probabilities are the entries of the stochastic transition matrix P, which completely defines the dynamics of a Markov chain, cf. Kemeny and Snell [3]. However, not all pairs of states define a feasible move of the process along the chain according to the rules of the system, i.e., not all pair of states are *adjacent.* Two states s_i and s_j are adjacent if $\Delta_{ij}s = (\Delta_{ij}s^l, \Delta_{ij}s^n) = s_j - s_i$ appears in the first column of the following table. The correspondent transition probability P_{ij} is given in the second column:

$(\Delta_{ij}s^l, \Delta_{ij}s^n)$	P_{ij}	Stage (1)	Stage (2)
$(-1,3)$	$p_i q_i(3-\Delta_{ij}s^n;3)$	A	3B
$(-1,2)$		A	A2B
$(0,1)$		A	2AB
$(0,0)$	$p_i q_i(3-\Delta_{ij}s^n;3)+(1-p_i)q_i(\lvert\Delta_{ij}s^n\rvert;0)$	A	3A
		B	3B
$(0,-1)$	$(1-p_i)q_i(\lvert\Delta_{ij}s^n\rvert;0)$	B	A2B
$(1,-2)$		B	2AB
$(1,-3)$		B	3A

Columns three and four provide the corresponding events observed in stages (1) and (2), respectively: the opinion of the robots in the next latent team finishing its latency period and the opinions of the robots that randomly form a new team in the set of non-latent robots. For values of $\Delta_{ij}s$ not included in column one, the transition probability is $P_{ij}=0$.

The probabilistic interpretation of P is straightforward: at any step ϑ, if the process is in state $s(\vartheta)=s_i$ it will move to state $s(\vartheta+1)=s_j$ with probability P_{ij}. It is worth noticing that, being the consensus states s_a and s_b two absorbing states, the probability mass of s_a and s_b is concentrated in the corresponding diagonal entries of P, that is: $P_{aa}=1$ and $P_{bb}=1$.

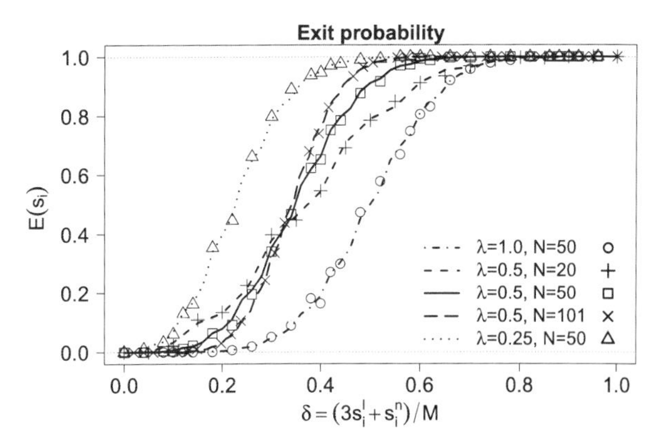

Fig. 79.1 Probability $E(s_i)$ to reach consensus on opinion A versus initial proportion δ of robots favoring that opinion for different settings of the system: ($M = 20$, $k = 6, \lambda = 0.5$), ($M = 50$, $k = 16, \lambda \in \{1, 0.5, 0.25\}$) and ($M = 101, k = 33, \lambda = 0.5$). *Lines* refer to the predictions of the Markov chain model, symbols refer to the average results of 1000 Monte Carlo simulations for each initial configuration of the system

79.2 Analysis of Opinion Dynamics

In order to analyze the dynamics of the majority rule with differential latency, we define, on the basis of P, the matrices Q, R, and N of Kemeny and Snell [3]. Q describes transitions between transient states, R gives the probability to move from a transient state to an absorbing state, and $N = (I - Q)^{-1}$ is the *fundamental* matrix with I being the identity matrix. From matrices Q, R, and N of the Markov chain model we study the behavior of the system. We validate the predictions of the model with the results of Monte Carlo simulations[2] averaged over 1000 independent runs for each choice of the parameters of the system.

First, we derive the exit probability $E(s_i)$, i.e., the probability that a system of M robots that starts in the initial configuration $s(\vartheta_0) = s_i$ reaches consensus on the opinion associated to the shortest latency—opinion A. This probability is given by the entries associated to the consensus state s_a of the product NR, which corresponds to the matrix of the absorption probabilities [3].

Figure 79.1 reports the predictions of the exit probability over the initial density $\delta = (3s_i^l + s_i^n)/M$ of robots favoring opinion A for several configurations of the system. As found by Scheidler [10], the larger is the expected latency period $1/\lambda$ associated to opinion B, the smaller is the initial number of preferences for opinion A such that the exit probability is $E(s_i) > 0.5$. Moreover, when the number of robots M increases, the exit probability approaches a step function around the critical density. It is worth noticing that the Markov chain model predicts the outcome of the simulations with great accuracy, regardless of the number of robots. A result that in general cannot be achieved using continuous approximations approaches.

[2]We simulated two sets of robots: latent teams characterized by an opinion and a latency, and non-latent robots described only by their opinions. The simulation proceeds as follow until consensus is reached: (1) the latent team having minimum latency is disbanded and its component robots are added to the set of non-latent robots, (2) 3 robots are randomly sampled from the set of non-latent robots and the majority rule is applied among them, (3) the new team is added to the set of latent teams and its latency is drawn from the exponential distribution according to the team's opinion.

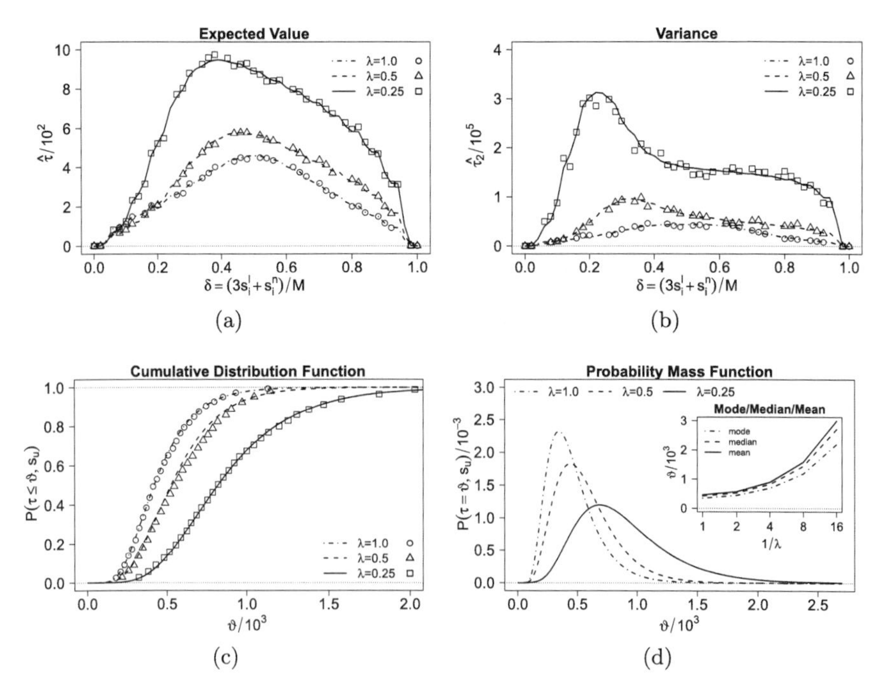

Fig. 79.2 Number τ of applications of the majority rule necessary to reach consensus for a system of $M = 50$ robots, $k = 16$ teams and $\lambda \in \{1, 0.5, 0.25\}$. (**a**) Expected value $\hat{\tau}$, (**b**) variance $\hat{\tau}_2$, (**c**) cumulative distribution function $P(\tau \leqslant \vartheta; s_u)$, (**d**) probability mass function $P(\tau = \vartheta; s_u)$ and details of mode, median and mean values. *Lines* refer to the predictions of the Markov chain model, symbols refer to the average results of 1000 Monte Carlo simulations for each initial configuration of the system

Next, we analyze the number τ of applications of the majority rule necessary to reach consensus. As stated above, we consider each step along the chain as one application on the decision-making mechanism. The expected value of τ is given by $\hat{\tau} = \xi N$, where ξ is a column vector of all 1s. The entries of $\hat{\tau}$ correspond to the row sums of the fundamental matrix N. In turn, N gives the mean sojourn time for each transient state of a Markov chain [3], that is, the expected number of times that a process started in state $s(\vartheta_0) = s_i$ passes from state s_j. The variance of τ is given by $\hat{\tau}_2 = (2N - I)\hat{\tau} - \hat{\tau}_{sq}$, where I is the identity matrix and $\hat{\tau}_{sq}$ is $\hat{\tau}$ with squared entries [3].

Figures 79.2(a) and 79.2(b) show the predictions of the expectation $\hat{\tau}$ and the variance $\hat{\tau}_2$ of the number of decisions necessary before consensus for a system with $M = 50$ robots. Again, the Markov chain model predicts the Monte Carlo simulations with great accuracy. Similarly to the findings of Scheidler [10] for the consensus time, the value of $\hat{\tau}$ is maximum near the critical density of the initial number of robots favoring opinion A. However, the expected number of decisions, which is related to the consensus time in Scheidler [10], is not a reliable statistics for this system. Indeed, the variance $\hat{\tau}_2$ is about three orders of magnitude larger than $\hat{\tau}$.

Finally, we derive the cumulative distribution function $P(\tau \leqslant \vartheta; s_i)$ of the number of decisions before consensus as well as its probability mass function $P(\tau = \vartheta; s_i)$. From a swarm robotics perspective, we are interested in the dynamics of a system initially unbiased, i.e., a system that starts with an equal proportion of preferences for the opinions A and B. Let $s(\vartheta_0) = s_u$ represents this initial unbiased configuration. Recalling that Q is the matrix of the transition probabilities for the transient states, we have that the entries Q^{ϑ}_{uj} of the ϑth power of Q give the probabilities to be in the transient state s_j at step ϑ when starting at s_u. Thus, the row sum of the uth row of Q^{ϑ} gives the probability to still be in one of the transient states. From this probability, we can derive the cumulative distribution function $P(\tau \leqslant \vartheta; s_u)$ simply by computing the series $\{1 - \sum_j Q^{\vartheta}_{uj}\}$ for values of ϑ such that $Q^{\vartheta} \to 0$.

Figure 79.2(c) shows the cumulative distribution function $P(\tau \leqslant \vartheta; s_u)$ for a system of $M = 50$ robots that starts unbiased. Obviously, the longer is the latency period $1/\lambda$ of opinion B, the larger is the number of applications of the majority rule necessary to reach consensus. Figure 79.2(d), provides the probability mass function $P(\tau = \vartheta; s_u)$, together with details of mode, median, and mean values of τ. As we can see, the values of the mode, median and mean statistics diverge for increasing values of the ratio $1/\lambda$ of the two expected latency periods. Moreover, when $1/\lambda \to \infty$ the shape of the distribution $P(\tau = \vartheta; s_u)$ tends to a flat function, thus revealing that the variance dominates the system.

79.3 Conclusion

We designed an absorbing Markov chain model for collective decisions in a system with a finite number of robots based on the majority rule with differential latency. Using our model, we derived: the probability that a system of M robots reaches consensus on the opinion associated to the shortest latency period, and the distribution of the number of applications of the majority rule necessary to reach consensus. This latter reveals that the system is characterized by a large variance of the number of decisions necessary before consensus, and thus, that its expected value, which was mainly adopted in previous studies, is a poor statistic for this system. In contrast to continuous approximations, we explicitly model the state space of the system—which is discrete—and the transition probabilities governing its dynamics. This approach allows us to always derive reliable predictions of a system regardless of its size.

Our contribution is relevant from a swarm robotics perspective not only for the reliability of its predictions; but also, because it allows us to perform a deeper analysis of the system. The analysis of our Markov chain model, with particular regard to the distribution of the number of decisions necessary to consensus, gives the possibility to perform statistical inference on certain interesting aspects of the system. Moreover, the approach can be easily extended to other voting schemata, allowing the comparison at design time of different choices for the decision-making mechanism.

In real-robot experiments, latency periods are unlikely to be exponentially distributed. Moreover, it is hard to ensure a constant number of teams in time. These assumptions represent hard constraints for a swarm robotics system. Massink et al. [6] propose to cope with these constraints modeling the latency period with an Erlang distribution. We plan to extend our approach in a similar way and to validate the resulting model with physics-based simulations and real-robot experiments.

Acknowledgements The research leading to the results presented in this paper has received funding from the European Research Council under the European Union's Seventh Framework Programme (FP7/2007–2013)/ERC grant agreement no. 246939. Mauro Birattari and Marco Dorigo acknowledge support from the F.R.S.-FNRS of Belgium's Wallonia-Brussels Federation.

References

1. Banisch S, Lima R, Araujo T (2011) Agent based models and opinion dynamics as Markov chains. arXiv:1108.1716v2
2. Galam S (1986) Majority rule, hierarchical structures, and democratic totalitarianism: a statistical approach. J Math Psychol 30(4):426–434
3. Kemeny JG, Snell JL (1976) Finite Markov chains. Springer, New York
4. Krapivsky PL, Redner S (2003) Dynamics of majority rule in two-state interacting spin systems. Phys Rev Lett 90:238701
5. Lambiotte R, Saramaki J, Blondel VD (2009) Dynamics of latent voters. Phys Rev E 79:046107
6. Massink M, Brambilla M, Latella D, Dorigo M, Birattari M (2012, in press) Analysing robot decision-making with bio-pepa. In: Proceedings of the eighth international conference on swarm intelligence, ANTS 2012. Lecture notes in computer science. Springer, Berlin
7. Montes de Oca M, Ferrante E, Scheidler A, Pinciroli C, Birattari M, Dorigo M (2011) Majority-rule opinion dynamics with differential latency: a mechanism for self-organized collective decision-making. Swarm Intell 5:305–327. doi:10.1007/s11721-011-0062-z
8. Nowak MA (2006) Five rules for the evolution of cooperation. Science 314(5805):1560–1563
9. Şahin E (2005) Swarm robotics: from sources of inspiration to domains of application. In: Şahin E, Spears W (eds) Swarm robotics. Lecture notes in computer science, vol 3342. Springer, Berlin, pp 10–20
10. Scheidler A (2011) Dynamics of majority rule with differential latencies. Phys Rev E 83:031116

Chapter 80
Computational Modeling of Collective Behavior of Panicked Crowd Escaping Multi-floor Branched Building

Dmitry Bratsun, Irina Dubova, Maria Krylova, and Andrey Lyushnin

Abstract The collective behavior of crowd leaving a room is modeled. The model is based on molecular dynamics approach with a mixture of socio-psychological and physical forces. The new algorithm for complicatedly branched space is proposed. It suggests that each individual develops its own plan of escape, which is stochastically transformed during the evolution. The algorithm includes also the separation of original space into rooms with possible exits selected by individuals according to their probability distribution. The model has been calibrated on the base of empirical data provided by fire case in the nightclub "Lame Horse" (Perm, 2009). The algorithm is realized as an end-user *Java* software. The code has been tested on a number of multi-level buildings with complicated geometry.

With the increasing size and frequency of mass events leading to more often crowd disasters, the study of collective behavior of panicked crowd has become important research area. However, even successful modeling approaches are hard to calibrate since the dynamics of crowd is sensitive to changes in the model. Probably, the only way to calibrate the model is to compare the results of computer simulations with scenarios of real events giving the most comprehensive empirical data.

In this paper, it is presented the computational modeling of the crowd panicking in closed space of multi-level branched building. The model is based on molecular

D. Bratsun (✉)
Theoretical Physics Department, Perm State Pedagogical University, Perm, Russia
e-mail: dmitribratsun@rambler.ru

I. Dubova · M. Krylova · A. Lyushnin
Department of Informatics, Perm State Pedagogical University, Perm, Russia

I. Dubova
e-mail: irina.dubova90@gmail.com

A. Lyushnin
e-mail: andry@pspu.ac.ru

T. Gilbert et al. (eds.), *Proceedings of the European Conference on Complex Systems 2012*, Springer Proceedings in Complexity, DOI 10.1007/978-3-319-00395-5_80,

Fig. 80.1 Pedestrian chooses his path randomly if there are several points of gathering

dynamics approach proposed by Helbing et al. [1]. Their model assumes a mixture of socio-psychological and physical forces influencing the behaviour in a crowd:

$$M_i \frac{d\mathbf{V}_i}{dt} = \mathbf{F}_i^1 + \sum_{i \neq m} \mathbf{F}_{im}^2 + \sum_{i \neq m} \mathbf{F}_{im}^3 + \sum_j \mathbf{F}_{ij}^{2W} + \sum_j \mathbf{F}_{ij}^{3W}, \qquad (80.1)$$

where $\mathbf{V}_i = d\mathbf{r}_i/dt$ is velocity of pedestrian i of mass M_i and radius R_i and

$$\mathbf{F}_i^1 = M_i \frac{\mathbf{U}_i - \mathbf{V}_i}{\Delta t_i}, \qquad (80.2)$$

$$\mathbf{F}_{im}^2 = A\mathbf{n}_{im} \exp \frac{D_{im}}{B}, \qquad (80.3)$$

$$\mathbf{F}_{im}^3 = D_{im} H(D_{im}) \big(k\mathbf{n}_{im} + K\big[(\mathbf{V}_m - \mathbf{V}_i) \cdot \boldsymbol{\tau}_{im}\big]\boldsymbol{\tau}_{im}\big), \qquad (80.4)$$

$$\mathbf{F}_{im}^{2W} = A^W \mathbf{n}_{im}^W \exp \frac{D_{im}^W}{B^W}, \qquad (80.5)$$

$$\mathbf{F}_{im}^{3W} = D_{im}^W H\big(D_{im}^W\big)\big(k\mathbf{n}_{im}^W + K\big[\mathbf{V}_i \cdot \boldsymbol{\tau}_{im}^W\big]\boldsymbol{\tau}_{im}^W\big), \qquad (80.6)$$

$$D_{im} \equiv R_i + R_m - |\mathbf{r}_i - \mathbf{r}_m|. \qquad (80.7)$$

Here $\mathbf{n}_{im}$ and $\boldsymbol{\tau}_{im}$ are normal and tangent unit vectors to the contact line between i and m pedestrians respectively; A, B, A^W, B^W, k, K are parameters. H stands for the function of Heaviside. The first force (80.2) is responsible for the state of panic of pedestrian. Each of pedestrians likes to move with a certain desired speed $\mathbf{U}_i$ in a certain direction, and therefore tends to correspondingly adapt his actual velocity $\mathbf{V}_i$ with a certain characteristic time Δt_i. Simultaneously, he tries to keep a velocity-dependent distance from other pedestrians j (80.3) and walls (80.5). The forces (80.4) and (80.6) describe the effects of counteracting body compression and a sliding friction impeding relative tangential motion.

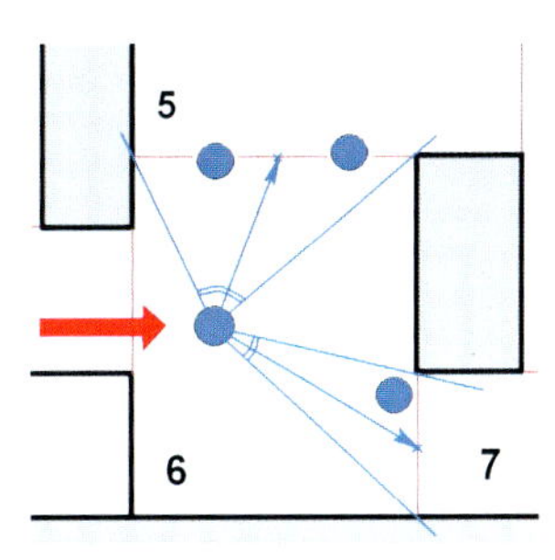

Fig. 80.2 Scheme illustrating the algorithm for selecting by the pedestrian a way out of the room

Keeping in mind the modeling in multi-floor buildings with a complicated structure, we have supplemented the model proposed in [1] by two important improvements [2]. First, we divide the initial space into separate rooms and define for each room the point (or points) of the gathering of pedestrians. Entering into a room everyone tends to reach this point. In the case when the number of gathering points more than one, the pedestrian must choose his path of the movement (Fig. 80.1). However, depending on how close pedestrian to a certain gathering point and how many other pedestrians seen in that direction the probability is higher for closer point and smaller number of people. The simplest algorithm provides a random selection based on distribution:

$$P_5 = \frac{N_7}{N_5 + N_7}, \tag{80.8}$$

where N_5 and N_7 are number of pedestrians in the direction 5 and 7 respectively (Fig. 80.2). As a result, each participant of crowd develops its own plan how to get out the building. This plan may be modified over time depending on the situation. Since the final result of stochastic system may vary, it should be averaged over the realisations.

The model has been calibrated using the empirical data provided by deadly Lame Horse fire [3]. The fire has occurred on December 5, 2009, around 1 a.m. in the nightclub "Lame Horse" located in Perm, Russia. A total of 282 people had reportedly been invited to the club's party. The fire started when sparks from fireworks ignited the low ceiling and its willow twig covering. The fire quickly spread to the walls and damaged the building's electrical wiring, causing the lights to fail. When the evacuation started, some people left via rear exits. The vast intake of oxygen turned the club's hall into a large fire tube and boosted the spread of fire. As fumes and smoke overtook the air, panic erupted and patrons stampeded toward the exit. According to witnesses, one leaf of the club's double doors was sealed shut, and the public was unaware of the backdoor exit behind the stage not shown by emergency lighting.

Subsequent events have shown that the people had only about 60 second to escape the club. But when fumes and smoke overtook the air, the panic has erupted. It is known that 153 people have died as a result of the fire and 62 people have become disabled. And only 70 persons have escaped the club itself within first minute.

In Fig. 80.3 we give an example of numerical simulation of panicked crowd of 283 pedestrians (143 men and 140 women) trying to escape the nightclub "Lame

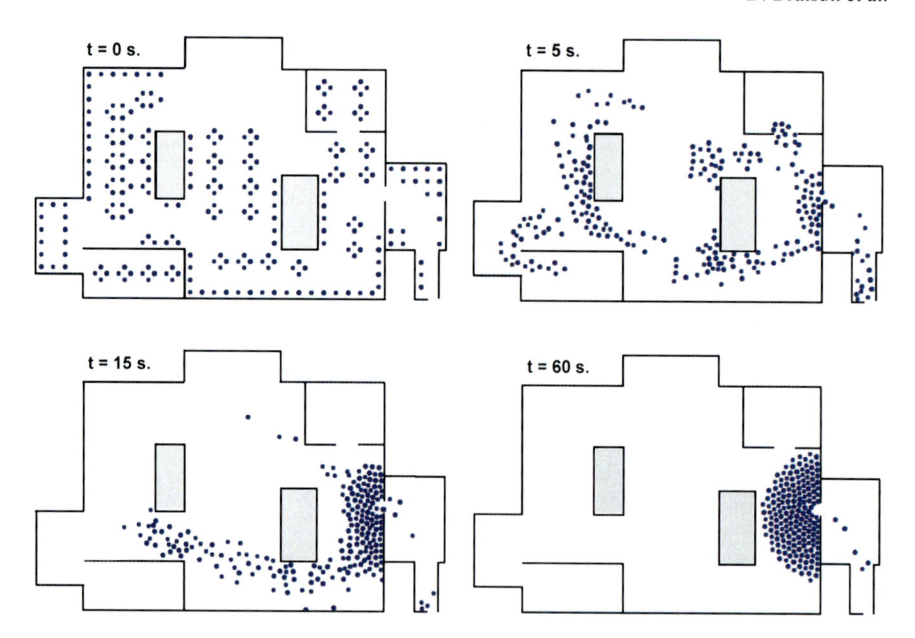

Fig. 80.3 Simulation of 283 pedestrians escaping the nightclub "Lame Horse"

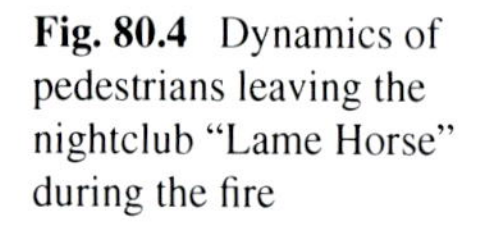
Fig. 80.4 Dynamics of pedestrians leaving the nightclub "Lame Horse" during the fire

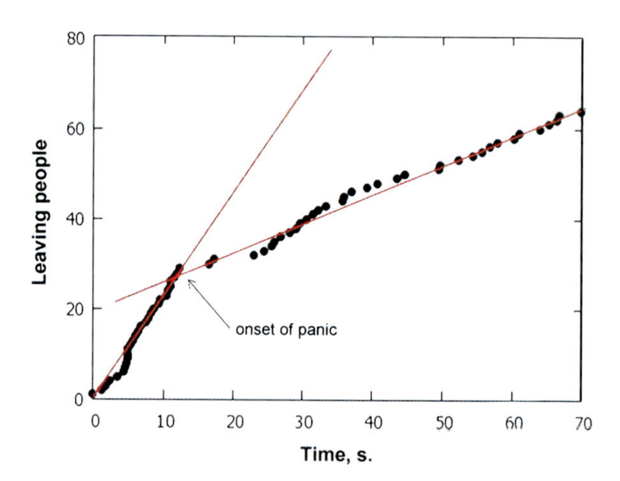

Horse". The plan of the club was taken from open sources. One can see that finally it is formed arch-like blocking of the exit. All people who were not able to leave the room within 60 second can be considered as victims of the fire. Figure 80.4 illustrates dynamics of number of leaving people in time. One can notice that the normal regime of the exit from the building extends approximately to 13 seconds. Then panic starts and the rate of leave falls sharply. By calibrating force functions and simulation parameters we have obtained the same rate as it was in real event: finally only 60 people have leaved the club within 60 second (Fig. 80.4).

The proposed algorithm has been applied to a number of multi-level buildings with complicated geometry. Such program could help to designers and architects to estimate potential dangers of internal structure of buildings for people in extraordinary situation and to minimize it on designing stage.

Acknowledgements The work was supported by the Department of Science and Education of Perm region (project C26/244), the Ministry of Science and Education of Russia (project 1.3103.2011) and Perm State Pedagogical University (project 031-F).

References

1. Helbing D, Farkas I, Vicsek T (2000) Simulating dynamical features of escape panic. Nature 407:487–490
2. Aptukov AM, Bratsun DA (2009) Modelling group dynamics of crowd panicking in the closed room. Vestn Perm Univ Fiz 3:18–23
3. BBC News 05.12.2009. http://news.bbc.co.uk/2/hi/europe/8396587.stm

Chapter 81
Spread of Disease During a Social Event

Lara Goscé and Anders Johansson

Abstract The aim of this work is to study how a disease can spread into a population during an event with a fixed duration, for example a congress, a concert, a religious event etc.

We will use the basic theory on networks and we'll make some new assumptions.

In the end we will try to find a boundary on the number of contacts so that the disease will eventually die out.

81.1 Introduction

We can start by imagining a multitude of individuals coming from different places in the world attending a single event with a limited duration, like a congress a few days long. We suppose the presence of a disease that could be spread by physical contact. Illnesses like the flu spread from person to person with droplets from the cough or sneeze of an infected person, the germs in these droplets can often live on someone hands for 2 hours or longer and can spread when people touch these by an handshake and then touch their eyes, mouth, and nose.

Since the limited duration of the event we are supposing that there are no removed individuals so we will consider only infected and susceptible compartments.

The individuals in this model are represented by vertex and the contacts between them by edges. For each contact between a susceptible individual and an infected one there is a probability that the infection will actually be transmitted. In this context, for simplicity, we are supposing that this probability is 1, so all contacts transmit infection.

81.2 The Model

We can now star to describe the model, we will follow the development of [1, 2] but we will make a new assumption.

L. Goscé (✉) · A. Johansson
Faculty of Engineering, University of Bristol, Bristol, UK

T. Gilbert et al. (eds.), *Proceedings of the European Conference on Complex Systems 2012*, Springer Proceedings in Complexity, DOI 10.1007/978-3-319-00395-5_81,

Consider the total population number N which is known because it is the number of people attending the conference. We imagine to divide the population in 4 subcategories:

1. Ph.D. students: n_1
2. Researchers: n_2
3. Professors: n_3
4. Keynote speakers and prizes winners: n_4

where n_i, $i = 1,2,3,4$ is the number of people belonging to each of this categories, this numbers are also known by the organizers of the conference. Obviously: $n_1 + n_2 + n_3 + n_4 = 1$.

For each category there is a parameter k_i, $i = 1,2,3,4$ which represent the average number of contacts of each category during the whole duration of the conference, this parameters are unknown. This mean that we are supposing that in an event the number of contacts of an individual depends of the role of that individual in that context. Of course the contacts depend also on the psychology and the sociality of each individual and it's possible that there would be PhD students more able than others to socialize with other people, but in the end the parameters k_i are average numbers and is a safe assumption to say that in a scientific context a generic student will have less contacts than a generic professor or Nobel prize winner. However we will come back on this numbers later on.

$\{p_{k_i}\}_{i=1,2,3,4}$ is the degree distribution of the graph, so each p_{k_i} is the fraction of the vertices having degree k_i. So:

$$p_{k_1} = \frac{n_1}{N}, \qquad p_{k_2} = \frac{n_2}{N}, \qquad p_{k_3} = \frac{n_3}{N}, \qquad p_{k_4} = \frac{n_4}{N}.$$

Of course:

$$\sum_{i=1}^{4} p_{k_i} = 1.$$

We can consider now the *generating function*:

$$G_0(z) = \sum_{i=1}^{4} p_{k_i} z^{k_i} = \frac{n_1}{N} z + \frac{n_2}{N} z^2 + \frac{n_3}{N} z^3 + \frac{n_4}{N} z^4$$

It's easy to prove that:

$$G_0(0) = 0, \qquad G_0(1) = 1, \qquad G_0'(z) > 0, \qquad G_0''(z) > 0$$

We define the *mean degree*:

$$\langle k \rangle = \sum_{i=1}^{4} k_i p_{k_i} = k_1 \frac{n_1}{N} + k_2 \frac{n_2}{N} + k_3 \frac{n_3}{N} + k_4 \frac{n_4}{N} = G_0'(1)$$

Suppose to reach a generic vertex by following a link, we will define the *excess degree* of that vertex the number of all of his links except the one we have used to reach it, so: $k_i - 1$, $i = 1,2,3,4$.

The probability of reaching a vertex with a given excess degree depends on the number of links of that vertex (k_i) and on his degree (that is number of vertex having that same excess degree, p_{k_i}), this must be weighted for a constant so that the sum of this probability is 1 (that is $\frac{1}{\langle k\rangle}$).

$$q_{k_i-1} = \frac{k_i p_{k_i}}{\langle k\rangle}, \quad i = 1, 2, 3, 4$$

We can now define a *generating function for the excess degree*:

$$G_1(z) = \sum_{i=1}^{4} q_{k_i-1} z^{k_i-1} = \frac{1}{\langle k\rangle}\sum_{i=1}^{4} k_i p_{k_i} z^{k_i-1} = \frac{1}{\langle k\rangle} G_0'(z)$$

The mean excess degree is:

$$\langle k_e\rangle = \sum_{i=1}^{4}(k_i - 1) q_{k_i-1} = \frac{1}{\langle k\rangle}\sum_{i=1}^{4} k_i(k_i-1) p_{k_i} = \frac{1}{\langle k\rangle}\sum_{i=1}^{4} k_i^2 p_{k_i} - \frac{1}{\langle k\rangle}\sum_{i=1}^{4} k_i p_{k_i}$$
$$= \frac{\langle k^2\rangle}{\langle k\rangle} - 1 = G_1'(1) = \mathcal{R}_0$$

This number is by definition the average number of secondary infection when a single infected individual is introduced in the population, that is the *basic reproduction number.*

Following the development of [1, 2] it is possible to show that if:

- $\mathcal{R}_0 < 1$, the probability that the infection will eventually die is 1
- $\mathcal{R}_0 > 1$, there is a probability that the infection will persist and will lead to an epidemic.

81.3 Control

Let's now try to find some conditions to gain control of the disease so that it won't become an epidemic.

The infection will die if:

$$\mathcal{R}_0 < 1 \iff \frac{1}{\langle k\rangle}\sum_{i=1}^{4} k_i(k_i-1) p_{k_i}$$
$$\iff \frac{k_1(k_1-1)\frac{n_1}{N} + k_2(k_2-1)\frac{n_2}{N} + k_3(k_3-1)\frac{n_3}{N} + k_4(k_4-1)\frac{n_4}{N}}{k_1\frac{n_1}{N} + k_2\frac{n_2}{N} + k_3\frac{n_3}{N} + k_4\frac{n_4}{N}} < 1$$

so, in order to have that fraction less than 1 we need that:

$$k_1(k_1-1)\frac{n_1}{N} + k_2(k_2-1)\frac{n_2}{N} + k_3(k_3-1)\frac{n_3}{N} + k_4(k_4-1)\frac{n_4}{N}$$
$$< k_1\frac{n_1}{N} + k_2\frac{n_2}{N} + k_3\frac{n_3}{N} + k_4\frac{n_4}{N}$$

this means that:

$$k_1(k_1-2)\frac{n_1}{N}+k_2(k_2-2)\frac{n_2}{N}+k_3(k_3-2)\frac{n_3}{N}+k_4(k_4-2)\frac{n_4}{N}<0$$

We are supposing that each individual interact with at least another one each $k_i > 0$, $i = 1, 2, 3, 4$ and, also $\frac{n_i}{N} > 0$, $i = 1, 2, 3, 4$ being ratios of positive quantities.

This means that we have to focus our attention on the quantities $k_i - 2$, $i = 1, 2, 3, 4$.

The simplest case is that:

$$\begin{aligned} k_1 &< 2 \\ k_2 &< 2 \\ k_3 &< 2 \\ k_4 &< 2 \end{aligned}$$

this means that each vertex has no more than one edge and so it's clear that at this point the infection is immediately eradicated since the infected individual infects the only other individual he is in contact with and the latter has no one to transmit the infection since his only edge is with the individual from which he received the infection.

This case is clearly not much realistic.

What we can do, instead, is trying to immunize one of the category so to not allow the disease to spread in the whole population. This mean that we are willing to make one of the quantities k_i less than one by control measures such as hand washing, face masks or vaccination. Which k_i will be more convenient to reduce will depend on the number of people belonging to each category. On the other hand, what we will do as a future work, is to consider a transmission probability less than 1, in this way not all the contacts will lead to a contagion and much more realistic cases will be studied.

81.4 A Few Considerations on the Number of Contacts

If we think about the parameters k_i, $i = 1, 2, 3, 4$ it is legitimate to suppose that $k_1 < k_2 < k_3 < k_4$, it is indeed plausible that a student (which is just entering into the world of research) has less contacts than a professor who has been in the field for years. We are saying that the number of contacts depend on the experience and the status of the individual himself.

We can so define a linear dependence between the smallest parameter and the others:

$$k_1, k_2 = s_1 k_1, \quad k_3 = s_2 k_1, \quad k_4 = s_3 k_1$$

where s_1, s_2, s_3 are positive real constants bigger than one and $s_1 < s_2 < s_3$.

This constants may be estimated by researches and inquiries on people in the academic field asking them personal experiences on the changes of their social interaction during their career.

So, if we go back at the basic reproductive number we have that:

$$\mathcal{R}_0 < 1 \iff k_1(k_1-2)\frac{n_1}{N} + s_1k_1(s_1k_1-2)\frac{n_2}{N} + s_2k_1(s_2k_1-2)\frac{n_3}{N} + s_3k_1(s_3k_1-2)\frac{n_4}{N} < 0$$

so by simple calculations

$$k_1^2\left(\frac{n_1}{N} + s_1^2\frac{n_2}{N} + s_2^2\frac{n_3}{N} + s_3^2\frac{n_4}{N}\right) - 2k_1\left(\frac{n_1}{N} + s_1\frac{n_2}{N} + s_2\frac{n_3}{N} + s_3\frac{n_4}{N}\right) < 0$$

so that

$$0 < k_1 < \frac{2(\frac{n_1}{N} + s_1\frac{n_2}{N} + s_2\frac{n_3}{N} + s_3\frac{n_4}{N})}{\frac{n_1}{N} + s_1^2\frac{n_2}{N} + s_2^2\frac{n_3}{N} + s_3^2\frac{n_4}{N}} = \theta$$

We have obtained a limitation θ on the number of contacts the Ph.D. students can make so that the disease will eventually die. In this simple case this boundary is still one. Goal of this research is to extent this procedure to much complicated cases in which disease control and transmissibility pays a central role.

References

1. Brauer F (2008) An introduction to networks in epidemic modelling. In: Brauer F, van den Driessche P, Wu J (eds) Mathematical epidemiology. Lecture notes in mathematics. Mathematical biosciences subseries, vol 1945. Springer, Berlin
2. Newman MEJ (2002) Spread of epidemic disease on networks. Physical Review E 66:016128

Chapter 82
A Collective Binomial Learning Methodology

Xiao Perdereau

Abstract In second-language learning, learners frequently have a poor environment for speaking and hearing the target language. Learning efficiency is thus limited. We propose a methodology involving the creation of temporary social structures. Collective interactions fed back among individuals and environment are constructed on a computer and practiced in a real world. A dynamic learning system which coherently ties together the practitioner's design, the learner's performance and the researcher's theories is possible. Our results call for language learning structures to include adaptive spoken structures, in contrast with existing educational systems.

Keywords Social structure · Social interaction · Feedback · Collective behavior · Learning system

82.1 Introduction

It is well known that people acquire spoken language quickly in favorable social and cultural environments. The most familiar example is children acquiring their mother tongue. Another example is foreign language acquisition in a country where the language is spoken. Both examples involve interactions between the learner and his or her social environment.

A more common situation for second-language learning is one in which the learners lack both an environment conducive to childlike language learning and any possibility to go to a country where the language is spoken. They have only one or two instructors in a classroom. These circumstances limit learning efficiency. In this paper, we propose to create appropriate social structures that foster social interactions, not only on paper or on a computer, but also in the real human world. Due to the

X. Perdereau (✉)
Centre d'études chinoises de l'Université de Bourgogne, Dijon, France
e-mail: xiao.chen-perdereau@u-bourgogne.fr

X. Perdereau
Laboratoire Interdisciplinaire Carnot de Bourgogne, UMR 6303, CNRS-Université de Bourgogne, 9, Av. A. Savary, BP 47870, 21078 Dijon Cedex, France

T. Gilbert et al. (eds.), *Proceedings of the European Conference on Complex Systems 2012*, Springer Proceedings in Complexity, DOI 10.1007/978-3-319-00395-5_82,

complexity of general learning systems, we need to construct new methodologies in which research and pedagogy guide each other.

82.2 Modelization

Each learner is considered an agent. The agents are classified in two equal categories, A-speakers and B-speakers. We define a binom as a pair of two agents, a A-speaker and a B-speaker. The two agents in a binom are partners.

We also construct a social structure, which involves a group and a number of spectators. The group is composed of a number (N) of binoms, and the spectators are external agents not belonging to the binoms. The group is then composed of $2N$ agents. N is an integer.

In the learning system, we consider 3 levels of interaction.

Level 1: Partner–Partner interaction
Level 2: Binom–Binom interaction
Level 3: Group–Spectators interaction

In a Partner–Partner interaction, the rules are based on cooperation. Each agent aims not only to improve his or her own proficiency, but also to have common improvement together with his or her partner. The agent's individual actions aid the welfare of the binom.

In a Binom–Binom interaction, rules are based on competition: the binoms should compete with each other, and if they observe differences indicating that certain binoms perform better than others, the poorer performers should reorganize their binomial learning procedures and cooperate more efficiently in order to give a better joint production.

In a Group–Spectator interaction, the rules are based on assessment. External agents (spectators) assess both partners in both languages and give the agents in the group a performance score.

82.3 Experiment

We built a temporary social structure in which all individuals were language learners. Approximately half of them were A-speakers learning language B, and the other half were B-speakers learning language A. A and B correspond to any two different languages. Every two different-language speakers temporarily formed a pair, which we call a binom, or more precisely an AB-binom. In a binom, the two individuals are partners.

In the learning sessions described, we recruited adult participants. Regardless of their main studies, they were all second-language learners. Half of them were native speakers of language A. We also included international students who were native speakers of language B. The participants were mostly at the beginner level in their second language. In this experiment, all learners of language A were partnered with learners of language B. Thus binoms were composed. These binoms form a group, which we call a binomial group.

We constructed a series of five sessions, which we called Binomial Learning. A session occurred every week. Each session was one hour and thirty minutes long. The time was managed as follows: the first session was guided by an instructor. Each language group had its own separate introductory session, with A-speakers together in one introductory session and B-speakers together in a separate introductory session.

The second, third and fourth sessions were devoted to binomial tasks: the learners met each other and chose partners in order to compose binoms. Participants were asked to learn a little about their partners, make their own speech fully understood, and present their partners and themselves to the spectators. This last task was done as a pair. The binoms could then change partners if they wished. Within each binom, the partners made agreements to make a speech in both languages (Chinese and French). A bilingual speaker was present at all sessions, but only gave translations when necessary. The fifth session was an organized party. An evaluation team examined the performance of each binom. Participants' evaluations were also collected.

Every individual was required to speak his or her partner's language. This rule guaranteed that for every learner, the language heard ("perceived language") was always L1, the learner's mother tongue, while the language spoken ("produced language") was always L2, the target language [1].

Typical phrases beginners might use and basic greetings [2] were provided to the participants at the start of a session. Inspired by Chomsky's definition of grammar [3] and the 'Learning by unlearning' method of Yang [4], we did not spend time learning or teaching grammar. Instead, we let the learners themselves explore their innate abilities for recognizing rules by comparing the syntax of the target language with that of their mother tongue.

We assumed that beginners would need materials they could use immediately. For this reason, the typical phrases we provided were adapted to each learner for immediate use in the organized events, also called "usage-events" [5]. The typical items comprised only a part of the speech used in each session. A second part was the native speakers' improvised productions. These productions were spontaneous and closely adapted to the events unfolding during the session. These cannot be replaced by any prefabricated language material. A third part of the speech was the learners' productions. These new utterances are also important.

82.4 Results

To assess a score for fluency, we chose to use rate of speech as a measurable marker. The participants showed surprising improvement in fluency and accuracy. Preliminary results suggest collective increase in learning efficiency. More analysis, including statistics, will come in the near future.

In measuring the time spent replicating the predetermined utterances during the five sessions, we saw that a learner in the binomial group spent less time than a learner in a traditional classroom learning the same language materials. We determined that the time was shorter because, according to the speaking rule, learners spent half of each session helping their partners. Despite the shorter time dedicated to each learner's target language, the level of improvement in fluency was higher than that seen in traditional classroom learning.

Due to the inclusion of native speakers, the binomial organization fosters feedback through interaction. There is an adaptation between partners; this is a part of the interaction. As pointed out by de Bot and Larsen-Freeman [6], co-adaptation is mutual, and the language resources of interlocutors are dynamically altered as each adapts to the other.

82.5 Conclusion

Binomial learning is favorable for language learning because it activates multi-leveled social interactions during the language-learning period. It should have a place in the landscape of the learning system diversity. The oral communication fostered by the interactions is invaluable to cognition. Therefore, spoken partnered communication should not be placed towards the end of the language-learning process, as is the case in many educational programs. Rather, spoken partnered communication belongs at the very beginning of the learning program, and it should continue in parallel with the learning progression. Educational systems should adapt language-learning programs to maximize the function of oral communication.

Acknowledgements We would like to thank the Presidency of Burgundy University for creating and supporting the Center of Chinese Studies. We also acknowledge the administrative support of Christoph Monny, Jean-Michel Dorlet and Maria Frich.

This work was financed by Conseil Régional de Bourgogne.

References

1. Perdereau X (2011) Mirror binomial language learning. In: Proceedings of fifteenth international conference on cognitive and neural systems, Boston, p 32
2. Matinrouge C (2009) Soirée zou. Yin et Langue, Paris
3. Chomsky N (1956) Three models for the description of language. IRE Trans Inf Theory 2(3):113–124
4. Yang CD (2006) The infinite gift: how children learn and unlearn the languages of the world. Scribner's, New York
5. Verspoor M, Behrens H (2011) Dynamic systems theory as a usage—based approach to second language development. In: Verspoor MH, de Bot K, Lowie W (eds) A dynamic approach to second language development. Benjamins, Amsterdam, pp 25–38
6. de Bot K, Larsen-Freeman D (2011) Researching second language development from a dynamic systems theory perspective. In: Verspoor MH, de Bot K, Lowie W (eds) A dynamic approach to second language development. Benjamins, Amsterdam, pp 5–23

Chapter 83
A Model for Social Network Evolution Affected by Individual Tolerance to Heterogeneity

Haoxiang Xia and Peng Liu

83.1 Extended Abstract

During the last decades, the evolution of social networks has attracted widespread attention; there are a lot of studies on models and mechanisms of social network evolution in statistical physics, social science, mathematics, computer science and other fields, and fruitful outcomes have been obtained, most notably the dynamical models on the formation of the small-world networks [9] and the scale-free networks [1], as well as tremendous subsequent contributions. Nevertheless, further explorations are still deserved to clarify the driving forces and mechanisms that underlie network evolution. In this work, we concern the possible influence of the social mechanism of "homophily" [5]. In essence, homophily indicates that individuals are likely to interact to others who are similar to themselves. Obviously this mechanism has prominent influence on the formation of social relationships and consequently the evolution of social network structure. Schelling [8] gave an early attempt to investigate the effect of homophily on the neighborhood structure in social communities in his well-noted segregation model on the regular lattice. With the integration of complex network theory and homophily research, some scholars gradually take homophily as the basis of micro-social relationship formation, to examine the social network structure evolution under different social contexts [2, 3]. A factor that may deserve further attention is the degree of homophily in the formation and maintenance of social relationships, in other word, individuals' tolerance to maintain a "heterogeneous" relationship. This work is therefore to study how such individuals' tolerance to heterogeneity affects the evolution of social network with a simple agent-based model.

In the presented model, a certain number of agents are interlinked to form a random network that is initially connected. The topological structure of the network then evolves over time through an edge-rewiring process. At each iteration step,

H. Xia (✉) · P. Liu
Institute of Systems Engineering, Dalian University of Technology, Dalian, Liaoning, China

T. Gilbert et al. (eds.), *Proceedings of the European Conference on Complex Systems 2012*, Springer Proceedings in Complexity, DOI 10.1007/978-3-319-00395-5_83,

a focal agent is arbitrarily chosen and this agent rewires one of its existing edges toward another non-neighboring agent that is randomly selected from the whole population. The decision of edge removal is based on the agent's tolerance to the heterogeneity of the focal agent in its neighborhood, which is measured by a threshold d in the model. The focal agent computes the differences of its "trait" or state with all its neighbors. If all the computed differences are not greater than the threshold d, it keeps all the current neighbors unchanged. Otherwise, the edge between the focal agent and the neighbor that is with largest difference in trait is removed and the focal agent then establishes a new edge to an arbitrarily-selected non-neighbor. In this edge-rewiring rule, the small d value indicates that agents can only maintain very homogenous relationships, while large d value indicates the agents are with high tolerance to maintain more heterogeneous relationships. The influence of the d value on the network structure is then tested by running computational simulations of this model under different d values.

In the simulations, we start with a random network of 100 agents and 280 edges for clarity, and the average degree $\langle k \rangle \approx 5.6 \gg \ln 100$ to ensure the initial connectivity of the randomly-generated network. The "trait" of each agent is assigned to a random real number that is uniformly distributed between 0.0 and 10.0 and parameter d ranges from 0.0 to 4.0. We observe significantly different topological structures of the network under different d values.

The entire network is basically connected when $0 \leq d \leq 0.2$. Meanwhile the clustering coefficient of the network (denoted as C) is significantly higher than that of the initial random network (denoted as $C0$), while the average shortest path length (denoted as L) is not significantly different from that of the initial random network (denoted as $L0$). In this sense, the network evolves into a *small-world* state as characterized by [9]. The network connectivity sharply declines when $0.2 < d < 0.4$, while it goes up again as d increases from 0.4 to 1.0. In both regions the network is fragmented into multiple isolated components. Further examination also shows that when $0.2 < d < 1.0$ each connected component is comprised by the agents with similar "trait" values and the overall network is highly clustered ($C/C0 > 8.0$). We can therefore assert that, when $0.2 < d < 1.0$, the network is in a "*segregation*" state that is comparable to the final state in Schelling's (1971) model. When d goes greater than 1.0, the overall network becomes connected again. However, two different states of the final network structure can be identified. When $1.0 \leq d \leq 2.0$ both the clustering coefficient and the average shortest path length are significantly greater than the counterparts of the initial random network (e.g. $C/C0 \approx 3.4$ and $L/L0 \approx 1.9$ when $d = 1.0$). In this state, a number of communities can be detected; and the agents within the same community have small differences in their trait values. What's more, two communities are interconnected by small amount of edges if the mean trait values of the two communities are not significantly different. For example, a community with the mean trait value 2.0 may be connected to another community whose mean trait value is 4.0; but a direct inter-community link seldom occurs between two communities with their respective mean trait values 2.0 and 8.0. In all, in this state (when $1.0 \leq d \leq 2.0$), the network topology shows a "*sequence*" structure which is close to a regular network. The regularity of the network structure

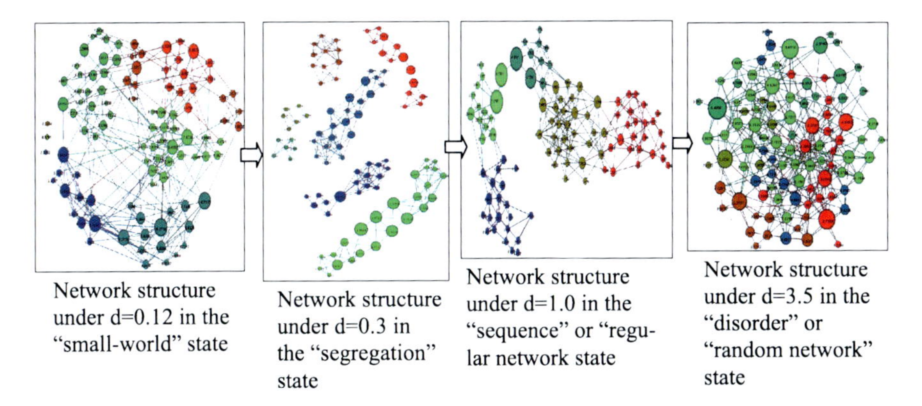

Network structure under d=0.12 in the "small-world" state

Network structure under d=0.3 in the "segregation" state

Network structure under d=1.0 in the "sequence" or "regular network state

Network structure under d=3.5 in the "disorder" or "random network" state

Fig. 83.1 Illustration of different network structures under different parameter values

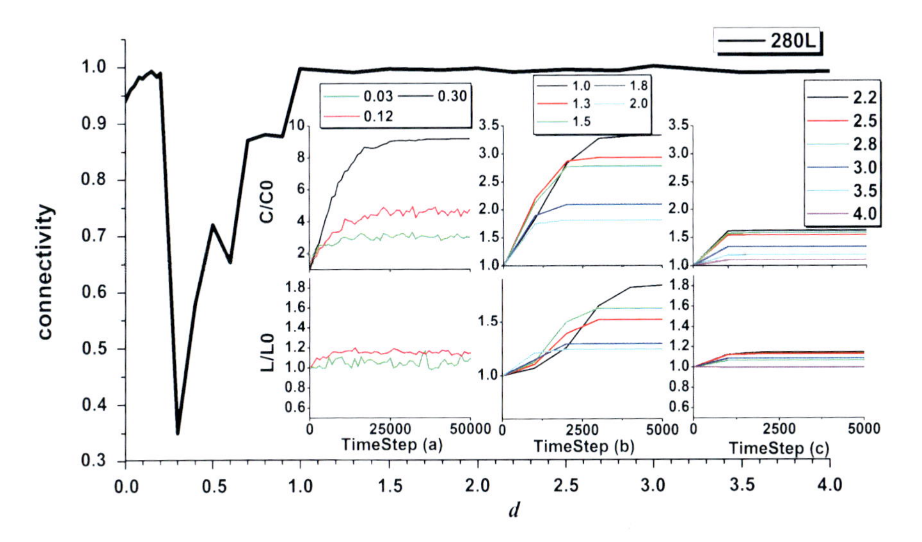

Fig. 83.2 Network properties under different parameter value

gradually diminishes as d increases. When d becomes greater than 2.0, the clustering coefficient and the average path length of the final network steadily approach to the counterparts of the initial random network. In particular, when $d = 3.5$, there is the observation of $C/C0 \approx 1.2$ and $L/L0 \approx 1.1$; When d reaches 4.0, the observation is then $C/C0 \approx 1.1$ and $L/L0 \approx 1.0$. From another aspect, the modularity of the network tremendously decreases and the network communities become hard to be detected when $d > 2.0$. We can basically assert that, as d increases from 2.0 to 4.0, the network steps into a random network state. Figure 83.1 illustrates the typical network structures under different parameter values, and Fig. 83.2 gives the network properties of typical states under different parameter values.

We give a partial explanation for the prior phenomenon observed in simulations. From the aspect of network structure, homophily is essentially a clustering mechanism to form cliques of homogenous agents. The scale of the cliques is determined by the value of parameter d. When d is small (in the simulations when $d < 0.2$), the formed cliques are small too; and the extra edges are randomly-distributed across the full population and the different cliques are basically interconnected through those random edges. Thus the small d value reflects the "*small-world*" state of the network. When d increases to the region of (0.2, 1.0), the scale of the cliques becomes big enough so that all the edges can be intra-clique edges. On the contrary, the cliques are so big enough that there are not adequate inter-clique edges to link different cliques. This causes the formation of the "*segregation*" state of the overall network. When d still goes larger to the region of (1.0, 2.0), the cliques become larger so that one clique can be connected to another clique and the boundaries between cliques are still clear. In this region, the network state is close to a "*regular network*". Finally, when d goes even larger, the agents can maintain relationships to very diverse agents and the only remaining clique become the whole population; correspondingly, the network goes into the "*random network*" state.

To sum up, the model presented in this work is in essence an extension of Schelling's segregation model in the context of social networks. By introducing the concept of individual tolerance, we find the social contact networks may evolve into multiple states, ranging from the random or disordered network to regular or sequenced network. The network evolution from random state to regular state also indicates that there may be a small world state between such two states. In other words, the reverse procedure of Watts & Strogatz's model exists in real social networks. As future work, we are going to investigate social network evolution by further incorporating the mechanisms of heterophily [4, 6] and social contagion [7].

References

1. Barabasi AL, Albert R (1999) Emergence of scaling in random networks. Science 286:509–512
2. Centola D, Gonzalez-Avella JC, Eguiluz VM, San Miguel M (2007) Homophily, cultural drift, and the co-evolution of cultural groups. J Confl Resolut 51:905–929
3. Holme P, Newman MEJ (2006) Nonequilibrium phase transition in the coevolution of networks and opinions. Phys Rev E 74
4. Liu WT, Duff RW (1972) The strength in weak ties. Public Opin Q 36:361–366
5. McPherson M, Smith-Lovin L, Cook JM (2001) Birds of a feather: homophily in social networks. Annu Rev Sociol 415–444
6. Rogers EM, Bhowmik DK (1970) Homophily-heterophily: relational concepts for communication research. Public Opin Q 34:523–538
7. Burt RS (1987) Social contagion and innovation: cohesion versus structural equivalence. Am J Sociol 92(6):1287–1335
8. Schelling TC (1971) Dynamic models of segregation. J Math Sociol 1:143–186
9. Watts DJ, Strogatz SH (1998) Collective dynamics of 'small-world' networks. Nature 393:440–442

Chapter 84
A Stochastic Lattice-Gas Model for Influenza Spreading

A. Liccardo and A. Fierro

Abstract We construct a stochastic SIR model for influenza spreading on a D-dimensional lattice, which represents the underlying contact network of individuals. An age distributed population is placed on the lattice and can move on it. The displacement from a site to a nearest neighbor empty site allows individuals to change the number and identities of their contacts. The model is validated against the age-distributed Italian epidemiological data for the influenza A(H1N1) during the 2009/2010 season, with sensible predictions for the epidemiological parameters.

Keywords Epidemic · Lattice-gas · H1N1

84.1 Introduction

There are two major approaches to model the spreading of infectious diseases in a space-structured population that are mostly used in recent literature: the Individual Based Models (IBM) and the Metapopualtion Models. The first ones [1–4] are obtained by coupling highly detailed socio-demographic models with a transmission model, while the second ones [5–11] are mainly focused on the role of human mobility, and thus are based on very accurate models of the mobility fluxes coupled with a transmission model. Both approaches require a huge amount of input

A. Liccardo (✉) · A. Fierro
Università degli Studi di Napoli Federico II, Complesso Universitario di Monte S. Angelo, Via Cintia, Edificio 6, 80126 Naples, Italy
e-mail: liccardo@na.infn.it

A. Fierro
e-mail: fierro@na.infn.it

A. Liccardo
INFN—Sezione Napoli, Complesso Universitario di Monte S. Angelo, Via Cintia, Edificio 6, 80126 Naples, Italy

T. Gilbert et al. (eds.), *Proceedings of the European Conference on Complex Systems 2012*, Springer Proceedings in Complexity, DOI 10.1007/978-3-319-00395-5_84,

data (census demographic information, daily commuting flows data, flight connections, etc.) necessary to reach a realistic description of the population in different geographic areas, and/or their connections at global level.

Following a complementary approach, we concentrate only on a few key factors, which we assume to be relevant for the spreading of an infectious disease, and try to explain the epidemic curves by constructing a model as simple as possible, involving only few parameters [12]. We use as a test case the Italian data on the H1N1 S-OIV during the season 2009/2010 furnished by the official site of the Italian Government (http://www.ministerosalute.it). The model is obtained by coupling a SIR model for the transmission process with a lattice-gas model, where the lattice represents the contact network of individuals. In spite of the regular structure of the lattice, the heterogeneity in the number of contacts is ensured by the existence of empty sites. Furthermore, the contact network is a dynamic network: by moving from a site to a nearest neighbor empty site, individuals change the number and the identities of their contacts. The mobility rules are fixed by imposing that young people prefer contacts with individual of similar age, as shown by many sociological analysis on mixing patterns (e.g.[13, 14]). The model allows to obtain sensible results for the epidemiological parameters of the influenza A(H1N1) during the 2009/2010 season, and theoretically interesting results for the generation time distribution.

84.2 The Model

84.2.1 The Contact Network

The model is essentially an attractive lattice-gas on a 3D cubic lattice which represents the contact network of individuals. The demographic structure is very simple: a population of N individuals is randomly distributed on the lattice, according to the age group distribution in Italy. We consider individuals of 4 different age classes, $\text{age} = 1, 2, 3, 4$, corresponding respectively to 0–4, 5–14, 15–64 and over 65 years old. Each site of the lattice is occupied at most by one individual, so that each person has at most 6 equidistant nearest neighbors. Periodic conditions are fixed on the lattice boundary. The dynamics of contacts among individuals is simulated using Monte Carlo techniques. We assume young people (0–4 and 5–14) to have an assortative behavior (i.e., to prefer contacts with individuals in the same age class), while adults stay indifferently with individuals of any age [14]. To encode this feature into the dynamics, we associate to each individual belonging to the age class "age" a nearest neighbor effective number, N^{age}, defined as the total number of nearest neighbors, if $\text{age} = 3, 4$, and as the number of nearest neighbors of the same age class, if $\text{age} = 1, 2$. More explicitly, the algorithm governing the dynamics is the following:

1. we choose at random an individual and a nearest neighbor site, on the lattice;
2. if the selected nearest neighbor site is occupied, we pass to the next step of dynamics and another individual is randomly chosen;

3. if the nearest neighbor site is empty, we try to move the individual from the initial site 1 to the destination site 2 with a transition probability:[1]

$$T(1 \to 2) = \min\left\{1, e^{\beta[N^{\mathrm{age}}(2) - N^{\mathrm{age}}(1)]}\right\}. \tag{84.1}$$

With the previous dynamics rules, after an initial transient, the system reaches a stationary state where the average of N^{age} on each age class is time independent, depending only on the age and on the parameter β. At low enough values of β the stationary state is also homogeneous, i.e. no large fluctuations in the local density are observed. We consider such a case and put $\beta = 1$.

84.2.2 The Transmission Model

The previous model for the population dynamic must be supplied with an epidemiological model. To this purpose, we construct the following SIR stochastic model: to each individual, we associate an internal degree of freedom for the healthy/infective status ($I_i = 0, 1$), and to healthy individuals, we attribute a further degree of freedom for susceptible/immune status ($\mathrm{anti}_i = 0, 1$). After a potentially contagious contact a susceptible ($I_i = 0, \mathrm{anti}_i = 0$) becomes infected ($I_i = 1, \mathrm{anti}_i = 0$) according to her/his specific age class susceptibility S_{age}. After the contagion process, the infected individual goes through an asymptomatic phase, lasting for a period T_s (where s stays for symptoms or stop). Then a symptomatic phase follows and lasts till the end of the infective period (T_{inf}).

Furthermore, when the epidemic is over, some disease adaptive rules are overimposed. Indeed infected individuals typically stay at home during the manifestation of symptoms and susceptible individuals tend to avoid contacts with them during the symptomatic phase. To encode these behaviors into the model we stop infected individuals during the symptomatic phase and interdict the empty sites that are nearest neighbor to them. Neighbors of symptomatic infected individuals can move without any restriction. Finally, each infected individual has her/his own stopping and infective period, while the infectivity is taken to be constant during the disease. A person no more infective, develops antibodies, thus her/his internal d.o.f. anti_i changes from 0 to 1, and then she/he starts to move again.

84.3 Results

We perform our simulations on a cubic lattice with linear size $L = 180$ with 20 % of occupation. The total number of individuals N is thus fixed to 1166400, and

[1] It is worth to notice that the probability to move an individual from the site 1 to a randomly chosen nearest neighbor site 2 is given by the probability that 2 is empty times $T(1 \to 2)$, where the first term favours spacing and the second one instead crowding.

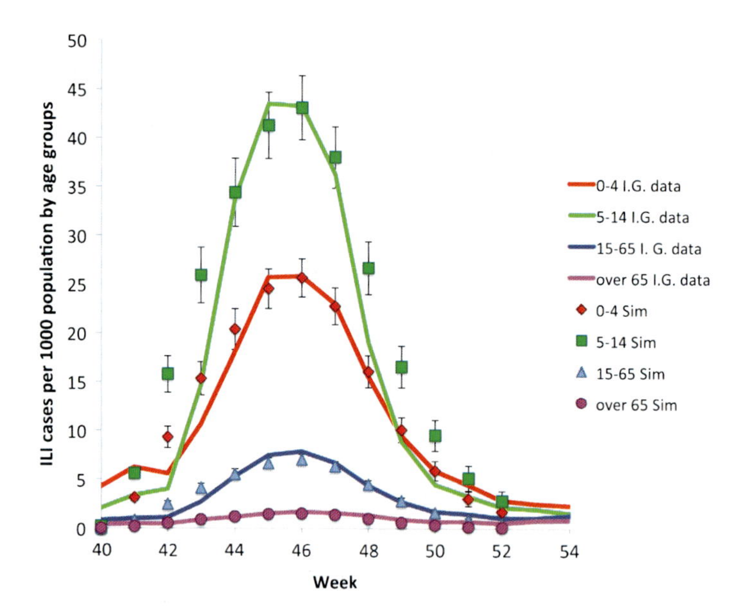

Fig. 84.1 Histogram of the observed illness prevalence (per thousand) of influenza cases in Italy from the 40th week of 2009 by age group, respectively 0–4 years old (*green*), 5–14 (*red*), 15–64 (*blue*), and over 65 (*violet*)

are distributed according to the age group densities of the Italian population, ρ_a ($a = 1, \ldots, 4$) given in Table 84.1. We assume both the infective period and the stop time to follow Gaussian distributions. The parameters for the infective period distribution are fixed to $\bar{T}_{\text{inf}} = 5$ days and $\sigma_{\text{inf}} = 1$ day, which correspond to the typical duration of influenza symptoms. For the stop time distribution, the average and the standard deviation, $\bar{T}_s = 0.5$ days, and $\sigma_s = 0.1$ days, have been fixed in order to obtain the best agreement between our data and the experimental ones.[2]

84.3.1 The Epidemic Curves

In Fig. 84.1, we compare the data on ILI cases during the H1N1 epidemic in 2009/2010, as reported by the I. G. government, with the simulated illness cases obtained with our model. The simulated illness cases and their errors have been evaluated respectively as mean values and standard deviations over the results of 64 processes, each characterized by a different random number generator. The transmission probabilities, given in Table 84.1, have been fixed in order to reproduce the

[2]Our data do not change significantly varying $\bar{T}_{\text{inf}}$ and σ_{inf}, provided that $\bar{T}_{\text{inf}} \gg \bar{T}_s$. This is consistent with the fact that the number of contagions during the stopping period is negligible with respect to those happening before the stopping (see the high frequency of contagions with generation time ≤ 0.5 days in Fig. 84.2).

Table 84.1 Age classes parameters

Age group (a)	0–4	5–14	15–64	+65
ρ_{age}	0.048	0.093	0.659	0.20
S_{age}	0.24	0.905	0.056	0.011

observed illness prevalence. The dissimilarity among the values of S_{age} reflects the fact that the H1N1 virus had a different incidence on different age classes, causing symptomatic disease mainly in younger population, probably as a consequence of a pre-existing partial immunity of older people carried over a previous pandemic [15, 16]. The smaller incidence on the 0–4 age class with respect to the 5–14 class is probably a consequence of a minor exposure due to the smaller percentage of children who attend day nursery in Italy.

The comparison between observed and simulated data shows that our model correctly reproduces the epidemiological data during the epidemic peak (from week 44-th to week 47-th) for all the age classes. For the age groups 0–4, 15–64 and +65 the model reproduces very accurately also the descendant phase of the epidemic. However the model fails to reproduce the first experimental points till week 43-th for each age class, as well as the descendant phase of the 5–14 years old age group. To interpret this mismatch, we first observe that the I. G. data we are comparing with, are those relative to the illness consultations and not to the ascertained cases. ILI cases are of course a reliable estimate of the real epidemic diffusion during the epidemic peak (when the alert of the population and public health system is utmost) but they are less significant at the beginning and at the end of the epidemic. Therefore, we interpret the disagreement at the beginning of the outbreak as due to an underestimation of the influenza diffusion at that time, due to the fact that not all the cases of influenza-like symptoms led to illness consultations, as instead it mostly happened during the peak. Moreover the disagreement in the descendant phase for the age group 5–14 can be naturally interpreted as due to self-initiated control measures adopted by many Italian families (e.g. reduction of scholar and extra scholar activities during the peak).

84.3.2 The Epidemic Parameters

Here we evaluate the generation time interval and the basic reproductive number. The generation interval ν is the time interval between the infection time of an infected person and that of her/his infector. Figure 84.2 shows the generation time distribution $g(\nu)$ obtained by our simulations. This distribution turns out to be well fitted by the function

$$g(\nu) \simeq A\left[1 - \operatorname{erf}\left(\frac{\nu - \bar{T}_s}{\sqrt{2}\sigma_s}\right)\right], \tag{84.2}$$

where $\bar{T}_s$ and σ_s coincide with the parameters of the Gaussian distribution chosen for the stop time T_s and A is a normalization constant. The functional form obtained in Eq. (84.2) is very interesting. In order to better understand it, we have done different simulations with different choices for the stop time distribution and the infective period distribution. We have found that $g(\nu)$ is strongly affected by the distribution law chosen for the stop time, while the dependence on the infective time distribution is weaker. We propose the following interpretation of this result: the generation interval distribution $g(\nu)$ may be seen as the product of 3 distinct probabilities $g(\nu) = g_1(\nu) \times g_2(\nu) \times g_3(\nu)$ where $g_1(\nu)$ is the probability that an individual infected at time 0 is still infective at time ν, $g_2(\nu)$ is the probability that a susceptible individual meets at time ν someone who was infected at time 0, and $g_3(\nu)$ is the contagion probability. The last one is given by the weighted average of the probability S_{age} on the age groups and is independent on ν. The probability $g_1(\nu)$ is by definition the complementary cumulative distribution of the infective period:

$$g_1(\nu) \propto P(T_{\text{inf}} > \nu) = \left[1 - \text{erf}\left(\frac{\nu - \bar{T}_{\text{inf}}}{\sqrt{2}\sigma_{\text{inf}}}\right)\right]. \tag{84.3}$$

In the time interval ($\nu < 1$ day), where the generation interval distribution $g(\nu)$ turns out to be significantly different from zero, the cumulative distribution of the infective period $g_1(\nu)$ is essentially equal to one, thus it trivially contributes to the generation interval distribution. This makes the generation time distribution $g(\nu)$ to reduce essentially to the probability $g_2(\nu)$, which conversely depends on ν and on the details of the dynamics, and is strongly affected by the distribution law chosen for the stop time. In particular $g_2(\nu)$ is expected to be proportional to the probability that the individual infected at time 0 is not recovered yet at time ν,

$$g_2(\nu) \propto P(T_s > \nu) = \left[1 - \text{erf}\left(\frac{\nu - \bar{T}_s}{\sqrt{2}\sigma_s}\right)\right], \tag{84.4}$$

in agreement with the distribution law obtained by our simulation in Eq. (84.2) (see Fig. 84.2). This very interesting result gives the explicit connection between generation time distribution, stop time distribution and infective period distribution.

The mean generation time corresponding to this distribution is $\bar{\nu} = (0.25 \pm 0.01)$ day. Such unrealistically low value is mainly due to the absence of any latent period in our model and to the fact that the contagions happen essentially during the dynamical phase (i.e. before the appearance of symptoms).

The reproductive number, R, is the Laplace transform inverse of the generation interval distribution, $g(\nu)$. For discrete distributions, it may be evaluated as [13]

$$R = \frac{r}{\sum_{i=1}^{n} g_i (e^{-r\nu_{i-1}} - e^{-r\nu_i})}, \tag{84.5}$$

where r is the exponential growth rate evaluated in the first week of the outbreak, $g_i = g(\nu_i)$ are the simulated frequencies of generation interval within ν_{i-1} and ν_i. With our simulation we get $R = 1.1 \pm 0.1$. This modest value of R is consistent with those found in literature for other countries [17, 18].

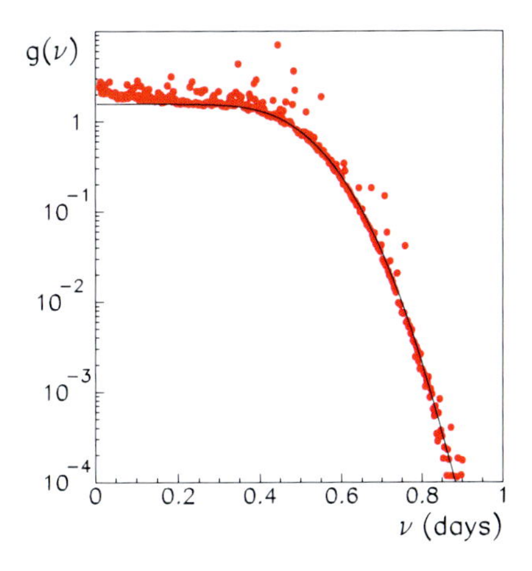

Fig. 84.2 Generation time distribution $g(\nu)$

In conclusion, in spite of its noticeable simplicity—few demographic information are required—our model correctly reproduces the behavior of the epidemiological Italian data at the peak of the H1N1 influenza for each age class, and also in the descendant phase for the 91 % of Italian population. The model has been generalized in order to predict the effect of limitation measures, such as vaccination or antiviral prophylaxis as well as limitation of population mobility.

References

1. Germann TC et al (2006) Proc Natl Acad Sci USA 103(15):5935
2. Ferguson NM et al (2005) Nature 437:209
3. Merler S, Ajelli M (2010) Proc - Royal Soc, Biol Sci 277:557
4. Ciofi degli Atti ML et al (2008) PLoS ONE 3:e1790
5. Rvachev LA, Longini IM (1985) Math Biosci 75:3
6. Watts D et al (2005) Proc Natl Acad Sci USA 102:11157–11162
7. Hufnagel L et al (2004) Proc Natl Acad Sci USA 101:15124
8. Colizza V et al (2007) PLoS Med 4:e13
9. Cooper BS et al (2006) PLoS Med 3:e12
10. Epstein JM et al (2007) PLoS ONE 2:e401
11. Colizza V et al (2006) Proc Natl Acad Sci USA 103:2015–2020
12. Fierro A, Liccardo A (2011) Eur Phys J E 34:11
13. Wallinga J, Lipsitch M (2007) Proc - Royal Soc, Biol Sci 274(1609):599
14. Del Valle SY et al (2007) Soc Netw 29:539
15. Khiabanian H et al (2009) PLoS ONE 4:8
16. Miller E et al (2010) Lancet 375:1100
17. Yang Y et al (2009) Science 326:729
18. Fraser C et al (2009) Science 324(5934):1557

Part VI
Social Systems, Economics and Finance

Chapter 85
CoopNet: A Social, P2P-Like Simulation Model to Explore Knowledge-Based Production Processes

Edoardo Mollona, Gian Paolo Jesi, and Matteo Vignoli

Abstract A prevalent claim is that we are in a knowledge economy, where firms can be viewed as networks of knowledge nodes interacting together.

In this work, we propose a social, P2P-like simulation model, CoopNet, to investigate how networking and organizational mechanisms interact to affect the emergent cooperation of workers. More specifically, we examine how (i) distinct reward mechanisms and (ii) distinct mobility assumptions on workers' combine to influence cooperation. As a result, we highlight the role of co-workers selection and the continuity of working relationships as alternative mechanisms to foster cooperation within intra-organizational network.

Keywords Knowledge economy · Agent-based · Self-emergent · Cooperation · Peer-to-peer

85.1 Introduction

A recent survey conducted by McKinsey [1] reports that, to facilitate knowledge sharing among co-workers (for example in product design), firms are investing in collaborative technology [2], peer-to-peer networks and social networks [3]. In addition, Mac-Duffie [4], quoting the Wall Street Journal [5] and data from the Gartner Group [6], reports that more than half of US companies with more than 5000 employees use virtual teams and that more than 60 % of professional employees surveyed reported working in a virtual team. Ultimately, management scholars describe the twenty-first-century firm as an intricately-woven web of dispersed operations

E. Mollona (✉) · G.P. Jesi
University of Bologna, Bologna, Italy
e-mail: emollona@cs.unibo.it

G.P. Jesi
e-mail: jesi@cs.unibo.it

M. Vignoli
University of Modena and Reggio Emilia, Modena, Italy
e-mail: matteo.vignoli@unimore.it

T. Gilbert et al. (eds.), *Proceedings of the European Conference on Complex Systems 2012*, Springer Proceedings in Complexity, DOI 10.1007/978-3-319-00395-5_85,

managed via collaborative technologies and organized around networks rather than rigid hierarchies [7]. Despite firms increasing investments in collaborative technologies such as peer-to-peer communities and social networks, the management problems that follow from the adoption of such technologies has been fairly neglected by received literature.

The aim of this paper is to analyze the circumstances under which cooperative behavior emerges in knowledge-based organizations when traditional hierarchical control mechanisms are weakened tools to prevent free-raiding.

We propose a model, CoopNet, that describes production processes as intra-organizational networks in which workers interact within local structures of exchange. We then explore unfolding aggregate behavior stemming from interaction among workers that respond to incentives crystallized in reward mechanisms.

85.2 Methodology

We adopted computer simulation in order to investigate the non obvious emergent consequences of changes in the reward mechanisms and in individual policies of social interaction. In this article, we operated with a computer simulation model as in a theoretical laboratory to explore the circumstances in which different patterns of cooperation among workers emerge.

The simulation approach allows us to explore the rich repertoire of different behaviors that potentially materialize in the setting that we designed [8, 9]. A minimal variation in any parameter, may lead to dramatically different emergent results as the dynamic of the interaction can be extremely complex and almost impossible to predict *a priori*. Essentially, the system behaves as a *Complex Adaptive System* (CAS) [3].

We designed and implemented an agent-based model [10]—the CoopNet model—to simulate interaction among employees. Increasingly social scientists are using the techniques of multi-agent based simulation [11] to explore complex dynamics in artificial social systems [12, 13].

Among similar lines, the CoopNet model should be viewed as an "artificial society" type model (i.e., similar to the SugarScape model [14]) that allows to represent in a stylized manner the process that may occur in a real organizations.

More specifically, in the CoopNet model, collaboration relationships among workers are represented by an undirected network where the employees are the nodes and each relation is a link among a couple of nodes. The communication between nodes is managed in a P2P-like fashion, using *message passing* and nodes have a *limited* (local) knowledge about the whole network.

85.3 Model

The simulation is "cycle" based, where at each cycle each node can run its protocol once. A node in the system holds a list of links to other nodes, which we call CACHE.

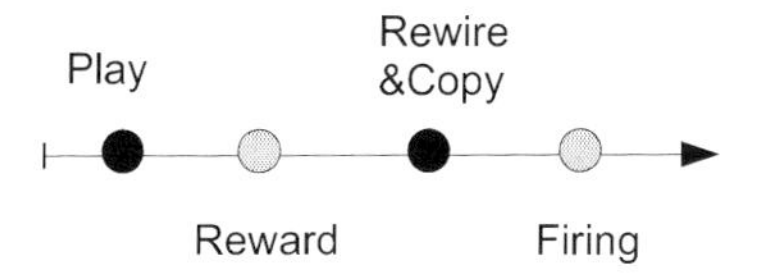

Fig. 85.1 The COOPNET model schema. *The dark circles* represent the *Individual Decision Making* phases, while *the light ones* represent the *Human Resource Management* phases

Therefore, an overlay network (the COOPNET overlay) defined by the relation "who knows whom" is induced by the cache's information.

Each node i holds a variable e_i representing its current *productivity effort* and a variable w_i representing its *wage*. With $e_i(j)$ we indicate the portion of the effort spent by node i reciprocating with j.

Each node i can act with a *pro-social* or *free rider* behavior; the behavior is assigned at bootstrap and may change during time (see Sect. 85.4.1). Pro-social nodes will try to invest more effort in their relations, while free rider nodes prefer to maximize the ratio between reward and effort spent. The system wide parameter MAX_{eff} represents the maximum allowed value for e_i $\forall$ node i.

The system rewards each node for its spent effort at regular time intervals. The wage w_i assigned to node i is calculated according to specific *reward policies* (see Sect. 85.4.2). In addition, we also consider the *utility* of a node i, which is defined as: $u_i = w_i - e_i$.

Finally, we consider that each node has the capability to contact a limited number of random nodes in the whole network independently of the COOPNET overlay. This limitation mimics the real world, as a person can hardly contact all the other persons in the company especially when the company is large. In P2P terms, this feature can be achieved by a sampling service [15].

The sequential schema followed by our model (shown in Fig. 85.1) is composed by four distinct phases: (a) PLAY, (b) REWARD, (c) REWIRE© and (d) FIRING phase. These phases respectively represent two core aspects highlighted by our model: the structure *of individual decision making* (i.e., phases (a), (c)) and the *human resource management* (i.e., phases (b), (d)).

85.4 Algorithms

In the following, we discuss the design of each phase according to the actual aspect it belongs to.

85.4.1 Individual Decision Making

Play Phase in this phase, each node tries to obtain the highest possible wage by interacting with its neighbors. If the interaction with one or more neighbors is not satisfactory, they are dropped from the current node's cache.

Each node i starts the simulation by committing a specific effort $e_i(j) = \frac{e_i}{degree_i}$ in the interaction with each neighbor j. In the next cycles, each node have to adjust its effort according to the other party's efforts. The basic idea is that when two nodes i and j plays distinct values, say x_i and x_j, they tend to close the difference gap between x_i and x_j.

How the gap is actually closed in associated to node's individual behavior. A pro-social node i will try to play the same amount of a neighbor j, but if j has played less than i, it will not modify its effort but will wait for extra effort from node j in the near future. The wait time is represented by t consecutive interactions (i.e., $t = 4$). A cooperative node j will likely fill—at least partially—the effort gap. Conversely, a free rider node i will not modify its effort if its neighbor j is playing more, but it will play less if its neighbor's effort is less than its own effort.

We consider the *willingness to cooperate* as the notion of satisfaction in a relation: it is sufficient that the party spending the lower amount of effort *partially* fills the effort gap thereby expressing its willingness to cooperate. In addition, we consider the issue where a neighbor is not able to reciprocate just because it is already busy in many, and possibly valuable, activities. This issue is managed by considering the total effort that i and j deploy in the interaction with their respective neighborhood: if $e_i < e_j$, j is considered a valuable resource and the link is preserved. Otherwise, node i drops the link and node j is forced to the same as the overlay is undirected. Our modeling here is strongly grounded on social exchange theory and the work about the relationships between reciprocity and perceived justice [16, 17].

Rewire&Copy Phase as in the PLAY phase, the rewiring process is related with the node's behavior. The basic idea is that nodes try to find a new neighbor with which to join and to improve the average effort of the cluster and hence its own wage. In addition, a node is likely to imitate a neighbor showing higher performances in the hope to achieve similar results.

The algorithmic steps performed by node i to manage this phase are the following:

1. *search*: if it has effort to spend, search for a set S of candidate nodes according to distinct *search policies*:
   - *Neighborhood Search Rewire* (NSR): the candidates set is composed of a random COOPNET neighbor s and all the neighbors of s
   - *Global Search Rewire* (GSR): the set is composed by k distinct nodes taken at random from the whole network, where k is a system wide constant
2. *select*: choose the "best" candidate in S according to a *selection policy*. Two alternative selection mechanism are provided. Nodes select partners either on (a) the basis of the *wage* that potential partners receive or (b) on the basis of the *utility* that accrue to potential partners. Node i checks if there is at least a candidate having an higher wage (or an higher utility); then, selects j that maximizes the difference $w_j - w_i > 0$ (or the difference $u_j - u_i > 0$) where $j \in \{1 \cdots \|S\|\}$.

3. *accept*: it is performed by i when receives a join request. If it has effort to spend, it accepts the newcomer.
4. *copy*: the decision the adopt the behavior of the high performing newcomer is designed as a stochastic process with probability p_{bcopy} (=90 %).

85.4.2 Human Resource Management

Reward Phase at the end of the PLAY phase, an organizational hierarchy (i.e., an oracle) measures the average productivity (*AP*) level: $AP = \sum_{i=1}^{N} \frac{e_i}{N}$, where e_i is the *effort* played by each node and N is the size of the network. If the COOPNET overlay is split into connected components (e.g., due to the rewiring process), the AP is computed on per-component basis: $\sum_{i=1}^{N_k} \frac{e_i}{N_k}$, where N_k is the size of the k-th connected component. For simplicity, we consider AP_i as the *AP* of component-i. Using the *AP* value, two distinct ways of computing the node wage are allowed:

- *WageBonus*: the basic wage of node i as follows: $w_i = \frac{\sum_{j=1}^{N} e_j}{N} \cdot (1 + \alpha)$, where α is an *added value* constant and N is the size of the cluster in which node i is embedded or the size of the whole network if the COOPNET overlay is not split. Then the system is able to measure the effective effort each node is investing, thus each node i will receive $w_i = w_i \cdot (1 + x)$, where $x \in [0 : 1]$ and has a stochastic nature if $e_i \geq$ BONUS_T, otherwise $x = 0$ (i.e., no bonus for node i).
- *WageNeighbor*: the basic idea is to associate each node's wage to the performance of a small group of nodes involved in a close relationship, rather than dispersing the individual responsibility in large groups. The small group is represented by a node's *neighborhood* of size d_{cnet} which can grow or shrink over time according to the node's behavior and the evolution of the system. Node's i wage is calculated as follows: $w_i = \frac{e_i + \sum_{j=1}^{c} e_j}{d_{\mathrm{cnet}}+1} \cdot (1 + \alpha)$, where α is an *added value* constant.

The idea here is to test different approaches to the design to group rewards [18].

Firing Phase when activated, the component managing that phase allows focal organization to fire underperforming employees. In the real world, knowing the actual productivity of an employee is a non trivial task and it is not fail-proof. The idea is that when a node's performance is at the lowest extreme (e.g., spending effort in the interval [0, 10]), it is easy to detect and it is fired with no uncertancy, but when its performance value is closer to the average performance (e.g., spending effort in the interval [10, 15[), then the system in unable to discriminate the difference with perfect accuracy an the firing process is stochastic according to a p_{fire} probability.

85.5 Experiments

We adopted PeerSim[1] as a simulation platform and we wrote a Java implementation of our model. The goal of our experiments is to generate hypotheses on the role that the context in which employees interact plays in facilitating, or hindering, cooperation.

We tested our model according to (a) the presence/absence of the firing feature, (b) the rewiring search policies (i.e., NSR, GSR), (c) the rewiring select policy (i.e., wage or utility-based) and (d) the different rewarding mechanism adopted in the simulated organization (i.e., WAGEBONUS, WAGENEIGHBOR).

We run two sets of experiments. In the first set, we assumed that the firm is able to asses individual productivity and to consequently assign a bonus to specific workers that show above-average productivity (i.e., WAGEBONUS rewarding scheme). In the second set, we assumed that the focal company is able to more accurately recognize smaller isles of productivity and to apply a fine-grained analysis of teams' productivity. Consequently, the firm assigns wages to workers on the basis of the average productivity of the neighborhood they belong to. On the other hand, the company is less efficient in assessing individual productivity (i.e., WAGENEIGHBOR rewarding scheme).

In both sets of experiments, we tested three different individual decision-making routine to select co-workers: (1) the GSR rewire policy allows to look for new suitable neighbors over a wider horizon, increasing the chance to find a good match, the (2) NSR policy allows employees to search only search over their neighborhood and the (3) UNSR policy considers the co-worker's utility rather than their wages.

Due to the space constraints, we just focus on a scenario involving 500 agents. The calibration of parameters adopted for our experiments was generated following an (i) extensive simulation campaigns and (ii) replication of similar setups found in literature [19–22] and (iii) general criteria of plausibility.

The network is randomly wired at bootstrap using $d_{\text{cnet}} = 5$ degree. The maximum effort a node can spend in a *time unit* is fixed to the constant $MAX_{\text{eff}} = 50$. The node's effort is assigned by a normal distribution having parameters: $\mu = 27.5$, $\sigma = 5.5$ and the initial behavior is assigned randomly to a 50–50 % proportion. The added value constant is $\alpha = 0.2$;

The probability to copy the behavior of a newly joined neighbor is $p_{\text{bcopy}} = 90\,\%$.

Each data represents an average value collected from 10 distinct simulation runs.

Figures 85.2(a) and (b) show the performance of the system when the WAGEBONUS reward policy is adopted. We performed experiments both activating and deactivating the firing mechanism. A benchmark is highlighted by a horizontal line in the upper part of each effort plot. It represents the ideal system performance when all nodes spend the maximum effort $e_i = 50\ \forall i = E_{\text{total}} = 25000$.

This first set of experiments presents two findings worth of being discussed. In Fig. 85.2, we notice that, when it is possible to detect and firing individuals that

[1] http://peersim.sf.net/.

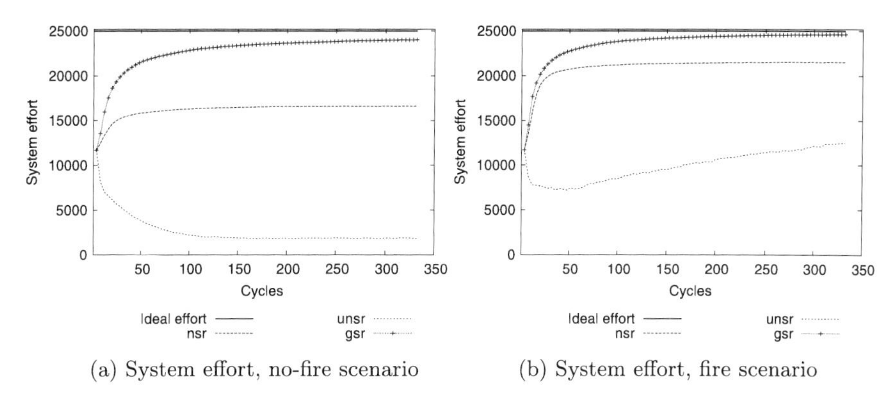

(a) System effort, no-fire scenario (b) System effort, fire scenario

Fig. 85.2 System effort achieved by each policy using WAGEBONUS reward. Both no-fire and fire scenarios are considered. Network size is 500

perform very low, performances of GSR and NSR are very similar. On the other hand, when detecting free-raiding and, consequently, firing becomes difficult, the two selection routine yield very different performances. In particular, the selection routine in which workers are allowed to search beyond the cluster of neighbors (GSR) yields much higher performances.

This finding suggests that, when organizational hierarchy is weakened by ambiguity of individual contribution to a shared endeavor, a firm ought to assign to individual workers larger autonomy in selecting co-workers all over the entire network of organizational employees. In this way, people willing to collaborate are given the opportunity to both punish free-riders, by severing specific working relationships, and to find and select co-workers willing to cooperate.

A second insight that flows from this first set of experiments regards the role of available information to individual decision-makers. As reported in Fig. 85.2, when the UNSR selection protocol is activated, the performance of the system are extremely low. Quickly, almost the entire population becomes free-rider and the firm's productivity collapses.

In the second set of experiment we applied the WAGENEIGHBOR policy. Figures 85.3(a) and (b) show that while the absolute performance is not as high as with the previous policy, the difference gap among GSR and NSR is much lower. Both achieves their result almost at the same time. This set of experiment elicits another important insight. In general, when a WAGENEIGHBOR policy is assumed, GSR protocol loses strength in comparison to NSR. The idea is that, when reward is assigned by evaluating productivity of small groups, it is important that co-workers within each group learn how to collaborate by both increasing their efforts or isolating free riders. Such learning process requires stability of groups of co-workers.

In this light, the NSR selection protocol, by constraining individual search for partners in the neighborhood, facilitate groups stability and intra-groups learning. On the other hand, the GSR protocol, which increase workers mobility, delays intra-group learning. Concluding, GSR and NSR illustrate virtues of two different mechanisms that facilitate emerging of cooperation.

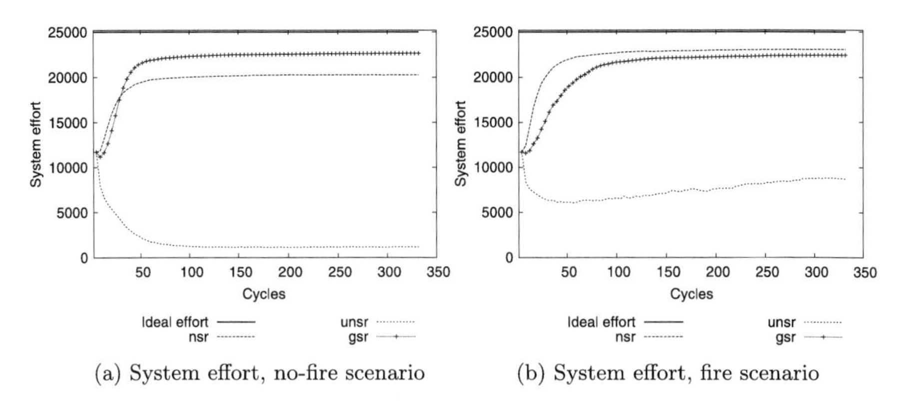

(a) System effort, no-fire scenario (b) System effort, fire scenario

Fig. 85.3 System effort achieved by each policy using WAGENEIGHBOR reward. Both no-fire and fire scenarios are considered. Network size is 500

85.6 Conclusions

We believe that concepts and points of view of different disciplines need to be integrated. More specifically, we suggest that organizational theory may gain a great deal from borrowing the concepts and techniques developed by computer scientists facing the complexity and emergent behaviors of large scale distributed systems (e.g., P2P)

In this article, we reported results from a research project in which we looked at organizations as hybrids that combine mechanisms that are typical of firms, such as rewarding mechanisms, and processes that characterize local interaction within P2P systems.

We developed our model by simulating the behavior of a large organization with an agent-based model that worked as a virtual laboratory for our analysis. We believe that our work discloses two avenues of future research.

Since the use of collaborative tools connecting virtual teams is increasing, which is the emergent consequence of their features over the specific organization structure and processes? It is possible to associate specific features of collaborative tools to specific types of organizations, or to specific types of products or services produced by an organization?

A second avenue of research is connected to the role that the information available to individual decision-makers. Our simulations conveyed a counterintuitive finding by suggesting that, in specific circumstances, less information is better than more. This outcome invites researchers to empirically test whether a correlation exists between information made available to co-workers by different collaborative tools, cooperative behavior and virtual teams' productivity. Concluding, we hope that our work was able to highlight the gains that may potentially stem from an interdisciplinary approach to the study of the role of information technology in firms' organization.

References

1. McKinsey & Company (2007) How businesses are using Web 2.0: a McKinsey's global survey. Technical report. McKinsey & Company
2. Guo J (2009) Collaborative conceptualization: towards a conceptual foundation of interoperable electronic product catalogue system design. Int J Enterp Inf Syst 3(1):59–94
3. Piao C, Han X, Wu H (2010) Research on e-commerce transaction networks using multi-agent modeling and open application programming interface. Int J Enterp Inf Syst 4(3):329–353
4. MacDuffie JP (2007) Hrm and distributed work. Acad Manag Ann 1(1):549–615
5. de Lisser E (1999) Update on small business: firms with virtual environments appeal to workers
6. Jones C (2004) Teleworking: the quiet revolution. Technical report, The Gartner Group
7. Cascio WF, Aguinis H (2008) Staffing twenty-first-century organizations. Acad Manag Ann 2(1):133–165
8. Gilbert G, Doran J (1994) Simulating societies: the computer simulation of social phenomena. UCL Press, London
9. Gilbert N, Troitzch K (2005) Simulation for the social scientist. J Manag Gov 12(2):225–231
10. Zhang Y, Bhattacharyya S (2007) Effectiveness of q-learning as a tool for calibrating agent-based supply network models. Int J Enterp Inf Syst 1(2):217–233
11. Li H, Wang H (2007) A multi-agent-based model for a negotiation support system in electronic commerce. Int J Enterp Inf Syst 1(4):457–472
12. Axelrod R (1997) Advancing the art of simulation in the social sciences. Complexity 3(2):16–22
13. Hales D, Edmonds B, Norling E, Rouchier J (2003) Multi-agent-based simulation III, MABS 2003. In: MABS. Lecture Notes in Computer Science, vol 2927. Springer, Berlin
14. Epstein J, Axtell R (1996) Growing artificial societies: social science from the bottom up. Brookings Institute Press, Washington
15. Jelasity M, Voulgaris S, Guerraoui R, Kermarrec AM, van Steen M (2007) Gossip-based peer sampling. ACM Trans Comput Syst 25(3):8
16. Molm LD, Quist TM, Wiseley PA (1993) Reciprocal justice and strategies of exchange. Soc Forces 72(1):19–44
17. Eckhoff T (1974) Justice: its determinants in social interaction. Rotterdam University Press, Rotterdam
18. Zenger TR, Hesterly WS (1997) The disaggregation of corporations: selective intervention, high-powered incentives, and molecular units. Organ Sci 8(3):209–222
19. Organ D (1988) Organizational citizenship behavior: the good soldier syndrome. Lexington Books, New York
20. Niehoff BP, Moorman RH (1993) Justice as a mediator of the relationship between methods of monitoring and organizational citizenship behaviour. Acad Manag J 36(3):527–556
21. Konosky MA, Pugh SD (1994) Citizenship behaviour and social exchange. Acad Manag J 37(3):656–669
22. Moorman RH, Blakely GL, Niehoff BP (1998) Does perceived organizational support mediate the relationship between procedural justice and organizational citizenship behaviour. Acad Manag J 41(3):351–357

Chapter 86
Analyses of Group Correlations in the KOSPI and the KOSDAQ

Jung Su Ko and Kyungsik Kim

Abstract We analyze the structure of cross-correlations in two Korean stock markets, the Korea Composite Stock Price Index (KOSPI) and the Korea Securities Dealers Automated Quotation (KOSDAQ). We investigate a remarkable agreement between the theoretical prediction and the empirical data concerning the density of eigenvalues in the KOSPI and the KOSDAQ. The research for the structure of group correlations undress the market-wide effect by using the Markowitz multi-factor model and network-based approach. In particular, the KOSPI has a dense correlation besides overall group correlations for stock entities, whereas both correlations are less for the KOSDAQ than for the KOSPI.

Keywords KOSPI · KOSDAQ · Group correlation · Random matrix theory

86.1 Introduction

There has been of considerable interest in the correlation-based approach among the various methodologies of the econophysics field in complex systems [1] for the last two decades. The empirical correlation matrices are very important for risk management and asset allocation. The probability of large losses for a certain portfolio or option is dominated by the correlated moves of its different constituents. Until now, the study of correlation matrices has a long history in finance and is one of the cornerstones in Markowitz's theory of optimal portfolios [2].

Financial market fluctuations may fundamentally result from the correlated decision making between buy and sell orders of various stock entities participating in the market. Because the concepts of the market are reflected in price fluctuations, we can obtain the structure of a given financial market by investigating financial cross-correlation data. Random matrix theory has provided highly meaningful information about the financial market structures of the US market [3]. It has been reported from previous works that the complicated structure of the stock market

---

J.S. Ko · K. Kim (✉)
Department of Physics, Pukyong National University, Busan 608-737, Republic of Korea
e-mail: kskim@pknu.ac.kr

T. Gilbert et al. (eds.), *Proceedings of the European Conference on Complex Systems 2012*, Springer Proceedings in Complexity, DOI 10.1007/978-3-319-00395-5_86,

cross-correlation was divided into three categories: a bulk random part, a market-wide part, and a group correlation between firms in the same business sector [4]. The investors want their stock portfolios to be optimized in relation to the structure of a particular business cluster. The decomposed group correlation is distinctly one of the key features of a mature stock market. No work has investigated the explicit structure of the group correlation, although papers studying the cross-correlation of the Korean stock market have been recently published [5, 6].

In this paper, we simulate and analyze the structures of business clusters by applying multifactor models in two Korean stock markets, the KOSPI and the KOSDAQ. For the KOSPI and the KOSDAQ, we ultimately test whether the structure of the group correlation in the two Korean financial markets is maintained consistently during a sufficiently long time or whether it shows a different pattern during the evolution of the system. Section 86.2 introduces the random matrix theory and the genuine information about the business sector among stocks. In Sect. 86.3, we describe the numerical calculation and the results of this research. The concluding remarks are given in last section.

86.2 Theoretical Background

Firstly, we defined the logarithmic return, $R_i(t)$, as follows,

$$R_i(t, \Delta t) = \log P_i(t + \Delta t) - \log P_i(t), \tag{86.1}$$

where $P_i(t)$ denotes the daily price of stock entity i at time t and Δt represents one day. In order to remove the effect of a magnitude of prices on the correlation coefficient, we used the normalized returns, $r_i(t)$, as follows:

$$r_i(t) = \frac{R_i(t) - \langle R_i \rangle}{\sigma_i}, \tag{86.2}$$

where σ_j is the standard deviation of the return and $\langle \ldots \rangle$ indicates a time average during L steps. The covariance matrix C is represented in terms of

$$C \equiv \langle r_i(t) r_j(t) \rangle = \begin{pmatrix} 1 & \cdots & c_{1N} \\ \vdots & \vdots & \vdots \\ c_{N1} & \cdots & 1 \end{pmatrix}. \tag{86.3}$$

The elements of the correlation matrix C_{ij} have a range of $-1 \leq C_{ij} \leq 1$. When the stocks i and j are perfectly-correlated, the value of C_{ij} is 1. On the other hand, $C_{ij} = -1$ if two stocks are perfectly anti-correlated. $C_{ij} = 0$ corresponds to the case where both stocks are uncorrelated. The cross-correlation between the stock entities provides significant and meaningful information about the intrinsic structure of the stock market.

Previous work published in recent years [7] suggested that the financial cross-correlation matrix C is separated into three groups, (a) a random part, (b) a market-wide effect part, and (c) a group correlation between entities included in the same industry, i.e., the so-called business sector. Therefore, for the latter, apart from the

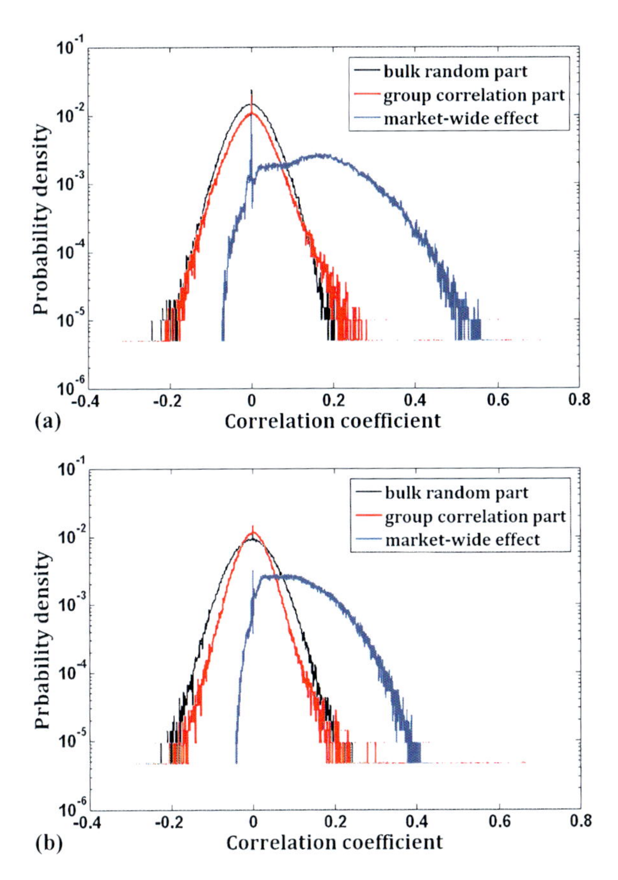

Fig. 86.1 Probability density of the correlation coefficient in the bulk random, the group correlation, and the market-wide effect for (**a**) the KOSPI (one-year periods 2006.1–2006.12) and (**b**) the KOSDAQ (one-year periods 2007.1–2007.12)

largest eigenvalue, genuine information about the business sector among stocks can be reconstructed by using several next-largest eigenvalues according to

$$C = \sum_{i=1}^{N_r} \lambda_i |\lambda_i\rangle\langle\lambda_i| + \sum_{j=1}^{N_g} \lambda_j |\lambda_j\rangle\langle\lambda_j| + \lambda_{largest} |\lambda_{largest}\rangle\langle\lambda_{largest}|, \tag{86.4}$$

where N_r and N_g represent the numbers of eigenvalues attributed to the random part and of the correlations based on the business sector, respectively.

86.3 Numerical Calculation and Result

We analyze the financial time-series data of Korean stock market securities from the KOSPI and the KOSDAQ. We respectively extracted the 629 and 650 largest stocks from the KOSPI and the KODAQ that were exchanged on the Korean stock market during the period 2006–2010, forming $L = 1244$ records of daily returns for Korean stocks.

Figure 86.1 shows the probability density of the correlation coefficient in the bulk random, the group correlation, and the market-wide effect for the KOSPI and

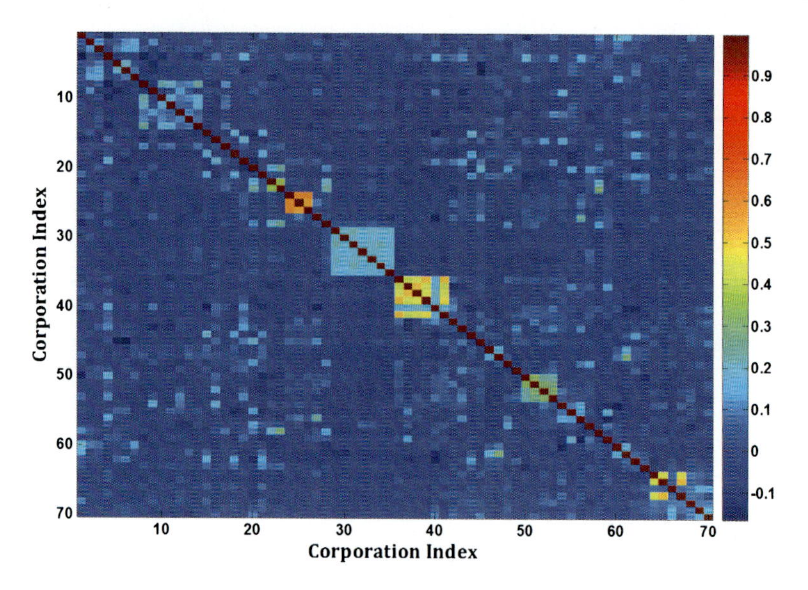

Fig. 86.2 Color map diagrams of group correlation matrix for $L = 1244$ in the KOSPI. If two firms are highly correlated, the representation approaches the color red, and if two firms are negatively correlated, the representation approaches *dark blue*. We show that we divided the diagrams into 10 block diagonals that correspond to business sectors: chemical, construction, distribution, electricity & electronics, finance, food, machinery, paper & plumber, service, and steel

the KOSDAQ. For the inverse participation ratio based on the daily return of the 629 KOSPI and the 650 KOSDAQ stock prices data after market-wide effects, a few large eigenvalues have distinguishable value, although most of the eigenvalues have uniformity for the inverse participation ratio value. This observation indicates that the regressed financial correlations contain two different kinds of information: (a) random noise and (b) genuine market information. In order to identify the genuine market information, we investigated the major eigenvector components that contributed significantly to a few large eigenvalues. Most notably, we find that there are inverse participation ratio values corresponding to the smallest eigenvalues, which suggests that the eigenvectors are localized, namely, only a few companies contribute to them. The small sizes of the corresponding eigenvalues show that these companies are uncorrelated with one another.

We investigate the group correlation between the stock entities included in the same business sector by using the color map representation of the group correlation matrix. We calculate the group correlation matrix in the following way. Firstly, we extract financial time series data and sort firms in the order of the business sector. Secondly, we calculate the cross-correlation matrix of these sorted time series data. Thirdly, we draw a color map according to the correlation coefficient of this group correlation matrix. Then, we obtain a correlation matrix with a size of 70 by 70. In Fig. 86.2, the indices 1–7 represent the 7 firms whose contributions are the largest, such as the 7 firms in the chemical business sector. The indices 8–14 represent the construction industry. In the same way, the indices 15–21, 22–28, and 29–35 mean the distribution, electricity & electronics, and finance, respectively. We show a color map diagram obtained using the method we mentioned. In the KOSPI, detection of the diagonal block map is a somewhat distinguished correction between the stock entities in the same business sector. However, in the KOSDAQ, detection of the diagonal block map is not clearly due to the distinctive correction between the stock entities. Each diagonal block corresponds to a business sector. In the KOSPI, there is an area that possesses un-correlation between the finance business sector and the others. However, the finance and food companies are each highly correlated with the returns of the financial market. On the other hand, in the KOSDAQ, medical companies are highly correlated with the returns of the financial market. In particular, some machinery industry is highly correlated with the returns of the financial market.

86.4 Conclusions

We have studied the group correlation of two Korean stock markets by applying the random matrix theory. As seen for the density distribution of the correlation coefficients, we have identified that both the KOSPI and the KOSDAQ have a positive correlation. We have found that both the KOSPI and the KOSDAQ appear to have high correlations in 2008 due to the global financial crisis. After comparing the market wide effect to the market removed correlation coefficient, we shows that the strong influence of the market wide effect on the KOSPI is similar to that on the KOSDAQ, and that the KOSPI is more affected than KOSDAQ. In other words, the KOSPI has more correlation than the KOSDAQ.

Observations from the color map show that the KOSPI diagram appears to be more distinct than the KOSDAQ diagram for the group correlation. Lastly, we examined the group by using network method, and we identified that the KOSPI had a certain correlation threshold of 0.105. Thus, we drew a group correlation network diagram for the KOSPI. However, in the KOSDAQ, we could not see a remarkable threshold value, so it was difficult to draw a network diagram. Eventually we ascertained that in the network diagram for the KOSDAQ, the threshold could be selected arbitrarily. Ultimately, we found that the KOSPI was larger than the KOSDAQ, not only in regards to correlations at large for stock entities (market-wide effect) but also in group correlations for the business sector. Other papers [5] detected a weak relationship between businesses and market capitalization for the KOSPI. However, when comparing developed markets such as the NYSE, they did not check for an exponential function of market capitalization when comparing developed markets such as the NYSE. From our results, we can logically assume that the Korean market has very many short-term portfolios and a conglomerate structure. Thus, we conclude that investors do not form a long-term investment strategy based on industrialized classifications.

Acknowledgements This work was supported by Center for Atmospheric Sciences and Earthquake Research (CATER 2012-6110).

References

1. Mantegna RN, Stanley HE (2000) An introduction to econophysics: correlation and complexity in finance. Cambridge University Press, Cambridge
2. Bouchaud JP, Potters M (1997) Theory of financial risk. Alea-Saclay, Paris
3. Plerou V, Gopikrishnan P, Rosenow H, Amaral LAN, Stanley HE (1999) Universal and non-universal properties of cross-correlations in financial time series. Phys Rev Lett 83:1471–1474
4. Plerou V, Gopikrishnan P, Rosenow H, Amaral LAN, Guhr T, Stanley HE (2002) Random matrix approach to cross-correlations in financial data. Phys Rev E 65:066126
5. Ahn S, Choi J, Lim G, Cha KY, Kim S, Kim K (2011) Identifying the structure of group correlation in the Korean financial market. Physica A 390:1991–2001
6. Kim DH, Jeong H (2005) Systematic analysis of group identification in stock markets. Phys Rev E 72:046133
7. Plerou V, Gopikrishnan P, Amaral LAN, Meyer M, Stanley HE (1999) Scaling of the distribution of price fluctuations of individual companies. Phys Rev E 60:6519–6529

Chapter 87
'Time is Money': An Heterogeneous Agent Model for the FX

Sophie Béreau

Abstract This paper deals with a Bayesian extension of a behavioral finance framework 'à la' De Grauwe and Grimaldi (The Exchange Rate in a Behavioural Finance Framework, Princeton University Press, Princeton, 2006) in which agents operating in the FX market differ in their forecasting time horizon for the exchange rate. In the short run, if we believe in the world described by Meese and Rogoff (J. Int. Econ., 14(1–2):3–24, 1983), this leads to a chartist rule, whereas in the long run, the PPP condition appears as a natural anchor. In between, i.e. in the medium run, we implement an APEER model using Bayesian tools, as an alternative to the FEER-BEER nexus. Our results show that the stabilizing impact of the intermediate rule depends on agents' good perception of the fundamentals.

87.1 Motivations

Exchange rate modelling [1, 5] remains one of the most challenging issues in International Macroeconomics and Finance, crystallizing many puzzles and sources of intellectual controversy within the scientific community. Among the various approaches that have been developed to address this issue, behavioural finance frame-

This is a long abstract for an article entitled 'Heterogeneous Agents in the FX Market: A Matter of Time Horizon?'. I would like to thank Jean-Yves Gnabo, Paul De Grauwe, Giulia Piccillo, Vincent Bouvatier, Guy Laroque, Jean-Michel Grandmont, Etienne Lehmann, Frédéric Bec, Vincent Blondel, and seminar participants at CORE, Large Graphs and Networks seminar, CREST, Thema—Université de Cergy-Pontoise, ADRES, University of Kent and Université Paris Ouest, for their very helpful comments and insights on this paper.

S. Béreau (✉)
Louvain School of Management, Université Catholique de Louvain, Louvain-la-Neuve, Belgium
e-mail: sophie.bereau@uclouvain.be

S. Béreau
CORE, Université Catholique de Louvain, Louvain-la-Neuve, Belgium

S. Béreau
EconomiX, Université Paris Ouest, Nanterre, France

T. Gilbert et al. (eds.), *Proceedings of the European Conference on Complex Systems 2012*, Springer Proceedings in Complexity, DOI 10.1007/978-3-319-00395-5_87,

work and heterogeneous agent models in particular stand out as promising tools to better understand underlying mechanisms at work.

The traditional opposition feature "fundamentalist" vs. "chartist" agents as defined by Frankel [4], but recent contributions such as [6, 7], or [8], document that rather than this kind of opposition, one can distinguish "FX dealers", who act on a daily basis from "fund managers", who are more sensitive to longer time horizons, which clearly paves the way to the investigation of an exchange rate model where the price dynamics would result from the opposition of different agents' trading strategies in terms of time horizon.

In this paper, we build on this stylized fact to derive a behavioural finance model for the exchange rate where long-run target values depend on the time horizon agents deal with, making this characteristics at core of agents' heterogeneity.

The remainder of the abstract is the following. Section 87.2 specifies the behavioural framework we rely on to model the exchange rate and Sect. 87.3 we draw and comment representative simulations. Finally, Section 87.4 concludes.

87.2 The Exchange Rate Model

The model we rely on for the exchange rate is derived from the behavioural finance framework described in Chap. 2 of [3] in which we include an intermediate forecasting rule.

The expression of the future exchange rate s_{t+1} depends on the current realization s_t plus the weighted expectations of exchange rate variations made by agents operating in the market, potentially of three type according to the time horizon they rely on for their forecasts, i.e. the short, medium and long run, plus a white noise assumed to follow a normal i.i.d. process.

$$s_{t+1} = s_t + \omega_{S,t}\beta(s_t - \bar{s}_{S,t}) - \omega_{M,t}\eta_M(s_t - \bar{s}_{M,t}) - \omega_{L,t}\eta_L(s_t - \bar{s}_{L,t}) + \tilde{\varepsilon}_{t+1} \quad (87.1)$$

with: $\tilde{\varepsilon}_{t+1} \sim \mathcal{N}(0, \sigma^2_{\tilde{\varepsilon}})$ representing the news in time $(t+1)$ that has not been incorporated in agents' forecasts.

In our framework, equilibrium values for the exchange rate vary with the corresponding time horizon. In the short run, a function of past realized values is used and in its simplest form, we get $\bar{s}_{S,t} = s_{t-1}$, whereas in the long run, the fundamental value $\bar{s}_{L,t}$ is known with certainty and follows a random walk without a drift. In between, i.e. in the medium run, the fundamental anchor $\bar{s}_{M,t}$ is estimated according to a state-space representation of a standard trend-cycle decomposition using Bayesian tools:

$$s_t = \bar{s}_{M,t} + m_{M,t} \quad (87.2)$$

$$\bar{s}_{M,t} = \bar{s}_{M,t-1} + \delta_{t-1} + \varepsilon_t, \quad \varepsilon_t \sim \mathcal{N}(0, \sigma^2_{\varepsilon}) \quad (87.3)$$

$$\delta_t = \delta_{t-1} + z_t, \quad z_t \sim \mathcal{N}(0, \sigma^2_z) \quad (87.4)$$

$$m_{M,t} = \phi_1 m_{M,t-1} + \phi_2 m_{M,t-2} + u_t, \quad u_t \sim \mathcal{N}(0, \sigma^2_u) \quad (87.5)$$

where $m_{M,t}$ denotes the medium-run misalignment, i.e. the gap between the current realized value for the exchange rate and its fundamental medium-run anchor.

Two sets of independent conjugate priors are assumed for (ϕ_1, ϕ_2) and $(\sigma_\varepsilon^{-2}, \sigma_z^{-2}, \sigma_u^{-2})$, i.e. the product of two normal distributions and three independent Gamma distributions respectively, which is rather standard in the context of Gaussian linear models. MCMC methods are then run in order to derive posterior conditional densities associated to our model.

At each period, agents may switch from one type to another depending on the *ex-post* performance of their forecasting rule. They use for that a fitness criterion inspired from [2] assumed to be as follows:

$$\omega_{i,t} = \frac{\exp(\gamma \pi'_{i,t})}{\sum\limits_{i \in \{S,M,L\}} \exp(\gamma \pi'_{i,t})}, \quad \forall i \in \{S, M, L\} \tag{87.6}$$

where $\omega_{i,t}$ is the share of the total population of agents relying on the rule of type i, $\gamma \geq 0$ the intensity with which agents revise their rules, $\pi'_{i,t}$ the risk-adjusted profit realized by means of the i's forecasting rule and S, M, L the short, medium and long-run respectively.

87.3 Simulation Results

Numerical simulations of the model are performed in order to measure how the confrontation of the different forecasting rules impact FX rate dynamics. Each set of stochastic simulations consists here in 10000 points. They correspond to, first, the draw of 5000 points considering only two forecasting rules, namely the short and the long-run rules, and then, to the draw of 5000 other points once the medium-run rule has been introduced, this for two reasons. First, because the medium-run approach needs the past of the exchange rate series to be implemented and second, because this allows us to study the differences in the exchange rate dynamics before and after the adoption of the medium-run rule.

Our results exhibited in Fig. 87.1 confirm previous findings when focusing on the short and long-run rules: in each case, the current exchange rate diverges from its fundamental value only when all the agents adopt the short-run (i.e. chartist) rule. On the contrary, when agents switch frequently between the short and the long-run (i.e. fundamentalist) rules, the observed dynamics is stable in the sense that the current exchange rate stays around its long-run fundamental value. There is thus a clear opposition between chartists and fundamentalists in terms of stabilizing characteristics. Whereas long-run agents' actions drive the exchange rate closed to its fundamental value, the chartists reproduce the past dynamics, which can lead to potential long-lasting disconnections of the exchange rate from its long-run equilibrium.

What happens now when we introduce a third rule in the game, i.e. a medium-run one, allowing agents to switch between three forecasting rules? The medium-run

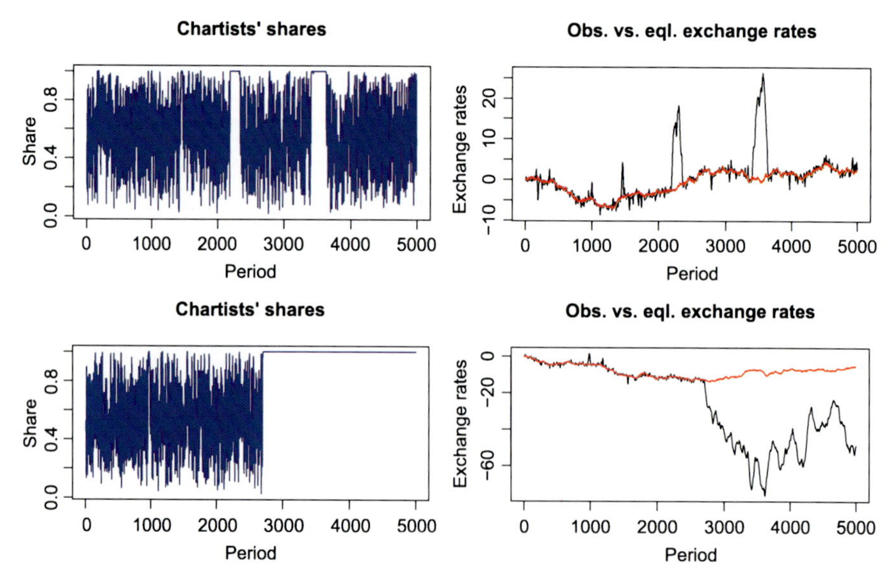

Fig. 87.1 Simulated data—convergence and divergence cases

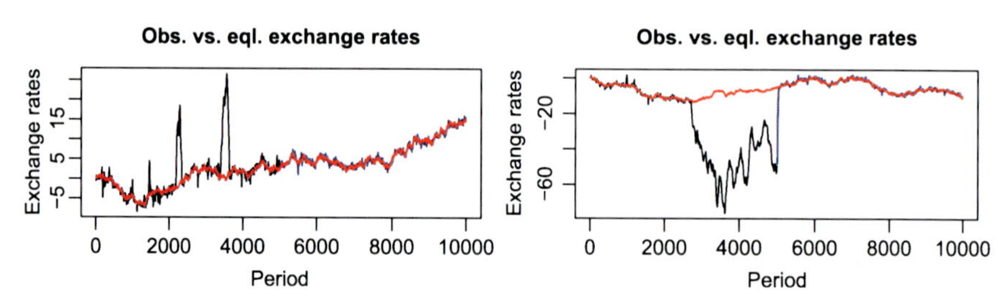

Fig. 87.2 What happens when medium-run agents are right?

rule is of a fundamentalist type, we expect it to be stabilizing, but is this characteristic conditioned by agents' good perception of the fundamentals' drivers? To give some insights on this issue, we propose to take a look at various outputs resulting from the three *scenarii*.

What happens when agents are right, i.e. when they rely on the MLE estimates for ψ?

As suggested on Fig. 87.2, we can see in both cases that the introduction of the medium-run rule leads to a drastic stabilization of the current rate closed to its PPP value. This correction stems from the fact that the filtered value, i.e. the medium-run equilibrium exchange rate, is very closed to the current one. Mechanically, the medium-run negative feedback rule is almost the exact opposite to the short-run positive one, corresponding to a "leaning against the wind strategy". Medium-run agents are thus reinforcing the mean-reverting dynamics in the market, thereby strengthening the hand of long-run agents at the expense of the short-run ones, see Chap. 9 from [3] or [11, 12] for more details.

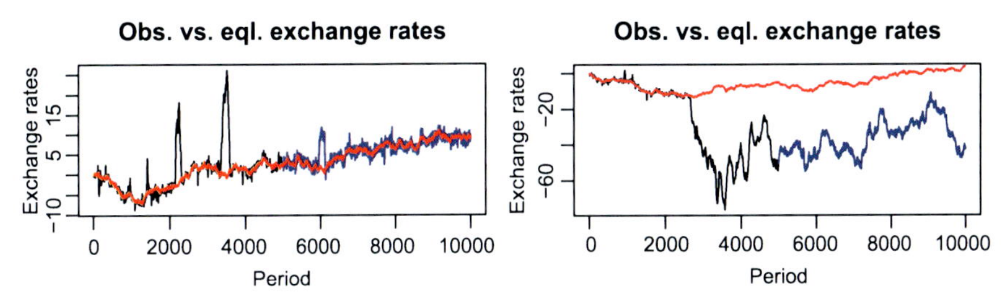

Fig. 87.3 What happens when medium-run agents are wrong?

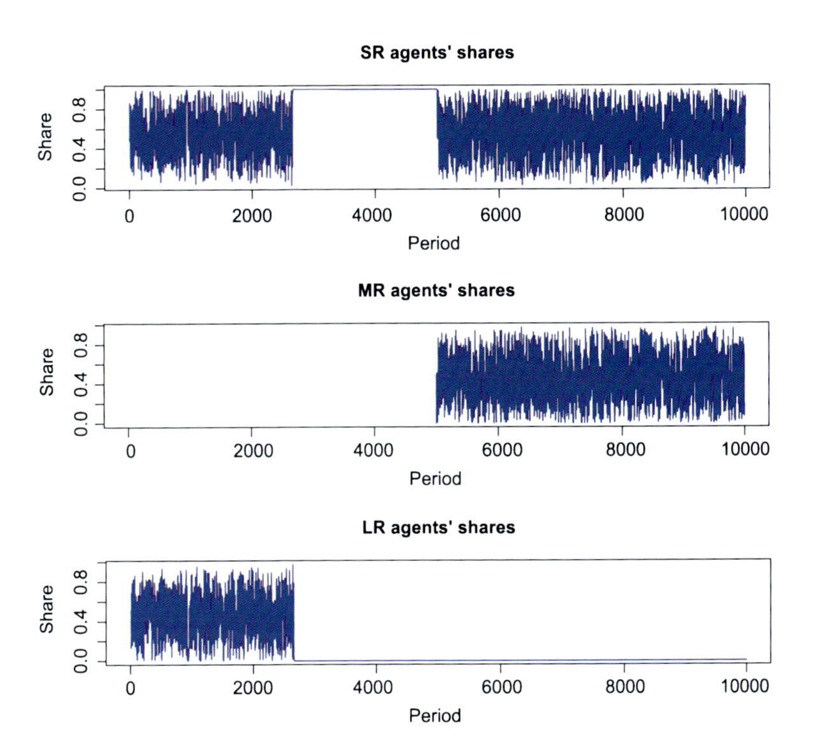

Fig. 87.4 Who is responsible for this divergence?

What happens now when agents are wrong, i.e. when they rely on values far from the MLE estimates for ψ?

We can see from Fig. 87.3 that the medium-run rule seems to stay stabilizing in the first case, although to the cost of greater volatility. Considering other sets of simulations, we show that this stabilizing characteristic is purely random. In addition, when we look at the second case in Fig. 87.3, i.e. when the exchange rate is already experiencing a long-lasting disconnection from its PPP value, we can see that the introduction of the medium-run rule cannot drive the current exchange rate towards its long-run fundamental rate. Since important switches can be observed from Fig. 87.4 between short and medium-run agents, this disconnection is due to

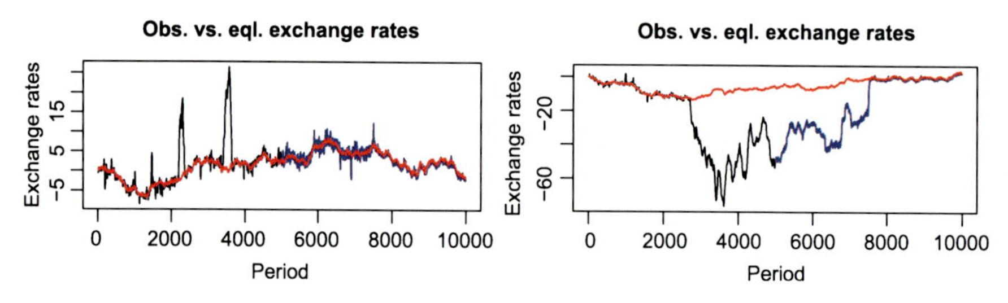

Fig. 87.5 What happens when medium-run agents are wrong but learn from the data?

both types. As whole, we conclude that the stabilizing properties of the medium-run rule seems to depend on agents' good perception of the fundamentals, summed up in ψ.[1]

As a last exercise, and to push further our analysis, we assume now that agents may update their perception of the fundamentals' drivers ψ through a Bayesian "learning process". In other words, we ask the following question: What happens when agents are wrong but then learn from the data and update their knowledge in a Bayesian manner? We can see from the graphs in Fig. 87.5 (derived from the same first two sets of simulations), that medium-run agents remain destabilizing until the revision step, in time $t = 7500$. After incorporating truncated information from the data,[2] the parameters become mechanically far much closer to the MLE estimates and allow to go back to "leaning against the wind strategy" as in the previous case.

87.4 Conclusions

To conclude, this piece of work proposes a Bayesian extension of the traditional behavioural finance model "à la" [3] in which agents differ in their forecasting time horizons as a way to re-explore Obstfeld and Rogoff's famous "exchange rate disconnect puzzle" [9]. As previously mentioned, those differences translate into the equilibrium values those agents target for the exchange rate. We then assume three rules corresponding to the short, the medium and the long run respectively. Whereas the first two rules stick to the usual representation, i.e. chartist vs. fundamentalist agents, we incorporate an intermediary regime, of a fundamental type, that relies on the assessment of a medium-run equilibrium value for the exchange rate supposed to potentially differ from the PPP and assessed here through an APEER model. In addition, we further extend the usual trial-and-error learning strategies on which agents rely for the sequential choice of their forecasting rules to the way they model the medium-run fundamental value, through a Bayesian sequential updating procedure. This modelling leads to two major conclusions: (i) In addition to the usual 'battle' between stabilizing long-run (fundamentalist) vs. destabilizing short-run (chartist)

---

[1] Other simulation sets confirm those conclusions. They are available upon request.

[2] In each case, Bayesian estimations have been performed on a sub-sample of 100 points.

agents, we show that medium-run agents may switch from one type to another depending on their ability to correctly perceive the drivers of exchange rate fundamentals. This means that even a fundamentalist type rule may lead to long-lasting disconnections between the exchange rate and its fundamental value; (ii) A good knowledge or perception of the fundamental drivers of the exchange rate is a precondition to make the median rule stabilizing. Further extensions discussed in the previous section are now to be done in order to test the robustness of those results in a more general context.

References

1. Bilson JF (1978) The monetary approach to the exchange rate: some empirical evidence. IMF Staff Pap 25:48–75
2. Brock WA, Hommes C (1998) A rational route to randomness. Econometrica 65:1059–1096
3. De Grauwe P, Gimaldi M (2006) The exchange rate in a behavioural finance framework. Princeton University Press, Princeton
4. Frankel JA, Froot KA (1990) Chartists, fundamentalists, and trading in the foreign exchange market. Am Econ Rev 80:181–185
5. Frenkel J (1976) A monetary approach to the exchange rate: doctrinal aspects and empirical evidence. Scand J Econ 78:200–224
6. Gehrig T, Menkhoff L (2003) Technical analysis in foreign exchange—the workhorse gains further ground. Working paper 278, Universität Hannover
7. Gehrig T, Menkhoff L (2003) The use of flow analysis in foreign exchange: explanatory evidence. J Int Money Financ 23:573–594
8. Gehrig T, Menkhoff L (2005) The rise of fund managers in foreign exchange: will fundamentalists ultimately dominate? World Econ 28:519–540
9. Lewis V, Markiewicz A (2009) Model misspecification, learning and the exchange rate disconnect puzzle. BE J Macroecon 9:13
10. Meese R, Rogoff K (1983) Empirical exchange rate models of the seventies: how well do they fit out of sample? J Int Econ 14(1–2):3–24
11. Westerhoof FH (2008) The use of agent-based financial market models to test the effectiveness of regulatory policies. J Econ Stat 228:195–227
12. Westerhoof FH (2006) The effectiveness of Keynes-Tobin transaction taxes when heterogeneous agents can trade in different markets: a behavioral finance approach. J Econ Dyn Control 30:293–322

Chapter 88
Anomalous Metastability and Fixation Properties of Evolutionary Games on Scale-Free Graphs

Michael Assaf and Mauro Mobilia

Abstract This contribution concerns the influence of scale-free graphs on the metastability and fixation properties of a set of evolutionary processes. In the framework of evolutionary game theory, where the fitness and selection are frequency-dependent and vary with the population composition, we analyze the dynamics of snowdrift games (characterized by a metastable coexistence state) on scale-free networks. Using an effective diffusion theory in the weak selection limit, we demonstrate how the scale-free structure affects the system's metastable state and leads to anomalous fixation. In particular, we analytically and numerically show that the probability and mean time of fixation are characterized by stretched exponential behaviors with exponents depending nontrivially on the network's degree distribution. Our approach is also shown to be applicable to models, like coordination games, not exhibiting metastability prior to fixation.

Keywords Metastability · Fixation · Complex networks · Evolutionary games · Diffusion theory

88.1 Introduction & Background

This contribution, based on Ref. [1], is concerned with the influence of complex spatial structure on the metastability and fixation properties of a set of evolutionary processes characterized by frequency-dependent selection.

The dynamics of systems where successful traits spread at the expense of others is naturally modeled in the framework of evolutionary game theory (EGT) [2–5].

M. Assaf (✉)
Racah Institute of Physics, Hebrew University of Jerusalem, Jerusalem 91904, Israel
e-mail: michael.assaf@mail.huji.ac.il

M. Mobilia (✉)
Department of Applied Mathematics, School of Mathematics, University of Leeds, Leeds LS2 9JT, UK
e-mail: M.Mobilia@leeds.ac.uk
url: http://www1.maths.leeds.ac.uk/~amtmmo/

T. Gilbert et al. (eds.), *Proceedings of the European Conference on Complex Systems 2012*, Springer Proceedings in Complexity, DOI 10.1007/978-3-319-00395-5_88,

While the EGT classic setting was originally proposed to describe the evolution of infinitely large and spatially homogeneous populations, it is known that evolutionary dynamics is affected by demographic noise and by the population's spatial arrangement [6–9]. Evolutionary dynamics is often characterized by the central notion of fixation [10, 11] that refers to the possibility that a "mutant type" takes over. In particular, one is interested in the probability that a given trait invades the entire population (*fixation probability*) and in the mean time for this event to occur (*mean fixation time*). In contrast to what happens in spatially-homogeneous (well-mixed) populations, the spatial arrangement of individuals can give rise to various very different scenarios [6–8]. In this context, evolutionary dynamics on networks [12–14] provides a general and unifying framework to describe the dynamics of both well-mixed and spatially-structured populations [15, 16]. In spite of their importance, the fixation properties of evolutionary processes on networks have been mostly studied in idealized situations, e.g. for two-state systems under a constant and weak selective bias [15–23]. In these works, it has been shown that the update rules and the complex network structure effectively renormalize the population size and affect the system's fixation properties. Furthermore, some properties of evolutionary games have been studied on scale-free networks by numerical simulations, and on regular graphs with mean field and perturbative treatments, see e.g. [24–26]. The models of Refs. [15–23] are of great interest but do *not* provide a general description of evolutionary dynamics on graphs. In particular, these references consider constant fitness and selection pressure, and thus can *not* describe systems possessing a long-lived *metastable* coexistence state prior to fixation [27–30]. In our work [1], within the EGT framework, we have studied metastability, which may arise as a consequence of frequency-dependent selection, and fixation on a class of scale-free networks. To the best of our knowledge, such an analytical study has not been conducted before. For concreteness, this contribution mainly focuses on "snowdrift games" (SGs) [2–5] that are the paradigmatic evolutionary games exhibiting metastability and characterized by an exponential dependence on the population size on complete graphs (well-mixed populations), see e.g. [29, 30]. Our findings are also directly relevant to various fields, e.g. to population genetics [31] and to the dynamics of epidemic outbreaks, for which a long-lived endemic state is often an intrinsic characteristic [27, 28, 32–36].

Our central result, of broad relevance and importance, is the demonstration that evolutionary dynamics on scale-free networks can lead to *anomalous* fixation and metastability characterized by a stretched exponential dependence on the population size, in stark contrast with their non-spatial counterparts. We also show that such a dependence characterizes fixation in models like coordination games which do *not* possess a long-lived metastable state prior to fixation [2–5].

88.2 The Model

We consider a network consisting of N nodes, each of which is either occupied by an individual of type C (cooperator) or D (defector). The occupancy of the node i

is encoded by the random variable η_i, with $\eta_i = 1$ if the node i is occupied by a C and $\eta_i = 0$ otherwise. The state of the system is thus described by $\{\eta\} = \{\eta_i\}^N$ and the density of cooperators present in the system is $\rho \equiv \sum_{i=1}^{N} \eta_i/N$. The adjacency matrix $\mathbf{A} = [A_{ij}]$ of this simple and undirected network is symmetric with elements $A_{ij} = 1$ if the nodes ij are connected and 0 otherwise [12–14]. The network is also characterized by its degree distribution $n_k = N_k/N$, where N_k is the number of nodes of degree k. For a generic two-strategy cooperation dilemma, the payoff of C against another C is denoted a and that of D playing against D is d. When C and D play against each other, the former gets payoff b and the latter gets c. For snowdrift games, in which we are chiefly interested, one has $c > a$ and $b > d$ and the (mean field) dynamics is characterized by two unstable absorbing states $\rho = 0$ and $\rho = 1$ separated by a (meta-)stable interior fixed point $\rho_* = (d - b)/(a - b - c + d)$. In a spatial setting, the interactions are among nearest-neighbor individuals and the species payoffs are defined locally: C and D players at node i interacting with a neighbor at site j respectively receive payoffs $\Pi_{ij}^C = a\eta_j + b(1 - \eta_j)$ and $\Pi_{ij}^D = c\eta_j + d(1 - \eta_j)$. In the spirit of the Moran model (with weak selection pressure) [9–11], each species local reproductive potential, or fitness, is given by the difference of $\Pi_{ij}^{C/D}$ relative to the population mean payoff $\bar{\Pi}_{ij}(t)$. Here, we make the mean-field-like choice $\bar{\Pi}_{ij}(t) = \rho(t)\Pi_{ij}^C + (1 - \rho(t))\Pi_{ij}^D$ to include what arguably is the simplest mechanism ensuring the formation of metastability. It is customary to introduce a selection strength $s > 0$ in the definition of the fitness to unravel the interplay between random fluctuations and selection [9–11]. By introducing a baseline contribution set to unity, the fitnesses of C/D at node i interacting with a neighbor at site j are $f_{ij}^C = 1 + s[\Pi_{ij}^C - \bar{\Pi}_{ij}]$ and $f_{ij}^D = 1 + s[\Pi_{ij}^D - \bar{\Pi}_{ij}]$, respectively. The system evolves according to the "link dynamics" (LD) [17–21]: a link is randomly selected at each time step and, if it connects a CD pair, one of the neighbors is randomly selected for reproduction with a rate proportional to its fitness while the other is replaced by the offspring. While various types of update rules are possible, we have used the LD for the sake of simplicity and to highlight the combined effects of the topology and frequency-dependent selection. However, we have checked that our conclusion is robust and holds for various other update rules leading to metastability [1].

88.3 Methodology

The system's evolutionary dynamics is analyzed in terms of $\{\rho_k\}$, where $\rho_k = \sum_i' \eta_i/N_k$ is the average number of cooperators on all nodes of degree k (the prime means that the sum is restricted to nodes of degree k). Hence, ρ_k is the *subgraph density* of C's on nodes of degree k. The analysis also relies on the mth−moment of the degree distribution, denoted $\mu_m \equiv \sum_k k^m n_k$, and the degree-weighted density of cooperators $\omega \equiv \sum_k (k/\mu_1) n_k \rho_k$. The analytical treatment thus proceeds in five steps, (a)–(e):

(a) **Birth-Death Process Formulation** In our model, the subgraph densities $\{\rho_k\}$ evolve according to a (continuous-time) birth-death process defined by the transition rates $T_k^+ = \sum_i' \sum_j A_{ij}\Psi_{ij}/\bar{N}$ and $T_k^- = \sum_i' \sum_j A_{ij}\Psi_{ji}/\bar{N}$, where $\Psi_{ij} = (1-\eta_i)\eta_j f_{ji}^C$, $\Psi_{ji} = (1-\eta_j)\eta_i f_{ji}^D$ and $\bar{N} \equiv N\mu_1$. At each time-increment $\delta t = N^{-1}$, the subgraph density ρ_k thus respectively changes by $\pm\delta\rho_k = \pm 1/N_k$ according to the transition rates $T_k^{\pm}$. In our analytical calculations, we have focused on degree-heterogeneous networks (of finite mean degree μ_1) with degree-uncorrelated nodes, see e.g. [38], yielding $A_{ij} = k_i k_j/\bar{N}$, while numerical simulations were performed using the "redirection algorithm" [39, 40].

(b) **Multivariate Fokker-Planck Equation** In the limit of weak selection intensity ($s \ll 1$) [29, 30], one can use the diffusion theory to investigate the properties of the above birth-death process [10, 11]. This yields a multivariate backward Fokker-Planck equation (FPE) whose generator reads [37]

$$\mathcal{G}(\{\rho_k\}) = \sum_k \left[\frac{(T_k^+ - T_k^-)}{n_k} \frac{\partial}{\partial \rho_k} + \frac{(T_k^+ + T_k^-)}{2Nn_k^2} \frac{\partial^2}{\partial \rho_k^2} \right], \quad \text{with}$$

$$T_k^+ = (n_k/\mu_1)\big[1 + s(b-d)(1-\rho)\big]k(1-\rho_k)\omega \quad \text{and} \tag{88.1}$$

$$T_k^- = (n_k/\mu_1)\big[1 - s(a-c)\rho\big]k\rho_k(1-\omega)$$

(c) **Time Scale Separation** While the multivariate FPE generated by (88.1) is generally not tractable, when the selection intensity is weak ($s \ll 1$), a time scale separation allows to greatly simplify the mathematical analysis [1, 19–23]. Indeed, when $t \ll s^{-1}$ the selection pressure is negligible and the quantity ρ is conserved [17, 18], before relaxing to its metastable value ρ_* on a time scale $t \sim s^{-1} \gg 1$. Furthermore, as shown in [1] and illustrated in Fig. 88.1, after a time of order $\mathcal{O}(1)$, one finds that on average all subgraph densities $\bar{\rho}_k$ and the mean degree-weighted density $\bar{\omega}$ approach the average density $\bar{\rho}$ and all these quantities approach ρ_* on a time scale $t \sim s^{-1}$.

(d) **Effective Single-Coordinate Fokker-Planck Equation** Since fixation occurs on much longer time scales than s^{-1}, when $\bar{\rho}_k \approx \bar{\omega} \approx \bar{\rho}$ (weak selection) and the fluctuations about the metastable state are small, one can use the approximation $\rho_k \approx \rho \approx \omega$ in the generator (88.1) of the backward FPE [1, 19–21]. Hence, with the change of variables $\rho_k \to \omega$ and $\rho_k \approx \rho \approx \omega$, Eq. (88.1) yields the *effective single-coordinate* FPE generator [1, 19–21]

$$\mathcal{G}_{\text{eff}}(\omega) = \frac{\omega(1-\omega)}{N_{\text{eff}}} \left[-\sigma(\omega - \rho_*) \frac{\partial}{\partial \omega} + \frac{1}{2} \frac{\partial^2}{\partial \omega^2} \right], \tag{88.2}$$

where $\sigma \equiv 2(b-d)N_{\text{eff}}s_{\text{eff}}/\rho_*$ [1]. The complex spatial structure renormalizes the population size and selection intensity, yielding the effective quantities $N_{\text{eff}} = N\,(\mu_1)^3/\mu_3$ and $s_{\text{eff}} \equiv s\ \mu_2/(\mu_1)^2$, whose expressions depend on the first three moments of the degree distribution. In the effective FPE generator (88.2), the drift term is thus proportional to $\sigma \propto sN\mu_1\mu_2/\mu_3$. For scale-free networks with degree distribution $n_k \sim k^{-\nu}$ and finite average degree (i.e. $\nu > 2$), the maximum degree is

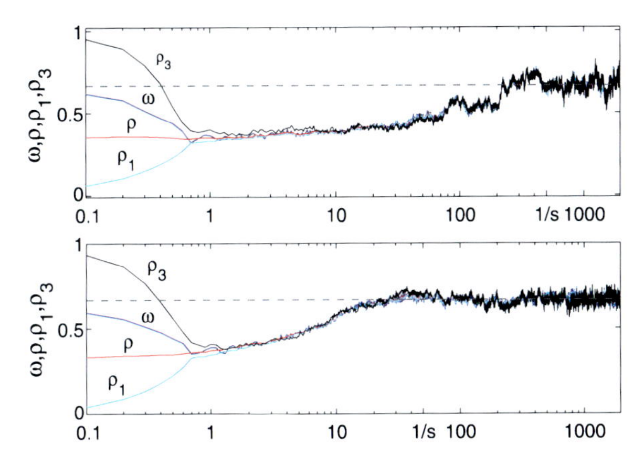

Fig. 88.1 (Adapted from [1]). Time scale separation in typical Monte Carlo trajectories of the densities ρ, ω, ρ_1, and ρ_3 on a scale-free network with $\nu = 3$ for a SG with $a = d = 1, b = 9, c = 5$ and $N = 10^4$. The selection intensity is $s = 0.002$ (*top*) and $s = 0.02$ (*bottom*), and initially $\rho_{k>\mu_1}(0) = 1$, $\rho_{k\leq\mu_1}(0) = 0$. One notices two time scales: after a time of order $\mathcal{O}(1)$, the trajectories of ρ, ω, ρ_1, and ρ_3 almost coincide and evolve together toward their common metastable value $\rho_* = 2/3$ (*dashed*) that is reached after a time of order $t \sim s^{-1}$ (indicated on the horizontal axis). Fixation occurs on a much longer time scale, see text and Ref. [1]

$k_{\max} \sim N^{1/(\nu-1)}$ [41]. From the moments μ_m, one thus obtains the nontrivial scaling of σ with the population size [1]:

$$\sigma \propto sN\frac{\mu_1\mu_2}{\mu_3} \sim \sigma_{\text{re}} = \begin{cases} sN, & \nu > 4 \\ sN/\ln N, & \nu = 4 \\ sN^{(2\nu-5)/(\nu-1)}, & 3 < \nu < 4 \\ s\sqrt{N}\ln N, & \nu = 3 \\ sN^{(\nu-2)/(\nu-1)}, & 2 < \nu < 3. \end{cases} \tag{88.3}$$

(e) **Calculation of the Fixation Properties** In the realm of the effective diffusion approximation (88.2), the fixation probability $\phi^C(\omega)$ that a system with initial degree-weighted density ω is taken over by cooperators is obtained analytically by solving the backward FPE $\mathcal{G}_{\text{eff}}(\omega)\phi^C(\omega) = 0$ with boundary conditions $\phi^C(0) = 1 - \phi^C(1) = 0$ [37]. Similarly, the unconditional mean (fixation) time τ to reach an absorbing boundary is given by the solution of $\mathcal{G}_{\text{eff}}(\omega)\tau(\omega) = -1$, with $\tau(0) = \tau(1) = 0$.

88.4 Results

The analytical results of this research have been corroborated by stochastic simulations and concern the thorough analysis of the influence of the scale-free network structure on the metastability and fixation properties of a class of evolutionary processes (snowdrift and coordination games).

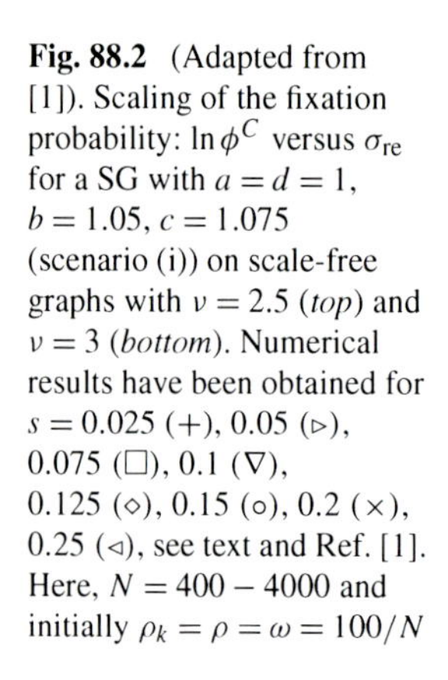

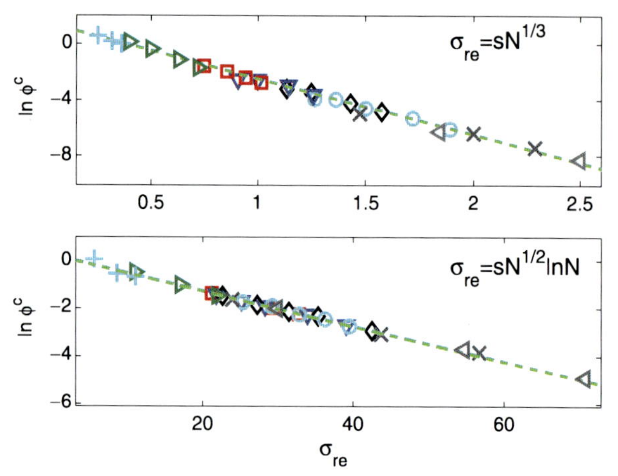

Fig. 88.2 (Adapted from [1]). Scaling of the fixation probability: $\ln \phi^C$ versus σ_{re} for a SG with $a = d = 1$, $b = 1.05$, $c = 1.075$ (scenario (i)) on scale-free graphs with $\nu = 2.5$ (*top*) and $\nu = 3$ (*bottom*). Numerical results have been obtained for $s = 0.025$ $(+)$, 0.05 $(\triangleright)$, 0.075 $(\square)$, 0.1 (∇), 0.125 $(\diamond)$, 0.15 $(\circ)$, 0.2 $(\times)$, 0.25 $(\triangleleft)$, see text and Ref. [1]. Here, $N = 400 - 4000$ and initially $\rho_k = \rho = \omega = 100/N$

88.4.1 Implications and Validity of the Effective Theory

We have shown that the scale-free structure is responsible for a nontrivial renormalization of the population size and selection intensity in an important class of evolutionary ("snowdrift") games. In the limit of weak selection, the properties of these models are accurately described by the effective diffusion theory (88.2)–(88.3). To appreciate the implications of the nontrivial scaling (88.3), it is instructive to consider the scale-free graphs with $2 < \nu < 3$ that are characterized by the divergence of μ_2 and μ_3 (when $N \to \infty$). For this class of complex networks that include nodes of high degree, we have found $N_{\mathrm{eff}} = N^{(2\nu-5)/(\nu-1)} \ll N$ and $s_{\mathrm{eff}} = sN^{(3-\nu)/(\nu-1)} \gg s$. In this case, the fluctuations intensity ($\propto N_{\mathrm{eff}}^{-1/2}$) and the drift strength ($\propto s_{\mathrm{eff}}$) are both enhanced by the complex topology. Yet, the product of these effective quantities yields $N_{\mathrm{eff}} s_{\mathrm{eff}} \sim sN^{(\nu-2)/(\nu-1)} \ll Ns$, which implies that the scale-free topology can drastically reduce the mean fixation time [1], see below.

We have checked that the predictions of the effective theory are accurate when $s_{\mathrm{eff}} \ll N_{\mathrm{eff}}^{-1}$ (e.g. when $s^2 \ll N^{-1/(\nu-1)}$ for $2 < \nu < 3$). Hence, when $2 < \nu < 4$, the effective diffusion approximation is applicable for a broader range of values of s than in a non-spatial setting (where the requirement is $s^2 \ll N^{-1}$ [29, 30]).

88.4.2 Fixation Probability in Snowdrift Games

Evolutionary dynamics is characterized by the fixation probability $\phi^C(\omega)$ that a system with initial degree-weighted density ω is taken over by cooperators.

As described in Sect. 88.3(e), we have analytically computed the fixation probability in the realm of the effective diffusion theory (88.2)–(88.3) and found

$$\phi^C(\omega) = \frac{\mathrm{erfi}[\rho_* \sqrt{\sigma}\,] - \mathrm{erfi}[(\rho_* - \omega)\sqrt{\sigma}\,]}{\mathrm{erfi}[\rho_* \sqrt{\sigma}\,] + \mathrm{erfi}[(1 - \rho_*)\sqrt{\sigma}\,]}, \tag{88.4}$$

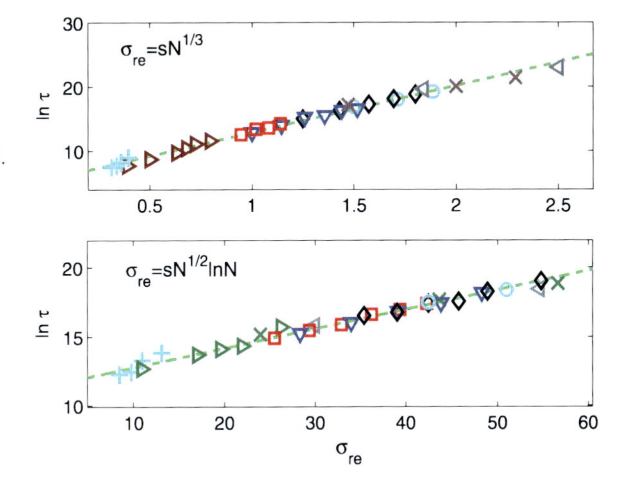

Fig. 88.3 (Adapted from [1]). Scaling of the mean fixation time: $\ln \tau$ versus σ_{re} for $\nu = 2.5$ (*top*) and $\nu = 3$ (*bottom*). Symbols and parameters are as in Fig. 88.2. Numerical results collapse along *the dashed line*, in agreement with (88.5). Initially $\rho_k = \rho = 0.5$, see text

where $\mathrm{erfi}(z) \equiv \frac{2}{\sqrt{\pi}} \int_0^z e^{u^2} du$. In Ref. [1], we have carefully discussed the biologically relevant case where $\omega \ll 1$ (low initial density of cooperators), the population size is large with $\sigma \gg 1$, and the selection pressure is weak [9–11]. In this situation, metastability (long-lived coexistence) is guaranteed and the fixation probability is characterized by a stretched-exponential dependence on the population size according to two main cases (when $\omega \ll 1$):

(i) $\ln \phi^C \simeq -(1 - 2\rho_*)\sigma$, when $\rho_* < 1/2$
(ii) $\ln(1 - \phi^C) \simeq -\omega(2\rho_* - \omega)\sigma$, if $\rho_* > 1/2$ and $\omega < 2\rho_* - 1$,

where the asymptotic dependence of $\sigma \propto sN\mu_1\mu_2/\mu_3$ is given by (88.3). The result (88.4) is corroborated by the stochastic simulations of Fig. 88.2 where results for various values of s and N have been rescaled to test the scaling relation (88.3). The linear data collapse and clear stretched-exponential dependence observed in Fig. 88.2 (see [1] for further results) confirm the predictions of (88.3) and (88.4). This demonstrates that the fixation probability of a few mutants may be drastically enhanced on a scale-free graph (with respect to a non-spatial setting).

88.4.3 Mean Fixation Time in Snowdrift Games

Metastability and evolutionary dynamics are also characterized by the mean fixation time (MFT) $\tau(\omega)$, i.e. the (unconditional) mean time necessary to reach an absorbing boundary.

As discussed in Sect. 88.3(e), we have solved the effective backward FPE for the unconditional mean fixation time τ and, to leading order in $\sigma \gg 1$, have found that τ exhibits a stretched-exponential dependence on N [1]:

$$\tau(\omega) \sim \begin{cases} (1 - \phi^C(\omega))e^{\sigma\rho_*^2}, & \text{when } \omega > \rho_* \\ \phi^C(\omega)e^{\sigma(1-\rho_*)^2}, & \text{otherwise.} \end{cases} \quad (88.5)$$

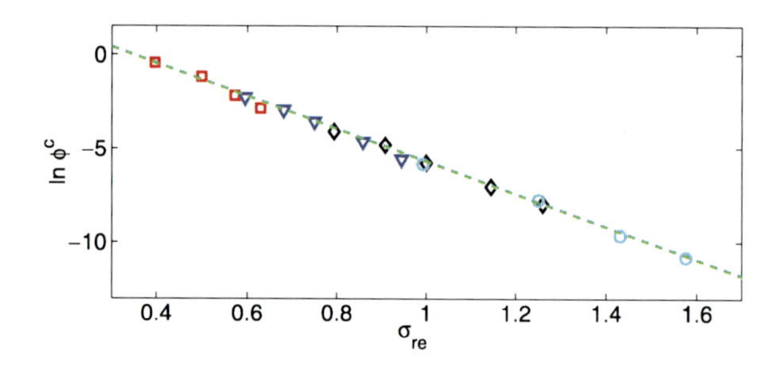

Fig. 88.4 Logarithm of the fixation probability $\ln \phi^C$ versus σ_{re} (88.3) for a CG with $a = 1.6, b = c = 1$ and $d = 1.2$ ($\rho_* = 0.75$) on a scale-free graph with $\nu = 2.5$. Here, $s = 0.05$ (□), 0.075 (△), 0.1 (◇) and 0.125 (○), $N = 1000$–4000 and $\sigma_k(0) = 0.5$. Numerics are consistent with the theoretical prediction $\ln \phi^C \sim -sN\mu_1\mu_2/\mu_3 \simeq -sN^{1/3}$, see text

With (88.4) and (88.5), one finds the leading-order contribution to the MFT: e.g., when $\sigma \gg 1$ and $\rho_* < 1/2$, τ is found to grow with N as the stretched exponential $\ln \tau \simeq \sigma \rho_*^2$. This prediction is confirmed by Fig. 88.3 (see Ref. [1] for further simulation results) and implies that on scale-free networks with degree distribution $n_k \sim k^{-\nu}$ and $2 < \nu < 4$, fixation occurs much more rapidly than on complete graphs. This phenomenon is called "hyperfixation" in population genetics [31].

88.4.4 Fixation Probability in Coordination Games

We have also studied the class of "coordination games" (CGs) for which $a > c$ and $d > b$. In CGs, the interior fixed point ρ_* is unstable, whereas the absorbing states $\rho = 0$ and $\rho = 1$ are stable (bistability) [2–5]. While there is no metastability in CGs, the fixation probability in coordination games evolving with the link dynamics on scale-free graphs has been found to have the same stretched-exponential dependence on N as in snowdrift games (see also [29, 30]), as shown in Fig. 88.4.

88.5 Conclusion

We have studied metastability and fixation of evolutionary processes on scale-free networks. For the sake of concreteness, we have focused on the paradigmatic case of "snowdrift games". The probability and mean fixation time have been computed from an effective Fokker-Planck equation derived by exploiting a time scale separation in the weak selection limit. These quantities exhibit a stretched-exponential dependence on the population size, in stark contrast with their non-spatial counterparts. We have checked with various update rules that the stretched-exponential behavior is a generic feature of metastability on scale-free graphs that also characterizes the fixation probability of coordination games. Important consequences of the stretched-exponential behavior are a drastic reduction of the mean fixation time, and the possible enhancement of the fixation probability of a few mutants with respect to a non-spatial setting.

References

1. Assaf M, Mobilia M (2012) Phys Rev Lett 109(18):188701
2. Smith JM (1982) Evolution and the theory of games. Cambridge University Press, Cambridge
3. Hofbauer J, Sigmund K (1998) Evolutionary games and population dynamics. Cambridge University Press, Cambridge
4. Nowak MA (2006) Evolutionary dynamics. Belknap Press, New York
5. Szabó G, Fáth G (2007) Phys Rep 446:97
6. Nowak MA, May RM (1992) Nature 359:826
7. Hauert C, Doebeli M (2004) Nature 428:643
8. Nowak M (2006) Science 314:1560
9. Nowak MA et al (2004) Nature 428:646
10. Crow JF, Kimura M (2009) An introduction to population genetics theory. Blackburn Press, New Jersey
11. Ewens WJ (2004) Mathematical population genetics. Springer, New York
12. Barabási AL, Albert R (1999) Science 286:509
13. Albert R, Barabási AL (2002) Rev Mod Phys 74:47
14. Newman MEJ (2010) Networks: an introduction. Oxford University Press, New York
15. Lieberman E et al (2005) Nature 433:312
16. Ohtsuki H et al (2006) Nature 441:502
17. Castellano C, Vilone D, Vespignani A (2003) Europhys Lett 63:153
18. Suchecki K, Eguiluz VM, San Miguel M (2005) Europhys Lett 69:228
19. Sood V, Redner S (2005) Phys Rev Lett 94:178701
20. Antal T et al (2006) Phys Rev Lett 96:188104
21. Sood V et al (2008) Phys Rev E 77:041121
22. Baxter GJ, Blythe RA, McKane AJ (2008) Phys Rev Lett 101:258701
23. Blythe RA (2010) J Phys A, Math Theor 43:385003
24. Santos FC, Pacheco JM (2005) Phys Rev Lett 95:098104
25. Ohtsuki H, Nowak MA (2006) J Theor Biol 243:86
26. Tarnita CE et al (2009) J Theor Biol 259:570
27. Assaf M, Meerson B (2006) Phys Rev Lett 97:200602
28. Assaf M, Meerson B (2010) Phys Rev E 81:155912
29. Mobilia M, Assaf M (2010) Europhys Lett 91:10002
30. Assaf M, Mobilia M (2010) J Stat Mech P09009:1–24
31. Wigham PA et al (2008) Theor Popul Biol 74:283
32. Pastor-Satorras R, Vespignani A (2001) Phys Rev Lett 86:3200
33. Nåsell I (2001) J Theor Biol 211:11
34. May RM, Lloyd AL (2001) Phys Rev E 64:066112
35. Newman MEJ (2002) Phys Rev E 66:016128
36. Durrett R (2010) Proc Natl Acad Sci USA 107:4491
37. Gardiner CW (2002) Handbook of stochastic methods. Springer, New York
38. Molloy M, Reed B (1995) Random Struct Algoritm 6:161
39. Dorogovtsev SN, Mendes JFF, Samukhin AN (2000) Phys Rev Lett 85:4633
40. Krapivsky PL, Redner S (2001) Phys Rev E 63:066123
41. Krapivsky PL, Redner S (2002) J Phys A 35:9517

Chapter 89
Constrained Graph Resampling for Group Assessment in Human Social Networks

Nicolas Tremblay, Pierre Borgnat, Jean-François Pinton, Alain Barrat, Mark Nornberg, and Cary Forest

Abstract The increasing availability of time—and space—resolved data of human activities and interactions gives insight into the study of both static and dynamic properties of human behavior. In practice, nevertheless, real-world datasets can often be considered as only one realisation of a particular event, giving rise to a key issue in social network analysis: the statistical significance of these properties. We focus in this work on features regarding groups of the networks and present a resampling—a.k.a. bootstrapping—method that enables us to add confidence intervals to such features. This in turn gives us the opportunity to compare groups' behaviors within any network. We apply this method to a new high resolution dataset of face-to-face proximity collected during two co-located scientific conferences, and it enables us to probe whether or not co-locating two conferences is an effective way of bringing together two different communities.

N. Tremblay (✉) · P. Borgnat · J.-F. Pinton
ENS Lyon, Laboratoire de Physique, CNRS UMR 5672, Université de Lyon, Lyon, France
e-mail: nicolas.tremblay@ens-lyon.fr

P. Borgnat
e-mail: pierre.borgnat@ens-lyon.fr

J.-F. Pinton
e-mail: jean-francois.pinton@ens-lyon.fr

A. Barrat
Centre de Physique Théorique de Marseille, CNRS UMR 6207, Marseille, France
e-mail: alain.barrat@ens-lyon.fr

A. Barrat
Data Science Laboratory, ISI Foundation, Torino, Italy

M. Nornberg · C. Forest
Physics Department, University of Wisconsin, Madison, WI, USA

M. Nornberg
e-mail: mark.nornberg@ens-lyon.fr

C. Forest
e-mail: cary.forest@ens-lyon.fr

T. Gilbert et al. (eds.), *Proceedings of the European Conference on Complex Systems 2012*, Springer Proceedings in Complexity, DOI 10.1007/978-3-319-00395-5_89,

Keywords Complex system · Dynamic network analysis · Graph resampling · Bootstrap

89.1 Introduction

High resolution experiments on face-to-face interactions between individuals in different social gatherings—such as scientific conferences, museums, schools, or hospitals—were made possible by the use of small radio sensors worn by participants, communicating with each other by bluetooth, wireless or active RFID (Radio Frequency Identification Device). These new data paved the way to many empirical investigations [2, 3, 6, 8] of human contacts, both static (existence of communities, clustering, distribution of degrees...) and dynamic (distribution of duration of contacts, of intercontacts, or of groups of different sizes...). An important issue regarding the analysis of these datasets is that each one of them can be considered as only one realisation of a particular event, it is therefore challenging to estimate confidence intervals to any of the measurable features. To this end, in the general non-network case, two methods based on constructing many random samples from the unique original data, have been widely used: the jackknife and the bootstrap methods [4]. In the case of networks, however, it is not clear how to directly transpose the classical bootstrap approach to graphs [5, 9]. In this paper, we focus on features of groups of a network, and formulate a resampling method that enables us to gain statistical significance by comparing the real data with random pseudosamples found under well-chosen constraints. The method is then applied to a dataset collected in two co-located conferences involving two distinct communities: it enables us to assess to what extent both communities mix together.

89.2 Resampling Method for Complex Human Contact Networks

There are admittedly several ways to model a human contact network by a graph, but one can always end up with a weighted graph where each node is an individual and where the strength of the interaction between two nodes is quantified by the weight of their associated link. In the following we consider such a weighted graph as well as a group of nodes within the graph that we call X^0, whose behavior we will compare to the behavior of random groups called bootstrap samples. Let us call R^0 the group of nodes that are not in X^0. We quantify X^0's "behavior" by looking at seven observables: N^0_{XX} the total number of links within X^0, N^0_{RR} the total number of links within R^0, N^0_{XR} the total number of links connecting the two groups, T^0_{XX} the total weight of intra-X^0 links, T^0_{RR} the total weight of intra-R^0 links, T^0_{XR} the total weight of the links connecting the two groups, and Q^0_X the modularity computed for the partitioning in two groups X^0 and R^0. The modularity, in this case of a partition in two groups, is a scalar between -0.5 and 0.5 and measures how well a particular partition of the nodes separates the network into distinct communities (a value tending to 0.5 denotes two strong communities) [7]. Depending on the specific issue addressed, other observables could be considered. The backbone of the resampling protocol is the following. First, formulate a Null Hypothesis regarding the behavior of X^0. Then, compute the behavior of a large number N of groups randomly chosen within the graph called bootstrap samples (we use X as a generic notation for the bootstrap samples) for which the Null Hypothesis is true. The novelty is that each bootstrap sample is a random group *under constraint* drawn with replacement. Finally, compare the behavior of X^0 to the statistical behavior of the bootstrap samples, and decide whether or not we can reject the hypothesis. If it is rejected, a measure d is proposed to compute to what extent X^0 differs from the bootstrap samples and hence the Null Hypothesis. The comparison between different groups' behavior now boils down to the comparison of the scalar d associated to each group.

89.3 The Data

We apply this method to a face-to-face proximity dataset collected in Salt Lake City in November 2011 during two co-located scientific conferences jointly organised by DPP (APS' Department of Plasma Physics) and GEC (Gaseous Electronics Conference) in an attempt to bring both communities—academic researchers and engineers respectively—together. In order to measure face-to-face proximity between the conference attendees wearing them, we use low power RFID tags embedded in conference badges, using the SocioPatterns sensing platform.[1] Two tags exchange packets only if they face each other (the human body acts as a shield at this frequency and power) within a distance of 1 to 1.5 meters. As soon as a tag receives a packet from another tag, it immediately uploads this information to RFID readers installed in the environment. By aggregating the five days of collected information, we obtain the overall network of contacts between the 320 participants of the experiment. Comparison with other similar experiments, and analysis of the contact patterns will be detailed in a later communication.

89.4 Is It Worthwhile to Co-locate Both Conferences?

89.4.1 Translating the Question

In terms closer to our graph approach, we translate this question in: how well do GEC nodes mix with DPP nodes? Of course, we cannot expect GEC to mix as well as any random group of the graph: it is a community. Hence, in order to answer to

[1] www.sociopatterns.org.

the question, we apply our method to assess the difference—or similarity—between GEC's behavior with three other known communities (that will act as a benchmark): the senior researchers from DPP (SEP), the juniors from DPP (JUP), and the students from DPP (STP). The group noted X^0 in the method will alternatively be GEC, SEP, STP, or JUP. We test those four groups to the same Null Hypotheses and compare the degree with which they reject them. To show that GEC's behavior is peculiar, we look for the appropriate Null Hypothesis—if it exists—that significantly discriminates GEC from the other groups. In the following, the aggregated graph is pre-processed by deleting links that have a total time of existence inferior to 1 minute (filtering threshold under which we consider the measurement to be noise).

89.4.2 Same Cardinal Constraint for GEC

Consider the following Null Hypothesis: GEC behaves like any group of $V_X = 39$ individuals in the conference. Here, the only constraint we impose to the bootstrap samples is to have a cardinal equal to V_X. We normalize each observable Z: $z = \frac{Z - \bar{Z}^*}{\sigma_Z^*}$ where $\bar{Z}^*$ is the expected value and σ_Z^* the standard deviation in a random graph with same total number of links and same weight sequence (this is done by randomly re-allocating the weights within the ensemble of possible links). Note that $\bar{Z}^*$ and σ_Z^* depend on V_X. We choose this mode of representation for its clarity (we can plot all 7 observables on the same figure) but also because it removes the effects due to the scale of the groups allowing us to compare the results between different groups. For each normalized observable z, we define d_z the distance between the actual measured value z^{X^0} and the interval $\bar{z}^b \pm 3\sigma_z^b$ ($d_z = 0$ if z^{X^0} is in the interval), where $\bar{z}^b$ and σ_z^b are computed on the bootstrap samples. This interval has the meaning of an acceptance interval for the Null Hypothesis. The sum d of the distances d_z computed for each observable is the measure we use to evaluate to what extent GEC rejects the Hypothesis: the larger is d, the higher is our confidence level to reject the Hypothesis.

For all the results presented in the following, we use $N = 1000$ bootstrap samples. In Fig. 89.1(a), we plot two histograms. The first one shows the number of times each node was chosen in a bootstrap sample. In the top right hand corner is its standard deviation σ_u: it indicates how uniformly the nodes were chosen. The second histogram shows how many nodes from GEC are in each bootstrap sample. The green dotted line represents the theoretical hypergeometric histogram computed for this same cardinal constraint. In the top right hand corner of the figure is the χ^2 distance between the observed and theoretical histograms. Each χ^2 value is computed with 10 bins that contain at least five realisations. An important point is that we do not use χ^2 for a goodness-of-fit test. In fact, we expect χ^2 to increase as soon as we impose stronger constraints on the bootstrap samples. The idea is that in the extreme case where we impose the boostraps to be exactly the GEC group, the distance d will obviously be null, σ_u will be larger than 300 and χ^2 larger than

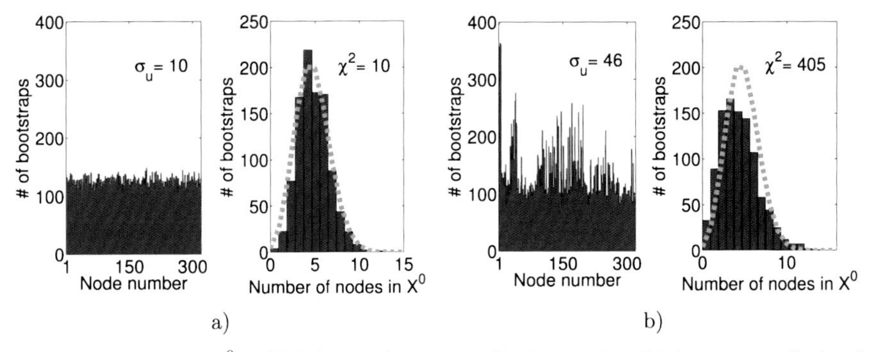

Fig. 89.1 Results for $X^0 = \text{GEC}$ for (**a**) the same cardinal constraint, (**b**) the same cardinal and same modularity constraint. *Left*: histogram of the number of occurrences of each node in the bootstrap samples and its standard deviation σ_u. *Right*: histogram of number of X^0-nodes in a bootstrap sample with its χ^2 distance from the theoretical hypergeometric histogram (*dotted line*)

10^{48} (the expected number of bootstrap samples having 39 GEC nodes is 10^{-48}), but we will have gained zero information. Therefore, we use χ^2 and σ_u as two control parameters of the "randomness" of the test, and make sure they stay reasonably small.

Finally, the top left figure of Fig. 89.2 summarizes the same cardinal constraint test for GEC: it compares the boxplots of the bootstrap samples with the measured behavior of GEC (big green crosses) and indicates d, σ_u and χ^2 in the bottom right hand corner of the figure.

89.4.3 Same Cardinal Constraint for All Four Groups

Fig. 89.2 shows the results for the four groups. If the graph was random, we would have boxplots centered around zero and whiskers between—typically: -3 and 3 (corresponding to more than 99 % coverage if the data was normally distributed). It is not the case (not for the whiskers) and this is an indirect proof that the graph is not random. Also, all groups have a non-null distance. This is not a surprise because those groups are known communities and behave as such: compared to the bootstrap samples, they tend to have high Q_X, N_{XX}, N_{RR}, T_{XX}, T_{RR} and low N_{XR}, T_{XR}. Interestingly, GEC's distance is clearly larger than the others: with this first naïve test, it already shows a peculiar behavior. However we can not say with statistical significance that its interaction with the rest of the conference is different from that of any specific group of people such as students or seniors (possibly also prone to discuss more with other people from their group).

89.4.4 Other Tests

To clearly show GEC's peculiar behavior, we need to find the appropriate test—if it exists—rejected by GEC but not by the others. To this end, we need to find a

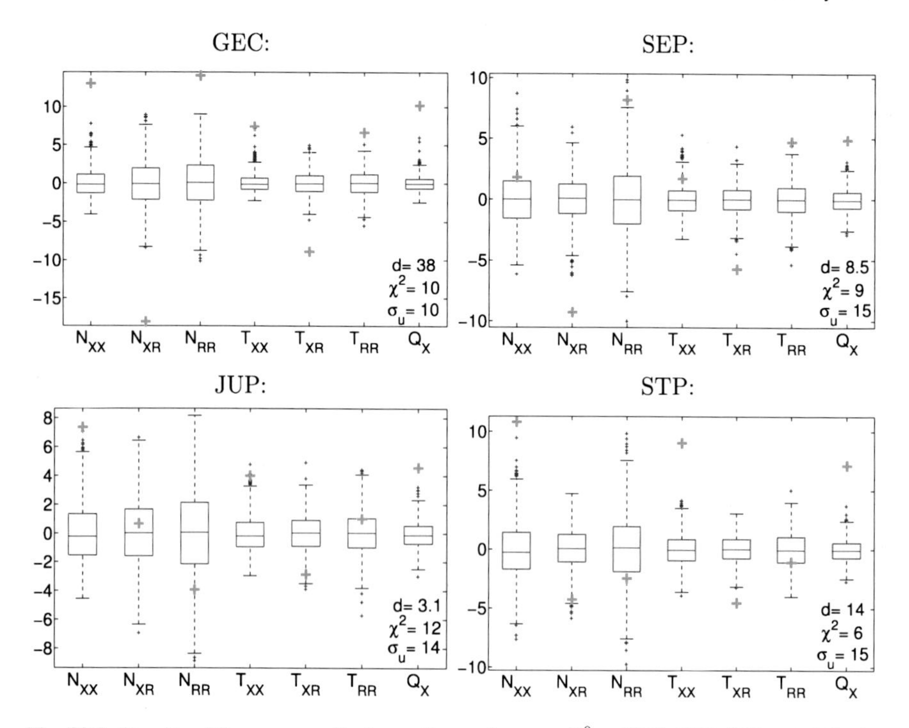

Fig. 89.2 Results of the same cardinal test. For each group X^0 = GEC, SEP, JUP and STP, the scalar d (*bottom right hand corner* of each figure) is an estimation of the distance between the statistical behavior of the bootstrap samples (boxplots) and the real data (*big green crosses*). χ^2 and σ_u are two control parameters of the "randomness" of the test—see text

compromise between strong enough constraints on the bootstrap samples to make the test more discriminative, but, as previously explained, loose enough so as to preserve the randomness of the test. In the first test, the only constraint we imposed on the boostraps was to have the same cardinal as X^0. We now refine the Null Hypothesis: X^0 behaves like any random group (with same cardinal) that has the same modularity, hence forming a community as strong as X^0. Requiring the exact same modularity is too strong a constraint and we relax it to: $Q_X^0(1-\delta) \leq Q_X \leq Q_X^0(1+\delta)$ with δ the error we tolerate. In the following, $\delta = 0.5$ %. We use a simulated annealing algorithm [1] to find such bootstrap samples and we plot the results for the four different groups in Fig. 89.3. First, we see that the boxplots are not centered around zero anymore, they indeed need to be in accordance with a high modularity. STP and JUP's distances are null. SEP's distance is almost ten times smaller than GEC's distance: this test is a satisfying confirmation that GEC behaves differently than the other groups. We plot in Fig. 89.1(b) the two same histograms as in Fig. 89.1(a) but for bootstrap samples under these new constraints (for X^0 = GEC). As expected, they show a higher σ_u and χ^2, yet not so large that the randomness of the bootstrap samples would be questionable.

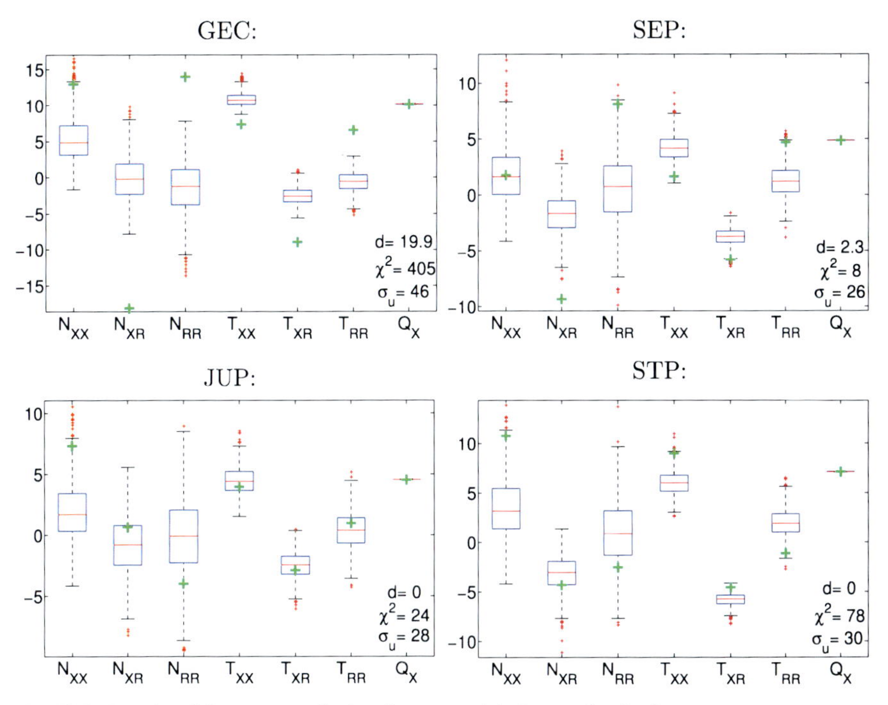

Fig. 89.3 Results of the same cardinal and same modularity test for the four groups

We also considered other kinds of constraints, for instance: keep N_{XX} constant, or keep the sum $T = 2 \times T_{XX} + T_{XR}$ constant (total time of conversation of nodes in X). Results are not plotted, but the distance to the bootstrap samples is always significantly bigger for GEC than for the other groups. Furthermore, these tests are robust with respect to the filtering threshold we choose: results are similar for a filtering of 1, 3 and 5 minutes.

89.5 Conclusion and On-going Work

We propose here a generic method to compare the behavior of different groups within a given graph. The method is inherently flexible: depending on the issue addressed in the data at hand, some observables and Null Hypotheses will be more appropriate than others. Furthermore, this method can be applied to any type of data that can be modelled by graphs. We are currently working on applying this general method to Null Hypotheses involving the dynamical behavior of groups, not only their aggregated behavior over time.

Acknowledgements We thank the SocioPatterns collaboration[2] for providing privileged access to the SocioPatterns sensing platform that was used in collecting the contact data.

[2]See footnote 1.

References

1. Brooks SP, Morgan BJT (1995) Optimization using simulated annealing. J R Stat Soc, Ser D, Stat 44(2):241–257
2. Cattuto C, Van den Broeck W, Barrat A, Colizza V, Pinton JF, Vespig nani A (2010) Dynamics of person-to-person interactions from distributed RFID sensor networks. PLoS ONE 5(7):e11596
3. Eagle N, Pentland A (2006) Reality mining: sensing complex social systems. Pers Ubiquitous Comput 10(4):255–268
4. Efron B (1982) The jackknife, the bootstrap, and other resampling plans, vol 38. SIAM, Philadelphia
5. Eldardiry H, Neville J (2008) A resampling technique for relational data graphs. In: Proceedings of the 2nd SNA workshop, 14th ACM SIGKDD conference on knowledge discovery and data mining
6. Hui P, Chaintreau A, Scott J, Gass R, Crowcroft J, Diot C (2005) Pocket switched networks and human mobility in conference environments. In: Proceedings of the 2005 ACM SIGCOMM workshop on delay-tolerant networking. ACM, New York, pp 244–251
7. Newman MEJ (2004) Analysis of weighted networks. Phys Rev E 70(5):056131
8. Salathe M, Kazandjieva M, Lee JW, Levis P, Feldman MW, Jones JH (2010) A high-resolution human contact network for infectious disease transmission. Proc Natl Acad Sci USA 107(51):22020–22025
9. Ying X, Wu X (2009) Graph generation with prescribed feature constraints. In: Proc of the 9th SIAM conference on data mining

Chapter 90
Automated Synthesis of Reliable and Efficient Systems Through Game Theory: A Case Study

Mickael Randour

Abstract Reactive computer systems bear inherent complexity due to continuous interactions with their environment. While this environment often proves to be uncontrollable, we still want to ensure that critical computer systems will not fail, no matter what they face. Examples are legion: railway traffic, power plants, plane navigation systems, etc. Formal verification of a system may ensure that it satisfies a given specification, but only applies to an already existing model of a system. In this work, we address the problem of synthesis: starting from a specification of the desired behavior, we show how to build a suitable system controller that will enforce this specification. In particular, we discuss recent developments of that approach for systems that must ensure Boolean behaviors (e.g., reachability, liveness) along with quantitative requirements over their execution (e.g., never drop out of fuel, ensure a suitable mean response time). We notably illustrate a powerful, practically usable algorithm for the automated synthesis of provably safe reactive systems.

90.1 Context

Nowadays, more and more aspects of our society depend on *critical reactive systems*, i.e., systems that continuously interact with their uncontrollable environment. Think about control programs of power plants, ABS for cars or airplane and railway traffic managing. Therefore, we are in dire need of systems capable of sustaining a safe behavior despite the nefarious effects of their environment.

Good developers know that testing do not capture the whole picture: never will it *proves* that no bug or flaw is present in the considered system. So for critical systems, it is useful to apply *formal verification*. That means using *mathematical tools* to prove that the system follows a given specification which models desired behaviors. While verification applies *a posteriori*, checking that the formal model of a system satisfies the needed specification, it is most of the time desirable to start *from* the specification and automatically build a system from it, in such a way that

Author supported by F.R.S.-FNRS. fellowship.

M. Randour (✉)
Institut d'Informatique, Université de Mons (UMONS), Mons, Belgium

T. Gilbert et al. (eds.), *Proceedings of the European Conference on Complex Systems 2012*, Springer Proceedings in Complexity, DOI 10.1007/978-3-319-00395-5_90,

desired properties are proved to be maintained in the process. This *a priori* process is known as *synthesis*.

The mathematical framework we use is *game theory*. It is a wide field with extensive formal bases and applications in numerous disciplines as diverse as economics, biology, operations research and, of course, computer science. Games model interactions between cooperating and/or competiting players who play to the best of their abilities in order to satisfy individual or common objectives. While interesting works of Borel [4] and even Cournot [12] precede them, von Neumann and Morgenstern are generally considered as the "Founding Fathers of (Modern) Game Theory" through their 1944 book titled *Theory of Games and Economic Behavior* [20].

Roughly speaking, we consider a reactive system as a player (player 1), and his uncontrollable environment as its adversary (player 2). We model their interactions in a game on a graph, where vertices model states of the system and its environment, and edges model their possible actions. Constructing a correct system controller then means devising a *strategy* (i.e., a succession of choices of actions) for player 1 such that, whatever the strategy of player 2, the outcome of the play satisfies the specification. Such game-theoretic formulations have proved useful for synthesis [11, 17, 18], verification [1], refinement [15], and compatibility checking [13] of reactive systems, as well as in analysis of emptiness of automata [19].

In this paper, we do not address the full theoretical deepness of such an approach but rather try to motivate and illustrate its usefulness towards an audience who may not be familiar with it. To that end, we discuss a motivating toy example. First, we present the informal description of a reactive system and the behavior it should enforce. Second, we show how to use the game-theoretic framework to model its relationship with its environment and formalize the desired specification. Third, we use the sound theory of synthesis and exhibit a suitable controller that ensures satisfaction of the specification. Our discussion is mostly high level and intuitive (see Fig. 90.1).

A wide variety of games (and thus system models) have been studied recently, with diverse enforceable behaviors [2, 5–9, 14, 16]. In this work, we will focus on systems that must satisfy *qualitative behaviors* (e.g., always eventually granting requests, never reaching a deadlock) along with multiple *quantitative requirements* (e.g., maintaining a bound on the mean response time, never running out of energy). In particular, we illustrate recent results of Chatterjee et al. [10] that are the first to provide a synthesis algorithm for such games, as well as a deep study of the complexity of the synthesized controllers.

90.2 Problem

Consider the following running example. We want to synthesize a controller for a robotized lawnmower. This lawnmower is automatically operated, without any human intervention. We present its informal specification, as well as the effects the environment can have on its operation.

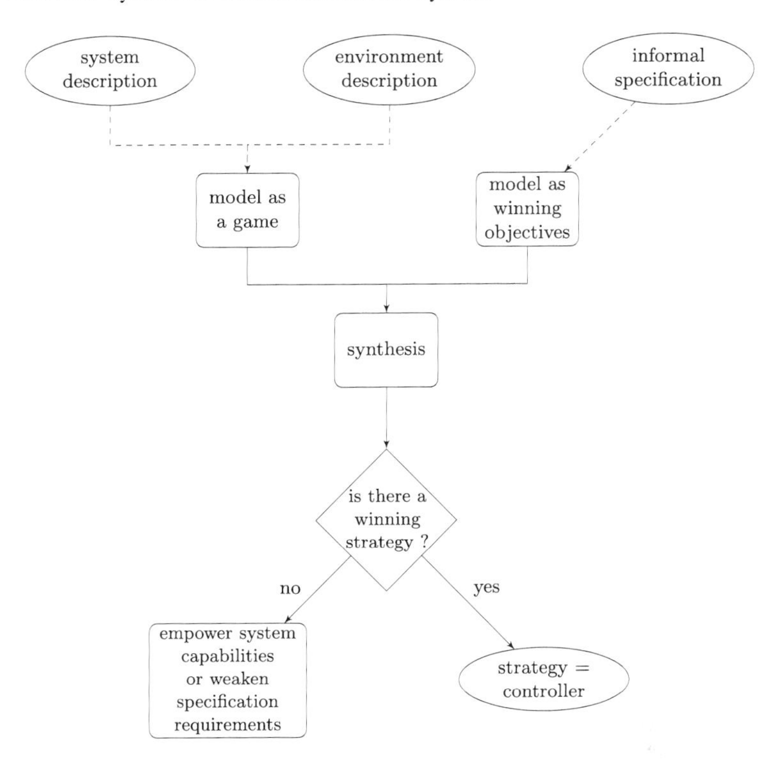

Fig. 90.1 Controller synthesis through game theory: process

- In this partial, simplified specification, the gardener do not ask for the lawnmower to satisfy any bound on the frequency of grass-cuttings. However, as he wants that the grass does not grow boundlessly, the lawnmower should cut the grass infinitely often in the future (as if it stops someday, the grass will not stop growing from then on).
- The lawnmower has an electric battery that can be recharged under sunshine thanks to solar panels, and a fuel tank that can only be filled when the lawnmower is back on its base. Both are considered unbounded to keep things simple.
- The weather can be cloudy or sunny.
- The lawnmower can refuel (2 fuel units) at its base under both weather conditions, but can only recharge its battery (2 battery units) when it is sunny. Resting at the base takes 20 time units.
- When cloudy, it can operate either under battery (1 battery unit) or using fuel (2 fuel units), both according to the same speed (5 time units). When sunny, the lawnmower may either cut the grass slowly, which always succeeds and consumes no energy (as the sun recharges the battery along the way), but takes 10 time units. Or it may cut the grass fast, which consumes both 1 unit of fuel and 1 unit of battery, but only takes 2 time units.

– When operating fast, the lawnmower makes considerably much noise, which may wake up the cat that resides in the garden and prompt it to attack the lawnmower. In that case, the grass-cutting is interrupted and the lawnmower goes back to its base, losing 40 time units as repair is needed. The cat does not go out if the weather is bad.
– As the gardener cannot benefit from his garden while the lawnmower is operating, he wants that the mean time required by actions of the lawnmower is less than 10 time units.

While simple, this toy example already involves qualitative requirements (i.e., the grass should be mown infinitely often), along with quantitative ones. There are indeed three quantities that have to be taken into account: battery and fuel are energy quantities, which should never be exhausted, and time per action is a quantity which mean over an infinite operating of the lawnmower should be less than a given bound.

Given this informal description of the capabilities of the system and its environment, as well as the specification the system should enforce, we need to build a system controller that guarantees satisfaction of the specification.

90.3 Modeling as a Game

Game We model the states and the interactions of the couple system/environment as a graph game where the system (here, the lawnmower) is player 1 and the environment is its adversary player 2. Formally, a *game structure* is a tuple $G = (S_1, S_2, s_{init}, E, k, w, p)$ where (i) S_1 and S_2 resp. denote the finite sets of *states* belonging to player 1 and player 2, with $S_1 \cap S_2 = \emptyset$; (ii) $s_{init} \in S = S_1 \cup S_2$ is the initial state; (iii) $E \subseteq S \times S$ is the set of *edges* s.t. for all $s \in S$, there exists $s' \in S$ s.t. $(s, s') \in E$; (iv) $k \in \mathbb{N}$ is the *dimension* of the weight vectors; (v) $w : E \to \mathbb{Z}^k$ is the multi-weight labeling function; and (vi) $p : S \to \mathbb{N}$ is the priority function.

The game starts at an initial state, and if the current state is a player 1 (resp. player 2) state, then player 1 (resp. player 2) chooses an outgoing *edge*. This choice is made according to a *strategy* of the player: given the sequence of visited states, a strategy chooses an outgoing edge. For this case study, we only consider strategies that operate this choice deterministically. This process of choosing edges is repeated forever, and gives rise to an outcome of the game, called a *play*, that consists of the infinite sequence of states that are visited. Formally, a *play* in G is an infinite sequence of states $\pi = s_0 s_1 s_2 \cdots$ s.t. $s_0 = s_{init}$ and for all $i \geq 0$, we have $(s_i, s_{i+1}) \in E$. The *prefix* up to the n-th state of play $\pi = s_0 s_1 \cdots s_n \cdots$ is the finite sequence $\pi(n) = s_0 s_1 \cdots s_n$. Such a prefix $\pi(n)$ belongs to player i, $i \in \{1, 2\}$, if $s_n \in S_i$. The set of plays of $G$is denoted by $\mathsf{Plays}(G)$. The set of prefixes that belong to player i is denoted by $\mathsf{Prefs}_i(G)$.

Applying this formalism, we represent the lawnmower problem as the game depicted on Fig. 90.2. Edges correspond to choices of the system or its environment and taking an edge implies a change on the three considered quantities, as denoted by the edge label. The *grass-cutting* state is special as the specification requires that it should be visited infinitely often by a suitable controller.

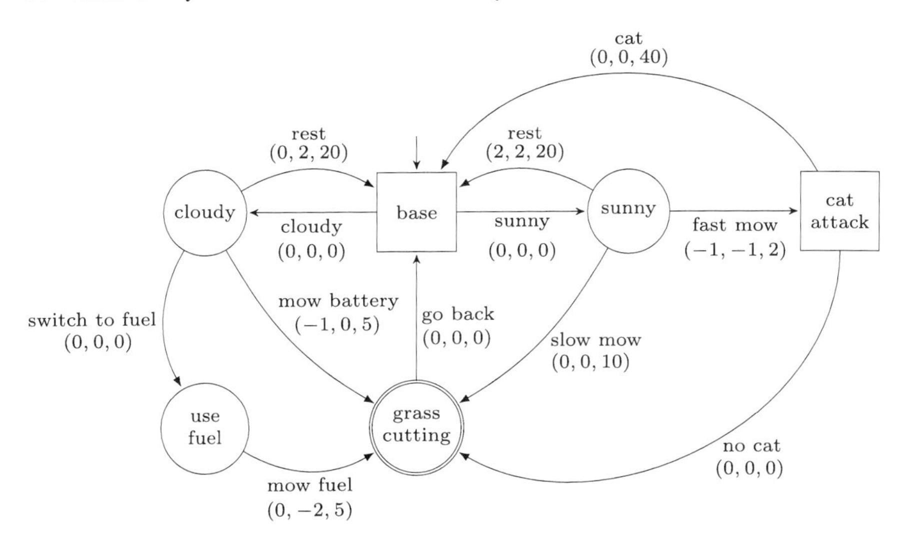

Fig. 90.2 Lawnmower game. *Edges* are fitted with tuples denoting changes in battery, fuel and time respectively

Strategies Formally, a *strategy* for $\mathcal{P}_i$, $i \in \{1, 2\}$, in G is a function λ_i : $\mathsf{Prefs}_i(G) \to S$ s.t. for all $\rho = s_0 s_1 \cdots s_n \in \mathsf{Prefs}_i(G)$, we have $(s_n, \lambda_i(\rho)) \in E$. The history of a play (i.e., the previously visited states and their order of appearance) may thus in general be used by a strategy to prescribe its choice. A strategy λ_i for $\mathcal{P}_i$ has *finite memory* if the history it needs to remember can be bounded. In that case, the strategy can be encoded by a deterministic Moore machine. As discussed earlier, a strategy of player 1 (the lawnmower) provides a complete description of a controller for the system, prescribing the actions to take in response to any situation. Therefore, our task is to build a strategy that satisfies the specification.

Objectives To devise such a strategy, it is needed to formalize the specification as objectives of the game. The conjunction of objectives yields a set of winning plays that endorse the specification. A strategy of player 1 is thus said to be *winning* if, *against every possible strategy of the adversary*, the play induced by following this strategy belongs to the winning set of plays.

The informal specification developed in Sect. 90.2 is encoded as the following objectives. We omit technical details for the sake of this case study.

- *Battery and fuel.* Both constitute energy types which quantities are never allowed to drop below zero. A play is thus winning for the *energy objective* if the running sum of the weights encountered along it (i.e., changes induced by the taken edges) never drops below zero on any of the first two dimensions.
- *Mean action time.* The specification asks that the lawnmower spends no more than 10 time units per action on average in the long run. That is, it is allowed to take more than 10 time units on some actions, but the long-run mean should be below this threshold. Therefore, the *mean-payoff objective* requires that the limit

of the mean of the third-dimension weights over the prefix of a play is lower than 10.
- *Infinitely frequent grass-cutting.* To satisfy this part of the specification, a strategy of player 1 must ensure that the grass-cutting state is visited infinitely often along the induced play. This is encoded as a *Büchi objective* (or as a *parity objective* via the priority function in the most general case).

90.4 Synthesis

Process Since our desire is to build practical real-world controllers, we are only interested in strategies that require *finite* memory. From a theoretical standpoint, there exist classes of games where infinite memory may help to achieve better results (see for example [8]), but infinite-memory strategies are of no practical use, as implementing a controller with infinite memory capabilities is obviously ruled out.

The core of the synthesis process depicted on Fig. 90.1 is thus to construct, if possible, a finite-memory strategy that ensures satisfaction of the previously defined objectives, as well as a corresponding initial value of the energy parameters, commonly referred to as *initial credit*. That is because for the energy objectives, it is allowed to start the game with some finite quantity in stock, before taking any action. Think about starting a race with some fuel in your tank.

While of importance for the analysis of systems with both qualitative and quantitative requirements, the synthesis problem for the class of games that is used to model the lawnmower problem, i.e., games with parity and multi energy or mean-payoff objectives, has only been considered recently [10]. In this paper, the complexity of synthesized controllers is studied and it is shown that for some systems, exponentially complex controllers are needed to enforce the specification. Moreover, exponential size controllers are always sufficient, i.e., if no exponential controller is able to enforce the specification, then implementing more complex controllers is no help.

Result 1 (Induced by [10, Theorem 1]) *Enforcing a specification combining both qualitative and quantitative aspects may require exponential size controllers in terms of memory requirements in the worst case.*

Interestingly, answering the question "does there exist a finite-memory controller that satisfies a given specification?" was shown to be coNP-complete in [8]. However, no deterministic algorithm was known to synthesize such a controller. Only quite recently, a practically implementable algorithm of *optimal complexity* for the synthesis of specification-wise suitable controllers was presented in [10]. This algorithm is both symbolic and incremental, and uses compact representations of data sets, thus being an ideal choice for implementation into synthesis tools.

Result 2 (Induced by [10, Theorem 2]) *The synthesis of controllers for systems with qualitative and quantitative requirements, such as the lawnmower, is in EXPTIME.*

This algorithm automatically builds suitable controllers with regard to the desired specification, if one is constructible. Therefore, it is the key tool in the synthesis process depicted on Fig. 90.1, and gives rise to an innovative and sound approach to the conception of provably safe reactive systems.

Lawnmower Controller To conclude our case study, we exhibit a synthesized controller that enforces the desired specification. Notice that there may exist other acceptable controllers. The one we present here is quite simple but already asks for some memory (in the form of bookkeeping of battery and fuel levels). The controller implements the following strategy:

- Start the game with empty battery and fuel levels.
- If the weather is sunny, mow slowly.
- If the weather is cloudy,
  - if there is at least one unit of battery, mow on battery,
  - otherwise, if there is at least two units of fuel, mow on fuel,
  - otherwise, rest at the base.

Notice that this strategy guarantees never running out of energy (which satisfies the energy objectives), induces infinitely frequent grass-cuttings (which satisfies the Büchi objective), and produces a play on which the mean time per action is less than 10 against any strategy of the adversary (which satisfies the mean-payoff objective). In this sample controller, the lawnmower never uses the "fast mow" action as the adversary could very well play "cat" and prevent visit of the grass-cutting state.

90.5 Conclusion

Through this case study, we have discussed how the game-theoretic framework can help in the synthesis of controllers. We have intuitively introduced some of the key underlying concepts such as games, strategies, qualitative and quantitative objectives. We have also discussed the recent development of a practically usable algorithm for the automated synthesis of valid controllers [10].

It is worthwhile noticing that automated synthesis suites for fragments of the presented formalism or similar logics are already in practical use, such as the LTL synthesis tool `Acacia+` for example [3] (which only applies to qualitative requirements). Thanks to the recent developments on the conjunction of qualitative and quantitative objectives [10], such tools could very well be extended to encompass all the needed complexity for the specification of real-world systems. Such an approach should be a leading trend for the analysis and synthesis of provably safe controllers for reactive systems in the near future. This discussion illustrates its interest, while abstracting the sound theory underneath.

References

1. Alur R, Henzinger TA, Kupferman O (2002) Alternating-time temporal logic. J ACM 49(5):672–713
2. Bloem R, Chatterjee K, Henzinger TA, Jobstmann B (2009) Better quality in synthesis through quantitative objectives. In: Proc of CAV. LNCS, vol 5643. Springer, Berlin, pp 140–156
3. Bohy A, Bruyère V, Filiot E, Jin N, Raskin J-F (2012) Acacia+, a tool for LTL synthesis. In: Proc of CAV. LNCS, vol 7358. Springer, Berlin, pp 652–657
4. Borel E, Ville J (1938) Applications aux jeux de hasard. Gauthier-Vilars, Paris
5. Bouyer P, Markey N, Olschewski J, Ummels M (2011) Measuring permissiveness in parity games: mean-payoff parity games revisited. In: Proc of ATVA. LNCS, vol 6996. Springer, Berlin, pp 135–149
6. Brázdil T, Jancar P, Kucera A (2010) Reachability games on extended vector addition systems with states. In: Proc of ICALP. LNCS, vol 6199. Springer, Berlin, pp 478–489
7. Chatterjee K, Doyen L (2010) Energy parity games. In: Proc of ICALP. LNCS, vol 6199. Springer, Berlin, pp 599–610
8. Chatterjee K, Doyen L, Henzinger TA, Raskin J-F (2010) Generalized mean-payoff and energy games. In: Proc of FSTTCS, LIPIcs 8, Schloss Dagstuhl, LZI, pp 505–516
9. Chatterjee K, Henzinger TA, Jurdzinski M (2005) Mean-payoff parity games. In: Proc of LICS. IEEE Comput Soc, Los Alamitos, pp 178–187
10. Chatterjee K, Randour M, Raskin J-F (2012) Strategy synthesis for multi-dimensional quantitative objectives. In: Proc of CONCUR. LNCS, vol 7454. Springer, Berlin, pp 115–131. Extended version on CoRR. arXiv:1201.5073 [cs.GT]
11. Church A (1962) Logic, arithmetic, and automata. In: Proceedings of the international congress of mathematicians, Institut Mittag-Leffler, pp 23–35
12. Cournot AA (1838) Recherches sur les principes mathématiques de la théorie des richesses/par Augustin Cournot. Hachette, Paris
13. de Alfaro L, Henzinger TA (2001) Interface theories for component-based design. In: Proc of EMSOFT. LNCS, vol 2211. Springer, Berlin, pp 148–165
14. Fahrenberg U, Juhl L, Larsen KG, Srba J (2011) Energy games in multiweighted automata. In: Proc of ICTAC. LNCS, vol 6916. Springer, Berlin, pp 95–115
15. Henzinger TA, Kupferman O, Rajamani S (2002) Fair simulation. Inf Comput 173(1):64–81
16. Martin DA (1998) The determinacy of Blackwell games. J Symb Log 63(4):1565–1581
17. Pnueli A, Rosner R (1989) On the synthesis of a reactive module. In: Proc of POPL, pp 179–190
18. Ramadge PJ, Wonham WM (1987) Supervisory control of a class of discrete-event processes. SIAM J Control Optim 25(1):206–230
19. Thomas W (1997) Languages, automata, and logic. In: Handbook of formal languages, beyond words, vol 3. Springer, Berlin, pp 389–455
20. Von Neumann J, Morgenstern O (1944) Theory of games and economic behavior. Princeton University Press, Princeton

Chapter 91
Evaluation of Latent Vocabularies Through Zipf's Law and Heaps' Law

Yukie Sano, Hideki Takayasu, and Misako Takayasu

Abstract We discuss about the number of latent distinct words through simulations by using Zipf's law and Heaps' law. From the standpoint of the number of latent distinct words which is estimated by our simulations, we can discuss about the difference among languages, author's properties such as professional and amateur authors and so on. In addition, Zipf's law and Heaps' law can be observed various field, thereby our approach has benefit not only for linguistic word occurrences but also various fields such as ecology and society to estimate hidden system size.

Keywords Zipf's law · Heaps' law · Language dynamics · Simulation

91.1 Introduction

Zipf's law is an empirical law stating that word frequency in documents is inversely proportional to word rank in descending order of occurrence [1]. Zipf's law is consistent with the following power law cumulative distribution of the number of word appearances:

$$P(\geq x) \propto x^{-\alpha}, \tag{91.1}$$

where the exponent α is a positive constant. Zipf's law is applicable not only to word frequencies in documents but also to incomes of firms and individuals which

Y. Sano (✉)
College of Science and Technology, Nihon University, Funabashi 274-8501, Japan
e-mail: sano.yukie@nihon-u.ac.jp

H. Takayasu
Sony Computer Science Laboratories, 3-14-13 Higashi-Gotanda, Tokyo 141-0022, Japan

H. Takayasu
Meiji Institute for Advanced Study of Mathematical Sciences, Tokyo, Japan

M. Takayasu
Tokyo Institute of Technology, Yokohama 226-8502, Japan

T. Gilbert et al. (eds.), *Proceedings of the European Conference on Complex Systems 2012*, Springer Proceedings in Complexity, DOI 10.1007/978-3-319-00395-5_91,

Table 91.1 Data description of the documents. Number of observed distinct words $D(N)$ in documents containing a total of N words and $D^*(N)$ optimized distinct words in simulation 2. Zipf's exponent α and Heaps' exponent β are also shown

	Language	Total N	Unique $D(N)$	Optimized $D^*(N)$	α	β
Pride and Prejudice	English	121464	6866	11122	0.84	0.51
Le Rouge et Noir	French	171729	14322	14332	0.97	0.58
Light and Darkness	Japanese	180623	10132	13100	1.04	0.49
Blog	Japanese	175136	10717	11057	0.94	0.51

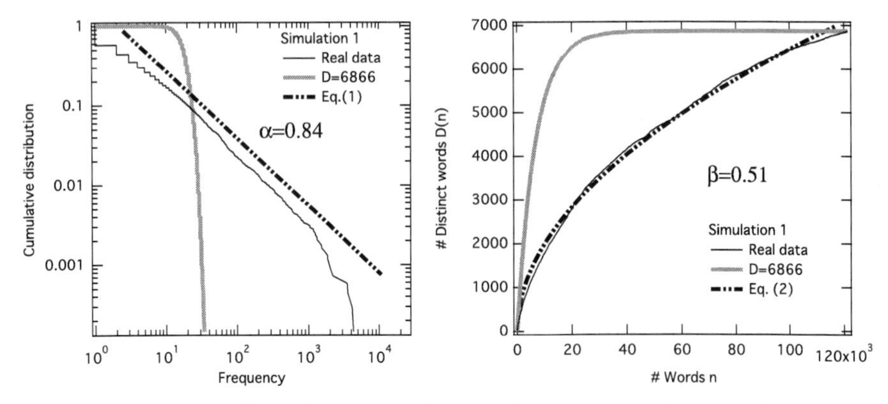

Fig. 91.1 Zipf's law and Heaps' law for English novel "Pride and Prejudice" (*thin line*) and artificially generated document by simulation 1 which introduced in Sect. 91.3.1 (*bold line*). Simulation 1 cannot reproduce the empirically observed results

is known as Pareto's law, the sizes of gypsum fragments, and the abundances of expressed genes [2].

Heaps' law states that the number of distinct words increases nonlinearly as the total number of words in a document increases [3]. Figure 91.1 shows typical result of Heaps' law which is observed in English novel "Pride and Prejudice". The number of distinct words $D(n)$ among the first n words of a document is approximated by the following power law:

$$D(n) \propto n^{\beta}, \tag{91.2}$$

where $0 < \beta < 1$. Note that while number of distinct words are limited in general, this Heaps' law curve is finally saturated in a very long document.

Heaps' law is also known as rarefaction curve and collector's curve in ecology. In this case, one can plot the number of species as a function of individuals sampled. Furthermore, when applied in ecology especially in island biogeography, Zipf's law is valid for species abundance distributions and Heaps' law describes species-area relationships in which the number of species found within an area increases nonlinearly for increasing area. Although the total number of species can be estimated based on the size of the area, species-area relationships are not fully equivalent.

However, it allows the calculation of the species richness for a given number of sampled individuals. Consequently, it is important issue to preserve biodiversity.

There are previous studies about relationship between Zipf's law and Heaps' law, and Heaps' law is often regarded as a derivation of Zipf's law. Cattuto et al. observed these two empirical laws in social bookmark data that focused on word tag co-occurrence distributions for Zipf's law and the Heaps' law curve for number of distinct word tags. Zipf's law and Heaps' law yielded the same exponents, i.e., $\beta = \alpha = 0.7$, for both empirical data and simple network simulations [4, 5]. Lü et al. showed by analytical and simple numerical simulation that, in the case of $\alpha \geq 1$, the value of β is equal to 1, while $\alpha = \beta$ in the case of $\alpha < 1$ [6].

91.2 Data Description

Here, we firstly confirm Zipf's law and Heaps' law using various types of real documents as follows;

- "Pride and Prejudice" was written in English by Jane Austen and published in 1813. It is one of most frequently downloaded documents from Project Gutenberg, which freely provides public domain book content on the web [7].
- "Le Rouge et le Noir" was written in French by Stendhal in 1830. It is the most frequently downloaded book in the French category of Project Gutenberg.
- "Light and Darkness (Meian)" was written in Japanese by Soseki Natsume who is one of the most popular authors in Japanese history.
- Blog entries posted by a single anonymous blogger in Japanese who was randomly selected from 20 thousand bloggers.

The details of the documents are listed in Table 91.1. In all of the documents, we removed all symbols, such as periods and colons, from the documents. Zipf's exponent α and Heaps' exponent β are shown in Table 91.1 by using the Gauss-Newton algorithm to minimize the sum of the squares of the errors in the whole area under the graph. As expected, Zipf's exponent α is approximately 1 in all cases, which is confirmed to be universal. On the other hand, Heaps' exponent β takes values in the range of from 0 to 1. Heaps' exponents β appear to be dependent on the observation size, the estimated values are scattered around 0.5. (Table 91.1).

91.3 Results

We reproduce observed Heaps' law by simple simulations. By following simulations 1 and 2, we generate artificial documents and confirm Zipf's law and Heaps' law.

91.3.1 Simulation 1

We set "word pool" that contains $D(N)$ distinct words and pick one word from the word pool. A word is selected by uniform probability and return to the word pool

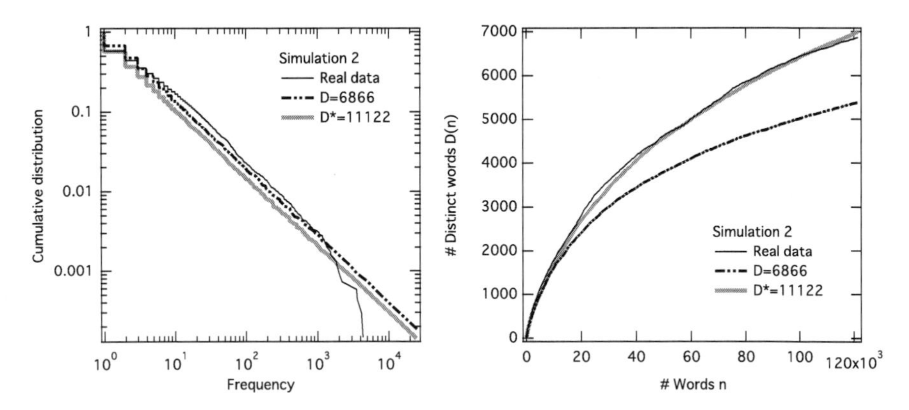

Fig. 91.2 Zipf's law and Heaps' law for English novel "Pride and Prejudice" (*thin line*) and artificially generated document by simulation 2 which introduced in Sect. 91.3.2. Simulation 2 reproduce Zipf's law by definition. Simulation 2 using a real number of distinct words $D(N) = 6866$ cannot reproduce Heaps' law (*dashed line*), while simulation using a number of optimized distinct words $D^*(N) = 11122$ can do (*bold line*)

after picking. We repeat this procedure N times and generate an artificial document that contains N words. Bold line in Fig. 91.1 shows the result of the simulation 1 compare to English novel "Pride and Prejudice". We set same number of total words $N = 121464$ and distinct words $D(N) = 6866$. It is obvious that Zipf's law cannot hold as words are selected by uniform probability. For Heaps' law, we can confirm clear deviation from the real data and simulation result reaches quickly to the upper limit of the distinct number $D(N)$.

91.3.2 Simulation 2

We also set word pool for simulation 2 and pick the words from the pool. However, we select a word from the word pool by weighted probability that follows empirically observed Zipf's law. For example, the most frequent word "the" appeared 4321 times while the least frequent word "abatement" appeared only once in "Pride and Prejudice", and Zipf's exponent α is 0.84. As a result, the probability of selecting word "the" is 3.7×10^4 times higher than "abatement".

Figure 91.2 shows result of the simulation 2 and confirm that Zipf's law holds by definition (Fig. 91.2(a)). For Heaps' law, the artificially generated documents using the empirically observed number of distinct words $D(N) = 6866$ cannot reproduce empirically observed Heaps' law (dashed line in Fig. 91.2(b)). We demonstrate that empirically observed Heaps' law holds for artificial documents in which a certain number of distinct words are added to empirically observed distinct words $D^*(N) = 11122$ (bold line in Fig. 91.2(b)).

Simulation using the optimized number of distinct words, $D^*(N)$, listed in Table 91.1 to minimize residual error and reproduce real data for all examples. It suggests that the vocabularies, thus the number of latent distinct words considered in the creation of a given document, can be predicted. We will also show results provided that total number of words N is the same for various documents with each estimated latent vocabularies $D^*(N)$.

Zipf's law and Heaps' law can be observed various field, thereby our approach has benefit not only for linguistic word occurrences but also various fields such as ecology and society to estimate hidden system size.

References

1. Zipf GK (1949) Human behavior and the principle of least effort. Addison-Wesley, Cambridge
2. Newman MEJ (2005) Contemp Phys 46:323
3. Heaps HS (1978) Information retrieval: computational and theoretical aspects. Academic Press, Orlando
4. Cattuto C, Loreto V, Pietronero L (2007) Proc Natl Acad Sci USA 104:1461
5. Cattuto C, Barrat A, Baldassarri A, Schehr G, Loreto V (2009) Proc Natl Acad Sci USA 106:10511
6. Lü L, Zhang Z-K, Zhou T (2010) PLoS ONE 5:e14139
7. Digital libraries providing public domain books. Gutenberg Project (http://www.gutenberg.org/) for various languages and Aozora bunko (http://www.aozora.gr.jp/) only for Japanese

Chapter 92
Complex Systems in Organizations and Their Influence on Human Resource Management

Tobias M. Scholz

Abstract Although complex systems have been part of organizational studies for several decades, they have witnessed significant changes in recent years. This research builds on existent work on complex systems by focusing on the shift toward power-law distributions. Evidentially, HRM is highly nested in those complex systems. Due to the advancements in the theoretical understanding of complexity and influenced by these, HRM is also shifting in new directions, especially towards a dynamic approach. In consequence, a new notion of a dynamic HRM is emerging, which could be characterized as "function follows process". Accordingly, understanding HRM as complex systems, HRM needs to observe the dynamic core of its processes as well as the outcomes of a changed governing distribution and derive from these its alternating functionality.

Keywords Complex systems · Human resource management · Organizational behavior

Recent years have amplified the environment of constant change and made it inevitable that organizations will interact with this volatile periphery [1]. It has become obvious that organizations nowadays are operating in a dynamic world. The reasons for that shift are globalization, decentralization, cultural interconnectedness and technological progress. Complicating the situation further, organizations are facing an increasing number of restrictions, e.g. the availability of natural resources or the "war for talents."

Those extreme changes in the environment [2] have led to an aggravation of the existing problems as well as unprecedented problems. The struggles with the high variability of issues and the aim of organizational stability have led to a system of constant reorganization and adaptation [3]. Organizations still tend to define and solve problems based on simplification, predictability, equilibrium and linearity [4]; however, current developments in the environment of organizations make it evident

T.M. Scholz (✉)
Chair for Human Resource Management and Organizational Behavior, University of Siegen, Siegen, Germany
e-mail: tobias.scholz@uni-siegen.de

T. Gilbert et al. (eds.), *Proceedings of the European Conference on Complex Systems 2012*, Springer Proceedings in Complexity, DOI 10.1007/978-3-319-00395-5_92,

that such an approach needs to be extended [5]. Barabási [6, p. 201] stated: "As companies face an information explosion and an unprecedented need for flexibility in a rapidly changing marketplace, the corporate model is in the midst of a complete makeover." To reinforce this, complex systems in organizations are predominantly influenced by the interaction amongst humans; organizations are forced to focus on the human factor [7].

This research addresses the question "How does HRM have to change in order to survive and thrive in today's environment?" Accordingly, our attention shifts to the forces (people) within an organization itself, away from purely outside influences [5], to deal with the unpredictability, non-equilibrium and non-linearity in a modern organization [1]. This paper tries to fill a research gap by examining HRM in an organization through the lens of complex systems. By applying the construct of Pareto distribution and the metaphors of organizational change [2], fractals, simple rules, self-organized criticality, emergence and adaptation, we can understand and research the effect of complex systems on HRM. HRM is facing a decreasing effect of fitting and common theoretical strategies [8]. Those alterations are reinforced by the existing shift to a dispersed environment, e.g. the war for talents, cultural diversity and heterogeneity. However, those changes in perception are still rooted in a static world. Similar to the dynamics in the environment, though, it becomes clear that HRM has to adapt to a dynamic approach. Complex systems with their terminology of dynamic, non-linearity and far-from equilibrium are becoming increasingly interesting for HRM. Complex systems deliver a setting for a description of such a dynamic world as that in which HRM now has to work.

Altering the perception of complex systems, HRM needs to rethink its position within an organization. Furthermore, it needs to analyze the influences of such fundamental changes. Therefore, HRM is facing a decision between two distributions, which are completely different and therefore lead to different paths. To emphasize, HRM is at a critical juncture [9], choosing a path for the future of its work and the processes within an organization.

HRM can choose amongst Gaussian distribution and Pareto distribution, or it can continue to follow the traditional approach by using the normal distribution. This means focusing on the average of phenomena. For HRM, this involves using methods to interact with the workforce in an organization based on the majority of people, and neglecting outliers [10]. Alternatively, HRM can adopt the power-law distribution and center its attention on the outliers. Contrary to normal distributions, such extremes can occur in two directions. Depending on the interaction, however, only one extreme, resulting in a long tail, will be interesting [11]. Essentially, outliers become the driving force behind power-law distributions.

Importantly, the logic of a complex system determines that a normal distribution can exist within a complex system. Based on the combination theory, several components of a complex system can have different distributions [12]. This theory allows us to compare the two distributions within the features of a complex system and enables us to administer the effects of both distributions on the HRM within an organization.

Gaussian vs. Paretian Fractals Gaussian fractals lead to centralization behavior. Based on normal distribution, different HR functions share a similar vision and strive to achieve this common vision. Even though such a vision is in constant flux in relation to the processes within an organization, all the different HR functions aspire to an average and common vision. In contrast, Paretian fractals state the drift towards decentralization. Decentralization means that a stated vision, as in the normal distribution, is not available and consequently not desirable. Therefore, several concurrent visions can emerge, according to the process and the function of HRM. Evidentially, through interaction, a set of key visions crystallizes, and thereby adapts to the necessary processes and combines, based on the visions, the HR functions that have similar handling processes.

Gaussian vs. Paretian Simple Rules Gaussian simple rules drift towards order. Those simple rules in an environment with a focus on the majority lead to a set of rules that state the core competence of HRM in the organization. This core competence also reveals the focus of HRM in general. Thereby, they confine HRM to an environmental setting for HR functions. Even though those HR functions are now well defined and HRM is organized, this implies an increase in observations of changes in the processes within the organization. The dynamic focus lies on the constant adaptation of the order towards the complex and chaotic system within an organization. Conversely, Paretian simple rules head towards disorder. Contrary to the Gaussian world, rules are simple in the way that they are general and minimalistic. Striving for disorder signifies that different parts of HRM are handling different HR functions. However, a general order emerges based on the necessity of providing all the essential functions for all the processes. This means that the main rule is: all processes need to exist. This, however, implies that a distribution within HRM is not specified in any way. This flexibility therefore increases the necessity of interaction amongst different parts of HRM, without which the general rule cannot be accomplished.

Gaussian vs. Paretian Self-organized Criticality Gaussian self-organized criticality veers towards an increase in attack tolerance. Based on the assumption that HRM and its sub-systems are aiming for the average and therefore similarity, the attack tolerance increases. In this case self-organized criticality focuses on the tipping point, below which a system remains intact and does not face unpredictable and uncontrollable system-wide changes. Research concerning those tipping points has already been carried out [13] and revealed that a system based on the Gaussian distribution has high attack tolerance. Attack means the random removal of parts of a system. For HRM this implies the removal of HR functions within an organization, but without replacement. As HR functions, in general, strive towards the average and similarity, another HR function can take over the spot of the removed HR function. Thus, HRM can tolerate many removals of that kind, without risking becoming overwhelmed by the processes it is trying to manage. Contrary to this, Paretian self-organized criticality targets the direction of HRM towards error tolerance. A system based on moving away from the average towards the extreme, however, means high error tolerance. Error thereby implies the random failure of HR functions. Other HR functions already have no mission to mimic other HR functions blindly. Therefore, any error will stay in the collapsing HR function, and other HR functions will try to adjust to the new situation. As a result of the failure of one HR function, related processes in the organization will not be handled. However, other HR functions will fill in and balance the system and the processes within an organization.

Gaussian vs. Paretian Emergence Gaussian emergence leads to an emergent behavior towards convergence. Through interaction with different parts of HRM, merging of its HR functions takes place. Based on that, different parts of HRM strive to align their functions and allow HRM to observe and manage the processes within an organization with a similar set of characteristics. Emergent processes in the function of HRM evolve within the complete HRM and spread quickly within the system. Contrarily, Paretian emergence is the process heading for divergence. This drift towards diversity exceeds diversity in the functionality of several HR tasks. It also means reaching divergence in every detail. Through interaction within such a divergent environment, HRM should trigger a variety of emergent processes. Contrary to the Gaussian world, spreading within the HRM is challenging, but this competition for successful emergence leads to better observation of the processes.

Gaussian vs. Paretian Adaptation Gaussian adaptation states the drift of HRM towards the average. It is the ambition of HRM to adapt to functions that can be applied to a majority of processes and thereby the similarity between HR functions increases. With the orientation towards alignment, HR functions can segment the necessary processes efficiently and adapt together towards the emergent processes within an organization. Adapting to the average, however, means neglecting the extreme emergent processes, which theoretically could be better. Paretian adaptation, however, is the quest of HRM regarding the outliers. Every HR task in the department needs to try to fit with the necessary processes within an organization. However, it is in constant interaction with parts of the inner sphere (other (HR) departments) and outer sphere (consultants and service partners) of an organization. Through improvements, HRM seeks to uphold the preferential attachment ability and therefore the necessity to serve the needs within an organization. Contrary to the Gaussian counterpart, balancing between processes leads to an evolutionary arms race. Those evolutionary processes, however, could lead to functions that fail to manage their actual processes.

In summary, the two distributions lead to different chances and obstacles. It becomes apparent for HRM that knowledge about the complex system of an organization is essential. The selection of the fitting distribution defines the function of HRM fundamentally and so influences the organization as well. In fact, the function of HRM therefore follows processes and even though both distribution paths describe the drift of HRM towards the average or extremes, the lock-in of the distribution does not describe the actual functions HRM implements on that information. The decision defines a drift in one direction, so it becomes necessary to observe the processes within an organization. Based on that information, the path can be narrowed down further.

Importantly, a lock-in into one of the distributions does not mean an eternal lock-in, but a path that defers all the following paths; evidently, it is a big change. HRM, however, needs to observe the complex system constantly, in order to realize the emergent processes contradicting the current distribution, as well as every other function of HRM.

It needs to be remembered that the paradigm shift behind complex system science is fundamental [10] and, if adapted, changes the work of HRM drastically. HRM approaches emergently such a shift towards the extremes and the power-law distribution in some distinct cases (e.g. cultural diversity, high potentials). Essentially HRM needs to understand the complex system of the organization; only through constant observation and interaction does it become apparent how to classify the organization. Especially through this classification, HRM can answer the distribution question economically and can initiate change that leads to stability [14].

Shifting towards the paradigm "function follows process", it allows us to review the HRM within an organization. Furthermore, the struggle of HRM reveals that the current methods do not describe the reality. Adopting complex systems and moving towards a dynamic approach reframes HRM fundamentally. The processes within an organization are constantly changing and, in a complex system, are non-linear and far from equilibrium. Therefore, they dictate that departments like HRM, in which the interaction is within the complex systems and its parts (people), need to observe, adopt and manage such systems. However, a temporary lock-in is probably inevitable; HRM should therefore try to minimize its lack towards emergent changes within an organization.

This paper tries to reveal that HRM needs to adapt to a new environment. Importantly, HRM needs to review this shift, but not abandon its history. As stated by Morin [15, p. 56], "Complex thought does not all reject clarity, order, or determinism. It knows they are insufficient, it knows that we cannot program discovery, knowledge, or action." Furthermore, Morin [15, p. 56] says: "Don't forget that reality is changing, don't forget that something new can (and will) spring up." Metaphorically the new equilibrium is changing and therefore dictates a shift from a static approach towards a dynamic approach in every facet of HRM. Even more, HRM has reached the limits of reductionism [16].

It is crucial that an adaption of complex system science and evidentially power-law distribution on the organizational level is unavoidable. Thus, we believe that HRM is in the midst of such a complex system and has the competence, objective and tools to understand, analyze and manage a complex system. However, such a role dictates a dynamic approach to HRM, in which function follows processes. Through such a dynamic and flexible approach, sustainable stability in an unpredictable, far from equilibrium and non-linear environment can be achieved.

References

1. Maguire S, Allen P, McKelvey B (2011) Complexity and management: introducing the SAGE handbook. In: Allen P, Maguire S, McKelvey B (eds) The SAGE handbook of complexity and management. Sage, Thousand Oaks, pp 1–26

2. Eoyang GH (2011) Complexity and the dynamics of organizational change. In: Allen P, Maguire S, McKelvey B (eds) The SAGE handbook of complexity and management. Sage, Thousand Oaks, pp 317–332
3. Maguire S, McKelvey B, Mirabeau L, Öztas N (2006) Complexity science and organization studies. In: Clegg SR, Hardy C, Lawrence TB, Nord WR (eds) The SAGE handbook of organization studies. Sage, Thousand Oaks, pp 165–214
4. Marion R (1999) The edge of organization: chaos and complexity theories of formal social organizations. Sage, Thousand Oaks,
5. Marion R (2008) Complexity theory for organizations and organizational leadership. In: Uhl-Bien M, Marion R (eds) Complexity leadership. Information Age, Charlotte, pp 1–16
6. Barabási A-L (2003) In: Linked. Plume, London
7. Pfeffer J (2010) Building sustainable organizations: the human factor. Acad Manag Perspect 24:34–45
8. Marion R, Uhl-Bien M (2011) Implications of complexity science for the study of leadership. In: Allen P, Maguire S, McKelvey B (eds) The SAGE handbook of complexity and management. Sage, Thousand Oaks, pp 385–399
9. Sydow J, Schreyögg G, Koch J (2009) Organizational path dependence: opening the black box. Acad Manag Rev 34:689–709
10. Andriani P, McKelvey B (2011) Managing in a Pareto world calls for new thinking. Management 14:89–118
11. Anderson C (2006) The long tail. Random House, London
12. Newman MEJ (2005) Power laws, Pareto distributions and Zipf' law. Contemp Phys 46:325–351
13. Alber R, Jeong H, Barabási A-L (2000) Error and attack tolerance of complex networks. Nature 406:378–382
14. Leana C, Barry B (2000) Stability and change as simultaneous experiences in organizational life. Acad Manag Rev 25:753–760
15. Morin E (2008) On complexity. Hampton, Cresskil
16. Barabási A-L (2012) The network takeover. Nat Phys 8:14–16

Chapter 93
Why First Movers May Fail: Global Versus Sequential Improvement of Complex Technological Artefacts

Adrien Querbes-Revier and Koen Frenken

Abstract We propose a new theory of late mover advantage where new entrants can leapfrog incumbents through introducing new functionality of an existing technology. Since first mover firms did not take into account new functionalities discovered after they entered, they limited their search on older functionalities and find it difficult to optimize functionalities once discovered later on. Late movers, in contrast, do not suffer from such technological irreversibilities, since they only start searching once all functionalities are known. Based on an agent-based model representing the innovation process of a complex technological artifact with a growing number of functionalities, we can conclude that, a first mover disadvantage can appear, particularly when the technology in question is large and complex, as for example in the case of current key technologies such as ICT, energy and mobility systems.

Keywords First mover advantage · Late mover advantage · Exaptation · Search · NK-model · Technological evolution · Complexity theory

93.1 Question and Background

In this paper, we analyse the economic and technological constraints in the development of new functionalities in complex technological artefacts (CTA). A firm introducing a new CTA has no certainty about the future evolution of this CTA. Particularly, the functions performed by this CTA may evolve over time thanks to scientific discoveries (technology push), or end user preferences (demand pull or mix & match) [4]. We argue that one key mechanism underlying this recurrent pattern in industrial dynamics is exaptation, which we understand as "the taking on of new functionality by existing structure" [8, p. 69]. As long as this evolution may last, the firm has then to make strategic decisions about how to improve its CTA

A. Querbes-Revier (✉) · K. Frenken
School of Innovation Sciences, Eindhoven University of Technology, P.O. Box 513, 5600 MB Eindhoven, The Netherlands
e-mail: A.Querbes@tue.nl

K. Frenken
e-mail: K.Frenken@tue.nl

T. Gilbert et al. (eds.), *Proceedings of the European Conference on Complex Systems 2012*, Springer Proceedings in Complexity, DOI 10.1007/978-3-319-00395-5_93,

relatively to its current use/functions, as well as to introduce new functions in line with the innovations of competitors or the expectation of users.

However, the design of complex systems is likely to produce technological irreversibility based on the interdependences between the different part and functions of the system. This is the reason why a first mover developing a CTA—with only limited knowledge about its functions—may suffer from the addition of new functionalities later on, more than later entrants will do. The principle of a first mover (dis)advantage has received a large attention over the past decades, but most of these researches have in common to explain this (dis)advantage by organizational or market-based reasons. Our aim is to propose an approach based on the growing technological irreversibilities of the CTA, complementary with the existing research on organizational inertia (e.g. [2, 3, 10]). Such dynamic appears for CTAs which are optimized function by function, since these functions are not initially foreseen by the first mover, but discovered by users or later entrants. Historical cases highlighting this dynamic include: the evolution of bicycles in the 1880s when the fast "Ordinary" bicycle was replaced by the "Safety" bicycle, very similar to our contemporary bicycles [9]; the transition of the steamboats in the 1830s from inland waters transportation to transatlantic transportation [6]; and the evolution of smartphones in the 2000s from mail-oriented devices to the boom of online application stores [11].

93.2 Model and Main Results

In order to capture this dynamic, we have built an agent-based model of simulation, using the representation of complex systems developed by Kauffman [7]: the NK model, later generalised by Altenberg [1]. This model comes from natural sciences, where it is used particularly to study the interaction between genes and traits of a biological organism. In social sciences, this model is implemented by many researchers in management and strategy, in order to study the structure of complex organizations (see [5] for a literature review in this field). From a technological perspective, this model is used to depict complex design of technological systems (e.g. [9] or [4]).

In this paper, the CTA is then depicted by N *technologies* interacting together to produce F *functions*, according to a *function-technology map*. This map is more or less complex (in terms of degree of technological interdependences between functions) depending on the level of K, the *pleiotropy* of these functions, i.e. the number of technologies affecting one function. The Fig. 93.1 shows an example of such a map with $N = 4$, $F = 2$ and $K = 3$.

This model makes the assumption that each technology can only have two states, producing then 2^N potential combinations of the technologies. The whole set of combinations forms what Kauffman calls a "rugged fitness landscapes", each com-

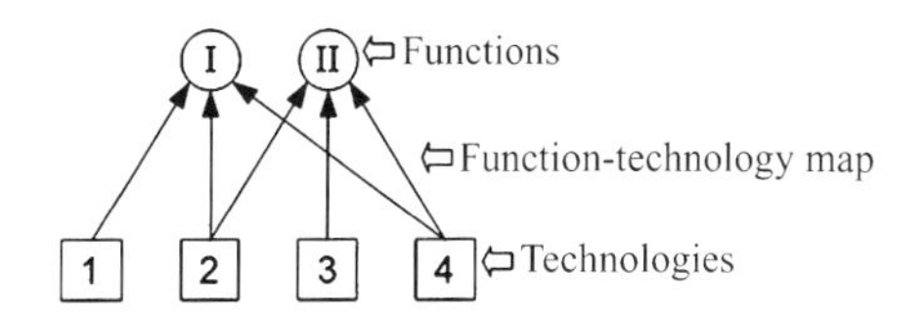

Fig. 93.1 An example of function-technology map

bination being a point of this landscape and receiving a particular payoff: the fitness. In line with Altenberg [1], the fitness is given by

$$\phi(x) = \frac{1}{F}\sum_{i=1}^{F}\phi_i(x_{j1(i)}, x_{j2(i)}, \ldots, x_{jK}) \tag{93.1}$$

with $\{x_{j1(i)}, x_{j2(i)}, \ldots, x_{jK}\} \subset \{x_1, \ldots, x_j, \ldots, x_N\}$ as the set of K technologies affecting the function i and $x = \{x_1, \ldots, x_j, \ldots, x_N\} \in \{0, 1\}^N$ is the binary string representing the whole technological system. Hence, the fitness of one function (ϕ_i) depends on the state of the K technologies affecting this function and the global fitness (ϕ) is the average of the fitness of every function.

In this paper, we compare two procedures dedicated to optimize this system by performing mutations at the level of technologies. The late mover looks at the system as a whole and then, at a given time t, one technology is mutated if this mutation results to $\phi_t(x) > \phi_{t-1}(x)$ (when many mutations give such a global fitness improvement, the best mutation is selected). The first mover looks at the system function by function, i.e. as long as exaptation co-opts new functionalities for this CTA. Hence, unlike the late mover (perfectly informed of the functions of the CTA), the first mover discovers the various functionalities of its system over time, function after function. Based on the example of Fig. 93.1, this first mover tries first to optimize function I, from mutations of technologies 1, 2 and 4 as long as $\exists\phi_{I,t}(x) > \phi_{I,t-1}(x)$. When a peak of the landscape is reached, one improves the next function (II) using the same principle, but from mutations of technology 3 only, since mutations of the others technologies would affect the state of the function I. Concretely, once a function has reached a peak, its affecting technologies are fixed.

In a nutshell, two results emerge from this agent-based model of simulation. On the one hand, the more the CTA combines a large number of technologies, a large number of functions or a high degree of complexity (K), more the late mover has a strong advantage to reach higher fitness (Fig. 93.2). On the other hand, this advantage has a cost: the necessity to test a larger amount of technological combination (Fig. 93.3). Hence, the first mover sustain its initial advantage (first entrant on the market) when $K < N$ to reach a satisfying fitness with a smarter search strategy.

93.3 Conclusion and Summary

We can generalize these results as follows: the late movers enjoy considerable advantages in technologies where new functionalities are discovered over time, a process known as exaptation. In particular, the more exaptation events take place after

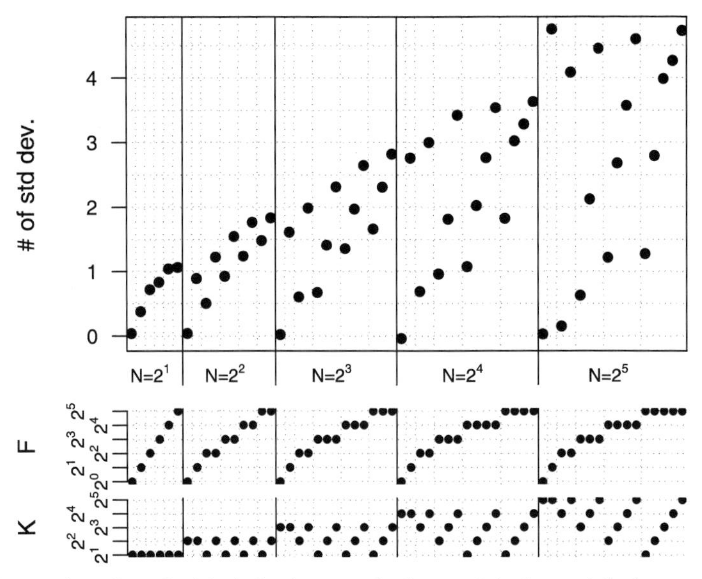

Fig. 93.2 Number of standard deviation between the fitness of the first and the late movers, over different combinations of the parameters N, F and K

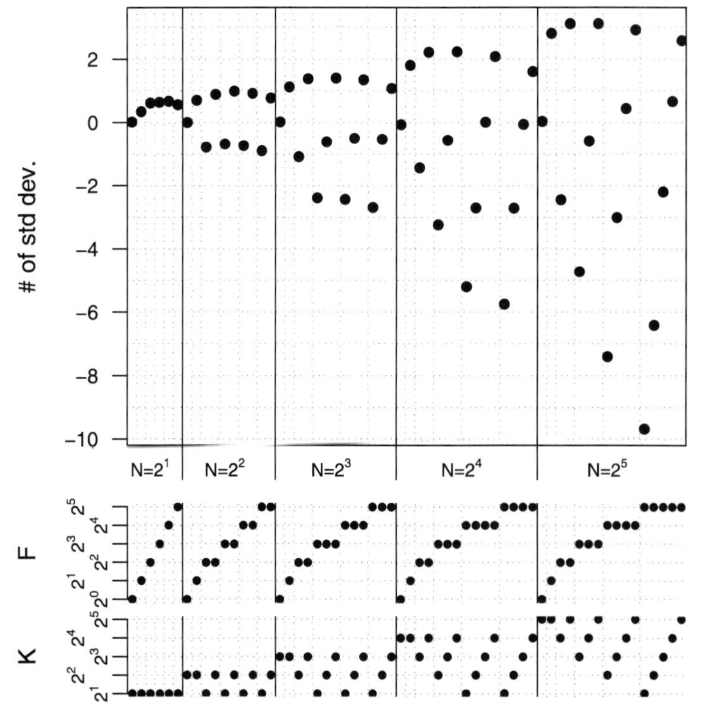

Fig. 93.3 Number of standard deviation between the fitness per screened technologies of the first and the late movers, over different combinations of the parameters N, F and K

a new technology has been introduced and the higher the complexity of the technology in question the more likely late movers will take over the leadership of first movers. The key lesson holds that the more complex a technology in terms of interdependencies between its underlying component technologies, the more difficulties first movers face to provide the new functionalities given the choices for component technologies made early on, and less likely their first mover advantage can be sustained.

References

1. Altenberg L (1997) NK fitness landscapes. In: Back T, Fogel D, Michalewicz Z (eds) Handbook of evolutionary computation. Oxford University Press, London
2. Brusoni S, Fontana R (2011) Incumbents' strategies for platform competition: shaping the boundaries of creative destruction. In: De Liso N, Leoncini R (eds) Internationalization, technological change and the theory of the firm. Routledge, New York, pp 66–88
3. Ethiraj SK, Levinthal D (2009) Hoping for A to Z while rewarding only A: complex organizations and multiple goals. Organ Sci 20(1):4–21
4. Frenken K (2006) Innovation, evolution and complexity theory. Edward Elgar, Cheltenham Glos
5. Ganco M, Hoetker G (2009) NK modeling methodology in the strategy literature: bounded search on a rugged landscape. In: Bergh D, Ketchen D (eds) Research methodology in strategy and management, vol 5. Emerald Group Publishing, Bingley, pp 237–268
6. Geels FW (2005) Technological transitions and system innovations: a co-evolutionary and socio-technical analysis. Edward Elgar, Cheltenham Glos
7. Kauffman SA (1993) The origins of order: self organization and selection in evolution. Oxford University Press, London
8. Lane DA (2011) Complexity and innovation dynamics. In: Antonelli C (ed) Handbook on the economic complexity of technological change. Edward Elgar, Cheltenham Glos, pp 63–80
9. van Nierop OA, Blankendaal ACM, Overbeeke CJ (1997) The evolution of the bicycle: a dynamic systems approach. J Des Hist 10(3):253–267
10. Tushman M, Smith WK, Wood RC, Westerman G, O'Reilly C (2010) Organizational designs and innovation streams. Ind Corp Change 19(5):1331–1366
11. West J, Mace M (2010) Browsing as the killer app: explaining the rapid success of Apple's iPhone. Telecommun Policy 34(5–6):270–286

Chapter 94
Market Opportunities, Customer Desires and Purchasing Selectiveness Modelling in Multi-layered Cellular Automata: A Study Case on Organizational Survivability

José V. Matos, Rui J. Lopes, and Yasmin Merali

Abstract The present work aims to contribute to a better understanding of the dynamics of organizational competition and survival in a supply chain network market context, while highlighting the potential of multi-layered cellular automata models as frameworks for accommodating increasing levels of complexity. More particularly, the implementation of inter-layer rules associated to k-bit words modelling of market opportunities, customer desires and purchasing selectiveness, and their impact on the dynamics of an evolutionary "ecology" of suppliers, competing organizations, and customers, following a complex adaptive systems approach is described and illustrated through a study case on organizational survivability. The implications of the study results—reflecting the interplay between market environment, competitors' strategic choice, and corresponding ability to succeed, survive crises and proliferate—are then discussed and the main aims of the work ahead highlighted.

Keywords Cellular automata · Complex adaptive systems · Crisis management · Organizational competition and survival · Supply-chain network modelling

J.V. Matos (✉) · R.J. Lopes
ISCTE-IUL Lisbon University Institute, Lisbon, Portugal
e-mail: jlvmatos@gmail.com

R.J. Lopes
e-mail: rui.lopes@iscte.pt

J.V. Matos · R.J. Lopes
Instituto de Telecomunicações, Lisbon, Portugal

J.V. Matos
Faculty of Sciences, University of Lisbon, Lisbon, Portugal

Y. Merali
Warwick Business School, University of Warwick, Warwick, UK
e-mail: yasmin.merali@wbs.ac.uk

T. Gilbert et al. (eds.), *Proceedings of the European Conference on Complex Systems 2012*, Springer Proceedings in Complexity, DOI 10.1007/978-3-319-00395-5_94,

94.1 Introduction

To succeed in a globally linked business market environment, companies are required to compete effectively and efficiently while simultaneously having the ability to survive unexpected events and potential supply chain disruptions. The increasing structural and dynamical complexity of business networks combined with the need to keep modelling at the simplest level required to provide meaningful and general insights, recommends the adoption of an incremental and comprehensive framework. In this context, the potential provided by multi-layered Cellular Automata (CA) models to provide such framework is highlighted and illustrated through a study case which aims to contribute to a better understanding of the organizational competition and survival dynamics in a supply chain network market context. While benefiting from the analysis advantages of simple CA models, the adopted approach, using multiple CA layers—corresponding to different agent types—combined with the implementation of inter-layer rules, provides a well-suited and versatile modelling methodology able to incorporate the properties and mechanisms of a complex adaptive system [1].

After providing a conceptual introduction to the competition and survival problem in the business world and addressing the modelling challenges involved, as well as the process leading to the adopted multi-layered CA base approach, complemented by the integration of fundamental market dynamics elements, the performed study case is described, and the obtained results presented. The implications of those results are then discussed, and the main aims of the work ahead highlighted.

94.2 Problem Domain Context, Components and Dynamics

To stay in business, companies need to simultaneously sustain a competitive advantageous market position while being able to avoid, escape, contain and recover from unexpected and threatening events that may lead to crisis and highly damaging situations. Organizational survivability requires an effective and efficient competitive strategy in normal times, as well as an adequate resilient strategy to handle potential threats and overcome crisis (i.e. the capability to anticipate, avoid, escape, contain, and recover from potential business continuity threats). In a resource constrained context, efficiency and resilience requirements are often at odds with each other.

Crises can be defined as an unstable time or state of affairs in which a decisive change is impending, with the distinct possibility of a highly undesirable outcome (adapted from Fink [2]). Organizational survivability and the effectiveness of firms' crisis preparedness and response measures may, however, depend on the specific competitive market environment in which those measures are implemented. That environment often includes a complex network of suppliers, competitors and customers, with different attributes (population densities, economic capacities, changing desires, etc.), linked through increasingly global and intricate supply chains.

To understand an organization's competitiveness and ability to avoid and survive crises we need to consider, not only its strategic choices, but also how its competitors' strategies affect the overall process dynamics. In this context, the underlying problem that the present work aims to help understand is how organizational strategic choices, in both normal and crisis times, affect the ability of a firm to successfully compete, overcome crises and survive under different market environments.

To address the present problem, the following components and dynamics of interest were considered: (1) supply-chain actors (suppliers, focal firms, and customers); (2) interactions between them, considering agent-centred limited knowledge about the market; (3) agents market-positioning behaviour reflecting Porter's five forces that shape industry competition (rivalry among existing competitors, bargaining power of buyers, bargaining power of suppliers, threat of new entrants, and threat of substitute products or services) [3]; (4) a set of generic strategies followed by competitors; (5) product/service opportunities and customer desires driving the demand and supply dynamics; (6) customers' purchasing behaviour based on product/service utility added-value; (7) downstream product/service and upstream revenues flows.

94.3 The Modelling Approach

94.3.1 Multi-layered Cellular Automata as a Supply Chain Modelling Framework

A three-layered CA was used as a supply-chain modelling framework, expanding the rule-based interactions from within to across populations (i.e. intra and inter-layer based rules). Aggregating the supply-chain actors in three individual population sets—suppliers, competing focal firms (or competitors), and customers (with different economic capacities and desires)—the rules driving their market position selection at each population layer are influenced, not only by the current attributes of the actor's neighbour sites on that layer, but also by the current attributes of the corresponding neighbour sites of the adjacent layers. As shown on Fig. 94.1, a competitor's functional (non-disrupted) supply chain implies the existence of at least one supplier and one customer on the same (x-axis) competitor's position, or adjacent positions, of the corresponding layers. A disrupted supply chain implies the interruption of the corresponding deliverable/revenue flows. In their pursuit of the corresponding delivery/revenue flows maximization through functional supply chains, the behaviour of suppliers (top layer) is influenced by the suppliers' and competitors' layers, the behaviour of competitors (middle layer) is influenced by all three layers, and the behaviour of customers is influenced by the competitors' (product/service deliverables) and customers' layers. In particular, the suppliers' positioning rules are driven towards the maximization of product/service demand from available supply chains' competitors—three at most, corresponding to the selected supplier's x-axis position X, and two ($X - 1$ and $X + 1$) neighbouring ones—while minimizing other (competing) suppliers in vicinity; competitors' positioning

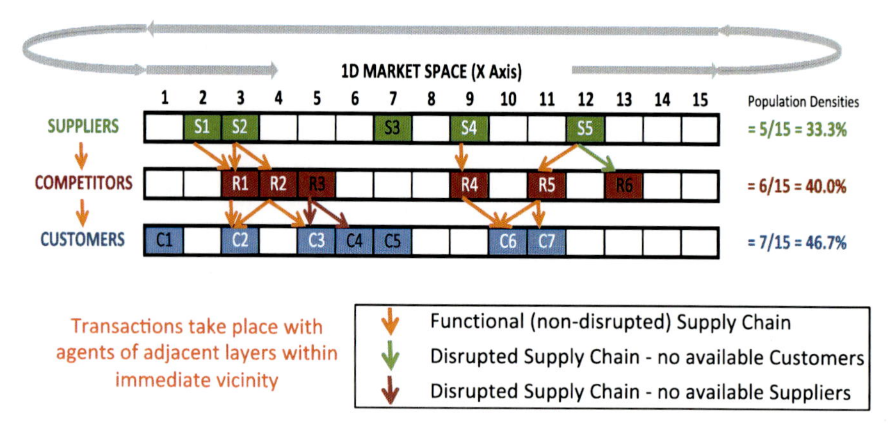

Fig. 94.1 A three-layered 1D cellular automata as a supply chain modelling framework

depends on their applicable strategy (as illustrated on Fig. 94.2); and customers' continuously seek the empty position on their layer that is best served by available competitors' deliverables.[1]

The adopted intra- and inter-layer neighbourhood rule-based CA model allows for the translation of the aspects related to Porter's five forces that shape industry competition [3] into the corresponding tension that drives suppliers, competitors and customers interests, as reflected in their market positioning selection rules. This modelling approach also provides a simple and comprehensive framework for the implementation of "structuralist" strategies. As defined by Kim and Maugborne [4], "structuralist" strategy types "assume the operational environment is given" (and have a bias to stay and defend their current positions). As such, their market positioning strategy is mainly driven by their market environment structural configuration, considering their current position, other competitors (particularly the most direct ones), potential customers, and business partners.

94.3.2 *Market Opportunities and Customer Desires Modelling as a Driver of Agents' Co-evolutionary Dynamics*

While the aforementioned model design can accommodate the implementation of a "structuralist" strategy type, it falls short on describing adequately both the demand/supply dynamics and to model "reconstructionist" strategy types, which seek to shape the operational environment, actively pursuing innovation and new opportunities [4]. In order to overcome these shortcomings the model should be enhanced with the introduction of product/service market opportunities and customer

[1]More specifically, with the introduction of market opportunities and customer desires, as later described in Sect. 94.3.2, customers will continuously seek the empty position that corresponds to the available deliverable that best matches its current desire.

desire features. Thus, product/service opportunities and customer desires are modelled using a similar k-bit word structure in which each bit represents a specific product/service attribute (that may, or not, match a customer's desire).

In this context, a k-bit opportunity word sub-model was implemented, associating a specific product/service opportunity to each site of the competitors CA layer (as represented in Fig. 94.2). These k-bit words distribution within that CA layer promotes opportunity diversity while allowing for competitors' possible exploitation of product/service similarities as well as local vicinity benefits. To accommodate those requirements, opportunity cardinal points—k-bit with words with Hamming distance of one between them—are regularly spaced on the corresponding CA layer,[2] and the intermediate words allowed to vary (based on Hamming distance criteria), with the level of variation defined as intended. The articulation between those k-bit opportunity words (translated to specific product/services offers when that opportunity is taken by a competitor) and the implementation of individual customer agents' k-bit desire words (which may evolve over time) fosters the model dynamics in terms of customers' desires and focal firms' products/services fitting objectives.

As illustrated on Fig. 94.2, each of the corresponding bits of the competitor offering (opportunity taken by the competitor) and customer desire words are compared. Customers have different economic capacities (providing different levels of attractiveness for the competitors) and seek offerings that best match their desires. Competitors following a "reconstructionist" strategy type steadily search for possible opportunity available market positions within their site neighbourhood that correspond to new and promising market niches (based on available opportunities, existing customers economic capacities and desires, and other competitors' offerings). Thus, the implementation of customer desires and market opportunities allows for distinct competitors' market positioning preferences, depending on their adopted strategy. As customers respond to changes in available offerings, and suppliers reposition themselves based on competitors' upstream demand and other suppliers positioning in their vicinity, downstream supplies and competitors' revenues change. This changing situation leads competitors to keep reassessing their current market positions in a continuous co-evolutionary process linking together suppliers', competitors', and customers' actions and outcomes.

94.3.3 Utility Added-Value and Customers' Purchasing Behaviour: Modelling Product/Service Selectiveness

As a last model feature implemented to drive the required supply and demand market dynamics, a competitor-customer inter-layer demand rule, based on an over-added utility parameter that characterizes customers' product/service selectiveness, is implemented. After comparing available product/services specific attributes'

[2] With all-zero and all-one k-bit opportunity words acting respectively as the "North" and "South" poles at the opposite positions of the (circular) one dimensional CA layer.

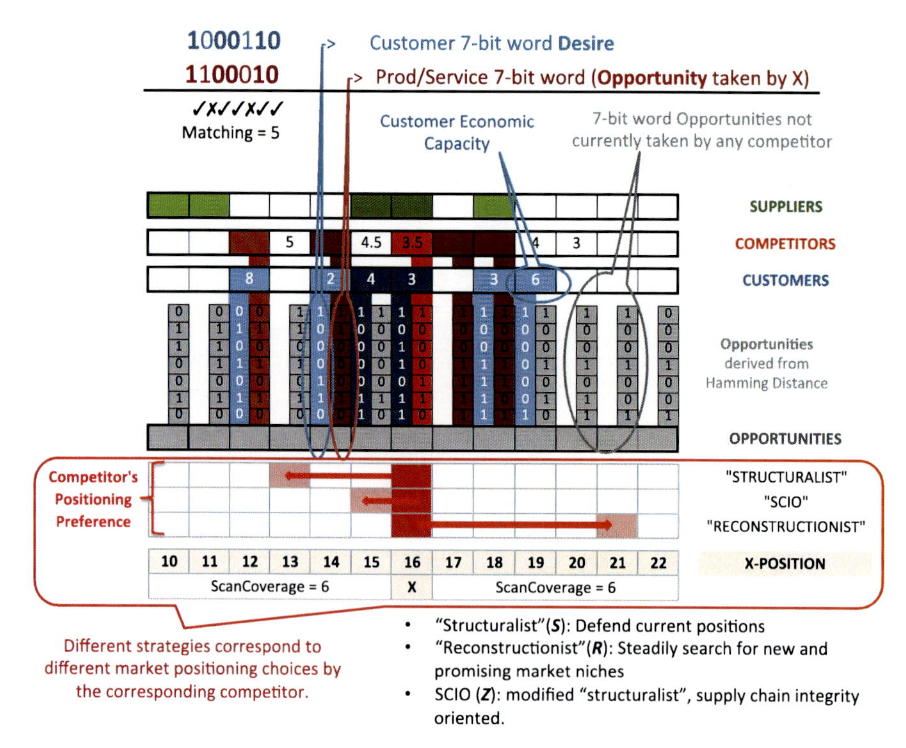

Fig. 94.2 Opportunities and desires as drivers of the co-evolutionary dynamics (excerpt representing positions 10 to 22 of the entire x-axis market space, corresponding to the local neighbourhood of competitor currently placed at position 16)

matching with customers' desire k-bit words, this over-added utility (OAU) parameter between -1 and $+1$ (-1 totally undervalues; 0 proportional to utility; 1 totally overvalues) is used to define the level of demand from the corresponding product/service providers.

Figure 94.3 provides an example illustrating a situation of a customer placed at position $X = 12$ with a 7-bit word desire "1000110". Considering a defined scanning coverage (SC) of 6, which allows the customer to check its x-axis neighbourhood from $X = 6$ to $X = 18$, the customer first assesses the available empty positions (as well as its current position) that are expected to allow it to obtain a product/service from a competitor that best matches its desire. In the provided example available positions at $X = 7$, 9 and 16 are expected to provide it with the best viable maximum utility (MAXUTIL) of 5, so it will choose one of these positions randomly.

Once at a specific position (in the example it is assumed that $X = 9$ was chosen) the customer can demand product/services from every available competitor (at $X - 1$, X, and $X + 1$) to a total amount corresponding to its economic capacity. In case $OAU = 0$ the customer will randomly demand product/services from the different available competitors with probabilities proportional to their relative desire matching. In Fig. 94.3 example it will demand with probability of $5/8$ from com-

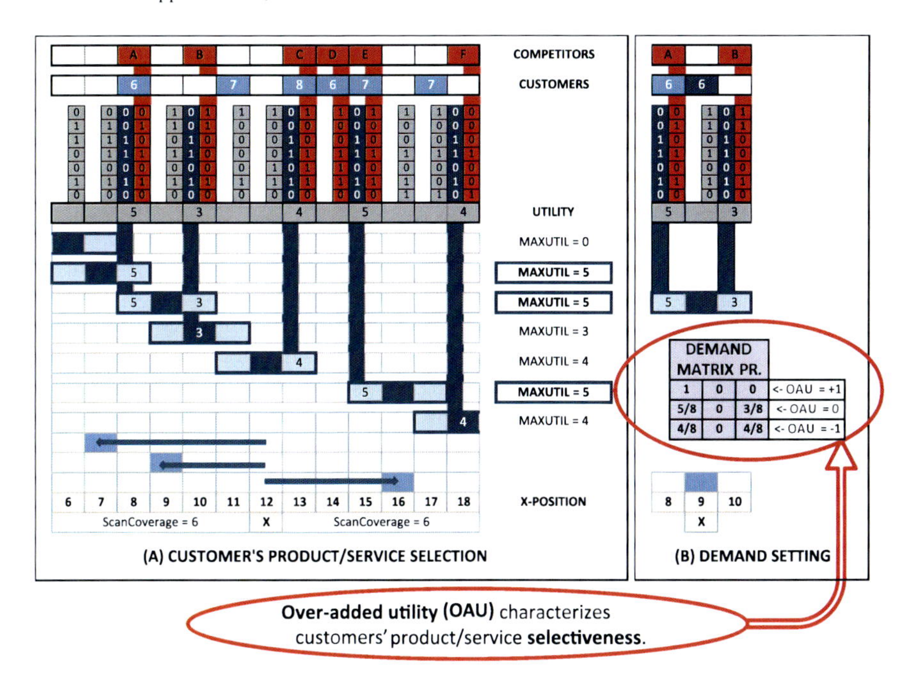

Fig. 94.3 Utility and selectiveness

petitor at $X = 8$ (desire matching $= 5$) and $3/8$ from competitor at $X = 10$ (desire matching $= 3$). If OAU $= 1$ it will demand the total amount only from the competitor that best matches its desire (in this case at $X = 8$). When OAU $= -1$ demand will be set with equal probability from all available competitors. In this way, OAU characterizes customers' product/service selectiveness.

94.4 Model Implementation: A Study Case on Organizational Survivability

94.4.1 The Organizational Survivability Study and Implemented Strategy Types

A model was developed and implemented—using Java with NetBeans IDE and the Repast [5] agent-based simulation platform—to better understand the relationship between strategic choice (in normal times and when in crisis) and organizational survivability under different market environment contexts. For this purpose, three agent type spaces (suppliers, competitors and customers) were implemented as individual layers, whose size reflects the common one-dimensional market, plus an underlying opportunity space. The implemented interaction between those spaces illustrates the integration of the multi-layered CA base solution with the additional modelling features previously described to ensure the competition and survival market co-evolutionary dynamics.

The added effort placed by organizations on ensuring the effectiveness of their supply chains, avoiding any disruption or disturbance that can perturb the swift flow of goods and services from their primary producers to the final customers, was also considered. In the present model, a modified "structuralist" strategy was also implemented to reflect these organizations that are willing to sacrifice a greater slice of their short-term operational results for the sake of a more robust position in terms of their business continuity objectives. Thus, besides a "reconstructionist" and a "structuralist" strategy, the model also uses a more "Supply-Chain Integrity Oriented" (SCIO) modified "structuralist" strategy.[3]

94.4.2 Competitors' Genetic Codes and Selection of the Fittest

To assess the impact of strategic choice on organizational survivability, each competitor agent is characterized by a genetic code that defines which strategy it follows in normal times and when in a crisis situation (which may be the same in both situations). This genetic code may be any of the nine possible outcomes from the permutation with repetition of the three ("reconstructionist", structuralist" and SCIO) strategies to the two applicable situations (normal times, and crisis).

Based on how good, or bad, a competitor is performing in terms of average revenues when compared to other competitors, its strategic genetic code may be selected (i.e. copied) by new market competing entrants,[4] or the competitor may be led to crisis (and eventual dismissal) respectively Competitors whose average revenues position them on the worst performers percentage, as defined by ICT (In-Crisis Threshold), are led into a crisis status. If agents in crisis are not able to improve their average revenues during a specified number of CE (Crisis Endurance) run steps, so that they become positioned above the defined OCT (Out-of-Crisis Threshold) percentage and return to a normal status, they will be dismissed and replaced by a new competitor. While a randomness factor (RSP)[5] is also incorporated in the selection of new entrants strategy defining genetic code, the selection of the fittest is implemented in an evolutionary process where the competitors' population genetic codes distribution tends to reflect the adaptation of the corresponding strategy pairs to the modelled market environment. As a specific genetic code population depends on the ratio of new entrants versus dismissed agents carrying that

[3]The way these three strategies may lead to different competitors' market positioning choices is illustrated on Fig. 94.2.

[4]A TPP (Top Performers Percentage) parameter defines the percentage of agents with best average revenues whose genetic code will be eligible for transmission to a new entrant (probability of selection depends on agent's average revenue).

[5]RSP (Random Strategy Probability) defines the probability that a new competitor's strategy defining genetic code is chosen randomly. Consequently, there is a probability of 1-RSP that its genetic code is selected from the top performing competitors.

code, its evolution depends on the average revenues of their agents (to increase their chances of being selected for reproduction, and avoid crises) as well as their ability to survive crises when they occur.

94.4.3 Model Flows and Parameterization

For the present work, the following main flows were considered: The upstream flow (from customers to suppliers) of product/service orders (demand), and the consequent downstream product/service flow (from suppliers to customers), with the corresponding upstream revenues flow.

Using different agents' population densities (PD), supplier/customer agents' replacement probabilities (RP) by new-borns, market scanning coverage (SC)(range of each agent local analysis neighbourhood), customers over-added utility (OAU) and desire change probabilities (DCP),[6] as well as other user-defined variables to characterize different market environments, the market positioning rules followed by each agent at each step (integrating the basic elements of Porter's five competitive forces) shape the overall process dynamics. After a random definition of competitors' genetic codes at each run initialization, the competitors' genetic code distribution along the co-evolutionary process is analysed.

94.5 Results, Analysis and Implications

Based on the populations co-evolution during multiple runs of 15000 steps each, the competitors were analysed by genetic code in terms of their average revenues and probability of being selected for possible genetic code transmission, percentage of agents in crisis, number of steps in crisis, crises survival ratio, dismissed agents ratio and total revenues. This analysis was performed under distinct PD ratios and different market environment volatilities, characterized in terms of different RP, SC, OAU, and DCP parameters (Table 94.1).

Analysis of the obtained results (Fig. 94.4) reveals that: (1) the survivability of each "genetic code" varies significantly with environment volatility and population density ratio changes; (2) while in low volatility environments the strategy employed in normal situations tends to be more determinant to survivability than the strategy adopted in crisis, as environment volatility increases the strategy employed in crisis

[6]These parameters are defined as follows: PD = Percentage of s occupied by suppliers, competitors and customers on their corresponding layers; RP = Probability that each supplier/customer is dismissed at the end of each run step and replaced by a new similar agent (but with a new random desire, in case of a customer); SC = Number of positions to each side of its current position that an agent is able to collect market information at its and adjacent layers; OAU = (see Sect. 94.3.3); DCP = Bit "mutation" probability (from "0" to "1", or from "1" to "0"), per run step, of each bit of customers' k-bit desire words.

Table 94.1 Simulation runs parameter values

PDR (Comp)	Competitors population density (ref.)	35 %
MEC (Cust)	Customers maximum economic capacity	8
K	Nr. of bits opportunity/desire k-bit words	7
CE	Crisis endurance	6
ICT	In-crisis threshold	10 %
OCT	Out-of-crisis threshold	15 %
TPP (Comp)	Competitors top performers percentage	60 %
RSP	Random strategy probability (new borns)	10 %

	Scenario's market volatility				
	Low	LM	Medium	MH	High
SC (Supp/Comp/Cust)	5	8	10	13	15
RP (Supp/Cust)	0.00	0.05	0.10	0.15	0.20
DCP	0.00	0.01	0.02	0.35	0.05
OAU	−1.0	−0.5	0.0	0.25	0.5

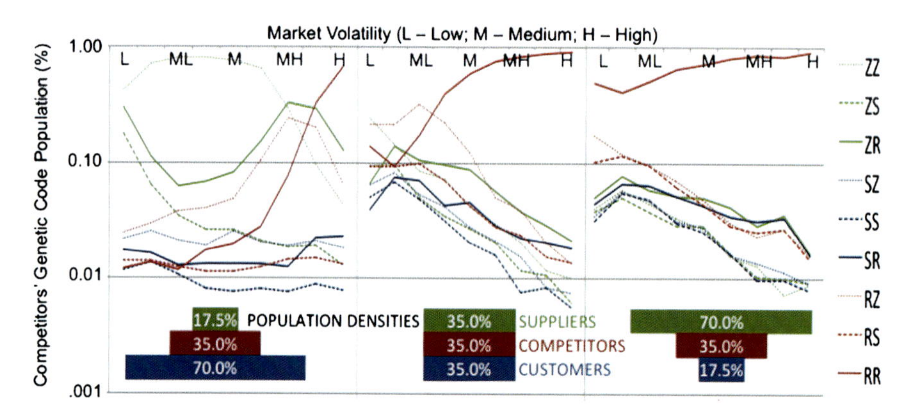

Fig. 94.4 Simulation results (supplier-to-competitor and competitor-to-customer density ratios of 1 : 2, 1 : 1, and 2 : 1)

tends to gain relevance in terms of relative survivability performance; (3) in low volatility and supply scarce/customer plenty environments, the highest levels of survivability are achieved by competitors using the "SCIO" as their normal strategy; (4) while the full "SCIO" (ZZ) strategy is very well adapted to low volatility and supply scarce contexts, and the full "reconstructionist" (RR) strategy dominates in high volatility or plenty supply scenarios, the "SCIO-reconstructionist" (ZR) strategy mix presents balanced results across the different scenarios and volatilities.

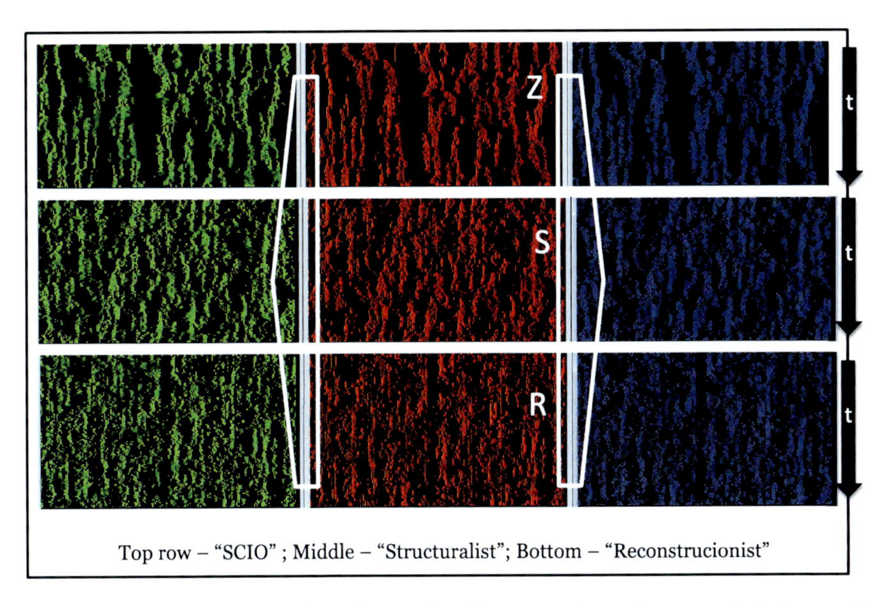

Fig. 94.5 Competitors' (agents in *red*) strategies effect on market environment, including suppliers (agents in *green*) and customers (agents in *blue*)

Scenarios with different environment volatilities and population densities were also run with competitors following only one strategy type. The effects induced on the business market by the competitors' strategy were then compared. The resulting patterns of suppliers, competitors and customers positioning evolution show that "SCIO" strategies tend to generate the more stable and concentrated market environments, while "reconstructionist" strategies present the more disperse and less stable ones (Fig. 94.5).

The obtained results suggest that: (1) as market environments become more volatile (more globally linked, with changing actors and customers' desires) the more critical to organizational survivability becomes the adoption of an effective crisis strategy; (2) as supply sources grow or market environments become more volatile, "reconstructionist" strategy types tend to have higher chances of success; (3) the proliferation of "reconstructionist" strategies, constantly pursuing new opportunities and shaping market trends, tend to induce higher instability in the market environment.

94.6 Conclusion and Future Work

The present study case on competition and survival modelling in a supply-chain market context illustrates: (1) multi-layered CA models potential as frameworks for accommodating increasing levels of complexity in terms of different agent populations' intra- and inter-layer (market) positioning-oriented rule-based modelling in a mutually shared space; (2) how the adopted approach does not hamper the clear, structured, and incremental implementation of those added levels of complexity, while withholding the desired CA modelling advantages in terms of a comprehensive understanding and analysis of the positioning dynamics at the individual and aggregate levels; (3) how the addition of market opportunities and changing customer desires, represented as k-bit words, in connection with an over-added utility algorithm to define the customer's purchasing selectiveness on the modelled supply-chain, plays a decisive role in the overall co-evolutionary market model dynamics.

The principal aims of the work ahead are as follows: (1) fine-tuned characterization of each variable effect on the obtained results; (2) applying information theory measures to analyse the model's phase-transition behaviour and quantify the system's sustainability [6]; (3) taking the model to empirical data.

Acknowledgements Special thanks to the support provided to this work by FCT project Pest-OE/EEI/LA0008/2011.

References

1. Holland JH (1995) Hidden order: how adaptation builds complexity. Addison-Wesley, Reading
2. Fink S (1986) Crisis management: planning for the inevitable. iUniverse, Lincoln
3. Porter ME (2008) The five competitive forces that shape strategy. Harv Bus Rev 86(1):79–93
4. Kim WC, Mauborgne R (2009) How strategy shapes structure. Harv Bus Rev 87(9):72–80
5. Repast. http://repast.sourceforge.net/
6. Ulanowicz RE, Goerner SJ, Lietaer B, Gomez R (2009) Quantifying sustainability: resilience, efficiency and the return of information theory. Ecol Complex 6:27–36

Chapter 95
When Pig Meets Pencil: The Beauty of Complexity in Industrial Networks

Andreas Ligtvoet

Abstract The making of pencils and the processing of pigs are complex industrial processes that are seemingly unrelated. We hypothesise about the interwovenness of these processes and argue that individuals wishing to influence these industrial networks need to deal with both detailed and dynamic complexity. This means being more humble about what can be measured and controlled, and being more open to learning and adapting policies.

95.1 Introduction

We live in a complex world in which many processes are dependent on other processes. This interdependence emerged from historical choices that were made by many different actors [2], most of them completely unaware of each other. In teaching and explaining complexity and industrial networks, it is sometimes difficult to convey the depth of the interwovenness [4]. In this article we introduce two stories (examples, archetypes) that in our opinion are already interesting on their own, but when told together really emphasise the multiple links that are at the core of *industrial networks*, but mostly remain unnoticed in our society.

95.2 I, Pencil

The short essay called "I, Pencil" was written by economist Leonard Read to explain the importance of free markets in producing an object as seemingly simple as a pencil [6]. He describes the production process of the "Mongol 482" type pencil produced by Eberhard Faber Pencil Company. Written from the first person point of

The author thanks Gerard P.J. Dijkema and Alex Ryan for their suggestions and comments and Chris B. Davis for sharing and discussing pencils and pigs. This work was supported by the Next Generations Infrastructures Foundation.

A. Ligtvoet (✉)
Faculty of Technology, Policy, and Management, Delft University of Technology, Delft, The Netherlands
e-mail: a.ligtvoet@tudelft.nl

T. Gilbert et al. (eds.), *Proceedings of the European Conference on Complex Systems 2012*, Springer Proceedings in Complexity, DOI 10.1007/978-3-319-00395-5_95,

view of a pencil, the story details the complexity of the pencil's creation, listing its components (wood, lacquer, graphite, ferrule, factice, pumice, wax, glue) and the numerous people involved.

The wood that is used for the pencil is harvested in Northern California and Oregon, shipped to a sawing mill, where it is cut into small, pencil-length slats less than 5 millimetres thick. This part of the process alone builds on the existence of transport modes (trucks, trains; see [1]), precision machinery and tools, iron ore smelting and processing, logging camps, lumberjacks, mechanical engineers, and the support of all the components by hardware stores, food stores and job markets.

Graphite is mined in Sri Lanka, using the appropriate tools, packaged and shipped to the USA. The graphite is mixed with clay from Mississippi and passed through numerous machines. To increase its strength and smoothness, the "lead" is then treated with a hot mixture which includes candelilla wax from Mexico.

At the top of the pencil is the eraser, made of an ingredient called factice. It is a rubber-like product made by reacting rape-seed oil from Indonesia with sulfur chloride. This is mixed with pumice from Italy; and the pigment which gives "the plug" its color is cadmium sulfide. The eraser is attached to the rest of the pencil by some metal—the ferrule—made of brass. Think of all the persons who mine zinc and copper and those who have the skills to make shiny sheet brass from these products of nature.

The wood receives six coats of lacquer, all made possible by the growers of castor beans and the refiners of castor oil. The label is a film formed by applying heat to carbon black mixed with resins. In the end it is all glued together, and the pencil is packed, shipped, and sold somewhere in a shop.

95.3 PIG05049

The book and art project "PIG05049" of designer Christien Meindertsma takes exactly the opposite approach: it starts with one pig randomly chosen from a commercial farm in the Netherlands and tracks its body parts through processing industries, ending up with a—sometimes surprising—array of 185 products [5]. The artist recalls a conversation with an old Italian farmer who explained that when in the olden days a pig was slaughtered, nearly all the parts of the animal were put into use by the villagers. Although the farmer assumes modern societies are more wasteful, Meindertsma's research shows that quite the opposite is the case.

By far the largest portion of the pig (some 52 %) is turned into pork meat (see Fig. 95.1). From the more delicate pieces: fillet or tenderloin, from the leg: ham and schnitzel, from the ribs: spare ribs, from the shoulder: chops, and from all the other pieces of meat a host of lower-grade products such as hamburgers, streaky bacon, salami, frankfurter, chorizo, ... Of course, the meat is also used in other food products such as sausage rolls, pizza, soup, cordon bleu, tortellini, egg rolls, pork pies, with as only boundary the creativity of the preprocessed food industry. Some of the other non-meat parts of the pig are turned into food: blood into blood sausage, fat into lard and diced bacon, liver into liver sausage, etc.

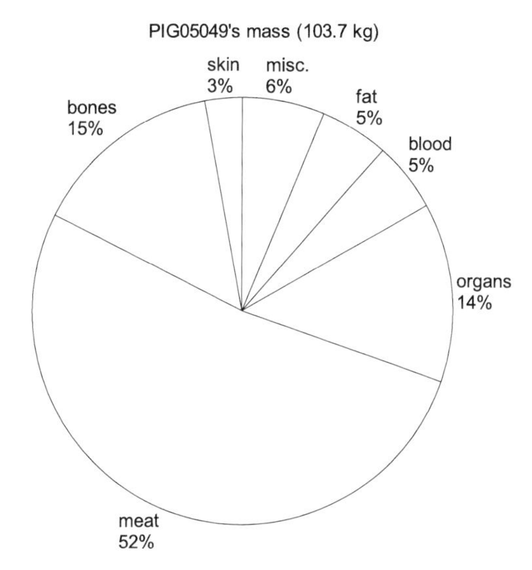

Fig. 95.1 Percentage of mass of PIG05049 [5]

The story turns more interesting when we follow the route of skin and bones to gelatins. These are used in a myriad of food products that seemingly have nothing to do with the original pig any more: liquorice, chewing gum, peppermint, marshmallows, cupcakes, vanilla pudding, chocolate mousse, ice cream, curd, aspic, ... Non-food uses are gelatine capsules for delivering medicine, paintballs, bath pearls, photographic paper, lubricants for ammunition, and ballistic gelatin for crime scene investigation.

Bones are very versatile in their use as calcium (extra calcium yogurt), bone protein (cellular concrete), and collagen (beauty masks and energy bars) are extracted. Fatty acids from bone fat is used in the production of anti-wrinkle cream, crayons, soap, washing powder, paint, shampoo, and body lotion. The glycerin from the same bone fat leads to toothpaste, antifreeze, and floor wax. The pancreas is harvested for insulin and the bladder is turned into a tambourine.

95.4 Pig Meets Pencil

When we look at the thousands of people involved in extracting value from the pig and the thousands that are involved in constructing the pencil, the question arises *whether, when, and how* these people and processes meet. Although this is hypothetical, there are a few interesting links between pigs, pencils, and their related industrial networks.

The first link is through one of the other products made of bone: glue. We could imagine that the glue that is derived from PIG05049 (some 0.4 % of its total mass) is actually used to glue together the wooden slats and the other components of the pencil. Second, animal fat is used as a wetting agent for the graphite that is to become the "lead"; also the candelilla wax for the lacquer is mixed with fat. Finally, another bone product is bone ash (0.3 %), which is turned into bone black pigment—the same pigment that can be used for the text on the side of the pencil.

Apart from this linear involvement of pigs and pencils (one could almost talk about a supply chain, although we prefer *supply network*; see Fig. 95.2), there are some other obvious and less obvious connections. The meat from the pig can feed the many labourers involved in the production process. We can safely assume that some pencils are used in processing the pig and in planning the pencil-making process. Also, some parts of the pig are turned into animal feed (giving rise to questions regarding dangers of prionic diseases).

95.5 Discussion

One can only marvel at the complexity of these supply and distribution networks of seemingly simple products. And the beauty is that most of these processes and their interlinks are not pre-planned or designed. In the described processes thousands and thousands of farmers, workers, and professionals are involved and not one of them could make anything as simple as a pencil without the help of many others. There is no mastermind at work, no one planning and designing the optimal pencil-production-path or the optimal pig-processing-path. It is one of the fascinating examples of the forces of demand and supply at work, interactively determining prices, determining choices, determining the final product. Some people would call this the "invisible hand", but it might be more precise to call this the emergent outcome of scores of invisible *hands*.

This is not a call for unbridled Austrian economics and a *laissez-faire* liberalised market. We believe that there need to be checks and balances lest the industrial ecology fully occupy all the other parts of our socio-technical systems like some invasive species. But we do agree with Read that no one individual, organisation, or government can fully understand the complexity of prices, property, profits, and incentives that are being created and destroyed every day on the millions of markets of which we only described a few in this article.

We should realise that life-cycle analyses and materials flow analyses provide a snapshot, like "I, Pencil" and "PIG05049", already outdated by the time they are written down, as someone somewhere in the network has probably figured out a more profitable way of processing pig fat. Nevertheless, these efforts are invaluable for understanding industrial complexity. At the same time, we need to try to understand the rules that guide actor behaviour; not only profit-seeking behaviour, but also higher-level societal goals like constitutionality, legality, accountability, transparency, justice, and equity [3], as these are equally constantly changing the shape of the network. This requires a balance between what can be called detailed complexity (intricate but static descriptions of all processes) and dynamic complexity (the forces that drive change) [7]. Often, when we try to understand these networks, we zoom in on a stable, measurable, narrowly bounded subsystem, collect and analyse

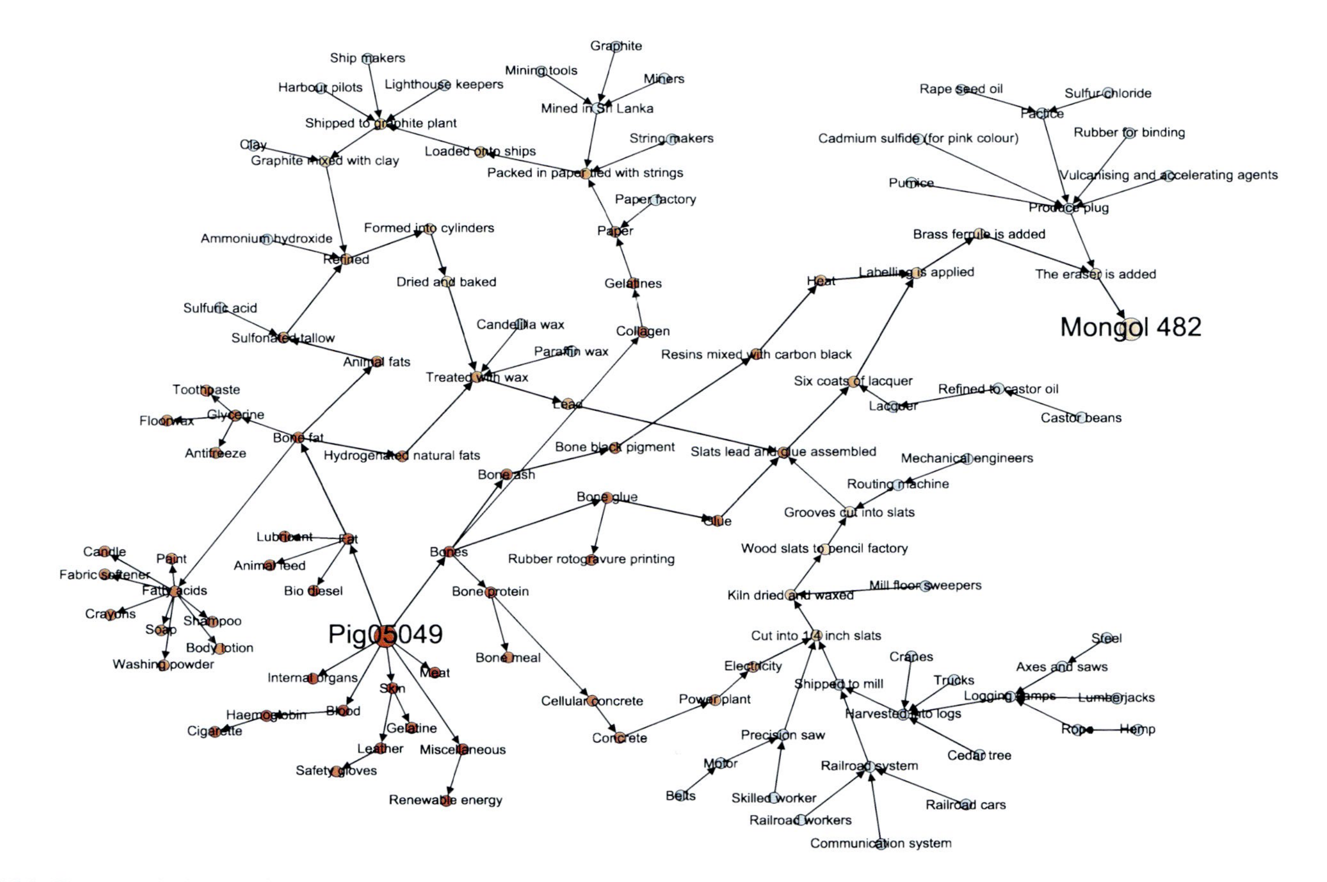

Fig. 95.2 The networked connections between pig and pencil

the data, build a model, and then confidently extrapolate our results to the whole system. Models of dynamic complexity, however, are rarely quantifiable, always contestable, and seem much less rigorous. However, they are more systemic, because they provide a synoptic perspective that connects disparate domains and specialised areas within the ecology.

Trying to regulate a network based on a static picture and simple linear chains will generate unintended consequences and (often unpleasant) surprises. The complexity, intransparency, and dynamic nature of these networks implies that policy makers need to approach the regulation of economic networks with a good deal more humility than they currently do. This means having to deal with the fact that we will never have complete understanding. When governments implement new policies, they often develop metrics to show that they are making progress. Instead, they should be using metrics that give early warning that the policies are generating unintended consequences, or have become irrelevant in a changing world. With this level of interdependence, we must assume from the outset that if we are not already wrong, we soon will be. Empirically grounded continuous learning is more important than getting the policy right up front.

References

1. Braden A (2009) The industrial ecology of emerging technologies. J Ind Ecol 13(2):168–183
2. Anderson P (1999) Complexity theory and organization science. Organ Sci 10(3):216–232
3. Bauer JM, Herder PM (2009) Designing socio-technical systems. In: Philosophy of technology and engineering sciences, vol 9. North-Holland, Amsterdam, pp 601–630
4. Dijkema GPJ, Basson L (2009) Complexity and industrial ecology. J Ind Ecol 13(2):157–164
5. Meindertsma C (2008) PIG05049. Flocks, Amsterdam
6. Read LE, I, pencil. Technical report, Foundation for Economic Education, New York (1958/2006). http://www.fee.org/library/books/i-pencil-2/
7. Senge PM (1990) The fifth discipline. Doubleday/Currency, New York

Chapter 96
Citation Networks Dynamics: A New Clustering Algorithm Using Recurrence Plots

F. Strozzi, C. Colicchia, A. Sorrenti, and J.M. Zaldívar

Abstract The development of science in various fields and the proliferation of the number of scientific studies have prompted researchers to represent the scientific literature as a network whose nodes are works and citations are the links (citation network). In this way, one can more easily detect milestones articles, and areas of greatest scientific activity. One major problem in big citation networks is to be able to carry out an efficient clustering which allows grouping the articles according to their similarities. This method should not be automatic and must not involve a careful reading of the selected works. From here one can see that the problem of finding a proper clustering is not easy to solve. There are several clustering methods with overlapping or not—for a complete review see Fortunato (Phys. Rep., 486:75–174, 2010). In this work we have developed a new clustering method for a citation network using a definition of distance introduced by Bommarito el al. (Physica A, 389:4201–4208, 2010). We have compared this method with the one developed by Pons and Latapy (Computer and information sciences, vol. 3733, pp. 284–293, 2005) applying both to a citation network on sustainability indicators in supply chains. The proposed method proves to be more efficient and, also, through the analysis of the adjacency matrix as a the Recurrence Plot, is able to view the dynamics of the various clusters in their evolution over time

Keywords Recurrence plot · Citation networks · Sustainability

96.1 Introduction

The necessity to face sustainability issues in terms of supply chain processes is becoming an argument of great relevance in Supply Chain Management (SCM). This is emphasized by the increasing pressures, both externally (e.g. legislative re-

F. Strozzi (✉) · C. Colicchia · A. Sorrenti
Cataneo University-LIUC, Castellanza, Italy

J.M. Zaldívar
European Commission, Joint Research Centre, Ispra, Italy

T. Gilbert et al. (eds.), *Proceedings of the European Conference on Complex Systems 2012*, Springer Proceedings in Complexity, DOI 10.1007/978-3-319-00395-5_96,

quirements) and internally (e.g. need for a more efficient use of resources), oriented toward the implementation of greener supply chains.

Sorrenti [17], using the Systematic Literature Network Analysis (SLNA) developed by Colicchia and Strozzi [4], built a citation network with the aim to classify sustainability indicators and papers. He found interesting clustering of the papers using the method, developed by Pons and Latapy [13], in accordance with the ISO 14031 and 14040.

Our contribution consists in developing new clustering techniques based on the distance matrix proposed by Bommarito et al. [1] and compare it with the Pons and Latapy method on the same citation network. The advantage of this new clustering technique is that it takes into account the time and allows a representation of the citation network as a time series. The time series obtained can then be studied using different time series analysis tools. In this work, we have tried to analyze the information that its Recurrence Plot representation adds. This new type of clustering can be a new tool for the SLNA methodology.

96.2 Citation Network

Three strings of keywords were adopted in Sorrenti [17] to build the citation network

- ((“environmental indicator*” OR “environmental performance indicator*”) AND “supply chain”)
- (“supply chain” AND environmental AND indicator* AND sustainab*)
- ((“supply chain” AND “environmental performance*”) AND indicator*)

The research was conducted by means Web of Science (WoS) database and the results are rearranged with HistCite which allowed the exportation of the file in Pajek [5], a software for the analysis and visualization of large networks. The citation network obtained has 84 nodes, and 20 connected components and its visualization using Pajek is showed in Fig. 96.1.

96.3 Main Connected Component (MCC)

For a more detailed analysis of individual components, please refer to Sorrenti [17]. In this work we will analyze only the Maximum Connected Component (MCC). From now on we will quote the items with corresponding serial numbers listed in the Table 96.1.

96.4 Cluster Analysis of MCC Using PL Algorithm

Clusters Analysis using Pons and Latapy (PL) algorithm [13] is carried out by means of R software. In network analysis, communities are created by defining a partition

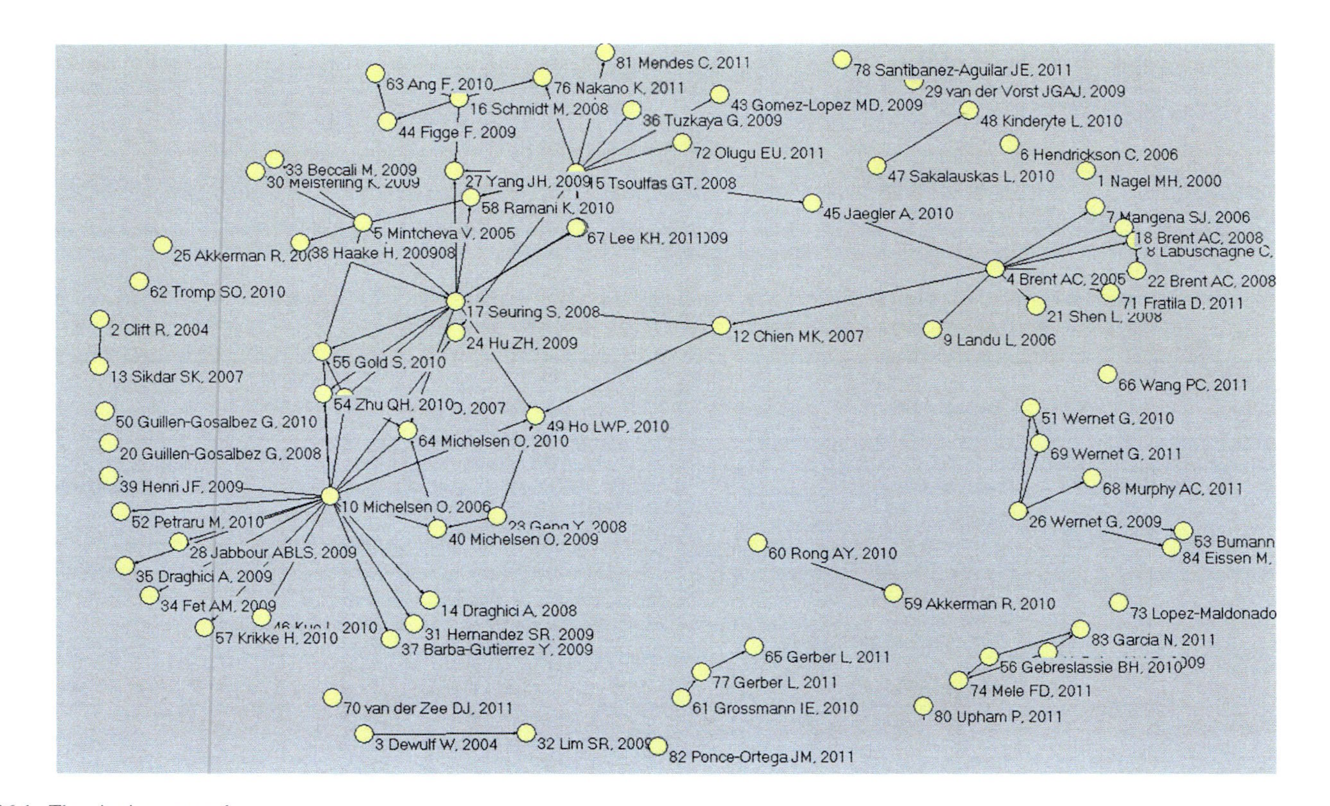

Fig. 96.1 The citation network

Table 96.1 The labels of the Maximum Connected Component (MCC) nodes

1	4 (Brent 2005)	17	23 (Geng 2008)	33	44 (Figge 2009)
2	5 (Mintcheva 2005)	18	24 (Hu 2009)	34	45 (Jaegler 2010)
3	7 (Mangena 2006)	19	27 (Yang 2009)	35	46 (Kuo 2010)
4	8 (Labuschagne 2006)	20	28 (Jabbour 2009)	36	49 (Ho 2010)
5	9 (Landu 2006)	21	30 (Meisterling 2009)	37	52 (Petraru 2010)
6	10 (Michelsen 2006)	22	31 (Hernandez 2009)	38	54 (Zhu 2010)
7	11 (Michelsen 2007)	23	33 (Beccali 2009)	39	55 (Gold 2010)
8	12 (Chien 2007)	24	34 (Fet 2009)	40	57 (Krikke 2010)
9	14 (Draghici 2008)	25	35 (Draghici 2009)	41	58 (Ramani 2010)
10	15 (Tsoulfas 2008)	26	36 (Tuzkaya 2009)	42	63 (Ang 2010)
11	16 (Schmidt 2008)	27	37 (Barba-Gutierrez 2009)	43	64 (Michelsen 2010)
12	17 (Seuring 2008)	28	38 (Haake 2009)	44	67 (Lee 2011)
13	18 (Brent 2008)	29	39 (Henri 2009)	45	71 (Fratila 2011)
14	19 (Seuring 2008)	30	40 (Michelsen 2009)	46	72 (Olugu 2011)
15	21 (Shen 2008)	31	42 (Cholette 2009)	47	76 (Nakano 2011)
16	22 (Brent 2008)	32	43 (Gomez-Lopez 2009)	48	81 (Mendes 2011)

that generates subsets, such that "the proportion of arcs within the subsets is high compared to the proportion of arcs between them" [12]. Their approach consists on performing random walks in the network and to analyze where they tend to be trapped, i.e. in densely connected parts. These provide the communities obtained by Pons and Latapy [13] algorithm. Figure 96.3 illustrates the presence of six clusters highlighted with different colors. By observing nodes disposition it is clear that the identified clusters have specific features.

- The green and red clusters are quite similar. Indeed within both clusters there is a "dominant" article [2, 11]. Moreover it is interesting to notice how the presence of a "dominant" paper involves a visualization of cluster nodes as a "fan".
- The yellow cluster is the biggest and most complex cluster. It is not clear how to identify one dominant node but there are two main centers: node 12 [16] and node 10 [19].
- The blue cluster highlights a triangular relationship based on node 11 [15].
- The remaining two clusters, i.e. the pink and the white one, are composed of only one paper.

96.5 Distance Matrix of the MCC

In [1] the authors considered, as a measure of the vicinity in a citation network, "the number of share sinks between two vertices". They stated that the direction of the arrows in the citation network from paper A to paper B means that A cites B. We have followed another notation, adopted for example by [5], in which the same arrow means that B cites A. Using this notation the direction of the arrow is associated to the flow of knowledge. This implies that the "sinks" for [1] are "sources" in this work. The distance defined by [1] becomes:

"Given a vertex i and its ancestor A_i, the *sources* of i are given by the set

$$S_i = \left\{x : \delta^+(x) = 0, x \in A_i\right\} = S \cap A_i$$

where $\delta^+(x)$ is the notation for the *in-degree* of vertex x and S in the set of all sources of the graph G".

The distance between vertices i and j is given by the proportion of *sources* they do not share:

$$D_{i,j} = 1 - \frac{|S_i \cap S_j|}{|S_i \cup S_j|}$$

where $|x|$ is the cardinality of set x''. The set S can be calculated using a shortest path algorithm.

96.6 From a Citation Network to a Time Series Using the Multidimensional Scaling

The transition from a time series to a network has been proposed recently in the literature [6, 9, 18]. In this work, we did the opposite: transform a network, the citation network, into a time series. The additional difficulty in this step is that in general a temporal order of the nodes does not exist, however, in a network of citations this happens.

The function *cmdscale* of Matlab performs the classical (metric) multidimensional scaling, also known as principal coordinates analysis. In practice, it consists on identifying the main components of the covariance of the distance matrix. We have applied the multidimensional scaling to the distance matrix calculated following the above procedure and, by plotting the principal eigenvector, we have obtained Fig. 96.2.

96.7 Clustering Using Time Series

Figure 96.2 shows how to bind to each paper a number that corresponds to its component along the principal eigenvector. We have assumed that the papers with similar values are related and we have obtained the clusters of the Table 96.2. In the following this clustering technique will be called Time Series (TS) clustering.

With Pajek we may overlap the partition of Table 96.2 to the MCC. The results are presented in Fig. 96.3.

96.8 Comparison of PL and TS Clustering

The first observation is that the two methods give very similar clusters with the exception of the yellow big cluster of PL that TS divides in two sub-clusters (see

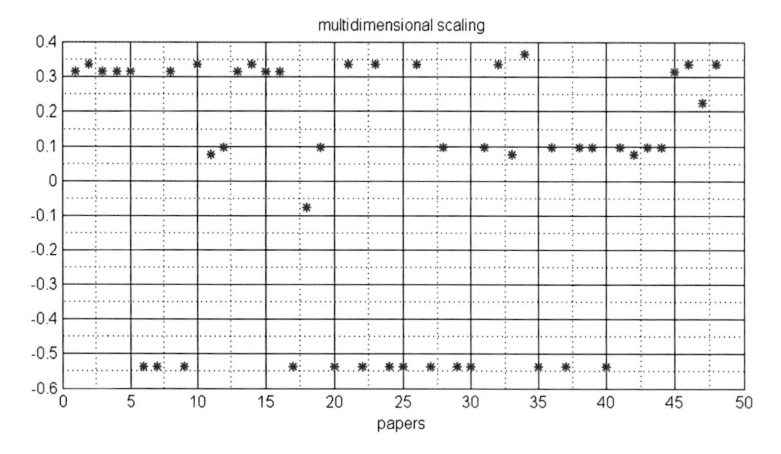

Fig. 96.2 Multidimensional scaling of the distance matrix

Table 96.2 TS Clustering

Cluster	Component value	TS papers	TS cluster color
1	0.3338	2, 10, 14, 21, 23, 26, 32, 46, 48	Yellow
2	0.3122	1, 3, 4, 5, 8, 13, 15, 16, 45	Green
3	0.2233	47	Red
4	0.09599	12, 19, 28, 31, 36, 38, 39, 41, 43, 44	Blu
5	0.07532	11, 33, 42	Pink
6	−0.07536	18	White
7	−0.5367	6, 7, 9, 17, 20, 22, 24, 25, 27, 29, 30, 35, 37, 40	Orange
8	0.3638	34	Violet

Table 96.3).The large yellow cluster identified by PL has more centers: node 12 [16] and node 10 [19] while the method TS divided it into two clusters. By analyzing the articles, it is possible to observe that this distinction is more appropriate. While node 10 is an overview of the various environmental indicators, node 12, emphasizes the lack of attention paid to social indicators of sustainability and the importance of sustainable suppliers. Indeed, it argues that "corporations are in turn being held responsible for the environmentally and social (or sustainability) performance of their suppliers" [16].

The papers of the cluster near to [16] show more attention to this aspect in contrast to those of cluster near to [19]. For example, in node 44, [10] the authors say: "...by evaluating and monitoring suppliers' performance CO_2 emissions, a focal company may avoid carbon-related risk and retain competitiveness based on its supply chain". Also in his short CV at the end of the paper one reads: "...He (Lee) has interests in all areas of Corporate Sustainability Management, but particularly in sustainable supply chain management, corporate social responsibility and stake-

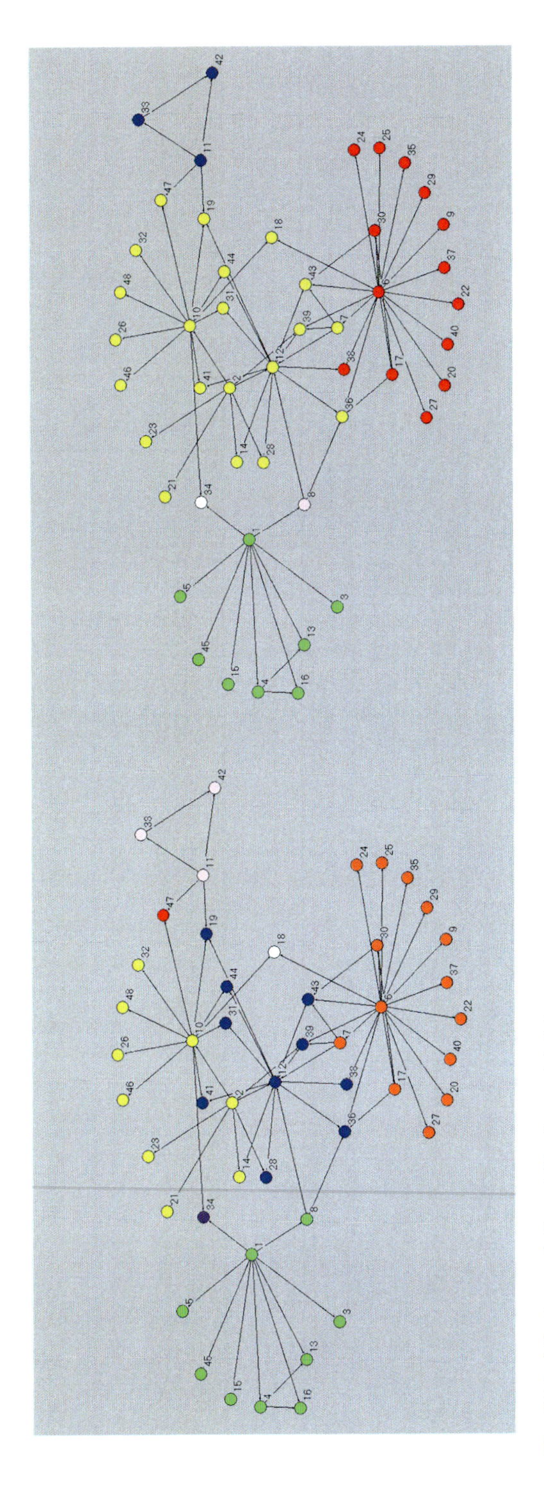

Fig. 96.3 TS cluster (*left*) and PL cluster (*right*)

Table 96.3 PL and TS clustering. In *bold* the papers belonging to the clusters using one method but not the other

PL clustering	TS clustering
1, 15, 13, 4, 16, 45, 3, 5	1, 15, 13, 4, 16, 45, 3, 5, **8**
6, 17, 30, 22, 24, 29, 27, 40, 35, 37, 9, 20, 25, **38**	6, 17, 30, 22, 24, 29, 27, 40, 35, 37, 9, 20, 25
12, 36, 43, **7**, 39, 28, 14, 19, 44, 31, 41, **18**	**38**, 12, 36, 43, 39, 28, 19, 44, 31, 41, **7**
2, 23, 21, 10, 26, 46, 48, 32, **47**	14, 2, 23, 21, 10, 26, 46, 48, 32
11, 42, 33	11, 42, 33
8	??
34	34

holder management". It is interesting to notice that, although the paper is linked to both the centers (nodes 10 and 12), the TS clustering method has put Lee in the cluster of node 12.

Other papers that use both papers are for example: node 41, [14] who says: "A company can select those suppliers that generated the least pollution in each individual phase". The node 31 [3] showed how their findings "...could be of use within an emission reduction program, as a component of an overall Corporate Social Responsibility (CSR) strategy".

Another way to compare the clusters of PL and TS methods is to observe the Table 96.3 in which we have underlined the papers that belong to the clusters using one method but not the other

The two methods give the same partitions for 90 % of the papers, but TS method is able to better differentiate the papers of the big yellow PL cluster.

96.9 Recurrence Plot from a Temporal Series and Cluster Evolution

Eckmann et al. [7] introduced a new graphical tool, which they called a recurrence plot (RP). The recurrence plot is based on the computation of the distance matrix between the reconstructed points in the phase space:

$$d_{ij} = \|S_i - S_j\|$$

where $S_i = \{s(t), s(t + \Delta t), s(t + 2\Delta t), \ldots, s(t + n\Delta t)\}$. This produces an $n \times n$ square matrix, **D**, n being the number of points under study. If this distance is lower that a predetermined cut-off, ε, the pixel located at specific (i, j) coordinates is darkened. To build the Recurrence Plot we use the following parameters: $\Delta t = 1$, $n = 1$ and $\varepsilon = 0.01$. Where n is the dimension of the time series and ε is the value that allows to differentiate the cluster (see Fig. 96.2).

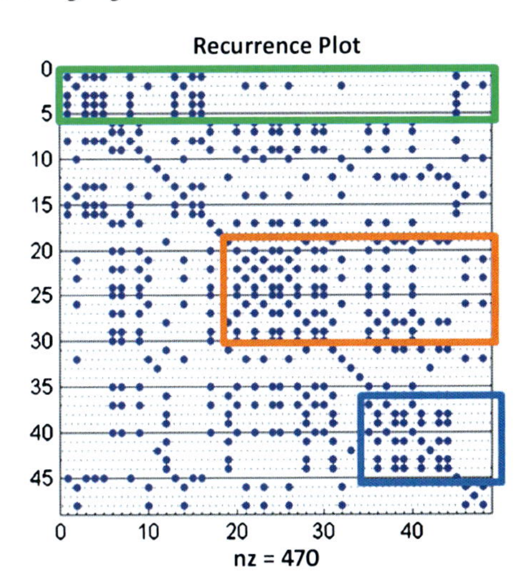

Fig. 96.4 Recurrence plot of the time series. In *green*, *red* and *blue* some big TS clusters and their evolution

With the Recurrence Plot we can identify again the same clusters of time series as one can see looking to the Fig. 96.4 in which the color of the boxes represents the cluster identified. For example, in the case of the green boxes we can observe that the bars with more points are corresponding to the papers 1, 3, 4, 5, 8, 13, 15, 16, 45, then exactly to the paper of green cluster (see Table 96.2). An advantage of this representation is that, since from left to right we find more recent papers, we can observe that the green cluster is started with the papers 1, 3, 4, 5 and then, after some time evolved adding papers 8, 13, 15, 16 and, only recently, 45.

In the same way the RP allows to analyze the evolution on one single paper regardless of the cluster to which it belongs.

96.10 Conclusions

In this paper we presented a new clustering method (TS) for citation networks. The method consists in building distance matrix where two papers are considered closer the more share the same sources. Subsequently to this matrix has been applied a method of multidimensional scaling and the projection along the main eigenvector of the resulting matrix has been considered. We have been grouped the papers according to their projection along the eigenvector. The eigenvector with respect to the number of its components is a time series as the component number is related to the year of papers publication. We see also that the Recurrence Plot of this time series allows you to see the temporal development of the different clusters. We applied TS clustering to a citation network on sustainability indicators in supply chains and we compared it with the method of [13] (PL). The TS clustering was more efficient than PL (with default parameters), managing to split the network into more subsets. Obviously, by changing its default parameters also PL could create more partitions, however, the parameters of PL are difficult to establish while in TS it was not necessary to define any. Another advantage of TS is that it allows the representation of the network through a Recurrence Plot and then the display of the time evolution of the clusters. It should be noted, however, a possible limitation of this method: the TS clustering is effective when the main eigenvalue is larger than the others and therefore when it is sufficient to describe in one dimension the nodes of the network. A future development of this work may be to assess the reliability of the method compared with the size of the main eigenvalue.

References

1. Bommarito MJ, Katz DM, Zelner JL, Fowler JH (2010) Distance measures for dynamic citation networks. Physica A 389:4201–4208
2. Brent AC, Visser JK (2005) An environmental performance resource impact indicator for life cycle management in the manufacturing industry. J Clean Prod 13:557–565
3. Cholette S, Venkat K (2009) The energy and carbon intensity of wine distribution: a study of logistical options for delivering wine to consumers. J Clean Prod 17:1401–1413
4. Colicchia C, Strozzi F (2012) Supply Chain Risk Management: a new methodology for a systematic literature review. Supply Chain Manag. doi:10.1108/13598541211246558
5. De Nooy W, Mrvar A, Batagelj V (2005) Exploratory social network analysis with Pajek. Cambridge University Press, Cambridge
6. Donner RV, Zou Y, Donges JF, Marwan N, Kurths J (2010) Recurrence networks—a novel paradigm for nonlinear time series analysis. New J Phys 12:033025
7. Eckmann JP, Kamphorst JO, Ruelle D (1987) Recurrence plots of dynamical systems. Europhys Lett 4:973–977
8. Fortunato S (2010) Community detection in graphs. Phys Rep 486:75–174
9. Hirata Y, Horai S, Kzuyuki A (2008) Reproduction of distance matrices and original time series from recurrence plots and their applications. Eur Phys J Spec Top 164:13–22
10. Lee KH, Cheong IM (2011) Measuring a carbon footprint and environmental practice: the case of Hyundai Motors Co (HMC). Ind Manag Data Syst 111:961–978
11. Michelsen O, Fet AM, Dahlsrud A (2006) Eco-efficiency in extended supply chains: a case study of furniture production. J Environ Manag 79:290–297
12. Newman ME, Girvan M (2004) Finding and evaluating community structure in networks. Phys Rev E 69:026113
13. Pons P, Latapy M (2005) Computing communities in large networks using random walks. In: Computer and information sciences. Lecture notes in computer science, vol 3733, pp 284–293
14. Ramani K, Ramanujan D, Bernstein WZ, Zhao F, Sutherland J, Handwerker C, Choi JK, Kim H, Thurston D (2010) Integrated sustainable life cycle design: a review. J Mech Des 132:091004
15. Schmidt M, Schwegler R (2008) A recursive ecological indicator system for the supply chain of a company. J Clean Prod 16:1658–1664
16. Seuring S, Muller M (2008) From a literature review to a conceptual framework for sustainable supply chain management. J Clean Prod 16:1699–1710
17. Sorrenti A (2012) Literature review on supply chain sustainability measures using citation network analysis. Thesis
18. Strozzi F, Poljansek K, Bono F, Gutierrez E, Zaldivar JM (2011) Recurrence networks: evolution and robustness. Int J Bifurc Chaos 21:1047–1063
19. Tsoulfas GT, Pappis CP (2008) A model for supply chains environmental performance analysis and decision making. J Clean Prod 16:1647–1657

Chapter 97
Bio-inspired Political Systems: Opening a Field

Nathalie Mezza-Garcia

Abstract In this paper we highlight the scopes of engineering *bio-inspired political systems*: political systems based on the properties of life that self-organize the increasing complexity of human social systems. We describe bio-inspired political systems and conjecture about various ways to get to them—most notably, metaheuristics, modeling and simulation and complexified topologies. Bio-inspired political systems operate with nature-based dynamics, inspired on the knowledge that has been acquired about complexity from natural social systems and life. Bio-inspired political systems are presented as the best alternative for organizing human sociopolitical interactions as computation and microelectronics-based technology profoundly modify the ways in which humans decide. Therefore, weakening classical political systems. For instance, dwindling top-down power structures, modifying the notion of geographical spatiality and augmenting the political granularity. We also argue that, more than a new theoretical proposal, bio-inspired political systems are coming to be the political systems of the future.

Keywords Metaheuristics · Modeling and simulation · Non-classical topologies · Complex network structures · Political granularity · Sociopolitical self-organization · Political regimes

97.1 Introduction

Human social systems are complexifying and it is becoming more difficult to frame and control them. At least, not through the traditional models of classical science. Bio-inspired political systems (BIPS) are an evolution of classical political systems. They are political systems based on the properties of life that self-organize the increasing complexity of the interactions among individuals and human social systems. In them, decision-making process follow metaheuristic algorithms inspired on nature, are tested via agent-based modeling and simulation, and are implemented by means of non-classical topologies. Bio-inspired political systems are an alternative to classical political systems because the latter are unable to handle the increasing complexity of the sociopolitical interactions in human social systems. We state, however, that more than an alternative, they will be, eventually, an emergence of the many shortcomings of classical political systems. Most of the limitations of the

T. Gilbert et al. (eds.), *Proceedings of the European Conference on Complex Systems 2012*, Springer Proceedings in Complexity, DOI 10.1007/978-3-319-00395-5_97,

latter relate to the global view that classical political systems pretend to have of the systems they try to control. Classical political systems base their decisions in the false assumption that it is possible to have a perfect knowledge about all the behaviors, individuals, elements and interactions that conform human social systems or that play a role in political systems. Apart from that, they are convinced that in problem-understanding, decision-making processes and decision-implementation, linear structures and mechanisms are sufficient enough to reflect, handle or describe the mentioned interactions This belief leads classical political systems to institutionalize human sociopolitical interactions by means of political regimes with tree topologies.

The institutionalization of politics is what the Greeks called *politiké* [1]. It is a narrowed conception of politics because it is directly linked with the problem of governability, so it leaves besides many aspects of the public space which are not necessary institutionalized and form part, beyond institutions, of the political dimension of human social systems. Among them are ethical, philosophical, economic, administrative, religious, scientific, educational, aesthetical or social aspects. The *politéia*, the sum of the latter, is politics as a worldview [1]. It is where this paper stands for formulating the critique to the characteristics and properties of classical political systems—whether representative democracy, monarchies, dictatorships or others-, and classical political regimes. The critique starts from stating that the topologies of classical political regimes do not reflect the complex nature of the topologies of human sociopolitical dynamics and neither their decision-making processes evolve in accord with sociopolitical interactions. This is not a surprise: institutionalizing the complexity of human social systems entails reducing normalizing and standardizing it by means of linear topologies and decision-making processes. This makes classical political systems incompatible with organizing complexity and, even less, with harnessing it. Harnessing complexity means to "explore how the dynamism of a complex adaptive system can be used for productive ends—instead of eliminating complexity" [2]. This could be done by means of political systems with more organic topological aspects and decentralized decision-making dynamics—biologically motivated.

A political system is the aggregate of decision-making processes in a social system—human social system. Political regimes are the institutional scaffolding and rules of a political system. Decision-making processes, individuals, elements and relations in political systems are so diverse, vast and non-linearly interconnected that there is no reason for political systems to be expressed and planned in such a non-complex way through classical political regimes—and, even less, through their tree topologies. The institutionalization of politics is, however, the current state of things. Such state is being left behind as technology complexifies the means in which humans interact at an accelerated rhythm. For instance, providing easier, faster and cheaper ways to trade, communicate and travel—physically or virtually. The aim of this paper is to open the quest to engineering bio-inspired political systems, using the same platforms that are making the interactions within and among human social systems more complex and, at the same time, uncontrollable by top-down political structures. Namely, via computers, computation and, ultimately, microelectronics-based technological advances. The type of engineering we refer here is complex systems engineering [3]. The latter is interested in uncertainty, evolvability, adaptability, resilience, robustness, self-organization—among others, instead of prediction, stability, reliability and centralized control [4].

Computation refers to information processing in computers (classical computation), computational systems (as Internet) and physical or biological systems (natural computing). Probably the most common example of computation among human social systems is Internet. Since the invention and later popularization of computers, there have been many social changes and advances related to how the processing of information among human social systems has evolved. On one side, more powerful computation has allowed getting people closer despite geographical distances, it has helped get faster and more comfortably from one place to another, and has accelerated the propagation of ideologies and ideas among groups and societies. On the other side, it has helped gain knowledge about the complex world in which we live, studying phenomena and behaviors that used to be a mystery, misunderstood or unknown—which is what spearhead science and engineering are doing nowadays. For instance, bio-inspired algorithms used in metaheuristics have helped find better ways to solve problems, thanks to the creation of models that imitate how living systems develop, evolve and interact with their environment [5]; farther more, modeling and simulation has widely benefit the learning about computation in emergent dynamics, dynamics of self-organization and collective intelligence in living systems [6]; and, besides, the discovery of fractal geometry [7] and complex networks has conducted to recognizing and understanding more clearly some of nature's structures and topological features. The three vias presented here (metaheuristics, agent-based modeling and simulation and complex topologies) are supported in the possibilities that computing provides. Computing is the main tool for engineering bio-inspired political systems, but it does not mean it should be the only one.

The claim that there is a tendency towards bio-inspired political systems and the description of the ways to get to them is studied into six sections. Firstly, the idea of bio-inspired models is contextualized. Secondly, bio-inspired political systems are introduced and some important remarks related to their design and engineering are marked as substantial. This leads to the study of the first, second and third order relationships between the three selected vias in which bio-inspired political systems can be engineered. Fourthly, some background elements of bio-inspired political systems are mentioned, showing how classical political systems and their structures are being affected, giving rise to a tendency towards more organic political systems. Fifthly, some possible implications of bio-inspired political systems are grasped. Finally, the paper concludes with several important remarks related to engineering bio-inspired political systems.

97.2 The Shift Towards Bio-inspired Models

A model is an abstraction (simplification) created to understand a system or phenomenon. Models should be as similar in structure to the *objects*, phenomena, be-

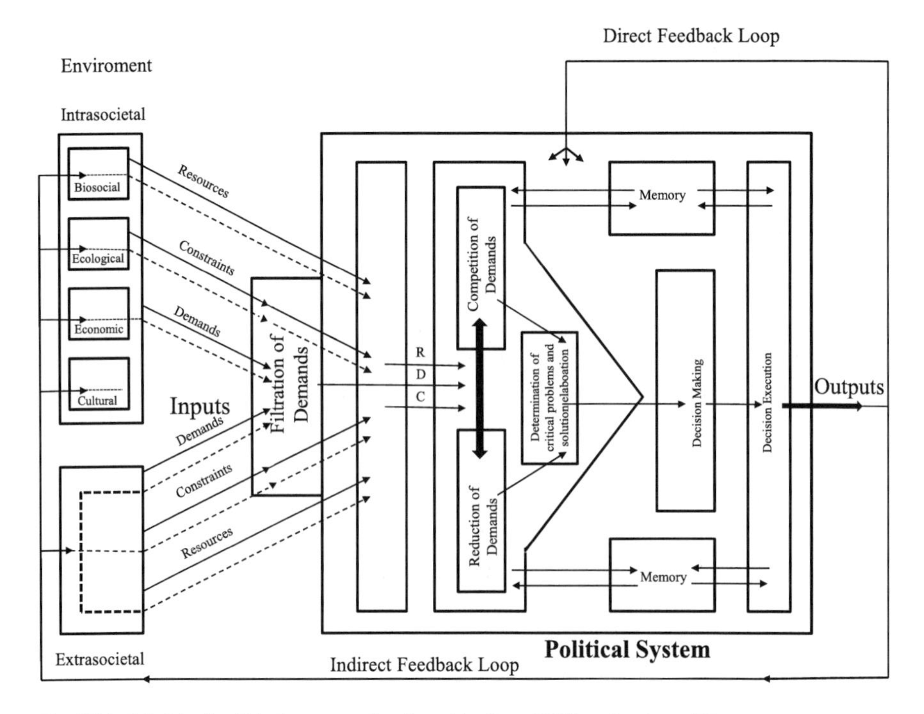

Fig. 97.1 Model of political systems developed by Jean-William Lapierre [9]

haviors, systems or problems that are being modeled [8]. For a long time, classical physics was the base for models in science—even political science. The result was simplistic models focused on analysis, control, predictability, rigidness, determinism, stability, equilibrium, certainty, centrality, reliability. That is, linearity. Classical political systems—and classical political regimes- are examples of physics-based models in political science because they strongly focus on the properties mentioned above. They both are reductionist approximations to the complexity of human social systems. This makes them to be designed with hierarchical centralized control mechanisms, cause-effect dynamics and top-down imposed normativity.

Figure 97.1 shows a model for political systems shared by the mainstream of political science. It was developed by Jean-William Lapierre [9], based on David Easton's model (see [10]). One of its many shortcomings is that it is conceived as a deterministic cause-effect system, where dynamics are understood as a linear sum of decision-making processes. With no doubt, we can claim that the model could have based on the theoretical implications of Newton's laws of motion. In general terms, the model errs in trying to schematize human sociopolitical dynamics from a non-complex point of view. A possible reason for this can be that the model was developed following a *general systems theory* perspective [9]. Correspondingly, it assumes a perfect knowledge about all the parts and interactions involved in the decision-making processes of political systems.

It is not a secret that if we are referring to complex systems not even knowing all the elements and interactions we can talk about determinism or perfect knowledge. When referring to political systems, this is also impossible. Political systems are imposed over human social systems and the interactions in them involve such complexity that not even when institutionalizing them by linear mechanisms trough classical regimes their complexity is eliminated. Maybe the farthest that this analytical and systemic models have reached has been to cooptate some of the traditional tools of the sciences of decision such as system dynamics, decision trees, real options and portfolio management [11]. Anyhow, understanding the *black box* of political systems assuming linear relations should never be tried to be done again.

Instead of physical-based models, and in contrast to the deterministic world showed above, human social systems—the systems that political systems try to organize trough political regimes—are feasible to be described by properties much closer to biology, such as evolvability, adaptation, uncertainty, emergence, self-organization, learning and synthesis. The paradigm of complexity are biological systems. More precisely, complexity has life as its core. For this reason, these properties are widely studied by the sciences of complexity. Therefore, for understanding and trying to organize human social systems, the best alternative is turning to complexity sciences. Notably, to bio-inspired models. This would benefit of the fact that complex systems engineering is interested at the organic properties mentioned above for generating close to real life models too. Complex systems engineering recognizes that bio-inspired engineering is a way to show how engineering is complexifying [4]—as our world complexifies as well. This, a union between political systems, bio-inspired models and complex systems engineering sustains in these motives.

The sciences of complexity have developed models, theories, concepts and tools for approaching non-linear—complex- behaviors, phenomena or systems. A great part of the most recognized models that complexity works with are more organic than those of classical science—without this meaning that they are more complicated. This allows a better comprehension about the systems that exhibit life-like behaviors. Among them, human social systems. We claim that the apprehension of complexity in political systems and political regimes is the best way (maybe the only one) to organize and harness the increasing complexity of human social systems and their interactions. It entails engineering bio-inspired models that replace classical ones and that reflect better the structures of human sociopolitical dynamics than those referenced the most in the study of politics and the political.

Every discipline among the human and social science studies complex systems. Many of them have already apprehended complexity and have given a certain shift towards bio-inspired models, although not necessary their mainstream. Economy recognized the chaotic, non-linear and self-similar nature of markets behaviors [12]; sociology acknowledged the complex adaptive nature of human social systems [13]; history recognized that it is not a sum of facts from the past, but a non-linear and open system that considers even facts that never happened [14] and, finally, the algorithmic complexity of art was recognized, enabling to think about the scopes of measures for the complexity of artistic pieces and their relation with subjective experiences [15].

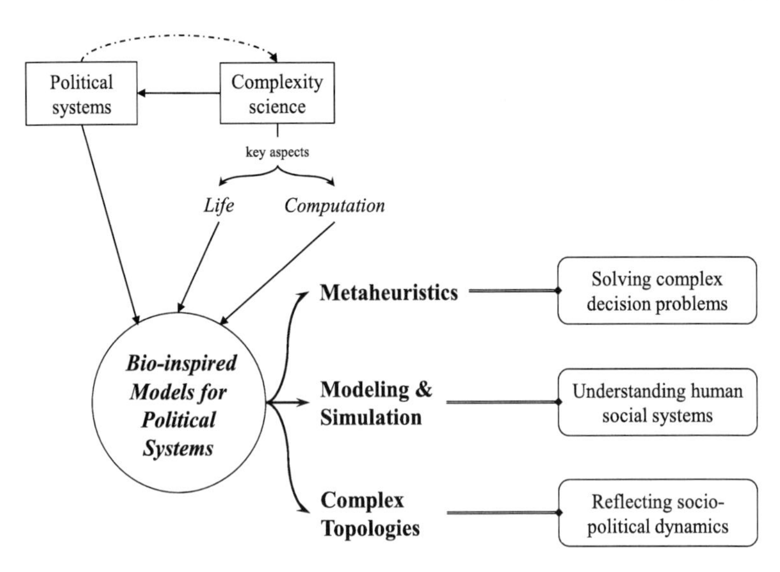

Fig. 97.2 Bio-inspired models

The most recent models they now use have life as the ground for building their explanations. In political science, however, this is not the case. Political Science studies some of the most complex systems that exist on earth—individuals, groups and human social systems—and, despite it, it is one of the disciplines among the human and social sciences that when working with complex systems has not completely recognized complexity. Hence its attachment to the classical realm of science. For approaching complex phenomena, classical models are non-viable anymore. In fact, they have largely demonstrated a wide spectrum of limitations for times of increasing complexity [16].

Figure 97.2 points to how political systems need be complexified turning to biologically-inspired models that present life-like behaviors—in this case, for understanding better the behavior of complex systems, finding better solutions to complex problems and designing political structures that reflect complexity. The reason is that life is the phenomenon that (i) presents most complexity, (ii) harmonically manages to self-organize, and (iii) harness complexity the most.

Bio-inspired models do not necessary have to be complex models. They can be very simple and still be complexity-based. The importance relies on the complexity of the dynamics they describe. In any case, it is important to bring up that the quality of life of a human social system largely depends on the complexity of its political system. Complex models are the best known road that can be selected for thinking about models for political systems—particularly bio-inspired models. As it will be shown, an advantage of the latter is that in the sciences of complexity, the latter are computational.

As explained in [17], the interest on life has been addressed from early philosophers to contemporary scientists, either interested in describing life, the nature of life or life's hallmarks. Originally, life was seen as a binary state: *something* was either alive or dead. Some contemporary approaches center their attention on whether the difference between the living and the non-living relies on composition, structure, function or a combination of the three. Ultimately, the differences between the living and the non-living are qualitative, in terms of degrees and of organization [18]. Computational models that study life-like behaviors have greatly contributed to this conception because in the middle of the living and the non-living there are computational bio-inspired models that have life-like behaviors. Additionally, they have taught us that the living properties of single individuals can be extrapolated to their social systems and, from a Darwinian point of view, we can state, for instance, that populations of human individuals—human social systems- present life-like behaviors. Therefore, it is valid to study them by using bio-inspired models that describe their life-like dynamics.

97.3 Engineering Bio-inspired Political

Engineering *bio-inspired political systems* (BIPS) implies to design *non-classical* and self-organized political systems where:

(a) Decisions are the result of metaheuristics processes.
(b) Comprehension and explication of sociopolitical phenomena are the result of (agent-based) modeling and simulation.
(c) Bio-inspired topologies of political regimes are the result of complex network structures.

Engineering these types of models is important because political systems face problems about human social systems. However, we must remember that in many situations, the problems involve humans but also other species in the planet or the biosphere itself. As a result, in some cases the scopes of the decisions taken by political systems entail negative bioethical, social, economic and political consequences. One explanation to this is that most of the time the decisions that are going to be implemented are not previously tested; the systems upon which political systems impose their decisions and environments are not well comprehended; and there is not a correspondence between the complexity of the affected systems and the linearity of the methods of decision-implementation. The best way to overcome these shortcomings is by approaching towards the engineering of models much closer to the complex nature of the systems affected by political systems. Bio-inspired models come as a substitute of classical models for political systems and regimes because there are some costs for maintaining the hegemony of them (lives, extinctions, economic or social consequences...) that should never be assumed anymore by any individual, species or population in the planet.

We decided to focus only in three ways for engineering BIPS: metaheuristics, modeling and simulation, and complex topologies. Their utility relies on their practical, theoretical and conceptual relevance, but it does not mean that new *teqniques*

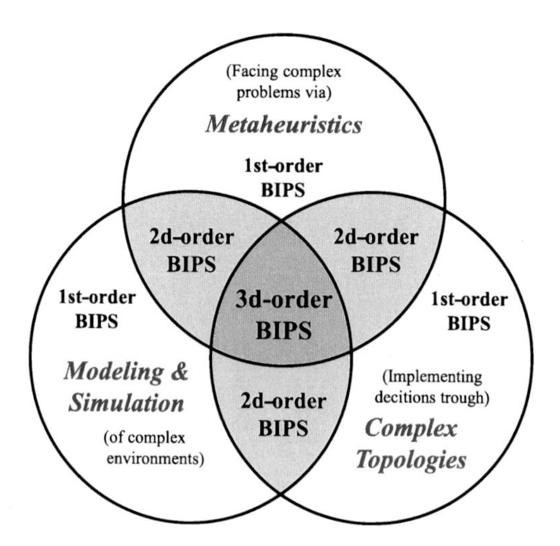

Fig. 97.3 Bio-inspired political systems: first, second and third order relationships

could not be incorporated to their logic in the coming years. The three vias mentioned above, when taken separately, are first-order models. The interaction between two of them, in any direction, (a • b, a • c, b • c) form second-order models, which we named basic hybrid models. And the interplay between the three—or more—conform third-order models and are the long-term desired scenario we think is needed for letting human social systems self-organize without the need of any imposed or elected ruler; or top-down system. Figure 97.3 shows the possible interactions of first-order, second-order and third-order bio-inspired models. In the following paragraphs we explain with more detail the role of each road for engineering BIPS and at the end of the section the possible relationships among them.

97.3.1 Using Metaheuristics for Decision-Making Processes

Political systems face problems by taking decisions upon phenomena that concern various kinds of complex systems, apart from human social systems. Consequently, the problems that political systems try to solve are complex problems. Metaheuristics are a tool for solving complex problems. They are crucial in bio-inspired political systems because of the complexity that characterizes the systems upon which they are imposed. Certainly, human social systems require better methods for problem-solving than those provided today by mere intuition of governors and based on analytic and reductionist methods.

Indeed, because "the main task of management (in political systems) consists of optimal decision-making" [19], the problems that political systems try to solve can be understood as optimization problems. I.e. problems that look for finding the best decision (which is not necessary the optimal) in a given moment.

Given the non-trivial nature of the problems, they should not be tried to be solved by decisions taken without rigor—theoretical, conceptual or ethical. However, decisions in political systems depend on the decision capacities of governors and the traditional methods they usually work with. That is, their own personal interests—which are not always oriented towards the general welfare—and limited tools of classical science. Theoretically, *they being there* means that they could solve and find the best alternative to any problem they face, thanks to the *unique* and in anyway incipient swarm-like collective intelligence of their voting and deliberation processes. In reality, having top-down methods for problem facing and individuals with their personal interests for deciding upon a global view of a system is not enough. Ergo the task and the role of governors are actually naive. This does not mean that complex problems in human social systems are always going to be faced by without rigor. Notably, the use of metaheuristics can help finding solutions close to the optimal.

Metaheuristics are computational tools for resolving complex problems regarding optimization and prediction—problems that cannot be elucidated by traditional analytical methods. Metaheuristics can be described as "general-purpose algorithms that can be applied to solve almost any optimization problem" [5]. They allow optimization under uncertainty contexts.

There are some metaheuristics that are physic-based, mathematics-based, biology-based and ethology-based [20]. The ones in which we are interested in are population-based and biologically motivated. They use algorithms inspired on natural phenomena or in the way in which some species solve problems, translating the processes into general frames that can be used for modeling various kind of complex phenomena. Metaheuristics start from considering that for any problem there is a defined space of multiple solutions. Population-based metaheuristics randomly search in the space of solutions and combines the best solutions between them, so in each generation the robustness of solution increases. It can be said that despite the different bio-inspired algorithms in metaheuristics, the basic metaphor is evolutionary.

This ways of finding solutions to problems is steps beyond how political systems do it nowadays. The following is a list with the main families of metaheuristics.

- Neural computation
- Evolutionary computation
- Swarm intelligence
- Inmune computing
- Membrane computing

The reason why complex problems should be faced by complex tools is because there are problems that (a) can have infinite solutions, (b) have dynamic solutions that exist in time-changing environments (c) are constrained and obey to restrictions, and (d) are based on contradictory principles because there can be many possibly conflicting objectives [21]. In other words: first, there are always missing pieces for the puzzle; second, there are pieces of the puzzle that fit in an x time, but not in a y time; and third, if a piece fits (in an x time), other pieces disengage (in the same x time o in a y time) [22]. These are complex problems. In them, finding *a* solution, *the* ultimate answer is impossible, or it would take millions of years if not an infinite computational time to be calculated. That is why the answer to life, the universe and everything is not 42.[1]

As metaheuristics will become more used by engineers and decision-makers [5], they should be promoted as one of the best tools there are for optimizing complex problems. We claim that in the future the magic of decision-makers will rely on whether they know how to translate a problem of political systems in terms of metaheuristics optimization.

97.3.2 Agent-Based Modeling and Simulating Bio-inspired Political Systems

"Agent-based modeling is a recent simulation modeling technique that consists on modeling a system from the bottom-up, capturing the interactions taking place between the system's constituent units" [23]. Agent-based modeling (ABMS) was born in the context of artificial life (AL), which creates synthetic life on computers that exhibit life-like properties and behaviors [24]. For the process of modeling, the system is understood as a collection of components (agents, parcels) that nonlinearly interact and give rise to emergent patterns and behaviors that cannot be directly traced back, simply, to the properties of the parts taken separately. ABMS can be for specific or general uses and can have strategic, tactical or operational domains [25]. By defining a set of basic rules, we can observe how patterns start to emerge bottom-up. By viewing the evolution of a system along generations, we can have deep insights to the comprehension of complex systems. This is something that classical modeling techniques lack because they fail in being able to work with nonlinearity [25].

In sum, according to [6], "in agent-based modeling, a system is modeled as a collection of autonomous decision-making entities called agents. Each agent individually assesses its situation and makes decisions on the basis of a set of rules". Agent-based modeling and simulation (ABMS) becomes a useful mindset when [6]:

- Agents exhibit complex behavior
- The interactions between the agents are nonlinear, discontinuous, or discrete.
- The topology of the interactions is heterogeneous and complex.
- The population is heterogeneous and each individual is (potentially) different
- Space is crucial and agents' positions are not fixed.

Political systems can benefit of ABMS because they are expressed in organizations and institutions—which are "often subject of operational risk [6]", and organizations are one area of application of ABMS. In this light, ABMS can help political systems by understanding the complexity of the sociopolitical dynamics

[1]Here we refer to the science fiction movie directed by Garth Jennings, The Hitchhiker's Guide to the Galaxy, where, in an ironic manner, a computer built by pan-dimensional beings calculates 42 as the answer to life, the universe and everything.

having place in and upon them. But it can also help with internal (organizational)—sometimes topological simulation. Thereby, lowering the impact of the operational risk of political systems due to the valuable information that ABMS provides about the behaviors of modeled agents and decisions in complex systems.

Political systems have always implemented—and are implementing, still—sets of decisions taken without even proving if they are good solutions. As a result, many times the decisions that are supposed to be favorable for a population result, instead, in negative outcomes and effects. This dues to the fact that nor the problem or the social system are fully comprehended. So decisions, in most cases, are arbitrarily imposed. For this we can say that almost every decision that classical political systems impose upon human and natural social systems are experiments with the real world, where governors are the scientists and the world is the laboratory.

Not testing the decisions that will further on be executed usually takes to two types of negative outcomes. Fist, aspects related to time or treasury and, second, and more importantly, bioethical results, such as the loss of lives, killing bio-diversity, augmenting poverty or polarizing more the world. With ABMS many of these negative outcomes can be avoided because of the gained comprehension about the systems that concern decisions and where decisions are implemented. ABMS is the best alternative so far for creating simulated environments that help testing solutions and decisions without having real-life individuals are guinea pigs.

The scopes of ABMS will bring questions about the role of future politicians as decision makers if ABMS continues to become more proficient at decision-making processes than governors. However, while this fully occurs, ABMS should be more used by public decision-makers, helping to anticipate potential outcomes or implementing—better informed- decisions [25]. Experimental proves in artificial life help narrow error margins when implementing a model or solution in real life.

In an on-going research using agent-based simulations, we manage to synthetize self-organized control mechanisms that adapt over time with changes in the environment, using only local information. Our quest is to find out whether human sociopolitical interactions, when they are not mediated by institutionalism, succeed on making coordinated patterns to emerge. Everything indicates that when defining basic elements in the base of the social system, it is plausible that human social systems give rise to adaptive and intelligent collective *swarm-like* behaviors. Unquestionably, ABMS is the best way to gaining comprehension about emergent behaviors in complex systems. Political systems need to appropriate of their use.[2]

97.3.3 Thinking Complex Topologies for Political Regimes

Topologies refer to the distribution of nodes in a network. Tree topologies (Fig. 97.4) are the structural models for institutions in classical political systems. That is, for

[2]Most of the cases where ABMS has been used in Political systems have been for activities related to military and war purposes.

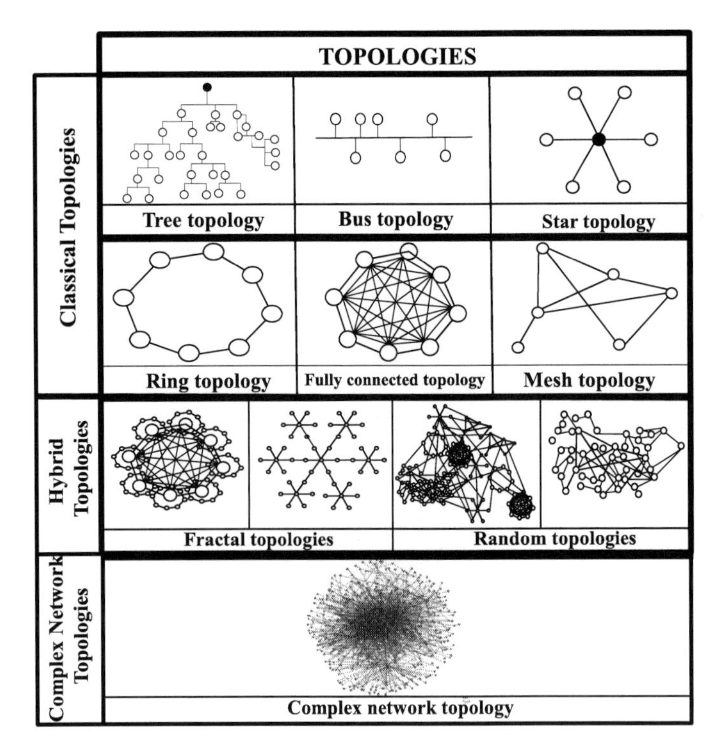

Fig. 97.4 Classical, hybrid and complex topologies

classical political regimes. Among the classical topologies presented in the Fig. 97.4 tree topologies are the ones that represent complexity the least. They are suitable for imposing restrictions to complexity because they have centralized control executed by means of a node in the top of the structure. This node *is* aware of all the information going throughout the system and it can be an emperor, a king, a queen, a prime minister, a president, a dictator, a parliament or a congress. Basically, any individual or group in charge of the direction of the decision-making processes in classical political systems.

Tree topologies for political regimes are obsolete for times of increasing complexity. It is not plausible for a single node in the top of a political structure to continue trying to have global information about all the dynamics of the complex systems over which it imposes upon. Therefore, organizing the complexity of human social systems should not be done by top-down methods, but bottom-up synthesis. In that way, complexity can be organized better and can be harnessed too. Nevertheless and despite that this is actually how sociopolitical interactions occur in human social systems (by bottom-up synthesis), the mainstream of science has been permeated with the idea that sociopolitical interactions must be top-down controlled.

Bio-inspired political systems recognize the importance of the topological properties of interactions for computational purposes, being some topologies more suitable for better information processing than others. We claim that the topologies of

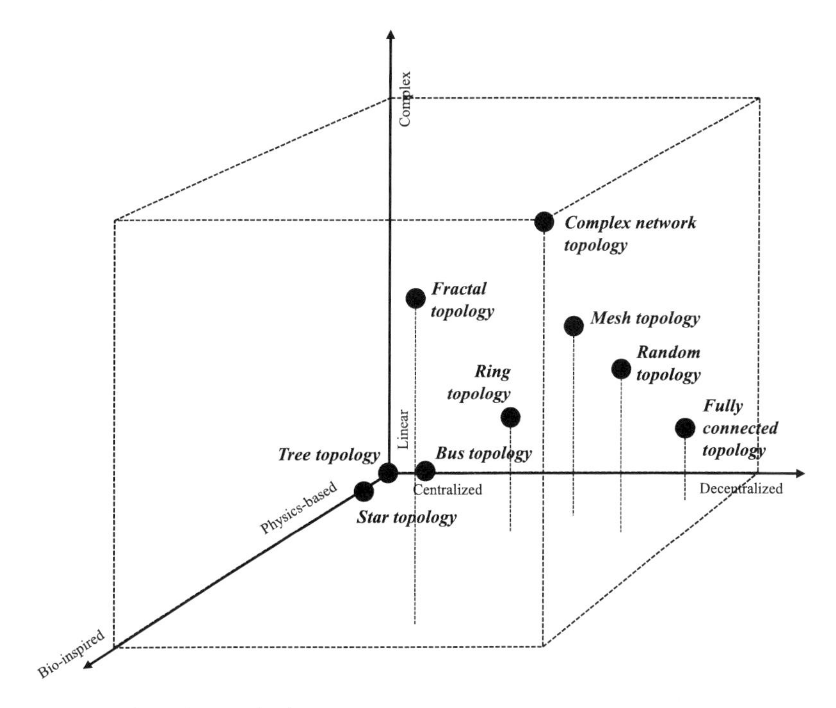

Fig. 97.5 Topology characterization

political regimes need be based on models of complexity because that is the nature of the systems they organize. A correspondence between the physical operations of the structures of political regimes and the logical structure of human social systems is needed, for them to reflect the computational structures (information processing) in human social systems. That is, they need to be isomorph or, even better, merge with sociopolitical interactions.

Figure 97.4 shows various kinds of models for topologies: classical, hybrid and complex network topologies. Among the classical models, we present basic structures for tree, bus, star, mesh, fully connected and ring topologies. We included within hybrid models those topologies with fractal structures and random ones, conformed either by single nodes stochastically distributed or models formed by other classical topologies, but different from tree topologies.

Figure 97.5 characterizes topologies. It has three axes. In one extreme of axis Z we located the property of being physical-based for topologies models and in the other extreme those more biologically motivated. In this case, tree topologies are characterized as the most physically-based, whereas complex network structures are presented as the more biologically-motivated. Axis X corresponds to how centralized or decentralized a topology is. Again, tree topologies, in this case, together with star topologies, are in one extreme—the most centralized. Axis Y goes from the most linear to the most complex. For this case, complex network structures are situated as the most complex of all and tree topologies are among the most linear.

Most of the times, in the middle of the axes we located hybrid models and the rest of classical topologies. Even the latter are preferable than tree topologies for the structuration of the regimes of political systems because, despite that their natural emergence is highly improbable, they are more decentralized than tree models.

The structures of the regimes in political systems must reflect the structures of the complex sociopolitical dynamics over which they are imposed upon. Complex political topologies are key in bio-inspired political systems because they are the result of sociopolitical dynamics synthesized bottom-up, which imply a better organization of the interactions between individuals, human social systems, among them and with their environments.

97.3.4 Second and Third-Order Bio-inspired Models

As Fig. 97.3 shows, second-order models correspond to the interactions between two of the three first-order models: (a) population-based metaheuristics, (b) agent-based modeling and simulation, (c) complex models for political topologies. That is, we can combine metaheuristics and modeling and simulation, metaheuristics and bio-inspired topologies or bio-inspired topologies and modeling and simulation. However, there will be times in which one might prevail over the other, which means that for second-order models, there are actually six combinations instead of three, as following.

(1) (a/b): When facing problems of optimization, the result of population-based metaheuristics can be tested in simulated environments before implementing them in the real world.
(2) (b/a): Agent-based models can be enriched by metaheuristics and, in that way, we can have models much closer to reality.
(3) (c/a): The more complex a topology is, the more it becomes a favorable environment for being receptive towards solutions and the logic itself of metaheuristics coming from non-mainframe power concentrators—or governors.
(4) (a/c): Metaheuristics can serve as a parameter for designing and deciding which topology to implement in which case for organizing certain sociopolitical dynamics.
(5) (b/c): The local information of ABMS dynamics would reinforce the processes of looking for topologies in congruence with the complexity of human social systems. Modeling and simulation could help the morphogenesis of political systems, finding topologies that actually reflect the structures of the sociopolitical dynamics over which they are imposed. It serves for seeing and proving the functioning of each topology.
(6) (c/b): Decentralized topologies are suitable spaces where the results of agent-based modeling can be taken into account because there would not be a central control deciding upon how a system behaves.

On the other hand, third-order models are relationships between the interactions of three—or more—models. Third-order bio-inspired models are the most plausible way for avoiding the deviations that occur when politics is institutionalized. They would be fully self-organized sociopolitical interactions with synthesized adaptive and self-organized bottom-up *control*. However, promoting self-organized models for organizing social systems in a world that has always been controlled top-down would imply non self-organized ways or anarchic spaces for it to occur.

Non self-organized vias for changing a political system or regime mean violent mediums, as the stories of most of political revolutions have proven. The other extreme, although pacific, would take too long. We would have to wait until the top-down structures of classical political systems continue to progressively weaken by the effects of technological advances in humans' exchanges of information. The price to pay would be that it will be extremely late for the wellbeing of some human groups, their habitats and for other species that inhabit them, and that are currently been affected by the decisions taken in the structures of classical political systems.

A good solution to accelerate the existence of third-order models would be to find an intermediate state between self-organization and design. Self-organization could be induced by means of a guided self-organization. I.e., engineering systems that tend to self-organize their dynamics. The, apparently, contradiction between both methods (design as planning and self-organization as non-determinism) was solved by Prokopenko [26], who found that combined both could lead to a point where self-organization is at the base of the desired (designed) dynamics. In the case of bio-inspired political systems, using metaheuristics for facing complex problems, modeling and simulating the environments where those problems take place and having complex topologies were decisions are bottom-up implemented would be the parameters that *guide* the self-organized dynamics of the system.

97.4 The Complexification of Human Social Systems: Bio-inspired Political Systems' Background

Bio-inspired political systems will be an emergence of the interconnection of some phenomena that are non-linearly transforming human social systems. Independently of the engineering of BIPS, there is, indeed, an increasing tendency from classical political systems towards more organic ones. The reason relates to some phenomena that are occurring in contemporary world and that are deeply transforming political systems by moving them away from how they have always been. Some examples are: the weakening of traditional power structures in top-down political systems, the reinvention of the local as the confluence centers for sociopolitical interactions or the propagation of ideas crossing artificial national boundaries. These facts serve as background for supporting the idea that, as the interactions between human, natural and artificial system get complexified, there is a tendency towards bio-inspired political systems. The following are some phenomena that, when combined, reinforce the tendency toward BIPS and the need of a field for their study.

97.4.1 The Network Society

As technology (microelectronics-based) complexifies the means trough which humans communicate—Internet, social networks, transportation, mobile phones, computers, and others, human communication modifies with it [27, 28]. This fact is in great part responsible for the small world phenomenon [29] that human social systems present. It is also the base of the concept of the network society [30], developed by Manuel Castells, according to which is "the social structure resulting from the interaction between the new technological paradigm and social organization at large" [31].

Network societies can be considered as an input for engineering BIPS because this phenomenon is decentralizing human interactions. Bounds are becoming more flexible and adaptive and are basing more on coordination than imposition. Of course, the network society implies technological basis for the communications of individuals, but not every individual—or political grain—in the world has access to them. However, we recognize that, along human history, the life conditions of individuals with less economic capacities has been increasing [32]. We hope that the same happens with the access to technology in the very long term.

There is something about sociopolitical relations in the network society that BIPS may influence in a positive way: "(the network society) excludes most of humankind, although all of humankind is affected by its logic and by the power relationships that interact in the global networks of social organization" [29].

This is supported by the fact that the network society brings up the phenomenon of *networked individualism* [31], which consists on how individuality becomes the center of social structures. Starting from there, individuals express their individuality by sociopolitical interactions facilitated by the connectivity in the network society. This leads towards public political spaces mediated by virtual interactions because technology provides more active and direct means for humans to connect to each other, to have voice, presence and action in the emerging non-geographical political arena [33, 34]. Thus, sociopolitical interactions facilitated by technology aim to work with less artificial geographic, migration, territorial, economic, ethnic, cultural, political or social restrictions. These types of societies claim for new political structures, better decision-making problems and better understanding of their problems.

97.4.2 Deterritorialization of Space

A great deal of the interactions taking place today in human social systems occur in non-geographical spaces; That is, by means of the Internet. An appropriate word is that they are *deterritorializing*, which means that every time it is becoming more difficult to link personal identities with a defined geographical territory or population. Identities are becoming more complex networks of experiences than clusters of traditions. This is explained by the fact that particular data about human social systems (religions, costumes, traditions, ideologies, fashions, styles, hobbies, etc.) is spreading more rapidly among them. Internet is a platform by means of which individuals can have access to *worlds beyond their own*. In that way, they can look for spaces and activities more related to their personal *wished* identities than to the geographical territory where they were born in. Two precisions must be brought out. First, complete deterritorialization will not happen fully unless Internet becomes affordable and accessible for underdeveloped and later-developed countries and not only for the so called developed nations. And, second, despite all the advantages, deterritorialization must not be a condition that should be pursued because it has many non-desired implications, for example, for the historic memory. Nevertheless, not pursuing it is not happening.

Deterritorialization comes after exponentially augmented information flows among human social systems. It implies that the group marks that used to facilitate tagging a population with certain symbols, shared features, generalized personal identities are no longer possible to be generalized. As a consequence, it becomes less easy to control human social systems by means of tree topologies because there are no generalized patterns that can serve as frames, such as adscription factors, symbols, anthems, flags, etc. or belonging to a defined geographical space. The relation between deterritorialization and bio-inspired political systems is that as the interactions in human social systems become more difficult to control by means of top-down mechanisms, they tend to self-organize and form complex sociopolitical networks or, at least, hybrid BIPS.

97.4.3 Finer Political Granularity

The political granularity measures the extension of the territorial parcels over which a political regime imposes its modes of organization upon the social systems that remain in a specific territory. The term *granularity* was extrapolated from molecular dynamics and engineering [35, 36]. For practical purposes, the parcels are called grains and their extension defines their granularity. The States' territories, provinces, regions, states and cities are examples of grains. Figure 97.5 summarizes three moments of the story of political parcels: empires, kingdoms and modern states. Every one of them exceeds its predecessor in energy consumption. The graphic shows how since the history of big empires, energy consumption has being in augment and, as this happens, the extensions of the territories over which the institutions of political systems impose their mandate have been decreasing. Thus, we can say that there is a non-linear tendency of progressively finer administrative and territorial political granularity going from coarse-grained parcels to fine-grained parcels.

Our *longue durée* approach [37] suggests that this counts even for the cases where political coarse grains are formed, at first, from political finer grained parcels; i.e. the EU conformed by states or big empires formed by smaller territorial organizations. Although it is possible that in the future some coarse-grained parcels that reunite finer-grained parcels will continue to emerge [38], it is highly unprobable that they

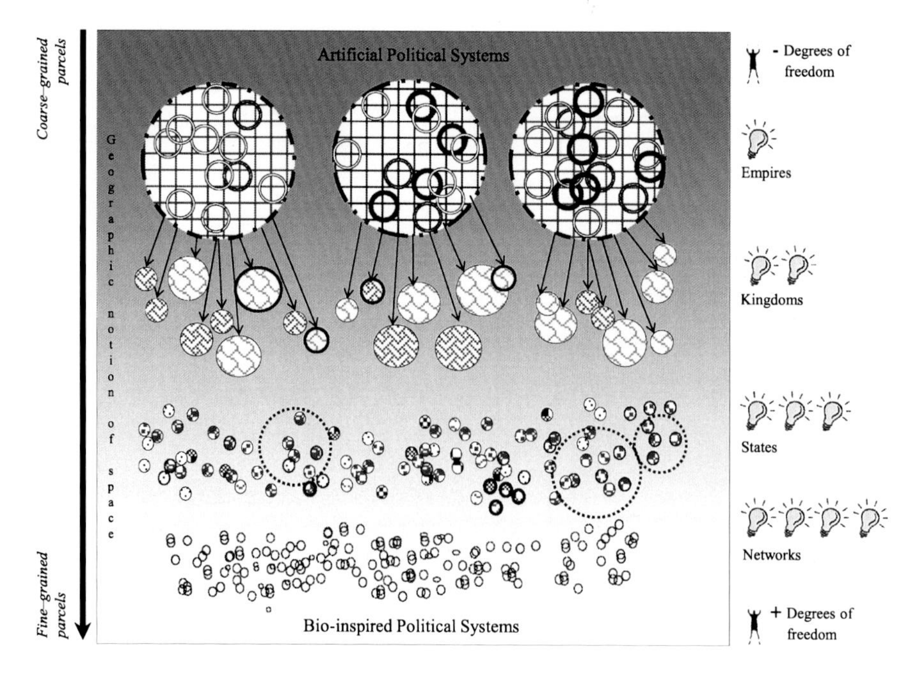

Fig. 97.6 The evolution of political granularity

will continue to have the control capacities they still have today due to the possibility of using physical violence—the basis of past and contemporary political power. This is explained by the idea that as human interactions will no longer only occur in a space with a geographical conception of territory, it will become more difficult to apply coercion over a population or territory, as it is becoming more difficult to frame them. In addition, parceling the globe's territory and management in finer-grained political parcels, reduces the capacities of mainframe power concentrators.

Finer political granularities lead towards bio-inspired political systems because as political grains become smaller, the importance of local information increases. Many contemporary examples of decentralized cooperation and trading among cities or between cities and national governments are proven to be highly effective, in comparison to nation-nation trading and cooperation [39]. Cities are political grains as well, and they cannot continue to be ruled by political institutions that operate in non-local levels, missing important information about the complex micro and local dynamics of the system.

When focusing on a geographical notion of space, finer political granularities mean non-centralized control. And, when it comes to non-geographical approaches for understanding territorial space, the idea of control, anyway, starts to vanish. Then, finer political granularities conduct towards BIPS, which are not control-focused. So the evolution towards finer political granularity leads to bio-inspired political systems. The link between the size of the grains and energy consumption shown in Fig. 97.6 suggests that there is a propensity towards finer political grains—formed by complex network structures.

97.4.4 Sociopolitical Self-organization

Sociopolitical self-organization consists on how political interactions among individuals bottom-up synthesize. This occurs without the need of any leaders, governors or top-down regimes. Sociopolitical self-organization is the *politéia* in its broadest sense. The contemporary expression of the phenomenon is being facilitated by the diffusion of technological means humans use to trade, travel or exchange, bypassing intermediaries, such as classical political systems and their regimes. Sociopolitical self-organization, besides being a continuously adaptive emergence of BIPS, as we previously stated, is also an antecessor of them. To the extent that humans can self-organize their sociopolitical interactions, hierarchical means and control are being left behind [27]. Together with this, the increasing diversity of human social systems, sociopolitical self-organization is pushing classical political systems to try to be more inclusive by becoming more open to diversity, which is one of the symptoms of a society that will collapse [40].

For instance, every time there are laws that try to regulate social aspects of individual's life, such as their sexual orientation or tendencies. With this, the hegemony of *unique* ways of thinking, living and manifesting individual identities are becoming a remembrance of the past. Regulating and legislating about such increased diversity in individuals personal and group identities by means of top-down structures is becoming more difficult—even more when individuals are recognizing their capacity to self-organize using social networks, not needing political intermediaries. Social changes are occurring at a rate so fast that when the formalisms in political systems are just starting to consider them, many new ones have already emerged.

In respect to social networks, the revolutions, insurgencies, marches and rebellion that are self-organized by using them are proving that the top-down structures of classical political systems cannot handle a world with increased complexity. Recent years provide many examples of sociopolitical self-organized dynamics, such as the Indignant Movement, the students' movement in South America and the Arab Spring [41]. The latter are emergencies that rose up against top-down political, economic and educative structures and regimes of classical political systems. There is nothing that points to the latter as contingencies. Rather, they must be considered as a tendency, as classical political systems continue to weaken and become openly incompatible with organizing and harnessing complexity.

Sociopolitical self-organization is also making political frontiers to become less rigid and more permeable. No longer a phenomenon taking place in a defined geographical space stays *in* there. It diffuses when it enters non geographical spaces (Internet), and afterwards it spreads again in geographical spaces. The international system is a space that feeds from self-organized geographical and non-geographical sociopolitical self-organized interactions. It is conformed by complex networks of international treaties and agreements, some of them voluntary and non-coercive, generated by diverse actors interested in cooperation, coordination, consensus and protection. The regimes act at micro or mezzo scales, such as cities, regions or states, but there are no generalized top-down regimes that apply globally for all the international system. The macro scale of it is increasing in importance and, as mentioned above, its contemporary basic inputs come from self-organized dynamics.

97.5 Implications of Engineering Bio-inspired Political Systems

A world where bio-inspired political systems have already been engineered would differ in many aspects from the world in which we currently live. Its economic structures would not be sustained by exploitation, being part of a collectiveness would not be defined by the adscription to a geographical territory, political regimes would not be something that is already set up before someone is born and to which individuals must be subjected, the *politéia* would not have to be subordinated to the *politiké* and, certainly, there would be better possibilities and hope for biodiversity on earth—at least for what remains of it. All these, however, require, first, for instance, a more generalized diffusion of means such as Internet—or its successors—along the world. We now turn to describing some of the implication of BIPS. We must clarify that the higher the order of the logic of the interactions between the vias for engineering BIPS, the more radical the implications mentioned here. For them to fully occur we have to wait some generations.

97.5.1 A Complex World Has Non-imposed Economic Structures

Through their regimes, political systems try to organize or, at least, to influence economy in political grains. They do so by means of laws, legislation, public policies, international treaties and agreements. Non-imposed economic structures will emerge, in the absence of legislation that regulate economic exchanges. The outstanding economist of the twentieth century, Friedrich Hayek, envisioned this spontaneous order [42] in economic catalaxies. Catalaxies are emergent economic structures which are the result of self-organized process of specialization and exchange among individuals in human social systems.

Internet will play a remarkable role in the emergence of no-imposed economic structures because it facilitates trading activities. Indeed, currently, internet is eliminating the need of having many intermediaries for trading and exchanging by helping individuals get closer to a more independent economic status [32], reducing the scenarios where they could be exploited. At the same time, this entails better bioethical labor conditions than those provided, imposed, guarantied or searched today by political regimes. Online crowdsourcing platforms and web pages where inventors can sell directly their work and ideas are examples that reaffirm the tendency [32]. The network society also plays an important role in this equation because the ability to work autonomously and be an active component of a network becomes paramount in the new economy: self-programmable economy. The same author who defined the concept also states [31]:

> Large corporations decentralize themselves as networks of semi-autonomous units; small and medium firms for business networks, keeping their autonomy and flexibility, while making possible to put together resources to attain critical mass small and medium business networks become providers and subcontractors to a variety of large corporations; large corporations and their ancillary networks engage in strategic partnerships on various projects

> concerning products, processes, markets, functions, resources, each one of this project being specific, and this building a specific network around such a project, so that at the end of the project, the network dissolves and its components form other networks around other projects. This, at any given point in time, economic activity is performed by networks of networks built around specific business projects.

In BIPS there is not a political structure that sets minimal conditions of payment or labor conditions, for example, for those who can benefit the least of the facilities that mediums such as Internet provide. However, it must be reminded that not even classical political systems can guarantee bioethical labor conditions, despite the fact that many of them have specific legislation for it. In bio-inspired political systems, individuals become *their own force of work*, which implies more diversity in the market. Ultimately, diversity reduces the possibilities of monopolies to be established. Hence, the self-organization of economic structures in BIPS would not point toward savage economic models—like today's. We hope that diversity in catalaxies entail that the influence of enterprises stops being feasible to be described by a power law.

97.5.2 A Complex World with Visa Policies Is Non-viable

Political regimes impose restrictions to international mobility by means of visa policies, international treats and agreements. Bio-inspired political systems would not forcibly try to maintain a population homogeneous, protecting it from *external perturbations*. That is, keeping it away from integrating it with citizens of different nationalities. On the contrary, BIPS would contribute to the mobility of individuals among human groups.

Three possible conditions could come from the phenomenon of traveling freely *abroad*. First, there could be a homogenization of religious, ethical, economic, cultural or philosophical aspects in human social systems. Second, the diversity of the latter could increase as a result of combining different religions, economies, products, cultures, subcultures, traditions, philosophies, etc., just as it happens with mutation and recombination in biology [43]. Third, both cases could occur simultaneously: the world could homogenize in some aspects and would become increasingly diverse in some others.

We state that the third is the most probable scenario—one that will boost decentralization and deterritorialization and will lead to a general wellbeing for individuals along the world. As economies would not be grounded anymore in the existence of currencies, economic wellbeing would not strictly base on the *opportunities and capabilities of choosing* [44]. That is, the wider is the network of degrees of freedom of a community, the better are the economic opportunities for individuals and groups. Being connected can be linked with economic wellbeing. The more connected a community is and can be, the better opportunities of economic growth it has. The more complex its network of citizens mobilization, the better its level of commodities and comfort because greater are the probabilities of profiting from increased diversity coming from catalaxies. Although it cannot be yet affirmed with

100 % certainty that there is casual direction between network diversity, connectivity factors and economic wellbeing, "social network diversity seem to be at the very least a strong structural signature for the economic development of a community" [45]. This correspondence can be confirmed by comparing economic conditions of the nations that are more connected today with those where citizens can barely travel abroad.

An additional argument that must be pointed out about the relation between diversity and complexity is that much diversity, whereas it is a variation, a distinction of species or a form of configuration, is not always a desired condition because the system can become inefficient or catastrophic [43]. On the other hand, diversity limited by some homogenization and interrelated with complex adaptive rules and other characteristics of complexity can produce robustness within a system. In this case, a variety of political system. Diversity changes the equilibrium of a system—which is dynamic in human social systems and this will bring robustness to the sociopolitical dynamics of the world because the more *emergent* diversity there is, the more complexity there exists as well [43].

97.5.3 True Democracy is only possible in a Politically Complex—Anarchist—World

Many of contemporary Political systems in western world are representative democracies with regimes structured by tree topologies. The idea of bio-inspired political systems sustains itself in the fact that there is not a real need of having centralized control mechanisms for organizing human social systems because complex systems tend to self-organize. A completely self-organized world is a world that has implemented third-order models. Third-order models are possible in anarchist contexts, based on the possibilities of democratic principles because we have to consider that decisions will continue to be taken—in complex network topologies, which makes our model an anarchic-democratic world.

Anarchy means having no principle or authority. The word is composed by the greek words *an* -ὰν- (without) and *arche* -ὰρχή- (principle), but there is nothing in is original meaning that links it with the absence of order. On the contrary, the great theorists of anarchism [46–49] described anarchy as a political system where order is synthesized bottom-up [50]. Anarchy is a social synergy rich-realm because it bases, mainly, in emergent cooperation networks in a dissipative self-organized complex system. Therefore, anarchy is compatible with networks of self-regulation [51], cooperation and consensus. On the other hand, the idea of democracy originated in the greeks, from the words demos -δῆμος- (people) and the word kratos -κράτος- (power). Then, whereas anarchy means the absence of government, democracy means that the power relies on the people. Together, both ideas are not incompatible.

Stating that bio-inspired political systems leads to an anarchic-democratic world is to state that people, *those who are today in the base of the pyramid*, will be their own governors; that is, that they will self-organize. An anarchic-democratic world is a world where complex networks of sociopolitical interactions dynamically self-organize human social systems. The interactions between this complex networks give rise to adaptive control mechanisms for every particular case. Anarchy is a consequence of BIPS because there are no individuals or groups directing or trying to organize human social systems. In respect to democracy, as there are no intermediates for the decision-making processes, every individual can be part of the conformation of sociopolitical dynamics, so we can refer to a truly democratic stadium of human history, despite the absence of coercive regimes that guide democratic actions. This dynamic is one of the most complex possibilities for living the political and politics. Going one step forward, we claim that the topology of a true democracy is the same as the topology of anarchy: complex network structures, which is the topology of third-order models for BIPS.

Unlike today, in an anarchic-democratic world there would be no discrimination among physical or virtual presence for problem-facing, decision-making and decision-implementation because the adscription of individuals to a defined geographical space will not matter for their sociopolitical interactions. Instead, this will be replaced by an interest or expression of a chosen collectivity or individuality. One feature that will make this phenomenon possible will be that there will be no regimes that concentrate the power and the rights over the decision-making processes. As a consequence, there will not exist centralized top-down power structures that some groups will try to bring down.

We intrinsically blame top-down power structures for the existence of political violence, guerrillas, insurrection, para-institutions, rebellion, strikes, civil violence, wars, and rebels in the form of academic embryos because the institutionalization of politics by top-down structures segregates groups that operate outside the defined boundaries of political regimes, which makes them try to bring down the establishment—by any means. Without having only one way for recognizing sociopolitical dynamics, there would be no establishment to dwindle, pervade, cooptate, change or eliminate with the use of coercive mechanisms.

In complex contexts, political adaptiveness is *much more* preferable than rigidness and sometimes citizen participation mechanisms are not flexible enough. A truly complex adaptive political system in a flexible environment can only be achieved in anarchic contexts where individuals, if they wish, can behave democratically. In the end, both, (direct) democracy and anarchy are expressions of politics as *politéia*.

97.5.4 A Complex World Much Closer to Bio-ethics

Since the industrial revolution, the resources on earth have been extracted in augment. This has caused a deterioration of the conditions for life on the planet and has killed and extinct thousands of species in such a brief period of time that it can hardly be said that extinctions obeyed to natural causes [52]. Classical political systems are responsible for great part of the harm committed towards our planet and the life that inhabits it: governors in turn sign international treaties and form internal policies about the extraction, production, deforestation and managing of the natural resources in their political parcels. They define the limits, or the absence of them, to every action that involves the national territory: the mining activities and the extraction of resources, human labor conditions, limits and permissions for hunting and frames for the economic activities of corporations and multinationals. Thus, they decide: scopes, limits and controls. That is why, in great part, classical political systems can be directly blamed for affecting negatively the conditions for life on earth, and life itself. However, blaming specifically governors is pointless because, anyway, independently of who is or are in turn concentrating political power, the structures of political regimes are suitable for expecting negative bioethical outcomes. Basically, the central node, hub or control core in the tree topology, basically, can take any decision it wants, even if it is not beneficial for the lives of some groups or species that inhabit the political grains or parcels in question. In sum, classical political systems are directly responsible for the reduction of bio-diversity on earth.

Bio-inspired political systems are a step closer towards a pacific interaction among individuals and human social systems because of their bottom-up synthesized dynamics. Biologically motivated models are the best alternative that humans can adopt in order to properly organize the biodiversity that is left. BIPS point towards the self-organization of complex networks that synthetize how the use of earth's resources are organized, defining the extraction, mining, deforestation, reforestation and hunting activities, using only local information and following decisions taken with the support of metaheuristics and modeling and simulation, and implemented by means of complex topologies. The combination of the three in BIPS might stop or reverse what can still be pulled back. For avoiding the disastrous situations currently experienced by the planet, subsequently of BIPS there would be synergies coming from the non-linear interactions of these complex networks. It is hoped that the idea of synthetizing bottom-up political systems permeates the mainstream of science and the political, in order not to return any longer to the type of top-down political structures that have harmed so profoundly life on earth. With no doubt, BIPS imply a network comprehension of living systems' relations and interactions, which entails bio-ethical ways for organizing social systems.

In addition, less antropocentric ways of organizing and structuring our human social systems will positively influence the way we exploit *Gaia*, evolving towards more dynamic equilibriums between human social systems, natural social systems and artificial systems. The road, however, is not yet paved. We still have to gain more knowledge about the complex computational dynamics of live, social systems and artificial life. Finding this type of adaptive balance is, with no doubt, the greatest challenge ever faced in human history [53]. Bio-inspired political systems are the only way in which, if political systems continue to exist, a political system can harness the increasing complexity of the interactions in our world.

Figure 97.7 conjugates the mentioned backgrounds with the implications of Bio-inspired political systems.

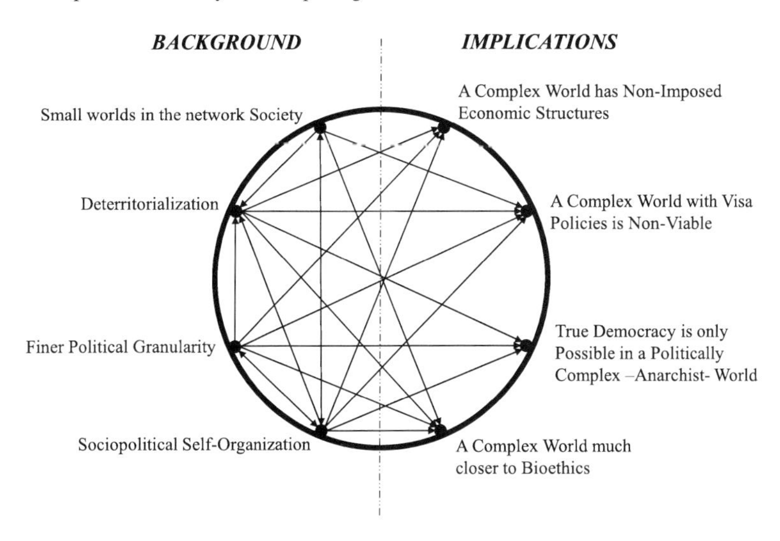

Fig. 97.7 Backgrounds and implications

97.6 Concluding Remarks

We stated that bio-inspired models for political systems mean materializing complexity. Agent-based modeling and simulation helps gaining comprehension about the systems modeled and testing decisions before implementing them. Metaheuristics are useful for problem-solving and exploring spaces of solutions that lead to solutions close to the optimal. And complex models for topologies permit generating congruence between the structure of political systems, the types of problems they must solve and the complex systems upon which they decide.

The complexity of BIPS would provide them with more degrees of freedom than classical political systems, for modifications to occur in their structures and dynamics *in accordance* with the evolution of human social systems. Their adaptability, based on the composition of the system and the types of relations between the nodes, would give them the capacity of pacifically generating variations on every temporal control parameter as the *politéia* synthesizes them.

This paper highlights the necessity of complexifying the means by which human social systems organize (political systems), since there is still a vacuum in visualizing the profound political implications of standing in complexity for thinking the political. For our case, political systems, imagining them in contexts of complexity leads necessary to think about ruptures, discontinuity, decentralization, evolvability, transformations and, of course, self-organization, which is really lacking in the mainstream of the studies of the political. Indeed, most of the theorists of politics that work in complexity and are interested in the study political systems are still theorizing in terms of voting dynamics, elections and governability. They assume that complexity is not being constrained by classical institutionalization mechanisms because human social systems present complex behaviors. They also assume that political systems are already self-organized [54].

Modeling and simulation added to metaheuristics plus complex topologies for political regimes can help us transform current political systems towards more organic ones, coping with the complexity of human sociopolitical dynamics. But they can also enlight us with an idea about the adjacent possible [55] of actual political systems. Based on this, the need of a field for the study of bio-inspired models for political systems is imperative. It is time for political science to underline the phenomenon that sooner or later will lead us to pass by the *social contract* era.

Acknowledgments: I sincerely thank Nelson Alfonso Gómez Cruz for the fructiferous conversations we had during the writing of this paper, his comments, suggestions and improvements on my graphics. I also acknowledge Carlos Eduardo Maldonado for his valuable corrections on the paper and for the one year internship I did with him, where I could work on developing my ideas. And I appreciate the precious time and suggestions coming from Beatriz Franco Cuervo.

References

1. Maldonado CE (2006) Política y sistemas no lineales: la biopolítica. Dilemas de la Política, Universidad Externado de Colombia, pp 91–142
2. Axelrod R, Cohen MD (2001) Harnessing complexity: organizational implications of a scientific frontier. Basic Books, New York
3. Braha D, Minai A, Bar-Yam Y (2007) Complex eingineered systems. Springer, Berlin
4. Maldonado CE, Gómez Cruz NA (2012) The complexification of eingineering. Complexity 17:8–15
5. Talbi E-G (2009) Metaheuristics. From design to implementation. Wiley, New York
6. Bonabeau E (2002) Agent-based modeling: methods and techniques for simulating human systems. Proc Natl Acad Sci USA 99:7280–7287
7. Mandelbrot BB (1983) The fractal geometry of nature. Freeman, New York
8. Rosenblueth A, Wiener N (1945) The role of models in science. Philos Sci 12:316–321
9. Lapierre J-L (1973) L'Analyse de systemes politiques. Presses Universitaires de France, Paris
10. Easton D (1957) An approach to the analysis of political systems. Q J Int Relat 9:383–400
11. Bonabeau E (2003) Don't trust your gut. Harv Bus Rev 81:116–123
12. Mandelbrot BB (1997) Fractals and scaling in finance. Springer, Berlin
13. Castellani B (2010) Sociology and complexity science: a new field of inquiry. Springer, Berlin
14. Maldonado CE (2012) Teoría de la historia, filosofía de la historia y complejidad. In: Maldonado CE (ed) Fronteras de la ciencia y complejidad. Universidad del Rosario, Bogotá, pp 17–48
15. Casti J (1998) Complexity and aesthetics. Complexity 3:11–16
16. Pahl-Wostl C (2007) The implications of complexity for integrated resources management. Environ Model Softw 22:561–569
17. Bedau MA, Cleland A (2010) The nature of life. Cambridge University Press, New York
18. Maldonado CE (2009) Significado y alcance de pensar sistemas vivos. Thelos E-journal. http://thelos.utem.cl/2009/09/15/significado-y-alcancede-pensar-en-sistemas-vivos/
19. Zäpfel G, Braune R, Bögl M (2010) Metaheuristic search concepts. Springer, Berlin
20. Dréo J, Pétrowski A, Siarry P, Taillard E (2006) Metaheuristics for hard optimization. Springer, Berlin
21. Michalewicz Z, Schmidt M, Michaelewicz M, Chiriac C (2007) Adaptive business intelligence. Springer, Berlin

22. Maldonado CE, Gómez Cruz NA (2011) El mundo de las ciencias de la complejidad. Universidad del Rosario, Bogotá
23. Bonabeau E, Hunt CW, Gaudiano P (2003) Agent-based modeling for testing and designing novel decentralizedcommand and control sstems paradigms. In: 8th international command and control research and technology symposium, Washington, DC
24. Macal C (2012) Agent-based modeling and artificial life. In: Meyers RA (ed) Computational complexity. Theory, techniques and applications. Springer, Berlin, pp 39–57
25. North M, Macal C (2007) Managing business complexity. Oxford University Press, London
26. Prokopenko M (2009) Guided self-organization. HSFP J 3:287–289
27. Bar-Yam Y (1997) Dynamics of complex systems. Westview Press, Boulder
28. Mainzer K (2007) Thinking in complexity. The computational dynamics of matter, mind an minkind. Springer, Berlin
29. Watts D (2003) Six degrees. The science of a connected age. Norton & Company, New York
30. Castells M (2010) The rise of the network society: the information age: economy, society, and culture. Wiley, New York
31. Castells M (2005) The network society: from knowledge to policy. In: Castells M, Cardoso G (eds) The network society: from knowledge to policy. Johns Hopkins University Press, Baltimore, pp 3–21
32. Ridley M (2010) The rational optimist: how prosperoty evolves. Harper Collins, New York
33. Cropf R, Krummenacher WS (2011) Information communication technologies and the virtual public sphere: impacts of network structures on civil society. IGI Global, Hershey
34. Ranerup A (2000) On-line Forums as a arena for political decisions. Lect Notes Comput Sci 1765:209–223
35. Goodman MA, Cowin SC (1972) A continuum theory for granular materials. Arch Ration Mech Anal 44:249–266
36. Pawlak Z (1982) Rough sets international. Int J Comput Sci 11:341–356
37. Braudel F (1958) Histoire et sciences sociales: la longue durée. Annales ESC 4:725–753
38. Axelrod R (1995) A model of the emergence of new political actors. In: Gilbert N, Conte R (eds) Artificial societies: the computer simulation of social life. University College Press, London
39. United Citied and Local Governments (2008) Decentralization and local democracy in the world: first global report. World Bank Publications, Barcelona
40. Brunk GE (2002) Why do societies collapse. J Theor Polit 14:195–230
41. Juris JS (2012) Reflections on #Occupy everywhere: special media, public space, and emerging logics of aggregation. Am Ethnol 39:259–279
42. Hayek F (1973) Law, legislation, and liberty. University of Chicago Press, Chicago
43. Page SE (2011) Diversity and complexity. Princeton University Press, Princeton
44. Sen A (2009) The idea of justice. Harvard University Press, Cambridge
45. Eagle N (2010) Network diversity and economic development. Science 328:1029–1031
46. Bakunin M (2012) Selected writings from Mikhail Bakunin: essays on anarchism. Red and Black Publishers, St Petesburg
47. Guerin D (1965) LÁnarchisme, de la doctrine a láction. Gallimard, Paris
48. Kropotkin P (1996) Mutual aid: a factor of evolution. Black Rose Books, Montreal
49. Proudhon P-J (2003) General idea of the revolution in the nineteenth century. Dover, New York
50. Baclay H (1996) People without government: an anthropology of anarchy. Kahn & Averill, London
51. Woodcock G (1962) Anarchism. A history of libertarian ideas and movements. World Publishing Company, New York
52. Leakey R, Lewin R (1995) The sixth extinction: patterns of life and the future of humankind. Doubleday, New York

53. Maldonado CE (2009) Ingenería de sistemas complejos. Panorama y oportunidades. In: Conferencias de la tercera asamblea de la red cartagenera de ingenería. El futuro de la educación en inenería y en gestión de la ingenería, Cartagena, Colombia. http://carlosmaldonado.org/articulos/Book02_RCI2009-Maldonado.pdf
54. Fuchs C (2004) The political system as a self-organizing information system. Cybern Syst 1:353–358
55. Kauffman S (2003) Investigations. Oxford University Press, Oxford

Chapter 98
The Family at the Center of Interdisciplinary Research in Complex Systems: A Call for Future Research Programs

Ana Teixeira de Melo and Madalena Alarcão

Abstract In this communication we explore different lines of inquiry holding the family at the center of interdisciplinary research programs in complexity science which aim not only to understand family dynamics with relevance for therapeutic or developmental interventions but also to other fields of science. We define three levels of research in family complexity. Assuming the family as a complex system as the departure level we propose the exploration of two additional research directions, one from the family down, into the dynamics of its microsystems and using them as windows to understand the biopsiconeurological worlds, and one from the family up, using it as a relevant sample of broader social complex system. We discuss possible lines of inquiry for each level of family complexity science.

Keywords Complexity science · Family science · Family interventions · Family complexity · Interdisciplinary research · Coordination dynamics

98.1 Introduction

Family science was soon informed by early systems sciences, particularly general systems theory and cybernetics, and rapidly incorporated its core concepts and metaphors. These influences gave an important impulse to the development of treatment approaches to human problems in the context of family interventions.

But the field of family systems research has long stagnated. Through today's complex adaptive nonlinear systems lenses the early conceptualizations of family functioning associated with main family therapy schools do not seem that complex after all. Contrary to the movement initiated in close areas in psychology [7] only with few exceptions (e.g. [6]) have family researchers and scholars kept up with systems' development and used the new conceptual, methodological and mathematical

A.T. de Melo (✉) · M. Alarcão
Faculty of Psychology and Education Sciences, University of Coimbra, Coimbra, Portugal
e-mail: anamelopsi@gmail.com

M. Alarcão
e-mail: madalena.alarcao@uc.pt

T. Gilbert et al. (eds.), *Proceedings of the European Conference on Complex Systems 2012*, Springer Proceedings in Complexity, DOI 10.1007/978-3-319-00395-5_98,

Fig. 98.1 The family and the complex world: a privileged domain to grasp complexity in micro, meso, and macro levels of complex living and social systems

tools of complexity to improve knowledge on family development and adaptation and to inform more effective interventions.

Understanding families as complex systems may contribute to the improvement of both research and practice in the fields of human services and, particularly, family interventions which are aimed at the promotion of the well-being and positive development of families, their members and their communities, both locally and globally.

A complexity approach can illuminate more effective and meaningful pathways to understand family change processes, with special relevance for the field of family interventions in general and for specialized areas such as the assessment of families facing multiple problems or in child protection cases.

A lot can be gained from studying families, developing and testing interventions under the scope of complexity sciences. On the other hand, exploring diverse lines of interdisciplinary inquiry with a focus on the families as complex systems can open avenues to deepen our knowledge of both the complex micro and macro human worlds (see Fig. 98.1).

98.2 The Proposal: Levels and Lines of Inquiry for Interdisciplinary Family Complexity Research

A Family Complexity Science may emerge from the integration of Family and Family Intervention Sciences and Complexity Science (Fig. 98.2).

The former can pose relevant specific research questions informed by the accumulated knowledge pertaining to family functioning, development and family interventions. The latter can guide theory and research advances by lending its core concepts and theories as well as its research paradigms and methods. Advances in family complexity science may then foster the review, reconstruction and expansion of both these fields separately and in different domains.

We propose there may be three main levels of research in family complexity research: (a) level 1 focuses on the family as a special domain of multiple forms of coordination and as a complex social system; (b) level 0 focuses the family and its numerous microsystems as well as their coordination dynamics and (3) level 2 focuses the family and its relation to macro social systems such as community and local networks or global society.

We next propose some lines of inquiry for each level of research.

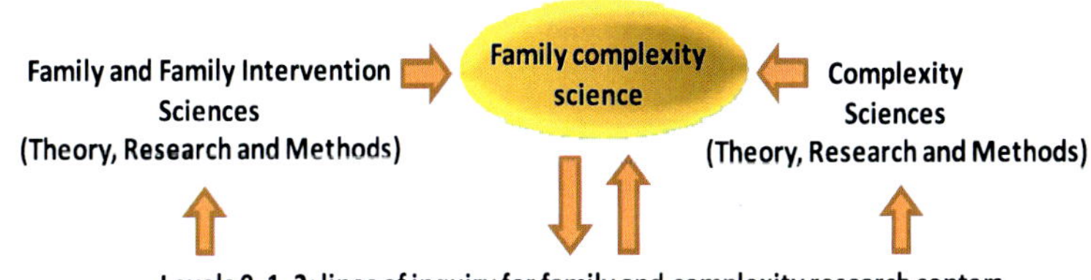

Fig. 98.2 Research integration in family complexity science

98.2.1 Level 1: The Family as a Complex System

This first line of research may contribute to ground Family Complexity Science in real data avoiding, therefore, a misleading importation of concepts from Complexity into Family Sciences. It aims at understanding the family in its complexity while possible contributing to clarify complex properties in other systems and the potential of Complexity to unravel, while unifying the mysteries of the social and natural worlds. The following lines of inquiry can be pursuit at this level:

- Description and comprehension of the family as a complex social [2];
- Identification, description and comprehension of relevant coordination pairs [11, 12] and their dynamics to understand family complexity;
- Exploration of properties of the families and family intervention systems as complex systems (e.g. features of self-organization; identification of emergent phenomena; self-similarity; type of dynamics) [4], including: (a) special groups of families (e.g. families with and without identified pathologies; families with simple patient-focuses symptoms vs. families with multiple symptoms and problems), (b) particular conditions of internal (e.g different family transitions; loss of family members; families with members with special needs; families with members with psychopathology) and external perturbations (e.g. families exposed to violence; families living in poverty), (c) different status in regard to interventions (e.g. families which benefit from treatment from those who don't; families in voluntary vs. mandatory treatments, etc.) and (d) different intervention settings (e.g. clinic vs. community-based treatments);
- Identification, description and understanding of core coordination variables/functional information (along with control parameters and boundary conditions) [12] to understand family functioning, development and change during interventions while attending to particular contexts; identification, description and explanation of relevant states, description of state spaces, coordination patterns and pattern dynamics associated with family positive development and adjustment as well as change during interventions.

98.2.2 Level 0: The Family and Micro Complex Systems

This line of research can contribute to deepen our understanding of families as emerging from the coordination of distinct elements, being them individuals (as wholes) or, at an even lower level of analysis, distinct neurobiological, emotional, cognitive and discursive subs-systems which interact in complex ways. On the other hand, this level of research can contribute with important insights to coordination science. Proposed lines of inquiry include the investigation of:

- Synchronization, interpersonal coordination (neurobiological, motor, emotional, behavioral, cognitive and discursive) [5, 13] and inter-brain coordination [3] between family members and between families and therapists [14] and their relation to family development, adaptation and change as well as intervention process and outcomes; coordination of individual and family change; neurobiological correlates and markers of coordination [12, 16];
- Intra-individual neurobiological coordination, self-organization and development [11] in relation to the quality and dynamics of the family context as possible control parameters and source of perturbation; constrains and opportunities for individual development in relation to family functioning, development and family coordination;
- Exploration of complementary pairs [12] of family dynamics (e.g. segregation–integration; competition–cooperation; opportunities–risks, stability–change) in relation to individual and family development, adaptation, change and intervention processes and outcomes;
- Coordination dynamics and coupling processes between families, professionals/trainees and trainers/supervisors in the context of family interventions during and after training programs in family interventions; learning, change and multilevel coupling processes.

98.2.3 Level 2: The Family and Macro Complex Systems

In this level of analysis the family is considered as window into the broader social complex human world inspected by fields such as sociology, economics, politics. Exploration of relevant issues for macro social systems can probably be performed with an exploratory focus on family processes. Some proposed lines of inquiry comprehend the exploration of the following:

- Evolutionary role of family processes and relation of within and between family dynamics with (positive) community and societal development [1, 10];
- Negotiation of individual vs. family goals, resolution of interpersonal dilemmas and decision-making processes in a family context: implications for larger groups and social interventions [1, 9, 15];
- Exploration of self-similarity and fractal-like phenomena or structures [4] in family and family intervention systems;
- Emergence, structure and dynamics of family support networks (informal and professional) and broader social contexts [17] in face of internal and external perturbations;

- Emergence, structure and dynamics [17] of complex multi-professionals/ interventions networks and local networks in “difficult cases” (e.g. families with multiple problems; poverty; child protection cases): relation to intervention process and outcomes and implications for interventions and policy;
- Informational regulation, decision-making and social influence processes in multi-professional networks in high pressure contexts

Different research methods and tools can be recruited, such as the diverse techniques integrated by the Sociology and Complexity Science Toolkit [2] (neural networking, social network analysis, grounded theory, cluster analysis). Also, a combination of real data and simulated data (agent-based modelling) can be well suited for these studies [8]. Through family intervention research studies changes in relevant parameters can be more easily experimented and their relation to collective variables tested than in larger groups. This kind of research could later inform experiments on larger systems which, on their turn, may help family sciences to move forward.

98.3 Conclusion

We propose the creation of interdisciplinary and cross-disciplinary family research centers under the scope of complexity in order to take advantage of the enormous potential of family-centered research to improve our knowledge of complex systems while using complexity theories and research tools to develop, apply and test new knowledge to promote family positive development and adaptation.

Researchers from diverse fields can participate in rich spaces of debate in family research centers looking for common variables, contexts and processes as studied in psychological, social, physical-chemical and neurobiological sciences and exploring new areas of research. In this context, complexity tools can not only be applied but also challenged for further developments calling fields as mathematics, physics, engineering and computer sciences to refine old methods and develop new ones for the exploration of the complex human world. We hope this communication can stimulate the approximation of family and complexity sciences and the development of an interdisciplinary space for family complexity science.

References

1. Axelrod R (1984) The evolution of co-operation. Penguin, London
2. Castellani B, Hafferty F (2009) Sociology and complexity science. Springer, Berlin
3. Dumas G, Nadela J, Soussignan R, Martinerie J, Garnero L (2010) Inter-brain synchronization during social interaction. PLoS ONE 5(8):1–10
4. Érdi P (2008) Complexity explained. Springer, Berlin
5. Fuchs A, Jirsa VK (2008) Coordination: neural, behavioral and social dynamics. Springer, Berlin

6. Gottman JM, Murray JD, Swanson CC, Tyson R, Swanson KR (2002) The mathematics of marriage. Dynamic nonlinear models. MIT Press, Cambridge
7. Guastello SJ, Koopmans M, Pincus D (eds) (2009) Chaos and complexity in psychology. The theory of nonlinear dynamical systems. Cambridge University Press, Cambridge
8. Helbing D, Balietti S (2011) How to do agent-based simulations in the future: from modeling social mechanisms to emergent phenomena and interactive systems design. Working paper 11-06-024. Available at http://www.santafe.edu/research/workingpapers/abstract/51b331dfecab44d50dc35fed2c6bbd7b/
9. Helbing D, Johansoon A (2009) Cooperation, norms and conflict: a unified approach. Santa Fe Institute working paper 09-09-040. Available at http://www.santafe.edu/research/workingpapers/abstract/11aa0aeb394147d8e75306ec3de3f141/
10. Lozada M, D'Adamo P, Fuentes MA (2010) Rewarding altruism. Working paper 10-07-014. Available at http://www.santafe.edu/research/workingpapers/10.07.014.pdf
11. Kelso JAS (1995) Dynamic patterns. The self-organization of brain and behavior. MIT Press, Cambridge
12. Kelso JAS, EngstrØm DA (2006) The complementary nature. MIT Press, Cambridge
13. Oullier O, de Guzman GC, Jantzen KJ, Kelso JAS (2008) Social coordination dynamics: measuring human bonding. Soc Neurosci 3(2):178–192
14. Ramsayer F, Tschacher W (2011) Nonverbal synchrony in psychotherapy: coordinated body movement reflects quality relationship and outcome. J Consult Clin Psychol 79(3):284–295
15. Sinha S, Raghavendra S (2004) Phase transition and pattern formation in a model of collective choice dynamics. Santa Fe working paper 04-09-028. Available at http://www.santafe.edu/media/workingpapers/04-09-028.pdf
16. Tognoli E, Lagarde J, DeGuzman GC, Kelso JAS (2007) The phi complex as a neuromarker of human coordination. Proc Natl Acad Sci USA 104(19):8190–8195
17. Watts DJ (2004) The "new" science of networks. Annu Rev Sociol 30:24–270

Chapter 99
Face-to-Face Discussions: Networking or Opinions Exchange?

Simone Righi and Timoteo Carletti

Abstract We use recent results of Cattuto et al. (PLoS ONE, 5(7):e11596, 2010) on face-to-face contact durations to try to answer the question: why do people engage in face-to-face discussions? In particular we focus on behavior of scientists in academic conferences. We show evidence that macroscopic measured data are compatible with two different micro-founded models of social interaction. We find that the first model, in which discussions are performed with the aim of introducing oneself (networking), explains the data when the group exhibits few well reputed scientists. On the contrary, when the reputation hierarchy is not strong, a model where agents' encounters are aimed at exchanging opinions explains the data better.

Keywords Face-to-face discussion · Opinion dynamics · Social interactions

Mathematical physics is today a well accepted framework where model and study social phenomena. After an initial phase relying mostly on qualitative models [5, 8, 13], scientists begun to obtain quantitative results [1, 6, 7, 12], mainly using quite abstract models able to capture universal patterns in human behavior without carefully adding complicating details that could blind a complete understanding of the relationships between the assumptions done and the outcomes observed.

Thanks to the advancements in technology a new phase recently started, where researchers have access to fine measurements about real world interactions. The huge amount of data (internet based applications, mobile network [2, 3, 9–11], face-to-face interactions [4]) is however often associated with a lack of a mathematical foundation of the emergent properties observed. Moreover not everything can be measured; understand why do people "internally" act in such a way as to determine the measured global properties, would necessitate a hierarchical model, whose levels

S. Righi (✉) · T. Carletti
Department of Economics, CRED and naXys, University of Namur (FUNDP), 8 Rempart de la Vierge, 5000 Namur, Belgium
e-mail: simone.righi@fundp.ac.be

T. Carletti
e-mail: timoteo.carletti@fundp.ac.be

T. Gilbert et al. (eds.), *Proceedings of the European Conference on Complex Systems 2012*, Springer Proceedings in Complexity, DOI 10.1007/978-3-319-00395-5_99,

will range from the "brain level", through the "individual level" to eventually arrive to the "group level". The complexity of such model would be too important to allow any useful analysis.

In the present paper we make a step forward and "measure the unmeasurable" using such kind of data. More precisely, we use recent results of face-to-face contact durations (see for instance [4]) to try to answer the question: why do people engage face-to-face discussions? While the present paper focus on interactions among scientists in conferences and workshops, cases to which data from [4] referred to, our setup is flexible enough to be applied to other situations and datasets.

Starting from a simple microscopic model we show evidence that macroscopic measured data, notably supra linear growth of total discussion time with respect to the agent connectivity, are compatible with two different scenarios based on the group structure. These can be summarized in two micro-founded models of social interaction. In the first one, we assume discussions among the agents to be performed about a neutral topic and thus the opinions exchanges do not influence the discussion times nor the selection of the partner, i.e. the goal of the discussions is to introduce himself to the partner (networking). On the other hand, in the second model we assume that agents' encounters are aimed to find a compromise on a subject of discussion and thus there could be an opinion update as consequence of the encounters.

We also assume that, in both models, each agent is characterized by a publicly known and accepted level of reputation and that agents aim to engage discussions with highly reputed peers. In the case of a scientific meeting this could be the h-factor of the individuals.

We find that when the group exhibits clearly identified well reputed people, then the outcomes of our model are consistent with the assumption that discussions are engaged mainly to share time with the most reputed person, namely agents perform *networking* actions, i.e. the first scenario does apply. On the other hand, when the reputation hierarchy is not strong, our findings are consistent with the hypothesis that people's discussions are finalized to *opinions exchange*, namely the second scenario holds.

We are aware that our models are very simple ones and that more variables could affect the discussion times among agents. However, as already stated, our goal is to reduce as much as possible the number of free parameters and thus, for instance, associate possible causes of long/short discussions to the experimental data.

The homogeneity assumption of the agent behavior, is also a strong one, thus in a second part of the paper we will relax this hypothesis by allowing the group to be composed by agents aimed at exchanging opinions and also agents doing networking. We will be thus interested in studying the robustness of our results as a fraction of the relative fraction of agents.

The paper is organized as follows. Section 99.1 will present the two main models, while Sect. 99.2 is aimed at reporting some numerical results. Finally Sect. 99.3 will contain the results of the non homogeneous model.

99.1 The Models

99.1.1 Networking (Model A)

To test the hypothesis that agents could be motivated, in their interactions, by the simple willingness to do networking, defined as introducing oneself and spending time to exchange personal informations with the more reputed persons, without the need of exchange valuable opinion, we introduce the following model (hereby named Model A).

Consider a closed group of N fixed agents, where social interactions last for discretized blocks of τ seconds; in the following we chose a value of $\tau = 20$ seconds to obtain simulations with the same time resolution of the experimental data from Barrat et al. [4]. As already stated each agent i is characterized by a level of reputation h_i, hereby defined by a positive real number distributed according to a power law distribution $\frac{1}{h^{\gamma_h}}$, where γ_h is a parameter of the model. As the dynamics unfolds, free agents[1] search for a partner to initiate a discussion. Our working assumption is that agents have a preference to engage in discussions with peers with high reputation, that is a publicly known value. Indeed, the probability of selecting an agent as partner is proportional to his reputation, properly renormalized. Namely the probability to select agent j is proportional to $h_j / \sum_{k=1}^{N} h_k$.

When a free agent meets another currently idle agent, the couple engages in a discussion for a time Δt, extracted from the power law distribution $\frac{1}{(\Delta t)^{\gamma}}$, where the parameter γ has been fixed to the value 1.5 that can be extracted from the experimental data of Barrat et al. [4]. However, the partner selection rule does not guarantee that each attempt will be *successful*, i.e. the wanted partner may not be free at that time, in this case the first agent needs to wait for Δt_w seconds before trying to initiate another conversation. The idea is to mimic the fact that an agent lose some time after an unfruitful approach by looking around or just taking a cup of coffee. We assume the distribution of the waiting times to be again a power law [14], $\frac{1}{(\Delta t_w)^{\gamma_w}}$. We assume moreover that the exponent γ_w varies for each agent and it is influenced by the degree of success of past interactions, more precisely $\gamma_w = 1 + x_i$, where x_i indicates the degree of frustration incurred, by the i-th agent that promotes the conversation, as a consequence of past interactions:

$$x_i = \frac{\text{Number of successful interactions of the } i\text{-th agent}}{\text{Total number of interactions attempted by the } i\text{-th agent}}. \tag{99.1}$$

The rationale of this assumption being the fact that an agent with many successful interactions will be more motivated to wait shorter times between one attempt and the next one than a peer with an history of many failed attempts. Our assumption is that the latter will have a higher probability to wait longer than the former before finding the "courage" of trying to approach someone else.

[1] An agent engaged in a discussion will be defined to be *occupied*, while an agent waiting for an available partner or proposing a conversation will be defined to be *free*.

99.1.2 Opinion Exchange (Model B)

The alternative hypothesis is that scientists interact, in the context of scientific conferences, in order to exchange opinions, for instance to try to convince their peers of some idea. In order to test this conjecture, we slightly modify the previous model by introducing the opinion exchange in the form proposed by [6] (this second model is hereby named Model B). The main novelty of this second setup is that all agents are also endowed with an opinion $(O_i)_{1\leq i\leq N}$, initially uniformly distributed in the interval $[0,1]$, and that this opinion has a feedback on the dynamics of the model. More precisely we assume that the distribution of the interaction times, $\frac{1}{(\Delta t)^\gamma}$, depends on the opinions difference among the two interacting agents: $\gamma = 1.5 + |O_i - O_j|$. The rationale being that the more affine the agents are, the higher is the probability their conversation will last longer: having similar opinions, the two agents can build a common ground and continue longer the discussion.

In line with [6], the social interaction can produce an opinions update if the agents are not too far in the opinion space, more precisely once agents i and j meet if $|O_i - O_j| \leq \sigma$, being $\sigma \in (0,1)$ a proxy for the openness of mind, their opinions will tend to converge:

$$O_i^{t+\Delta t} = O_i^t + \mu(\Delta t)(O_j^t - O_i^t) \tag{99.2}$$

$$O_j^{t+\Delta t} = O_j^t + \mu(\Delta t)(O_i^t - O_j^t). \tag{99.3}$$

We assumed that the convergence parameter, μ, depends on the duration of the discussion:

$$\mu(\Delta t) = \frac{1}{2}\tanh(\beta\Delta t), \tag{99.4}$$

this results in a *effectiveness* of the conversation: the longer lasts the interaction, i.e. the greater Δt, the stronger will be the convergence of the opinions to their average. This choice is made to mimic the fact that during a long conversation one has comparatively more chances of changing idea, or to convince the partner, than during a short chat.

99.2 The Results

We simulate both models for a time duration of $T = 12$ hours, once again to compare our results with [4]. At the end of the simulations, each agent is characterized by the number of distinct contacts he had, that can be called the degree k_i of the interaction network, and by the total discussion time he had, namely the strength s_i. The latter relates to the total discussion time, w_{ij} (weight), agents i and j had, by $s_i = \sum' w_{ij}$, being the sum restricted to agents j that had at least a contact with i.

In this preliminary work we focused on the relationship between the degree k and the average node strength $\langle s(k)\rangle$ as a function of the degree:

$$\langle s(k)\rangle = \frac{\sum_{i:k_i=k} s_i}{N_k}, \tag{99.5}$$

where N_k denotes the total number of agents with degree k.

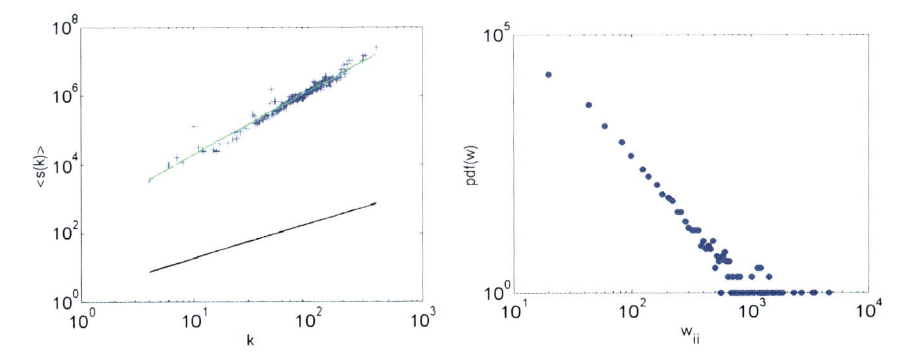

Fig. 99.1 Average strength versus degree and weight distribution (Model A). *On the left panel*, the log-log plot of the average strength for a generic simulation with: $N = 575$ and $\gamma_h = 3.0$, together with a best linear fit (*lighter line*), *the black line* denotes the linear growth. On *the right panel*, the probability distribution function of agents weights for the same simulation

Table 99.1 Super-linear growth of the average strength vs the degree : $\langle s(k)\rangle \sim k^{\rho}$, for different values of γ_h for Model A and Model B

ρ	Model A (Networking)	Model B (Opinion exchange)
$\gamma_h = 3$	~ 1.8	~ 1.6
$\gamma_h = 1.1$	~ 1.6	~ 1.8

As it can be observed from Fig. 99.1 a generic simulation of our model generates a pretty well defined super-linear relationship between the degree and the average strength. The data extracted from Barrat et al. [4] show that the relationship between the degree and the average node strength is given by $\langle s(k)\rangle \sim k^{1.73}$. We are able to tune the model parameters, in particular γ_h responsible for the group organization, in order to obtain results which are significantly close to those obtained experimentally. Moreover the matching between the fit on empirical data and the synthetic ones obtained via our models, allows us to gain insights into which kind of activity is dominant in a certain dataset.

Both models are, in principle, capable of generating outcomes similar to the experimental ones; however our aim is to give a social interpretation to the parameters values, allowing us to unravel the mechanism at play : which is the social force, networking or opinion exchange, responsible for the measured observables, average strength? In particular we are interested in studying different scale free distributions of agents reputation. We observe that, in a group where agents with very high reputation are present, people tend to do networking, in fact the Model A generates a slope much closer to the experimental one than the Model B. On the contrary when the group of interacting agents is constituted mainly by equally reputed agents, the Model B seems to fit better the experimental data. These results are summarized in Table 99.1, for a given set of parameters values.

99.3 Extension: Mixed Population Hypothesis

In the analyses performed in the previous section we showed that it is possible to fit the data observed by [4] with two different models. Both are based on the assumption that each member of the observed population share the same objective (networking or opinion exchange).

It could be argued that, in reality, not every participant to a conference has the same objective. Some people may be willing to share and possibly change opinions, while others may just want to do networking.

In order to make our model more realistic we now relax the homogeneity assumption of group's objective, allowing for a mixed population, where pN agents are willing to discuss while $(1-p)N$ agents do networking. The similar structure of Models A and B makes the task of merging them relatively straightforward. When two agents of the same type are selected to initiate an interaction we can simply assume that the same rules outlined in Sect. 99.1, for the homogenous population case, apply. The only complication comes when two agents of different types meet. In this case there are two conflicting rules for the extraction of the interaction time. To solve this problem we observe that a conversation can last only as long as both the participants are willing to pursue it. Thus, we assume that each of the agents extracts an interaction time corresponding to his own rule of behavior (a networker as in model A and an opinion exchanger as in model B) and the actual length of the conversation is the shortest of the two. Moreover, the agent interested in exchanging opinions, may change his own idea while the agent only interested in knowing someone more important will keep the same opinion as before.

With this unified modeling setup, we are able to study the effect of the population' composition on the relationship between agents' connectivity and the average strength of their interactions, $\langle s(k)\rangle \sim k^{\rho}$. In Fig. 99.2 we show the dependence of the exponent ρ on the fraction, p, of agents interested into discussing, for different values of the parameters γ, γ_h and γ_w.

One can immediately observe that the distribution of h-indexes in the population (driven by γ_h) does not have a large influence on ρ. Moreover, from the data one can also deduce that p should be large enough, i.e. larger than 0.4, to have an appreciable influence on the value of ρ, for instance to make it to decrease. Finally, this trend is stronger the larger is γ_w (i.e. the less likely is for an agent to wait for long times). We propose the following explanation for this phenomenon. When γ_w is small, agents are more likely to wait for longer amounts of time after a failed interaction and thus they can be considered available once a second free agent tries to contact them. Whenever a large amount of persons are not engaged in conversations, there is an higher probability to have positive interactions (and thus less "frustration"). The consequence is that the presence of a significative number of opinion exchangers increases the length of interactions between them thus leading to lower ρ. When γ_w increases the coefficient ρ decreases even more because the effect of the introduction of opinion exchanges on their relative interactions is stronger.

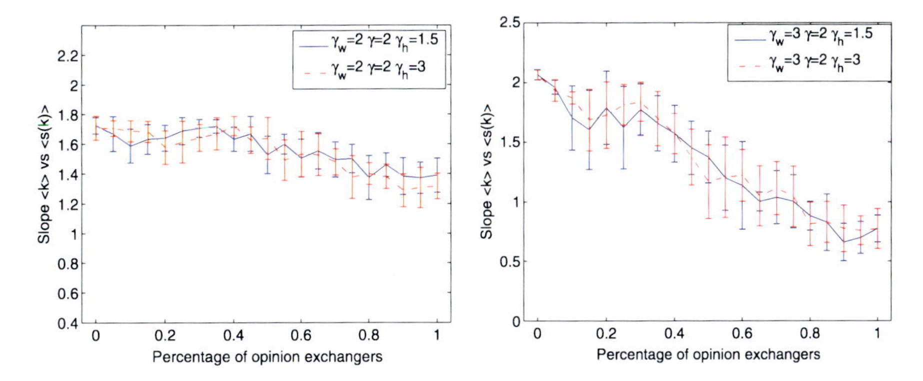

Fig. 99.2 Dependence of ρ on p, where $\langle s(k)\rangle \sim k^{\rho}$ and p is the fraction of agents aiming at exchanging opinion. On *the left panel*, we report numerical results for the set of parameters: $N = 575$, $\gamma = 2$, $\gamma_w = 2$ and $\gamma_h = \{1.5, 3.0\}$, while on *the right panel*: $N = 575$, $\gamma = 2$, $\gamma_w = 3$ and $\gamma_h = \{1.5, 3.0\}$

99.4 Conclusions

In this paper we presented and studied a simple Agent Based Model, grounded on experimental data, with the aim of reproducing some macroscopic measured features, for instance the super linear growth of the total discussion times as a function of the number of distinct contacts agents had. Moreover we are able to provide a sociological interpretation of the parameters values allowing us to make a distinction between the case where individuals engage discussions with a networking goal, with respect to the case where individuals meet to exchange opinions and try to convince their partners to modify prior beliefs.

The model is constructed upon quite simple social interaction rules and strongly use some experimentally measured data, notably the distribution of contact times. The model is voluntarily simple, relying on few parameters, whose role can be completely understood. This choice allows us to provide a clear social interpretation of the latter.

We agree that the model could be improved by allowing several discussion topics, while in the present form only one subject is at play. This modification could be easily introduced in a forthcoming paper, however we stress that even under the proposed very simple scenario, the model is able to reproduce some real features according to the social structure of the group. We believe that this fact can be ascribed to the use of measured data that contain important hidden informations about the individuals and their actions, that otherwise could be very difficulty modeled and/or need the introduction of several unknown parameters. For instance, in the reality agents are not homogeneous and each one has his own "discussion strategy" or his own "use of time strategy" that are very difficult to evaluate; however this information is captured by the distribution of contacts times and thus we avoid the introduction of a complicated modeling for each agent.

We believe that such strategy, where models are built using simple social rules together with experimental data, could be very fruitful in the future to gain new insights into social dynamics.

References

1. Axelrod R (1997) The dissemination of culture: a model with local convergence and global polarization. J Confl Resolut 41(2):203
2. Blondel V, Krings G, Thomas I (2010) Regions and borders of mobile telephony in Belgium and in the Brussels metropolitan zone. Brussels Studies 42(4)
3. Blondel V (2008) Fast unfolding of communities in large networks. J Stat Mech 2008:P10008
4. Cattuto C et al (2010) Dynamics of person-to-person interactions from distributed RFID sensor networks. PLoS ONE 5(7):e11596
5. Chamberlin EH (1948) An experimental imperfect market. J Polit Econ 56:95
6. Deffuant G et al (2000) Mixing beliefs among interacting agents. Adv Complex Syst 3:87–98
7. Galam S (1986) Majority rule, hierarchical structures and democratic totalitarism: a statistical approach. J Math Psychol 30:426
8. Gerard HG, Orive R (1987) The dynamics of opinion formation. Adv Exp Soc Psychol 20:171
9. Krings G et al (2009) Urban gravity: a model for inter-city telecommunication flows. Journal of Statistical Mechanics L07003. doi:10.1088/1742-5468/2009/07/L07003
10. Isella L et al (2011) What's in a crowd? Analysis of face-to-face behavioral networks. J Theor Biol 271:166
11. Lambiotte R et al (2008) Geographical dispersal of mobile communication networks. Physica A 387:5317–5325
12. Schelling TC (1971) Dynamic models of segregation. J Math Sociol 1:143
13. van Dijk SJ, van Winden F (2006) On the dynamics of social ties structures in groups. J Pers Soc Psychol 43(1):78
14. Zhao K et al (2011) Social network dynamics of face to face interactions. Phys Rev E 83:056109

Chapter 100
Evolution of Fairness and Conditional Cooperation in Public Goods Dilemmas

Sven Van Segbroeck, Jorge M. Pacheco, Tom Lenaerts, and Francisco C. Santos

Abstract Cooperation prevails in many collective endeavours. To ensure that co-operators are not exploited by free riders, mechanisms need to be put into place to protect them. Direct reciprocity, one of these mechanisms, relies on the facts that individuals often interact more than once, and that they are capable of retaliating when exploited. Yet in groups, strategies targeting retaliation against specific group members may be unfeasible, because individuals may not be able to identify clearly who contributed and who did not. Still, they may assess what constitutes a fair income from a collective endeavour. We discuss here how conditional cooperation in group interactions emerges naturally and how natural selection leads populations to evolve towards a specific level of fairness (Van Segbroeck et al., Phys. Rev. Lett., 108:158104, 2012), contingent on the nature and size of the collective dilemma faced by individuals.

Keywords Reciprocity · N-Player games · Repeated games · Fairness · Evolutionary game theory

Darwinian evolution dictates that cooperation, which is the act of helping someone at a personal cost, is not evolutionary viable as a behavioral strategy since others may profit directly from this act and are not required to behave in the same way.

S. Van Segbroeck (✉) · T. Lenaerts
MLG, Université Libre de Bruxelles, Brussels, Belgium

J.M. Pacheco
Departamento de Matemática e Aplicações, Universidade do Minho, Braga, Portugal

J.M. Pacheco · F.C. Santos
ATP-group, CMAF, Complexo Interdisciplinar, Lisbon, Portugal

T. Lenaerts
AI-lab, Vrije Universiteit Brussel, Brussels, Belgium

F.C. Santos
DEI & INESC-ID, Instituto Superior Técnico, TU Lisbon, Lisbon, Portugal

T. Gilbert et al. (eds.), *Proceedings of the European Conference on Complex Systems 2012*, Springer Proceedings in Complexity, DOI 10.1007/978-3-319-00395-5_100,

Yet, cooperation is prevalent at all biological levels and has been identified as an essential complexity inducing mechanism in different contexts. In light of this paradox, theoretical biologists have postulated a number of mechanisms that explain the evolution of cooperation in the competitive world defined by Darwin's natural selection [1].

One of the most prominent mechanisms is Robert Trivers' *direct reciprocity* [2], which became famous through the invention, by Anatol Rapaport, of the tit-for-tat strategy within the game theoretical tournaments setup by Robert Axelrod [3]. Direct reciprocity states that two players in a strategic interaction may prefer to cooperate if there is a high chance that they meet again, capturing the essence of "*If you scratch my back, I will scratch yours*". Trivers showed mathematically for pairs of players that if the probability of interacting again (w) is higher than the cost-to-benefit ratio of the game ($w > c/b$), then mutual cooperation would evolve.

Given that a lot of strategic situations involve more than two players, as for instance in Social Welfare or Climate Change related problems, one can wonder how these results extend to those situations. In that context, tit-for-tat can no longer be used as a direct retaliation strategy by each player since it is no longer obvious against whom to retaliate. In addition, retaliating blindly may send the wrong signal to the other participants, triggering them to defect in turn. In [4] we examined whether there exists a strategy like tit-for-tat that leads to cooperation in groups, providing at the same time a mechanism to respond towards good or bad "group behavior"? One solution could be to decide to cooperate if a sufficient number of members cooperated in the previous round, as is the case in the two-player situation. Yet, how many players should cooperate before one decides to do the same? In other words, what is acceptable, or even, fair?

To answer these questions we analyzed an evolutionary model (for both infinite and finite populations) in which N individuals interact in the context of a public goods dilemma, the (repeated) N-persons prisoner's dilemma (NPD) [1, 5]. In this game all players have the chance to contribute an amount c to the public good. The sum of the effective contributions is then multiplied by a factor F and this result is then shared equally among all the N group members, irrespective of their contribution. This entire process repeats itself with a probability w, resulting in an average number of $\langle r \rangle = (1-w)^{-1}$ rounds per group. The outcome of the game may differ from round to round, as individuals can base their decision to contribute on the result of the previous round. We distinguished N different types of reciprocators, encoded in terms of the strategies $R_M (M \in \{1, \ldots, N\})$, where R_M players always contribute in the first round and subsequently contribute to the public good in the next round only if at least M players did the same in the previous round. In addition to these N different reciprocator types, we also included the strategy always defect (AD) to account for unconditional defectors.

In infinitely large populations, using a replicator equation [6] to describe the evolutionary dynamics between a particular reciprocator type R_M and AD, our results identify that (when $F < N$) cooperation may be enforced by the reciprocators when a sufficient fraction, called the coordination equilibrium x_L, of these reciprocators is present. When the number of individuals belonging to the type R_M is lower than

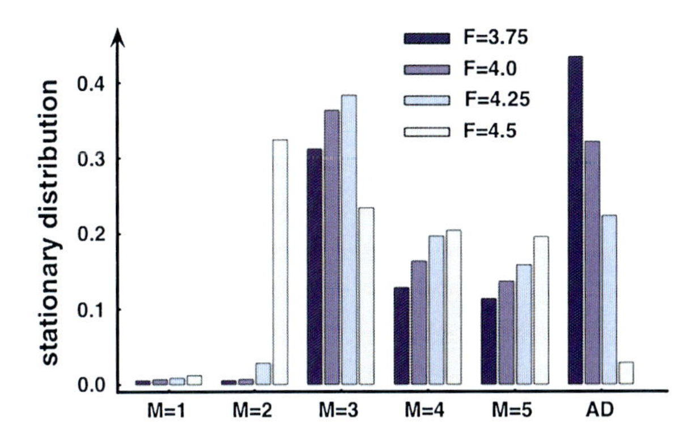

Fig. 100.1 The stationary distributions (prevalence in time of each strategy) for different values of the multiplication factor F ($w = 0.9$, $N = 5$, $Z = 100$, $C = 1$). This figure was reproduced from [4]

this threshold x_L, R_M players are unsuccessful against the AD players since they receive insufficient benefits from cooperating always in the first round. Yet, even when the number of R_M individuals is higher than x_L, they will not take over the entire population. There is always a fraction of AD players remaining, defining a coexistence equilibrium x_R . Not only do the values of these two equilibria depend on the probability of repeating the game, they are also different for each value of M: the bigger M, i.e. the more group members are required to contribute to the public good, the less often the game needs to be repeated for cooperation to become viable in the population. Still for an insufficient number of repetitions, AD will dominate the population. In general, we show [4] how the presence of conditional cooperators transforms the nature of the dilemma from defection dominance, towards N-person coordination games [7, 8].

This first analysis only takes into account the selection dynamics between two types of players, i.e. a single type of reciprocators R_M and AD. There may be an evolutionary preference for a particular value M. As such, one needs to explore the viability of all the strategies together. Moreover, as populations are finite, certain stochastic effects influence the evolutionary equilibrium obtained in this game. Consequently, we examined analytically a stochastic, finite population analogue of the deterministic evolutionary dynamics defined above, in which strategies evolve according to a mutation-selection process defined in discrete time. We assumed furthermore a limit in which mutations are rare, allowing us to compute the stationary distributions of the six different strategies. We could also show through numerical simulations that our results hold for a wider interval of mutation regimes.

As shown in Fig. 100.1, there is a specific concept of fairness, associated with an aspiration level M^*, whose corresponding strategy is most favored by evolution for a given value of F, being most prevalent among all strategies. Moreover, several values of M, corresponding to more stringent requirements in terms of the number of contributions ($M > M^*$), may coexist in the population. Their abundance is determined by the harshness of the dilemma, which is in the NPD defined by the value of the multiplication factor F: As F decreases, the fractions of the other reciprocators decreases in favor of R_M and AD, until at a certain point when the current M is no longer viable and more stringent conditions (higher M) are required to enforce cooperation.

This evolutionary dynamics becomes clearer when one considers the different evolutionary flows among strategies. By adhering to conditional reciprocation towards groups, individuals find a way towards widespread cooperation. Yet, if by chance, the population ends up adopting less demanding conditions (low M), then defection may prosper again, as it increases the temptation for this behavior to spread. Hence, stochastic effects may lead to cyclic behaviors corresponding to the oscillations between cooperation and defection akin to those observed both in nature and human history.

In summary, we have shown in [4] that even in repeated group interactions cooperation triggered by reciprocal strategies becomes viable when the probability of repeating the game is sufficiently high. Besides, this probability is dependent on what individuals perceive as a fair collective effort M that defines each reciprocal strategy. Furthermore, we show that this process leads to the emergence particular levels of fairness, which depend on the dilemma at stake.

References

1. Sigmund K (2010) The calculus of selfishness. Princeton series in theoretical and computational biology. Princeton University Press, Princeton
2. Trivers RL (1971) The evolution of reciprocal altruism. Q Rev Biol 46:35–57
3. Axelrod R (1984) The evolution of cooperation. Basic Books, New York
4. Van Segbroeck S, Pacheco JM, Lenaerts T, Santos FC (2012) Emergence of fairness in repeated group interactions. Phys Rev Lett 108:158104
5. Hardin G (1968) The tragedy of the commons. Science 162:1243–1248
6. Hofbauer J, Sigmund K (1998) Evolutionary games and population dynamics. Cambridge University Press, Cambridge
7. Pacheco JM, Santos FC, Souza MO, Skyrms B (2009) Evolutionary dynamics of collective action in N-person stag hunt dilemmas. Proc - Royal Soc, Biol Sci 276(1655):315–321
8. Santos FC, Pacheco JM (2011) Risk of collective failure provides an escape from the tragedy of the commons. Proc Natl Acad Sci USA 108(26):10421

Chapter 101
Patterns in the Occupational Mobility Network of thc Higher Education Graduates. Comparative Study in 12 EU Countries

Eliza-Olivia Lungu, Ana-Maria Zamfir, and Cristina Mocanu

Abstract The article investigates the properties of the occupational mobility network (OMN) in 12 EU countries. Using REFLEX database we construct for each country an empirical OMN that reflects the job movements of the university graduates, during the first five years after graduation (1999–2005). The nodes are represented by the occupations coded at 3 digits according to ISCO-88 and the links are weighted with the number of graduates switching from one occupation to another. We construct the networks as weighted and directed. This comparative study allows us to see what are the common patterns in the OMN over different EU labor markets.

Keywords Occupational mobility network · Social network · Econophysics

101.1 Introduction

Job mobility represents an important characteristic of the first years after collage graduation, as the rate of job change declines with age and experience [1]. According to human capital theory, youth inexperience lowers the cost of occupational mobility. This period is called the shopping and thrashing stage and is characterized by exploration and testing of the labour market.

Empirical evidences suggest that matching takes place at occupational level as information obtained by individuals working in a job is used to predict the quality of the match at other jobs within the same occupation. Thus, those working their first job are more likely to leave the current job than those working their second, third etc. job in the same occupation.

E.-O. Lungu (✉) · A.-M. Zamfir · C. Mocanu
National Research Institute for Labour and Social Protection, 6-8 Povernei Street, Sector 1, Bucharest 010643, Romania
e-mail: eliza.olivia.lungu@gmail.com

A.-M. Zamfir
e-mail: anazamfir@incsmps.ro

C. Mocanu
e-mail: mocanu@incsmps.ro

T. Gilbert et al. (eds.), *Proceedings of the European Conference on Complex Systems 2012*, Springer Proceedings in Complexity, DOI 10.1007/978-3-319-00395-5_101,

Investigations on wage formation showed that there are occupation-specific skills that are transferable across employers. This means that when a worker switches the employer or the sector, he or she loses less human capital than when changing occupation [2]. Higher loss of human capital represents higher costs for mobile workers. However, some occupations are linked to each other due to the transferability of skills. Such occupations in which skills and experience can be partially or fully transferred form career paths.

By representing the career switches in terms of networks (occupations → nodes, people changing jobs → links) we have first a visual representation of the job mobility and secondly access to all the methods developed lately by the complex network community to better understand this social and economic phenomena.

101.2 Data

The empirical occupational mobility network (OMN) is build using REFLEX database (The Flexible Professional in the Knowledge Society New Demands on Higher Education in Europe). The database contains information regarding the employment history of ISCED5 graduates from fifteen countries (Austria, Belgium/Flanders, Czech Republic, Estonia, Finland, France, Germany, Italy, Japan, the Netherlands, Norway, Portugal, Spain, Switzerland and UK). For each country it was drawn a representative sample of individuals which graduated in the academic year 1999/2000, while the data collection took place in 2005.

We restricted our study to EU countries for which we have data regarding the occupations at 3 digits of the higher education graduates (we excluded Japan, Spain and Switzerland). We employ the records about their career mobility to built for each EU country, an occupational mobility network (OMN) that covers their job shifts in the considered period. The generation under study shows a high mobility, 59 % of them changing their job in the considered time period at least once, while 30 % changed it just once.

101.3 Occupational Mobility Network

The first step of the analysis is to generate the occupational mobility network for each country, from the transition matrix between the first and the current job after graduation. We construct the networks as directed and weighted. The nodes represent the occupations coded at 3 digits according to ISCO-88[1] and the links are weighted with the number of persons moving from one occupation to another. The

[1]The International Standard Classification of Occupations (ISCO-88) can be consulted at the International Labour Office web page http://laborsta.ilo.org/applv8/data/isco88e.html (Accessed on October 5, 2012).

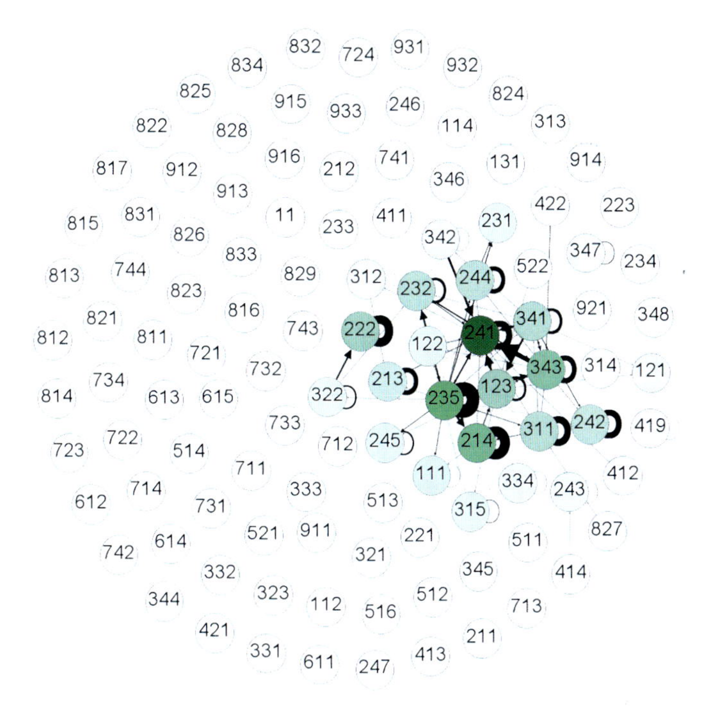

Fig. 101.1 OMN of Belgium. The thickness of the links is proportional to their weights and the color of the nodes reflects their in-strength (*light green*—isolated node, *dark green*—highly connected node). The labels of the vertexes represent the occupational codes according to ISCO-88

occupation change is defined as a modification in the occupational code at 3 digits of a graduate between the two time moments (first job and current job). We represent as self-loops the job switches in the same occupation.

As an example, the occupational mobility network of Belgium is plotted in Fig. 101.1. The vertexes are labeled with the occupational codes and their color reflects the node in-strength (light green—isolated node, dark green—highly connected node). The in-strength is calculated at the total weight of the links that point directly to a node i and it is interpreted as the number of individuals that an occupation i attracted in the considered time span. In the case of Belgium, the node with highest in-strength centrality is 241—Business professionals.

Further we calculate preliminary node specific statistics: node degree (total, in-degree and out-degree), node strength (total, in-strength and out-strength) and aggregated network statistics: network density, percentage of self-loops and the size of the giant component (see Table 101.1).

The degree of a node i is the number of links connected with it. In the case of directed networks, there are also defined in-degree, the number of links that point to node i and out-degree, the number of links that leave from node i. In this context, the in-degree of a node (occupation) i is interpreted as the total number of occupations

Table 101.1 General network properties by country (122—Production and operations department managers; 123—Other department managers; 214—Architects, engineers and related professionals; 223—Nursing and midwifery professionals; 231—College, university and higher education teaching professionals; 235—Other teaching professionals; 241—Business professionals; 323—Nursing and midwifery associate professionals; 331—Primary education teaching associate professionals; 341—Finance and sales associate professionals; 343—Administrative associate professionals; 419—Other office clerks)

Country	Density	Self loops (%)	Diameter	Giant component	Highest in-degree	Highest out-degree	Highest in-strength	Highest out-strength
Italy	0.030	11.7	5	50 %	341	343	214	214
France	0.017	19.4	9	54 %	241	343	241	123
Austria	0.017	13.9	6	45 %	241	241	241	241
Germany	0.016	13.7	7	48 %	241	214	214	214
Netherlands	0.043	9.4	6	63 %	241	343	241	123
United Kingdom	0.030	10.6	6	62 %	241	419	241	419
Finland	0.032	11.3	7	59 %	241	343	214	214
Norway	0.022	13.9	8	58 %	122	331	223	323
Czech Republic	0.035	9.5	5	55 %	343	343	241	214
Portugal	0.009	25.0	7	34 %	231	241	241	241
Belgium	0.018	12.9	5	52 %	241	235	241	235
Estonia	0.020	11.9	7	51 %	241	343	241	343

from which it can attract people and the out-degree as the number of occupations where the labour force can switch from node i.

For the majority of the countries the occupation that attracts labour force from the highest number of occupations is 241—Business professionals. In the case of out-degree, the highest value is encounter for 343—Administrative associate professionals. In terms of transferable skills theory, it implies that both occupations use such skills in a high extend.

The node strength (total, in-strength, out-strength) is calculated in the same manner as the node degree, only that this time we sum the weights of the direct neighbors of a node i. As in the case of in-degree, the node with the highest in-strength is 241—Business professionals, in most of the countries.

The network density quantifies the proportion of actual links with respect to the all possible ones. The occupational mobility network is sparse in all the countries, having a low network density (see Table 101.1). Our result confirms the importance of education in labour market allocation, as higher education graduates reach just some of the occupations, especially those requiring higher qualifications.

We also noticed a correlation between EPL (Employment Protection Legislation) and network's density, in the sense that the more flexible is the national labour market the more diverse is the occupational mobility of the collage graduates.

101.3.1 Community Structure

Communities are blocks of nodes that present dense connections within them and sparse ones with the other communities. In the context of OMN, we interpret a network community as group of occupations that actively exchange labour force between them, thus share a common set of required skills.

For this analysis we consider the OMN as binary and undirected and apply modularity optimization method. Th basic idea of the algorithm is to maximize the modularity function Q, defined as the difference between the fraction of edges within communities and expected fraction of edges in a random network.

We estimate the community structure for each country and after that we overlap them in order to see the common patterns. On average we found 5, 6 distinct communities in each country.

We calculate an index of similarity between the occupations community structure along the EU countries. The more countries in which two occupations share the same community, the higher is the value of the index. We plot the results in a heap map that has on the axis the occupation codes at 3 digits according to ISCO-88 and in each square the value of the similarity index (Fig. 101.2). The redder is the cell, then in more countries two occupations share the same community. A dark blue cell means that two occupations do not share the same community in all the countries while a dark red one means that they share the same community in all the countries. So, the blue lines represent isolated occupations (nodes) over all the countries. We can noticed the red patterns are concentrated in the first half of the

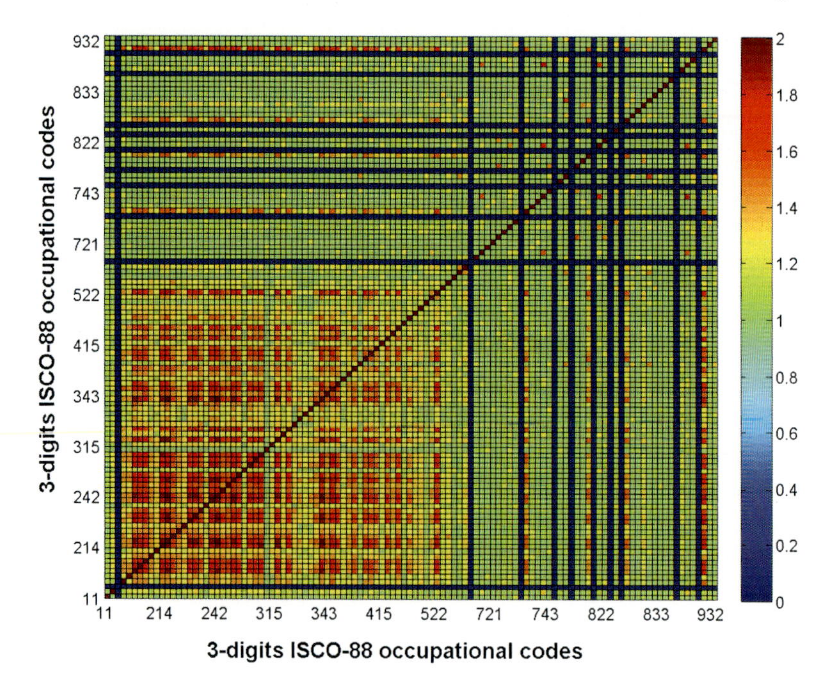

Fig. 101.2 Heap map of the similarity index between occupational communities in EU countries (0—isolated occupation in all the countries and 2—connected occupation in all the countries)

ISCO-88 classification, which makes sense because here are the occupations that requires a bachelor degree. We also notice that the occupations from this area are less connected with the rest of the network.

101.3.2 3-Node Network Motifs & Anti-motifs

The concept of network motifs had been introduced by R. Milo and his collaborators to denote "patterns interconnections occuring in complex networks at numbers that are significantly higher than those in randomized networks" [3]. For this study we are interested in the identification of the statistically significant three-nodes connected motifs (Fig. 101.3). For detecting them we employ Onnela's algorithm and compute the motif intensity $I(g)$ of a subgraph g with links l_g:

$$I(g) = \left(\prod_{(ij) \in l_g} w_{ij} \right)^{\frac{1}{|l_g|}} \qquad (101.1)$$

where $|l_g|$ is the number of links in subgraph g. The total intensity I_M of a three-nodes motif is calculated as the sum of the its subgraph intensities I_g. We preferred this method because we can take into account the weights.

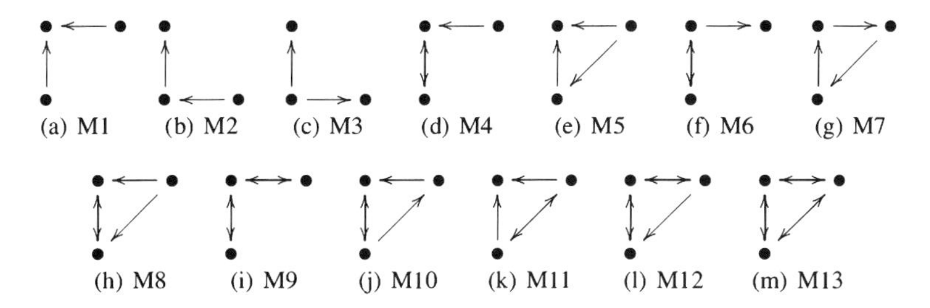

Fig. 101.3 All types of 3-node network motifs

Table 101.2 Motif fingerprint by country (1—motif, 0—absent and −1—anti-motif)

Country	M1	M2	M3	M4	M5	M6	M7	M8	M9	M10	M11	M12	M13
Italy	−1	−1	−1	1	0	0	0	0	0	0	0	0	1
France	−1	−1	−1	0	−1	0	−1	0	0	0	0	1	0
Austria	−1	−1	−1	0	−1	−1	−1	0	0	0	0	0	1
Germany	−1	−1	−1	1	0	0	0	0	1	0	0	1	0
Netherlands	0	−1	−1	1	0	0	0	0	0	0	0	1	1
UK	0	0	−1	0	0	0	0	0	0	0	0	0	1
Finland	0	−1	−1	0	0	0	0	1	0	0	0	1	1
Norway	0	0	0	1	0	0	0	0	0	0	0	1	0
Czech Republic	0	−1	−1	−1	−1	0	−1	0	0	0	0	1	1
Portugal	0	−1	−1	0	0	0	0	0	0	0	0	1	0
Belgium	−1	−1	−1	0	0	0	0	1	0	0	0	1	1
Estonia	0	0	−1	0	0	0	0	0	0	0	0	1	0

Further, we calculate the motif intensity score in order to test their statistical significance.

$$\tilde{z}_M = \frac{I_M - \langle i_M \rangle}{(\langle i_M^2 \rangle - \langle i_M \rangle^2)^{1/2}} \tag{101.2}$$

where i_M is the total intensity of motif M in the reference networks. The null-model network is constructed as an ensemble of random networks generated by reshuffling the empirical weights [4].

Table 101.2 summarizes the statistically significant motifs, by country. We also include in the table the motifs that are under represented in the OMN, as anti-motifs. We notice that M3 is poorly represented in all the countries, beside Norway, which means that there are not occupations from where the labour force just moves away. Rather, we could say that it is more a reciprocal exchange between occupations (M12, M13).

101.4 Conclusions

We investigate patterns of occupational mobility by employing a network based approach in which the nodes are occupations are 3 digits and the links are weighted with the flows of individuals moving from one occupation to another. Such an approach helps us to better visualize paths of mobility and calculate network indicators in order to understand models of connectivity between different occupations.

References

1. Topel RH, Ward MP (1992) Job mobility and the careers of young men. Q J Econ 107(2):439–479
2. Kambourov G, Manovskii I (2009) Occupational specificity of human capital. Int Econ Rev 50(1):63–115
3. Milo R, Shen-Orr S, Itzkovitz S, Kashtan N, Chklovskii D, Alon U (2002) Network motifs: simple building blocks of complex networks. Science 5594:824–827
4. Rubinov M, Sporns O (2010) Complex network measures of brain connectivity: uses and interpretations. NeuroImage 52(3):1059–1069

Part VII
Satellite Meeting: Complexity in Spatial Dynamics

Chapter 102
Modeling Urban Patterns Across Geographical Scales by a Fractal Diffusion-Aggregation Approach

Roberto Murcio and Suemi Rodríguez-Romo

Abstract An integral urban growth model is introduced. We developed this model to capture the different spatial morphologies and urban dynamics observed when more than one city are interacting on a specific region creating large metropolitan areas at different geographical scales. For small scales (1:1500000) our model is based on two well-known fractal growth processes, diffusion and percolation, in order to represent two of the main urban growth drivers: people migration and economics of agglomeration respectively. Morphology at large scales (1:50000) is derived from a Self-Organized Criticality (SOC) model, adapted to urban interactions to explore the possible relations between "avalanches" and city redensification processes.

102.1 Assumptions for the Model

The approach presented here is based on three previously reported urban models:

- Colored diffusion-limited aggregation for urban migration [1],
- Modeling large Mexican urban metropolitan areas by a Vicsek-Szalay approach [2], and
- Modeling Mexican urban metropolitan areas by a self-organized criticality approach [3].

The model is based on the following assumptions:

1. Cities and systems of cities, at certain scales, reflect fractal growth patterns.
2. Cities and systems of cities exhibit Self-Organization capabilities.
3. Cities and systems of cities can be thought of as complex systems with certain Self-Organization Criticality characteristics.

R. Murcio (✉) · S. Rodríguez-Romo
Centro de Investigaciones Teóricas, Facultad de Estudios Superiores Cuautitlán, Universidad Nacional Autónoma de México, Estado de México, Mexico
e-mail: rmurcio@unam.mx

T. Gilbert et al. (eds.), *Proceedings of the European Conference on Complex Systems 2012*, Springer Proceedings in Complexity, DOI 10.1007/978-3-319-00395-5_102,

Fig. 102.1 Great metropolitan area of central Mexico 12044.8 km^2 (7528 square miles) 22 million people

Fig. 102.2 Vector image of the actual central México metropolitan areas

Figures 102.1 and 102.2 depict two different views of the Central Mexico Metropolitan Area (CMMA). Notice that it is composed primarily of three large metropolitan areas, México, Puebla and Toluca.

Our ambition is to model metropolitan areas from first principles, at two different geographical scales, 1:1000000 and 1:200000. To accomplish this, different theoretical models will be used:

(i) Diffusion Limited Aggregation,
(ii) Percolation, and
(iii) Self-Organized Criticality.

102.2 Urban Migration—Diffusion Limited Aggregation

Why people move? Is there a reason why a group of persons would change its way of life in one geographical place to begin a new one in another? Economics, land and government changes changes are never simple, whether in type or intensity. New opportunities emerge in different places. Things could just not have been working any more at home, or, maybe, are getting better somewhere else. It is the difference

Fig. 102.3 Classic configuration obtained with the DLA urban model. This model can be consulted in Colored diffusion-limited aggregation for urban migration can be consulted in Ref. [1]

between the present and the feeling of other opportunities which push people's decisions in motion. When this feeling grows bigger and bigger and there are no barriers in the way, a migration takes place. For migration to occur we need three important features:

- Complementarity—In order to migrate between places, an offer must exist in one place and a need in another
- Intervention opportunity—Complementarity between places can generate exchange just in the absence of intervention opportunities
- Migration Cost—Is a weight on real time and cost. If the time and cost of moving a distance is too big, movement will not take place, regardless of perfect complementarity and the absence of intervention opportunities

Figure 102.3 depicts a typical Diffusion Limited Agreggation growth pattern [4], observed under those conditions [1].

102.3 Urban Growth Through Economics of Agglomeration—Vicsek-Szalay-Batty

Tamás Vicsek and Alexander S. Szalay [5] proposed a cellular automaton model (VS) to study the fractal distribution of galaxies. In their lattice model, a cell i, which represents a mass element, would become part of a galaxy based on two parameters; the potential of belonging to a galaxy at position i, and a given threshold that regulates such potential.

Economic agglomerations theory states that

> "Just as matter in the solar system is concentrated in a small number of bodies (the planets and their satellites); economic life is concentrated in a fairly limited number of human settlements (cities and clusters). Furthermore, paralleling large and small planets, there are large and small settlements with very different combinations of firms and households [6]."

Fig. 102.4 Example of a configuration obtained by our model set to ($\Phi = 4.5$, $t = 500$). The three main clusters from the actual CMMA are clearly present, along with several satellite clusters. This model can be consulted in Modeling large Mexican urban metropolitan areas by a Vicsek Szalay approach can be consulted in Ref. [2]

The use of a model developed for galaxies to understand urban settlements is thus a reasonable approach. Batty [7] took VS and adapted it to an urban context, the VSB model. Letting $P_i(t)$ denote the potential of population growth at site i and time t, we have the following evolution:

$$P_i(t+1) = \sum_{j=\Omega_i} \frac{P_j(t)}{5} + \varepsilon_i(t) \tag{102.1}$$

$$P_i(t) > \Phi \quad \text{and} \quad D_i(t-1) = 0 \quad \text{then } D_i(t) = 1, \quad \text{otherwise } D_i(t) = 0 \tag{102.2}$$

In this system, Ω_i denotes the four cells neighboring i on a two-dimensional square grid, $\varepsilon_i(t)$ is a random variable which take values ± 1 with equal probabilities, Φ is a threshold parameter, and $D_i(t)$ is an index indicating whether or not cell i has undergone development.

Using this approach, we found important and, to some extent, surprising similarities with the actual digitized area. These include Box Counting dimension, Zipf's law, morphology, and possible 2nd-order phase transitions between sprawl and compact configurations depending on the threshold parameter. These configurations are Independent of "time".

Figure 102.4 illustrates the type of configurations derived by the model [2].

102.4 Urban Redensification—Self-organized Criticality

Like a pile of sand, cities change, it is an undeniable fact. Cities change from one state to another all the time due the aggregation of new activities, such as births, migration, new real estate developments, etc. Most of the activities change through

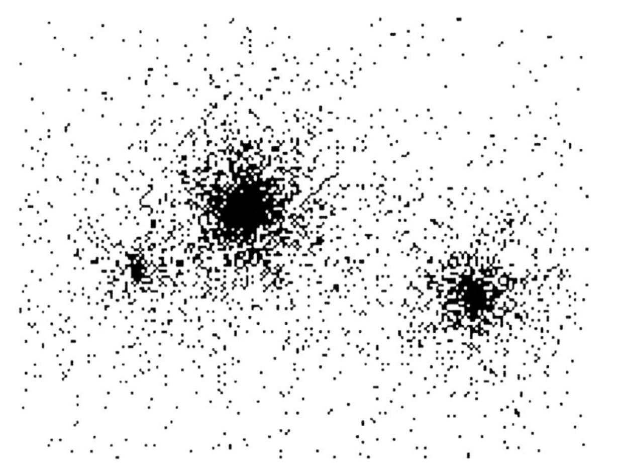

Fig. 102.5 Example of a configuration obtained by our model at $t = 5000$. The three main clusters from the actual CMMA are clearly present, along with many isolated points, each one representing an urban town. This model can be consulted in Ref. [3]

processes of redistribution. Each time an activity changes its location, it triggers a chain reaction in which other activities are motivated to move, because economic agents that make up these activities readjust to the new circumstances. Empirical evidence tell us that the city continues to exist with basically the same morphology while these chain reactions occurs, and that such reactions do not continue indefinitely. These processes can be modeled using self-organized criticality [8–10]. In first instance, assume that a population is distributed in an urban area according to a empirical power law that relates population density $\rho(r)$ at a distance r from the city's Central Business District (CBD) as follows:

$$\rho(r) \sim r^{-\alpha} \tag{102.3}$$

where α is a density gradient parameter, which in Ref. [7] is held constant over time, even though it appears to decrease gradually as the city grows. This parameter affects the density of the population, not the SOC process itself, so for practical purposes, we follow this recommendation in this paper, and α will remain constant through all simulations. In terms of the sand pile model, the critical slope of the distribution of the population over a city is controlled by α. Certainly avalanches occur, but do not follow a power law.

Figure 102.5 illustrates a configuration obtained by this model.

102.5 Integrated Model

We combine the initial parameters of Vicsek Szalay Batty and SOC in the same layout

- Initial Potential—1:1000000
- Maximum Capacity MC—1:200000
- Current Capacity CC—1:200000

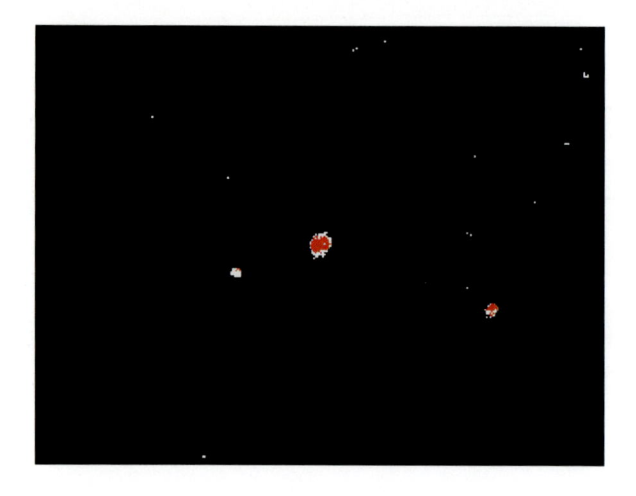

Fig. 102.6 Urban configuration at $t < 50$. Threshold equals to 4.5. *The red portions* represent the avalanches

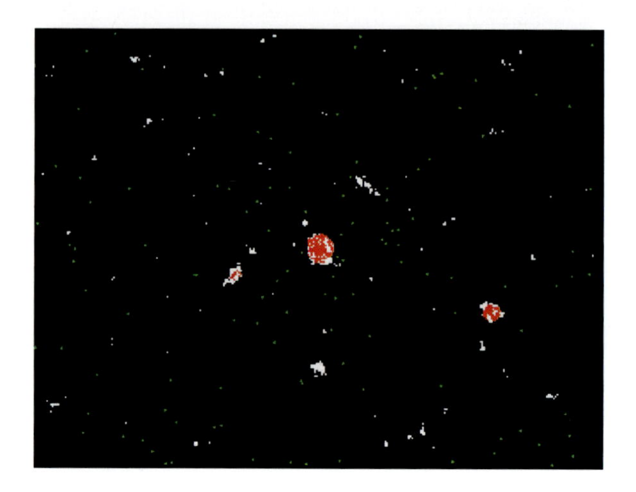

Fig. 102.7 Urban configuration at t = 150. Threshold equals to 4.5

The local interactions provoke local avalanches, which increase CC, which, at some point, affects the global potential of a zone. The potential is acting at 1:1000000, a scale at which we can actually observe urban growth.

Figures 102.6, 102.7 and 102.8 summarize the configuration observed in the course of time.

At $t = 0$ the process is initiated.

At $t = 50$ a series of migration waves begins to take place. These migrating units (agents) follow a direct walk towards the central cluster, as seen in Figs. 102.7 and 102.8.

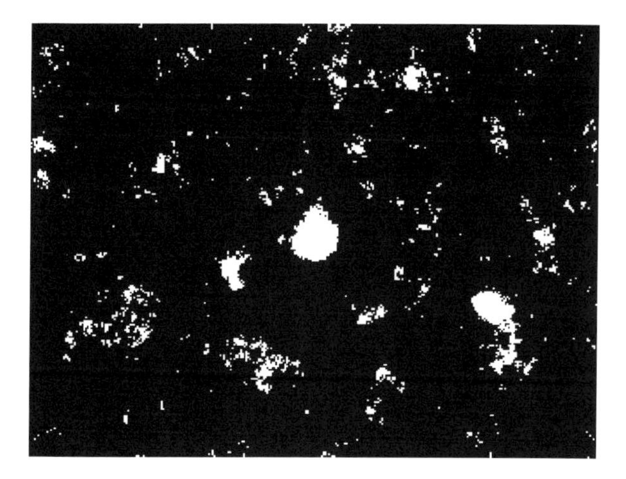

Fig. 102.8 Urban configuration at t = 300. Threshold equals to 4.5

102.6 Comments

- At the 1:1000000 scale we have not obtained significantly different morphologies than with VSB.
- Avalanches do not follow a power law distribution at a Metropolitan Area scale
- But Zipf's law is still fulfilled—Fuzzy Clustering
- Emerging patters obtained are consistent with the actual urban patterns.
- The influence of the Current Capacity parameter needs further investigation.
- With only VSB and Migration, we have obtained a possible second order transition.
- This situation will probably would repeat with the inclusion of the SOCs parameters.

References

1. Murcio R, Rodriguez-Romo S (2009) Physica A 388:2689–2698
2. Murcio R, Rodríguez-Romo S (2011) Physica A 390(16):2895–2903
3. Murcio R, Rodriguez-Romo S (2011) Modeling Mexican urban metropolitan area by a self-organized approach. In: Sayama H, Minai A, Braha D, Bar-Yam Y (eds) Unifying themes in complex systems volume VIII: Proceedings of the eighth international conference on complex systems. New England complex systems institute series on complexity. NECSI Knowledge Press, Cambridge, pp 630–642. ISBN 978-0-9656328-4-3
4. Witten TA, Sander LM (1981) Phys Rev Lett 47:1400
5. Vicsek T, Szalay A (1987) Fractal distribution of galaxies modeled by a cellular-automaton-type stochastic process. Phys Rev Lett 58:2818–2821
6. Fujita M, Thisse JF (2002) Economics of agglomeration. Cities, industrial location, and regional growth. Cambridge University Press, Cambridge

7. Batty M (2007) Cities and complexity: understanding cities with cellular automata, agent-based models, and fractals. MIT Press, Cambridge
8. Allen PM (1997) Cities and regions as self-organizing systems: models of complexity. Gordon & Breach, New York
9. Bak P, Tang C, Wiesenfeld K (1987) Phys Rev Lett 59:381
10. Bak P, Tang C, Wiesenfeld K (1988) Phys Rev A 38:364

Chapter 103
Generating Individual Behavioural Routines from Massive Social Data for the Simulation of Urban Dynamics

Nick Malleson and Mark Birkin

103.1 Introduction

This paper presents recent methodological developments aimed at establishing the daily spatio-temporal behaviour of individual people from their activity on social-networking services. Ultimately, the methods will be used to provide supplementary data to a complex agent-based model of urban dynamics. This work will review recent developments in the use of 'crowd-source' data for understanding society (Sect. 103.2), outline novel methods for capturing the spatio-temporal behaviour of individual users (Sect. 103.3) and discuss how the this information can be incorporated into a model of urban dynamics (Sect. 103.4). Finally, Sect. 103.5 will conclude the article with a discussion about current challenges and future research.

103.2 The Social Data Deluge

Recent work recognising the character of cities as complex systems suggests that aggregate models have the effect of smoothing out a city's underlying dynamism [1]. Hence individual-level modelling approaches (such as agent-based modelling) have become a popular means of representing various urban phenomena such as disease spread [12], crime [15] and land-use [16]. However, individual-level models are often criticised because they lack proper validation, which stems from a shortage of suitable data. Models often use population censuses and social surveys that have four major drawbacks:

1. they tend to deal with aggregate groups rather than individuals;
2. they occur infrequently;

N. Malleson (✉) · M. Birkin
School of Geography, University of Leeds, Leeds LS2 9JT, UK
e-mail: n.malleson06@leeds.ac.uk

T. Gilbert et al. (eds.), *Proceedings of the European Conference on Complex Systems 2012*, Springer Proceedings in Complexity, DOI 10.1007/978-3-319-00395-5_103,

3. they are usually focused on the attributes and characteristics of the population, rather than their attitudes and behaviours;
4. they provide a snapshot view rather than a dynamic and continuous perspective.

New sources of social data—commonly referred to as "crowd-sourced data" [18] and "volunteered geographical information" [10]—are becoming available that resolve many these drawbacks. Services such as Facebook, FourSquare, Flikr and Twitter provide large-scale social network data that have the potential to revolutionise the study of social phenomena and our approach to social simulation. These sources potentially contain a wealth of information about peoples' spatio-temporal behaviour, if relevant information can be extracted from the vast amounts of noise that are also present. Numerous researchers are developing new methods for data collection [5, 17, 20] and making use of crowd sourced data in diverse fields such as social network analysis [6], crisis management/evaluation [8], disease spread [9] and election forecasts [21]. However, the application of crowd-sourced data to understanding urban dynamics is a relatively under-researched field.

The use of crowd-sourced data in the social sciences shares a number of similarities with the "fourth paradigm" [2] data intensive activities that are usually limited to the physical and biological sciences. Under this new paradigm, the vast amounts of data that are often generated by physical or biological experiments can be accessed through publicly available interfaces which effectively provides a much larger number of researchers with the opportunity to use the data. This new paradigm is not an unrealistic goal for the social sciences.

The growing use of online social media also has the potential to lead to a paradigm shift in the design and delivery of traditional social surveys. For example, the Mapiness [13] project has used a mobile phone application to collect data on peoples' perceived levels of happiness. Importantly, the geographical location of the respondent is known (using the phone's GPS equipment) so the researchers are able to relate their happiness to environmental conditions. At the time of writing, over 50000 people had installed the application and the project had collected approximately 3.5 million individual data points. Clearly this amount of data would be extremely difficult to collect using a traditional survey; suggesting the potential for a more concerted effort in developing applications that are attractive to end users. In a similar vein, Birkin et al. [3] utilised data collected through a large online survey which was publicised on UK local news. The survey sought peoples' views on how their travel behaviour would change following the implementation of a road charging scheme and Birkin et al. used this information to calibrate a model of future traffic flows. Again, the number of responses to the survey (more than 15000) would have been difficult to gather using traditional methods.

103.3 Establishing Individual Behaviours

The overarching aim of the research is to build an accurate agent-based model of the daily ebb and flow of a city. At present, the research focusses on commuting

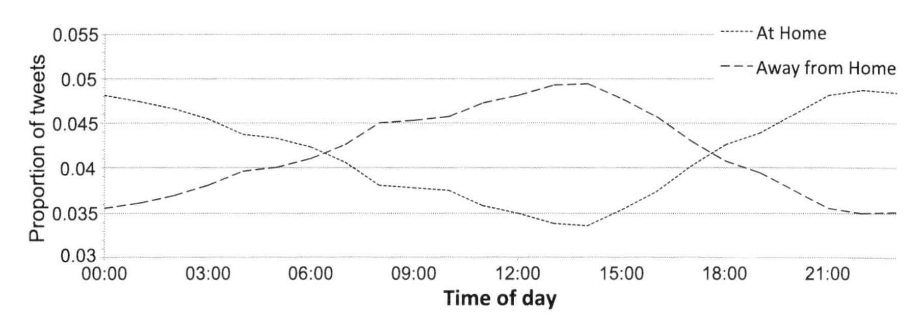

Fig. 103.1 The proportions of tweets corresponding to 'home' or 'work' activities for an average weekday (all Monday–Thursday tweets) for all users

behaviour, but future work will incorporate additional key behaviours (shopping, leisure activities, school, etc.). Therefore in the current research iteration we attempt to identify two specific behavioural states: 'at home' or 'at work'. This section will discuss how data from Twitter can automatically be classified into these two categories.

Initially, a person's base location (referred to as their 'home') is identified by calculating the most dense location of all their tweets. In future, it will be possible to generate more accurate estimates of behaviour through the incorporation of the textual content of the message as well as spatial location—see, for example, Eisenstein et al. [7]. The Kernel Density Estimation (KDE) algorithm, as formulated by Silverman [19], was applied to every distinct tweet location to identify the point with the highest density which was assumed to be the user's home. This process can be repeated for every user in the data.

Following the identification of 'home' locations, it is possible identify where each tweet was sent from relative to the location of the user's home. At present, it is assumed that all tweets within 50 meters of a user's home indicate that they are 'at home', otherwise they are assumed to be 'away from home'. By temporally aggregating this information for every user into a single day, it is possible to perform a cursory validation of the behaviour that has been identified. To this end, Fig. 103.1 presents the proportions of each behaviour in each hourly time period for all users. Tweets on days Friday to Sunday have been ignored at this stage in order to capture likely commuting behaviour. It is clear that there are more 'away from home' behaviours during the day and a greater proportion of 'at home' tweets in the evening which is an encouraging preliminary result.

103.4 Towards an Individual-Level Model

The identification of messages that originate from 'home', or elsewhere, has laid the groundwork for later use in an agent-based model of urban dynamics (which is currently under development). The aim is to use these data in conjunction with other sources, such as population censuses, to provide novel ways of seeding, calibrating and validating the model.

The first stage of the modelling process is to use *microsimulation* to disaggregate the 2001 UK census (and the 2011 census when it becomes available) to generate a synthetic population of individuals who will occupy the model [4, 11]. Each individual is created with number of personal attributes which can be used in later research to tailor their behaviour. The individuals are grouped into households during the microsimulation process and these are geo-referenced to the resolution of the smallest census area boundary (called an 'output area'). With the incorporation of additional building data they can be assigned to distinct houses [14, e.g.], although this has not been attempted at present. Individuals' destinations (their workplaces) are also estimated from UK census.

Once the synthetic population has been initialised, an agent-based model is used to simulate their behaviour. At present, individuals travel once from home to work each day. A simplifying assumption at this stage is that agents travel at a constant speed and in a straight line from home to work, although constraining and allocating the flows to transport networks is planned as a future objective. At each iteration, agents decide whether to go to their home or to their work depending on the simulated time. Agent's sample from normal probability distributions to determine the times that they will start to travel home or to work. Thus the model can be configured by altering the mean and standard deviation of the probability distributions. There are four parameters overall: the mean and standard deviation of the time the agents leave their homes (μ_h and σ_h) and the times that the agents leave their work (μ_w and σ_w). Figure 103.2 illustrates the distribution of agents in the running model with default (uncalibrated) values for μ_h, σ_h, μ_w and σ_w. Agents travel from within Leeds and the outskirts of the city towards the central business district for work in the morning and then return home again in the afternoon.

Crowd-sourced data will then used for model calibration. A genetic algorithm will be applied in order to search for optimal values for the model parameters and model error will be calculated by comparing the overall times that agents spend at work or at home to the data from Twitter (e.g. the data presented in Fig. 103.1). Hence the research will generate a temporally realistic, spatially-explicit model of individual commuting behaviour, calibrated using novel crowd-sourced data, which will help us to understand the spatio-temporal dynamics of urban areas.

103.5 Challenges, Conclusions and Ongoing Research

Crowd-sourced data have the potential to revolutionise the ways that social scientists collect data and how simulation models make use of data. However, there are considerable challenges that must be addressed before the data will be truly useful.

Bias There are potentially insurmountable biases associated with crowd-sourced data. For this research, only a small percentage of all tweets—found to be 1 % by

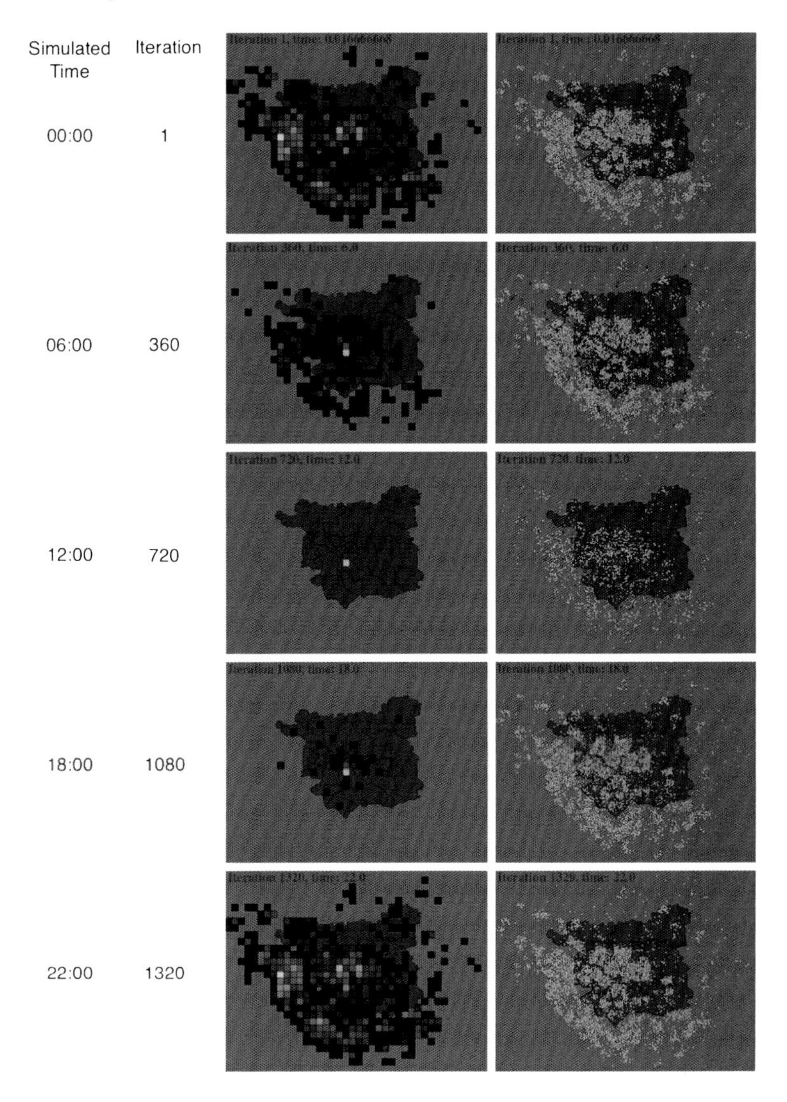

Fig. 103.2 The prototype model running over a single simulated day (uncalibrated). Agents travel from within Leeds and the outskirts of the city towards the central business district. *The left maps* show density (*white cells* are the most densely populated). *On the right maps*, *white points* indicate stationary agents, *red* signifies agents travelling to work and *green* shows agents travelling home

[8]—that are sent from mobile devices also have accurate GPS coordinates associated with them. In addition, approximately only a 1 % sample of all activity is actually revealed by Twitter without charge. Perhaps more substantially, there are large sections of society that do not use social-networking services and hence will not be captured in the analysis. However, this does not need to be an insurmountable drawback if precautions are taken. By analysing crowd-sourced data with other sources,

such as population censuses and geo-demographic products, it will be possible to estimate which sections of society are not captured in the data. Supplementary data can then be used to capture the behaviour of these missing populations. In essence, the crowd-sourced data can be applied to populations that are well represented—young people visiting night clubs might be an example of a situation where this will work—and replaced with alternative data for phenomena that involve people who do not participate in social networking.

Reliability Accepting biases in the data, there are still questions over how reliable the process of generating behavioural profiles can be. Although the behavioural estimates appear to be reliable in aggregate (see Fig. 103.1) there are numerous individuals for whom nonsensical behaviour is generated (e.g. individuals who appear to be away from home for 24 hours per day). However, this suggests that an analysis of the locations of twitter messages *in isolation* is insufficient to establish a person's behaviour, rather than a crippling drawback with the approach in general. It can be expected that the coupling of text mining techniques with traditional spatial analysis (e.g."spatio-temporal text mining") will provide a means of more accurately classifying a person's behaviour.

Ethical Considerations Finally, ethical considerations must play a large part in any future research. The concept of crowd-sourced/volunteered data is relatively new and ethical frameworks for using the data in research are still under-developed. Messages posted on public forums (including Twitter) are inherently accessible to the general public and hence do not fall under the remit of traditional human-subject research. However, it is possible that users are naively unaware of the amount of information they reveal about themselves when they use social networking services. Although early work in this area suggests that users do not need to be asked for consent and university ethics committee consideration need not be sought [22], more work in this area is clearly needed.

103.6 Conclusion

This paper has presented a novel effort to classify individual behaviour from the spatial location of twitter messages and use this information to improve the accuracy of an agent-based model of urban dynamics. Although current behaviours are limited to commuting to work, the intention is to extend this to a broader range of behaviour including shopping, socialising, education, etc. (see [23] for example). Immediate future work will focus on methods to generate more reliable behavioural estimates (e.g. taking the textual content into of messages into account as well as their spatial location) including machine-learning algorithms and necessary modelling improvements such as the incorporation of a transport network. Although there are considerable barriers to the use of these data in earnest, this preliminary research has shown, at least, that there is utility in continuing to use crowd-sourced data. This type of data is only going to become more prolific hereafter so the tools developed now will undoubtedly be useful, after some refinements, in the future.

References

1. Batty M (2005) Agents, cells, and cities: new representational models for simulating multi-scale urban dynamics. Environ Plan A 37:1373–1394
2. Bell G, Hey T, Szalay A (2009) Beyond the data deluge. Science 323:1297–1298
3. Birkin M, Malleson N, Hudson-Smith A, Gray S, Milton R (2011) Calibration of a spatial simulation model with volunteered geographical information. Int J Geogr Inf Sci 25(8):1221–1239
4. Birkin M, Turner A, Wu B (2006) A synthetic demographic model of the UK population: Methods, progress and problems. In: Proceedings of the second international conference on e-social science, Manchester, UK
5. Cheng Z, Caverlee J, Lee K (2010) You are where you tweet: a content-based approach to geo-locating twitter users. In: Proceedings of the 19th ACM international conference on Information and knowledge management, CIKM'10. ACM Press, New York, pp 759–768
6. Davis CA Jr, Pappa GL, de Oliveira DRR, de F, Arcanjo L (2011) Inferring the location of twitter messages based on user relationships. Trans GIS 15(6):735–751
7. Eisenstein J, O'Connor B, Smith NA, Xing EP (2010) A latent variable model for geographic lexical variation. In: Proceedings of the 2010 conference on empirical methods in natural language processing, EMNLP'10. ACL, Stroudsburg
8. Gelernter J, Mushegian N (2011) Geo-parsing messages from microtext. Trans GIS 15(6):753–773
9. Gomide J, Veloso A, Meira W, Almeida V, Benevenuto F, Ferraz F, Teixeira M (2011) Dengue surveillance based on a computational model of spatio-temporal locality of Twitter. In: Proceedings of the ACM, Koblenz, Germany, pp 1–8
10. Goodchild M (2007) Citizens as sensors: the world of volunteered geography. GeoJournal 69:211–221
11. Harland K, Heppenstall A, Smith D, Birkin M (2012) Creating realistic synthetic populations at varying spatial scales: a comparative critique of population synthesis techniques. J Artif Soc Soc Simul 15(1):1. http://jasss.soc.surrey.ac.uk/15/1/1.html
12. Johansson A, Batty M, Hayashi K, Al Bar O, Marcozzi D, Memish Z (2012) Crowd and environmental management during mass gatherings. Lancet Infect Dis 12(2):150–156
13. MacKerron G (2012) Happiness and environmental quality. PhD thesis, The London School of Economics and Political Science, London
14. Malleson N, Birkin M (2011) Towards victim-oriented crime modelling in a social science e-infrastructure. Philos Trans R Soc, Math Phys Eng Sci 369(1949):3353–3371
15. Malleson N, Heppenstall A, See L (2010) Crime reduction through simulation: an agent-based model of burglary. Comput Environ Urban Syst 34(3):236–250
16. Matthews R, Gilbert N, Roach A, Polhill J, Gotts N (2007) Agent-based land-use models: a review of applications. Landsc Ecol 22(10):1447–1459
17. Russell M (2011) Mining the social web. In: Analyzing data from Facebook, Twitter, LinkedIn, and other social media sites. O'Reilly Media, Sebastopol
18. Savage M, Burrows R (2007) The coming crisis of empirical sociology. Sociology 41(5):885–899
19. Silverman BW (1986) Density estimation for statistics and data analysis. Chapman & Hall, New York
20. Stefanidis A, Crooks A, Radzikowski J (2011) Harvesting ambient geospatial information from social media feeds. GeoJournal 1–20. doi:10.1007/s10708-011-9438-2
21. Tumasjan A, Sprenger TO, Sandner PG, Welpe IM (2011) Election forecasts with Twitter: how 140 characters reflect the political landscape. Soc Sci Comput Rev 29(4):402–418
22. Wilkinson D, Thelwall M (2011) Researching personal information on the public web: methods and ethics. Soc Sci Comput Rev 29(4):387–401
23. Yang Y, Atkinson P, Ettema D (2008) Individual space–time activity-based modelling of infectious disease transmission within a city. J R Soc Interface 5(24):759–772

Chapter 104
Spatial Externalities Approach to Modelling the Preferential Attachment Process in Urban Systems

Igor Lugo

Abstract We studied the application of spatial externalities and complex network analysis to model a spatial network and to explore a preferential attachment mechanism in urban systems. Nodes in the spatial network represented cities (urban polygons) with hierarchical attributes computed by complex network measures, and edges corresponded to road segments (highways) related to their length. City attributes consisted of city size, accessibility, and a new measure that we called city-road infrastructure. To identify rules of connectivity among cities, we used the Cobb-Douglas function to correlate such attributes and spatial econometrics models to test its functional form and statistical significance. The case of Mexico illustrated the application of this approach. The results of this study indicated that not only hierarchical attributes but also spatial dependences of cities have to be specified in the preferential attachment mechanism.

Keywords Spatial externalities · Preferential attachment · Urban systems

104.1 Introduction

The analysis of the structure and evolution of networks is closely related to the study of urban systems, in particular to understand basic rules that govern the growth and connectivity of networks. However, there has been little attention to model and explore a large-scale spatial system where nodes are related to the hierarchical differentiation of cities, and road infrastructures are associated with edge attributes. For this reason, we investigate the application of spatial externalities, in economics, and complex network analysis to generate a planar spatial network that provides the basic system configuration and to study rules of connectivity among cities based on a preferential attachment mechanism.

I would like to thank Denise Pumain for her comments and suggestion related to this document.

I. Lugo (✉)
Centro Regional de Investigaciones Multidisciplinarias, Universidad Nacional Autónoma de México, Av. Universidad S/N, Cto. 2, Col. Chamilpa 62210, Cuernavaca, Morelos, Mexico
e-mail: igorlugo@correo.crim.unam.mx

T. Gilbert et al. (eds.), *Proceedings of the European Conference on Complex Systems 2012*, Springer Proceedings in Complexity, DOI 10.1007/978-3-319-00395-5_104,

One of the first discussions and analyses of the basic rules of connectivity in urban systems emerged during the 1970s with the document of Burghardt [9]. He investigated the formation of road and city networks in the Roman Pannonia, suggesting that military objectives explained the generation of roads between cities. On the other hand, the work of Irwin and Huges [17] pointed out that centrality measures in social networks analysis could be applied to study spatial networks. More recently, a large volume of published documents have described either the economic perspective of spatial dependences among cities, for example Black and Henderson [8], Duranton and Puga [10], and Ioannides and Skouras [16], or the structure and dynamics of roads in spatial network, for example Yamis et al. [24], and Xie and Levinson [22, 23].

Two important methods have been developed to analyze urban and spatial systems: spatial econometrics and complex networks. The former is related to the field of spatial externalities that study the causality between economic variables associated with geospatial information, for example Anselin [4], and the latter analyzes the structure and change of networks based on their topological and geometrical properties, for example Barthélemy [6]. Because both methods have investigated different aspects of the same subject, the purpose of the analysis is to link them generating a planar spatial network, where nodes represent cities (urban polygons) with hierarchical attributes computed by complex network measures, and edges correspond to road segments (highways) related to their length, and proposing a preferential attachment process based on a functional form commonly used in econometrics. City attributes consist of city size, accessibility, and a new measure that we called city-road infrastructure. Therefore, a Cobb-Douglas function is used to correlate such attributes, and spatial econometrics models were estimated to test its statistical significance.

To illustrate the application of this approach, we study the case of Mexico based on geospatial data of urban polygons and road lines in the year of 2005 (INEGI, [15]). The result to emerge from the analysis is that not only hierarchical attributes but also spatial dependences of cities have to be specified in the preferential attachment mechanism. This finding suggests that geography and road infrastructures matter for explaining the generation of new road sections among cities.

104.2 Methods

104.2.1 Spatial Externalities

Spatial externalities can be defined as the neighboring configuration effect among spatial objects. It encompasses the concepts of spatial spillovers and dependence (Abbot [1], Sampson et al. [19]). The economics view of such externalities is related to the notion of economies of agglomeration, sources of strong geographic concentration of consumption and production (Fujita and Thisse, [12]), and they are measured by spatial econometrics models.

These models require variables and their causality to specify and estimate spatial dependences. We were interested in three data that can be related to the hierarchical differentiation of cities: city size, accessibility, and city-road infrastructure. The city size measured the number of habitants, and it was related to the Pareto-Zipf distribution suggesting a network structure with high levels of interactions in few places (Pumain, [18]; Sarabia and Prieto, [20]; Eeckhout, [11]). Accessibility was used to represent proximity of one geospatial object with respect to all others (Wilson, [21]; Haynes and Fotheringham, [14]; Getis, [13]). In particular, we were interested in the type 2 accessibility because it is related to physical infrastructures (Batty, [7]). Such accessibility can be defined as the total distance from one place to all others based on the shortest routes in a planar graph. The city-road infrastructure reflected the level of coordination and mediation of a city into the flow of resources in a spatial network, that is, the concentration of roads in a city. We computed it constraining the betweenness centrality to a group of urban areas that contains a number of road segments.

To determine whether the measure of the city-road infrastructure provided fundamental information about hierarchy in urban systems, we used a Cobb-Douglas function and defined the city size and accessibility as explanatory variables:

$$CRI_i = K S_i^{\alpha} A_i^{\beta} \tag{104.1}$$

where CRI_i is the city-road infrastructure of city i, S_i is the city size, A_i is the accessibility, K is a proportionality constant, and α and β are parameters related to the participation of each explanatory data in the response variable. Taking logarithms to the left side of (104.1), we specified the statistical model as following:

$$CRI_i = \ln(K) + \alpha \ln(S_i) + \beta \ln(A_i) + \varepsilon_i \tag{104.2}$$

where ε_i represented the error term. Therefore, this model exemplified a linear representation of city attributes and corresponded to a constrained model in spatial econometrics (Anselin, [2]).

104.2.2 Network Analysis

The creation of the planar spatial network incorporated polygon data into road segments and associated node attributes (road intersections and dead ends) with inside and outside polygon characteristics. The geoprocessing operation for creating such a network was divided into three steps: identifying lines inside polygons, transforming line segments to individual lines, and defining node attributes.

In addition, we used five types of complex network measures that help to identify representative nodes inside urban areas, compute topological and geometrical characteristics of the planar graph, and determine the level of accessibility and the city-road infrastructure. These measures were the following: the weighted average nearest-neighbors degree, shortest path, diameter, circuity, and edge betweenness centrality.

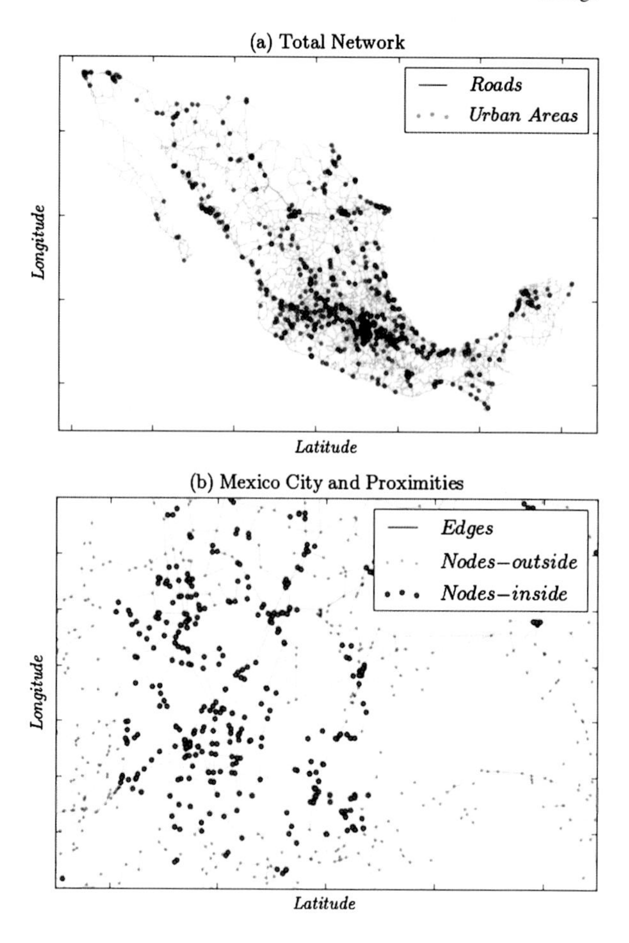

Fig. 104.1 Visualization of the planar spatial network

Therefore, in urban systems, where cities are connected by road, the preferential attachment process can be defined as the probability that city i is connected to j by a new set of road sections. Based on the causality, spatial configuration, and hierarchical measures of cities, we proposed to test the following mechanism:

$$P_{i \to j} = \frac{CRI_i}{\Sigma_j CRI_j} \tag{104.3}$$

where the probability depended on the city-road infrastructure measure, which is a function of the city size and accessibility.

104.3 Results

The result obtained from the geoprocessing operation is presented in Fig. 104.1. The planar spatial network had 891 urban areas, where the total number of nodes

Table 104.1 Model estimations (CRI_i response variable)

Predictors	Models		
	OLS	Lag (ML)	Error (ML)
K	−0.0031	−0.0058	−0.0017
	(0.005)	(0.006)	(0.006)
$\ln(S_i)$	0.0015	0.0016	0.0016
	(0.0005)	(0.0005)	(0.0005)
$\ln(A_i)$	−0.0135	−0.0122	−0.0197
	(0.0065)	(0.0066)	(0.0086)
$\mathbf{W}\ln(CRI_i)$		0.2196	
		(0.0979)	
$\mathbf{W}\varepsilon_i$			0.2686
			(0.0944)
N	526	526	526
Log likelihood	1412.97	1415.27	1416.17

Note: Values in parenthesis correspond to standard errors, and p-values of coefficients are statistical significant at 0.05. **W** is the spatial weighted matrix with a minimum threshold value of 147.9 km. The number of observation was less than the original database because some of them did not report their value of habitant, and the computation of accessibility did not take into account urban polygons without a road line inside it

Table 104.2 Spatial dependence tests

Test	Value
Moran's I	0.0440
Lagrange Multiplier (lag)	5.3566847 [0.0206430]
Robust LM (lag)	7.5892495 [0.0058717]
Lagrange Multiplier (error)	7.0555872 [0.0079018]
Robust LM (error)	9.2881519 [0.0023064]

Note: Values in square brackets correspond to p-values

inside and outside them were 2696 and 21948 respectively. The number of individual edges, corresponding to the road network, was 27026. Figure 104.1(a) presents a large-scale system (complete network), where nodes correspond to points inside urban areas, and edges consist of road segments. Figure 104.1(b) shows a local system (Mexico City and proximities) where nodes were characterized by inside and outside attributes.

In order to assess (104.3), we used two types of statistical methods: ordinary least square (OLS) and maximum likelihood (ML). The OLS method worked as explorative analysis, testing the model specification and the level of spatial autocorrelation (Anselin et al., [5]). On the other hand, ML incorporated spatial dependences: spatial lag and error (Anselin, [3]). To distinguish the model with an appropriate functional form, we looked into the log-likelihood test (Table 104.1) and the spatial dependence tests (Table 104.2).

Table 104.1 compared estimations of three model specifications. All models presented significant and expected values of S_i and A_i coefficients. An increased change in S_i produced a positive effect in CRI_i, that is, more people in a city improve the flow of resources. On the other hand, a decreased change in A_i (higher levels of accessibility) created a positive effect in CRI_i. Then, A_i impacted more than S_i the level of CRI_i. However, the OLS model presented a spatial dependence indicated in the Moran's I test. The lag (ML) model corrected the spatial autocorrelation problem, Lagrange Multiplier (lag) and Robust LM (lag), in OLS but the log-likelihood test was lower than the error (ML) model. Such an error model rectified even more the spatial dependence, Lagrange Multiplier (error) and Robust LM (error), and presented the higher value of the log-likelihood test. Therefore, this model represented the best functional form.

From these data we could specify the preferential attachment process in (104.3) as following:

$$P_{i \to j} = \frac{S_i^{0.0016} A_i^{-0.0197} \mathbf{W}\varepsilon_i^{0.2686}}{\Sigma_j S_j^{0.0016} A_j^{-0.0197} \mathbf{W}\varepsilon_j^{0.2686}} \tag{104.4}$$

where $\mathbf{W}\varepsilon$ solved the spatial autocorrelation problem and improved the precision of the mechanism.

104.4 Conclusion

The results of this study indicate that not only hierarchical attributes but also spatial dependences of cities have to be specified in the preferential attachment mechanism. The probability of generating a new set of road sections among cities depends on first place in the distance, which incorporates the effect of neighbors in the error term. In addition, accessibility is the second important attribute, and the last one is the city size.

Contrary to expectations, the city size attribute had small effect in the preferential attachment process because of the model specification. The configuration of the spatial system gave more attention to city attributes related to the road network, for example geospatial, geometrical, and topological characteristics. Therefore, different functional forms and more historical geodata of cities and roads can produce variations in our results. Future studies on this observation are therefore recommended.

These findings have important theoretical and empirical implications in the study of urban systems. The spatial externality approach enhances the study of the structure and evolution of networks in urban systems if cities and roads are modeled as a planar spatial network, and spatial econometrics and complex networks are complements rather than substitutes methods.

References

1. Abbot A (1997) Of time and space: the contemporary relevance of the Chicago school. Soc Forces 75:1149–1182
2. Anselin L (1988) Spatial econometrics: methods and models. Kluwer Academic, Dordrecht

3. Anselin L (2003) Spatial externalities, spatial multipliers, and spatial econometrics. Int Reg Sci Rev 26(2):153–166
4. Anselin L (2006) Spatial econometrics. In: Mills TC, Patterson K (eds) Palgrave handbook of econometrics. Volume 1. Econometric Theory. Palgrave Macmillan, Basingstoke, UK, pp 901–969
5. Anselin L, Bera AK, Florax R, Yoon MJ (1996) Simple diagnostic test for spatial dependence. Reg Sci Urban Econ 26(1):77–104
6. Barthélemy M (2011) Spatial networks. Phys Rep 499:1–101
7. Batty M (2009) Accessibility: in search of a unify theory. Environ Plan B, Plan Des 36:191–194
8. Black D, Henderson V (1999) A theory of urban growth. J Polit Econ 107:252–284
9. Burghardt AF (1979) The origin of the road and city network of Roman Pannonia. J Hist Geogr 5(1):1–20
10. Duranton G, Puga D (2004) Micro-foundations of urban agglomeration economies. In: Henderson JV, Thisse JF (eds) Handbook of regional and urban economics, 1st edn, vol 4. Elsevier, Amsterdam, pp 2063–2117
11. Eeckhout J (2009) Gibrats law for (all) cities: reply. Am Econ Rev 99:1679–1683
12. Fujita M, Thisse JF (2002) Economies of agglomeration. Cambridge University Press, Cambridge
13. Getis A (1991) Spatial interaction and spatial autocorrelation: across-product approach. Environ Plan A 23:1269–1277
14. Haynes KA, Fotheringham AS (1984) Gravity and spatial interaction models. Sage, Thousand Oaks
15. Instituto Nacional de Estadística y Geografía (INEGI). http://www.inegi.org.mx/geo/contenidos/topografia/InfoEscala.aspx
16. Ioannides Y, Skouras S (2009) Gibrat's Law for (all) cities: a rejoinder. Discussion papers series, 0740, Department of Economics, Tufts University
17. Irwin MD, Hughes HL (1992) Centrality and the structure of urban interaction: measures, concepts, and applications. Soc Forces 71(1):17–51
18. Pumain D (2006) Alternative explanation of hierarchical differentiation in urban systems. In: Pumain D (ed) Hierarchy in nature and social sciences, vol 64, pp 169–221
19. Sampson RJ, Morenoff J, Earls F (1999) Beyond social capital: spatial dynamics of collective efficacy for children. Am Sociol Rev 64:633–660
20. Sarabia JM, Prieto F (2009) The Pareto-positive stable distribution: a new descriptive model for city size data. Physica A 388(19):4179–4191
21. Wilson AG (1971) A family of spatial interaction models and associated developments. Environ Plann A 3:1–32
22. Xie F, Levinson D (2009) Topological evolution of surface transportation networks. Comput Environ Urban Syst 33:211–223
23. Xie F, Levinson D (2007) Measuring the structure of road networks. Geogr Anal 39:336–356
24. Yamis D, Rasmussen S, Fogel D (2003) Growing urban roads. Netw Spat Econ 3:69–85

Part VIII
Satellite Meeting: Space-Time Phases

Chapter 105
Some Properties of Persistent Mutual Information

Peter Gmeiner

Abstract Persistent mutual information is an information measure suggested to measure the amount of information which survive for a long time in a dynamical system. We consider systems which are given as stochastic processes and investigate properties and relations of the persistent mutual information to other known complexity measures for stochastic processes. In particular we show that the excess entropy is an upper bound for the persistent mutual information. We also calculate the persistent mutual information explicitly for some simple examples.

Keywords Excess entropy · Persistent mutual information · Emergence

105.1 Introduction

The persistent mutual information (PMI) is a complexity measure for stochastic processes. It is related to well-known complexity measures like excess entropy or statistical complexity. Essentially it is a variation of the excess entropy so that it can be interpreted as a specific measure of system internal memory. It was first introduced in 2010 by Ball, Diakonova and MacKay as a measure for (strong) emergence [1]. We define the PMI mathematically and investigate the relation to excess entropy and statistical complexity. In particular we prove that the excess entropy is an upper bound for the PMI. Furthermore we show some properties of the PMI and calculate it explicitly for some example processes.

105.2 Preliminaries

Let $(\Omega, \mathcal{F}, P)$ be a probability space with a metric space Ω, a σ-algebra $\mathcal{F}$ and a probability measure P. For random variables $X, Y : \Omega \to \mathcal{A}$ mapping to a finite

P. Gmeiner (✉)
Department Mathematik, Friedrich-Alexander-Universität Erlangen-Nürnberg, Cauerstr. 11, 91058 Erlangen, Germany
e-mail: gmeiner@mi.uni-erlangen.de

T. Gilbert et al. (eds.), *Proceedings of the European Conference on Complex Systems 2012*, Springer Proceedings in Complexity, DOI 10.1007/978-3-319-00395-5_105,

alphabet $\mathcal{A}$ the Shannon entropy is denoted as $H(X)$ and the conditioned Shannon entropy as $H(X|Y)$. The mutual information between two random variables is $I(X;Y) := H(X) - H(X|Y)$.

We consider a time-discrete stationary stochastic process $\overleftrightarrow{S} := (S_t)_{t\in\mathbb{Z}}$ with random variables $S_t : \Omega \to \mathcal{A}$ for all $t \in \mathbb{Z}$. We define the semi-infinite processes $\overleftarrow{S} := (S_{-t})_{t\in\mathbb{N}}$ interpreted as past and $\overrightarrow{S} := (S_t)_{t\in\mathbb{N}_0}$ interpreted as future respectively. Blocks of random variables with finite length are denoted by $S_a^b := (S_k)_{k\in[a,b]\cap\mathbb{Z}}$ for $-\infty < a \le b < \infty$ and the corresponding block entropy is $H(L) := H(S_1^L) = H(S_1, \dots, S_L)$. The one-sided sequence space is $\mathcal{A}^{\mathbb{N}} := \times_{i\in\mathbb{N}}\mathcal{A}$ and in the same way the two-sided sequence space $\mathcal{A}^{\mathbb{Z}}$ is defined. We introduce the shift function $\sigma : \mathcal{A}^{\mathbb{Z}} \to \mathcal{A}^{\mathbb{Z}}$ by $\sigma(x)_i := x_{i+1}$. At any time $t \in \mathbb{Z}$ we have random variables $S_{-\infty}^t := (S_k)_{k\le t}$ and $S_{t+1}^{\infty} := (S_k)_{k\ge t+1}$ that govern the systems observed behaviour respectively in the shifted past and the shifted future. The mutual information between these two variables is the well-known *excess entropy* [3, 5]

$$E := \lim_{L\to\infty} I\left(S_0^{L-1}; S_{-L}^{-1}\right). \tag{105.1}$$

In general, it is not clear if the limit in (105.1) exists (for Markov processes of finite order one can prove the existence). With the assumption that the limit in (105.1) exists as a finite number the following equality holds: $E = I(\overleftarrow{S}; \overrightarrow{S})$, see Chap. 2.2 in [15].

105.3 Conceptualization

The definition of the excess entropy (105.1) allows a concrete information theoretic interpretation. In particular the excess entropy can be seen as a specific measure of *system internal memory*. We will take this as a basis to define a term, first suggested in [1], which will capture the structural behavior of a dynamical system on the whole time-domain. In particular it should be possible to detect some existing *inherent structure* of the system which will survive for all times. In order to achieve this goal we adapt the mutual information-based representation of the excess entropy and introduce the following expression

$$E_{t,\tau}^L := I\left(S_t^{t+L-1}; S_{-\tau-L+1}^{-\tau}\right).$$

For $t = 0$ and $\tau = 1$ we have $E = \lim_{L\to\infty} E_{0,1}^L$. For arbitrary t and τ-values we get a family of similar terms $E_{t,\tau} := \lim_{L\to\infty} E_{t,\tau}^L$. Every expression $E_{t,\tau}$ is the excess entropy with a time-gap of size $|t-\tau|$ between a random variable block of the past and the future and we write $E_k^L := E_{0,k}^L = I(S_0^{L-1}; S_{-k-L+1}^{-k})$.

Next we consider the convergence of the double sequence $(E_k^L)_{L,k\in\mathbb{N}}$ (this is only the case if and only if $\lim_{n\to\infty} E_{k(n)}^{L(n)}$ converges to the same value for all subsequences with $\lim_{n\to\infty} k(n) = \lim_{n\to\infty} L(n) = \infty$). In particular then it holds that $\lim_{k\to\infty}\lim_{L\to\infty} E_k^L = \lim_{L\to\infty}\lim_{k\to\infty} E_k^L$, and the time gap between the past and future goes to infinity.

Definition 105.1 Let a stochastic process with values in a finite alphabet $\mathcal{A}$ be given. The *persistent mutual information* of such a process is defined by

$$PMI := \lim_{k,L\to\infty} E_k^L.$$

If the *PMI* exists, it is enough to consider the iterated limits

$$PMI = \lim_{L\to\infty}\lim_{k\to\infty} E_k^L = \lim_{k\to\infty}\lim_{L\to\infty} E_k^L.$$

This expression was first proposed by Ball and collaborators in [1]. Similar to Definition 105.1 we can define the *PMI* for semi-infinite (or one-sided) processes.

Remark 105.2 In this paper we only consider stationary stochastic processes. The existence of *PMI* is a priori not clear. Nevertheless we can show that it exists for Markov processes of finite order or for periodic processes (see Sect. 105.6). One sufficient condition for existence is the almost everywhere convergence of the random variables of the stochastic process. Then the *PMI* can be seen as the excess entropy of a process with constant past. Thus the *PMI* can be understood as the amount of information which is communicated from a very far past to the future.

105.4 Relation to Statistical Complexity

We now pick up the sketched ideas in [1], to express the *PMI* with so called *causal states*. In particular one can show that the *statistical complexity* (internal entropy of the causal states) is an upper bound for the *PMI*. In the rest of this section we assume that the *PMI* exist. We consider shifted blocks of random variables $\overleftarrow{S}_\tau := (S_{-\tau-t})_{t\in\mathbb{N}}$, $\overrightarrow{S}_\tau := (S_{\tau+t})_{t\in\mathbb{N}_0}$, for $\tau \in \mathbb{N}_0$. The sets of realisations[1] are denoted by $\overleftarrow{\mathbf{S}}_\tau$, $\overrightarrow{\mathbf{S}}_\tau$ and the sub-σ-algebras which are generated by cylinder sets are denoted with $\mathcal{C}_{\tau,-\mathbb{N}}, \mathcal{C}_{\tau,\mathbb{N}_0}$. On the set $\mathcal{A}^{\mathbb{N}}$ of all shifted past trajectories of the process $\overleftrightarrow{S}$ we define an equivalence relation

$$\overleftarrow{s} \sim \overleftarrow{s}' :\Leftrightarrow \Pr(\overrightarrow{S} = \overrightarrow{s} \mid \overleftarrow{S}_\tau = \overleftarrow{s}) = \Pr(\overrightarrow{S} = \overrightarrow{s} \mid \overleftarrow{S}_\tau = \overleftarrow{s}'), \quad \forall \overrightarrow{s} \in \mathcal{C}_{\mathbb{N}_0},$$

where $\overleftarrow{s}, \overleftarrow{s}' \in \mathcal{A}^{\mathbb{N}}$ and $\Pr(\overrightarrow{S} = \overrightarrow{s} \mid \overleftarrow{S}_\tau = \overleftarrow{s})$ is a regular version of the conditional expectation. The equivalence classes

$$S_\tau^+(\overleftarrow{s}) := \left\{\overleftarrow{s}' \in \mathcal{A}^{\mathbb{N}} : \overleftarrow{s}' \sim \overleftarrow{s}\right\} \subset \mathcal{A}^{\mathbb{N}}$$

of this relation are called *shifted causal states*. The set of all shifted causal states is denoted by $\mathcal{S}_\tau^+ := \{S_\tau^+(s) \mid s \in \mathcal{A}^{\mathbb{N}}\}$.

[1] For every $\omega \in \Omega$ the mapping $R_\omega : t \mapsto S_{-t}(\omega)$ is called a realisation of the process $\overleftarrow{S}$. The set of all realisations is defined as $\overleftarrow{\mathbf{S}} := \{(R_\omega(t))_{t\in\mathbb{N}} : \omega \in \Omega\}$.

In the same sense we define (future) shifted causal states $S_\tau^-(\overrightarrow{s})$ and $\mathcal{S}_\tau^-$.

For sake of simplicity we are only considering stochastic processes with a finite set of shifted causal states $\mathcal{S}_\tau^+ = \{S_1^+, \ldots, S_n^+\}$ and $\mathcal{S}_\tau^- = \{S_1^-, \ldots, S_m^-\}$. Given a past observation of infinite length $s_{-\infty}^t \in \mathcal{A}^{\mathbb{Z}}$ at time $t \in \mathbb{Z}$ using stationarity we identify this shifted past with a shifted causal state $S_\tau^+(\sigma^{-t-\tau-1}(s_{-\infty}^t)) \in \mathcal{S}_\tau^+$. Together with the next symbol s_{t+1} generated by the process the next shifted causal state $S_\tau^+(\sigma^{-t-\tau-2}(s_{-\infty}^t s_{t+1})) \in \mathcal{S}_\tau^+$ is uniquely determined and the shifted causal states are Markov [14, 17]. We define the Markov kernels between two shifted causal states $S_i^+, S_j^+ \in \mathcal{S}_\tau^+$ emitting an output symbol $r \in \mathcal{A}$ for any $t \in \mathbb{Z}$ as follows

$$\begin{aligned} T_{\tau,i,j}^{+(r)} &:= T(S_i^+)(S_j^+, r) \\ &= \Pr\big(S\big(\sigma^{-t-\tau-2}\big(s_{-\infty}^t s_{t+1}\big)\big) = S_j^+ \quad \text{and} \\ S_{t+1}^+ &= r \mid S\big(\sigma^{-t-\tau-1}\big(s_{-\infty}^t\big)\big) = S_i^+\big). \end{aligned}$$

The set of transition matrices is denoted with $T_\tau^+ := \{(T_{\tau,i,j}^{+(r)})_{i,j=1}^n : r \in \mathcal{A}\}$. The probability of a shifted causal state $S_i^+ \in \mathcal{S}_\tau^+$ is denoted by $p_i^+ := P(S_i^+)$. The ordered pair $M_\tau^+ := (T_\tau^+, (p_1^+, \ldots, p_n^+))$ is called *shifted (past) ϵ-machine*. In the same way we can define a *shifted (future) ϵ-machine* $M_\tau^- := (T_\tau^-, (p_1^-, \ldots, p_m^-))$.

The shifted ϵ-machines has internal state entropies

$$C_{P,\tau}^+ := H\big(\mathcal{S}_\tau^+\big) = -\sum_{j=1}^n p_j^+ \log p_j^+,$$

and

$$C_{P,\tau}^- := H\big(\mathcal{S}_\tau^-\big) = -\sum_{j=1}^m p_j^- \log p_j^-,$$

which are also known as (shifted) *statistical complexities* [13, 17]. We can write the *PMI* as follows.

Proposition 105.3 *Assume that PMI exists, then* $PMI = \lim_{\tau\to\infty} I(\overleftarrow{S}; \overrightarrow{S}_\tau)$.

Proof Since *PMI* exists we can change the limits and write them as

$$PMI = \lim_{\tau\to\infty} \lim_{L\to\infty} I\big(S_{-L}^{-1}; S_{\tau-1}^{\tau+L-2}\big). \tag{105.2}$$

We can decompose the limits in (105.2) into two independent limits and get with a limit property of the mutual information (see [15, Chap. 2.2]) applied two times

$$PMI = \lim_{\tau\to\infty} \lim_{L\to\infty} I\big(S_{-L}^{-1}; S_{\tau-1}^{\tau+L-2}\big) = \lim_{\tau\to\infty} I(\overleftarrow{S}; \overrightarrow{S}_\tau). \qquad \square$$

Similar to the fact that the excess entropy can be expressed via causal states [8, 11, 12] we can also express the *PMI* via shifted causal states.

Proposition 105.4 *Assume that PMI exists, then we have*

$$PMI = \lim_{\tau\to\infty} I\big(\mathcal{S}_0^+; \mathcal{S}_\tau^-\big) = \lim_{\tau\to\infty} I\big(\mathcal{S}_\tau^+; \mathcal{S}_0^-\big).$$

Proof We take $\overrightarrow{S}_\tau$ instead of $\overrightarrow{S}$ and $\mathcal{S}_0^+, \mathcal{S}_\tau^-$ instead $\mathcal{S}^+, \mathcal{S}^-$, then the proof is with Proposition 105.3 analogous to the proof of Proposition B.2 in [12]. We obtain the second equality with the symmetry of the mutual information and the stationarity of the stochastic process. □

With this expression we get the following inequalities.

Corollary 105.5 *The statistical complexities are upper bounds for the PMI, if it exists,*

$$PMI \le C_P^-, \qquad PMI \le C_P^+,$$

with equality if and only if $\lim_{\tau\to\infty} H(\mathcal{S}_0^+|\mathcal{S}_\tau^-) = 0$ *or* $\lim_{\tau\to\infty} H(\mathcal{S}^-|\mathcal{S}_\tau^+) = 0$.

Proof With Proposition 105.4, the stationarity and the definition of the statistical complexity we get

$$PMI = H\big(\mathcal{S}_0^+\big) - \lim_{\tau\to\infty} H\big(\mathcal{S}_0^+|\mathcal{S}_\tau^-\big) \le C_P^+.$$

With the symmetry of the mutual information we get $PMI \le C_P^-$. □

105.5 Relation to Excess Entropy

One might expect that the *PMI* coincide with the excess entropy as soon as the structure of the past coincide with the structure of the future. The next proposition shows that this is indeed the case for processes with zero metric entropy. The metric entropy is defined as the following limit $h_P := \lim_{L\to\infty} \frac{H(L)}{L}$ which exists for all stationary stochastic processes. With an elementary investigation one can show the following proposition.

Proposition 105.6 *Assume that the excess entropy E and PMI exists, then it holds that*

$$PMI = E \iff h_P = 0.$$

Proof $\Rightarrow$: Since E is finite we have with Proposition B.1 in [12] that $H(L) \sim E + Lh_P$ as $L \to \infty$, see also [13]. Furthermore we get

$$PMI = \lim_{L\to\infty}\Big(2H(L) - \lim_{k\to\infty} H\big(S_0^{L-1}, S_{-k-L+1}^{-k}\big)\Big) \tag{105.3}$$

$$= \lim_{L\to\infty}\Big(2E + 2Lh_P - \lim_{k\to\infty} H\big(S_0^{L-1}, S_{-k-L+1}^{-k}\big)\Big). \tag{105.4}$$

Hence we get with $PMI = E$

$$\lim_{L\to\infty}\Big(\lim_{k\to\infty} H\big(S_0^{L-1}, S_{-k-L+1}^{-k}\big) - 2Lh_P\Big) = 2E - PMI = E$$
$$= \lim_{L\to\infty}\big(H(L) - H\big(S_0^{L-1}|S_{-L}^{-1}\big)\big).$$

Since $1 \le \frac{\lim_{k\to\infty} H(S_0^{L-1}, S_{-k-L+1}^{-k})}{H(L)}$ and $\frac{H(L)}{\lim_{k\to\infty} H(S_0^{L-1}, S_{-k-L+1}^{-k})} \le 1$, it follows that

$$\lim_{k\to\infty} H\big(S_0^{L-1}, S_{-k-L+1}^{-k}\big) \sim H(L) \quad \text{as } L \to \infty,$$

which leads with (105.3) to

$$PMI = \lim_{L\to\infty} H(L) = E.$$

Finally this implies because of Proposition B.1 in [12] that $h_P = 0$.

$\Leftarrow$: Due to $h_P = 0$ it holds that $E = \lim_{L\to\infty} H(L)$. Furthermore it follows that

$$H(2L+k) \ge H\big(S_0^{L-1}, S_{-k-L+1}^{-k}\big) \ge H(L),$$

using $H(2L+k) \overset{k\to\infty}{\to} E$ and $H(L) \overset{L\to\infty}{\to} E$ leads to

$$\lim_{L\to\infty}\lim_{k\to\infty} H\big(S_0^{L-1}, S_{-k-L+1}^{-k}\big) = E.$$

Together we get

$$PMI = \lim_{L\to\infty}\Big(2H(L) - \lim_{k\to\infty} H\big(S_0^{L-1}, S_{-k-L+1}^{-k}\big)\Big) = 2E - E = E. \qquad \square$$

More generally with the representation of the *PMI* in terms of causal states as given in Sect. 105.4, we can show that the *PMI* is bounded from above by the excess entropy.

Proposition 105.7 *Assume that the PMI exists, then we have PMI* $\le E$.

Proof With Proposition 105.4, Proposition B.2. in [12] and the rule $H(X|Y) \le H(X)$ for two random variables X, Y, we get

$$PMI = \lim_{\tau\to\infty} I\big(\mathcal{S}^-; \mathcal{S}_\tau^+\big) = \lim_{\tau\to\infty}\big(H\big(\mathcal{S}^-\big) - H\big(\mathcal{S}^-|\mathcal{S}_\tau^+\big)\big)$$
$$= \lim_{\tau\to\infty}\big(H\big(\mathcal{S}^-\big) - H\big(\mathcal{S}^-|\overleftarrow{S}_\tau\big)\big)$$
$$= \lim_{\tau\to\infty}\lim_{L\to\infty}\big(H\big(\mathcal{S}^-\big) - H\big(\mathcal{S}^-|S_{-\tau-L+1}^{-\tau-1}\big)\big)$$
$$\le \lim_{\tau\to\infty}\lim_{L\to\infty}\big(H\big(\mathcal{S}^-\big) - H\big(\mathcal{S}^-|S_{-\tau-L+1}^{-1}\big)\big)$$

$$= H(\mathcal{S}^{-}) - H(\mathcal{S}^{-} \mid \overleftarrow{S})$$
$$= H(\mathcal{S}^{-}) - H(\mathcal{S}^{-} \mid \mathcal{S}^{+})$$
$$= I(\mathcal{S}^{-}; \mathcal{S}^{+}) = E. \qquad \square$$

Remark 105.8 Proposition 105.7 says that the *PMI* do not care about some random variables which are considered from the excess entropy. It forgets this information and the excess entropy use the full information available from the realisations of the process. In this sense the *PMI* is a coarser complexity measure than the excess entropy. With that we get a graduation of the considered complexity measures from a coarse to a fine one, i.e.

$$PMI \leq E \leq C_P^{+}. \tag{105.5}$$

105.6 Explicit Representations

With elementary investigations we can show a series of explicit representations of the *PMI* for simple processes. First we consider periodic processes.[2] For that case the following corollary of Proposition 105.6 give us the result.

Corollary 105.9 *Let a periodic process with period L be given. Then the PMI amounts to $PMI = H(L)$, in particular it holds $PMI = E$.*

Proof With Proposition 105.6 and the fact that $h_P = 0$ hold for periodic processes, the claim follows with the fact that $E = H(L)$ for periodic processes. $\square$

For Markov-processes the *PMI* vanishes, since the dependencies between the past and future blocks disappear in finite time.

Proposition 105.10 *Let a Markov-process of order R be given. Then $PMI = 0$.*

Proof With the Markov-property and the abbreviation $\tilde{S}_k := S_{-k-L+1}^{-k}$ it follows for $-k < R$

$$H(S_0^{L-1} \mid \tilde{S}_k)$$
$$= - \sum_{\sigma,\xi \in \mathcal{A}^L} \Pr(S_0^{L-1} = \sigma, \tilde{S}_k = \xi) \log(\Pr(S_0^{L-1} = \sigma \mid \tilde{S}_k = \xi))$$
$$= - \sum_{\sigma,\xi \in \mathcal{A}^L} \Pr(S_0^{L-1} = \sigma \mid \tilde{S}_k = \xi) \Pr(\tilde{S}_k = \xi) \log(\Pr(S_0^{L-1} = \sigma \mid \tilde{S}_k = \xi))$$

[2] A process is called *periodic* with period L if $S_t = S_{t+L}$ for all $t \in \mathbb{Z}$ and $S_t \neq S_{t+k}$ for $k < L$.

$$\stackrel{-k<R}{=} - \sum_{\sigma,\xi\in\mathcal{A}^L} \Pr\bigl(S_0^{L-1}=\sigma\bigr)\Pr(\tilde{S}_k=\xi)\log\bigl(\Pr\bigl(S_0^{L-1}=\sigma\bigr)\bigr)$$

$$= H(L).$$

Hence the persistent mutual information is

$$PMI = \lim_{L\to\infty}\Bigl(H(L) - \lim_{k\to\infty} H\bigl(S_0^{L-1} \mid S_{-k-L+1}^{-k}\bigr)\Bigr) = \lim_{L\to\infty}\bigl(H(L) - H(L)\bigr)$$

$$= 0. \qquad \square$$

105.7 Example Processes

In the following we calculate the *PMI* for concrete examples of stochastic processes.

Independent, Identically, Distributed (i.i.d.) Process A stochastic process is called independent, identical distributed if the finite dimensional distributions are independent and all distributions are equal, i.e. if for finite times $t_1 < \cdots < t_n \in \mathbb{Z}$ it holds that $\Pr(S_{t_1}, \ldots, S_{t_n}) = \Pr(S_{t_1}) \cdots \Pr(S_{t_n})$ and $\Pr(S_{t_i}) = \Pr(S_{t_j})$ for all $i, j \in \{1, \ldots, n\}$. The probability distributions are not depending on the time distance because they are identical distributed. Hence it holds that

$$H\bigl(S_t^{t+L-1}, S_{-\tau-L+1}^{-\tau}\bigr) = H\bigl(S_0^{L-1}, S_{-L}^{-1}\bigr).$$

Hence the *PMI* coincide with the excess entropy E by definition. Furthermore we have

$$\Pr\bigl(S_0^{L-1}, S_{-L}^{-1}\bigr) = \Pr\bigl(S_0^{L-1}\bigr)\Pr\bigl(S_{-L}^{-1}\bigr).$$

With the definition of the mutual information we get for every $L \in \mathbb{N}$

$$I\bigl(S_0^{L-1}; S_{-L}^{-1}\bigr) = 0,$$

and finally

$$PMI = E = \lim_{L\to\infty} I\bigl(S_0^{L-1}; S_{-L}^{-1}\bigr) = 0.$$

Thue-Morse Process The Thue-Morse sequence consists of a two symbol alphabet $\mathcal{A} = \{0, 1\}$ and is constructed via the substitution $\mathcal{G} : \mathcal{A} \to \mathcal{A}^2$, with $\mathcal{G}(0) = 01, \mathcal{G}(1) = 10$. The Thue-Morse sequence is defined as the fixed point of $\mathcal{G}$ as

$$u := \mathcal{G}^\infty(0) = 011010011001\ldots = \mathcal{G}(u).$$

The stochastic process generating the Thue-Morse sequence is called *Thue-Morse process*. The used probability measure is the counting measure. Using spectral analytical methods [10, 16] we can explicitly calculate the frequencies of symbol-blocks of length n in u and we can also show that the Thue-Morse process is

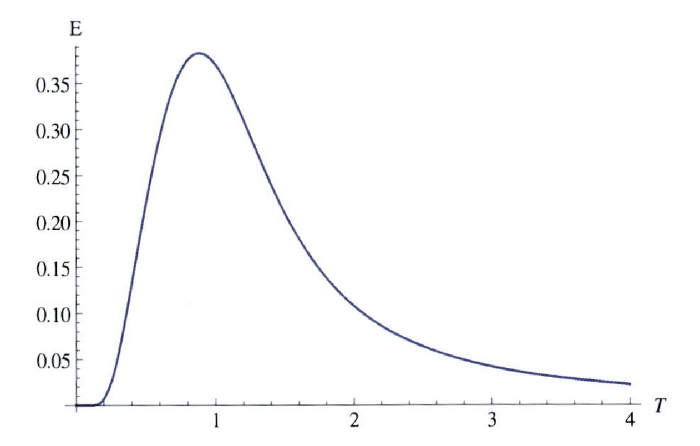

Fig. 105.1 Excess entropy for a one-dimensional Ising model depending on the temperature T

uniquely ergodic, see also [11, 12]. In [2] it is proven that for all $k \geq 1$ the first derivative of the block entropy $\triangle H(n) := H(n) - H(n-1)$ for the Thue-Morse process can be written as $\triangle H(n) = \frac{4}{3 \cdot 2^k}$, if $2^k + 1 \leq n \leq 3 \cdot 2^{k-1}$, and $\triangle H(n) = \frac{2}{3 \cdot 2^k}$, if $3 \cdot 2^{k-1} + 1 \leq n \leq 2^{k+1}$. With that the metric entropy vanishes $h_P = \lim_{n \to \infty} \triangle H(n) = 0$ and with some further technical lemmas from [6, 7] and [16] one can show that $PMI = \infty$. In particular one can also show that $E = \infty$. A detailed calculation is given in [12] and [11].

Persistent Mutual Information for an One-Dimensional Ising-Spin Chain For the one-dimensional Ising-spin chain the PMI is zero due to the fact that the spin chain is a Markov-process of first order and with Proposition 105.10 we have $PMI = 0$.

Remark 105.11 Crutchfield et al. calculated in [4] and in [9] an explicit expression of the excess entropy E for that example. It turns out that depending on the temperature the excess entropy attains a maximum at some critical temperature and get close to zero for very low and very high temperatures, see Fig. 105.1.

It is well known that in the one-dimensional Ising model no phase-transition appears (we consider a phase-transition as an example for *weak emergence*). Nevertheless the fact that E attains a nontrivial expression in that case and PMI is zero shows that E seems to measure complex structure at a finer level. Thus for detecting emergent structures the PMI seems to be a more useful complexity measure. To confirm this intuition more concrete calculation examples are needed. For a detailed discussion on complexity measures and their relation to emergence, see [11, 12].

Acknowledgements I would like to thank Andreas Knauf for motivating me to work on this project, for his constant support and many useful discussions.

References

1. Ball RC, Diakonova M, MacKay RS (2010) Quantifying emergence in terms of persistent mutual information. Adv Complex Syst 13(03):327

2. Berthé V (1994) Conditional entropy of some automatic sequences. J Phys A 27:7993–8006
3. Crutchfield J, Packard NH (1983) Symbolic dynamics of noisy chaos. Physica D 7(1–3):201
4. Crutchfield J, Feldman D (1997) Statistical complexity of simple 1D spin systems. Phys Rev E 55(2):1239R–1243R
5. Crutchfield J, Feldman D (2003) Regularities unseen, randomness observed: levels of entropy convergence. Chaos 15:25–54
6. Dekking FM (1992) On the Prouhet-Thue-Morse measure. Acta Univ Carol, Math Phys 33:35–40
7. de Luca A, Varrichio S (1989) Some combinatorical properties of the Thue-Morse sequence. Theor Comput Sci 63:333–348
8. Ellison C, Mahoney J, Crutchfield J (2009) Prediction, retrodiction and the amount of information stored in the present. J Stat Phys 136(6):1005–1034
9. Feldman D, Crutchfield J (1998) Discovering noncritical organization: statistical mechanical, information theoretic and computational views of patterns in simple one-dimensional spin systems. Santa Fe Institute working paper 98-04-026
10. Fogg NP (2008) Substitutions in dynamics, arithmetics and combinatorics. Springer, Berlin
11. Gmeiner P (2010) Komplexitätsmaße und Emergenz. Diploma-Thesis
12. Gmeiner P (2012) Properties of persistent mutual information and emergence. arXiv:1210.5058
13. Grassberger P (1986) Toward a quantitative theory of self-generated complexity. Int J Theor Phys 25(9):907–938
14. Löhr W (2010) Models of discrete-time stochastic processes and associated complexity measures. PhD thesis, Leipzig
15. Pinsker MS (1964) Information and information stability of random variables and processes. Holden-Day, Oakland
16. Queffélec M (1987) Substitution dynamical systems—spectral analysis. Springer, Berlin
17. Shalizi CR (2001) Causal architecture, complexity and self-organization in time. PhD thesis

Part IX
Satellite Meeting: Complex Dynamics in Cellular Systems

Chapter 106
Demographic Fluctuations and Inherent Time Scales in a Genetic Circuit

Hildegard Meyer-Ortmanns and Darka Labavić

Abstract We review results on a genetic circuit made out of a self-activating species A that activates its own repressor B in a negative feedback loop. We consider this motif in three descriptions: a deterministic coarse-grained one from the start, its stochastic pendant, and a stochastic version with an improved time resolution. We study the conditions under which we can derive the deterministic coarse-grained from the stochastic time-resolved version. As it can be shown from the time-resolved version, the regular oscillations which are found in a number of realizations of this motif, fade away for slow binding rates of the transcription factors to the promoter regions of the genes. Results of our Gillespie simulations match well with mean-field predictions if the averaging over states accounts for the inherent time scales. The occurrence of quasi-cycles in the stochastic descriptions raises the question as to which oscillations in natural systems are of mere demographic origin.

Keywords Genetic circuits · Quasi-cycles · Coarse-graining

106.1 Introduction

A frequently found motif in biological networks is the combination of a self-activating species A that also activates its own repressor, a second species B, in a negative feedback loop [1–7]. The very motif should be understood only as an effective coarse-grained description in the sense that the concrete realizations may differ in the number of intermediate steps before the loops close and in the biological realization of the positive and negative feedback, e.g. on the transcriptional level or via direct repression between proteins, for example. It is sketched in Fig. 106.1.

We termed the motif bistable frustrated unit (BFU), as the bistability is inherent in the self-activation loop of A, while either of the connecting bonds between A and

H. Meyer-Ortmanns (✉) · D. Labavić
School of Engineering and Science, Jacobs University, Campus Ring 8, 28759 Bremen, Germany
e-mail: h.ortmanns@jacobs-university.de

D. Labavić
e-mail: d.labavic@jacobs-university.de

T. Gilbert et al. (eds.), *Proceedings of the European Conference on Complex Systems 2012*, Springer Proceedings in Complexity, DOI 10.1007/978-3-319-00395-5_106,

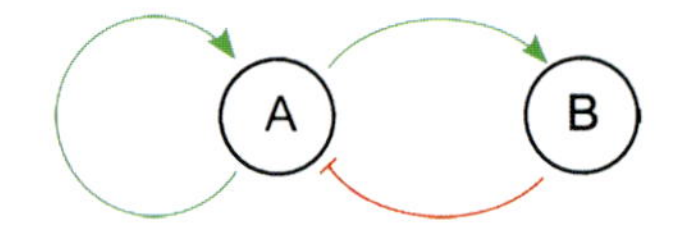

Fig. 106.1 Basic motif of a self-activating species A, activating also its own repressor B. *Pointed arrows* denote activation, *blunt arrow* denotes repression

itself or between B and A is frustrated. The frustration results from the fact that A receives a conflicting input from its own activation and the repression via B at the same time. The term points on the analogy to spin systems and was first used in [8]. From the physics' point of view it is therefore of interest whether common dynamical features of this motif exist independently of its realization and description. In this contribution we review three descriptions of the BFU referring to the same realization with repression on the transcriptional level, but differing in the details which are taken into account. We start in Sect. 106.2 with the simplest description in terms of two deterministic ordinary differential equations for concentrations of proteins A and B in which underlying processes on the genetic and the mRNA level are subsumed in rate constants. In Sect. 106.3 we consider a fully stochastic version of the very same system, now described in terms of biochemical reactions, or, equivalently, in terms of a master equation so that demographic fluctuations in the number of proteins A and B are taken into account. In Sect. 106.4 we zoom into the time resolution of the motif including different ratios of binding rates of genes with respect to the protein decay rates. Here we start from a fully stochastic formulation, based on reactions between the promoter regions of genes and proteins as well as production and decay rates of proteins. We then formulate an equivalent set of master equations. Here we illustrate how the appropriated averaging procedure over fast processes depends on the inherent time scales and leads to distinct deterministic sets of equations that differ in their bifurcation patterns. Only in the special case of so-called fast genes we recover the former deterministic coarse-grained description of Sect. 106.2. Therefore the simple description cannot be claimed to capture the dynamical performance of this motif.

106.2 The Bistable Frustrated Unit in a Deterministic Description

In its simplest formulation the BFU is described by the following set of differential equations for the time evolution of protein concentrations Φ_A and Φ_B:

$$\frac{d\phi_A}{dt} = \frac{\alpha}{1+\phi_B/K}\,\frac{b+\phi_A^2}{1+\phi_A^2} - \phi_A, \tag{106.1}$$

$$\frac{d\phi_B}{dt} = \gamma(\phi_A - \phi_B), \tag{106.2}$$

where γ is the ratio of the half-life of A to that of B. Here we focussed on the case $\gamma \ll 1$, that is when the protein B has a much longer half-life than A with a slow reaction on changes in A, while A has a fast response to changes in B. The parameter K sets the strength of repression of A by B. We assumed $K \ll 1$, so that already a small concentration of B will inhibit the production of A. The parameter b determines the basal expression level of A. In units where the production rate of B is equal to its degradation rate, K plays the role of a Michaelis constant that sets the strength of the repression of A by K. The choice of the Hill coefficients (here as $h = 1$ in $(B/K)^h$, activation of B by A with Hill coefficient $h = 0$ and self activation of A involving powers of $h = 2$)), used in the repression of A by B, may influence the very bifurcation pattern. The parameter α is the maximal rate of production of A for full activation ($\phi_A^2 \gg b$) and no repression ($\phi_B \approx 0$). It served as our bifurcation parameter.

Results In [12] we analyzed the phase structure of this model as a function of α. For small α, ($\alpha < 31.1$), and large α, ($\alpha > 97.9$), keeping $K = 0.02$, $b = 0.01$, $\gamma = 0.01$ fixed in both cases, we observed excitable behavior: For small perturbations below a certain threshold the values of the concentrations directly return to their fixed point values, while for perturbations above this threshold they return after a long excursion in phase space. In the intermediate α-regime, separated from the fixed-point regimes by subcritical Hopf bifurcations, we found regular limit cycles. It is these oscillations, which are supposed to describe the oscillatory behavior in a number of genetic systems, sharing the motif of the BFU.

106.3 Stochastic Version of the Bistable Frustrated Unit

Our next goal was to formulate a stochastic counterpart of the model (106.1)–(106.2) by introducing individual molecules of A and B and their corresponding numbers N_A and N_B to account for the demographic fluctuations in N_A, N_B. The concentrations become $\Phi_A = N_A/N_0$, $\Phi_B = N_B/N_0$, where N_0 plays the role of the system size, that is the average number of molecules A and B. It allows to control the size of the demographic fluctuations. Rather than adding noise terms to Eqs. (106.1)–(106.2) (directly "expanding" about the deterministic limit), the fully stochastic description amounts to a set of biochemical reactions, given as

$$\text{production of } A \quad N_A \to N_A + 1 \quad \text{with rate } N_0 f(N_A/N_0, N_B/N_0) \tag{106.3}$$

$$\text{decay of } A \quad N_A \to N_A - 1 \quad \text{with rate } N_A \tag{106.4}$$

$$\text{production of } B \quad N_B \to N_B + 1 \quad \text{with rate } \gamma N_A \tag{106.5}$$

$$\text{decay of } B \quad N_B \to N_B - 1 \quad \text{with rate } \gamma N_B \tag{106.6}$$

where "rate" denotes the transition rate of a specific process, and $f(\phi_A, \phi_B)$ is given by

$$f(\phi_A, \phi_B) = \frac{\alpha}{1 + \phi_B/K} \frac{b + \phi_A^2}{1 + \phi_A^2}. \tag{106.7}$$

This set of reactions is simulated with the Gillespie algorithm [10] to yield trajectories in the (N_A, N_B)-phase space. Equivalently the time evolution of our system can be described by the following master equation for the probability $P(N_A, N_B; t)$ for finding N_A proteins of type A and N_B proteins of type B at time t:

$$\begin{aligned}\frac{\partial P(N_A, N_B)}{\partial t} &= -\big(N_0 f(N_A/N_0, N_B/N_0) + N_A + \gamma N_A + \gamma N_B\big) P(N_A, N_B) \\ &\quad + (N_A + 1) P(N_A + 1, N_B) \\ &\quad + N_0 f\big((N_A - 1)/N_0, N_B/N_0\big) P(N_A - 1, N_B) \\ &\quad + \gamma(N_B + 1) P(N_A, N_B + 1) + \gamma N_A P(N_A, N_B - 1). \quad (106.8)\end{aligned}$$

On the right-hand-side we have as loss terms for $P(N_A, N_B)$ the production of A with rate $N_0 f(N_A/N_0, N_B/N_0)$, the decay of A proportional to N_A, the production of B proportional to γN_A, and its deletion proportional to γN_B. Gain terms to $P(N_A, N_B)$ result from the decay of A out of a state with $N_A + 1$ proteins, its production from a state with $N_A - 1$ proteins with rate $N_0 f((N_A - 1)/N_0, N_B/N_0)$, the decay of B with rate $\gamma(N_B + 1)$ and its production with rate γN_A from $N_B - 1$ proteins of type B. When we numerically integrate Eq. (106.8), the resulting probability distribution should match with the histogram which can be derived from the Gillespie measurements of the trajectories in phase space. To treat Eq. (106.8) in an analytical approximation, we applied the van Kampen expansion in $1/\sqrt{N_0}$ [11]. For $N_0 \to \infty$, we obtain Eqs. (106.1)–(106.2) in terms of $\langle N_S \rangle$ with $\langle N_S \rangle = \sum_{N_S} N_S P(N_A, N_B; t)$, $S = A, B$. To next order we obtain a Fokker-Planck equation for the probability to observe fluctuations in N_A, N_B at time t. As outlined in [11], the solution of the Fokker-Planck equation is a Gaussian distribution. It is therefore sufficient to express the first and second moments in our case in terms of the parameters of the Eqs. (106.3)–(106.7), evaluated either at the deterministic fixed-point solution, when the parameters are chosen in the fixed-point regime, or at the stationary limit-cycle solution for parameters out of the limit-cycle regime. Furthermore we can calculate the variance of the fluctuations and the autocorrelation functions and compare the results with those derived from the Gillespie simulations.

Results For a transient time in the Gillespie simulations, the numbers N_A and N_B seem to converge to a neighborhood of the deterministic fixed points and the deterministic limit cycles, respectively. For simplicity we keep the characterization as fixed-point and limit-cycle regime also for the parameter ranges in the stochastic formulation, although these terms become meaningful in a strict sense only in the deterministic limit. So the stochastic trajectories roughly follow the deterministic ones with fluctuations being the larger the smaller the system size. This result corresponds to the naive expectation, and if it would also reflect the long-time behavior, we would not report on it here. For a larger number of Gillespie steps, however, we see cycles also deeply in both fixed-point regimes, in particular for $\alpha = 15$ and $\alpha = 150$, see Fig. 106.2(top) [9].

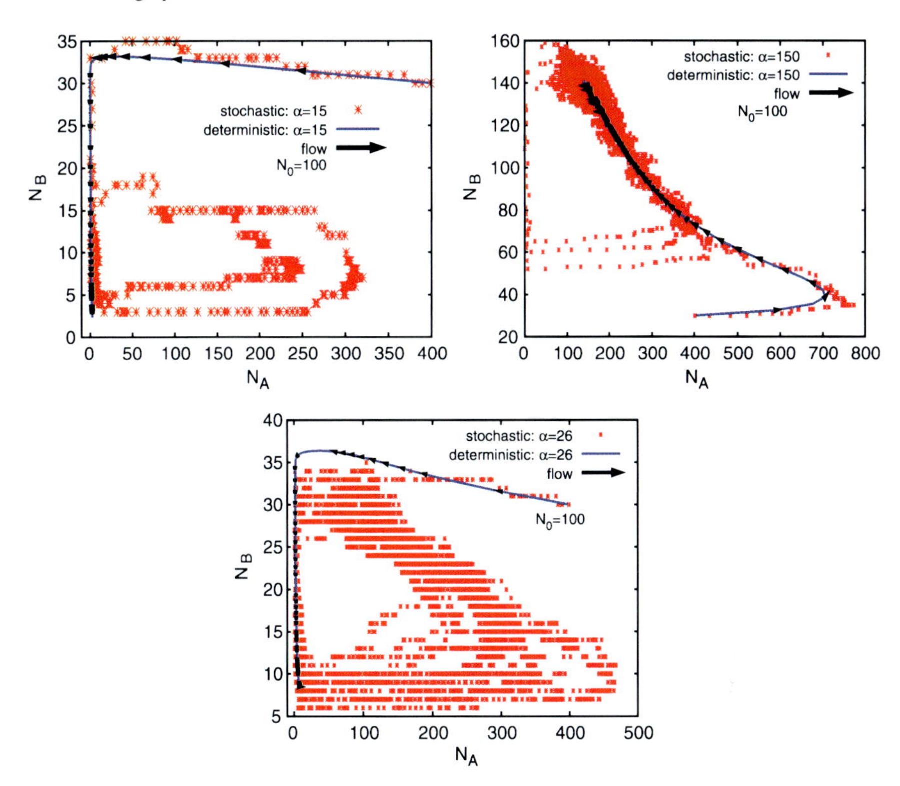

Fig. 106.2 Trajectories in phase space for Gillespie steps $T_G = 10^4, \alpha = 15$ (*top left*), and $T_G = 4 \cdot 10^5, \alpha = 150$ (*top right*). *Bottom*: for $\alpha = 26$, close to the transition region, we see cycles for the stochastic simulations (*red points*) and a trajectory (*blue* (*full*) *line*) converging to the fixed point as deterministic solution. $N_0 = 100$ in all cases

These cycles are called quasi-cycles due to their absence in the limit $N_0 \to \infty$. In Fig. 106.2(bottom) α is close to, but still below the value of the deterministic bifurcation point, accordingly the deterministic trajectory converges to a fixed point, but the stochastic trajectories show large fluctuations and approximately proceed along cycles. The figure shows that in the vicinity of the transition region between fixed-point and limit-cycle behavior it is inherently difficult to disentangle regular limit cycles from quasi-cycles. Better suited than plots of the phase trajectories are measurements of the autocorrelations to identify the origin of the oscillatory behavior. Autocorrelations decay faster for quasi-cycles than for the stochastic version of limit cycles. In [9] we measured the autocorrelations and the power spectrum in the different regimes as a function of α, both from the Gillespie simulations and via the van Kampen expansion. In particular we could reproduce the strong variation in the size of the variance along the noisy limit cycles. The van Kampen expansion leads to results that agree with the Gillespie simulations whenever the fluctuations are not large compared to the mean-field level, that is not too close to the transition region and not too deep in the fixed-point regime. It should be mentioned that the impact

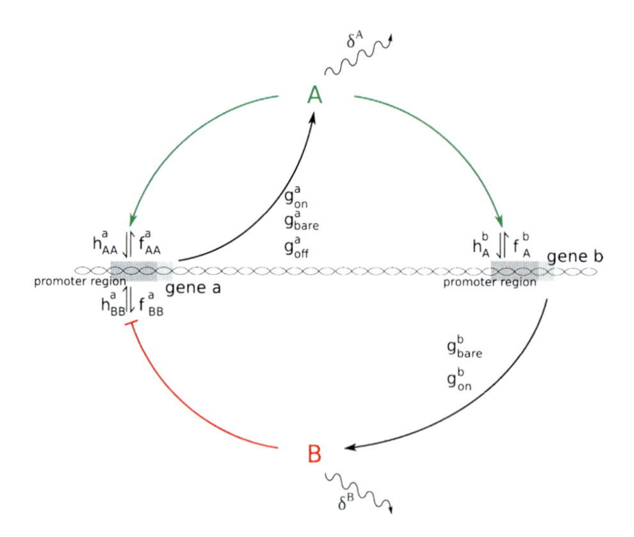

Fig. 106.3 Zoom into the motif of Fig. 106.1 with a realization via genes a and b leading to the production of proteins A and B with rates $g^a_{on,bare,off}$ and $g^b_{bare,on}$, respectively, depending on the bound transcription factors to the promoter region of a and b. Transcription factors $A(B)$ bind with rate $h^a_{AA}(h^a_{BB})$ to the promoter region of a, and unbind with rates $f^a_{AA}(f^a_{BB})$, respectively. Transcription factor A also binds to the promoter region of gene b with rate h^a_{A} and unbinds with rate f^b_A. Proteins A and B decay with rates δ^A and δ^B, respectively

of fluctuations on the power spectrum also depends on the feature of how spiky the dynamics is. This feature can be tuned via the ratio of half-life of protein A to protein B. Very spiky dynamics leads to strong fluctuations in the physical time it takes the system to perform one cycle in the protein numbers. For further details we refer to [9].

106.4 Caveats in Deriving Models on the Coarse-Grained Scale

In the former deterministic and stochastic versions of the model, the origin of activation and repression was hidden in the rate constants. Now we consider a realization of the motif as shown in Fig. 106.3.

It is the genes a and b that lead to the production of proteins A and B with rates $g^a_{on,bare,off}$ and $g^b_{on,bare}$, respectively, depending on which transcription factors are bound to the promoter region of a and b. Transcription factors $A(B)$ bind with rate h^a_{AA} (h^a_{BB}) to the promoter region of a, turning gene a into the on- (off-) state, respectively, and unbind with rates f^a_{AA} (f^a_{BB}), respectively. If no transcription factor is bound, we call the gene states "bare". Transcription factor A also binds to the promoter region of b with rate h^a_{A} (inducing the on-state of gene b) and unbinds with rate f^b_{A} (leaving b in a bare state). Proteins A and B decay with rates δ^A and δ^B, respectively. To focus on the effect of binding rates we drop the intermediate mRNA production steps and only zoom into a better time resolution of the binding rates with respect to the decay rates of the proteins, the fast one (A) and the slow one (B). Therefore we directly start from a fully stochastic description in terms of the following protein production and decay reactions:

$$\begin{aligned}
a_{on} &\xrightarrow{g^a_{on} N_0} A + a_{on} \\
a_{bare} &\xrightarrow{g^a_{bare} N_0} A + a_{bare} \\
a_{off} &\xrightarrow{g^a_{off} N_0} A + a_{off} \\
A &\xrightarrow{\delta^A} \phi \\
b_{on} &\xrightarrow{g^b_{on} N_0} B + b_{on} \\
b_{bare} &\xrightarrow{g^b_{bare} N_0} B + b_{bare} \\
B &\xrightarrow{\delta^B} \phi .
\end{aligned} \tag{106.9}$$

The reactions should be read according to "gene a in the on-state produces protein A with rate $g^a_{on} N_0$" etc.. Depending on which transcription factors are bound to the promoter region, we distinguish between the three states of gene a, a_i with ($i =$ on, bare, off) and the two states of gene b, b_j with ($j =$ on, bare). The binding/unbinding reactions of transcription factors to the promoter regions of genes are chosen as

$$\begin{aligned}
a_{bare} + 2A &\xrightarrow{h^a_{AA}/N_0^2} a_{on} \\
a_{on} &\xrightarrow{f^a_{AA}} a_{bare} + 2A \\
a_{bare} + 2B &\xrightarrow{h^a_{BB}/N_0^2} a_{off} \\
a_{off} &\xrightarrow{f^a_{BB}} a_{bare} + 2B \\
b_{bare} + A &\xrightarrow{h^b_A/N_0} b_{on} \\
b_{on} &\xrightarrow{f^b_A} b_{bare} + A .
\end{aligned} \tag{106.10}$$

The values of the effective binding rates and the corresponding unbinding rates are chosen to be of the same order according to

$$\frac{h^a_{AA} N_A^2}{N_0^2} = \frac{h^a_{BB} N_B^2}{N_0^2} = \frac{h^b_A N_A}{N_0} \sim f^a_{AA} = f^a_{BB} = f^b_A . \tag{106.11}$$

Table 106.1 Parameters kept fixed

g^a_{bare}	g^a_{off}	g^b_{on}	g^b_{bare}	δ^A	δ^B
25	0	2.5	0.025	1	0.01

Table 106.2 Binding and unbinding parameters

Genes	N_0	N_A	N_B	$f^a_{AA} = f^a_{BB} = f^b_A$	$\frac{h^a_{AA}}{N_0^2} N_A^2 = \frac{h^a_{BB}}{N_0^2} N_B^2$	$\frac{h^b_A}{N_0} N_A$	$\delta^A \gg \delta^B$
Fast	1	100	100	100	100	100	$\gg \delta^A \gg \delta^B$
Slow	1	100	100	1	1	1	$\sim \delta^A \gg \delta^B$
Ultra-slow	1	100	100	0.01	0.01	0.01	$\sim \delta^B \ll \delta^A$

The final choice of our parameters is displayed in Table 106.1 for the production and decay parameters that are kept fixed and in Table 106.2 for the binding and unbinding parameters.

In our numerical approach we used again Gillespie simulations of the reactions (106.9) and (106.10), later displayed in Figs. 106.4 and 106.5 for two extreme cases of fast and ultra-slow genes. The ultimate goal now is to reproduce the observed features of the stochastic simulations in the deterministic limit by appropriate sets of differential equations. These differential equations should be derived from the set of master equations corresponding to the reactions (106.9) and (106.10). In particular the question arises as to whether Eqs. (106.1) and (106.2) from Sect. 106.2 can be recovered and under what conditions. The explicit form of the master equations can be found in [13]. They determine the probability $P_{ij}(N_A, N_B; t)$ for finding N_A proteins of type A, N_B proteins of type B at time t under the condition that gene a is in state i, ($i =$ on, bare, off), and gene b is in state j, ($j =$ on, bare).

Our aim is to derive equations on the mean-field level for the first moments of N_A, N_B by averaging over all processes that are fast compared to the decay rates of both proteins. First of all we assume that we can factorize the probability $P_{ij}(N_A, N_B, t)$ according to

$$P_{ij}(N_A, N_B, t) = a_i b_j P(N_A, N_B, t) \tag{106.12}$$

with a_i the probability of finding gene a in the i-state and b_j the probability of finding gene b in the j-state, while $P(N_A, N_B, t)$ is the probability of finding the respective protein numbers whatever states the genes are in. As a further approximation we shall replace higher order moments $\langle N_S^k \rangle$, $k \geq 2$, by $\langle N_S \rangle^k$, $S = A, B$, where $\langle N_S \rangle = \sum_{ij} \langle N_S \rangle a_i b_j$. Both approximations, the factorization of the probability and the drop of higher order correlations in the moments, are in principle crude and can be justified only à posteriori, if the resulting equations reflect the deterministic limit of the Gillespie simulations.

Within these approximations we next derive three equations for $\frac{d(\langle N_A \rangle a_i)}{dt}$, ($i =$ on, bare, off) and two equations for $\frac{d(\langle N_B \rangle b_i)}{dt}$, ($i =$ on, bare). These equations follow from averages over the protein numbers as well as over the states of gene a

for protein B and over the states of gene b for protein A, since the change in N_A is assumed to be independent of the states of gene b and vice versa for N_B. To extract the equations for the moments N_A and N_B, we therefore have also to derive the time evolution of the probabilities to find the genes in one of their allowed states, that is equations for $\frac{da_i}{dt}$, $\frac{db_j}{dt}$ that follow from $a_i = \sum_{N_A,N_B,j} P_{ij}(N_A, N_B; t)$ and $b_j = \sum_{N_A,N_B,i} P_{ij}(N_A, N_B; t)$. At this stage we are left with five equations for the first moments of proteins and five equations for the probabilities for finding the genes in their specific states. Details of the derivation can be found in [13].

For the averaging so far we have not used any assumption on the binding rates and their relation to the decay rates. For the next steps, however, in which we want to further reduce the set of equations, we shall distinguish the three cases of fast, slow and ultra-slow genes with parameters chosen according to Table 106.2.

Fast Genes Here we assume that proteins A (B) see only average values of their own genes a (b), respectively. Therefore we sum $\sum_i d/dt(\langle N_A\rangle a_i)$, ($i$ = on, bare, off) and $\sum_j d/dt(\langle N_B\rangle b_j)$, ($j$ = on, bare), and insert for the probabilities a_i and b_j their stationary values which follow from the solutions of $\frac{da_i}{dt} = 0 = \frac{db_j}{dt}$. The result is

$$\frac{d\Phi_A}{dt} = \frac{\gamma^a_{bare} + \gamma^a_{on} x^a_{AA} \Phi^2_A + \gamma^a_{off} x^a_{BB} \Phi^2_B}{1 + x^a_{AA}\Phi^2_A + x^a_{BB}\Phi^2_B} - \Phi_A \tag{106.13}$$

$$\frac{d\Phi_B}{dt} = \frac{\delta^B}{\delta^A}\left(\frac{\gamma^b_{bare} + \gamma^b_{on} x^b_A \Phi_A}{1 + x^b_A \Phi_A} - \Phi_B\right) \tag{106.14}$$

with the following definitions: $\langle N_S\rangle/N_0 =: \Phi_S$, $S = A, B$, $x^m_n := \frac{h^m_n}{f^m_n}$ with ($m = a, b$), ($n = A, B$ or AA, BB), respectively, $\gamma^a_{on} = \frac{g^a_{on}}{\delta^A}$, $\tau = t\delta^A$, and the like. Equations (106.13)–(106.14) correspond to Eqs. (106.1)–(106.2) of the first section, so that we have recovered the former deterministic description apart from some minor differences. In particular the replacement of the dynamical probabilities a_i, b_j by their stationary values and the averaging over the remaining gene states reduced the number of dynamical equations to two, for $\Phi_A = \langle N_A\rangle/N_0$ and $\Phi_B = \langle N_B\rangle/N_0$, as compared to ten at the start.

Figure 106.4 shows which features of the Gillespie trajectories in (N_A, N_B)-space and their corresponding probability distributions (PDFs) are reflected by the deterministic equations. The locations of the maxima in the PDFs in the upper and lower right figure exactly correspond to the fixed-point values of the deterministic equations, indicated by vertical lines. In the limit-cycle regime (second row of the figure) the vertical lines mark the maximal and minimal extension of the limit cycles in Φ_A and Φ_B when determined by the deterministic equations; their values are in reasonable agreement with the PDF as derived from the Gillespie simulations. So the overall phase structure of two fixed-point regimes and one intermediate limit-cycle regime is preserved in the coarse-grained description of Eqs. (106.13)–(106.14).

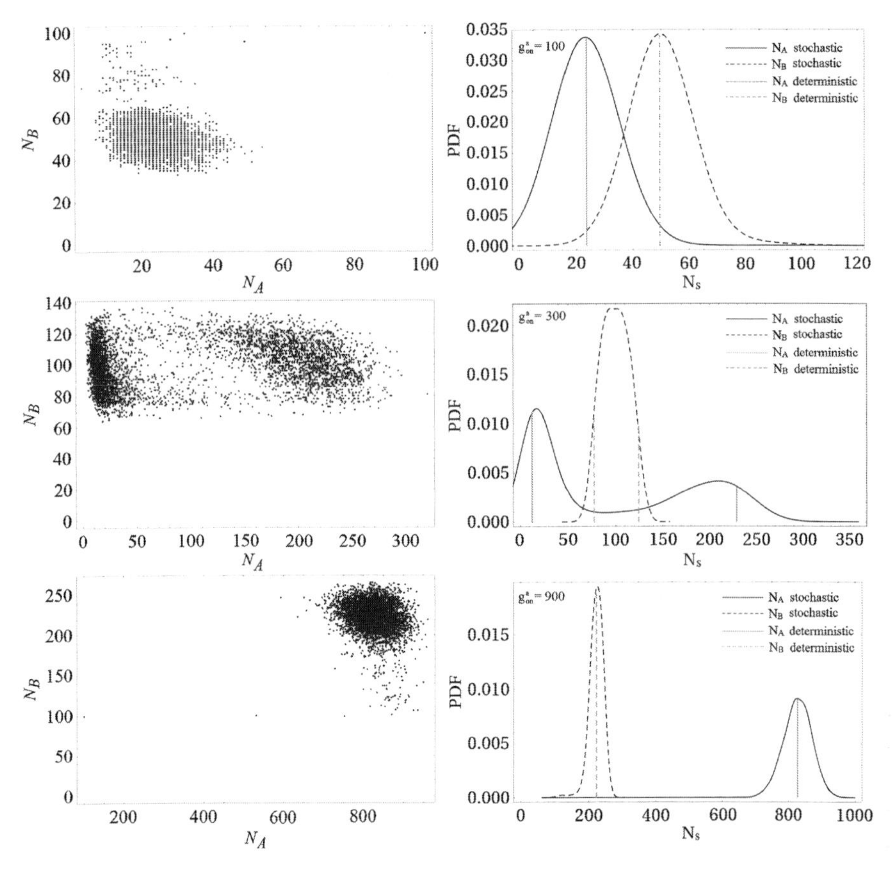

Fig. 106.4 Gillespie simulations for fast genes, that is $h^a_{AA} = h^a_{BB} = 0.01$, $h^b_A = 1.0$, $f^a_{AA} = f^a_{BB} = f^b_A = 100$. Phase portraits of the number of proteins N_B versus N_A within Gillespie time $T_G = 5000$ (*left column*) and corresponding probability density functions (PDF) (*right column*) of N_A (*full line black*) and N_B (*dashed line black*), while *the gray full and dashed vertical lines* indicate the position of the fixed points in the first and third row. In *the first and third row* $g^a_{on} = 100$ and $g^a_{on} = 900$, respectively, in *the left panels* we see the stochastic pendant of the fixed points observed in the deterministic case. For clarity of the figure we do not plot every Gillespie step, but only 5000 of them. The maxima of the PDFs agree well with the fixed points in the deterministic description. In *the second row* we see the stochastic version of limit cycles for $g^a_{on} = 300$. *The vertical lines* here mark the maximal and minimal extension of the limit cycles in Φ_A and Φ_B when integrated as solutions of the deterministic equations (106.13)–(106.14). These plots confirm our former model (106.1)–(106.2) as a suitable coarse-grained description

Slow Genes From the corresponding Gillespie simulations (not displayed here) one can see that switches between the on- and the bare states of gene a are still fast, but those between on- and off-states are less frequent. Accordingly we sum now $\frac{d}{dt}(\langle N_A \rangle a_{on} + \langle N_A \rangle a_{bare})$ and $\frac{d}{dt}(\langle N_A \rangle a_{bare} + \langle N_A \rangle a_{off})$. It then depends on the initial conditions whether protein A is either described by the dynamics of the first sum or by that of the second sum, reflecting the fact that protein A sees two different average values of the gene states, either over "on-bare" or over "bare-off". If we assume that the switches between the on- and bare states of gene b, which are of the same order as the decay time of protein A, are still fast as compared to protein B, we average $\frac{d}{dt}(\langle N_B\rangle b_j)$ further over ($j=$ on, bare), independently of which alternative (I or II) describes the dynamics of protein A. This way we obtain two sets of two differential equations each, either $\frac{d\langle N_A^I\rangle}{dt}$ and $\frac{d\langle N_B\rangle}{dt}$ or $\frac{d\langle N_A^{II}\rangle}{dt}$ and $\frac{d\langle N_B\rangle}{dt}$. If we still insert the stationary values of a_i and b_j, a linear stability analysis reveals that the first set of equations has up to three fixed points, depending on the choice of parameters, two stable and one saddle, the second set of equations has one stable fixed point, but none of the two sets allows for a regime of stable limit cycles. A comparison with the results of the Gillespie simulations (not displayed here) works best for large values of the bifurcation parameter, where the first fixed point, predicted by $\frac{d\langle N_A^I\rangle}{dt}$ and the second fixed point predicted by $\frac{d\langle N_A^{II}\rangle}{dt}$ are visible in the two maxima of the PDF derived from the Gillespie trajectories. These trajectories show, however, stripes of N_A, N_B events, well localized in N_A, corresponding to the two noisy versions of the fixed points in N_A, but smeared out over a large range of N_B values. This result is at odds with the fixed-point predicted by $\frac{d\langle N_B\rangle}{dt}$ and shows that in contrast to protein A, protein B is not fast enough to resolve the different states, here of gene b, and to adapt to them in time before the gene state changes.

Ultra-Slow Genes This limit refers to a situation, in which the binding (unbinding) rates of transcription factors are of the order of the slow protein B. So the time which genes a and b spend in one of their possible states is long as compared to $1/\delta^A$, the lifetime of protein A. It then does no longer make sense to further sum the equations for $\frac{d\langle N_A\rangle a_i}{dt}$ over any states of gene a, nor to sum $\frac{d\langle N_B\rangle b_j}{dt}$ over any states of gene b, neither to insert stationary values for a_i, b_j. Instead we insert da_i/dt, db_j/dt from the formerly derived expressions and end up with five uncoupled equations for the first moments $\Phi_S=\langle N_S\rangle/N_0$, $S=A,B$, namely

$$\begin{aligned}\frac{d\Phi_A}{dt}&=g_i^a-\delta^A\Phi_A, \quad i=\text{on, bareoff}\\ \frac{d\Phi_B}{dt}&=g_i^b-\delta^A\Phi_A, \quad i=\text{on, bare}\end{aligned} \tag{106.15}$$

with solutions that for $t\to\infty$ exponentially decay to the fixed points g_i^s/δ^S with $(S=A,B)$, $(s=a,b)$, ($i=$ on, bare, off) for $s=a$, and ($i=$ on, bare) for $s=b$, leading to six fixed points. Depending on the initial conditions, the system converges to one of them. For this choice of binding rates, no reduction to a smaller set of differential equations for the moments $\langle N_A\rangle$, $\langle N_B\rangle$ is therefore possible, neither can the probabilities for gene states be absorbed in effective rate constants.

In the Gillespie simulations of this limit we expect the system to switch between three possible states with respect to N_A and two with respect to N_B, so between six fixed points in the deterministic limit. The former oscillations in the limit-cycle regime are clearly gone. For N_A we see both in the phase portraits (localized stripes

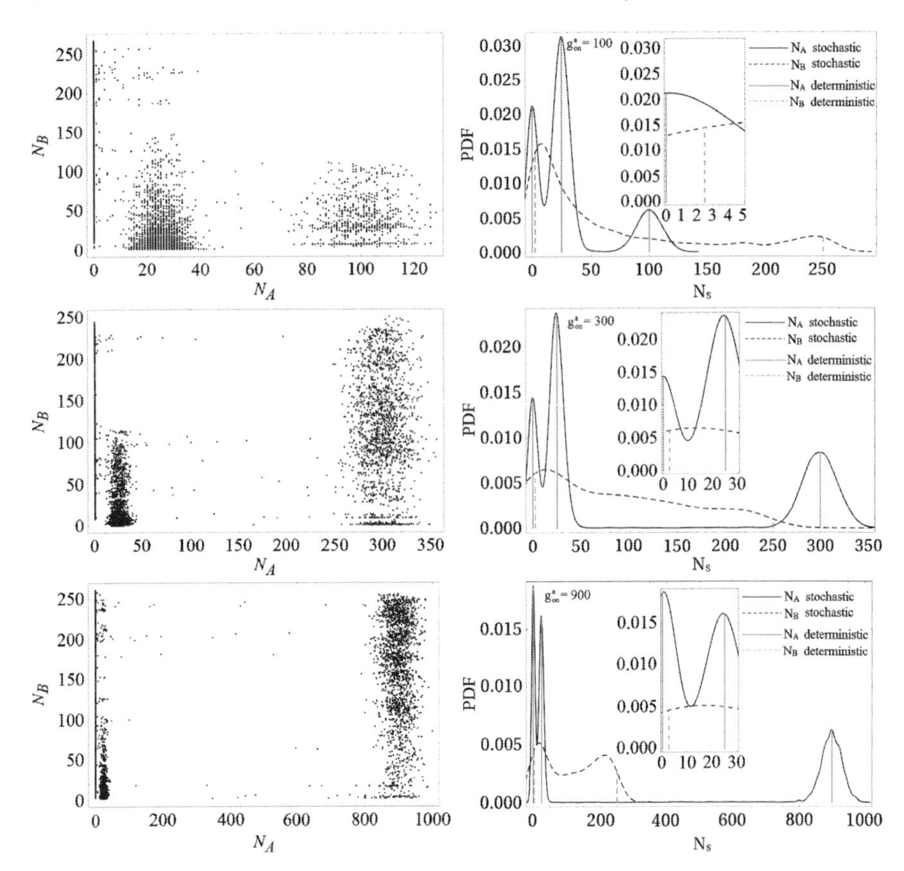

Fig. 106.5 Same as Fig. 106.4, but for ultra-slow genes, that is $h^a_{AA} = h^a_{BB} = 0.000005$, $h^b_A = 0.0001$, $f^a_{AA} = f^a_{BB} = f^b_A = 0.01$. In *all three rows, left column*, we see remnants of three fixed points in the deterministic limit with respect to values of N_A, while the values of N_B are broadly spread, since B is too slow to follow the different states of gene b. *The vertical lines* from the deterministic prediction of the fixed point values (*right column*) match well the maxima of the PDFs for N_A and only roughly for N_B. *The insets* zoom into the (N_A, N_B) values of the two fixed points for lower N_A-values

in N_A) and in the probability density functions (pronounced maxima in N_A) remnants of three distinct fixed-point values of N_A (cf. Fig. 106.5), while the remnants of two possible fixed-point values of N_B are only vaguely visible as two broad maxima in the probability distribution. Obviously the ultra-slow genes are still not slow enough to allow protein B to adjust to the different states of gene b.

Our description of Sect. 106.3 in terms of Eqs. (106.1)–(106.2) would therefore completely fail to predict the time evolutions of Φ_A and Φ_B. Last but not least the regime of noisy limit cycles is also gone for ultra-slow genes as for the case of slow genes.

106.5 Summary and Conclusions

We have studied the motif of a bistable frustrated unit from the physics' perspective, independently of whether our parameter choice is realized in one of the natural systems where it is found. A question of main interest concerned the existence of regular oscillations for an intermediate range of parameters, independently of the realization of the motif. The answer is negative. No regular oscillations are found for cases in which the binding rates of genes are not fast compared to the protein decay rates. In general one should therefore account for caveats in finding the appropriate averaging procedure over intermediate states of the system and over subsets of variables that usually serve to reduce the number of degrees of freedom for a coarse-grained description.

The occurrence of quasi-cycles in both stochastic versions of Sects. 106.3 and 106.4 is not new from the physics' point of view. Quasi-cycles in ecological systems are known for example from [14, 15]. In our context of genetic systems the observation of quasi-cycles raises a question which is of interest from the biological point of view: Are there genetic circuits whose oscillatory behavior is exclusively a result of demographic fluctuations, e.g. in the number of proteins? Since oscillations can be created in a number of different ways, both in models and in nature, it may be rather involved to trace back the true origin of oscillations in a concrete case.

Acknowledgements We would like to thank our collaborators A. Garai (UC San Diego), W. Janke and H. Nagel (University of Leipzig) as well as B. Waclaw (University of Edinburgh) for their contributions to different parts of the work.

References

1. Martiel J, Goldbeter A (1987) A model on receptor desensitization for cyclic AMP signaling in dictyostelium cells. Biophys J 52:807–828
2. Novak B, Tyson JJ (1993) Numerical analysis of a comprehensive model of M-phase control in xenopus oocyte extracts and intact embryos. J Cell Sci 106:1153–1168
3. Pomerening JR, Kim SY, Ferrell JE Jr. (2005) Systems-level dissection of the cell-cycle oscillator: bypassing positive feedback produces damped oscillations. Cell 122:565–578
4. Tyson JJ (1991) Modeling the cell division cycle: cdc2 and cyclin interactions. Proc Natl Acad Sci USA 88:7328–7332
5. Qiao L, Nachbar RB, Kevrekidis IG, Shvartsman SY (2007) Bistability and oscillations in the Huang-Ferrell model of MAPK signaling. PLoS Comput Biol 3:1819–1826
6. Vilar JMG, Kueh HY, Barkai N, Leibler S (2002) Mechanisms of noise-resistance in genetic oscillators. Proc Natl Acad Sci USA 99:5988–5992
7. Ingolia NT, Murray AW (2004) The ups and downs of modeling the cell cycle. Curr Biol 14:R771–R777
8. Krishna S, Semsey S, Jensen M (2009) Frustrated bistability as a means to engineer oscillations in biological systems. Phys Biol 6:036009. 8pp
9. Garai A, Waclaw B, Nagel H, Meyer-Ortmanns H (2012) Stochastic description of a bistable frustrated unit. J Stat Mech 2012:P01009 (28 pp)
10. Gillespie DT (1977) Exact stochastic simulation of coupled chemical reactions. J Phys Chem 81:2340–2361

11. Van Kampen NG (2005) Stochastic processes in physics and chemistry. North-Holland, Amsterdam
12. Kaluza P, Meyer-Ortmanns H (2010) On the role of frustration in exciatble systems. Chaos 20:043111 (11 pp)
13. Labavić D, Nagel H, Janke W, Meyer-Ortmanns H (2012) Coarse-grained modeling of genetic circuits as a function of the inherent time scales. arXiv:1209.0581v1
14. Butler T, Goldenfeld N (2009) Robust ecological pattern formation induced by demographic noise. Phys Rev E 80:030902(R) (4 pp)
15. Butler T, Goldenfeld N (2011) Fluctuation-driven Turing patterns. Phys Rev E 84:011112 (12 pp)

Part X
Satellite Meeting: Information Processing with Recurrent Dynamical Systems: Theory and Experiment

Chapter 107
Memory and Nonlinear Mapping in Reservoir Computing with Two Uncoupled Nonlinear Delay Nodes

Silvia Ortín, Luis Pesquera, and José Manuel Gutiérrez

Abstract A novel architecture based on a single nonlinear node with delayed feedback has been recently proposed for Reservoir Computing (Appeltant et al., Nat. Commun., 2:468–472, 2011). We analyze the interplay of memory and nonlinear mapping for a single sigmoid delay node and for two uncoupled delay nodes with different parameters for the sigmoid function. When two nodes are used, the memory capacity increases and the performance for low nonlinearity task degree is clearly improved.

Keywords Reservoir computing · Time-delay systems

107.1 Introduction

Standard Reservoir Computing (RC) utilizes a large network of randomly connected nonlinear dynamical nodes with fixed weights [1]. Recently, it has been demonstrated that RC can also be realized by replacing the complex network by a single nonlinear node subject to delayed feedback [2]. The reservoir with delay systems is constructed in a completely deterministic manner as in the case of simple cyclic reservoirs [3]. The reduction to a single delay node has allowed hardware implementation in electronic [3] and photonic [4, 5] systems.

In this paper we first show that a single delay node with a sigmoid function exhibits short-term memory. However, real-world tasks require both memory and a nonlinear mapping. We use a recently introduced task (delayed continuous XOR [6]) to study the interplay between nonlinearity of the mapping and memory in the reservoir. In order to increase memory capacity (MC), we consider two uncoupled

S. Ortín (✉) · L. Pesquera · J.M. Gutiérrez
Instituto de Física de Cantabria (CSIC-Universidad de Cantabria), Santander, Spain
e-mail: ortin@ifca.unican.es

L. Pesquera
e-mail: pesquerl@ifca.unican.es

J.M. Gutiérrez
e-mail: gutierjm@unican.es

T. Gilbert et al. (eds.), *Proceedings of the European Conference on Complex Systems 2012*, Springer Proceedings in Complexity, DOI 10.1007/978-3-319-00395-5_107,

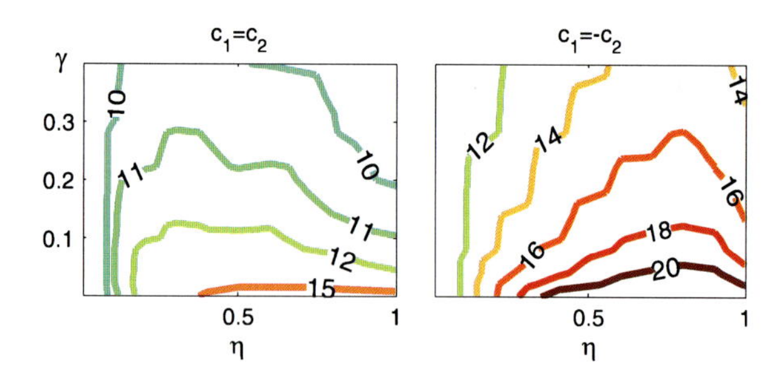

Fig. 107.1 Quality memory capacity m_q. *Left*: Single delay node with sigmoid function exponent $c = 4$, $N = 400$ and $\tau = 80$. *Right*: Two uncoupled delay nodes with $c_1 = 4$ and $c_2 = -4$. Each node has $N = 200$ virtual nodes and $\tau = 40$

delay nodes with different parameters for the sigmoid function. It is found that performance for low nonlinearity task degree is clearly improved.

107.2 One Single Nonlinear Delay Node

The reservoir is an input-driven delay dynamical system with a single nonlinear node [1]

$$dx(t)/dt = -x(t) + \eta f\bigl(x(t-\tau) + \gamma J(t)\bigr), \qquad (107.1)$$

where J being the input, τ the delay time, η the feedback strength and γ the input scaling. Input is multiplexed in time across τ by using a random binary mask (values $+1$ and -1) of N bits. The delay interval is divided into N pieces of length $\theta = \tau/N = 0.2$ with their endpoints representing virtual nodes from which linear readouts learn to extract information and perform computation through linear regression (see [1] for details). We consider a sigmoid function $f(u) = 1/(1+\exp(-cu)) - 0.45$ with an exponent $c = 4$, such that the system operates in a stable fixed point without input ($\gamma = 0$) when $\eta < 1$.

The input $u(t)$ consists of a random signal with values in $[-0.5, 0.5]$. We first compute the quality MC, m_q, that is given by the sum of the normalized correlation $m(d)$ between the output and the delayed input $u(t-d)$ over d, without taking into account low values in the tail of $m(d)$ [7]. The results plotted in Fig. 107.1(left) for $N = 400$ virtual nodes and $\tau = 80$ show that m_q is greater than 12 for low γ and a large range of values for η. The maximum value achieved for m_q is 21.

We now evaluate the performance of the single delay node for a delayed continuous XOR-task [6]. This task offers control of the amount of nonlinearity and memory required from the reservoir. The task is to reconstruct the following delayed nonlinear function of the input $u(t)$

$$y_{d,p}[t] = \operatorname{sign}\bigl[u(t-d)u(t-d-1)\bigr]\bigl|u(t-d)u(t-d-1)\bigr|^{p}, \qquad (107.2)$$

where the delay d is the required memory and the power p determines the degree of task nonlinearity. The maximum delay $d_{\max}$ with a normalized root mean square error ($\mathrm{NRMSE_{XOR}}$) smaller than 0.1 is shown in Fig. 107.2(top) for $p = 1$ and $p = 2$. For low nonlinearity ($p = 1$) good performance ($\mathrm{NRMSE_{XOR}} < 0.1$) is obtained with $d = 6$ for low γ and a large range of values for η. The maximum value

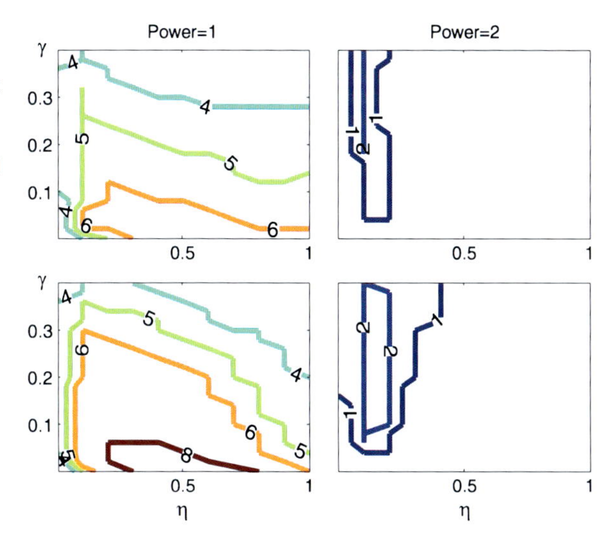

Fig. 107.2 Maximum delay with $\mathrm{NRMSE}_{\mathrm{XOR}} < 0.1$ for continuous delayed XOR-task for powers p = 1 (*left*) and 2 (*right*). *Top*: Single delay node. *Bottom*: Two uncoupled delay nodes. Same parameters as in Fig. 107.1

achieved for $d_{\max}$ is 7. When nonlinearity is increased ($p = 2$) we obtain $d_{\max} = 2$ for a small range of parameter values.

107.3 Two Uncoupled Nonlinear Delay Nodes

We now consider two uncoupled delay nodes with different exponents, c_1 and c_2, for the sigmoid function. Using more delay nodes can increase the diversity of the reservoir. We have found that an increase in the MC is obtained when c_1 and c_2 have opposite signs.

We show in Fig. 107.1(right) m_q for two uncoupled delay nodes with exponents $c_1 = 4$ and $c_2 = -4$. We have considered $N = 200$ and $\tau = 40$ for each node. Quality MC m_q is greater than 20 for low γ and a large range of values for η. The maximum value achieved for m_q is 30.5. This represents a clear increase of MC with respect to the values obtained with a single node for the same number of virtual nodes (see Fig. 107.1(left)).

We now evaluate the performance of two uncoupled delay nodes for the delayed continuous XOR-task. The maximum delay with $\mathrm{NRMSE}_{\mathrm{XOR}} < 0.1$, $d_{\max}$, obtained with two uncoupled delay nodes is shown in Fig. 107.2(bottom) for powers $p = 1$ and $p = 2$. For low nonlinearity ($p = 1$) we obtain $d_{\max} = 8$ for low γ and a large range of values for η.

The maximum value achieved for $d_{\max}$ is 9. The performance is clearly improved with respect to the system with one single delay node. When nonlinearity is increased ($p = 2$) a value $d_{\max} = 2$ is obtained. The range of parameter values for $p = 2$ with $d_{\max} = 2$ shows a clear increase with respect to the one obtained for the single node system. These performance levels are comparable to or even better than those obtained with conventional RC. The $\mathrm{NRMSE}_{\mathrm{XOR}}$ obtained for $N = 100$ neu-

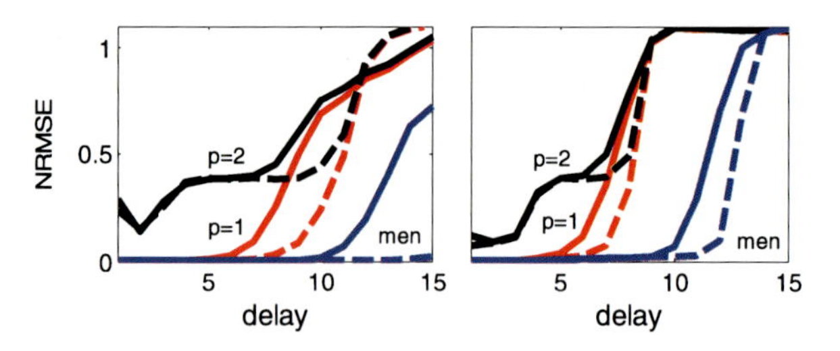

Fig. 107.3 NRMSE for continuous delayed XOR task with $p = 1$ (*red*), 2 (*black*) and memory task (*blue*). *Left* (*right*): $\gamma = 0.02$ (0.14) and $\eta = 0.4$ (0.1). *Solid* (*dashed*) *lines*: Single (two uncoupled) delay node with the same parameters as in Fig. 107.1

rons with standard RC [7] for power $p = 1$ (107.2) is greater than 0.1 when $d > 3$ ($d > 1$), that is $d_{\max} = 3$ ($d_{\max} = 1$).

Finally, we compare the performance for delayed continuous XOR task, $\text{NRMSE}_{\text{XOR}}$, and memory task (target: delayed input $u(t-d)$), $\text{NRMSE}_{\text{mem}}$, in Fig. 107.3 for both single and two uncoupled node systems. Two different set of parameter values are considered: $\gamma = 0.02$, $\eta = 0.4$ and $\gamma = 0.14$, $\eta = 0.1$, that correspond to the optimal parameter regions for delayed continuous XOR task with $p = 1$ and $p = 2$, respectively. Very low errors for memory task are obtained for the delay range that corresponds to $\text{NRMSE}_{\text{XOR}} < 0.5$. For $p = 1$ we obtain a $\text{NRMSE}_{\text{mem}} < 0.009$ when $\text{NRMSE}_{\text{XOR}} < 0.5$ in all the cases shown in Fig. 107.3. Then $\text{NRMSE}_{\text{XOR}}$ is not determined by the error for memory task, even for low nonlinearity ($p = 1$).

107.4 Conclusions and Discussion

We have shown that a single delay node with sigmoid nonlinearity exhibits short-term memory. However, the maximum MC reached by this type of system is limited. This can limit the performance for tasks that require high MC. It is shown that MC increases when two uncoupled delay nodes with different parameters for the sigmoid function are used.

We have analyzed the interplay of memory and nonlinear mapping by using a delayed continuous XOR task. The performance has a faster degradation with delay when the nonlinearity degree is increased. Performance for low nonlinearity task ($p = 1$) is clearly improved by using two uncoupled nodes. The achieved performance levels are comparable to or even better than those obtained with conventional RC. We have found that the performance is not only determined by the error of memory task, even for low nonlinearity.

We have considered two nodes to increase the diversity of the reservoir. This can be also achieved by using a single node with a nonlinear function that has different operation points. Preliminary results show that MC increases by using a single node with a $\cos^2$ function.

The reduction to a single node has allowed a hardware implementation of RC [2, 4, 5]. A crucial point in experimental realizations is the evaluation of noise effects on the computational properties. The effect of noise will be important for tasks that require low values of the input scaling, as for long delays.

Acknowledgements This work has been funded by EC Project PHOCUS (FP7-ICT-2009-C-240763).

References

1. Lukosevicius M, Jaeger H (2009) Reservoir computing approaches to recurrent neural network training. Comput Sci Rev 3:127–149
2. Appeltant L, Soriano MC, Van der Sande G, Danckaert J, Massar S, Dambre J, Schrauwen B, Mirasso CR, Fischer I (2011) Information processing using a single dynamical node as complex system. Nat Commun 2:468–472
3. Rodan A, Tino P (2011) Minimum complexity echo state network. IEEE Trans Neural Netw 22(1):131–144
4. Larger L, Soriano MC, Brunner D, Appeltant L, Gutierrez JM, Pesquera L, Mirasso CR, Fischer I (2012) Photonic information processing beyond Turing: an optoelectronic implementation of reservoir computing. Opt Express 20:3241–3249
5. Paquot Y, Duport F, Smerieri A, Dambre J, Schrauwen B, Haelterman M, Massar S (2012) Optoelectronic reservoir computing. Sci Rep 2:287
6. Verstraeten D, Dambre J, Dutoit X, Schrauwen B (2010) Memory versus non-linearity in reservoirs. In: The 2010 international joint conference on neural networks (IJCNN), pp 1–8
7. Hermans M, Schrauwen B (2010) Memory in linear recurrent neural networks in continuous time. Neural Netw 23:341–355

Part XI
Satellite Meeting: Complexity in the Real World—From Policy Intelligence to Intelligent Policy

Chapter 108
What Networks to Support Innovation? Evidence from a Regional Policy Framework

Annalisa Caloffi, Federica Rossi, and Margherita Russo

Abstract We explore how the implementation of a set of policy programmes over a period of six years induced some "emergent" learning effects which had not originally been envisaged by policymakers. This way, we show how policy evaluation can be used not only to assess the expected impact of policy interventions but also to discover their unexpected behavioural effects, and therefore provides an important instrument to guide the design of future interventions.

Keywords Policy evaluation · Social network analysis · Behavioural effects · Policy design

108.1 Introduction

Complexity-based approaches to innovation have emphasized the role of interactions among heterogeneous agents as key sources of innovation [1] highlighting the elements of such interactions that are associated with greater likelihood to generate innovations and to foster long-lasting relationships giving rise to innovation cascades. In management theory, it has been recognized that, as technologies become more complex and economic environments more uncertain, firms increasingly rely upon external sources of knowledge for their innovation processes, leading their innovation activities to become more open and distributed [2]. Firms' ability to access

A. Caloffi (✉)
Department of Economics, University of Padova, Padova, Italy
e-mail: annalisa.caloffi@unipd.it

F. Rossi
School of Business, Economics and Informatics, Birkbeck College, University of London, London, UK
e-mail: f.rossi@bbk.ac.uk

M. Russo
Department of Economics, University of Modena and Reggio Emilia, Modena, Italy
e-mail: margherita.russo@unimore.it

T. Gilbert et al. (eds.), *Proceedings of the European Conference on Complex Systems 2012*, Springer Proceedings in Complexity, DOI 10.1007/978-3-319-00395-5_108,

knowledge through interactions with other organizations, including universities, is increasingly recognized as a source of competitive advantage.

Hence, the ability to effectively access external knowledge through networking is a very important competence for firms wishing to innovate successfully. However, not all organizations are equally able to engage in effective networking. Small firms, for example, may find it difficult to distract resources from their main activities in order to engage in the search for external partners, to interact with organizations that are cognitively very distant (like universities or large multinational corporations) and even to identify the appropriate social channels through which contacts with potential partners could be made.

While policies directed at improving the education of the workforce may increase the networking capabilities of organizations in the long term (it has been shown that a higher share of highly qualified personnel increases an organization's absorptive capacity and hence its ability to search for and absorb external knowledge) another more immediate approach could be to encourage organizations to gain experience in networking with external partners by promoting the set up of innovation networks.

Policies fostering inter-organizational collaborations have been undertaken for a very long time (in Europe, at least since the launch of the first collaborative research programmes in the 1980s) but usually their stated objective is to promote joint R&D or technology transfer—promoting participants' networking skills is only incidental. Only a few programmes in the EU have had networking per se as a specific objective, and even these promote the formation of networks, not the strengthening of the participants' ability to network with others.

This points to the need to investigate what instruments can be used to strengthen organizations' networking skills. In this paper we explore whether policies sponsoring the formation of innovation networks may have as a significant "emergent effect" the strengthening of the participants' networking abilities, and if so which characteristics of these policies may be particularly conducive to enhancing networking skills. We do this thanks to the empirical analysis of a set of nine policy programmes in support of innovation networks implemented in the same region (Tuscany) between 2002 and 2008. The time dimension allows us to investigate whether agents' repeated participation to these policies enhances their ability to form "better" innovation networks. This approach fits with the recent debate in policy analysis on the need to investigate whether policies have "behavioural effects" in terms of stimulating learning processes on the part of the participants [3, 4].

108.2 What Makes a "Good" Innovation Network?

By not only promoting the set up of innovation networks but also by imposing constraints on their characteristics, the policymaker may be able to facilitate learning processes within and between the networks and thus to stimulate to a greater or lesser extent the development of the participants' networking capabilities. In our empirical analysis, we focus on three types of policy constraints which could promote learning and which were actually present in some of the policy interventions that we investigated.

1. *Heterogeneity*. By requiring the networks to include a certain degree of heterogeneity, the policymaker may facilitate learning processes thanks to which organizations improve their ability to interact with diverse organizations and hence to form heterogeneous networks in the future. The experience of engaging in and managing relationships with agents characterized by different cognitive frames and modes of operation is likely not only to facilitate the emergence of novelty [1, 5] but also to teach organizations how to improve their ways of interacting with others.
2. *Stability*. Stable relationships are important in order to promote knowledge spillovers and innovation diffusion to agents who are cognitively very distant and hence may need more time and greater interaction in order to absorb external knowledge. The policymaker may facilitate the consolidation of stable relationships by providing continuity in the policy framework. In fact, participants to policy-supported innovation networks have highlighted that the different time scales at which innovation processes and innovation policy interventions unfold (the former develop along a much longer time scale than the latter) can be problematic [6].
3. *Intermediaries*. The policymaker may want to ensure that networks include new participants, even in the presence of a stable core, in order to avoid "lock in" into communities of stable collaborators which can lead to undesirable effects like closure to outsiders, inward-looking attitudes, dependence on a partner, and the emergence of lobbying behaviour. The involvement of innovation intermediaries might provide support to achieve this aim. Intermediaries are organizations that play a mediating role in innovation processes, facilitating connections between other organizations that are engaged in the invention, development and production of new products, processes and services. They ensure interaction and communication among heterogeneous participants, which differ in language, systems of incentives and objectives, etc. [7].

In the policy practice, it can be very difficult to find a balance between fostering efficient and effective teamwork (allowing the time to create mutual understanding and routines) and favouring the creation of ruptures and novelty. The tension between temporary and stable relationships could be solved by considering the specific objectives of the network: that is, networks that explicitly prioritize innovation diffusion processes or the absorption of spillovers resulting from established innovations may be more effective when built around relatively stable communities of innovators that include either small and large firms or enterprises and universities; while networks aimed at the production of radical innovations may be more effective when new relationships play a prominent role.

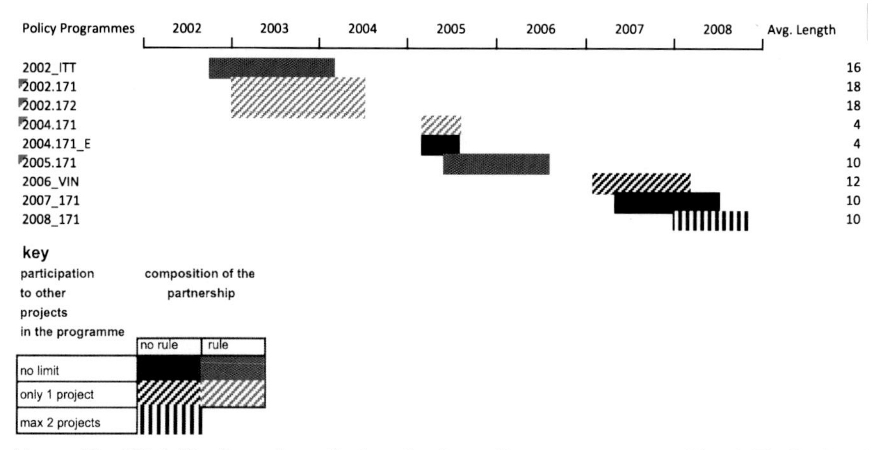

Note to Fig. 108.1: *The first column* displays the nine policy programmes considered. The Regional Programmes of Innovative Action are identified with the following labels: 2002_ITT (Regional Programme of Innovative Action issued in 2002, whose acronym was ITT—Tuscany Technological Innovation) and 2006_VIN (acronym: Virtual INnovation and Cooperative Integration, issued in 2006). The different calls of the two lines 1.7.1 and 1.7.2 included in the Single Programming Document are identified with the name of the line and of the reference year, as identified by the administrative documents we have analysed

Fig. 108.1 The time profile and rules of the different programmes

108.3 The Regional Policy Programmes

In the programming period 2000–2006, Tuscany's regional government promoted nine consecutive programmes aimed at supporting innovative projects carried out by networks of heterogeneous economic agents. The set of policy programmes can be divided into two major periods. The first period, which included the majority of programmes and participants, ran from 2002 to 2005 (the last projects were completed towards the end of 2006). It included six programmes (2002_ITT, 2002_171, 2002_172, 2004_171, 2004_171E, 2005_171). In the vision of policymakers, these programmes would have led to the development and strengthening of innovation clusters made of SMEs and large companies working together with innovation service providers and other agents supporting innovation. Strongly inspired by the regional innovation system framework (which was dominant in the European innovation strategies of the time) the regional policy maker considered the emergence of such clusters as the first step towards the formation of Tuscany's innovation system. These programmes were characterized by the imposition of numerous constraints—on the size and composition of the partnership and on the number of projects in which each organization could participate, as shown in Fig 108.1.

The second period started in 2006, and ended with the last intervention implemented in 2008. It included three programmes (2006_VIN, 2007_171 and 2008_171). The policymaker's goal was to consolidate the networks formed in the previous period.[1] No constraints were present in this period. This allows us to test, in our empirical analysis, whether the policy constraints imposed in the first period had some impact on the participants' learning processes influencing the development of their networking abilities in the second period.

Overall, the nine programmes were assigned almost €37 million, representing around 40 % of the total funds spent on innovation policies, and sponsored 168 projects. The total number of different organizations involved in the nine programmes was 1127, a subset of which (348) took part in more than one project. We classified the organizations involved in the programmes into nine categories: firms (60.3 % of all organizations involved),[2] business service providers (7.6 %); private research companies (2 %); local business associations (7.5 %); universities and other public research providers; service centres (generally publicly funded or funded via public-private partnerships; 3 %); chambers of commerce (1 %); local governments (6.8 %); and other public bodies (3.5 %).

The various programmes addressed a set of technology/industry targets. A large share of funds was devoted to ICT and multimedia (48.2 %), with the objective to widen their adoption in traditional industries and SMEs. Projects in opto-electronics, an important competence network in the region, received 16.4 % of funds. The third targeted area, projects in mechanics, received 7.5 % of funds. The remaining technological fields included organic chemistry (5 %), biotech (4 %), and others (new materials, nanotechnologies and a combination of the previously mentioned technologies).

108.4 Assessing Learning Effects/1: The Heterogeneity of Project Networks

We assessed the heterogeneity of each project by measuring the diversity of the types of participants (using the reciprocal of the Herfindahl index computed on the shares of participants belonging to each of nine categories outlined earlier). The average heterogeneity index is not too dissimilar across different programmes. The only exception is the RPIA programme launched in 2006, which had lower average heterogeneity and low dispersion of these values around the mean. Remarkably, in the first period, there was very little difference in the mean and dispersion of the

---

[1] Interestingly, these interventions had not been planned at the beginning of the programming period. Rather the region was able to procure additional funds that enabled it to implement a further RPIA and two more waves of the SPD line supporting innovation networks (programme 171).

[2] In terms of economic activity (based on Nace Rev. 1.1 codes) and size, the largest share of participating enterprises were manufacturing companies (68 %): of these, 21.8 % were micro and small firms in the traditional industries of the region (marble production and carving, textiles, mechanics, jewelery), while the remaining share were micro firms in the service sector (Nace Rev. 1.1:72). The latter were an active group, with 1.8 projects per agent on average. The share of participating enterprises varied in the different programmes, ranging from a minimum of 37.1 % in programme 172_2002 to a maximum of 100 % in the smallest programme (171_2004).

heterogeneity index between the five programmes that imposed a minimum heterogeneity constraint and the programme that did not; nor there was lower average heterogeneity in the second period, when no constraints were imposed. We also find that greater project size was associated with greater heterogeneity, and that project networks funded within programmes where a minimum heterogeneity constraint was present were generally much larger than those funded within programmes without such constraint, and very often much larger than the minimum size required to fulfill the heterogeneity constraint. A possible explanation for this is that the imposition of a mandatory heterogeneity constraint forced projects coordinators to include organizations that were not strictly necessary to the project's success and required them to increase the network size to include all the desired participants; while the elimination of such constraints allowed the partnership to be designed according to the effective project requirements and to economise on the number of partners without necessarily reducing heterogeneity. This would therefore recommend caution in imposing arbitrary heterogeneity constraints without taking into account the actual partnership needs of the different projects.

Computing the heterogeneity index at the level of the entire programme, rather than at the level of individual project networks, provides a different outlook. The heterogeneity index in terms of participants' fluctuated around a stable trend, and programmes with a minimum heterogeneity constraint were no more heterogeneous than the others. Instead, the heterogeneity index in terms of participants' technology areas was increasing over time, indicating that the programmes progressively involved a wider range of technologically diverse organizations.

To detect the learning effects of the policy interventions on the organizations' networking abilities, we consider the 205 organizations that took part in projects in both periods, 2002–2005 and 2006–2008, and we test whether an organization's participation in policy interventions in the first period (and the features of that participation) had an impact on its ability to engage in heterogeneous partnerships in the second period.

We measure heterogeneity of an agent's networks in period 2002–2005 using the average of the heterogeneity index of all the project networks that the organization was involved during that period (*avgdiversity_20068*). We then regress this variable on a set of variables that capture the involvement of the agent in previous and current policy programmes:

- *p2002ITT*, *p2002171*, *p2002172*, *p2004171*, *p2004171E*: set of five dummy variables capturing which policy programmes the organization was involved in during period 2002–2005;
- *avgdiversity_20025*: average heterogeneity index of the projects the organization was involved in during period 2002–2005;
- *avgfunding_20025*: average funding obtained by the organization in projects during period 2002–2005;
- *avgpctSC_20025*: average share of partners that were service centres in the projects the organization was involved in during period 2002–2005;
- *avgp_20025*: average number of partners in the projects the organization was involved in during period 2002–2005;

Table 108.1 Regression explaining average heterogeneity of organization's networks in 2006–2008

	Coefficient	Robust standard error	Sign.
Dependent variable	*avgdiversity_20068*		
Number of obs	197		
p2002ITT	−0.007	0.089	
p2002171	−0.391	0.230	*
p2002172	0.080	0.221	
p2004171	0.318	0.222	
p2004171E	0.448	0.225	**
p2005171	−0.088	0.039	**
avgdiversity_20025	0.122	0.103	
avgfunding_20025	0.000	0.000	
avgpctSC_20025	−3.289	1.773	*
avgp_20025	0.004	0.018	
avgfunding_20068	0.000	0.000	*
avgpctSC_20068	0.110	0.072	
avgp_20068	0.094	0.024	***
_cons	3.040	0.592	***

Note: * 0.1, ** 0.01, *** 0.001. F(31, 165): 7.51, Prob > F: 0.0000, R-squared: 0.4229, Root MSE: 0.94736

- *avgfunding_20068*: average funding obtained by the organization in projects during period 2006–2008;
- *avgpctSC_20068*: average share of partners that were service centres in the projects the organization was involved in during period 2006–2008;
- *avgp_20068*: average number of partners in the projects the organization was involved in during period 2006–2008;

We also consider a set of control variables capturing the type of agent, its size and the share of projects it engaged in each technological area. We use OLS with robust standard errors to control for possible correlation among the errors.[3] Due to some missing observations, the overall number of observations is 197.

These result displayed in Table 108.1 suggest that the constraints imposed by the policy did not have the expected impact on the participants' behaviour. In fact, participation in two of the programmes with minimum heterogeneity constraints (171_2002 and 171_2005) had a significantly negative effect on the heterogeneity of networks in the second period, while participation in the only programme *without* a minimum heterogeneity constraint (2004_171E) had a significantly *positive*

[3]To check whether the "learning effects" induced by the policy were effectively due to the policy participation rather than to joint participation to other projects, we experimented with including a dummy variable equal to 1 if the organization had already collaborated with another participant in the policy programmes but outside of the set of regional policies, in both regressions. The inclusion of this variable reduced the number of observations to 182 due to missing values, did not change the sign and significance of the coefficients, and was itself not significant. Hence we did not include it in the final analysis.

effect. This may suggest (negative) policy learning: as the imposition of a minimum heterogeneity constraint forced organizations to form partnerships that were larger and more heterogeneous than was necessary, this negative experience may have led these organizations to limit the heterogeneity of their later partnerships in order to avoid inefficiencies. Hence, the constraint was not effective, maybe because the type of heterogeneity devised by the policymaker did not match the actual needs of the participants. Vice versa, participation in a programme where no such constraint was present seemed to have encouraged partners to experiment with more heterogeneous networks in the second period.

A greater share of relationships with the types of agents that the policymaker had envisaged could play the role of innovation intermediaries, the service centres, has a significantly *negative* effect on heterogeneity in the second period. As service centres are generally focused on specific technological areas, this may indicate that relationships with service centres did not encourage the encounter with organizations in different fields but rather only promoted relationships within the same area. This is not to say that service centres were not instrumental in facilitating relationships, but rather they did not seem to promote the ability of organizations to form relationships with heterogeneous partners (at least in the very aggregate terms we have measured it).

Greater average funding and larger networks in the second period were associated with greater heterogeneity in the same period. This suggests that the organizations that have the resources to obtain and manage more funds and to engage in larger projects also have better networking competences that enable them to organize heterogeneous partnerships.

108.5 Assessing Learning Effects/2: The Stability of Relationships

By definition, the first programme included participants and relationships that were new to the programme. Then, as time went by, there was a progressive increase in the number of agents that have already benefited from these policies. Nonetheless, continuous participation (that is, having been continuously active in all the previous programmes) and relatively stable participation (that is, having been present in at least one of the previous programmes) were associated with new relationships among new and old participants, as the share of new relationships remained high across all programmes; in particular, it remained constant and near 100 % in the first period, while it declined (non monotonically) in the second period, remaining however above 80 %.

This is consistent with the general policy objectives which, as we discussed earlier, were focused on the construction of new networks in the first period and on the consolidation of existing relationships in the second period.

We also find that the programmes that attracted the largest share of new participants were (besides the first) those which required project networks to have a min-

Table 108.2 Regression explaining stability of relationships of participants in 2006–2008

	Coefficient	Robust standard error	Sign.
Dependent variable	*Pctrepeated20068*		
Number of obs	197		
p2002ITT	0.025	0.032	
p2002171	0.220	0.099	**
p2002172	−0.004	0.058	
p2004171	0.100	0.072	
p2004171E	0.035	0.055	
p2005171	0.069	0.017	***
avgdiversity_20025	0.034	0.028	
avgfunding_20025	0.000	0.000	
avgpctSC_20025	0.103	0.466	
avgp_20025	0.001	0.004	
avgfunding_20068	0.000	0.000	
avgpctSC_20068	0.022	0.025	
avgp_20068	0.015	0.007	**
_cons	0.042	0.172	

Note: * 0.1, ** 0.01, *** 0.001. F(31, 165): 9.20, Prob > F: 0.000, R-squared: 0.4853, Root MSE: 0.24836

imum number of participants (172_2002 and 171_2005). Therefore, one of the effects of the presence of a high minimum number of participants was the involvement of a large number of agents that were new to the policy. On the contrary, broadening the range of target sectors/technology areas—as implemented in the programmes after 2004—did not appear to have the same effect.

Our results show that around 86 % of the total number of relationships was repeated over at least two years. Very often, such relationships developed between firms, between firms and universities, or between firms and service providers (service centres or private business service providers) indicating that repeated relationships developed among organizations that have a common research or technological focus.

We then test whether an organization's participation in policy interventions in the first period (and the features of that participation) had an impact on its ability to engage in stable partnerships in the second period, by regressing the stability of links in 2006–2008 (measured as the percentage of relationships of each agent in 2006–2008 which already existed in 2002–2005, *Pctrepeated20068*) on the same regressors and control variables as in the regression used to study heterogeneity. We consider the set of 205 agents that participated in projects in the two periods and we run a OLS regression with robust standard errors. Due to some missing observations, the overall number of observations is 197.

The results are displayed in Table 108.2.

Participation in the two programmes that provided funds only to projects that had a minimum number of participants (172_2002 and 171_2005) had a significantly positive effect on the stability of relationship in the subsequent period. We have already noted how this constraint seems to have encouraged the involvement of new participants in the programme; this result seems to suggest that these participants have also gone on to form relationships that were repeated in the second period. There was a positive effect of average number of partners in 2006–2008 on the stability of an organization's relationships in the same period, suggesting that organizations building larger networks relied to a greater extent on partners they had already collaborated with. This may be explained on the basis of the need to be able to rely on trusted partners with whom communication and knowledge exchange are easier, when managing the complexities of larger networks.

When considering the control variables (not shown), we find that local governments tend to have a greater share of stable relationships, and hence do not appear as playing a role of brokers of new relationships in the networks. Organizations involved in projects in certain technological areas, especially those can be characterized as "high tech", are less likely to have a greater share of stable partnerships. This provides some (weak) support for our suggestion that projects that entail greater technological complexity and that may have the potential for more radical innovation aim for greater novelty in the partnership's composition.

108.6 Conclusions

In this paper we have shown, using some simple econometric tools, how the imposition of constraints on network formation in the context of policy interventions supporting innovation networks may have some learning effects, stimulating the participants' ability to form heterogeneous and stable partnerships, although not always in the direction envisaged by the policymaker. This analysis represents one step in a wider research programme focused on the exploration of innovative analytical tools in order to investigate the behavioural effects of policy interventions, which involves the use of qualitative research, econometric analysis and static and dynamic social network analysis.

References

1. Lane DA, Maxfield R (1997) Foresight complexity and strategy. In: Arthur WB, Durlauf S, Lane DA (eds) The economy as an evolving complex system II. Addison-Wesley, Reading
2. Chesbrough H (2003) Open innovation: the new imperative for creating and profiting from technology. Harvard Business School Press, Boston
3. Autio E, Kanninen S, Gustaffson R (2008) First- and second-order additionality and learning outcomes in collaborative R&D programs. Res Policy 37(1):59–76
4. Clarysse B, Wright M, Mustar P (2009) Behavioural additionality of R&D subsidies: a learning perspective. Res Policy 38:1517–1533
5. Nooteboom B (2000) Learning by interaction: absorptive capacity, cognitive distance and governance. J Manag Gov 4(1–2):69–92
6. Russo M, Rossi F (2009) Cooperation partnerships and innovation. A complex system perspective to the design, management and evaluation of an EU regional innovation policy programme. Evaluation 15(1):75–100
7. Howells J (2006) Intermediation and the role of intermediaries in innovation. Res Policy 35:715–728

Chapter 109
Computational Complete Economy Models: A Model Class that Bridges the Gap Between Conventional Economic Modeling and Agent-Based Models

Davoud Taghawi-Nejad and Samuel G. Asfaha

Abstract Computational Complete Economy models are an agent-based model class that is based on Computable General Equilibrium models. Individual firms and households are modeled based on calibrated utility and sector specific production functions using the techniques inherited from CGE. Unemployment, that emerges from non-equilibrium markets, is explained as a result of the interaction of the individual agents in the macroeconomy. This agent-based model builds on previous knowledge and expertise on CGE within the ILO, the UN and other institutions. Building the CCE on a model already existing in these organizations allows us to overcome organisational resistance.

Keywords ABM · Agent-based computational economics · Labor economics · Policy · Macroeconomics · CGE · Computable general equilibrium

In the aftermath of the economic crisis, the inadequacy of conventional tools of analysis have become clear to policy-makers and development practitioners. Policy-makers are often not acquainted with heterodox academic environments and lack the time to learn new tools. Let alone applying them to their daily work. We therefore develop an agent-based model class that is carefully designed to be easily comprehensible to policy-makers who are trained in conventional economics. These computational complete economy (CCE)[1] models capture economy-wide interactions among economic agents, analogous to the widely used computable general equilibrium (CGE) models. Like other traditional macro-economic models CGE models assume representative consumers and firms; actors are rational and have perfect

[1]The current working paper can be found at the Computational Economics and Finance 2012 website: http://editorialexpress.com/conference/CEF2012/program/CEF2012.html.

D. Taghawi-Nejad (✉)
Department of Economics "S. Cognetti de Martiis", University of Turin, Turin, Italy
e-mail: davoud@taghawi-nejad.de

S.G. Asfaha
International Training Center of the ILO (United Nations), Turin, Italy

T. Gilbert et al. (eds.), *Proceedings of the European Conference on Complex Systems 2012*, Springer Proceedings in Complexity, DOI 10.1007/978-3-319-00395-5_109,

foresights. CCEs in contrast explicitly reflect the complexity of the real world and the heterogeneity of economic agents in particular consumers and firms.

Just like CGE models, CCE models analyze the effects of policy changes or shocks in the economy. Contrary to CGE models, CCE models take into account the interaction of individual players in the macro-economy. What is more CCE models do not assume equilibrium. In CCE models each person, firm and institution is individually simulated in a computer program. These millions of economic agents who produce, trade and consume together represent the complete economy.

In the simplest case, individual firms may be modeled on the basis of calibrated production functions, taking firm size and capacity constraints into accounts. People are modeled on calibrated utility functions and wealth. One can describe individual agents in as much detail as necessary to study a policy or phenomenon. One could, for example, model people as having bounded rationality and even individual traits like different skills, age and family relationships. The modeling of agents is not limited by the constraints of equilibrium mathematics. For example, agents may put in varying amounts of effort to search for a job, acquire new skills on the job or between jobs, or create new institutions such as unions. Firms and employees can have explicit labor contracts. What is more institutions can be modeled as agents. Trade unions, for example, could be negotiating collective labor contracts for their members.

Conventional CGE models can be seen as a special case of a simple CCE model. But CCE models allow us to observe short-term and non equilibrium effects. In CGE models all actions of the agents are at equilibrium prices. In CCE models agents produce, invest, trade, consume and save given the current prices. They update their prices by learning from past actions. The development of prices toward equilibrium is explicit. Short term impacts of policies are therefore observable.

We could create a CCE model without additional features like heterogeneity and institutions. If this model was simulated for hundreds of years the last year would be equivalent to the corresponding CGE model. But the CCE model would show us all the previous years, when the simulated world was not in equilibrium and genuine unemployment existed. In this case, a CGE model could be regarded as a very special case of CCE models.

Our current project is an illustration of a CCE model: we use Indonesian data, to calibrate an economy. We augment the agents representing workers, by giving them different sector specific skills. When we now run a trade policy simulation, there is a skill mismatch and the transition of workers between sectors and firms takes time: in the transition we observe out of equilibrium unemployment.

The advantages of CCEs over CGEs are fourfold. First, because the economy in CCE models is by definition in disequilibrium, we can observe such common non-equilibrium phenomena as unemployment, business cycles and crisis. Second, heterogeneity of firms and people can be explicitly modeled in CCE models. With this capability, CCE models could capture complex human and business behaviors, human capital differences as well as knowledge, social status constraints and other personal and cultural attributes. Third, we can model institutions such as trade unions. Last, we can be confident that the equilibrium assumption does not interfere with the result of a simulation. For instance, one can never exclude that the positive effects of trade in trade simulations are just a result of the equilibrium assumption.

We thus have a model class which in its base-line is similar to the widely used and popular CGE models but which in reality is more robust and realistic, because it captures real life heterogeneity and behavioral complexity. In today's world of rapidly changing economic structure and growing instability, academics and policy makers are increasingly skeptical of equilibrium economics. We, as does the Economist in its article 'Agents of change' (22nd July 2010), argue that agent-based models are the alternative.

Part XII
Satellite Meeting: Data-Driven Modeling of Contagion Processes

Chapter 110
Malaria Incidence Forecasting and Its Implication to Intervention Strategies in South East Asia Region

Ankit Bansal, Sarita Azad, and Pietro Lio

Abstract Forecasting an epidemic is a complex task because of its dependence on multiple parameters. The challenges pose by sparse and error-prone data is addressed by stochastic data assimilation model. A two-step algorithm based on ensemble Kalman filter is applied to forecast malaria incidence. The temporal dependence of the data is modelled using simple Markov process and the time series is cast into a state space model.

The risk perception of various demographic regions is assessed based on the availability of prevention and control measures. The ranking of different countries in SEAR region is prepared on the basic of multiple attribute decision making approach. Our results show that India, Myanmar, Indonesia are in low ranking in availability of control measures, though number of malaria cases are highest in these countries.

Forecasting cases in next year depends on how much population is covered under these schemes. It is vital to monitor malaria trends to see if malaria control campaigns are being effective, and to make improvements.

Keywords Malaria · SEAR · Forecasting · Control measures

110.1 Introduction

Despite more than five decades of intensive control efforts, malaria still remains one of the most serious problems faced by the countries of WHO South-East Asia Region (SEAR) [1–3]. The disease is endemic in all the countries of the Region except the Maldives, which has remained free of indigenous cases since 1984. Every year, nearly 100 million cases are estimated in SEAR, which is highest after African region. However these estimates of malaria burden are uncertain [4]. It is crucial

A. Bansal · S. Azad (✉)
Indian Institute of Technology Mandi, Mandi, India
e-mail: sarita@iitmandi.ac.in

P. Lio
Computer Laboratory, University of Cambridge, Cambridge, UK

T. Gilbert et al. (eds.), *Proceedings of the European Conference on Complex Systems 2012*, Springer Proceedings in Complexity, DOI 10.1007/978-3-319-00395-5_110,

to provide accurate estimates in defining intervention strategies against malaria. It has been noticed that having an efficient statistical methodology will contribute to a more focussed approach for control, and have a positive impact on the resource allocation for malaria control over space and time. In SEAR, India, Indonesia and Myanmar have an estimated 94 % of all the cases contributed from these countries alone, but still have very poor record of malaria preventions and control coverage. Malaria has re-emerged in India after 1970's due to reduction in supply of DDT and the resistance developed by mosquitoes to DDT and increased resistance of malaria parasite to chloroquine. According to WHO estimates less than even of 25 % of high risk population is covered under indoor residual spray (IRS), whereas less than 2 % population is covered under insecticide treated bednet (ITN) [5].

The rate of malaria incidence is increasing in SEAR due to changing climate, increased parasite resistance to anti-malarial drugs, and poor risk perception in the communities. Accurate prediction and early detection of malaria cases are key factors in the containment of the disease. A good forecasting model will inform health officials about the situation and prepare them take preventive measures timely. The malaria transmission can be studied using time series analysis approaches. Time series analysis originally developed for economic forecasting and geophysical signal processing has been shown relevant and valuable in the public health context [6]. The methodology to forecast disease incidence can be classified into three distinct approaches: statistical, empirical and dynamical. Most commonly used statistical models are based on seasonal climate forecasts which investigate the association between climatic variability and the number of malaria cases [7–9]. Dynamical model couples the dynamics of the disease in both the mosquito vector and the human host using SIR (susceptible-infectious-removal) differential equations [10–12]. We follow empirical method which is based on a time series analysis of past disease data and do not use any predictors. We are particularly interested in time series based stochastic models [13–15]. Ensemble Kalman filter has been widely used to provide climate forecast [16], and to study disease transmission [17–19]. We adopt a two-step algorithm based on ensemble Kalman filter to forecast malaria incidence. The temporal dependence of the data is modelled using simple Markov process and the time series is cast into a state space model.

110.2 Data and Methods

This section describes the data used for the analyses, methods for pre-processing of the data, types of models tested and the criteria for model selection.

The WHO SEAR includes eleven countries: Bangladesh, Bhutan, DPR Korea, India, Indonesia, Maldives, Myanmar, Nepal, Sri Lanka, Thailand, and Timor-Leste. The annual malaria incidence data (1985–2010) for SEAR individual countries are obtained from the World Health Organization (WHO). Various methods to control malaria in this region have been adopted including integrated vector control through selective spraying of residual insecticides to interrupt transmission by reducing vector longevity and the use of insecticide-impregnated bed nets. The data of control measures in SEAR is obtained from WHO.

110.2.1 Ensemble Kalman Filter (EnKf)

In this paper we employ a two step forecasting model. The basic dynamic of the disease is modelled based on a time series method. The time series models such as autoregressive moving average (ARMA) have an advantage that it can handle the temporal dependence of data over large period of time. We cast the time series of malaria infected cases into a state space model which represents a Markov process. The coefficients a_t and b_t in Eq. (110.1) of ARMA model represent the dynamics state of the disease:

$$\begin{aligned} x_t &= a_t x_{t-1} + b_t w_{t-1} + w_t \\ \begin{bmatrix} a_t \\ b_t \end{bmatrix} &= \begin{bmatrix} a_{t-1} \\ b_{t-1} \end{bmatrix} + v_t \end{aligned} \tag{110.1}$$

Once the outcome of the next measurement (data mixed with random noise) is available the forecast are updated using a weighted average, with more weights given to estimates with higher certainty.

$$x_t = a_t^f x_{t-1} + b_t^f w_{t-1} + w_t \tag{110.2}$$

The measurement model is given in Eq. (110.2). Model and measurement noise matrices $(v_{t,} w_t)$ are assumed to be known (Gaussian white noise), and are time invariant. In present we consider, we consider only one parameter model. The goal of the model is to find the best estimate of ARMA parameter a_t^a, given the forecast a_t^f and measurement of infected individuals at x_t. The best estimate of infected population is based on assimilated estimate a_t^a. The Kalman filter equations are expressed in two steps, the forecast step, where information from the measurements is used in time series model, and the analysis step, where this information is used to obtain assimilated value using Kalman gain matrix [20].

Forecast step is defined as

$$a_t^{f_i} = a_{t-1} + v_t^i \tag{110.3}$$

$$x_t = a_t x_{t-1} + w_t \tag{110.4}$$

Analysis steps are defined as

$$a_t^{a_i} = a_t^{f_i} + K\left(x_t + w_t^i - a_t^{f_i} x_{t-1}\right) \tag{110.5}$$

$$x_t^{f_i} = a_t^{a_i} x_{t-1} \tag{110.6}$$

The error is calculated using root mean square error

$$\text{RMSE} = \sqrt{\frac{1}{n}\sum_{t=1}^{n}\left(\frac{x_t^f - x_t}{x_t}\right)^2} \tag{110.7}$$

Fig. 110.1 Ensemble Kalman filter prediction of malaria cases in India

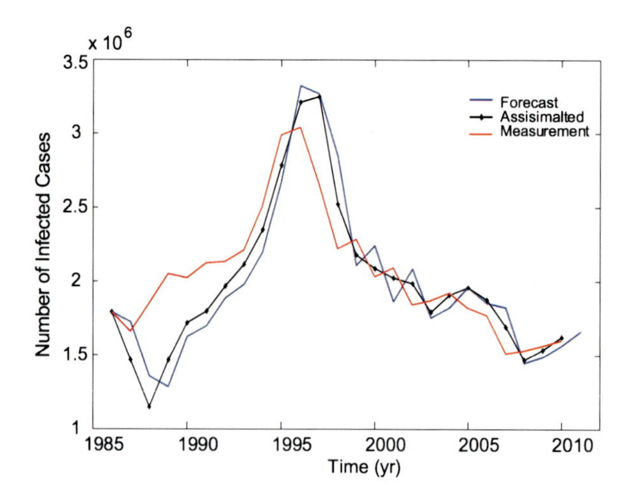

110.2.2 Ranking Approach

In order to rank countries based on their performance in malaria control measures, we employ Multiple Attribute Decision Making (MADM) technique. This kind of ranking can help in monitoring and evaluation of the effectiveness of control measures for different countries. A number of MADM methods are reported in the literature [21]. In the present study we have used TOPSIS method [22]. TOPSIS implies that a decision matrix having 'm' alternatives and 'n' attributes can be assumed to be problem of 'n' dimensional hyperplane having 'm' points whose location is given by the value of their attributes. The methodology consists of evaluating the Euclidean distance between given alternative and the positive ideal solution (best possible case) and the negative ideal solution (worst possible case) respectively. The ideology is that the best possible alternative will be the one having the least distance from the positive ideal solution and the most distance from the negative ideal solution.

110.3 Results and Discussions

The time series model using EnKf is applied to forecast the malaria cases in SEA Region. Initially we define our model as lag-one autoregressive process. The parameters of this process are then estimated using EnKf using procedure defined in Sect. 110.2. We assume constant values of model noise variance ($v_t = 0.01$) and measurement noise variance (wt $= 2\ \%$ of the observed number of cases). Figure 110.1 shows the measured, forecasted and assimilated malaria cases for India during 1985–2010. Also, forecasted malaria cases for 2011 are shown in Table 110.1. It is seen that the prediction improves as data is assimilated over the years. The relative root mean square error is calculated between measured and assimilated values over 1985–2010 and is 0.07 which shows that the model is able

Table 110.1 Predicted values of malaria cases for year 2010 and 2011

Country	Measured value for 2010	Assimilated value for 2010	Forecasted value for 2011	RRMSE
Bangladesh	55873	53137	48395	0.3651
Bhutan	436	534	232	0.3469
Korea	13520	11508	10347	0.3800
India	1600000	1482300	1559000	0.0754
Indonesia	229819	303360	147580	0.3103
Myanmar	420808	426230	458990	0.1713
Nepal	3335	2971	3050	0.1781
Sri Lanka	736	369	480	0.4920
Thailand	32502	21004	31186	0.1426
Timor-Leste	48137	59476	25361	0.4807

to predict malaria cases for India very well. Also, our model forecast 1.5 million cases of malaria for 2011, whereas 1.53 million cases were reported by WHO. In Table 110.1, rest of the SEAR countries assimilated cases for 2010 and forecasted cases for 2011 are listed. It is seen from the Table 110.1 that forecasted values for other SEAR countries show large errors. For example, prediction error for Sri Lanka is large as 49 % for one year ahead predictions. It is expected that the forecast estimates would improve made by adding more parameters in our state space model.

Such forecasts can be useful to understand real situation of malaria in SEAR countries. Table 110.2 shows data of control measures for ten SEAR countries. Among SEAR countries, India has largest population and also has large population at malaria risk. These countries have different control strategies which are largely based on their economy, policy and resources availability. In order to compare these countries with respect to their control measures and performance against malaria interventions, we have used a MADM technique (TOPSIS). MADM techniques have been used for ranking objects based on multiple attributes in various scientific and engineering fields. We consider seven attributes to prepare our ranking, these are: population at risk, annual parasite index, cumulative availability of long lasting nets (ITNS and LLIN), population covered by indoor residuals(IRS), population covered under, IRS/ IITNs/LLIN, and percentage of population at risk. Our results show that Bhutan and Myanmar stand on first and last rank among the SEAR countries under study. It is also be noted that TOPSIS index is very close for the countries Timor Laste and DPRK (0.4); Thailand, Nepal and Bangladesh (0.3); Indonesia and Myanmar (0.1). It indicates that these groups of countries have almost similar situation in malaria interventions.

110.4 Conclusions

Enkf approach is used to forecast malaria incidence for ten SEAR countries. This study provides a quantitative framework for future and strategic planning. The re-

Table 110.2 Ranking of SEAR countries based on the malaria control measures

Country	Population at risk	PAR (API > 1)	Cum. avail. of ITNs+LLINs	Persons covered under ITNs+LLNs	Pop. covered by IRS	Pop. covered under IRS/ITNs/LLINs	% pop. at risk covered under all this	TOPSIS Index	Ranks
Bhutan	512705	284512	131984	263968	140503	404471	100	0.803990555	1
Sri Lanka	4876833	0	1327000	2654000	314146	2968146	60.9	0.634801274	2
Timor Leste	1149029	1149029	245831	491662	58825	550487	47.9	0.438448475	3
DPRK	14992000	3401546	206600	413200	2000000	2413200	70.9	0.406392935	4
Thailand	45616927	4596611	992043	1984086	568799	2552885	55.5	0.327253686	5
Nepal	20357013	611597	1090001	2108002	76835	2256837	11.1	0.322743467	6
Bangladesh	55155059	11355453	2896943	5793886	0	5793886	51	0.315035977	7
India	1.025E+09	261124000	4752000	9504000	53432930	62936930	24.1	0.27540743	8
Indonesia	117351457	70410874	4408840	8817680	0	8817680	12.5	0.162130495	9
Myanmar	37390892	27811803	999325	1998650	12709	2011359	7.2	0.105933397	10

sults show that the combined autoregressive and Enkf predicts the disease reasonably well. However, multivariate model may improve the accuracy of the forecast. The forecasted estimates suggest that a locally defined malaria control strategy would be inefficient in future as number of malaria cases tend to increase in countries especially India. This study interprets that India has maximum number of malaria cases and stand at a lower rank among SEAR countries. India requires more effort in controlling malaria in near future.

References

1. Jain S, Chugh TD (2011) An overview of malaria burden in India and road-blocks in its control. J Clin Diagn Res 5(5):915–916
2. Kondrashin AV, Rooney W (1992) Overview: epidemiology of malaria and its control in countries of the WHO South-East Asia region. Southeast Asian J Trop Med Public Health 23(4):13–22
3. Kondrashin AV (1992) Malaria in the WHO Southeast Asia region. Indian J Malariol 29(3):129–160
4. Farooquia HH, Hussain MA, Zodpey S (2012) Malaria control in India: has sub-optimal rationing of effective interventions compromised programme efficiency? WHO South-East Asia J Public Heal 1(2):128–132
5. Trig P (2004) Malaria epidemics: forecasting, prevention, early detection and control. WHO report
6. Diggle P (1990) Time series: a biostatistical introduction. Clarendon, Oxford
7. Patz JA, Olson SH (2006) Malaria risk and temperature. Influences from global climate change and local land use practices. Proc Natl Acad Sci USA 103:5635–5636
8. Mostashari F, Kulldorff M, Hartman JJ, Miller JR, Kulasekera V (2003) Dead bird clusters as an early warning system for West Nile virus activity. Emerg Infect Dis 9:641–646
9. Thomson MC, Connor SJ (2001) The development of malaria early warning systems for Africa. Trends Parasitol 17(9):438–445
10. Filipe JA, Riley EM, Drakeley CJ, Sutherland CJ, Ghani AC (2007) Determination of the processes driving the acquisition of immunity to malaria using a mathematical transmission model. PLoS Comput Biol 3:2569–2579
11. Maire N, Smith T, Ross A, Owusu-Agyei S, Dietz K, Molineaux L (2006) A model for natural immunity to asexual blood stages of Plasmodium falciparum malaria in endemic areas. Am J Trop Med Hyg 75:19–31
12. Koenraadt JM, Paaijmans KP, Schneider P, Githeko AK, Takken W (2006) Low larval vector survival explains unstable malaria in the western Kenya highlands. Trop Med Int Health 11:1195–1205
13. Laneri K, Bahdra A, Ionides E, Bouma M, Dhiman R, Rajpal Y, Pascual M (2010) Forcing vs feedback: epidemic malaria and monsoon rains in NW India. PLoS Comput Biol 6:e1000898
14. Reuman DC, Desharnais RA, Costantino RF, Ahmad OS, Cohen JE (2006) Power spectra reveal the influence of stochasticity on nonlinear populationdynamics. Proc Natl Acad Sci USA 103:18860–18865
15. Codeço CT, Lele S, Pascual M, Bouma M, Ko AI (2008) A stochastic model for ecological systems with strong nonlinear response. J R Soc Interface 5:247–252
16. Kalnay E (2003) Atmospheric modeling, data assimilation, and predictability. Cambridge University Press, New York
17. Schiff SJ (2010) Towards model-based control of Parkinson's disease. Phil Trans Royal Soc A 368(1918):2269–2308
18. Lehnertz K, Mormann F, Osterhage H, Muller A, Prusseit J (2007) State-of-the-art of seizure prediction. Clin Neurophysiol 24:147–153

19. Mormann F, Andrzejak RG, Elger CE, Lehnertz K (2007) Seizure prediction: the long and winding road. Brain 130:314–333
20. Gillijns S, Mendoza OB, Chandrasekar J, De Moor BLR, Bernstein DS, Ridley A (2006) What is the ensemble Kalman filter and how well does it work? In: Proceedings of the American control conference Minneapolis, Minnesota, USA, June 14–16
21. Tzeng G-H, Huang JJ (2011) Multiple attribute decision making: methods and applications. CRC Press, Boca Raton
22. Chauhan A, Vaish R (2013) Pareto-optimal microwave dielectric materials. Ad Sci, Eng and Medic 5:149–159

Chapter 111
Studying Disease Dynamics Under Diverse Population Structures and Contagion Scenarios

Iris N. Gomez-Lopez, Olivia Loza, and Armin R. Mikler

Abstract Disease dynamics are strongly affected by the structure of the population. According to demographics and geographic characteristics, the population naturally segregates into clusters of people with similar characteristics. Specific distributions of the population create unique environments for disease progression. To understand diverse epidemic patterns, it is necessary to ascertain its dynamics in differentiated communities. Synthetic communities reconstruction with computer simulations are convenient to conduct epidemiological investigations under different scenarios. In this work, a synthetic population is constructed, considering demographic and geographic attributes, to identify different population structures. By clustering population, the path the epidemic follows during its progression can be ascertained. Consequently, assumptions for identifying clusters with specific demographics and geographics are to be made to identify risk groups. A methodology that hierarchisizes the population's attributes is utilized to produce different assortments of individuals into groups. Therefore diversified synthetic scenarios are produced to facilitate the experimentation and observation of the disease dynamics.

Keywords Population · Dynamics · Demographics · Geographics · Epidemic

111.1 Introduction

Emerging infectious diseases have been a major concern for Public Health since the extent of strategies to anticipate or mitigate a contagion are limited [1]. In this regard, methodologies and theories have been utilized to implement insightful computational models to conduct public health and epidemiological investigations by studying epidemics and their associated factors [2]. By constructing models to simulate epidemic scenarios, it is possible to study the dynamics of both populations and diseases. Experimentation allows to effectively analyze strategic approaches that lessen the effects of the spread of an infectious disease on a population. Strategies,

I.N. Gomez-Lopez (✉) · O. Loza · A.R. Mikler
University of North Texas, Denton, TX, USA

T. Gilbert et al. (eds.), *Proceedings of the European Conference on Complex Systems 2012*, Springer Proceedings in Complexity, DOI 10.1007/978-3-319-00395-5_111,

such as monitoring, vaccination, prophylaxis and social distancing, entirely rely on the understanding of the factors that steer the disease dynamics. For instance, factors such as disease pathogenesis or social affinity of the population influence differently an infectious disease progression within the same community [3]. Disease dynamics are strongly affected by the population characteristics. Demographic and geographic characteristics exert people to prefer individuals to interact with others of similar characteristics. Hence, the correlation among people's characteristics and their preferences is expressed by distributing individuals into clusters of similar characteristics [4]. In addition, a particular distribution of the population into clusters depicts a community structure of clusters with differentiated characteristics. It can be noticed that, within a set of clusters, a clustering not only there exists local interaction in the group but also globally from their group to other groups of similar characteristics. The Center for Diseases Control has reported that attributes, such as age, gender, ethnicity, and race are associated with the emergence or prevalence of certain diseases. In current work, the dynamics of both population and disease are studied and a methodology is proposed for experimentation and analysis of synthetic epidemic scenarios.

Current work produces health-seeking behavior data under different assumptions [5]. First, a synthetic population of individuals with diverse demographic and geographic attributes has to be constructed by disaggregating US Census 2010 data. Characteristics of the synthesized population such as geographic granularity and number of demographic features can be customized. After disaggregation has been performed, the geographic assignment of coordinates to synthetic individuals has to be matched with real geographic distribution data. Next, the *Syntheticum* database is created with synthetic individuals with 11 demographic attributes such as *age, ethnicity, school grade, income, gender, household income,* and *household size*, and five geographic location levels, such as *state, county, census blockgroup, school attendance zone*, and *school* are included, in this regard, different criteria for hierarchization of the *Syntheticum* attributes are stated and a methodology to produce different distributions of the population is proposed. Finally, a clustering of the population is generated and a contagion network is constructed to study various behaviors of an infectious disease spread. As the case may require, relationships of the population arrangements along with their characteristics and the disease dynamics are unified in a framework in order to identify correlations of attribute hierarchization and the epidemic behavior.

111.2 Methodology

In this section, the methodology of current approach is depicted in Fig. 111.1. This flow chart shows the actual process to produce the final contagion scenario for a particular structure of the population.

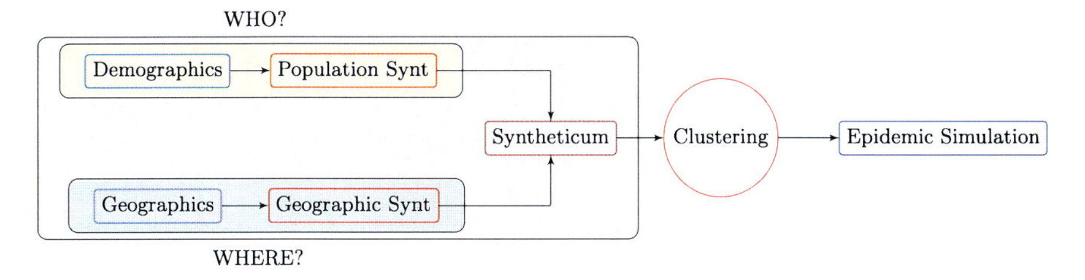

Fig. 111.1 Methodology in a nutshell

111.2.1 Databases Processing

The first stages of the process relay in the preprocessing of the Demographic and Geographic data of the population. This is accomplished by mining two database sources: *Census 2000* and the *School District* database of the State of Texas *Texas Education Agency (TEA)*. These databases serve the purpose of integrating the Demographic and the Geographic characteristics of the final synthetic population database: *The Syntheticum*.

Demographic Attributes The module *WHO?* builds a non-homogeneous population, see Fig. 111.1, making use of the *Census 2000* online database. This process is divided into two steps, the information extraction and the population synthesis. The Population Synthesizer, *PopGen* achieves close matching with US Census marginal distributions by making use of some Iterative Proportional Fitting (IPF-based) procedures along with an Iterative Proportional Update(IPU) algorithm. This tool is capable of synthesizing populations with approximations that are close to Census data distributions of demographic and some geographic attributes [6]. In conclusion, this framework divides the task in four steps: *Project setup, Data import, Set Corresponding variables and Synthesizer run*. In current work, the synthesized population corresponds to Denton County, TX, with a geographic granularity of census *blockgroups*. The number of individuals is 426583 with 28 demographic and geographic attributes; nonetheless, 1500 individuals are extracted and 10 attributes such as *age, gender, race, educational level and grade, work type, income*, and *household* are selected to construct a sample population and conduct visualizable examples.

Geographic Attributes The previous module provides geographic distribution for the synthetic individuals. However, they can only be mapped with the precision of the census *blockgroup* distribution (approx. 600–3000 ind. per blockgroup) In the *WHERE?* module, the fidelity of the geographic location assignment for individuals is improved. By taking advantage of the fact that education is compulsory in US and that enforces people to affiliate to a particular School facility, the geographic location precision is enhanced. The geographic boundaries of the *School Districts* and the *School attendance zones* contributes to a higher and more realistic geographic granularity. As a result, the fraction of the population who attends school and their families are highly biased towards a real catchment area; nonetheless the influence can be extended to those families who are not attached to any school but live within its boundary limits [7, 8]. To accomplish this, the synthetic population form *WHO?* and *WHERE?* modules are projected together in the same geographic space. This computation results in a new database called *Syntheticum*. The contribution of the new database is the geographic distribution of the individuals which now can be mapped spatially with precision of *School attendance*, *School attendance zone* and *School District* boundary. In this matter, geographic attributes such as diverse school zone overlappings (multiple zone belongingness), families tied to various schools (siblings attending different schools) or families falling within certain school/zone boundary can be known.

111.2.2 Clustering

In this module structures of the population are constructed by making use of the *Syntheticum* comprehensive geographic and demographic attribute database. For this purpose, a hierarchy of the *Syntheticum*'s attributes is produced. After this, a *attribute weight* assignment is conducted according to the hierarchy dendrogram. Thus, the attributes in the top of the dendrogram will be granted a higher weight than those in the lower layers. After this, the *Syntheticum* individuals' attributes and the *attribute-hierarchy* are cross-referenced so a final *weighted Syntheticum* is produced,

Vector Space Model In order to compute similarity between individuals, *The Syntheticum* is represented as $P = \{p_1, p_2, p_3, \ldots, p_n\}$, where $p_i = \{a_1, a_2, a_3, \ldots, a_m\}$ is an array of weighted demographic and geographic attributes that represents the i individual in a m-dimensional space. Making use of the m-dimensional vectors, the similarity between any two individuals by computing their *Euclidean distance* can be computed; the same manner as in a set of documents represented as a collection of vectors with various dimensions, Thus, a matrix of distances among individuals in the population can be constructed.

Clustering Implementation Finally, in this section the partitioning of the population into clusters is conducted. Current approach makes use of k-*means* clustering to obtain a partition of the population into clusters. However, this approach is rather naive if an assessment of the k number of clusters and the resulting clustering is not made. In this regard, hierarchical clustering is utilized to estimate a value of k by deriving a partitioning of the population and analyzing the resulting dendrogram. An automated method determines the number of k-groups by learning from the dendrogram properties [9]. As a consequence, the value of k is utilized to compute *k-means* for clustering. Finally, *goodness* metrics such as within-cluster compactness metrics and between-cluster distances are applied to the resulting clustering so that the partitioning is assessed as well. Methods such as *Hubberts gamma* coefficient and the *Dunn* index are utilized to determine cluster compactness, ratio of the between-cluster and within-cluster similarity and also to show a correlation of within-cluster compactness and well separated clusters [10–12].

111.3 Population Structure: Graphs

Once *The Syntheticum* is partitioned into clusters,a graph representation is utilized to map the resulting clustering into a network of clusters. In this context, every cluster is depicted as a node with an associated weight according to its characteristics. Also the relationships among the clusters are represented with links connecting them.

111.4 Results and Discussion

Experiments on the sample population are described in this section. Two distinct clustering of the population are considered so the effects of a specific structure of the population on the epidemic can be studied. The first clustering is made under the assumption of a heterogeneous population with uniform *attribute-weight* distribution. Conversely, the second clustering considers a heterogeneous population with non-uniform *attribute-weight* distribution. As a result, not only the number of clusters and the distribution of individuals per cluster are different for each partitioning, but also the disease progression expresses differently. In Fig. 111.2 the progression of the epidemic by quantifying the number of susceptible, infectious and recovered individuals can be observed.

The following are correlations between specific characteristics of the population structure and the epidemic shown on the results:

Using an uniform weight distribution for the 1500 population sample:

- The number of clusters is six.
- The people is distributed into clusters with a non-uniform fashion.
- The disparity of size between clusters is alike.
- Disease onset-time is shifted with one to two days increment in time.
- Disease is exported by every cluster.
- Disease is spread sequentially by infecting first the larger clusters.

The non-uniform weight distribution for the 1500 population sample:

- The number of clusters is five.
- The people is distributed into clusters with a quasi-uniform fashion.
- There is a major disparity of size between the largest and smallest cluster.
- Disease onset-time is shifted with two to three days increment in time.
- Disease is exported mainly faster by major populated clusters.
- Disease is spread sequentially by infecting first the larger clusters.

Finally, it can be said that every *attribute-weighting* scheme produces a specific structure of the population and exerts specific effects on an epidemic.

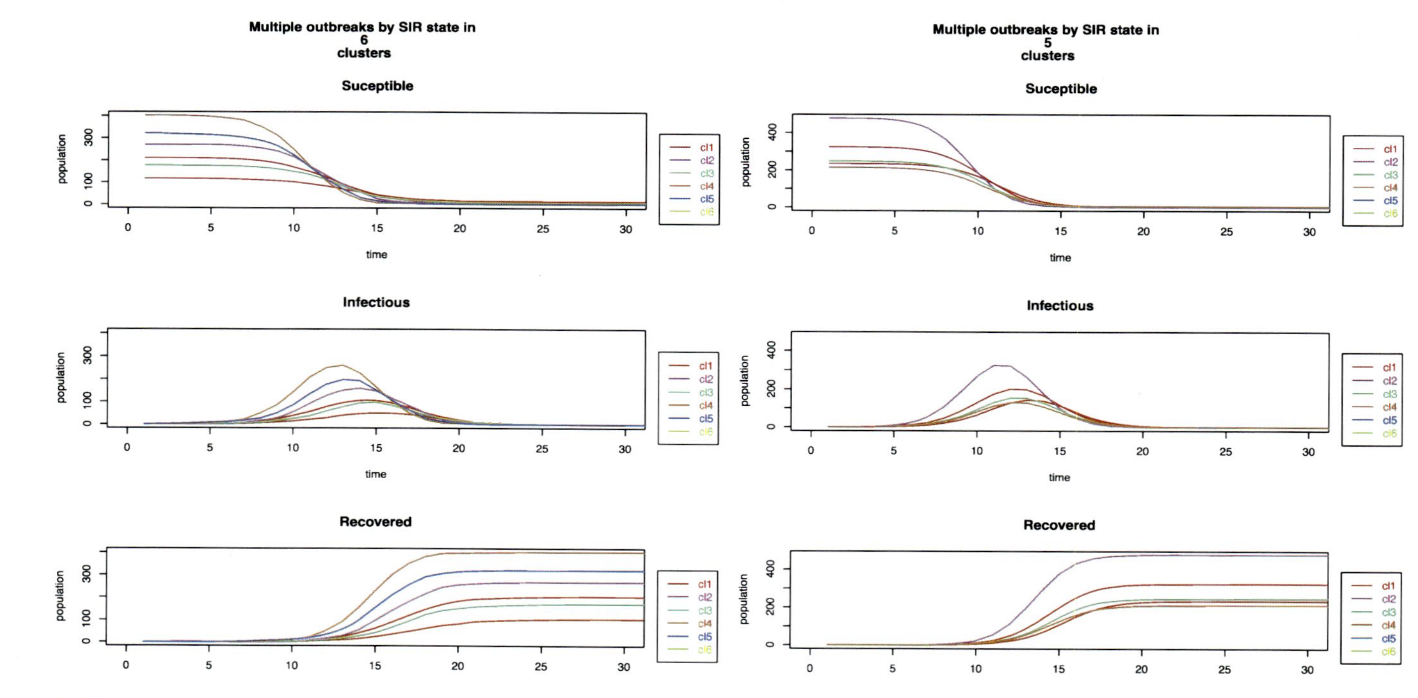

Fig. 111.2 Multiple outbreaks in two different structures of population. Non-weighted in *the left* and weighted in *the right*

References

1. Berkelhamer J (2007) Pandemic influenza: warning, children at-risk. Technical report, American Academy of Pediatricsand Trust for America's Health
2. Wang D, Xiong SJ (2008) Effects of disease characteristics and population distribution on dynamics of epidemic spreading among residential sites. Physica A 387(13):3155–3161
3. Mcpherson M (2001) Birds of a feather: homophily in social networks. Annu Rev Sociol 27:415–444
4. Christensen C, Albert I, Grenfell B, Albert R (2010) Disease dynamics in a dynamic social network. Physica A 389(13):2663–2674
5. Barrett CL, Eubank SG, Smith JP (2005) If smallpox strikes Portland. Sci Am 292:54–61
6. Ye X, Konduri K, Pendyala RM, Sana B, Wadell P (2009) A methodology to match distributions of both households and person attributes in the generation of synthetic populations. In: 88th annual meeting of the transportation research board
7. Salathé M, Kazandjieva M, Lee JW, Levis P, Feldman MW, Jones JH (2010) A high-resolution human contact network for infectious disease transmission. Proc Natl Acad Sci USA 107(51):22020–22025
8. Yang Y, Sugimoto JD, Halloran ME, Basta NE, Chao DL, Matrajt L, Potter G, Kenah E, Longini IM (2009) The transmissibility and control of pandemic influenza a (H1N1) virus. Science 326:729–733
9. Langfelder P, Zhang B, Horvath S (2007) Defining clusters from a hierarchical cluster tree: the Dynamic Tree Cut library for R. Bioinformatics 24(5):719–720
10. Hubert L, Schultz J (1976) Quadratic assignment as a general data analysis strategy. Br J Math Stat Psychol 29(2):190–241
11. Dunn JC (1974) Well separated clusters and optimal fuzzy-partitions. J Cybern 4:95–104
12. Rousseeuw P (1987) Silhouettes: a graphical aid to the interpretation and validation of cluster analysis. J Comput Appl Math 20:53–65

Chapter 112
Stochastic Computational, Thermal, and Vertical Transmission Models to Simulate Dengue Persistence in Vector and Human Populations

Angel Bravo-Salgado, Armin R. Mikler, and Thiraphat Meesumrarn

Abstract In this paper, a stochastic-contact computational model (GSCM) is combined with thermal (TM) and vertical transmission (VT) models. Furthermore, the simulation of the reemergence of dengue virus outbreaks in a vector and human population during seasonal cycles and periods of unfavorable weather was conducted in a specific region in Thailand. Five years of regional data from Thailand was used to corroborate the simulation results.

112.1 Introduction

Emerging vector-borne diseases have triggered considerable concerns around the world. Public health experts have estimated that the number of cases increases in tropical and subtropical regions every year. For instance, the World Health Organization (WHO) and Pediatric Dengue Vaccine Initiative (PDVI) have estimated that approximately 2.5 billion to 3.6 billion people are at risk to contract Dengue [1]. Moreover, it is been hypothesized that climate change will broaden the natural limits of the disease, thus increasingthe already raised number of cases.

Coping with yearly outbreaks requires the intervention of public health authorities. The available health resources are used to embrace interventions that prevent, promote, cure, or rehabilitate people from the disease effects. From 2004 to 2007, the total expenditure of eight countries to fight dengue was estimated to exceed I$587 million [2].

The use of analytic tools such as mathematical or computational models permit both quantitative and qualitative analysis of virtual contagion scenarios. In order to anticipate and establish an adequate preparedness plan, an effective policy requires a rapid intervention and optimal utilization of public health resources. The systematic study of disease scenarios through models and simulations facilitate the development of the necessary capabilities of public health experts to cope with real sanitary emergencies.

In this paper, a framework, the GSCM-DEN, is utilized to study dengue disease dynamics in vector and human population. The constituent parts of the GSCM-DEN

A. Bravo-Salgado (✉) · A.R. Mikler · T. Meesumrarn
University of North Texas, Denton, TX, USA

T. Gilbert et al. (eds.), *Proceedings of the European Conference on Complex Systems 2012*, Springer Proceedings in Complexity, DOI 10.1007/978-3-319-00395-5_112,

framework are the stochastic contact (GSCM), thermal (TM) and vertical transmission (VT) models. GSCM-DEN delivers a number of practical uses. First, it simulates the reemergence of dengue virus in a vector and human population during seasonal cycles. GSCM-DEN is a practical tool for public health field agents to simulate and observe the course of the vector-borne diseases within a particular region. Finally, approximation models that describe the biological processes of the vector population and the development of the pathogen within the host are implemented.

112.1.1 Background

Equation-based and computational vector-borne disease models commonly consider horizontal disease transmission. Horizontal disease transmission occurs either from an infectious human to a susceptible vector or from an infectious vector to a susceptible human. In both cases, transmission takes place via the vector's bite. Another form of transmission is the vertical transmission which happens when an infectious female vector transmits the dengue disease to its offspring. Nonetheless, it has been hypothesized that dengue vertical transmission could be the reason for the persistence of dengue cases through out the year, though little is known about this contagion process.

Temperature affects multiple facets of the biology of the vector including development, survival of eggs, larvae, pupae, and more importantly the life span of the vector. It is also known that the speed of development of arthropods' pathogens are largely determined by the changes in temperature.

Computational models using stochastic processes for the simulation of disease outbreak scenarios can rapidly maximize the benefits compared with those implemented using equation-based models. Since high-fidelity computational models should provide outcomes in a timely manner during a sanitary emergency, the necessity of constant improvement in the simulation overhead has become important.

112.2 Methodology

112.2.1 Stochastic Contact Model

During an epidemic and following a compartment-based paradigm Eq. (112.1),

$$|V| = |S_v| + |L_v| + |I_v| \tag{112.1}$$

represents the size of the mosquito population. Hence, $|S_v|$, $|L_v|$, and $|I_v|$ represent the number of vectors in the susceptible, latent, and infectious compartments. Respectively, Eq. (112.2) represents the size of the human population. Hence, $|S_p|$, $|L_p|$, $|I_p|$, and $|R_p|$ represent the size of susceptible, latent, infectious, and recover compartments.

$$|P| = |S_p| + |L_p| + |I_p| + |R_p| \tag{112.2}$$

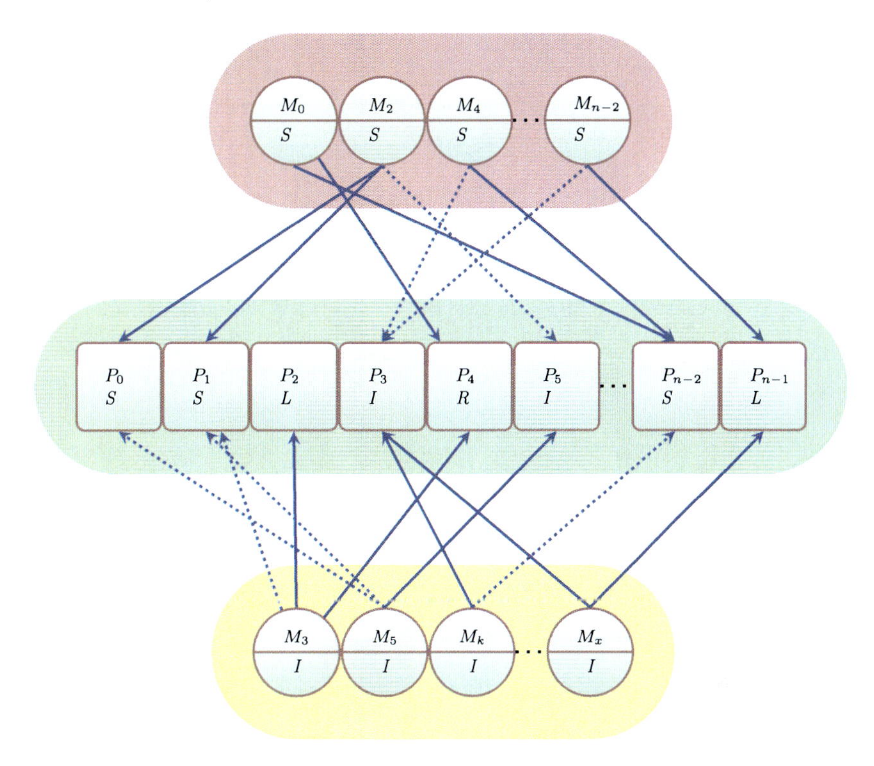

Fig. 112.1 Interactions between population members within the stochastic-contact model

β is the number of bites per day of a vector. From the computational perspective, each bite is considered an interaction. Assuming an uniform distribution to generate interactions, a naive-contact interaction model will generate a total number of interactions per day equal to,

$$Naive_{int} = \beta|V||P| \tag{112.3}$$

or

$$Naive_{int} = \beta|S_v + L_v + I_v||P| \tag{112.4}$$

In this model, it should be noted that the interactions that can effectively transmit the disease to any of the two populations are of type (S_v, I_p) and (I_v, S_p). Figure 112.1 shows these interactions with a subtle difference. The difference is that the interactions are of the kind $(S_v, \{S_p, L_p, I_p, R_p\})$ and $(I_v, \{S_p, L_p, I_p, R_p\})$ since uniformity has to be preserved. Formally,

$$GSCM_{int} = \beta|S_v||P| + |I_v||P| \tag{112.5}$$

or

$$GSCM_{int} = \beta|S_v + I_v||P| \tag{112.6}$$

Clearly, Eq. (112.6) shows an improvement over the number of interaction that have to be generated during simulation by the naive interaction model.

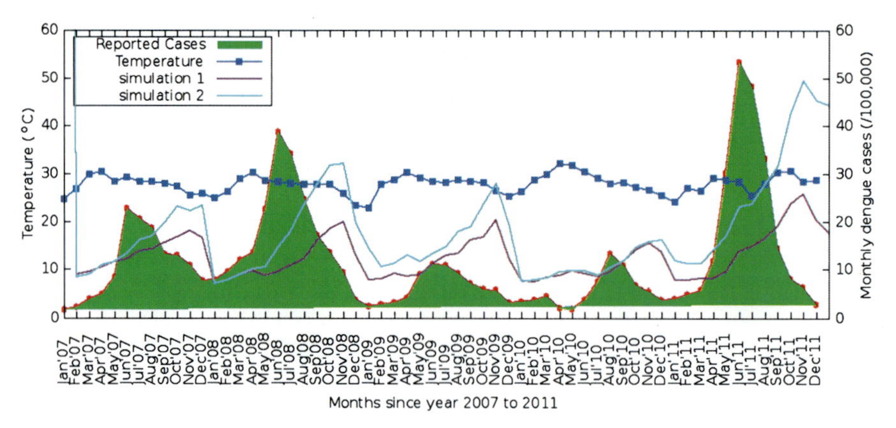

Fig. 112.2 Number of dengue cases compared to the results of two simulations

112.2.2 Thermal Model

Temperature affects the biology of the vector such as its development and the survival of its eggs, larvae, and pupae (*gonotrophic cycle*). Moreover, temperature affects the life span of the vector. Other factors such as presence of dengue virus in the salivary cells in arthropods are also determined by the changes in temperature [3]. Furthermore, studies in [4] provide the field work necessary for the modeling of vector biological processes with nine approximation models to determine the dynamics of the gonotrophic cycle of the *Aedes Aegypti* mosquitoes. The polynomial degree of each model is determined and assessed by choosing the one with the smallest normal error. Thus, the thermal model becomes an essential part in the study of disease dynamics in regard of the available daily average temperature.

112.2.3 Vertical Transmission

Studies have shown that vector eggs infected with dengue virus can carry it to adulthood and later pass it on to future generations. It is also known that the dengue causes an induced-birth control over the vector infected eggs. This effect increases the proportion of vector fatality. Vertical transmission data reported from experiments on infectious vectors were used to design a computational vertical transmission model (VT). The VT facilitates the reemergence of the vector-borne disease after periods of unfavorable-vector weather conditions [5]. Using the studies in [6], we compute the average proportion of infected female eggs that survive and remain infectious until they reach adulthood.

112.3 Results and Discussion

Preliminary experiments show the results of our efficient and high-fidelity computational model. In Fig. 112.2 the number of dengue cases and the monthly average temperature in Thailand are depicted. In addition, the results of two executions of the GSCM-DEN outbreak are presented over a time span of five years. In Fig. 112.2, the similarity between simulation (simulation 1 and 2) and the actual monthly reported cases of dengue in a region in Thailand is presented. These simulated trajectories present a shift over the original reported cases. It is important to note that only temperature data was used to produce the preliminary experiments. Future work will consider the integration of a rain model that makes use of rain data to improve the current results.

References

1. Who—dengue in the western pacific region (2010). http://www.wpro.who.int/health_topics/dengue/
2. Suaya JA, Shepard DS, Siqueira JAB, Martelli CT, Lum LCS, Tan LH, Kongsin S, Jiamton S, Garrido F, Montoya R, Armien B, Huy R, Castillo L, Caram M, Sah BK, Sughayyar R, Tyo KR, Halstead SB (2009) Cost of dengue cases in eight countries in the Americas and Asia: a prospective study. Am J Trop Med Hyg 80(5):846–855
3. Watts D, Burke D, Harrison B, Whitmire R, Nisalak A (1987) Effect of temperature on the vector efficiency of aedes aegypti for dengue 2 virus. Am J Trop Med Hyg 36(1):143
4. Beserra EB, Castro FPD Jr, Santos JWD, Santos TDS, Fernandes CRM (2006) Biologia e exigências térmicas de aedes aegypti (l.) (diptera: Culicidae) provenientes de quatro regiões bioclimáticas da paraíba. Neotrop Entomol 35(6):853–860
5. Adams B, Boots M (2010) How important is vertical transmission in mosquitoes for the persistance of dengue? insights from a mathematical model. Epidemics 2. doi:10.1016/j.epidem.2010.01.001
6. Vinod J, Sharma RC (2001) Impact of vertically-transmitted dengue virus on viability of eggs of virus-inoculated aedes aegypti. Dengue Bull 25:103–106

Part XIII
Satellite Meeting: Complex Behavior in Discrete Dynamical Systems

Chapter 113
Biham-Middleton-Levine Traffic Model in Two-Dimensional Hexagonal Lattice

J. Carlos García Vázquez, Salvador Rodríguez Gómez, and Fernando Sancho Caparrini

Abstract In this paper, a Biham-Middleton-Levine traffic model on a 2D hexagonal lattice is studied by computer simulations and its behavior is compared with the BML model on a 2D square lattice and a 3D cubic lattice (Supported by Excellence project TIC-6064 of Junta de Andalucía cofinanced with FEDER founds).

113.1 Introduction

In [1] Biham, Middleton and Levine (BML) introduced one of the most studied models about traffic flow in recent years. Based on a two-dimensional cellular automaton, this extremely simple model exhibits self-organization, pattern formation and phase transitions.

BML model considers the motions of two species of particles, north- and east-bound particles, in a two dimensional square lattice with $N \times N$ sites with periodic boundary conditions in such a way that the geometric model corresponds to the discrete torus $\mathbb{Z}_N \times \mathbb{Z}_N$. Several interesting generalizations of the model have been considered, such as free boundary conditions [6], non-square aspect ratio of the underlying square lattice [5], four-directional traffic [3], and an extension to three dimensional cubic lattice with periodic boundary conditions [2].

Supported by Excellence project TIC-6064 of Junta de Andalucía cofinanced with FEDER founds.

J.C.G. Vázquez (✉)
Departamento de Análisis Matemático, Universidad de Sevilla, Sevilla, Spain
e-mail: garcia@us.es

S.R. Gómez
Departamento de Física, Química y Ciencias Naturales, Universidad Pablo de Olavide, Sevilla, Spain
e-mail: salrodgom@upo.es

F.S. Caparrini
Departamento de Ciencias de la Computación e Inteligencia Artificial, Universidad de Sevilla, Sevilla, Spain
e-mail: fsancho@us.es

T. Gilbert et al. (eds.), *Proceedings of the European Conference on Complex Systems 2012*, Springer Proceedings in Complexity, DOI 10.1007/978-3-319-00395-5_113,

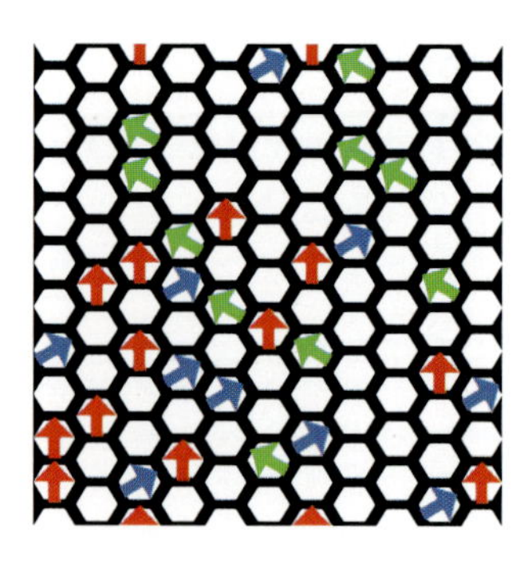

Fig. 113.1 BML model on a two dimensional hexagonal tesselation with periodic boundary conditions

In this paper we consider a BML model on a two-dimensional hexagonal lattice and we pose the question of compare this model with the 2D square lattice [1] and the 3D cubic lattice case [2].

113.2 BML on a Two-Dimensional Hexagonal Lattice

We consider an hexagonal lattice tessellation of the torus with three families of particles (red, blue and green) that try to move in north, north-east and north-west direction. In order to consider the hexagonal lattice with periodic boundary conditions we need the system size be an even natural number (see Fig. 113.1).

Initially, particles are placed randomly according to a parameter $0 \leq p \leq 1$ which is the probability of a lattice site to contain a particle. Each hexagonal cell has a probability $p/3$ of having a blue particle, $p/3$ of having a red particle, $p/3$ of having a green particle and $1 - p$ of being empty. In this way p is the particle density in the system.

The discrete time dynamics is determined by the following rules: (a) first, all red particle synchronously try to move forward one site in north direction. If the site northward of a red particle is currently empty, it advances. Otherwise that red particle stays in the present location, even if the northward site is to become empty during the current time step; (b) second, blue particles follow the same rules but they try to move in north-east direction; (c) finally, green particles try to move in north-west direction following the same rules as before. This marks the end of a time step and the above particle moving process is repeated over and over again. Let us observe that the dynamics is fully deterministic and the only random event in the model occurs in the initial condition.

In every discrete time step t, the instantaneous speed of the system v_t is computed as the ratio between the number of particles that succeed to move and the total number of them. If $v_t = 0$ then no particle has moved in the time step t; if $v_t = 1$ then all particles have moved. The asymptotic speed v is the speed v_t averaged over the asymptotic cycle and $\langle v \rangle$ is the average asymptotic speed over the random initial configurations. We want to study the average asymptotic speed $\langle v \rangle$ as a function of p.

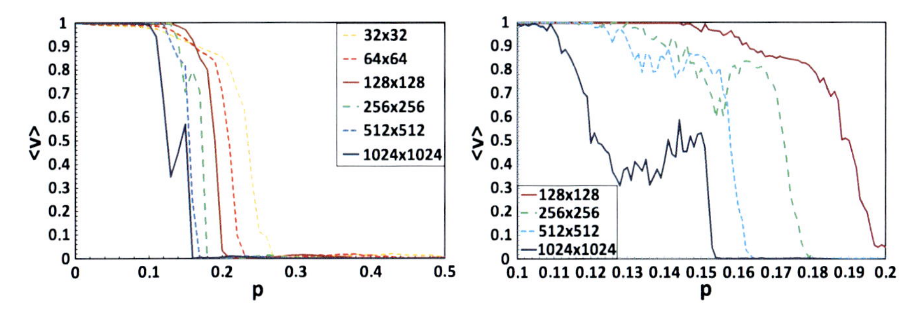

Fig. 113.2 $\langle v \rangle$ as a function of p for different system sizes

113.3 Simulation Results

We have implemented the BML model on hexagonal lattices of finite size $N \times N$, for $N = 32, 64, 128, 256, 512$ and 1024.

In the left panel of Fig. 113.2, $\langle v \rangle$ is seen as a function of p. These curves have been obtained as an average of 1000 random initial configurations for every value of p. The asymptotic speed v for every initial configuration is approximated by the average of v_t between time steps $t = 50.000$ and $t = 51.000$. The step increment of the parameter p has been of 0.01 units. Since we have a special interest in the sudden drop of $\langle v \rangle$, we have completed the numerical study repeating the experiments moving p from 0.1 to 0.2 with increments of 0.001 and with the same conditions as before. Results can be seen in the right panel of Fig. 113.2.

113.4 Analysis of the Results

Similar to the results in [1] for the 2D square lattice case and to [2] for the 3D cubic lattice case, in the 2D hexagonal lattice case there are two qualitatively different asymptotic states separated by a sharp dynamical transition. Below the transition all particles move freely and the system reaches asymptotic speed 1 or very close to 1 (the free flow phase) and above the transition the particles get stuck and the system gets speed 0 (the global jam phase). It is interesting to observe the geometric patterns of the system. Below the transition the system is able to self-organize and to form very well defined geometric patterns that can be seen in the left panel of Fig. 113.3: ordered monochromatic clusters of particles almost freely flowing. Above the transition the system forms one or more percolating clusters of jammed particles (right panel of Fig. 113.3).

Following [1] we define the critical region as the range of densities where both asymptotic states can be found with a non 0 probability. The size of this region depends on the system size and the critical probability is taken as the center of the critical region. We have made an estimation of the value of the critical probability for different system sizes. We have considered the critical region as the region of

Fig. 113.3 Typical geometric patterns of the free flow and global jam phases in the hexagonal lattice BML model

Table 113.1 Critical region as a function of N

N	$\langle v \rangle < 0.99$	$\langle v \rangle > 0.01$	Critical density
128	$p > 0.14$	$p < 0.2$	0.17
256	$p > 0.13$	$p < 0.18$	0.155
512	$p > 0.12$	$p < 0.17$	0.145
1024	$p > 0.1$	$p < 0.16$	0.13

densities for which speed lies between 0.01 and 0.99, and the critical probability as the center of the interval. We see, as in [1], that the critical probability slightly decrease as the system size increases (see Table 113.1).

In [1] Biham, Middleton and Levine reported the existence of two phases in the two dimensional square lattice model separated by a sharp dynamical transition, and the general belief was that the system showed a first order transition. However, D'Souza [4, 5] found a region of metastable intermediate states with an asymptotic speed around $v = 2/3$, and with a very well defined geometry. Thus, the BML model does not necessarily exhibit a sharp phase transition from free flow to global jam, but instead has a range of intermediate states with regions of free flow intersecting at jammed wave-fronts. In the 2D hexagonal lattice case, as in the 3D cubic lattice case, we report that we have not found intermediate states: asymptotic speed is always very close to 1 or to 0. Although we have observed a few results with v around 0.84 the number of that cases is decreasing when the system size grows.

The curves for the hexagonal lattice model are more similar to the 3D cubic lattice than to the 2D square lattice case. From the results we believe that the BML model on the hexagonal lattice may show a first order phase transition. To support this hypothesis we have plotted the hysteresis curve (Fig. 113.4) obtained in the following way: starting from $p = 0$, the density is increased by a small increment by randomly introducing particles to the empty sites of the system. Then the system evolves until it reaches a recurrent state and speed is measured. This is done again and again until $p = 1$. The blue curve represents this evolution. Now, we decrease the density by the same increment by removing randomly particles from the system, following the same process. The red line represents this process. In this way the hysteresis loop, typical of first order phase transitions, is obtained.

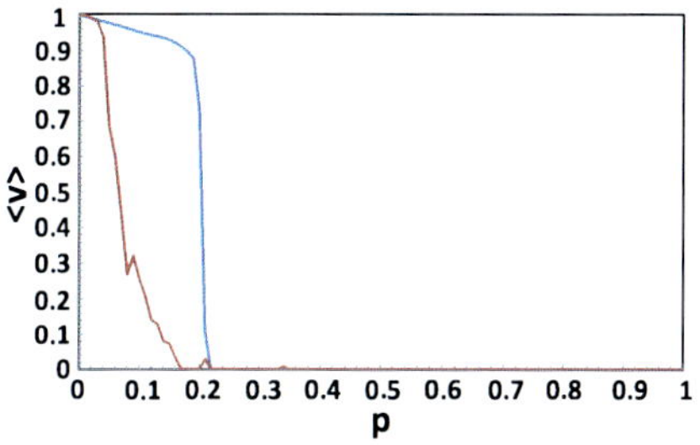

Fig. 113.4 Hysteresis curve: the blue and red curves represent the evolution of the system when particles are slowly added to or removed from the system, respectively

Fig. 113.5 Typical geometric pattern of the slow speed phase in the hexagonal lattice BML model

In [2] the authors also point out one interesting difference with respect the 2D square lattice case. Besides the free flow phase and the global jam phase, they observed an extensive region of densities with a low but non 0 asymptotic speed. According to the authors this phase is a result of the formation of spatially limited-extended percolating cluster of particles, where most particles are jammed by colliding into the percolating cluster and a small number of residual particles freely move. Since almost all particles are jammed the speed is very slow. This is a difference with the 2D square lattice case, because in that case all particles will eventually merge into the percolating backbone leading to a completely jamming configuration. Completely similar to the 3D cubic lattice case, in the 2D hexagonal lattice case there exists a slow speed phase where speed is very close to 0, but greater than 0. The system is able to form percolating clusters of jammed particles letting a residual small number of particles flow freely. In Fig. 113.5 we can see a typical geometric pattern of this phase.

One of the complications of the BML model from a theoretical point of view is that it is not a monotonous model in the following sense: adding particles to a configuration that is known to jam can actually change the sequence of particle interactions and result a configuration going to free flowing instead of jamming. In this model we can see this kind of behavior, where increasing the density causes increased speed. This can be seen clearly in the critical region for system sizes 256 and 1024 (Fig. 113.2) and to a lesser extent in the slow speed phase, where we can see a peak in the plots (Fig. 113.6). Our model can be a tool to study this behavior that does not appear in the other two models.

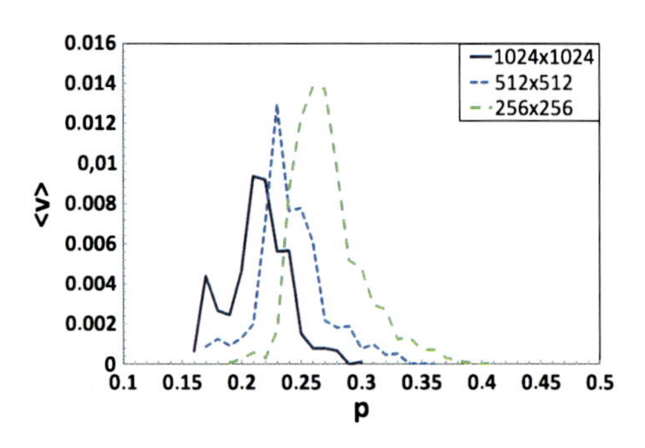

Fig. 113.6 $\langle v \rangle$ is not decreasing as a function of p in the slow speed phase

113.5 Conclusions

From the numerical results we conclude that the phase diagram of the 2D hexagonal lattice model is more similar to the 3D cubic lattice than to the 2D square lattice case. We can reproduce most of the features of the 3D cube lattice case which do not appear in the 2D square lattice case, by changing the topology of the lattice that supports the 2D model.

References

1. Biham O, Middleton AA, Levine D (1992) Self organization and a dynamical transition in traffic flow models. Phys Rev A 46:6124–6127
2. Chau HF, Wan KY (1999) Phase diagram of the Biham-Middleton-Levine traffic model in three dimensions. Phys Rev E 60:5301–5304
3. Huang D, Huang W (2006) Biham-Midleton-Levine model with four-directional traffic. Physica A 370:747–755
4. Linesch NJ, D'Souza RM (2008) Periodic states, local effects and coexistence in the BML traffic jam model. Physica A 387:6170–6176
5. D'Souza RM (2005) Coexisting phases and lattice dependence of a cellular automata model for traffic flow. Phys Rev E 71:066112
6. Tadaki S (1994) Two dimensional cellular automaton model of traffic flow with open boundaries. Phys Rev E 54:2409–2413

Chapter 114
Pesin's Relation for Weakly Chaotic One-Dimensional Systems

Alberto Saa and Roberto Venegeroles

Abstract We explore a recent rigorous result due to Zweimüller in order to propose an extension, for the case of weakly chaotic systems, of the usual Pesin's relation between the Lyapunov exponent and the Kolmogorov-Sinai entropy for one-dimensional systems. We show, furthermore, that Zweimüller's result does provide an efficient prescription for the evaluation of the algorithm complexity for systems exhibiting subexponential instabilities. Our results are confirmed by exhaustive numerical simulations. We also compare our proposal with a recent one base on the Krengel entropy.

Keywords Weak chaos · Kolmogorov-Sinai entropy · Lyapunov exponents

114.1 Introduction

The main purpose of the present work [1] is to extend the well-known Pesin relation [2] for the case of weakly chaotic one-dimensional systems. For usual one-dimensional chaotic systems, the Pesin relation is given simply by $h = \lambda$, with h and λ standing, respectively, for the Kolmogorov-Sinai (KS) entropy and the usual Lyapunov exponent. We will show that adequate subexponential generalizations of the KS entropy and of the Lyapunov exponent will obey exactly the same Pesin-type relation, for almost all trajectories.

We consider here the general class of Pomeau-Manneville (PM) maps [3] $x_{t+1} = f(x_t)$, with $f : [0, 1] \to [0, 1]$ such that

$$f(x) \sim x\left(1 + ax^{z-1}\right), \tag{114.1}$$

for $x \to 0$, with $a > 0$ and $z > 1$. Systems of the type (114.1) exhibit exactly the kind of subexponential instability for nearby trajectories that we are concerned

A. Saa (✉)
Departamento de Matemática Aplicada, UNICAMP, 13083-859 Campinas, SP, Brazil
e-mail: asaa@ime.unicamp.br

R. Venegeroles
Centro de Matemática, Computação e Cognição, UFABC, 09210-170 Santo André, SP, Brazil
e-mail: roberto.venegeroles@ufabc.edu.br

T. Gilbert et al. (eds.), *Proceedings of the European Conference on Complex Systems 2012*, Springer Proceedings in Complexity, DOI 10.1007/978-3-319-00395-5_114,

here: $\delta x_t \sim \delta x_0 \exp(\lambda_\alpha t^\alpha)$, with $0 < \alpha < 1$. For intermittent systems like (114.1), the statistics of a given observable ϑ for randomly distributed initial conditions has some peculiar properties. For ergodic systems, the time average $t^{-1} \sum_{k=0}^{t-1} \vartheta(f^k(x))$ converges to the spatial average $\int \vartheta \, d\mu$, with $d\mu = \omega(x)dx$. On the other hand, if the system has a diverging invariant measure, the time average will typically depend on the chosen trajectory. Nevertheless, the ADK theorem [4] ensures in this case that a suitable time-weighted average does converge in distribution terms towards a Mittag-Leffler distribution with unit first moment. For the subexponential regime ($z > 2$), the ADK theorem assures the convergence in distribution terms towards a Mittag-Leffler distribution of the subexponential finite-time Lyapunov exponent

$$\lambda_t^{(\alpha)}(x) = \frac{1}{t^\alpha} \sum_{k=0}^{t-1} \ln\left| f'\left(f^k(x)\right)\right|, \tag{114.2}$$

for $t \to \infty$. The generalized Lyapunov exponent (114.2) plays for intermittent systems the same role did by the usual exponent (corresponding to $\alpha = 1$ in (114.2)) for one-dimensional chaotic systems.

114.2 Pesin's Relation

In order to investigate the connection between subexponential instability and the corresponding degree of randomness of an intermittent dynamical system like (114.1), we will consider the Kolmogorov-Chaitin concept of complexity [5]. Let us assume that the phase space of the map (114.1) is partitioned and completely covered by a set of non overlapping ordered cells. A given trajectory $\{x_t\}$ generated by the map (114.1) can be represented by a sequence of symbols $\{s_t\}$, which we assume to be integers such that s_t corresponds to the cell where x_t belongs. The next step in the analysis consists in eliminating redundancies that may appear in $\{s_t\}$ by performing a compression of information. This can be done, for instance, by introducing the so-called algorithmic complexity function $C_t(\{s_t\})$, which is defined as the length of the shortest possible program able to reconstruct the sequence $\{s_t\}$ on a universal Turing machine [5]. Systems that exhibit some degree of regularity are able to generate sequences of symbols at a rate higher than needed for recording their programs. For example, a periodic sequence can be recreated by replaying the periodic pattern over the total length. Typically, for these cases, one has $C_t \sim \ln t$. On the other hand, if the trajectory is completely random, there is no way of reproducing it other than memorizing the whole trajectory, resulting in a sequence length that increases linearly in time, i.e., $C_t \sim t$. The finite time KS entropy is defined simply as $h_t = C_t/t$. An important recent rigorous result due to Zweimüller [6] unveils the relation between KS entropy and the Lyapunov exponent for systems exhibiting subexponential instability. According to this result, we have, for almost all initial conditions,

$$\frac{C_t}{\sum_{k=0}^{t-1} \vartheta(f^k(x))} \to \frac{h_\mu(f)}{\int \vartheta d\mu}, \tag{114.3}$$

for $t \to \infty$, for any observable function ϑ, where $h_\mu(f)$ stands for the Krengel entropy, which can be expressed by the so-called Rohlin's formula

$$h_\mu(f) = \int \ln|f'| d\mu. \qquad (114.4)$$

By choosing again $\vartheta = \ln|f'|$, we get from (114.3) the surprisingly simple relation

$$h_t^{(\alpha)} \to \lambda_t^{(\alpha)}, \qquad (114.5)$$

for $t \to \infty$ and for almost all initial conditions, where

$$h_t^{(\alpha)} = \frac{C_t}{t^\alpha} \qquad (114.6)$$

is the subexponential generalization of the finite-time KS entropy. The relation (114.5) is the most natural generalization of the Pesin relation for systems of the type (114.1). From the ADK theorem and (114.5), we have that both $h_t^{(\alpha)}$ and $\lambda_t^{(\alpha)}$ converge in distribution terms toward the same Mittag-Leffler distribution. Nevertheless, Zweimüller's result is indeed stronger, assuring that, for almost all trajectories, $h_t^{(\alpha)}$ coincides with $\lambda_t^{(\alpha)}$ in the limit $t \to \infty$. In this way, the relation (114.3) does provide an efficient prescription for evaluating the algorithmic complexity of a given trajectory for one-dimensional maps, namely

$$C_t \to \sum_{k=0}^{t-1} \ln\left|f'\left(f^k(x)\right)\right|, \qquad (114.7)$$

for large t. The power of the prescription (114.7) resides in the fact that it does provide, for the systems in question, a computable way for the calculation of the algorithmic complexity function C_t, a well-known non-computable function in general [5].

114.3 Connection with the Krengel Entropy

We close with some final remarks on the connection between the Krengel entropy and the Lyapunov exponent. The ADK theorem gives (see, for instance, [7])

$$\langle\lambda^{(\alpha)}\rangle_{\text{ADK}} = \frac{1}{ba}\left(\frac{a}{\alpha}\right)^\alpha \frac{\sin(\pi\alpha)}{\pi\alpha} \int \ln|f'(x)|\omega(x)dx, \qquad (114.8)$$

where b is given by the asymptotics of the invariant density near the origin

$$\omega(x) \sim bx^{-\frac{1}{\alpha}}. \qquad (114.9)$$

From (114.8), Rohlin's formula (114.4) for the Krengel entropy implies immediately that

$$\frac{1}{b}h_\mu(f) = a\left(\frac{\alpha}{a}\right)^\alpha \frac{\pi\alpha}{\sin(\pi\alpha)}\langle\lambda^{(\alpha)}\rangle_{\text{ADK}}, \qquad (114.10)$$

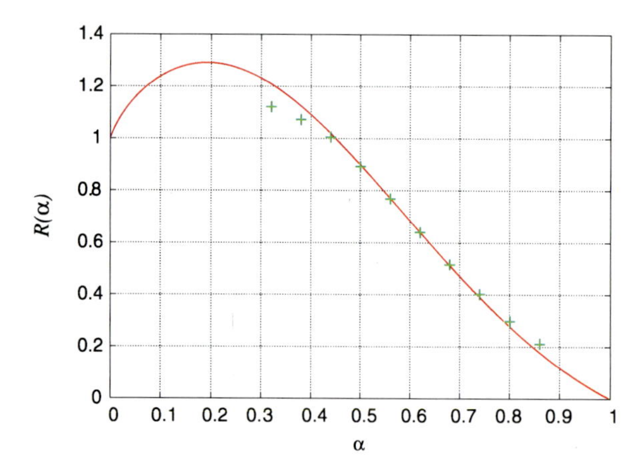

Fig. 114.1 The ratio $R(\alpha)$ for the Thaler map (114.11). The solid line corresponds to the expression (114.14). The points correspond to the numerical evaluation of (114.13) for 25×10^3 random initial conditions with 6×10^5 iterations of (114.11). A good concordance of the ADK prediction (114.14) with the numerical simulations is observed. Notice that the (b-dependent) relation proposed in [8] is $R(\alpha) = \alpha^{-1}$, which is incompatible with our numerical results. For the details on the numerics, see [7]

which is the correct relation (compare with that one proposed recently in [8]) between Krengel entropy and a dynamically meaningful average of subexponential Lyapunov exponents for maps of the type (114.1). The ADK average is not only a dynamically meaningful average, it is essentially the dynamically meaningful average for these systems. For instance, the average of the subexponential Lyapunov exponents (114.2) calculated for randomly chosen (with any absolutely continuous measure with respect to the usual Lebesgue measure on the interval [0, 1]) initial conditions x will converge to the ADK average for large t, see [7] for some recent applications of this important fact. Notice also that both sides of (114.10) are invariant under the symmetry $\omega \to \xi\omega$.

The so-called Thaler map [9]

$$f(x) = x\left[1 + \left(\frac{x}{1+x}\right)^{z-2} - x^{z-2}\right]^{-1/(z-2)}, \tag{114.11}$$

mod 1, provides a convenient setup to test numerically such results. Its invariant density is given, up to a multiplicative factor b, by

$$\omega(x) = x^{-1/\alpha} + (1+x)^{-1/\alpha}, \tag{114.12}$$

where $\alpha = (z-1)^{-1}$. From (114.11) one has $a = 1$. Figure 114.1 depicts the b-invariant ratio

$$R(\alpha) = (h_\mu/b)^{-1}\langle\lambda^{(\alpha)}\rangle_{\text{ADK}}, \tag{114.13}$$

which for the Thaler map (114.11) is given by

$$R(\alpha) = \frac{\sin\pi\alpha}{\pi\alpha^{\alpha+1}}. \tag{114.14}$$

A good concordance with the numerical simulations is observed. For the details about the convergence of the averages for this kind of map, see [7].

Acknowledgements This work was supported by the Brazilian agencies CNPq and FAPESP. The authors wish to thank Professor Leon Brenig for the warm hospitality at the Université Libre de Bruxelles, where part of this work was done.

References

1. Saa A, Venegeroles R (2012) J Stat Mech P03010. arXiv:1109.5419
2. Pesin YaB (1977) Russ Math Surv 32:55
3. Pomeau Y, Manneville P (1980) Commun Math Phys 74:189
4. Aaronson J (1997) An introduction to infinite ergodic theory. AMS, Providence
5. Gaspard P (1998) Chaos, scattering and statistical mechanics. Cambridge University Press, Cambridge
6. Zweimüller R (2006) Discrete Contin Dyn Syst 15:353
7. Pires CJA, Saa A, Venegeroles R (2011) Phys Rev E 84:066210
8. Korabel N, Barkai E (2009) Phys Rev Lett 102:050601
9. Thaler M (2001) Stud Math 143.103

Chapter 115
An Agent-Based Sorting Model for City Size and Wealth Distributions

Steffen Eger

Abstract In the current work, we undertake to propose a new model for explaining both city size and wealth distributions in an economy. Our model is, first, based upon agents with preferences over neighborhoods; rich/wealthy neigborhoods are more attractive than poorer ones and agents generally want to 'sort' into the neighborhood whose wealth level is largest. A second feature is that neighborhoods have an impact upon the members of their community, which we define in terms of the neighborhood's average wealth level. Finally, agents are inert in the sense that they are unwilling to leave their current neighborhood without reason and generally incur costs of relocation; and agents are boundedly rational in the sense that they perform local instead of global relocation decisions and that they do not anticipate/predict other agents' behavior. We show by simulation that under reasonable parametrizations, this model generates Zipfean city size distributions with coefficient alpha close to 1. It does not, however, by itself, generate Pareto wealth distributions. To this end, we add a stochastic component to individual agents' wealth levels, which we specify such that it either entails linear or exponential average growth. Nontrivially, this seems to lead to the 'correct shape' of the wealth distribution function for both the linear and exponential growth paradigms, but with 'suitable' coefficient beta only under the exponential growth implication.

115.1 Introduction

One version of *Zipf's law* for city sizes states that if one ranks cities by size and plots city size versus rank in log-log-scale one obtains, roughly, a *straight line* with slope between $-\alpha = -0.8$ and $\alpha = -1.5$ [22, 29], where we refer to α as the *Zipf coefficient*. In Fig. 115.1(left) we illustrate the law, using city size data from the United States for the year 2009. One sees that the fit is not perfect, with some deviation particularly for the largest cities New York, Los Angeles, and Chicago, but overall seemingly pretty good and an excellent rule of thumb. Remarkably, a related law apparently holds for the distribution of wealth among subjects in economies. If

S. Eger (✉)
Goethe University Frankfurt, Frankfurt, Germany

T. Gilbert et al. (eds.), *Proceedings of the European Conference on Complex Systems 2012*, Springer Proceedings in Complexity, DOI 10.1007/978-3-319-00395-5_115,

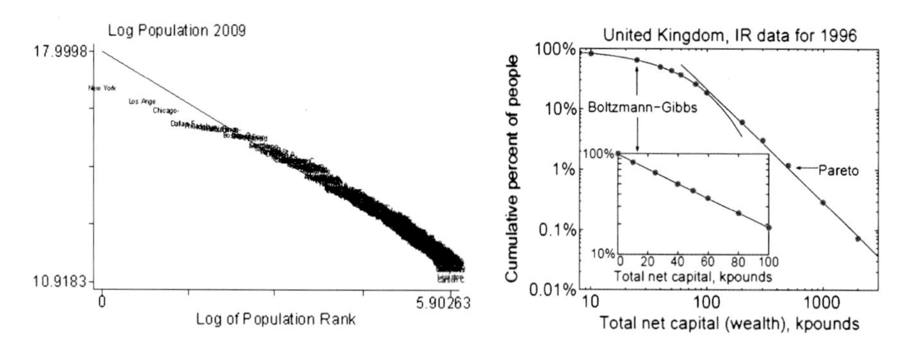

Fig. 115.1 *Left*: City size distribution of the United States for 2009, from http://economix.blogs.nytimes.com/2010/04/20/a-tale-of-many-cities/. *Right*: Wealth distribution of the United Kingdom for 1996, reprinted from [5]

one plots the probability that an individual has at least wealth level w against w in log-log-scale one obtains for the 'rich' tail of the distribution, again, a straight line, this time with slope between $-\beta = -1$ and $-\beta = -3$, where we call β the *Pareto coefficient*. This relationship, which we exemplify in Fig. 115.1(right), is termed *Pareto's law* for wealth distributions.

Both laws were discovered around the turn of the 19th century and, although *prima facie* having different interpretations, can both be phrased as 'rank size' rules; for Pareto's law we have that if one ranks richest subjects by wealth and plots wealth level versus rank in log-log-scale one obtains a straight line with slope between $-1/3$ and $-1/1 = -1$ (see below). The 'rank size' formulation allows a simple conceptualization; assuming a Zipf coefficient of 1 and a Pareto coefficient of 2, then, the largest city in an economy has about double the size of the second largest, three times the size of the third largest, etc. Accordingly, the wealthiest individual in an economy is about $\sqrt{2}$ times wealthier than the second wealthiest, $\sqrt{3}$ times wealthier than the third wealthiest, etc.

The two laws have received considerable attention ever since their discovery, particularly from physisicts, economists, and statisticians, where the leading question is about the mechanism(s) generating such outcomes. Two stochastic explanations for the Zipf and Pareto phenomena are as follows. For city sizes, Simon [26] found that a 'preferential attachment' rule can imply the observed regularity; cities grow proportionally to their current size, that is, larger cities obtain more new inhabitants. For wealth, a distribution process in analogy to those observed in statistical physics has been suggested (cf. [2]) where at each time step two randomly selected individuals 'gamble' about an individually determined share of their current wealth, leaving total wealth unchanged. Other stochastic models include [6, 11, 20, 34] and [3, 7, 13, 21, 23, 27, 28]. While these models may be at least partly convincing and undoubtedly appealing in their abstractness and simplicity, they do not tell us about *instrinsic motivations* and preferences of the agents involved in their setups. Some other models address this issue, for example, by defining *optimizing* (economic) agents that, in the case of cities, relocate e.g. on the basis of preferences defined over population density; e.g. agents may rejoice in companionship, but certainly want to avoid overcrowding (cf. [18, 19]). Further agent-based models include [9, 12, 24, 32, 33].

In the current work, we undertake to propose a new model for explaining both city size and wealth distributions in an economy. Our model is, first, based upon agents with preferences over *neighborhoods*; rich/wealthy neigborhoods are more attractive than poorer ones and agents generally want to 'sort' into the neighborhood whose wealth level is largest.[1] A second, separate but related, feature is that neighborhoods have an impact upon the members of their community, which we define in terms of the neighborhood's average wealth level. Such neighborhood or peer effects are well-known and well-documented in economics (cf. [4, 10], etc.) and, for example, in e.g. businesses it is estimated that income and productivity of individual workers increase considerably as the average income and productivity of co-workers increase (cf. [15, 25]). Finally, agents are *inert* in the sense that they are unwilling to leave their current neighborhood without reason and generally incur costs of relocation;[2] and agents are *boundedly rational* in the sense that they perform local instead of global relocation decisions and that they do not anticipate/predict other agents' behavior. Probably as remarkable as Zipf's law itself, we show by simulation that under reasonable parametrizations, this model generates Zipfean city size distributions with coefficient α close to 1. It does not, however, by itself, generate Pareto wealth distributions. To this end, we add a stochastic component to individual agents' wealth levels, which we specify such that it either entails linear or exponential average growth. Nontrivially, while not infringing upon the Zipfean city size distribution law, this seems to lead to the 'correct shape' of the wealth distribution function (straight line for the rich tail in log-log-scale, exponential regime for the poor tail, see below) for both the linear and exponential growth paradigms, but with 'suitable' coefficient β only under the exponential growth implication.

While we believe our model to be abstract and general enough to potentially apply to the migratory behavior of many living organisms, we note that it is, at the same time, rooted in economic theory. For example, the classical Tiebout sorting model [31] holds that individuals sort into neighborhoods based upon the latters' attractiveness (in terms of taxes, public goods, etc.). Moreover, *neighborhood effects* (or more specificially *peer effects*) have been well-studied in economics and may include such phenomena as status formation motives, aversion to pay inequality (sometimes also referred to by the phrase 'Keeping up with the Joneses'), learning, and exogenous effects due to environmental characteristics of an area (cf. [17]). Our

[1]The fact that human agents, or biological organisms, undertake to settle in rich ecological neighborhoods seems self-evident to us. We emphasize with a few examples. First, immigration from third-world countries such as Africa to the industrialized countries such as the United States, Europe, etc. Secondly, in the animal world, usually rivers, food-rich habitats, etc., exogenous sources of wealth, attract multitudes of organisms. Thirdly, the 'poor chasing the rich' literature (cf. [1, 30, 31]) has as key insight that people care for the average income or wealth in the community in which they live for at least three reasons: status, peer groups, and taxes.

[2]This may capture physical restrictions (e.g. aversion or inability to cover long distances) or social restrictions (e.g. unwillingness to give up one's social ties, etc.).

approach is, to the best of our knowledge, one of the few (general) agent-based approaches to modeling Zipfean city size distributions and the first one based on the concept of neighborhoods. Moreover, it is apparently the first such model based on *wealth* as a decision variable and the first model to discuss city size and wealth distributions in a unified framework.

115.2 The Model

115.2.1 Setup

Agents (or *players*) $i = 1, \dots, n$, $n \geq 2$, inhabit a *world* (which we also refer to as *grid* or *lattice*) $X \subseteq \mathbb{N}^k$, $k \geq 1$. Only a fraction $s \in [0, 1]$ of all places (or points) in X are inhabited, the remaining are empty. Each agent i has *payoff* (which we refer to as his *wealth*) $Y_{i,t}$ in *periods* $t = 0, 1, \dots, T$. Payoffs are determined by the agent's current payoff and his environment's payoffs according to

$$Y_{i,t+1} = Y_{i,t} + \delta(\bar{Y}_{i,t} - Y_{i,t}) + \epsilon_{i,t+1}, \tag{115.1}$$

where $\delta \in (0, 1)$ is the *adaption rate*, $\bar{Y}_{i,t}$ is some 'average' payoff in agent i's environment at time t, and $\epsilon_{i,t+1}$ is a random component. Generally, the *average payoff* $\bar{Y}_{i,t}$ is determined according to

$$\bar{Y}_{i,t} = \sum_j w_{ij} Y_{j,t}, \tag{115.2}$$

where w_{ij} is the (distance-dependent/dependent on their respective positions) *weight* assigned to agent j with regard to agent i, with $w_{ij} \geq 0$ and $\sum_j w_{ij} = 1$.

115.2.2 Relocation Dynamics

All agents receive random initial wealth $Y_{i,0}$, for $i = 1, \dots, n$. Then, for all periods $t = 1, \dots, T$ they solve the following maximization problem

$$\max: \quad \mathbb{E}_t[Y_{i,t+1}], \tag{115.3}$$

by playing a best response to the world as it is at time t: of the $|X|(1 - s)$ free positions in world X, agent i chooses the one with the highest utility (i.e. expected next period wealth) assuming that the world remains unchanged, i.e. assuming that all other agents stay at their position.[3] This choice has an obvious solution since $Y_{i,t+1}$

[3] Agents assume that the world stays as it is except for their own hypothetical movements. Otherwise, if this was not considered, agents would avoid 'empty regions' as these have low average income values.

has only one component which is not determined exogenously at time t, namely $\bar{Y}_{i,t}$.[4] Thus, in period t agent i chooses position $p \in X \backslash \mathcal{O}$—where $\mathcal{O}$ is the set of occupied places in X—with the highest average wealth value. Importantly, the order in which agents are allowed to choose their position in period t is determined randomly; agents who move later make more 'informed' decisions.

To account for individuals' 'inertia' we introduce positive moving costs $c : X^2 \to \mathbb{R}^{\geq 0}$. This changes the optimization conditions only slightly; agents now choose $p \in X \backslash \mathcal{O}$ such that

$$\mathbb{E}_t\big[Y_{i,t+1} - c(p_i, p)\big] \tag{115.4}$$

is maximized, where p_i is agent i's position prior to his relocation decision. Moreover, to model 'bounded rationality' we generally restrict agents to conduct a local search for optimal grid positions instead of a global search; i.e. they may only choose positions in the vicinity of their current habitat. When all n agents have moved (some may have remained at the position they were occupying before), their wealth is updated using Eq. (115.1).

115.2.3 Discussion of the Model

To specify the agent's world $X \subseteq \mathbb{N}^k$, usually a one or two-dimensional grid is assumed, as in the agent-based models discussed in Sect. 115.1. Each 'place' (or point) $p \in X$ can then be occupied by a *single* agent or is otherwise empty.

As to the agents' interactions, Eq. (115.1) implements the neighborhood influences discussed above. Note that, disregarding the error, $\delta \to 1$ implies that $Y_{i,t+1} \to \bar{Y}_{i,t}$ and $\delta \to 0$ implies $Y_{i,t+1} \to Y_{i,t}$, so that the adaption rate determines how strongly an agent's wealth is determined by the average wealth and his last period wealth, respectively. Also note that Eq. (115.1) is a variant of the updating rule in self-organzing maps (cf. [16]), neural networks (cf. [14]), etc. and thus related to the literature on unsupervised self-organizing adaptive systems, etc.

The random component in Eq. (115.1) need not necessarily be realized as white noise, i.e. as zero-mean, independent random variables. On the contrary, as discussed in Sect. 115.1, we might want to design our model in such a way that the *rich are getting richer*. To do so, the conditional expectation $\mathbb{E}[\epsilon_{i,t+1} \mid Y_{i,t}]$ could be specified as an increasing function of $Y_{i,t}$. A simple choice, leading to exponential growth and implementing 'preferential attachment', would be to let $\mathbb{E}[\epsilon_{i,t+1} \mid Y_{i,t}] = \mu Y_{i,t}$ for $\mu > 0$.

115.3 Results

We consider the following parametrization. The world is a one-dimensional grid of size $1 \cdot N$, with places $p \in X = \{1, \ldots, N\}$, where we choose $N = 1000$. Moreover,

[4] The variable $\bar{Y}_{i,t}$ is not exogenously determined at time t since the weights w_{ij} may depend on the position of agent i relative to agent j.

Table 115.1 Model calibration. By $\tilde{f}(x;\mu,\sigma^2,a,b)$ we denote the (adequately normalized and truncated) normal density function with mean μ, variance σ^2, and truncation parameters a and b

Parameter	Value	Meaning
N	1000	grid size
n	300	number of agents
χ	0.001	inertia/moving cost parameter
x_0	50	$Y_{a,0} \sim U(x_0, x_1)$
x_1	60	$Y_{a,0} \sim U(x_0, x_1)$
δ	$\in [0, 1]$	adaption rate
ρ	$\in \{5, 10, 20, 30, 100, 400, N\}$	spatial reach
T	$\in \{5000, 10000\}$	number of periods
$w_{i,j}$	$\tilde{f}(p_j; p_i, 1, p_i - 5, p_i + 5)$	weight for agent j with respect to agent i

we set the number of agents, n, to 300. Agents conduct a local search for optimal grid positions by considering the ρ, $\rho \in \mathbb{N}$, places 'around' their current position for potential relocation. Agents a have moving costs of

$$c(p_a, q) = \begin{cases} \chi |p_a - q| & \text{if } |p_a - q| \leq \rho \\ \infty & \text{else,} \end{cases}$$

where $p_a, q \in X$ (p_a being agent's a current position), and where we set χ to the 'low' value of 0.001. Each agent a is initially assigned a random place $p_a \in X$ and a random wealth level $Y_{a,0} \sim U(x_0, x_1)$, where $U(x_0, x_1)$ is the continuous uniform distribution on the interval $[x_0, x_1]$. As to determining the average wealth at any given grid point $p \in X$, we use the (normalized and truncated) normal density function centered at p with the 'low' variance of 1; we set the density weights to zero for agents at positions $q \in X$, with $|p - q| > 5$. Finally, we define a city as a continuous set of occupied grid points with no empty spaces in between. We summarize the model calibration in Table 115.1.

In the subsequent analyses, all values discussed are averages over 10 runs, and single numerical results are taken after $T = 5000$ iterations, unless stated otherwise. We estimate β on the basis of the 10 % richest, mainly due to our small size of n (e.g. it would not be wise to estimate a regression on the basis of $1\ \% \cdot n = 3$ data points).

115.3.1 Linear Growth

First, consider the case where the random component in (115.1) is close to zero and iid across agents. We set

$$\epsilon_{i,t+1} \sim \mathcal{N}(0.2, 1),$$

Table 115.2 Calibration as in Table 115.1. From left to right: $\rho = 100$, $\delta = 0.015$, $\delta = 0$. Linear growth

	δ					
	0.001	0.005	0.015	0.05	0.1	0.5
α size	1.662	1.198	0.994	0.961	0.942	0.938
R^2	0.883	0.898	0.920	0.941	0.920	0.905
β size	142.85	90.90	45.45	15.62	14.92	15.62
R^2	0.976	0.956	0.938	0.876	0.834	0.767

	ρ				
	5	10	20	400	10, π_a
α size	0.651	0.847	0.892	1.563	0.888
R^2	0.770	0.909	0.918	0.893	0.830
β size	142.86	125.0	90.90	50.0	166.67
R^2	0.972	0.939	0.921	0.978	0.965

	ρ		
	5	10	20
α size	0.705	1.448	1.483
R^2	0.570	0.893	0.890
β size	38.46	33.33	25.64
R^2	0.963	0.968	0.977

where the mean of 0.2 is chosen so as to avoid that agents get poorer, on average, from period to period due to positive moving costs. Note also that if δ was zero (and ignoring moving costs) this would entail that $Y_{i,t}$, defined in (115.1), follows, in expectation, an affine-linear growth process,

$$\mathbb{E}[Y_{i,t}] = \mathbb{E}[Y_{i,t-1} + \mu] = \mu \cdot t + Y_{i,0},$$

where, in our case, $\mu = 0.2$. By simulation we find that such a process, by itself, leads to a large Pareto coefficient β for the top 10 % wealthiest of around 36.69 after $T = 5000$ periods, where the fit is very good, with R^2 value of about 98 %.

We outline our main results in Table 115.2, and Figs. 115.2 and 115.3. From Table 115.2, we deduce that for many different calibrations of ρ and δ, our model entails city size distributions with Zipf parameter α in the 'right' range between 0.8 and 1.5. As a reference, we found by simulation that distributing $n = 300$ agents randomly among $N = 1000$ grid points implies a coefficient α of around 2.128 with R^2 value of around 0.839. Our model performs much 'better', even with a value of δ equal to 0; for example, for $\delta = 0$ and $\rho \in \{10, 20\}$, α is on average smaller than 1.5 after $T = 5000$ periods and the R^2 value is close to 90 %. We obtain 'best' results for δ above/around 1 % and ρ relatively small, in the range between $\rho = 20$ and $\rho = 100$; the R^2 fit is then usually larger than 90 % and α is close to unity.

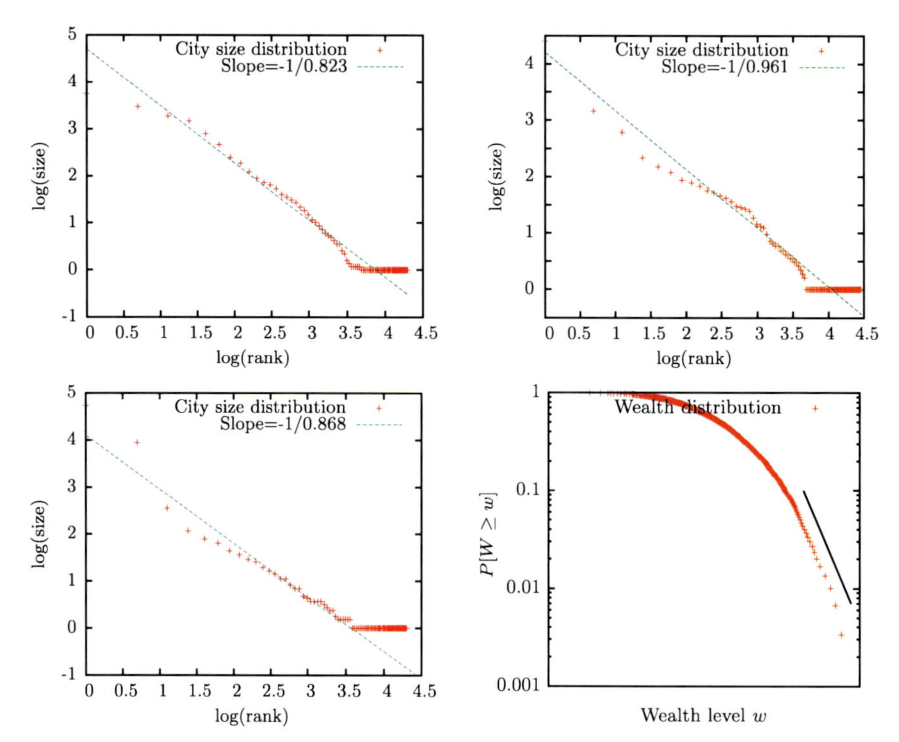

Fig. 115.2 City size and wealth distributions after $T = 5000$ periods. Calibration as in Table 115.1, and $\delta = 0.05$ throughout. From left to right: $\rho = 10$ and π_a (ρ is a random variable here, define as $\rho = N$ with probability π_a and $\rho = 10$ with probability $1 - \pi_a$, where π_a is proportional to agent a's wealth), $\rho = 100$, $\rho = 20$, $\rho = 5$. The marked slope in the wealth distribution curve has value -200. Linear growth

Moreover, Fig. 115.3 shows that, at least for specific parameter settings of ρ and δ, α is quite stable over time and it usually takes fewer than 1000 periods for it to settle within a narrowly defined band around its asymptotic (as it seems) value.

Concerning the Pareto wealth coefficient β, its size is generally (magnitudes) too large under the given calibration and the linear wealth growth process. While the fit is usually good (R^2 above 90 % for small δ)—note also that the 'correct' *form* of the wealth distribution function is usually reproduced by our model (cf. Fig. 115.2), i.e., a Boltzmann-Gibbs type distribution for 'small' wealth levels w and a Pareto distribution for 'large' w, the lowest value recorded in the simulations summarized in Table 115.2 is more than five times larger than observed in real economies. This is a rather unsurprising result, given the high level of β under $\delta = 0$ and ignoring moving costs (see above), the system's inherent tendency toward convergence (proven in an extended version of the paper), and the thus implied assimilation of agents' wealth levels. Moreover, in Fig. 115.3, we see that β is much less stable than α under the given calibration, displaying considerable fluctuation over time.

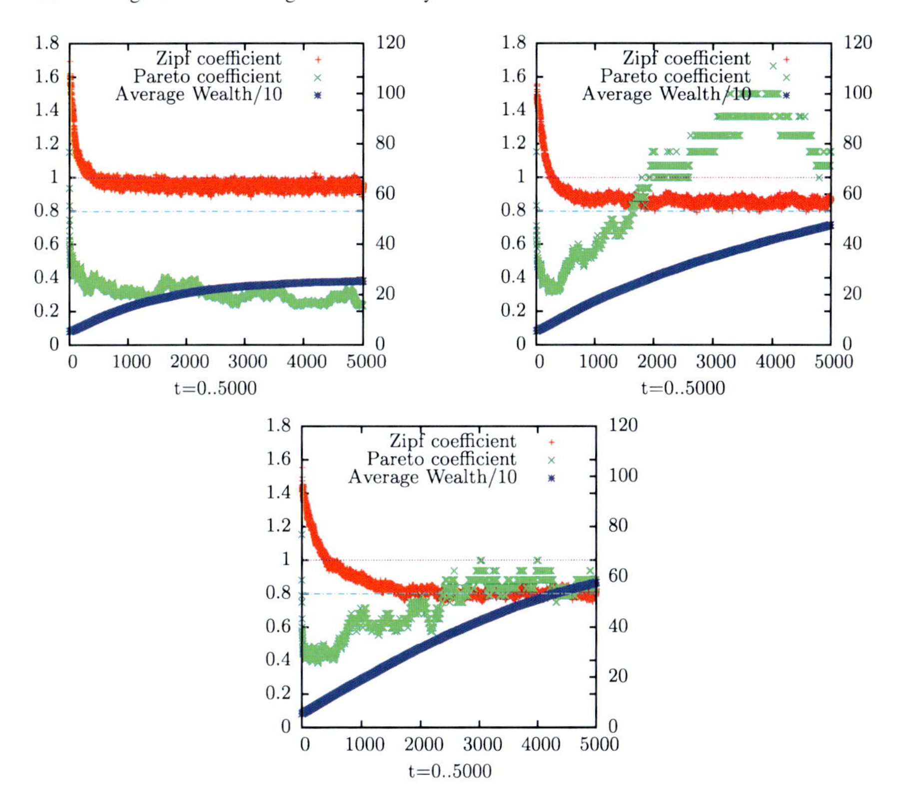

Fig. 115.3 Parameter evolution of α, β, average wealth, and Gini coefficients over time. Average wealth and β are drawn with respect to the y_2-axis indicated on the right side of each plot. Calibration as in Table 115.1, and $\delta = 0.05$. From left to right: $\rho = 100$, $\rho = 20$, $\rho = 10$ and π_a. Linear growth

115.3.2 Exponential/Proportional Growth

Next, consider the following specification of $\epsilon_{i,t+1}$,

$$\epsilon_{i,t+1} \sim \mathcal{N}(\mu Y_{i,t}, \kappa |Y_{i,t}|), \tag{115.5}$$

where $\mu \in (0, 1)$ and $\kappa \in (0, 1)$. If δ were zero and ignoring moving costs, this would entail the following evolution of $Y_{i,t+1}$, defined in (115.1),

$$Y_{i,t+1} = (1 + \mu) Y_{i,t} + \nu_{i,t+1}, \qquad \mathbb{E}[Y_{i,t}] = (1 + \mu)^t Y_{i,0}. \tag{115.6}$$

where $\nu_{i,t+1} \sim \mathcal{N}(0, \kappa |Y_{i,t}|)$; i.e., this implies in expectation, an exponential growth process. Note also that (115.5) and (115.6) can be interpreted as an instantiation of 'proportional attachment'; the increase in an agent's wealth level between two periods is proportional, in expectation, to the size of the agent's current wealth level—the richer experience a larger increment. That we choose the variance of $\epsilon_{i,t+1}$ to be

Table 115.3 Calibration as in Table 115.1. From left to right: $\rho = 30$, $\delta = 0.015$, $\delta = 0$. Exponential growth

	δ			
	0.001	0.005	0.015	0.02
α size	1.490	1.500	1.157	1.147
R^2	0.890	0.891	0.927	0.938
βsize	3.401	6.667	2.873	2.994
R^2	0.960	0.947	0.921	0.813

	ρ			
	5	10	20	10, π_a
α size	0.664	0.969	0.988	1.281
R^2	0.852	0.928	0.938	0.891
β size	12.821	11.765	6.172	15.385
R^2	0.930	0.930	0.939	0.955

	ρ		
	5	10	20
α size	0.850	1.295	1.321
R^2	0.566	0.855	0.875
β size	2.81	2.73	2.403
R^2	0.948	0.947	0.947

a function of $|Y_{i,t}|$ seems plausible and entails a constant relative standard deviation of $\epsilon_{i,t+1}$ across agents. In the following, we set $\mu = 0.0005$ and $\kappa = 0.1$. By simulation, we find that this parametrization leads to a coefficient β of around 3.02, with R^2 value of around 0.950, under $\delta = 0$ and no moving costs.

We summarize results in Table 115.3 and Fig. 115.4. Contrasting with the results under linear growth, we see that the growth process may also affect city size distributions. For example, under $\delta = 0.015$, α is closer to unity under exponential growth than under linear growth, with no worse R^2 values. The best result is, again, obtained for δ around 1 % and ρ relatively small, this time around 30.[5] In that case, α is very close to unity and β is smaller than 3 after 5000 periods, with R^2 values larger than 90 %. Figure 115.4 shows that β is in the range between 2 and 3 after about 4500 periods and seems to remain stable, although still fluctuating more heavily than α.

115.4 Conclusion

A few remarks on our results must be made. First, the fact that exponential growth is implied (or apparently required) by our model seems to be a too specific assump-

[5] For larger ρ we frequently noticed a 'poverty trap', in which agents become poorer initially—due to higher average moving costs—and can then not recover due to the small size of μ.

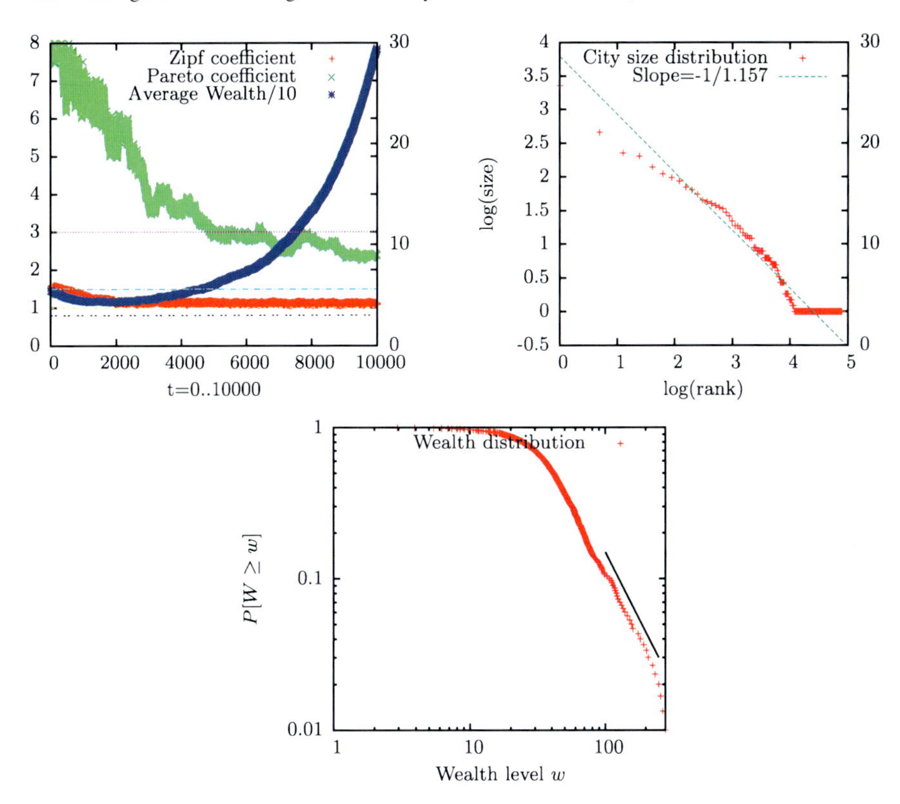

Fig. 115.4 Calibration as in Table 115.1, and $\delta = 0.015$ and $\rho = 30$. Parameter evolution of α, β, average wealth, and Gini coefficients over time, and city size and wealth distributions after $T = 5000$ periods. Average wealth is drawn with respect to the y_2-axis indicated on the right side of the respective plot. The marked slope in the wealth distribution curve has value -2.873. Exponential growth

tion, on the one hand, since exponential economic growth is presumably a feature of the last few hundred years exclusively, in effect only since the industrial revolution, whereas most of human history was, by all appearances, characterized by virtually no growth at all, at least on average. On the other hand, it evidently opposes the wealth distribution models proposed in econophysics, which assume zero-sum wealth processes. Two things might be answered to this; namely, that a) the universal applicability of Pareto's law is apparently still not fully ascertained to date, and some authors call it a property of capitalist societies (cf. [8]), for which exponential growth is valid, allegedly. Moreover, b) it must be said that our model does not necessarily require exponential growth but could presumably do with a proportional attachment rule without such implications (e.g. due to high enough moving costs, some agents having large negative net wealth and thus clearing the balance, etc.); also note that in Fig. 115.4, there is in fact practically no growth at all for the first 5000 periods and still the Pareto coefficient approaches 3.

Next, the result that Zipf's law is apparently 'best' reproduced for small values of δ around 1 % appears to be a plausible outcome. It is known that, for example, peer effects at the work-place are usually between 5 and 15 % (cf. [15, 25]), and neighborhood effects in a community should certainly be lower, as the degree of interaction between individuals, there, is presumably considerably lower.

For future work, it might be of interest to find an 'agent-based'—instead of a stochastic—solution for generating Pareto wealth distributions within our Tiebout-like sorting model. Potentially, the inclusion and adequate weighting of further variables such as density or inequality aversion might be helpful here, but we think this unlikely. Presumably, unless a process is defined whereby agents veritably lose and gain wealth beyond the rather small neighborhood effects implemented—be it through 'gambling' or other mechanisms—a wealth distribution that is unequal 'enough' in the rich tail will not be achieved. To check for the robustness of our results, implementing our model in a two-dimensional scenario might be a further aspect of concern.

References

1. Bucovetsky S, Glazer A (2010) Peer group effects, sorting, and fiscal federalism. Working papers 091006, University of California-Irvine, Department of Economics
2. Chatterjee A, Chakrabarti BK, Manna SS (2003) Money in gas-like markets: Gibbs and Pareto laws. Phys Scr T 106:36
3. Coelho R, Richmond O, Barry J, Hutzler S (2008) Double power laws in income and wealth distributions. Physica A 387(3847):3847–3851
4. Dietz R (2001) Estimation of neighborhood effects in the social sciences: an interdisciplinary approach. Working paper
5. Drăgulescu AA (2003) Applications of physics to economics and finance: money, income, wealth, and the stock market. PhD thesis, Department of Physics, University of Maryland, USA. arXiv:cond-mat/0307341
6. Drossel B, Schwabl F (1992) Self-organized critical forest-fire model. Phys Rev Lett 69:1629–1632
7. Dorogovtsev SN, Mendes JFF (2003) Evolution of networks: from biological nets to the internet and WWW. Oxford University Press, London
8. Düring B, Matthes D, Toscani G (2008) Kinetic equations modelling wealth distributions: a comparison of approaches. Phys Rev E 78:056103
9. Epstein JM, Axtell RL (1996) Growing artificial societies: social science from the bottom up. MIT Press, Cambridge
10. Falk A, Ichino A (2006) Clean evidence on peer effects. J Labor Econ 24(1):39–57
11. Gabaix X (1999) Zipf's law for cities: an explanation. Q J Econ 114:739–767
12. Gardner M (1970) Mathematical games—the fantastic combinations of John Conway's new solitaire game "life". Sci Am 223:120–123
13. Gupta AK (2006) Models of wealth distributions—a perspective. In: Chakrabarti BK, Chakraborti A, Chatterjee A (eds) Econophyiscs and sociophysics: trends and perspectives. Wiley, New York
14. Hopfield JJ (1982) Neural networks and physical systems with emergent collective computational abilities. Proc Natl Acad Sci USA 79:2554–2558
15. Ichino A, Maggi G (2000) Work environment and individual background: explaining regional shirking differentials in a large Italian firm. Q J Econ 115(3):1057–1090

16. Kohonen T (1984) Self-organization and associative memory. Springer, Berlin
17. Manski CF (1993) Economic analysis of social interactions. J Econ Perspect 14(3):115–136
18. Mansury Y, Gulyás L (2007) The emergence of Zipf's Law in a system of cities: an agent-based simulation approach. J Econ Dyn Control 31:2438–2460
19. Page SE (1998) On the emergence of cities. J Urban Econ 45:184–208
20. Reed WJ, Hughes BD (2002) From gene families and genera to incomes and internet files: why power laws are so common in nature. Phys Rev E 66:067103
21. Richmond P, Solomon S (2001) Power laws are Boltzmann laws in disguise. Int J Mod Phys C 12(3). doi:10.1142/S0129183101001754
22. Rosen K, Resick M (1980) The size distribution of cities: an examination of the Pareto law and primacy. J Urban Econ 8(2):165–186
23. Santos MA, Coelho R, Hegyi G, Néda Z, Ramasco J (2007) Wealth distribution in modern and medieval societies. Eur Phys J Spec Top 143:81–85
24. Schelling T (1978) Micromotives and macrobehavior. Norton, New York
25. Shvydko T (2007) Interactions at the workplace: peer effects in earnings. Working paper
26. Simon HA (1955) On a class of skew distribution functions. Biometrika 42:425–440
27. Slanina F (2004) Inelastically scattering particles and wealth distribution in an open economy. Phys Rev E 69:046102
28. Solomon S, Richmond P (2002) Stable power laws in variable economies; Lotka–Volterra implies Pareto–Zipf. Eur Phys J B 27:257–261
29. Soo KT (2005) Zipf's law for cities: a cross country investigation. Reg Sci Urban Econ 35(3):239–263
30. Strahilevitz LJ (2006) Exclusionary amenities in residential communities. Viriginia Law Rev 92:437–499
31. Tiebout C (1956) A pure theory of local expenditures. J Polit Econ 64(5):416–424
32. Wilensky U (1998) NetLogo wealth distribution model. Center for Connected Learning and Computer-Based Modeling, Northwestern University, Evanston, IL. http://ccl.northwestern.edu/netlogo/models/WealthDistribution
33. Wilensky U (1999) NetLogo. Center for Connected Learning and Computer-Based Modeling, Northwestern University, Evanston, IL. http://ccl.northwestern.edu/netlogo/
34. Yule GU (1957) A mathematical theory of evolution based on the conclusions of Dr JC Willis. Philos Trans R Soc Lond B 213:21–87

Chapter 116
Characteristic Features of the Sustainable Strategies in the Evolvable Iterated Prisoners' Dilemma

Mieko Tanaka-Yamawaki and Ryota Itoi

Abstract In the realm of iterated prisoners' dilemma equipped with evolutional generation of strategies, a model has been proposed by Lindgren that allows elongation of genes, represented by one-dimensional binary arrays, by three kinds of mutations: duplications, splittings, and point mutations, and the strong strategies are set to survive according to their performance at the change of generations. The actions that the players can choose are assumed to be either cooperation represented by C or defection represented by D. It is convenient to use {0, 1} instead of {D, C}. Each player has a strategy that determines the player's action based on the history of actions chosen by both players in each game. Corresponding to the history of actions, represented by a binary tree of depth m, a strategy is represented by the leaves of that tree, an one-dimensional array of length 2^m.

We have performed extensive simulations until many long genes appear, and by evaluating those genes we have discovered that the genes of high scores are constructed by 3 common quartet elements, [1001[, [0001], and [0101]. Furthermore, we have determined the strong genes commonly have the element [1001000100010001] that have the following four features:

(1) never defects under the cooperative situation, represented by having '1' in the fourth element of the quartet such as [* * * 1],

(2) retaliates immediately if defected, represented by having '0' in the first element and the third element in the quartet such as [0 * 0 *],

(3) volunteers a cooperative action after repeated defections, represented by '1' in the first element of the genes,

(4) exploits the benefit whenever possible, represented by having '0' in the quartet such as [* 0 * *].

This result is stronger and more specific compared to [1**1 0*** 0*** *001] reported in the work of Lindgren as the structure of strong genes.

M. Tanaka-Yamawaki (✉) · R. Itoi
Department of Information and Knowledge Engineering, Graduate School of Engineering, Tottori University, 101-4 Koyamacho-Minami, Tottori 680-8552, Japan
e-mail: mieko@ike.tottori-u.ac.jp

T. Gilbert et al. (eds.), *Proceedings of the European Conference on Complex Systems 2012*, Springer Proceedings in Complexity, DOI 10.1007/978-3-319-00395-5_116,

116.1 Introduction

A game theory is a tool to study the strategic interactions developed between individuals that mutually influence each other, which can be is used to understand human behavior and economics. The prisoners' dilemma (abbreviated as PD, hereafter) is a well-known example of the two-person-game models in which the most rational choice for individuals results in the less beneficial choice for the society that those individuals belong to. The PD has been studied for a long time in discussing the issues of CO2 emission, nuclear power possession, and price competitions, etc. [1–3] It is well known that the strategy called Tit-For-Tat was the most powerful strategy in the historical experiment performed by Axelrod [4]. However, it has also been known that the strongest strategy changes under different environments.

A question arises whether the Tit-For-Tat strategy can be developed from any initial condition and any historical paths, or it is generated from a specific conditions. We are also interested in whether this strategy eventually disappear under certain circumstances, or not. In order to answer to those question, we take a view of the artificial life to study evolution of genes in a wider framework of time and space to study the problem of automatic generation of strategies in the multi-agent simulations.

Murakami and one of the authors have considered a simple model to generate the Tit-For-Tat strategy according to the rule of survival of the fittest, and observed how this specific strategy is generated [5].

Lindgren considered a model of evolutional genes in the multi-agent model that allows elongation of genes, represented by one-dimensional binary arrays, by three kinds of mutations: duplications, splittings, and point mutations, and the strong strategies are set to survive according to their performance at the change of generations [6].

In this paper, we report our result of long-term simulations that clarified the existence of common elements found in strong and sustainable strategies and the three specific common quartet elements contained in such strategies and the features represented by those elements.

116.2 Iterated Prisoners' Dilemma

The prisoners' dilemma is defined by the payoff structure of both players shown in Table 116.1. We assume the players have only two actions to choose, to cooperate (C, hereafter) or to defect (D, hereafter). There are four parameters R, P, S, T which are set tp satisfy $S < P < R < T$ and $S + T < 2R$. The key point of the situation under which the two players are set in this model is the better choice for individual results in the worst choice of both. For example, if we assume B cooperates, A's rational choice is to defect because $R < T$. However, even if we assume B defects, A's rational choice is still to defect because $S < P$. Thus A is supposed to defect whatever B chooses. The situation is the same for B. Thus both A and B end up

Table 116.1 The payoff table of the prisoners' dilemma provided $S < P < R < T$ and $S + T < 2R$

(A's payoff, B's payoff)	B's action is C	B's action is D
A's action is C	(R, R)	(S, T)
A's action is D	(T, S)	(P, P)

Table 116.2 Actions of the TFT pair and the PAV pair when the second player commits an error at $t = 2$

Time t	(TFT, TFT)	(PAV, PAV)
1	(C, C)	(C, C)
2	(C, 'D')	(C, 'D')
3	(D, C)	(D, D)
4	(C, D)	(C, C)
5	(D, C)	(C, C)

with choosing to defect. However, the payoff P is smaller than R. How can they choose the better option of mutual cooperation?

The poor solution (P, P) is inevitable for a single game, unless they promise to start with the cooperative actions. When they repeat the game by starting with the cooperative actions, then the best choice for both of them is to continue to cooperate except the last match. Because once each of them defects, then the opponent will retaliate in the next match. Therefore if they know the time to end the repeated game, they will defect at the last match. For this reason, the iterated prisoners dilemma game (abbreviated as IPD, hereafter) is played without fixing the time to end. In such a game, a particular strategy called Tit-For-Tat (TFT, hereafter) wins over the other strategies. In general, good strategies including the TFT, share the following three features:

(1) to cooperate as long as the opponent cooperates
(2) to retaliate immediately if defected
(3) to offer cooperation after continuous defections.

However, it has been known that the Pavlov strategy (PAV, hereafter) is better than the TFT under a certain condition. The PAV keeps the same action after getting T or R which are the good payoff, and changes the action from the previous one after getting S or P which are the poor payoff. This strategy is stronger than the TFT in a model allowing errors in actions in which the player chooses an opposite action from the one chosen by the strategy [7].

This situation is depicted in an example shown in Table 116.2. In this case, both (TFT,TFT) and (PAV, PAV) begin the game from the cooperative relationships at the time $t = 1$. Suppose if an error occurs at $t = 2$ in the second player, then the TFT pair immediately fall into pose if an error occurs at $t = 2$ in the second player, then the TFT pair immediately fall into (C, D) a series of (C, D) and (D, C), while the PSV pair can recover the original cooperative situation of (C, C). Thus the TFT is not always the best under errors.

116.3 Evolvable Strategies in the IPD

In the framework of the artificial life (ALIFE), a new scheme of searching for the better strategies was presented in Ref. [6] in a multi-agent model of evolvable strategies, in which the strategies grow like genes. Here the strategies are represented by one-dimensional binary strings.

The two actions, the cooperation and the defection, {D, C}, are represented by {0, 1}. Each player has a strategy that determines the player's action based on the history of actions chosen by both players in each game. Corresponding to the history of actions, represented by a binary tree of depth m, a strategy is represented by the leaves of that tree, an one-dimensional array of length 2^m. It is convenient to set the two edges of the binary tree to have 0 in the left edges and 1 in the right edges.

For example there are four strategies represented by [00], [01], [10], [11], for $m = 1$ corresponding to a model to simply count the opponent's previous action as the history. The strategy [00] is called as ALLD because only D is chosen irrelevant to the opponent's past action. Likewise, [11] is called as ALLC. The strategy [01] is the TFT because D is chosen only when the opponent's action of the immediate past is D. Likewise the strategy [10] is called as anti-TFT (abbreviated as ATFT).

If we count the actions of both players as the history, that is the case of $m = 2$ and the corresponding strategy becomes a binary string of length 4. For example, a strategy [1001] chooses C if the past actions of both players are the same, i.e., both C or both D, and chooses D if the past actions of both players were not the same, i.e., when one player's action was C, the other player's action was D. This corresponds to the PAV. A strategy [0101] is the same as [01] because D is chosen for the opponent's defective action and C is chosen for the opponent's cooperative action irrelevant to the past action of the other side. Likewise, the strategy [0000] is the same as [00]. A strategy represented by [0001] chooses C only when the past actions of both sides were C. We call this strategy as the retaliation-oriented TFT (abbreviated by RTFT).

For larger m, the history and the corresponding strategy can be written as $h_m = (a_m, \ldots, a_2, a_1)_2$ and $S_m = [A_1 A_2 \cdots A_n]$ for $n = 2^m$. An example of the strategy for the case of $m = 3$ represented by a string of 10010001 is shown in Fig. 116.1. Out of all the possible strategies, good ones are chosen by employing the genetic algorithm. The typical job-flow of this mechanism is illustrated in Fig. 116.2.

Starting from the initial population of agents, which could be the entire set of possible strings or a randomly sampled subset of the entire set, pairs of agents play the IPD of indefinite length. After all the agents playing the game with all the other agents, their total payoff are counted and their population is renewed according to the population dynamics explained below. Subsequently the mechanism of three types of mutation, (1) point mutation (2) doubling, and (3) fission, are applied in order to grow the strategy strings to create new patterns and the new lengths that the previous generation did not know.

We have followed the scenario written by Lindgren [6], except for the two points: the first point is the stochastic ending of IPD, and the second point is that we have performed extensive amount of simulations. As a result, we have discovered the

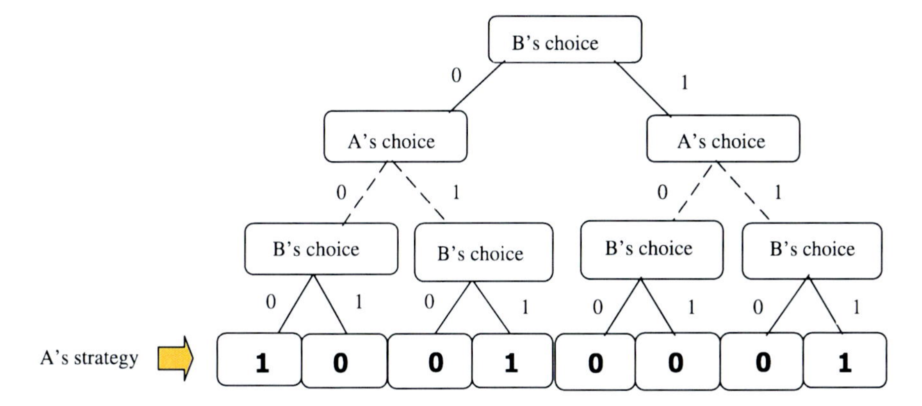

Fig. 116.1 An strategy and the binary tree of the history

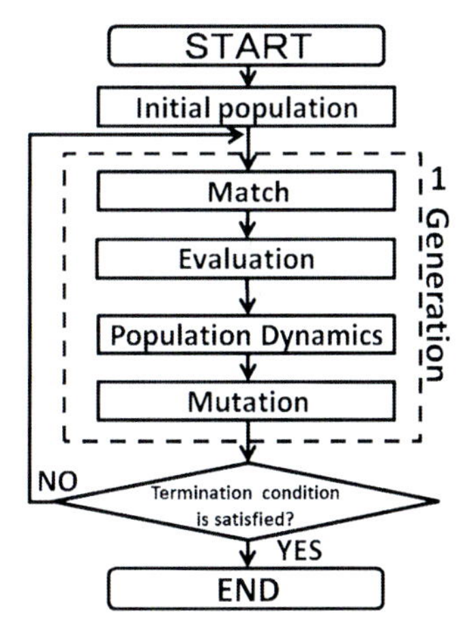

Fig. 116.2 The Job-flow of the evolvable IPD simulation

type of gene structure of sustainable strategies in more specific manner compared to [1**1 0*** 0*** *001] suggested in Ref. [6].

116.4 Simulation Result

We have run our program by the following conditions. We have tried two different initial conditions to start the simulation. The first type consists of the four $m = 1$ strategies, [00], [01], [10], and [11] with equal populations of 250 each, and the second type consists of 1000 random sequences of length 32. Either case, the total population of agents is kept unchanged from the initial value of 1000 throughout the simulation. The number of simulations are 50 for the first type and 40 for the second type. The rate of point mutation, the duplication rate, and the split rate are set to be 2×10^{-5}, 10^{-6}, 10^{-6}, the same as in Ref. [6]. We also assume the rate

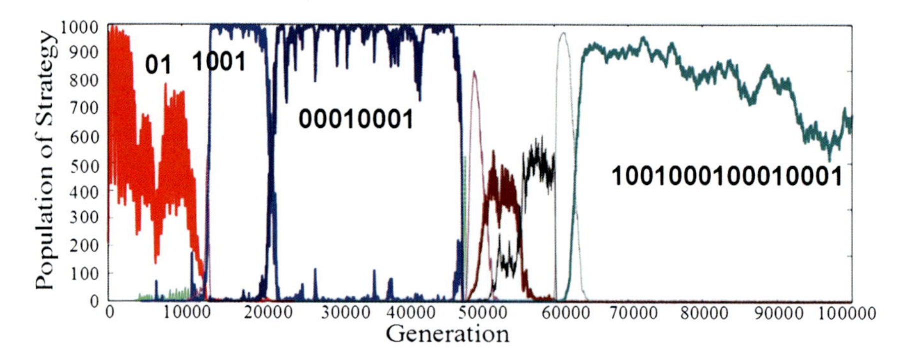

Fig. 116.3 The triplet of TFT-PAV-RTFT is observed in this example of simulation starting from the Type I initial population having four $m = 1$ strategies in the equal weight

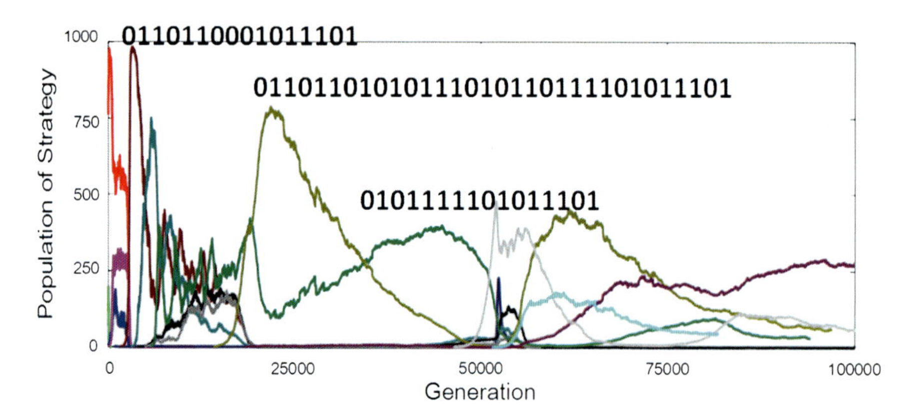

Fig. 116.4 A pattern of changing strategies starting from random strategies

of error, i.e., with which the opposite action prescribed by the gene is executed, to be 0.01. The payoff parameters in Table 116.1 are also chosen to be S = 0, P = 1, R = 3, and T = 5.

The length of each game is not fixed in order to avoid the convergence to the ALLD dominance, but the end of the game is announced with the probability of 0.005.

We show simulation results of the Type I and the Type II initial populations in Figs. 116.3 and 116.4, respectively, in which the horizontal axis shows the generation and the vertical axis shows the population of strategies. Both cases exhibit drastic changes of dominating strategies as the generation increases.

An interesting feature is observed in Fig. 116.3. Namely, the [01] (=TFT) dominance followed by the [1001] (=PAV) dominance, then the [0001] (RTFT) dominance comes and the [01] dominance. This particular pattern is observed in 37 examples out of 50 independent runs of the first type initial condition, and this triplet pattern of TFT => PAV => RTFT is sometimes repeated for many generations.

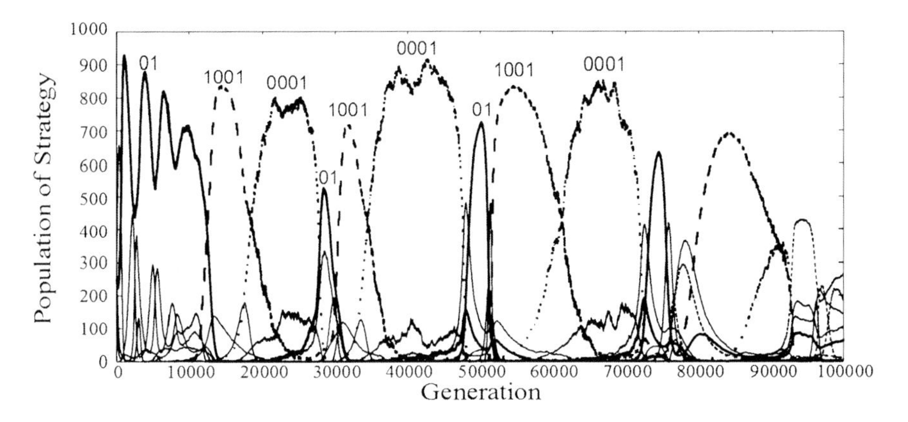

Fig. 116.5 An example of the triplet TFT-PAV-RTFT repeated for three cycles

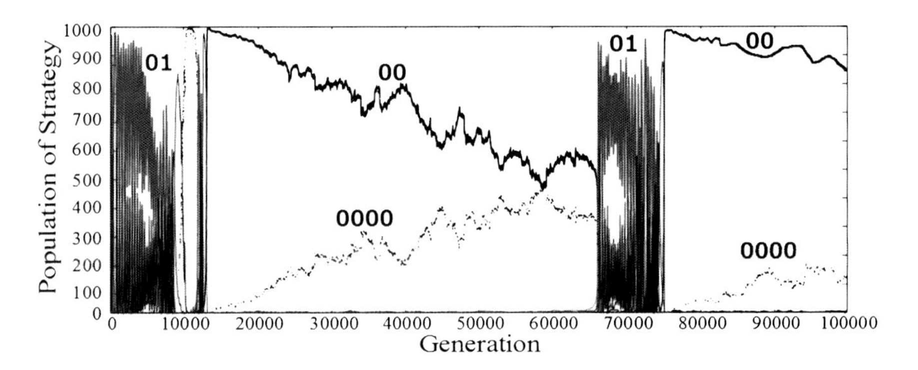

Fig. 116.6 An example of the triplet pattern collapsed by the ALLD dominance

However, as the length of the genes reaches the size of 16 or 32, this triplet pattern disappear and the [1001000100010001] dominates.

Figure 116.5 is an example of the triplet pattern of TFT-PAV-RTFT repeated for three cycles. Figure 116.6 is an example of the triplet pattern collapsed after one cycle, due to the strong dominance of ALLD strategy. Figure 116.7 shows a case of the triplet pattern washed away by the emergence of the longer and the stronger strategies.

116.5 Evaluation of the Strategies

We try to quantify the degree of sustainability of those strategies by means of a fitness parameter W_i defined by the accumulated sum of population throughout the total generation. The 8153 strategies emerged in the 45 simulations of Type I initial condition and the 11753 strategies emerged in the 50 simulations of Type II initial

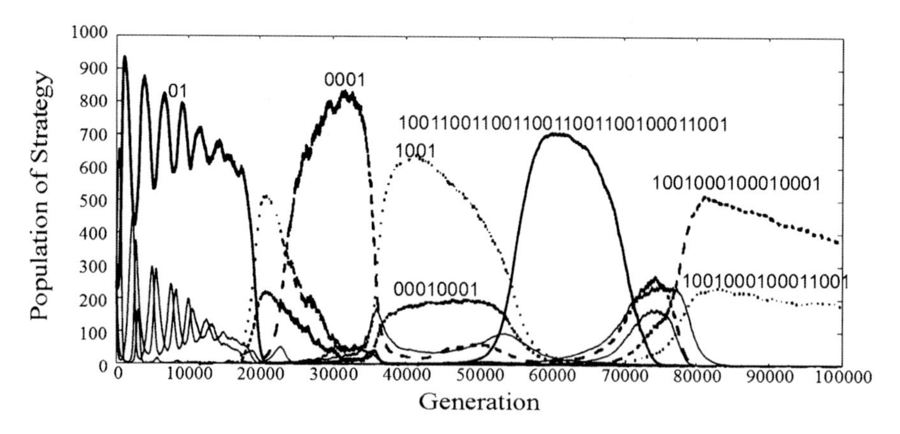

Fig. 116.7 An example of the triplet pattern collapsed by the emergence of the longer strategies

Table 116.3 Evaluation of the strategies

Type I initial strategies (fixed)		Type II initial strategies (random)	
Strategy	W_i	Strategy	W_i
1101 1001	0.123	1011	0.078
0101 1001	0.077	0000 0011	0.070
1101 0110	0.064	1101 1010	0.059
1010 0011	0.050	1001 1001	0.049
1101 0100	0.047	1101 1011 1101 1011	0.045
1001 0001 0001 0001	0.041	0001 0011	0.038
0001 1011	0.040	1101 0101 0001 1001	0.036
0100 1001	0.039	1101 1101 0000 0111	0.032
1101 0111	0.029	1000 0000 0100 0001	0.029
1001 1011 1001 1011	0.028	1111 0101 0101 1110	0.027

condition are sorted in the descending order of W_i in Table 116.3. The strategies having positive values of fitness are chosen as 'good' strategies and selected for further analysis. The number of 'good' strategies, satisfying the of the positive fitness condition, was 340 out of 8153 for the case of Type I initial condition, and 785 out of 11753 for the case of Type II initial condition.

We search for a possible characteristic feature of those strategies selected by using the goodness criterion. We first set the length of all those strategies to the equal length (= 32), by doubling and count the frequency of symbol '1' at each site, as illustrated in Fig. 116.8. The rates of '1' for all the 32 sites are shown in Fig. 116.9 for the Type I initial population and in Fig. 116.10 for the Type II initial population. This structure can be assumed to be the prototype strategy. The result shows that both Type I and Type II derived the same structure of [1001000100010001].

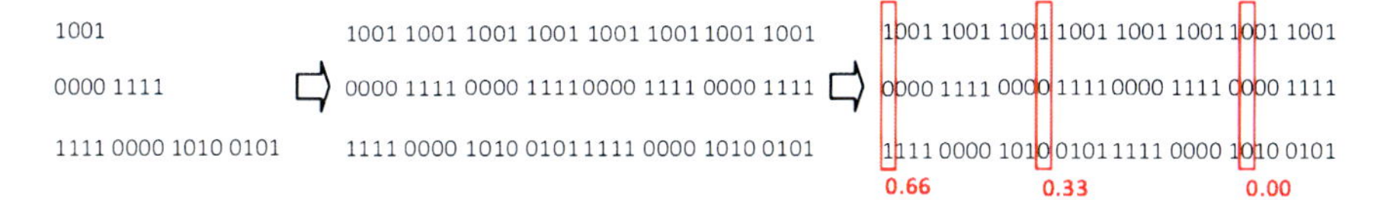

Fig. 116.8 Compute the rate of occurrence of '1' at each site

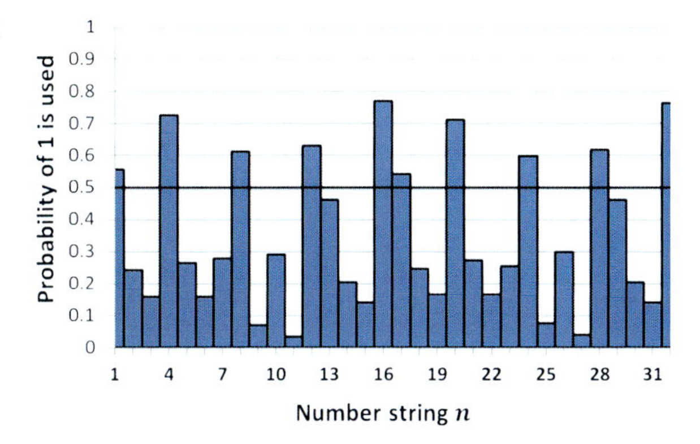

Fig. 116.9 The rates of '1' at each site of total 32 sites for Type I initial population

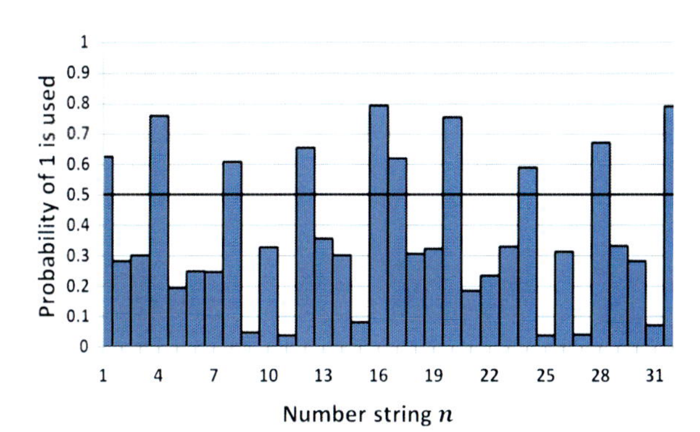

Fig. 116.10 The rates of '1' at each site of total 32 sites for Type II initial population

116.6 Discussion

Based on the result of our simulations, 'good' strategies who survive longer with larger population compared to the others have a common prototype gene structure of [1001000100010001]. Moreover, this result was irrelevant to the initial population. This gene structure is characterized by the following 4 features.

(1) cooperate if the opponent cooperates (This feature is seen in common to [***1 ***1 ***1 ***1], TFT, PAV, and [1**1 0*** 0*** *001], etc.)
(2) immediately retaliate if defected (This feature is seen in common to [0*0* 0*0* 0*0* 0*0* 0*0*], TFT, PAV, but not in [1**1 0*** 0*** *001].)
(3) generous (This feature is seen in common to [1*** **** **** ****], PAV, and [1**1 0*** 0*** *001], but not in TFT.)

Also, the structure [1*** **** **** ****] has an advantage over PAV for being more robust against ALL-D due to longer term of patience.

(4) coolness (This feature is in common to [*0** *0** *0** *0**] having 0 against the opponent's cooperative action. TFT, [1**1 0*** 0*** *001] do not have such a feature.)

116.7 Conclusion

We have performed extensive simulations of IPD and analyzed to determine the prototype structure of 'good' genes having a structure of [1001000100010001]. Although this is a specific example of the structure of strong gene, [1**1 0*** 0*** *001], suggested in Ref. [6], our analysis have reached much stronger specification of the gene structure of the strategy 'better' than TFT. This prototype consists of two types of quartets corresponding to PAV and RTFT. In other words, this strategy acts like the Pavlov when the actions of both players were 'Defect' at the game before the last, but acts like RTFT for the other three cases. This strategy has stronger tendency of retaliation against the opponent's defection compared to the Pavlov strategy. The advantage of this strategy compared to TFT is based on the structure of starting with '1', which helps to offer cooperation under defective situations, which is considered to be a key to solve the dilemma structure of many social problems.

References

1. Novak MA, Sigmund K (1998) Evolution of indirect reciprocityby image scoring. Nature 393:573–576
2. Roberts G, Sherratt TN (1998) Development of cooperative relationship through increasing investment. Nature 394:175–178
3. Yao X, Darwen P (1999) How important is your reqtutation in a multiagent environment. In: IEEE-SMC1999, pp 575–580
4. Axelrod R (1984) The evolution of cooperation
5. Tanaka-Yamawaki M, Murakami T (2009) Effect of reputation on the formation of cooperative network of prisoners. In: New advances in intelligent decision technologies, SCI199. Springer, Berlin, pp 615–623
6. Lindgren K (1990) Evolutionary phenomena in simple dynamics. In: Artificial life II. Addison-Wesley, Reading, pp 295–312
7. Nowak MA, Sigmund K (1993) A strategy of win-stay, lose-shift that outperforms tit-for-tat in Prisoner's dilemma. Nature 364:56–58

Chapter 117
Lyapunov Exponent: A Qualitative Ranking of Block Cipher Modes of Operation

Jeaneth Machicao, Anderson Marco, and Odemir Bruno

Abstract In Cryptography, a mode of operation is a technique to improve a block cipher effect. A challenge question of how to "qualitatively compare" among block cipher modes of operation, remains poorly studied. To overcome this lack we propose a methodology through some analogies to discrete dynamical systems (DDS). The method consists in the measure of chaos on the modes of operation by estimations of the Lyapunov exponent (LE). We opted to exploit the LE measures to compare among: ECB, CBC, OFB, CFB and CTR modes of operation. Results showed an effectively tool to qualitatively rank modes of operation. These novel findings represent an important advance to cryptography.

Keywords Cipher block · Modes of operation · Lyapunov exponent

117.1 Introduction

In cryptography many encryption algorithms have been actively developed to keep information as safe as possible [1]. A block cipher divides these information into blocks (plaintext) and transform them into an unreadable output (ciphertext) depending on a secret key [1] and usually a mode of operation is combined with an underlying block cipher to enhance its effect.

Unfortunately, we identify some issues that have paid little attention in literature: "How to choose the best block cipher modes of operation among many?", "How to compare among ciphers?" or even "can we extract measures from an encryption algorithm?". We notice that these are hard questions because of the nature of encryption algorithm should not follow any pattern.

Some researchers have studied techniques to classify block ciphers using techniques from retrieval information [2], artificial intelligence [3], support vector ma-

J. Machicao gratefully acknowledges the financial support of FAPESP (The State of São Paulo Research Foundation) (Proc. ♯ 2011-05461-0).

J. Machicao (✉) · A. Marco · O. Bruno
Instituto de Física de São Carlos (IFSC), Universidade de São Paulo, São Carlos, São Paulo, Brazil

T. Gilbert et al. (eds.), *Proceedings of the European Conference on Complex Systems 2012*, Springer Proceedings in Complexity, DOI 10.1007/978-3-319-00395-5_117,

chines (SVM) [4] and including histogram methods and prediction block [5]. But those cryptography issues remain poorly studied.

There is considerable evidence that appoint analogies between chaos properties and cryptography [6–8]. In dynamical systems (DS), the Lyapunov exponent (LE) is a measure to quantify chaos and to determine whether a DS is chaotic or not. In this manuscript, we report the results of a new method specifically designed to estimate the LE of the five modes of operation: ECB, CBC, OFB, CFB and CTR [9], which is capable of qualitatively rank them according to their respective LE "measures". We considered these modes of operation as DDS, e.g., cellular automata (CA). Therefore, we quantify the LE based on a discrete calculation, which was first applied to CA [10].

117.2 Background: NIST Modes of Operation

In 2001 the NIST [9] standardized five modes of operation: ECB, CBC, OFB, CFB and CTR. A *mode of operation* is a technique to enhance a block cipher effect, it defines a process to encrypt data (divided into blocks), and it usually takes additional inputs, *e.g.*, initialization vector(IV) or random functions [1]. Consider the *plaintext* P and *ciphertext* C as Boolean vectors of size N, grouped in blocks $P_1 P_2 \cdots P_b$ of n bits, where n is the block-length and $N = b \times n$. A n-block cipher is defined as $C = E_K(P)$, to encrypt a single block with key K. We study an isolated mode of operation M_{mode}, this means without the block cipher influence, and define $C = M(P)$, where M transforms the utter P into C without the block cipher influence. Table 117.1 presents both: classical modes of operation and its isolated adaptation.

117.3 Estimation of Lyapunov Exponent λ

We trace some analogies between the modes of operation and DDS, for instance we considered them as CA. Thus, we define our variables as a phase space, represented as a tuple $\langle C, s, P, M_{mode} \rangle$. The *ciphertext* C is a one-dimensional space composed of N bits c_i. The output function $s(c_i, t)$ maps the states of bit c_i at time t. The *plaintext* P assigns to every c_i an initial state, $s(\cdot, 0) = P$. The transition function M_{mode} describes a mode of operation, that transforms the state of each bit c_i under $s(c_i, t+1) = M_{mode}(s(c_i, t))$.

Then, we measure the rate of divergence λ of two random *plaintexts* P and P^*, initially separated by only one bit (the rightmost bit), while is encrypted by the isolated mode of operation. Note that, once the mode of operation is encrypting, it behaves as a DDS and is endowed with previous "state" in time, i.e., the *IV* or the rand function. For that reason, is important to track the previous state of the mode to encrypt at next iteration. To measure the LE we quote the damage spread theory [11]. Thus, we employ a damage vector $h(\cdot, t) = s(\cdot, t) \oplus s^*(\cdot, t)$, so that $h(\cdot, 0)$ contains only one defective bit.

Table 117.1 Comparison between the NIST modes of operation and its isolated adaptation

NIST modes of operation		Isolated modes of operation M_{mode}	
Electronic Codebook ECB	$C_i = E_K(P_i)$	M_{ECB}	$C_i = P_i$
Cipher Block Chaining CBC	$C_i = E_K(C_{i-1} \oplus P_i)$	M_{CBC}	$C_i = C_{i-1} \oplus P_i$
	$C_0 = IV$		$C_0 = IV$
Output Feedback OFB	$C_i = P_i \oplus O_i$	M_{OFB}	$C_i = P_i \oplus O_i$
	$O_i = E_K(O_{i-i})$		$O_i = O_{i-1}$
	$O_i = E_K(IV)$		$O_i = IV$
Cipher Feedback CFB	$C_i = P_i \oplus E_K(C_{i-1})$	M_{CFB}	$C_i = P_i \oplus C_{i-1}$
	$C_1 = P_1 \oplus E_K(IV)$		$C_1 = P_1 \oplus IV$
Counter CTR	$C_i = P_i \oplus O_i$	M_{CTR}	$C_i = P_i \oplus O_i$
	$P_i = C_i \oplus O_i$		$P_i = C_i \oplus O_i$
	$O_i = E_K(ctr_i)$		$O_i = ctr_i$
	$ctr_i = rand(ctr_{i-1})$		$ctr_i = rand(ctr_{i-1})$
	$ctr_i = rand(IV)$		$ctr_i = rand(IV)$

1. Consider P and P^*, initially $s(\cdot, 0)$ and $s^*(\cdot, 0)$ such that $\sum_{c_i} h(c_i, 0) = 1$.
2. Lets encrypt P using $M_{mode}(s(\cdot, 0))$. In similar way but separately, lets encrypt P^* with $M_{mode}(s^*(\cdot, 0))$. Calculate $s(\cdot, 1)$ and $s^*(\cdot, 1)$. The damage vector $h(\cdot, 1)$ takes one when $s^*(c_j, 1) \neq s(c_j, 1)$, otherwise takes zero.
3. For every c_j for which $h(c_j, 1) = 1$, create a replica r^j containing the Boolean complement of $s(c_j, 1)$, such that $r^j(c_j, 1) = \bar{s}(c_j, 1)$.
4. Lets encrypt $M_{mode}(s(c_i, 2))$. On the other hand, lets encrypt each of the replicas r^j with previous M_{mode}. The damage vector $h(c_j, 2) = \sum_{r^j} \sum_{c_i} h(c_i, 1)$, contains the sum of flipped bits of all replicas.
5. For every c_j and $r^j(\cdot, 2)$ for which $r(c_j, 2) \neq s(c_j, 2)$, create a replica r^j, such that $r^j(c_j, 2) = \bar{s}(c_j, 2)$.
6. Repeat steps (4–5) for every iteration $t + 1$.
7. Finally, the LE λ is defined as the total of the defective bits while iteration.

$$\lambda(t) = \frac{1}{t} \log \left(\frac{\sum_{r^j} \sum_{c_i} h(c_i, t)}{\sum_{c_i} h(c_i, 0)} \right) \tag{117.1}$$

117.4 Results and Discussion

For experimental purpose we employed *plaintexts* and *ciphertexts* of size $N = 640$ bits and divided into five blocks of $n = 128$ bits, which represent a typical block-length in cryptography. Both parameters, P and IV were randomly generated at each test.

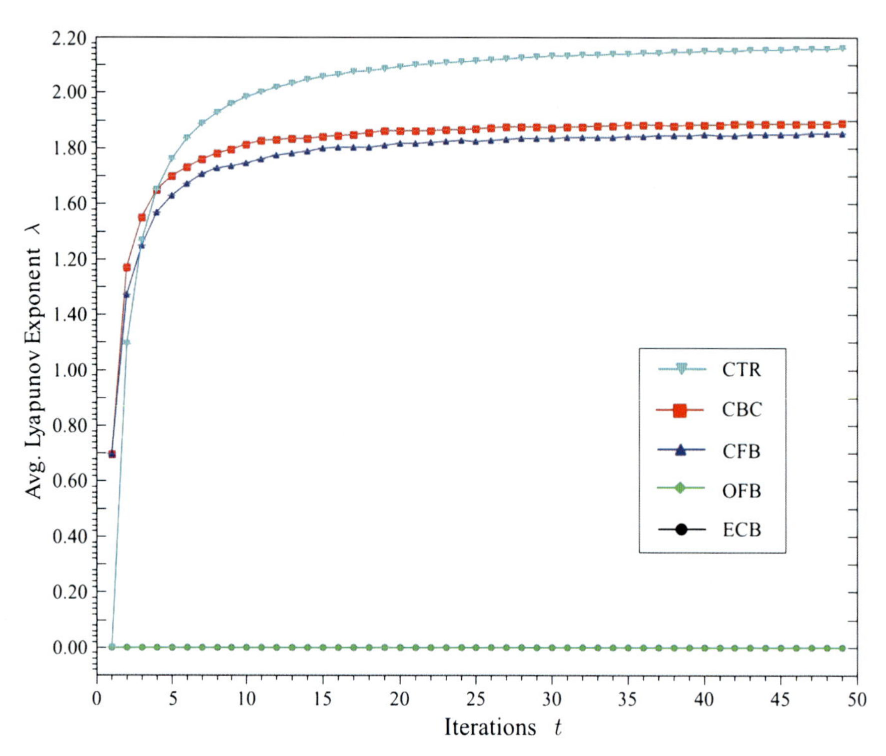

Fig. 117.1 The spectrum LE generated by the fives modes of operations. *The inset figure* shows the global view, while the main figure shows the zoom view of the curves

The calculation of LE has been performed on $t = 50$ iterations for each mode. Moreover, this test was repeated 200 times and the average LE with its related standard deviation were estimated in terms of the number of 200 repetitions.

Fig. 117.1 shows the spectrum LE generated by the fives modes of operations. The horizontal axis indicates the number of iterations t and vertical axis shows the average LE $\bar{\lambda}$. Noted that, all curves showed standard deviation $\sigma = 0$ (error bar), this statistically means that each modes is represented by an unique curve, where the number of repetitions do not influence in the calculation.

The straight lines at $\bar{\lambda} = 0$, corresponding to ECB and OFB modes, which theoretically means a neutral system, where the trajectories of initial P and P^* posses slow relative change. While the modes of operation: CTR, CBC and CFB, got positive LE $\lambda > 0$, which is defined to be chaotic, because its dynamics is directly affected by it is sensitive dependence on initial conditions.

117.4.1 Ranking Evaluation

We found that, based on the position placed of the curves from top to bottom, the rank list in order of quality is: CTR, CBC, CFB, OFB and ECB (in descending

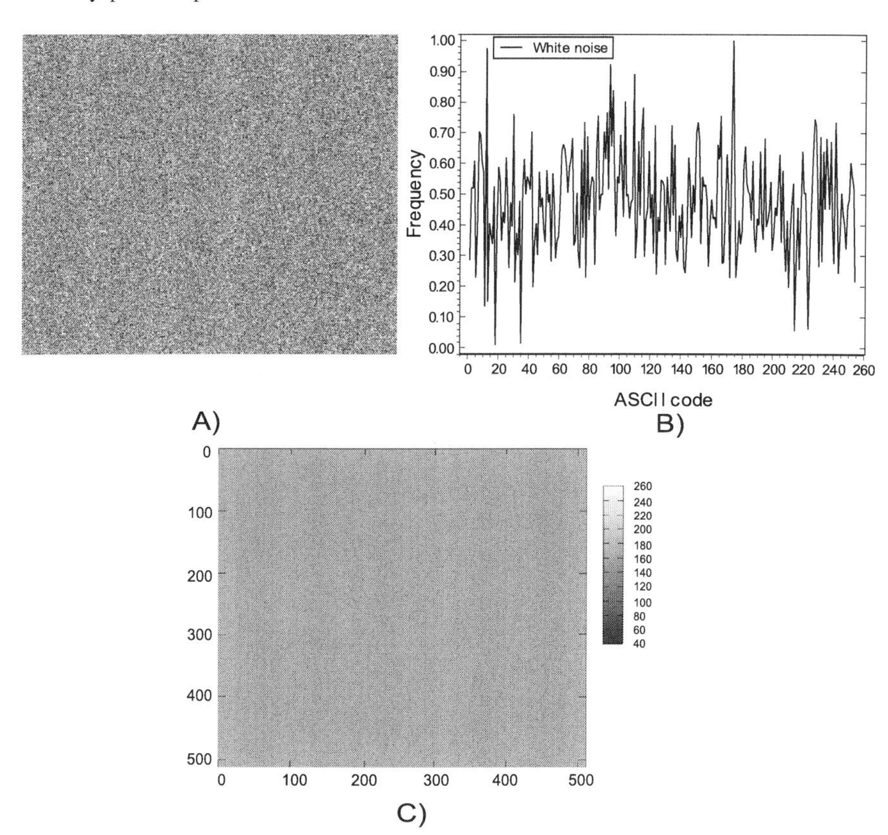

Fig. 117.2 An image analysis composed of (**A**) white-noise image, (**B**) related histogram and (**C**) Fourier power spectrum. First, the original image is a random signal, which, by definition, contains no information [13]. Second, its histogram presents the ASCII characters frequency which seems uniformly distributed. Third, there is a dot at the center of the power spectrum, which is the origin of the frequency coordinate system and, represent any frequency information from the original image

order) for the NIST block cipher modes of operation. Furthermore, this affirmation can be validated with a visual criteria test.

We used the 2D Fourier power spectrum analysis to validate our results [12]. The 2D FFT converts an image from the spatial domain into the frequency domain, in this manner; the 2D FFT contains all the information enclosed in the original image. We intuitive understand an FFT image by visually interpreting its coordinate system. For example, in Fig. 117.2 we see an image analysis of white-noise composed of the original image, histogram and 2D FFT, we interpreted that the white-noise contains no information [13].

Based on this allegation, we can analyze the modes of operation in a similar way as shown in the previous example. We processed an input image with each isolated mode of operation, calculated its frequency analysis and its 2D Fourier

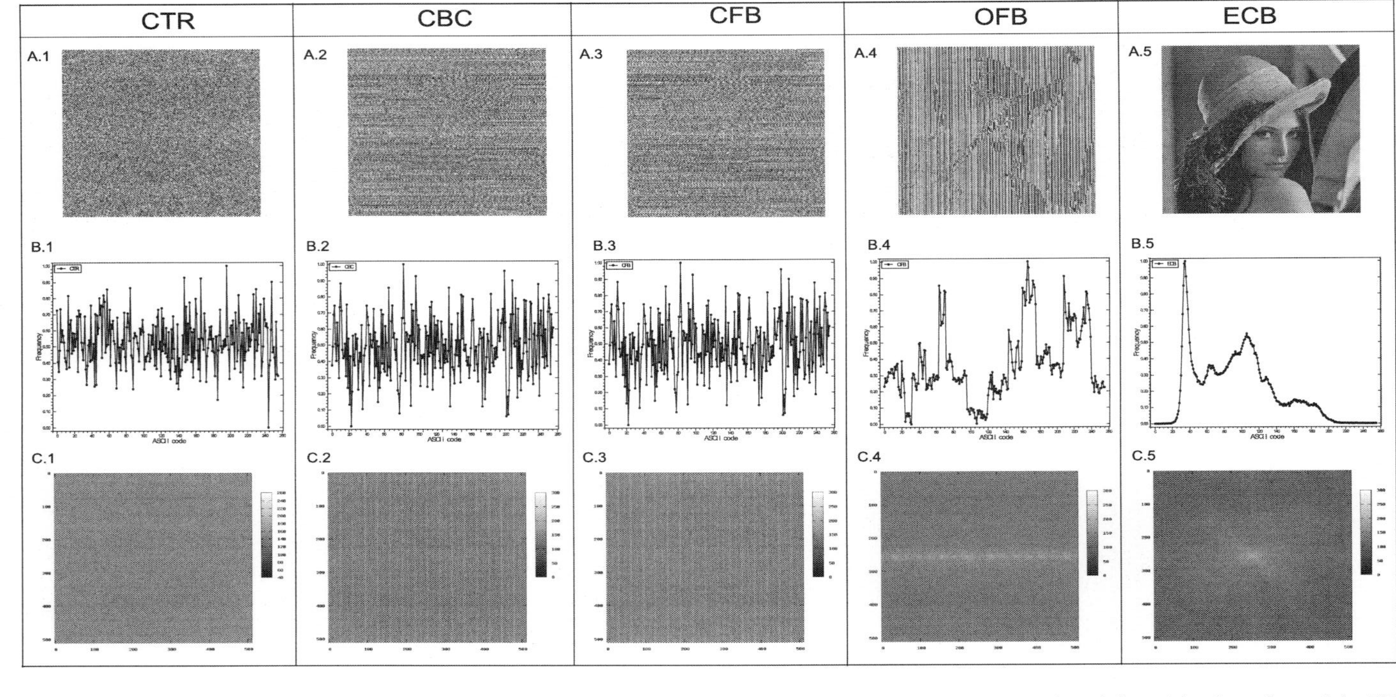

Fig. 117.3 Comparison of five modes of operation: (**A**) Input image, (**B**) histogram and (**C**) Fourier power spectrum, from left to right, the order rank is: CTR, CBC, CFB, OFB and ECB

power spectrum. We visually demonstrated in Fig. 117.3 the rank order analysis before mentioned, because it was expected a nearly flat spectrum comparable to a flat white-noise spectrum.

117.5 Conclusions

We have described the implementation details of the Lyapunov exponent approach. The application of LE to NIST modes of operation, effectively, can compare in a qualitative manner. We found that: CTR, ECB, CFB, OFB and ECB, in this particular order, as the quality ranking of the modes of operation. These results show an important advance to cryptography, because it has never been reported a qualitative comparison research in public literature. Furthermore, we corroborate the strong relationship between chaos and cryptography, as an active source of study.

References

1. Menezes AJ, Vanstone SA, Van Oorschot PC (1996) Handbook of applied cryptography, 1st edn. CRC Press, Boca Raton
2. de Carvalho CAB (2006) O uso de técnicas de recuperação de informações em criptoanálise. Instituto Militar de Engenharia
3. de Souza WAR (2007) Identificação de padrões em criptogramas usando técnicas de classificação de textos. Instituto Militar de Engenharia
4. Dileep AD, Sekhar CC (2006) Identification of block ciphers using support vector machines. In: IJCNN. IEEE Press, New York, pp 2696–2701
5. Nagireddy S (2008) A pattern recognition approach to block cipher identification. Indian Institute of Technology Madras
6. Amigó JM, Szczepanski J, Kocarev L (2005) Discrete chaos and cryptography. In: International symposium on nonlinear theory and its applications (NOLTA2005), pp 461–464
7. Álvarez G, Li S (2006) Some basic cryptographic requirements for chaos-based cryptosystems. Int J Biol Chem 16:2129–2151
8. Millerioux G, Amigó JM, Daafouz J (2008) A connection between chaotic and conventional cryptography. IEEE Trans Circuits Syst I, Regul Pap 55(6):1695–1703
9. Dworkin M (2001) Recommendation for block cipher modes of operation. National Institute of Standards and Technology
10. Baetens JM, De Baets B (2010) Phenomenological study of irregular cellular automata based on Lyapunov exponents and Jacobians. Chaos 20(3):033112
11. Vichniac GY (1990) Boolean derivatives on cellular automata. Physica D 45(1–3):63–74
12. Rukhin A, Soto J, Nechvatal et al. (2010) A statistical test suite for random and pseudorandom number generators for cryptographic applications. National Institute of Standards and Technology
13. Wu Z, Huang NE (2004) A study of the characteristics of white noise using the empirical mode decomposition method. Proc R Soc A, Math Phys Eng Sci 460(2046):1597–1611

Part XIV
Satellite Meeting: Self-organization, Management and Control

Chapter 118
Improving Individual Accessibility to the City

Arnaud Banos, Nicolas Marilleau, and MIRO Team

Abstract In this work we address the issue of sustainable cities by focusing on one of its very central components, daily mobility. Indeed, if cities can be interpreted as spatial organisations favouring social interactions, the number of daily movements needed to reach this goal is continuously increasing. Therefore, improving urban accessibility merely results in increasing the traffic and its negative externalities (congestion, accidents, pollution, noise...), while reducing at the end the accessibility of people to the city. We therefore propose to investigate this issue from the complex systems point of view. The real spatio-temporal urban accessibility of citizens can not be approximated just by focusing on space and implies to take into account the space-time activity patterns of individuals, in a more dynamic way. However, given the importance of local interactions in such a perspective, an agent based approach seems to be a relevant solution. This kind of individual based and "interactionist" approach allows exploring the possible impact of individual behaviours on the global behaviour of city but also the possible impact of global measures on individual behaviours.

Keywords Accessibility · Agent-based modelling · Cities · Urban modeling · Urban planning

118.1 Extended Abstract

Our approach is largely based on the conceptual framework of "time geography" defined by Torsten Hagerstrand (see [1]) in the 1970's. This underlying approach can be summarized with a few broad unifying principles:

A. Banos (✉)
Géographie-Cités, CNRS/Paris1, Paris, France
e-mail: arnaud.banos@parisgeo.cnrs.fr

N. Marilleau
UMI 209 UMMISCO, IRD/UPMC, Bondy, France
e-mail: nicolas.marilleau@ird.fr

MIRO Team
http://miro.csregistry.org/tiki-index.php

T. Gilbert et al. (eds.), *Proceedings of the European Conference on Complex Systems 2012*, Springer Proceedings in Complexity, DOI 10.1007/978-3-319-00395-5_118,

- movement should be treated as a demand deriving from demands for other activities independent of mobility itself,
- we should focus no longer on simple and segmented movements but upon sequences, chaining of trips. Therefore, we should examine closely the chaining and duration of activities and trips,
- spatial, temporal and social constraints should be incorporated, and similarly, the interactions between individuals and between individuals and their environment,
- it is crucial to recognise interdependencies between events separated in space and in time,
- finally, activities and movements should be analysed in a fundamentally dynamic manner.

In such perspective, the connection with agent-based modelling is quite natural and provides both an elegant and efficient way to explore various issues related to sustainable cities. One of the key issues relates closely to the following question: under which conditions will non coordinated but interdependent local actions eventually lead to the emergence of global spatial configurations of interest? That issue, we develop a model based on an agent based framework: NetLogo [2]. This toy model allows us exploring city configurations (see Fig. 118.1) to evaluate for example service accessibility or using.

Let's imagine a very simplified city composed of a regular network. Each node of this network may remains unoccupied or host, with a given probability, one of the following "reservoirs": a residence, a workplace, a public service, a commercial service. A population of virtual citizens is then generated. For each one of them, a residence and workplace are randomly, and the "citizen" is initially located on the node corresponding to its residence (a residence may host several "citizens"). Each "citizen" then chooses a public service and a commercial service, such that the total trip chain minimises the travel time of the "citizen". Each of these agents will have to perform the same trip chain, driving: Residence–Workplace–Public Service–Commercial Service–Residence.

$$V_i = V_{fi} e^{-\alpha(\frac{n_i}{K_i})} \tag{118.1}$$

Each edge of the network is valued by a maximum speed and the speed of each agent is defined during the simulation with a deterministic Single-Regime speed-Density function specified by Eq. (118.1) where [4]:

- V_i the speed of vehicles driving on edge i,
- V_{fi} the free-flow speed (maximum speed) of edge i,
- K_i the maximum traffic capacity of edge i,
- n_i the number of vehicles driving of edge i and
- α a congestion impact factor

At the end of the simulation (i.e. when every "citizen" is back home), public and commercial services (initially located at random on the regular network) may change node, especially if two few "citizens" included them in their trip chain. Two distinct strategies are then implemented:

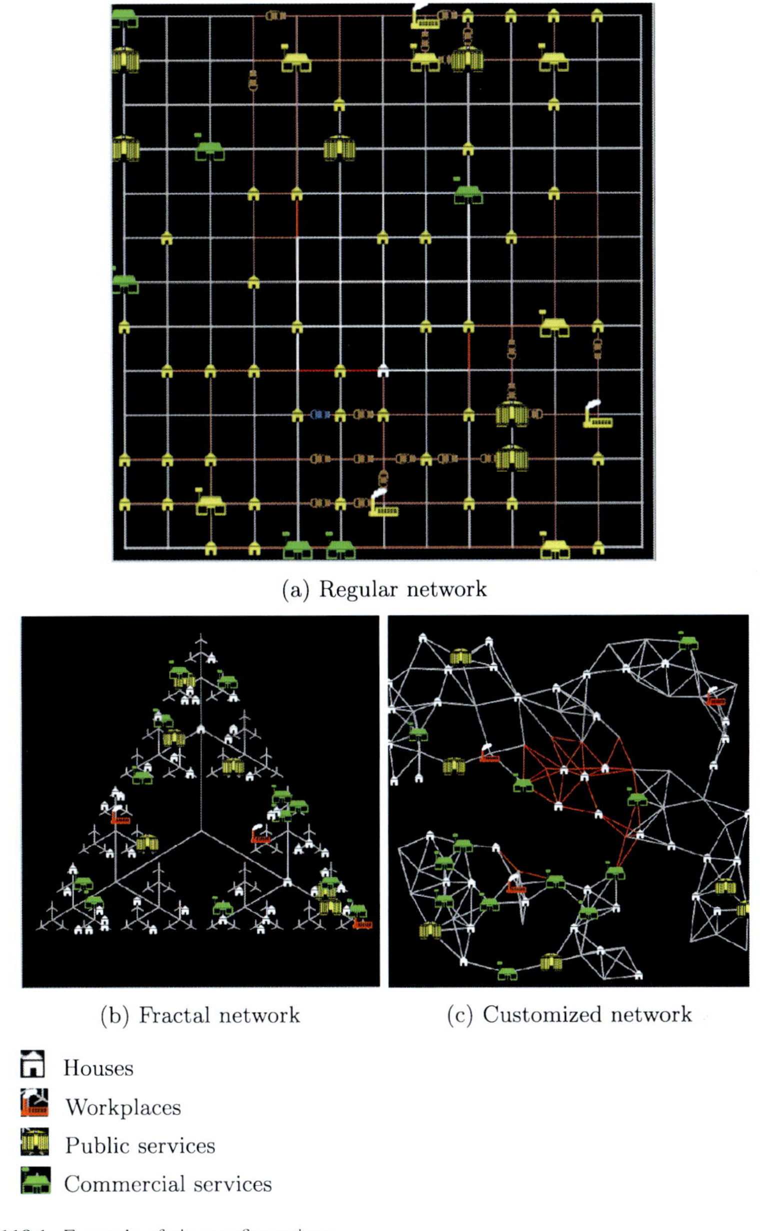

Fig. 118.1 Example of city configurations

– non satisfied public services will "listen" to non satisfied "citizens" (i.e. those "citizens" for whom the travelling time corresponding to the public service part of their trip chain exceeds a given threshold) and will get closer to it, under the constraint of a minimum population coverage threshold,

– non satisfied commercial services will try to find a free node characterised by a high traffic, in order to increase their chance to be included in "citizens" trip chains.

Given that "citizens" residences and workplaces are fixed and that no capacity constraints (road + services) are taken into account, it seems quite obvious that public and commercial services will tend to concentrate, under the constraint of residences and workplaces locations. However, introducing constraints (road + services) will amplify the feedback between "citizens" and services and will inevitably increase the complexity of both processes and patterns. Our presentation will focus on this specific point

This Netlogo toys model is designed in an experimental way while we are creating and developing a large scale model based on GIS and enquiries done at two middle sized towns (Dijon and Grenoble, France). This second model based on GAMA [3] applies to real case studies hypothesis that were established thanks to the Netlogo toy model.

Acknowledgements The research for this work was supported by the French ANR "Sustainable Cities".

References

1. Hagerstrand T (1970) What about people in regional science. Pap Reg Sci 24:7–21
2. Sklar E (2011) NetLogo, a multi-agent simulation environment. Artif Life 13(3):303–311
3. Taillandier P, Drogoul A, Vo DA, Amouroux E (2012) GAMA: a simulation platform that integrates geographical information data, agent-based modeling and multi-scale control. In: The 13th international conference on principles and practices in multi-agent systems (PRIMA'2011), India, vol 7057, pp 242–258
4. Underwood R (1961) Speed, volume, and density relationship quality and theory of traffic flow. Yale Bureau of Highway Traffic, New Haven

Chapter 119
Passification Based Controlled Synchronization of Complex Networks

Alexander Fradkov, Ibragim Junussov, and Anton Selivanov

Abstract In the paper an output synchronization problem for a networks of linear dynamical agents is examined based on passification method and recent results in graph theory. The static output feedback and adaptive control are proposed and sufficient conditions for synchronization are established ensuring synchronization of agents under incomplete measurements and incomplete control. The results are extended to the networks with sector bounded nonlinearities in the agent dynamics and information delays.

Keywords Complex networks · Passification · Synchronization

119.1 Introduction

Controlled synchronization of networks has a broad area of important applications: control of power networks, cooperative control of mobile robots, control of lattices, control of biochemical, ecological networks, etc. [1–5]. However most existing papers deal with control of the networks of dynamical systems (agents) with full state measurements and full control (vectors of agent input, output and state have equal dimensions). In the case of synchronization by output feedback additional dynamical systems (observers) are incorporated into network controllers.

The work was supported by Russian Foundation for Basic Research (project 11-08-01218) and Russian Federal Program "Cadres" (contract 16.740.11.0042).

A. Fradkov (✉)
Institute for Problems of Mechanical Engineering, St. Petersburg, Russia
e-mail: fradkov@mail.ru

A. Fradkov · I. Junussov · A. Selivanov
St. Petersburg State University, St. Petersburg, Russia

I. Junussov
e-mail: dxdtfxut@gmail.com

A. Selivanov
e-mail: antonselivanov@gmail.com

T. Gilbert et al. (eds.), *Proceedings of the European Conference on Complex Systems 2012*, Springer Proceedings in Complexity, DOI 10.1007/978-3-319-00395-5_119,

In this paper the synchronization problem for networks of linear agents with arbitrary numbers of inputs, outputs and states by static output neighbor-based feedback is solved based on passification method [6, 7] and recent results in graphs theory. The results are extended to the networks with sector bounded nonlinearities in the agent dynamics and information delays.

119.2 Problem Statement

Let the network S consist of d agents $S_i, i = 1, \ldots, d$. Each agent S_i is modeled as a controlled system

$$\dot{x}_i = Ax_i + B_0 f(x_i) + Bu_i, \qquad y_i = C^{\mathrm{T}} x_i, \tag{119.1}$$

where $x_i \in \mathbb{R}^n$ is a state vector, $u_i \in \mathbb{R}^1$ is a controlling input (control), $y_i \in \mathbb{R}^l$ is a vector of measurements (output). Let $\mathcal{G} = (\mathcal{V}, \mathcal{E})$ be the digraph with the set of vertices $\mathcal{V}$ and the set of arcs $\mathcal{E} \subseteq \mathcal{V} \times \mathcal{V}$ such that for $i = 1, \ldots, d$ the vertex v_i is associated with the agent S_i.

Let the control goal be:

$$\lim_{t\to\infty} \big(x_i(t) - x_j(t)\big) = 0, \quad i, j = 1, \ldots, d. \tag{119.2}$$

119.3 Static Control

Let control law for S_i be

$$u_i(t) = K \sum_{j\in\mathcal{N}_i} \big(y_i(t-\tau) - y_j(t-\tau)\big) = KC^{\mathrm{T}} \sum_{j\in\mathcal{N}_i} \big(x_i(t-\tau) - x_j(t-\tau)\big), \tag{119.3}$$

where $K \in \mathbb{R}^{1\times l}$, $\mathcal{N}_i = \{k = 1, \ldots, d | (v_i, v_k) \in \mathcal{E}\}$ is the set of neighbor vertices to v_i, $\tau \geq 0$ is communication delay.

The problem is to find K from (119.3) such that the goal (119.2) holds.

The problem is first analyzed for linear agent dynamics ($B_0 = 0$) without delays ($\tau = 0$) under the following assumptions:

(A1) *There exists a vector* $g \in \mathbb{R}^l$ *such that the function* $g^{\mathrm{T}} W(s)$ *is hyper-minimum-phase, where* $W(s) = C^{\mathrm{T}}(sI - A)^{-1}B$. (Recall that the rational function $\chi(s) = \beta(s)/\alpha(s)$ is called *hyper minimum phase*, if its numerator $\beta(s)$ is a Hurwitz polynomial and its highest coefficient β_{n-1} is positive [7].)

(A2) *The interconnection graph is undirected and connected.*

(A2D) *The interconnection graph is directed and has the directed spanning tree.*

Let $A(\mathcal{G})$ denote adjacency matrix of the graph $\mathcal{G}$. For digraph $\mathcal{G}$ consider the graph $\widehat{\mathcal{G}}$ such that $A(\widehat{\mathcal{G}}) = A(\mathcal{G}) + A(\mathcal{G})^{\mathrm{T}}$. Laplacian $L(\widehat{\mathcal{G}}) = D(\widehat{\mathcal{G}}) - A(\widehat{\mathcal{G}})$ of the graph $\widehat{\mathcal{G}}$ is symmetric and has the eigenvalues: $0 = \lambda_1 < \lambda_2 \leq \cdots \leq \lambda_d$, [1, 3]. The main result is as follows.

Theorem 119.1 *Let assumptions* A1 *and either* A2 *or* A2D *hold and* $k \geq 2\kappa/\lambda_2$, *where*

$$\kappa = \sup_{\omega \in \mathbb{R}^1} \mathrm{Re}\big(g^{\mathrm{T}} W(i\omega)\big)^{-1}. \tag{119.4}$$

Then the control law (119.3) *with feedback gain* $K = -k \cdot g^{\mathrm{T}}, k \in \mathbb{R}^1$ *ensures the goal* (119.2).

Similar results are obtained for undirected and balanced directed communication graphs.

119.4 Adaptive Control

Let agent S_i be able to adjust its control gain, i. e. each local controller is adaptive. Let each controller have the following form:

$$u_i(t) = \theta_i(t) y_i(t), \tag{119.5}$$

where $\theta_i(t) \in \mathbb{R}^{1 \times l}$—tunable parameter which is tuned based on the measurements from the neighbors of i-th agent.

Denote:

$$\overline{y}_i = \sum_{j \in \mathcal{N}_i} (y_i - y_j), \quad i = 1, \ldots, N$$

and consider the following adaptation algorithm:

$$\theta_i(t) = -g^{\mathrm{T}} \cdot k_i(t), \qquad \dot{k}_i(t) = y_i(t)^{\mathrm{T}} g g^{\mathrm{T}} \overline{y}_i. \tag{119.6}$$

Adaptive synchronization conditions are formulated as follows.

Theorem 119.2 *Let assumptions* A1, A2 *hold. Then adaptive controller* (119.5)–(119.6) *ensures achievement of the goal* (119.2).

The above results are extended to the networks with sector bounded nonlinearities in the agent dynamics and information delays.

119.5 Conclusions

The control algorithm for synchronization of networks based on static output feedback (119.3) to each agent from the neighbor agents is proposed. Since the number of inputs and outputs of the agents are less than the number of agent state variables, synchronization of agents is achieved under incomplete measurements and incomplete control. Synchronization conditions include passifiability (hyper-minimum-phase property) for each agent and some connectivity conditions for interconnection graph: existence of the directed spanning tree in case of directed graph and connectivity in case of undirected graph. Similar conditions are obtained for adaptive passification-based control of network with undirected interconnection graph, for sector bounded nonlinearities in the agent dynamics and information delays.

The proposed solution for output feedback synchronization unlike those of [4, 5] does not use observers. Compared to static output feedback result of [4, Theorem 4] the proposed synchronization conditions relax passivity condition for agents to their passifiability that allows for unstable agents. The paper [4], however, deals with time-varying network topology. The presented results extend our previous results [8–10].

Simulation results for the networks of double integrators and Chua circuits illustrate the theoretical results.

References

1. Olfati-Saber R, Murray RM (2004) Consensus problems in networks of agents with switching topology and time-delays. IEEE Trans Autom Control 49(9):1520–1533
2. Boccaletti S, Latora V, Moreno Y, Chavez M, Hwang DU (2006) Complex networks: structure and dynamics. Phys Rep 424(4–5):175–308
3. Bullo F, Cortez J, Martinez S (2009) Distributed control of robotic networks. Princeton University Press, Princeton
4. Scardovi L, Sepulchre R (2009) Synchronization in networks of identical linear systems. Automatica 45(11):2557–2562
5. Li Z, Duan Z, Chen G, Huang L (2010) Consensus of multiagent systems and synchronization of complex networks: a unified viewpoint. IEEE Trans Circuits Syst I 57(1):213–224
6. Fradkov AL, Miroshnik IV, Nikiforov VO (1999) Nonlinear and adaptive control of complex systems. Kluwer, Dordrecht
7. Fradkov AL (2003) Passification of nonsquare linear systems and feedback Yakubovich-Kalman-Popov lemma. Eur J Control 6:573–582
8. Fradkov AL, Junussov IA (2011) Output feedback synchronization for networks of linear agents. In: 7th European nonlinear dynamics conference (ENOC 2011), Rome
9. Dzhunusov IA, Fradkov AL (2011) Synchronization in networks of linear agents with output feedbacks. Autom Remote Control 72(8):1615–1626
10. Fradkov AL, Grigoriev GK, Selivanov AA (2011) Decentralized adaptive controller for synchronization of dynamical networks with delays and bounded disturbances. In: Proc 50th IEEE conf dec contr, Orlando, pp 1110–1115

Part XV
Satellite Meeting: Complex Multiphase Systems

Chapter 120
Inertia and Hydrodynamic Interactions in Dynamical Density Functional Theory

Benjamin D. Goddard, Andreas Nold, Nikos Savva, Grigorios A. Pavliotis, and Serafim Kalliadasis

Abstract We study the dynamics of a colloidal fluid in the full position-momentum phase space, including hydrodynamic interactions, which strongly influence the non-equilibrium properties of the system. For large systems, the number of degrees of freedom prohibits direct simulation and a reduced model is necessary. Under standard assumptions, we derive a dynamical density functional theory (DDFT), which is a generalisation of many existing DDFTs, and shows good agreement with stochastic simulations.

Keywords Dynamical density functional theory · Colloids · Hydrodynamic interactions · Inertia

120.1 Dynamics of Colloidal Fluids

Consider a large number N of identical, spherically symmetric colloidal particles of mass m suspended in a bath of many more much smaller and much lighter particles. This separation of scales allows the effects of the bath to be treated in a probabilistic way, leading to stochastic Newton's equations for the particles. For the $3N$-dimensional vectors $\mathbf{r}$ and $\mathbf{p}$, containing all particle positions and momenta

B.D. Goddard (✉) · S. Kalliadasis
Department of Chemical Engineering, Imperial College London, London SW7 2AZ, UK
e-mail: b.goddard@imperial.ac.uk

A. Nold
Center of Smart Interfaces, TU Darmstadt, Petersenstr. 32, 64287 Darmstadt, Germany

N. Savva
School of Mathematics, Cardiff University, Cardiff CF24 4AG, UK

G.A. Pavliotis
Department of Mathematics, Imperial College London, London SW7 2AZ, UK

G.A. Pavliotis
Institut für Mathematik, Freie Universität Berlin, Arnimallee 6, 14195 Berlin, Germany

T. Gilbert et al. (eds.), *Proceedings of the European Conference on Complex Systems 2012*, Springer Proceedings in Complexity, DOI 10.1007/978-3-319-00395-5_120,

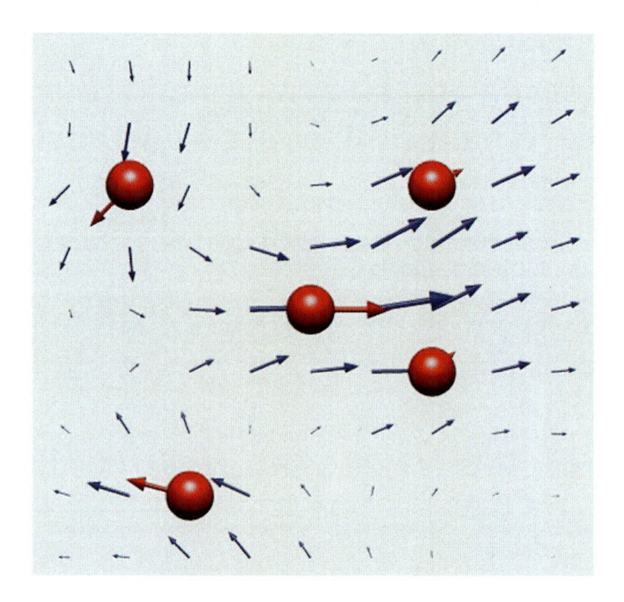

Fig. 120.1 Bath fluid flows (*blue*) caused by colloidal motion (*red*)

respectively, and potential V:

$$\frac{\mathrm{d}\mathbf{r}}{\mathrm{d}t} = \frac{\mathbf{p}}{m}, \qquad \frac{\mathrm{d}\mathbf{p}}{\mathrm{d}t} = -\nabla_{\mathbf{r}} V(\mathbf{r}, t) - \boldsymbol{\Gamma}(\mathbf{r})\mathbf{p} + \mathbf{A}(\mathbf{r})\mathbf{f}(t). \tag{120.1}$$

The final two terms in (120.1) model the effects of the bath. The motion of colloidal particles cause flows in the bath, which in turn cause forces on the colloidal particles, called hydrodynamic interactions (HI), see Fig. 120.1. The momenta and forces are related by the $3N \times 3N$ positive-definite friction tensor $\boldsymbol{\Gamma}$. Collisions of bath particles with colloidal particles are described by the stochastic forces $\mathbf{f}$. The strength of these collisions is related to $\boldsymbol{\Gamma}$ via a generalized fluctuation-dissipation theorem, giving $\mathbf{A} = (mk_B T\boldsymbol{\Gamma})^{1/2}$, where T is the temperature and k_B is Boltzmann's constant. If an explicit form for $\mathbf{A}$ is known (e.g. if HI are neglected and $\boldsymbol{\Gamma} = \gamma\mathbf{1}$, with γ the friction felt by a single isolated particle), then the numerical solution of (120.1) takes order N operations at each timestep. If $\mathbf{A}$ must be determined from $\boldsymbol{\Gamma}$, this is increased to order N^3. Hence HI are often neglected for numerical reasons, whilst physically they are important in all but very dilute systems as they decay only with the inverse of particle separation.

When N is large, we are interested not in particular realizations of $\mathbf{r}$ and $\mathbf{p}$, given by (120.1), but in their joint probability distribution $f^{(N)}(\mathbf{r}, \mathbf{p}, t)$. Averaging (120.1) over the noise leads to the Kramers equation, a high-dimensional PDE:

$$\begin{aligned} &\partial_t f^{(N)} + \frac{1}{m}\mathbf{p} \cdot \nabla_{\mathbf{r}} f^{(N)} - \nabla_{\mathbf{r}} V(\mathbf{r}, t) \cdot \nabla_{\mathbf{p}} f^{(N)} \\ &\quad = \nabla_{\mathbf{p}} \cdot \left[\boldsymbol{\Gamma}(\mathbf{r})(\mathbf{p} + mk_B T\nabla_{\mathbf{p}}) f^{(N)}\right]. \end{aligned} \tag{120.2}$$

It is known rigorously that $f^{(N)}$ is a functional of the one-body position distribution $\rho(\mathbf{r}_1) = N\int \mathrm{d}\mathbf{p}\mathrm{d}\mathbf{r}' f^{(N)}(\mathbf{r}, \mathbf{p}, t)$, where $\mathrm{d}\mathbf{r}'$ denotes integration over all but $\mathbf{r}_1$, the position of the first particle. This motivates the derivation of a *dynamical density functional theory (DDFT)*, i.e. an equation of the form $\partial_t \rho(\mathbf{r}_1, t) = -\nabla_{\mathbf{r}_1} \cdot \mathbf{a}(\mathbf{r}_1, t, [\rho])$, where $\mathbf{a}$ is a functional of ρ. Existing DDFTs have been formulated in the overdamped (high friction) regime [1, 2], where inertial effects may be neglected, or ignore HI [3]. We present a general DDFT, of which existing formulations [1–3] are special cases.

120.2 Derivation of a Dynamical Density Functional Theory

We denote the position and momentum of the jth particle by $\mathbf{r}_j$ and $\mathbf{p}_j$ and let $\boldsymbol{\Gamma}(\mathbf{r}) = \gamma[\mathbf{1} + \tilde{\boldsymbol{\Gamma}}(\mathbf{r})]$, where the HI tensor $\tilde{\boldsymbol{\Gamma}}$ is decomposed into 3×3 blocks $\tilde{\boldsymbol{\Gamma}}_{ij}$ [4]. By taking momentum moments of (120.2), we obtain an infinite hierarchy of equations. Truncating at the second equation gives a continuity equation (120.3), and the time evolution of the current [5]:

$$0 = \partial_t \rho(\mathbf{r}_1, t) + \nabla_{\mathbf{r}_1} \cdot \big(\rho(\mathbf{r}_1, t)\mathbf{v}(\mathbf{r}_1, t)\big), \tag{120.3}$$

$$\begin{aligned}
0 &= \partial_t \big(\rho(\mathbf{r}_1, t)\mathbf{v}(\mathbf{r}_1, t)\big) + \gamma \rho(\mathbf{r}_1, t)\mathbf{v}(\mathbf{r}_1, t) \\
&\quad + \frac{1}{m} \int \mathrm{d}\mathbf{r}' \nabla_{\mathbf{r}_1} V(\mathbf{r}, t) \rho^{(N)}(\mathbf{r}, t) && \text{(V)} \\
&\quad + \frac{N\gamma}{m} \sum_{j=1}^{N} \int \mathrm{d}\mathbf{p}\mathrm{d}\mathbf{r}' \tilde{\boldsymbol{\Gamma}}_{1j}(\mathbf{r}) \mathbf{p}_j f^{(N)}(\mathbf{r}, \mathbf{p}, t) && \text{(H)} \\
&\quad + \nabla_{\mathbf{r}_1} \cdot \int \mathrm{d}\mathbf{p}_1 \frac{\mathbf{p}_1 \otimes \mathbf{p}_1}{m^2} f^{(1)}(\mathbf{r}_1, \mathbf{p}_1, t). && \text{(K)}
\end{aligned}$$

To close the second equation, it is necessary to deal with terms arising from many-body potentials (V), HI (H), and 'kinetic pressure' (K) effects.

For (V), at equilibrium, there exists an exact functional identity

$$\int \mathrm{d}\mathbf{r}' \nabla_{\mathbf{r}_1} V(\mathbf{r}) \rho^{(N)}(\mathbf{r}) = \rho(\mathbf{r}_1, t) \nabla_{\mathbf{r}_1} \frac{\delta \mathcal{F}[\rho]}{\delta \rho} - k_B T \nabla_{\mathbf{r}_1} \rho(\mathbf{r}_1),$$

where $\mathcal{F}[\rho] = k_B T \int \mathrm{d}\mathbf{r}_1 \rho(\mathbf{r}_1)[\ln(\Lambda^3 \rho(\mathbf{r}_1)) - 1] + \mathcal{F}_{\mathrm{exc}}[\rho] + \int \mathrm{d}\mathbf{r}_1 \rho(\mathbf{r}_1) V_1(\mathbf{r}_1)$ is the Helmholtz free energy functional. In general, $\mathcal{F}_{\mathrm{exc}}$ is unknown but has been well-studied at equilibrium and good approximations exist, e.g. [6, 7]. We thus assume that this identity also holds out of equilibrium, in particular obtaining the correct equilibrium behaviour.

Since HI vanish at equilibrium there exists no analogous identity for (H). For ease of exposition, we assume that the HI are two-body:

$$\tilde{\boldsymbol{\Gamma}}_{ij}(\mathbf{r}) = \delta_{ij} \sum_{\ell \neq i} \mathbf{Z}_1(\mathbf{r}_i, \mathbf{r}_\ell) + (1 - \delta_{ij}) \mathbf{Z}_2(\mathbf{r}_i, \mathbf{r}_j).$$

We also assume that

$$f^{(2)}(\mathbf{r}_1,\mathbf{r}_2,\mathbf{p}_1,\mathbf{p}_2,t) \approx f^{(1)}(\mathbf{r}_1,\mathbf{p}_1,t) f^{(1)}(\mathbf{r}_1,\mathbf{p}_1,t) g\big(\mathbf{r}_1,\mathbf{r}_2,[\rho]\big)$$

for a *known* functional g. Again, this has been well-studied at equilibrium [8, 9].

As (K) is analogous to the kinetic pressure tensor, there is no reason to expect it to be a simple functional of ρ and $\mathbf{v}$. However, for general $f^{(1)}$, we have $f^{(1)}(\mathbf{r}_1,\mathbf{p}_1,t) = f^{(1)}_{\text{le}}(\mathbf{r}_1,\mathbf{p}_1,t) + f^{(1)}_{\text{neq}}(\mathbf{r}_1,\mathbf{p}_1,t)$, where $f^{(1)}_{\text{le}}(\mathbf{r}_1,\mathbf{p}_1,t) = (2\pi m k_B T)^{-3/2}\rho(\mathbf{r}_1,t)\exp(-|\mathbf{p}_1 - m\mathbf{v}(\mathbf{r}_1,t)|^2/(2mk_BT))$ is the local-equilibrium part. The non-equilibrium part $f^{(1)}_{\text{neq}}$ satisfies $\int d\mathbf{p}_1 \alpha(\mathbf{p}_1) f^{(1)}_{\text{neq}}(\mathbf{r}_1,\mathbf{p}_1,t) = 0$ for $\alpha(\mathbf{p}_1) \in \{1, \mathbf{p}_1, |\mathbf{p}_1|^2\}$. Combining these three approximations gives

$$\begin{aligned}
&\partial_t \mathbf{v}(\mathbf{r}_1,t) + \big(\mathbf{v}(\mathbf{r}_1,t)\cdot\nabla_{\mathbf{r}_1}\big)\mathbf{v}(\mathbf{r}_1,t) + \frac{1}{m}\nabla_{\mathbf{r}_1}\frac{\delta\mathcal{F}[\rho]}{\delta\rho} \\
&\quad + \frac{1}{\rho(\mathbf{r}_1,t)}\nabla_{\mathbf{r}_1}\cdot\int d\mathbf{p}_1 \frac{\mathbf{p}_1\otimes\mathbf{p}_1}{m^2} f^{(1)}_{\text{neq}}(\mathbf{r}_1,\mathbf{p}_1,t) + \gamma\mathbf{v}(\mathbf{r}_1,t) \\
&\quad + \gamma\int d\mathbf{r}_2 \rho(\mathbf{r}_2,t) g\big(\mathbf{r}_1,\mathbf{r}_2,[\rho]\big)\big[\mathbf{Z}_1(\mathbf{r}_1,\mathbf{r}_2)\mathbf{v}(\mathbf{r}_1,t) + \mathbf{Z}_2(\mathbf{r}_1,\mathbf{r}_2)\mathbf{v}(\mathbf{r}_2,t)\big] = 0
\end{aligned} \tag{120.4}$$

which, along with the continuity equation (120.3), and under the additional assumption that the term containing $f^{(1)}_{\text{neq}}$ may be neglected or approximated as a functional of ρ and $\mathbf{v}$, gives a DDFT. Neglecting this term can be rigorously motivated both close to local equilibrium, giving a generalised Navier-Stokes equation with non-local terms, and in the overdamped limit [4], recovering a Smoluchowski equation with a novel diffusion tensor. The terms involving $\mathbf{Z}_j$ in (120.4) are absent in previous DDFTs. The term involving $\mathbf{Z}_1$ combines with $\gamma\mathbf{v}$ to give an effective, density dependent friction coefficient, whilst that involving $\mathbf{Z}_2$ couples the velocities in a non-local way.

120.3 Numerical Simulations

We study a 3D radially symmetric system where, for $r = |\mathbf{r}|$, $V_1(r,t) = 0.1r^2(1 - h(r,t)) - \beta\exp[(r - r_0(t))^2/\alpha^2]$ where $h(r,t) = 1/2[\text{erf}[(r + r_0(t))/\alpha] - \text{erf}[(r - r_0(t))/\alpha]]$ is a smooth cutoff function. As such, the density depends only on the radial coordinate and the DDFT may be reduced to a 1D calculation. We consider 50 identical hard spheres of diameter 1 with $\alpha = 4$, $\beta = 10$ and $r_0(t) = 3 + \sin(t\pi)$, where the particles are initially at equilibrium under potential $V_1(r,0)$. All simulations have $m = k_BT = 1$ and $\gamma = 8$. Figure 120.2 shows the evolution of the mean radial position and velocity for eight simulations. DDFTs are shown by lines and stochastic simulations are shown by symbols. Red, denotes the solution of (120.3) and (120.4) (long dashes) and the Euler-Maruyama solution of (120.1) without HI,

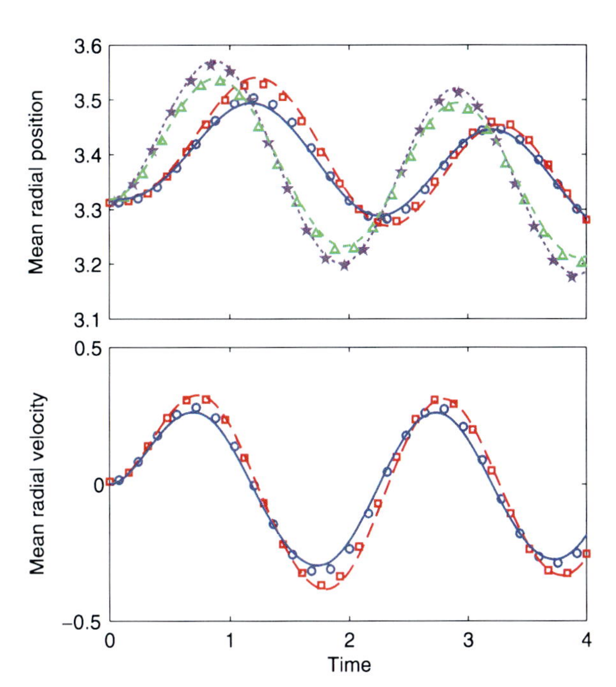

Fig. 120.2 Comparison of DDFT and stochastic simulations; see text for details

i.e. $\mathbf{\Gamma} = \gamma \mathbf{1}$ (squares) (see also [3]). Purple is the corresponding overdamped DDFT [1, 2] (dots) and Ermak-McCammon stochastic simulation [10] (stars). Blue is as red, but the HI term is given by the inverse of the Rotne-Prager tensor for the stochastic simulation (circles) and by the 11-term Jeffrey-Onishi expansion for the DDFT [11, 12] (solid). Whilst these are not strictly equivalent and thus we still obtain excellent agreement. Finally, green is as purple but with the HI term given by the Rotne-Prager tensor for both DDFT (short dashes) and stochastic simulation (triangles).

The agreement in all cases is very good and clearly demonstrates the need to include both inertial effects and HI. In particular, it is clear that, in such a system, the HI damp the dynamics. Including inertia results in a slower initial motion of the centre of mass and also affects the period of the dynamics.

References

1. Marconi UMB, Tarazon P (1999) J Chem Phys 110:8032
2. Rex M, Löwen H (2009) Eur Phys J E 28:139
3. Archer AJ (2009) J Chem Phys 130:014509
4. Goddard BD, Pavliotis GA, Kalliadasis S (2012) Multiscale Model Simul 10:633
5. Goddard BD, Nold A, Savva N, Pavliotis GA, Kalliadasis S (2012) Phys Rev Lett 109:120603

6. Rosenfeld Y (1989) Phys Rev Lett 63:980
7. Roth R, Evans R, Lang A, Kahl G (2002) J Phys Condens Matter 14:12063
8. Evans R (1979) Adv Phys 28:143
9. Trokhymchuk A, Nezbeda I, Jirsák J, Henderson D (2005) J Chem Phys 123:024501
10. Ermak DL, McCammon JA (1978) J Chem Phys 69:1351
11. Rotne J, Prager S (1969) J Chem Phys 50:4831
12. Jeffrey DJ, Onishi Y (1984) J Fluid Mech 139:261

Chapter 121
Effective Macroscopic Stokes-Cahn-Hilliard Equations for Periodic Immiscible Flows in Porous Media

Markus Schmuck, Grigorios A. Pavliotis, and Serafim Kalliadasis

Abstract Using thermodynamic and variational principles we study a basic phase field model for the mixture of two incompressible fluids in strongly perforated domains. We rigorously derive an effective macroscopic phase field equation under the assumption of periodic flow and a sufficiently large Péclet number with the help of the multiple scale method with drift and our recently introduced splitting strategy for Ginzburg-Landau/Cahn-Hilliard-type equations (Schmuck et al., Proc. R. Soc. A, 468:3705–3724, 2012). As for the classical convection-diffusion problem, we obtain systematically diffusion-dispersion relations (including Taylor-Aris-dispersion). In view of the well-known versatility of phase field models, our study proposes a promising model for many engineering and scientific applications such as multiphase flows in porous media, microfluidics, and fuel cells.

Keywords Homogenization · Diffusion-dispersion relations · Porous structures · Stokes-Cahn-Hilliard equations

121.1 Introduction and Results

We describe an arbitrary interface between two fluids by the total energy density,

$$e\big(\mathbf{x}(\mathbf{X},t),t\big) := \frac{1}{2}\left|\frac{\partial \mathbf{x}(\mathbf{X},t)}{\partial t}\right|^2 - \frac{\lambda}{2}\big|\nabla_{\mathbf{x}}\phi\big(\mathbf{x}(\mathbf{X},t),t\big)\big|^2 - \frac{\lambda}{2}F\big(\phi\big(\mathbf{x}(\mathbf{X},t),t\big)\big), \tag{121.1}$$

M. Schmuck (✉) · S. Kalliadasis
Department of Chemical Engineering, Imperial College London, South Kensington Campus, London SW7 2AZ, UK
e-mail: m.schmuck@imperial.ac.uk

S. Kalliadasis
e-mail: s.kalliadasis@imperial.ac.uk

M. Schmuck · G.A. Pavliotis
Department of Mathematics, Imperial College London, South Kensington Campus, London SW7 2AZ, UK

G.A. Pavliotis
e-mail: g.pavliotis@imperial.ac.uk

T. Gilbert et al. (eds.), *Proceedings of the European Conference on Complex Systems 2012*, Springer Proceedings in Complexity, DOI 10.1007/978-3-319-00395-5_121,

where ϕ is a conserved order-parameter that evolves between different fluid phases represented as the minima of a homogeneous free energy F. The parameter λ represents the surface tension effect, i.e. $\lambda \propto$ (surface tension) $\times$ (capillary width). We allow for free energies F which represent polynomials of the form

$$F(\phi) := \int_0^{\phi} f(s)ds, \quad \text{and} \quad f(s) := a_3 s^3 + a_2 s^2 + a_1 s. \tag{121.2}$$

We introduce the Lagrangian coordinate $\mathbf{X}$ for the (initial) material configuration and we denote by $\mathbf{x}(\mathbf{X}, t)$ the Eulerian (reference) coordinate. The last two terms in (121.1) represent the well-known Cahn-Hilliard/Ginzburg-Landau free energy density adapted to the flow map $\mathbf{x}(\mathbf{X}, t)$ defined by

$$\begin{cases} \frac{\partial \mathbf{x}}{\partial t} = \mathbf{u}(\mathbf{x}(\mathbf{X}, t), t), \\ \mathbf{x}(\mathbf{X}, 0) = \mathbf{X}. \end{cases} \tag{121.3}$$

By the kinetic energy, i.e., the first term in (121.1), we can account for fluid flow of incompressible materials with the viscosity μ, i.e.,

$$\begin{cases} \frac{\partial \mathbf{u}}{\partial t} + (\mathbf{u} \cdot \nabla)\mathbf{u} - \mu \Delta \mathbf{u} + \nabla p = \mathbf{g}, \\ \operatorname{div} \mathbf{u} = 0, \end{cases} \tag{121.4}$$

where $\mathbf{g}$ is a driving force acting on the fluid. We consider mixtures of two incompressible and immiscible fluids of the same viscosity μ which is satisfied in many practical situations.

Suppose that the fluid initially occupies a domain $\Omega \subset \mathbb{R}^d$, with $d > 0$ the dimension of space. For an arbitrary length of time $T > 0$ we then define the total energy by

$$E(\mathbf{x}) := \int_0^T \int_{\Omega} e\big(\mathbf{x}(\mathbf{X}, t), t\big) d\mathbf{X} dt. \tag{121.5}$$

Equation (121.5) combines an action functional for the flow map $\mathbf{x}(\mathbf{X}, t)$ and a free energy for the order parameter ϕ and hence combines mechanical and thermodynamic energies [2–6]. We will focus our studies on quasi-stationary, i.e., $\mathbf{u}_t = \mathbf{0}$ and $\mathbf{g} \neq \mathbf{0}$, and low-Reynolds number flows, i.e., $(\mathbf{u} \cdot \nabla)\mathbf{u} = \mathbf{0}$. By using calculus of variations [7] and the theory of gradient flows together with the imposed boundary condition $\int_{\partial\Omega} w(\mathbf{x}) do(\mathbf{x})$ for $w(\mathbf{x}) \in H^{3/2}(\partial\Omega)$, where do is the surface measure, we derive the following set of equations

$$\text{(Homogeneous case)} \quad \begin{cases} -\mu \Delta \mathbf{u} + \nabla p = \mathbf{g} & \text{in } \Omega_T, \\ \phi_t + \mathrm{Pe}(\mathbf{u} \cdot \nabla)\phi = \lambda \operatorname{div}(\nabla(f(\phi) - \Delta\phi)) & \text{in } \Omega_T, \\ \nabla_n \phi := \mathbf{n} \cdot \nabla \phi = w(\mathbf{x}) & \text{on } \partial\Omega_T, \\ \nabla_n \Delta \phi = 0 & \text{on } \partial\Omega_T, \\ \phi(\mathbf{x}, 0) = h(\mathbf{x}) & \text{on } \Omega, \end{cases} \tag{121.6}$$

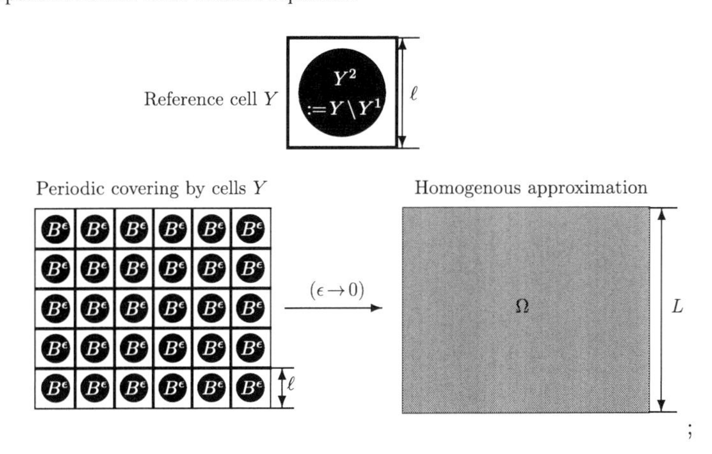

Fig. 121.1 *Left*: Porous medium $\Omega^\epsilon := \Omega \setminus B^\epsilon$ as a periodic covering of reference cells $Y := [0, \ell]^d$. *Top*: Reference cell $Y = Y^1 \cup Y^2$. *Right*: "Homogenization limit"

where $\operatorname{div}\mathbf{u} = 0$ in Ω_T, $\mathbf{u} = \mathbf{0}$ on $\partial\Omega_T$, $\Omega_T := \Omega \times]0, T[$, $\partial\Omega_T := \partial\Omega^1 \times]0, T[$, λ represents the elastic relaxation time of the system, and the driving force $\mathbf{g}$ accounts for the elastic energy [5]

$$\mathbf{g} = -\gamma \operatorname{div}(\nabla\phi \otimes \nabla\phi), \tag{121.7}$$

where γ corresponds to the surface tension [8] and hence we set $\gamma = \lambda$ for simplicity as in [9]. We denote by $\mathrm{Pe} := \frac{k\tau L\mathrm{U}}{D}$ the dimensionless Péclet number for a reference fluid velocity $\mathrm{U} := |\mathbf{u}|$, L is the characteristic length of the porous medium, and via Einstein's relation we obtain the diffusion constant $D = k\tau M$ from the mobility M for the Boltzmann constant k and temperature τ. Our restriction to the Stokes equation (in difference to [5]) is motivated here by the fact that such flows turn into Darcy's law in porous media, e.g. [10, 11]. The main objective of our study is to derive effective macroscopic equations describing (121.6) in the case of perforated domains $\Omega^\epsilon \subset \mathbb{R}^d$ instead of a homogeneous $\Omega \subset \mathbb{R}^d$. A well-accepted approach is to represent a porous medium $\Omega = \Omega^\epsilon \cup B^\epsilon$ periodically with pore space Ω^ϵ and solid phase B^ϵ. We define $I^\epsilon := \partial\Omega^\epsilon \cap \partial B^\epsilon$ where $\epsilon > 0$ defines the heterogeneity $\epsilon = \frac{\ell}{L}$ for a characteristic pore size ℓ and the characteristic length of the porous medium L, see Fig. 121.1. Then, we define the porous medium by a periodic covering with a reference cell $Y := [0, \ell_1] \times [0, \ell_2] \times \cdots \times [0, \ell_d]$ which represents a single, characteristic pore. For simplicity, we set $\ell_1 = \ell_2 = \cdots = \ell_d = 1$. The periodicity assumption allows, by passing to the limit $\epsilon \to 0$, for the derivation effective macroscopic porous media equations, as depicted in Fig. 121.1. The pore and the solid phase of the medium are defined as usual by, $\Omega^\epsilon := \bigcup_{\mathbf{z}\in\mathbb{Z}^d} \epsilon(Y^1 + \mathbf{z}) \cap \Omega$, and $B^\epsilon := \bigcup_{\mathbf{z}\in\mathbb{Z}^d} \epsilon(Y^2 + \mathbf{z}) \cap \Omega = \Omega \setminus \Omega^\epsilon$, where the subsets $Y^1, Y^2 \subset Y$ are defined such that Ω^ϵ is a connected set. More precisely, Y^1 denotes the pore phase (e.g. liquid or gas phase in wetting problems), see Fig. 121.1.

For a feasible and reliable upscaling, we restrict ourselves to periodic flows defined on a single reference cell Y, i.e.,

$$\text{(Periodic flow)} \quad \begin{cases} -\mu\Delta_{\mathbf{y}}\mathbf{u} + \nabla_{\mathbf{y}} p = -\gamma\,\mathrm{div}_{\mathbf{y}}(\nabla_{\mathbf{y}}\psi \otimes \nabla_{\mathbf{y}}\psi) & \text{in } Y^1, \\ \quad \mathbf{u} \text{ is } Y^1\text{-periodic}, & \\ \mathrm{Pe}(\mathbf{u}\cdot\nabla_{\mathbf{y}})\psi = \lambda\,\mathrm{div}_{\mathbf{y}}(\nabla_{\mathbf{y}}(f(\psi) - \Delta_{\mathbf{y}}\psi)) & \text{in } Y^1, \\ \nabla_n\psi := (\mathbf{n}\cdot\nabla_{\mathbf{y}})\psi = w & \text{on } \partial Y^2, \\ \nabla_n\Delta_{\mathbf{y}}\psi = 0 & \text{on } \partial Y^2, \end{cases} \tag{121.8}$$

where ψ is Y^1-periodic $\mathrm{div}_{\mathbf{y}}\,\mathbf{u} = 0$ in Y^1 and $\mathbf{u} = \mathbf{0}$ on ∂Y^2.

Motivated by [12, 13], we study the case of large Péclet number and consider the following distinguished limit.

Assumption (LP) *The Péclet number scales with respect to the characteristic pore size* $\epsilon > 0$ *as follows*: $\mathrm{Pe} \sim \frac{1}{\epsilon}$.

Let us first discuss Assumption (LP). If one introduces the microscopic Péclet number $\mathrm{Pe}_{mic} : - = \frac{k\tau\ell U}{D}$, then it follows immediately that $\mathrm{Pe} = \frac{\mathrm{Pe}_{mic}}{\epsilon}$. Since we introduced a periodic flow problem on the characteristic length scale $\ell > 0$ of the pores by problem (121.8), it is obvious that we have to apply the microscopic Péclet number in a corresponding microscopic formulation,

$$\text{(Microscopic problem)} \quad \begin{cases} \frac{\partial}{\partial t}\phi_\epsilon + \frac{\mathrm{Pe}_{mic}}{\epsilon}(\mathbf{u}(\mathbf{x}/\epsilon)\cdot\nabla)\phi_\epsilon & \\ \quad = \lambda\,\mathrm{div}(\nabla(f(\phi_\epsilon) - \Delta\phi_\epsilon)) & \text{in } \Omega_T^\epsilon, \\ \nabla_n\phi_\epsilon := \mathbf{n}\cdot\nabla\phi_\epsilon = w(\mathbf{x}/\epsilon) & \text{on } I_T^\epsilon, \\ \nabla_n\Delta\phi_\epsilon = 0 & \text{on } I_T^\epsilon, \end{cases} \tag{121.9}$$

for the initial condition $\phi_\epsilon(\cdot, 0) = \psi(\cdot)$ on Ω^ϵ and the definition $I_T^\epsilon := I^\epsilon \times]0, T[$. The microscopic system (121.9) leads to a high-dimensional problem even under the assumption of periodicity since the space discretization parameter needs to be chosen to be much smaller than the characteristic size ϵ of the heterogeneities of the porous structure, e.g. left-hand side of Fig. 121.1.

Obviously, the systematic and reliable derivation of practical, convenient, and low-dimensional approximations is the key to feasible numerics of problems posed in porous media and provides a basis for computationally efficient schemes. We further note that physically, the periodic fluid velocity defined by (121.8) can be considered as the spatially periodic velocity of a moving frame. Hence, the periodic fluid velocity $\mathbf{u}(\mathbf{x}/\epsilon) := \mathbf{u}(\mathbf{y})$ enters the microscopic phase field problem as follows

$$\text{(Microscopic problem)} \quad \begin{cases} \frac{\partial}{\partial t}\phi_\epsilon + \frac{\mathrm{Pe}_{mic}}{\epsilon}(\mathbf{u}(\mathbf{x}/\epsilon)\cdot\nabla)\phi_\epsilon & \\ \quad = \lambda\,\mathrm{div}(\nabla(f(\phi_\epsilon) - \Delta\phi_\epsilon)) & \text{in } \Omega_T^\epsilon, \\ \nabla_n\phi_\epsilon := \mathbf{n}\cdot\nabla\phi_\epsilon = w(\mathbf{x}/\epsilon) & \text{on } I_T^\epsilon, \\ \nabla_n\Delta\phi_\epsilon = 0 & \text{on } I_T^\epsilon, \\ \phi_\epsilon(\cdot, 0) = \psi(\cdot) & \text{on } \Omega^\epsilon. \end{cases} \tag{121.10}$$

The main result of our study is the systematic derivation of upscaled immiscible flow equations which effectively account for pore geometries starting from the microscopic system (121.8)–(121.10) by passing to the limit $\epsilon \to 0$, i.e.,

$$\text{(Upscaled equation)} \quad \begin{cases} p\frac{\partial \phi_0}{\partial t} = \operatorname{div}([pf'(\phi_0)\hat{M} - (2\frac{f(\phi_0)}{\phi_0} - f'(\phi_0))\hat{M}_v - \hat{\Theta}(\mathbf{x},t) \\ \qquad - \hat{C}(\mathbf{x},t)]\nabla\phi_0) - f'(\phi_0)\operatorname{div}((\hat{M}_v + \hat{K})\nabla\phi_0) \\ \qquad + \frac{\lambda^2}{p}\operatorname{div}(\hat{M}_w\nabla(\operatorname{div}(\hat{D}\nabla\phi_0) - \tilde{w}_0)), \end{cases} \tag{121.11}$$

where $\hat{\Theta}(\mathbf{x},t) := \{\theta_{kl}\}_{1\leq k,l\leq d}$ and $\hat{C}(\mathbf{x},t) := \{c_{ik}\}_{1\leq k,l\leq d}$ take the fluid convection into account, i.e.,

$$\theta_{kl} := \frac{\mathrm{Pe}_{mic}}{|Y|}\int_{Y^1}(\mathbf{u}\cdot\nabla_{\mathbf{y}})\zeta^{kl}(\mathbf{y})d\mathbf{y}, \qquad c_{ik} := \frac{\mathrm{Pe}_{mic}}{|Y|}\int_{Y^1}\left(u^i - v^i\right)\delta_{ik}\xi_v^k(\mathbf{y})d\mathbf{y}. \tag{121.12}$$

These two tensors account for the so-called diffusion-dispersion relations (e.g. Taylor-Aris-dispersion [14–16]). The tensors $\hat{M}$, $\hat{M}_v$ and $\hat{M}_w$ are derived and defined in [17]. The result (121.11) makes use of the recently proposed splitting strategy for homogenization of fourth order problems in [1] and an asymptotic multiscale expansion with drift introduced in [18, 19]. We note that the nonlinear problem (121.11) is characterized by a complex coupling between the micro- and the macroscale. As a consequence, the reference cell problems need to be computed for each macroscopic degree of freedom now and seems to be an intrinsic feature of upscaling nonlinearly coupled problems [1, 17, 20].

121.2 Conclusion

The main new result here is the extension of the results in the study by Schmuck et al. in the absence of flow [1] to include a periodic fluid flow in the case of sufficiently large Péclet number. The resulting new effective porous media approximation (121.11) of the microscopic Stokes-Cahn-Hilliard problem (121.8)–(121.10) reveals interesting physical characteristics such as diffusion-dispersion [15] relations by (121.12). The homogenization methodology serves as a systematic tool for the reliable and rigorous derivation of effective macroscopic porous media equations starting with the fundamental work on Darcy's law [10, 11]. A qualitative quantification of the new equations by error estimates [21] would be very interesting.

Acknowledgements We acknowledge financial support from EPSRC Grant No. EP/H034587, EU-FP7 ITN Multiflow and ERC Advanced Grant No. 247031.

References

1. Schmuck M, Pradas M, Pavliotis GA, Kalliadasis S (2012) Upscaled phase-field models for interfacial dynamics in strongly heterogeneous domains. Proc R Soc A 468:3705–3724

2. de Gennes PG, Prost RL (1993) The physics of liquid crystals. Oxford University Press, London
3. Doi M (1986) The theory of polymer dynamics. Oxford Science Publication, London
4. Lin FH, Liu C (1995) Nonparabolic dissipative systems, modeling the flow of liquid crystals. Commun Pure Appl Math XLVIII:501–537
5. Liu C, Shen J (2003) A phase field model for the mixture of two incompressible fluids and its approximation by a Fourier-spectral method. Physica D 179:211–228
6. Lowengrub J, Truskinovsky L (1998) Quasi-incompressible Cahn-Hilliard fluids and topological transitions. Proc R Soc A 454:2617–2654
7. Struwe M (2008) Variational methods: applications to nonlinear partial differential equations and Hamiltonian systems. Springer, Dordrecht
8. Liu C, Walkington NJ (2001) An Eulerian description of fluids containing visco-hyperelastic particles. Arch Ration Mech Anal 159:229–252
9. Abels H (2011) Double obstacle limit for a Navier-Stokes/Cahn-Hilliard system. In: Progress in nonlinear differential equations and their applications, vol 43. Springer, Berlin, pp 1–20
10. Carbonell R, Whitaker S (1983) Dispersion in pulsed systems—II: Theoretical developments for passive dispersion in porous media. Chem Eng Sci 38(11):1795–1802
11. Hornung U (1997) Homogenization and porous media. Springer, Berlin
12. Mei CC (1992) Method of homogenization applied to dispersion in porous media. Transp Porous Media 9:261–274. doi:10.1007/BF00611970
13. Rubinstein J, Mauri R (1986) Dispersion and convection in periodic porous media. SIAM J Appl Math 46(6):1018–1023
14. Aris R (1956) On the dispersion of a solute in a fluid flowing through a tube. Proc R Soc A 235(1200):67–77
15. Brenner H (1980) Dispersion resulting from flow through spatially periodic porous media. Philos Trans R Soc Lond A 297(1430):81–133
16. Taylor G (1953) Dispersion of soluble matter in solvent flowing slowly through a tube. Proc R Soc A 219(1137):186–203
17. Schmuck M, Berg P (2013) Homogenization of a catalyst layer model for periodically distributed pore geometries in PEM fuel cells. Appl Math Res Express. doi:10.1093/amrx/abs011
18. Allaire G, Brizzi R, Mikelić A, Piatnitski A (2010) Two-scale expansion with drift approach to the Taylor dispersion for reactive transport through porous media. Chem Eng Sci 65:2292–2300
19. Marušic-Paloka E, Piatnitski AL (2005) Homogenization of a nonlinear convection-diffusion equation with rapidly oscillating coefficients and strong convection. J Lond Math Soc 72(02):391
20. Schmuck M (2012) A new upscaled Poisson-Nernst-Planck system for strongly oscillating potentials. Preprint. arXiv:1209.6618v1
21. Schmuck M (2012) First error bounds for the porous media approximation of the Poisson-Nernst-Planck equations. Z Angew Math Mech 92(4):304–319

Chapter 122
Bound State Formation and Self-organization in Interfacial Turbulence

Marc Pradas, Serafim Kalliadasis, Phuc-Khanh Nguyen, and Vasilis Bontozoglou

Abstract We study pulse interactions in falling liquid films by means of both an analytical low-dimensional model and direct numerical simulations (DNS) of the full Navier-Stokes equations and associated wall and free-surface boundary conditions. We observe a rich dynamics between pulses, including a monotonic and oscillatory interaction in a binary system, and a pulse self-organization process in a large multi-pulse system that brings the pulses to be separated by well-defined distances. We find very good agreement between weak interaction theory and numerical computations of both the low-dimensional model and DNS.

122.1 Introduction

The dynamics and interactions of coherent structures in nonlinear systems is a very active topic that finds applications in many different areas such as optics, quantum mechanics, or fluid mechanics. An example is the so-called dissipative or interfacial turbulence [1] where localized coherent structures arise as a result of the competition between energy production, internal dissipation–energy accumulation and nonlinearity. These dissipative structures are in continuous weak interaction and may eventually self-organize to exhibit some sort of complexity such as bound-state formation [2–5], i.e. groups of two or more structures behaving as a single object, or synchronisation [6]. In this sense, interfacial turbulence is low-dimensional spatio-temporal chaos where despite the apparent complexity of the system it is still possible to develop a coherent-structure theory to study interactions between localized states. Very well-know examples are thermal convection [7], granular media [8], the Faraday parametric instability [9], or falling liquid films [10–12].

We present here a detailed analysis of pulse dynamics and interactions in a liquid film falling over a vertical wall which is an example of a convectively unstable open-flow hydrodynamic system with a sequence of spatio-temporal transitions which are

M. Pradas (✉) · S. Kalliadasis
Department of Chemical Engineering, Imperial College London, London SW7 2AZ, UK

P.-K. Nguyen · V. Bontozoglou
Department of Mechanical Engineering, University of Thessaly, Pedion Areos 38334, Volos, Greece

T. Gilbert et al. (eds.), *Proceedings of the European Conference on Complex Systems 2012*, Springer Proceedings in Complexity, DOI 10.1007/978-3-319-00395-5_122,

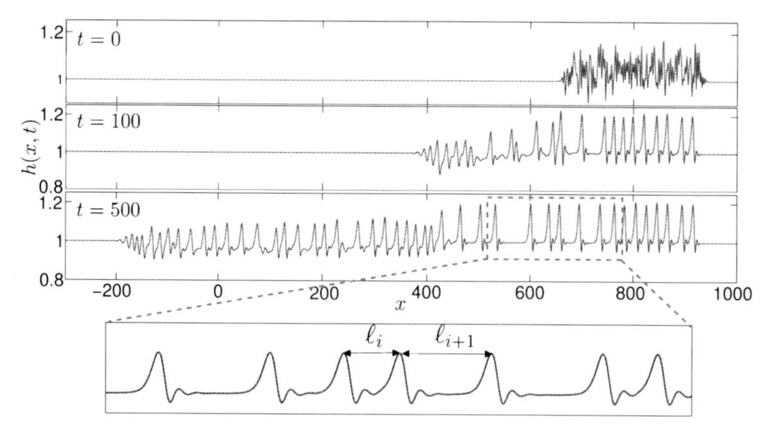

Fig. 122.1 Spatio-temporal evolution of an initial disturbance obtained with the second-order model (122.1a)–(122.1b)

generic to a large class of hydrodynamic and other nonlinear systems [12]. It is well-known that for small-to-moderate values of the Reynolds number (defined typically as the ratio of the inlet flow rate to the kinematic viscosity), the initial growth of small disturbances at the inlet is rapidly followed by a nearly spanwise-uniform wavy regime where localised coherent structures which are sufficiently stable and robust are separated by relatively large portions of nearly flat films [13–15]. Each of these structures resembles an (infinite-domain) solitary pulse consisting of a non-linear hump with a long tail at the back and a steep front edge preceded by small capillary ripples (see Fig. 122.1 for a spatio-temporal numerical evolution of an initial localized disturbance).

122.1.1 Interfacial Turbulence in Falling Liquid Films

The analytical study of falling liquid films is based on a low-dimensional second-order formulation which was derived in [16–18] and describes the evolution of both the interface thickness $h(x,t)$ and local flow rate $q(x,t)$:

$$\delta q_t = \frac{5}{6}h - \frac{5}{2}\frac{q}{h^2} + \delta\left(\frac{9}{7}\frac{q^2}{h^2}h_x - \frac{17}{7}\frac{q}{h}q_x\right) + \frac{5}{6}hh_{xxx} + \eta\left[4\frac{q}{h^2}(h_x)^2 - \frac{9}{2h}q_x h_x - 6\frac{q}{h}h_{xx} + \frac{9}{2}q_{xx}\right], \tag{122.1a}$$

$$h_t = -q_x, \tag{122.1b}$$

where δ is a reduced Reynolds number and η a viscous dispersion number that controls the strength of the long-wave second-order terms. The above model was obtained by combining the long-wave expansion up to second order with a weighted residual technique based on a Galerkin projection in which the velocity field is expanded onto a basis with polynomial test functions [16–18].

122.1.2 Coherent-Structure Theory

Bound-state formation and pulse interactions in the above model (122.1a)–(122.1b) are studied by making use of a coherent-structure theory that has been recently developed in [19]. The main idea of the theory is to assume that at long times and for sufficiently large domains, pulse dynamics is dominated mainly by weak interactions so that the solution is given as a superposition of N pulses and a small correction function, i.e., $\mathbf{\Phi} = \mathbf{\Phi}_N + \sum_i \mathbf{\Phi}_0(x - x_i(t)) + \hat{\mathbf{\Omega}}$, for $i = 1, \ldots, N$ written in vectorial form, where $\mathbf{\Phi} = (q\ h)^T$, $\mathbf{\Phi}_N$ corresponds to the Nusselt flat film solution, $\mathbf{\Phi}_0$ is the steady-state solution of a single solitary wave located at $x_i(t)$, and $\hat{\mathbf{\Omega}}$ is the small correction. Via appropriate asymptotic expansions, the following dynamic equation for the correction function is then obtained:

$$\hat{\mathbf{\Omega}}_t - \dot{x}_i(t)\mathbf{\Phi}_{ix} = \mathcal{L}_i\hat{\mathbf{\Omega}} + \mathcal{J}_i\mathbf{\Phi}_{i-1} + \mathcal{J}_i\mathbf{\Phi}_{i+1}, \qquad (122.2)$$

where $\mathcal{L}_i$ and $\mathcal{J}_i$ correspond to linear differential operators. By rigorously scrutinising the spectral properties of $\mathcal{L}_i$, it is shown that the location of each pulse is given by the following dynamical system:

$$\dot{x}_i(t) = \alpha_i S_1(x_{i+1} - x_i) + \beta_i S_2(x_i - x_{i-1}), \qquad (122.3)$$

with $\alpha_i = \beta_i = 1$ for $i = 2, \ldots, N-1$, $\alpha_1 = \beta_N = 1$, and $\alpha_N = \beta_1 = 0$. The S_1 and S_2 functions are given by the projection of the terms $\mathcal{J}_i\mathbf{\Phi}_{i+1}$ and $\mathcal{J}_i\mathbf{\Phi}_{i-1}$, respectively, to the null space of $\mathcal{L}_i$.

122.2 Results

122.2.1 Bound State Formation

We use water as working liquid throughout the whole study with viscosity $\mu = 10^{-3}$ kg/ms, density $\rho = 1000$ kg/m^3, and surface tension $\sigma = 72.01$ mN/m.

In the simplest scenario of a system compound of two pulses, Eq. (122.3) reduces to a dynamical system for the separation length $\ell = x_2 - x_1$ between both pulses:

$$\dot{\ell} = -\partial_\ell V_c(\ell), \qquad (122.4)$$

where we have defined the potential function $V_c(\ell) \equiv \int_\ell [S_1(\ell) - S_2(\ell)]$ with $V_c(\ell \to \infty) = 0$. It is important to note that functions S_1 and S_2 are actually related to the interaction from the capillary ripples and monotonic tails of the pulses, respectively. Moreover, as it has been shown in [20], this potential function can be

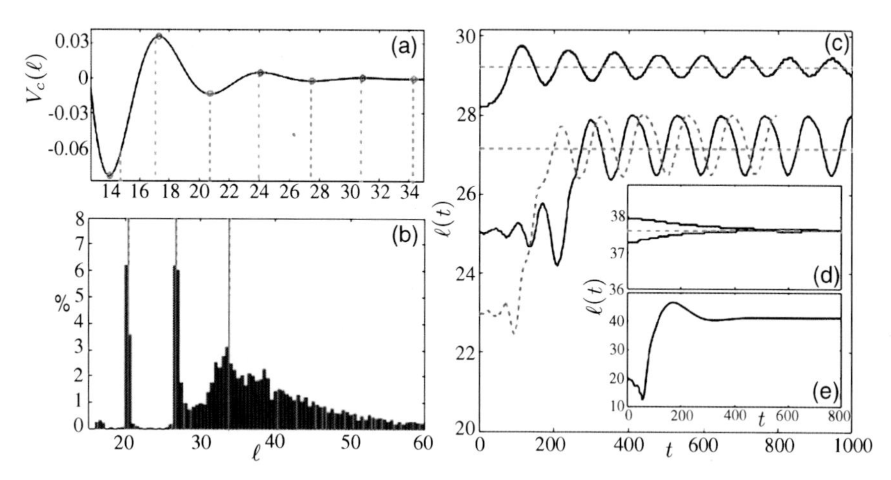

Fig. 122.2 (**a**) $V_c(\ell)$ obtained from Eq. (122.4) for $\delta = 1$. *Dashed* and *dot-dashed lines* correspond to the bound states obtained from DNS. (**b**) Hystogram of separation lengths in the numerical experiment shown in Fig. 122.1. (**c**)–(**e**) Separation length dynamics in a binary system obtained with DNS

physically related to the net capillary force between both pulses, and its local maxima and minima correspond to nominally unstable and stable bound states, i.e. both pulses travel at the same speed ($\dot{\ell} = 0$).

Figure 122.2(a) shows the potential function $V_c(\ell)$ for $\delta = 1$ obtained from coherent-structure theory (Eq. (122.4)). We can observe the existence of stable (local minima) and unstable (local maxima) bound states. On the other hand, we have also performed direct numerical simulations (DNS) of the full Navier-Stokes equations and associated wall and free-surface boundary conditions (see [21, 22] for references on the numerical method). Our DNS results show the existence of stable and unstable bound states the location of which is marked as dashed and dot-dashed lines, respectively, in Fig. 122.2. We can see there is very good agreement between the theoretically predicted bound state separation distances and the DNS computations. It is important to note however that for the shortest separation lengths ($\ell < 20$), strong pulse interactions start to be important and some deviations between weak-interaction theory (i.e. well-separated pulses assumption) and DNS results are observed.

122.2.2 Pulse dynamics and self-organization

Pulse dynamics in a binary system is studied by means of DNS. The reader is referred to [20] for a full study of numerical computations of the second-order model (122.1a)–(122.1b). In the present study we start by placing both pulses separated with an initial length ℓ_0 and then follow the time evolution of the separation length $\ell(t)$. The results are presented in Figs. 122.2(c), (d), and (e). For large initial separation lengths (Fig. 122.2(d)) we observe both pulses to monotonically attract ($\dot{\ell} < 0$) or repeal ($\dot{\ell} > 0$) to each other approaching a stable predicted bound-state (see horizontal dashed line). For shorter separation lengths ($\ell_0 \sim 29$ in Fig. 122.2(c)), our DNS results show that the pulse separation length is characterized now with an oscillatory dynamics which is eventually damped, approaching again a theoretically predicted stable bound state. Interestingly, when we decrease even more the initial separation length ($\ell_0 \sim 25$) this oscillatory dynamics becomes self-sustained and the separation length oscillates with time at constant frequency and around a stable bound state predicted by the theory. For even shorter initial separation lengths (Fig. 122.2(e)), strong pulse interactions start to dominate the dynamics which is characterized by a strong initial repulsion until both pulses approach a stable bound state located at a much larger distance. It is important to emphasize that our DNS results are in fully agreement with numerical computations of the second-order model (see [20]), indicating hence the reliability of using low-dimensional models to study pulse dynamics and interactions.

Finally, we also study large domain systems containing several pulses interacting with each other. Note that DNS computations are actually computationally very expensive to study such large domains, and we therefore make use of the low-dimensional model (122.1a)–(122.1b) instead. Figure 122.1 shows the numerical experiment of an initial disturbance evolving into a wave packet which in turn gives birth to a number of pulses. We performed 1000 different realisations on the initial random condition and calculated the separation distances ℓ from the first 12 pulses for $\delta = 1.0$. Figure 122.2(b) depicts the histogram of ℓ, where we can observe the presence of peaks which correspond to the predicted two-pulse bound state distances. This is a clear indication that many interacting pulses self-organize to be separated by well-defined distances.

122.3 Conclusions

We have performed a detailed and systematic investigation of pulse interactions in falling liquid films via both DNS and analysis–numerical simulations of the low-dimensional model derived in [16–18] that leads to a second-order model for the local flow rate and interface thickness. We made use of the coherent-structure theory developed in [19] for a system of N pulses to study the case of a two-pulse system.

Our DNS results have shown the existence of two-pulse bound-states the distance of which has been found to be in excellent agreement with the theoretical predictions of coherent-structure theory. We have also observed that two-pulse dynamics is very complex, observing monotonic, oscillatory dynamics or strong repulsions depending on the initial conditions. We have also found very good agreement between DNS computations and numerical results of the low-dimensional model. Our results therefore show the reliability of using low-dimensional models to study pulse dynamics.

By performing large domain computations of the second-order model, we have investigated the interaction of many pulses. We have observed the emergence of a self-organized state where the pulses appear to be separated by well-defined distances which correspond to the theoretically predicted distances for two-pulse bound-states.

References

1. Manneville P (1990) Dissipative structure and weak turbulence. Academic Press/Elsevier, San Diego/Amsterdam
2. Elphick C, Ierley GR, Regev O, Spiegel EA (1991) Phys Rev A 44:1110–1122
3. Duprat C, Giorgiutti-Dauphiné F, Tseluiko D, Saprykin S, Kalliadasis S (2009) Phys Rev Lett 103:234501
4. Tseluiko D, Saprykin S, Duprat C, Giorgiutti-Dauphiné F, Kalliadasis S (2010) Physica D 239:2000–2010
5. Tseluiko D, Kalliadasis S (2012) IMA J Appl Math. doi:10.1093/imamat/hxs064
6. Turaev D, Vladimirov AG, Zelik S (2012) Phys Rev Lett 108:263906
7. Krishnamurti R (1970) J Fluid Mech 42:295–307
8. Umbanhowar PB, Melo F, Swinney HL (1990) Nature (London) 382:793
9. Miles J, Henderson D (1990) Annu Rev Fluid Mech 22:143
10. Kapitza PL, Kapitza SP (1949) Zh Eksp Teor Fiz 19:105–120
11. Chang H-C (1994) Annu Rev Fluid Mech 26:103–136
12. Kalliadasis S, Ruyer-Quil C, Scheid B, Velarde MG (2012) Falling liquid film. Springer, London
13. Liu J, Paul JD, Gollub JP (1993) J Fluid Mech 250:69–101
14. Chang H-C (1994) J Fluid Mech 294:123–154
15. Demekhin EA, Kalaidin EN, Kalliadasis S, SYa V (2007) Phys Fluids 19:114103
16. Ruyer-Quil C, Manneville P (1998) Eur Phys J B 6:277–292
17. Ruyer-Quil C, Manneville P (2000) Eur Phys J B 15:357–369
18. Ruyer-Quil C, Manneville P (2002) Phys Fluids 14:170–183
19. Pradas M, Tseluiko D, Kallidasis S (2011) Phys Fluids 23:044104
20. Pradas M, Tseluiko D, Kallidasis S (2012) IMA J Appl Math 77:408–419
21. Vlachogiannis M, Bontozoglou V (2001) J Fluid Mech 435:191–215
22. Malamataris NA, Vlachogiannis M, Bontozoglou V (2002) Phys Fluids 14:1082–1094

Part XVI
Satellite Meeting: Information Processing in Complex Systems

Chapter 123
Dynamics of Artificial Markets on Irregular Topologies

Ranaivo Mahaleo Razakanirina and Bastien Chopard

Abstract We investigate the factors that influence the stability of artificial markets. We consider particularly the cases of goods, labour and money markets. On the one hand, we show that all the agents belonging to circuits of goods and labour markets remain available during the whole simulation. These particular topologies are stable. On the other hand, in the contrary, we show that the money market is sensitive to the existence of circuits. Systems with circuits crash and become chaotic with short-term loan duration.

Keywords Economical model · Complex system · Multilayer cellular automata on a graph · Complex network · Strongly connected components · Lend-redeem

123.1 Introduction

Economic systems are complex systems [1–4] built upon the complex interactions between heterogenous agents. From the microscopical dynamics of these interactions the high level abstraction laws are stated, laws that help understanding the behaviors of such systems. These systems are naturally defined using a graph due to the irregularities of the interactions.

Multilayer Cellular Automata on Graph (MCAG) [5–8] is an effective framework to simulate such complex systems. If offers an unified formalism to deal with multi-phase dynamics, multi-layer topologies, distinct topologies at each layer level and global information sharing. The artificial markets that we simulate in this paper are defined naturally using MCAG formalism.

Our goal in this paper is to investigate the dynamics of some abstract markets based on three main wealth ingredients which are goods, working hours and money. Particularly, we focus on the stability of such markets. Stability in terms of state

R.M. Razakanirina (✉) · B. Chopard
University of Geneva, Geneva, Switzerland
e-mail: ranaivo.razakanirina@unige.ch

B. Chopard
e-mail: bastien.chopard@unige.ch

T. Gilbert et al. (eds.), *Proceedings of the European Conference on Complex Systems 2012*, Springer Proceedings in Complexity, DOI 10.1007/978-3-319-00395-5_123,

and also in terms of topology. We build the local homogeneous dynamics of these markets from a certain economic reality describing the local interaction between agents. For instance, such an interaction can be defined by "an agent buys more from a seller which proposes a lower unit price of goods".

This paper is organized as follows. In Sect. 123.2, we describe the construction of a goods market dynamics where agents buy and sell goods. In Sect. 123.3, we enrich the goods market by adding a labour market which considers the production and the consumption of goods. In Sect. 123.4, we describe the money market in standalone, a market which simulates the lending redeeming process between agents.

123.2 Give and Take Model (GT-Model)

This model simulates an artificial goods market dynamics where agents interact by buying-selling goods. The interactions are modeled by a directed graph denoted G_g where edge (i, j) means that source agent i buys goods form target agent j. Cash goes from source to target and goods flow in the reverse direction. The role of each agent, either buyer or seller or both, is completely defined by the graph structure. We assume that agents do not buy goods from themselves. Thus G_g is a simple directed graph.

The wealth of each agent i at any time t consists of its cash $c_i(t)$ and its goods $g_i(t)$ (both real positive numbers and infinitely divisible). The initial value $p_i(t)$ of a unit price of goods sold by agent i is chosen arbitrary. In terms of the MCAG formalism, the GT-model consists of three layers which are the cash, the goods, and the unit price layers. The last two layers are the perfect symmetry of the cash layer.

At each time iteration t, the dynamics consist of four steps

- During the first step, each buyer i takes a fraction λ of its cash and distribute is to every seller j it is connected to. The amount c_{ij} of cash i gives to j is inversely proportional to p_j.
- During the second step, each seller j take a fraction μ of its goods and distribute its to each buyer who gave money in the first phase. The amount g_{ji} of goods agent i receives from j is proportional to c_{ij}.
- During the third step, the unit price is updated to reflect the above transaction. The price $p_j(t+1)$ is computed as the amount of cash divided by the amount of goods received for that cash

$$p_j(t+1) = \frac{c_{ij}}{g_{ji}} \tag{123.1}$$

  Due to the proportionality of g_{ji} to c_{ij}, the price only depends on the seller j, not the buyer i.
- The last step of the dynamics updates the topology so as to only maintain interesting trading relations. A buyer may decide to stop the interaction with a seller if the offered price is too high compared to the price of its other sellers. A buyer also decides to stop trading with a seller if the quantity of goods received is too small.

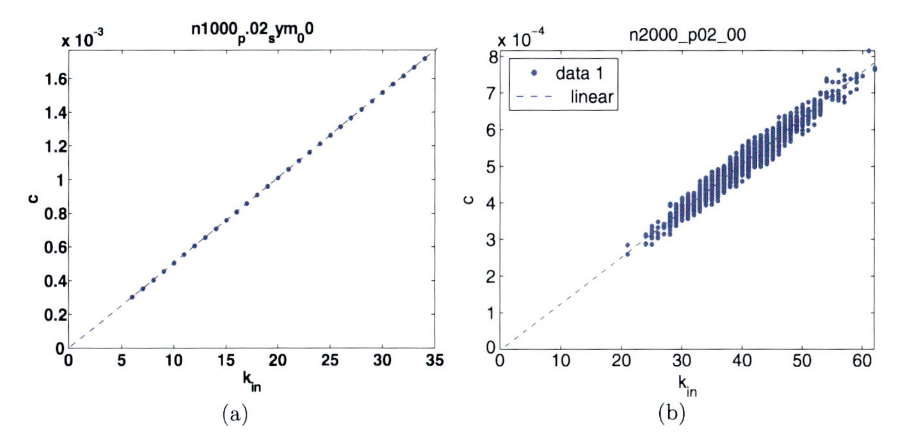

Fig. 123.1 *The dots* show the distribution of cash of a GT-model: (**a**) for a symmetric random graph (with $n = 1000$ and pairewise links created with probability $p_{link} = 0.02$) versus the indegree k^{in} at the steady state. *The dotted line* is the distribution of cash calculated analytically. (**b**) For an asymmetric Erdos Rényi random graph versus indegree k^{in} at the steady state ($n = 2000$ and $p_{link} = 0.02$. *The dotted lines* are the linear estimations of the distribution, in a least-squares sense, given by Eq. (123.4)

From the above rule, it is easy to show that the total cash c_{tot} and the total quantity of goods g_{tot} are conserved.

The time evolution of this model is characterized by a transient regime where edges may be cut. The effect of the edge dynamics is the emergence of the strongly connected components [9] (SCC) of the initial topology. The link between two agents survives only if there is a path back to where it came even the path is long.

Each SCC converges to a stationary state and becomes a sub-market where the unit price of goods converges to unique equilibrium price p_e. When this steady state is established, the total incoming cash c_i^{in} is equal to total outgoing cash c_i^{out}, for each agent. Similarly, g_i^{in} the total incoming amount of goods equals g_i^{out}, the total outgoing amount of goods. Thus we have $c_i^{in} = c_i^{out} = \lambda c_i$ and $g_i^{in} = g_i^{out} = \mu g_i$. Therefore, $p_e = \frac{\lambda c_i}{\mu g_i}$. Since, in the steady state, p_e is the same for all agents, we obtain, by summing over i,

$$p_e = \frac{\lambda c_{tot}}{\mu g_{tot}}, \tag{123.2}$$

which is the supply and demand relation.

Another result is that the cash of a particular agent tends to be proportional to its total number of clients. Particularly if G_g is symmetric, we obtain a linearity (Figs. 123.1(a) and 123.1(b)).

Using the stationary assumption, this distribution of cash can be computed mathematically by the following homogeneous system of equations [5]

$$c_i - \sum_{\forall j} \frac{a_{ji}}{k_j^{out}} c_j = 0, \tag{123.3}$$

where a_{ji} are the elements of the adjacency matrix of G_g, k_j^{out} is the total number of outgoing neighbors of j. Equation (123.3) corresponds to an eigenvector centrality measure commonly used in complex networks. The division by k_j^{out} makes this measure similar to the so-called page rank index.

In case of a symmetric graph ($a_{ij} = a_{ji}$), Eq. (123.3) can be solved analytically

$$c_i = \frac{k_i^{in}}{m} c_{tot}, \tag{123.4}$$

where m is the total number of edges in G_g and k_i^{in} the number of clients connected to i. Here we confirm the results depicted in Fig. 123.1(a) This relation shows that the cash is inversely proportional to the market size m and that it is proportional to k_i^{in}. Thus, the cash distribution follows Zipf law if the market topology is a scale free graph [10].

For an asymmetrical graph, Eq. (123.3) can be solved numerically using the following matrix representation

$$(I - B)C = 0, \tag{123.5}$$

where the elements b_{ij} of B are $b_{ij} = \frac{a_{ji}}{k_j^{out}}$ and $C = (c_0, c_1, \ldots, c_{n-1})^T$, n is the order of G_g. The existence of solution depends on the rank of $(I - B)$. Particularly, if $rank(I - B) = n - 1$, we can use the conservation of cash $\sum_{\forall \ell} c_\ell = c_{tot}$ as additional constraint to obtain the non trivial solution of (123.5). These solutions fit well the values depicted in Fig. 123.1(b) in case of asymmetric graph.

Knowing p_e, the distribution of goods can be computed as $g_i = \frac{\lambda}{\mu p_e} c_i$.

123.3 Labour and Goods Market Model (LGM-Model)

In this section, we enrich the above model by adding a mechanism for creating goods out of the labor of the agents. Thus g_{tot} is no longer conserved, but total cash c_{tot} still is.

We consider a model consisting of n_f firms and n_e employees with $n_f \ll n_e$. Firms produce goods by transforming the labour furnished by its employees and these goods are bought by any employees in the system. Firms do not work with other firms and employees buy only from firms. This interaction creates two coupled markets (Fig. 123.2). The first one is the labour market that we denote G_ℓ where firms and employees exchange labour against salary (cash). The second one is the goods market that we denote G_g where firms and employees exchange cash against goods. The Labour and Goods Market Model (LGM-Model) is built upon this scheme.

With LGM, firms play both the role of employer and seller respectively in the labour market G_ℓ and in goods market G_g. Employees play both the role of worker and buyer respectively in G_ℓ and in G_g.

From now on, we denote the firms (also sellers) by i and the employees (also buyers) by j. G_ℓ and G_g are two distinct bipartite directed graphs. The directed edge

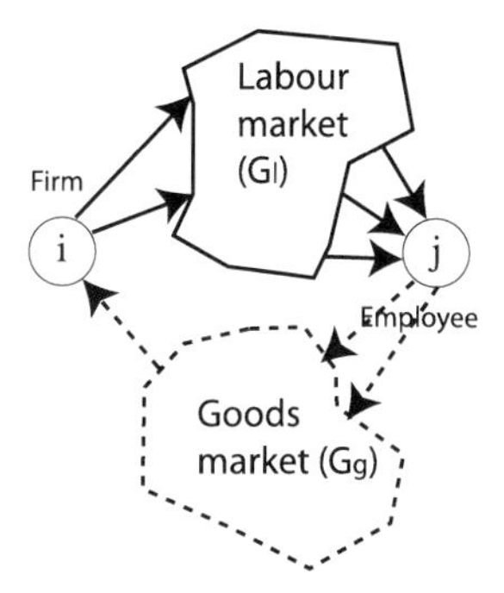

Fig. 123.2 Firm i is an employer in the labour market and a seller in the goods market. Employee j is a worker in the labour market and a buyer in the goods market

in G_ℓ which connects two agents means that the target agent works for the source agent. The flow of cash (salary) goes from source to target and the flow of working hours from target to source. We assume that the agents are not self-employed. Thus, G_ℓ is a simple directed graphs. The definition of G_g is equivalent to the definition given in GT-model (described in the Sect. 123.2).

The amount of working hours $h_i(t)$ available to each firm i at any time t becomes a part of its wealth, in addition to its cash $c_i(t)$ and its quantity of goods $g_i(t)$. The cost of working hours is expressed by the hourly wage $w_i(t)$ offered to the employees. We start the simulations with homogeneous values of these wealth and without goods. Goods are created and consumed during the transition rule.

Firms transform at any time the working hours that are only given by its employees using its production factor π. We choose that each firm produces three units of goods per person hour. Thus $\pi = 3$. Employees do not produce goods.

Buyers consume at any time a fraction ϕ_j of its goods to balance the total quantity of goods produced in the system. Sellers consume goods less that the buyers ($\phi_i < \phi_j$). We choose $\phi_i = 0.1$ and $\phi_j = 0.9$.

We assume also that firms are not a self-employed. The working hours cannot be conserved, therefore, at the end of the transition rule, each employee j resets its working hours to a value δ. Here we consider that each iteration corresponds to a working day, therefore, we choose $\delta = 8$. The hours of labor available to firms is always set to 0.

The dynamics consists of two main steps which are the labour dynamics and the goods dynamics.

1. During the labour dynamics,
   - Firstly each employee gives the totality of its working hours to its employers. An employee split its daily δ hours among the employers it is connected to in proportion to the hourly wage which is offered.
   - Secondly, each firm pays its workers proportionally to the furnished labour, investing a fraction γ of its cash. The hourly wage $w_i(t+1)$ at the next time iteration is the ratio between the cash and the quantity of hours transiting through the edge (i, j)

$$w_i(t+1) = \frac{c_{ij}}{w_{ji}}. \tag{123.6}$$

The topology of G_ℓ evolves according to the level of hourly wage proposed by the firms. Employee j may cut its cooperation with the firm which has the cheapest hourly wage.

- Thirdly, the employers transform its available labour to goods.
  At the end of the labour dynamics (superscript *), the balance of hours, cash and goods of each firm i are respectively $h_i^* = h^{in}$, where h^{in} is the total quantity of working hours received from the workers, $c_i^* = c_i(t) - c^{out}$ where $c^{out} = \gamma c_i(t)$ is the total cash to pay workers and $g_i^* = g_i(t) + \pi h^{in}$. πh^{in} is the production of the firm.
  The balance of wealth of each employee j are $h_j^* = 0$, $c_j^* = c_j(t) + c^{in}$ where c^{in} is the total salary given by its employers and $g_j^* = g_j(t)$.

2. The goods market dynamics is similar to that described in GT-Model but adding after the last step the goods consumption phase.
   Therefore, the balance of firm wealth becomes $h_i(t+1) = 0$, $c_i(t+1) = c_i^* + c^{*in}$, where c^{*in} is the total cash given by employers after selling goods, $g_i(t+1) = (1-\phi_i)(g_i^* - g_i^{*out})$ where $g_i^{*out} = \mu g_i^*$ is the total quantity of goods furnished to buyers.
   The balance of employee wealth is $h_j(t+1) = \delta$, $c_j(t+1) = c_j^* - c_j^{*out}$ where $c_j^{*out} = \lambda c_j^*$ the total cash that employee uses to buy goods and $g_j(t+1) = (1-\phi_j)(g_j^* + g_j^{*in})$ where g_j^{*in} is the total quantity of goods obtained by the employee after buying operation.

The values of γ, λ, μ are set to 0.9, values with which we obtain the following interesting behaviors.

During a transient regime, the edges of G_ℓ and G_g may be cut. We observe that the SCC of the whole market $G = G_\ell \cup G_g$ emerge initially due to the edge dynamics of labour and goods markets. The dynamics tend to create inside each SCC one sub-market where the value of unit price of goods is homogeneous and the value of hourly wage is homogeneous. This fact creates competition (competition of unit-price and hourly wage) between the firms in the SCC. The level of competition is measured by the difference of unit price and hourly wage between the firms inside one SCC. One SCC is split to several little parts that we call sub-SCC if the level of competition is higher than the threshold defined in the edge dynamics of labour and goods markets (Fig. 123.3). Then, each sub-SCC tends to stationary state. Otherwise, if the level of competition that cannot create sub-SCC, the behavior of SCC becomes chaotic.

The distribution of cash inside one SCC at the stationary state (Fig. 123.4) shows that a large number of individuals are poor and that they correspond to the employees. On the other hand, a few individuals are rich and they correspond to the firms. The cash distribution tends to be proportional to the number of clients connected to each firm (Fig. 123.5).

We estimate the linear approximation of this distribution and we find that [8]

$$c_i = \left(\frac{c_{tot} - n\bar{c_e}}{m^*}\right)k_i^* + \bar{c_e}, \qquad (123.7)$$

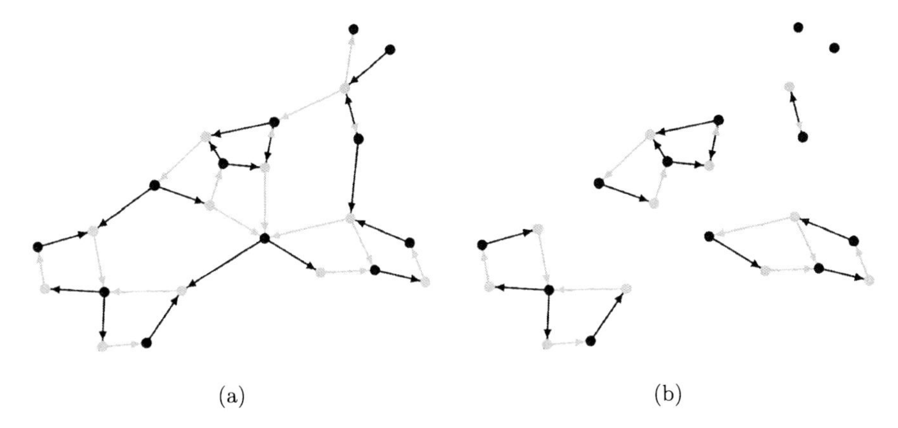

Fig. 123.3 (**a**): Initial topology of LGM-model with 11 firms (*light grey nodes*) and 12 employees (*black nodes*). *The light grey edges* are the flows of cash for the labour market G_ℓ and *the black edges* are the flows of cash for the goods market G_g. (**b**): The topology at $t = 100$ of the LGM–model described in Fig. 123.3(a). The SCC of the initial topology emerge. Each SCC is then split to little parts called sub-SCC when the competition between the firms/sellers tends to have uniform wage and uniform price of goods in each SCC

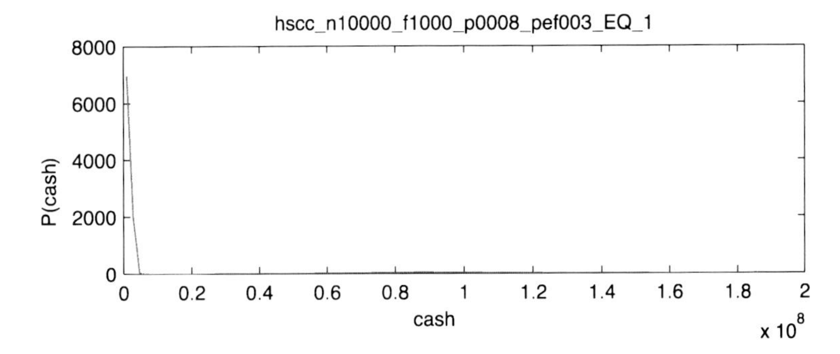

Fig. 123.4 The distribution of cash at the stationary state of LGM-model with 1000 firms, 9000 employees which shows that a large number of individuals are poor and a few individuals are rich

where $n = n_f + n_e$ is the total number of agents, m^* is the size of the goods market (G_g), k_i^* is the number of clients of agent i (G_g) and $\bar{c_e}$ is the mean value of cash of employees at the stationary state. We have $\bar{c_e} = \frac{c_e}{n_e}$, where n_e is the number of employees and $c_e = \sum_j c_j$, the total cash of employees j.

Equation (123.7) gives us the rank of each agent which depends on its number of customer in goods market. In particular, the employees do not have customers. Consequently, they have the lowest rank.

At the stationary state, the hourly wages and the unit price of goods inside one sub-SCC becomes homogeneous. Thus one can describe the cash distribution mathematically [8].

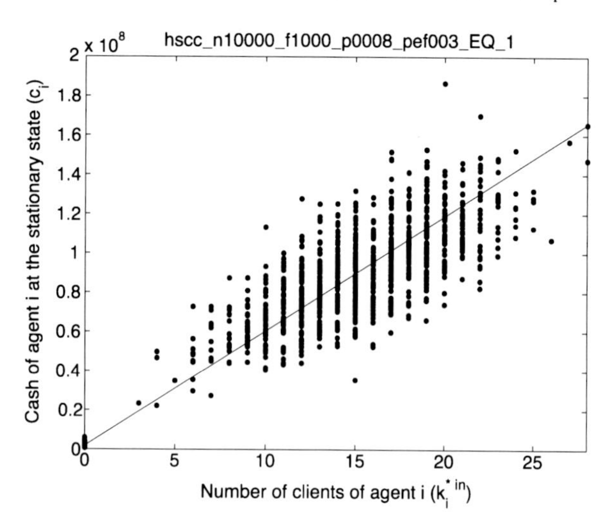

Fig. 123.5 *Dots* are the cash versus the number of connected buyers. *The line* is the mean distribution of cash versus the number of clients computed with Eq. (123.7)

123.4 Lend Redeem Model (LR-Model)

Lending process is really important in economy because it gives to firms the financial tools to prosper by borrowing cash, transforming it to production factors and redeem the borrowed cash from the benefits. The time delay from accepting to redeeming the loan is fundamental in this process. If the duration of the loan is not sufficient to get sufficient benefits, the reimbursement is not possible and the system can crash. Our goal in this section is to play with this time duration and with the topology of lending agreement to identify the factors that can bring down the system.

Thus, we consider a system called Lend Redeem Model (LR-Model) in standalone. This model simulates the evolution of cash between agents which lend and redeem the borrowed cash when the debt becomes mature. The structure of the lending agreements is defined as a directed graph G_{LA}. The directed edge which connects two agents means that the source agent lends cash to the target agent. Lending agreement means that target agent should accept the amount of loan decided by source agent. The duration of the loan is defined by the borrower. In exchange, when the loan reaches the maturation date, the target agent should redeem the corresponding debt. According to the structure of G_{LA}, an agent may be either a lender or a borrower or both.

The wealth of each agent i at any time t is its cash $c_i(t)$. This quantity is a real positive value and infinitely divisible. We denote $\tau_i > 0$ the duration of the loans of agent i. This quantity is constant during the simulation and is expressed in terms of the number of iterations until maturity.

The dynamics consist of two sequential steps. The first one is the redeem step and the second one is the lending step. The maturation dates of debts previously agreed decrease systematically at each time iteration t.

1. During the redeem step, all matured debts (debts that maturation date reaches 0) are redeemed to the corresponding lender. The remaining cash is computed as

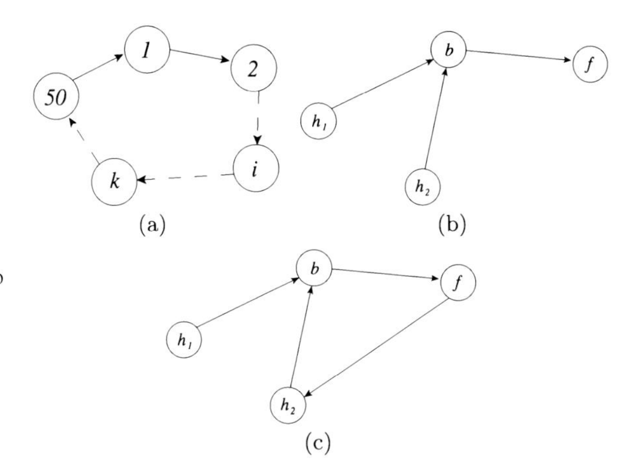

Fig. 123.6 LR topologies used during the simulations. (**a**): RING50, a ring with 50 agents and a unique value of loan duration τ. (**b**): B01H02F01 with one bank b, two households h_1 and h_2 and one firm f. (**c**): B01H02F01_RING with the edge (f, h_2) in addition to B01H02F01. $\tau_b = 10$ and $\tau_f = 5$

$c_i = c_i(t) + r_i^{in} - r_i^{out}$ where r_i^{in} is the total amount of cash redeemed to i from its borrowers and r_i^{out} is the total amount of cash redeemed by i to its lender. This balance may be negative.

2. During the lending step, each agent i lends a fraction λ of its cash to its borrowers j proportionally to the solvability of the borrower. The solvability factor is given by $s_j = \frac{c_j}{c_{tot}}$. More solvable agents obtain a higher amount of cash. The borrowers should accept the loan and the duration is set as τ_j. The balance of cash at the next time iteration is $c_i(t+1) = c_i + \ell_i^{in} - \ell_i^{out}$ where ℓ_i^{in} is the total debt accepted by i and $\ell_i^{out} = \lambda c_i$ is the total loan given by i during the current iteration.

We start the simulations with arbitrary values of cash and arbitrary values of loan duration. We choose $\lambda = 0.6$. The topologies depicted in Fig. 123.6 are used during the simulations.

Figure 123.7 depicts the time evolution of cash on RING50 and B01H02F01_RING topologies. We show that according to the value of τ or τ_{h_2}, the system converges to equilibrium or to a chaotic state. In these topologies, only a loan duration of 1 insures the stability of such systems. The instability is caused by the topology of the lending agreement. A circuit in G_{LA} tends to accumulate the debts of each agent that belongs to this circuit. Figure 123.8 shows that from $t = 0$ to $t = 50$, each agent of RING50 accumulates debt and from $t = 50$, the system starts to break down due to the high level of debt previously accumulated (Fig. 123.7(b)).

As opposed to the previous case, the evolution of B01H02F01, that does not contain circuit, tends to a stationary state (Fig. 123.9). From these observations, we assume that the lend agreement topologies without circuit are more robust to the change of loan duration.

123.5 Conclusion

This paper investigated the behaviors of artificial markets based on three main ingredients which are cash, working hours and goods.

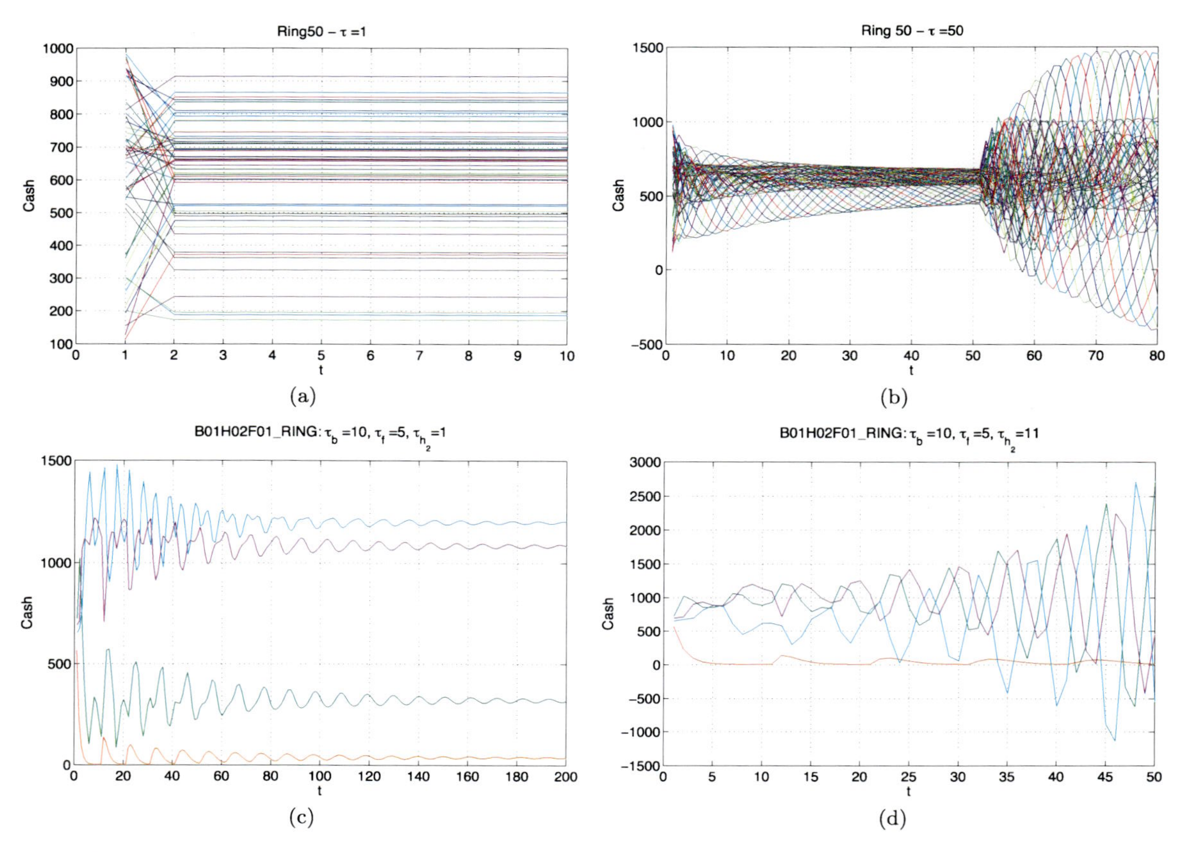

Fig. 123.7 Evolution of cash of RING50 (**a**) when $\tau = 1$, (**b**) when $\tau = 50$. Evolution of cash of B01H02F01_RING (**c**) when $\tau_{h_2} = 1$, (**d**) when $\tau_{h_2} = 11$

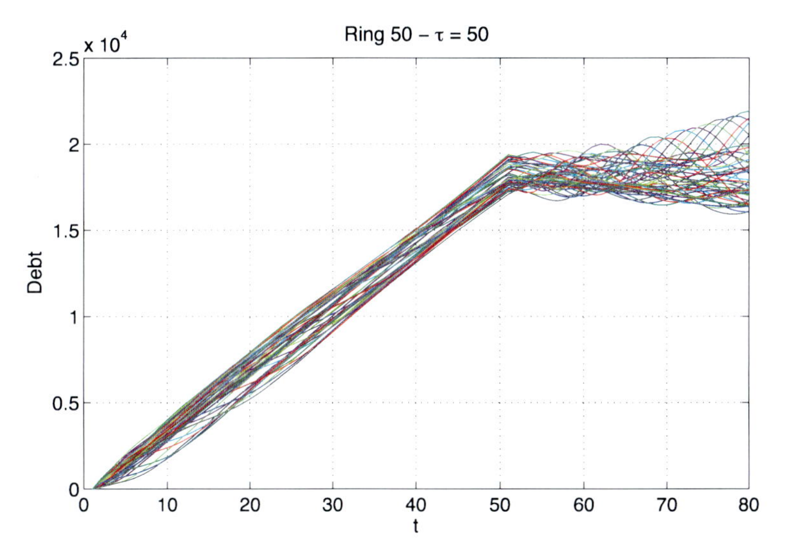

Fig. 123.8 Accumulation of debt with RING50 when $\tau = 50$

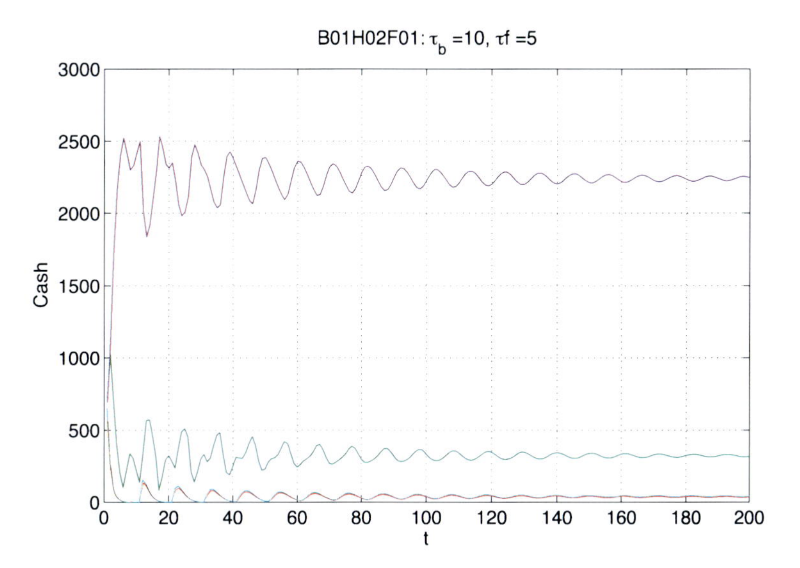

Fig. 123.9 Evolution of cash of B01H02F01

We started our analysis with the dynamics of supply and demand of goods using the Give and Take Model (GT-Model) where the state dynamics are built upon the exchanges of cash against goods between agents in goods market (G_g) and the edge dynamics depends on the mutual interests between agents. The beneficial relations in terms of flow are conserved and the interactions with sellers that propose very high unit price are cut. We found that the Strongly Connected Components (SCC) of G_g emerge. Each SCC becomes a sub-market where the unit price of goods is homogeneous. It appears that all the circuits of G_g remain available during the simulation.

Secondly, we merge the goods market G_g with a new market G_ℓ called labour market. The resulting Labour and Goods Market model (LGM) adds the transformation of working hours into goods. Here we focused on a particular topology where there are only a few firms and a high number of employees. Firms are not employed by other firms and employees are not employed by other employees. Employees buy only goods to firms and not to other employees. Firms sell only goods to employees and not to other firms. The SCC of $G = G_\ell \cup G_g$ emerge and are possibly split to little parts called sub-SCC, depending on the competition between firms inside each SCC.

Here we found that the edge dynamics of both GT and LGM models tends to conserve the circuits of the topology of whole market. Circuit means stability. The distribution of cash in both GT and LGM at the stationary state is proportional to the number of clients connected to each agent. As the number of clients is equal to the number of incoming edges, this property is similar to the so-called page rank. If the structure of the market is a scale free graph, the distribution of cash follows Zipf law. Consequently, the firms of LGM are richer than the employees.

Finally, we analyze independently the dynamics of lending-redeeming cash process. This model simulates the exchanges of cash and the redeeming processes after a certain duration. We found in contradiction with the results described with goods and labour markets that in such model, the existence of circuit increases the risk of crash. Circuit in this case means instability due to the accumulation of debt inside the circuit. The only way that ensures the stability is to choose a very short-term loan duration. In our case, it corresponds to one time step.

Acknowledgements We acknowledge financial support from the Swiss National Science Foundation.

References

1. Di Matteo T, Aste T, Hyde ST (2003) Exchanges in complex networks: income and wealth distributions. In: The physics of complex systems (New advances and perspectives), pp 435–442
2. Serrano M, Boguñá M, Vespignani A (2007) Patterns of dominant flows in the world trade web. J Econ Interact Coord 2:111–124
3. Bhattacharya K, Mukherjee G, Saramäki J, Kaski K, Manna SS (2008) The international trade network: weighted network analysis and modelling. J Stat Mech Theory Exp 2008(02):P02002
4. Glattfelder JB (2010) Ownership networks and corporate control: mapping economic power in a globalized world. PhD thesis, ETH Zurich, Department of Management, Technology, and Economics
5. Razakanirina RM (2007) Cellular automata over graph applied on financial and goods flows simulation. Master's thesis, University of Geneva, Computer Science Department

6. Chopard B, Falcone J-L, Razakanirina R, Hoekstra A, Caiazzo A (2008) On the collision-propagation and gather-update formulations of a cellular automata rule. In: ACRI proceedings, pp 144–151
7. Razakanirina RM, Chopard B (2011, in press) Multilayer cellular automata on a graph applied to the exchanges of cash and goods. J Cell Autom
8. Razakanirina RM, Chopard B (2012) Labour and goods market dynamics using an abstract microeconomical model. Acta Phys Pol B, Proc Suppl 5(1):131–143
9. Hartmann AK, Weigt M (2005) Phase transitions in combinatorial optimization problems. Wiley, New York
10. Caldarelli G (2007) Scale-free networks: complex webs in nature and technology (Oxford finance). Oxford University Press, London

Chapter 124
Multiple Levels in Self-adaptive Complex Systems: A State-Based Approach

Luca Tesei, Emanuela Merelli, and Nicola Paoletti

Abstract This work introduces a general multi-level model for self-adaptive systems. A self-adaptive system is seen as composed by two levels: the lower level describing the actual behaviour of the system and the upper level accounting for the dynamically changing environmental constraints on the system. In order to keep our description as general as possible, the lower level is modelled as a state machine and the upper level as a second-order state machine whose states have associated formulas over observable variables of the lower level. Thus, each state of the second-order machine identifies the set of lower-level states satisfying the constraints. Adaptation is triggered when a second-order transition is performed; this means that the current system no longer can satisfy the current high-level constraints and, thus, it has to adapt its behaviour by reaching a state that meets the new constraints. The semantics of the multi-level system is given by a flattened transition system that can be statically checked in order to prove the correctness of the adaptation model. To this aim we formalize two concepts of weak and strong adaptability providing both a relational and a logical characterization. We report that this work gives a formal computational characterization of multi-level self-adaptive systems, evidencing the important role that (theoretical) computer science could play in the emerging science of complex systems.

Keywords Self-adaptive systems · Multi-level model · Finite state machines · Adaptability relations

L. Tesei (✉) · E. Merelli · N. Paoletti
School of Science and Technology, Computer Science Division, University of Camerino, 62032 Camerino, Italy
e-mail: luca.tesei@unicam.it

E. Merelli
e-mail: emanuela.merelli@unicam.it

N. Paoletti
e-mail: nicola.paoletti@unicam.it

T. Gilbert et al. (eds.), *Proceedings of the European Conference on Complex Systems 2012*, Springer Proceedings in Complexity, DOI 10.1007/978-3-319-00395-5_124,

124.1 Introduction

Self-adaptive systems are a particular kind of systems able to modify their own behaviour according to their environment and to their current configuration. They learn from the environment and develop new strategies in order to fulfil an objective, to better respond to problems, or more generally to maintain desired conditions. Self-adaptiveness is an intrinsic property of the living matter. Complex biological systems naturally exhibit auto-regulative mechanisms that continuously trigger internal changes according to external stimuli. Moreover, self-adaptation drives both the evolution and the development of living organisms.

Recently there has been an increasing interest in self-adaptive properties of software systems. In [19] the following definition is given: "*Self-adaptive software evaluates its own behaviour and changes behaviour when the evaluation indicates that it is not accomplishing what the software is intended to do, or when better functionality or performance is possible*".

As a matter of fact, software systems are increasingly resembling complex systems and they need to dynamically adapt in response to changes in their operational environment and in their requirements/goals. Two different types of adaptation are typically distinguished:

– *Structural adaptation*, which is related to architectural reconfiguration. Examples are addition, migration and removal of components, as well as reconfiguration of interaction and communication patterns.
– *Behavioural adaptation*, which is related to functional changes, e.g. changing the program code or following different trajectories in the state space.

Several efforts have been made in the formal modelling of self-adaptive software, with particular focus on verifying the correctness of the system after adaptation. Zhang et al. give a general state-based model of self-adaptive programs, where the adaptation process is seen as a transition between different non-adaptive regions in the state space of the program [27]. In order to verify the correctness of adaptation they define a new logic called A-LTL (an adapt-operator extension to LTL) and model-checking algorithms [28] for verifying adaptation requirements. In PobSAM [15, 16] actors expressed in Rebeca are governed by managers that enforce dynamic policies (described in an algebraic language) according to which actors adapt their behaviour. Different adaptation modes allow to handle events occurring during adaptation and ensuring that managers switch to a new configuration only once the system reaches a safe state. Another example is the work by Bruni et al. [7] where adaptation is defined as the run-time modification of the control data and the approach is instantiated into a formal model based on labelled transition systems. In [6], graph-rewriting techniques [20] are employed to describe different characterizations of dynamical software architectures. Meseguer and Talcott [21] characterize adaptation in a model for distributed object reflection based on rewriting logic and nesting of configurations. Theorem-proving techniques have also been used for assessing the correctness of adaptation: in [18] a proof lattice called transitional invariant lattice is built to verify that an adaptive program satisfies global invariants before and after adaptation. In particular it is proved that if it is possible to build that lattice, then adaptation is correct.

There are several other works worth mentioning, but here we do not aim at presenting an exhaustive state-of-the-art in this widening research field. We address the interested reader to the surveys [9, 24] for a general introduction to the essential aspects and challenges in the modelling of self-adaptive software systems.

124.1.1 A Multi-level View of Self-adaptation

examples are Harel's statecharts [13], a visual formalism for hierarchical state machines. In the field of concurrency and process-algebra, we can mention the action refinement [1] and state refinement [26] techniques, as well as Cardelli's mobile ambients [8]. In addition, membrane systems [22] are one of the leading hierarchical models in the field of unconventional and biologically-inspired computing. Complex systems can be regarded as multi-level systems, where two fundamental levels can be distinguished: a *behavioural level* B accounting for the dynamical behaviour of the system; and a higher *structural level* S accounting for the global and more persistent features of the system. These two levels affect each other in two directions: *bottom-up*, e.g. when a collective global behaviour or new emergent patterns are observed; and *top-down*, e.g. when constraints, rules and policies are superimposed on the behavioural level. These two fundamental levels and their relationships are the base to scale-up to multi-level models. In a generic multi-level model, any n-th level must resemble the behavioural level, the corresponding $n+1$-level has to match with the structural level and the relationships between them will have to show the same characteristics. We discuss how this scale-up is implemented in our setting in Sect. 124.5

Multiple levels arise also when software systems are concerned. For instance, in [11] Corradini et al. identify and formally relate three different levels: the *requirement level*, dealing with high-level properties and goals; the *architectural level*, focusing on the component structure and interactions between components; and the *functional level*, accounting for the behaviour of a single component. Furthermore, Kramer and Magee [17] define a three-level architecture for self-managed systems consisting of a *component control level* that implements the functional behaviour of the system by means of interconnected components; a *change management level* responsible for changing the lower component architecture according to the current status and objectives; and a *goal management level* that modifies the lower change management plans according to high-level goals. Hierarchical finite state machines and Statecharts [13] have also been employed to describe the multiple architectural levels in self-adaptive software systems [14, 25].

In this work we introduce $S[B]$-*systems*: a general state-based model for self-adaptive systems where the lower behavioural level describes the actual dynamic behaviour of the system and the upper structural level accounts for the dynamically changing environmental constraints imposed on the lower system. The B-level is modelled as a state machine B. The upper level is also described as a state machine where each state has associated a set of constraints (logical formulas) over variables resulting from the observation of the lower-level states, so that each S-state identifies the set of B-states satisfying the constraints. Therefore, a set of dynamically changing constraints underlies a *second-order structure* S whose states are sets of B-states and, consequently, transitions relate sets of B-states.

We focus on *behavioural and top-down adaptation*: the B-level adapts itself according to the higher-level rules. In other words the upper level affects and constrains the lower level. Adaptation is expressed by firing a higher-order transition, meaning that the S-level switches to a different set of constraints and the B-level has adapted its behaviour by reaching a state that meets the new constraints. Our idea is broadly inspired by Zhang et al. [27], i.e. the state space of an adaptive program can be separated into a number of regions exhibiting a different *steady-state behaviour* (behaviour without reconfiguration). However, in our model the steady-state regions are represented in a more declarative way using constraints associated to the states of the S-level. Moreover, in $S[B]$-systems not only the behavioural level, but also the adaptation model embedded in the structural level is dynamic. Adaptation of the B-level is not necessarily instantaneous and during this phase the system is left unconstrained but an invariant condition that is required to be met during adaptation. Differently to [27], the invariants are specific for every adaptation transition making this process controllable in a finer way. The semantics of the multi-level system is given by a flattened transition system that can be statically checked in order to prove the correctness of the adaptation model. To this aim we also formalize the notion of adaptability, i.e. the ability of the behavioural level to adapt to a given structural level. We distinguish between weak and strong adaptability, providing both a relational and a logical characterization for each of them.

$S[B]$-systems has been inspired by some of the authors' recent work in the definition of a spatial bio-inspired process algebra called *Shape Calculus* [4, 5]. In that case, a process $S[B]$ is characterized by a reactive behaviour B and by a shape S that imposes a set of geometrical constraints on the interactions and on the occupancy of the process. This idea is shifted in a more general context in the $S[B]$-systems where, instead, we consider sets of structural constraints on the state space of the B-level. We want to underline that previous work and, mainly, this work have been conceived as contributions not only in the area of adaptive software system, but also in the area of modelling complex natural systems.

The notion of multiple levels that characterizes our approach for computational adaptive systems is something well-established in the science of complex systems. As pointed out by Baianu and Poli [3] "*All adaptive systems seem to require at least two layers of organization: the first layer of the rules governing the interactions of the system with its environment and with other systems, and a higher-order layer that can change such rules of interaction*". $S[B]$-systems are similarly built on two levels: the B-level describes the state-based behaviour of the system and the S-level regulates the dynamics of the lower level. In our settings, communication and interactions are not explicitly taken into account. Indeed the behavioural finite state machine can describe the semantics of a system made by several interacting components.

Another accepted fact is that higher levels in complex adaptive systems lead to higher-order structures. Here the higher S-level is described by means of a second order state machine (i.e. a state machine over the powerset of the B-states). Similar notions have been formalized by Baas [2] with the *hyperstructures* framework for multi-level and higher-order dynamical systems; and by Ehresmann and Vanbremeersch with their *memory evolutive systems* [12], a model for hierarchical autonomous systems based on category theory.

The paper is organized as follows. Section 124.2 introduces the formalism and the syntax of $S[B]$-systems, together with an ecological example that will be used also in the following. In Sect. 124.3 we give the operational semantics of a $S[B]$-system by means of a flattened transition system. In Sect. 124.4 we formalize the concepts of weak and strong adaptability both in a relational and in a logical form. Finally, conclusions and possible future developments of the model are discussed in Sect. 124.5.

124.2 A Multi-level State-Based Model

An $S[B]$-system encapsulates the behavioural (B) and the structural/adaptive (S) aspects of a system. The behavioural level is classically described as a finite state machine of the form $B = (Q, q_0, \rightarrow_B)$. In the following, the states $q \in Q$ will also be referred to as B-states and the transitions as B-transitions.

The structural level is modelled as a finite state machine $S = (R, r_0, \rightarrow_S, L)$ (R set of states, r_0 initial state, $\rightarrow_S$ transition relation and L state labelling function). In the following, the states $r \in R$ will be also referred to as S-states and the transitions as S-transitions. The function L labels each S-state with a set of formulas (the constraints) over an *observation* of the B-states in the form of a set of variables X. Therefore an S-state r uniquely identifies the set of B-states satisfying $L(r)$ and S gives rise to a second-order structure $(R \subseteq 2^Q, r_0, \rightarrow_S \subseteq 2^Q \times 2^Q, L)$.

In this way, behavioural adaptation is achieved by switching from an S-state imposing a set of constraints to another S-state where a (possibly) different set of constraints holds. During adaptation the behavioural level is no more regulated by the structural level, except for a condition, called *transition invariant*, that must be fulfilled by the system undergoing adaptation. We can think of this condition as a minimum requirement to which the system must comply to when it is adapting and, thus, it is not constrained by any S-state.

Note that an $S[B]$-system *dynamically* adapts and reconfigures its behaviour, thus both the behavioural level and the structural level are dynamic.

Definition 124.1 ($S[B]$-System Behaviour) The behaviour of an $S[B]$-system $S[B]$ is a tuple $B = (Q, q_0, \rightarrow_B)$, where

- Q is a finite set of states and $q_0 \in Q$ is the initial state; and
- $\rightarrow_B \subseteq Q \times Q$ is the transition relation.

In general, we assume no reciprocal internal knowledge between the S- and the B-level. In other words, they see each other as *black-box systems*. However, in order to realize our notion of adaptiveness, there must be some information flowing bottom-up from B to S and some information flowing top-down from S to B. In particular, the bottom-up flow is modelled here as a set of variables $X = \{x_1, \ldots, x_n\}$ called *observables* of the S-level on the B-level. The values of these variables must always be *derivable* from the information contained in the B-states, which can possibly hold more "hidden" information related to internal activity. This keeps our approach black-box-oriented because the S-level has not the full knowledge of the B-level, but only some derived (e.g. aggregated, selected or calculated) information. Concerning the top-down flow, the B-system only knows whether its current state satisfies the current constraint or not. If not, we can assume that the possible target S-states and the relative invariants are outputted by the S-system and given in input to the B-system.

Definition 124.2 ($S[B]$-System Structure) The structure of an $S[B]$-system $S[B]$ is a tuple $S = (R, r_0, \rightarrow_S, L)$, where

- R is a finite set of states and $r_0 \in R$ is the initial state;
- $\rightarrow_S \subseteq R \times \Phi(X) \times R$ is a transition relation, labelled with a formula called *invariant*; and
- $L : R \rightarrow \Phi(X)$ is a function labelling each state with a formula over a set of *observables* $X = \{x_1, \ldots, x_n\}$.

Thus, an $S[B]$-system has associated a finite set $X = \{x_1, \ldots, x_n\}$ of typed variables over finite domains $\{D_1, \ldots, D_n\}$ whose values must be completely determined in each state of Q. More formally,

Definition 124.3 (Observation Function) Given an $S[B]$-system $S[B]$ with a set $X = \{x_1, \ldots, x_n\}$ of observables, an *observation function* $\mathcal{O}\colon Q \rightarrow \prod_{i=1}^{n} D_i$ is a total function that maps each B-state q to the tuple of variable values $(v_1, \ldots, v_n) \in D_1 \times \cdots \times D_n$ observed at q.

Note that we do not require this function to be bijective. This means that some different states can give the same values to the observables. In this case, the difference is not visible to S, but it is internal to B.

We indicate with $\Phi(X)$ the set of formulas over the variables in X. We assume that constraints are specified with a first-order logic-like language.

Definition 124.4 (Satisfaction Relation) Let $S[B]$ be a $S[B]$-system with a set $X = \{x_1, \ldots, x_n\}$ of observables and with an observation function $\mathcal{O}$. A state $q \in Q$ *satisfies* a formula $\varphi \in \Phi(X)$, written $q \models \varphi$, iff φ is satisfied applying the substitution $\{v_1/x_1, \ldots, v_n/x_n\}$, where $\mathcal{O}(q) = (v_1, \ldots, v_n)$, using the interpretation rules of the logic language.

Let us also define an evaluation function $[\![\_]\!] : \Phi(X) \rightarrow 2^Q$ mapping a formula $\varphi \in \Phi(X)$ to the set of B-states $Q' = \{q \in Q \mid q \models \varphi\}$, i.e. those satisfying φ.

Let us now give an intuition of the adaptation semantics. Let the active S-state be r_i and $r_i \xrightarrow{\varphi}_S r_j$. Assume that the behaviour is in a steady state (i.e. not adapting) q_i and therefore $q_i \models L(r_i)$. If there are no B-transitions $q_i \rightarrow_B q_j$ such that $q_j \models L(r_i)$ the system starts adapting to the target S-state r_j. In this phase, the B-level is no more constrained, but during adaptation the invariant φ must be met. Adaptation ends when the behaviour reaches a state q_k such that $q_k \models L(r_j)$.

The following definition determines when the structure S of a $S[B]$-system is well formed, that is: it must no contain inconsistencies w.r.t. all possible variable observations and the initial B-state must satisfy the initial S-state.

Definition 124.5 (Well-Formed Structure) Let $S[B]$ be an $S[B]$-system. The structural level S is well-formed iff the following conditions hold:

- for all S-states $r \in R$, $L(r)$ must be *satisfiable*, in the sense that there must be a variable observation under which $L(r)$ holds ($\exists q \in Q.\ q \models L(r)$) and
- the initial B-state must satisfy the constraints in the initial S-state, i.e. $q_0 \models L(r_0)$.

In the remainder of the paper we assume to deal with well-formed structures without explicitly mentioning it.

124.2.1 An Example from Ecology

In this part we introduce a case study in the field of ecology and population biology: *the adaptive 1-predator 2-prey food web*. This system describes a variant of classical prey-predator dynamics where in normal conditions the predator consumes its favourite prey p_0. When the availability of p_0 is no longer sufficient for the survival of the predator, it has to adapt its diet to survive and it consequently starts consuming another species p_1. For the sake of showing the features of our model, here we present an oversimplified version of this system that omits quantitative aspects like predation rates and growth of prey. We assume that the predator initially consumes the prey p_0 (variable $p = 0$) and that prey may be available (variable $a_i = 1,\ i = 0, 1$) or not (variable $a_i = 0,\ i = 0, 1$). The effect of consuming an available prey is to make that prey unavailable, as expected. The predator may also decide not to eat and change its diet (variable $p = 1$). A boolean variable tells whether in the current state the predator has eaten some prey (variable *eat*). At each step the predator can do one of the following:

- eat the currently favourite prey p_i, if available ($a_i \leftarrow a_i - 1$ and $eat \leftarrow$ true);
- do not eat and switch its favourite prey ($p \leftarrow |1 - p|$ and $eat \leftarrow$ false); or
- do not eat.

Finally, if the predator does not feed itself for two consecutive times, it migrates to a more suitable habitat (variable $moved =$ true) and no further actions are possible. The attentive reader may notice that under these restrictions the system will inevitably lead to a state where the predator moves to a different habitat. This is due

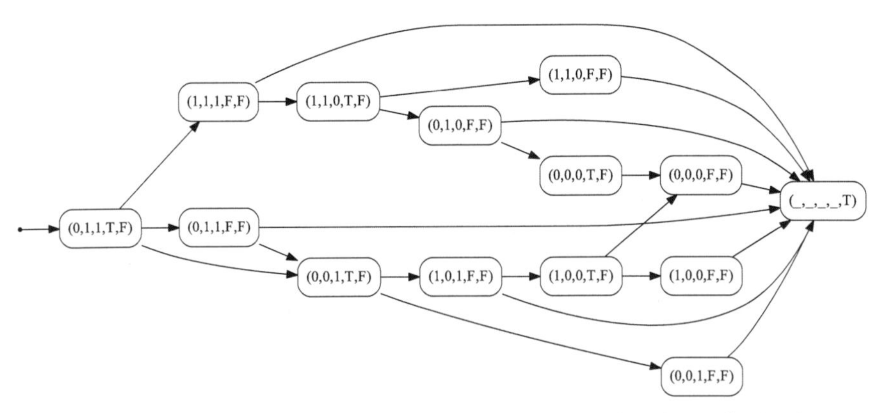

Fig. 124.1 The behavioural state machine B for the adaptive 1-predator 2-prey food web example. Each state is characterized by a different combination of the variables $(p, a_0, a_1, eat, moved)$ (favourite prey, availability of p_0, availability of p_1, has the predator eaten?, has the predator migrated?). The initial state is $(0, 1, 1, \mathsf{true}, \mathsf{false})$. All the states where $moved = \mathsf{true}$ has been grouped for simplicity to a single state $(\_, \_, \_, \_\mathsf{true})$

to the fact that prey growth is not modelled here and it is always the case that the system eventually reaches a state where the predator cannot feed because of the unavailability of both prey. Each state of the behavioural level (depicted in Fig. 124.1) is described by a different evaluation of the involved variables:

$$(p, a_0, a_1, eat, moved) \in \{0, 1\} \times \{0, 1\} \times \{0, 1\} \times \{\mathsf{false}, \mathsf{true}\} \times \{\mathsf{false}, \mathsf{true}\}.$$

In this example we consider two different S-levels (represented in Fig. 124.3): S_0 and S_1, but with the same set of S-states. More specifically S_0 is given by:

$$\begin{aligned} R &= \{r_0, r_1, r_2\}, \\ \rightarrow_S &= \left\{r_0 \xrightarrow{\neg moved} r_1, r_0 \xrightarrow{\neg eat} r_2, r_1 \xrightarrow{\neg moved} r_0, r_1 \xrightarrow{\neg eat} r_2\right\} \\ L(r) &= \left\{p == 0 \wedge (\neg eat \implies a_0 > 0) \wedge \neg moved\right\} \quad \text{if } r = r_0 \\ &\quad \left\{p == 1 \wedge (\neg eat \implies a_1 > 0) \wedge \neg moved\right\} \quad \text{if } r = r_1 \\ &\quad \{moved\} \quad \text{if } r = r_2. \end{aligned}$$

On the other hand, S_1 differs from S_0 only in the transition function, that is:

$$\rightarrow_S = \left\{r_0 \xrightarrow{p==1} r_1, r_1 \xrightarrow{\neg eat} r_2\right\}$$

The three different S-states model three different stable regions in the prey-predator dynamics:

- r_0: the predator consumes p_0. More precisely, the constraints require that the favourite prey must be p_0 ($p == 0$); that the predator has not moved to another habitat ($\neg moved$); and that if the predator is not currently feeding, the prey p_0

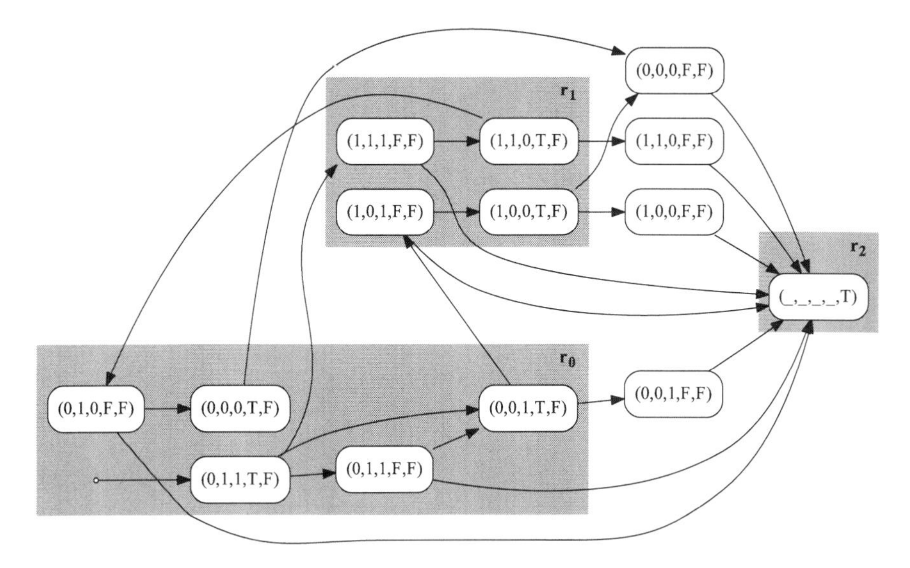

Fig. 124.2 S-states determining stable regions in the adaptive 1-predator 2-prey system

must be available so that the predator can eat in the following step ($\neg eat \implies a_0 > 0$).

- r_1: the predator consumes p_1; the constraints are the same as r_0, but referred to prey p_1.
- r_2: the predator has migrated.

Figure 124.2 shows how the structural constraints identify different stable regions in the behavioural level. The adaptation dynamics, regulated by the transitions in S_0, allow the predator to adapt from r_0 to r_1, under the invariant $\neg moved$ indicating that during adaptation the predator cannot migrate. The equivalent S-transition is defined from r_1 to r_0, so that the predator is able to return to its initially favourite prey. Both from r_0 and r_1 a S-transition to r_2 is allowed under the invariant $\neg eat$. In this way, the predator can adapt itself and migrate to a different habitat under starvation conditions. On the other hand, the transition relation in S_1 has been defined in a simpler way, which makes the predator adapt deterministically from r_0 to r_1 and finally to r_2. In this case, the adaptation invariant from r_0 to r_1 requires that the predator has changed its diet to prey p_1.

The following section will show the operational rule for deriving the transitional semantics of the $S[B]$-system as a whole and the semantics of $S_0[B]$ and $S_1[B]$ in the adaptive 1-predator 2-prey system will be given as well.

124.3 Operational Semantics

In this part, we give the operational semantics of an $S[B]$-system as a transition system resulting from the flattening of the behavioural and of the structural levels.

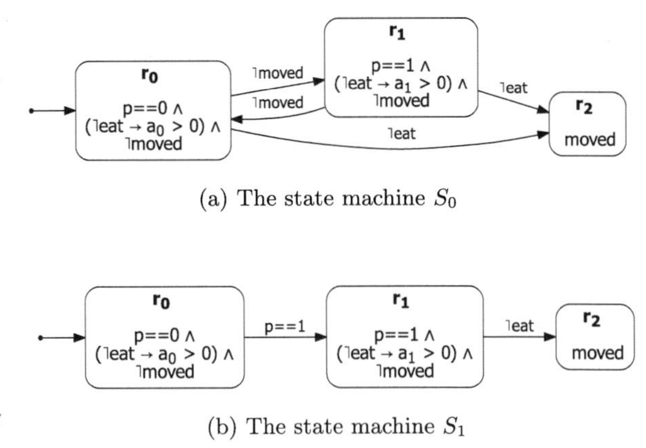

Fig. 124.3 The two different structural levels S_0 and S_1 in the adaptive 1-predator 2-prey food web example. In each S-state r_i the constraint imposed to the behavioural level are represented. Transition labels indicate adaptation invariants. S_0 allows the predator to adapt its diet and migrate due to starvation anytime. In S_1 adaptation is guided from r_0 (consume prey p_0), to r_1 (consume prey p_1) and finally to r_2 (migration)

We obtain a Labelled Transition System (LTS) over states of the form (q, r, ρ), where

- $q \in Q$ and $r \in R$ are the active B-state and S-state, respectively; and
- ρ keeps the target S-state that can be reached during adaptation and the invariant that must be fulfilled during this phase. Therefore ρ is either empty (no adaptation is occurring), or a singleton $\{(\varphi, r')\}$, with $\varphi \in \Phi(X)$ a formula and $r' \in R$ an S-state.

Definition 124.6 (Flat $S[B]$-System) Let $S[B]$ be an $S[B]$-system. A flat $S[B]$-system is a LTS $F(S[B]) = (F, f_0, \xrightarrow{r} \cup \xrightarrow{r,\varphi,r'})$ where

- $F \subseteq Q \times R \times 2^{\Phi(X) \times R}$ is the set of states;
- $f_0 = (q_0, r_0, \emptyset)$ is the initial state;
- $\xrightarrow{r} \subseteq F \times F$, with $r \in R$, is a family of transition relations between non-adapting states satisfying $L(r)$; and
- $\xrightarrow{r,\varphi,r'} \subseteq F \times F$, with $r, r' \in R$ and $\varphi \in \Phi(X)$, is a family of transition relations between states during the adaptation determined by the S-transition $r \xrightarrow{\varphi}_S r'$. As a consequence it holds that for all r, r', φ, $\xrightarrow{r} \cap \xrightarrow{r,\varphi,r'} = \emptyset$.

Table 124.1 lists the set of rules characterizing the flattened transitional semantics of an $S[B]$-system:

- Rule STEADY describes the steady (i.e. non-adapting) behaviour of the system. If the system is not adapting and the B-state q can perform a transition to a q' that satisfies the current constraints $L(r)$, then the flat system can perform a non-adapting transition $\xrightarrow{r}$ of the form $(q, r, \emptyset) \xrightarrow{r} (q', r, \emptyset)$.
- Rule ADAPTSTART regulates the starting of an adaptation phase. Adaptation occurs when none of the next B-states satisfy the current specification ($\forall q''.(q \rightarrow_B q'' \implies q'' \not\models L(r))$, or more compactly $(q, r, \emptyset) \not\xrightarrow{r}$). In this case, for each

Table 124.1 Operational semantics of a $S[B]$-system

$$\text{STEADY}\ \frac{q \rightarrow_B q' \quad q' \models L(r)}{(q, r, \emptyset) \xrightarrow{r} (q', r, \emptyset)}$$

$$\text{ADAPTSTART}\ \frac{\forall q''.(q \rightarrow_B q'' \implies q'' \not\models L(r)) \quad q \rightarrow_B q' \quad r \xrightarrow{\varphi}_S r' \quad q' \models \varphi}{(q, r, \emptyset) \xrightarrow{r,\varphi,r'} (q', r, \{(\varphi, r')\})}$$

$$\text{ADAPT}\ \frac{q \rightarrow_B q' \quad q' \models \varphi \quad q \not\models L(r')}{(q, r, \{(\varphi, r')\}) \xrightarrow{r,\varphi,r'} (q', r, \{(\varphi, r')\})}$$

$$\text{ADAPTEND}\ \frac{q \models L(r')}{(q, r, \{(\varphi, r')\}) \xrightarrow{r,\varphi,r'} (q, r', \emptyset)}$$

S-transition $r \xrightarrow{\varphi}_S r'$ an adaptation towards the target state r' under the invariant φ starts and the flat system performs an adapting transition $\xrightarrow{r,\varphi,r'}$ of the form $(q, r, \emptyset) \xrightarrow{r,\varphi,r'} (q', r, \{(\varphi, r')\})$.

- Rule ADAPT describes the evolution during the actual adaptation, leading to transitions of the form $(q, r, \{(\varphi, r')\}) \xrightarrow{r,\varphi,r'} (q', r, \{(\varphi, r')\})$. During adaptation the behaviour is not regulated by the specification and it must not satisfy the target constraints $L(r')$ ($q \not\models L(r')$). We also require that the invariant $\varphi \in \Phi(X)$ must always hold during this phase. Note that the semantics does not immediately assure that a state where the target formula holds is eventually reached. Formulations of the adaptability requirement are given in Sect. 124.4.
- Rule ADAPTEND describes the end of the adaptation phase, i.e. a transition $\xrightarrow{r,\varphi,r'}$ from an adapting state $(q, r, \{(\varphi, r')\})$ where q satisfies the set of target constraints ($q' \models L(r')$), to the steady non-adapting state $(q, r', \emptyset)$.

Note that rules STEADY+ADAPTSTART ensure that there cannot exist a non-adapting state with both an outgoing non-adapting transition $\xrightarrow{r}$ and an outgoing adapting transition $\xrightarrow{r,\varphi,r'}$. Conversely, rules ADAPT+ADAPTEND ensure that there cannot exist an adapting state with both an outgoing non-adapting transition and an adapting transition.

The flattened transitional semantics of the two systems $S_0[B]$ and $S_1[B]$ in the adaptive 1-predator 2-prey food web example presented in Sect. 124.2.1 is depicted in Fig. 124.4. First, we observe that the flat $S_0[B]$ system has a larger state space than the flat $S_1[B]$, due to the higher number of S-transitions in S_0. In both cases two different adaptation phases can be noticed, the first starting from the flat state $((0, 0, 1, \mathsf{true}, \mathsf{false}), r_0, \emptyset)$ and the second starting from $((1, 0, 0, \mathsf{true}, \mathsf{false}), r_1, \emptyset)$. While in $S_0[B]$ it is possible to adapt to the migration region also in the first phase, in $S_1[B]$ this is possible only in the second phase, i.e. when both prey become unavailable. Moreover in $S_0[B]$, we notice that in each adaptation phase there always exists an adaptation path leading to a target stable region, but some adaptation paths

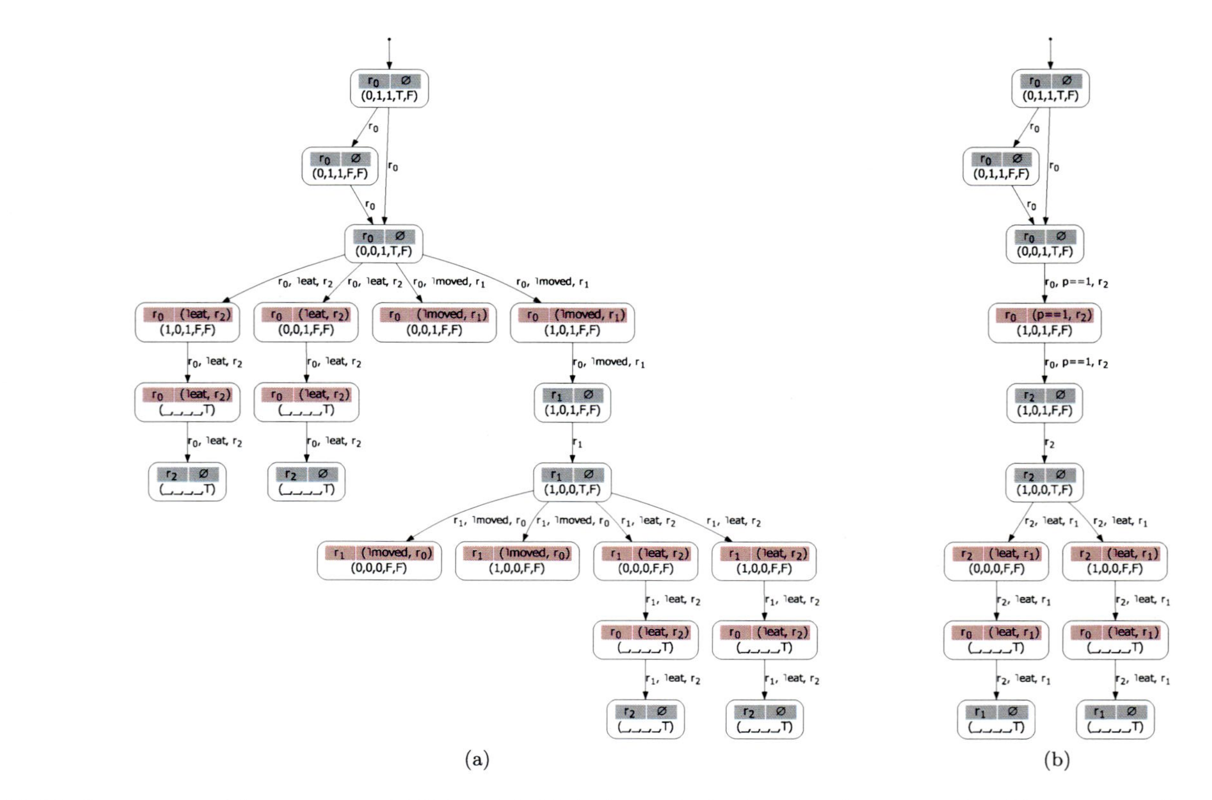

Fig. 124.4 The flat semantics of the two systems $S_0[B]$ (**a**) and $S_1[B]$ (**b**) in the adaptive 1-predator 2-prey example. Different structural levels lead to different adaptation capabilities. Two adaptation phases (*light red* marked ones) can be recognized: the first occurs when the predator stops consuming the prey p_0, the second when it stops consuming p_1. In both systems there always exists an adaptation path leading to a target stable region, but in $S_0[B]$ some paths violate the invariant and cannot proceed. In $S_1[B]$ every adaptation path leads to a target S-state

cannot proceed because they violate the invariant. Conversely, in $S_1[B]$ every adaptation path leads to a target S-state. Therefore the same behavioural level B possesses different adaptation capabilities, depending on the structure S it is embedded in. These two different kinds of adaptability are formalized in Sect. 124.4.

Although, depending on the structure S, the flat semantics could possibly lead to a model larger than the behavioural model B, the flat $S[B]$-system lends itself quite naturally to on-the-fly representation techniques. Indeed, during non-adapting phases it would be necessary to keep in memory just the subsystem restricted to the set $[\![L(r)]\!] \subseteq B$ of B-states that satisfy the current constraints $L(r)$. On the other hand, as soon as an adaptation of the form $(q, r, \emptyset) \xrightarrow{r,\varphi,r'} (q', r, \{(\varphi, r')\})$ takes place, it would be sufficient to store those B-states q'' such that $q'' \models \varphi \wedge q'' \not\models L(r')$, i.e. those state where the invariant is met, but the target constraints are not.

124.4 Adaptability Relations

The above described transitional semantics for $S[B]$-systems does not guarantee that an adaptation process always leads to a state satisfying the target constraints, or that the system can always start adapting when the current constraints are not met. We characterize this requirements on the adaptability of an $S[B]$-system by means of two binary relations over the set of B-states and the set of S-states, namely the *weak adaptability relation* $\mathcal{R}_w$ and the *strong adaptability relation* $\mathcal{R}_s$.

Informally, B is *weak adaptable* to S if any active B-state q satisfies the constraints imposed by the active S-state r, or it can start adapting and there exists a finite path reaching a B-state q' satisfying the constraints dictated by a target S-state r'. On the other hand, B is *strong adaptable* to S if any active B-state q satisfies the constraints imposed by the active S-state r, or it can start adapting towards a target S-state r' and all paths reach a B-state q' satisfying the constraints $L(r')$ in a finite number of transitions.

In the following definitions the notation $\rightarrow^i$ with $i \in \mathbb{N}$ indicates the exponentiation of the transition relation $\rightarrow$, i.e. $\rightarrow^i = (\rightarrow)^i = \rightarrow (\rightarrow)^{i-1}$. We use this notation to remark that adaptation paths must be of finite length.

Definition 124.7 (Weak Adaptability) *Weak-adaptability* is a binary relation $\mathcal{R}_w \subseteq Q \times R$ defined as follows. Let $q \in Q$ be a B-state and $r \in R$ be an S-state. Then, $q\mathcal{R}_w r$ iff

- $q \models L(r)$ and
- for all $q' \in Q$, whenever $q \rightarrow_B q'$, it holds that either
  - $q'\mathcal{R}_w r$, or
  - there exists $q'' \in Q, \varphi \in \Phi(X), r' \in R, i \in \mathbb{N}$, $(q, r, \emptyset) \xrightarrow{r,\varphi,r'} (q', r, \{(\varphi, r')\}) \xrightarrow{r,\varphi,r'}{}^{i} (q'', r', \emptyset)$ and $q''\mathcal{R}_w r'$.

Let $S[B]$ be an $S[B]$-system. Then B is weak adaptable to S if their initial states are weak adaptable, i.e. $q_0 \mathcal{R}_w r_0$.

Definition 124.8 (Strong Adaptability) *Strong-adaptability* is a binary relation $\mathcal{R}_s \subseteq Q \times R$ defined as follows. Let $q \in Q$ be a B-state and $r \in R$ be an S-state. Then, $q \mathcal{R}_s r$ iff

- $q \models L(r)$ and
- for all $q' \in Q$, whenever $q \rightarrow_B q'$, it holds that either
  - $q' \mathcal{R}_s\, r$, or
  - $(q, r, \emptyset) \xrightarrow{r,\varphi,r'} (q', r, \{(\varphi, r')\})$ for some $\varphi \in \Phi(X), r' \in R$ *and* every path starting from $(q', r, \{(\varphi, r')\})$ leads, in a finite number of consecutive $\xrightarrow{r,\varphi,r'}$ transitions, to a state $(q'', r', \emptyset)$ such that $q'' \mathcal{R}_s r'$.

Let $S[B]$ be an $S[B]$-system. Then B is strong adaptable to S if their initial states are strong adaptable, $q_0 \mathcal{R}_s r_0$.

In the remainder of the paper we will alternatively say that a system $S[B]$ is weak (strong) adaptable, in the sense that B is weak (strong) adaptable to S. It is straightforward to see that strong adaptability implies weak adaptability, since the strong version of the relation requires that every adaptation path reaches a target S-state, while the weak version just requires that at least one adaptation path reaches a target S-state. Now that a relational characterization of adaptability has been given, a concept of equivalence between B-states that are adaptable to the same S-states naturally arises. Therefore we define the weak adaptation equivalence and the strong adaptation equivalence over the set of B-states as follows.

Definition 124.9 (Weak Adaptation Equivalence) Two B-states $q_1, q_2 \in Q$ are said to be equivalent under weak adaptation, written $q_1 \approx_w q_2$, iff for each S-state $r \in R$, $q_1 \mathcal{R}_w r \iff q_2 \mathcal{R}_w r$.

Definition 124.10 (Strong Adaptation Equivalence) Two B-states $q_1, q_2 \in Q$ are said to be equivalent under strong adaptation, written $q_1 \approx_s q_2$, iff for each S-state $r \in R$, $q_1 \mathcal{R}_s r \iff q_2 \mathcal{R}_w r$.

As discussed in Sect. 124.3, the adaptive 1-predator 2-prey system possesses different adaptation capabilities depending on the structural level S. In particular we notice that the system $S_0[B]$ is *weak adaptable*, since in each adaptation phase there always exists an adaptation path leading to a target S-state. Nevertheless, it is not strong adaptable because there are adaptation paths that violate the invariant and consequently cannot end adapting. On the other hand, $S_1[B]$ is *strong adaptable*, because every adaptation path leads to a target S-state.

124.4.1 A Logical Characterization for Adaptability

In this part we formulate the above introduced adaptability requirements in terms of temporal formulae that can be statically checked on the flat $S[B]$-system. To this purpose we describe such properties in the well known *CTL (Computational Tree Logic)* [10], a branching-time logic whose semantics is defined in term of states. The set of well-formed CTL formulas are given by the following grammar:

$$\phi ::= \mathsf{false} \mid \mathsf{true} \mid p \mid \neg\phi \mid \phi \wedge \phi \mid \phi \vee \phi \mid \mathbf{AX}\phi \mid \mathbf{EX}\phi \mid \mathbf{AF}\phi \mid \mathbf{EF}\phi \mid$$
$$\mathbf{AG}\phi \mid \mathbf{EG}\phi \mid \mathbf{A}[\phi\mathbf{U}\phi] \mid \mathbf{E}[\phi\mathbf{U}\phi],$$

where p is an atomic proposition, logical operators are the usual ones ($\neg, \wedge, \vee$) and temporal operators (**X** next, **G** globally, **F** finally, **U** until) are preceded by the universal path quantifier **A** or the existential path quantifier **E**. Starting from a state s, CTL operators are interpreted as follows. $\mathbf{AX}\phi$: for all paths, ϕ holds in the next state; $\mathbf{EX}\phi$: there exists a path s.t. ϕ holds in the next state; $\mathbf{AF}\phi$: for all paths, ϕ eventually holds; $\mathbf{EF}\phi$: there exists a path s.t. ϕ eventually holds; $\mathbf{AG}\phi$: for all paths, ϕ always holds; $\mathbf{EG}\phi$: there exists a path s.t. ϕ always holds; $\mathbf{A}[\phi_1\mathbf{U}\phi_2]$: for all paths, ϕ_1 holds until ϕ_2 holds; and $\mathbf{E}[\phi_1\mathbf{U}\phi_2]$: there exists a path s.t. ϕ_1 holds until ϕ_2 holds).

In the following we provide the CTL formulas characterizing a weak adaptable and a strong adaptable $S[B]$-system. Formulas are evaluated over the flat semantics and we employ the proposition *adapt* to denote an adapting state. More formally, given a flat $S[B]$-system F and a state $s = (q_s, r_s, \rho_s)$, $\langle F, s\rangle \models adapt$ if and only if $\rho_s \neq \emptyset$. Additionally, the connective $\phi_1 \implies \phi_2$ has the usual meaning: $\neg\phi_1 \vee \phi_2$.

- *Weak adaptability*: for all paths, it always holds that as soon as adaptation starts, there exists at least one path for which the system eventually ends the adaptation phase leading to a target S-state.

$$\mathbf{AG}\big((\neg adapt \wedge \mathbf{EX}adapt) \implies \mathbf{EF}\neg adapt\big) \quad (124.1)$$

- *Strong adaptability*: for all paths, it always holds that whenever the system is in an adapting state, for all paths it eventually ends the adaptation phase leading to a target S-state.

$$\mathbf{AG}(adapt \implies \mathbf{AF}\neg adapt) \quad (124.2)$$

Proposition 124.1 (Equivalent Formulations of Weak Adaptability) *Let $S[B]$ be an $S[B]$-system. Then, $S[B]$ is weak adaptable if and only if $S[B]$ satisfies the weak adaptability CTL formula (Eq.* 124.1*). Formally, $q_0\mathcal{R}_s r_0 \iff \langle F, f_0\rangle \models$ $\mathbf{AG}((\neg adapt \wedge \mathbf{EX}\, adapt) \implies \mathbf{EF}\neg adapt)$, where F is the flat semantics of $S[B]$, q_0, r_0 and f_0 are the initial states of the behavioural level B, of the structural level S and of the flattened system F, respectively.*

Proposition 124.2 (Equivalent Formulations of Strong Adaptability) *Let $S[B]$ be an $S[B]$-system. Then, $S[B]$ is strong adaptable if and only if $S[B]$ satisfies the strong adaptability CTL formula (Eq.* 124.2). *Formally,* $q_0 \mathcal{R}_w r_0 \iff \langle F, f_0 \rangle \models \mathbf{AG}(adapt \implies \mathbf{AF} \neg adapt)$, *where F is the flat semantics of $S[B]$, q_0, r_0 and f_0 are the initial states of the behavioural level B, of the structural level S and of the flattened system F, respectively.*

Note that since we assume that the behavioural and the structural state machines are finite state, then the CTL adaptability properties can be model checked. This means that the defined notions of weak and strong adaptability are decidable.

124.5 Discussion and Conclusion

In this work we presented $S[B]$-systems, a general multi-level model for self-adaptive systems, where the lower B-level is a state machine describing the behaviour of the system and the upper S-level is a second-order state machine accounting for the dynamical constraints with which the system has to comply. Higher-order S-states identify stable regions that the B-level may reach by performing adaptation paths. An intriguing (but here simplified) case study from ecology has been provided to demonstrate the capabilities of $S[B]$-systems: the adaptive 1-predator 2-prey system. The semantics of the multi-level system is given by a flattened transition system and two different concepts of adaptability (namely, weak and strong adaptability) have been formalized, both in a relational flavour and with CTL formulas that can be model checked. We report that this work gives a formal computational characterization of self-adaptive systems, based on concepts like multiple levels and higher-order structures that are well-established in the science of complex systems.

Note also that in this work we defined in details just two levels, namely the S-level and the B-level. However, our approach can be easily extended in order to consider multiple levels arising from the composition of multiple $S[B]$-systems. Let $\{S^n[B^n]_i \mid i \in I\}$ be a set of $S[B]$-systems at a certain level n. Their parallel composition would be defined as $\|_{i\in I} S^n[B^n]_i$. Then, if we let $B^{n+1} = \|_{i\in I} S^n[B^n]_i$ be the behavioural state machine at level $n+1$, an higher-level $S[B]$-system $S^{n+1}[B^{n+1}]$ can be built by defining a structure S^{n+1} at level $n+1$, together with a set of observable variables X^{n+1} and with an observation function $\mathcal{O}^{n+1}$.

The present work is just an initial attempt and several extensions can be integrated into the model in the next future. First, the definition of a higher-level algebraic language for specifying $S[B]$-systems would be useful in order to handle more complex and larger models of adaptive systems. Additionally, we are currently investigating further adaptability relations and different models for the structural level, where adaptation can occur not only when no possible future behaviours satisfy the current constraints, but also when stability conditions are met. Then, another possible research direction would be embedding quantitative aspects into the two levels of an $S[B]$-system. In this way, an S-transition would have associated a measure of its cost/propensity, for distinguishing the adaptation paths more likely to occur (e.g. in the 1-predator 2-prey example, the predator adapting its diet), to those less probable (e.g. the predator migrating even under prey availability conditions).

Finally we assume that the reciprocal knowledge between the two levels is limited: they see each other as black-box systems. However, this approach could be extended in order that the structure S has a more comprehensive knowledge of the behaviour B. Under the *white-box assumption*, the structure could act as a sort of monitor that is able to statically check the behavioural model for properties of safe adaptation. In this way, the system will know in advance if an adaptation path eventually leads to a target S-state and if not, it will avoid that path. In other words, runtime model checking techniques allows the system to behave in an *anticipatory way*. Anticipation is a crucial property in complex self-adaptive systems, since it makes possible to adjust present behaviour in order to address future faults. A well-know definition is given by Rosen [23]: "*An anticipatory system is a system containing a predictive model of itself and/or its environment, which allows it to change state at an instant in accord with the model's predictions pertaining to a later instant*". In the settings of $S[B]$-systems, the predictive model of the system could be the behavioural level itself, or a part of it if we assume that S does not have a complete knowledge of B and is able to "look ahead" only at a limited number of future steps. The verdict of runtime model checking would be what Rosen refers to as model's predictions.

Acknowledgements The authors thank TOPDRIM: *Topology Driven Methods for Complex Systems* funded by the European Commission (FP7 ICT FET Proactive—N. 318121). The authors thank Marianna Taffi for helping in the definition of the case study.

References

1. Aceto L, Hennessy M (1993) Towards action-refinement in process algebras. Information and Computation 103:204
2. Baas N (1994) Emergence, hierarchies, and hyperstructures. In: Langton C (ed) Artificial life III, vol 17. Addison-Wesley, Reading, pp 515–537
3. Baianu I, Poli R (2010) From simple to super-and ultra-complex systems: a paradigm shift towards non-abelian emergent system dynamics. In: Theory and applications of ontology, vol 2
4. Bartocci E, Cacciagrano D, Di Berardini M, Merelli E, Tesei L (2010) Timed operational semantics and well-formedness of shape calculus. Scientific Annals of Computer Science 20:33–52
5. Bartocci E, Corradini F, Di Berardini M, Merelli E, Tesei L (2010) Shape calculus. A spatial mobile calculus for 3D shapes. Scientific Annals of Computer Science 20:1–31
6. Bruni R, Bucchiarone A, Gnesi S, Melgratti H (2008) Modelling dynamic software architectures using typed graph grammars. Electronic Notes in Theoretical Computer Science 213(1):39–53
7. Bruni R, Corradini A, Gadducci F, Lluch-Lafuente A, Vandin A (2012) A conceptual framework for adaptation. In: Proceedings of the 15th international conference on fundamental approaches to software engineering (FASE 2012). Springer, Berlin
8. Cardelli L, Gordon A (1998) Mobile ambients. In: Foundations of software science and computation structures. Springer, Berlin, pp 140–155

9. Cheng B, de Lemos R, Giese H, Inverardi P, Magee J, Andersson J, Becker B, Bencomo N, Brun Y, Cukic B et al (2009) Software engineering for self-adaptive systems: a research roadmap. In: Software engineering for self-adaptive systems, pp 1–26
10. Clarke E, Emerson E, Sistla A (1986) Automatic verification of finite-state concurrent systems using temporal logic specifications. ACM Transactions on Programming Languages and Systems (TOPLAS) 8(2):244–263
11. Corradini F, Inverardi P, Wolf A (2006) On relating functional specifications to architectural specifications: a case study. Science of Computer Programming 59(3):171–208
12. Ehresmann A, Vanbremeersch J (2007) Memory evolutive systems: hierarchy, emergence, cognition, vol 4. Elsevier, Amsterdam
13. Harel D (1987) Statecharts: a visual formalism for complex systems. Science of computer programming 8(3):231–274
14. Karsai G, Ledeczi A, Sztipanovits J, Peceli G, Simon G, Kovacshazy T (2003) An approach to self-adaptive software based on supervisory control. Self-adaptive software: applications 77–92. doi:10.1007/3-540-36554-0_3
15. Khakpour N, Jalili S, Talcott C, Sirjani M, Mousavi M (2010) PobSAM: policy-based managing of actors in self-adaptive systems. Electronic Notes in Theoretical Computer Science 263:129–143
16. Khakpour N, Khosravi R, Sirjani M, Jalili S (2010) Formal analysis of policy-based self-adaptive systems. In: Proceedings of the 2010 ACM symposium on applied computing. ACM, New York, pp 2536–2543
17. Kramer J, Magee J (2007) Self-managed systems: an architectural challenge. In: Future of software engineering, FOSE'07. IEEE Press, New York, pp 259–268
18. Kulkarni S, Biyani K (2004) Correctness of component-based adaptation. Component-Based Software Engineering 3054:48–58
19. Laddaga R (1997) Self-adaptive software. Technical report 98-12, DARPA BAA
20. Le Métayer D (1998) Describing software architecture styles using graph grammars. IEEE Transactions on Software Engineering 24(7):521–533
21. Meseguer J, Talcott C (2006) Semantic models for distributed object reflection. In: ECOOP 2002—Object-oriented programming, pp 1637–1788
22. Paun G (2003) Membrane computing. In: Fundamentals of computation theory. Springer, Berlin, pp 177–220
23. Rosen R (1985) Anticipatory systems. Pergamon, Elmsford
24. Salehie M, Tahvildari L (2009) Self-adaptive software: landscape and research challenges. ACM Transactions on Autonomous and Adaptive Systems (TAAS) 4(2):14
25. Shin M (2005) Self-healing components in robust software architecture for concurrent and distributed systems. Science of Computer Programming 57(1):27–44
26. Uselton A, Smolka S (1993) State refinement in process algebra. In: Proceedings of the North American process algebra workshop, Ithaca, New York
27. Zhang J, Cheng B (2006) Model-based development of dynamically adaptive software. In: Proceedings of the 28th international conference on software engineering. ACM, New York, pp 371–380
28. Zhang J, Goldsby H, Cheng B (2009) Modular verification of dynamically adaptive systems. In: Proceedings of the 8th ACM international conference on aspect-oriented software development. ACM, New York, pp 161–172

Chapter 125
Information Filtering and Learning: From Heuristics to Social Eudaimonia

Pietro Liò, Luce Jacovella, Lucia Bianchi, and Viet Nguyen

Abstract Big data pose many question on how our mental structures are prepared to analyse and use them to solve problems in the real world and to generate wellbeing and prosperity for our society. There are decision-making practices and analytical tools that we, as humans, have evolved to deal with simple problems of survival. can they help to efficiently take decisions moving across the complexity of Big data? Here we discuss how, following the intuitions of David Bohm and other authors, we can think of new approaches relaying on intuition and collective awareness to extract collective solutions. From the 'ghetto perspective' based on a limited number of data and simple heuristics to solve social demands problems, social media offer the opportunity to move towards new perspectives to confront societal challenge, promote sustainable development, social, economic and systemic change. Here we present some suggestions for a new organization of data based on collective awareness and introduces the concept of social eudaimonia. This new approach on data based on a collective response might lead to converging and harmonized, though not 'absolute', responses that provide a dynamic and value-based relationship between data and problem solving.

125.1 The Impact of Big Data in Our Evolution

Humans, because of their cognitive limitations, are often unable to perform rational calculations and information processing and instead rely on error—prone heuristics; moreover, even when people could optimise, that is to compute the best decision, they often rely on heuristics to save efforts at the price of sacrificing some accuracy.

P. Liò (✉) · V. Nguyen
Computer Laboratory, University of Cambridge, Cambridge, UK
e-mail: pietro.lio@cl.cam.ac.uk

V. Nguyen
e-mail: Viet.Nguyen@cl.cam.ac.uk

L. Jacovella
Department of Law, Queen Mary University, London, UK
e-mail: l.jacovella@qmul.ac.uk

T. Gilbert et al. (eds.), *Proceedings of the European Conference on Complex Systems 2012*, Springer Proceedings in Complexity, DOI 10.1007/978-3-319-00395-5_125,

These concepts are based on the principle of an accuracy—effort trade—off: the less information, computation, or time one uses, the less accurate one's judgments will be. This trade—off is believed to be one of the few general laws of the mind that Humans have evolved in the interaction with their environment to improve their conditions and survive. Such heuristics are also at the basis of our risk perception. However, as humans we struggle to grasp the value of information and data and to pin down what value is or whether value is absolute value (true value), or its value depends on other factors, such as its utility. We start from analyzing some common heuristics used by Humans to show how they are short-term and contingent and therefore inadequate to behave as adequate decision making tools to deal with Big data. A memory based recognition of the world would already help to develop value-based solution. However, we argue that Social media could provide a social memory and elaboration tool, based on collective intelligence to lead to a new organization of information, decision making and acting that we term "social eudaimonia". Big Data is definitely a new territory for our brains. We do very well in spotting something moving in a fixed environmental background, but recent works in physiology have shown that we do not do as well in crowded moving environment. There is a widespread belief that social media has increased the amount and quality of information but also background rumors. Big data is built on a huge number of contributions that have different quality and context properties and history and keeps changing. We need to keep track of the good quality bits and discard data with poor value. In doing this operation we attribute value to information on the basis of a scale of parameters or categories of value and we operate a value judgment. Discarding valuable information will result in decreasing the overall quality of the data we got. To avoid this, we often build metadata such indexes, scales of values, and other parameters to account for all changes and additions. Meta data, however, occupies large memory and consume energy to process it. In order to analyse Big data, it is fundamental therefore, to understand the process by which value can be created. Gasping to get the complete picture we may be able to do small bites and see small parts of the data in a detailed way and the bulk only with summary statistics. Observers, i.e. users, i.e. small biters from different sides of the Big Data may sample differently the Big data and have different views and arrive to very different conclusions. This would happen despite the advances in computational power, given that value judgment depends on the relationship between the evaluating subject and the evaluated object. This subject-object relationship is unique and subjective . Today, multi-scale and complex social media data are gathered and analysed in a rather simple way that completely misses the opportunity to uncover combinations of predictive profiles. Moreover, even if we could observe what happens at more or less all scales, from the single data point to the whole network, putting things together in order to obtain real understanding is much more difficult and much less developed in our evolutionary past. In using Big data, we need to keep in mind that we are too keen to apply heuristics that we developed for a different environment. We also need to carefully consider the history of additions and changes when we trade quality and quantity and also give high consideration for the information diversity. Finally, we need to have new statistics-based judgement, i.e. a sort of statistical sense.

125.2 The Heuristics We Acquired in Our Evolution

Through evolution we developed the capability of learning and using cognitive heuristics that operate fast, frugal and adaptive filtering strategies, i.e. effective, simple rules, requiring little estimation time and working under incomplete knowledge of the problem solution space. How do people make decisions when optimization is out of reach? Examples of the "embedded heuristics toolbox" we use at individual or social levels are: the Recognition heuristic, which states that if one of the two alternatives is recognized, one will infer that it has the higher value on the criterion (less-is-more effect is often detected); The $1/N$ equality heuristic, which allocates resources equally to each of N alternatives; Tit-for-Tat, in which one cooperates first and then imitates her/his partner's last behavior. Other widely used heuristics are the imitation of the behavior of majority, and the imitation of the successful person. The last two heuristics are recognized as a driving force in bonding and group identification and therefore play an important role in our everyday choices. In 1950 Herbert Simon first proposed that the people satisfice rather than maximise. Maximisation means optimisation, the process of finding the best solution for a problem, whereas satisficing (satisfying and sacrificing at the same time) means finding a good enough solution. This corresponds to a well known heuristics: in order to select a good alternative (e.g. a house or a spouse) from a series of options encountered sequentially, a person sets an aspiration level, chooses the first one that meets the aspiration and then terminates the search. The aspiration level can be fixed or adjusted following experience. Among these heuristics, Goldstein and Gigerenzer [1] have studied the recognition heuristic. This heuristic has proved to be not only fast and frugal, but it is also ecologically rational, in that it exploits structures of information coming from the environment in order to work. How well our heuristics are doing with Big Data such as social media. Is this a good or bad thing? Probably both. Because it so hard to grasp a full understanding of complex phenomena and wisdom seems to escaping from every day's meanings, the heuristic decision-making is short-term. Decisions are taken without their long term effects sufficiently elaborated in order to safe energy and respond to immediate problems or threats. We often hold the belief that a given analysis or decision holds a true value, until our 'truth' is challenged by new facts and data or by changing circumstances. What bridges the gap between the overload of data and the capacity of reading the data is often the purpose for which such analysis has been carried out. Any value judgment is open to mistakes and the attribution of value to data can change depending on what is under observation. Makiguchi for example, defines value as the quantitative expression of the relationship between the evaluating subject and the evaluated object [3]. Because value is the measure of the relationship between the object and the evaluating subject, it is essential that the subject base her judgment upon a correct scale of values. Makiguchi pointed that: "Value arises from the relationship between the evaluating subject and the object of evaluation". If either changes relative to the other, it is only obvious that the perceived value will change [3]. Any value judgment is only relative to circumstances and it is not absolute. Whatever the chosen criteria are, they have to allow sufficient flexibility to adapt to changing circumstances and they have

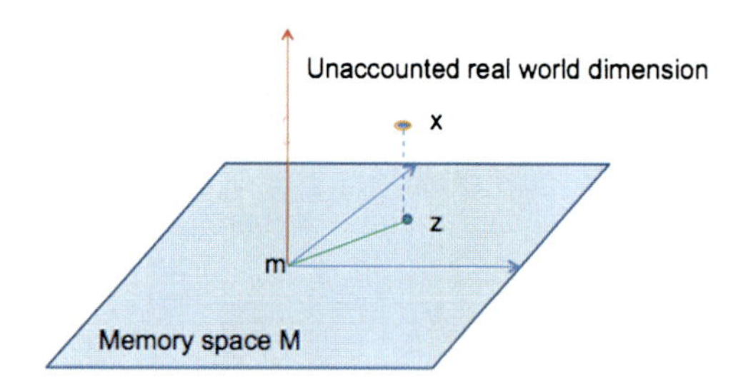

Fig. 125.1 From real world to cognition memory (see text)

to reflect their temporal viability. Moreover, the value cannot be separated from the subject since the value judgment depends upon who is the judge. The value changes with the changing of the subject, of the object and of their relationship. Makiguchi was probably trying to expose the complexity of the value judgments to overcome the simplistic dichotomies that are at the basis of heuristic thinking. He said: "In comparison with the complexity of reality, the criterion of mere utility, therefore the dichotomy yes/no, is no longer applicable". The object of the value judgment, knowledge, can become obsolete due to further developments in a given field or the priorities of the subject can shift, altering the relationship and the subsequent value. For this reasons, any viable value judgment has to reflect the complexity of the phenomena or objects involved, as well as the shifting priorities of the valuing subject. It is a constant challenge for meaning where individual and society dynamically reflect upon each other in the pursuing of value creation.

125.3 Cognition Memory and the Bayesian Brain

An important step would be to model the heuristics into a cognitive memory and behavioral model. A minimal cognition memory is constructed from two orthogonal basis, and the real world observation could is characterised by a three-dimensional space. As in Fig. 125.1 an observation about one object in real world, available as feature vector x, could be mapped to the hidden memory space as lower-dimensional representation vector z. The memory or mental space is defined by the mean m, and orthogonal basis U, which set the main parameters for calculating the likelihood function. The likelihood function of observation x, or the level of recognition given the current memory construct, are assumed to be computable from such parameters by addressing two components. The first component, which compare the position of z to m in the memory space, could be seen as an internal distance function of the mental space. The second component measures the distance of the same object between two spaces: the real world description of an object (x), and the perceived account of the same object into a personal mental space (z). This part could be seen as the discrepancy or the level of subjectivity of that particular person. We use the epsilon function to emphasise this.

Babies of about 24 month old are able starting to grasp the general structure of the surrounding world. From our framework, this coincides with the accumulation

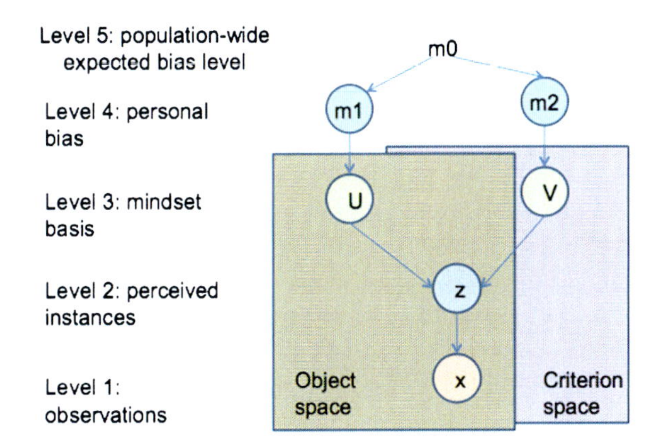

Fig. 125.2 A hierarchical Bayesian framework for cognition memory construction: here the distinction is between object and criterion space

threshold of reasonable amount of encounters i.e. It is now possible for the brain to derive a reasonable representation of the outside world, and use that basis to accelerate learning. (e.g. Once a child know that new words (dogs, cats) often associate with new shapes, not colours, the speed of learning new words start to steep up). Then the model could be extended to a cognitive representation of the world as in Fig. 125.2 with different levels of description and then including the five personality traits: Openness (inventive/curious vs. consistent/cautious), Conscientiousness (efficient/organized vs. easy-going/ careless) Extraversion (outgoing/energetic vs. solitary/reserved), Agreeableness (friendly/ compassionate vs. cold/unkind), Neuroticism (sensitive/ nervous vs. secure/ confident). A further extension may provide a representation of the parameter space of the cognitive behavior ranging from collaboration social value to competition/individual interest and distributed versus centrally controlled which, at population level, would synthesise the collective awareness and the social eudaimonia.

125.4 Social Eudaimonia Leads to Collective Intelligence Awareness

This daimon, is an intrinsic quality of the human nature that requires a constant dialectic exercise and the refusal of any dogma or pre-ordinate assumptions with one's self and with society. It is import to continue the journey of discovering based on principles, intellect, rather than demonstrative knowledge. The development of knowledge or the accumulation of data as a value per se, without any regard for the wider implications, is one of the current wrong assumptions of modern society. We assume that having a lot of data is per se a positive thing. Social media can generate data, even an excess of data, but can also help developing collective intelligence tools and improve our problem solving approach. In other words, social media produce a lot of noise, but can also expand the range of variables that we can elaborate collectively and still save energy. Drawing from Bohm implicate order [2], the solution to the dilemma is an inward order that manifests spontaneously in creativity. Social media order could develop from a multiscale order in which the analysis, interpretation and use of user generated data that is done at level of network or community level is replicated at larger scale, to an overarching collective order. Bohm assimilates the generative order to the process of art creation. When different levels are integrated by a general comprehensive order, we can produce a masterpiece; otherwise, the result is just mediocre work. This suggests a new notion of hierarchy, in which the more general principle, is immanent, that is, actively pervading and indwelling, not only in the less general, but ultimately in the reality as a whole. Emerging in this fashion, hierarchies are no longer fixed and rigid structures, involving domination of lower level by the higher. Rather they develop out of an immanent generative principle, from the more general to the less general. This new order arises or comes into play spontaneously as new organization of information. An interesting aspect of Bohm's analogy with a paint or work of art is the interpretation of the viewer [2]. In the context of social media the user is at the same time the generator of content and data and the viewer. This is unprecedented in modern history and requires a different schema or paradigm that allows the interpretation of data and the consequent decision-making process and action. When the schema changes, until it is absorbed, it might receive a negative reaction by the viewer. This fresh new order made of a hierarchy of parameters, or meta-order, requires the viewer to respond to new forms that are unfamiliar and therefore, disturbing. This represent the move from an implicate order to a super-implicate order that organizes the implicate order. Noteworthy, Aristotle links eudaimonia to the social capacities of human nature as contributing to the community is the highest fulfillment of human nature. Perhaps, we need to evolve new paradigms to be capable of living in a world in which the appreciation of complexity and collective intelligence easy our need for absolute certainty, a world without truths, but rich in understanding and wisdom, in which we can live pursuing a collective, social eudaimonia.

Acknowledgements This study was supported by the FP7 EU project RECOGNITION: Relevance and Cognition for Self-Awareness in a Content-Centric Internet (257756).

References

1. Gigerenzer G, Goldstein DG (2002) Models of ecological rationality: the recognition heuristic. Psychological Review 109(1):75–90
2. Bohm D, Peat FD (1987) Science, order and creativity. Routledge, London, p 163
3. Bethel DM (1973) Makiguchi the value creator: revolutionary Japanese educator and founder of the Soka Gakkai. John Weatherhill, Inc, New York, p 169

Part XVII
Satellite Meeting: Genomic Complexity

Chapter 126
Modelling the Genetic and Epigenetic Signals in Colon Cancer Using a Bayesian Network

Irina A. Roznovăţ and Heather J. Ruskin

Abstract Cancer, the unregulated growth of cells, has been a major area of focus of research for years due to its impact on human health. Cancer development can be traced back to aberrant modifications in genetic and epigenetic mechanisms within the body over time. Given time and cost implications of human genome experimentation, computational modeling is increasingly being employed to improve understanding of mechanisms which determine cancer initiation and progression. Here, we introduce a network-based model for genetic and epigenetic signals in colorectal cancer, with the focus on the gene level and tumor pathways. The current framework also considers the influence of ageing for micromolecular events in cancer development.

Keywords Genetic and epigenetic events · Colon cancer · Bayesian network · Ageing

126.1 Introduction

During recent decades, a novel direction in cancer research has been the identification and study of the *epigenetic* events that affect gene expression in tumor pathways. Defined as non-genetic hereditable modifications in chromatin structure [1], epigenetic signals have been linked to cancer development due to their impact on the function of both *tumor suppressor genes* and *oncogenes*. Detected in the earliest stages of different neoplastic diseases [2], epigenetic events are considered markers for cancer initiation. Additionally, they are already exploited in some cancer treatments, due to their *reversibility* property. Different genetic and epigenetic events have been reported as being progressive over time in cancer phenotype [3, 4]. For example, the mutation rate within the body increases due to the defects in the DNA

I.A. Roznovăţ (✉) · H.J. Ruskin
Centre for Scientific Computing & Complex Systems Modelling (SCI-SYM), School of Computing, Dublin City University, Dublin, Ireland
e-mail: iroznovat@computing.dcu.ie

H.J. Ruskin
e-mail: hruskin@computing.dcu.ie

T. Gilbert et al. (eds.), *Proceedings of the European Conference on Complex Systems 2012*, Springer Proceedings in Complexity, DOI 10.1007/978-3-319-00395-5_126,

repair system, which accumulate as ageing takes place. Additionally, specific genes (such as *estrogen receptor* (ER), *insulin-like growth factor II* (IGF2)) have been identified to be more sensitive to age-related methylation [5].

It is well known that cancer development is linked to the micromolecular events that affect the regulation of key cellular mechanisms. A network-based approach to model these systems can be an important tool in understanding the tumor pathways. In this paper, we describe in brief a computational model that is being developed to study the relationships between genetic and epigenetic signals in disease conditions. The focus here is on the influence of ageing over these micromolecular events during colorectal cancer development.

126.2 Colon Cancer Model

The current model is being built using a Bayesian network approach for describing gene relationships at different cancer stages. The external input of the framework is based on statistical data for the interdependencies between the micromolecular events (genetic and epigenetic) observed in colon cancer development. The model to date integrates data from *StatEpigen* database [6], and will be extended further to use information also from the other databases, such as *PubMeth* [7]. StatEpigen is a specifically targeted to colon cancer, manually curated and annotated database, developed at Sci-Sym, Dublin City University, Ireland and contains information on gene relationships in different pathology phenotype levels.

Hyper and *hypo*methylation are epigenetic events associated with oncogene activation, tumor suppressor gene inactivation, and with chromosomal instability in cancer development. They have a direct influence on the DNA methylation[1] (DNAm) level of a gene. Based on this important role in tumor development, DNAm is the main feature of the colon cancer model to date. The DNAm level is updated mainly by means of the data extracted from StatEpigen, which includes gene mutations and *hyper* and *hypo*methylation for different genes (known to have significant impact in colon cancer). In addition, based on the recently established dependence between the post-transcriptional histone[2] modifications (HM) and DNAm patterns, information about the core histone methylation and acetylation (the addition of acetyl group to a chemical compound) is incorporated, in order to compute a new gene DNAm level [8, 9]. These authors have reported that HM are more naturally unstable than DNAm, and that the relationships between HM and DNAm therefore must be described using different dynamics. The entire process of updating the DNAm level is referred into the model as a *methylation cycle*.

Considered a major risk factor in cancer development, *ageing* is an integrated factor for updating the DNAm level and HM in the current model. To analyze the

[1] The addition of a methyl group to cytosine ring.

[2] Histones are the proteins that pack DNA in nucleosomes. They are grouped into two categories: the core histones (H2A, H2B, H3, H4) and the linker histones (H1 and H5).

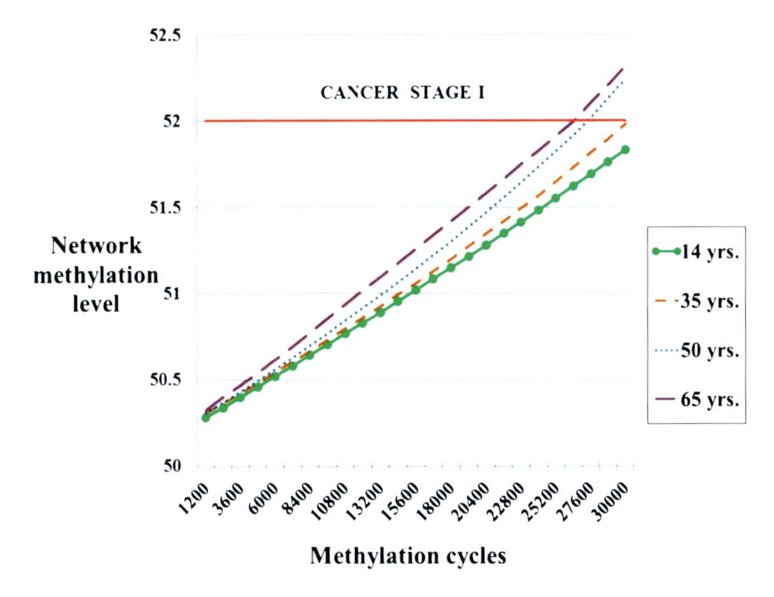

Fig. 126.1 Comparison between the gene methylation level for four patients (aged 14, 35, 50 and 65 respectively) accumulated over a 3×10^4-methylation cycles period

relationship between ageing and DNA methylation level, a case study was utilized including four patients of different ages (14, 35, 50 and 65 respectively). This experiment was started using empirical data on gene relationships extracted from StatEpigen and initially, tumors of all patients were considered to be at Stage 0 of cancer. The gene network in each case was allowed to evolve over time and the methylation level was recorded for all four, during $\sim 3 \times 10^4$ methylation cycles, as illustrated in Fig. 126.1.

At the end of the simulation, we need to know if the tumors of the patients were in the same cancer stage or not. In order to study the tumor progression, we analyzed two attributes for each network, namely the average methylation level and the highly-methylated genes. The results highlighted that methylation level updates follow different patterns for the four gene networks, with updating rate lower for younger than for older patients. Consequently, at the end of this simulation, it was observed that the tumor of the 50 and 65 year old patients had advanced to the next cancer stage (Stage 1).

126.3 Conclusions and Future Work

In this paper, a computational model for colorectal cancer dynamics was described in brief. Its focus is to study the impact of different molecular events on specific genes during cancer development, in order to predict tumor initiation and progression. The gene framework also considers the influences of ageing on genetic and epigenetic signals in updating the gene DNAm level. The results showed that the rate of update of the DNAm level was higher for older than for younger patients. This implies that the tumor was more aggressive for the former than for the latter and that ageing does influence the molecular events underlying cancer development. A limitation of the model to date is the fact that the relationship between gender and ageing has not yet been incorporated. For example, the analysis described in [10] highlighted a 5-year difference between genders in respect of colon cancer initiation. In addition to inclusion of this information in future development, a parallel programing approach will be designed and implemented to better accommodate the complexity of the molecular events in gene framework.

Acknowledgements This project acknowledges financial support from CIESCI ERA-Net Complexity Project, (EU/IRCSET). Access to the StatEpigen database is also gratefully acknowledged.

References

1. Allis CD, Jenuwein T, Reinberg D, Caparros ML (2007) Epigenetics. Cold Spring Harbor Laboratory Press, New York
2. Herman JG, Baylin SB (2003) Gene silencing in cancer in association with promoter hypermethylation. N Engl J Med 349(21):2042–2054
3. Fraga MF, Esteller M (2007) Epigenetics and aging: the targets and the marks. Trends Genet 23(8):413–418
4. Fraga MF, Agrelo R, Esteller M (2007) Cross-talk between aging and cancer. Ann NY Acad Sci 1100(o):60–74
5. Ahuja N, Li Q, Mohan AL, Baylin SB, Issa JPJ (1998) Aging and DNA methylation in colorectal mucosa and cancer. Cancer Res 58(23):5489–5494
6. Barat A, Ruskin HJ (2010) A manually curated novel knowledge management system for genetic and epigenetic molecular determinants of colon cancer. Open Color Cancer J 3:36–46
7. Ongenaert M, Van Neste L, De Meyer T, Menschaert G, Bekaert S, Van Criekinge W (2008) PubMeth: a cancer methylation database combining text-mining and expert annotation. Nucleic Acids Res 36(1):D842–D846
8. Cedar H, Bergman Y (2009) Linking DNA methylation and histone modification: patterns and paradigms. Nat Rev Genet 10(5):295–304
9. Raghavan K, Ruskin HJ (2011) Computational epigenetic micromodel-framework for parallel implementation and information flow. In: The eighth international conference on complex systems, vol 8. NECSI Knowledge Press, Boston, pp 340–353
10. Brenner H, Hoffmeister M, Arndt V, Haug U (2007) Gender differences in colorectal cancer: implications for age at initiation of screening. Br J Cancer 96(5):828–831

Chapter 127
The Role of the Genome in the Evolution of the Complexity of Metabolic Machines

Claudio Angione, Giovanni Carapezza, Jole Costanza, Pietro Lió, and Giuseppe Nicosia

Abstract In this positional paper, we consider the metabolic network of a bacterium as a living computer, and we are able to program it in order to obtain desired outputs. Furthermore, we discuss how the mutation-recombination-fixation events in the genome could lead to an increased computational capability, as well as to an increased complexity of the metabolic machine associated with the bacterium.

127.1 Introduction

Gene duplication events are important sources of novel biochemical and metabolic functions. After gene duplication, mutations cause the gene copies to diverge. The classical model predicts that these mutations will generally lead to the loss of function of one gene copy; rarely, new functions will be created and both duplicate genes are conserved. Most known genes belong to large families with extensive DNA sequence similarities. New proteins evolve from existing proteins by mutation. The neofunctionalisation model suggests that after a gene duplicates, since the two resulting copies are functionally redundant, one can accumulate mutations leading to a new function, while the other copy remains conserved [1]. The duplication, degeneration, complementation model proposes that if the ancestral gene possesses several

C. Angione (✉) · P. Lió
Computer Laboratory, University of Cambridge, Cambridge, UK
e-mail: costanza@dmi.unict.it

P. Lió
e-mail: pietro.lio@cl.cam.ac.uk

G. Carapezza · J. Costanza · G. Nicosia
Department of Mathematics and Computer Science, University of Catania, Catania, Italy

G. Carapezza
e-mail: carapezza@dmi.unict.it

J. Costanza
e-mail: claudio.angione@cl.cam.ac.uk

G. Nicosia
e-mail: nicosia@dmi.unict.it

T. Gilbert et al. (eds.), *Proceedings of the European Conference on Complex Systems 2012*, Springer Proceedings in Complexity, DOI 10.1007/978-3-319-00395-5_127,

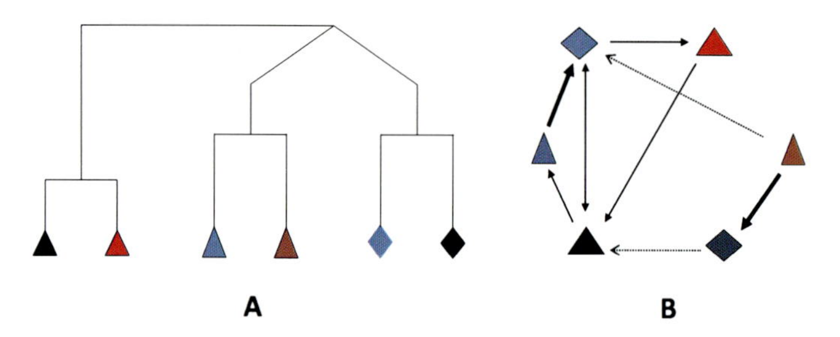

Fig. 127.1 In (**A**), the tree describes the events of the duplication. Recent events are separated by short distances in the tree. In (**B**), the duplicated genes conserve the reaction patterns in both metabolic pathways

subfunctions, these could be alternatively distributed among its duplicate descendants by neutral mutations. Although both models assume that the duplication itself has no intrinsic advantage, clearly an increased gene dosage (i.e. increased concentrations of a protein) after duplication can be beneficial in itself and alter the robustness and sensitivity of the biosystem [2]. There is a clear relation between gene duplication and metabolic complexity as shown in Fig. 127.1. Many, if not most, enzymes can promiscuously catalyze reactions, or act on substrates, other than those for which they evolved. The process of gene duplication could lead to structural changes resulting in the modification of the infidelity of molecular recognition.

Let us now turn into the relation between computation and metabolism inspired by Turing [3], who proposed computational processes in the morphogenesis. Turing states that an organism, most of the time, develops from one pattern into another. Many years later, Bray [4] argued that a single protein is able to transform one or multiple input signals into an output signal, thus it can be viewed as a computational or information carrying element. Following this line of thought, we provide a framework to show that bacteria could have computational capability and act as molecular machines. This relationship is based on the mapping between the metabolism and a RM (equivalent to a TM). Specifically, we think the reactions in the bacterium as increment/decrement instructions of the RM, where the RM registers count the number of molecules of each metabolite.

It is well known that a von Neumann architecture is composed of a processing unit, a control unit, a memory to store both data and instructions, and input-output mechanisms. We propose an effective formalism to map the von Neumann architecture to an entire bacterial cell, which becomes a molecular machine. We model the processing unit of the bacterium as the collection of all its chemical reactions, so as to associate the chemical reaction network of bacteria with a TM [5]. Here we use GDMO [6] to obtain Pareto fronts representing multi-objective optimisations in the metabolism. Each point of the Pareto front provided by GDMO is a molecular machine to execute a particular task. Pareto optimality allows to obtain not only a wide range of Pareto optimal solutions, but also the *best trade-off design*. In Fig. 127.2 we show a Pareto front obtained with GDMO when optimising acetate and succinate.

Optimal genetic interventions in cells, framed as optimal programs to be run in a molecular machine, can be exploited to extend and modify the behaviour of

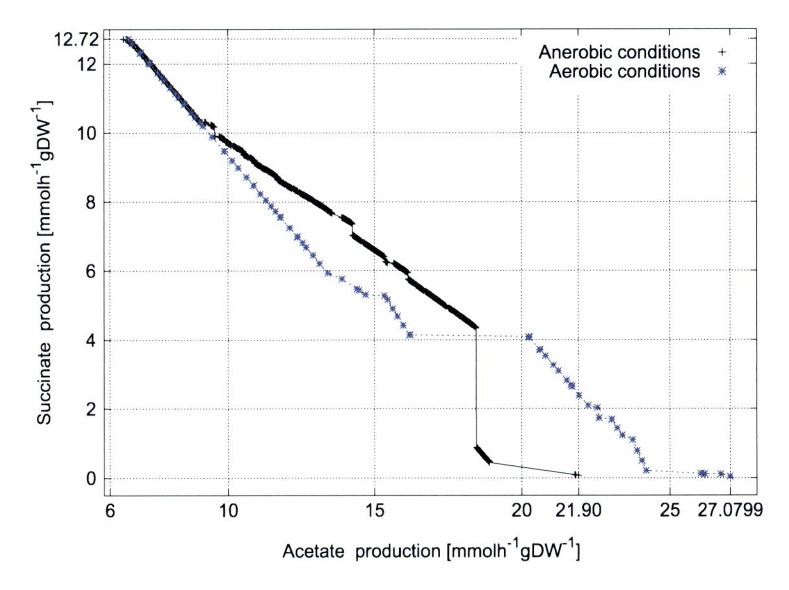

Fig. 127.2 Pareto fronts for the simultaneous maximisation of succinate and acetate production obtained by GDMO in anaerobic and aerobic conditions ($O_2 = 10$ mmol h^{-1} gDW^{-1}), with glucose feed equal to 10 mmol h^{-1} gDW^{-1}

cells and cell aggregates. For instance, programs can instruct cells to make logic decisions according to environmental factors, current cell state, or a specific user-imposed aim, with reliable and reproducible results.

127.2 Computing with Bacteria

Inspired by Brent and Bruck [7], who studied similarities and differences between biological systems and von Neumann computers, in Fig. 127.3 we propose a mapping between the von Neumann architecture and bacteria. Specifically, the metabolism of a bacterium can be viewed as a Turing Machine (TM).

The bacterium takes as input the substrates required for its growth and, thanks to its chemical reaction network, produces desired metabolites as output. The string y acts as a program stored in the RAM [5]. Let us consider the multiset Y of the bits of y. A partition Π of the multiset $Y = \{y_1, y_2, \dots, y_L\}$ is a collection $\{b_1, b_2, \dots, b_p\}$ of submultisets of Y that are nonempty, disjoint, and whose union equals Y. The elements $\{b_s\}_{s=1,\dots,p}$ of a partition are called blocks. We denote by $P(Y;p)$ the set of all partitions of Y with p blocks. $P(Y;p)$ has a cardinality equal to the Stirling number, namely $|P(Y;p)| = S_{L,p}$.

In order to formalise the control unit behaviour, let us define the function:

$$g_\Phi : \{0,1\}^L \longrightarrow \bigcup_{y \in \{0,1\}^L} P(Y;p), \qquad \bar{y} \in \{0,1\}^L \longmapsto \Pi \in P(\bar{Y};p),$$

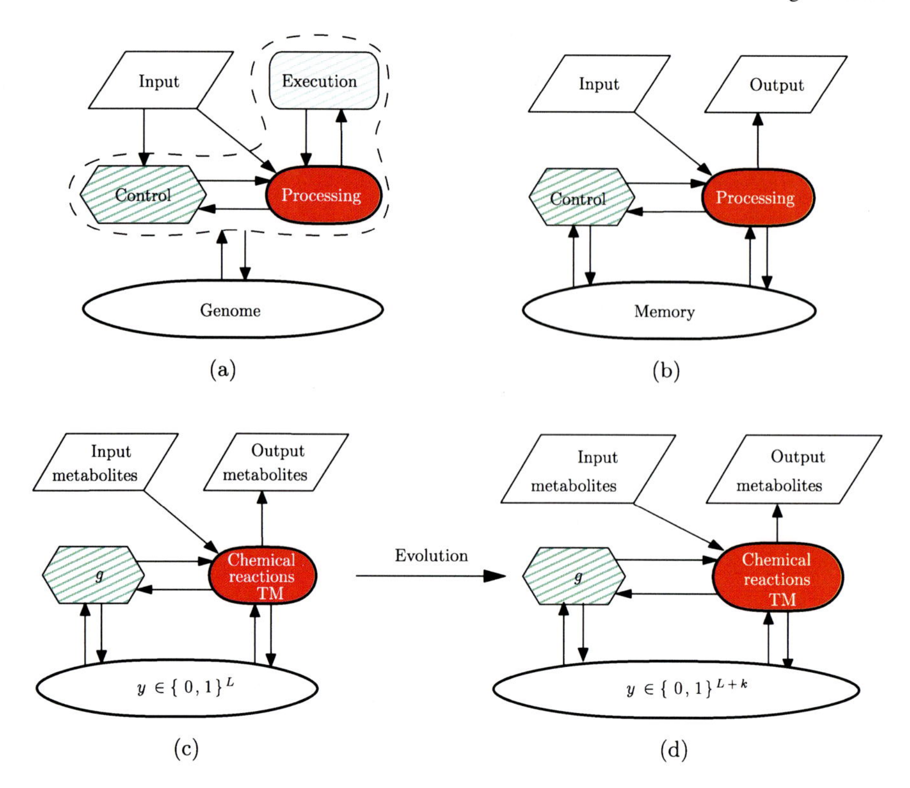

Fig. 127.3 Comparison inspired by [7] among biological systems (**a**), von Neumann architecture (**b**), and evolving bacteria (**c**)–(**d**). The string y is a program stored in the RAM. The function g represents the control unit: it interprets the binary string y and turns gene sets off. The processing unit is the metabolism of bacteria, composed of all the chemical reactions taking place in it. The goal is to produce desired metabolites as output of the molecular machine. The evolution and the growth of a bacterium is the consequence of duplication and mutation occurring in the genome, which increases its length (**d**)

where the partition Π is uniquely determined by the pathway-based clustering of the chemical reaction network. We can formalise this clustering as a p-blocks partition Φ of the set of the bit indexes in the string y. In particular, if we denote by $[L]$ the set of the first L natural numbers, we have $\Phi \in P([L]; p)$ [5]. The partition Φ allows the control function g_Φ to partition the multiset Y associated with the string y (see Fig. 127.4).

The function g_Φ turns syntax into semantics [5], i.e. it works as a control unit that translates the binary string y and makes use of it to turn gene sets on and off, according to the pathways in the metabolism. Each element of the partition Π is the submultiset b_s of all the gene sets related to reactions in the s-th pathway. The processing unit of the bacterium could be modelled as the collection of all its chemical reactions. Therefore, the chemical reaction network of bacteria can be associated with a TM [8].

Let us consider the *Minsky's register machine* (RM), i.e. a finite state machine augmented with a finite number of registers. Formally, a Minsky machine $\mathcal{M} = (D, i_0, i_1, \varphi)$ is composed of a finite set D of states, a finite set $H = \{H_r\}_r$ of registers, and a multivalued mapping $\varphi : D\backslash\{i_0\} \longrightarrow \{(H_r, i), (H_r, j, k) \mid H_r \in H, j, k \in D\}$. The set D has two distinguished elements $i_0, i_1 \in D$ representing the initial state and the halting state respectively. Each register H_r of the RM stores a non-negative integer. The instruction $inc(i, r, j)$ increments register r by 1 and causes the machine to move from state i to state j through the mapping $\varphi(i) = j$. Conversely, the instruction $dec(i, r, j, k)$, given that $H_r > 0$, decrements register r by 1 and causes the machine to move from state i to state j ($\varphi(i) = j$); if $H_r = 0$, the machine moves from state i to state k ($\varphi(i) = k$). The Minsky's RM has been proven to be equivalent to the TM [9]. Indeed, a RM is a multitape TM with the tapes restricted to act like simple registers (i.e. "counters"). A register is represented by a left-handed tape that can hold only positive integers by writing stacks of marks on the tape; a blank tape represents the count '0'.

The chemical reaction network of a bacterium can be mapped to the RM by defining [8]: (i) the set of state species $\{D_i\}$, where each D_i is associated with the state i of the RM; (ii) the set of register species $\{H_r\}$, where each H_r is associated with the register r of the RM, and therefore represents the molecular count of species r. The instruction $inc(i, r, j)$ represents the chemical reaction $D_i \rightarrow D_j + H_r$, while the instruction $dec(i, r, j, k)$ represents either $D_i + H_r \rightarrow D_j$ or $D_i \rightarrow D_k$ depending on whether $H_r > 0$ or $H_r = 0$ respectively. The molecular machine performs the "test for zero" by executing the reaction $D_i \rightarrow D_k$ only when H_r is over, since the r-th register cannot be decreased and the reaction $D_i + H_r \rightarrow D_j$ cannot take place. In the FBA approach coupled with the metabolic machine, the variables are the fluxes of the chemical reactions, therefore a high flux corresponds to both a high rate of reaction and a high mass of products. Hence, given the increment reaction $inc(i, r, j)$, the value of H_r is positively correlated with the reaction flux; conversely, in the decrement reaction $dec(i, r, j, k)$, when $H_r > 0$ the value of H_r is negatively correlated with the reaction flux. In a fixed volume V in which the reactions occur, given two reactions *inc* and *dec* with fluxes v_1 and v_2 respectively, the metabolism of the bacterium has a probability of error per step equal to $\epsilon = v_2/(v_1/V + v_2)$.

Since the simulated TM can be universal, the correspondence between metabolism and TM allows to perform any kind of computation through a set of species and chemical reactions characterised by their flux. As a result, bacteria can carry out at least any computation performed by a computer. A program embedded in a bacterium, whose metabolism works like a TM, could be able to implement the robust knockout strategy found by GDMO [6].

127.3 The Complexity of an Organism

Formally, let y be the array representing the sequence of the L genes of the organism. During the evolution process, a gene or a subsequence of genes (e.g. an

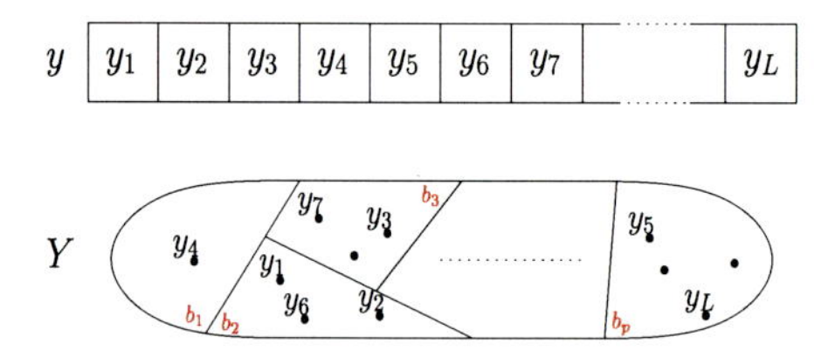

Fig. 127.4 The multiset Y associated with y is partitioned by Π in p blocks. The elements of Π are submultisets of Y, since y is a string of bits, thus 0 and 1 may occur more than once in the same subset. In this example, $\Pi = \{\{y_4\}, \{y_1, y_6, y_2\}, \dots, \{y_5, \dots, y_L\}\}$, $\Phi = \{\{4\}, \{1, 6, 2\}, \dots, \{5, \dots, L\}\}$

operon) can be duplicated and inserted in the sequence. Without loss of generality, let us assume that the last k genes are duplicated:

$$y = (y_1, \dots, y_L) \longrightarrow y = (y_1, \dots, y_L, y_{L+1}, \dots, y_{L+k}).$$

This process is called gene amplification or gene duplication. Initially, the following *condition of duplication* holds: $y_l = y_{l-k}, \forall l = L+1, \dots, L+k$. In fact, since the duplication is a stochastic process, the condition of duplication is not always guaranteed. However, after the duplication, mutations occur on new and existing genes, thus we obtain the final string $y = (y_1, \dots, y_{L'})$, where $L' = L + k$. Let us suppose that the gene y_L was responsible for the reaction $D_i \to D_j + H_r$, and therefore for the instruction $inc(i, r, j)$ in the RM. After the duplication, both y_L and y_{L+k} will code for the same reaction $D_i \to D_j + H_r$. Conversely, after the mutation, y_{L+k} will code for another reaction, say $D_{i'} \to D_{j'} + H_{r'}$. As a result, the complexity of the metabolic machine has increased, since a new reaction $inc(i', r', j')$ is now operating in the RM.

Starting from an ancestor, each duplication followed by a mutation shapes the computational capability of the metabolic machines represented by its metabolism. The mutation is a stochastic process that creates the possibility of a new instruction of the metabolic machine, while the natural selection can keep or discard this new instruction. Hence, the complexity of an organism evolves on the basis of stochastic processes and natural selection. The genome amplification allows the organism to increase the range of chemical reactions available, increasing also the range of increment and decrement instruction in the RM associated with the metabolism. As a result, the metabolic machine increases its computational power.

Acknowledgements P. Lió acknowledges the MIUR Flagship (PB05) INTEROMICS Initiative. G. Carapezza acknowledges HPC Edinburgh.

References

1. Sikosek T, Chan HS, Bornberg-Bauer E (2012) Escape from adaptive conflict follows from weak functional trade-offs and mutational robustness. Proc Natl Acad Sci USA 109(37):14888–14893

2. Massingham T, Davies LJ, Lió P (2001) Analysing gene function after duplication. BioEssays 23(10):873–876
3. Turing AM (1952) The chemical basis of morphogenesis. Philos Trans R Soc Lond B, Biol Sci 237(641):37–72
4. Bray D et al (1995) Protein molecules as computational elements in living cells. Nature 376(6538):307–312
5. Angione C, Carapezza G, Costanza J, Lió P, Nicosia G (2012) Computing with metabolic machines. In: Voronkov A (ed) Turing-100. EPiC Series, vol 10, pp 1–15
6. Costanza J, Carapezza G, Angione C, Lió P, Nicosia G (2012) Robust design of microbial strains. Bioinformatics. doi:10.1093/bioinformatics/bts590
7. Brent R, Bruck J (2020) computing: can computers help to explain biology? Nature 440(7083):416–417. 2006
8. Soloveichik D et al (2008) Computation with finite stochastic chemical reaction networks. Nat Comput 7(4):615–633
9. Minsky ML (1967) Computation. Prentice Hall, New York

Chapter 128
Can We Understand Parameter Values in the Human Genome?

Wentian Li

Abstract We aim at compiling a list of human genome parameters with the following questions in mind: Are these parameters accurately estimated? How do the values in human genome compare to those in other primates? Can we understand the evolutionary origin or cause of these values?

128.1 Introduction

Mathematics has its set of unitless (dimensionless) constants, such as π, e, golden ratio, Euler constant, etc. (http://en.wikipedia.org/wiki/Mathematical_constant). Similarly, physics has its own set of fundamental constants that can be measured in a physics experiment (http://physics.nist.gov/cuu/Constants/). Examples include the speed of light, gravitational constant, electron charge, etc. Then why don't we see discussions on constants in the field of biology? It is because biology is essentially about evolution, on how genotype and phenotype change with species, with environment, and with time. Though some features remain constant for very long time, such as the genetic code, or the number of gender in sexual species, these are the exceptions instead of rules. However, if we limit to one species in a given period of time, we may use a set of parameter values to define that species, for example, us the *homo sapient*.

Collecting a set of human genome parameters can be useful for both theoretical and practical investigations. We can have a better idea of where human stands in the space of all parameter values. We can compare human genome parameters with those from other closest primates' genomes, such as chimpanzee [4] and gorilla [35]. We will also carry out a back-of-envelop genomic estimation more easily.

We do not discuss human parameters that are not related to the genome; for example, total number of cells in an adult ($10–100 \times 10^{12}$), number of different cell types (between 200 and 400), etc. Also not discussed are derived quantities, such as the nearest-neighbor-base correlation, as a correlation can be measured in a variety

W. Li (✉)
The Robert S. Boas Center for Genomics and Human Genetics, The Feinstein Institute for Medical Research, North Shore LIJ Health System, 350 Community Drive, Manhasset, NY 11030, USA

T. Gilbert et al. (eds.), *Proceedings of the European Conference on Complex Systems 2012*, Springer Proceedings in Complexity, DOI 10.1007/978-3-319-00395-5_128,

Table 128.1 Some human genome parameters (*: not discussed in [19])

Parameter name	Range of values	Unit	References
Genome size	3×10^9	base	
% repeats (*)	50 % (up to 69 %)	%	[6]
Chromosomes	23	pairs	
Genes	20000–100000	num.	[12, 34]
Exon length (*)	100–150	base	
Transcript length (*)	2100–3000	base	[20]
Transcription factor	2000	num.	[41]
(G+C) content	40 %	%	[23]
(G+C)/gene-rich regions	120	num.	[19]
SNP	$3\text{–}30 \times 10^6$	num.	[30]
New deleterious mutations	0.5–2	num./generation	[28]
Crossovers	50–100 (F), 30–70 (M)	num.	[3, 15, 37]
Crossover hotspots (*)	15000–25000	number	[24, 27]

of ways (observed-over-expected, observed-subtract-expected, etc.). For one human genome parameter, we would like to know whether it is accurately measured, how does its value compare to other primates; and whether there are any theoretical justifications of its value. Some parameters are intrinsically linked to the issue of genome complexity. Most of this presentation were already in [19]. Here some parameters not discussed before are added.

128.2 A List of Human Genome Parameters

Some human genome parameters are listed in Table 128.1.

Genome Size The change of genome size from the common ancestor of primates and human is thought to be mainly caused by transposable elements. In the current human population, two individual's genome sizes may not be identical due to "structural variation" such as copy-number-variations (large-scale insertions and deletions). If traced even early to fish, then whole genome duplication could be an important factor [5, 25, 32]. Allowing size to increase is a key feature in modelling dynamical systems with certain property [17, 18], and genome size increase may well be a cause of large-scale correlation in genomes [22].

Proportion of Transposable Elements/Repetitive Sequences Closely related to the genome size is the percentage of human genome that are transposable elements. In a standard procedure, this information is obtained by running the RepeatMasker program (http://www.repeatmasker.org/) which aligns the known list of transposable elements with the genome. However, a recent article argues that ancestral transposable elements may experience much more mutation, thus the "repeat-derived" sequences may not be detected by alignment [6]. Using an alternative approach ("oligo cloud"), they have reached a value of 66–69 % as the proportion of repetitive or repetitive-derived sequences.

Number of Chromosome Pairs The number of 23 pairs of chromosomes (22 autosomal pairs plus either XX for female or XY for male) is a rather stable number within the human species. Missing one copy of a chromosome in a pair (monosomy) is fatal for an autosomal chromosome and leads to Turner syndrome for the sex chromosome (XO). Possessing an extra copy of a chromosome (trisomy) leads to Down syndrome (chr21), or Edwards syndrome (chr18), or Patau syndrome (chr13), but fatal for other chromosomes except for the sex chromosome. Most other primates have 24 pairs of chromosome, indicating a fusion event leading to *homo sapient* [8].

Number of Genes If the number refers to the protein-coding loci, it is around 20000. If the number refers to the gene-product, considering the possibility of alternative splicing, it can be as high as 10^5. There are two major collections of human genes. One is the NCBI RefSeqGene [34] which is not human-specific. Another is the GENCODE [12] that include protein-coding loci with all alternative transcribed products as well as non-protein-coding loci with transcript evidence (e.g. long non-coding RNA [7]) and pseudogenes. On the theoretical front, there were discussions on a minimum number of genes needed to make a functional cell [16], though this is more relevant to bacteria than to human. Another line of thought is that the more genes, the higher the chance for an individual to experience a lethal mutation, thus the maximum number of genes should be limited [14, 33].

Exon Lengths Using the RefSeqGene data, it can be shown that the histogram of exon-length has a single peak (mode) in the range of 100–150 bp. The distribution is very long-tailed, though can be made more symmetric if exon-length is log-transformed. Since neither exon-lengths nor log-exon-lengths are normally distributed, mean and median are not expected to the same with the mode. Interestingly, the value of mode is close to the 147 bp—the length of DNA that wraps around histone for form nucleosome. Somewhat related, a previous study points to a length of 205 bp for the exon-intron pair [1], and there are other papers exploring the relationship between exon-intron pattern and chromatin structure [36, 40].

Transcript Lengths This is the length that adds up all exons in a gene, i.e., the DNA sequence that is transcribed to mRNA. Very similar to that of exon distribution, the histogram of transcript length is more symmetric when the length is log-transformed [20]. Since not all RNA sequences are translated (there are 5' and 3' untranslated regions (UTR)), the mode of transcript length may not match the mode of protein sequence length (multiplied by 3).

Number of Transcription Factors (TF) The number of TFs is roughly 10 % of the number of genes [41]. It is tempting to conclude that one TF has 10 targets to regulate, but in fact, the number of targets follow a power-law distribution [19].

Using ChIP-seq technology, TF binding sites can be systematically studied, which have been done for 42 selected TFs [9]. An approximate alternative technology is to combine the chromatin open sites, determined by the DNase I hypersensitive sites, with the sequence motif prediction [39]. In the 2012 Nobel-Prize-winning work, four TFs were shown to be sufficient to convert mouse fibroblasts cells to pluripotent stem cells [38], highlighting the crucial role played by TF in cell differentiation. TFs are hypothesized to be the fast evolving part of a genome, and may contribute the most to the complexity of a genome.

Guanine-Cytosine Content (G+C-Content) (G+C)-content of human genome (40 %) is higher than expected by a neutral biased mutation model (33 %) [19, 23]. In fact, human genome is not homogeneously high in (G+C)-content, but mainly in regions with high gene contents and high recombination rates. There could be many debates on the cause-effect arrow of these correlated quantities [10], but no conclusive evidence in sight.

Number of (G+C)-Rich and Gene-Rich Regions A preliminary analysis indicates there are 120 (G+C)-rich and gene-rich regions [19]. This number is comparable to that of a recently proposed "super-structure" of human genome [2]

Number of Single-Nucleotide-Polymorphisms (SNPs) We calculate the number of neutral-mutation caused polymorphism [30, 31, 43]: $N_{snp} = 4N_e M(0.577 + \log(n))$, where N_e is the ancestral population size at the bottleneck (3100 for European/Asian, 7500 for African), M is the number of newly generated mutation per haploid-genome per generation (33), n is the current world population size in haploid unit ($2 \times 6.8 \times 10^9$). The more people re-sequenced, the larger the number of SNPs. It would be interesting to see whether the number predicted by this formula is correct.

Number of New Deleterious Mutation per Generation We partition the mutations first into unobserved (because these are lethal) and observed ones; then in the observed group, partition them into neutral, deleterious, and advantageous ones. Using the estimation of mutation rate from [28], it can shown that the new non-lethal deleterious mutation per generation is between 0.5 and 2. The interest in this number is due to Müller [26], usually refer to as Müller's ratchet—one-way decay towards lower fitness caused by the accumulation of deleterious mutations [29, 42]. A recent work claims that Müller's ratchet can be halted by advantageous mutations [11].

Number of Crossovers During Meiosis There are two lines of evidences in measuring the number of crossovers. One is through genetic analysis of pedigree data [15]. Another is by an immunocytogenetic assay to mark the mismatch repair protein MLH1 [3, 37]. While for male data, the two techniques lead to similar conclusion, there are discrepancies for female data. The number of crossovers in a chromosome is proportional to the chromosome length [21], and one may also estimate the number of crossovers from the genome size.

Number of Crossover (Recombination) Hotspots Although 15000 [24] and 25000 [27] recombination hotpots in the human genome were claimed, there are many questions remain. For example, why the locations of hotspots are not conserved in chimpanzee [44]? Could it be an artifact of a population demographic history [13]? It is not impossible that this number is unstable thus wouldn't be considered as a parameter after all.

References

1. Beckmann JS, Trifonov EN (1991) Splice junctions follow a 205-base ladder. Proc Natl Acad Sci USA 88:2380–2383
2. Carpena P, Oliver JL, Hackenberg M, Coronado AV, Barturen G, Bernaola-Galván P (2011) High-level organization of isochores into gigantic superstructures in the human genome. Phys Rev E 83:031908
3. Cheng EY et al (2009) Meiotic recombination in human oocytes. PLoS Genet 5:e1000661
4. Cheng EY et al (2005) Chimpanzee sequencing and analysis consortium: initial sequence of the chimpanzee genome and comparison with the human genome. Nature 437:69–87
5. Dehal P, Boore JL (2005) Two rounds of whole genome duplication in the ancestral vertebrate. PLoS Biol 3:e314
6. De Koning APJ, Gu W, Castoe TA, Batzer MA, Pollock DD (2011) Repetitive elements may comprise over two-thirds of the human genome. PLoS Genet 7:e1002384
7. Derrien T et al (2012) The GENCODE v7 catalog of human long noncoding RNAs: analysis of their gene structure, evolution, and expression. Genome Res 22:1775–1789
8. Dutrillaux B, Rethoré MO, Lejeune J (1975) Comparison du caryotype de l'orang-outang (Pongo pygmaeus) á celui de l'homme, du chimpanzé et du gorille. Ann Génét 18:153–161
9. ENCODE Project Consortium (2012) An integrated encyclopedia of DNA elements in the human genome. Nature 489:57–74
10. Freudenberg J, Wang M, Yang Y, Li W (2009) Partial correlation analysis indicates causal relationships between GC-content, exon density and recombination rate in the human genome. BMC Bioinform 10(1):S66
11. Goyal S, Balick DJ, Jerison ER, Neher RA, Shraiman BI, Desai MM (2011) Rare beneficial mutations can halt Muller's ratchet. arXiv:1110.2939
12. Harrow J, et al (2012) GENCODE: the reference human genome annotation for the ENCODE project. Genome Res 22:1760–1774
13. Johnston HR, Cutler DJ (2012) Population demographic history can cause the appearance of recombination hotspots. Am J Hum Genet 90:774–783
14. King JL, Jukes TH (1969) Non-Darwinian evolution. Science 164:788–798
15. Kong A et al (2002) A high-resolution recombination map of the human genome. Nat Genet 31:241–247
16. Koonin EV (2000) How many genes can make a cell: the minimal-gene-set concept. Annu Rev Genomics Hum Genet 1:99–116
17. Li W (1989) Spatial 1/f spectra in open dynamical systems. Europhys Lett 10:395–400
18. Li W (1991) Expansion-modification systems: a model for spatial 1/f spectra. Phys Rev A 43:5240–5260
19. Li W (2011) On parameters of the human genome. J Theor Biol 288:92–104
20. Li W (2012) Menzerath's law at the gene-exon level in the human genome. Complexity 17:49–53
21. Li W, Freudenberg J (2009) Two-parameter characterization of chromosome-scale recombination rate. Genome Res 19:2300–2307
22. Li W, Kaneko K (1992) Long-range correlation and partial 1/f spectrum in a non-coding DNA sequence. Europhys Lett 17:655–660

23. Lynch M (2010) Rate, molecular spectrum, and consequences of human mutation. Proc Natl Acad Sci USA 107:961–968
24. McVean G, Myers S, Hunt S, Deloukas P, Bentley DR, Donnelly P (2004) The fine-scale structure of recombination rate variation in the human genome. Science 304:581–584
25. Meyer A, Van de Peer Y (2005) From 2R to 3R: evidence for a fish-specific genome duplication (FSGD). BioEssays 27:937–945
26. Müller HJ (1950) Our load of mutations. Am J Hum Genet 2:111–176
27. Myers C, Bottolo L, Freeman C, McVean G, Donnelly P (2005) A fine-scale map of recombination rates and hotspots across the human genome. Science 310:321–324
28. Nachman MW, Crowell SL (2000) Estimate of the mutation rate per nucleotide in human. Genetics 156:297–304
29. Neher RA, Shraiman BI (2012) Fluctuations of fitness distributions and the rate of Muller's ratchet. Genetics 191:1283–1293
30. Nei M, Li WH (1979) Mathematical model for studying genetic variation in terms of restriction endonucleases. Proc Natl Acad Sci USA 76:5269–5273
31. Nei M (1987) Molecular evolutionary genetics. Columbia University Press, New York
32. Ohno S (1970) Evolution by gene duplication. Springer, Berlin
33. Ohno S (1972) So much 'junk' DNA in our genome. Brookhaven Sysp Biol 23:366–370
34. Pruitt KD, Tatusova T, Brown GR, Maglott DR (2012) NCBI reference sequences (RefSeq): current status, new features and genome annotation policy. Nucleic Acids Res 40(Database issue), D130–D135
35. Scally A et al (2012) Insights into hominid evolution from the gorilla genome sequence. Nature 483:169–175
36. Schwartz S, Meshorer E, Ast G (2009) Chromatin organization marks exon-intron structure. Nat Struct Mol Biol 16:990–995
37. Sun F et al (2006) Variation in MLH1 distribution in recombination maps for individual chromosomes from human males. Hum Mol Genet 15:2376–2391
38. Takahashi K, Yamanaka S (2006) Induction of pluripotent stem cells from mouse embryonic and adult fibroblast cultures by defined factors. Cell 126:663–676
39. Thurman RE et al (2012) The accessible chromatin landscape of the human genome. Nature 489:75–82
40. Tilgner H, Nikolaou C, Althammer S, Sammeth M, Beato M, Valcárcel J, Guigó R (2009) Nucleosome positioning as a determinant of exon recognition. Nat Struct Mol Biol 16:996–1001
41. Vaquerizas JM, Kummerfeld SK, Teichmann SA, Luscombe NM (2009) A census of human transcription factors: function, expression and evolution. Nat Rev Genet 10:252–263
42. Wardlaw AM, Agrawal AF (2012) Temporal variation in selection accelerates mutational decay by Muller's ratchet. Genetics 191:907–916
43. Watterson GA (1975) On the number of segregating sites in genetical models without recombination. Theor Popul Biol 7:256–276
44. Winckler W et al (2005) Comparison of fine-scale recombination rates in humans and chimpanzees. Science 308:107–111

Part XVIII
Satellite Meeting: Critical Phenomena and Collective Behavior of Multi-particle Systems

Chapter 129
Kinetic Theory of Two-Species Coagulation

Carlos Escudero

Abstract We study the stochastic process of two-species coagulation. This process consists in the aggregation dynamics taking place in a ring. Particles and clusters of particles are set in this ring and they can move either clockwise or counterclockwise. They have a probability to aggregate forming larger clusters when they collide with another particle or cluster. We study the stochastic process both analytically and numerically. Analytically, we derive a kinetic theory which approximately describes the process dynamics and determine its asymptotic behavior. In particular we answer the question of how the system gets ordered, with all particles and clusters moving in the same direction, in the long time.

Keywords Stochastic processes · Coagulation dynamics · Kinetic equations

The theoretical study of coagulation and its kinetic description is of broad interest because of its vast applicability in diverse topics such as aerosols [20], polymerization [22, 27], Ostwald ripening [5, 15], galaxies and stars clustering [21], and population biology [17] among many others. We propose a generalization of the stochastic process of coagulation. We consider the coagulation process among two different species: the aggregation takes place only when one element of one of the species interacts with an element of the other species. In particular, we place particles and clusters of particles in a ring, where they move with constant speed, either clockwise or counterclockwise. When two clusters (or two particles or one particle and one cluster) meet they have the chance to aggregate and form a cluster containing all particles involved in the collision. The direction of motion of the newborn cluster is chosen following certain probabilistic rules. We are interested in the properties of the realizations of such a stochastic process, and in particular in their long time behavior. Our main theoretical technique is the use of kinetic equations, an approach we have outlined in [10]. This is of course just one possible extension of the theory of coagulation. We have designed it getting inspiration from self-organizing systems and in particular from collective organism behavior. Let us note that this

C. Escudero (✉)
Departamento de Matemáticas & ICMAT (CSIC-UAM-UC3M-UCM), Universidad Autónoma de Madrid, 28049 Madrid, Spain

T. Gilbert et al. (eds.), *Proceedings of the European Conference on Complex Systems 2012*, Springer Proceedings in Complexity, DOI 10.1007/978-3-319-00395-5_129,

is a field that has been studied using a broad range of different theoretical techniques [2, 6, 7, 23, 25]. Another field which has inspired ourselves is the study of the dynamics of opinion formation and spreading [3, 8, 24], which is represented for instance by the classical voter model [4, 13, 16]. As a final influence, we mention that clustering has been previously studied in population dynamics models [12] including swarming systems [14], and coagulation equations have been used in both swarming [18] and opinion formation models [19]. Despite of its simplicity, the two-species coagulation model could be related to some of these systems.

A particular system that has influenced the current developments is the collective motion of locusts. The experiment performed in [1] revealed that locusts marching on a (quasi one dimensional) ring presented a coherent collective motion for high densities; low densities were characterized by a random behavior of the individuals and intermediate densities showed coherent displacements alternating with sudden changes of direction. The models that have been introduced to describe this experiment assume that the organisms behave like interacting particles [1, 11, 26]. Related interacting particle models have been used to describe the collective behavior of many different organisms and analyzing the mathematical properties of such models has been a very active research area [2, 6, 7]. The two-species coagulation model could be thought of as a particular limit of some of these models or as a simplified version of them which still retains some desirable features.

The goal of our current work is not to describe the detailed behavior of any specific system. Instead, we explore the mathematical properties of a stochastic process which has been designed by borrowing inspiration from different self-organizing systems. Therefore the focus is on mathematical tractability. We determine under which conditions consensus is reached and what form it adopts. We introduce the kinetic theory that approximately describes the stochastic process and concentrate on its mathematical analysis. We also study the system by means of direct numerical simulations of the stochastic process. We use them to check the predictions of our kinetic theory and explore the stochastic process beyond the kinetic level. Kinetic approximations neglect many sources of fluctuations and thus numerical simulations are required in order to describe many properties of the individual realizations of the stochastic process which are not reflected at the kinetic level.

Our analysis is limited to the one-dimensional spatial situation with periodic boundary conditions (the dynamics is taking place in a circumference). Our approach is based on coagulation equations. This kinetic description is in general invalid for one-dimensional systems because it is known that spatial correlations do propagate in this dimensionality. So we assume a collision takes place when two clusters meet with a very small probability. The probability should be so small that all the particles travel the whole system several times before one collision happens on average. This way the system becomes well-stirred, and so we can neglect spatial correlations and treat the system as if it were zero dimensional, what allows mathematical tractability. The content of this summary is based on the developments reported in a paper that is currently being considered for its possible publication [9].

References

1. Buhl J, Sumpter DJT, Couzin ID, Hale JJ, Despland E, Miller ER, Simpson SJ (2006) From disorder to order in marching locusts. Science 312:1402–1406
2. Carrillo JA, D'Orsogna MR, Panferov V (2009) Double milling in self-propelled swarms from kinetic theory. Kinet Relat Models 2:363–378
3. Castellano C, Fortunato S, Loreto V (2009) Statistical physics of social dynamics. Rev Mod Phys 81:591–646
4. Clifford P, Sudbury A (1973) A model for spatial conflict. Biometrika 60:581–588
5. Conti M, Meerson B, Peleg A, Sasorov PV (2002) Phase ordering with a global conservation law: Ostwald ripening and coalescence. Phys Rev E 65:046117
6. Czirok A, Barabasi A-L, Vicsek T (1999) Collective motion of self-propelled particles: kinetic phase transition in one dimension. Phys Rev Lett 82:209–212
7. D'Orsogna MR, Chuang YL, Bertozzi AL, Chayes LS (2006) Self-propelled particles with soft-core interactions: patterns, stability and collapse. Phys Rev Lett 96:104302
8. Deffuant G, Neu D, Amblard F, Weisbuch G (2000) Mixing beliefs among interacting agents. Adv Complex Syst 3:87–98
9. Escudero C, Macia F, Toral R, Velazquez JJL Kinetic theory and numerical simulations of two-species coagulation. Preprint
10. Escudero C, Macia F, Velazquez JJL (2010) Two-species coagulation approach to consensus by group level interactions. Phys Rev E 82:016113
11. Escudero C, Yates CA, Buhl J, Couzin ID, Erban R, Kevrekidis IG, Maini PK (2010) Ergodic directional switching in mobile insect groups. Phys Rev E 82:011926
12. Hernandez-Garcia E, Lopez C (2004) Clustering, advection and patterns in a model of population dynamics. Phys Rev E 70:016216
13. Holley RA, Liggett TM (1975) Ergodic theorems for weakly interacting infinite systems and the voter model. Ann Probab 3:643–663
14. Huepe C, Aldana M (2004) Intermittency and clustering in a system of self-driven particles. Phys Rev Lett 92:168701
15. Lifshitz IM, Slyozov VV (1961) The kinetics of precipitation from supersaturated solid solutions. J Phys Chem Solids 19:35–50
16. Liggett TM (1985) Interacting particle systems. Springer, New York
17. Niwa HS (1998) School size statistics of fish. J Theor Biol 195:351–361
18. Peruani F, Deutsch A, Bar M (2006) Nonequilibrium clustering of self-propelled rods. Phys Rev E 74:030904(R)
19. Pineda M, Toral R, Hernandez-Garcia E (2009) Noisy continuous-opinion dynamics. J Stat Mech P08001. doi:10.1088/1742-5468/2009/08/P08001
20. Seinfeld J (1986) Atmospheric chemistry and physics of air polution. Wiley, New York
21. Silk J, White SD (1978) The development of structure in the expanding universe. Astrophys J 223:L59–L62
22. Sintes T, Toral R, Chakrabarti A (1994) Reversible aggregation in self-associating polymer systems. Phys Rev E 50:2967–2976
23. Toral R, Marro J (1987) Cluster kinetics in the lattice gas model: the Becker-Doring type of equations. J Phys C, Solid State Phys 20:2491–2500
24. Toral R, Tessone CJ (2007) Finite size effects in the dynamics of opinion formation. Commun Comput Phys 2:177–195
25. Vazquez F, Lopez C (2008) Systems with two symmetric absorbing states: Relating the microscopic dynamics with the macroscopic behavior. Phys Rev E 78:061127
26. Yates CA, Erban R, Escudero C, Couzin ID, Buhl J, Kevrekidis IG, Maini PK, Sumpter DJT (2009) Inherent noise can facilitate coherence in collective swarm motion. Proc Natl Acad Sci USA 106:5464–5469
27. Ziff RM (1980) Kinetics of polymerization. J Stat Phys 23:241–263

List of Participants[1]

Syed Muhammad Ali, Abbas ali@cfpm.org
Tapio, Ala-Nissila alanissi@gmail.com
Rodrigo, Aldecoa raldecoa@ibv.csic.es
Peter, Allen p.m.allen@cranfield.ac.uk
Pierpaolo, Andriani pier2paolo@gmail.com
Chris, Antonopoulos chantonopoulos@teimes.gr
Andrea, Apolloni abu.apolloni@gmail.com
Tom, Arbuckle tom.arbuckle@ul.ie
Elsa, Arcaute e.arcaute@ucl.ac.uk
Gil, Ariel arielg@math.biu.ac.il
Alberto, Armoni alberto@armoni.com.ar
Peter, Arndt arndt@molgen.mpg.de
Defne, ASKAR defne.askar@gmail.com
M., Aziz-Alaoui aziz.alaoui@univ-lehavre.fr
Sergio, Bacelar sergio.bacelar@gmail.com
Claude, Baesens claude.baesens@warwick.ac.uk
Jan, Baetens jan.baetens@ugent.be
Paolo, Bajardi paolo.bajardi@isi.it
Duygu, Balcan duygu.balcan@isi.it
Sven, Banisch sven.banisch@universecity.de
Martine, Barons m.barons@warwick.ac.uk
Vasileios, Basios vbasios@ulb.ac.be
Peter, Baudains p.baudains@ucl.ac.uk
Declan, Baugh declan.baugh2@mail.dcu.ie
Gareth, Baxter gjbaxter@ua.pt
Itzhak, Benenson bennya@post.tau.ac.il
Charles H., Bennett bennetc@us.ibm.com
Sophie, Béreau sophie.bereau@uclouvain.be

[1]This list is limited to the 389 participants who requested their names to be published.

T. Gilbert et al. (eds.), *Proceedings of the European Conference on Complex Systems 2012*, Springer Proceedings in Complexity, DOI 10.1007/978-3-319-00395-5,

Pedro, Bernaola-Galvan rick@uma.es
Daniel, Bernardes daniel.bernardes@polytechnique.org
Cyrille, Bertelle cyrille.bertelle@gmail.com
Maria Letizia, Bertotti marialetizia.bertotti@unibz.it
Katrien, Beuls katrien@ai.vub.ac.be
Vincent, Blondel vincent.blondel@uclouvain.be
Peter, Boait p.boait@dmu.ac.uk
Jean-Pierre, Boon jpboon@ulb.ac.be
Christian, Borghesi christian.borghesi@u-cergy.fr
Roumen, Borissov roumen.borissov@ec.europa.eu
Melania, Borit melania.borit@uit.no
Wojciech, Borkowski wborkowsk@poczta.onet.pl
Benny, Bornfeld bennybornfeld@gmail.com
Arianna, Bottinelli ariannabottinelli@gmail.com
Jean, Boulton jboulton@claremont-mc.co.uk
Tassos, Bountis bountis@math.upatras.gr
Marc, Bourgois marc.bourgois@eurocontrol.int
J. Javier, Brey brey@us.es
Arnaud, Browet arnaud.browet@uclouvain.be
David, Bryngelsson david.bryngelsson@chalmers.se
Marcello, Budroni mbudroni@ulb.ac.be
Luminita Manuela, Bujorianu lmbujorianu@gmail.com
Domenico, Bullara dbullara@ulb.ac.be
Grégory, Bulnes Cuetara gbulnesc@ulb.ac.be
Petre, Caraiani petre.caraiani@gmail.com
Miriam, Carbon-Mangels miriam.carbon-mangels@pei.de
Timoteo, Carletti timoteo.carletti@fundp.ac.be
Enrico, Carlon enrico.carlon@fys.kuleuven.be
Maria Carmela, Catone mariacarmela.catone@unifi.it
Yassin, Chaffi ychaffi@ulb.ac.be
Aviel, Chaimovich avielchaimovich@umail.ucsb.edu
Lock Yue, Chew lockyue@ntu.edu.sg
Jean-Christophe, Chiem jean-christophe.chiem@uclouvain.be
Matteo, Chinazzi matteo.chinazzi@gmail.com
Tommaso, Ciarli t.ciarli@sussex.ac.uk
Pierre, Colinet pcolinet@ulb.ac.be
Vittoria, Colizza contagion.eccs2012@isi.it
Pierre, Collet pierre.collet@unistra.fr
Matthew, Cook matthew.cook@open.ac.uk
Rebecca, Cotton-Barratt r.langham@warwick.ac.uk
Daniel, Czamanski danny@czamanski.com
Agnieszka, Czaplicka agaczapl@if.pw.edu.pl
Joni, Dambre joni.dambre@ugent.be
Alexandre, Dauphin adauphin@ulb.ac.be
Toby, Davies toby.davies.09@ucl.ac.uk

Pedro, de Mendonca g.p.a.de-mendonca@warwick.ac.uk
Sarah, de Nigris denigris.sarah@gmail.com
Anne, de Wit adewit@ulb.ac.be
Adeline, Decuyper adeline.decuyper@uclouvain.be
Marcelo, del Castillo-Mussot marcelodlcstll@yahoo.com
Amene, Deljoo a.deljoo@tudelft.nl
Antoine, Delmotte antoine.delmotte09@imperial.ac.uk
Jean-Charles, Delvenne jean-charles.delvenne@uclouvain.be
Jonathan, Demaeyer jodemaey@ulb.ac.be
Basak, Demires Ozkul basakdemires@gmail.com
Jacques, Demongeot jacques.demongeot@imag.fr
Jean-Louis, Deneubourg jldeneub@ulb.ac.be
Peter, Dick peter.dick@dh.gsi.gov.uk
Alexandra, Diem alexandra.diem@uni-jena.de
Irene, Donato irene.donato@polito.it
Johan, Dubbeldam j.l.a.dubbeldam@tudelft.nl
Marco-Antonio, Duenas-Esterling maduenase@gmail.com
Ralph, Dum ralph.dum@ec.europa.eu
Geneviève, Dupont gdupont@ulb.ac.be
François, Duport francois.duport@ulb.ac.be
Miguel A., Durán-Olivencia migduroli@gmail.com
Antoine, Dutot antoine.dutot@gmail.com
Erik, Edlund edlunde@gmail.com
Bruce, Edmonds b.edmonds@mmu.ac.uk
Andrew, Edwards aedwards@maths.man.ac.uk
Manfred, Eigen julia.freiberg@mpibpc.mpg.de
Marion, Erpelding marion.erpelding@fys.uio.no
Thomas, Evans thomas.evans.11@ucl.ac.uk
Qingyi, Feng qingqingwin@gmail.com
Lucas, Fernandes ldiasfis@ifi.unicamp.br
Jose, Fernandez-Villacanas jose.fernandez-villacanas@ec.europa.eu
Eliseo, Ferrante eferrant@ulb.ac.be
Alessandro, Filisetti alessandro.filisetti@unibo.it
Ingo, Fischer ingo@ifisc.uib-csic.es
Greg, Fisher greg.fisher@synthesisips.net
Francisca, Flinterman franciscacaron@hotmail.com
Santo, Fortunato santo.fortunato@aalto.fi
Alexander, Fradkov fradkov@mail.ru
Gaihua, Fu gaihua.fu@ncl.ac.uk
Giuseppe, Futia giuseppe.futia@polito.it
Riccardo, Gallotti rgallotti@gmail.com
Viviane, Galvao vivgalvao@gmail.com
Cornelia, Gamst gamst@math.fu-berlin.de
Jordi, Garcia-Ojalvo jordi.g.ojalvo@upc.edu
Juan Carlos, García-Vázquez garcia@us.es

Philip, Garnett philip.garnett@durham.ac.uk
Pierre, Gaspard gaspard@ulb.ac.be
Marcos, Gaudiano marcosgaudiano@gmail.com
Theo, Geisel geisel@ds.mpg.de
Lendert, Gelens lendert.gelens@gmail.com
Claude, Gerard cgerard1@ulb.ac.be
Carlos, Gershenson cgg@unam.mx
Fakhteh, Ghanbarnejad fakhteh@bioinf.uni-leipzig.de
Marco, Gherardi gocram@gmail.com
Thomas, Gilbert tgilbert@ulb.ac.be
Nigel, Gilbert n.gilbert@surrey.ac.uk
Vladimir, Gligorijevic vgligorijevic@gmail.com
Peter, Gmeiner gmeiner@mi.uni-erlangen.de
Benjamin, Goddard b.goddard@imperial.ac.uk
Albert, Goldbeter agoldbet@ulb.ac.be
Natasa, Golo natasa.golo@gmail.com
Didier, Gonze dgonze@ulb.ac.be
Lara, Gosce cexlg@bristol.ac.uk
Willy, Govaerts willy.govaerts@ugent.be
Peter, Grassberger peter.grassberger@ucalgary.ca
Andrej, Grebenc andrej.grebenc@ec.europa.eu
Patrick, Grosfils pgrosfi@ulb.ac.be
Roberto, Gutiérrez betuchas3@gmail.com
Konrad, Halupka halupka@biol.uni.wroc.pl
Mark, Hardman mark.hardman@canterbury.ac.uk
Rosemary, Harris rosemary.harris@qmul.ac.uk
Tomonori, Hasegawa tomonori.hasegawa2@mail.dcu.ie
Florence, Haudin fhaudin@ulb.ac.be
Holger, Hennig holgerh@nld.ds.mpg.de
Niel, Hens niel.hens@gmail.com
Burkhard, Hense burkhard.hense@helmholtz-muenchen.de
Hans-Peter, Herzel h.herzel@biologie.hu-berlin.de
Ryohei, Hisano em072010@yahoo.co.jp
Jaroslav, Hlinka hlinka@cs.cas.cz
Petter, Holme petter.holme@physics.umu.se
Janusz, Holyst jholyst@if.pw.edu.pl
Cars H., Hommes c.h.hommes@uva.nl
Ágnes, Horvát h_emoke_a@yahoo.de
Mihnea R., Hristea mihnea.hristea@ilsr.at
Akira, Ishii ishii.akira.t@gmail.com
Gabriel, Istrate gabriel.istrate@gmail.com
Magdalena, Jagielska magda.jagielska@hotmail.com
Ryan, James rgjames@ucdavis.edu
Ljubomir, Jankovic lubo.jankovic@bcu.ac.uk
Andrzej, Jarynowski andrzej.jarynowski@uj.edu.pl

Cristian, Jimenez Romero cjr_personal@yahoo.es
Hang-Hyun, Jo h2jo23@gmail.com
Jeffrey, Johnson j.h.johnson@open.ac.uk
Emma, Jonson emma.jonson@chalmers.se
Marco, Jose marcojose@biomedicas.unam.mx
Remi, Joubaud remi.joubaud@gmail.com
Jaap, Kaandorp j.a.kaandorp@uva.nl
Maria, Kalimeri maria.kalimeri@ibpc.fr
Christel, Kamp christel.kamp@pei.de
Mirko, Kämpf mirko.kaempf@gmail.com
Giorgos, Kanellopoulos giorgoskanellopoulos@gmail.com
Raymond, Kapral rkapral@chem.utoronto.ca
Benjamin, Kaube bk308@imperial.ac.uk
Hans, Keune hans.keune@inbo.be
Kyungsik, Kim kskim@pknu.ac.kr
Beom Jun, Kim beomjun@skku.edu
Markus, Kirkilionis mak@maths.warwick.ac.uk
Alexander, Klauer alexander.klauer@itwm.fraunhofer.de
Eugene, Korotkov bioinf@yandex.ru
Robert, Kosinski kosinski@if.pw.edu.pl
Nikos, Kouvaris nkoub@fhi-berlin.mpg.de
Enrique, Kremers kremers@eifer.org
Ursula, Kummer ursula.kummer@bioquant.uni-heidelberg.de
Jorge, Kurchan jorge@pmmh.espci.fr
Robin, Lamarche-Perrin robin.lamarche-perrin@imag.fr
Renaud, Lambiotte renaud.lambiotte@fundp.ac.be
Felipe, Lara-Rosano flararosano@gmail.com
Dukhee, Lee dhlnexys@kaist.ac.kr
Marc, Lefranc marc.lefranc@univ-lille1.fr
Robert, Legenstein legi@igi.tugraz.at
Jean-Marie, Lehn lehn@unistra.fr
Jean-Christophe, Leloup jleloup@ulb.ac.be
Lorena, Lemaigre llemaigr@ulb.ac.be
Antonella, Liccardo liccardo@na.infn.it
Oskar, Lindgren lindgreo@chalmers.se
Hartmut, Loewen hlowen@thphy.uni-duesseldorf.de
Ricardo, Lopez-Ruiz rilopez@unizar.es
András, Lörincz lorincz@inf.elte.hu
René, Lozi rlozi@unice.fr
Ihor, Lubashevsky i-lubash@u-aizu.ac.jp
Pablo, Lucas pablo.lucas@ucd.ie
Igor, Lugo igorlugo@correo.crim.unam.mx
Liv, Lundberg livlundberg@gmail.com
Eliza Olivia, Lungu eliza.olivia.lungu@gmail.com
James F., Lutsko jlutsko@ulb.ac.be

Marina Jeaneth, Machicao Justo mj.machicao@gmail.com
Suman Kumar, Maji sumankumar.maji@inria.fr
Nick, Malleson n.malleson06@leeds.ac.uk
Piero, Manfredi manfredi@ec.unipi.it
Ed, Manley ucesejm@ucl.ac.uk
Thanos, Manos thanos.manos@gmail.com
Gaetan, Marceau Caron gaetan.marceau-caron@inria.fr
Elio, Marchione e.marchione@ucl.ac.uk
Ignacio, Marin imarin@ibv.csic.es
Serge, Massar smassar@ulb.ac.be
Anthony, Masys anthony.masys@drdc-rddc.gc.ca
José, Matos jlvmatos@gmail.com
Barry, McMullin barry.mcmullin@dcu.ie
Jean-Luc, Mercier jlm@unistra.fr
Hildegard, Meyer-Ortmanns h.ortmanns@jacobs-university.de
Nathalie, Mezza-Garcia mezzagarcia@gmail.com
Piotr, Migdal piotr.migdal@icfo.es
Armin, Mikler mikler@unt.edu
Mai, Minoura maichi_mino@yahoo.co.jp
Pedro, Miramontes pmv@ciencias.unam.mx
Eve, Mitleton-Kelly e.mitleton-kelly@lse.ac.uk
Marija, Mitrovic marija.mitrovic@aalto.fi
Mauro, Mobilia m.mobilia@leeds.ac.uk
Joanna, Moir j.moir@warwick.ac.uk
Joon-Young, Moon joon.young.moon@gmail.com
Alfredo, Morales alfredo.moralesg@alumnos.upm.es
Fumito, Mori mori.fumito@ocha.ac.jp
Stefan C., Müller tsuji-mueller@t-online.de
Roberto, Murcio rmurcio@unam.mx
Moustafa, Nakechbandi moustafa.nakechbandi@univ-lehavre.fr
Jan, Naudts jan.naudts@ua.ac.be
Izaak, Neri izaakneri@gmail.com
Stamatios, Nicolis snicolis@math.uu.se
Grégoire, Nicolis gnicolis@ulb.ac.be
Francesco, Niglia francesco@fnstudio.net
Martin, Nilsson Jacobi mjacobi@chalmers.se
Susu, Nousala s.nousala@gmail.com
Sylvie, Occelli occelli@ires.piemonte.it
Jeremi, Ochab jeremi.ochab@uj.edu.pl
Paul, Ormerod pormerod@volterra.co.uk
Bruno, Pace bruno.pace@gmail.com
Evangelia, Panagakou epanagakou@chem.demokritos.gr
Michael B., Paradowski m.b.paradowski@uw.edu.pl
Diane, Payne diane.payne@ucd.ie
Luis, Pesquera pesquerl@ifca.unican.es

Giovanni, Petri giovanni.petri@isi.it
Benjamin, Pfeuty benjamin.pfeuty@univ-lille1.fr
de Buyl, Pierre pdebuyl@ulb.ac.be
Draga, Pihler-Puzovic draga.pihler-puzovic@manchester.ac.uk
Leonid, Pismen pismen@technion.ac.il
Dariusz, Plewczynski darman@icm.edu.pl
Chiara, Poletto chiara.poletto@isi.it
Yves, Pomeau pomeau@lps.ens.fr
Adrien, Poncelet adrien.poncelet@uclouvain.be
Lise, Ponselet lise.ponselet@uclouvain.be
Oriol, Pont oriol@ffn.ub.es
Jefferson, Portela stafusa@gmail.com
Márton, Pósfai posfaim@gmail.com
Astero, Provata aprovata@chem.demokritos.gr
Gunnar, Pruessner g.pruessner@imperial.ac.uk
Ivan, Puga-Gonzalez ivanpuga@gmail.com
Rick, Quax r.quax@uva.nl
Adrien, Querbes-Revier a.querbes@tue.nl
Matti, Raasakka matti.raasakka@aei.mpg.de
Giovanni, Rabino giovanni.rabino@polimi.it
Filippo, Radicchi f.radicchi@gmail.com
Shama, Rahman rahman.shama@gmail.com
Mickael, Randour mickael.randour@gmail.com
Jonathan, Reades j.reades@ucl.ac.uk
Frank, Redig f.h.j.redig@tudelft.nl
Carl Henning, Reschke carlhenning.reschke@imfk.de
Mauricio, Ribeiro ribeiro@cbpf.br
Simone, Righi simone.righi@fundp.ac.be
Luis, Riolfo lriolfo@ulb.ac.be
Marianne, Robert mars.robert@hotmail.fr
Martin, Robert mrobert@ttck.keio.ac.jp
Juan, Rocha juan.rocha@stockholmresilience.su.se
Tarcisio, Rocha Filho marciano@fis.unb.br
Karina, Rojas karinarojas10@gmail.com
Andrea, Roli andrea.roli@unibo.it
Laurence, Rongy lrongy@ulb.ac.be
David A., Rosenblueth drosenbl@servidor.unam.mx
Bridget, Rosewell brosewell@volterra.co.uk
Magda, Roszczynska magda.roszczynska@gmail.com
Charles, Rougé charles.rouge@irstea.fr
Tarik, Roukny troukny@ulb.ac.be
Catherine, Rouvas cnicolis@oma.be
Irina, Roznovat iroznovat@computing.dcu.ie
Nicolas, Rubido n.rubido.obrer@abdn.ac.uk
Alberto, Saa asaa@ime.unicamp.br

Marcel, Salathé salathe@psu.edu
Alfons, Salden alfons@almende.org
Jose, Salgado portosalgado@gmail.com
Leonidas, Sandoval lsandovaljr@hotmail.com
Yukie, Sano yukie.sano@gmail.com
Juan Luis, Santos juanluis.santos@iaes.es
Pier Paolo, Saviotti pier-paolo.saviotti@wanadoo.fr
Antonio, Scala antonio.scala.phys@gmail.com
Christoph, Schmal christoph.schmal@web.de
Markus, Schmuck m.schmuck@imperial.ac.uk
Max, Schneider max.schneider15@gmail.com
Tobias, Scholz tobias.scholz@uni-siegen.de
Simone, Schopper simone.schopper@biol.lu.se
Mantas, Sekmokas chiupt@yahoo.com
Marcelo, Serrano Zanetti mszanetti@yahoo.com
Hassan, Shafiey h.shafiey@gmail.com
Adi, Shklarsh adi.shklarsh@gmail.com
Jean-Christophe, Sibel jeanchristophesibel@gmail.com
Alina, Sirbu alina.sirbu@isi.it
Haris, Skokos hskokos@auth.gr
Piotr, Slowinski slowinski.piotr@gmail.com
Stephen, Soehnlen stephen.soehnlen@springer.com
Charlotte, Sonck charlotte.sonck@ugent.be
Kohei, Sonoda koheisonoda@gmail.com
Konstantin, Starkov konproz@gmail.com
Daniel, Strombom strombom@math.uu.se
Fernanda, Strozzi fstrozzi@liuc.it
Dalibor, Stys stys@jcu.cz
Cédric, Sueur cedric.sueur@iphc.cnrs.fr
Yulia, Suvorova suvorovay@gmail.com
Charlotte, Szostek charlotte.szostek@bris.ac.uk
Andrea, Tacchella andreatacchella@gmail.com
Taro, Takaguchi taro4001@gmail.com
Hisa-Aki, Tanaka htan@ee.uec.ac.jp
Mieko, Tanaka-Yamawaki mieko@ike.tottori-u.ac.jp
Ana, Teixeira de Melo anamelopsi@gmail.com
Luca, Tesei luca.tesei@gmail.com
Ruediger, Thul ruediger.thul@nottingham.ac.uk
Mario Vincenzo, Tomasello mtomasello@ethz.ch
Claude, Tomberg ctomberg@ulb.ac.be
Petter, Törnberg pettert@chalmers.se
Thomas, Toulias t.toulias@teiath.gr
Emmanouil, Tranos e.tranos@vu.nl
Maguy, Trefois maguy.trefois@uclouvain.be
Nicolas, Tremblay nicolas.tremblay@ens-lyon.fr

Miquel, Trias mtrias@iac3.eu
Kolbjorn, Tunstrom tunstrom@princeton.edu
Francesco, Vaccarino francesco.vaccarino@gmail.com
Eugenio, Valdano eugenio.valdano@gmail.com
Andrea, Valente val.andrea@hotmail.it
Gabriele, Valentini gabriele.valentini@ulb.ac.be
Devaraj, van der Meer d.vandermeer@utwente.nl
Ko, van der Weele weele@math.upatras.gr
Aernout, van Enter avanenter@gmail.com
Carlo, Vanderzande carlo.vanderzande@uhasselt.be
Stéphane, Vannitsem svn@meteo.be
Liz, Varga liz.varga@cranfield.ac.uk
Daniel, Vasata daniel.vasata@fit.cvut.cz
Viktoras, Veitas vveitas@gmail.com
Roberto, Venegeroles roberto.venegeroles@ufabc.edu.br
Vilhelm, Verendel vive@chalmers.se
Clement, Vidal clement.vidal@philosophons.com
Jan, Viebahn jan.viebahn@zmaw.de
Bruno, Vieira Ribeiro bruno.vieiraribeiro@univ-amu.fr
Marco, Villani marco.villani@unimore.it
Daniele, Vilone daniele.vilone@gmail.com
Cristina, Vinas cristinavinas@gmail.com
Stefania, Vitali svitali82@gmail.com
Burton, Voorhees burt@athabascau.ca
Sylvia, Walby s.walby@lancaster.ac.uk
Huijuan, Wang h.wang@tudelft.nl
Iain, Weaver isw1g10@soton.ac.uk
Raoul, Weiler raoul.weiler@telenet.be
David, Weinbaum david.weinbaum@vub.ac.be
Pieter, Wellens pieter@ai.vub.ac.be
Lander, Willem lander.willem@ua.ac.be
Ruthild, Winkler-Oswatitsch rome@gwdg.de
Martin, Wirz mail@wirzm.ch
Florian, Wodlei florian.wodlei@ilsr.at
Marta, Wronikowska marta.wronikowska@yahoo.com
Lidia, Yamamoto lidia.yamamoto@unistra.fr
Tatsuo, Yanagita yanagita@isc.osakac.ac.jp
Il Gu, Yi superstring19@gmail.com
Nicky, Zachariou nnz04@ic.ac.uk
Marcin, Zagórski marcin.zagorski@uj.edu.pl
Massimiliano, Zanin mzanin@innaxis.org
Anna, Zaytseva annza944@gmail.com
Hector, Zenil hzenilc@gmail.com
Arkady, Zgonnikov arkadiy.zgonnikov@gmail.com
Aleksandra, Zhukova sasha.mymail@gmail.com
Wolfgang, zu Castell castell@helmholtz-muenchen.de

Author Index

T. Gilbert et al. (eds.), *Proceedings of the European Conference on Complex Systems 2012*, Springer Proceedings in Complexity, DOI 10.1007/978-3-319-00395-5,

Printed by Publishers' Graphics LLC